Lecture Notes in Computer Science   13568

Advanced Research in Computing and Software Science
Subline of Lecture Notes in Computer Science

More information about this series at https://link.springer.com/bookseries/558

Armando Castañeda ·
Francisco Rodríguez-Henríquez (Eds.)

LATIN 2022:
Theoretical Informatics

15th Latin American Symposium
Guanajuato, Mexico, November 7–11, 2022
Proceedings

 Springer

Editors
Armando Castañeda (iD)
Universidad Nacional Autónoma de México
Mexico City, Mexico

Francisco Rodríguez-Henríquez
Centro de investigación y de Estudios
Avanzados
Mexico City, Mexico

Cryptography Research Centre of the
Technology Innovation Centre
Abu Dhabi, United Arab Emirates

ISSN 0302-9743 ISSN 1611-3349 (electronic)
Lecture Notes in Computer Science
ISBN 978-3-031-20623-8 ISBN 978-3-031-20624-5 (eBook)
https://doi.org/10.1007/978-3-031-20624-5

This Springer imprint is published by the registered company Springer Nature Switzerland AG
The registered company address is: Gewerbestrasse 11, 6330 Cham, Switzerland

Preface

This volume contains the papers presented at the 15th Latin American Theoretical Informatics Symposium (LATIN 2022) held during November 7–11, 2022, in Guanajuato, Mexico. Previous editions of LATIN took place in São Paulo, Brazil (1992), Valparaíso, Chile (1995), Campinas, Brazil (1998), Punta del Este, Uruguay (2000), Cancún, Mexico (2002), Buenos Aires, Argentina (2004), Valdivia, Chile (2006), Búzios, Brazil (2008), Oaxaca, Mexico (2010), Arequipa, Perú (2012), Montevideo, Uruguay (2014), Ensenada, Mexico (2016), Buenos Aires, Argentina (2018), and São Paulo, Brazil (2021).

The symposium received 114 submissions from around the world. Each submission was reviewed by four Program Committee members, and carefully evaluated on quality, originality, and relevance to the conference. Committee members often reviewed the submissions with the help of additional external referees. Based on an extensive electronic discussion, the committee selected 46 papers. In addition to the accepted contributions, the symposium featured keynote talks by David Eppstein (University of California, Irvine, USA), Mauricio Osorio (Universidad de las Américas, Mexico), Merav Parter (Weizmann Institute of Science, Israel), and Jeffrey D. Ullman (Stanford University, USA).

LATIN 2022 featured two awards: the 2022 Imre Simon Test-of-Time Award and the Alejandro López-Ortiz Best Paper Award. In this edition, the Imre Simon Test-of-Time Award winner was Johannes Fischer for his paper "Optimal Succinctness for Range Minimum Queries," which appeared in LATIN 2010. For the Alejandro López-Ortiz Best Paper Award, the Program Committee selected the paper "Theoretical Analysis of Git Bisect," by Julien Courtiel, Paul Dorbec, and Romain Lecoq. We thank Springer for supporting both awards.

A round table to honor the research and legacy of Héctor García-Molina was held as part of the LATIN 2022 program. The panel comprised Carlos Coello Coello (Cinvestav, Mexico), Jeffrey D. Ullman (Stanford University, USA), and Gio Wiederhold (Stanford University, USA). The round table was moderated by Mariano Rivera (CIMAT, Mexico).

The program of the symposium included tutorial sessions devoted mainly to theory students and young researchers. Edgar Chávez (CICESE, Mexico) chaired the Tutorial Session Committee.

The main organizer of the conference was the Centro de Investigación en Matemáticas (CIMAT), located in Guanajuato, Mexico. Mariano Rivera chaired the Local Arrangements Committee.

Many people helped to make LATIN 2022 possible. First, we would like to recognize the outstanding work of the members of the Program Committee. Their commitment contributed to a very detailed discussion on each of the submitted papers. The LATIN Steering Committee offered valuable advice and feedback; the conference benefitted immensely from their knowledge and experience. We would also like to recognize Conrado Martínez, Jacques Sakarovitch, and Yoshiko Wakabayashi for their work in the 2022 Imre Simon Test-of-Time Award Committee.

Finally, the conference would not have been possible without our generous sponsors, Springer, the Cryptography Research Center of the Technology Innovation Institute, Abu Dhabi, United Arab Emirates, and the Centro de Investigación en Matemáticas (CIMAT), Guanajuato, Mexico. We are also grateful for the facilities provided by EasyChair for paper evaluation and the preparation of this volume.

November 2022

Armando Castañeda
Francisco Rodríguez-Henríquez

The Imre Simon Test-of-Time Award

The winner of the 2022 Imre Simon Test-of-Time Award, considering papers up to the 2012 edition of the Latin American Theoretical Informatics Symposium (LATIN), is

Optimal Succinctness for Range Minimum Queries by Johannes Fischer,
LATIN 2010, LNCS 6034, 158–169, 2010.

Range Minimum Query (RMQ) is used on arrays to find the position of an element with the minimum value between two specified indices. This simple problem—in its formulation—has many different applications including the fundamental problem of finding the *least common ancestor* (LCA) of two nodes in a tree or the *longest common prefix problem* (LCP), as well as other exact and approximate string matching problems. A witness of the relevance of these problems is the first Imre Simon Test-of-Time award won in 2012 by the LATIN 2000 paper *The LCA Problem Revisited*, by Martin Farach-Colton and Michael Bender.

In order to make RMQs very efficient, there has been a long quest for preprocessing algorithms and data structures with which RMQs could later be answered very efficiently, ideally in constant time. The first non-trivial solution to the problem, presented by Berkman and Vishkin (SIAM J. Computing, 1993), required linear time for preprocessing and linearithmic space ($\Theta(n \log n)$ bits, for an array of n items). Many authors have thus been looking for succinct data structures; in this case, data structures using a linear number of bits, without sacrificing the preprocessing or the query times. The first solution which does not need to keep the original input to answer RMQs (*non-systematic*, in the terminology of the awarded paper) was by Sadakane (J. Discrete Algorithms, 2007), using $4n + o(n)$ bits for the balanced-parentheses-encoding of the Cartesian tree. Besides the query and preprocessing times, the space used in the final data structure **and** during the preprocessing phase has been of concern. The LATIN paper of 2010 presented the first scheme achieving $O(1)$ time for queries, $O(n)$ preprocessing time, using only $2n + o(n)$ bits in the final succinct data structure to answer queries—thus meeting the information-theoretic bound—, and only $n + o(n)$ additional bits during construction time.

To achieve the space and time efficiency above, the paper introduced *2d-min-heaps*, which are equivalent to the also well-known LRM-trees (left-to-right minima trees) of Navarro and Sadakane (ACM Trans. Algorithms, 2014). The 2d-min-heaps were originally intended to efficiently support RMQs, but they have proved also very useful for several other applications, e.g., the succinct representation of ordinal trees.

Fischer's contributions in the LATIN 2010 paper made their way into the journal article *Space-Efficient Preprocessing Schemes for Range Minimum Queries on Static Arrays*, published in *SIAM Journal on Computing* 40(2):465–492, together with Volker Heun, in 2011. That paper became very influential in the area of compressed and succinct data structures, and it is a milestone in the quest for the best solutions in time and space to RMQs, with numerous quotations and references too.

The relevance of the problem addressed, the originality of the technique used to solve it, the clarity of presentation, and the continued and widespread recognition of this contribution throughout the years since its publication heavily weighed in the committee's choice.

The committee for the 2022 Imre Simon Test-of-Time Award.

Conrado Martínez
Jacques Sakarovitch
Yoshiko Wakabayashi

Organization

Conference Chair

Francisco Rodríguez
  Henríquez

Centro de investigación y de Estudios Avanzados,
  Mexico, and Cryptography Research Centre, TII,
  United Arab Emirates

Program Committee Chair

Armando Castañeda

Universidad Nacional Autónoma de México, Mexico

Steering Committee

Michael A. Bender
Cristina G. Fernandes
Joachim von zur Gathen

Evangelos Kranakis
Conrado Martínez
Flávio Keidi Miyazawa

Stony Brook University, USA
Universidade de São Paulo, Brazil
Bonn-Aachen International Center for Information
  Technology, Germany
Carleton University, Canada
Universitat Politècnica de Catalunya, Spain
Universidade Estadual de Campinas, Brazil

Program Committee

Martin Aumüler
Jérémy Barbay
Leonid Barenboim
Frédérique Bassino
Luciana Buriol
Witold Charatonik
Min Chih Lin
Amalia Duch Brown
Leah Epstein
Martín Farach-Colton
Esteban Feuerstein
David Flores-Peñaloza
Fedor Fomin
Jesper Jansson
Gabriela Jeronimo
Christos Kaklamanis
Shuji Kijima
Gregory Kucherov
François Le Gall

University of Copenhagen, Denmark
Universidad de Chile, Chile
The Open University of Israel, Israel
Université Sorbonne Paris Nord, France
Amazon, USA
University of Wrocław, Poland
Universidad de Buenos Aires, Argentina
Universitat Politècnica de Catalunya, Spain
University of Haifa, Israel
Rutgers University, USA
Universidad de Buenos Aires, Argentina
Universidad Nacional Autónoma de México, Mexico
University of Bergen, Norway
Kyoto University, Japan
Universidad de Buenos Aires, Argentina
University of Patras and CTI "Diophantus", Greece
Kyushu University, Japan
CNRS/LIGM, France
Nagoya University, Japan

Jérémy Ledent	University of Strathclyde, UK
Reut Levi	IDC Herzliya, Israel
Giovanni Manzini	Università di Pisa, Italy
Andrea Marino	Universitè degli Studi di Firenze, Italy
Elvira Mayordomo	Universidad de Zaragoza, Spain
Marco Molinaro	Pontifical Catholic University of Rio de Janeiro, Brazil
Guilherme Oliveira Mota	Universidade de São Paulo, Brazil
Lucia Moura	University of Ottawa, Canada
Gonzalo Navarro	Universidad de Chile, Chile
Rafael Oliveira	University of Waterloo, Canada
Daniel Panario	Carleton University, Canada
Gopal Pandurangan	University of Houston, USA
Seth Pettie	University of Michigan, USA
Miguel A. Pizaña	Universidad Autónoma Metropolitana, Mexico
Igor Potapov	University of Liverpool, UK
Svetlana Puzynina	Sobolev Institute of Mathematics, Russia
Pablo Rotondo	Université Gustave Eiffel, France
Jared Saia	University of New Mexico, USA
Rodrigo I. Silveira	Universitat Politècnica de Catalunya, Spain
Mohit Singh	Georgia Tech, USA
José A. Soto	Universidad de Chile, Chile
Frank Stephan	National University of Singapore, Singapore
Martin Strauss	University of Michigan, USA
Subhash Suri	University of California, Santa Barbara, USA
Dimitrios M. Thilikos	Université de Montpellier, CNRS, France
Christopher Thraves	Universidad de Concepción, Chile
Denis Trystram	Université de Grenoble Alpes, France
Seeun William Umboh	University of Sydney, Australia
Jorge Urrutia	Universidad Nacional Autónoma de México, Mexico
Alfredo Viola	Universidad de la República, Uruguay
Mikhail V. Volkov	Ural Federal University, Russia
Sebastian Wild	University of Liverpool, UK
Georg Zetzsche	Max Planck Institute for Software Systems, Germany

Imre Simon Test-of-Time Award Committee

Conrado Martínez (Chair)	Universitat Politècnica de Catalunya, Spain
Jacques Sakarovitch	IRIF CNRS and Telecom ParisTech, France
Yoshiko Wakabayashi	Universidade de São Paulo, Brazil

Organizing Committee

Tutorial Chair

Edgar Chávez	CICESE, Ensenada, Mexico

Local Arrangements Chair

Mariano Rivera   CIMAT, Guanajuato, Mexico

Publicity Chair

Tássio Naia   Universidade de São Paulo, Brazil

Student Scolarships Chair

Joel A. Trejo-Sánchez   CIMAT, Mérida, Mexico

Additional Reviewers

Mikkel Abrahamsen	Varsha Dani	Peter Gyorgyi
Federico Albanese	Pablo De Caria	Sariel Har-Peled
Jean-Paul Allouche	Paloma de Lima	Atsuya Hasegawa
José Alvarado Morales	Elie de Panafieu	Meng He
Avah Banerjee	Holger Dell	Juan Heguiabehere
Kevin Bello	Diego Diaz	Shuichi Hirahara
Rémy Belmonte	Daniel A. Díaz-Pachón	Carlos Hoppen
Shalev Ben-David	Michael Dinitz	Khalid Hourani
Pascal Bergsträßer	Henk Don	Clemens Huemer
Giulia Bernardini	Gaëtan Douéneau-Tabot	Ullrich Hustadt
Dario Bezzi	Philippe Duchon	Tomohiro I.
Hans L. Bodlaender	Fabien Dufoulon	Emmanuel Iarussi
Richard Brewster	Bruno Escoffier	Allen Ibiapina
Adam Brown	Vincent Fagnon	Christian Ikenmeyer
Boris Bukh	Wenjie Fang	Shunsuke Inenaga
Andrei Bulatov	Andreas Emil Feldmann	Mohammad Abirul Islam
Laurent Bulteau	Moran Feldman	Haitao Jiang
Xavier Bultel	Michael Fuchs	Vincent Jugé
Sergiy Butenko	Martin Fürer	Nikolay Kaleyski
Sergio Cavero	Eric Fusy	Haim Kaplan
Sofia Celi	Joaquim Gabarro	Roman Karasev
María Chara	Anahi Gajardo	Jarkko Kari
Abhranil Chatterjee	Guilhem Gamard	Maximilian Katzmann
Vincent Chau	Moses Ganardi	Dominik Kempa
Juhi Chaudhary	Peter Gartland	Ali Keramatipour
Bhadrachalam Chitturi	Christian Glasser	Kamil Khadiev
Julien Clément	Eric Goles	Eun Jung Kim
Alessio Conte	Petr Golovach	Bjørn Kjos-Hanssen
Alfredo Costa	Lev Gordeev	Hartmut Klauck
Jonas Costa	Inge Li Gørtz	Yasuaki Kobayashi
Marcio Costa Santos	Claude Gravel	Yusuke Kobayashi
Ágnes Cseh	Daniel Grier	Ekkehard Köhler
Luís Cunha	Luciano Grippo	Dominik Köppl
Konrad K. Dabrowski	Mauricio Guillermo	Tuukka Korhonen

William Kretschmer
Matjaž Krnc
Michal Kunc
O-Joung Kwon
Jacques-Olivier Lachaud
Hung Le
Thierry Lecroq
Asaf Levin
Jason Li
Bingkai Lin
Alexei Lisitsa
Markus Lohrey
Sylvain Lombardy
Bruno Lopes
Matteo Loporchio
Jérémie Lumbroso
Marten Maack
Pasin Manurangsi
Vladislav Makarov
Andreas Maletti
Javier Marenco
Nestaly Marín
Conrado Martínez
Moti Medina
Wilfried Meidl
Sihem Mesnager
Félix Miravé
Benjamin Monmege
Fabrizio Montecchiani
Peter Mörters
William K. Moses Jr.
Phablo Moura
Tuan Anh Nguyen

Yakov Nekrich
Joachim Niehren
Hugo Nobrega
Loana Nogueira
Pascal Ochem
Alexander Okhotin
Francisco Olivares
Juan M. Ortiz de Zarate
Ferruh Ozbudak
Anurag Pandey
Roberto Parente
Lehilton L. C. Pedrosa
Pablo Pérez-Lantero
Kévin Perrot
Dominik Peters
Duy Pham
Marta Piecyk
Aditya Pillai
Solon Pissis
Madhusudhan R. Pittu
Thomas Place
Adam Polak
Nicola Prezza
Evangelos Protopapas
Saladi Rahul
Santiago Ravelo
Kilian Risse
Javiel Rojas-Ledesma
Puck Rombach
Hamza Safri
Nicolás Sanhueza
  Matamala
Ignasi Sau

David Saulpic
Luke Schaeffer
Marinella Sciortino
Louisa Seelbach Benkner
Maria Serna
Jeffrey Shallit
Mark Siggers
Jefferson Elbert Simões
Matias Pavez Signe
Patrick Solé
Mauricio Soto
Uéverton Souza
Raphael Steiner
Benedikt Stufler
Dániel Szabó
Marek Szykuła
Guido Tagliavini
Claudio Telha
Alexander Tiskin
Gabriel Tolosa
Alfredo Torrico
Konstantinos Tsakalidis
Cristian Urbina
André van Renssen
Nikolay Vereshchagin
Stéphane Vialette
Giorgio Vinciguerra
Yukiko Yamauchi
Maxwell Young
Raphael Yuster
Pascal Weil
Viktor Zamaraev

Sponsors

Springer
Centro de Investigación en Matemáticas, Guanajuato, Mexico
Cryptography Research Center of the Technology Innovation, Institute, Abu Dhabi,
   United Arab Emirates

Contents

Cryptography

Social Choice Theory

Theoretical Machine Learning

Automata Theory and Formal Languages

Combinatorics and Graph Theory

Complexity Theory

Computational Geometry

Algorithms and Data Structures

Cutting a Tree with Subgraph Complementation is Hard, Except for Some Small Trees

Dhanyamol Antony[1]([✉]), Sagartanu Pal[2]([✉]), R. B. Sandeep[2], and R. Subashini[1]

[1] National Institute of Technology Calicut, Kozhikode, India
{dhanyamol_p170019cs,suba}@nitc.ac.in
[2] Indian Institute of Technology Dharwad, Dharwad, India
{183061001,sandeeprb}@iitdh.ac.in

Abstract. For a graph property Π, Subgraph Complementation to Π is the problem to find whether there is a subset S of vertices of the input graph G such that modifying G by complementing the subgraph induced by S results in a graph satisfying the property Π. We prove that the problem of Subgraph Complementation to T-free graphs is NP-Complete, for T being a tree, except for 41 trees of at most 13 vertices (a graph is T-free if it does not contain any induced copies of T). This result, along with the 4 known polynomial-time solvable cases (when T is a path on at most 4 vertices), leaves behind 37 open cases. Further, we prove that these hard problems do not admit any subexponential-time algorithms, assuming the Exponential Time Hypothesis. As an additional result, we obtain that Subgraph Complementation to paw-free graphs can be solved in polynomial-time.

Keywords: Subgraph complementation · Graph modification · Trees · Paw

1 Introduction

A graph property is hereditary if it is closed under vertex deletions. It is well known that every hereditary property is characterized by a minimal set of forbidden induced subgraphs. For example, for chordal graphs, the forbidden set is the set of all cycles on at least four vertices, for split graphs, the forbidden set is $\{2K_2, C_4, C_5\}$, for cluster graphs it is $\{P_3\}$, and for cographs it is $\{P_4\}$. The study of structural and algorithmic aspects of hereditary graph classes is central to theoretical computer science.

A hereditary property is called H-free if it is characterized by a singleton set $\{H\}$ of forbidden subgraphs. Such hereditary properties are very interesting for their rich structural and algorithmic properties. For example, triangle-free

This work is partially supported by SERB Grants "SRG/2019/002276" and "MTR/2021/000858".

A. Castañeda and F. Rodríguez-Henríquez (Eds.): LATIN 2022, LNCS 13568, pp. 3–19, 2022.
https://doi.org/10.1007/978-3-031-20624-5_1

graphs could be among the most studied graphs classes. There is an extensive list of structural studies of H-free graphs, for examples, see [1] for claw-free graphs, [2] for cographs, and [3] for paw-free graphs. There are many important hard problems, such as Independent Set [4–8], which admit polynomial-time algorithms for H-free graphs, for various graphs H.

Graph modification problems refer to problems in which the objective is to transform the input graph into a graph with some specific property Π. The constraints on the allowed modifications and the property Π define a graph modification problem. For an example, the objective of the Chordal Vertex Deletion problem is to check whether it is possible to transform the input graph by deleting at most k vertices so that the resultant graph is a chordal graph. Graph modification problems, where the target property is H-free, have been studied extensively for the last four decades under various paradigms - exact complexity [9–22], parameterized complexity [23–25], kernelization complexity [13,15,26–34], and approximation complexity [12,16,17,35]. We add to this long list by studying the exact complexity of a graph modification problem known as Subgraph Complementation, where the target property is H-free.

A *subgraph complement* of a graph G is a graph G' obtained from G by flipping the adjacency of pairs of vertices of a subset S of vertices of G. The operation is known as *subgraph complementation* and is denoted by $G' = G \oplus S$. The operation was introduced by Kamiński et al. [36] in relation with clique-width of a graph. For a class $\mathcal{G}$ of graphs, *subgraph complementation to $\mathcal{G}$* is the problem to check whether there is a set of vertices S in the input graph G such that $G \oplus S \in \mathcal{G}$. A systematic study of this problem has been started by Fomin et al. [37]. They obtained polynomial-time algorithms for this problem for various classes of graphs including triangle-free graphs and P_4-free graphs. A superset of the authors of this paper studied it further [11] and settled the complexities of this problem (except for a finite number of cases) when $\mathcal{G}$ is H-free, for H being a complete graph, a path, a star, or a cycle. They proved that subgraph complementation to H-free graphs is polynomial-time solvable if H is a clique, NP-Complete if H is a path on at least 7 vertices, or a star graph on at least 6 vertices, or a cycle on at least 8 vertices. Further, none of these hard problems admit subexponential-time algorithms, assuming the Exponential-Time Hypothesis. Very recently, an algebraic study of *subgraph complementation distance* between two graphs – the minimum number of subgraph complementations required to obtain one graph from the other – has been initiated by Buchanan, Purcell, and Rombach [38].

We study subgraph complementation to H-free graphs, where H is a tree. We come up with a set $\mathcal{T}$ of 41 trees of at most 13 vertices such that if $T \notin \mathcal{T}$, then subgraph complementation to T-free graphs is NP-Complete. Further, we prove that, these hard problems do not admit subexponential-time algorithms, assuming the Exponential-Time Hypothesis. These 41 trees include some paths, stars, bistars (trees with 2 internal vertices), tristars (trees with 3 internal vertices), and some subdivisions of claw. Among these, for four paths (P_ℓ, for $1 \leq \ell \leq 4$), the problem is known to be polynomial-time solvable. So, our result leaves behind only 37 open cases, which are listed in Fig. 1. Additionally, we prove that the problem is hard when H is a 5-connected non-self-complementary prime graph with at least 18 vertices. As a separate result, we obtain that the problem can be solved in polynomial-time when H is a paw (the unique connected graph on 4 vertices having a single triangle).

We use 9 reductions to obtain our results on Subgraph Complementation to T-free graphs (SC-TO-$\mathcal{F}(T)$) - each of them is either from 3-SAT (or from a variant of it) or from SC-TO-$\mathcal{F}(T')$, where T' is an induced subgraph of T. Due to space constraints we have moved all proofs, except that of the main Theorem (Theorem 2) and that of two reductions (in Sect. 3) which are representatives of the two types of reductions that we use. The omitted proofs have been moved to a full version of this paper due to space constraints.

#	Name(s)	Tree	#	Name(s)	Tree	#	Name(s)	Tree	#	Name(s)	Tree
1	P_5 ($T_{1,0,1}$)		11	$T_{2,2}$		20	$T_{1,1,3}$		29	$T_{1,3,2}$	
2	$K_{1,3}$ ($C_{1,1,1}$)		12	$T_{2,3}$		21	$T_{1,1,4}$		30	$T_{1,3,3}$	
3	$K_{1,4}$		13	$T_{2,4}$		22	$T_{1,1,5}$		31	$T_{1,3,4}$	
4	$C_{1,2,3}$		14	$T_{3,3}$		23	$T_{1,2,1}$		32	$T_{1,3,5}$	
5	$C_{1,3,3}$		15	$T_{3,4}$		24	$T_{1,2,2}$		33	$T_{1,4,1}$	
6	$C_{2,2,2}$		16	$T_{4,4}$		25	$T_{1,2,3}$		34	$T_{1,4,2}$	
7	$C_{2,2,3}$		17	$T_{1,0,2}$ ($C_{1,1,3}$)		26	$T_{1,2,4}$		35	$T_{1,4,3}$	
8	$T_{1,2}$ ($C_{1,1,2}$)		18	$T_{1,1,1}$ ($C_{1,2,2}$)		27	$T_{1,2,5}$		36	$T_{1,4,4}$	
9	$T_{1,3}$		19	$T_{1,1,2}$		28	$T_{1,3,1}$		37	$T_{1,4,5}$	
10	$T_{1,4}$										

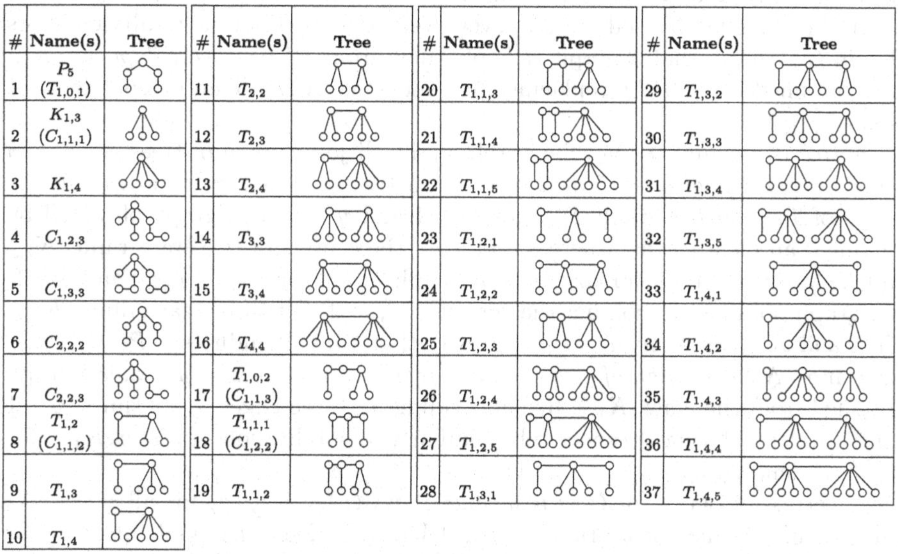

Fig. 1. The trees T for which the complexity of SC-TO-$\mathcal{F}(T)$ is open

2 Preliminaries

For a graph G, the vertex set and edge set are denoted by $V(G)$ and $E(G)$ respectively. A graph G is *H-free* if it does not contain H as an induced subgraph. By $\mathcal{F}(H)$ we denote the class of H-free graphs. The vertex connectivity, $\mathcal{K}(G)$, of a graph G is the minimum number of vertices in G whose removal either causes G disconnected or reduces G to a graph with only one vertex. A graph G is said to be k-connected if $\mathcal{K}(G) \geq k$. By K_n, nK_1, $K_{1,n-1}$, C_n, and P_n, we denote the complete graphs, empty graphs, star graphs, cycles, and paths on

n vertices respectively. A graph G which is isomorphic to its complement $\overline{G}$ is called a *self-complementary* graph. If G is not isomorphic to $\overline{G}$, then it is called *non-self-complementary*. By $G + H$, we denote the *disjoint union* of two graphs G and H. By rG, we denote the disjoint union of r copies of a graph G. For a subset X of vertices of G, by $G - X$ we denote the graph obtained from G by removing the vertices in X.

The open neighborhood of a vertex $v \in V(G)$, denoted by $N(v)$, is the set of all the vertices adjacent to v, and the closed neighborhood of v, denoted by $N[v]$, is defined as $N(v) \cup \{v\}$. Let u be a vertex and X be a vertex subset of G. By $N_X(u)$ and $N_{\overline{X}}(u)$, we denote the neighborhood of u inside the sets X and $V(G) \setminus X$, respectively. We extend the notion of adjacency to sets of vertices as: two sets A and B of vertices of G are *adjacent* (resp., *non-adjacent*) if each vertex of A is adjacent (resp., non-adjacent) to each vertex of B. By *replacing* a vertex u with a graph H in G, we mean the graph obtained by deleting u from G and introducing H, and making every vertex in H adjacent to all neighbors of u in G. We say that a graph H is obtained from H' by *vertex duplication*, if H is obtained from H' by replacing each vertex v_i in H' by an independent set of size $r_i \geq 1$.

A *tree* is a connected acyclic graph, and a disjoint union of trees is called a *forest*. The *internal tree* T' of a tree T is a tree obtained by removing all the leaves of T. A *bistar* graph $T_{x,y}$, for $x \geq 1$ and $y \geq 1$, is a graph obtained by making a and b adjacent, where a and b are the centers of two star graphs $K_{1,x}$ and $K_{1,y}$ respectively. Similarly, *tristar* graph $T_{x,y,z}$, for $x \geq 1$, $y \geq 0$ and $z \geq 1$, is a graph obtained by joining the centers a, b, and c of three star graphs $K_{1,x}$, $K_{1,y}$, and $K_{1,z}$ respectively in such a way that $\{a, b, c\}$ induces a P_3 with b as the center. A *subdivision of claw*, denoted by $C_{x,y,z}$ for $1 \leq x \leq y \leq z$, is a graph obtained from the claw, $K_{1,3}$, by subdividing its three edges $x - 1$ times, $y - 1$ times, and $z - 1$ times respectively. Similarly, a subdivision of a star $K_{1,a}$, for $a \geq 3$, is denoted by $C_{x_1, x_2, \ldots, x_a}$, where $1 \leq x_1 \leq x_2 \leq \ldots \leq x_a$.

A vertex subset X of G is a *module* if $N_{\overline{X}}(u) = N_{\overline{X}}(v)$ for all $u, v \in X$. The *trivial* modules of a graph G are $\emptyset, V(G)$, and all the singletons $\{v\}$ for $v \in V(G)$. A graph is *prime* if it has at least 3 vertices and all its modules are trivial, and *nonprime* otherwise. A nontrivial module M is a *strong module* of a graph G if for every other module M' in G, if $M \cap M' \neq \emptyset$, then either $M \subseteq M'$ or $M' \subseteq M$. A module which induces an independent set is called *independent module* and a module which induces a clique is called *clique module*. Let G be a nonprime graph such that both G and $\overline{G}$ are connected graphs. Then there is a unique partitioning $\mathcal{P}$ of $V(G)$ into maximal strong modules. The quotient graph Q_G of G has one vertex for each set in $\mathcal{P}$ and two vertices in Q_G are adjacent if and only if the corresponding modules are adjacent in G.

In a 3-SAT formula, every clause contains exactly three literals of distinct variables and the objective of the 3-SAT problem is to find whether there exists a truth assignment which assigns TRUE to at least one literal per clause. The Exponential-Time Hypothesis (ETH) and the Sparsification Lemma imply that 3-SAT cannot be solved in subexponential-time, i.e., in time $2^{o(n+m)}$, where n

is the number of variables and m is the number of clauses in the input formula. To prove that a problem does not admit a subexponential-time algorithm, it is sufficient to obtain a linear reduction from a problem known not to admit a subexponential-time algorithm, where a linear reduction is a polynomial-time reduction in which the size of the resultant instance is linear in the size of the input instance. All our reductions are trivially linear and we may not explicitly mention the same. We refer to the book [39] for a detailed description of these concepts. In a k-SAT formula, every clause contains exactly k literals. The objective of the k-SAT$_{\geq 2}$ problem is to find whether there is a truth assignment for the input k-SAT formula such that at least two literals per clause are assigned TRUE. For every $k \geq 4$, by two simple standard linear reductions from 3-SAT to 4-SAT$_{\geq 2}$ and then to k-SAT$_{\geq 2}$, one can prove the hardness of k-SAT$_{\geq 2}$.

Proposition 1 (folklore). *For $k \geq 4$, k-SAT$_{\geq 2}$ is NP-Complete. Further, the problem cannot be solved in time $2^{o(n+m)}$, assuming the ETH.*

By $G \oplus S$, for a graph G and $S \subseteq V(G)$, we denote the graph obtained from G by flipping the adjacency of pairs of vertices in S. The problem that we deal with in this paper is given below.

SC-TO-$\mathcal{F}(H)$: Given a graph G, find whether there is a set $S \subseteq V(G)$ such that $G \oplus S$ is H-free.

Proposition 2 ([11]). *Let T be a path on at least 7 vertices or a star on at least 6 vertices. Then SC-TO-$\mathcal{F}(T)$ is NP-Complete. Further, the problem cannot be solved in time $2^{o(|V(G)|)}$, unless the ETH fails.*

We say that two problems A and B are linearly equivalent, if there is a linear reduction from A to B and there is a linear reduction from B to A.

Proposition 3 ([11]). *SC-TO-$\mathcal{F}(H)$ and SC-TO-$\mathcal{F}(\overline{H})$ are linearly equivalent.*

3   Reductions for General Graphs

In this section, we introduce two reductions which will be used in the next section to prove hardness for SC-TO-$\mathcal{F}(H)$, when H is a tree. We believe that these reductions will be useful in an eventual dichotomy for the problem for general graphs H. The first reduction is a linear reduction from SC-TO-$\mathcal{F}(H')$ to SC-TO-$\mathcal{F}(H)$ where H is obtained from H' by vertex duplication. The second reduction proves that for every 5-connected non-self-complementary prime graph H with a clique or independent set of size 4, SC-TO-$\mathcal{F}(H)$ is NP-Complete and does not admit a subexponential-time algorithm, assuming the ETH.

Graphs with Duplicated Vertices. Here, with the help of a linear reduction, we prove that the hardness results for a prime graph H' translate to that for H, where H is obtained from H' by vertex duplication.

Lemma 1. *Let H' be a prime graph with vertices $V(H') = \{v_1, v_2, \ldots, v_t\}$. Let H be a graph obtained from H' by replacing each vertex v_i in H' by an independent set I_i of size r_i, for some integer $r_i \geq 1$. Then there is a linear reduction from SC-TO-$\mathcal{F}(H')$ to SC-TO-$\mathcal{F}(H)$.*

Let H' and H be graphs mentioned in Lemma 1. Let r be the maximum integer among the r_is, i.e., $r = \max_{i=1}^{i=t} r_i$. We note that H' is the quotient graph of H.

Construction 1. *Given a graph G' and an integer $r \geq 1$, the graph G is constructed from G' as follows: for each vertex u of G', replace u with a set W_u which induces an rK_r. The so obtained graph is G.*

Lemma 2. *If $G' \oplus S' \in \mathcal{F}(H')$ for some $S' \subseteq V(G')$, then $G \oplus S \in \mathcal{F}(H)$, where S is the union of vertices in W_u for every vertex $u \in S'$.*

Proof. Let an H be induced by A (say) in $G \oplus S$. Recall that G is constructed by replacing each vertex u in G' with a module W_u which induces an rK_r. If $A \subseteq W_u$ for some vertex u in G', then H is an induced subgraph of either rK_r (if $u \notin S'$) or $\overline{rK_r}$ (if $u \in S'$). Then H', the quotient graph of H, is either an independent set or a complete graph. This is not true as H' is a prime graph. Therefore, A has nonempty intersection with more than one W_us. For a vertex u in G', either W_u is a subset of S (if $u \in S'$) or W_u has empty intersection with S (if $u \notin S'$). Therefore, if A has nonempty intersection with W_u, then $A \cap W_u$ is a module of the H induced by A. Therefore, $A \cap W_u \subseteq I_i$ for some $1 \leq i \leq t$. Let U_i be the set of vertices u in G' such that I_i (in the H induced by A) has a nonempty intersection with W_u. Arbitrarily choose one vertex from U_i. Let A' be the set of such chosen vertices for all $1 \leq i \leq t$. We claim that A' induces an H' in $G' \oplus S'$. Let u_i and u_j be the vertices chosen for I_i and I_j respectively, for $i \neq j$. Since $A \cap W_{u_i} \subseteq I_i$ and $A \cap W_{u_j} \subseteq I_j$, and $i \neq j$, we obtain that $u_i \neq u_j$. It is enough to prove that u_i and u_j are adjacent in $G' \oplus S'$ if and only if v_i and v_j are adjacent in H'. If u_i and u_j are adjacent in $G' \oplus S'$, then W_{u_i} and W_{u_j} are adjacent in $G \oplus S$. This implies that I_i and I_j are adjacent in H. Hence v_i and v_j are adjacent in H'. For the converse, assume that v_i and v_j are adjacent in H'. This implies that I_i and I_j are adjacent in H. Therefore, W_{u_i} and W_{u_j} are adjacent in $G \oplus S$. Hence u_i and u_j are adjacent in $G' \oplus S'$.

Lemma 3. *If $G \oplus S \in \mathcal{F}(H)$ for some $S \subseteq V(G)$, then $G' \oplus S' \in \mathcal{F}(H')$, where S' is a subset of vertices of G' obtained in such a way that whenever all vertices of a K_r from a module W_u (which induces an rK_r) are in S, then the corresponding vertex u in G' is included in S'.*

Proof. Suppose $G' \oplus S'$ contains an H' induced by a set $A' = \{v_1, v_2, \ldots, v_t\}$. If a vertex u in G' is in S', then all vertices of a K_r from W_u in G are in S. Therefore, there is an independent set of size r in $W_u \cap S$ in $G \oplus S$. Similarly, if $u \notin S'$, then there is an independent set of size r in $W_u \setminus S$ in $G \oplus S$ formed by one vertex, which is not in S, from each copy of K_r in W_u which is not in S. We construct A as follows: for each vertex $v_i \in A'$, if $v_i \in S'$, include in A an independent set $I_i \subseteq W_{v_i} \cap S$ such that $|I_i| = r_i$, and if $v_i \notin S'$, include in A an independent set $I_i \subseteq W_{v_i} \setminus S$ such that $|I_i| = r_i$. We claim that A induces an H in $G \oplus S$. Note that each chosen I_i is a module in $G \oplus S$. Since $I_i \subseteq S$ if and only if $v_i \in S'$, we obtain that I_i and I_j are adjacent in $G \oplus S$ if and only if v_i and v_j are adjacent in the H' induced by A'. This completes the proof.

Lemma 1 follows directly from Lemma 2 and 3. When the lemma is applied on trees, we get the following corollary. We note that the quotient tree Q_T of a tree is prime if and only if T is not a star graph.

Corollary 1. *Let T be a tree which is not a star graph, and let Q_T be its quotient tree. Then there is a linear reduction from* SC-TO-$\mathcal{F}(Q_T)$ *to* SC-TO-$\mathcal{F}(T)$.

5-Connected Graphs. Here, we obtain hardness results for SC-TO-$\mathcal{F}(H)$, where H is a 5-connected graph satisfying some additional constraints.

Theorem 1. *Let H be a 5-connected non-self-complementary prime graph with an independent set of size 4 or with a clique of size 4. Then* SC-TO-$\mathcal{F}(H)$ *is NP-Complete. Further, the problem cannot be solved in time $2^{o(|V(G)|)}$, unless the ETH fails.*

Let H be a 5-connected graph satisfying the constraints mentioned in Theorem 1. Let H have t vertices and let $V' \subseteq V(H)$ induce either a K_4 or a $4K_1$ in H. We use Construction 2 for a reduction from 4-SAT$_{\geq 2}$ to prove Theorem 1.

Construction 2. *Let Φ be a 4-SAT formula with n variables $X_1, X_2, \cdots, X_n$, and m clauses $C_1, C_2, \cdots, C_m$. We construct the graph G_Φ as follows. For each variable X_i in Φ, the variable gadget also named as X_i consists of the union of two special sets $X_{i1} = \{x_i\}$ and $X_{i2} = \{\overline{x_i}\}$, and $t - 2$ other sets $X_{i3}, X_{i4} \ldots X_{it}$ such that each X_{ij}, for $3 \leq j \leq t$ induces an $\overline{H}$. Make the adjacency between these $X_{ij}s$ in such a way that taking one vertex each from these sets induces an H, where X_{i1} and X_{i2} correspond to two non-adjacent vertices, if V' forms a K_4, and correspond to two adjacent vertices, if V' forms a $4K_1$. If V' forms a clique then add an edge between X_{i1} and X_{i2}, and if V' forms an independent set, then remove the edge between X_{i1} and X_{i2}. The vertices $x_i s$ and $\overline{x_i}s$ are called literal vertices denoted by a set L, which induces a clique, if V' is a clique, and induces an independent set, if V' is an independent set.*

For each clause C_i of the form $(\ell_{i1} \vee \ell_{i2} \vee \ell_{i3} \vee \ell_{i4})$ in Φ, the clause gadget also named as C_i consists of $t - 4$ copies of $\overline{H}$ denoted by C_{ij}, for $1 \leq j \leq (t - 4)$. Let the four vertices introduced (in the previous step) for the literals $\ell_{i1}, \ell_{i2}, \ell_{i3}$, and ℓ_{i4} be denoted by $L_i = \{y_{i1}, y_{i2}, y_{i3}, y_{i4}\}$. The adjacency among each of these $C_{ij}s$ and the literal vertices L_i is in such a way that, taking one vertex from each $C_{ij}s$ and the vertices in L_i induces an H. This completes the construction.

An example of the construction is shown in Fig. 3 for a graph H given in Fig. 2. Keeping a module isomorphic to $\overline{H}$ guarantees that not all vertices in the module is present in a solution S of G_Φ (i.e., $G_\Phi \oplus S$ is H-free). The purpose of variable gadget X_i is to make sure that both x_i and $\overline{x_i}$ are not placed in a solution S, so that we can assign TRUE to all literals corresponding to literal vertices placed in S, to get a valid truth assignment for Φ. On the other hand, any truth assignment assigning TRUE to at least two literals per clause makes sure that the set S formed by choosing literal vertices corresponding to TRUE literals destroys copies of H formed by clause gadgets C_i and the corresponding sets L_i of literal vertices. Now, to prove Theorem 1, we use Lemma 4 and Lemma 5.

Fig. 2. An example of a 5-connected non-self-complementary prime graph with a K_4 (formed by the lower four vertices)

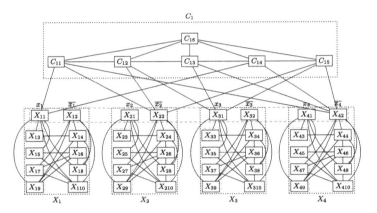

Fig. 3. An example of Construction 2 for the formula $\Phi = C_1$ where $C_1 = x_1 \vee \overline{x_2} \vee x_3 \vee \overline{x_4}$ corresponding to the graph H shown in Fig. 2 with a K_4. The lines connecting two rectangles indicate that each vertex in one rectangle is adjacent to all vertices in the other rectangle. If there is no line shown between two rectangles, then the vertices in them are non-adjacent, with the exceptions – (i) all the vertices in a red rectangle (dashed) together form a clique; (ii) the rectangles in each green rectangle (dashed) are adjacent.

Lemma 4. *Let Φ be a yes-instance of 4-SAT$_{\geq 2}$ and ψ be a truth assignment satisfying Φ. Then $G_\Phi \oplus S$ is H-free where S is the set of literal vertices whose corresponding literals were assigned TRUE by ψ.*

Proof. Let $G_\Phi \oplus S$ contain an H induced by A (say). Since H is a prime graph and $\overline{H}$ is not isomorphic to H, $|A \cap Y| \leq 1$ where Y is a module isomorphic to $\overline{H}$. Thus, $|A \cap X_{ij}|$ is at most one. Therefore, since $\{x_i, \overline{x_i}\}$ is not a subset of S, we obtain that X_i does not have an induced H in $G_\Phi \oplus S$. Recall that, the vertices in X_{ij} (for $3 \leq j \leq t$) are non-adjacent to $V(G) \setminus X_i$, and H is 5-connected. This implies that $A \cap (X_i \setminus \{x_i, \overline{x_i}\}) = \emptyset$.

Since C_i contains $t - 4$ sets of $\overline{H}$s, $|C_i \cap A| \leq t - 4$. Now assume that A contains vertices from two clause gadgets C_i and C_j. Since the vertices in C_i are only adjacent to the four literal vertices corresponding to the clause C_i, and H is 5-connected, removing the four literal vertices corresponding to C_i disconnects the graph which is not possible –note that C_i and C_j are non-adjacent. Hence, A contains vertices from at most one clause gadget C_i.

Note that L induces a $K_n \times nK_1$ in $G_\Phi \oplus S$, if V' induces a clique, and induces a $K_n + nK_1$ in $G_\Phi \oplus S$, if V' induces an independent set. Therefore, H is not an induced subgraph of the graph induced by L in $G_\Phi \oplus S$. Recall that the vertices in $A \cap C$ are from at most one clause gadget C_i, and at most one vertex from each of the sets C_{ij} in C_i is in $A \cap C_i$. We know that C_i is non-adjacent to all literal vertices corresponding to the literals not in the clause C_i, and H is 5-connected. Therefore, $A \cap L = \{y_{i1}, y_{i2}, y_{i3}, y_{i4}\}$. Since at least two vertices in $A \cap L$ are in S, the graph induced by A in $G \oplus S$ is not isomorphic to H.

Lemma 5. *Let Φ be an instance of 4-SAT$_{\geq 2}$. If $G_\Phi \oplus S$ is H-free for some $S \subseteq V(G_\Phi)$, then there exists a truth assignment satisfying Φ.*

Proof. Let $G_\Phi \oplus S$ be H-free for some $S \subseteq V(G_\Phi)$. We want to find a satisfying truth assignment of Φ. Since each of the C_{ij}s in C_i, for $1 \leq i \leq m$ and $1 \leq j \leq t - 4$, induces an $\overline{H}$, there is at least one vertex in each C_{ij} which is not in S. Then, if at least two vertices from L_i are not in S, then there is an induced H by vertices in L_i and one vertex each from $C_{ij} \setminus S$, for $1 \leq j \leq t - 4$. Therefore, at least two vertices from L_i are in S. Next we prove that $\{x_i, \overline{x_i}\}$ is not a subset of S. For each X_{ij} (for $3 \leq j \leq t$), since each of them induces an $\overline{H}$, at least one vertex is not in S. Then, if both x_i and $\overline{x_i}$ are in S, then there is an H induced by $x_i, \overline{x_i}$, and one vertex each from $X_{ij} \setminus S$, for $3 \leq j \leq t$. Now, it is straight-forward to verify that assigning TRUE to every literal x_i such that $x_i \in S$, is a valid satisfying truth assignment of Φ.

4   Trees

By $\mathcal{T}$ we denote the set $\mathcal{P} \cup \mathcal{T}_1 \cup \mathcal{T}_2 \cup \mathcal{T}_3 \cup \mathcal{C}$, where $\mathcal{P} = \{P_x \mid 1 \leq x \leq 5\}$, $\mathcal{T}_1 = \{K_{1,x} \mid 1 \leq x \leq 4\}$, $\mathcal{T}_2 = \{T_{x,y} \mid 1 \leq x \leq y \leq 4\}$, $\mathcal{T}_3 = \{T_{1,0,1}, T_{1,0,2}\} \cup \{T_{x,y,z} \mid x = 1, 1 \leq y \leq 4, 1 \leq z \leq 5\}$, and $\mathcal{C} = \{C_{1,1,1}, C_{1,1,2}, C_{1,1,3}, C_{1,2,2}, C_{1,2,3}, C_{1,3,3}, C_{2,2,2}, C_{2,2,3}\}$. These sets denote the paths, stars, bistars, tristars, and subdivisions of claw not handled by our reductions.

We note that $|\mathcal{P}| = 5$, $|\mathcal{T}_1| = 4$, $|\mathcal{T}_2| = 10$, $|\mathcal{T}_3| = 22$, and $|\mathcal{C}| = 8$. However, a star graph $K_{1,x}$ is a path in $\mathcal{P}$ if $x \leq 2$, the bistar graph $T_{1,1}$ is the path P_4, the tristar graphs $T_{1,0,1}$ is the path P_5, and the subdivision of claw $C_{1,1,1}$ is the star graph $K_{1,3}$, $C_{1,1,2}$ is the bistar graph $T_{1,2}$, $C_{1,1,3}$ is the tristar graph $T_{1,0,2}$, and $C_{1,2,2}$ is the tristar graph $T_{1,1,1}$. Therefore, $|\mathcal{T}| = 41$, and the tree of maximum order in $\mathcal{T}$ is $T_{1,4,5}$ with 13 vertices. We prove the following theorem in this section, which is the main result of the paper.

Theorem 2. *Let T be a tree not in $\mathcal{T}$. Then* SC-TO-$\mathcal{F}(T)$ *is NP-Complete. Further, the problem cannot be solved in time $2^{o(|V(G)|)}$, unless the ETH fails.*

This task is achieved in three parts. In the first part, we give two general reductions for trees. First we prove that there is a linear reduction from SC-TO-$\mathcal{F}(T')$ to SC-TO-$\mathcal{F}(T)$, where T is a prime tree and T' is its internal tree. Then we deal with trees with at least 4 leaves and at least 3 internal vertices, and satisfying some additional constraints. In the second part, we deal with various subclasses of trees - bistars, tristars, paths, and subdivisions of claw. We combine all these results in the third part to prove Theorem 2.

General Reductions for Trees. Here, with a very simple reduction, we obtain that the hardness transfers from T' to T, where T is a prime tree and T' is its internal tree.

Lemma 6. *Let T be a prime tree and let T' be its internal tree. Then there is a linear reduction from* SC-TO-$\mathcal{F}(T')$ *to* SC-TO-$\mathcal{F}(T)$.

Next, we obtain hardness results for SC-TO-$\mathcal{F}(T)$, when T is a tree with at least 4 leaves and at least 3 internal vertices, and satisfying some additional constraints. The reduction is from k-SAT$_{\geq 2}$, where k is the number of leaves in T.

Theorem 3. *Let T be a tree with at least 4 leaves and at least 3 internal vertices. Let T' be the internal tree of T. Assume that the following properties are satisfied.*

(i) If T' is a star graph, then at least one of the following conditions are satisfied:
  (a) every leaf of T' has at least two leaves of T as neighbors, or
  (b) the center of the star T' has no leaf of T as neighbor, or
  (c) T is either a $C_{1,2,2,2}$, or a $C_{1,2,2,2,2}$.
(ii) There are no two adjacent vertices of degree 2 in T such that neither of them is adjacent to any leaf of T.

Then SC-TO-$\mathcal{F}(T)$ *is NP-Complete. Further, the problem cannot be solved in time $2^{o(|V(G)|)}$, unless the ETH fails.*

Corollary 2. *Let x, y, z be integers such that $1 \leq x \leq z$, $y \geq 0$ and either of the following conditions is satisfied.*

(i) $x = 1, y = 0, z \geq 3$, or
(ii) $x \geq 2$

Then SC-TO-$\mathcal{F}(T_{x,y,z})$ *is NP-Complete. Further, the problem cannot be solved in time $2^{o(|V(G)|)}$, unless the ETH fails.*

Bistars, Tristars, Paths, and Subdivisions of Claw. It turns out that, given the general reductions obtained in Sect. 3 and the general reductions for trees, it is sufficient to handle the following subclasses of trees - paths, stars, bistars, tristars, and subdivisions of claw. By Proposition 2, we know that SC-TO-$\mathcal{F}(P_t)$ is hard for $t \geq 7$. Here, we resolve the case of P_6, which is in fact essential to reduce the unsolved cases to a finite set. Stars, except for claw and $K_{1,4}$, have already been handled in [11] - See Proposition 2. Here we prove the hardness for bistars and tristars, except for 10 bistars and 22 tristars. We also resolve the cases of subdivisions of claw, except for 8 subdivisions of claw. These results are listed below. Theorem 4 along with Proposition 2 implies Corollary 3, and a linear reduction from SC-TO-$\mathcal{F}(K_{1,y})$ is used to prove the hardness of SC-TO-$\mathcal{F}(T_{x,y})$ in Theorem 5.

Theorem 4. SC-TO-$\mathcal{F}(P_6)$ *is NP-Complete. Further, the problem cannot be solved in time* $2^{o(|V(G)|)}$, *unless the ETH fails.*

Corollary 3. *Let* $\ell \geq 6$ *be an integer. Then* SC-TO-$\mathcal{F}(P_\ell)$ *is NP-Complete. Further, the problem cannot be solved in time* $2^{o(|V(G)|)}$, *unless the ETH fails.*

Theorem 5. *Let* x, y *be two integers such that* $1 \leq x \leq y$ *and* $y \geq 5$. *Then* SC-TO-$\mathcal{F}(T_{x,y})$ *is NP-Complete. Further, the problem cannot be solved in time* $2^{o(|V(G)|)}$, *unless the ETH fails.*

Recall that, as a corollary (Corollary 2) of the theorem (Theorem 3) which deals with trees with at least 4 leaves and three internal vertices, we have resolved some cases of tristar graphs: we proved that SC-TO-$\mathcal{F}(T_{x,y,z})$ is hard if $z \geq x \geq 2$ or if $x = 1, y = 0, z \geq 3$. In this section, we handle the rest of the cases when $x = 1$ and $y \geq 1$, except for a finite number of cases. Firstly, a linear reduction from SC-TO-$\mathcal{F}(T_{y,z-1})$ is used to prove the hardness of SC-TO-$\mathcal{F}(T_{x,y,z})$. This will take care of the cases when $y \geq 5$ or $z \geq 6$ (recall that the problem for $T_{y,z-1}$ is hard if $y \geq 5$ or $z \geq 6$). But, for the reduction to work, there is an additional constraint that $z \geq 3$. So, to handle the case when $z \leq 3$, we use another reduction which is from SC-TO-$\mathcal{F}(K_{1,y})$ and does not have any constraint on z. Thus, we have the following result, which along with Corollary 2 implies Theorem 6.

Lemma 7. *Let* $1 \leq x \leq z$, *and* $y \geq 0$ *be integers such that* $y \geq 5$ *or* $z \geq 6$. *Then* SC-TO-$\mathcal{F}(T_{x,y,z})$ *is NP-Complete. Further, the problem cannot be solved in time* $2^{o(|V(G)|)}$, *unless the ETH fails.*

Theorem 6. *Let* $1 \leq x \leq z$ *and* $y \geq 0$ *be integers such that at least one of the following conditions is satisfied: (i)* $x \geq 2$, *or (ii)* $y \geq 5$, *or (iii)* $z \geq 6$, *or (iv)* $x = 1, y = 0, z \geq 3$. *Then* SC-TO-$\mathcal{F}(T_{x,y,z})$ *is NP-Complete. Further, the problem cannot be solved in time* $2^{o(|V(G)|)}$, *unless the ETH fails.*

A subdivision of a claw has exactly three leaves. Due to this, we cannot handle them using the reduction used to handle trees with 4 leaves (Theorem 3). Let $T = C_{x,y,z}$ be a subdivision of claw, where $x \leq y \leq z$. If $x = y = 1$, then T is obtained from P_{z+2} by duplicating a leaf. Therefore, we can use Lemma 1 and Corollary 3 to prove the hardness, when $z \geq 4$. If $y > 1$, then T is prime and if T has at least 9 vertices, then $\overline{T}$ is 5-connected and has an independent set of size 4. Then the hardness results for 5-connected prime graphs (Theorem 1) can be used to prove the hardness for $\overline{T}$ and hence for T (Proposition 3). But, there is a particular subdivision of claw, $C_{1,2,4}$, which is not handled by any of these reductions. Further, there is an infinite family of trees, which is obtained by duplicating the leaf adjacent to the center of the claw in $C_{1,2,4}$, not handled by Theorem 3, as each tree in the family violates condition (ii) of Theorem 3. This requires us to handle $C_{1,2,4}$ separately.

Theorem 7. SC-TO-$\mathcal{F}(C_{1,2,4})$ *is NP-Complete. Further, the problem cannot be solved in time* $2^{o(|V(G)|)}$, *unless the ETH fails.*

We observe that, for any integer $t \geq 4$, the subdivision of claw $C_{1,1,t-2}$ is obtained by introducing a false-twin for a leaf of a P_t. Then, Observation 8 follows directly from Lemma 1.

Observation 8. There is a linear reduction from SC-TO-$\mathcal{F}(P_t)$ to SC-TO-$\mathcal{F}(C_{1,1,t-2})$.

Theorem 9. *Let* $x \leq y \leq z$ *be integers such that at least one of the following conditions are satisfied.*

(i) $x = 1, y = 2, z = 4$, *or*
(ii) $x = y = 1$, *and* $z \geq 4$, *or*
(iii) $x + y + z \geq 8$.

Then SC-TO-$\mathcal{F}(C_{x,y,z})$ *is NP-Complete. Further, the problem cannot be solved in time* $2^{o(|V(G)|)}$, *unless the ETH fails.*

Corollary 4. *Let* T *be a subdivision of claw not in* $\{C_{1,1,1}, C_{1,1,2}, C_{1,1,3}, C_{1,2,2}, C_{1,2,3}, C_{1,3,3}, C_{2,2,2}, C_{2,2,3}\}$. *Then* SC-TO-$\mathcal{F}(T)$ *is NP-Complete. Further, the problem cannot be solved in time* $2^{o(|V(G)|)}$, *unless the ETH fails.*

Putting Them Together. Here, we prove the main result (Theorem 2) of this paper by using the results proved so far. The proof makes use of the following observation, which essentially says that if a prime tree T does not satisfy condition (ii) of Theorem 3, then T is handled by Theorem 1, Corollary 3, or Theorem 7.

Observation 10. Let T be a prime tree such that there are two adjacent internal vertices u, v which are not adjacent to any leaf of T. Then either of the following conditions is satisfied.

(i) T has an independent set of size 4 and $\overline{T}$ is 5-connected, or
(ii) T is either a P_6, or a P_7, or the subdivision of claw $C_{1,2,4}$.

Now, we are ready to prove the main result of the paper.

Proof. (Proof of Theorem 2). Let p be the number of internal vertices of T. If $p = 1$, then T is a star graph and the statements follow from Proposition 2. If $p = 2$, then T is a bistar graph and the statements follow from Theorem 5. If $p = 3$, then T is a tristar graph and the statements follow from Theorem 6. Assume that $p \geq 4$. If T has only two leaves, then T is isomorphic to P_ℓ, for

$\ell \geq 6$. Then the statements follow from Corollary 3. If T has exactly three leaves, then T is a subdivision of claw. Then the statements follow from Corollary 4. Assume that T has at least four leaves.

Let Q_T be the quotient tree of T. If Q_T has two adjacent internal vertices which are not adjacent to any leaves, then by Observation 10, either (i) Q_T has an independent set of size 4 and $\overline{Q_T}$ is 5-connected or (ii) Q_T is either a P_6, or a P_7, or a $C_{1,2,4}$. If (i) is true, then by Theorem 1, SC-TO-$\mathcal{F}(\overline{Q_T})$ is NP-Complete and cannot be solved in subexponential-time (assuming the ETH). Then, so is for SC-TO-$\mathcal{F}(Q_T)$, by Proposition 3. Then the statements follow from Lemma 1. If (ii) is true, then SC-TO-$\mathcal{F}(Q_T)$ is hard by Corollary 3 and Theorem 7. Then the statements follow from Lemma 1. Therefore, assume that Q_T has no two adjacent internal vertices not adjacent to any leaves of Q_T. Hence, T has no two adjacent internal vertices not adjacent to any leaves of T. Then, if T', the internal tree of T, is not a star graph, then the statements follow from Theorem 3. Assume that T' is a star graph. If the condition (i) of Theorem 3 is satisfied, then we are done. Assume that the condition (i) of Theorem 3 is not satisfied, i.e., the center of T' has at least one leaf of T as a neighbor, one leaf of T' has exactly one leaf of T as a neighbor, and T is neither $C_{1,2,2,2}$ nor $C_{1,2,2,2,2}$. Assume that T has exactly 4 internal vertices. Then T' is a claw and Q_T is $C_{1,2,2,2}$. Then by Theorem 3, SC-TO-$\mathcal{F}(Q_T)$ is NP-Complete and cannot be solved in subexponential-time (assuming the ETH). Then the statements follow from Corollary 1. Similarly, when T has exactly 5 internal vertices, we obtain that Q_T is $C_{1,2,2,2,2}$ and then the statements follow from Theorem 3 and Corollary 1. Assume that T has at least 6 internal vertices. Then, T' is a $K_{1,a}$, for some $a \geq 5$. Then by Lemma 6, there is a linear reduction from SC-TO-$\mathcal{F}(K_{1,a})$ to SC-TO-$\mathcal{F}(Q_T)$. By Proposition 2, SC-TO-$\mathcal{F}(K_{1,a})$ is NP-Complete and cannot be solved in subexponential-time (assuming the ETH). Then the statements follow from Corollary 1.

5   Polynomial-Time Algorithm

In this section, we prove that SC-TO-$\mathcal{F}(paw)$ can be solved in polynomial-time, where paw is the unique graph on 4 vertices having exactly one vertex with degree 1. We use a result by Olariu [3] that every component of a paw-free graph is either triangle-free or complete mutitipartite.

A graph is complete multipartite if and only if it does not contain any $K_2 + K_1$ as an induced subgraph. It is known that SC-TO-$\mathcal{F}(K_3)$ and SC-TO-$\mathcal{F}(K_2 + K_1)$ can be solved in polynomial-time. The former is proved in [37] and the latter is implied by another result from [37] that Subgraph Complementation problems admit polynomial-time algorithms if the target graph class is expressible in MSO_1 and has bounded clique-width.

Algorithm for SC-TO-$\mathcal{F}(paw)$

Input: A graph G.

Output: If G is a yes-instance of SC-TO-$\mathcal{F}(paw)$, then returns YES; returns NO otherwise.

Step 1 : If G is paw-free, then return YES.

Step 2 : If G is a yes instance of SC-TO-$\mathcal{F}(K_3)$, or a yes-instance of SC-TO-$\mathcal{F}(K_2 + K_1)$, then return YES.

Step 3 : For every triangle uvw in G, if $(N(u) \cap N(v)) \cup (N(u) \cap N(w)) \cup (N(v) \cap N(w))$ is a solution, then return YES.

Step 4 : For every ordered pair of adjacent vertices (u, v), do the following:
  (i) Compute R_u and R_v, the lists of component partitions of $N(u) \setminus N[v]$ and $N(v) \setminus N[u]$ respectively.
  (ii) For every (X_u, Y_u) in R_u, and for every (X_v, Y_v) in R_v, if $Y_u \cup Y_v \cup (N[u] \cap N[v])$ is a solution, then return YES.
  (iii) Let N_{uv} be $N(u) \cap N(v)$.
  (iv) For every (X_v, Y_v) in R_v, and for every subset S_1' of N_{uv} such that $|S_1'| \geq |N_{uv}| - 2$, and for every set V_2' of at most three mutually non-adjacent vertices in G, and for every set S_2' of at most three mutually adjacent vertices in G, do the following:
     (a) $S_1'' = Y_v \cup S_1' \cup \{u\}$
     (b) $V_2'' = X_v \cup V_2'$
     (c) Let Z_2 be the set of vertices such that every vertex in Z_2 is adjacent to every vertex in S_1'' and at least one vertex in V_2''.
     (d) $S_2'' = S_2' \cup Z_2$
     (e) If $S_1'' \cup S_2''$ is a solution, then return YES.
     (f) Let S_2'' be the set of vertices part of clique components K in the graph $G - S_1''$ such that every vertex in K is adjacent to every vertex S_1''.
     (g) If $S_1'' \cup S_2''$ is a solution, then return YES.

Step 5 : Return NO.

Let G be an input graph of SC-TO-$\mathcal{F}(paw)$. We define a *component partition* of a graph G as a partition of its vertices into two sets P, Q such that P induces a single component or an independent set of size at most 3, and Q contains the remaining vertices. We observe that all component partitions of a graph can be found in polynomial-time.

Before running the algorithm, we preprocess the input graph by removing paw-free components and by replacing independent (clique) modules of size at least 4 in G by independent (clique) modules of size 3. The correctness of the former rule is trivial and that of the latter rule follows from the fact that a paw does not contain an independent or clique module of size 3. Therefore, we can assume that the input graph does not have any paw-free components and that there are no independent (clique) modules of size at least 4 present in the graph.

The algorithm works as follows. First we check whether G is a trivial yes-instance by checking whether G is a paw-free graph or not (Step 1). Next we check whether G can be transformed into a triangle-free graph or a complete multipartite graph by Subgraph Complementation, using the algorithms from [37] (Step 2). If yes, then the instance is a yes-instance and we are done. If not, then every solution of G transforms G into a graph having multiple components, at least one of it is guaranteed to be a complete multipartite component (disjoint union of two triangle-free graphs is triangle-free). Let S be a solution of G and let $G_1, G_2, \ldots, G_t$ be the components of $G \oplus S$. Let S_i be $S \cap V(G_i)$ and V_i be $V(G_i) \setminus S_i$, for $1 \leq i \leq t$. Clearly, S intersects with each G_i, otherwise G_i is a paw-free component in G, which contradicts with our assumption that G has no paw-free components. Further, every vertex in S_i is adjacent to every vertex in S_j for $i \neq j$. Assume that $G \oplus S$ has at least three components. Let $u \in S_1, v \in S_2$, and $w \in S_3$. Then, S is the union of common neighbors of u and v, common neighbors of u and w, and the common neighbors of v and w (Step 3). The only case left is when $t = 2$, i.e., $G \oplus S$ has exactly two components for every solution S. Let $u \in S_1$ and $v \in S_2$. Assume that G_1 is a complete multipartite component. Consider $W = N(u) \setminus N[v]$. The vertices in $W \cap V_1$ is not adjacent to the vertices in $W \cap S_2$. Further, since $W \cap V_1$ induces a complete multipartite graph, we obtain that W induces either a connected graph or an independent set of size at most 3 (every independent set of the complete multipartite graph induced by V_1 is a module in G and hence cannot have size more than 3). Therefore, by enumerating all component partitions (X_u, Y_u) of the graph induced by W, we obtain the one in which $X_u = W \cap V_1$ and $Y_u = W \cap S_2$.

As we noted already, at least one component must be complete multipartite. Then there are two cases to consider - in the first case both G_1 and G_2 are complete multipartite graphs and in the second case G_1 is triangle-free and G_2 is complete multipartite. The algorithm returns YES in Step 4(ii) in the first case and in Step 4(vi) in the second case.

Theorem 11. SC-TO-$\mathcal{F}(paw)$ *can be solved in polynomial-time.*

6   Concluding Remarks

In this paper, we resolved the computational complexity of SC-TO-$\mathcal{F}(T)$, for all trees T, except for 37 trees listed in Fig. 1. Among these open cases, we would like to highlight the tree $C_{1,2,2}$. If we can prove that SC-TO-$\mathcal{F}(C_{1,2,2})$ is hard, then the list of open cases reduces to 17 trees, i.e., all the trees numbered 18 to 37 in the list vanishes due to Corollary 1. The tree resisted all our attempts to cut it down. Among other open cases, we believe that those with 5 vertices ($P_5, K_{1,4}$, and $T_{1,2}$) are the most challenging - we do not have any result so far on non-trivial 5-vertex graphs. The case of P_5 was stated as an open problem in [37]. We also believe that the claw may admit a polynomial-time algorithm, similar to paw - the difficulty in getting such a result seems to reside in the intricacies of the structure theorem for claws.

To get a complete P versus NP-Complete dichotomy for SC-TO-$\mathcal{F}(H)$, for general graphs H, one major hurdle is to tackle the graphs which are self-complementary. Introducing $\overline{H}$ in a reduction for SC-TO-$\mathcal{F}(H)$ helps us to make sure that at least one vertex in a set of vertices is untouched by any solution. We find it very difficult to find alternate reductions which do not use $\overline{H}$ – a reason why we do not have hardness results so far for any self-complementary graphs.

References

1. Chudnovsky, M., Seymour, P.D.: The structure of claw-free graphs. Surv. Combinat. **327**, 153–171 (2005)
2. Corneil, D.G., Lerchs, H., Stewart Burlingham, L.: Complement reducible graphs. Discret. Appl. Math. **3**(3), 163–174 (1981)
3. Olariu, S.: Paw-fee graphs. Inf. Process. Lett. **28**(1), 53–54 (1988)
4. Minty, G.J.: On maximal independent sets of vertices in claw-free graphs. J. Comb. Theo. Ser. B **28**(3), 284–304 (1980)
5. Gartland, P., Lokshtanov, D.: Independent set on P_k-free graphs in quasi-polynomial time. In: Irani, S., (ed.) 61st IEEE Annual Symposium on Foundations of Computer Science, FOCS 2020, Durham, NC, USA, November 16–19, 2020, pp. 613–624. IEEE (2020)
6. Lokshantov, D., Vatshelle, M., Villanger, Y.: Independent set in P_5-free graphs in polynomial time. In: Chekuri, C., (ed) Proceedings of the Twenty-Fifth Annual ACM-SIAM Symposium on Discrete Algorithms, SODA 2014, Portland, Oregon, USA, January 5–7, 2014, pp. 570–581. SIAM (2014)
7. Pilipczuk, M., Pilipczuk, M., Rzążewski, P.: Quasi-polynomial-time algorithm for independent set in P_t-free graphs via shrinking the space of induced paths. In: Viet Le H., King, V., (eds.), 4th Symposium on Simplicity in Algorithms, SOSA 2021, Virtual Conference, January 11–12, 2021, pp. 204–209. SIAM (2021)
8. Grzesik, A., Klimosová, T., Pilipczuk, M., Pilipczuk, M.: Polynomial-time algorithm for maximum weight independent set on P_6-free graphs. ACM Trans. Algor. **18**(1), 4:1-4:57 (2022)
9. Aravind, N.R., Sandeep, R.B., Sivadasan, N.: Dichotomy results on the hardness of H-free edge modification problems. SIAM J. Discret. Math. **31**(1), 542–561 (2017)
10. Yannakakis, M.: Edge-deletion problems. SIAM J. Comput. **10**(2), 297–309 (1981)
11. Antony, D., Garchar, J., Pal, S., Sandeep, R.B., Sen, S., Subashini, R.: On subgraph complementation to H-free graphs. In: Kowalik, Ł., Pilipczuk, M., Rzążewski, P. (eds.) WG 2021. LNCS, vol. 12911, pp. 118–129. Springer, Cham (2021). https://doi.org/10.1007/978-3-030-86838-3_9
12. Alon, N., Stav, U.: Hardness of edge-modification problems. Theor. Comput. Sci. **410**(47–49), 4920–4927 (2009)
13. Brügmann, D., Komusiewicz, C., Moser, H.: On generating triangle-free graphs. Electron. Notes Discret. Math. **32**, 51–58 (2009)
14. El-Mallah, E.S., Colbourn, C.J.: The complexity of some edge deletion problems. IEEE Trans. Circ. Syst. **35**(3), 354–362 (1988)
15. Komusiewicz, C., Uhlmann, J.: Cluster editing with locally bounded modifications. Discret. Appl. Math. **160**(15), 2259–2270 (2012)
16. Shamir, R., Sharan, R., Tsur, D.: Cluster graph modification problems. Discret. Appl. Math. **144**(1–2), 173–182 (2004)
17. Sharan, R.: Graph modification problems and their applications to genomic research. PhD thesis, Tel-Aviv University (2002)
18. Jelínková, E., Kratochvíl, J.: On switching to H-free graphs. J. Graph Theo. **75**(4), 387–405 (2014)

19. Hage, J., Harju, T., Welzl, E.: Euler graphs, triangle-free graphs and bipartite graphs in switching classes. Fundam. Informat. **58**(1), 23–37 (2003)
20. Hayward, R.B.: Recognizing P_3-structure: a switching approach. J. Comb. Theo. Ser. B **66**(2), 247–262 (1996)
21. Hertz, A.: On perfect switching classes. Discret. Appl. Math. **94**(1–3), 3–7 (1999)
22. Kratochvíl, J., Nešetřil, J., Zýka, O.:. On the computational complexity of Seidel's switching. In: Annals of Discrete Mathematics, vol. 51, pp. 161–166. Elsevier (1992)
23. Gramm, J., Guo, J., Hüffner, F., Niedermeier, R.: Graph-modeled data clustering: fixed-parameter algorithms for clique generation. In: Petreschi, R., Persiano, G., Silvestri, R. (eds.) CIAC 2003. LNCS, vol. 2653, pp. 108–119. Springer, Heidelberg (2003). https://doi.org/10.1007/3-540-44849-7_17
24. Drange, P.: Parameterized graph modification algorithms. PhD thesis (2015)
25. Drange, P.G., Dregi, M., Sandeep, R.B.: Compressing bounded degree graphs. In: Kranakis, E., Navarro, G., Chávez, E. (eds.) LATIN 2016. LNCS, vol. 9644, pp. 362–375. Springer, Heidelberg (2016). https://doi.org/10.1007/978-3-662-49529-2_27
26. Marx, D., Sandeep, R.B.: Incompressibility of H-free edge modification problems: towards a dichotomy. J. Comput. Syst. Sci. **125**, 25–58 (2022)
27. Kratsch, S., Wahlström, M.: Two edge modification problems without polynomial kernels. Discret. Optim. **10**(3), 193–199 (2013)
28. Cai, L., Cai, Y.: Incompressibility of H-free edge modification problems. Algorithmica **71**(3), 731–757 (2015)
29. Cai, Y.: Polynomial kernelisation of H-free edge modification problems. Mphil thesis, Department of Computer Science and Engineering, The Chinese University of Hong Kong, Hong Kong SAR, China (2012)
30. Cao, Y., Chen, J.: Cluster editing: kernelization based on edge cuts. Algorithmica **64**(1), 152–169 (2012)
31. Cao, Y., Rai, A., Sandeep, R.B., Ye, J.: A polynomial kernel for diamond-free editing. Algorithmica **84**(1), 197–215 (2022)
32. Guillemot, S., Havet, F., Paul, C., Perez, A.: On the (non-)existence of polynomial kernels for P_ℓ-free edge modification problems. Algorithmica **65**(4), 900–926 (2013)
33. Yuan, H., Ke, Y., Cao, Y.: Polynomial kernels for paw-free edge modification problems. Theor. Comput. Sci. **891**, 1–12 (2021)
34. Eiben, E., Lochet, W., Saurabh, S.: A polynomial kernel for paw-free editing. In: 15th International Symposium on Parameterized and Exact Computation, IPEC 2020, vol. 180 of LIPIcs, pp. 10:1–10:15. Schloss Dagstuhl - Leibniz-Zentrum für Informatik (2020)
35. Bliznets, I., Cygan, M., Komosa, P., Pilipczuk, M.: Hardness of approximation for H-free edge modification problems. ACM Trans. Comput. Theory **10**(2), 91–932 (2018)
36. Kamiński, M., Lozin, V.V., Milanič, M.: Recent developments on graphs of bounded clique-width. Discret. Appl. Math. **157**(12), 2747–2761 (2009)
37. Fomin, F.V., Golovach, P.A., Strømme, T.J.F., Thilikos, D.M.: Subgraph complementation. Algorithmica **82**(7), 1859–1880 (2020)
38. Buchanan, C., Purcell, C., Rombach, P.: Subgraph complementation and minimum rank. Electron. J. Comb., 29(1) (2022)
39. Cygan, M., et al.: Parameterized Algorithms. Springer, Cham (2015). https://doi.org/10.1007/978-3-319-21275-3

Elastic-Degenerate String Matching with 1 Error

Giulia Bernardini[1], Esteban Gabory[2], Solon P. Pissis[2,3,4],
Leen Stougie[2,3,4], Michelle Sweering[2], and Wiktor Zuba[2(✉)]

[1] University of Trieste, Trieste, Italy
`giulia.bernardini@units.it`
[2] CWI, Amsterdam, The Netherlands
{`esteban.gabory,solon.pissis,leen.stougie,michelle.sweering,`
`wiktor.zuba`}`@cwi.nl`
[3] Vrije Universiteit, Amsterdam, The Netherlands
[4] INRIA-Erable, Villeurbanne, France

Abstract. An elastic-degenerate (ED) string is a sequence of n finite sets of strings of total length N, introduced to represent a set of related DNA sequences, also known as a *pangenome*. The ED string matching (EDSM) problem consists in reporting all occurrences of a pattern of length m in an ED text. The EDSM problem has recently received some attention by the combinatorial pattern matching community, culminating in an $\tilde{\mathcal{O}}(nm^{\omega-1}) + \mathcal{O}(N)$-time algorithm [Bernardini et al., SIAM J. Comput. 2022], where ω denotes the matrix multiplication exponent and the $\tilde{\mathcal{O}}(\cdot)$ notation suppresses polylog factors. In the k-EDSM problem, the approximate version of EDSM, we are asked to report all pattern occurrences with at most k errors. k-EDSM can be solved in $\mathcal{O}(k^2 mG + kN)$ time under edit distance, where G denotes the total number of strings in the ED text [Bernardini et al., Theor. Comput. Sci. 2020]. Unfortunately, G is only bounded by N, and so even for $k = 1$, the existing algorithm runs in $\Omega(mN)$ time in the worst case. Here we make progress in this direction. We show that 1-EDSM can be solved in $\mathcal{O}((nm^2 + N)\log m)$ or $\mathcal{O}(nm^3 + N)$ time under edit distance. For the decision version of the problem, we present a faster $\mathcal{O}(nm^2\sqrt{\log m} + N\log\log m)$-time algorithm. Our algorithms rely on non-trivial reductions from 1-EDSM to special instances of classic computational geometry problems (2d rectangle stabbing or range emptiness), which we show how to solve efficiently.

Keywords: String algorithms · Approximate string matching · Edit distance · Degenerate strings · Elastic-degenerate strings

The work in this paper is supported in part by: the Netherlands Organisation for Scientific Research (NWO) through project OCENW.GROOT.2019.015 "Optimization for and with Machine Learning (OPTIMAL)" and Gravitation-grant NETWORKS-024.002.003; the PANGAIA and ALPACA projects that have received funding from the European Union's Horizon 2020 research and innovation programme under the Marie Skłodowska-Curie grant agreements No 872539 and 956229, respectively; and the MUR - FSE REACT EU - PON R&I 2014–2020.

© Springer Nature Switzerland AG 2022
A. Castañeda and F. Rodríguez-Henríquez (Eds.): LATIN 2022, LNCS 13568, pp. 20–37, 2022.
https://doi.org/10.1007/978-3-031-20624-5_2

1   Introduction

String matching (or pattern matching) is a fundamental task in computer science, for which several linear-time algorithms are known [18]. It consists in finding all occurrences of a short string, known as the *pattern*, in a longer string, known as the *text*. Many representations have been introduced over the years to account for unknown or uncertain letters in the pattern or in the text, a phenomenon that often occurs in real data. In the context of computational biology, for example, the IUPAC notation [26] is used to represent locations of a DNA sequence for which several alternative nucleotides are possible. Such a notation can encode the consensus of a population of DNA sequences [1,2,22,32] in a gapless multiple sequence alignment (MSA).

Iliopoulos et al. generalized these representations in [25] to also encode insertions and deletions (gaps) occurring in MSAs by introducing the notion of elastic-degenerate strings. An *elastic-degenerate* (ED) string $\tilde{T}$ over an alphabet Σ is a sequence of finite subsets of Σ^* (which includes the empty string ε), called *segments*. The number of segments is the *length* of the ED string, denoted by $n = |\tilde{T}|$; and the total number of letters (including symbol ε) in all segments is the *size* of the ED string, denoted by $N = \|\tilde{T}\|$. Inspect Fig. 1 for an example.

In Table 1, m is the length of the pattern, n is the length of the ED text, N is its size, and ω is the matrix multiplication exponent. These algorithms are also *on-line*: the ED text is read segment-by-segment and occurrences are reported as soon as the last segment they overlap is processed. Grossi et al. [24] presented an $\mathcal{O}(nm^2 + N)$-time algorithm for EDSM. This was later improved by Aoyama et al. [5], who employed fast Fourier transform to improve the time complexity of EDSM to $\mathcal{O}(nm^{1.5}\sqrt{\log m}+N)$. Bernardini et al. [8] then presented a lower bound conditioned on Boolean Matrix Multiplication suggesting that it is unlikely to solve EDSM by a combinatorial algorithm in $\mathcal{O}(nm^{1.5-\epsilon} + N)$ time, for any $\epsilon > 0$. This was an indication that fast matrix multiplication may improve the time complexity of EDSM. Indeed, Bernardini et al. [8] presented an $\mathcal{O}(nm^{1.381} + N)$-time algorithm, which they subsequently improved to an $\tilde{\mathcal{O}}(nm^{\omega-1}) + \mathcal{O}(N)$-time algorithm [9], both using fast matrix multiplication, thus breaking through the conditional lower bound for EDSM.

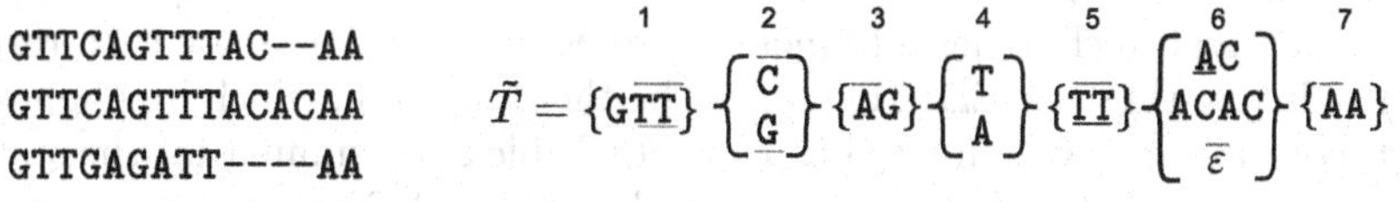

Fig. 1. An MSA of three sequences and its (non-unique) representation $\tilde{T}$ as an ED string of length $n = 7$ and size $N = 20$. The only two *exact* occurrences of $P = $ TTA in $\tilde{T}$ end at positions 6 (black underline) and 7 (blue overline); a *1-error* occurrence of P in $\tilde{T}$ ends at position 2 (green underline); and another *1-error* occurrence of P in $\tilde{T}$ ends at position 3 (red overline). Note that other 1-error occurrences of P in $\tilde{T}$ exist (e.g., ending at positions 1 and 5). (Color figure online)

Table 1. The upper-bound landscape of the EDSM problem.

EDSM	Features	Running time
Grossi et al. [24]	Combinatorial	$\mathcal{O}(nm^2 + N)$
Aoyama et al. [5]	Fast Fourier transform	$\mathcal{O}(nm^{1.5}\sqrt{\log m} + N)$
Bernardini et al. [8]	Fast matrix multiplication	$\mathcal{O}(nm^{1.381} + N)$
Bernardini et al. [9]	Fast matrix multiplication	$\tilde{\mathcal{O}}(nm^{\omega-1}) + \mathcal{O}(N)$

Table 2. The state of the art result for k-EDSM and our new results for $k = 1$. Note that $n \leq G \leq N$. All algorithms underlying these results are combinatorial and the reporting algorithms are all on-line.

k-EDSM	Features	Running time
Bernardini et al. [10]	k errors	$\mathcal{O}(k^2 mG + kN)$
This work	1 error	$\mathcal{O}(nm^3 + N)$
This work	1 error	$\mathcal{O}((nm^2 + N)\log m)$
This work	1 error (decision)	$\mathcal{O}(nm^2\sqrt{\log m} + N\log\log m)$

Our Results and Techniques. In string matching, a single extra or missing letter in a potential occurrence results in missing (many or all) occurrences. Hence, many works are focused on approximate string matching for standard strings [4,13,17, 23,27,28]. For approximate EDSM (k-EDSM), Bernardini et al. [7,10] gave an on-line $\mathcal{O}(k^2 mG + kN)$-time algorithm under edit distance and an on-line $\mathcal{O}(kmG + kN)$-time algorithm under Hamming distance, where k is the maximum allowed number of errors (edits) or mismatches, respectively, and G is the total number of strings in all segments. Unfortunately, G is only bounded by N, and so even for $k = 1$, the existing algorithms run in $\Omega(mN)$ time in the worst case.

Let us remark that the special case of $k = 1$ is not interesting for approximate string matching on standard strings: the existing algorithms have a polynomial dependency on k and a linear dependency on the length n of the text, and thus for $k = 1$ we trivially obtain $\mathcal{O}(n)$-time algorithms under edit or Hamming distance. However, this is not the case for other string problems, such as text indexing with errors, where the first step was to design a data structure for 1 error [3]. The next step, extending it to k errors, required the development of new highly non-trivial techniques and incurred some exponential factor with respect to k [16]. Interestingly, k-EDSM seems to be the same case, which highlights the main theoretical motivation for this paper. In Table 2, we summarize the state of the art result for k-EDSM and our new results for $k = 1$. Note that the reporting algorithms underlying our results are also *on-line*.

Indeed, to arrive at our main results, we design a rich combination of algorithmic techniques. Our algorithms rely on non-trivial reductions from 1-EDSM to special instances of classic computational geometry problems (2d rectangle stabbing or 2d range emptiness), which we show how to solve efficiently.

The combinatorial algorithms we develop here for approximate EDSM are good in the following sense. First, the running times of our algorithms do not depend on G, a highly desirable property. Specifically, all of our results replace $m \cdot G$ by an $n \cdot \text{poly}(m)$ factor. Second, our $\tilde{\mathcal{O}}(nm^2 + N)$-time algorithms are at most one $\log m$ factor slower than $\mathcal{O}(nm^2 + N)$, the best-known bound obtained by a combinatorial algorithm (not employing fast Fourier transforms) for *exact* EDSM [24]. Last, our $\mathcal{O}(nm^3 + N)$-time algorithm has a linear dependency on N, another highly desirable property (at the expense of an extra m-factor).

Paper Organization. In Sect. 2, we provide the necessary definitions and notation, we describe the basic layout of the developed algorithms, and we formally state our main results. In Sect. 3, we present our solutions under edit distance. In Sect. 4, we conclude with some basic open questions for future work.

2 Preliminaries

We start with some basic definitions and notation following [18]. Let $X = X[1] \ldots X[n]$ be a *string* of length $|X| = n$ over an ordered alphabet Σ whose elements are called *letters*. The *empty string* is the string of length 0; we denote it by ε. For any two positions i and $j \geq i$ of X, $X[i \mathinner{.\,.} j]$ is the *fragment* of X starting at position i and ending at position j. The fragment $X[i \mathinner{.\,.} j]$ is an *occurrence* of the underlying *substring* $P = X[i] \ldots X[j]$; we say that P occurs at *position* i in X. A *prefix* of X is a fragment of the form $X[1 \mathinner{.\,.} j]$ and a *suffix* of X is a fragment of the form $X[i \mathinner{.\,.} n]$. By XY or $X \cdot Y$ we denote the *concatenation* of two strings X and Y, i.e., $XY = X[1] \ldots X[|X|]Y[1] \ldots Y[|Y|]$. Given a string X we write $X^R = X[|X|] \ldots X[1]$ for the *reverse* of X.

An *elastic-degenerate string* (ED string) $\tilde{T} = \tilde{T}[1] \ldots \tilde{T}[n]$ over an alphabet Σ is a sequence of $n = |\tilde{T}|$ finite sets, called *segments*, such that for every position i of $\tilde{T}$ we have that $\tilde{T}[i] \subset \Sigma^*$. By $N = ||\tilde{T}||$ we denote the total length of all strings in all segments of $\tilde{T}$, which we call the *size* of $\tilde{T}$; more formally, $N = \sum_{i=1}^{n} \sum_{j=1}^{|\tilde{T}[i]|} |\tilde{T}[i][j]|$, where by $\tilde{T}[i][j]$ we denote the jth string of $\tilde{T}[i]$. (As an exception, we also add 1 to account for empty strings: if $\tilde{T}[i][j] = \varepsilon$, then we have that $|\tilde{T}[i][j]| = 1$.) Given two sets of strings S_1 and S_2, their *concatenation* is $S_1 \cdot S_2 = \{XY \mid X \in S_1, Y \in S_2\}$. For an ED string $\tilde{T} = \tilde{T}[1] \ldots \tilde{T}[n]$, we define the *language* of $\tilde{T}$ as $\mathcal{L}(\tilde{T}) = \tilde{T}[1] \cdot \ldots \cdot \tilde{T}[n]$. Given a set S of strings we write S^R for the set $\{X^R \mid X \in S\}$. For an ED string $\tilde{T} = \tilde{T}[1] \ldots \tilde{T}[n]$ we write $\tilde{T}^R$ for the ED string $\tilde{T}[n]^R \ldots \tilde{T}[1]^R$.

Given a string P and an ED string $\tilde{T}$, we say that P *matches* the fragment $\tilde{T}[j \mathinner{.\,.} j'] = \tilde{T}[j] \ldots \tilde{T}[j']$ of $\tilde{T}$, or that an *occurrence* of P *starts* at position j and *ends* at position j' in $\tilde{T}$ if there exist two strings U, V, each of them possibly empty, such that $P = P_j \cdot \ldots \cdot P_{j'}$, where $P_i \in \tilde{T}[i]$, for every $j < i < j'$, $U \cdot P_j \in \tilde{T}[j]$, and $P_{j'} \cdot V \in \tilde{T}[j']$ (or $U \cdot P_j \cdot V \in \tilde{T}[j]$ when $j = j'$). Strings U, V and P_i, for every $j \leq i \leq j'$, specify an *alignment* of P with $\tilde{T}[j \mathinner{.\,.} j']$. For each occurrence of P in $\tilde{T}$, the alignment is, in general, not unique. In Fig. 1, $P = \texttt{TTA}$ matches $\tilde{T}[5 \mathinner{.\,.} 6]$ with two alignments: both have $U = \varepsilon$, $P_5 = \texttt{TT}$, $P_6 = \texttt{A}$, and V is either $\texttt{C}$ or $\texttt{CAC}$.

We will refer to P as the *pattern* and to $\tilde{T}$ as the *ED text*. We want to accept matches with edit distance at most 1.

Definition 1. *Given two strings P and Q over an alphabet Σ, we define the edit distance $d_E(P, Q)$ between P and Q as the length of a shortest sequence of letter replacements, deletions, and insertions, to obtain P from Q.*

Lemma 1 ([18]). *The function d_E is a distance on Σ^*.*

We define the main problem considered in this paper as follows:

1-Error EDSM
Input: A string P of length m and an ED string $\tilde{T}$ of length n and size N.
Output: All positions j' in $\tilde{T}$ such that there is at least one string P' with an occurrence ending at position j' in $\tilde{T}$, and with $d_E(P, P') \leq 1$ (reporting version); or YES if and only if there is at least one string P' with an occurrence in $\tilde{T}$, and with $d_E(P, P') \leq 1$ (decision version).

Let P' be a string starting at position j and ending at position j' in $\tilde{T}$ with $d_E(P, P') = 1$. We call this *an occurrence of P with 1 error* (or a *1-error occurrence*); or equivalently, we say that *P matches $\tilde{T}[j \mathinner{.\,.} j']$ with 1 error*. Let $U P'_j, \ldots, P'_{j'} V$ be an alignment of P' with $\tilde{T}[j \mathinner{.\,.} j']$ and $i \in [j, j']$ be an integer such that the single replacement, insertion, or deletion required to obtain P from $P' = P'_j \cdot \ldots \cdot P'_{j'}$ occurs on P'_i. We then say that the alignment (and the occurrence) *has the 1 error in $\tilde{T}[i]$*. (It should be clear that for one alignment we may have multiple different i.) We show the following theorem.

Theorem 1. *Given a pattern P of length m and an ED text $\tilde{T}$ of length n and size N, the reporting version of 1-Error EDSM can be solved on-line in $\mathcal{O}(nm^2 \log m + N \log m)$ or $\mathcal{O}(nm^3 + N)$ time. The decision version of 1-Error EDSM can be solved off-line in $\mathcal{O}(nm^2 \sqrt{\log m} + N \log \log m)$ time.*

Definition 2. *For a string $P = P[1 \mathinner{.\,.} m]$, an ED string $\tilde{T} = \tilde{T}[1] \ldots \tilde{T}[n]$, and a position $1 \leq i \leq n$, we define three sets:*

- *$AP_i \subseteq [1, m]$, such that $j \in AP_i$ if and only if $P[1 \mathinner{.\,.} j]$ is an active prefix of P in $\tilde{T}$ ending in the segment $\tilde{T}[i]$, that is, a prefix of P which is also a suffix of a string in $\mathcal{L}(\tilde{T}[1] \ldots \tilde{T}[i])$.*
- *$AS_i \subseteq [1, m]$, such that $j \in AS_i$ if and only if $P[j \mathinner{.\,.} m]$ is an active suffix of P in $\tilde{T}$ starting in the segment $\tilde{T}[i]$, that is, a suffix of P which is also a prefix of a string in $\mathcal{L}(\tilde{T}[i] \ldots \tilde{T}[n])$.*
- *$1\text{-}AP_i \subseteq [1, m]$, such that $j \in 1\text{-}AP_i$ if and only if $P[1 \mathinner{.\,.} j]$ is an active prefix with 1 error of P in $\tilde{T}$ ending in the segment $\tilde{T}[i]$, that is, a prefix of P which is also at edit distance at most 1 from a suffix of a string in $\mathcal{L}(\tilde{T}[1] \ldots \tilde{T}[i])$.*

For convenience we also define $AP_0 = AS_{n+1} = 1\text{-}AP_0 = \emptyset$.

The following lemma shows that the computation of active suffixes can be easily reduced to computing the active prefixes for the reversed strings.

Lemma 2. *Given a pattern $P = P[1 .. m]$ and an ED text $\tilde{T} = \tilde{T}[1 .. n]$, a suffix $P[j .. m]$ of P is an active suffix in $\tilde{T}$ starting in the segment $\tilde{T}[i]$ if and only if the prefix $P^R[1 .. m - j + 1] = (P[j .. m])^R$ of P^R is an active prefix in $\tilde{T}^R$, ending in the segment $\tilde{T}^R[n - i + 1] = (\tilde{T}[i])^R$.*

Proof. If $P[j .. m]$ is a prefix of $S \in \mathcal{L}(\tilde{T}[i .. n])$, then $P^R[1 .. m - j + 1]$ is a suffix of $S^R \in \mathcal{L}(\tilde{T}[1 ... n]^R)$. From the definition of $\tilde{T}^R$ we have $\tilde{T}[i .. n]^R = (T[n])^R ... (T[i])^R = \tilde{T}^R[1 .. n - i + 1]$, hence $S^R \in \mathcal{L}(\tilde{T}^R[1 .. n - i + 1])$. This proves the forward direction of the lemma; the converse follows from symmetry.

□

The efficient computation of active prefixes was shown in [24], and constitutes the main part of the combinatorial algorithm for exact EDSM. Similarly, computing the sets 1-AP plays the key role in the reporting version of our algorithm for 1-ERROR EDSM (see Fig. 2). Finding active prefixes (and, by Lemma 2, suffixes) reduces to the following problem, formalized in [8].

ACTIVE PREFIXES EXTENSION (APE)
Input: A string P of length m, a bit vector U of size m, and a set $\mathcal{S}$ of strings of total length N.
Output: A bit vector V of size m with $V[j] = 1$ if and only if there exists $S \in \mathcal{S}$ and $i \in [1, m]$, such that $P[1 .. i] \cdot S = P[1 .. j]$ and $U[i] = 1$.

Lemma 3 ([24]). *The APE problem for a string P of length m and a set $\mathcal{S}$ of strings of total length N can be solved in $\mathcal{O}(m^2 + N)$ time.*

Given an algorithm for the APE problem working in $f(m) + N$ time, we can find *all* active prefixes for a pattern P of length m in an ED text $\tilde{T} = \tilde{T}[1] ... \tilde{T}[n]$ of size N in $\mathcal{O}(nf(m) + N)$ total time:

Corollary 1 ([24]). *For a pattern P of length m and an ED text $\tilde{T} = \tilde{T}[1] ... \tilde{T}[n]$ of total size N, computing the sets AP_i for all $i \in [1, n]$ takes $\mathcal{O}(nm^2 + N)$ time.*

As depicted in Fig. 2, the computation of active prefixes with 1 error (1-AP_i) and the reporting of occurrences with 1 error reduce to a problem where the error can only occur in a single, fixed $\tilde{T}[i]$. In particular, this problem decomposes into 4 cases, which we formalize in the following proposition.

Proposition 1. *Let $\tilde{T} = \tilde{T}[1] ... \tilde{T}[n]$ be an ED text and P be a pattern that has an occurrence with 1 error in $\tilde{T}$. For each alignment corresponding to such occurrence, at least one of the following is true:*

Easy Case: *P matches $\tilde{T}[i]$ with 1 error for some $1 \le i \le n$.*
Anchor Case: *P matches $\tilde{T}[j .. j']$ with 1 error in $\tilde{T}[i]$ for some $1 \le j < i < j' \le n$. $\tilde{T}[i]$ is called the anchor of the alignment.*
Prefix Case: *P matches $\tilde{T}[j .. i]$ with 1 error in $\tilde{T}[i]$ for some $1 \le j < i \le n$, implying an active prefix of P which is a suffix of a string in $\mathcal{L}(\tilde{T}[j .. i - 1])$.*

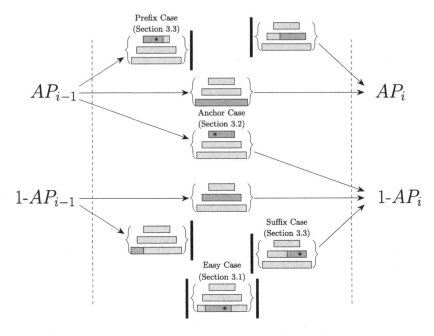

Fig. 2. The layout of the algorithms for computing AP_i, $1\text{-}AP_i$, and reporting occurrences. The green areas correspond to the (partial) matches in $\tilde{T}[i]$, and the symbol $*$ indicates the position of the error. The vertical bold lines indicate the beginning/the end of an occurrence or a 1-error occurrence. The cases without a label allow only exact matches and were already solved by Grossi et al. in [24]. (Color figure online)

Suffix Case: *P matches $\tilde{T}[i \mathinner{.\,.} j']$ with 1 error in $\tilde{T}[i]$ for some $1 \leq i < j' \leq n$, implying an active suffix of P which is a prefix of a string in $\mathcal{L}(\tilde{T}[i+1 \mathinner{.\,.} j'])$.*

Proof. Suppose P has a 1-error occurrence matching $\tilde{T}[j \mathinner{.\,.} j']$ with $1 \leq j \leq j' \leq n$. If $j = j'$ we are in the Easy Case. Otherwise, each alignment has an error in some $\tilde{T}[i]$ for $j \leq i \leq j'$. If $j < i < j'$, we are in the Anchor Case; if $j < i = j'$, we are in the Prefix Case; and if $j = i < j'$, we are in the Suffix Case. $\qquad\square$

3   1-Error EDSM

In this section, we present algorithms for finding all 1-error occurrences of P given by each type of possible alignment described by Proposition 1 (inspect Fig. 3). The Prefix and Suffix Cases are analogous by Lemma 2; the only difference is in that, while the Suffix Case computes new $1\text{-}AP$, the Prefix Case is used to actually report occurrences. They are jointly considered in Sect. 3.3.

We follow two different procedures for the decision and reporting versions. For the decision version, we precompute sets AP_i and AS_i, for all $i \in [1, n]$, using Corollary 1, and we simultaneously compute possible exact occurrences of P. Then we compute 1-error occurrences of P by grouping the alignments

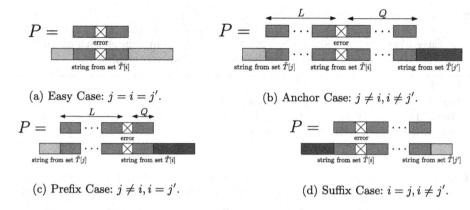

(a) Easy Case: $j = i = j'$.

(b) Anchor Case: $j \neq i, i \neq j'$.

(c) Prefix Case: $j \neq i, i = j'$.

(d) Suffix Case: $i = j, i \neq j'$.

Fig. 3. Possible alignments of 1-error occurrences of P in $\tilde{T}$. Each occurrence starts at segment $\tilde{T}[j]$, ends at $\tilde{T}[j']$, and the error occurs at $\tilde{T}[i]$

depending on the segment i in which the error occurs, and using AP_i and AS_i. For the reporting version, we consider one segment $\tilde{T}[i]$ at a time (on-line) and extend partial exact or 1-error occurrences of P to compute sets AP_i and $1\text{-}AP_i$ using just sets AP_{i-1} and $1\text{-}AP_{i-1}$ computed at the previous step. We design different procedures for the 4 cases of Proposition 1. We can sort all letters of P, assign them rank values from $[1, m]$, and construct a perfect hash table over these letters supporting $\mathcal{O}(1)$-time look-up queries in $\mathcal{O}(m \log m)$ time [30]. Any letter of $\tilde{T}$ not occurring in P can be replaced by the same special letter in $\mathcal{O}(1)$ time. In the rest we thus assume that the input strings are over $[1, m + 1]$.

Two problems from computational geometry have a key role in our solutions. We assume the word RAM model with coordinates on the integer grid $[1, n]^d = \{1, 2, \dots, n\}^d$. In the *2d rectangle emptiness* problem, we are given a set $\mathcal{P}$ of n points to be preprocessed, so that when one gives an axis-aligned rectangle as a query, we report YES if and only if the rectangle contains a point from $\mathcal{P}$. In the "dual" *2d rectangle stabbing* problem, we are given a set $\mathcal{R}$ of n axis-aligned rectangles to be preprocessed, so that when one gives a point as a query, we report YES if and only if there exists a rectangle from $\mathcal{R}$ containing the point.

Lemma 4 ([11,21]). *After $\mathcal{O}(n\sqrt{\log n})$-time preprocessing, we can answer 2d rectangle emptiness queries in $\mathcal{O}(\log \log n)$ time.*

Lemma 5 ([15,31]). *After $\mathcal{O}(n \log n)$-time preprocessing, we can answer 2d rectangle stabbing queries in $\mathcal{O}(\log n)$ time.*

In Sect. 3.4, we note that the 2d rectangle stabbing instances arising from 1-ERROR EDSM have a special structure. We show how to solve them efficiently thus shaving logarithmic factors from the time complexity.

3.1 Easy Case

The Easy Case can be reduced to approximate string matching with at most 1 error (1-SM), for which we have the following well-known results.

1-SM
Input: A string P of length m and a string T of length n.
Output: All positions j in T such that there is at least one string P' ending at position j in T with $d_E(P, P') \leq 1$.

Lemma 6 ([17,28]). *Given a pattern P of length m, a text T of length n, and an integer $k > 0$, all positions j in T such that the edit distance of $T[i\,..\,j]$ and P, for some position $i \leq j$ on T, is at most k, can be found in $\mathcal{O}(kn)$ time or in $\mathcal{O}(\frac{nk^4}{m} + n)$ time.[1] In particular, 1-SM can be solved in $\mathcal{O}(n)$ time.*

We find occurrences of P with at most 1 error that are in the Easy Case for segment $\tilde{T}[i]$ in the following way: we apply Lemma 6 for $k = 1$ and every string of $\tilde{T}[i]$ whose length is at least $m - 1$ (any shorter string is clearly not relevant for this case) as text. If, for any of those strings, we find an occurrence of P, we report an occurrence at position i (inspect Fig. 3a). The time for processing a segment $\tilde{T}[i]$ is $\mathcal{O}(N_i)$, where N_i is the total length of all the strings in $\tilde{T}[i]$.

3.2 Anchor Case

Let $\tilde{T}$ be an ED text and P be a pattern with a 1-error occurrence and an alignment in the Anchor Case with anchor $\tilde{T}[i]$. Further let $L = P[1\,..\,\ell]S'$ and $Q = S''P[q\,..\,m]$ be a prefix and a suffix of P, respectively, for some $\ell \in AP_{i-1}, q \in AS_{i+1}$, where S', S'' are a prefix and a suffix of some $S \in \tilde{T}[i]$, respectively (strings S', S'' can be empty). By definition of the edit distance, a pair L, Q gives a 1-error occurrence of P if one of the following holds:

1 mismatch: $|L| + |Q| + 1 = m$ *and* $|S'| + |S''| + 1 = |S|$ (inspect Fig. 3b).
1 deletion in P: $|L| + |Q| = m - 1$ *and* $|S'| + |S''| = |S|$.
1 insertion in P: $|L| + |Q| = m$ *and* $|S'| + |S''| + 1 = |S|$.

We show how to find such pairs with the use of a geometric approach. For convenience, we only present the Hamming distance (1 mismatch) case. The other cases are handled similarly.

Let $\lambda \in AP_{i-1}$ be the length of an active prefix, and let ρ be the length of an active suffix, that is, $m - \rho + 1 \in AS_{i+1}$. Note that AP_{i-1} and AS_{i+1} can be precomputed, for all i, in $\mathcal{O}(nm^2 + N)$ total time by means of Corollary 1. (In particular, AS_{i+1} is required only for the decision version; for the reporting version, we explain later on how to avoid the precomputation of AS_{i+1} to obtain

---

[1] Charalampopoulos et al. have announced an improvement on the exponent of k from 4 to 3.5; specifically they presented an $\mathcal{O}(\frac{nk^{3.5}\sqrt{\log m \log k}}{m} + n)$-time algorithm [14].

an on-line algorithm.) We will exhaustively consider all pairs (λ, ρ) such that $\lambda + \rho < m$. Clearly, there are $\mathcal{O}(m^2)$ such pairs.

Consider the length $\mu = m - (\lambda + \rho) > 0$ of the substring of P still to be matched for some prefix and suffix of P of lengths (λ, ρ), respectively. We group together all pairs (λ, ρ) such that $m - (\lambda + \rho) = \mu$ by sorting them in $\mathcal{O}(m^2)$ time. We construct, for each such group μ, the compacted trie T_μ of the fragments $P[\lambda + 1 .. m - \rho]$, for all (λ, ρ) such that $m - (\lambda + \rho) = \mu$, and analogously the compacted trie T_μ^R of all fragments $P^R[\rho + 1 .. m - \lambda]$. For each group μ, this takes $\mathcal{O}(m)$ time [29]. We enhance all nodes with a perfect hash table in $\mathcal{O}(m)$ total time to access edges by the first letter of their label in $\mathcal{O}(1)$ time [20].

We also group all strings in segment $\tilde{T}[i]$ of length less than m by their length μ. The group for length μ is denoted by G_μ. This takes $\mathcal{O}(N_i)$ time. Clearly, the strings in G_μ are the only candidates to extend pairs (λ, ρ) such that $m - (\lambda + \rho) = \mu$. Note that the mismatch can be at any position of any string of G_μ: its position determines a prefix S' of length h and a suffix S'' of length k of the same string S, with $h + k = \mu - 1$, that must match a prefix and a suffix of $P[\lambda + 1 .. m - \rho]$, respectively. We will consider all such pairs of positions (h, k) whose sum is $\mu - 1$ (intuitively, the minus one is for the mismatch). This guarantees that $L = P[1 .. \lambda]S'$ and $Q = S''P[m - \rho + 1 .. m]$ are such that $|L| + |Q| + 1 = m$. The pairs are $(0, \mu - 1), (1, \mu - 2), \ldots, (\mu - 1, 0)$. This guarantees that L and Q are *one position apart* $(|S'| + |S''| + 1 = |S|)$.

The number of these pairs is $\mathcal{O}(\mu) = \mathcal{O}(m)$. Consider one such pair (h, k) and a string $S \in G_\mu$. We treat every such string S separately. We spell $S[1 .. h]$ in T_μ. If the whole $S[1 .. h]$ is successfully spelled ending at a node u, this implies that all the fragments of P corresponding to nodes descending from u share $S[1 .. h]$ as a prefix. We also spell $S^R[1 .. k]$ in T_μ^R. If the whole of $S^R[1 .. k]$ is successfully spelled ending at a node v, then all the fragments of P corresponding to nodes descending from v share $(S^R[1 .. k])^R$ as a suffix. Nodes u and v identify an interval of leaves in T_μ and T_μ^R, respectively. We need to check if these intervals both contain a leaf corresponding to the same fragment of P. If they do, then we obtain an occurrence of P with 1 mismatch (see Fig. 4). We now have two different ways to proceed, depending on whether we need to solve the off-line decision version or the on-line reporting version.

Decision Version. Recall that T_μ, T_μ^R by construction are ordered based on lexicographic ranks. For every pair (T_μ, T_μ^R), we construct a data structure for 2d rectangle emptiness queries on the grid $[1, \ell]^2$, where ℓ is the number of leaves of T_μ (and of T_μ^R), for the set of points (x, y) such that x is the lexicographic rank of the leaf of T_μ representing $P[\lambda + 1 .. m - \rho]$ and y is the rank of the leaf of T_μ^R representing $P^R[\rho + 1 .. m - \lambda]$ for the same pair (λ, ρ). This denotes that the two leaves correspond to *the same fragment* of P. For every (T_μ, T_μ^R), this preprocessing takes $\mathcal{O}(m\sqrt{\log m})$ time by Lemma 4, since ℓ is $\mathcal{O}(\mu) = \mathcal{O}(m)$. For all $\mu \leq m$ groups, the whole preprocessing thus takes $\mathcal{O}(m^2\sqrt{\log m})$ time.

We then ask 2d range emptiness queries that take $\mathcal{O}(\log \log m)$ time each by Lemma 4. Note that all rectangles for S can be collected in $\mathcal{O}(|S|) = \mathcal{O}(\mu)$ time

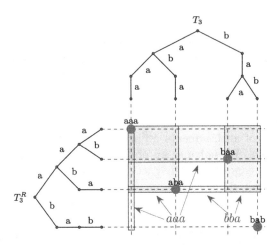

Fig. 4. An example of points and rectangles (solid shapes) for the decision version of the Anchor Case with 1 mismatch. Here $P = bbaaaabababb$, $AP_{i-1} = \{1, 2, 4, 7, 8, 9\}$, $AS_{i+1} = \{5, 6, 9, 11, 12\}$, $\mu = 3$, and $\tilde{T}[i] = \{aaa, bba\}$. T_3 and T_3^R are built for 4 strings: $P[2..4] = baa, P[3..5] = aaa, P[8..10] = aba, P[9..11] = bab$; the 5 rectangles correspond to pairs $(\varepsilon, aa), (a, a), (aa, \varepsilon), (\varepsilon, ab), (b, a)$, namely, the pairs of prefixes and reversed suffixes of aaa and bba (rectangle (bb, ε) does not exist as T_3 contains no node bb).

by spelling S through T_μ and S^R through T_μ^R, one letter at a time. Thus the total time for processing all G_μ groups of segment i is $\mathcal{O}(m^2\sqrt{\log m} + N_i \log \log m)$. If any of the queried ranges turns out to be non-empty, then P' such that $d_H(P, P') \leq 1$ appears in $\mathcal{L}(\tilde{T})$ with anchor in $\tilde{T}[i]$; we do not have sufficient information to output its ending position however.

Reporting Version. For this version, we do the dual. We construct a data structure for 2d rectangle stabbing queries on the grid $[1, \ell]^2$ for the set of rectangles collected for all strings $S \in G_\mu$. By Lemma 5, for all μ groups, the whole preprocessing thus takes $\mathcal{O}(N_i \log N_i)$ time.

For every (T_μ, T_μ^R), we then ask the following queries: (x, y) is queried if and only if x is the rank of a leaf representing $P[\lambda + 1..m - \rho]$ and y is the rank of a leaf representing $P^R[\rho + 1..m - \lambda]$. For every (T_μ, T_μ^R), this takes $\mathcal{O}(m \log N_i)$ time by Lemma 5 and by the fact that for each group G_μ there are $\mathcal{O}(m)$ pairs (λ, ρ) such that $m - (\lambda + \rho) = \mu$. For all groups G_μ (they are at most m), all the queries thus take $\mathcal{O}(m^2 \log N_i)$ time. Thus the total time for processing all G_μ groups of segment i is $\mathcal{O}((m^2 + N_i) \log N_i)$.

We are not done yet. By performing the above algorithm for active prefixes and active suffixes, we find out which pairs can be completed to a full occurrence of P with at most 1 error. This information is not sufficient to compute where such an occurrence ends (and storing additional information together with the active suffixes may prove costly). To overcome this, we use some ideas from the decision algorithm, appropriately modified to preserve the on-line nature of the reporting algorithm. Instead of iterating ρ over the lengths of precomputed active suffixes, we iterate it over *all* possible lengths in $[0, m]$ (including 0 because we may want to include m in 1-AP_i). A suffix of P of length ρ completes a partial occurrence computed up to segment i exactly when $m - \rho \in$ 1-AP_i (a pair $x \in$ 1-$AP_i, x + 1 \in AS_{i+1}$ corresponds to an occurrence). We thus use the reporting algorithm to compute the part of 1-AP_i coming from the extension of AP_{i-1} (see Fig. 2), and defer the reporting to the no-error version of the Prefix Case for the right j'; which was solved by Grossi et al. [24] in linear time.

3.3   Prefix Case

Let $\tilde{T}$ be an ED text and P be a pattern with a 1-error occurrence and an alignment in the Prefix Case with active prefix ending at $\tilde{T}[i - 1]$. Let $L = P[1 .. \ell]S'$, with $\ell \in AP_{i-1}$, be a prefix of P that is extended in $\tilde{T}[i]$ by S'; and Q be a suffix of P occurring in some string of $\tilde{T}[i]$ (strings S', Q can be empty). By definition of the edit distance, we have 3 possibilities for any alignment of a 1-error occurrence of P in the Prefix Case:

1 mismatch: $|L| + |Q| + 1 = m$, S' is a prefix of the same string in which Q occurs, and they are one position apart (inspect Fig. 3c).

1 deletion in P: $|L| + |Q| = m - 1$, S' is a prefix of the same string in which Q occurs, and they are consecutive.

1 insertion in P: $|L| + |Q| = m$, S' is a prefix of the same string in which Q occurs, and they are one position apart.

For convenience, we only present the method for Hamming distance (1 mismatch). The other possibilities are handled similarly. The techniques are similar to those for the Anchor Case (Sect. 3.2). We group the prefixes of all strings in $\tilde{T}[i]$ according to their length $\mu \in [1, m)$. The total number of these prefixes is $\mathcal{O}(N_i)$. The group for length μ is denoted by G_μ. We construct the compacted trie T_{G_μ} of the strings in G_μ, and the compacted trie $T_{G_\mu}^R$ of the reversed strings in G_μ. This can be done in $\mathcal{O}(N_i)$ total time for all compacted tries. To achieve this, we employ the following lemma by Charalampopoulos et al. [12]. (Recall that we have already sorted all letters of P. In what follows, we assume that $N_i \geq m$; if this is not the case, we can sort all letters of $\tilde{T}[i]$ in $\mathcal{O}(m + N_i)$ time.)

Lemma 7 ([12]). *Let X be a string of length n over an integer alphabet of size $n^{\mathcal{O}(1)}$. Let I be a collection of intervals $[i, j] \subseteq [1, n]$. We can lexicographically sort the substrings $X[i .. j]$ of X, for all intervals $[i, j] \in I$, in $\mathcal{O}(n + |I|)$ time.*

We concatenate all the strings of $\tilde{T}[i]$ to obtain a single string X of length N_i, to which we apply, for each μ, Lemma 7, with a set I consisting of the intervals over X corresponding to the strings in G_μ. By sorting, in this way, all strings in G_μ (for all μ), and by constructing [19] and preprocessing [6] the generalized suffix tree of the strings in $\tilde{T}[i]$ in $\mathcal{O}(N_i)$ time to support answering lowest common ancestor (LCA) queries in $\mathcal{O}(1)$ time, we can construct all T_{G_μ} in $\mathcal{O}(N_i)$

total time. We handle $T_{G_\mu}^R$, for all μ, analogously. Similar to the Anchor Case we enhance all nodes with a perfect hash table within the same complexities [20].

In contrast to the Anchor Case, we now only consider the set AP_{i-1}: namely, we do not consider AS_{i+1}. Let $\lambda \in AP_{i-1}$ be the length of an active prefix. We treat every such element separately, and they are $\mathcal{O}(m)$ in total. Let $\mu = m - \lambda > 0$ and consider the group G_μ whose strings are all of length μ. The mismatch being at position $h + 1$ in one such string S determines a prefix S' of S of length h that must extend the active prefix of P of length λ, and a fragment Q of S of length $k = \mu - h - 1$ that must match a suffix of P. We will consider all such pairs (h, k) whose sum is $\mu - 1$. The pairs are again $(0, \mu - 1), (1, \mu - 2), \ldots, (\mu - 1, 0)$, and there are clearly $\mathcal{O}(\mu) = \mathcal{O}(m)$ of them.

Consider (h, k) as one such pair. We spell $P[\lambda + 1 \mathinner{.\,.} \lambda + h]$ in T_{G_μ}. If the whole $P[\lambda + 1 \mathinner{.\,.} \lambda + h]$ is spelled successfully, this implies an interval of leaves of T_{G_μ} corresponding to strings from $\tilde{T}[i]$ that share $P[\lambda + 1 \mathinner{.\,.} \lambda + h]$ as a prefix. We spell $P^R[1 \mathinner{.\,.} k]$ in $T_{G_\mu}^R$. If the whole $P^R[1 \mathinner{.\,.} k]$ is spelled successfully, this implies an interval of leaves of $T_{G_\mu}^R$ corresponding to strings from $\tilde{T}[i]$ that have the same fragment $(P^R[1 \mathinner{.\,.} k])^R$. These two intervals form a rectangle in the grid implied by the leaves of T_{G_μ} and $T_{G_\mu}^R$. We need to check if these intervals both contain a leaf corresponding to the same prefix of length μ of a string in $\tilde{T}[i]$. If they do, then we have obtained an occurrence with 1 mismatch in $\tilde{T}[i]$.

To do this we construct, for every $(T_{G_\mu}, T_{G_\mu}^R)$, a 2d range data structure for the set of points (x, y) such that x is the rank of a leaf of T_{G_μ}, y is the rank of a leaf of $T_{G_\mu}^R$, and the two leaves correspond to *the same prefix* of length μ of a string in $\tilde{T}[i]$. For every $(T_{G_\mu}, T_{G_\mu}^R)$, this takes $\mathcal{O}(|G_\mu|\sqrt{\log |G_\mu|})$ time by Lemma 4. For all G_μ groups, the whole preprocessing takes $\mathcal{O}(N_i\sqrt{\log N_i})$ time.

We then ask 2d range emptiness queries each taking $\mathcal{O}(\log \log |G_\mu|)$ time by Lemma 4. Note that all rectangles for λ can be collected in $\mathcal{O}(m)$ time by spelling $P[\lambda + 1 \mathinner{.\,.} \lambda + \mu - 1]$ through T_{G_μ} and $P^R[1 \mathinner{.\,.} \mu - 1]$ through $T_{G_\mu}^R$, one letter at a time. This gives a total of $\mathcal{O}(m^2 \log \log N_i + N_i\sqrt{\log N_i})$ time for processing all G_μ groups of $\tilde{T}[i]$, because $\sum_\mu |G_\mu| \leq N_i$.

To solve the Suffix Case (compute active prefixes with 1 error starting in $\tilde{T}[i]$) we employ the mirror version of the algorithm, but iterating λ over the whole $[0, m]$ instead of AS_{i+1} (like in the reporting version of the Anchor Case).

3.4   Shaving Logs Using Special Cases of Geometric Problems

Anchor Case: Simple 2d Rectangle Stabbing

Lemma 8. *We can solve the Anchor Case (i.e., extend AP_{i-1} into 1-AP_i) in $\mathcal{O}(m^3 + N_i)$ time.*

Proof. By Lemma 5, 2d rectangle stabbing queries can be answered in $\mathcal{O}(\log n)$ time after $\mathcal{O}(n \log n)$-time preprocessing.

Notice that in the case of the 2d rectangle stabbing used in Sect. 3.2 the rectangles and points are all in a predefined $[1, m] \times [1, m]$ grid. In such a case we can also use an easy folklore data structure of size $\mathcal{O}(m^2)$, which after an $\mathcal{O}(m^2 + |\text{rectangles}|)$-time preprocessing answers such queries in $\mathcal{O}(1)$ time.

Namely, the data structure consists of a $[1, m + 1]^2$ grid Γ (a 2d-array of integers) in which for every rectangle $[u, v] \times [w, x]$ we add 1 to $\Gamma[u][w]$ and $\Gamma[v+1][x+1]$ and -1 to $\Gamma[u][x+1]$ and $\Gamma[v+1][w]$. Then we modify Γ to contain the 2d prefix sums of its original values (we first compute prefix sums of each row, and then prefix sums of each column of the result). After these modifications, $\Gamma[x][y]$ stores the number of rectangles containing point (x, y), and hence after $\mathcal{O}(m^2 + |\text{rectangles}|)$-time preprocessing we can answer 2d rectangle stabbing queries in $\mathcal{O}(1)$ time. In our case we have a total of $\mathcal{O}(m)$ such grid structures, each of $\mathcal{O}(m^2)$ size, and ask $\mathcal{O}(m^2)$ queries, and hence obtain an $\mathcal{O}(m^3 + N_i)$-time and $\mathcal{O}(m^2)$-space solution for computing 1-AP_i from AP_{i-1}. $\qquad\square$

Prefix Case: A Special Case of 2d Rectangle Stabbing. Inspect the example of Fig. 4 for the Anchor Case. Note that the groups of rectangles for each string have the special property of being composed of *nested intervals*: for each dimension, the interval corresponding to a given node is included in the one corresponding to any of its ancestors. Thus for the Prefix Case, where we only spell fragments of the same string P in both compacted tries, we consider the following special case of off-line 2d rectangle stabbing.

Lemma 9. *Let $p_1, \ldots, p_h$ and $q_1, \ldots, q_h$ be two permutations of $[1, h]$. We denote by Π the set of h points $(p_1, q_1), (p_2, q_2), \ldots, (p_h, q_h)$ on $[1, h]^2$. Further let R be a collection of r axis-aligned rectangles $([u_1, v_1], [w_1, x_1]), \ldots, ([u_r, v_r], [w_r, x_r])$, such that*

$$[u_r, v_r] \subseteq [u_{r-1}, v_{r-1}] \subseteq \cdots \subseteq [u_1, v_1] \quad \text{and} \quad [w_1, x_1] \subseteq [w_2, x_2] \subseteq \cdots \subseteq [w_r, x_r].$$

Then we can find out, for every point from Π, if it stabs any rectangle from R in $\mathcal{O}(h + r)$ total time.

Proof. Let H be a bit vector consisting of h bits, initially all set to zero. We process one rectangle at a time. We start with $([u_1, v_1], [w_1, x_1])$. We set $H[p] = 1$ if and only if $(p, q) \in \Pi$ for $p \in [u_1, v_1]$ and any q. We collect all p such that $(p, q) \in \Pi$ and $q \in [w_1, x_1]$, and then search for these p in H: if for any p, $H[p] = 1$, then the answer is positive for p. Otherwise, we remove from H every p such that $p \in [u_1, v_1]$ and $p \notin [u_2, v_2]$ by setting $H[p] = 0$. We proceed by collecting all p such that $(p, q) \in \Pi$, $q \in [w_2, x_2]$ and $q \notin [w_1, x_1]$, and then search for them in H: if for any p, $H[p] = 1$, then the answer is positive for p. We repeat this until H is empty or until there are no other rectangles to process.

The whole procedure takes $\mathcal{O}(h + r)$ time, because we set at most h bits on in H, we set at most h bits back off in H, we search for at most h points in H, and then we process r rectangles. $\qquad\square$

Lemma 10. *We can solve the Prefix (resp. Suffix) Case, that is, report 1-error occurrences ending in $\tilde{T}[i]$ (resp. compute active prefixes with 1 error starting in $\tilde{T}[i]$) in $\mathcal{O}(m^2 + N_i)$ time.*

Proof. We employ Lemma 9 to get rid of the 2d range data structure. The key is that for every length-μ suffix $P[\lambda + 1 \mathinner{.\,.} m]$ of the pattern we can afford to pay $\mathcal{O}(\mu + |G_\mu|)$ time plus the time to construct T_{G_μ} and $T_{G_\mu}^R$ for set G_μ. Because the grid is $[1, |G_\mu|]^2$, we exploit the fact that the intervals found by spelling $P[\lambda + 1 \mathinner{.\,.} \lambda + \mu - 1]$ through T_{G_μ} and $P^R[1 \mathinner{.\,.} \mu - 1]$ through $T_{G_\mu}^R$, one letter at a time, are subset of each other, and querying μ such rectangles is done in $\mathcal{O}(\mu + |G_\mu|)$ time by employing Lemma 9. Since we process at most m distinct length-μ suffixes of P, the total time is $\mathcal{O}(m^2 + N_i)$, because $\sum_\mu |G_\mu| \leq N_i$. □

3.5 Wrapping-up

To obtain Theorem 1 for the decision version of the problem we first compute AP_i and AS_i, for all $i \in [1, n]$, in $\mathcal{O}(nm^2 + N)$ total time (Corollary 1). We then compute all the occurrences in the Easy Cases using $\mathcal{O}(N)$ time in total (Sect. 3.1); and we finally compute all the occurrences in the Prefix and Suffix Cases in $\sum_i \mathcal{O}(m^2 + N_i) = \mathcal{O}(nm^2 + N)$ total time (Lemma 10).

Now, to solve the decision version of the problem, we solve the Anchor Cases with the use of the precomputed AP_{i-1} and AS_{i+1} for each $i \in [2, n-1]$ in $\mathcal{O}(m^2\sqrt{\log m} + N_i \log \log m)$ time (Sect. 3.2), which gives $\mathcal{O}(nm^2\sqrt{\log m} + N \log \log m)$ total time for the whole algorithm.

For the reporting version we proceed differently to obtain an on-line algorithm; note that this is possible because we can proceed without AS_i (see Fig. 2). We thus consider one segment $\tilde{T}[i]$ at the time, for each $i \in [1, n]$, and do the following. We compute 1-AP_i, as the union of three sets obtained from:

- The Suffix Case for $\tilde{T}[i]$, computed in $\mathcal{O}(m^2 + N_i)$ time (Lemma 10).
- Standard APE with 1-AP_{i-1} as the input bit vector, computed in $\mathcal{O}(m^2 + N_i)$ time (Lemma 3).
- Anchor Case computed from AP_{i-1} in $\mathcal{O}((m^2 + N_i) \log N_i)$ (Sect. 3.2) or $\mathcal{O}(m^3 + N_i)$ time (Lemma 8).

If $N_i \geq m^3$, the algorithm of Lemma 8 works in the optimal $\mathcal{O}(m^3 + N_i) = \mathcal{O}(N_i)$ time, hence we can assume that the $\mathcal{O}((m^2 + N_i) \log N_i)$-time algorithm is only used when $N_i \leq m^3$, and thus it runs in $\mathcal{O}((m^2 + N_i) \log m)$ time. Therefore over all i the computations require $\mathcal{O}((nm^2 + N) \log m)$ or $\mathcal{O}(nm^3 + N)$ total time. For every segment i we can also check whether an active prefix from 1-AP_{i-1} or from AP_{i-1} can be completed to a full match in $\tilde{T}[i]$ using the algorithms of Grossi et al. from [24] and Prefix Case, respectively, in $\mathcal{O}(m^2 + N_i)$ extra time.

By summing up all these we obtain Theorem 1.

4 Open Questions

We leave the following basic questions open for future investigation:

1. Can we design an $\mathcal{O}(nm^2 + N)$-time algorithm for 1-EDSM?
2. Can our techniques be efficiently generalized for $k > 1$ errors?

References

1. t al Alzamel, M., e.: Degenerate string comparison and applications. In: Parida, L., Ukkonen, E. (eds.) 18th International Workshop on Algorithms in Bioinformatics, WABI 2018, Helsinki, Finland, 20–22 August 2018, LIPIcs, vol. 113, pp. 21:1–21:14. Schloss Dagstuhl - Leibniz-Zentrum für Informatik (2018). https://doi.org/10.4230/LIPIcs.WABI.2018.21

2. Alzamel, M., et al.: Comparing degenerate strings. Fundam. Informaticae **175**(1–4), 41–58 (2020). https://doi.org/10.3233/FI-2020-1947

3. Amir, A., Keselman, D., Landau, G.M., Lewenstein, M., Lewenstein, N., Rodeh, M.: Text indexing and dictionary matching with one error. J. Algorithms **37**(2), 309–325 (2000). https://doi.org/10.1006/jagm.2000.1104

4. Amir, A., Lewenstein, M., Porat, E.: Faster algorithms for string matching with k mismatches. J. Algorithms **50**(2), 257–275 (2004). https://doi.org/10.1016/S0196-6774(03)00097-X

5. Aoyama, K., Nakashima, Y., Inenaga, T., Inenaga, S., Bannai, H., Takeda, M.: Faster online elastic degenerate string matching. In: Navarro, G., Sankoff, D., Zhu, B. (eds.) Annual Symposium on Combinatorial Pattern Matching, CPM 2018, Qingdao, China, 2–4 July 2018, LIPIcs, vol. 105, pp. 9:1–9:10. Schloss Dagstuhl - Leibniz-Zentrum für Informatik (2018). https://doi.org/10.4230/LIPIcs.CPM.2018.9

6. Bender, M.A., Farach-Colton, M.: The LCA problem revisited. In: Gonnet, G.H., Viola, A. (eds.) LATIN 2000. LNCS, vol. 1776, pp. 88–94. Springer, Heidelberg (2000). https://doi.org/10.1007/10719839_9

7. Bernardini, G., Pisanti, N., Pissis, S., Rosone, G.: Pattern matching on elastic-degenerate text with errors. In: 24th International Symposium on String Processing and Information Retrieval (SPIRE), pp. 74–90 (2017). https://doi.org/10.1007/978-3-319-67428-5_7

8. Bernardini, G., Gawrychowski, P., Pisanti, N., Pissis, S.P., Rosone, G.: Even faster elastic-degenerate string matching via fast matrix multiplication. In: Baier, C., Chatzigiannakis, I., Flocchini, P., Leonardi, S. (eds.) 46th International Colloquium on Automata, Languages, and Programming, ICALP 2019, Patras, Greece, 9–12 July 2019, LIPIcs, vol. 132, pp. 21:1–21:15. Schloss Dagstuhl - Leibniz-Zentrum für Informatik (2019). https://doi.org/10.4230/LIPIcs.ICALP.2019.21

9. Bernardini, G., Gawrychowski, P., Pisanti, N., Pissis, S.P., Rosone, G.: Elastic-degenerate string matching via fast matrix multiplication. SIAM J. Comput. **51**(3), 549–576 (2022). https://doi.org/10.1137/20M1368033

10. Bernardini, G., Pisanti, N., Pissis, S.P., Rosone, G.: Approximate pattern matching on elastic-degenerate text. Theor. Comput. Sci. **812**, 109–122 (2020). https://doi.org/10.1016/j.tcs.2019.08.012

11. Chan, T.M., Larsen, K.G., Patrascu, M.: Orthogonal range searching on the RAM, revisited. In: Hurtado, F., van Kreveld, M.J. (eds.) Proceedings of the 27th ACM Symposium on Computational Geometry, Paris, France, 13–15 June 2011, pp. 1–10. ACM (2011). https://doi.org/10.1145/1998196.1998198

12. Charalampopoulos, P., Iliopoulos, C.S., Liu, C., Pissis, S.P.: Property suffix array with applications in indexing weighted sequences. ACM J. Exp. Algorithmics **25**, 1–16 (2020). https://doi.org/10.1145/3385898

13. Charalampopoulos, P., Kociumaka, T., Wellnitz, P.: Faster approximate pattern matching: a unified approach. In: Irani, S. (ed.) 61st IEEE Annual Symposium on

Foundations of Computer Science, FOCS 2020, Durham, NC, USA, 16–19 November 2020, pp. 978–989. IEEE (2020). https://doi.org/10.1109/FOCS46700.2020.00095

14. Charalampopoulos, P., Kociumaka, T., Wellnitz, P.: Faster pattern matching under edit distance. CoRR **abs/2204.03087** (2022). https://doi.org/10.48550/arXiv.2204.03087, (announced at FOCS 2022)

15. Chazelle, B.: A functional approach to data structures and its use in multidimensional searching. SIAM J. Comput. **17**(3), 427–462 (1988). https://doi.org/10.1137/0217026

16. Cole, R., Gottlieb, L., Lewenstein, M.: Dictionary matching and indexing with errors and don't cares. In: Babai, L. (ed.) Proceedings of the 36th Annual ACM Symposium on Theory of Computing, Chicago, IL, USA, 13–16 June 2004, pp. 91–100. ACM (2004). https://doi.org/10.1145/1007352.1007374

17. Cole, R., Hariharan, R.: Approximate string matching: a simpler faster algorithm. SIAM J. Comput. **31**(6), 1761–1782 (2002). https://doi.org/10.1137/S0097539700370527

18. Crochemore, M., Hancart, C., Lecroq, T.: Algorithms on Strings. Cambridge University Press, Cambridge (2007). https://doi.org/10.1017/CBO9780511546853

19. Farach, M.: Optimal suffix tree construction with large alphabets. In: 38th Annual Symposium on Foundations of Computer Science, FOCS 1997, Miami Beach, Florida, USA, 19–22 October 1997, pp. 137–143. IEEE Computer Society (1997). https://doi.org/10.1109/SFCS.1997.646102

20. Fredman, M.L., Komlós, J., Szemerédi, E.: Storing a sparse table with 0(1) worst case access time. J. ACM **31**(3), 538–544 (1984). https://doi.org/10.1145/828.1884

21. Gao, Y., He, M., Nekrich, Y.: Fast preprocessing for optimal orthogonal range reporting and range successor with applications to text indexing. In: Grandoni, F., Herman, G., Sanders, P. (eds.) 28th Annual European Symposium on Algorithms, ESA 2020, Pisa, Italy (Virtual Conference), 7–9 September 2020, LIPIcs, vol. 173, pp. 54:1–54:18. Schloss Dagstuhl - Leibniz-Zentrum für Informatik (2020). https://doi.org/10.4230/LIPIcs.ESA.2020.54

22. Gawrychowski, P., Ghazawi, S., Landau, G.M.: On indeterminate strings matching. In: Gørtz, I.L., Weimann, O. (eds.) 31st Annual Symposium on Combinatorial Pattern Matching, CPM 2020, Copenhagen, Denmark, 17–19 June 2020, LIPIcs, vol. 161, pp. 14:1–14:14. Schloss Dagstuhl - Leibniz-Zentrum für Informatik (2020). https://doi.org/10.4230/LIPIcs.CPM.2020.14

23. Gawrychowski, P., Uznanski, P.: Towards unified approximate pattern matching for Hamming and L1 distance. In: Chatzigiannakis, I., Kaklamanis, C., Marx, D., Sannella, D. (eds.) 45th International Colloquium on Automata, Languages, and Programming, ICALP 2018, Prague, Czech Republic, 9–13 July 2018, LIPIcs, vol. 107, pp. 62:1–62:13. Schloss Dagstuhl - Leibniz-Zentrum für Informatik (2018). https://doi.org/10.4230/LIPIcs.ICALP.2018.62

24. Grossi, R., Iliopoulos, C.S., Liu, C., Pisanti, N., Pissis, S.P., Retha, A., Rosone, G., Vayani, F., Versari, L.: On-line pattern matching on similar texts. In: Kärkkäinen, J., Radoszewski, J., Rytter, W. (eds.) 28th Annual Symposium on Combinatorial Pattern Matching, CPM 2017, Warsaw, Poland, 4–6 July 2017, LIPIcs, vol. 78, pp. 9:1–9:14. Schloss Dagstuhl - Leibniz-Zentrum für Informatik (2017). https://doi.org/10.4230/LIPIcs.CPM.2017.9

25. Iliopoulos, C.S., Kundu, R., Pissis, S.P.: Efficient pattern matching in elastic-degenerate strings. Inf. Comput. **279**, 104616 (2021). https://doi.org/10.1016/j.ic.2020.104616

26. IUPAC-IUB Commission on Biochemical Nomenclature: Abbreviations and symbols for nucleic acids, polynucleotides, and their constituents. Biochemistry **9**(20), 4022–4027 (1970). https://doi.org/10.1016/0022-2836(71)90319-6

27. Landau, G.M., Vishkin, U.: Efficient string matching with k mismatches. Theor. Comput. Sci. **43**, 239–249 (1986). https://doi.org/10.1016/0304-3975(86)90178-7

28. Landau, G.M., Vishkin, U.: Fast string matching with k differences. J. Comput. Syst. Sci. **37**(1), 63–78 (1988). https://doi.org/10.1016/0022-0000(88)90045-1

29. Na, J.C., Apostolico, A., Iliopoulos, C.S., Park, K.: Truncated suffix trees and their application to data compression. Theor. Comput. Sci. **304**(1–3), 87–101 (2003). https://doi.org/10.1016/S0304-3975(03)00053-7

30. Ružić, M.: Constructing efficient dictionaries in close to sorting time. In: Aceto, L., Damgård, I., Goldberg, L.A., Halldórsson, M.M., Ingólfsdóttir, A., Walukiewicz, I. (eds.) ICALP 2008. LNCS, vol. 5125, pp. 84–95. Springer, Heidelberg (2008). https://doi.org/10.1007/978-3-540-70575-8_8

31. Shi, Q., JáJá, J.F.: Novel transformation techniques using q-heaps with applications to computational geometry. SIAM J. Comput. **34**(6), 1474–1492 (2005). https://doi.org/10.1137/S0097539703435728

32. The Computational Pan-Genomics Consortium: Computational pan-genomics: status, promises and challenges. Brief. Bioinf. **19**(1), 118–135 (2018). https://doi.org/10.1093/bib/bbw089

Median and Hybrid Median
K-Dimensional Trees

Amalia Duch[1] , Conrado Martínez[1](✉) , Mercè Pons[2],
and Salvador Roura[1]

[1] Department of Computer Science, Universitat Politècnica de Catalunya,
Barcelona, Spain
{duch,conrado,roura}@cs.upc.edu

[2] Departament d'Ensenyament, Generalitat de Catalunya, Barcelona, Spain

Abstract. We consider here two new variants of K-dimensional binary
search trees (K-d trees): median K-d trees and hybrid-median K-d trees.
These two kinds of trees are designed with the aim to get a tree as
balanced as possible. This goal is attained by heuristics that choose for
each node of the K-d tree the appropriate coordinate to discriminate. In
the case of median K-d trees, the chosen dimension to discriminate at
each node is the one whose point value at that node is the most centered
one. In hybrid-median K-d trees, the heuristic is similar except that it
should be followed in a cyclic way, meaning that, at every path of the
tree, no dimension can be re-selected to discriminate unless all the other
dimensions have already been selected. We study the expected internal
path length (IPL) and the expected cost of random partial match (PM)
searches in both variants of K-d trees. For both variants, we prove that
the expected IPL is of the form $c_K \cdot n \log_2 n +$ lower order terms, and the
expected cost of PM is of the form $\Theta(n^\alpha)$ with $\alpha = \alpha(s, K)$. We give
the explicit equations satisfied by the constants c_K and the exponents α
which we can then numerically solve. Moreover, we prove that as $K \to \infty$
the trees in both variants tend to be perfectly balanced ($c_K \to 1$) and we
also show that $\alpha \to \log_2(2 - s/K)$ for median K-d trees when $K \to \infty$.
In the case of hybrid median K-d trees we conjecture that $\alpha \to 1 - s/K$,
when $K \to \infty$, which would be optimal.

Keywords: K-d trees · Multidimensional data structures · Partial
match queries · Analysis of algorithms

1 Introduction

In this work we study two variants of K-dimensional binary search trees [1,14]
(K-d trees, for short): *median K-d tree* and *hybrid median K-d tree*; both were
introduced by Pons [12] in 2010. When built from uniformly distributed input

This work has been supported by funds from the MOTION Project (Project
PID2020-112581GB-C21) of the Spanish Ministry of Science & Innovation
MCIN/AEI/10.13039/501100011033.

A. Castañeda and F. Rodríguez-Henríquez (Eds.): LATIN 2022, LNCS 13568, pp. 38–53, 2022.
https://doi.org/10.1007/978-3-031-20624-5_3

data sets, these two simple variants of K-d trees achieve better costs for exact searches and insertions than other variants of K-d trees. They also perform better with respect to *partial match queries* which in turn implies better performance in other *associative queries* like *orthogonal range* or *nearest neighbour* queries.

Recall that a K-d tree is a binary search tree that stores a collection F of items, each endowed with a K-dimensional key $\mathbf{x} = (x_0, \ldots, x_{K-1})$. In addition to the data point key $\mathbf{x}$, each node $\langle \mathbf{x}, j \rangle$ of a K-d tree stores a *discriminant*, a value j, $0 \leq j < K$, which is the coordinate that will be used to split the inserted keys into the left and right subtrees rooted at $\langle \mathbf{x}, j \rangle$: the data points with a key $\mathbf{y}$ such that $y_j < x_j$ are recursively inserted into the left subtree, whereas those with a key $\mathbf{z}$ such that $x_j < z_j$ are recursively inserted into the right one[1].

The original K-d trees —we will refer to these as *standard K-d tree*— were introduced by Bentley in the mid 70s [1] with a rule that assigns discriminants in a cyclic way. Thus, a node at level $\ell \geq 0$ has discriminant $\ell \bmod K$. Several variants of K-d trees differ in the way in which the discriminants are assigned to nodes, whereas other variants apply local (for example, Kdt trees [2]) or global rebalancing rules (for example, *divided K-d tree* [8]). Among the variants that use alternative rules to assign discriminants we have *relaxed K-d tree* [4], which assign discriminants uniformly and independently at random, and *squarish K-d tree* [3], which try to get a partition as balanced as possible of the data space.

Median K-d trees and hybrid median K-d trees also aim to build a more balanced tree. In median K-d trees the rule is to choose as discriminant at each node the coordinate that would presumably divide the forthcoming elements as evenly as possible into the two subtrees of the node. While this can be easily accomplished if we have the collection of n items beforehand, median K-d trees achieve a similar outcome using a heuristic based on the usual assumption that the keys from which the tree is built are drawn uniformly at random in $[0,1]^K$. Besides, hybrid median K-d trees combine the heuristics of standard and median K-d trees: at every node the coordinate used to discriminate is chosen using the median K-d tree heuristic but, in a cyclic way as in standard K-d trees.

We use here the *internal path length* (IPL)[2] [7] of median K-d trees and hybrid median K-d trees as a measure of their degree of balance and of the cost of building the tree and of exact (successful) searches. As general purpose data structures, K-d trees provide efficient (not necessarily optimal and only on expectation) support for dynamic insertions, exact searches and several associative queries. In particular, we focus here on *random partial match* queries (random PM queries), first because of their own intrinsic interest and second because their analysis is a fundamental block in the analysis of other kind of associative queries such as orthogonal range and nearest neighbour queries.

---

[1] We have omitted on purpose what to do with elements $\mathbf{v}$ such that $x_j = v_j$; several alternatives exist to cope with such situation, but in the random model which we will use for the analysis such event does not occur and hence the strategy used to cope with such situation becomes unimportant.

[2] The internal path length of a binary search tree is the sum, over all its internal nodes, of the paths from the root to every node of the tree.

A random PM query is a pair $\langle \mathbf{q}, \mathbf{u} \rangle$, where $\mathbf{q} = (q_0, \ldots, q_{K-1})$ is a K-dimensional point independently drawn from the same continuous distribution as the data points, and $\mathbf{u} = (u_0, \ldots, u_{K-1})$ is the *pattern* of the query; each $u_i = S$ (the i-th attribute of the query is *specified*) or $u_i = *$ (the i-th attribute is *unspecified*). The goal of the PM search is to report all data points $\mathbf{x} = (x_0, \ldots, x_{K-1})$ in the tree such that $x_i = q_i$ whenever $u_i = S$ where s is the number of specified coordinates; the interesting cases arise when $0 < s < K$.

Our main tool for the analysis of the expected IPL and the expected cost of random PMs is the continuous master theorem (CMT, for short) [13] and some "extensions" developed here to cope with systems of divide-and-conquer recurrences. In particular, we give the main order term of the expected IPL of median K-d trees and hybrid median K-d trees: in both cases it is of the form $\sim c_K n \log_2 n$ for a constant c_K depending on K and on the variant of K-d tree considered (Theorems 1 and 3); median K-d trees and hybrid median K-d trees perform better than other variants, for all $K \geq 2$, since $c_K < 2$ —while $c_K = 2$ for all K in standard, relaxed and squarish K-d trees. Moreover, in median K-d trees and hybrid median K-d trees $c_K \to 1$ as $K \to \infty$, which is optimal for data structures built using key comparisons. We also show that the expected cost of random PM searches will be always $\Theta(n^\alpha)$ for an exponent α which depends on the variant of K-d trees, the dimension K and the number s of coordinates which are specified in the PM query. We give the equations satisfied by the exponent α in each case (Theorems 2 and 4). Although in general these equations are not analytically solvable, it is possible to provide accurate numerical approximations. In the case of median K-d trees, the expected cost of PM queries lies somewhere between that of standard K-d trees and that of relaxed K-d trees, and $\alpha \to \log_2(2 - s/K)$ as $K \to \infty$. For hybrid median K-d trees the expected PM cost outperforms that of relaxed and standard K-d trees for all $K \geq 2$, and we conjecture that $\alpha \to 1 - s/K$ as $K \to \infty$, which is optimal. Table 1 summarizes our results comparing them to other variants of K-d trees.

Table 1. An abridged comparison of median K-d trees and hybrid median K-d trees with other families of K-d trees, giving the coefficient of $n \log n$ for IPL and the exponent α for PM where $*$ indicates conjectured.

Family	IPL		Partial match $s = 1, s = K/2$,	
	$K = 2$	$K \to \infty$	$K = 2$	$K \to \infty$
Standard K-d trees [1,6]	2	2	0.56155	0.56155
Relaxed K-d trees [4,10]	2	2	0.618	0.618
Squarish K-d trees [3]	2	2	0.5	0.5
Median K-d trees [**this paper**]	1.66	$\to 1.443$	0.602	$\to 0.585$
Hybrid median K-d trees [**this paper**]	1.814	$\to 1.443$	0.546	$\to 0.5^*$

This paper is organized as follows. In Sect. 2 we give the definition of random median K-d trees as well as the previous known results on them and we present the analysis of their expected IPL (Subsec. 2.1) and the expected cost of random PMs (Subsec. 2.2). In Sect. 3 we proceed as in the preceding section, now with the analysis of random hybrid median K-d trees. Finally, in Sect. 4 we give our conclusions and guidelines for future work.

2   Median K-d Trees

Median K-d trees were introduced by Pons [12] and they are a simple variant of standard K-d trees: the only difference lies in the way to choose the dimension used to discriminate at each node.

As happens in plane binary search trees, in K-d trees the insertion of an item creates a new node that replaces a leaf of the current tree. It is worth noting that every leaf of a K-d tree corresponds to a region of the space from which the elements are drawn and hence the whole tree induces a partition of the space —$[0,1]^K$ in our case. The region delimited by the leaf that a new node replaces at the moment of its insertion into the tree is known as its *bounding box*.

In median K-d trees, when a new data point $\mathbf{x} = (x_0, \ldots, x_{K-1})$ is inserted in the bounding box $R = [\ell_0, u_0] \times \cdots [\ell_{K-1}, u_{K-1}]$ the discriminant j is chosen as follows,

$$j = \arg\min_{0 \leq i < K} \left\{ \left| \frac{x_i - \ell_i}{u_i - \ell_i} - \frac{1}{2} \right| \right\}.$$

An example of median K-d tree together with its induced partition of the space is shown in Fig. 1.

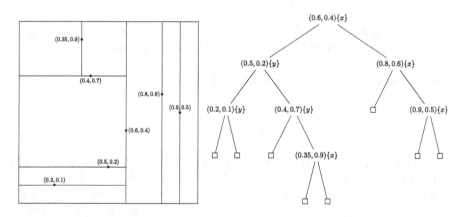

Fig. 1. Example of a median K-d tree built from 2-dimensional points.

In the analysis of the expected IPL and the expected cost of random PM in a median K-d tree of size n, we will assume, as usual in the literature, that the tree is randomly built. That is, that the n points are random and independently drawn from $[0,1]^K$, with each coordinate x_i of a data point $\mathbf{x}$ independently and uniformly drawn from $[0,1]$.

In [12] it is shown that (i) the expected IPL of random median K-d trees is $I_n \sim c_K n \log_2 n$ for a constant c_K depending on K; it is also stated there without formal proof that $c_K \to 1$ as $K \to \infty$; and (ii) that, for $K = 2$ and $K = 3$, the expected cost of a random PM is $\Theta(n^\alpha)$ with $\alpha(1,2) \approx 0.60196\ldots$, $\alpha(2,3) \approx 0.74387\ldots$ and $\alpha(1,3) \approx 0.42756\ldots$.

Here, using the CMT, we obtain the same results for the expected IPL and extend the analysis of the expected cost of random PM to any value of K and s proving also that $\alpha \to \log_2(2 - s/K)$ as K grows (and s/K remains constant).

In order to proceed with the analysis, we need to compute the probability that the left subtree of a random median K-d tree is of size j, given that the tree is of size n. This is crucial in order to set up the recurrences for the expected IPL and the expected cost of partial matches in the next subsections, and it enables the systematic application of the CMT (see [13] or Appendix A of [5]) to solve the recurrences, instead of the ad-hoc arguments given in [12].

Let $\mathbf{x} = (x_0, x_1, \ldots, x_{K-1})$ be the key stored at the root of a median K-d tree T that contains n data points. We can define the *rank vector* of $\mathbf{x}$ as $\mathbf{r} = \mathbf{r}(\mathbf{x}, T) = (r_0, r_1, \ldots, r_{K-1})$ where r_i is the number of data points in T with i-th coordinate smaller or equal to x_i. If the root of T discriminates with respect to the i-th coordinate then —because we assume that the tree is randomly built— the size of the left subtree L of T will be $r_i - 1$ and the size of the right subtree will be $n - r_i$. In an idealization of median K-d trees the chosen discriminating coordinate will be i if r_i is the closest rank to $\lfloor (n+1)/2 \rfloor$ —ties are resolved in favor of the coordinate with smallest index. It follows that

$$\mathbb{P}\{|L| = j \mid |T| = n\} = \mathbb{P}\{Z_{n,K} = j+1\}, \qquad 0 \le j < |T|,$$

where $Z_{n,K}$ denotes the closest integer to $\lfloor (n+1)/2 \rfloor$ (equivalently the closest integer to $\lceil n/2 \rceil$) in a set of K given integers independently and uniformly drawn from $\{1, \ldots, n\}$.

For example, let $K = 2$ and $Z := Z_{n,2}$. Then we have

$$\mathbb{P}\{Z = j\} = \begin{cases} \frac{4j-1}{n^2} & \text{if } j \le \lfloor \frac{n}{2} \rfloor, \\ \frac{4(n-j)+1}{n^2} & \text{if } j > \lfloor \frac{n}{2} \rfloor. \end{cases} \tag{1}$$

To see why, suppose that $n = 2\lambda + 1$ and $j \le \lambda = \lfloor n/2 \rfloor$. Then $Z = j$ will occur if (1) both ranks are equal to j, this happens with probability $1/n^2$ or (2) one rank is j the other is $< j$ or $\ge n+1-j$, which will happen with probability $2 \cdot (1/n) \cdot (j-1+j)/n = (4j-2)/n^2$. Hence the probability of $Z = j$ when $j \le \lambda$ is $(4j-1)/n^2$. The case for $j > \lambda + 1$ is similar except that ties in the distance to the center are resolved in favor of the smallest rank: thus if $j > \lambda + 1$ then $n - j + 1$ will be at the same distance to the center but smaller than j hence $Z = j$ requires one rank to be j and the other be smaller than $n + 1 - j$.Thus, the probability that $Z = j$ when $j > \lambda + 1$ is

$1/n^2 + 2 \cdot 2(n-j)/n^2 = (4(n-j)+1)/n^2$. On the other hand, if $j = \lambda + 1$ then we will have $Z = j$ no matter what the other rank is; we have that the probability of $Z = \lambda + 1$ is $1/n^2 + 2(n-1)/n^2 = (2n-1)/n^2 = (4\lambda+1)/n^2 = (4(n-j)+1)/n^2$. Therefore, we can write that the probability of $Z = j$ when $j \geq \lambda + 1 > \lfloor n/2 \rfloor$ is $(4(n-j)+1)/n^2$. For even n, when $n = 2\lambda$, the arguments are identical and Eq. (1) holds too.

For the general case, we can reason in an analogous way, assuming that $\ell \geq 1$ of the K ranks are j and $K - \ell$ are either smaller that j or greater than $n - j$. If $j \leq \lfloor n/2 \rfloor$ then

$$\mathbb{P}\{Z = j\} = \frac{1}{n^K} \cdot \left[(2j)^K - (2j-1)^K \right],$$

and if $j > \lfloor n/2 \rfloor$ then the analysis is analogous but we need a small correction as we cannot allow any coordinate to be $n + 1 - j$, hence in that case

$$\mathbb{P}\{Z = j\} = \frac{1}{n^K} \cdot \left[(2(n-j)+1)^K - (2(n-j))^K \right].$$

2.1   Internal Path Length

Let us start writing down the recurrence for the expected IPL I_n of a random median K-d tree T of size n, for $n > 0$. For that, we condition on the size of the left subtree L, thus

$$
\begin{aligned}
I_n &= n - 1 + \sum_{j=0}^{n-1} \pi_{n,j}(I_j + I_{n-1-j}) = n - 1 + \sum_{j=0}^{n-1} \pi_{n,j} I_j + \sum_{j=0}^{n-1} \pi_{n,n-1-j} I_j \\
&= n - 1 + \sum_{j=0}^{n-1} (\pi_{n,j} + \pi_{n,n-1-j}) I_j,
\end{aligned}
\tag{2}
$$

with $\pi_{n,j} = \mathbb{P}\{|L| = j \mid |T| = n\} = \mathbb{P}\{Z_{n,K} = j+1\}$ and $I_0 = 0$. Indeed, the IPL of T is the sum of the IPL of its subtrees L and R, and we add $+1$ for every internal node other than the root. In order to apply the continuous master theorem we identify $\omega_{n,j} = \pi_{n,j} + \pi_{n,n-1-j}$ as the weights sequence in the divide-and-conquer recurrence. Substituting j by $z \cdot n$, multiplying by n and taking the limit when $n \to \infty$ we get the *shape function*

$$
\omega_K(z) = \lim_{n \to \infty} n \cdot \omega_{n,z \cdot n} =
\begin{cases}
2K(2z)^{K-1} = K2^K z^{K-1}, & \text{if } z \leq 1/2, \\
2K(2(1-z))^{K-1} = K2^K (1-z)^{K-1}, & \text{if } z \geq 1/2.
\end{cases}
$$

When $n \to \infty$, the shape function derived for the idealization using ranks is the actual shape function for median K-d trees, where we would have had to compute the probability that, given a random set of K points $X_0, \ldots, X_{K-1}$ independently and uniformly drawn from $[0,1]$, we have $Z'_{n,K} = j$ with

$$Z'_{n,K} = \#\{X_i \mid X_i < X_\ell\},$$

where $\ell = \arg\min_{0 \le i < K}\{|X_i - 1/2|\}$.

Once we have the shape function for the divide-and-conquer recurrence, we can get the const-entropies for all $K \ge 1$:

$$\mathcal{H}_K = 1 - \int_0^1 z\,\omega_K(z)\,dz = 0.$$

As they all are zero, we need to compute the log-entropies:

$$\mathcal{H}'_K = -\int_0^1 z\ln(z)\omega_K(z)\,dz. \tag{3}$$

No easy closed form for $\mathcal{H}'_K$ is available; but we can compute any value of $\mathcal{H}'_K$ and thus of the expected IPL (see Table 2).

Theorem 1 (Pons, 2010). *The expected IPL of random median K-d tree of size n is*

$$I_n = c_K n \ln n + o(n \log n)$$

where

$$c_K^{-1} = \mathcal{H}'_K = -K2^K\left[A_K + \sum_{0 \le i < K}\binom{K-1}{i}(-1)^i B_{i+1}\right],$$

with $B_j = -(A_j + 1/(j+1)^2)$ and

$$A_j = \int_0^{1/2} z^j \ln z\,dz = -\frac{1 + (j+1)\ln 2}{2^{j+1}(j+1)^2},$$

The IPL gives a measure of the cost of building the K-d tree in the first place, but also of the cost of exact successful searches. Indeed, $\frac{I_n}{n} = c_K \cdot \ln n + o(\log n)$ is the expected depth of a random node. We can use the definition of $\mathcal{H}'_K$ to show that $\mathcal{H}'_K < \mathcal{H}'_{K+1}$ and thus the coefficients $c_K = (\mathcal{H}'_K)^{-1}$ are monotonically decreasing with K. It is also easy to prove that $c_K \to 1/\ln 2$ which implies that median K-d trees tend to get perfectly balanced, as $K \to \infty$ (see Fig. 3 on page 12). Indeed, from the definition (3) of $\mathcal{H}'_K$, if we let $K \to \infty$ the shape function under the integral sign degenerates to a Dirac's delta distribution at $z = 1/2$ and thus

$$\mathcal{H}'_K \to -\int_0^1 \ln z\,\delta_{1/2}(z)\,dz = -\ln(1/2) = \ln 2.$$

2.2   Random Partial Match

Consider a random partial match with s specified coordinates, $0 < s < K$. Because of the symmetries of the problem all the coordinates are equivalent with respect to the query pattern and thus we can assume without loss of generality that the query is of the form $\mathbf{q} = (q_0, \ldots, q_{s-1}, *, \ldots, *)$ with q_i a uniformly

Table 2. Coefficient of the first order term in the expected IPL of random median K-d trees.

K	$\mathcal{H}'_K$	$\mathbb{E}\{I_n\}/(n\ln n) \sim c_K = 1/\mathcal{H}'_K$
1	$1/2$	2
2	$5/6 - 1/3\ln 2 \approx 0.6023$	1.660
3	$4/3 - \ln 2 \approx 0.6402$	1.562
4	$131/60 - 11/5\ln 2 \approx 0.6584$	1.519
...	...	...
∞	$\ln 2 \approx 0.6931$	$1/\ln 2 \approx 1.443$

drawn real number in $[0, 1]$. Then the recurrence for the expected cost $P_n := P_n^{(K,s)}$ of the PM is

$$P_n = 1 + \frac{s}{K}\sum_{j=0}^{n-1}\pi_{n,j}\left(\frac{j+1}{n+1}P_j + \frac{n-j}{n+1}P_{n-1-j}\right) + \frac{K-s}{K}\sum_{j=0}^{n-1}\pi_{n,j}(P_j + P_{n-1-j})$$

$$= 1 + \frac{s}{K}\sum_{j=0}^{n-1}(\pi_{n,j} + \pi_{n,n-1-j})\frac{j+1}{n+1}P_j + \frac{K-s}{K}\sum_{j=0}^{n-1}(\pi_{n,j} + \pi_{n,n-1-j})P_j. \quad (4)$$

To derive the recurrence above, we condition on the size of the left subtree, and consider two possibilities: with probability s/K the discriminating coordinate of the root is specified, and we have to continue recursively in the left or the right subtree with probability proportional to their number of leaves of each subtree. On the other hand, with probability $(K - s)/K$ the discriminating coordinate of the root is not specified and the PM must continue in both subtrees. We have thus that the shape function is

$$\omega_K(z) = \begin{cases} K2^K z^{K-1}(\rho z + 1 - \rho), & \text{if } z \leq 1/2, \\ K2^K(1-z)^{K-1}(\rho z + 1 - \rho), & \text{if } z \geq 1/2, \end{cases}$$

with $\rho := s/K \in (0, 1)$. Then the const-entropy is

$$\mathcal{H}_K = 1 - \int_0^1 \omega_K(z)\, dz = \rho - 1,$$

which is always negative, since $\rho < 1$. In this situation the CMT tells us that the expected PM cost will be $P_n = \Theta(n^\alpha)$, where α is the unique root in $[0, 1]$ of the equation

$$\int_0^1 z^\alpha \omega_K(z)\, dz - 1 = 0,$$

Theorem 2. *The expected cost of a random partial match with s specified coordinates out of K, $0 < s < K$, in a random median K-d tree of size n is $P_n = \Theta(n^\alpha)$, where $\alpha \in [0, 1]$ is the unique real solution of*

$$2^{-\alpha} \left(\frac{K(1-\rho)}{K+\alpha} + \frac{K\rho}{2(K+\alpha+1)} \right)$$
$$+ K2^K \left\{ \rho\, B(1/2; K+1, \alpha+1) + (1-\rho)\, B(1/2; K, \alpha+1) \right\} = 1, \quad (5)$$

with $B(z; a, b) = \int_0^z t^{a-1}(1-t)^{b-1}\, dt$ denoting the incomplete Beta function [11, Ch. 8] and $\rho = s/K$.

While we cannot give a closed form for α in terms of K and ρ, Eq. (5) can be used to compute numerical approximations with a high degree of accuracy.

We can also find the value of α as K grows and $\rho = s/K$ remains constant. For very large K, known asymptotic expansions of the incomplete Beta function (see for instance [9] or [11, Ch. 8, pp. 183–184]) yield that α must satisfy

$$2^{-\alpha} \left(1 - \rho + \frac{\rho}{2} \right) + K2^K \left(\frac{1}{2} \right)^{\alpha} \frac{1}{K2^K} (\rho/2 + 1 - \rho) \; = \; 2^{-\alpha}(2 - \rho) \; = \; 1,$$

and hence $\alpha = \log_2(2 - \rho)$. In it is interesting to note that it coincides with the exponent of the expected cost of random PM in relaxed K-d tries [10].

Figure 2 plots the excess $\vartheta(x) := \alpha(x) - (1 - x)$ in the exponent of the cost of random PM of median K-d trees for various values of K (and $x \equiv s/K$), and, for comparison, we also plot the excess $\vartheta(x)$ for relaxed K-d trees [4,10], standard K-d trees [6] and the limit curve $\log_2(2 - x) - 1 + x$ that corresponds to the excess in the exponent for relaxed K-d tries [10].

3   Hybrid Median K-d Trees

Hybrid. K-d trees, also introduced in [12], combine two different rules to choose discriminants. In particular, the hybridization of median K-d trees with standard K-d trees are the so called hybrid median K-d trees, where, for an arbitrary dimension $K \geq 2$, the rule to assign the discriminants is the following:

1. Nodes at levels $\ell \equiv 0 \pmod{K}$ discriminate with respect to the median rule applied to all K coordinates
2. Nodes at levels $\ell \equiv j \pmod{K}$, $0 < j < K$, discriminate with respect the median rule applied to all the coordinates not used as discriminant by any of its $j - 1$ immediate ascendants.

The above implies that, in such a tree, in any path from the root to a leaf, looking at the discriminants of the nodes along the path we will find a sequence of permutations of order K (except for the last part of the path, which will eventually contain only $j < K$ distinct discriminants).

The analysis of the IPL and random partial match in hybrid median K-d trees now becomes more complicated as it requires considering a system of divide-and-conquer recurrences instead of a single divide-and-conquer recurrence as we had when analyzing median K-d trees.

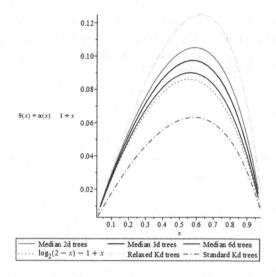

Fig. 2. The excess $\vartheta(x) = \alpha(x) - 1 - x$ for various median K-d trees and other K-d trees.

3.1 Internal Path Length

Let us consider first the IPL of an hybrid median K-d tree. Let $I_n^{(\ell)}$ denote the expected IPL of an hybrid median K-d tree of size n where there are only ℓ available choices for the discriminant at the root (because the other $K - \ell$ discriminants have been already used for the immediate ancestors), then the probability that the left subtree is of size j is given by $\pi_{n,j}^{(\ell)} = \mathbb{P}\{Z_{n,\ell} = j + 1\}$ and if $\ell > 1$ we have

$$I_n^{(\ell)} = n - 1 + \sum_{j=0}^{n-1} \pi_{n,j}^{(\ell)} \left(I_j^{(\ell-1)} + I_{n-1-j}^{(\ell-1)} \right), \qquad 1 < \ell \leq K \text{ and } n > 0,$$

and

$$I_n^{(1)} = n - 1 + \sum_{j=0}^{n-1} \pi_{n,j}^{(1)} \left(I_j^{(K)} + I_{n-1-j}^{(K)} \right), \qquad n > 0.$$

Define now the sequences of vectors $\boldsymbol{F}_n = (I_n^{(K)}, \ldots, I_n^{(1)})^T$ and $\boldsymbol{t}_n = (n - 1, \ldots, n - 1)^T$, and the sequence of weight matrices $\boldsymbol{\Omega}_{n,k} = \left(\omega_{n,k}^{(i,j)} \right)_{K \times K}$, where $\omega_{n,k}^{(i,i+1)} = \pi_{n,k}^{(K+1-i)} + \pi_{n,n-1-k}^{(K+1-i)}$ if $i < K$, $\omega_{n,k}^{(K,1)} = \pi_{n,k}^{(1)} + \pi_{n,n-1-k}^{(1)}$ and all other $\omega_{n,k}^{(i,j)} = 0$. Then we can compactly express the system for the IPL as

$$\boldsymbol{F}_n = \boldsymbol{t}_n + \sum_{0 \leq k < n} \boldsymbol{\Omega}_{n,k} \cdot \boldsymbol{F}_k.$$

Let us suppose that we substitute in the recurrences above each $F_k^{(i)} \equiv I_k^{(i)}$ by its corresponding "row" in the system. This substitution can be expressed in terms of the following operation between weight sequences $\{\omega_{n,k}\}$ and $\{\omega'_{n,k}\}$, giving a new sequence $\{\omega''_{n,k}\}$ defined by

$$\omega''_{n,k} = (\omega \otimes \omega')_{n,k} := \sum_{k<j<n} \omega_{n,j} \cdot \omega'_{j,k}.$$

The operation can be naturally extended to sequences of square $d \times d$ matrices ($d = K$ in our instance). The (i,j) component of each matrix in the sequence $\{\tilde{\boldsymbol{\Omega}}_{n,k}\} := \{(\boldsymbol{\Omega} \otimes \hat{\boldsymbol{\Omega}})_{n,k}\} = \{\boldsymbol{\Omega}_{n,k}\} \otimes \{\hat{\boldsymbol{\Omega}}_{n,k}\}$ is given by

$$\tilde{\boldsymbol{\Omega}}_{n,k}^{(i,j)} = (\boldsymbol{\Omega} \otimes \hat{\boldsymbol{\Omega}})_{n,k}^{(i,j)} = \sum_{\ell} \left(\omega^{(i,\ell)} \otimes \hat{\omega}^{(\ell,j)} \right)_{n,k}.$$

Then we can write one substitution step as

$$\boldsymbol{F}_n = \boldsymbol{t}_n + \sum_{0 \le k < n} \boldsymbol{\Omega}_{n,k} \cdot \boldsymbol{t}_k + \sum_{0 \le k < n} (\boldsymbol{\Omega} \otimes \boldsymbol{\Omega})_{n,k} \cdot \boldsymbol{F}_k$$

The substitution process can be iterated repeatedly:

$$\boldsymbol{F}_n = \boldsymbol{t}_n + \sum_{0 \le k < n} \boldsymbol{\Omega}_{n,k} \cdot \boldsymbol{t}_k + \sum_{0 \le k < n} \boldsymbol{\Omega}_{n,k}^{[2]} \cdot \boldsymbol{t}_k + \cdots + \sum_{0 \le k < n} \boldsymbol{\Omega}_{n,k}^{[\ell-1]} \cdot \boldsymbol{t}_k$$
$$+ \sum_{0 \le k < n} \boldsymbol{\Omega}_{n,k}^{[\ell]} \cdot \boldsymbol{F}_k,$$

where $\boldsymbol{\Omega}^{[1]} \equiv \boldsymbol{\Omega}$ and $\boldsymbol{\Omega}^{[\ell]} = \boldsymbol{\Omega} \otimes \boldsymbol{\Omega}^{[\ell-1]}$, for $\ell > 1$. This new operation $\otimes$ —let us call it *substitution product*— of weight sequences is associative and commutative, and distributes respect to the sum. Its extension to matrices is associative but not commutative, exactly as ordinary matrix products. In the case of the IPL of hybrid K-d trees it turns out that the matrix $\boldsymbol{\Omega}_{n,k}^{[K]}$ is diagonal. This is a very lucky circumstance since then we obtain a set of K independent divide-and-conquer recurrences, and each one can be readily solved using the CMT. To that end, we would only need to compute the weight matrix $\boldsymbol{\Omega}_{n,k}^{[K]}$ and the new toll function

$$\hat{\boldsymbol{t}}_n = \boldsymbol{t}_n + \sum_{0 \le k < n} \left(\boldsymbol{\Omega}_{n,k} + \boldsymbol{\Omega}_{n,k}^{[2]} + \cdots + \boldsymbol{\Omega}_{n,k}^{[K-1]} \right) \cdot \boldsymbol{t}_k.$$

Rather than computing $\boldsymbol{\Omega}_{n,k}^{[\ell]}$ for all $\ell > 1$, the special structure of the problem can be further exploited to obtain our final result (Theorem 3 below, whose proof is given in Appendix B of [5]). In particular, to prove the theorem we introduce the *shape matrix* $\boldsymbol{\Omega}(z)$ in which the (i,j) entry is the shape function for the sequence $\{\omega_{n,k}^{(i,j)}\}$ and the matrices

$$\boldsymbol{\Phi}_\ell(x) = \left(\int_0^1 (\boldsymbol{\Omega}^{[\ell]}(z))^{(i,j)} z^x \, dz \right)_{K \times K}, \quad \boldsymbol{\Phi}'_\ell(x) = \left(-\int_0^1 (\boldsymbol{\Omega}^{[\ell]}(z))^{(i,j)} z^x \ln z \, dz \right)_{K \times K}$$

which are the K-dimensional analogous of the const- and log-entropies of the CMT. Properties of $\otimes$ (such as those proven in Appendix C of [5]) are used to simplify the calculation and show that $\boldsymbol{F}_n \sim (\boldsymbol{\Phi}'_K(1))^{-1}\hat{\boldsymbol{t}}_n \ln n + o(\boldsymbol{1}n \log n)$, where $\hat{\boldsymbol{t}}_n = (Kn, Kn, \ldots, Kn)^T + o(\boldsymbol{1})$. We also show that $\boldsymbol{\Phi}'_K(1)$ is a diagonal matrix where all non-null entries are equal to $\mathcal{H}'_1 + \cdots + \mathcal{H}'_K$, with $\mathcal{H}'_i$ the log-entropy for the expected IPL in median i-dimensional trees.

Theorem 3. *The expected IPL of a random hybrid median K-d tree of size n is*

$$I_n = c_K^{[hm]} n \ln n + o(n \log n)$$

where

$$c_K^{[hm]} = \frac{K}{\mathcal{H}'_1 + \ldots + \mathcal{H}'_K},$$

and the values of $\mathcal{H}'_i$ are those given in Theorem 1.

To conclude, let us observe that for all K, $c_K^{[hm]} \geq c_K^{[med]} = \frac{1}{\mathcal{H}'_K}$ and also that $c_K^{[hm]} \to 1/\ln 2$, albeit the convergence speed is slower than for median K-d trees (as can be seen in Fig. 3).

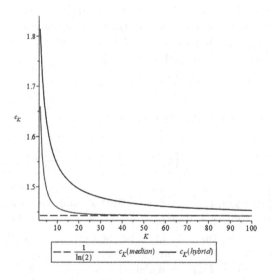

Fig. 3. The coefficient of $n \ln n$ in the average IPL of median K-d trees (red) and hybrid median K-d trees (blue). (Color figure online)

3.2 Random Partial Match

Let $P_n^{(i,\ell)}$ denote the expected cost of a random PM in a hybrid median K-d tree of size n in which there are only $i \geq 1$ coordinates to choose as discriminants — the remaining $K - i$ have been used in the immediate ancestors— and $0 \leq \ell \leq i$ of them are specified in the query. We are interested in $P_n^{(K,s)}$ with $0 < s < K$.

Suppose $i > \ell \geq 1$. With probability ℓ/i the discriminant coordinate — chosen by the median rule among i choices— is specified and thus we will either continue in the left subtree of size j or the right subtree of size $n - 1 - j$ with probability $\pi_{n,j}^{(i)} \frac{j+1}{n+1}$ or $\pi_{n,j}^{(i)} \frac{n-j}{n+1}$, respectively, but now in the next level we will be paying the expected cost of a random PM with only $i - 1$ available coordinates of which only $\ell - 1$ are specified. On the other hand, with probability $(i - \ell)/i$ the discriminant won't be specified and the recursion will continue in both subtrees with only $i - 1$ available coordinates to chose from to discriminate but still ℓ coordinates specified. If $i = \ell > 0$ the reasoning above applies with only branching to either the left or the right subtrees; and if $i > \ell = 0$ then we will continue in both subtrees as no specified coordinate is among those that can be used as discriminants. Hence, if $i > 1$ and $0 \leq \ell \leq i$ we have

$$P_n^{(i,\ell)} = 1 + \frac{\ell}{i} \sum_{j=0}^{n-1} \left(\pi_{n,j}^{(i)} + \pi_{n,n-1-j}^{(i)} \right) \frac{j+1}{n+1} P_j^{(i-1,\ell-1)}$$

$$+ \frac{i - \ell}{i} \sum_{j=0}^{n-1} (\pi_{n,j}^{(i)} + \pi_{n,n-1-j}^{(i)}) P_j^{(i-1,\ell)}$$

The special cases are thus: (1) when $i = 1$ and $\ell = 1$, then the recursion follows in the appropriate subtree but all the K discriminants become available in the next level; and (2) when $i = 1$ and $\ell = 0$, then the recursion follows in both subtrees but with all K coordinates again usable to discriminate. That is,

$$P_n^{(1,0)} = 1 + \sum_{j=0}^{n-1} (\pi_{n,j}^{(1)} + \pi_{n,n-1-j}^{(1)}) P_j^{(K,s)}$$

$$P_n^{(1,1)} = 1 + \sum_{j=0}^{n-1} \left(\pi_{n,j}^{(1)} + \pi_{n,n-1-j}^{(1)} \right) \frac{j+1}{n+1} P_j^{(K,s)}$$

The resulting call graph is more complicated than the one of IPL, and the system of D&C recurrences will involve $d = (K - s + 1)(s + 1) - 1$ "algorithms" with costs $P_n^{(i,\ell)}$, see for example Fig. 4 for the case $K = 3$ and $s = 2$.

Once we have set up the system of divide-and-conquer recurrences we can construct a shape matrix $\boldsymbol{\Omega}(z)$ in which the entry (u, v) is the shape function $\omega^{(u,v)}(z)$ corresponding to weight sequence $\omega_{n,k}^{(u,v)}$; vertices u and v correspond to partial match algorithms with parameters (i, ℓ) and (i', ℓ'). Many entries will be null as algorithm u (or (i, ℓ)) does not call algorithm v (or (i', ℓ')). We can think of this shape matrix as the adjacency matrix for the call digraph in which each edge (u, v) is labelled by $\omega^{(u,v)}(z)$. Likewise we can define the matrix $\boldsymbol{\Phi}(x)$ in which the entries are the definite integrals $\int_0^1 \omega^{(u,v)}(z) z^x \, dz$. Then we can find the expected cost $P_n^{(K,s)}$ thanks to the following result.

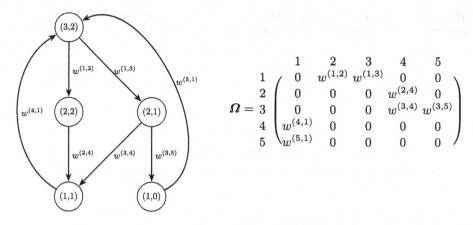

Fig. 4. Call graph for the system of D&C recurrences of the PM costs in hybrid median K-d trees for $K = 3$ and $s = 2$.

Theorem 4. *The expected cost of a random partial match with s specified coordinates out of K, $0 < s < K$, in a random hybrid median K-d tree of size n is $P_n^{(K,s)} = \Theta(n^\alpha)$, where $\alpha \in [0, 1]$ is the unique real solution of $\det(\boldsymbol{I} - \boldsymbol{\Phi}(x)) = 0$, where $\boldsymbol{\Phi}(x) = \int_0^1 \boldsymbol{\Omega}(z)\, z^x\, dz$ and $\boldsymbol{\Omega}(z)$ is the shape matrix corresponding to the system of d divide-and-conquer recurrences, with $d = (K - s + 1)(s + 1) - 1$.*

The proof of this result can be found in Appendix D of [5]. It is based in the properties of iterated substitution matrices $\boldsymbol{\Omega}^{[K]}(z)$ (and $\boldsymbol{\Phi}_K(x)$), and those of the determinant of $\boldsymbol{\Phi}(x) - \boldsymbol{I}$ once we see it as the (weighted) adjacency matrix of the call graph in which we add self-loops to every vertex of the call graph.

We report the values of α for $K \leq 6$ in Table 3. Next to each entry we give inside parentheses the corresponding values of α for standard K-d trees. All values have been rounded to three significant figures. These values suggest that random PM in hybrid median K-d trees perform better on average than in standard K-d trees. Namely, we conjecture that $\alpha^{[\text{hyb}]}(s, K) < \alpha^{[\text{std}]}(s, K)$ for all s and K. Moreover, we conjecture that as $K \to \infty$, $\alpha^{[\text{hyb}]}(s, K) \to 1 - s/K$, which is optimal. One argument in favor of this conjecture is that hybrid median K-d trees get increasingly balanced as K grows, but the hybridization guarantees that we cycle over all K coordinates as we follow paths down the tree—any path from a node at level $r \cdot K$ to a node at level $(r+1) \cdot K - 1$ has used all coordinates as discriminants. Hence partial match in hybrid median K-d trees should behave as in standard K-d tries [6], for which $\alpha(s, K) = 1 - s/K$.

Table 3. Values of α for the expected cost of PM in hybrid median K-d trees. In parentheses, the corresponding values of α for standard K-d trees.

K	s				
	1	2	3	4	5
2	0.546 (0.562)	–	–	–	–
3	0.697 (0.716)	0.368 (0.395)	–	–	–
4	0.771 (0.79)	0.53 (0.562)	0.275 (0.306)	–	–
5	0.815 (0.833)	0.624 (0.656)	0.425 (0.463)	0.218 (0.25)	–
6	0.845 (0.862)	0.685 (0.716)	0.522 (0.562)	0.354 (0.395)	0.181 (0.211)

4   Conclusions and Final Remarks

Throughout this work we have considered two variants of K-d trees: median K-d trees and hybrid median K-d trees. Both are simple and easy to implement, and neither requires significant extra space. We show that both variants are more balanced than most other well known variants of K-d trees based on key comparisons, such as standard, relaxed and squarish K-d trees. This is due to the fact that their expected IPL is $\sim c_K n \ln n$ with $c_K < 2$ for all $K \geq 2$, and $c_K \to 1/\ln 2$ as $K \to \infty$, while for the other mentioned variants $c_K = 2$. We have also shown that their expected cost for random PM is $\Theta(n^\alpha)$, where $\alpha = \alpha(s, K)$. For median K-d trees this expected cost is better than that of relaxed K-d trees but not than that for standard K-d trees. In contrast, hybrid median K-d trees outperfom standard and relaxed K-d trees and we conjecture that they approach the optimal exponent —only attained by squarish K-d trees- $\alpha = 1 - s/K$ as K gets larger. In view of these results, good choices would be hybrid median K-d trees if the efficiency of insertions and exact searches were to be prioritized — while not deviating too much from the optimal performance in partial matches— or squarish K-d trees if the priority were the efficiency of partial match, with slightly worse expected costs for insertions and exact searches.

To derive analytic results, our main tool has been the continuous master theorem —the CMT. For the analysis of median K-d trees the most challenging step was to find the probability that a random median K-d tree of size n has a left subtree of size j, but once computed an almost direct application of the CMT provides the sought answers. Hybrid median K-d trees have posed an entirely new challenge as we have had to cope with systems of divide-and-conquer recurrences that can not be solved directly using the CMT. Nevertheless, we have been able to exploit the special structure of the systems corresponding to the IPL and the random PM in hybrid median K-d trees to find the constants c_K and the equations satisfied by the exponents $\alpha(s, K)$ by developing a limited extension of the CMT to cope with systems of recurrences.

Last but not least, our work constitutes a new example of the power of the CMT as a fundamental tool in the analysis of algorithms: without its help the analysis of median K-d trees would be a daunting task. It would have been desirable to have a full developed set of results and tools in the spirit of the CMT to cope with systems of divide-and-conquer recurrences such as those arising in the analysis of hybrid median K-d trees. Indeed, the extensions of the CMT that we have developed in this work could constitute a first step towards this goal.

References

1. Bentley, J.L.: Multidimensional binary search trees used for associative searching. Commun. ACM **18**(9), 509–517 (1975)
2. Cunto, W., Lau, G., Flajolet, P.: Analysis of kdt-trees: Kd-trees improved by local reorganisations. In: Dehne, F., Sack, J.-R., Santoro, N. (eds.) WADS 1989. LNCS, vol. 382, pp. 24–38. Springer, Heidelberg (1989). https://doi.org/10.1007/3-540-51542-9_4
3. Devroye, L., Jabbour, J., Zamora-Cura, C.: Squarish k-d trees. SIAM J. Comput. **30**, 1678–1700 (2000)
4. Duch, A., Estivill-Castro, V., Martínez, C.: Randomized K-dimensional binary search trees. In: Chwa, K.-Y., Ibarra, O.H. (eds.) ISAAC 1998. LNCS, vol. 1533, pp. 198–209. Springer, Heidelberg (1998). https://doi.org/10.1007/3-540-49381-6_22
5. Duch, A., Martínez, C., Pons, M., Roura, S.: The analysis of median and hybrid median K-dimensional trees: two heuristically balanced variants of K-dimensional trees. Available from ResearchGate (2022). https://doi.org/10.13140/RG.2.2.12891.44322, Preprint with the full version of this extended abstract
6. Flajolet, P., Puech, C.: Partial match retrieval of multidimensional data. J. ACM **33**(2), 371–407 (1986)
7. Knuth, D.E.: The Art of Computer Programming: Sorting and Searching, vol. 3, 2nd edn. Addison-Wesley, Boston (1998)
8. van Kreveld, M.J., Overmars, M.H.: Divided k-d-trees. Algorithmica **6**, 840–858 (1991)
9. López, J.L., Sesma, J.: Asymptotic expansion of the incomplete beta function for large values of the first parameter. Integral Transf. Spec. Funct. **8**(3), 233–236 (1999)
10. Martínez, C., Panholzer, A., Prodinger, H.: Partial match queries in relaxed multidimensional search trees. Algorithmica **29**(1–2), 181–204 (2001)
11. Paris, R.B.: Incomplete Gamma and related functions. In: Olver, F.W.J., Lozier, D.W., Boisvert, R.F., Clark, C.W. (eds.) NIST Handbook of Mathematical Functions, vol. 8. Cambridge University Press (2010)
12. Pons, M.: Design, Analysis and Implementation of New Variants of Kd-trees. Master's thesis, Universitat Politècnica de Catalunya (2010)
13. Roura, S.: Improved master theorems for divide-and-conquer recurrences. J. ACM **48**(2), 170–205 (2001)
14. Samet, H.: Foundations of Multidimensional and Metric Data Structures. Morgan Kaufmann, Burlington (2006)

Weighted Connected Matchings

Guilherme C. M. Gomes[1], Bruno P. Masquio[2(✉)], Paulo E. D. Pinto[2],
Vinicius F. dos Santos[1], and Jayme L. Szwarcfiter[2,3]

[1] Universidade Federal de Minas Gerais (UFMG), Belo Horizonte, MG, Brazil
{gcm.gomes,viniciussantos}@dcc.ufmg.br
[2] Universidade do Estado do Rio de Janeiro (UERJ), Rio de Janeiro, RJ, Brazil
{brunomasquio,pauloedp}@ime.uerj.br
[3] Universidade Federal do Rio de Janeiro (UFRJ), Rio de Janeiro, RJ, Brazil
jayme@nce.ufrj.br

Abstract. A matching M is a $\mathscr{P}$-matching if the subgraph induced by the endpoints of the edges of M satisfies property $\mathscr{P}$. As examples, for appropriate choices of $\mathscr{P}$, the problems INDUCED MATCHING, UNIQUELY RESTRICTED MATCHING, CONNECTED MATCHING and DISCONNECTED MATCHING arise. For many of these problems, finding a maximum $\mathscr{P}$-matching is a knowingly NP-hard problem, with few exceptions, such as CONNECTED MATCHING, which has the same time complexity as the usual MAXIMUM MATCHING problem. The weighted variant of MAXIMUM MATCHING has been studied for decades, with many applications, including the well-known ASSIGNMENT problem. Motivated by this fact, in addition to some recent research in weighted versions of acyclic and induced matchings, we study the MAXIMUM WEIGHT CONNECTED MATCHING. In this problem, we want to find a matching M such that the endpoints of its edges induce a connected subgraph and the sum of the edge weights of M is maximum. Unlike the unweighted CONNECTED MATCHING problem, which is in P for general graphs, we show that MAXIMUM WEIGHT CONNECTED MATCHING is NP-hard even for bounded diameter bipartite graphs, starlike graphs, planar bipartite graphs, and subcubic planar graphs, while solvable in linear time for trees and graphs having degree at most two. When we restrict edge weights to be non-negative only, we show that the problem turns out to be polynomially solvable for chordal graphs, while it remains NP-hard for most of the other cases. In addition, we consider parameterized complexity. On the positive side, we present a single exponential time algorithm when parameterized by treewidth. As for kernelization, we show that, even when restricted to binary weights, WEIGHTED CONNECTED MATCHING does not admit a polynomial kernel when parameterized by vertex cover number under standard complexity-theoretical hypotheses.

Keywords: Algorithms · Complexity · Induced subgraphs · Matchings

V.F. dos Santos—Partially supported by FAPEMIG and CNPq
J. L. Szwarcfiter—Partially supported by FAPERJ and CNPq.

A. Castañeda and F. Rodríguez-Henríquez (Eds.): LATIN 2022, LNCS 13568, pp. 54–70, 2022.
https://doi.org/10.1007/978-3-031-20624-5_4

1   Introduction

Matching problems have a long history and a vast literature in both structural and algorithmic graph theory [13, 22, 24–28]. A matching is a subset $M \subseteq E$ of the edges of a graph $G = (V, E)$ that do not share any endpoint. A $\mathscr{P}$-matching is a matching such that $G[M]$, the subgraph of G induced by the endpoints of edges of M, satisfies property $\mathscr{P}$. One of the most natural and important properties in network applications is the connectivity of a graph. In this paper we focus on matchings whose endpoints induce a connected graph.

The problem of deciding whether or not a graph admits a $\mathscr{P}$-matching of a given size has been investigated for many different properties $\mathscr{P}$ over the years. One of the most well-known examples is the INDUCED MATCHING problem, where $\mathscr{P}$ is the property of being 1-regular; in [8] it was shown to be NP-complete even for bipartite graphs. Other NP-hard matching problems include ACYCLIC MATCHING [17], k-DEGENERATE MATCHING [1], UNIQUELY RESTRICTED MATCHING [18], and DISCONNECTED MATCHING [7, 19]. One of the few exceptions of a $\mathscr{P}$-matching problem that is solvable in polynomial time is CONNECTED MATCHING [17], in which $G[M]$ has to be connected.

It is worth mentioning that the name CONNECTED MATCHING is also used for a different problem, where the aim is to find a matching M in a given graph G such that every pair of edges in M has a common adjacent edge [9]. We adopt the more recent meaning of CONNECTED MATCHING, given by Goddard et al. [17], who also proved that the sizes of a maximum matching and a maximum connected matching in a connected graph coincide. In [7], it was shown that, given a maximum matching and a graph G, a maximum connected matching of G can be obtained in linear time.

Recently, some $\mathscr{P}$-matching concepts were extended to edge-weighted problems, where, in addition to the matching to have a certain property $\mathscr{P}$, the sum of the weights of the matching edges must be sufficiently large. It was shown that MAXIMUM WEIGHT INDUCED MATCHING can be solved in linear time for convex bipartite graphs [21] and in polynomial time for circular-convex and triad-convex bipartite graphs [29]. In [14], Fürst et al. showed that MAXIMUM WEIGHT ACYCLIC MATCHING is solvable in polynomial time for P_4-free graphs and $2P_3$-free graphs.

Motivated by these studies, we consider connected matchings on edge-weighted graphs. In particular, we investigate both the decision and optimization versions, which we formally define as follows; note that the decision problem is in NP.

WEIGHTED CONNECTED MATCHING
Instance: An edge-weighted graph G and an integer k.
Question: Is there a connected matching M of weight at least k?

MAXIMUM WEIGHT CONNECTED MATCHING
Instance: An edge-weighted graph G.
Task: Find a connected matching M of G of maximum weight.

In some cases, we approach WEIGHTED CONNECTED MATCHING separately when negative weights are allowed or not. Note that, unlike some weighted matching problems, such as MAXIMUM WEIGHT MATCHING, negative weighted edges are relevant and may even be required to be in an optimal solution.

Our Results. In this paper, we study the complexity of WEIGHTED CONNECTED MATCHING under different constraints.

Our investigation begins with a starlike graphs, a subclass of chordal graphs, where we show that the problem is NP-complete when arbitrary weights are allowed. On the other hand, we present a polynomial-time algorithm for chordal graphs if all weights are non-negative.

Afterwards, we turn our attention to bipartite graphs, first showing that even for non-negative weights, WEIGHTED CONNECTED MATCHING remains NP-complete, even if the graph is planar or its diameter is bounded. If negative weights are allowed, we prove that the problem remains hard for subcubic planar graphs.

We then prove the existence of polynomial-time algorithms for: (i) graphs of maximum degree two, establishing a complexity dichotomy based on the maximum degree, (ii) graphs of bounded treewidth, and (iii) trees; the latter is an improvement upon the treewidth algorithm, since it runs in linear time. The treewidth algorithm implies fixed-parameter tractability for the treewidth parameterization, which leads us to our final result, where we prove that no polynomial kernel exists when parameterizing by vertex cover unless NP ⊆ coNP/poly. We summarize our main results in Table 1.

Table 1. Summary of our results for WEIGHTED CONNECTED MATCHING.

Graph class		Complexity	
		Weights ≥ 0	Any weights
General		NP-complete (Theorem 3)	
Bipartite having diameter at most 4			
Chordal		P (Theorem 2)	NP-complete (Theorem 1)
Starlike			
Planar	bipartite	NP-complete (Theorem 4)	
	subcubic	?	NP-complete (Theorem 5)
$\Delta \leq 2$		P (Theorem 6)	
Tree		P (Theorem 8)	

Preliminaries. For an integer k, we define $[k] = \{1, \ldots, k\}$. We use standard graph theory notation and nomenclature as in [5,6], and refer to [10] for parameterized complexity. Let G be a graph, $M \subseteq E(G)$, and $V(M)$ be the set of endpoints of edges of M, which are also called M-saturated vertices, or just saturated. Let $\Delta(G)$ be the maximum vertex degree of G. For $W \subseteq V(G)$, we denote by $G[W]$ the subgraph of G induced by W; in abuse of notation, we define $G[M] = G[V(M)]$. A matching is said to be maximum if there is no other matching of G with greater cardinality and to be maximum weight if there is no other matching of G having greater sum of edge weights. A matching is perfect if $V(M) = V(G)$. Also, M is said to be connected if $G[M]$ is connected. Let uv be an edge of G. We denote by $w(uv)$ the weight of the edge uv and define $w(M) = \sum_{uv \in M} w(uv)$. The operations $G - uv$ and $G - v$ result, respectively, in the graphs $G' = (V, E \setminus \{uv\})$ and $G[V \setminus \{v\}]$. We denote by $K_{a,b}$ the complete bipartite graph with a vertices in one part and b vertices in the other. A star is a graph isomorphic to $K_{1,b}$, for some b. The graphs P_n and C_n are path and cycle graphs having n vertices. A graph G is H-free if G has no copy of H as an induced subgraph; G is chordal if it has no induced cycle with more than three vertices. A clique tree of G is a tree T representing G in which vertices and edges of T correspond, respectively, to maximal cliques and minimal separators of G. A graph is a starlike graph if it is chordal and has a clique tree that is a star graph. A graph is planar if it can be embedded in the plane without edge crossings.

2   Chordal Graphs

We begin our study on chordal graphs and one of its subclasses, starlike graphs. WEIGHTED CONNECTED MATCHING for these classes has different time complexities depending on the admittance of negative weights on the input graph; more specifically, we show that the problem is NP-complete if negative weights are allowed and is in P otherwise. To reach this result, we prove the NP-completeness for starlike graphs having weights in $\{-1, +1\}$ in Sect. 2.1 and a polynomial-time algorithm for chordal graphs having non-negative weights in Sect. 2.2.

2.1   Starlike Graphs

Our first result is a proof that WEIGHTED CONNECTED MATCHING is NP-complete on starlike graphs having edge weights in $\{-1, +1\}$. Our reduction is from the NP-complete problem 3SAT [16], with input given by the pair $(X, \mathcal{C})$, where X is the set of variables and $\mathcal{C}$ is the set of clauses; w.l.o.g. we assume that each clause contains exactly three literals. We set the input of WEIGHTED CONNECTED MATCHING as $k = |X| + |\mathcal{C}|$ and $G_{X,\mathcal{C}}$ built by the following rules:

(I) For each variable $x_i \in X$, add a copy of C_3 whose vertices are labeled x_i, x_i^+ and x_i^-. Set weight -1 to edge $x_i^+ x_i^-$ and $+1$ to the other edges.

(II) For each pair of variables $x_i, x_j \in X$, add all possible edges between vertices of $\{x_i^-, x_i^+\}$ and $\{x_j^-, x_j^+\}$ and set its weights to -1.

(III) For each clause $C_i \in \mathcal{C}$, add a copy of K_2 whose edge weight is $+1$ and label its endpoints as c_i^+, and c_i^-. Also, for each literal x_j of C_i, connect both c_i^- and c_i^+ to x_j^- if x is negated, or x_j^+ otherwise; in both cases, the added edges have weight -1.

This graph is indeed starlike, as its clique tree is a star, having as center the maximal clique containing the vertices $\{x_i^+, x_i^- \mid x_i \in X\}$, one leaf clique $\{x_i, x_i^+, x_i^-\}$ for each $x_i \in X$, and one leaf clique for each clause. A connected matching of weight at least k in $G_{X,\mathcal{C}}$ corresponds to an assignment of X such that $x_i \in X$ is set to true if and only if x_i^+ is M-saturated.

Lemma 1. *Given a solution for the* 3SAT *instance* $(X, \mathcal{C})$, *we can obtain a connected matching* M *in* $G_{X,\mathcal{C}}$ *having weight* $|X| + |\mathcal{C}|$.

Proof. We show how to obtain the matching M. (i) For each clause $C_i \in \mathcal{C}$, add the edge $c_i^- c_i^+$ to M. Also, (ii) for each variable $x_i \in X$, if x_i is set to true, we saturate the edge $x_i^+ x_i$; otherwise, $x_i^- x_i$.

Next, we show that M is connected. Edges from (ii) are connected as each one is incident to a vertex x_i, which is part of the center clique. Each edge from (i), obtained by clause $C_i \in \mathcal{C}$, having x_j as the variable related to the literal that resolves to true in C_j, is connected. This holds because, if x_j is negated, then $c_i^+ x_j^- \in E(G_{X,\mathcal{C}})$ and x_j^- is saturated. Otherwise, $c_i^+ x_j^+ \in E(G_{X,\mathcal{C}})$ and x_j^+ is saturated.

Lemma 2. *Given an input* $(X, \mathcal{C})$ *for* 3SAT *and a connected matching* M *in* $G_{X,\mathcal{C}}$ *having weight* $|X| + |\mathcal{C}|$, *we can obtain an assignment* R *of* X *that solves* 3SAT.

Proof. Denote by W_{-1} and W_1 the edge sets from $G_{X,\mathcal{C}}$ whose weights are, respectively, -1 and 1. First, we show that a matching having weight $|X| + |\mathcal{C}|$ contains exactly $|X| + |\mathcal{C}|$ edges from W_1 and no edges from W_{-1}.

Note that there can be at most $|X| + |\mathcal{C}|$ edges from W_1. This holds because, for each variable $x_i \in X$, a matching contains at most one edge of $\{x_i^+ x_i, x_i^- x_i\}$, since both have an endpoint in vertex x_i. Also, for each clause $C_i \in \mathcal{C}$, the edge $c_i^- c_i^+$ can also be contained in the matching.

Since all edges contained in W_{-1} have negative weights, if there is a matching with $|X| + |\mathcal{C}|$ vertices from W_1 and no vertices from W_{-1}, then it is maximum. Since M is a matching whose weight is $|X| + |\mathcal{C}|$, then $|M \cap W_1| = |X| + |\mathcal{C}|$ and $|M \cap W_{-1}| = 0$. Also, as M is connected, then, for each edge $c_i^+ c_i^- \in M$, there is a saturated vertex adjacent to c_i^+, either x_j^+ or x_j^-, for $x_j \in C_i$ and $C_i \in \mathcal{C}$. Those vertices are exactly the ones representing a literal in clause C_i. So, to obtain R, for each variable $x_i \in X$, we set x_i to true if and only if x_i^+ is saturated.

Theorem 1. WEIGHTED CONNECTED MATCHING *is* NP-*complete even for starlike graphs whose edge weights are in* $\{-1, +1\}$.

Proof. Note that the problem is in NP. According to the correspondence between WEIGHTED CONNECTED MATCHING and 3SAT solutions described in Lemmas 1 and 2, the 3SAT problem, which is NP-complete, can be reduced to WEIGHTED CONNECTED MATCHING using a starlike graph whose edge weights are either -1 or $+1$. Therefore, WEIGHTED CONNECTED MATCHING is NP-complete even for starlike graphs whose weights are in $\{-1, +1\}$.

As an example, let an input of 3SAT be $(x_1 \lor \overline{x_2} \lor \overline{x_4}) \land (x_1 \lor \overline{x_3} \lor x_5) \land$ $(\overline{x_1} \lor \overline{x_2} \lor x_4) \land (x_2 \lor x_3 \lor x_5)$. The related graph is illustrated in Fig. 1, as well as a connected matching having weight $|X| + |\mathcal{C}| = 9$, corresponding to the assignment (F, T, F, F, T) of the variables $(x_1, x_2, x_3, x_4, x_5)$, in this order.

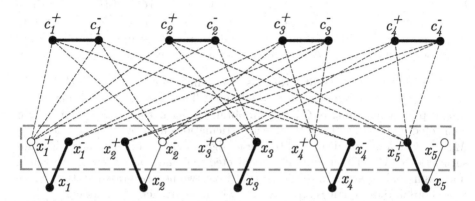

Fig. 1. Example of a starlike graph built by the reduction and a connected matching. Dashed edges have weight -1 and solid edges have weight 1. Vertices in the dashed rectangle induce a clique in which the omitted edges have weight -1.

2.2 Chordal Graphs for Non-negative Weights

Theorem 1 directly implies that WEIGHTED CONNECTED MATCHING is also NP-complete for chordal graphs. Interestingly, if the weights are restricted to be non-negative, we can solve WEIGHTED CONNECTED MATCHING in polynomial time in this class. To this end, we present a polynomial-time reduction to the MAXIMUM WEIGHT PERFECT MATCHING problem, which can be solved in polynomial time [11,12]. In this problem, we are given a graph G and we want to find a perfect matching M whose sum of the edge weights is maximum.

For the reduction, let $G = (V, E)$ be the input graph to MAXIMUM WEIGHT CONNECTED MATCHING. We build the input graph $G_p = (V_p, E_p)$ to MAXIMUM WEIGHT PERFECT MATCHING as follows:

(I) Set $V_p = V$. If $|V|$ is odd, add a vertex h to V_p.
(II) Set $E_p = E$. Now, for each pair of non articulation vertices $v_1, v_2 \in V \cup \{h\}$, if $v_1v_2 \notin E$, add the 0-weight edge v_1v_2 to E_p.

In the next lemma, we show that there is always a maximum weight connected matching that saturates all articulations of a graph.

Lemma 3. *Let G be a connected graph having no negative weight edge. There is a maximum weight connected matching M that saturates all articulations.*

Proof. Let M be a maximum weight connected matching such that an articulation v is not saturated. We show that we can saturate v only by adding edges to M, in a way M is still connected. Let $\mathcal{C} = \{C_1, \ldots, C_{|\mathcal{C}|}\}$ be the connected components of $G - v$. Note that $|\mathcal{C}| \geq 2$ and the edges of M are contained in exactly one component $C_i \in \mathcal{C}$, because, otherwise, $G[M]$ would not be connected. Let $U = ((\bigcup_{u \in V(M)} N(u)) \setminus V(M)) \setminus \{v\}$. Note that $U \subseteq V(C_i)$. If $U \neq \emptyset$ and there is a path $P = (p_1, \ldots, p_q)$ between $p_1 \in U$ and $p_q = v$ in $G - V(M)$, saturate $p_i p_{i+1}$ for every possible i even. If q is odd, then v is saturated. If P does not exist, or q is even or $U = \emptyset$, saturate v with any vertex of $C_j \neq C_i$ of $\mathcal{C}$.

Without loss of generality, there is also a maximum weight matching in a chordal graph that saturates all articulations. Therefore, if M_p is a maximum weight perfect matching in G_p, we can obtain in linear time a maximum weight connected matching which saturates all articulations in G by the union of $M^* = M_p \cap E$ with a maximal set S of 0-weight edges having endpoints in vertices not saturated by M^*. This results in a connected matching, as otherwise we would find a minimal separator of G that has at least two non-saturated vertices; since G is chordal, these vertices are adjacent, which contradicts the maximality of S. Hence, WEIGHTED CONNECTED MATCHING can be solved in polynomial time, as stated in the following theorem.

Theorem 2. MAXIMUM WEIGHT CONNECTED MATCHING *for chordal graphs whose edge weights are all non-negative can be solved in polynomial time.*

3   Bipartite Graphs

The hardness result stated in Theorem 1 is highly dependent on the existence of several non-trivial cliques, which are forbidden in some graph classes, including bipartite graphs. This raises the question of whether or not the absence of these structures makes the problem easier; in this section, we answer this in the negative by showing that WEIGHTED CONNECTED MATCHING remains hard on bipartite graphs having only binary weights.

Our proof is also based on a reduction from 3SAT, whose input is $(X, \mathcal{C})$. Let the input of WEIGHTED CONNECTED MATCHING be $k = |X| + |\mathcal{C}| + 1$ and the graph $G_{X,\mathcal{C}}$ obtained by the following rules.

(I)   Add two vertices, h^+ and h^-, connected by a 1-weight edge.
(II) For each variable $x_i \in X$, add a copy of P_3 whose edge weights are 1, and label its endpoints as x_i^+ and x_i^-. Moreover, connect the other vertex, labeled x_i, to h^+ and set this edge weight to 0.
(III) For each clause $C_i \in \mathcal{C}$, add a copy of K_2 whose edge weight is 1 and label its vertices as c_i^+ and c_i^-. Also, for each literal x_j of C_i, add a 0-weight edge $c_i^+ x_j^-$ if x_j is negated, or $c_i^+ x_j^+$ otherwise.

A connected matching M of weight at least k of $G_{X,\mathcal{C}}$ corresponds to an assignment of X such that variable x_i is set to true if and only if x_i^+ is M-saturated.

Figure 2 presents an example where the input formula of 3SAT is $(x_1 \vee \overline{x_2} \vee \overline{x_4}) \wedge (x_1 \vee \overline{x_3} \vee x_5) \wedge (\overline{x_1} \vee \overline{x_2} \vee x_4) \wedge (x_2 \vee x_3 \vee x_5)$. The illustrated connected matching corresponds to the assignment (F, T, F, F, T), in this order, of the variables $(x_1, x_2, x_3, x_4, x_5)$.

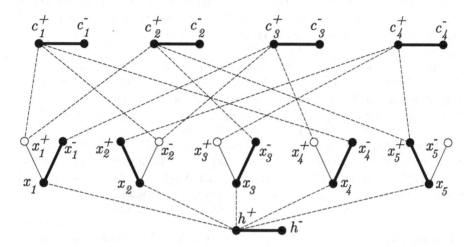

Fig. 2. Example of a bipartite graph built by the reduction and a connected matching of size $|X| + |\mathcal{C}| + 1 = 10$. Dashed and solid edges represent weights 0 and 1, respectively.

It is also possible to strengthen our NP-completeness proof in terms of the graph diameter, adding a vertex u and the 0-weight edges defined by $\{c_i^+ u \mid C_i \in \mathcal{C}\} \cup \{x_i^+ h^-, x_i^- h^- \mid x_i \in X\} \cup \{x_i x_j^+, x_i x_j^-, x_j x_i^+, x_j x_i^- \mid x_i, x_j \in X\}$. Connected matchings of weight k have the same properties and patterns as previously stated.

In next lemmas, we show that, given an input $(X, \mathcal{C})$ for 3SAT, it is possible to obtain in linear time a connected matching M in $G_{X,\mathcal{C}}$, $w(M) = k$, if we have a solution for $(X, \mathcal{C})$, and vice versa. Denote W_0 and W_1 by the edge sets from $G_{X,\mathcal{C}}$ whose weights are, respectively, 0 and 1.

Lemma 4. *Given a solution for the 3SAT instance $(X, \mathcal{C})$, we can obtain a connected matching M having weight $|X| + |\mathcal{C}| + 1$ in $G_{X,\mathcal{C}}$.*

Proof. We show how to obtain the matching M. (i) For each clause $C_i \in \mathcal{C}$, add the edge $c_i^- c_i^+$ to M. Also, (ii) for each variable $x_i \in X$, if x_i is set to true, we saturate the edge $x_i^+ x_i$; otherwise, $x_i^- x_i$. Moreover, (iii) saturate the edge $h^+ h^-$.

Now, we prove that this matching is connected. Edges from (ii) are connected as they are connected to the edge $h^+ h^-$ of (iii). Each edge from (i), obtained by clause $C_i \in \mathcal{C}$, having x_j as the variable related to a literal that resolves to true in C_i, is connected. This holds because, if x_j is negated, then $c_i^+ x_j^- \in E(G_{X,\mathcal{C}})$ and x_j^- is saturated. Otherwise, $c_i^+ x_j^+ \in E(G_{X,\mathcal{C}})$ and x_j^+ is saturated.

Lemma 5. *Given an input* $(X, \mathcal{C})$ *for* 3SAT *and a connected matching* M *in* $G_{X,\mathcal{C}}$ *having weight* $|X| + |\mathcal{C}| + 1$, *we can obtain an assignment of* X *that solves* 3SAT *in polynomial time.*

Proof. First, we show that a matching M having weight $|X| + |\mathcal{C}| + 1$ contains exactly $|X| + |\mathcal{C}| + 1$ edges from W_1 and no edges from W_0. Note that $|M \cap W_1| \leq |X| + |\mathcal{C}| + 1$ because, for each variable $x_i \in X$, there is at most one matched edge of $\{x_i^+ x_i, x_i^- x_i\}$, since both have an endpoint in vertex x_i. Also, M can contain the edges $\{h^+ h^-\} \cup \{c_i^- c_i^+ \mid C_i \in \mathcal{C}\}$.

Observe that each edge from W_0 has an endpoint in either $\{h^+\}$ or $S_1 = \{c_i^+ \mid C_i \in \mathcal{C}\}$. Saturating any of these vertices by a W_0 edge will decrease the number of possibly matched edges of W_1 and the weight of M, resulting in $w(M) < k$. Namely, if we saturate h^+ or $c_i^+ \in S_1$, we are not able to saturate, respectively, $h^+ h^-$ or $c_i^+ c_i^-$. Thus, $|M \cap W_1| = |X| + |\mathcal{C}| + 1$ and $|M \cap W_0| = 0$.

If the matching M is connected, then, for each edge $c_i^+ c_i^- \in M$, there is an M-saturated vertex adjacent to c_i^+, either x_j^+ or x_j^-, for a variable x_j contained in clause C_i. Also, for each variable $x_i \in X$, the vertex x_i is saturated, which is connected to the edge $h^+ h^-$. Hence, to obtain an assignment to X, for each variable $x_i \in X$, we set x_i to true if and only if x_i^+ is saturated.

Theorem 3. WEIGHTED CONNECTED MATCHING *is* NP-*complete on bipartite graphs of diameter* 4 *even if all edge weights are in* $\{0, 1\}$.

Proof. Note that the problem is in NP. According to the correspondence between WEIGHTED CONNECTED MATCHING and 3SAT solutions described in Lemmas 4 and 5, 3SAT, which is NP-complete, can be reduced to WEIGHTED CONNECTED MATCHING using a bipartite graph whose diameter is 4 and the edge weights are either 0 or 1. Hence, WEIGHTED CONNECTED MATCHING is NP-complete even for bipartite graphs whose weights are in $\{0, 1\}$ and diameter is 4.

A follow-up question is if there are subclasses of bipartite graphs that admit polynomial-time algorithms. As we show in Sect. 5, there exists such algorithms for trees and graphs of maximum degree two. Nevertheless, we would like to study non-trivial classes, such as chordal bipartite graphs and planar bipartite graphs. We leave the former as an open problem, but proceed to study the latter, and other subclasses of planar graphs, in our next section.

4   Planar Graphs

Aside from the planar bipartite case, which is shown to be NP-complete in Sect. 4.1, we investigate the complexity of WEIGHTED CONNECTED MATCHING in planar graphs under degree constraints in Sect. 4.2, proving that the problem remains hard in the subcubic case.

4.1   Planar Bipartite Graphs

In this section, we prove the NP-completeness of WEIGHTED CONNECTED MATCHING on planar bipartite graphs having weights either 0 or 1. We use a polynomial-time reduction from a SAT variant that we explain in the following.

Let B be a conjunctive formula where $\mathcal{C} = \{C_1, \ldots, C_q\}$ and $X = \{x_1, \ldots, x_m\}$ are the sets of clauses and variables of B, respectively. Let $X_c = (x_1, \ldots, x_m)$ be an ordering of X. Let $G(B) = (V, E)$ be the graph in which there is a vertex for each clause and each variable of B, namely $V = \{x_i \mid x_i \in X\} \cup \{c_j \mid C_j \in \mathcal{C}\}$. The edge set E is partitioned in A_1, A_2 where $A_1 = \{x_i c_j \mid \{x_i, \overline{x_i}\} \cap C_j \neq \emptyset\}$ and $A_2 = \{x_i x_{i+1} \mid 1 \leq i < m\} \cup \{x_m x_1\}$; note that A_2 induces a cycle containing all variable vertices and follows the ordering X_c.

In [23], Lichtenstein defined a conjunctive formula B as planar if $G(B)$ is planar; he showed that, for every instance of SAT, we can build in polynomial time an equivalent planar formula as well as its planar embedding. The problem of finding a true assignment to a planar formula was named as PLANAR SAT, and was proven to be NP-complete. Later in the paper, the author also proved the NP-completeness of a PLANAR SAT variant of our interest. In this problem, the input planar boolean formula B is monotone, that is, each of its clauses consists only of positive literals or only of negative literals. Also, there is a planar embedding of $G(B)$ such that each edge referencing a positive(negative) literal is connected to the top(bottom) of the variable vertices. We refer to it as PLANAR MONOTONE SAT. We assume that X_c is also part of its input, since it was shown in [23] how to obtain such embedding of $G(B)$ in polynomial time.

We use a reduction from PLANAR MONOTONE SAT to WEIGHTED CONNECTED MATCHING, for which the input is defined as $k = 2|X| + |\mathcal{C}|$ and the planar bipartite graph, that we call $H(B)$, obtained from $G(B)$, with the addition of the following rules.

(I)   For each variable $x_i \in X$, generate four vertices, x_i^+, x_i^-, v_i, u_i. Also, add the 1-weight edges $\{v_i u_i, x_i x_i^+, x_i x_i^-\}$.

(II) For each edge having an endpoint in x_i, $1 \leq i \leq |X|$, representing a positive(negative) literal, we connect this edge to $x_i^+(x_i^-)$ instead of x_i and set its weight to 1.

(III) Remove all the edges from the variable vertices cycle, that is, from A_2, and add the 0-weight edges $\{v_i x_i, v_i x_{i+1} \mid 1 \leq i < |X|\} \cup \{v_m x_m, v_m x_1\}$.

(IV) For each clause $C_i \in \mathcal{C}$, we label the corresponding vertex as c_i^+ and connect it to a new vertex, c_i^-, by a 1-weight edge.

Similarly to the reduction described in Sect. 3, a connected matching M having weight at least k in $H(B)$ corresponds to an assignment of B in such a way that variable $x_i \in X$ is set to true if and only if x_i^+ is M-saturated.

Theorem 4. WEIGHTED CONNECTED MATCHING *is* NP-*complete on planar bipartite graphs whose edge weights are in* $\{0, 1\}$.

4.2   Subcubic Planar Graphs

Now, we approach subcubic planar graphs, showing the NP-completeness of
WEIGHTED CONNECTED MATCHING for this class where edge weights in
$\{-1, +1\}$. Our proof is very similar to another reduction, made by Marzio De
Biasi [2].

Let us define a polynomial-time reduction from STEINER TREE, one of Karp's
original 21 NP-complete problems [20]. In this problem, we are given a graph
$H = (V_H, E_H)$, a subset $R \subseteq V_H$ and an integer $k' > 0$; we want to know if there
is a subgraph $T = (V_T, E_T)$ of H such that T is a tree, $R \subseteq V_T$ and $|E_T| \le k'$.
Our reduction is from STEINER TREE in which the input graph is planar; Garey
and Johnson showed in [15] that this problem is NP-complete.

Let $(H = (V_H, E_H), R, k')$ be the input of STEINER TREE such that H is
planar. Also, let $q = \Delta(H)$, $p = q(|V_H| - |R|) + 1$ and $r = p|E_H| + 1$. For
WEIGHTED CONNECTED MATCHING input (G, k), we define $k = r|R| - pk'$ and
the graph G built by the following procedures:

(I) For each vertex $w \in V_H$, add a copy of a cycle with $2r$ vertices if $w \in R$, or
    $2q$ otherwise. If this number is less than 3, instead, add a copy of a path with
    the same number of vertices. Set the weights of all these edges to 1. Now,
    for each $u \in N(w)$, arbitrarily label one of the vertices of this subgraph as
    v_{wu}. We denote this subgraph by C_w.
(II) For each edge wu in E_H, generate a copy of P_{2p} whose edge weights are
    -1 and make one of its endpoints vertices adjacent to v_{wu} and the other to
    v_{uw}. We denote this subgraph by P_{wu} or P_{uw}.

A connected matching of weight at least k in G corresponds to a tree $T =
(V_T, E_T)$ where $V_T = \{w \mid w \in V_H, V(C_w) \cap V(M) \ne \emptyset\} \subseteq R$ and $E_T = \{wu \mid
wu \in E_H, V(P_{wu}) \subseteq V(M)\}$.

As an example, consider the input for STEINER TREE as $k' = 1$, $R = \{a, b\}$
and $H = (V_H, E_H)$, isomorphic to C_3, with $V_H = \{a, b, c\}$. So, $q = \Delta(H) = 2$,
$p = q(|V_H| - |R|) + 1 = 3$, $r = p|E_H| + 1 = 10$. For the WEIGHTED CONNECTED
MATCHING input, we set $k = 17$, and the graph as illustrated in Fig. 3.

Theorem 5. WEIGHTED CONNECTED MATCHING *is* NP-*complete even for subcubic planar graphs having edge weights in* $\{-1, +1\}$.

5   Polynomial-Time Algorithms

So far, we have proven that WEIGHTED CONNECTED MATCHING is NP-complete
even when some constraints are imposed, such as limits on the weights, planarity
and degree bounds. In this section, we turn our attention to other tractable
cases, presenting algorithms for graphs of maximum degree two, trees and, more
generally, graphs of bounded treewidth.

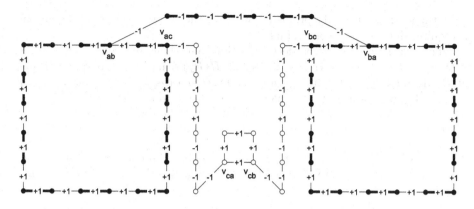

Fig. 3. Example of a subcubic planar graph G generated from a cycle of length three. The connected matching represents the tree subgraph $G[\{a, b\}]$ of G.

5.1 Graphs Having Degree at Most 2

Due to Theorem 5, we know that MAXIMUM WEIGHT CONNECTED MATCHING is NP-hard for graphs of maximum degree three. As such, we turn our attention to graphs of maximum degree two, i.e. the disjoint unions of paths and cycles, and prove that they allow for a linear time algorithm; in fact, for the case of paths, we reuse the algorithm for trees previously described.

As for cycles, we use following method. Given graph $G = (V, E)$, take two arbitrary edges $E' = \{uw, wv\} \subseteq E$. A maximum connected matching in G contains one or none elements of E'. So, we compare three maximum connected matchings, defined in $G - uw$, $G - wv$, and $G - w$; they can be obtained in linear time, by using a dynamic programming if $|E' \cap M| = 1$, or by our linear algorithm for trees(see Sect. 5.2) otherwise. Among these three, the matching having the largest weight is a maximum weight connected matching.

Theorem 6. MAXIMUM WEIGHT CONNECTED MATCHING *can be solved in linear time for cycles. Furthermore, the problem can be solved in linear time for graphs of maximum degree at most* 2.

5.2 Bounded Treewidth Graphs

A tree decomposition of a graph G is a pair $\mathbb{T} = (T, \mathcal{B} = \{B_j \mid j \in V(T)\})$, where T is a tree and $\mathcal{B} \subseteq 2^{V(G)}$ is a family where: $\bigcup_{B_j \in \mathcal{B}} B_j = V(G)$; for every edge $uv \in E(G)$ there is some B_j such that $\{u, v\} \subseteq B_j$; for every $i, j, q \in V(T)$, if q is in the path between i and j in T, then $B_i \cap B_j \subseteq B_q$. Each $B_j \in \mathcal{B}$ is called a *bag* of the tree decomposition. G has treewidth at most t if it admits a tree decomposition such that no bag has more than $t + 1$ vertices. For further properties of treewidth, we refer to [30]. Our algorithm relies on the rank based approach of Bodlaender et al. [3] for optimizing dynamic programming algorithms for connectivity problems; we omit the several additional definitions it requires due to space constraints.

For each node x of a tree decomposition, our algorithm constructs a table $f_x(S, U) \subseteq \Pi(B_x) \times \mathbb{R}$, with $S, U \subseteq B_x$ and $\Pi(B_x)$ being the set of all partitions of B_x. Intuitively, each entry $(p, w) \in f_x(S, U)$ corresponds to a matching M of the subgraph induced by the bags of the subtree rooted at x with weight w, where each block $p \in \Pi(S \cup U)$ is part of a distinct connected component of $G[M]$.

Theorem 7. *Given a tree decomposition of width t of the n-vertex input graph,* MAXIMUM WEIGHT CONNECTED MATCHING *can be solved in $2^{\mathcal{O}(t)} n^{\mathcal{O}(1)}$ time.*

Trees. The algorithm described in Sect. 5.2 implies that WEIGHTED CONNECTED MATCHINGS can be solved in $n^{\mathcal{O}(1)}$ time on trees. We strengthen this result by a linear time algorithm for this class.

Given a tree T, we begin by rooting it in some vertex $r \in V(T)$. Then, we traverse this rooted tree in post-order such a way that, when visiting vertex v, we find the weight of a maximum weight connected matching in the subtree defined by v and its descendants and such that v is saturated. The matching having the largest weight is a maximum weight connected matching in T.

In Theorem 8, we give details about this algorithm, proving its correctness and analyzing the running time.

Theorem 8. MAXIMUM WEIGHT CONNECTED MATCHING *on trees can be solved in linear time.*

Proof. We describe a linear algorithm that solves WEIGHTED CONNECTED MATCHING for trees. Let $T = (V, E)$ be a tree. We denote by T^r the tree T rooted in r. Moreover, for $v \in V$, T_v^r is the subtree of T^r that contains v and all its descendants, and $S(r, v)$ is the set of children of v in T^r. Also, consider $B_{r,v}$ as the weight of a maximum weight connected matching in T_v^r such that, if v is not a leaf, then v is saturated. Moreover, $\overline{B}_{r,v}$ is the weight of a matching M defined as the union of the maximum connected matchings in T_u^r, for each $u \in S(r, v)$, such that $G[V(M) \cup \{v\}]$ is connected.

Next, we describe a dynamic programming that, for given a root $r \in V$, can be used to obtain $\overline{B}_{r,v}$ and $B_{r,v}$ for every $v \in V$. For the base case, the vertex v is a leaf in T_v^r, and then $B_{r,v} = \overline{B}_{r,v} = 0$. Otherwise, v is not a leaf, and we can obtain $B_{r,v}, \overline{B}_{r,v}$ as follows.

$$f(r, vu) = \overline{B}_{r,u} + w(vu) + \sum_{s \in S(r,v) \setminus \{u\}} \max\{B_{r,s}, 0\}$$

$$\overline{B}_{r,v} = \sum_{u \in S(r,v)} \max\{B_{r,u}, 0\}$$

$$B_{r,v} = \max_{u \in S(r,v)} f(r, vu)$$

Next, we show that we can run this dynamic programming in linear time. Clearly, summations $\overline{B}_{r,v} = \sum_{u \in S(r,v)} \max\{B_{r,u}, 0\}$ for all vertices $v \in V$ can be calculated in linear time. For $B_{r,v}$, note that it can also be written as follows.

$$B_{r,v} = \overline{B}_{r,v} + \max_{u \in S(r,v)} (w(uv) + \overline{B}_{r,u} - B_{r,u})$$

This leads to a linear time procedure to obtain $B_{r,v}$ and $\overline{B}_{r,v}$, for every $v \in V$. Now, we can find a maximum weight connected matching in T by reconstructing the matching that generated weight $B_{r,h}$, for the vertex h that maximizes $B_{r,h}$.

6 Kernelization

Theorem 7 implies that MAXIMUM WEIGHT CONNECTED MATCHING parameterized by treewidth is in FPT, which immediately prompts an investigation into whether its decision version admits a polynomial kernel under the same parameterization. We answer this negatively by showing that WEIGHTED CONNECTED MATCHING parameterized by vertex cover number does not admit a polynomial kernel, unless NP $\subseteq$ coNP/poly, even if the input is restricted to bipartite graphs of bounded diameter and the allowed weights are in $\{0,1\}$, which implies the same result when parameterizing by treewidth, since treewidth is upper bounded by the vertex cover number.

We prove our result through an OR-cross-composition [4] from the 3SAT problem. Our construction is heavily inspired by the proof described in Sect. 3. Formally let, $\mathcal{H} = \{(X_1, \mathcal{C}_1), \ldots, (X_t, \mathcal{C}_t)\}$ be a set of t 3SAT instances such that $X_i = X = \{x_1, \ldots, x_n\}$ for every $i \in [t]$. Also, let $\mathcal{C} = \bigcup_{i \in [t]} \mathcal{C}_i$. Finally, let (G, k) be the WEIGHTED CONNECTED MATCHING instance we are going to build.

We begin our construction by adding to G a pair of vertices c_j, c'_j for each $C_j \in \mathcal{C}$ and a unit weight edge between them. Then, for each $x_i \in X$, we add vertices x_i^-, x_i^*, x_i^+ and edges $x_i^- x_i^*, x_i^* x_i^+$, each of weight 1. Now, for each $C_j \in \mathcal{C}$ and $i \in [n]$, if $x_i \in C_j$, we add the 0-weight edge $x_i^+ c_j$ to G, otherwise, if $\overline{x_i} \in C_j$, we add the weight 0 edge $x_i^- c_j$. We conclude this first part of the construction by adding a pair of vertices h, h' to G, making them adjacent with an edge of weight 1, and adding an edge of weight 0 between h and x_i^* for every $x_i \in X$. At this point, we have an extremely similar graph to the one constructed in Sect. 3.

For the next part of the construction, we add a copy of $K_{1,t}$, where the vertex on the smaller side is labeled q and, the vertices on the other side are each assigned a unique label from the set $Y = \{y_1, \ldots, y_t\}$, with each edge having weight 1. Now, for each $y_\ell \in Y$ and $C_j \in \mathcal{C} \setminus \mathcal{C}_\ell$, we add the 0-weight edges $c'_j y_\ell$ and $h y_\ell$. Finally, we set $k = |\mathcal{C}| + |X| + 2$, i.e. we must pick one edge in each clause gadget and vertex gadget plus the edge hh' and one edge between q and Y. Note that $|V(G)| = 3|X| + 2|\mathcal{C}| + |Y| + 3 \leq 3|X| + 2|X|^3 + |Y| + 3$, which implies that $V(G) \setminus Y$ is a vertex cover of G of size $\mathcal{O}(|X|^3)$, as required by the cross-composition framework. Moreover, note that G is bipartite, as we can partition it as follows: $L = \{q, h\} \cup \{x_i^+, x_i^- \mid i \in [n]\} \cup \{c'_j \mid C_j \in \mathcal{C}\}$ and $R = V(G) \setminus L$, where both L and R are independent sets.

Theorem 9. *Unless* NP ⊆ coNP/poly, WEIGHTED CONNECTED MATCHING *does not admit a polynomial kernel when parameterized by vertex cover number and required weight even if the input graph is bipartite and edge weights are in* $\{0, 1\}$.

7   Conclusions and Future Work

Motivated by previous works on weighted $\mathscr{P}$-matchings, such as WEIGHTED INDUCED MATCHING [21,29] and WEIGHTED ACYCLIC MATCHING [14], in this paper we introduced and studied WEIGHTED CONNECTED MATCHING problem.

We begin our investigation on the complexity of the problem by imposing restrictions on the input graphs and weights. In particular, we showed that the problem is NP-complete on planar bipartite graphs and bipartite graphs of diameter 4 for binary weights, and on subcubic planar graphs and starlike graphs when weights are restricted to $\{-1, +1\}$. On the positive side, we presented polynomial-time algorithms for MAXIMUM WEIGHT CONNECTED MATCHING on chordal graphs with non-negative weights, graphs having maximum degree at most two with arbitrary weights, on trees and, more generally, on graphs of bounded treewidth. The latter algorithm implies that WEIGHTED CONNECTED MATCHING is fixed-parameter tractable under the treewidth parameterization. This prompted our study of the problem from the kernelization point of view; our inquiry showed that no polynomial kernel exists when parameterized by vertex cover and the minimum required weight unless NP ⊆ coNP/poly.

Possible directions for future work include determining the complexity of the problem for different combinations of graph classes and allowed edge weights. In particular, we would like to know the complexity of WEIGHTED CONNECTED MATCHING for diameter 3 bipartite graphs when weights are non-negative, chordal bipartite graphs, and subcubic planar graphs under the same constraint. Other graph classes of interest include cactus graphs and block graphs.

We are also interested in the parameterized complexity of the problem. In terms of natural parameterizations, we see two possible directions: parameterizing by the number of edges in the matching or by the weight of the matching; while we have some negative kernelization results for these parameters, tractability is still unknown. Other possibilities include the study of other structural parameterizations, with the main open question being tractability for the cliquewidth parameterization.

References

1. Baste, J., Rautenbach, D.: Degenerate matchings and edge colorings. Discrete Appl. Math. **239**, 38–44 (2018). https://doi.org/10.1016/j.dam.2018.01.002
2. Biasi, M.D.: Max-weight connected subgraph problem in planar graphs. Theor. Comput. Sci. Stack Exch. https://cstheory.stackexchange.com/q/21669

3. Bodlaender, H.L., Cygan, M., Kratsch, S., Nederlof, J.: Deterministic single exponential time algorithms for connectivity problems parameterized by treewidth. Inf. Comput. **243**, 86–111 (2015). https://doi.org/10.1016/j.ic.2014.12.008, 40th International Colloquium on Automata, Languages and Programming (ICALP 2013)

4. Bodlaender, H.L., Jansen, B.M.P., Kratsch, S.: Cross-composition: a new technique for kernelization lower bounds. In: Proceedings of the 28th International Symposium on Theoretical Aspects of Computer Science (STACS), LIPIcs, vol. 9, pp. 165–176 (2011). https://doi.org/10.4230/LIPIcs.STACS.2011.165

5. Bondy, J.A., Murty, U.S.R.: Graph Theory. Springer, London (2008). https://www.springer.com/gp/book/9781846289699

6. Brandstädt, A., Le, V.B., Spinrad, J.P.: Graph Classes: A Survey. Society for Industrial and Applied Mathematics, Philadelphia (1999)

7. de C. M. Gomes, G., Masquio, B.P., Pinto, P.E.D., dos Santos, V.F., Szwarcfiter, J.L.: Disconnected matchings. CoRR abs/2112.09248 (2021). https://arxiv.org/abs/2112.09248

8. Cameron, K.: Induced matchings. Discrete Appl. Math. **24**(1), 97–102 (1989). https://doi.org/10.1016/0166-218X(92)90275-F

9. Cameron, K.: Connected matchings. In: Jünger, M., Reinelt, G., Rinaldi, G. (eds.) Combinatorial Optimization — Eureka, You Shrink! LNCS, vol. 2570, pp. 34–38. Springer, Heidelberg (2003). https://doi.org/10.1007/3-540-36478-1_5

10. Cygan, M., et al.: Parameterized Algorithms. Springer, Cham (2015). https://doi.org/10.1007/978-3-319-21275-3

11. Duan, R., Pettie, S., Su, H.H.: Scaling algorithms for weighted matching in general graphs. ACM Trans. Algorithms **14**(1) (2018). https://doi.org/10.1145/3155301

12. Edmonds, J.: Maximum matching and a polyhedron with 0, 1-vertices. J. Res. Natl Bureau Stand. B **69**, 125–130 (1965). https://doi.org/10.6028/jres.069B.013

13. Edmonds, J.: Paths, trees, and flowers. Can. J. Math. **17**, 449–467 (1965). https://doi.org/10.4153/CJM-1965-045-4

14. Fürst, M., Rautenbach, D.: On some hard and some tractable cases of the maximum acyclic matching problem. Ann. Oper. Res. **279**(1), 291–300 (2019). https://doi.org/10.1007/s10479-019-03311-1

15. Garey, M.R., Johnson, D.S.: The rectilinear Steiner tree problem is NP-complete. SIAM J. Appl. Math. **32**(4), 826–834 (1977). https://doi.org/10.1137/0132071

16. Garey, M.R., Johnson, D.S.: Computers and Intractability: a Guide to the Theory of NP-Completeness. W. H. Freeman & Co., New York (1979)

17. Goddard, W., Hedetniemi, S.M., Hedetniemi, S.T., Laskar, R.: Generalized subgraph-restricted matchings in graphs. Discrete Math. **293**(1), 129–138 (2005). https://doi.org/10.1016/j.disc.2004.08.027

18. Golumbic, M.C., Hirst, T., Lewenstein, M.: Uniquely restricted matchings. Algorithmica **31**(2), 139–154 (2001). https://doi.org/10.1007/s00453-001-0004-z

19. Gomes, G.C.M., Masquio, B.P., Pinto, P.E.D., dos Santos, V.F., Szwarcfiter, J.L.: Disconnected matchings. In: Chen, C.-Y., Hon, W.-K., Hung, L.-J., Lee, C.-W. (eds.) COCOON 2021. LNCS, vol. 13025, pp. 579–590. Springer, Cham (2021). https://doi.org/10.1007/978-3-030-89543-3_48

20. Karp, R.: Reducibility among combinatorial problems. In: Miller, R., Thatcher, J. (eds.) Complexity of Computer Computations, pp. 85–103. Plenum Press (1972). https://doi.org/10.1007/978-1-4684-2001-2_9

21. Klemz, B., Rote, G.: Linear-time algorithms for maximum-weight induced matchings and minimum chain covers in convex bipartite graphs. Algorithmica **84**(4), 1064–1080 (2022). https://doi.org/10.1007/s00453-021-00904-w

22. Kobler, D., Rotics, U.: Finding maximum induced matchings in subclasses of claw-free and P5-free graphs, and in graphs with matching and induced matching of equal maximum size. Algorithmica **37**(4), 327–346 (2003). https://doi.org/10.1007/s00453-003-1035-4
23. Lichtenstein, D.: Planar formulae and their uses. SIAM J. Comput. **11**(2), 329–343 (1982). https://doi.org/10.1137/0211025
24. Lozin, V.V.: On maximum induced matchings in bipartite graphs. Inf. Process. Lett. **81**(1), 7–11 (2002). https://doi.org/10.1016/S0020-0190(01)00185-5
25. Masquio, B.P.: Emparelhamentos desconexos. Master's thesis, Universidade do Estado do Rio de Janeiro (2019). http://www.bdtd.uerj.br/handle/1/7663
26. Micali, S., Vazirani, V.V.: An $O(\sqrt{|V|}|E|)$ algorithm for finding maximum matching in general graphs. In: 21st Annual IEEE Symposium on Foundations of Computer Science, pp. 17–27, October 1980. https://doi.org/10.1109/SFCS.1980.12
27. Moser, H., Sikdar, S.: The parameterized complexity of the induced matching problem. Discrete Appl. Math. **157**(4), 715–727 (2009). https://doi.org/10.1016/j.dam.2008.07.011
28. Panda, B.S., Chaudhary, J.: Acyclic matching in some subclasses of graphs. In: Gąsieniec, L., Klasing, R., Radzik, T. (eds.) IWOCA 2020. LNCS, vol. 12126, pp. 409–421. Springer, Cham (2020). https://doi.org/10.1007/978-3-030-48966-3_31
29. Panda, B.S., Pandey, A., Chaudhary, J., Dane, P., Kashyap, M.: Maximum weight induced matching in some subclasses of bipartite graphs. J. Comb. Optim. **40**(3), 713–732 (2020). https://doi.org/10.1007/s10878-020-00611-2
30. Robertson, N., Seymour, P.D.: Graph minors. II. algorithmic aspects of tree-width. J. Algorithms **7**(3), 309–322 (1986). https://doi.org/10.1016/0196-6774(86)90023-4

Space-Efficient Data Structure for Next/Previous Larger/Smaller Value Queries

Seungbum Jo[1]([✉])[iD] and Geunho Kim[2][iD]

[1] Chungnam National University, Daejeon, South Korea
sbjo@cnu.ac.kr
[2] Chungbuk National University, Cheongju, South Korea
gnho@chungbuk.ac.kr

Abstract. Given an array of size n from a total order, we consider the problem of constructing a data structure that supports various queries (range minimum/maximum queries with their variants and next/previous larger/smaller queries) efficiently. In the encoding model (i.e., the queries can be answered without the input array), we propose a $(3.701n + o(n))$-bit data structure, which supports all these queries in $O(\log^{(\ell)} n)$ time, for any positive integer ℓ (here, $\log^{(1)} n = \log n$, and for $\ell > 1$, $\log^{(\ell)} n = \log(\log^{(\ell-1)} n)$). The space of our data structure matches the current best upper bound of Tsur (Inf. Process. Lett., 2019), which does not support the queries efficiently. Also, we show that at least $3.16n - \Theta(\log n)$ bits are necessary for answering all the queries. Our result is obtained by generalizing Gawrychowski and Nicholson's $(3n - \Theta(\log n))$-bit lower bound (ICALP, 15) for answering range minimum and maximum queries on a permutation of size n.

Keywords: Range minimum queries · Encoding model · Balanced parenthesis sequence

1 Introduction

Given an array $A[1, \ldots, n]$ of size n from a total order and an interval $[i, j] \subset [1, n]$, suppose there are k distinct positions $i \leq p_1 \leq p_2 \ldots \leq p_k \leq j$ where $p_1, p_2, \ldots, p_k$ are the positions of minimum elements in $A[i, \ldots, j]$. Then, for $q \geq 1$, *range q-th minimum query* on the interval $[i, j]$ ($\mathsf{RMin}(i, j, q)$) returns the position p_q (returns p_k if $q > k$), and *range minimum query* on the interval $[i, j]$ ($\mathsf{RMin}(i, j)$) returns an arbitrary position among $p_1, p_2, \ldots, p_k$. One can also analogously define *range q-th maximum query* (resp. *range maximum query*) on the interval $[i, j]$, denoted by $\mathsf{RMax}(i, j, q)$ (resp. $\mathsf{RMax}(i, j)$).

In addition to the above queries, one can define next/previous larger/smaller queries as follows. When the position i is given, the *previous smaller value query* on the position i ($\mathsf{PSV}(i)$) returns the rightmost position $j < i$, where $A[j]$ is smaller than $A[i]$ (returns 0 if no such j exists), and the *next smaller value query* on the position i ($\mathsf{NSV}(i)$) returns the leftmost position $j > i$ where $A[j]$

© Springer Nature Switzerland AG 2022
A. Castañeda and F. Rodríguez-Henríquez (Eds.): LATIN 2022, LNCS 13568, pp. 71–87, 2022.
https://doi.org/10.1007/978-3-031-20624-5_5

is smaller than $A[i]$ (returns $n + 1$ if no such j exists). The *previous (resp. next) larger value query* on the position i, denoted by $\mathsf{PLV}(i)$ (resp. $\mathsf{NLV}(i)$)) is also defined analogously.

In this paper, we focus on the problem of constructing a data structure that efficiently answers all the above queries. We consider the problem in the *encoding model* [15], which does not allow access to the input A for answering the queries after prepossessing. In the encoding data structure, the lower bound of the space is referred to as the *effective entropy* of the problem. Note that for many problems, their effective entropies have much smaller size compared to the size of the inputs [15]. Also, an encoding data structure is called *succinct* if its space usage matches the optimal up to lower-order additive terms. The rest of the paper only considers encoding data structures and assumes a $\Theta(\log n)$-bit word RAM model, where n is the input size.

Previous Work. The problem of constructing an encoding data structure for answering range minimum queries has been well-studied because of its wide applications. It is well-known that any two arrays have a different set of answers of range minimum queries if and only if their corresponding Cartesian trees [19] are distinct. Thus, the effective entropy of answering range minimum queries on the array A of size n is $2n - \Theta(\log n)$ bits. Sadakane [17] proposed the $(4n + o(n))$-bit encoding with $O(1)$ query time using the balanced-parenthesis (BP) [11] of the Cartesian tree on A with additional nodes. Fisher and Heun [7] proposed the $(2n + o(n))$-bit data structure (hence, succinct), which supports $O(1)$ query time using the depth-first unary degree sequence (DFUDS) [2] of the *2d-min heap* on A. Here, a 2d-min heap of A is an alternative representation of the Cartesian tree on A. By maintaining the encodings of both 2d-min and max heaps on A (2d-max heap can be defined analogously to 2d-min heap), the encoding of [7] directly gives a $(4n + o(n))$-bit encoding for answering both range minimum and maximum queries in $O(1)$ time. Gawrychowski and Nicholson [8] reduced this space to $(3n + o(n))$-bit while supporting the same query time for both queries. They also showed that the effective entropy for answering the range minimum and maximum queries is at least $3n - \Theta(\log n)$ bits.

Next/previous smaller value queries were motivated from the parallel computing [3], and have application in constructing compressed suffix trees [14]. If all elements in A are distinct, one can answer both the next and previous smaller queries using Fischer and Heun's encoding for answering range minimum queries [7]. For the general case, Ohlebusch et al. [14] proposed the $(3n + o(n))$-bit encoding for supporting range minimum and next/previous smaller value queries in $O(1)$ time. Fischer [6] improved the space to $2.54n + o(n)$ bits while maintaining the same query time. More precisely, their data structure uses the *colored 2d-min heap* on A, which is a 2d-min heap on A with the coloring on its nodes. Since the effective entropy of the colored 2d-min heap on A is $2.54n - \Theta(\log n)$ bits [10], the encoding of [6] is succinct. For any $q \geq 1$, the encoding of [6] also supports the range q-th minimum queries in $O(1)$ time [9].

From the above, the encoding of Fischer [6] directly gives a $(5.08n + o(n))$-bit data structure for answering the range q-th minimum/maximum queries and

Table 1. Summary of the upper and lower bounds results of encoding data structures for answering q-th minimum/maximum queries and next larger/smaller value queries on the array $A[1, \ldots, n]$, for any $q \geq 1$ (here, we can choose ℓ as any positive integer). Note that all our upper bound results also support the range minimum/maximum and previous larger/smaller value queries in $O(1)$ time (the data structures of [6,9] with $O(1)$ query time also support these queries in $O(1)$ time).

Array type	Space (in bits)	Query time	Reference
Upper bounds			
$A[i] \neq A[i+1]$ for all $i \in [1, n-1]$	$4n + o(n)$	$O(1)$	[9]
	$3.585n$	$O(n)$	[18]
	$3.585n + o(n)$	$O(\log^{(\ell)} n)$	This paper
General array	$5.08n + o(n)$	$O(1)$	[6,9]
	$4.088n + o(n)$	$O(n)$	[9]
	$4.585n + o(n)$	$O(1)$	
	$3.701n$	$O(n)$	[18]
	$3.701n + o(n)$	$O(\log^{(\ell)} n)$	This paper
Lower bounds			
Permutation	$3n - \Theta(\log n)$		[8]
General array	$3.16n - \Theta(\log n)$		This paper

next/previous larger/smaller value queries in $O(1)$ time by maintaining the data structures of both colored 2d-min and max heaps. Jo and Satti [9] improved the space to (i) $4n + o(n)$ bits if there are no consecutive equal elements in A and (ii) $4.585n + o(n)$ bits for the general case while supporting all the queries in $O(1)$ time. They also showed that if the query time is not of concern, the space of (ii) can be improved to $4.088n + o(n)$ bits. Recently, Tsur [18] improved the space to $3.585n$ bits if there are no consecutive equal elements in A and $3.701n$ bits for the general case. However, their encoding does not support the queries efficiently ($O(n)$ time for all queries).

Our Results. Given an array $A[1, \ldots, n]$ of size n with the interval $[i, j] \subset [1, n]$ and the position $1 \leq p \leq n$, we show the following results:

(a) If A has no two consecutive equal elements, there exists a $(3.585n + o(n))$-bit data structure, which can answer (i) RMin(i, j), RMax(i, j), PSV(p), and PLV(p) queries in $O(1)$ time, and (ii) for any $q \geq 1$, RMin(i, j, q), RMax(i, j, q), NSV(p), and NLV(p) queries in $O(\log^{(\ell)} n)$ time[1], for any positive integer ℓ.
(b) For the general case, the data structure of (a) uses $3.701n + o(n)$ bits while supporting the same query time.

Our results match the current best upper bounds of Tsur [18] up to lower-order additive terms while supporting the queries efficiently.

---

[1] Throughout the paper, we denote $\log n$ as the logarithm to the base 2.

The main idea of our encoding data structure is to combine the BP of colored 2d-min and max heap of A. Note that all previous encodings in [8,9,18] combine the DFUDS of the (colored) 2d-min and max heap on A. We first consider the case when A has no two consecutive elements (Sect. 3). In this case, we show that by storing the BP of colored 2d-min heap on A along with its color information, there exists a data structure that uses at most $3n + o(n)$ bits while supporting range minimum, range q-th minimum, and next/previous smaller value queries efficiently. The data structure is motivated by the data structure of Jo and Satti [9] which uses DFUDS of colored 2d-min heap on A. Compared to the data structure of [9], our data structure uses less space for the color information. Next, we show how to combine the data structures on colored 2d-min and max heap on A into a single structure. The combined data structure is motivated by the idea of Gawrychowski and Nicholson's encoding [8] to combine the DFUDS of 2d-min and max heap on A.

In Sect. 4, we consider the case that A has consecutive equal elements. In this case, we show that by using some additional auxiliary structures, the queries on A can be answered efficiently from the data structure on the array A', which discards all the consecutive equal elements from A.

Finally, in Sect. 5, we show that the effective entropy of the encoding to support the range q-th minimum and maximum queries on A is at least $3.16n - \Theta(\log n)$ bits. Our result is obtained by extending the $(3n - \Theta(\log n))$-bit lower bound of Gawrychowski and Nicholson [8] for answering the range minimum and maximum queries on a permutation of size n. We summarize our results in Table 1.

2   Preliminaries

This section introduces some data structures used in our results.

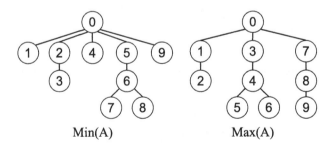

Fig. 1. Min(A) and Max(A) on the array $A = $ 5 4 5 3 1 2 6 3 1.

2D Min-Heap and Max-Heap. Given an array $A[1, \ldots, n]$ of size n, the 2d min-heap on A (denoted by Min(A)) [6] is a rooted and ordered tree with $n + 1$ nodes, where each node corresponds to the value in A, and the children are ordered from left to right. More precisely, Min(A) is defined as follows:

1. The root of $\mathsf{Min}(A)$ corresponds to $A[0]$ ($A[0]$ is defined as $-\infty$).
2. For any $i > 0$, $A[i]$ corresponds to the $(i+1)$-th node of $\mathsf{Min}(A)$ according to the preorder traversal.
3. For any non-root node corresponds to $A[j]$, its parent node corresponds to $A[\mathsf{PSV}(j)]$.

In the rest of the paper, we refer to the node i in $\mathsf{Min}(A)$ as the node corresponding to $A[i]$ (i.e., the $(i+1)$-th node according to the preorder traversal). One can also define the 2d-max heap on A (denoted as $\mathsf{Max}(A)$) analogously. More specifically, in $\mathsf{Max}(A)$, $A[0]$ is defined as ∞, and the parent of node $i > 0$ corresponds to the node $\mathsf{PLV}(i)$ (see Fig. 1 for an example). In the rest of the paper, we only consider $\mathsf{Min}(A)$ unless $\mathsf{Max}(A)$ is explicitly mentioned. The same definitions, and properties for $\mathsf{Min}(A)$ can be applied to $\mathsf{Max}(A)$.

For any $i > 0$, $\mathsf{Min}(A)$ is the *relevant tree* of the node i if the node i is an internal node in $\mathsf{Min}(A)$. From the definition of $\mathsf{Min}(A)$, Tsur [18] showed the following lemma.

Lemma 1 ([18]). *For any $i \in \{1, 2, \ldots, n-1\}$, the following holds:*

(a) *If $\mathsf{Min}(A)$ is a relevant tree of the node i, then the node $(i+1)$ is the leftmost child of the node i in $\mathsf{Min}(A)$.*
(b) *If A has no two consecutive equal elements, $\mathsf{Min}(A)$ (resp. $\mathsf{Max}(A)$) is a relevant tree of the node i if and only if the node i is a leaf node in $\mathsf{Max}(A)$ (resp. $\mathsf{Min}(A)$).*

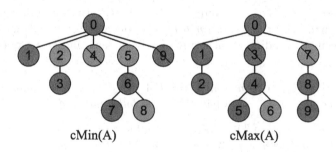

cMin(A)   cMax(A)

Fig. 2. $\mathsf{cMin}(A)$ and $\mathsf{cMax}(A)$ on the array $A = $ 5 4 5 3 1 2 6 3 1. The nodes with slash line indicate the valid nodes. (Color figure online)

Colored 2D Min-Heap and Max-Heap. The colored 2d-min heap of A (denoted by $\mathsf{cMin}(A)$) [6] is $\mathsf{Min}(A)$ where each node is colored *red* or *blue* as follows. The node i in $\mathsf{cMin}(A)$ is colored red if and only if i is not the leftmost child of its parent node, and $A[i] \neq A[j]$, where the node j is the node i's immediate left sibling. Otherwise, the node i is colored blue. One can also define the colored 2d-max heap on A (denoted by $\mathsf{cMax}(A)$) analogously (see Fig. 2 for an example). The following lemma says that we can obtain the color of some nodes in $\mathsf{cMin}(A)$ from their tree structures.

Lemma 2. *For any node i in $\mathsf{cMin}(A)$, the following holds:*

(a) If the node i is the leftmost child of its parent node, the color of the node i is always blue.

(b) If A has no two consecutive equal elements, the color of the node i is always red if its immediate left sibling is a leaf node.

Proof. (a) is directly proved from the definition of $\mathsf{cMin}(A)$. Also, if the immediate left sibling j of the node i is a leaf node, j is equal to $i - 1$ (note that the preorder traversal of $\mathsf{cMin}(A)$ visits the node i immediately after visiting the node j). Thus, if A has no two consecutive equal elements, the color of the node i is red.   □

We say the node i in $\mathsf{cMin}(A)$ is *valid* if the node i is a non-root node in $\mathsf{cMin}(A)$ which is neither the leftmost child nor the immediate right sibling of any leaf node. Otherwise, the node i is *invalid*. By Lemma 2, if A has no two consecutive equal elements, the color of the invalid nodes of $\mathsf{cMin}(A)$ can be decoded from the tree structure.

Rank and Select Queries on Bit Arrays. Given a bit array $B[1, \ldots, n]$ of size n, and a pattern $p \in \{0,1\}^{+}$, (i) $\mathsf{rank}_p(i, B)$ returns the number of occurrence of the pattern p in $B[1, \ldots, i]$, and (ii) $\mathsf{select}_p(j, B)$ returns the first position of the j-th occurrence of the pattern p in B. The following lemma shows that there exists a succinct encoding, which supports both rank and select queries on B efficiently.

Lemma 3 ([12,16]). *Given a bit array $B[1, \ldots, n]$ of size n containing m 1s, and a pattern $p \in \{0,1\}^{+}$ with $|p| \leq \frac{\log n}{2}$, the following holds:*

- *There exists a $(\log \binom{n}{m} + o(n))$-bit data structure for answering both $\mathsf{rank}_p(i, B)$ and $\mathsf{select}_p(j, B)$ queries in $O(1)$ time. Furthermore, the data structure can access any $\Theta(\log n)$-sized consecutive bits of B in $O(1)$ time.*
- *If one can access any $\Theta(\log n)$-sized consecutive bits of B in $O(1)$ time, both $\mathsf{rank}_p(i, B)$ and $\mathsf{select}_p(j, B)$ queries can be answered in $O(1)$ time using $o(n)$-bit auxiliary structures.*

Balanced-Parenthesis of Trees. Given a rooted and ordered tree T with n nodes, the balanced-parenthesis (BP) of T (denoted by $\mathsf{BP}(T)$) [11] is a bit array defined as follows. We perform a preorder traversal of T. We then add a 0 to $\mathsf{BP}(T)$ when we first visit a node and add a 1 to $\mathsf{BP}(T)$ after visiting all nodes in the subtree of the node. Since we add single 0 and 1 to $\mathsf{BP}(T)$ per each node in T, the size of $\mathsf{BP}(T)$ is $2n$. For any node i in T, we define

$f(i, T)$ and $s(i, T)$ as the positions of the 0 and 1 in $\mathsf{BP}(T)$ which are added when the node i is visited, respectively. When T is clear from the context, we write $f(i)$ (resp. $s(i)$) to denote $f(i, T)$ (resp. $s(i, T)$). If T is a 2d-min heap, $f(i, T) = \mathsf{select}_0(i + 1, BP(T))$ by the definition of 2d-min heap.

3   Data Structure on Arrays with No Consecutive Equal Elements

In this section, for any positive integer ℓ, we present a $(3.585n + o(n))$-bit data structure on $A[1, \ldots, n]$, which supports (i) range minimum/maximum and previous larger/smaller queries on A in $O(1)$ time, and (ii) range q-th minimum/maximum and next larger/smaller value queries on A in $O(\log^{(\ell)} n)$ time for any $q \geq 1$, when there are no two consecutive equal elements in A. We first describe the data structure on $\mathsf{cMin}(A)$ for answering the range minimum, range q-th minimum, and next/previous smaller value queries on A. Next, we show how to combine the data structures on $\mathsf{cMin}(A)$ and $\mathsf{cMax}(A)$ in a single structure.

Encoding Data Structure on cMin(A). We store $\mathsf{cMin}(A)$ by storing its tree structure along with the color information of the nodes. To store the tree structure, we use $\mathsf{BP}(\mathsf{cMin}(A))$. Also, for storing the color information of the nodes, we use a bit array c_{min}, which stores the color of all valid nodes in $\mathsf{cMin}(A)$ according to the preorder traversal order. In c_{min} we use 0 (resp. 1) to indicate the color blue (resp. red). It is clear that $\mathsf{cMin}(A)$ can be reconstructed from $\mathsf{BP}(\mathsf{cMin}(A))$ and c_{min}. Since $\mathsf{BP}(\mathsf{cMin}(A))$ and c_{min} takes $2(n + 1)$ bits and at most n bits, respectively, the total space for storing $\mathsf{cMin}(A)$ takes at most $3n + 2$ bits. Note that a similar idea is used in Jo and Satti's *extended DFUDS* [9], which uses the DFUDS of $\mathsf{cMin}(A)$ for storing the tree structure. However, extended DFUDS stores the color of all nodes other than the leftmost children, whereas c_{min} does not store the color of all invalid nodes. The following lemma shows that from $\mathsf{BP}(\mathsf{cMin}(A))$, we can check whether the node i is valid or not without decoding the entire tree structure.

Lemma 4. *The node i is valid in cMin(A) if and only if $f(i) > 2$ and both $BP(\mathsf{cMin}(A))[f(i) - 2]$ and $BP(\mathsf{cMin}(A))[f(i) - 1]$ are 1.*

Proof. If both $BP(\mathsf{cMin}(A))[f(i) - 2]$ and $BP(\mathsf{cMin}(A))[f(i) - 1]$ are 1, the preorder traversal of $\mathsf{cMin}(A)$ must complete the traversal of two subtrees consecutively just before visiting the node i for the first time, which implies the node i's immediate left sibling is not a leaf node (hence the node i is valid).

Conversely, if $BP(\mathsf{cMin}(A))[f(i) - 1] = 0$, $\mathsf{cMin}(A)$ is a relevant tree of the node $i - 1$. Thus, the node i is the leftmost child of the node $(i - 1)$ by Lemma 1. Next, if $BP(\mathsf{cMin}(A))[f(i) - 2] = 0$ and $BP(\mathsf{cMin}(A))[f(i) - 1] = 1$, the node $(i-1)$ is the immediate left sibling of the node i since $f(i) - 2$ is equal to $f(i-1)$. Also the node $(i - 1)$ is a leaf node since $f(i) - 1$ is equal to $s(i - 1)$. Thus, the node i is invalid in this case. $\square$

Now we describe how to support range minimum, range q-th minimum, and next/previous smaller value queries efficiently on A using $\mathsf{BP}(\mathsf{cMin}(A))$ and c_{min} with $o(n)$-bit additional auxiliary structures. Note that both the range minimum and previous smaller value query on A can be answered in $O(1)$ time using $BP(\mathsf{cMin}(A))$ with $o(n)$-bit auxiliary structures [5,13]. Thus, it is enough to consider how to support a range q-th minimum and next smaller value queries on A. We introduce the following lemma of Jo and Satti [9], which shows that one can answer both queries with some navigational and color queries on $\mathsf{cMin}(A)$.

Lemma 5 ([9]). *Given $\mathsf{cMin}(A)$, suppose there exists a data structure, which can answer (i) the tree navigational queries (next/previous sibling, subtree size, degree, level ancestor, child rank, child select, and parent[2]) on $\mathsf{cMin}(A)$ in $t(n)$ time, and the following color queries in $s(n)$ time:*

- *color(i): return the color of the node i*
- *PRS(i): return the rightmost red sibling to the left of the node i.*
- *NRS(i): return the leftmost red sibling to the right of the node i.*

Then for any $q \geq 1$, range q-th minimum, and the next smaller value queries on A can be answered in $O(t(n) + s(n))$ time.

Since all tree navigational queries in Lemma 5 can be answered in $O(1)$ time using $\mathsf{BP}(\mathsf{cMin}(A))$ with $o(n)$-bit auxiliary structures [13], it is sufficient to show how to support $\mathsf{color}(i)$, $\mathsf{PRS}(i)$, and $\mathsf{NRS}(i)$ queries using $\mathsf{BP}(\mathsf{cMin}(A))$ and c_{min}. By Lemma 3 and 4, we can compute $\mathsf{color}(i)$ in $O(1)$ time using $o(n)$-bit auxiliary structures by the following procedure: We first check whether the node i is valid using $O(1)$ time by checking the values at the positions $f(i) - 1$ and $f(i) - 2$ in $\mathsf{BP}(\mathsf{cMin}(A))$. If the node i is valid (i.e., both the values are 1), we answer $\mathsf{color}(i)$ in $O(1)$ time by returning $c_{min}[j]$ where j is $\mathsf{rank}_{110}(f(i), \mathsf{BP}(\mathsf{cMin}(A)))$ (otherwise, by Lemma 4, we answer $\mathsf{color}(i)$ as blue if and only if the node i is the leftmost child of its parent node). Next, for answering $\mathsf{PRS}(i)$ and $\mathsf{NRS}(i)$, we construct the following ℓ'-level structure (ℓ' will be decided later):

- At the first level, we mark every $(\log n \log \log n)$-th child node and maintain a bit array $M_1[1, \ldots, n]$ where $M_1[t] = 1$ if and only if the node t is marked (recall that the node t is the node in $\mathsf{Min}(A)$ whose preorder number is t). Since there are $n/(\log n \log \log n) = o(n)$ marked nodes, we can store M_1 using $o(n)$ bits while supporting rank queries in $O(1)$ time by Lemma 3 (in the rest of the paper, we ignore all floors and ceilings, which do not affect to the results). Also we maintain an array P_1 of size $n/(\log n \log \log n)$ where $P_1[j]$ stores both $\mathsf{PRS}(s)$ and $\mathsf{NRS}(s)$ if s is the j-th marked node according to the preorder traversal order. We can store P_1 using $O(n \log n/(\log n \log \log n)) = o(n)$ bits.
- For the i-th level where $1 < i \leq \ell'$, we mark every $(\log^{(i)} n \log^{(i+1)} n)$-th child node. We then maintain a bit array M_i which is defined analogously to M_1. We can store M_i using $o(n)$ bits by Lemma 3.

---

[2] refer to Table 1 in [13] for detailed definitions of the queries.

Now for any node p, let $cr(p)$ be the child rank of p, i.e., the number of left siblings of p. Also, let $pre_{(i-1)}(p)$ (resp. $next_{(i-1)}(p)$) be the rightmost sibling of p to the left (resp. leftmost sibling of p to the right) which is marked at the $(i-1)$-th level. Suppose s is the j-th marked node at the current level according to the preorder traversal order. Then we define an array P_i of size $n/(\log^{(i)} n \log^{(i+1)} n)$ as $P_i[j]$ stores both (i) the smaller value between $cr(s) - cr(PRS(s))$ and $cr(s) - cr(pre_{(i-1)}(s))$, and (ii) the smaller value between $cr(NRS(s)) - cr(s)$ and $cr(next_{(i-1)}(s)) - cr(s)$. Since both (i) and (ii) are at most $\log^{(i-1)} n \log^{(i)} n$, we can store P_i using $O(n \log^{(i)} n/(\log^{(i)} n \log^{(i+1)} n)) = o(n)$ bits. Therefore, the overall space is $O(n/\log^{(\ell'+1)} n) = o(n)$ bits in total for any positive integer ℓ'.

To answer $PRS(i)$ (the procedure for answering $NRS(i)$ is analogous), we first scan the left siblings of i using the previous sibling operation. Whenever the node i_1 is visited during the scan, we check whether (i) $color(i_1) = red$, or (ii) $M_{\ell'}[i_1] = 1$ in $O(1)$ time. If i_1 is neither the case (i) nor (ii), we continue the scan. If i_1 is in the case (i), we return i_1 as the answer. If i_1 is in the case (ii), we jump to the i_1's left sibling i_2 whose child rank is $cr(i_1) - P_{\ell'}[j]$, where $j = \mathrm{rank}_1(i_1, M_{\ell'-1})$. Since the node i_2 always satisfies one of the following: $color(i_2) = red$ or $M_{\ell'-1}[i_2] = 1$, we can answer $PRS(i)$ by iteratively performing child rank and rank operations at most $O(\ell')$ times after finding i_2. Thus, we can answer $PRS(i)$ in $O(\ell')$ time in total (we scan at most $O(\ell')$ nodes to find i_2). By choosing ℓ as $\ell' + 2$, we obtain the following theorem.

Theorem 1. *Given an array $A[1, \ldots, n]$ of size n and any positive integer ℓ, we can answer (i) range minimum and previous smaller value queries in $O(1)$ time, and (ii) range q-th minimum and next smaller value queries for any $q \geq 1$ in $O(\log^{(\ell)} n)$ time, using $BP(cMin(A))$ and c_{min} with $o(n)$-bit auxiliary structures.*

Theorem 1 implies that there exists a data structure of $cMax(A)$ (composed to $BP(cMax(A))$ and c_{max} with $o(n)$-auxiliary structures), which can answer (i) range maximum and previous larger value queries in $O(1)$ time, and (ii) range q-th maximum and next larger value queries for any $q \geq 1$ in $O(\log^{(\ell)} n)$ time.

Combining the Encoding Data Structures on cMin(A) and cMax(A). We describe how to combine the data structure of Theorem 1 on $cMin(A)$ and $cMax(A)$ using $3.585n + o(n)$ bits in total. We first briefly introduce the idea of Gawrychowski and Nicholson [8] to combine the DFUDS of $Min(A)$ and $Max(A)$. In DFUDS, any non-root node i is represented as a bit array $0^{d_i}1$ where d_i is the degree of i [2]. The encoding of [8] is composed of (i) a bit array $U[1, \ldots, n]$, where $U[i]$ indicates the relevant tree of the node i, and (ii) a bit array $S = s_1 s_2 \ldots s_n$ where s_i is the bit array, which omits the first 0 from the DFUDS of the node i on its relevant tree. To decode the DFUDS of the node i in $Min(A)$ or $Max(A)$, first check whether the tree is the relevant tree of the node i by referring to $U[i]$. If so, one can decode it by prepending 0 to s_i. Otherwise, the decoded sequence is simply 1 by Lemma 1(b). Also, Gawrychowski and Nicholson [8] showed that

U and S take at most $3n$ bits in total. The following lemma shows that a similar idea can also be applied to combine $BP(cMin(A))$ and $BP(cMax(A))$ (the lemma can be proved directly from Lemma 1).

Lemma 6. *For any node $i \in \{1, 2, \ldots, n-1\}$, if $cMin(A)$ is a relevant tree of the node i, $f(i+1, cMin(A)) = f(i, cMin(A)) + 1$, and $f(i+1, cMax(A)) = f(i, cMax(A)) + k$, for some $k > 1$. Otherwise, $f(i+1, cMax(A)) = f(i, cMax(A)) + 1$, and $f(i+1, cMin(A)) = f(i, cMin(A)) + k$, for some $k > 1$.*

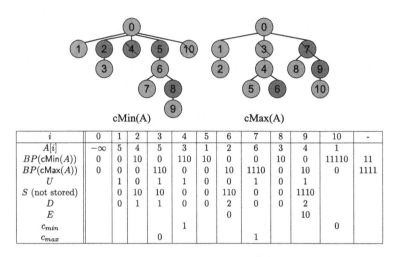

i	0	1	2	3	4	5	6	7	8	9	10	-
$A[i]$	$-\infty$	5	4	5	3	1	2	6	3	4	1	
$BP(cMin(A))$	0	0	10	0	110	10	0	0	10	0	11110	11
$BP(cMax(A))$	0	0	0	110	0	0	10	1110	0	10	0	1111
U		1	0	1	0	0	1	0	1	0		
S (not stored)		0	10	10	0	0	110	0	0	1110		
D		0	1	1	0	0	2	0	0	2		
E							0			10		
c_{min}					1						0	
c_{max}				0				1				

Fig. 3. Combined data structure of $cMin(A)$ and $cMax(A)$. i-th column of the table shows (i) the substring of $BP(cMin(A))$ and $BP(cMax(A))$ begin at position $f(i-1)+1$ and end at position $f(i)$ (shown in the second and the third row, respectively), and (ii) s_i for each i (shown in the fourth row).

We now describe our combined data structure of $cMin(A)$ and $cMax(A)$. We first maintain the following structures to store $BP(cMin(A))$ and $BP(cMax(A))$:

1. The same bit array $U[1, \ldots, n-1]$ as in the encoding of [8]. We define $U[i] = 0$ (resp. $U[i] = 1$) if $cMin(A)$ (resp. $cMax(A)$) is a relevant tree of the node i. For example, $U[6] = 0$ since $cMin(A)$ is a relevant tree of the node 6.
2. For each node $i \in \{1, 2, \ldots, n-1\}$, suppose the tree $T \in \{cMin(A), cMax(A)\}$ is not a relevant tree of i, and let k_i be the number of ones between $f(i, T)$ and $f(i+1, T)$. Now let $S = s_1 s_2 \ldots s_{n-1}$ be a bit array, where s_i is defined as $1^{k_i - 1} 0$. For example, since there exist three 1's between $f(6)$ and $f(7)$ in $cMax(A)$, $s_6 = 110$. Then, S is well-defined by Lemma 6 ($k_i \geq 1$ for all i). Also, since there are at most $n-1$ ones and exactly $n-1$ zeros by Lemma 1(b), the size of S is at most $2(n-1)$. We maintain S using the following two arrays:

(a) An array $D[1, \ldots n-1]$ of size n where $D[i] = 0$ if s_i contains no ones, $D[i] = 1$ if s_i contains a single one, and $D[i] = 2$ otherwise. For example, $D[6] = 2$, since s_6 has two ones. We maintain D using the data structure of Dodis et al. [4], which can decode any $\Theta(\log n)$ consecutive elements of D in $O(1)$ time using $\lceil (n-1) \log 3 \rceil$ bits. Now let k and ℓ be the number of 1's and 2's in D, respectively.

(b) Let i_2 be the position of the i-th 2 in D. Then, we store a bit array $E = e_1 e_2, \ldots, e_\ell$ where e_i is a bit array defined by omitting the first two 1's from s_{i_2}. For example, since the 6 is the first position of D whose value is 2 and $s_6 = 110$, e_1 is defined as 0. The size of E is at most $2(n-1) - (n-1) - (k+\ell) = n - k - \ell$.

3. We store both $f(n, \mathsf{cMin}(A))$, and $f(n, \mathsf{cMax}(A))$ using $O(\log n)$ bits.

To store both c_{min} and c_{max}, we simply concatenate them into a single array c_{minmax}, and store the length of c_{min} using $O(\log n)$ bits. Then, by Lemma 4, the size of c_{minmax} is $k+\ell$. Thus, our encoding of $\mathsf{cMin}(A)$ and $\mathsf{cMax}(A)$ takes at most $(n-1) + (n-1) \log 3 + (n-k-\ell) + (k+\ell) + O(\log n) = (2 + \log 3)n + O(\log n) < 3.585n + O(\log n)$ bits in total [18]. An overall example of our encoding is shown in Fig. 3. Now we prove the main theorem in this section.

Theorem 2. *Given an array $A[1, \ldots, n]$ of size n and any positive integer ℓ, suppose A has no two consecutive equal elements. Then there exists a $(3.585n + o(n))$-bit encoding data structure which can answer (i) range minimum/maximum and previous larger/smaller value queries in $O(1)$ time, and (ii) range q-th minimum/maximum and next larger/smaller value queries in $O(\log^{(\ell)} n)$ time, for any $q \geq 1$.*

Proof. We show how to decode any $\log n$ consecutive bits of $\mathsf{BP}(\mathsf{cMin}(A))$, which proves the theorem. Note that the auxiliary structures and the procedure for decoding $\mathsf{BP}(\mathsf{cMax}(A))$ are analogous. Let $B[1, \ldots, f(n) - 1]$ be a subarray of $\mathsf{BP}(\mathsf{cMin}(A))$ of size $f(n) - 1$, which is defined as $\mathsf{BP}(\mathsf{cMin}(A))[2, \ldots, f(n)]$. Then it is enough to show how to decode $\log n$ consecutive bits of B in $O(1)$ time using $o(n)$-bit auxiliary structures (note that $\mathsf{BP}(\mathsf{cMin}(A))$ is $0 \cdot B \cdot 1^{2n+2-f(n)}$). We also denote $f(n) - 1$ by $f'(n)$ in this proof.

We first define correspondences between the positions of B and D, and between the positions of B and E as follows. For each position $j \in \{1, \ldots, f'(n)\}$ of B, let $\alpha(j)$ and $\beta(j)$ be the corresponding positions of j in D and E, respectively. We define both $\alpha(1)$ and $\beta(1)$ as 1, and for each $j \in \{2, \ldots, f'(n)\}$, we define $\alpha(j)$ as $\mathsf{rank}_0(j-1, B)$. Next, let k be the number of 2's in $D[1, \ldots, \alpha(j)]$ and j' be the number of 1's in B between $B[j]$ and the and the leftmost 0 in $B[j, \ldots, f(\alpha(j+1))]$. Then $\beta(j)$ is defined as (i) 1 if $k = 0$, (ii) $\mathsf{select}_0(k, E) + 1$ if $k > 0$ and $D[\alpha(j)] \neq 2$, and (iii) $\mathsf{select}_0(k, E) - \max(j' - 3, 0)$ otherwise. Then any subarray of B starting from the position j can be constructed from the subarrays of U, D and E starting from the positions $\mathsf{rank}_0(j, B)$, $\alpha(j)$ and $\beta(j)$, respectively.

Now, for $i \in \{1, 2, \ldots, \lceil (f'(n))/\log n \rceil\}$, let the i-th block of B be $B[\lceil (i-1) \log n + 1 \rceil, \ldots, \min(\lceil i \log n \rceil, f'(n))]$. Then, it is enough to decode at most two consecutive blocks of B to construct any $\log n$ consecutive bits of B. Next, we define the i-th block of U, D, and E as follows:

- i-th block of U is defined as a subarray of U whose starting and ending positions are $\mathsf{rank}_0(\lceil (i-1)\log n \rceil, B)$, and $\mathsf{rank}_0(\min(\lceil i \log n \rceil - 1, f'(n)), B)$, respectively. To decode the blocks of U without B, we mark all the starting positions of the blocks of U using a bit array U_1 of size $f'(n)$ where $U_1[i] = 1$ if and only if the position i is the starting position of the block in U. Then, since U_1 contains at most $O(f'(n)/\log n) = o(n)$ 1's, we can store U_1 using $o(n)$ bits while supporting rank and select queries in $O(1)$ time by Lemma 3.
- i-th block of D is defined as a subarray of D whose starting and ending positions are $\alpha(\lceil (i-1)\log n \rceil + 1)$ and $\alpha(\min(\lceil i \log n \rceil, f'(n)))$, respectively. Then, the size of each block of D is at most $\log n$, since any position of D has at least one corresponding position in B. We maintain a bit array D_1 analogous to U_1 using $o(n)$ bits. Also, to indicate the case that two distinct blocks of D share the same starting position, we define another bit array D_2 of size $\lceil f'(n)/\log n \rceil$ where $D_2[i] = 1$ if and only if i-th block of D has the same starting position as the $(i-1)$-th block of D. We store D_2 using the data structure of Lemma 3 using $o(n)$ bits to rank and select queries in $O(1)$ time. Then, we can decode any block of D in $O(1)$ time using rank and select operations on D_1 and D_2.
- i-th block of E is defined as a subarray of E whose starting and ending positions are $\beta(\lceil (i-1)\log n \rceil + 1)$ and $\beta(\min(\lceil i \log n \rceil, f'(n)))$, respectively. To decode the blocks of E, we maintain two bit arrays E_1 and E_2 analogous to D_1 and D_2, respectively, using $o(n)$ bits.

  Note that, unlike D, the size of some blocks in E can be arbitrarily large since some positions in E do not have the corresponding positions in B. To handle this case, we classify each block of E as *bad block* and *good block* where the size of bad block is at least at $c \log n$ for some constant $c \geq 9$, whereas the size of good block is less than $c \log n$. If the i-th block of E is good (resp. bad), we say it as i-th good (resp. bad) block.

  For each i-th bad block of E, let F_i be a subsequence of the i-th bad block, which consists of all bits at the position j where $\beta^{-1}(j)$ exists. We store F_i explicitly, which takes $\Theta(n)$ bits in total (the size of F_i is at most $\log n$). However, we can apply the same argument used in [8] to maintain *min-bad block* due to the fact that each position in E corresponds to at least one position in either $\mathsf{BP}(\mathsf{cMin}(A))$ or $\mathsf{BP}(\mathsf{cMax}(A))$. The argument says that for each i-th bad block of E, one can save at least $\log n$ bits by maintaining it in a compressed form. Thus, we can maintain F_i for all i-th bad blocks of E without increasing the total space.

Next, let $g(u, d, e, b)$ be a function, which returns a subarray of B from the subarrays of U, and D, and E as follows (suppose $u = u[1] \cdot u'$ and $d = d[1] \cdot d'$):

$$g(u, d, e, b) = \begin{cases} \epsilon & \text{if } u = \epsilon \text{ or } d = \epsilon \\ 0 \cdot g(u', d', e, b) & \text{if } u[1] \neq b \text{ and } d[1] \neq 2 \\ 0 \cdot g(u', d', e', b) & \text{if } u[1] \neq b, \ d[1] = 2, \text{ and } e = 1^t 0 \cdot e' \\ 10 \cdot g(u', d', e, b) & \text{if } u[1] = b \text{ and } d[1] = 0 \\ 110 \cdot g(u', d', e, b) & \text{if } u[1] = b \text{ and } d[1] = 1 \\ 1^t \cdot g(u, d, e', b) & \text{if } u[1] = b, \ d[1] = 2, \text{ and } e = 1^t \cdot e' \\ 1^{t+3} 0 \cdot g(u', d', e', b) & \text{if } u[1] = b, \ d[1] = 2, \text{ and } e = 1^t 0 \cdot e' \end{cases}$$

We store a precomputed table that stores $g(u, d, e, b)$ for all possible u, d, and e of sizes $\frac{1}{4} \log n$ and $b \in \{0, 1\}$ using $O(2^{\frac{1}{4} \log n + \frac{3}{2} \cdot \frac{1}{4} \log n + \frac{1}{4} \log n} \log n) = O(n^{\frac{7}{8}} \log n) = o(n)$ bits.

To decode the i-th block of B, we first decode the i-block of U and D in $O(1)$ time using rank and select queries on U_1, D_1, and D_2. Let these subarrays be b_u and b_d, respectively. Also, we decode the i-th block of E using rank and select queries on E_1 and E_2. We then define b_e as F_i if the i-th block of E is bad. Otherwise, we define b_e as the i-th good block of E. Next, we compute $g(b_u, b_d, b_e, 0)$ in $O(1)$ time by referring to the precomputed table $O(1)$ times, and prepend 0 if we decode the first block of B. Finally, note that there are at most $q \leq 4$ consecutive positions from p to $p + q - 1$ of B whose corresponding positions are the same in both D and E. Because such a case can only occur when $B[p] = B[p+1] = \cdots = B[p+q-2] = 1$ and $B[p+q-1] = 0$, we maintain an array R of size $O(n/\log n)$, which stores the four cases of the number of consecutive 1's (0, 1, 2, or at least 3) from the beginning of the i-th block of B. Then, if the number of consecutive 1's from the beginning of $g(b_u, b_d, b_e, 0)$ is at most 3, we delete some 1s from the beginning of $g(b_u, b_d, b_e, 0)$ by referring to R as the final step. □

4   Data Structure on General Arrays

In this section, we present a $(3.701n + o(n))$-bit data structure to support the range q-th minimum/maximum and next/previous larger/smaller value queries on the array $A[1 \ldots, n]$ without any restriction. Let $C[1, \ldots, n]$ be a bit array of size n where $C[1] = 0$, and for any $i > 1$, $C[i] = 1$ if and only if $C[i-1] = C[i]$. If C has k ones, we define an array $A'[1, \ldots, n-k]$ of size $n-k$ that discards all consecutive equal elements from A. Then from the definition of colored 2d-min and max heap, we can observe that if $C[i] = 1$, (i) the node i is a blue-colored leaf node, and (ii) i's immediate left sibling is also a leaf node, both in cMin(A) and cMax(A). Furthermore, by deleting all the bits at the positions $f(i, \text{cMin}(A)) - 1$, and $f(i, \text{cMin}(A))$ from BP(cMin(A)) we can obtain BP(cMin(A')). We can also obtain BP(cMax(A')) from BP(cMax(A')) analogously. Now we prove the following theorem.

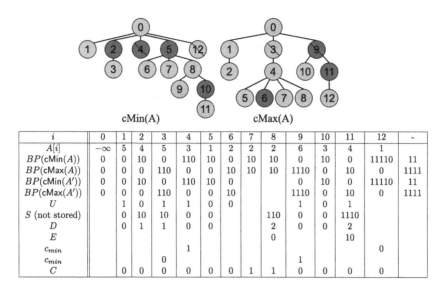

cMin(A)   cMax(A)

i	0	1	2	3	4	5	6	7	8	9	10	11	12	-
$A[i]$	$-\infty$	5	4	5	3	1	2	2	2	6	3	4	1	
$BP(\mathsf{cMin}(A))$	0	0	10	0	110	10	0	10	10	0	10	0	11110	11
$BP(\mathsf{cMax}(A))$	0	0	0	110	0	0	10	10	10	1110	0	10	0	1111
$BP(\mathsf{cMin}(A'))$	0	0	10	0	110	10	0			0	10	0	11110	11
$BP(\mathsf{cMax}(A'))$	0	0	0	110	0	0	10			1110	0	10	0	1111
U		1	0	1	1	0	0			1	0	1		
S (not stored)		0	10	10	0	0			110	0	0	1110		
D		0	1	1	0	0			2	0	0	2		
E									0			10		
c_{min}					1								0	
c_{min}				0						1				
C		0	0	0	0	0	0	1	1	0	0	0	0	

Fig. 4. Combined data structure of $\mathsf{cMin}(A)$ and $\mathsf{cMax}(A)$. Note that A' is the same array as the array in Fig. 3.

Theorem 3. *Given an array $A[1,\ldots,n]$ of size n and any positive integer ℓ, there exists a $(3.701n + o(n))$-bit encoding data structure which can answer (i) range minimum/maximum and previous larger/smaller value queries in $O(1)$ time, and (ii) range q-th minimum/maximum and next larger/smaller value queries in $O(\log^{(\ell)} n)$ time, for any $q \geq 1$.*

Proof. The data structure consists of C and the data structure of Theorem 2 on A', which can answer all the queries on A' in $O(\log^{(\ell)} n)$ time (see Fig. 4 for an example). By maintaining C using the data structure of Lemma 3, the data structure takes at most $(2 + \log 3)(n - k) + \binom{n}{k} + o(n) \leq 3.701n + o(n)$ bits in total [18] while supporting rank and select queries on C in $O(1)$ time. For any node i in $\mathsf{cMin}(A)$ and $\mathsf{cMax}(A)$, we can compute the color of the node i in $O(1)$ time as follows. If $C[i] = 0$, we return the color of the node $(\mathsf{rank}_0(i, C) - 1)$ in $\mathsf{cMin}(A')$ and $\mathsf{cMax}(A')$, respectively. Otherwise, we return blue. Now we describe how to decode any $\log n$ consecutive bits of $BP(\mathsf{cMin}(A))$ in $O(1)$ time using $o(n)$-bit auxiliary structures, which proves the theorem (the auxiliary structures and the procedure for decoding $BP(\mathsf{cMax}(A))$ are analogous). In the proof, we denote $BP(\mathsf{cMin}(A))$ and $BP(\mathsf{cMin}(A'))$ as B and B', respectively.

For each position j of B, we say j is *original* if $B[j]$ comes from the bit in B', and *additional* otherwise. That is, the position j is additional if and only if j is $f(j') - 1$ or $f(j')$ where $C[j'] = 1$. For each original position j, let $b'(j)$ be its corresponding position in B'.

Now we divide B into the blocks of size $\log n$ except the last block, and let s_i be the starting position of the i-th block of B. We then define a bit array $M_{B'}$ of size $2(n-k)$ as follows. For each $i \in 1, \ldots, \lceil (2(n+1)/\log n \rceil$, we set the $b'(s_i)$-th position of $M_{B'}$ as one if s_i is original. Otherwise, we set the $b'(s_i')$-th position of $M_{B'}$ as one where s_i' is the leftmost original position from s_i to the right in B.

All other bits in $M_{B'}$ are 0. Also, let $M'_{B'}$ a bit array of size $\lceil(2(n+1)/\log n\rceil$ where $M'_{B'}[i]$ is 1 if and only if we mark the same position for s_i and s_{i-1}. Since $M_{B'}$ has at most $\lceil(2(n+1)/\log n\rceil = o(n)$ ones, we can maintain both $M_{B'}$ and $M'_{B'}$ in $o(n)$ bits while supporting rank and select queries in $O(1)$ time by Lemma 3. Similarly, we define a bit array M_C of size n as follows. If s_i is original, we set the $(\text{rank}_0(s_i-1,B))$-th position of M_C as one. Otherwise, we set the $(\text{rank}_0(s_i-1,B))$-th (resp. $(\text{rank}_0(s_i,B)$-th) position of M_C as one if $B[s_i]$ is 0 (resp. 1). We also maintain a bit array M'_C analogous to $M'_{B'}$. Again, we can maintain both M_C and M'_C using $o(n)$ bits while supporting rank and select queries on them in $O(1)$ time.

Next, let $h(b,c)$ be a function, which returns a subarray of B from the subarrays of B' and C, defined as follows (suppose $c = c[1] \cdot c'$):

$$h(b,c) = \begin{cases} 1^t & \text{if } b = 1^t \text{ and } c[1] = 0 \\ 1^t 0 \cdot h(b',c') & \text{if } b = 1^t 0 \cdot b' \text{ and } c[1] = 0 \\ 10 \cdot h(b,c') & \text{if } c[1] = 1 \end{cases}$$

We store a precomputed table, which stores $h(b,c)$ for all possible b, c of size $\frac{1}{4}\log n$ using $O(2^{\frac{1}{2}\log n}\log n) = O(\sqrt{n}\log n) = o(n)$ bits.

To decode the i-th block of B, we first decode $\log n$-sized subarrays of B' and C, $b_{b'}$ and b_c, whose starting positions are $\text{select}_1(\text{rank}_0(i, M'_{B'}), M_{B'})$ and $\text{select}_1(\text{rank}_0(i, M'_C), M_C)$, respectively. We then compute $h(b_{b'}, b_c)$ in $O(1)$ time by referring to the precomputed table $O(1)$ times. Finally, we store a bit array of size $o(n)$, which indicates whether the first bit of the i-th block of B is 0 or not. As the final step, we delete the leftmost bit of $h(b_{b'}, b_c)$ if the i-th block of B starts from 0, and s_i is additional (this can be done by referring to the bit array). □

5   Lower Bounds

This section considers the effective entropy to answer range q-th minimum and maximum queries on an array of size n, for any $q \geq 1$. Note that for any $i \in \{1, \ldots, n\}$, both PSV(i) and PLV(i) queries can be answered by computing q-th range minimum and maximum queries on the suffixes of the substring $A[1, \ldots, i]$, respectively. Similarly, both NSV(i) and NLV(i) queries can be answered by computing q-th range minimum and maximum queries on the prefixes of the substring $A[i, \ldots, n]$, respectively.

Let $\mathcal{A}_n$ be a set of all arrays of size $n \geq 2$ constructed from the following procedure:

1. For any $0 \leq k \leq n-1$, pick arbitrary k positions in $\{2, \ldots, n\}$, and construct a *Baxter Permutation* [1] π_{n-k} of size $n-k$ on the rest of $n-k$ positions. Here, a Baxter permutation is a permutation that avoids the patterns 2–41–3 and 3–14–2.
2. For k picked positions, assign the rightmost element in π_{n-k} to the left.

Since the number of all possible Baxter permutations of size $n - k$ is at most $2^{3(n-k)-\Theta(\log n)}$ [8], the effective entropy of $\mathcal{A}_n$ is at least $\log|\mathcal{A}_n| \geq \log(\sum_{k=0}^{n-1} 2^{3(n-k)-\Theta(\log n)} \cdot \binom{n-1}{k})) \geq \max_k(3n - 3k + \log\binom{n}{k} - \Theta(\log n)) \geq n \log 9 - \Theta(\log n) \geq 3.16n - \Theta(\log n)$ bits [18]. The following theorem shows that the effective entropy of the encoding to support the range q-th minimum and maximum queries on an array of size n is at least $3.16n - \Theta(\log n)$ bits.

Theorem 4. *Any array A in $\mathcal{A}_n$ for $n \geq 2$ can be reconstructed using range q-th minimum and maximum queries on A.*

Proof. We follow the same argument used in the proof of Lemma 3 in [8], which shows that one can reconstruct any Baxter permutation of size n using range minimum and maximum queries.

The proof is induction on n. the case $n = 2$ is trivial since only the possible cases are $\{1, 1\}$ or $\{1, 2\}$, which can be decoded by range first and second minimum queries. Now suppose the theorem statement holds for any size less than $n \geq 3$. Then, both $A_1 = A[1, \ldots, n-1]$ and $A_2 = A[2, \ldots, n]$ from $\mathcal{A}_{n-1}$ can be reconstructed by the induction hypothesis. Thus, to reconstruct A from A_1 and A_2, it is enough to compare $A[1]$ and $A[n]$.

If any answer of $\mathsf{RMax}(1, n, q)$ and $\mathsf{RMin}(1, n, q)$ contains the position 1 or n, we are done. Otherwise, let x and y be the rightmost positions of the smallest and largest element in $[2, n-1]$, which can be computed by $\mathsf{RMax}(1, n, q)$ and $\mathsf{RMin}(1, n, q)$, respectively. Without a loss of generality, suppose $x < y$ (other case is symmetric). In this case, [8] showed that (i) there exists a position $i \in [x, y]$, which satisfies $A[1] < A[i] < A[n]$ or $A[1] > A[i] > A[n]$, or (ii) $A[1] < A[n]$, which proves the theorem (note that $A[1]$ cannot be equal to $A[n]$ in this case since the same elements in A always appear consecutively). $\qquad\square$

6   Conclusion

This paper proposes an encoding data structure that efficiently supports range (q-th) minimum/maximum queries and next/previous larger/smaller value queries. Our results match the current best upper bound of Tsur [18] up to lower-order additive terms while supporting the queries efficiently.

Note that the lower bound of Theorem 4 only considers the case that the same elements always appear consecutively, which still gives a gap between the upper and lower bound of the space. Improving the lower bound of the space for answering the queries would be an interesting open problem.

Acknowledgements. This work was supported by the National Research Foundation of Korea (NRF) grant funded by the Korea government (MSIT) (No. NRF-2020R1G1A1101477). We would like to thank to Srinivasa Rao Satti for helpful discussions.

References

1. Baxter, G.: On fixed points of the composite of commuting functions. Proc. Am. Math. Soc. **15**(6), 851–855 (1964)
2. Benoit, D., Demaine, E.D., Munro, J.I., Raman, R., Raman, V., Rao, S.S.: Representing trees of higher degree. Algorithmica **43**(4), 275–292 (2005)
3. Berkman, O., Schieber, B., Vishkin, U.: Optimal doubly logarithmic parallel algorithms based on finding all nearest smaller values. J. Algorithms **14**(3), 344–370 (1993)
4. Dodis, Y., Patrascu, M., Thorup, M.: Changing base without losing space. In: STOC 2010, pp. 593–602. ACM (2010)
5. Ferrada, H., Navarro, G.: Improved range minimum queries. In: 2016 Data Compression Conference, DCC 2016, pp. 516–525. IEEE (2016)
6. Fischer, J.: Combined data structure for previous- and next-smaller-values. Theor. Comput. Sci. **412**(22), 2451–2456 (2011)
7. Fischer, J., Heun, V.: Space-efficient preprocessing schemes for range minimum queries on static arrays. SIAM J. Comput. **40**(2), 465–492 (2011)
8. Gawrychowski, P., Nicholson, P.K.: Optimal encodings for range top-k, selection, and min-max. In: Halldórsson, M.M., Iwama, K., Kobayashi, N., Speckmann, B. (eds.) ICALP 2015. LNCS, vol. 9134, pp. 593–604. Springer, Heidelberg (2015). https://doi.org/10.1007/978-3-662-47672-7_48
9. Jo, S., Satti, S.R.: Simultaneous encodings for range and next/previous larger/smaller value queries. Theor. Comput. Sci. **654**, 80–91 (2016)
10. Merlini, D., Sprugnoli, R., Verri, M.C.: Waiting patterns for a printer. Discret. Appl. Math. **144**(3), 359–373 (2004)
11. Munro, J.I., Raman, V.: Succinct representation of balanced parentheses and static trees. SIAM J. Comput. **31**(3), 762–776 (2001)
12. Munro, J.I., Raman, V., Rao, S.S.: Space efficient suffix trees. J. Algorithms **39**(2), 205–222 (2001)
13. Navarro, G., Sadakane, K.: Fully functional static and dynamic succinct trees. ACM Trans. Algorithms **10**(3), 16:1–16:39 (2014)
14. Ohlebusch, E., Fischer, J., Gog, S.: CST++. In: Chavez, E., Lonardi, S. (eds.) SPIRE 2010. LNCS, vol. 6393, pp. 322–333. Springer, Heidelberg (2010). https://doi.org/10.1007/978-3-642-16321-0_34
15. Raman, R.: Encoding data structures. In: Rahman, M.S., Tomita, E. (eds.) WALCOM 2015. LNCS, vol. 8973, pp. 1–7. Springer, Cham (2015). https://doi.org/10.1007/978-3-319-15612-5_1
16. Raman, R., Raman, V., Satti, S.R.: Succinct indexable dictionaries with applications to encoding k-ary trees, prefix sums and multisets. ACM Trans. Algorithms **3**(4), 43 (2007)
17. Sadakane, K.: Succinct data structures for flexible text retrieval systems. J. Discrete Algorithms **5**(1), 12–22 (2007)
18. Tsur, D.: The effective entropy of next/previous larger/smaller value queries. Inf. Process. Lett. **145**, 39–43 (2019)
19. Vuillemin, J.: A unifying look at data structures. Commun. ACM **23**(4), 229–239 (1980)

Near-Optimal Search Time in δ-Optimal Space

Tomasz Kociumaka[1], Gonzalo Navarro[2], and Francisco Olivares[2($\boxtimes$)]

[1] Max Planck Institute for Informatics, Saarbrücken, Germany
[2] CeBiB — Center for Biotechnology and Bioengineering, Department of Computer Science, University of Chile, Santiago, Chile
`folivares@uchile.cl`

Abstract. Two recent lower bounds on the compressiblity of repetitive sequences, $\delta \leq \gamma$, have received much attention. It has been shown that a string $S[1..n]$ can be represented within the optimal $O(\delta \log \frac{n}{\delta})$ space, and further, that within that space one can find all the *occ* occurrences in S of any pattern of length m in time $O(m \log n + occ \log^\epsilon n)$ for any constant $\epsilon > 0$. Instead, the near-optimal search time $O(m + (occ + 1) \log^\epsilon n)$ was achieved only within $O(\gamma \log \frac{n}{\gamma})$ space. Both results are based on considerably different locally consistent parsing techniques. The question of whether the better search time could be obtained within the δ-optimal space was open. In this paper, we prove that both techniques can indeed be combined in order to obtain the best of both worlds, $O(m + (occ + 1) \log^\epsilon n)$ search time within $O(\delta \log \frac{n}{\delta})$ space.

1 Introduction

The amount of data we are expected to handle has been growing steadily in the last decades [20]. The fact that much of the fastest-growing data is composed of highly repetitive sequences has raised the interest in text indexes whose size can be bounded by some measure of repetitiveness [17], and in the study of those repetitiveness measures [16]. Since statistical compression does not capture repetitiveness well [13], various other measures have been proposed for this case. Two recent ones, which have received much attention because of their desirable properties, are the size γ of the smallest string attractor [9] and the substring complexity δ [3,10]. It holds that $\delta \leq \gamma$ for every string [3] (with $\delta = o(\gamma)$ in some string families [11]), and that γ asymptotically lower-bounds a number of other measures sensitive to repetitiveness [9] (e.g., the size of the smallest Lempel–Ziv parse [14]). On the other hand, any string $S[1..n]$ can be represented within $O(\delta \log \frac{n}{\delta})$ space, and this bound is tight for every n and δ [10,11,18].

A more ambitious goal than merely representing S in compressed space is to *index* it within that space so that, given any pattern $P[1..m]$, one can efficiently find all the *occ* occurrences of P in S. Interestingly, it has been shown that, for any constant $\epsilon > 0$, one can index S within the tight $O(\delta \log \frac{n}{\delta})$ space, so as to

Funded in part by Basal Funds FB0001, Fondecyt Grant 1-200038 and a Conicyt Doctoral Scholarship, ANID, Chile.

© Springer Nature Switzerland AG 2022
A. Castañeda and F. Rodríguez-Henríquez (Eds.): LATIN 2022, LNCS 13568, pp. 88–103, 2022.
https://doi.org/10.1007/978-3-031-20624-5_6

search for P in time $O(m \log n + occ \log^\epsilon n)$ time [10, 11]. If one allows the higher $O(\gamma \log \frac{n}{\gamma})$ space, the search time can be reduced to $O(m + (occ + 1) \log^\epsilon n)$ [3], which is optimal in terms of the pattern length and near-optimal in the time per reported occurrence. Within (significantly) more space, $O(\gamma \log \frac{n}{\gamma} \log n)$, one obtains truly optimal search time, $O(m + occ)$.

The challenge of obtaining the near-optimal search time $O(m + (occ + 1) \log^\epsilon n)$ within tight space $O(\delta \log \frac{n}{\delta})$ was posed [3, 10, 11], and this is what we settle on the affirmative in this paper. Both previous results build a convenient context-free grammar on S and then adapt a classical grammar-based index on it [4, 5]. The index based on attractors [3] constructs a grammar from a locally consistent parsing [15] of S that forms blocks in S ending at every minimum of a randomized mapping on the alphabet, collapsing every block into a nonterminal and iterating. The smaller grammar based on substring complexity [11] uses another locally consistent parsing called recompression [7], which randomly divides the alphabet into "left" and "right" symbols and combines every left-right pair into a nonterminal, also iterating. The key to obtaining δ-bounded space is to pause the pairing on symbols that become too long for the iteration where they were formed [10, 11]. We show that the pausing idea can be applied to the first kind of locally consistent grammar as well and that, although it leads to possibly larger grammars, it still yields the desired time and space complexities. The next theorem summarizes our result.

Theorem 1.1. *For every constant $\epsilon > 0$, given a string $S[1..n]$ with measure δ, one can build in $O(n)$ expected time a data structure using $O(\delta \log \frac{n}{\delta})$ words of space such that, later, given a pattern $P[1..m]$, one can find all its occ occurrences in S in time $O(m + \log^\epsilon \delta + occ \log^\epsilon(\delta \log \frac{n}{\delta})) \subseteq O(m + (occ + 1) \log^\epsilon n)$.*

2 Notation and Basic Concepts

A string is a sequence $S[1..n] = S[1] \cdot S[2] \cdots S[n]$ of symbols, where each symbol belongs to an alphabet $\Sigma = \{1, \ldots, \sigma\}$. We denote as $\Sigma(S)$ the subset of Σ consisting of symbols that occur in S. The length of S is denoted $|S| = n$. We assume that the alphabet size is a polynomial function of n, that is, $\sigma = n^{O(1)}$. The concatenation of strings S and S' is denoted $S \cdot S' = SS'$. A string S' is a substring of S if S' is the empty string ε or $S' = S[i..j] = S[i] \cdots S[j]$ for some $1 \le i \le j \le n$. We also use "(" and ")" to denote non-inclusive intervals: $S(i..j) = S[i + 1..j - 1]$, $S(i..j] = S[i + 1..j]$, and $S[i..j) = S[i..j - 1]$. With the term *fragment*, we refer to a particular occurrence $S[i..j]$ of a substring in S (not just the substring content). We use S^{rev} to denote the reverse of S, that is, $S^{rev} = S[n] \cdot S[n - 1] \cdots S[1]$.

We use the RAM model of computation with word size $w = \Theta(\log n)$ bits. By default, we measure the space in words, which means that $O(x)$ space comprises of $O(x \log n)$ bits.

A *straight line program* (SLP) is a context-free grammar where each nonterminal appears once at the left-hand side of a rule, and where the nonterminals can be sorted so that the right-hand sides refer to terminals and preceding nonterminals. Such an SLP generates a single string. Furthermore, we refer to a *run-length straight line program* (RLSLP) as an SLP that, in addition, allows rules of the form $A \to A_1^m$, where A, A_1 are nonterminals and $m \in \mathbb{Z}_{\geq 2}$, which means that A is a rule composed by concatenating m copies of A_1.

A *parsing* is a way to decompose a string S into non-overlapping *blocks*, $S = S_1 \cdot S_2 \cdots S_k$. A *locally consistent parsing (LCP)* [1] is a parsing where, if two fragments $S[i..j] = S[i'..j']$ appear inside equal long enough contexts $S[i - \alpha..j + \beta] = S[i' - \alpha..j' + \beta]$, then the same blocks are formed inside $S[i..j]$ and $S[i'..j']$. The meaning of "long enough" depends on the type of LCP [1,3,6].

3   A New δ-Bounded RLSLP

The measure δ was originally introduced in a stringology context [18], but it was formally defined later [3] as a way to construct a grammar of size $O(\gamma \log \frac{n}{\gamma})$ without knowing γ. For a given string $S[1..n]$, let $d_k(S)$ be the number of distinct length-k substrings in S. The sequence of all values $d_k(S)$ is known as the *substring complexity* of S. Then, δ is defined as

$$\delta = \max \left\{ \frac{d_k(S)}{k} : k \in [1..n] \right\}.$$

An RLSLP of size $O(\delta \log \frac{n}{\delta})$ was built [11] on top of the recompression method [7]. In this section, we show that the same can be achieved on top of the block-based LCP [15]. Unlike the previous construction, ours produces an RLSLP with $O(\delta \log \frac{n}{\delta})$ rules in $O(n)$ deterministic time, though we still need randomization in order to ensure that the total grammar size is also $O(\delta \log \frac{n}{\delta})$.

We adapt the preceding construction [11], which uses the so-called *restricted recompression* [12]. This technique pauses the processing for symbols whose expansion is too long for the current stage. A similar idea was used [2,8] for adapting another LCP, called *signature parsing* [19]. We apply restriction (the pausing technique) to the LCP of [15] that forms blocks ending at local minima of a randomized bijective function, which is interpreted as an alphabet permutation. This LCP will be used later to obtain near-optimal search time, extending previous work [3]. We call our parsing *restricted block compression*.

3.1   Restricted Block Compression

Given a string $S \in \Sigma^+$, our restricted block compression builds a sequence of strings $(S_k)_{k \geq 0}$ over the alphabet $\mathcal{A}$ defined recursively to contain symbols in Σ, pairs formed by a symbol in $\mathcal{A}$ and an integer $m \geq 2$, and sequences of at least two symbols in $\mathcal{A}$; formally, $\mathcal{A}$ is the least fixed point of the expression

$$\mathcal{A} = \Sigma \cup (\mathcal{A} \times \mathbb{Z}_{\geq 2}) \cup \bigcup_{i=2}^{\infty} \mathcal{A}^i.$$

In the following, we denote $\bigcup_{i=2}^{\infty} \mathcal{A}^i$ with $\mathcal{A}^{\geq 2}$.

Symbols in $\mathcal{A} \setminus \Sigma$ are *non-terminals*, which are naturally associated with productions $(A_1, \ldots, A_j) \rightarrow A_1 \cdots A_j$ for $(A_1, \ldots, A_j) \in \mathcal{A}^{\geq 2}$ and $(A_1, m) \rightarrow A_1^m$ for $(A_1, m) \in \mathcal{A} \times \mathbb{Z}_{\geq 2}$. Setting any $A \in \mathcal{A}$ as the starting symbol yields an RLSLP. The string generated by this RLSLP is $\exp(A)$, where $\exp : \mathcal{A} \rightarrow \Sigma^+$ is the *expansion* function defined recursively:

$$\exp(A) = \begin{cases} A & \text{if } A \in \Sigma, \\ \exp(A_1) \cdots \exp(A_j) & \text{if } A = (A_1, \ldots, A_j) \text{ for } A_1, \ldots, A_j \in \mathcal{A}, \\ \exp(A_1)^m & \text{if } A = (A_1, m) \text{ for } A_1 \in \mathcal{A} \text{ and } m \in \mathbb{Z}_{\geq 2}. \end{cases}$$

Then, for every string $(S_k)_{k \geq 0}$ generated using restricted block compression, if the expansion function is extended homomorphically to $\exp : \mathcal{A}^* \rightarrow \Sigma^*$, with $\exp(A_1 \cdots A_m) = \exp(A_1) \cdots \exp(A_m)$ for $A_1 \cdots A_m \in \mathcal{A}^*$, then it must hold that $\exp(S_k) = S$ for every $k \in \mathbb{Z}_{\geq 0}$. Starting from $S_0 = S$, the strings $(S_k)_{k \geq 1}$ are built by the alternate applications of two functions, both of which decompose a string $T \in \mathcal{A}^+$ into *blocks* (by placing *block boundaries* between some characters) and then collapse blocks of length $m \geq 2$ into individual symbols in $\mathcal{A}$. In Definition 3.1, the blocks are maximal *runs* of the same symbol in a subset $\mathcal{B} \subseteq \mathcal{A}$, and they are collapsed to symbols in $\mathcal{A} \times \mathbb{Z}_{\geq 2}$.

Definition 3.1 (Run-length encoding). *Given $T \in \mathcal{A}^+$ and a subset of symbols $\mathcal{B} \subseteq \mathcal{A}$, we define $rle_{\mathcal{B}}(T) \in \mathcal{A}^+$ as the string obtained by decomposing T into blocks and collapsing these blocks as follows:*

1) For every $i \in [1..|T|)$, place a block boundary between $T[i]$ and $T[i+1]$ if $T[i] \notin \mathcal{B}$, $T[i+1] \notin \mathcal{B}$, or $T[i] \neq T[i+1]$.
2) For each block $T[i..i+m)$ of $m \geq 2$ equal symbols A, replace $T[i..i+m) = A^m$ with the symbol $(A, m) \in \mathcal{A}$.

In Definition 3.3, the blocks boundaries are determined by local minima of a permutation on $\mathcal{A}$, and the blocks are collapsed to symbols in $\mathcal{A}^{\geq 2}$.

Definition 3.2 (Local minima). *Given $T \in \mathcal{A}^+$ and a bijective function $\pi : \Sigma(T) \rightarrow [1..|\Sigma(T)|]$, we say that $j \in (1..|T|)$ is a local minimum if*

$$\pi(T[j-1]) > \pi(T[j]) \text{ and } \pi(T[j]) < \pi(T[j+1]).$$

Definition 3.3 (Restricted block parsing). *Given $T \in \mathcal{A}^+$, a bijective function $\pi : \Sigma(T) \rightarrow [1..|\Sigma(T)|]$, and a subset of symbols $\mathcal{B} \subseteq \mathcal{A}$, we define $bc_{\pi,\mathcal{B}}(T) \in \mathcal{A}^+$ as the string obtained by decomposing T into blocks and collapsing these blocks as follows:*

1) For every $i \in [1..|T|)$, place a block boundary between $T[i]$ and $T[i+1]$ if $T[i] \notin \mathcal{B}$, $T[i+1] \notin \mathcal{B}$, or i is a local mimimum with respect to π.
2) For each block $T[i..i+m)$ of length $m \geq 2$, replace $T[i..i+m)$ with a symbol $(T[i], \ldots, T[i+m-1]) \in \mathcal{A}$.

Note that $\mathcal{B}$ consists of *active* symbols that can be combined into larger blocks; we say that the other symbols are *paused*. The idea of our restricted block compression is to create successive strings S_k, starting from $S_0 = S$. At the odd levels k we perform run-length encoding on the preceding string S_{k-1}. On the even levels k, we perform block parsing on the preceding string S_{k-1}. We pause the symbols whose expansions have become too long for that level.

Definition 3.4 (Restricted block compression). *Given a string $S \in \Sigma^+$, the strings S_k for $k \in \mathbb{Z}_{\geq 0}$ are constructed as follows, where $\ell_k := \left(\frac{4}{3}\right)^{\lceil k/2 \rceil - 1}$, $\mathcal{A}_k := \{A \in \mathcal{A} : |\exp(A)| \leq \ell_k\}$, and $\pi_k : \Sigma(S_{k-1}) \to [1..|\Sigma(S_{k-1})|]$ is a bijection satisfying $\pi_k(A) < \pi_k(B)$ for every $A \in \Sigma(S_{k-1}) \setminus \mathcal{A}_k$ and $B \in \Sigma(S_{k-1}) \cap \mathcal{A}_k$:*

- *If $k = 0$, then $S_k = S$.*
- *If $k > 0$ is odd, then $S_k = rle_{\mathcal{A}_k}(S_{k-1})$.*
- *If $k > 0$ is even, then $S_k = bc_{\pi_k, \mathcal{A}_k}(S_{k-1})$.*

Note that $\exp(S_k) = S$ holds for all $k \in \mathbb{Z}_{\geq 0}$.

3.2   Grammar Size Analysis

Our RLSLP will be built by performing restricted block compression as long as $|S_k| > 1$. Although the resulting RLSLP has infinitely many symbols, we can remove those having no occurrences in any S_k. To define the actual symbols in the grammar, for all $k \in \mathbb{Z}_{\geq 0}$, denote $\mathcal{S}_k := \{S_k[j] : j \in [1..|S_k|]\}$ and $\mathcal{S} := \bigcup_{k=0}^{\infty} \mathcal{S}_k$.

We first prove an upper bound on $|S_k|$ which, in particular, implies that $|S_k| = 1$ holds after $O(\log n)$ iterations.

Lemma 3.5. *For every $k \in \mathbb{Z}_{\geq 0}$, we have $|S_k| < 1 + \frac{4n}{\ell_{k+1}}$.*

Proof. We proceed by induction on k. For $k = 0$, we have $|S_0| = n < 1 + 4n = 1 + \frac{4n}{\ell_1}$. If k is odd, we note that $|S_k| \leq |S_{k-1}| < 1 + \frac{4n}{\ell_k} = 1 + \frac{4n}{\ell_{k+1}}$. If k is even, let us define

$$J = \{j \in [1..|S_{k-1}|] : S_{k-1}[j] \notin \mathcal{A}_k\}.$$

Since $A \notin \mathcal{A}_k$ implies $|\exp(A)| > \ell_k$, we have $|J| < \frac{n}{\ell_k}$. Then, since no two consecutive symbols can be local minima, we have

$$|S_k| \leq 2|J| + 1 + \frac{|S_{k-1}| - (2|J| + 1)}{2} = \frac{1 + |S_{k-1}|}{2} + |J| < 1 + \frac{2n}{\ell_k} + \frac{n}{\ell_k} = 1 + \frac{3n}{\ell_k}$$
$$= 1 + \frac{4n}{\ell_{k+1}}. \qquad \square$$

Our next goal is to prove that restricted block compression is a locally consistent parsing. For this, we associate S_k with a decomposition of S into *phrases*.

Definition 3.6 (Phrase boundaries). *For every $k \in \mathbb{Z}_{\geq 0}$ and $j \in [1..|S_k|]$, we define the level-k phrases of S induced by S_k as the fragments*

$$S(|\exp(S_k[1..j))|..|\exp(S_k[1..j])|] = \exp(S_k[j]).$$

We also define the set B_k of phrase boundaries induced by S_k:

$$B_k = \{|\exp(S_k[1..j])| : j \in [1..|S_k|]\}.$$

Lemma 3.7. *Consider integers $k, m, \alpha \geq 0$ with $\alpha \geq 8\ell_k$, as well as positions $i, i' \in [m + 2\alpha..n - \alpha]$ such that $S(i - m - 2\alpha..i + \alpha] = S(i' - m - 2\alpha..i' + \alpha]$.*

1) If $i \in B_k$, then $i' \in B_k$.
2) If $S(i - m..i]$ is a level-k phrase, then $S(i' - m..i']$ is a level-k phrase corresponding to the same symbol in S_k.

Proof. We proceed by induction on k, with a weaker assumption $\alpha \geq 7\ell_k$ for odd k. In the base case of $k = 0$, the claim is trivial because $B_k = [1..n)$ and $S_k = S$. Next, we prove that the claim holds for integers $k > 0$ and $\alpha > \ell_k$ assuming that it holds for all $k - 1$ and $\alpha - \lfloor \ell_k \rfloor$. This is sufficient for the inductive step: If $\alpha \geq 8\ell_k$ for even $k > 0$, then $\alpha - \lfloor \ell_k \rfloor \geq 7\ell_k = 7\ell_{k-1}$. Similarly, if $\alpha \geq 7\ell_k$ for odd k, then $\alpha - \lfloor \ell_k \rfloor \geq 6\ell_k = 8\ell_{k-1}$.

We start with the first item, where we can assume $m = 0$ without loss of generality. For a proof by contradiction, suppose that $S(i - 2\alpha..i + \alpha] = S(i' - 2\alpha..i' + \alpha]$ and $i \in B_k$ yet $i' \notin B_k$ for some $i, i' \in [2\alpha..n - \alpha]$. By the inductive assumption (applied to positions i, i'), $i \in B_k \subseteq B_{k-1}$ implies $i' \in B_{k-1}$. Let us set $j, j' \in [1..|S_{k-1}|)$ so that $i = |\exp(S_{k-1}[1..j])|$ and $i' = |\exp(S_{k-1}[1..j'])|$. By the assumptions on i, i', the parsing of S_{k-1} places a block boundary between $S_{k-1}[j]$ and $S_{k-1}[j+1]$, but it does not place a block boundary between $S_{k-1}[j']$ and $S_{k-1}[j' + 1]$. By Definitions 3.1 and 3.3, the latter implies $S_{k-1}[j'], S_{k-1}[j'+1] \in \mathcal{A}_k$. Consequently, the phrases $S(i' - \ell..i'] = \exp(S_{k-1}[j'])$ and $S(i'..i' + r] = \exp(S_{k-1}[j' + 1])$ around position i' are of length at most $\lfloor \ell_k \rfloor$. Since $i' - \lfloor \ell_k \rfloor \leq i' - \ell \leq i' + r \leq i' + \lfloor \ell_k \rfloor$, the inductive assumption applied to positions i', i and $i' + r, i + r$ implies that $S(i - \ell..i]$ and $S(i..i + r]$ are parsed into $S_{k-1}[j] = S_{k-1}[j']$ and $S_{k-1}[j + 1] = S_{k-1}[j' + 1]$, respectively.

If k is odd, then a boundary between two symbols in $\mathcal{A}_k$ is placed if and only if the two symbols differ. Consequently, $S_{k-1}[j'] = S_{k-1}[j' + 1]$ and $S_{k-1}[j] \neq S_{k-1}[j + 1]$. This contradicts $S_{k-1}[j] = S_{k-1}[j']$ and $S_{k-1}[j + 1] = S_{k-1}[j' + 1]$.

Thus, it remains to consider the case of even k. Since the block parsing places a boundary between $S_{k-1}[j], S_{k-1}[j + 1] \in \mathcal{A}_k$, we conclude from Definition 3.3 that j must be a local minimum with respect to π_k, i.e., $\pi_k(S_{k-1}[j - 1]) > \pi_k(S_{k-1}[j]) < \pi_k(S_{k-1}[j + 1])$. Due to $S_{k-1}[j] \in \mathcal{A}_k$, the condition on π_k imposed in Definition 3.4 implies $S_{k-1}[j - 1] \in \mathcal{A}_k$. Consequently, the phrase $S(i - \ell'..i - \ell] = \exp(S_{k-1}[j - 1])$ is of length at most $\lfloor \ell_k \rfloor$. Since $i' - 2\lfloor \ell_k \rfloor \leq i' - \ell' \leq i' - \ell \leq i'$, the inductive assumption, applied to positions $i - \ell, i' - \ell$ implies that $S(i' - \ell'..i' - \ell]$ is parsed into $S_{k-1}[j' - 1] = S_{k-1}[j - 1]$. Thus, $\pi_k(S_{k-1}[j' - 1]) = \pi_k(S_{k-1}[j - 1]) > \pi_k(S_{k-1}[j']) = \pi_k(S_{k-1}[j]) < \pi_k(S_{k-1}[j' + 1]) = \pi_k(S_{k-1}[j + 1])$, which means that j' is a local minimum with respect to π_k and, by Definition 3.3, contradicts $i' \notin B_k$.

Let us proceed to the proof of the second item. Let $S_{k-1}(j - m'..j]$ be the block corresponding to the level-k phrase $S(i - m..i]$. By the inductive assumption, $S(i' - m..i']$ consists of level-$(k - 1)$ phrases that, in S_{k-1}, are collapsed into a fragment $S_{k-1}(j' - m'..j']$ matching $S_{k-1}(j - m'..j]$. Moreover, by the first item, the parsing of S_{k-1} places block boundaries before $S_{k-1}[j' - m']$ and after $S_{k-1}[j']$, but nowhere in between. Consequently, $S_{k-1}(j - m'..j]$ and

$S_{k-1}(j' - m'..j']$ are matching blocks, which means that they are collapsed into matching symbols of S_k, Thus, the level-k phrases $S(i - m..i]$ and $S(i' - m..i']$ are represented by matching symbols in S_k. □

Our next goal is to prove that $|\mathcal{S}| = O(\delta \log \frac{n}{\delta})$ (Corollary 3.12). As a first step, we show that $|\mathcal{A}_{k+1} \cap \mathcal{S}_k| = O(\delta)$ (Lemma 3.9). The idea for this proof is to consider the leftmost occurrence of all symbols of S_k and then bound the set of those occurrences in relation to δ (Claims 3.10 and 3.11). At a high level, we build on the arguments of [11], where the same bound was proved in expectation, but we obtain worst-case results with our parsing. We start by generalizing Lemma 3.5.

Lemma 3.8. *For every $k \in \mathbb{Z}_{\geq 0}$ and every interval $I \subseteq [1..n]$, we have*

$$|B_k \cap I| < 1 + \frac{4|I|}{\ell_{k+1}}.$$

Proof. We proceed by induction on k. For $k = 0$, we have $|B_k \cap I| = |I| < 1 + 4|I| = 1 + \frac{4|I|}{\ell_1}$. If k is odd, we note that $B_k \subseteq B_{k-1}$ and therefore $|B_k \cap I| \leq |B_{k-1} \cap I| < 1 + \frac{4|I|}{\ell_k} = 1 + \frac{4|I|}{\ell_{k+1}}$. If k is even, let us define

$$J = \{j \in [1..|S_{k-1}|] : S_{k-1}[j] \notin \mathcal{A}_k\},$$
$$J_I = \{j \in J : |\exp(S_{k-1}[1..j))| \in I\} \subseteq B_{k-1} \cap I.$$

Since $A \notin \mathcal{A}_k$ implies $|\exp(A)| > \ell_k$, we have $|J_I| < \frac{|I|}{\ell_k}$. Then, since no two consecutive symbols can be local minima, we have

$$|B_k \cap I| \leq 2|J_I| + 1 + \frac{|B_{k-1} \cap I| - (2|J_I| + 1)}{2} = \frac{1 + |B_{k-1} \cap I|}{2} + |J_I|$$
$$< 1 + \frac{2|I|}{\ell_k} + \frac{|I|}{\ell_k} = 1 + \frac{3|I|}{\ell_k} = 1 + \frac{4|I|}{\ell_{k+1}}. \quad □$$

The following result is used to bound both the number of symbols $|\mathcal{S}|$ (where we only care about $|\mathcal{S}_k \cap \mathcal{A}_{k+1}|$, i.e., the number of substrings with $m = 1$ active symbol) and the size of the RLSLP resulting from restricted block compression.

Lemma 3.9. *If the string S has measure δ, then, for all integers $k \geq 0$ and $m \geq 1$, the string S_k contains $O(m\delta)$ distinct length-m substrings in $\mathcal{A}_{k+1}^*$.*

Proof. Denote $\alpha := \lceil 8\ell_k \rceil$ and $\ell := 3\alpha + \lfloor m\ell_{k+1} \rfloor$, and let L be the set of positions in S covered by the leftmost occurrences of substrings of S of length at most ℓ, as well as the trailing ℓ positions in S. We first prove two auxiliary claims.

Claim 3.10. *The string S_k contains at most $|L \cap B_k|$ distinct length-m substrings in $\mathcal{A}_{k+1}^*$.*

Proof. Let us fix a length-m substring $T \in \mathcal{A}_{k+1}^*$ of S_k and let $S_k(j - m..j]$ be the leftmost occurrence of T in S_k. Moreover, let $p = |\exp(S_k[1..j - m])|$ and $q = |\exp(S_k[1..j])|$ so that $S(p..q]$ is the expansion of $S_k(j - m..j]$. By $S_k(j - m..j] \in \mathcal{A}_{k+1}^*$, we have $q - p \leq m\lfloor \ell_{k+1} \rfloor \leq \ell - 3\alpha$.

We shall prove that $q \in L$; for a proof by contradiction, suppose that $q \notin L$. Due to $(0..\ell] \cup (n - \ell..n] \subseteq L$, this implies that $q \in (\ell..n - \ell]$ is not covered by the leftmost occurrence of any substrings of length at most ℓ. In particular, $S(p - 2\alpha..q + \alpha]$ must have an earlier occurrence $S(p' - 2\alpha..q' + \alpha]$ for some $p' < p$ and $q' < q$. Consequently, Lemma 3.7, applied to subsequent level-k phrases comprising $S(p..q]$, shows that $S(p'..q']$ consists of full level-k phrases and the corresponding fragment of S_k matches $S_k(j - m..j] = T$. By $q' < q$, this contradicts the assumption that $S_k(j - m..j]$ is the leftmost occurrence of T in S_k, which completes the proof that $q \in L$.

A level-k phrase ends at position q, so we also have $q \in B_k$. Since the position q is uniquely determined by the substring T, this yields an upper bound of $|L \cap B_k|$ on the number of choices for T. □

Claim 3.11. *The set L forms $O(\delta)$ intervals of total length $O(\delta\ell)$.*

Proof. Each position in $L \cap (0..n - \ell]$ is covered by the leftmost occurrence of a substring of length exactly ℓ, and thus L forms at most $\lfloor \frac{1}{\ell}|L|\rfloor$ intervals of length at least ℓ each. Hence, it suffices to prove that the total length satisfies $|L| = O(\delta\ell)$. For this, note that, for each position $j \in L \cap [\ell..n - \ell]$, the fragment $S(j - \ell..j + \ell]$ is the leftmost occurrence of a length-2ℓ substring of S; this because any length-ℓ fragment covering position j is contained within $S(j - \ell..j + \ell]$. Consequently, $|L| \leq d_{2\ell}(S) + 2\ell = O(\delta\ell)$ holds as claimed. □

By Claim 3.10, it remains to prove that $|L \cap B_k| = O(\delta m)$. Let $\mathcal{I}$ be the family of intervals covering L. For each $I \in \mathcal{I}$, Lemma 3.8 implies $|B_k \cap I| \leq 1 + \frac{4|I|}{\ell_{k+1}}$. By the bounds on $\mathcal{I}$ following from Claim 3.11, this yields the announced result:

$$|B_k \cap L| \leq |\mathcal{I}| + \frac{4}{\ell_{k+1}} \sum_{I \in \mathcal{I}} |I| = O(\delta + \tfrac{\delta\ell}{\ell_{k+1}}) = O(\delta m). \qquad \square$$

The proof of our main bound $|\mathcal{S}| = O(\delta \log \frac{n}{\delta})$ combines Lemmas 3.5 and 3.9.

Corollary 3.12. *For every string S of length n and measure δ, we have $|\mathcal{S}| = O(\delta \log \frac{n}{\delta})$.*

Proof. Note that $|\mathcal{S}| \leq 1 + \sum_{k=0}^{\infty} |\mathcal{S}_k \setminus \mathcal{S}_{k+1}|$. We combine two upper bounds on $|\mathcal{S}_k \setminus \mathcal{S}_{k+1}|$, following from Lemmas 3.5 and 3.9, respectively.

First, we observe that Definition 3.4 guarantees $\mathcal{S}_k \setminus \mathcal{S}_{k+1} \subseteq \mathcal{S}_k \cap \mathcal{A}_{k+1}$. Moreover, each symbol in $\mathcal{S}_k \cap \mathcal{A}_{k+1}$ corresponds to a distinct length-1 substring of S_{k+1}, and thus $|\mathcal{S}_k \setminus \mathcal{S}_{k+1}| \leq |\mathcal{S}_k \cap \mathcal{A}_{k+1}| = O(\delta)$ holds due to Lemma 3.9. Secondly, we note that $|\mathcal{S}_k \setminus \mathcal{S}_{k+1}| = 0$ if $|S_k| = 1$ and $|\mathcal{S}_k \setminus \mathcal{S}_{k+1}| \leq |S_k| \leq 2(|S_k| - 1)$ if $|S_k| \geq 2$. Hence, Lemma 3.5 yields

$$|\mathcal{S}_k \setminus \mathcal{S}_{k+1}| \leq 2(|S_k| - 1) \leq \tfrac{8n}{\ell_{k+1}} = O((\tfrac{3}{4})^{k/2} n).$$

We apply the first or the second upper bound on $|\mathcal{S}_k \setminus \mathcal{S}_{k+1}|$ depending on whether $k < \lambda := 2\lfloor \log_{4/3} \frac{n}{\delta} \rfloor$. This yields

$$\sum_{k=0}^{\infty} |\mathcal{S}_k \setminus \mathcal{S}_{k+1}| = \sum_{k=0}^{\lambda-1} O(\delta) + \sum_{k=\lambda}^{\infty} O((\tfrac{3}{4})^{k/2} n)$$

$$= 2\lfloor \log_{4/3} \tfrac{n}{\delta} \rfloor \cdot O(\delta) + \sum_{i=0}^{\infty} O((\tfrac{3}{4})^{i/2}\delta) = O(\delta \log \tfrac{n}{\delta}).$$

Overall, we conclude that $|\mathcal{S}| = 1 + O(\delta \log \frac{n}{\delta}) = O(\delta \log \frac{n}{\delta})$ holds as claimed. $\square$

Next, we show that the total expected grammar size is $O(\delta \log \frac{n}{\delta})$.

Theorem 3.13. *Consider the restricted block compression of a string $S[1..n]$ with measure δ, where the functions $(\pi_k)_{k \geq 0}$ in Definition 3.4 are chosen uniformly at random. Then, the expected size of the resulting RLSLP is $O(\delta \log \frac{n}{\delta})$.*

Proof. Although Corollary 3.12 guarantees that $|\mathcal{S}| = O(\delta \log \frac{n}{\delta})$, the remaining problem is that the size of the resulting grammar (i.e., sum of production sizes) can be larger. Every symbol in $\Sigma \cup (\mathcal{A} \times \mathbb{Z}_{\geq 2})$ contributes $O(1)$ to the RLSLP size, so it remains to bound the total size of productions corresponding to symbols in $\mathcal{A}^{\geq 2}$. These symbols are introduced by restricted block parsing, i.e., they belong to $\mathcal{S}_{k+1} \setminus \mathcal{S}_k$ for odd $k > 0$. In order to estimate their contribution to grammar size, we shall fix $\pi_0, \dots, \pi_k$ and compute the expectation with respect to the random choice of π_{k+1}. In this setting, we prove the following claim:

Claim 3.14. *Let $k > 0$ be odd and $T \in \mathcal{A}_k^m$ be a substring of S_k. Restricted block parsing $bc_{\pi_{k+1},\mathcal{A}_{k+1}}(S_k)$ creates a block matching T with probability $O(2^{-m})$.*

Proof. Since $S_k = rle_{\mathcal{A}_k}(S_{k-1})$ and $\mathcal{A}_{k+1} = \mathcal{A}_k$, every two subsequent symbols of T are distinct. Observe that if T forms a block, then there is a value $t \in [1..m]$ such that $\pi_{k+1}(T[1]) < \cdots < \pi_{k+1}(T[t]) > \cdots > \pi_{k+1}(T[m])$; otherwise, there would be a local minimum within every occurrence of T in S_{k-1}. In particular, denoting $h := \lfloor m/2 \rfloor$, we must have $\pi_{k+1}(T[1]) < \cdots < \pi_{k+1}(T[h+1])$ (when $t > h$) or $\pi_{k+1}(T[m-h]) > \cdots > \pi_{k+1}(T[m])$ (when $t \leq h$). However, the probability that the values $\pi_{k+1}(\cdot)$ for $h+1$ consecutive characters form a strictly increasing (or strictly decreasing) sequence is at most $\frac{1}{(h+1)!}$: either exactly $\frac{1}{(h+1)!}$ (if the characters are distinct) or 0 (otherwise); this is because π_{k+1} shuffles $\Sigma(S_k) \cap \mathcal{A}_{k+1}$ uniformly at random. Overall, we conclude that the probability that T forms a block does not exceed $\frac{2}{(h+1)!} \leq 2^{-\Omega(m \log m)} \leq O(2^{-m})$. $\square$

Next, note that every symbol in $\mathcal{S}_{k+1} \setminus \mathcal{S}_k$ is obtained by collapsing a block of m active symbols created within $bc_{\pi_{k+1},\mathcal{A}_{k+1}}(S_k)$ (with distinct symbols obtained from distinct blocks). By Lemma 3.9, the string S_k has $O(\delta m)$ distinct substrings $T \in \mathcal{A}_{k+1}^m$. By Claim 3.14, any fixed substring $T \in \mathcal{A}_{k+1}^m$ yields a symbol in $\mathcal{S}_{k+1} \setminus \mathcal{S}_k$ with probability $O(2^{-m})$. Consequently, the total contribution of symbols in $\mathcal{S}_{k+1} \setminus \mathcal{S}_k$ to the RLSLP size is, in expectation, $\sum_{m=2}^{\infty} O(m \cdot \delta m \cdot 2^{-m}) = O(\delta)$.

At the same time, $\mathcal{S}_{k+1} \setminus \mathcal{S}_k = \emptyset$ if $|\mathcal{S}_k| = 1$ and, if $|\mathcal{S}_k| \geq 2$, the contribution of symbols in $\mathcal{S}_{k+1} \setminus \mathcal{S}_k$ to the RLSLP size is most $|\mathcal{S}_k| \leq 2(|\mathcal{S}_k| - 1) \leq \frac{8n}{\ell_{k+1}} = O((\frac{3}{4})^{k/2}n)$, where the bound on $|\mathcal{S}_k|$ follows from Lemma 3.5. This sums up to $O(\delta)$ across all odd levels $k > \lambda := 2\lfloor \log_{4/3} \frac{n}{\delta} \rfloor$. Overall, we conclude that the total expected RLSLP size is $O(\delta \log \frac{n}{\delta} + (\lambda + 1)\delta) = O(\delta \log \frac{n}{\delta})$. □

We are now ready to show how to build an RLSLP of size $O(\delta \log \frac{n}{\delta})$ in linear expected time.

Corollary 3.15. *Given $S[1..n]$ with measure δ, we can build an RLSLP of size $O(\delta \log \frac{n}{\delta})$ in $O(n)$ expected time.*

Proof. We apply Definition 3.4 on top of the given string S, with functions π_k choices uniformly at random. It is an easy exercise to carry out this construction in $O(\sum_{k \geq 0} |\mathcal{S}_k|) = O(n)$ worst-case time.

The expected size of the resulting RLSLP is $c \cdot \delta \log \frac{n}{\delta}$ for some constant c; we can repeat the construction (with fresh randomness) until it yields an RLSLP of size at most $2c \cdot \delta \log \frac{n}{\delta}$. By Markov's inequality, we succeed after $O(1)$ attempts in expectation. As a result, in $O(n)$ expected time, we obtain a grammar of total worst-case size $O(\delta \log \frac{n}{\delta})$. □

Remark 3.16 (Grammar height). In the algorithm of Corollary 3.15, we can terminate restricted block compression after $\lambda := 2\lfloor \log_{4/3} \frac{n}{\delta} \rfloor$ levels and complete the grammar with an initial symbol rule $A_\lambda \rightarrow S_\lambda[1] \cdots S_\lambda[|S_\lambda|]$ so that $\exp(A_\lambda) = S$. Lemma 3.5 yields $|S_\lambda| = O(1 + (\frac{3}{4})^{\lambda/2}n) = O(\delta)$, so the resulting RLSLP is still of size $O(\delta \log \frac{n}{\delta})$; however, the height is now $O(\log \frac{n}{\delta})$. □

4 Local Consistency Properties

We now show that the local consistency properties of our grammar enable fast indexed searches. Previous work [3] achieves this by showing that, thanks to the locally consistent parsing, only a set $M(P)$ of $O(\log |P|)$ pattern positions need be analyzed for searching. To use this result, we now must take into account the pausing of symbols. Surprisingly, this modification allows for a much simpler definition of $M(P)$.

Definition 4.1. *For every non-empty fragment $S[i..j]$ of S, we define*

$$B_k(i,j) = \{p - i : p \in B_k \cap [i..j]\}$$

and

$$M(i,j) = \bigcup_{k \geq 0} (B_k(i,j) \setminus [2\alpha_{k+1}..j - i - \alpha_{k+1}] \cup \{\min(B_k(i,j) \cap [2\alpha_{k+1}..j - i - \alpha_{k+1}))\}),$$

where $\alpha_k = \lceil 8\ell_k \rceil$ and $\{\min \emptyset\} = \emptyset$.

Intuitively, the set $B_k(i,j)$ lists (the relative locations of) all level-k phrase boundaries inside $S[i..j]$. For each level $k \geq 0$, we include in $M(i,j)$ the phrase boundaries that are close to either of the two endpoints of $S[i..j]$ (in the light of Lemma 3.7, it may depend on the context of $S[i..j]$ which of these phrase boundaries are preserved in level $k+1$) as well as the leftmost phrase boundary within the remaining internal part of $S[i..j]$.

Lemma 4.2. *The set $M(i,j)$ satisfies the following properties:*

1) For each $k \geq 0$, if $B_k(i,j) \neq \emptyset$, then $\min B_k(i,j) \in M(i,j)$.
2) We have $|M(i,j)| = O(\log(j-i+2))$.
3) If $S[i'..j'] = S[i..j]$, then $M(i',j') = M(i,j)$.

Proof. Let us express $M(i,j) = \bigcup_{k \geq 0} M_k(i,j)$, setting

$$M_k(i,j) := B_k(i,j) \setminus [2\alpha_{k+1}..j-i-\alpha_{k+1}] \cup \{\min(B_k(i,j) \cap [2\alpha_{k+1}..j-i-\alpha_{k+1}))\}.$$

As for Item 1, it is easy to see that $\min B_k(i,j) \in M_k(i,j)$: we consider two cases, depending on whether $\min B_k(i,j)$ belongs to $[2\alpha_{k+1}..j-i-\alpha_{k+1})$ or not.

As for Item 2, let us first argue that $|M_k(i,j)| = O(1)$ holds for every $k \geq 0$. Indeed, each element $q \in B_k(i,j) \cap [0..2\alpha_{k+1})$ corresponds to $q + i \in B_k \cap [i..i+2\alpha_{k+1})$ and each element $q \in B_k(i,j) \cap [j-i-\alpha_{k+1}..j-i)$ corresponds to $q + i \in B_k \cap [j - \alpha_{k+1}..j)$. By Lemma 3.8, we conclude that $|M_k(i,j)| \leq 1 + (1 + \frac{8\alpha_{k+1}}{\ell_{k+1}}) + (1 + \frac{4\alpha_{k+1}}{\ell_{k+1}}) = O(1)$. Moreover, if $\ell_k > 4(j-i)$, then Lemma 3.8 further yields $|B_k(i,j)| = |B_k \cap [i..j]| \leq 1$. Since $M_k(i,j)$ and $B_{k+1}(i,j)$ are both subsets of $B_k(i,j)$, this means that $\left| \bigcup_{k:\ell_k>4(j-i)} M_k(i,j) \right| \leq 1$. The number of indices k satisfying $\ell_k \leq 4(j-i)$ is $O(\log(j-i+2))$, and thus

$$|M(i,j)| \leq O(1) \cdot O(\log(j-i+2)) + 1 = O(\log(j-i+2)).$$

As for Item 3, we shall prove by induction on k that $M_k(i,j) \subseteq M(i',j')$. This implies $M(i,j) \subseteq M(i',j')$ and, by symmetry, $M(i,j) = M(i',j')$. In the base case of $k = 0$, we have

$$M_0(i,j) = ([0..2\alpha_1] \cup [j-i-\alpha_1..j-i)) \cap [0..j-i) = M_0(i',j').$$

Now, consider $k > 0$ and $q \in M_k(i,j)$. If $q \in B_k(i,j) \setminus [2\alpha_k..j-i-\alpha_k)$, then $q \in M_{k-1}(i,j)$, and thus $q \in M(i',j')$ holds by the inductive assumption. As for the remaining case, $M_k(i,j) \cap [2\alpha_k..j-i-\alpha_k) = M_k(i',j') \cap [2\alpha_k..j'-i'-\alpha_k)$ is a direct consequence of $B_k(i,j) \cap [2\alpha_k..j-i-\alpha_k) = B_k(i',j') \cap [2\alpha_k..j'-i'-\alpha_k)$, which follows from Lemma 3.7. $\square$

Definition 4.3. *Let P be a substring of S and let $S[i..j]$ be its arbitrary occurrence. We define $M(P) := M(i,j)$; by item 3 of Lemma 4.2, this does not depend on the choice of the occurrence.*

By Lemma 4.2, the set $M(P)$ is of size $O(\log |P|)$, yet, for every level $k \geq 0$ and every occurrence $P = S[i..j]$, it includes the leftmost phrase boundary in $B_k(i,j)$. Our index exploits the latter property for the largest k with $B_k(i,j) \neq \emptyset$.

5   Indexing with Our Grammar

In this section, we adapt the results on attractors [3, Sec. 6] to our modified parsing, so as to obtain our main result.

Definition 5.1 ([3]). *The grammar tree of a RLCFG is obtained by pruning its parse tree: all but the leftmost occurrences of each nonterminal are converted into leaves and their subtrees are pruned. We treat rules $A \rightarrow A_1^s$ (assumed to be of size 2) as $A \rightarrow A_1 A_1^{[s-1]}$, where the node labeled $A_1^{[s-1]}$ is always a leaf (A_1 is also a leaf unless it is the leftmost occurrence of A_1).*

Note that the grammar tree has exactly one internal node per distinct non-terminal and its total number of nodes is the grammar size plus one. We identify each nonterminal A with the only internal grammar tree node labeled A. We also sometimes identify terminal symbols a with grammar tree leaves.

The search algorithm classifies the occurrences of a pattern $P[1..m]$ in S into "primary" and "secondary", according to the partition of S induced by the grammar tree leaves.

Definition 5.2 ([3]). *The leaves of the grammar tree induce a partition of S into phrases. An occurrence of $P[1..m]$ at $S[t..t + m)$ is primary if the lowest grammar tree node deriving a range of S that contains $S[t..t+m)$ is internal (or, equivalently, the occurrence crosses the boundary between two phrases); otherwise it is secondary.*

The general idea of the search is to find the primary occurrences by looking for prefix-suffix partitions of P and then find the secondary occurrences from the primary ones [5].

5.1   Finding the Primary Occurrences

Let nonterminal A be the lowest (internal) grammar tree node that covers a primary occurrence $S[t..t + m)$ of $P[1..m]$. Then, if $A \rightarrow A_1 \cdots A_s$, there exists some $i \in [1..s]$ and $q \in [1..m]$ such that (1) a suffix of $\exp(A_i)$ matches $P[1..q]$, and (2) a prefix of $\exp(A_{i+1}) \cdots \exp(A_s)$ matches $P[q..m]$. The idea is to index all the pairs $(\exp(A_i)^{rev}, \exp(A_{i+1}) \cdots \exp(A_s))$ and find those where the first and second component are prefixed by $(P[1..q])^{rev}$ and $P[q..m]$, respectively. Note that there is exactly one such pair per border between two consecutive phrases (or leaves in the grammar tree).

Definition 5.3 ([3]). *Let v be the lowest (internal) grammar tree node that covers a primary occurrence $S[t..t + m)$ of P. Let v_i be the leftmost child of v that overlaps $S[t..t + m)$. We say that node v is the parent of the primary occurrence $S[t..t + m)$ of P and node v_i is its locus.*

The index [3] builds a two-dimensional grid data structure. It lexicographically sorts all the components $\exp(A_i)^{rev}$ to build the x-coordinates, and all the components $\exp(A_{i+1}) \cdots \exp(A_s)$ to build the y-coordinates; then, it fills the grid with points $(\exp(A_i)^{rev}, \exp(A_{i+1}) \cdots \exp(A_s))$, each associated with the locus A_i. The size of this data structure is of the order of the number of points, which is bounded by the grammar size, $g = O(\delta \log \frac{n}{\delta})$ in our case. The structure can find all the p points within any orthogonal range in time $O((p + 1) \log^\epsilon g)$, where $\epsilon > 0$ is any constant fixed at construction time.

Given a partition $P = P[1..q] \cdot P(q..m]$ to test, they search for $P[1..q]^{rev}$ in a data structure that returns the corresponding range in x, search for $P(q..m]$ in a similar data structure that returns the corresponding range in y, and then perform the corresponding range search on the geometric data structure.

They show [3, Sec. 6.3] that the x- and y-ranges of any τ cuts of P can be computed in time $O(m + \tau \log^2 m)$, within $O(g)$ space. All they need from the RLCFG to obtain this result is that (1) one can extract any length-ℓ prefix or suffix of any $\exp(A)$ in time $O(\ell)$, which is proved for an arbitrary RLCFG; and (2) one can compute a Karp–Rabin fingerprint of any substring of S in time $O(\log^2 \ell)$, which is shown to be possible for any locally contracting grammar, which follows from our Lemma 3.8.

In total, if we have identified τ cuts of P that suffice to find all of its occurrences in S, then we can find all the $occ_p \leq occ$ primary occurrences of P in time $O(m + \tau(\log^\epsilon g + \log^2 m) + occ_p \log^\epsilon g)$.

5.2 Parsing the Pattern

The next step is to set a bound for τ with our parsing and show how to find the corresponding cuts.

Lemma 5.4. *Using our grammar of Sect. 3, there are only $\tau = O(\log m)$ cuts $P = P[1..q] \cdot P(q..m]$ yielding primary occurrences of $P[1..m]$. These positions belong to $M(P) + 1$ (see Definition 4.3).*

Proof. Let A be the parent of a primary occurrence $S[t..t+m)$, and let k be the round where A is formed. There are two possibilities:

(1) $A \rightarrow A_1 \cdots A_s$ is a block-forming rule, and for some $i \in [1..s)$, a suffix of $\exp(A_i)$ matches $P[1..q]$, for some $q \in [1..m)$. This means that $q - 1 = \min B_{k-1}(t, t + m - 1)$.
(2) $A \rightarrow A_1^s$ is a run-length nonterminal, and a suffix of $\exp(A_1)$ matches $P[1..q]$, for some $q \in [1..m)$. This means that $q - 1 = \min B_{k-1}(t, t + m - 1)$.

In either case, $q \in M(P) + 1$ by Lemma 4.2. Further, $|M(P)| = O(\log m)$.   □

The parsing is done in $O(m)$ time almost exactly as in previous work [3, Sec. 6.1], with the difference that we have to care about paused symbols. Essentially, we store the permutations π_k drawn when indexing S and use them to parse P in the same way, level by level. We then work for $O(\log m)$ levels on exponentially decreasing sequences, in linear time per level, which adds up to $O(m)$. There are a few differences with respect to previous work, however [3]:

1) In the parsing of [3], the symbols are disjoint across levels, so the space to store the permutations π_k is proportional to the grammar size. In our case, instead, paused symbols exist along several consecutive levels and participate in several permutations. However, by Lemma 3.9, we have $|\mathcal{S}_k \cap \mathcal{A}_{k+1}| = O(\delta)$ active symbols in S_k. We store store the values of π_{k+1} only for these symbols and observe that the values π_{k+1} for the remaining symbols do not affect the placement of block boundaries in Definition 3.3: If $S_k[j], S_k[j+1] \in \mathcal{A}_{k+1}$, then, due condition imposed on π_{k+1} in Definition 3.4, j may only be a local minimum if $S_k[j-1] \in \mathcal{A}_{k+1}$. When parsing P, we can simply assume that $\pi_{k+1}(A) = 0$ on the paused symbols $A \in \Sigma(S_k) \setminus \mathcal{A}_{k+1}$ and obtain the same parsing of S. By storing the values of π_k only for the active symbols, we use $O(\delta \log \frac{n}{\delta})$ total space.

2) They use that the number of symbols in the parsing of P halve from a level to the next in order to bound the number of levels in the parse and the total amount of work. While this is not the case in our parsing with paused symbols, it still holds by Lemmas 3.5 and 3.8 that the number of phrases in round k is less than $1 + \frac{4m}{\ell_{k+1}}$, which gives us, at most, $h = 12 + 2\lfloor \log_{4/3} m \rfloor = O(\log m)$ parsing rounds and a total of $\sum_{k=0}^{h}(1 + \frac{4m}{\ell_{k+1}}) = O(m)$ symbols processed along the parsing of P.

5.3 Secondary Occurrences and Short Patterns

The occ_s secondary occurrences can be obtained in $O(occ_s)$ time given the primary ones, with a technique that works for any arbitrary RLCFG and within $O(g)$ space [3, Sec. 6.4]. Plugged with the preceding results, the total space of our index is $O(\delta \log \frac{n}{\delta})$ and its search time is $O(m + \tau(\log^\epsilon g + \log^2 m) + occ \log^\epsilon g) = O(m + \log^\epsilon g \log m + occ \log^\epsilon g)$. This bound exceeds $O(m + (occ + 1) \log^\epsilon g)$ only when $m = O(\log^\epsilon g \log \log g)$. In that case, however, the middle term is $O(\log^\epsilon g \log \log g)$, which becomes $O(\log^\epsilon g)$ again if we infinitesimally adjust ϵ.

The final touch is to reduce the $O(m + \log^\epsilon g + occ \log^\epsilon g)$ complexity to $O(m + \log^\epsilon \delta + occ \log^\epsilon g)$. This is relevant only when $occ = 0$, so we need a way to detect in time $O(m + \log^\epsilon \delta)$ that P does not occur in S. We already do this in time $O(m + \log^\epsilon g)$ by parsing P and searching for its cuts in the geometric data structure. To reduce the time, we note that $\log^\epsilon g \in O(\log^\epsilon(\delta \log \frac{n}{\delta})) \subseteq O(\log^\epsilon \delta + \log \log \frac{n}{\delta})$, so it suffices to detect in $O(m)$ time the patterns of length $m \leq \ell = \log \log \frac{n}{\delta}$ that do not occur in S. By definition of δ, there are at most $\delta\ell$ strings of length ℓ in S, so we can store them all in a trie using total space $O(\delta\ell^2) \subseteq O(\delta \log \frac{n}{\delta})$. By implementing the trie children with perfect hashing, we can verify in $O(m)$ time whether a pattern of length $m \leq \ell$ occurs in S. We then obtain Theorem 1.1.

6 Conclusions and Future Work

We have obtained the best of two worlds [3,10] in repetitive text indexing: an index of asymptotically optimal size, $O(\delta \log \frac{n}{\delta})$, with nearly-optimal search time, $O(m + (occ + 1) \log^\epsilon n)$, which is built in $O(n)$ expected time. This closes a question open in those previous works.

Our result could be enhanced in various ways, as done in the past with γ-bounded indexes [3]. For example, is it possible to search in optimal $O(m + occ)$ time within $O(\delta \log \frac{n}{\delta} \log^\epsilon n)$ space? Can we count the number of pattern occurrences in $O(m + \log^{2+\epsilon} n)$ time within our optimal space, or in $O(m)$ time within $O(\delta \log \frac{n}{\delta} \log n)$ space? We believe the answer to all those questions is affirmative and plan to answer them in the extended version of this article.

References

1. Batu, T., Sahinalp, S.C.: Locally consistent parsing and applications to approximate string comparisons. In: De Felice, C., Restivo, A. (eds.) DLT 2005. LNCS, vol. 3572, pp. 22–35. Springer, Heidelberg (2005). https://doi.org/10.1007/11505877_3
2. Birenzwige, O., Golan, S., Porat, E.: Locally consistent parsing for text indexing in small space. In: 31st Annual ACM-SIAM Symposium on Discrete Algorithms, SODA 2020, pp. 607–626. SIAM (2020). https://doi.org/10.1137/1.9781611975994.37
3. Christiansen, A.R., Ettienne, M.B., Kociumaka, T., Navarro, G., Prezza, N.: Optimal-time dictionary-compressed indexes. ACM Trans. Algorithms **17**(1), 8:1–8:39 (2021). https://doi.org/10.1145/3426473
4. Claude, F., Navarro, G.: Improved grammar-based compressed indexes. In: Calderón-Benavides, L., González-Caro, C., Chávez, E., Ziviani, N. (eds.) SPIRE 2012. LNCS, vol. 7608, pp. 180–192. Springer, Heidelberg (2012). https://doi.org/10.1007/978-3-642-34109-0_19
5. Claude, F., Navarro, G., Pacheco, A.: Grammar-compressed indexes with logarithmic search time. J. Comput. Syst. Sci. **118**, 53–74 (2021). https://doi.org/10.1016/j.jcss.2020.12.001
6. Cole, R., Vishkin, U.: Deterministic coin tossing and accelerating cascades: micro and macro techniques for designing parallel algorithms. In: 18th Annual ACM Symposium on Theory of Computing, STOC 1986, pp. 206–219 (1986). https://doi.org/10.1145/12130.12151
7. Jeż, A.: A really simple approximation of smallest grammar. Theor. Comput. Sci. **616**, 141–150 (2016). https://doi.org/10.1016/j.tcs.2015.12.032
8. Kempa, D., Kociumaka, T.: Dynamic suffix array with polylogarithmic queries and updates. In: 54th Annual ACM SIGACT Symposium on Theory of Computing, STOC 2022, pp. 1657–1670 (2022). https://doi.org/10.1145/3519935.3520061
9. Kempa, D., Prezza, N.: At the roots of dictionary compression: string attractors. In: 50th Annual ACM SIGACT Symposium on Theory of Computing, STOC 2018, pp. 827–840 (2018). https://doi.org/10.1145/3188745.3188814
10. Kociumaka, T., Navarro, G., Prezza, N.: Towards a definitive measure of repetitiveness. In: Kohayakawa, Y., Miyazawa, F.K. (eds.) LATIN 2021. LNCS, vol. 12118, pp. 207–219. Springer, Cham (2020). https://doi.org/10.1007/978-3-030-61792-9_17

11. Kociumaka, T., Navarro, G., Prezza, N.: Towards a definitive compressibility measure for repetitive sequences, October 2021. https://arxiv.org/pdf/1910.02151
12. Kociumaka, T., Radoszewski, J., Rytter, W., Waleń, T.: Internal pattern matching queries in text and applications (2021). Unpublished manuscript
13. Kreft, S., Navarro, G.: On compressing and indexing repetitive sequences. Theor. Comput. Sci. **483**, 115–133 (2013). https://doi.org/10.1016/j.tcs.2012.02.006
14. Lempel, A., Ziv, J.: On the complexity of finite sequences. IEEE Trans. Inf. Theor. **22**(1), 75–81 (1976). https://doi.org/10.1109/TIT.1976.1055501
15. Mehlhorn, K., Sundar, R., Uhrig, C.: Maintaining dynamic sequences under equality tests in polylogarithmic time. Algorithmica **17**(2), 183–198 (1997). https://doi.org/10.1007/BF02522825
16. Navarro, G.: Indexing highly repetitive string collections, part I: repetitiveness measures. ACM Comput. Surv. **54**(2), 29:1–29:31 (2021). https://doi.org/10.1145/3434399
17. Navarro, G.: Indexing highly repetitive string collections, part II: compressed indexes. ACM Comput. Surv. **54**(2), 26:1–26:32 (2021). https://doi.org/10.1145/3432999
18. Raskhodnikova, S., Ron, D., Rubinfeld, R., Smith, A.D.: Sublinear algorithms for approximating string compressibility. Algorithmica **65**(3), 685–709 (2013). https://doi.org/10.1007/s00453-012-9618-6
19. Sahinalp, S.C., Vishkin, U.: On a parallel-algorithms method for string matching problems (overview). In: Bonuccelli, M., Crescenzi, P., Petreschi, R. (eds.) CIAC 1994. LNCS, vol. 778, pp. 22–32. Springer, Heidelberg (1994). https://doi.org/10.1007/3-540-57811-0_3
20. Stephens, Z.D., et al.: Big data: Astronomical or genomical? PLoS Biology **13**(7), e1002195 (2015). https://doi.org/10.1371/journal.pbio.1002195

Computing and Listing Avoidable Vertices and Paths

Charis Papadopoulos$^{(\boxtimes)}$ and Athanasios E. Zisis

Department of Mathematics, University of Ioannina, Ioannina, Greece
charis@uoi.gr, athanas.zisis@gmail.com

Abstract. A simplicial vertex of a graph is a vertex whose neighborhood is a clique. It is known that listing all simplicial vertices can be done in $O(nm)$ time or $O(n^\omega)$ time, where $O(n^\omega)$ is the time needed to perform a fast matrix multiplication. The notion of avoidable vertices generalizes the concept of simplicial vertices in the following way: a vertex u is avoidable if every induced path on three vertices with middle vertex u is contained in an induced cycle. We present algorithms for listing all avoidable vertices of a graph through the notion of minimal triangulations and common neighborhood detection. In particular we give algorithms with running times $O(n^2 m)$ and $O(n^{1+\omega})$, respectively. Additionally, based on a simplified graph traversal we propose a fast algorithm that runs in time $O(n^2 + m^2)$ and matches the corresponding running time of listing all simplicial vertices on sparse graphs with $m = O(n)$. Moreover, we show that our algorithms cannot be improved significantly, as we prove that under plausible complexity assumptions there is no truly subquadratic algorithm for recognizing an avoidable vertex. To complement our results, we consider their natural generalizations of avoidable edges and avoidable paths. We propose an $O(nm)$-time algorithm that recognizes whether a given induced path is avoidable.

1 Introduction

Closely related to chordal graphs is the notion of a simplicial vertex, that is a vertex whose neighborhood induces a clique. In particular, Dirac [11] proved that every chordal graph admits a simplicial vertex. However not all graphs contain a simplicial vertex. Due to their importance to several algorithmic problems, such as finding a maximum clique or computing the chromatic number, it is natural to seek for fast algorithms that list all simplicial vertices of a graph. For doing so, the naive approach takes $O(nm)$ time, whereas the fastest algorithms take advantage of computing the square of an $n \times n$ binary matrix and run in $O(n^\omega)$ and $O(m^{2\omega/(\omega+1)})$ time [17]. Hereafter we assume that we are given a graph G on n vertices and m edges; currently, $\omega < 2.37286$ [2].

C. Papadopoulos is supported by Hellenic Foundation for Research and Innovation (H.F.R.I.) under the "First Call for H.F.R.I. Research Projects to support Faculty members and Researchers and the procurement of high-cost research grant", Project FANTA (eFficient Algorithms for NeTwork Analysis), number HFRI-FM17-431.

A. Castañeda and F. Rodríguez-Henríquez (Eds.): LATIN 2022, LNCS 13568, pp. 104–120, 2022.
https://doi.org/10.1007/978-3-031-20624-5_7

A natural way to generalize the concept of simplicial vertices is the notion of an avoidable vertex. A vertex u is avoidable if either there is no induced path on three vertices with middle vertex u, or every induced path on three vertices with middle vertex u is contained in an induced cycle. Thus every simplicial vertex is avoidable, however the converse is not necessarily true. As opposed to simplicial vertices, it is known that every graph contains an avoidable vertex [1,5,7,20]. Extending the notion of avoidable vertices is achieved through avoidable edges and, more general, avoidable paths. This is accomplished by replacing the middle vertex in an induced path on three vertices by an induced path on arbitrary $k \geq 2$ vertices, denoted by P_k. Beisegel et al. [3] proved first that every non-edgeless graph contains an avoidable edge, considering the case of $k = 2$. Regarding the existence of an avoidable induced path of arbitrary length, Bonamy et al. [9] settled a conjecture in [3] and showed that every graph is either P_k-free or contains an avoidable P_k. Gurvich et al. [13] strengthened the later result by showing that every induced path can be shifted in an avoidable path, in the sense that there is a sequence of neighboring induced paths of the same length. Although the provided proof in [13] is constructive and identifies an avoidable path given an induced path, the proposed algorithm was not settled whether it runs in polynomial time.

Since avoidable vertices generalize simplicial vertices, it is expected that avoidable vertices find applications in further algorithmic problems. Indeed, Beisegel et al. [3] revealed new polynomially solvable cases of the maximum weight clique problem that take advantage of the notion of avoidable vertices. Similar to simplicial vertices, the complexity of a problem can be reduced by removing avoidable vertices, tackling the problem on the reduced graph. It is therefore of interest to list all avoidable vertices efficiently. If we are only interested in computing two avoidable vertices this can be done in linear time by using fast graph searches [3,5]. However, an efficient elimination process, such as deleting or removing avoidable vertices, is not enough to recursively compute the rest of the avoidable vertices. Thus, computing the set of all avoidable vertices requires to decide for each vertex of the graph whether it is avoidable and a usual graph search cannot guarantee to test all vertices.

Concerning lower bounds, it is known [18] that the problem of finding a triangle in an n-vertex graph can be reduced in $O(n^2)$ time to the problem of counting the number of simplicial vertices. Moreover, Ducoffe proved that under plausible complexity assumptions computing the diameter of an AT-free graph is at least as hard as computing a simplicial vertex [12]. For general graphs, the quadratic time complexity of diameter computation cannot be improved by much [22]. We note that the currently fastest algorithms for detecting a triangle run in time $O(nm)$ and $O(n^\omega)$ [16]. Notably, we show a similar lower bound for recognizing an avoidable vertex. In particular, via a reduction form the Orthogonal-Vector problem, we prove that under SETH, there is no truly subquadratic algorithm for deciding whether a given vertex is avoidable. This gives a strong evidence that our $O(nm)$- and $O(n^\omega)$-recognition algorithms upon which are based our listing algorithms cannot be improved significantly.

A naive approach that recognizes a single vertex u of a graph G of whether it is avoidable or not, needs to check if all neighbors of u are pairwise connected in an induced subgraph of G. Thus the running time of recognizing an avoidable vertex is $O(n^3 + n^2 m)$ or, as explicitly stated in [3], it can be expressed as $O(\overline{m} \cdot (n + m))$ where $\overline{m}$ is the number of edges in the complement of G. Inspired by both running times, we first show that we can reduce in linear time the listing problem on a graph G having $m \geq n$ and $\overline{m} \geq n$. In a sense such a result states that graphs that are sparse ($m < n$) or dense ($\overline{m} < n$) can be decomposed efficiently to smaller connected graphs for which their complement is also connected. Towards this direction, we give an interesting connection with the avoidable vertices on the complement of G. As a result, the naive algorithms for listing all avoidable vertices take $O(n^3 \cdot m)$ and $O(n \cdot \overline{m} \cdot m)$ time, respectively.

Our main results consist of new algorithms for listing all avoidable vertices in running times comparable to the ones for listing simplicial vertices. More precisely, we propose three main approaches that result in algorithms for listing all avoidable vertices of a graph G with the following running times:

- $O(n^2 \cdot m)$, by using a minimal triangulation of G. A close relationship between avoidable vertices and minimal triangulation was already known [3]. However, listing all avoidable vertices through the proposed characterization is inefficient, since one has to produce *all* possible minimal triangulations of G. Here we strengthen such a characterization in the sense that it provides an efficient recognition based on one particular minimal triangulation of G. More precisely, we take advantage of vertex-incremental minimal triangulations that can be computed in $O(nm)$ time [8].
- $O(n^2 + m^2)$, by exploring structural properties on each edge of G. This approach is based on a modified, traditional breadth-first search algorithm. Our task is to construct search trees rooted at a particular vertex that reach all vertices of a predescribed set S, so that every non-leaf vertex does not belong to S. If such a tree exists then every path from the root to a leaf that belongs to S is called an S-excluded path. It turns out that S-excluded paths can be tested in linear time and we need to make $2m$ calls of a modified breadth-first search algorithm.
- $O(n^{1+\omega})$, where $O(n^\omega)$ is the running time for matrix multiplication. For applying a matrix multiplication approach, we contract the connected components of G that are outside the closed neighborhood of a vertex. Then we observe that a vertex u is avoidable if the neighbors of u are pairwise in distance at most two in the contracted graph. As the distance testing can be encapsulated by the square of its adjacency matrix, we deduce an algorithm that takes advantage of a fast matrix multiplication.

We should note that each of the stated algorithms is able to recognize if a given vertex u of G is avoidable in time $O(nm)$, $O(d(u)(n + m))$, and $O(n^\omega)$, respectively, where $d(u)$ is the degree of u in G. Further, all of our proposed algorithms consist of basic ingredients that avoid using sophisticated data structures.

In addition, we consider the natural generalizations of avoidable vertices, captured within the notions of the avoidable edges and avoidable paths. A naive algorithm that recognizes an avoidable edge takes time $O(n^2 \cdot m)$ or $O(\overline{m} \cdot m)$. Here we show that recognizing an avoidable edge of a graph G can be done in $O(n \cdot m)$ time. This is achieved by taking advantage of the notions of the S-excluded paths and their efficient detection by the modified breadth-first search algorithm. Also notice that an avoidable edge is an avoidable path on two vertices. We are able to reduce the problem of recognizing an avoidable path of arbitrary length to the recognition of an avoidable edge. In particular, given an induced path we prove that we can replace the induced path by an edge and test whether the new added edge is avoidable or not in a reduced graph. Therefore our recognition algorithm for testing whether a given induced path is avoidable takes $O(n \cdot m)$ time. As a side remark of the later algorithm, we partially resolve an open question raised in [13].

2   Preliminaries

All graphs considered here are finite undirected graphs without loops and multiple edges. We refer to the textbook by Bondy and Murty [10] for any undefined graph terminology. We use n to denote the number of vertices of a graph and use m for the number of edges. Given $x \in V_G$, we denote by $N_G(x)$ the neighborhood of x. The closed neighborhood of x, denoted by $N_G[x]$, is defined as $N_G(x) \cup \{x\}$. For a set $X \subset V(G)$, $N_G(X)$ denotes the set of vertices in $V(G) \setminus X$ that have at least one neighbor in X. Analogously, $N_G[X] = N_G(X) \cup X$. The induced path on $k \geq 2$ vertices is denoted by P_k and the induced cycle on $k \geq 3$ vertices is denoted by C_k. For an induced path P_k, the vertices of degree one are called *endpoints*. Notice that for any two vertices x and y of a connected graph there is an induced path having x and y as endpoints. Given two vertices u and v of a connected graph G, a set $S \subset V_G$ is called (u, v)-*separator* if u and v belong to different connected components of $G - S$. We say that S is a separator if there exist two vertices u and v such that S is a (u, v)-separator. A graph G is *co-connected* if its complement $\overline{G}$ is connected and a *co-component* of G is a connected component of $\overline{G}$.

Given an edge $e = xy$, the *contraction* of e removes both x and y and replaces them by a new vertex w, which is made adjacent to those vertices that were adjacent to at least one of the vertices x and y, that is $N(w) = (N(x) \cup N(y)) \setminus \{x, y\}$. Contracting a set of vertices S is the operation of substituting the vertices of S by a new vertex w with $N(w) = N(S)$.

A vertex v is called *simplicial* if the vertices of $N_G(v)$ induce a clique. Listing all simplicial vertices of a graph can be done $O(nm)$ time. The fastest algorithm for listing all simplicial vertices takes time $O(n^\omega)$, where $O(n^\omega)$ is the time needed to multiply two $n \times n$ binary matrices [17]. Avoidable vertices and edges generalize the concept of simplicial vertices in a natural way.

Definition 2.1. *A vertex v is called* avoidable *if every P_3 with middle vertex v is contained in an induced cycle. Equivalently, v is avoidable if $d_G(v) \leq 1$ or for every pair $x, y \in N_G(v)$ the vertices x and y belong to the same connected component of $G - (N_G[u] \setminus \{x, y\})$.*

Every simplicial vertex is avoidable, however the converse is not necessarily true. It is known that every graph contains an avoidable vertex [1,7,20]. Omitted proofs of statements can be found in a preliminary full version [21].

Observation 2.1. *Let G be a graph and let u be a vertex of G. Then u is non-avoidable if and only if there is an (x,y)-separator S that contains u such that $S \subset N_G[u]$ for some vertices $x, y \in N_G(u)$.*

3   A Lower Bound For Recognizing an Avoidable Vertex

In the forthcoming sections, we give algorithms for recognizing an avoidable vertex in $O(nm)$ time and $O(n^\omega)$ time. Here we show that, under plausible complexity assumptions, a significant improvement on the stated running times is unlikely, as we show that there is no truly subquadratic algorithm for deciding whether a given vertex is avoidable. By *truly subquadratic*, we mean an algorithm with running time $O(n^{2-\epsilon})$, for some $\epsilon > 0$ where n is the size of its input.

More precisely, the Strong Exponential-Time Hypothesis (SETH) states that for any $\epsilon > 0$, there exists a k such that the k-SAT problem on n variables cannot be solved in $O((2 - \epsilon)^n)$ time [15]. The Orthogonal-Vector problem (OV) takes as input two families A and B of n sets over a universe C, and asks whether there exist $a \in A$ and $b \in B$ such that $a \cap b = \emptyset$. An instance of OV is denoted by $OV(A, B, C)$. It is known that under SETH, for any $\epsilon > 0$, there exists a constant $c > 0$ such that $OV(A, B, C)$ cannot be solved in $O(n^{2-\epsilon})$, even if $|C| \leq c \cdot \log n$ [25]. For deciding whether a given vertex is avoidable, we give a reduction from OV.

Theorem 3.1. *The OV problem with $|A| = |B| = n$ and $|C| = O(\log n)$ can be reduced in $O(n \log n)$ time to the problem of deciding whether a particular vertex of an $O(n)$-vertex graph is avoidable.*

Proof. Let $OV(A, B, C)$ be an instance of OV. We construct a graph G as follows. The vertex set of G consists of $A \cup B \cup C$ and three additional vertices u, c_A, c_B. For the edges of G, we have:

- u is adjacent to every vertex of $A \cup B$;
- c_A is adjacent to every vertex of A and c_B is adjacent to every vertex of B;
- for every $a \in A$ and every $c \in C$, $ac \in E(G)$ if and only if $c \in a$;
- for every $b \in B$ and every $c \in C$, $bc \in E(G)$ if and only if $c \in b$.

These are exactly the edges of G. In particular notice that $G[C \cup \{u, c_A, c_B\}]$ is an independent set. Moreover, observe that G has $2n + |C| + 3$ vertices and the number of edges is $O(n \log n)$. We claim that $OV(A, B, C)$ is a yes-instance if and only if u is non-avoidable in G.

Assume that there are sets $a \in A$ and $b \in B$ such that $a \cap b = \emptyset$. Let $x \in A$ and $y \in B$ be the vertices of A and B that correspond to a and b, respectively. By construction, x and y are non-adjacent in G. Moreover, by construction, x and y have no common neighbor in C, as $a \cap b = \emptyset$. Now notice that all neighbors of x and y that do not belong to $N_G[u] = A \cup B \cup \{u\}$ are in $C \cup \{c_A, c_B\}$ and $G[C \cup \{c_A, c_B\}]$ is an edgeless graph. Thus x and y belong to different components in $G - (N_G[u] \setminus \{x, y\})$ and u is non-avoidable in G.

For the converse, assume that u is non-avoidable in G. Since $N(u) = A \cup B$ there are vertices $x, y \in A \cup B$ such that x and y lie in different components in $G - (N_G[u] \setminus \{x, y\})$. If both x and y belong to A, then they have a common neighbor c_A in $G - (N_G[u] \setminus \{x, y\})$ which is not possible. Similarly, both x and y do not belong to B due to vertex c_B. Thus $x \in A$ and $y \in B$. As there are no edges in $G[C \cup \{c_A, c_B\}]$, we deduce that x and y have no common neighbor in C. Hence there are sets in A and B that correspond to the vertices x and y, respectively, that have no common element and $OV(A, B, C)$ is a yes-instance. $\qquad\square$

4   Avoidable Vertices in Sparse or Dense Graphs

Here we show how to compute efficiently all avoidable vertices on sparse or dense graphs. In particular, for a graph G on n vertices and m edges, we consider the cases in which $m < n$ (sparse graphs) or $\overline{m} < n$ (dense graphs), where $\overline{m} = |E(\overline{G})|$. Our main motivation comes from the naive algorithm that lists all avoidable vertices in $O(n \cdot \overline{m} \cdot (n+m))$ time that takes advantage of the non-edges of G [3]. We will show that we can handle the non-edges in linear time, so that the running time of the naive algorithm can be written as $O(n^3 \cdot m)$. For doing so, we consider the behavior of avoidable vertices on the complement of a graph by considering the connected components in both G and $\overline{G}$.

It is not difficult to handle sparse graphs. Observe that $m < n$ implies that G is disconnected or G is a tree. The connectedness assumption of the input graph G follows from the fact that a vertex u is avoidable in G if and only if u is avoidable in the connected component containing u, since there are no paths between vertices of different components. Moreover, trees have a trivial solution as the leaves are exactly the set of avoidable vertices. We include both properties in the following statement.

Observation 4.1. *Let u be a vertex of G and let $C(u)$ be the connected component of G containing u. Then u is avoidable if and only if u is avoidable in $G[C(u)]$. Moreover, if G is a tree then u is avoidable if and only if u is a leaf.*

We can follow almost the same approach on the complement of G. For doing so, we first prove the following result which interestingly relates avoidability on G and $\overline{G}$. Note, however, that the converse is not necessarily true (e.g., consider the C_5 in which all vertices are avoidable in both G and $\overline{G}$).

Lemma 4.1. *Let u be a non-avoidable vertex of G. Then, u is avoidable in $\overline{G}$.*

Proof. Since u is a non-avoidable vertex in G, there is a separator S that contains u such that $S \subset N_G[u]$ by Observation 2.1. Let $C_1, \ldots, C_k$ be the connected components of $G - S$, with $k \geq 2$. Notice that at least two components of

$C_1, \ldots, C_k$ contain a neighbor of u. Without loss of generality, assume that $C_1 \cap N_G(u) \neq \emptyset$ and $C_2 \cap N_G(u) \neq \emptyset$. Consider the complement $\overline{G}$ and let x, y be two neighbors of u in $\overline{G}$. Observe that both x and y do not belong to S, since $S \subset N_G[u]$. Thus $x \in C_i$ and $y \in C_j$, for $1 \leq i, j \leq k$. We show that either $xy \in E(\overline{G})$ or there is a path in $\overline{G}$ between x and y that avoids vertices of $N_{\overline{G}}(u)$. If $i \neq j$ then $xy \in E(\overline{G})$, because every vertex of C_i is adjacent to every vertex of C_j in $\overline{G}$. Suppose that $x, y \in C_i$. If $C_i \neq C_1$ then there is a vertex $w_1 \in C_1 \cap N_G(u)$ such that $w_1 u \notin E(\overline{G})$ and $w_1 x, w_1 y \in E(\overline{G})$. If $C_i = C_1$ then there is a vertex $w_2 \in C_2 \cap N_G(u)$ such that $w_2 u \notin E(\overline{G})$ and $w_2 x, w_2 y \in E(\overline{G})$. Thus in both cases there is a path of length two between x and y that avoids vertices $N_{\overline{G}}(u)$. Therefore, u is avoidable in $\overline{G}$.   □

We next deal with the case in which $\overline{G}$ is disconnected. Notice that if $G = K_n$ then every vertex of G is simplicial and thus avoidable.

Lemma 4.2. *Let $G \neq K_n$, $u \in V(G)$, and let $\overline{C}(u)$ be the co-component containing u. Then, u is avoidable in G if and only if $|\overline{C}(u)| > 1$ and u is avoidable in $G[\overline{C}(u)]$.*

In general, avoidability is not a hereditary property with respect to induced subgraphs, even when restricted to the removal of non-avoidable vertices. However, the removal of universal vertices does not affect the rest of the graph.

Lemma 4.3. *Let G be a graph and let w be a universal vertex of G. Then w is avoidable if and only if G is a complete graph. Moreover, any vertex $u \in V(G) \setminus \{w\}$ is avoidable in G if and only if u is avoidable in $G - w$.*

To conclude the cases for which $\overline{m} < n$, we consider graphs whose complement is a tree. By Observation 4.1 we restrict ourselves on connected graphs.

Lemma 4.4. *Let G be a connected graph such that $\overline{G}$ is a tree T. A vertex u of G is avoidable if and only if u is a non-leaf vertex in T.*

Based on the previous results, we can reduce our problem to a graph G that is both connected and co-connected and neither G nor $\overline{G}$ are isomorphic to trees. To achieve this in linear time we apply known techniques that avoid computing explicitly the complement of G, since we are mainly interested in recursively detecting the components and co-components of G. Such a decomposition, known as the *modular decomposition*, can be represented by a tree structure, denoted by $T(G)$, of $O(n)$ size and can be computed in linear time [19, 24]. More precisely, the leaves of $T(G)$ correspond to the vertices of G and every internal node w of $T(G)$ is labeled with three distinct types according to whether the subgraph of G induced by the leaves of the subtree rooted at w is (i) not connected, or (ii) not co-connected, or (iii) connected and co-connected. Moreover the connected components and the co-components of types (i) and (ii), respectively, correspond to the children of w in $T(G)$. Let $\mathcal{G}$ be a collection of maximal vertex-disjoint induced subgraphs of G that are both connected and co-connected. Then $T(G)$ determines all graphs of $\mathcal{G}$ in linear time. In addition, we call $\mathcal{G}$, *typical collection* of G if for each graph $H \in \mathcal{G}$: H is connected and co-connected, $|V(H)| \leq |E(H)|$, $|V(H)| \leq |E(\overline{H})|$, and every avoidable vertex in H is an avoidable vertex in G. The results of this section deduce the following algorithm.

Theorem 4.1. *Let G be a graph and let $A(G)$ be the set of avoidable vertices in G. There is a linear-time algorithm, that computes a typical collection $\mathcal{G}$ of maximal vertex-disjoint induced subgraphs of G and for every vertex $v \in V(G) \setminus V(\mathcal{G})$, decides if $v \in A(G)$.*

5 Computing Avoidable Vertices Directly From G

Here we give two different approaches for computing all avoidable vertices. Both of them deal with the input graph itself without shrinking any unnecessary information, as opposed to the algorithms given in forthcoming sections. Our first algorithm makes use of notions related to minimal triangulations of G and runs in time $O(n^2 m)$. The second algorithm runs in time $O(n^2 + m^2)$ and is based on a modified, traditional breadth-first search algorithm.

Let us first explain our algorithm through a minimal triangulation of G. We first need some necessary definitions. A graph is *chordal* if it does not contain an induced cycle of length more than three. In different terminology, G is chordal if and only if G is $(C_4, C_5, \ldots)$-free graph. A graph $H = (V, E \cup F)$ is a *minimal triangulation* of $G = (V, E)$ if H is chordal and for every $F' \subset F$, the graph $(V, E \cup F')$ is not chordal. The edges of F in H are called *fill edges*. Several $O(nm)$-time algorithms exist for computing a minimal triangulation [4,6,14,23]. In connection with avoidable vertices, Beisegel et al. [3] showed the following.

Theorem 5.1 ([3]). *Let u be a vertex of G. Then u is avoidable in G if and only if u is a simplicial vertex in some minimal triangulation of G.*

Although such a characterization is complete, it does not lead to an efficient algorithm for deciding whether a given vertex is avoidable, since one has to produce *all* possible minimal triangulations of G. Here we strengthen such a characterization in the sense that it provides an efficient recognition based on a particular, *nice*, minimal triangulation of G.

Lemma 5.1. *Let u be a vertex of a graph $G = (V, E)$ and let $H = (V, E \cup F)$ be a minimal triangulation of G such that u is not incident to any edge of F. Then u is avoidable in G if and only if u is simplicial in H.*

Proof. If u is simplicial in H then by Theorem 5.1 we deduce that u is avoidable in G. Suppose that u is non-simplicial in H. Then there are two vertices $x, y \in N_G(u)$ that are non-adjacent in H. Since G is a subgraph of H, we have $xy \notin E(G)$. We claim that there is no path in G between x and y that avoids any vertex of $N_G[u] \setminus \{x, y\}$. Assume for contradiction that there is such a path P. Then $V(P) \setminus \{x, y\}$ is non-empty and contains vertices only from $V \setminus N[u]$. This means that x, y belong to the same connected component of H induced by $(V \setminus N[u]) \cup \{x, y\}$. As u is non-adjacent to any vertex of $V \setminus N[u]$ in H, the vertices of $(V \setminus N[u]) \cup \{x, y, u\}$ induce an induced cycle of length at least four in H. Then we reach a contradiction to the chordality of H. Therefore, there is no such path between x and y, which implies that u is non-avoidable in G. □

Next we show that such a minimal triangulation with respect to u, always exists and can be computed in $O(nm)$ time. Our approach for computing a nice minimal triangulation of G is *vertex incremental*, in the following sense. We take the vertices of G one by one in an arbitrary order $(v_1, \ldots, v_n)$, and at step i we compute a minimal triangulation H_i of $G_i = G[\{v_1, \ldots, v_i\}]$ from a minimal triangulation H_{i-1} of G_{i-1} by adding only edges incident to v_i. This is possible thanks to the following result.

Lemma 5.2 ([8]). *Let G be an arbitrary graph and let H be a minimal triangulation of G. Consider a new graph $G' = G + v$, obtained by adding to G a new vertex v. There is a minimal triangulation H' of G' such that $H' - v = H$.*

We denote by $H(v_1, \ldots, v_n)$ a vertex incremental minimal triangulation of G which is obtained by considering the vertex ordering $(v_1, \ldots, v_n)$ of G. Computing such a minimal triangulation of G, based on any vertex ordering, can be done in $O(nm)$ time [8].

Lemma 5.3. *Let u be a vertex of G and let $X = N_G(u)$ and $A = V(G) \setminus N_G[u]$. In any vertex incremental minimal triangulation $H(A, u, X)$ of G, no fill edge is incident to u.*

Proof. Let $H(A, u, X) = (V, E \cup F)$ be a vertex incremental minimal triangulation of G. Consider the vertex ordering (A, u, X). Observe that when adding u to $H[A]$ no fill edge is required, as the considered graph $H[A] + u$ is already chordal. Moreover u is adjacent in G to every vertex appearing after u in the described ordering (A, u, X). Thus u is non-adjacent to any vertex of A in $H(A, u, X)$ which means that no edge of F is incident to u. □

A direct consequence of Lemmas 5.1 and 5.3 is an $O(nm)$-time recognition algorithm for deciding whether a given vertex u is avoidable. For every vertex u, we first construct a vertex incremental minimal triangulation $H(A, u, X)$ of G by applying the $O(nm)$-time algorithm given in [8]. Then we simply check whether u is simplicial in the chordal graph $H(A, u, X)$ by Lemma 5.1, which means that the overall running time is $O(nm)$. The details are given in Algorithm 1. By applying Algorithm 1 on each vertex, we obtain the following result.

Theorem 5.2. *Listing avoidable vertices by using Algorithm 1 takes $O(n^2 m)$ time.*

5.1   A Fast Algorithm For Listing Avoidable Vertices

Our second approach is based on the following notion of *protecting* that we introduce here. Given a set of vertices $S \subseteq V$, an *S-excluded path* is a path in which no internal vertex belongs to S. Observe that an edge is an S-excluded

---

Algorithm 1: Using a vertex incremental minimal triangulation

---

Input : A graph G, a minimal triangulation H of G, and a vertex u
Output: True iff u is avoidable in G
1 Let $X = N_G(u)$ and $A = V(G) \setminus N_G[u]$
2 Initialize a new graph $H' = H[A \cup \{u\}]$
3 Add the vertices of X in H' in an arbitrary order and maintain a minimal triangulation H' of G by applying the $O(nm)$-time algorithm given in [8]
4 **if** u *is simplicial in* H' **then return** *true*
5 **else return** *false*

Algorithm 2: Detecting whether there is an S-excluded path.

Input : A graph G, a vertex x, and a target set $S \subseteq V(G)$
Output: True iff there is an S-excluded path between x and every vertex of S
1 Initialize a queue $Q = \{x\}$, set $T = \emptyset$, and mark x
2 **while** Q *is not empty* **do**
3     $s = Q.pop()$
4   **for** $v \in N(s)$ **do**
5   **if** v *is unmarked* **then**
6   **if** $v \in S$ **then** $T = T \cup \{v\}$
7   **else** $Q.add(v)$
8   Mark v

9 **return** $T == S$

---

path, for any choice of S. By definition a single vertex is connected to itself by the trivial path. Whenever there is an S-excluded path in G between vertices a and b, notice that a can reach b through vertices of $V(G) \setminus S$.

Definition 5.1 (Protecting). *Let x and y be two vertices of G. We say that x protects y if there is a $N_G[y]$-excluded path between x and every vertex of $N_G(y)$. In other words, x protects y if for any $z \in N_G(y) \setminus \{x\}$, either $xz \in E(G)$ or x can reach z through vertices of $V(G) \setminus N_G[y]$.*

Let us explain how to check if x protects y in linear time, that is in $O(n+m)$ time. We consider the graph $G' = G - y$ and run a slight modification of a breadth-first search algorithm on G' starting from x. In particular, we try to reach the vertices of $N_G(y) \setminus \{x\}$ (target set) from x in G'. Every time we encounter a vertex v of the target set, we include v in a set T of discovered target vertices and we do not continue the search from v by avoiding to place v within the search queue. Consequently, no vertex of the target set is a non-leaf node of the constructed search tree. Algorithm 2 shows in detail the considered modification of a breadth-first search.

Lemma 5.4. *Algorithm 2 is correct and runs in $O(n + m)$ time.*

Algorithm 3: Detecting whether the neighbors of u protect u

    Input   : A graph G and a vertex u
    Output: True iff u is avoidable in G
1 Let $X = N_u$ and $G' = G - u$
2 **for** $x \in X$ **do**
3 | Set $S = X \setminus \{x\}$
4 |_ **if** *Algorithm 2(G', x, S) is not true* **then** **return** *false*

5 **return** *true*

Proof. For the correctness, let T be the search tree discovered by the algorithm when the search starts from x. Observe that the basic concepts of the breadth-first search are maintained, so that the key properties with the shortest paths between the vertices of G and the search tree T are preserved. If there is a leaf vertex v in the constructed tree T such that $v \in S$ then the unique path in T is an S-excluded path in G between x and v, since no vertex of S is a non-leaf vertex of T. On the other hand, assume that there is an S-excluded path in G between x and every vertex of S. For every $v \in S$, among such S-excluded paths between x and v, choose $P(v)$ to be the shortest. Let $p(v)$ be the neighbor of v in $P(v)$. Clearly x and every vertex $p(v)$ belong to the same connected component of G. Consider the graph $G - S$. Notice that every vertex $p(v)$ belongs to the same connected component with x in $G - S$, since for otherwise some vertices of S separate x and a vertex v of S which implies that there is no S-excluded path in G between x and v in G. Now let T_x be a breadth-first search tree of $G - S$ that contains x. Then the distance between x and $p(v)$ in T_x corresponds to the length of their shortest path in $G - S$. Construct T by attaching every vertex v of S to be a neighbor of $p(v)$ in T_x. Therefore T is a tree that contains the shortest S-excluded paths between x and the vertices of S. Regarding the running time, notice that no additional data structure is required compared to the classical implementation of the breadth-first search. Hence the running time of Algorithm 2 is bounded by the breadth-first search algorithm. □

Therefore we can check whether x protects y by running Algorithm 2 on the graph $G - y$ with target set $S = N_G(y) \setminus \{x\}$. The connection to the avoidability of a vertex, can be seen with the following result.

Lemma 5.5. *Let u be a vertex of a graph $G = (V, E)$. Then u is avoidable in G if and only if x protects u for every vertex $x \in N_G(u)$.*

Theorem 5.3. *Listing all avoidable vertices by using Algorithm 3 takes $O(n^2 + m^2)$ time.*

6   Avoidable Vertices via Contractions

Here we show how to compute all avoidable vertices of a graph G through contractions. Given a graph $G = (V_G, E_G)$ and a vertex $u \in V_G$, we denote by G_u the graph obtained from G by contacting every connected component of $G - N_G[u]$. We partition the vertices of $G_u - u$ into (X, C), such that $X = N_G(u)$ and C contains the contracted vertices of $G - N_G[u]$. We denote by $G_u(X, C)$ the contracted graph where (X, C) is the vertex partition with respect to G_u. Observe that $G_u[X \cup \{u\}] = G[X \cup \{u\}]$ and $G_u[C \cup \{u\}]$ is an independent set.

Observation 6.1. *Given a vertex u of $G = (V, E)$, the construction of $G_u(X, C)$ can be done in $O(n + m)$ time.*

Next we show that $G_u(X, C)$ holds all necessary information of important paths of G with respect to the avoidability of u.

Lemma 6.1. *Let u be a vertex of a graph $G = (V, E)$. Then u is avoidable in G if and only if u is avoidable in $G_u(X, C)$.*

Lemma 6.1 implies that we can apply all of our algorithms given in the previous section in order to recognize an avoidable vertex. Although such an approach does not lead to faster theoretical time bounds, in practice the contracted graph has substantial smaller size than the original graph and may lead to practical running times. We next show that the contracted graph results in an additional algorithm with different running time.

Let $G_u(X, C)$ be the contracted graph of a vertex u. The *filled-contracted graph*, denoted by $H_u(X, C)$, is the graph obtained from $G_u(X, C)$ by adding all necessary edges in order to make every neighborhood of $C_i \in C$ a clique. That is, for every $C_i \in C$, $N_{H_u}(C_i)$ is a clique. The following proof resembles the characterization given through minimal triangulations in Lemma 5.1. However $H_u(X, C)$ is not necessarily a chordal graph, because $X \nsubseteq N_{G_u}(C)$.

Lemma 6.2. *A vertex u is avoidable in G if and only if $H_u[X]$ is a clique.*

We take advantage of Lemma 6.2 in order to recognize whether u is avoidable. The naive construction of $H_u(X, C)$ requires $O(n^3)$ time, since $|X| \leq n$ and $|C| \leq n$. Instead of constructing $H_u(X, C)$, we are able to check $H_u[X]$ in an efficient way through matrix multiplication. To do so, we consider the graph G' obtained from $G_u(X, C)$ by removing u and deleting every edge with both endpoints in X. Observe that the resulting graph G' is a bipartite graph with bipartition (X, C), as $G_u[C \cup \{u\}]$ is an independent set. It turns out that it is enough to check whether two vertices of X are in distance two in G' which can be encapsulated by the square of its adjacency matrix. Algorithm 4 shows in details our proposed approach.

Theorem 6.1. *Listing all avoidable vertices by using Algorithm 4 takes $O(n^{1+\omega})$ time.*

Proof. We apply Algorithm 4 on each vertex of G. Let us first discuss on the correctness of Algorithm 4. By Lemma 6.2, it is enough to show that $H_u[X]$ is a clique if and only if $M_3[X]$ has non-zero entries in its non-diagonal positions. Let G_1 and G_2 be the two constructed graphs in Algorithm 4. Observe that the

---

Algorithm 4: Testing if u is avoidable by using matrix multiplication

---
Input : A graph G and a vertex u
Output: True iff u is avoidable in G
1 Construct the contracted graph $G_u(X, C)$ of u
2 Let $G_1 = G_u(X, C) - u$
3 Construct the adjacency matrix M_1 of G_1
4 Let G_2 be the bipartite graph obtained from G_1 by removing every edge having both endpoints in X
5 Construct the adjacency matrix M_2 of G_2
6 Compute the square of M_2, i.e., $M_2^2 = M_2 \cdot M_2$
7 Construct the matrix $M_3 = M_1 + M_2^2$
8 **for** $x, y \in X$ **do**
9 $\quad\lfloor$ **if** *the entry $M_3[x, y]$ is zero* **then return** *false*

10 **return** *true*

---

square of G_2, denoted by G_2^2, is the graph obtained from the same vertex set of G_2 and two vertices u, v are adjacent in G_2^2 if the distance of u and v is at most two in G_2. Thus the matrix M_2^2 computed by Algorithm 4 corresponds to the adjacency matrix of G_2^2. Now it is enough to notice that two vertices x, y of X are adjacent in $H_u[X]$ if and only if $xy \in E(G_1) \cup E(G_2^2)$. In particular observe that if x and y have a common neighbor w in G_2 then w is a vertex of C since there is no edge between vertices of X in G_2 and $u \notin V(G_2)$. Therefore $M_3[x, y]$ has a non-zero entry if and only if x and y are adjacent in $H_u[X]$.

Regarding the running time, notice that the construction of G_u take linear time by Observation 6.1. All steps besides the computation of M_2^2 can be done in $O(n^2)$ time. The most time-consuming step is the matrix multiplication involved in computing M_2^2, which can be done in $O(n^\omega)$ time. Hence the total running time for recognizing all n vertices takes $O(n^{1+\omega})$ time. $\square$

7   Recognizing Avoidable Edges and Paths

Natural generalizations of avoidable vertices are avoidable edges and avoidable paths. Here we show how to efficiently recognize an avoidable edge and an avoidable path. Recall that the two vertices having degree one in an induced path P_k on $k \geq 2$ vertices are called *endpoints*. Moreover, the edge obtained after removing the endpoints from an induced path P_4 is called *middle edge*.

Definition 7.1 (Simplicial and avoidable edge). *An edge uv is called simplicial if there is no P_4 having uv as a middle edge. An edge uv is called avoidable if either uv is simplicial, or every P_4 with middle edge uv is contained in an induced cycle.*

Given two vertices x and y of G, we define the following sets of their neighbors: $B(x, y)$ contains the common neighbors of x and y; i.e., $B(x, y) = N_G(x) \cap N_G(y)$; A_x contains the private neighbors of x; i.e., $A_x = N_G(x) \setminus (B(x, y) \cup \{y\})$; A_y

contains the private neighbors of y; i.e., $A_y = N_G(y) \setminus (B(x,y) \cup \{x\})$. Under this terminology, observe that $A_x \cap A_y = \emptyset$ and $N_G(\{x,y\})$ is partitioned into the three sets $B(x,y), A_x, A_y$.

Observation 7.1. *An edge xy of G is simplicial if and only if $A_x = \emptyset$ or $A_y = \emptyset$ or every vertex of A_x is adjacent to every vertex of A_y.*

By Observation 7.1, the recognition of a simplicial edge can be achieved in $O(n+m)$ time: consider the bipartite subgraph $H(A_x, A_y)$ of $G[A_x \cup A_y]$ which is obtained by removing every edge having both endpoints in either A_x or A_y. Then it is enough to check whether $H(A_x, A_y)$ is a complete bipartite graph.

We show that the more general concept of an avoidable edge can be recognized in $O(nm)$ time. For doing so, we will take advantage of Algorithm 2 and the notion of protecting given in Definition 5.1.

Definition 7.2. *An edge xy is* protected *if there is an $(N_G[x] \cup N_G[y])$-excluded path between every vertex of $N_G(x)$ and every vertex of $N_G(y)$.*

We note that if an edge xy is protected then x protects y and y protects x in accordance to Definition 5.1. However, the reverse is not necessarily true.

Lemma 7.1. *Let xy be an edge of G. Then xy is an avoidable edge in G if and only if xy is a protected edge in $G - B(x,y)$.*

Based on Lemma 7.1 and Algorithm 2, we deduce the following running time for recognizing an avoidable edge. Notice that the stated running time is comparable to the $O(d(u)(n+m))$-time algorithm for recognizing an avoidable vertex u implied by Theorem 5.3.

Theorem 7.1. *Recognizing an avoidable edge can be done in $O(n \cdot m)$ time.*

Proof. Let xy be an edge of G. We first collect the vertices of $B(x,y)$ in $O(n)$ time. By Lemma 7.1 we need to check whether xy is protected in $H = G - B(x,y)$. If xy is simplicial edge then xy is avoidable and, by Observation 7.1, this can be tested in $O(n+m)$ time. Otherwise, both sets A_x, A_y are non-empty. Without loss of generality, assume that $|A_x| \leq |A_y|$. In order to check if xy is protected, we run $|A_x|$ times Algorithm 2: for every vertex $a \in A_x$, run Algorithm 2 on the graph $(H - ((A_x \setminus \{a\}) \cup \{x,y\})$ started at vertex a with a target set A_y. In particular, we test whether there is an A_y-excluded path between a and every vertex of A_y without considering the vertices of $(A_x \setminus \{a\}) \cup \{x,y\}$, that is on the graph $H - ((A_x \setminus \{a\}) \cup \{x,y\})$. If all vertices of A_x have an A_y-excluded path with all the vertices of A_y on each corresponding graph, then such paths do not contain any internal vertex from $A_x \cup A_y \cup \{y\}$. Since $N_H[x] = A_x \cup \{x,y\}$ and $N_H[y] = A_y \cup \{x,y\}$, we deduce that xy is a protected edge, and thus, xy is avoidable in G. Regarding the running time, observe that we make at most $n \geq |A_x|$ calls to Algorithm 2 on induced subgraphs of G. Therefore, by Lemma 5.4, the total running time is $O(nm)$. $\qquad\square$

Let us now show how to extend the recognition of an avoidable edge towards their common generalization of avoidable induced paths. The *internal path* of a non-edgeless induced path P is the path obtained from P without its endpoints and its vertex set is denoted by $in(P)$.

Definition 7.3 (Simplicial and avoidable path). *An induced path P_k on $k \geq 2$ vertices is called* simplicial *if there is no induced path on $k + 2$ vertices that contains P_k as an internal path. An induced path P_k on $k \geq 2$ vertices is called* avoidable *if either P_k is simplicial, or every induced path on $k + 2$ vertices that contains P_k as an internal path is contained in an induced cycle.*

Let P_k be an induced path on k vertices of a graph G with $k \geq 3$ having endpoints x and y. We denote by $I[P_k]$ the vertices of $N_G[in(P_k)] \setminus \{x, y\}$ and we denote by $G + xy$ the graph obtained from G by adding the edge xy.

Theorem 7.2 *Let P_k be an induced path on $k \geq 3$ vertices of a graph G having endpoints x and y. Then P_k is an avoidable path in G if and only if xy is an avoidable edge in $G + xy - I[P_k]$. Moreover, testing whether P_k is avoidable can be done in $O(n \cdot m)$ time.*

8   Concluding Remarks

The running times of our algorithms for listing all avoidable vertices are comparable to the corresponding ones for listing all simplicial vertices. The notion of protecting and the relative S-excluded paths seem to tackle further problems concerning avoidable structures. Our recognition algorithm for avoidable edges results in an algorithm for listing avoidable edges with running time $O(nm^2)$ which is comparable to the $O(m^2)$-algorithm for listing avoidable vertices. Regarding avoidable paths on k vertices, one needs to detect first with a naive algorithm a path P_k in $O(n^k)$ time and then test whether P_k being avoidable or not. As observed in [9], such a detection is nearly optimal, since we can hardly avoid the dependence of the exponent in $O(n^k)$. Therefore by Theorem 7.1 we get an $O(n^{k+1} \cdot m)$-algorithm for listing all avoidable paths on k vertices.

An interesting direction for further research along the avoidable paths is to reveal problems that can be solved efficiently by taking advantage the list of all avoidable paths in a graph. For instance, one could compute a minimum length of a sequence of shifts transforming an induced path P_k to an avoidable induced path. Gurvich et al. [13] proved that each induced path can be transformed to an avoidable one by a sequence of shifts, where two induced paths on k vertices are shifts of each other if their union is an induced path on $k + 1$ vertices. To compute efficiently a minimum length of shifts, one could construct a graph H that encodes all neighboring induced paths on k vertices of G. In particular, the nodes of H correspond to all induced paths on k vertices in G and two nodes in H are adjacent if and only if their union is an induced path on $k + 1$ vertices in G. Note that H contains $O(n^k)$ nodes and can be constructed in $n^{O(k)}$ time.

Having the list of avoidable paths on k vertices, we can mark the nodes of H that correspond to such avoidable paths. Now given an induced path P_k on k vertices in G we may ask the shortest path in H from the node that corresponds to P_k towards a marked node that corresponds to an avoidable path. Such a path always exists from the results of [13] and can be computed in time linear in the size of H. Therefore, for fixed k, our algorithm computes a minimum length of sequence of shifts in polynomial time answering an open question given in [13].

References

1. Aboulker, P., Charbit, P., Trotignon, N., Vuskovic, K.: Vertex elimination orderings for hereditary graph classes. Discret. Math. **338**(5), 825–834 (2015)
2. Alman, J., Williams, V.V.: A refined laser method and faster matrix multiplication. In: Proceedings of SODA 2021, pp. 522–539. SIAM (2021)
3. Beisegel, J., Chudnovsky, M., Gurvich, V., Milanic, M., Servatius, M.: Avoidable vertices and edges in graphs. In: Proceedings of WADS 2019. vol. 11646, pp. 126–139 (2019)
4. Berry, A.: A wide-range efficient algorithm for minimal triangulation. In: Proceedings of SODA 1999, pp. 860–861. ACM/SIAM (1999)
5. Berry, A., Blair, J.R.S., Bordat, J.P., Simonet, G.: Graph extremities defined by search algorithms. Algorithms **3**(2), 100–124 (2010)
6. Berry, A., Blair, J.R.S., Heggernes, P., Peyton, B.W.: Maximum cardinality search for computing minimal triangulations of graphs. Algorithmica **39**(4), 287–298 (2004)
7. Berry, A., Bordat, J.P.: Separability generalizes dirac's theorem. Discret. Appl. Math. **84**(1–3), 43–53 (1998)
8. Berry, A., Heggernes, P., Villanger, Y.: A vertex incremental approach for maintaining chordality. Discret. Math. **306**(3), 318–336 (2006)
9. Bonamy, M., Defrain, O., Hatzel, M., Thiebaut, J.: Avoidable paths in graphs. Electron. J. Comb. **27**(4), P4.46 (2020)
10. Bondy, J.A., Murty, U.S.R.: Graph Theory. Springer (2008)
11. Dirac, G.A.: On rigid circuit graphs. Abhandlungen aus dem Mathematischen Seminar der Universitat Hamburg **25**(1), 71–76 (1961)
12. Ducoffe, G.: The diameter of at-free graphs. J. Graph Theory **99**, 594–614 (2022)
13. Gurvich, V., Krnc, M., Milanic, M., Vyalyi, M.N.: Shifting paths to avoidable ones. J. Graph Theory **100**, 69–83 (2022)
14. Heggernes, P.: Minimal triangulations of graphs: a survey. Discret. Math. **306**(3), 297–317 (2006)
15. Impagliazzo, R., Paturi, R.: On the complexity of k-sat. J. Comput. Syst. Sci. **62**, 367–375 (2001)
16. Itai, A., Rodeh, M.: Finding a minimum circuit in a graph. SIAM J. Comput. **7**, 413–423 (1978)
17. Kloks, T., Kratsch, D., Müller, H.: Finding and counting small induced subgraphs efficiently. Inf. Process. Lett. **74**(3–4), 115–121 (2000)
18. Kratsch, D., Spinrad, J.P.: Between $O(nm)$ and $o(n^{alpha})$. SIAM J. Comput. **36**, 310–325 (2006)
19. McConnell, R.M., Spinrad, J.P.: Modular decomposition and transitive orientation. Discrete Math. **201**, 189–241 (1999)

20. Ohtsuki, T., Cheung, L.K., Fujisawa, T.: Minimal triangulation of a graph and optimal pivoting order in a sparse matrix. J. Math. Anal. Appl. **54**(3), 622–633 (1976)
21. Papadopoulos, C., Zisis, A.: Computing and listing avoidable vertices and paths. CoRR abs/2108.07160 (2021)
22. Roditty, L., Williams, V.V.: Fast approximation algorithms for the diameter and radius of sparse graphs. In: Proceedings of STOC 2013, pp. 515–524 (2013)
23. Rose, D.J., Tarjan, R.E., Lueker, G.S.: Algorithmic aspects of vertex elimination on graphs. SIAM J. Comput. **5**(2), 266–283 (1976)
24. Tedder, M., Corneil, D., Habib, M., Paul, C.: simpler linear-time modular decomposition via recursive factorizing permutations. In: Aceto, L., Damgård, I., Goldberg, L.A., Halldórsson, M.M., Ingólfsdóttir, A., Walukiewicz, I. (eds.) ICALP 2008. LNCS, vol. 5125, pp. 634–645. Springer, Heidelberg (2008). https://doi.org/10.1007/978-3-540-70575-8_52
25. Williams, R.: A new algorithm for optimal 2-constraint satisfaction and its implications. Theor. Comput. Sci. **348**, 357–365 (2005)

Klee's Measure Problem Made Oblivious

Thore Thießen[(✉)] and Jan Vahrenhold

Department of Computer Science, University of Münster, Münster, Germany
{t.thiessen,jan.vahrenhold}@uni-muenster.de

Abstract. We study Klee's measure problem — computing the volume of the union of n axis-parallel hyperrectangles in $\mathbb{R}^d$ — in the oblivious RAM (ORAM) setting. For this, we modify Chan's algorithm [12] to guarantee memory access patterns and control flow independent of the input; this makes the resulting algorithm applicable to privacy-preserving computation over outsourced data and (secure) multi-party computation.

For $d = 2$, we develop an oblivious version of Chan's algorithm that runs in expected $\mathcal{O}(n \log^{5/3} n)$ time for perfect security or $\mathcal{O}(n \log^{3/2} n)$ time for computational security, thus improving over optimal general transformations. For $d \geq 3$, we obtain an oblivious version with perfect security while maintaining the $\mathcal{O}(n^{d/2})$ runtime, i.e., without any overhead.

Generalizing our approach, we derive a technique to transform divide-and-conquer algorithms that rely on linear-scan processing into oblivious counterparts. As such, our results are of independent interest for geometric divide-and-conquer algorithms that maintain an order over the input. We apply our technique to two such algorithms and obtain efficient oblivious counterparts of algorithms for inversion counting and computing a closest pair in two dimensions.

Keywords: Klee's measure problem · Oblivious RAM · Data-oblivious algorithms · Data-oblivious divide-and-conquer algorithms

1 Introduction

First introduced by Klee [20] in 1977, *Klee's measure* problem is a well-known problem in computational geometry:

> Given a set B of n axis-parallel hyperrectangles (for short: *boxes*) in $\mathbb{R}^d$, compute $\|\bigcup(B)\|$. Here, $\|\cdot\|$ is the d-dimensional volume of a (measurable) subset of $\mathbb{R}^d$.

On the theoretical side, the problem is related to other geometric problems, e.g., the *depth* and *coverage* problems [12]. There has been a series of works improving the upper bounds for Klee's measure problem [11,12,16,23]. The currently best runtime of $\mathcal{O}(n \log n + n^{d/2})$ for any (fixed) number d of dimensions is obtained by Chan's algorithm [12].

A special case of Klee's measure problem, computing the so-called *hypervolume indicator*, is used in, e.g., evaluating multi-objective optimizations [19].

A. Castañeda and F. Rodríguez-Henríquez (Eds.): LATIN 2022, LNCS 13568, pp. 121–138, 2022.
https://doi.org/10.1007/978-3-031-20624-5_8

For the hypervolume indicator, the boxes are restricted to have a common lower bound in all dimensions [12,19]. For this and other special cases, faster algorithms than for the general problem are known [12].

Oblivious Algorithms. We design a *data-oblivious* (for short: *oblivious*) algorithm for Klee's measure problem. In the random access machine (RAM) context, the notion of (data-)obliviousness was introduced by Goldreich and Ostrovsky [17]. Informally, the requirement is that the *probe sequence*, i. e., the sequence of memory operations and memory access locations, must be independent of the input. This guarantees that no information about the input can be derived from observing the memory access patterns. Oblivious algorithms — in combination with encryption — can be used to perform privacy-preserving computation over outsourced data. Additionally, oblivious algorithms can be transformed into efficient protocols for multi-party computation [15,25].

We distinguish two notions of obliviousness: Let x and y be two inputs of equal length n. Then the probe sequences for x and y must be identically distributed (*perfect security*) or indistinguishable by any polynomial-time algorithm except for negligible probability in n (*computational security*).[1] In line with standard assumptions for oblivious algorithms [6,15], we include the control flow in the probe sequence. Access to a constant number of *private memory* cells (registers) as well as the memory contents are not considered to be part of the probe sequence (since we may assume that the memory is encrypted [17]).

As the underlying model of computation, we assume the *word RAM* model (in line with standard assumptions for oblivious algorithms). However, we note that our results with perfect security also hold in the *real RAM* model usually assumed in computational geometry. We use oblivious sorting as a building block. There are oblivious sorting algorithms with $\mathcal{O}(n \log n)$ runtime and perfect security, e. g., due to asymptotically optimal sorting networks [1].

In recent years, there has been a lot of progress regarding oblivious algorithms. It is known that — in general — transforming a RAM program into an oblivious program incurs an $\Omega(\log N)$ (multiplicative) overhead [17,21] (where N is the space used by the program). On the constructive side, there is an ORAM construction matching this lower bound [4]. This provides — for programs with at least linear runtime — a black-box transformation to achieve obliviousness with only a logarithmic overhead in runtime. Even with such general transformations, there are still some limitations: Optimal general ORAM transformations currently entail high constant runtime factors that make them unsuitable for practical application [4]. Additionally, all known optimal transformations only satisfy the weaker requirement of computational security [7]. The state-of-the-art general transformation with perfect security is due to Chan et al. [9] and achieves a runtime overhead of $\mathcal{O}(\log^3 N / \log \log N)$.

To address these issues and overcome the $\Omega(\log n)$ lower bound associated with the general transformation, oblivious algorithms for specific problems have

---

[1] See the definitions by Asharov et al. [3, Section 3] for a more formal introduction of oblivious security applicable to oblivious algorithms.

been proposed: examples include sorting algorithms [2,22], graph algorithms [6], and some algorithms for fundamental geometric problems [15]. To the best of our knowledge, oblivious algorithms for Klee's measure problem and its variants have not been considered in the literature so far.

Algorithm 1. Chan's RAM algorithm [12] to compute Klee's measure for $d \geq 2$. The coordinates of the boxes B are sorted in a pre-processing step.

1: **function** MEASURE(B, Γ)
2:     **if** $|B| \leq$ some suitably chosen constant **then**
3:         **return** $\|\Gamma \setminus \bigcup(B)\|$ (computed using a brute-force approach) $\triangleright \mathcal{O}(d)$
4:     SIMPLIFY(B, Γ) $\triangleright \mathcal{O}(d \cdot n)$
5:     $\langle \Gamma_{\mathrm{L}}, \Gamma_{\mathrm{R}} \rangle \leftarrow$ CUT(B, Γ) $\triangleright \mathcal{O}(d \cdot n)$
6:     $B_{\mathrm{L}} \leftarrow$ all $b \in B$ intersecting Γ_{L}; $B_{\mathrm{R}} \leftarrow$ all $b \in B$ intersecting Γ_{R} $\triangleright \mathcal{O}(d \cdot n)$
7:     **return** MEASURE($B_{\mathrm{L}}, \Gamma_{\mathrm{L}}$) + MEASURE($B_{\mathrm{R}}, \Gamma_{\mathrm{R}}$)

---

2 Warm-Up: Shaving Off a $\log \log n$ Factor

As a warm-up, we first show how to improve over general and naive transformations of Chan's algorithm for $d = 2$ by a $\mathcal{O}(\log \log n)$ factor in runtime. We also introduce the representation of the boxes used throughout the paper. In Sect. 3, we build on this approach when showing how to improve the runtime for $d = 2$ to our main result. For completeness, we first describe Chan's algorithm as presented in the original paper [12]. We will also briefly sketch how to obtain an oblivious modification for $d \geq 3$.

2.1 Original Algorithm

The input for Klee's measure problem is a set B of n axis-parallel hyperrectangles. Chan's algorithm, shown in Algorithm 1, first computes the measure $\overline{m}$ of the complement of $\bigcup(B)$ relative to some box Γ (*domain*) containing all boxes B. For the final result, $\overline{m}$ is subtracted from the measure of Γ. The algorithm to compute $\overline{m}$ is — at its core — a simple divide-and-conquer algorithm with two helper functions: SIMPLIFY and CUT. In each call, a set of boxes intersecting the current domain Γ is processed. Throughout the algorithm, the coordinates of the boxes B are maintained in sorted order for each dimension.

The main idea of SIMPLIFY is to remove the area covered by the boxes $D \subseteq B$ that cover the domain Γ in all but one dimension: For a given dimension $i \in \{1, \dots, d\}$, let D_i denote the boxes covering Γ in all but dimension i. The simplification can be performed by first computing the union of the boxes D_i and then adjusting the x_i-coordinates of all other boxes in B so that the volume of all connected components of $\bigcup(D_i)$ is reduced to 0. Since the coordinates are sorted in each dimension, this simplification can be realized with a linear scan in each dimension. To maintain the measure of the complement, the extent of Γ in dimension i (*height*) is reduced accordingly, i.e., by the height of $\bigcup(D_i)$.

Intuitively, the simplification reduces the problem complexity by reducing the number of sub-domains a box intersects with. Consider the problem for $d = 2$: When cutting the domain Γ for the recursion, the simplification guarantees that — on every level of the recursion — each box only has to be considered in a constant number of sub-domains; this is since the area covered by the box is removed from the covered sub-domains in-between.

For the CUT-step, a weighted median is computed that splits Γ into two sub-domains Γ_L and Γ_R. Like in k-d-tree constructions, the algorithm cycles through the dimensions in a round-robin manner. In the analysis, Chan shows that cutting reduces the weight of each sub-domain by a factor of $2^{2/d}$. Since the weight is related to the number of boxes by a constant factor (depending on d), this also provides an upper bound for the number of boxes in each sub-domain. The runtime for Algorithm 1 is thus bounded by the recurrence $T(n) = 2 \cdot T(n \,/\, 2^{2/d}) + \mathcal{O}(n)$. With the time $\mathcal{O}(n \log n)$ for pre-sorting the coordinates this implies a runtime of $\mathcal{O}(n^{d/2})$ for $d \geq 3$ and a runtime of $\mathcal{O}(n \log n)$ for $d = 2$ overall.

2.2 High Dimensions

Before presenting an oblivious modification of Chan's algorithm for $d = 2$, we will briefly consider higher dimensions $d \geq 3$. Here, we note that Chan's algorithm can easily be transformed into an oblivious algorithm while maintaining the runtime complexity. This leaves the planar case $d = 2$ as the "hard case".

To see why the case $d \geq 3$ is actually easier, consider again the recurrence bounding the runtime: As we will discuss in Sect. 2.5, the main difficulty for the oblivious algorithm is to maintain the sorted coordinates for an efficient simplification. If we solve this naively by sorting, we have costs of $\mathcal{O}(n \log n)$ instead of $\mathcal{O}(n)$ in each recursive call. This results in a runtime bounded by the recurrence $T(n) = 2 \cdot T(n \,/\, 2^{2/d}) + \mathcal{O}(n \log n)$. A naive transformations thus leads to a runtime of $T(n) \in \mathcal{O}(n \log^2 n)$ for $d = 2$, yet for $d \geq 3$ this immediately solves to $T(n) \in \mathcal{O}(n^{d/2})$, maintaining the runtime of the original algorithm.

There still remain challenges for an oblivious implementation of Chan's algorithm for $d \geq 3$. We will briefly sketch how to address them: For the representation, it suffices to store the individual boxes with their bounds in each dimension; the boxes can be duplicated and sorted as needed. An oblivious implementation also needs to ensure that the input size for the recursive calls does only depend on the problem size n, otherwise the runtimes for the recursive calls might leak information about the input. For this, we note that the recurrence given by Chan already bounds the number of boxes in the recursive calls. By carefully padding the input with additional (empty) boxes we can always recursively call with the worst-case size. Splitting the n boxes for the recursive calls can be done by duplication and oblivious routing in $\mathcal{O}(n \log n)$ time [18].

2.3 Oblivious Box Representation

For the planar case $d = 2$, Chan's algorithm can be simplified to only cutting the domain in the x-dimension [12]. Consequently, simplification is only performed

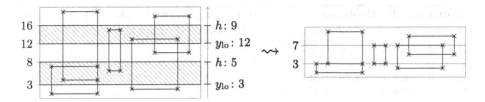

Fig. 1. Simplification of a sub-domain Γ with 5 boxes to remove the area covered by the hatched slabs. The result of the simplification is shown on the right.

for the y-dimension. We say a box b *covers* the domain Γ iff b spans over Γ in the x-dimension; in this case we call b a *slab*. For $d = 2$ the cutting can also be performed without weights, balancing the number of vertices in each sub-domain. Note that since the initial domain contains all vertices and cuts are only performed in the x-dimension, we only need to check the x-dimension to see if a vertex is contained in a given sub-domain Γ.

We represent each box by its four vertices. The sequence of vertices is not maintained in a fixed order; instead, we explicitly sort the sequence when needed. Since we need information about the horizontal extent of the boxes when processing the vertices, we keep the full x-interval in each vertex. We also store the relative location in both dimensions, i.e., whether the vertex is a lower (lo) or upper (hi) vertex. So overall, each vertex is represented by a tuple

$$b \in \underbrace{\mathbb{R}^2 \times \{\mathrm{lo}, \mathrm{hi}\}}_{x\text{-interval}} \times \underbrace{\mathbb{R} \times \{\mathrm{lo}, \mathrm{hi}\}}_{y\text{-coordinate}} .$$

For a vertex b, we write b_x, b_y to denote the coordinates and $b_{[x]}$ to denote the x-interval. We write box(b) to refer to the box the vertex b (partially) represents.

Since our algorithms for the planar case are based on processing the vertices in y-order, we simplify terminology by referring to the lower y-vertices as *start*- and to the upper y-vertices as *endpoints*. Similarly, we will refer to the lower x-vertices as *left* and to the upper x-vertices as *right* vertices. Storing the x-interval in each vertex allows us to represent the box by either both left or both right vertices. We will leverage this when cutting the box at sub-domain boundaries.

2.4   Simplification in One Sub-domain

For an oblivious modification of Chan's algorithm we need a way to *obliviously* simplify a given domain Γ, i.e., to remove the area covered by the slabs D. To do so, we reverse the order of cutting and simplification: We remove the area covered by slabs in a sub-domain (while maintaining the measure of the complement) before recursing. This exchange of the two algorithmic steps means that in each recursive call, the number of boxes (and vertices) remains unchanged — an important requirement for not leaking any information about the input.

For the simplification, Chan's algorithm first computes the connected components of $\bigcup(D)$ and then adjusts the coordinates of the remaining boxes in a

Algorithm 2. Algorithm to remove the area covered by slabs D in Γ. The return value is the combined height of all connected components of $\bigcup(D)$.

1: **procedure** OBLSIMPLIFY(B, Γ)
2:     $y_{\mathrm{lo}} \leftarrow$ **undef**; $c \leftarrow 0$; $h \leftarrow 0$ ▷ y-anchor; overlap; height
3:     **for** each vertex $b \in B$ **do**   ▷ in y-order
4:         **if** box(b) covers Γ **then** UPDATE($\langle y_{\mathrm{lo}}, c, h \rangle, b$)   ▷ update the state
5:         **if** b is in Γ **then** ADJUST($\langle y_{\mathrm{lo}}, c, h \rangle, b$)   ▷ adjust the vertex
6:     **return** h
7: **procedure** UPDATE($\langle y_{\mathrm{lo}}, c, h \rangle, b$)
8:     **if** b is a startpoint $\wedge$ $c = 0$ **then** $y_{\mathrm{lo}} \leftarrow b_y$ ▷ start of a component
9:     **if** b is a startpoint **then** $c \leftarrow c + 1$ **else** $c \leftarrow c - 1$
10:     **if** b is an endpoint $\wedge$ $c = 0$ **then** $h \leftarrow h + (b_y - y_{\mathrm{lo}})$ ▷ end of the component
11: **procedure** ADJUST($\langle y_{\mathrm{lo}}, c, h \rangle, b$)
12:     $\Delta \leftarrow h$
13:     **if** $c > 0$ **then** $\Delta \leftarrow \Delta + (b_y - y_{\mathrm{lo}})$ ▷ in the component
14:     $b_y \leftarrow b_y - \Delta$

---

synchronized traversal. Unfortunately, such traversal is infeasible in the oblivious context since doing so might leak information about the coordinates' distribution. To address this, we redesign the subroutine to perform a single linear scan: The vertices B are processed in y-order (start- before endpoints); each vertex $b \in B$ contained in the sub-domain Γ ($b_x \in \Gamma_{[x]}$) is adjusted to remove the area covered by the slabs D. Specifically, b_y is reduced by the height of $\bigcup(D)$ below b_y.

The resulting Algorithm 2 is the key ingredient to realize the subroutine SIMPLIFY; we can maintain the measure of the complement by subtracting the obtained value h from the height of Γ. An example of a simplification is shown in Fig. 1. It is essential that the algorithm performs exactly one linear scan over the entire input — processing (and potentially modifying) each vertex with a constant number of operations — and uses only a constant amount of space for the state $\langle y_{\mathrm{lo}}, c, h \rangle$. This immediately implies that this algorithm can be implemented as a linear time, oblivious program.

Regarding the correctness of OBLSIMPLIFY for a set B of vertices and a domain Γ, we let $B_\Gamma := \{ \text{box}(b) \mid b \in B \wedge b_x \in \Gamma_{[x]} \}$ be the boxes represented by vertices in Γ and let $D_\Gamma := \{ \text{box}(b) \mid b \in B \wedge b_{[x]} \supseteq \Gamma_{[x]} \}$ be the boxes covering Γ. Let B'_Γ be the boxes B_Γ after running the algorithm. By considering the connected components of D_Γ when projected on the y-axis we can easily prove:

Lemma 1. *Given a sequence B of vertices sorted by their y-coordinate and a domain Γ, OBLSIMPLIFY adjusts the y-coordinates and returns a value h so that*

$$\left\| \bigcup(B'_\Gamma) \right\| = \left\| \bigcup(B_\Gamma) \setminus \bigcup(D_\Gamma) \right\| \quad \text{and} \quad \left\| \bigcup(D_\Gamma) \right\| = h \cdot \text{width}(\Gamma)$$

where all measures are restricted to the domain Γ.

There are two properties of this simple algorithm that are worth pointing out as they are required for the correctness: Firstly note that, for a slab $b \in D$, it does not matter whether we process the left, the right, or both pairs of vertices. Processing both pairs temporarily increases the overlap, but — since both start- and both endpoints have the same y-coordinate — effects neither the anchor point y_{lo} nor the height h. This property ensures the correctness of the complete algorithm since we process both the left and the right vertices of a box before eventually separating them by cutting at a sub-domain boundary.

Secondly, it may be that we adjust a vertex of a slab $b \in D$ in ADJUST, i. e., that we adjust a box that previously caused an update in UPDATE. In this case the height of b is reduced to 0. This implies that changes to y_{lo} and c due to the vertices of b in any subsequent invocation of OBLSIMPLIFY have no effect; for the analysis we can thus assume that the vertices b with $b_x \in \Gamma_{[x]} \subseteq b_{[x]}$ are removed. We stress that no vertices are ever actually removed.

2.5   Oblivious Algorithm

While OBLSIMPLIFY realizes an oblivious simplification, it still requires the vertices to be sorted by their y-coordinate. This is necessary in each recursive call of the divide-and-conquer algorithm. Chan's RAM algorithm pre-sorts the coordinates in each dimension and maintains these orders, e. g., with a doubly-linked list over the boxes for each dimension. This is possible since both SIMPLIFY and CUT do not effect the relative order. For $d = 2$, another option would be to stably partition the vertices for the recursive calls.

Neither can be done efficiently in the ORAM model: The transformation lower bound mentioned in the introduction implies a $\Omega(\log n)$ lower bound on random access to n elements, barring us from efficiently maintaining links. This makes a direct implementation of Chan's approach inefficient. Unfortunately, there also is an $\Omega(n \log n)$ lower bound on stable partitioning n elements (assuming indivisibility) [22]; this implies that — unlike in the RAM model — stable partitioning is no faster than sorting.

As we will show in Sect. 4, this challenge is not unique to this algorithm and in fact arises for many geometric divide-and-conquer algorithms. Our approach — instead of maintaining the order — is to re-sort the vertices by y-coordinate in each recursive call. To make up for at least some part of this runtime overhead, we increase the number of recursive calls from 2 to m (a value to be determined below). Algorithm 3 shows our general outline for the combined cutting and simplification with $m \geq 2$ recursive calls. Remember that in contrast to Chan's algorithm, we simplify immediately after cutting.

To determine the boundaries of the m sub-domains $\Gamma_1, \ldots, \Gamma_m$, we first sort the vertices by their x-coordinate. We ensure that both left vertices of a box always end up in the same sub-domain, same for the right vertices. To avoid leaking information about the input via the runtime of the recursive calls, the number of vertices in each sub-domain must only depend on n: Since there may be several vertices with identical x- and y-coordinates, we assign an additional unique identifier to each box. By establishing a total order this allows us to ensure an even distribution of the $n := |B|$ vertices to the m sub-domains, independent

Algorithm 3. Procedure for cutting the given vertices B into m sub-domains and simplifying each sub-domain.

```
1: procedure CUTANDSIMPLIFY(B, Γ)
2:     sort B by x-coordinate                                              ▷ O(n log n)
3:     det. sub-domains Γ₁, ..., Γₘ (boundaries: B[⌈(i·|B|)/m⌉]]ₓ for 1 ≤ i < m)   ▷ O(m)
4:     sort B by y-coordinate                                              ▷ O(n log n)
5:     for each sub-domain Γᵢ (simultaneously) do                          ▷ O(n · m)
6:         hᵢ ← OBLSIMPLIFY(B, Γᵢ)
7:         height(Γᵢ) ← height(Γᵢ) − hᵢ
8:     sort B by x-coordinate                                              ▷ O(n log n)
9:     return ⟨Γ₁, ..., Γₘ⟩
```

of the input configuration. The value for m only depends on n and we only access the sub-domain representations in a non–input-dependent manner; thus we can store the sub-domain representations in $\Theta(m)$ successive memory cells and access them directly without violating the oblivious security requirement.

For the planar case $d = 2$ we consider, it suffices to partition the vertices, there is no need to duplicate or explicitly remove vertices. This follows from the properties of Algorithm 2: The initial domain Γ is recursively divided into sub-domains. Any box b can — on any level of the recursion — partially cover no more than two sub-domains, one for each of its vertical sides. The sub-domains in-between are fully covered, and through the simplification the area covered by b is removed from these sub-domains. If a left or right pair of vertices coincides with the bounds of a sub-domain (and thus no longer forms a left or right box), that box is implicitly removed by reducing its height to 0.

After determining the sub-domains, we can sort the vertices by their y-coordinates to apply OBLSIMPLIFY for simplification. Afterwards, we reduce the height of Γ_i by the height h_i removed in that sub-domain. Sequentially processing the sub-domains might reduce the height of a slab covering the next sub-domain, thus resulting in an incorrect simplification. To avoid this, we can perform all simplifications simultaneously: We process all vertices $b \in B$, keeping a separate state for each of the m sub-domains. Again, we can store the states in $\Theta(m)$ successive memory cells. In each iteration, we process the current vertex b for each sub-domain and keep track of the single adjustment to the vertex (which we can apply before processing the next vertex). The final sorting step partitions the vertices according to their sub-domain for the recursive calls.

To balance the costs of sorting and simplifying, we set $m := \max(\lceil \log_2 n \rceil, 2)$. The recursion tree then has $\mathcal{O}(\log_m n) = \mathcal{O}(\log n / \log \log n)$ height; this yields $\mathcal{O}(n \log^2 n / \log \log n)$ total runtime. Not only does this improve over an optimal general transformation by a $\mathcal{O}(\log \log n)$ factor, our construction also guarantees (deterministic) identical memory access patterns, implying perfect security.

3   Improved Processing Using Oblivious Data Structures

When processing the m sub-domains individually, the algorithm presented in the previous section spends considerable time processing boxes not intersecting the current sub-domain and repeatedly adjusting vertices in different sub-domains to remove the area covered by the same slab. Here, we show how to avoid both of these, resulting in further improvement over the general transformation.

For our improved algorithm, we apply Algorithm 2 to m slabs simultaneously in a hierarchical manner. As a prerequisite, we first note on how to construct the necessary data structures with perfect and computational security. We then describe a novel, tree-based data structure; using this data structure we construct the improved algorithm and state its runtime and security properties.

3.1 Oblivious Data Structures

The conceptual idea of our oblivious data structure is to model the recursion tree of a divide-and-conquer algorithm up to a certain height. For this, we first sketch how to construct oblivious static binary trees from known oblivious primitives. We then describe how to use the static binary trees to model the divide-and-conquer recursion tree; for this, we will explicitly state some necessary conditions.

Oblivious Static Binary Tree. As observed by Wang et al. [27], it is possible to build efficient oblivious data structures from *position-based* ORAM constructions. Position-based ORAMs are often used as a building block to construct a "full" ORAM [7,24,26] since they are efficient in practice [24,26] and suitable to build ORAM constructions with perfect security [7,9].

To access the elements efficiently, position-based ORAMs assign a (temporary) label to each element. Accessing an element reveals its label, so to hide the access pattern it is necessary to assign a new label every time an element is accessed. The labels for all elements are maintained in a *position map* which — in general — does not fit in the private memory. *Recursive* position-based ORAM schemes address this by recursively storing the position map in a (smaller) position-based ORAM until it fits in private memory [7,9,24,26]. Depending on the construction, this recursion usually leads to an additional logarithmic runtime overhead [7,24,26]. For data structures where the access pattern graph exhibits some degree of predictability the recursion (with the associated overhead) is not necessary [27]. An example are rooted trees with a bounded degree; here, the labels of the $\mathcal{O}(1)$ direct successors can be encoded in the current element.

We build on this insight to construct oblivious static binary trees with perfect security. Wang et al. use *Path ORAM* [24] to represent oblivious binary trees [27]: Their representation allows traversing a rooted path in a (static) binary tree with N nodes and height h in $\mathcal{O}(h \log N)$ time. However, due to the constraints of the employed ORAM construction, they only achieve the (weaker) statistical security and require a super-constant number of private memory cells [24].[2] To

---

[2] Most statistically-secure ORAM constructions incur a $\mathcal{O}(\log n)$ time overhead to achieve a security failure probability negligible in n [9]. This excludes many ORAM construction with good performance in practice [24,26] for our range of parameters.

achieve perfect security, we apply their technique to a position-based ORAM construction with perfect security, e. g., the construction of Chan et al. [7].[3]

In conclusion, by leveraging the technique of Wang et al. [27], we can obtain perfectly-secure oblivious static binary tree data structures. The runtime properties follow from the ORAM construction of Chan et al. [7].

Lemma 2. *For a static binary tree $\mathcal{T}$ with N nodes and height h, both public parameters, we can construct an oblivious representation with perfect security. Then $\mathcal{T}$ requires $\mathcal{O}(N)$ space, $\mathcal{O}(1)$ private memory cells, and can be constructed in $\mathcal{O}(N \log N)$ time. A rooted path in $\mathcal{T}$ can be traversed in expected (over the random choices) and amortized (over the path traversals) $\mathcal{O}(h \log^2 N)$ time.*

We will refer to $\mathcal{T}$ as an *oblivious binary tree*. Note that for complete binary trees $h = \lceil \log_2(N + 1) \rceil$, thus N is effectively the only public parameter.

By relaxing the security requirement to computational security (and assuming the existence of a family of pseudorandom functions with negligible security failure probability), the query complexity stated in Lemma 2 can be reduced by a $\mathcal{O}(\log N)$ factor. For this, the tree nodes are stored as elements in an asymptotically optimal ORAM construction, e. g., *OptORAMa* [4]. This results in an amortized $\mathcal{O}(h \log N)$ time for traversing a rooted path. As will become clear later, we have (for a constant $c > 1$) some small number $N \in \Theta(2^{\log^{1/c} n})$ of nodes and a large number $T \in \mathcal{O}(n^{\mathcal{O}(1)})$ of accesses. For these parameters, we explicitly consider the parameter λ guarding the security failure probability for OptORAMa: Choosing $\lambda \in \Theta(n)$, we can apply a theorem due to Asharov et al. [4, Theorem 7.2] and simultaneously achieve an (amortized) $\mathcal{O}(h \log N)$ runtime and a negligible security failure probability.

Oblivious Processing Tree. We use the oblivious binary tree as a tool to model the recursion tree of what we will refer to as a *scan-based* divide-and-conquer algorithm: an algorithm that — in each invocation — performs a linear processing scan over the input using $\mathcal{O}(1)$ additional space and then stably partitions the input in preparation for some (constant) number a of recursive calls. One example for such an algorithm is the closest-pair algorithm of Bentley and Shamos [5] (see Sect. 4). Overall, these restrictions imply that the algorithm is order-preserving, i. e., the processing order remains the same in all recursive calls.

The challenge when obliviously implementing a scan-based divide-and-conquer algorithm is the $\Omega(n \log n)$ lower bound for stably partitioning n elements [22]. Thus, although conceptually simple, a naive oblivious computation of such a divide-and-conquer algorithm up to some depth h would incur a runtime of $\mathcal{O}(h \cdot n \log n)$ for a given input of size n. To avoid this, we explicitly construct an oblivious tree $\mathcal{T}$ where the nodes correspond to the recursive calls

---

[3] The state-of-the-art perfectly-secure ORAM construction [8] further improves the runtime by a $\mathcal{O}(\log \log n)$ factor; this improvement is due to a reduction in the recursion depth and therefore does not benefit our application.

of the algorithm, i.e., the nodes of the recursion tree. Each node $v \in T$ stores the state of the processing step (state(v)) in the corresponding recursive call as well as routing information, i.e., information on how the elements are partitioned for the recursive calls. We then individually trace each element e through the tree, guided by the routing information. For each node v on the path we update state(v) by simulating the processing step with e as the current element.

This approach correctly simulates the recursion since we require the processing step to be linear and sequence of elements processed in each node corresponds to the sequence of elements processed in the corresponding recursive call. Since we traverse paths of length $h \in \mathcal{O}(\log m)$ in a tree with $N \in \mathcal{O}(m)$ nodes, the oblivious binary trees described above yield:

Lemma 3. *Using an oblivious binary tree we can obliviously simulate a scan-based divide-and-conquer algorithm up to depth h, resulting in $m := a^h$ leaves. The processing steps can be applied to $n \in \Omega(m) \cap \mathcal{O}(2^{\log^c m})$ elements, for $c > 1$, in expected $\mathcal{O}(n \log^3 m)$ time for perfect security or $\mathcal{O}(n \log^2 m)$ time for computational security, $\mathcal{O}(m)$ additional space, and $\mathcal{O}(1)$ private memory cells.*

3.2 Simplification in m Sub-domains

We now show how to simplify the m given sub-domains by processing all vertices and sub-domains simultaneously in a single scan. Rather than concentrating on the geometric details, we will derive this from Algorithm 2 in a general way by applying the above processing technique. We use an oblivious binary tree with m leaves as described above: This recreates the recursion structure of the original algorithm (except we simplify before recursing). As in Sect. 2 the sub-domains $\Gamma_1, \ldots, \Gamma_m$ are determined in advance by sorting the vertices.

Each of the m sub-domains Γ_i corresponds to the i-th leftmost leaf ℓ_i in the oblivious binary tree T. Then $\Gamma_{\ell_i} = \Gamma_i$ is the domain of the leaf ℓ_i; the domain Γ_v of an inner node v is exactly the union of sub-domains below v. To efficiently traverse T, we annotate each inner node v with their left and right sub-domains $\Gamma_L = \Gamma_{\text{left}(v)}$ and $\Gamma_R = \Gamma_{\text{right}(v)}$. This allows us to easily identify whether a vertex is contained in or a box covers the left or right sub-domain.

The algorithm is shown in Algorithm 4; see Fig. 2 for an example. To simulate the recursion of the original algorithm, we trace each vertex b through the explicit recursion tree T, adjusting b to remove the area covered by slabs. To simplify the left and right sub-domains before recursing, we keep two separate states per inner node v of the binary recursion tree: s_L for simplification in the left and s_R for simplification in the right sub-domain of v. After processing all vertices it remains to determine the combined height h_i of the areas removed in each sub-domain Γ_i. For this, we can traverse the paths to each leaf ℓ_i and sum the heights h in the respective states (in s_L for $\Gamma_{\ell_i} \subseteq \Gamma_L$ and in s_R for $\Gamma_{\ell_i} \subseteq \Gamma_R$).

We now state the correctness of OBLCOMBINEDSIMPLIFY. For each $i \in \{1, \ldots, m\}$, let $B_i := \{\text{box}(b) \mid b \in B \wedge b_x \in (\Gamma_i)_{[x]}\}$ be the set of boxes represented by the vertices in Γ_i and let $D_i := \{\text{box}(b) \mid b \in B \wedge b_{[x]} \supseteq (\Gamma_i)_{[x]}\}$ be the slabs over Γ_i. Let each B_i' be the boxes B_i after running the algorithm:

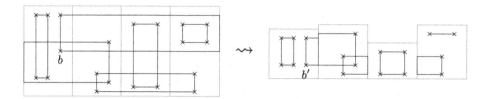

Fig. 2. Simplification of $\Gamma_1, \Gamma_2, \Gamma_3, \Gamma_4$ with 20 vertices. The boundaries are drawn in gray and the vertices are indicated by the marks. Consider processing the vertex b: In the root, b updates the state s_R for the domain $\Gamma_3 \cup \Gamma_4$ and is adjusted according to the initial state s_L. Then b is processed in the left successor where b updates the state s_R for Γ_2 and is adjusted according to s_L for Γ_1; this results in the processed b'.

Algorithm 4. Algorithm to simplify in all sub-domains Γ_i simultaneously. The procedures UPDATE and ADJUST correspond to the sub-routines in Algorithm 2.

1: **procedure** OBLCOMBINEDSIMPLIFY($B, \Gamma_1, \ldots, \Gamma_m$)
2:    $\mathcal{T} \leftarrow$ (complete) oblivious binary tree with m leaves   ▷ recursion tree
3:    **for** each vertex $b \in B$ **do**   ▷ in y-order
4:        $v \leftarrow \text{root}(\mathcal{T})$
5:        **do**
6:            $\Gamma_L \leftarrow$ domain for left(v); $\Gamma_R \leftarrow$ domain for right(v)
7:            $\langle s_L, s_R \rangle \leftarrow \text{state}(v)$ ▷ read the simplification states
8:            **if** box(b) covers Γ_L **then** UPDATE(s_L, b)   ▷ update the states for
9:            **if** box(b) covers Γ_R **then** UPDATE(s_R, b)   covered sub-domains
10:           **if** b is in Γ_L **then** ADJUST(b, s_L); $v \leftarrow$ left(v)   ▷ adjust according to
11:           **if** b is in Γ_R **then** ADJUST(b, s_R); $v \leftarrow$ right(v)   the respective state
12:           state(v) $\leftarrow \langle s_L, s_R \rangle$ ▷ write back the simplification states
13:        **while** v is not a leaf
14:    **return** heights $\langle h_1, \ldots, h_m \rangle$ of the boxes removed from each sub-domain

---

Lemma 4. *Given a sequence B of vertices sorted by their y-coordinate and m disjoint sub-domains $\Gamma_1, \ldots, \Gamma_m$, OBLCOMBINEDSIMPLIFY adjusts the y-coordinates and returns values $h_1, \ldots, h_m$ so that*

$$\left\| \bigcup(B_i') \right\| = \left\| \bigcup(B_i) \setminus \bigcup(D_i) \right\| \quad \text{and} \quad \left\| \bigcup(D_i) \right\| = h_i \cdot \text{width}(\Gamma_i)$$

for each $i \leq m$ where all measures are restricted to the respective domain Γ_i.

Proof (sketch). Let $\widehat{B}$ be a sorted sequence of vertices and $\widehat{B}'$ those vertices adjusted to remove the area covered by some slabs $\widehat{D}$. For the sub-procedures UPDATE and ADJUST we first show that orderly processing UPDATE(s, b) for the vertices $b \in \widehat{B}'$ maintains the height of $\bigcup(\{\text{box}(b) \mid b \in \widehat{B}\}) \setminus \bigcup(\widehat{D})$ in the state s.

We then proceed by induction over the levels of $\mathcal{T}$, starting with the root: We maintain the invariant that, for any node $v \in \mathcal{T}$ and any vertex $b \in \Gamma_v$, $b_y^v = b_y - h_{\leq b_y}(\bigcup(\bigcap_{\Gamma_i \subseteq \Gamma_v} D_i))$ (*). Here, b_y^v is defined to be the y-coordinate of b before the iteration of the `do-while` loop for v and $h_{\leq y}(S)$ is the height of $S_{\leq y} := S \cap \mathbb{R} \times (-\infty, y]$, i.e., the one-dimensional measure of $S_{\leq y}$ when projected onto the y-axis. This invariant together with the above properties of UPDATE and ADJUST implies that, for any $v \in \mathcal{T}$ with direct successor w, UPDATE$(s, \cdot)$ maintains the height of $\bigcup(D_w) \setminus \bigcup(D_v)$ in the state s for the sub-domain Γ_w $(\star)$.

The first equality of Lemma 4 then follows from $(*)$ for the respective leaf $v = \ell_i$ by considering the bounds of the connected components of $\bigcup(B_i')$. For the second equality, consider the state s corresponding to each non-root node w (with parent v) after processing the last vertex b: Clearly $s.c = 0$ and $s.h = \text{height}(\bigcup(D_w)) - \text{height}(\bigcup(D_v))$ according to $(\star)$. By summing the heights $s.h$ on the path to each leaf ℓ_i we obtain $h_i = \text{height}(\bigcup(D_i))$.   □

With Lemma 3 and since the inner loop in Lines 6 to 12 only accesses a constant number memory cells in the current node, the following properties hold:

Lemma 5.   OBLCOMBINEDSIMPLIFY *obliviously simplifies* $m \in \Omega(2^{\log^{1/c} n}) \cap \mathcal{O}(n)$ *sub-domains, for* $c > 1$, *with a total of* n *vertices in expected* $\mathcal{O}(n \log^3 m)$ *time for perfect security or* $\mathcal{O}(n \log^2 m)$ *time for computational security,* $\mathcal{O}(m)$ *space, and* $\mathcal{O}(1)$ *private memory cells.*

3.3   Putting Everything Together

To obtain a faster algorithm for Klee's measure problem, we replace the one-slab-at-a-time simplification (Algorithm 2) with the multi-slab simplification (Algorithm 4) in Algorithm 3. As before, the returned values h_i can be used to update the heights of the sub-domains Γ_i.

Since adjusting the y-coordinates for m sub-domains can be done in expected $\mathcal{O}(n \log^c m)$ time (with $c = 2$ for computational security and $c = 3$ for perfect security), we can balance the cost of sorting and updating the coordinates by choosing $m := \max(2^{\lceil \log_2^{1/c} n \rceil}, 2)$. This leads to a recursion tree height of $\mathcal{O}(\log_m n) = \mathcal{O}(\log^{1-1/c} n)$ for the complete algorithm. With the time required for sorting in each recursive call this yields a $\mathcal{O}(n \log^{2-1/c} n)$ runtime overall.

The security of Algorithm 4 immediately follows from the security of the oblivious binary tree $\mathcal{T}$: The algorithm repeatedly traverses rooted paths in $\mathcal{T}$, performing a constant number of operations for each node. Algorithm 3 is secure due to the security of oblivious sorting and the fact that the sub-domain access is independent of the input. With a total order over the boxes, the input sizes for the recursive calls solely depend on the problem size n. The base case for $\mathcal{O}(1)$ boxes with runtime $\mathcal{O}(1)$ can trivially be transformed into an oblivious algorithm. Since each recursive call processes the boxes in an oblivious manner, the obliviousness of the full divide-and-conquer algorithm follows.

Theorem 1. *There is an oblivious algorithm solving Klee's measure problem for* $d = 2$ *in expected* $\mathcal{O}(n \log^{5/3} n)$ *time for perfect security or* $\mathcal{O}(n \log^{3/2} n)$ *time for computational security,* $\mathcal{O}(n)$ *additional space, and* $\mathcal{O}(1)$ *private memory cells.*

4 General Technique

Above we showed how to use the oblivious binary tree to construct an efficient algorithm for Klee's measure problem. The technique is not specific to Chan's algorithm and can be applied to other geometric divide-and-conquer algorithms as well. In this section, we will outline the necessary conditions for the application of our technique. As an illustration, we will also sketch the application to two other problems, namely inversion counting and the closest-pair problem for $d = 2$. For both problems we will focus on the linear processing step; the runtime — both for perfect and computational security — then immediately follows as for Klee's measure problem above.

Our technique for transforming divide-and-conquer algorithms into oblivious counterparts requires that the following conditions are met:

(a) The input sizes for the recursive calls must only depend on the problem size.
(b) Each element must be contained in the input to at most one recursive call.
(c) Within each recursive call, the elements are processed with a linear scan using $\mathcal{O}(1)$ additional memory cells.

The first condition (a) is necessary for security, otherwise the algorithm might leak information via the runtime of the recursive calls. It is often possible to accommodate small size variations without affecting the runtime by padding the inputs with dummy elements. The conditions (b) and (c) are necessary to individually trace the elements through the recursion tree. This individual tracing also limits the information that can be passed to the recursive calls.

Inversion Counting. The *inversion counting* problem — for a sequence A of length n — is to determine the number of indices i, j so that $i < j$ and $A[i] > A[j]$. In the RAM model, this problem can be solved in $\mathcal{O}(n \log n)$ time by an augmented merge-sort algorithm [14, Exercise 2-4 d].[4] Since merging is a linear-time operation in the RAM model, but has an $\Omega(n \log n)$ lower bound in the oblivious RAM model (assuming indivisibility) [22], a direct implementation of this approach results in an $\mathcal{O}(n \log^2 n)$ time oblivious algorithm.

To improve over this, we interpret this algorithm as scan-based divide-and-conquer algorithm: Since the RAM algorithm is a divide-and-conquer algorithm, it remains to describe how to obliviously count inversion pairs using linear scans. For this, note that it is possible to separate the merging and counting steps by marking each element. Counting inversions can then be done in a linear scan by counting the elements from the second half and adding the current count to the number of inversions for every encountered element from the first half.

---

[4] In the word RAM model of computation there is an $\mathcal{O}(n \log^{1/2} n)$ time algorithm due to Chan and Pătraşcu [13]. For computational security, it is thus also possible to obtain an $\mathcal{O}(n \log^{3/2} n)$ time oblivious algorithm through optimal oblivious transformation.

For more efficient processing we split the input into m parts and recurse for each part. We then annotate each element with its part, sort the elements, and count inversions between the m parts using an oblivious binary tree: The annotated parts identify the leafs and each node has a counter for the elements belonging to a leaf in the right subtree. For the final number of inversions we sum the inversions counted in all nodes and in the recursive calls.

Closest Pair. For the planar *closest-pair* problem, we are given a set of n points $P \subset \mathbb{R}^2$ and want to determine a pair $p, q \in P$ with minimal distance according to some metric d. For an oblivious algorithm, we apply our technique to the divide-and-conquer algorithm of Bentley and Shamos [5].[5]

As for the inversion counting above, we begin by describing the necessary modifications to obtain a scan-based divide-and-conquer algorithm: After partitioning the points into P_L and P_R according to the median x-coordinate and recursing, we have minimal distances δ_L and δ_R; let $\delta := \min\{\delta_L, \delta_R\}$. It remains to check whether there is a pair in $P_L \times P_R$ with a smaller distance: While the original algorithm performs a synchronized traversal over P_L and P_R, we need to slightly modify this to adhere to the requirement that we may only perform a linear scan over the complete input. We thus sort the points by increasing y-coordinate and perform a linear scan over the sorted sequence. Using a standard packing argument [5], we maintain a queue Q of constant size that stores all points within distance at most δ from the median x-coordinate and within distance of at most δ below the current point. Whenever we encounter a point p with $|s_x - p_x| \geq \delta$ where s_x is the median x-coordinate, we simply ignore this point; otherwise, we check p against all points currently in Q to see whether $d(p, q) < \delta$ for any point q on the other side of s_x. If we find such a point, we update δ to $d(p, q)$. We then put p into Q while pruning all points too far below p or away from s_x.

Again, we can utilize an oblivious binary tree by sorting the input according to the x-coordinate and splitting evenly into m parts before recursing. We additionally store the separating x-coordinate s_x in each node of the binary tree; then we sort the elements by y-coordinate and process them so that for each node v the state consists of the queue Q (of constant size) as well as the current closest pair. Finally, we iterate over all nodes, updating the closest pair as needed.

5   Conclusion and Future Work

We gave an efficient oblivious modification of Chan's algorithm [12] for Klee's measure problem for $d = 2$, both for perfect and computational security.

---

[5] Eppstein et al. [15] discuss how to obliviously compute a closest pair in the plane in $\mathcal{O}(n \log n)$ time through an efficient construction of a well-separated pair decomposition [10]. For this, the input points need to have integer coordinates or a bounded spread. In contrast, our algorithm works without any such assumptions.

For $d \geq 3$, we sketched how to maintain the runtime of the original algorithm. Our oblivious algorithms only require $\mathcal{O}(1)$ private memory cells and can be used to construct a protocol for multi-party computation. We constructed our results with a general technique for oblivious divide-and-conquer algorithms and demonstrated its generality by applying it to the inversion counting and closest-pair problems.

Some open problems remain: Most notably, is it possible to obliviously solve Klee's measure problem for $d = 2$ in $\Theta(n \log n)$ time? From a more general perspective, faster oblivious binary tree implementations would not only improve our algorithm, but also a more general class of divide-and-conquer algorithms using our technique. Designing efficient oblivious tree data structures thus remains an interesting open problem.

Acknowledgments. We thank the reviewers for their constructive comments that helped to improve the presentation.

References

1. Ajtai, M., Komlós, J., Szemerédi, E.: An $\mathcal{O}(n \log n)$ sorting network. In: Proceedings of the Fifteenth Annual ACM Symposium on Theory of Computing, pp. 1–9 (1983). https://doi.org/10.1145/800061.808726
2. Asharov, G., Chan, T.H.H., Nayak, K., Pass, R., Ren, L., Shi, E.: Bucket oblivious sort: an extremely simple oblivious sort. In: Symposium on Simplicity in Algorithms (SOSA), pp. 8–14 (2020). https://doi.org/10.1137/1.9781611976014.2
3. Asharov, G., Komargodski, I., Lin, W.K., Nayak, K., Peserico, E., Shi, E.: OptORAMa: optimal oblivious RAM. Cryptology ePrint Archive, Report 2018/892 (2018). https://ia.cr/2018/892
4. Asharov, G., Komargodski, I., Lin, W.K., Nayak, K., Peserico, E., Shi, E.: OptORAMa: optimal oblivious RAM. In: Advances in Cryptology - EUROCRYPT 2020, pp. 403–432 (2020). https://doi.org/10.1007/978-3-030-45724-2_14
5. Bentley, J.L., Shamos, M.I.: Divide-and-conquer in multidimensional space. In: Proceedings of the Eighth Annual ACM Symposium on Theory of Computing, pp. 220–230 (1976). https://doi.org/10.1145/800113.803652
6. Blanton, M., Steele, A., Alisagari, M.: Data-oblivious graph algorithms for secure computation and outsourcing. In: Proceedings of the 8th ACM SIGSAC Symposium on Information, Computer and Communications Security, pp. 207–218 (2013). https://doi.org/10.1145/2484313.2484341
7. Chan, T.H.H., Nayak, K., Shi, E.: Perfectly secure oblivious parallel RAM. In: Theory of Cryptography, pp. 636–668 (2018). https://doi.org/10.1007/978-3-030-03810-6_23
8. Chan, T.H.H., Shi, E., Lin, W.K., Nayak, K.: Perfectly oblivious (parallel) RAM revisited, and improved constructions. Cryptology ePrint Archive, Report 2020/604 (2020). https://ia.cr/2020/604
9. Chan, T.H.H., Shi, E., Lin, W.K., Nayak, K.: Perfectly oblivious (parallel) RAM revisited, and improved constructions. In: 2nd Conference on Information-Theoretic Cryptography (ITC 2021), pp. 8:1–8:23 (2021). https://doi.org/10.4230/LIPIcs.ITC.2021.8

10. Chan, T.M.: Well-separated pair decomposition in linear time? Inf. Process. Lett. **107**(5), 138–141 (2008). https://doi.org/10.1016/j.ipl.2008.02.008
11. Chan, T.M.: A (slightly) faster algorithm for Klee's measure problem. Comput. Geom. **43**(3), 243–250 (2010). https://doi.org/10.1016/j.comgeo.2009.01.007
12. Chan, T.M.: Klee's measure problem made easy. In: 2013 IEEE 54th Annual Symposium on Foundations of Computer Science, pp. 410–419 (2013). https://doi.org/10.1109/FOCS.2013.51
13. Chan, T.M., Pătraşcu, M.: Counting inversions, offline orthogonal range counting, and related problems. In: Proceedings of the 2010 Annual ACM-SIAM Symposium on Discrete Algorithms (SODA), pp. 161–173 (2010). https://doi.org/10.1137/1.9781611973075.15
14. Cormen, T.H., Leiserson, C.E., Rivest, R.L., Stein, C.: Introduction to Algorithms, 3rd edn. MIT Press, Cambridge (2009)
15. Eppstein, D., Goodrich, M.T., Tamassia, R.: Privacy-preserving data-oblivious geometric algorithms for geographic data. In: Proceedings of the 18th SIGSPATIAL International Conference on Advances in Geographic Information Systems, pp. 13–22 (2010). https://doi.org/10.1145/1869790.1869796
16. Fredman, M.L., Weide, B.: On the complexity of computing the measure of $\bigcup [a_i, b_i]$. Commun. ACM **21**(7), 540–544 (1978). https://doi.org/10.1145/359545.359553
17. Goldreich, O., Ostrovsky, R.: Software protection and simulation on oblivious RAMs. J. ACM **43**(3), 431–473 (1996). https://doi.org/10.1145/233551.233553
18. Goodrich, M.T.: Data-oblivious external-memory algorithms for the compaction, selection, and sorting of outsourced data. In: Proceedings of the Twenty-Third Annual ACM Symposium on Parallelism in Algorithms and Architectures, pp. 379–388 (2011). https://doi.org/10.1145/1989493.1989555
19. Guerreiro, A.P., Fonseca, C.M., Paquete, L.: The hypervolume indicator: computational problems and algorithms. ACM Comput. Surv. **54**(6), 1–42 (2021). https://doi.org/10.1145/3453474
20. Klee, V.: Can the measure of $\bigcup_1^n [a_i, b_i]$ be computed in less than $\mathcal{O}(n \log n)$ steps? Am. Math. Mon. **84**(4), 284–285 (1977). https://doi.org/10.1080/00029890.1977.11994336
21. Larsen, K.G., Nielsen, J.B.: Yes, there is an oblivious RAM lower bound! In: Advances in Cryptology - CRYPTO 2018, pp. 523–542 (2018). https://doi.org/10.1007/978-3-319-96881-0_18
22. Lin, W.K., Shi, E., Xie, T.: Can we overcome the $n \log n$ barrier for oblivious sorting? In: Proceedings of the 2019 Annual ACM-SIAM Symposium on Discrete Algorithms (SODA), pp. 2419–2438 (2019). https://doi.org/10.1137/1.9781611975482.148
23. Overmars, M.H., Yap, C.K.: New upper bounds in Klee's measure problem. SIAM J. Comput. **20**(6), 1034–1045 (1991). https://doi.org/10.1137/0220065
24. Stefanov, E., et al.: Path ORAM: an extremely simple oblivious RAM protocol. In: Proceedings of the 2013 ACM SIGSAC Conference on Computer & Communications Security, pp. 299–310 (2013). https://doi.org/10.1145/2508859.2516660
25. Wang, G., Luo, T., Goodrich, M.T., Du, W., Zhu, Z.: Bureaucratic protocols for secure two-party sorting, selection, and permuting. In: Proceedings of the 5th ACM Symposium on Information, Computer and Communications Security, pp. 226–237 (2010). https://doi.org/10.1145/1755688.1755716

26. Wang, X., Chan, H., Shi, E.: Circuit ORAM: on tightness of the Goldreich-Ostrovsky lower bound. In: Proceedings of the 22nd ACM SIGSAC Conference on Computer and Communications Security, pp. 850–861 (2015). https://doi.org/10.1145/2810103.2813634
27. Wang, X.S., et al.: Oblivious data structures. In: Proceedings of the 2014 ACM SIGSAC Conference on Computer and Communications Security, pp. 215–226 (2014). https://doi.org/10.1145/2660267.2660314

Approximation Algorithms

A Parameterized Approximation Algorithm for the Multiple Allocation k-Hub Center

Marcelo P. L. Benedito$^{(\boxtimes)}$, Lucas P. Melo , and Lehilton L. C. Pedrosa

Institute of Computing, University of Campinas, Campinas, Brazil
{mplb,lehilton}@ic.unicamp.br

Abstract. In the MULTIPLE ALLOCATION k-HUB CENTER, we are given a connected edge-weighted graph G, sets of clients $\mathcal{C}$ and hub locations $\mathcal{H}$, where $V(G) = \mathcal{C} \cup \mathcal{H}$, a set of demands $\mathcal{D} \subseteq \mathcal{C}^2$ and a positive integer k. A solution is a set of hubs $H \subseteq \mathcal{H}$ of size k such that every demand (a, b) is satisfied by a path starting in a, going through some vertex of H, and ending in b. The objective is to minimize the largest length of a path. We show that finding a $(3 - \epsilon)$-approximation is NP-hard already for planar graphs. For arbitrary graphs, the approximation lower bound holds even if we parameterize by k and the value r of an optimal solution. An exact FPT algorithm is also unlikely when the parameter combines k and various graph widths, including pathwidth. To confront these hardness barriers, we give a $(2 + \epsilon)$-approximation algorithm parameterized by treewidth, and, as a byproduct, for unweighted planar graphs, we give a $(2 + \epsilon)$-approximation algorithm parameterized by k and r. Compared to classical location problems, computing the length of a path depends on non-local decisions. This turns standard dynamic programming algorithms impractical, thus our algorithm approximates this length using only local information.

Keywords: Parameterized approximation algorithm · Hub location problem · Treewidth

1 Introduction

In the classical location theory, the goal is to select a set of centers or facilities to serve a set of clients [10,12,25,26]. Usually, each client is simply connected to the closest selected facility, so that the transportation or connection cost is minimized. In several scenarios, however, the demands correspond to connecting a set of pair of clients. Rather than connecting each pair directly, one might select a set of hubs that act as consolidation points to take advantage of economies

Research supported by São Paulo Research Foundation (FAPESP), grants #2015/11937-9 and #2019/10400-2 and National Council for Scientific and Technological Development (CNPq), grants #422829/2018-8 and #312186/2020-7.

A. Castañeda and F. Rodríguez-Henríquez (Eds.): LATIN 2022, LNCS 13568, pp. 141–156, 2022.
https://doi.org/10.1007/978-3-031-20624-5_9

of scale [8, 22, 30, 31]. In this case, each origin-destination demand is served by a path starting at the origin, going through one or more selected hubs and ending at the destination. Using consolidation points reduces the cost of maintaining the network, as a large number of goods is often transported through few hubs, and a small fleet of vehicles is sufficient to serve the network [9].

Many hub location problems have emerged through the years, that vary depending on the solution domain, whether it is discrete or continuous; on the number of hub stops serving each demand; on the number of selected hubs, and so on [1, 16]. Central to this classification is the nature of the objective function: for *median* problems, the objective is to minimize the total length of the paths serving the demands, while, for *center* problems, the objective is to find a solution whose maximum length is minimum. In this paper, we consider the MULTIPLE ALLOCATION k-HUB CENTER (MAkHC), which is a center problem in the one-stop model [29, 42], where clients may be assigned to multiple hubs for distinct demands, and whose objective is to select k hubs to minimize the worst connection cost of a demand.

Formally, an instance of MAkHC is comprised of a connected edge-weighted graph G, sets of clients $\mathcal{C}$ and hub locations $\mathcal{H}$, where $V(G) = \mathcal{C} \cup \mathcal{H}$, a set of demand pairs $\mathcal{D} \subseteq \mathcal{C}^2$ and a positive integer k. The objective is to find a set of hubs $H \subseteq \mathcal{H}$ of size k that minimizes $\max_{(a,b) \in \mathcal{D}} \min_{h \in H} d(a, h) + d(h, b)$, where $d(u, v)$ denotes the length of a shortest path between vertices u and v. In the decision version of MAkHC, we are also given a non-negative number r, and the goal is to determine whether there exists a solution of value at most r.

This problem is closely related to the well-known k-CENTER [23, 26], where, given an edge-weighted graph G, one wants to select a set of k vertices, called centers, so that the maximum distance from each vertex to the closest center is minimized. In the corresponding decision version, one also receives a number r, and asks whether there is a solution of value at most r. By creating a demand (u, u) for each vertex u of G, one reduces k-CENTER to MAkHC, thus MAkHC can be seen as a generalization of k-CENTER. In fact, MAkHC even generalizes the k-SUPPLIER [27], that is a variant of k-CENTER whose vertices are partitioned into clients and locations, only clients need to be served, and centers must be selected from the set of locations.

For NP-hard problems, one might look for an α-approximation, that is a polynomial-time algorithm that finds a solution whose value is within a factor α of the optimal. For k-CENTER, a simple greedy algorithm already gives a 2-approximation, that is the best one can hope for, since finding an approximation with smaller factor is NP-hard [23]. Analogously, there is a best-possible 3-approximation for k-SUPPLIER [27]. These results have been extended to MAkHC as well, which also admits a 3-approximation [39]. We prove this approximation factor is tight, even if the input graph is unweighted and planar.

An alternative is to consider the problem from the perspective of parameterized algorithms, that insist on finding an exact solution, but allow running times with a non-polynomial factor that depends only on a certain parameter of the input. More precisely, a decision problem with parameter w is *fixed-parameter tractable* (FPT) if it can be decided in time $f(w) \cdot n^{O(1)}$, where n is the size of the input and f is a function that depends only on w. Feldmann and Marx [19] showed that k-CENTER is W[1]-hard for planar graphs of constant doubling dimension when the parameter is a combination of k, the highway dimension and the pathwidth of the graph. Blum [5] showed that the hardness holds even if we additionally parameterize by the skeleton dimension of the graph. Under the assumption that FPT $\neq$ W[1], this implies that k-CENTER does not admit an FPT algorithm for any of these parameters, even if restricted to planar graphs of constant doubling dimension.

Recently, there has been interest in combining techniques from parameterized and approximation algorithms [18,36]. An algorithm is called a parameterized α-approximation if it finds a solution within factor α of the optimal value and runs in FPT time. The goal is to give an algorithm with improved approximation factor that runs in super-polynomial time, where the non-polynomial factors of the running time are dependent on the parameter only. Thus, one may possibly design an algorithm that runs in FPT time for a W[1]-hard problem that, although it finds only an approximate solution, has an approximation factor that breaks the known NP-hardness lower bounds.

For k-CENTER, Demaine *et al.* [14] give an FPT algorithm parameterized by k and r for planar and map graphs. All these characteristics seem necessary for an exact FPT algorithm, as even finding a $(2 - \epsilon)$-approximation with $\epsilon > 0$ for the general case is W[2]-hard for parameter k [17]. If we remove the solution value r and parameterize only by k, the problem remains W[1]-hard if we restrict the instances to planar graphs [19], or if we add structural graph parameters, such as the vertex-cover number or the feedback-vertex-set number (and thus, also treewidth or pathwidth) [32].

To circumvent the previous barriers, Katsikarelis *et al.* [32] provide an efficient parameterized approximation scheme (EPAS) for k-CENTER with different parameters w, i.e., for every $\epsilon > 0$, one can compute a $(1 + \epsilon)$-approximation in time $f(\epsilon, w) \cdot n^{O(1)}$, where w is either the cliquewidth or treewidth of the graph. More recently, Feldmann and Marx [19] have also given an EPAS for k-CENTER when it is parameterized by k and the doubling dimension, which can be a more appropriate parameter for transportation networks than r. For constrained k-CENTER and k-SUPPLIER, Goyal and Jaiswal [24] give parameterized approximations for variants such as capacitated, fault-tolerant, outlier and ℓ-diversity.

Our Results and Techniques. We initiate the study of MA*k*HC under the perspective of parameterized algorithms. We start by showing that, for any $\epsilon > 0$, there is no parameterized $(3 - \epsilon)$-approximation for MA*k*HC when the parameter is k, the value r is bounded by a constant and the graph is unweighted, unless FPT = W[2]. For unweighted planar graphs, finding a good constant-factor approximation remains hard in the polynomial sense, as we show that it is NP-hard to find a $(3 - \epsilon)$-approximation for MA*k*HC in this case.

To challenge the approximation lower bound, one might envisage an FPT algorithm by considering an additional structural parameter, such as vertex-cover and feedback-vertex-set numbers or treewidth. However, this is unlikely to lead to an exact FPT algorithm, as we note that the hardness results for k-CENTER [5,19,32] extend to MAkHC. Namely, we show that, unless FPT = W[1], MAkHC does not admit an FPT algorithm when parameterized by a combination of k, the highway and skeleton dimensions as well as the path-width of the graph, even if restricted to planar graphs of constant doubling dimension; or when parameterized by k and the vertex-cover number. Instead, we aim at finding an approximation with factor strictly smaller than 3 that runs in FPT time.

In this paper, we present a $(2+\epsilon)$-approximation for MAkHC parameterized by the treewidth of the graph, for $\epsilon > 0$. The running time of the algorithm is $\mathcal{O}^*((\text{tw}/\epsilon)^{\mathcal{O}(\text{tw})})$, where polynomial factors in the size of the input are omitted. Moreover, we give a parameterized $(2+\epsilon)$-approximation for MAkHC when the input graph is planar and unweighted, parameterized by k and r.

Our main result is a non-trivial dynamic programming algorithm over a tree decomposition, that follows the spirit of the algorithm by Demaine et al. [14]. We assume that we are given a tree decomposition of the graph and consider both k and r as part of the input. Thus, for each node t of this decomposition, we can guess the distance from each vertex in the bag of t to its closest hub in some (global) optimal solution H^*. The subproblem is computing the minimum number of hubs to satisfy each demand in the subgraph G_t, corresponding to t.

Compared to k-CENTER and k-SUPPLIER, however, MAkHC has two additional sources of difficulty. First, the cost to satisfy a demand cannot be computed locally, as it is the sum of two shortest paths, each from a client in the origin-destination pair to some hub in H^* that satisfies that pair. Second, the set of demand pairs $\mathcal{D}$ is given as part of the input, whereas every client must be served in k-CENTER or in k-SUPPLIER. If we knew the subset of demands D_t^* that are satisfied by some hub in $H^* \cap V(G_t)$, then one could solve every subproblem in a bottom-up fashion, so that every demand would have been satisfied in the subproblem corresponding to the root of the decomposition.

Guessing D_t^* leads to an FPT algorithm parameterized by tw, r and $|\mathcal{D}|$, which is unsatisfactory as the number of demands might be large in practice. Rather, for each node t of the tree decomposition, we compute deterministically two sets of demands $D_t, S_t \subseteq \mathcal{D}$ that enclose D_t^*, that is, that satisfy $D_t \subseteq D_t^* \subseteq D_t \cup S_t$. By filling the dynamic programming table using D_t instead of D_t^*, we can obtain an algorithm that runs in FPT time on parameters tw and r, and that finds a 2-approximation.

The key insight for the analysis is that the minimum number of hubs in G_t that are necessary to satisfy each demand in D_t by a path of length at most r is a lower bound on $|H^* \cap V(G_t)|$. At the same time, the definition of the set of demands S_t ensures that each such demand can be satisfied by a path of length at most $2r$ using a hub that is close to a vertex in the bag of t. This is the main technical contribution of the paper, and we believe that these ideas might find usage in algorithms for similar problems whose solution costs have non-local components.

Using only these ideas, however, is not enough to get rid of r as a parameter, as we need to enumerate the distance from each vertex in a bag to its closest hub. A common method to shrink a dynamic programming table with large integers is storing only an approximation of each number, causing the solution value to be computed approximately. This eliminates the parameter r from the running time, but adds a term ϵ to the approximation factor. This technique is now standard [34] and has been applied multiple times for graph width problems [4, 14, 20, 32].

Specifically, we employ the framework of approximate addition trees [34]. For some $\delta > 0$, we approximate each value $\{1, \ldots, r\}$ of an entry in the dynamic programming table by an integer power of $(1+\delta)$, and show that each such value is computed by an addition tree and corresponds to an approximate addition tree. By results in [34], we can readily set δ appropriately so that the number of distinct entries is polynomially bounded and each value is approximated within factor $(1 + \epsilon)$.

Related Work. The first modern studies on hub location problems date several decades back, when models and applications were surveyed [37,38]. Since then, most papers focused on integer linear programming and heuristic methods [1,16]. Approximation algorithms were studied for the single allocation median variant, whose task is to allocate each client to exactly one of the given hubs, minimizing the total transportation cost [2,21,28]. Later, constant-factor approximation algorithms were given for the problem of, simultaneously, selecting hubs and allocating clients [3]. The analog of MA*k*HC with median objective was considered by Bordini and Vignatti [7], who presented a (4α)-approximation algorithm that opens $\left(\frac{2\alpha}{2\alpha-1}\right)k$ hubs, for $\alpha > 1$.

There is a single allocation center variant that asks for a two-level hub network, where every client is connected to a single hub and the path satisfying a demand must cross a given network center [35,41]. Chen *et al.* [11] give a $\frac{5}{3}$-approximation algorithm and showed that finding a $(1.5 - \epsilon)$-approximation, for $\epsilon > 0$, is NP-hard. This problem was shown to admit an EPAS parameterized by treewidth [4] and, to our knowledge, is the first hub location problem studied in the parameterized setting.

Organization. The remainder of the paper is organized as follows. Section 2 introduces basic concepts and describes the framework of approximate addition trees. Section 3 shows the hardness results for MA*k*HC in both classical and parameterized complexity. Section 4 presents the approximation algorithm parameterized by treewidth, which is analyzed in Sect. 5. Section 6 presents the final remarks.

2   Preliminaries

An α-*approximation algorithm* for a minimization problem is an algorithm that, for every instance I of size n, runs in time $n^{\mathcal{O}(1)}$ and outputs a solution of value at most $\alpha \cdot \mathrm{OPT}(I)$, where $\mathrm{OPT}(I)$ is the optimal value of I. A *parameterized algorithm* for a parameterized problem is an algorithm that, for every instance (I, k), runs in time $f(k) \cdot n^{\mathcal{O}(1)}$, where f is a computable function that depends only on the parameter k, and decides (I, k) correctly. A parameterized problem that admits a parameterized algorithm is called *fixed-parameter tractable*, and the set of all such problems is denoted by FPT. Finally, a *parameterized α-approximation algorithm* for a (parameterized) minimization problem is an algorithm that, for every instance I and corresponding parameter k, runs in time $f(k) \cdot n^{\mathcal{O}(1)}$ and outputs a solution of value at most $\alpha \cdot \mathrm{OPT}(I)$. For a complete exposition, we refer the reader to [13,36,40].

We adopt standard graph theoretic notation. Given a graph G, we denote the set of vertices and edges as $V(G)$ and $E(G)$, respectively. For $S \subseteq V(G)$, the subgraph of G induced by S is denoted as $G[S]$ and is composed by the vertices of S and every edge of the graph that has both its endpoints in S.

A *tree decomposition* of a graph G is a pair $(\mathcal{T}, X)$, where $\mathcal{T}$ is a tree and X is a function that associates a node t of $\mathcal{T}$ to a set $X_t \subseteq V(G)$, called *bag*, such that:

(i) $\cup_{t \in V(\mathcal{T})} X_t = V(G)$;
(ii) for every $(u, v) \in E(G)$, there exists $t \in V(\mathcal{T})$ such that $u, v \in X_t$;
(iii) for every $u \in V(G)$, the set $\{t \in V(\mathcal{T}) : u \in X_t\}$ induces a connected subtree of $\mathcal{T}$.

The width of a tree decomposition is $\max_{t \in V(\mathcal{T})} |X_t| - 1$ and the *treewidth* of G is the minimum width of any tree decomposition of the graph. Also, for a node $t \in V(\mathcal{T})$, let $\mathcal{T}_t$ be the subset of nodes that contains t and all its descendants, and define G_t as the induced subgraph of G that has $\bigcup_{t' \in \mathcal{T}_t} X_{t'}$ as the set of vertices.

Dynamic programming algorithms over tree decompositions often assume that the decomposition has a restricted structure. In a *nice tree decomposition* of G, $\mathcal{T}$ is a binary tree and each node t has one of the following types:

(i) *leaf node*, which has no child and $X_t = \emptyset$;
(ii) *introduce node*, which has a child t' with $X_t = X_{t'} \cup \{u\}$, for $u \notin X_{t'}$;
(iii) *forget node*, which has a child t' with $X_t = X_{t'} \setminus \{u\}$, for $u \in X_{t'}$;
(iv) *join node*, which has children t' and t'' with $X_t = X_{t'} = X_{t''}$.

Given a tree decomposition $(\mathcal{T}, X)$ of width tw, there is a polynomial-time algorithm that finds a nice tree decomposition with $\mathcal{O}(\text{tw} \cdot |V(G)|)$ nodes and the same width [33]. Moreover, we may assume without loss of generality that our algorithm receives as input a nice tree decomposition of G whose tree has height $\mathcal{O}(\text{tw} \cdot \log |V(G)|)$, using the same arguments as discussed in [4,6].

2.1 Approximate Addition Trees

An *addition tree* is an abstract model that represents the computation of a number by successively adding two other previously computed numbers.

Definition 1. *An* addition tree *is a full binary tree such that each leaf u is associated to a non-negative integer input y_u, and each internal node u with children u' and u'' is associated to a computed number $y_u := y_{u'} + y_{u''}$.*

One can replace the sum with some operator $\oplus$, which computes each such sum only approximately, up to an integer power of $(1 + \delta)$, for some parameter $\delta > 0$. The resulting will be an *approximate addition tree*. While the error of the approximate value can pile up as more operations are performed, Lampis [34] showed that, for some $\epsilon > 0$, as long as δ is not too large, the relative error can bounded by $1 + \epsilon$.

Definition 2. *An* approximate addition tree *with parameter $\delta > 0$ is a full binary tree, where each leaf u is associated to a non-negative integer input z_u, and each internal node u with children u' and u'' is associated to a computed value $z_u := z_{u'} \oplus z_{u''}$, where $a \oplus b := 0$ if both a and b are zero, and $a \oplus b := (1 + \delta)^{\lceil \log_{1+\delta}(a+b) \rceil}$, otherwise.*

For simplicity, here we defined only a deterministic version of the approximate addition tree, since we can assume that the height of the tree decomposition is bounded by $\mathcal{O}(\mathrm{tw} \cdot \log |V(G)|)$. For this case, Lampis showed the following result.

Theorem 1 ([34]). *Given an approximate addition tree of height ℓ, if $\delta < \frac{\epsilon}{2\ell}$, then, for every node u of the tree, we have $\max \left\{ \frac{z_u}{y_u}, \frac{y_u}{z_u} \right\} < 1 + \epsilon$.*

2.2 Preprocessing

For an instance of MAkHC and a demand $(a, b) \in \mathcal{D}$, define G_{ab} as the induced subgraph of G with vertex set $V(G_{ab}) = \{v \in V(G) : d(a, v) + d(v, b) \leq r\}$.

Notice that if a solution H has a hub $h \in V(G_{ab})$, then the length of a path serving (a, b) that crosses h is at most r. In this case, we say that demand (a, b) is *satisfied* by h with cost r. Thus, in an optimal solution H^* of MAkHC, for every $(a, b) \in \mathcal{D}$, the set $H^* \cap V(G_{ab})$ must be non-empty.

Also, if there is $v \in V(G)$ such that $d(a, v) + d(v, b) > r$ for every $(a, b) \in \mathcal{D}$, then v does not belong to any (a, b)-path of length at most r, and can be safely removed from G. From now on, assume that we have preprocessed G in polynomial time, such that for every $v \in V(G)$,

$$\min_{(a,b) \in \mathcal{D}} d(a, v) + d(v, b) \leq r.$$

Moreover, we assume that each edge has an integer weight and that the optimal value, OPT, is bounded by $\mathcal{O}(\frac{1}{\epsilon} |V(G)|)$, for a given constant $\epsilon > 0$. If not, then we solve another instance for which this holds and that has optimal value $\mathrm{OPT}' \leq (1 + \epsilon)\mathrm{OPT}$ using standard rounding techniques [40]. It suffices finding a constant-factor approximation of value $A \leq 3\mathrm{OPT}$ [39], and defining a new distance function such that $d'(u, v) = \left\lceil \frac{3|V(G)|}{\epsilon A} d(u, v) \right\rceil$.

3 Hardness

Next, we observe that approximating MAkHC is hard, both in the classical and parameterized senses. First, we show that approximating the problem by a factor better than 3 is NP-hard, even if the input graph is planar and unweighted. This result strengthens the previous known lower bound and matches the approximation factor of the greedy algorithm [39].

Theorem 2. *For every $\epsilon > 0$, if there is a $(3 - \epsilon)$-approximation for MAkHC when G is an unweighted planar graph, then $\mathrm{P} = \mathrm{NP}$.*

To find a better approximation guarantee, one might resource to a parameterized approximation algorithm. The natural candidates for parameters of MAkHC are the number of hubs k and the value r of an optimal solution. The next theorem states that this choice of parameters does not help, as it is W[2]-hard to find a parameterized approximation with factor better than 3, when the parameter is k, the value r is bounded by a constant and G is unweighted.

Theorem 3. *For every $\epsilon > 0$, if there is a parameterized $(3 - \epsilon)$-approximation for MAkHC with parameter k, then $\mathrm{FPT} = \mathrm{W}[2]$. This holds even for the particular case of MAkHC with instances I such that $\mathrm{OPT}(I) \le 6$.*

Due to the previous hardness results, a parameterized algorithm for MAkHC must consider different parameters, or assume a particular case of the problem. In this paper, we focus on the treewidth of the graph, that is one of the most studied structural parameters [13], and the particular case of planar graphs. This setting is unlikely to lead to an (exact) FPT algorithm, though, as the problem is W[1]-hard, even if we combine these conditions. The next theorem follows directly from a result of Blum [5], since MAkHC is a generalization of k-CENTER.

Theorem 4. *Even on planar graphs with edge lengths of constant doubling dimension, MAkHC is W[1]-hard for the combined parameter $(k, \mathrm{pw}, h, \kappa)$, where* pw *is the pathwidth, h is the highway dimension and κ is the skeleton dimension of the graph.*

Note that MAkHC inherits other hardness results of k-CENTER by Katsikarelis *et al.* [32], thus it is W[1]-hard when parameterized by a combination of k and the vertex-cover number.

Recall that the treewidth is a lower bound on the pathwidth, thus the previous theorem implies that the problem is also W[1]-hard for planar graphs when parameterized by a combination of k and tw. To circumvent these hardness results, we give a $(2 + \epsilon)$-approximation algorithm for MAkHC for arbitrary graphs parameterized by tw, breaking the approximation barrier of 3.

4   The Algorithm

In this section, we give a $(2 + \epsilon)$-approximation parameterized only by the treewidth. In what follows, we assume that we receive a preprocessed instance of MAkHC and a nice tree decomposition of the input graph G with width tw and height bounded by $\mathcal{O}(\mathrm{tw} \cdot \log |V(G)|)$. Also, we assume that G contains all edges connecting pairs $u, v \in X_t$ for each node t. Moreover, we are given an integer r bounded by $\mathcal{O}((1/\epsilon)|V(G)|)$. Our goal is to design a dynamic programming algorithm that computes the minimum number of hubs that satisfy each demand with a path of length r. The overall idea is similar to that of the algorithm for k-CENTER by Demaine *et al.* [14], except that we consider a tree decomposition, instead of a branch decomposition, and that the computed solution will satisfy demands only approximately.

Consider some fixed global optimal solution H^* and a node t of the tree decomposition. Let us discuss possible candidates for a subproblem definition. The subgraph G_t corresponding to t in the decomposition contains a subset of H^* that satisfies a subset D_t^* of the demands. The shortest path serving each demand with a hub of $H^* \cap V(G_t)$ is either completely contained in G_t, or it must cross some vertex of the bag X_t. Thus, as in [14], we guess the distance i from each vertex u in X_t to the closest hub in H^*, and assign "color" $\downarrow i$ to u if the corresponding shortest path is in G_t, and color $\uparrow i$ otherwise.

Since the number of demands may be large, we cannot include D_t^* as part of the subproblem definition. For k-CENTER, if the shortest path serving a vertex in G_t crosses a vertex $u \in X_t$, then the length of this path can be bounded locally using the color of u, and the subproblem definition may require serving all vertices. For MAkHC, however, there might be demands (a, b) such that a is in G_t, while b is not, thus the coloring of X_t is not sufficient to bound the length of a path serving (a, b).

Instead of guessing D_t^*, for each coloring c of X_t, we require that only a subset $D_t(c)$ must be satisfied in the subproblem, and they can be satisfied by a path of length at most $2r$. Later, we show that the other demands in D_t^* are already satisfied by the hubs corresponding to the coloring of X_t. More specifically, we would like to compute $A_t(c)$ as the minimum number of hubs in G_t that satisfy each demand in $D_t(c)$ with a path of length at most $2r$ and that respect the distances given by c.

Since we preprocessed the graph in Sect. 2, there must be a hub in H^* to each vertex of X_t at distance at most r. Thus, the number of distinct colorings to consider for each t is bounded by $r^{\mathcal{O}(\mathrm{tw})}$. To get an algorithm parameterized only by tw, we need one more ingredient: in the following, the value of each color is stored approximately as an integer power of $(1 + \delta)$, for some $\delta > 0$. Later, using the framework of approximate addition trees, for any constant $\epsilon > 0$, we can set δ such that the number of subproblems is bounded by $\mathcal{O}^*((\mathrm{tw}/\epsilon)^{\mathcal{O}(\mathrm{tw})})$, and demands are satisfied by a path of length at most $(1 + \epsilon)2r$.

The set of approximate colors is

$$\Sigma = \{\downarrow 0\} \cup \{\uparrow i, \downarrow i \, : \, j \in \mathbb{Z}_{\geq 0} \, , \, i = (1 + \delta)^j \, , \, i \leq (1 + \epsilon)r \}.$$

A coloring of X_t is represented by a function $c : X_t \to \Sigma$. For each coloring c, we compute a set of demands that are "satisfied" by c.

Definition 3. *Define $S_t(c)$ as the set of demands (a, b) for which there exists $u \in X_t$ with $c(u) \in \{\uparrow i, \downarrow i\}$ and such that $d(a, u) + 2i + d(u, b) \leq (1 + \epsilon)2r$.*

The intuition is that a demand $(a, b) \in S_t(c)$ can be satisfied by a hub close to u by a path of length at most $(1 + \epsilon)2r$. Also, we compute a set of demands that must be served by a hub in G_t by the global optimal solution.

Definition 4. *Define $D_t(c)$ as the set of demands (a, b) such that $(a, b) \notin S_t(c)$ and either: (i) $a, b \in V(G_t)$;   or   (ii) $a \in V(G_t)$, $b \notin V(G_t)$ and there is $h \in V(G_{ab}) \cap V(G_t)$ such that $d(h, V(G_{ab}) \cap X_t) > r/2$.*

We will show in Lemmas 4 and 5 that $D_t(c) \subseteq D_t^* \subseteq D_t(c) \cup S_t(c)$, thus we only need to take care of demands in $D_t(c)$ in the subproblem. Formally, for each node t of the tree decomposition and coloring c of X_t, our algorithm computes a number $A_t(c)$ and a set of hubs $H \subseteq \mathcal{H} \cap V(G_t)$ of size $A_t(c)$ that satisfies the conditions below.

(C1) For every $u \in X_t$, if $c(u) = \downarrow i$, then there exists $h \in H$ and a shortest path
    P from u to h of length at most i such that $V(P) \subseteq V(G_t)$;
(C2) For every $(a, b) \in D_t(c)$, $\min_{h \in H} d(a, h) + d(h, b) \leq (1 + \epsilon)2r$.

If the algorithm does not find one such set, then it assigns $A_t(c) = \infty$. We describe next how to compute $A_t(c)$ for each node type.

For a *leaf node* t, we have $V(G_t) = \emptyset$, then $H = \emptyset$ satisfies the conditions, and we set $A_t(c_\emptyset) = 0$, where $c_\emptyset$ denotes the empty coloring.

For an *introduce node* t with child t', let u be the introduced vertex, such that $X_t = X_{t'} \cup \{u\}$. Let $I_t(c)$ be the set of colorings c' of $X_{t'}$ such that c' is the restriction of c to $X_{t'}$ and, if $c(u) = \downarrow i$ for some $i > 0$, there is $v \in X_{t'}$ with $c'(v) = \downarrow j$ such that $i = d(u, v) \oplus j$. Note that this set is either a singleton or is empty. If $I_t(c)$ is empty, discard c. Define:

$$A_t(c) = \min_{\substack{c' \in I_t(c): \\ D_t(c) \subseteq D_{t'}(c')}} \begin{cases} A_{t'}(c') + 1 & \text{if } c(u) = \downarrow 0, \\ A_{t'}(c') & \text{otherwise.} \end{cases}$$

If H' is the solution corresponding to $A_{t'}(c')$, we output $H = H' \cup \{u\}$ if $c(u) = \downarrow 0$, or $H = H'$ otherwise.

For a *forget node* t with child t', let u be the forgotten vertex, such that $X_t = X_{t'} \setminus \{u\}$. Let $F_t(c)$ be the set of colorings c' of $X_{t'}$ such that c is the restriction of c' to X_t and, if $c'(u) = \uparrow i$, then there is $v \in X_t$ such that $c(v) = \uparrow j$ and $i = d(u, v) \oplus j$. If $F_t(c)$ is empty, discard c. Define:

$$A_t(c) = \min_{\substack{c' \in F_t(c): \\ D_t(c) \subseteq D_{t'}(c') \cup S_{t'}(c')}} A_{t'}(c').$$

We output as solution the set $H = H'$, where H' corresponds to the solution of the selected subproblem in t'.

For a *join node* t with children t' and t'', we have $X_t = X_{t'} = X_{t''}$. Let $J_t(c)$ be the set of pairs of colorings (c', c'') of X_t such that, for every $u \in X_t$, when

$c(u)$ is $\downarrow 0$ or $\uparrow i$, then $c'(u) = c''(u) = c(u)$; else, if $c(u)$ is $\downarrow i$, then $(c'(u), c''(u))$ is either $(\uparrow i, \downarrow i)$ or $(\downarrow i, \uparrow i)$. If $J_t(c)$ is empty, discard c. Define:

$$A_t(c) = \min_{\substack{(c', c'') \in J_t(c): \\ D_t(c) \subseteq D_{t'}(c') \cup D_{t''}(c'')}} A_{t'}(c') + A_{t''}(c'') - h(c),$$

where $h(c)$ is the number of vertices u in X_t such that $c(u) = \downarrow 0$. We output a solution $H = H' \cup H''$, where H' and H'' are the solutions corresponding to t' and t'', respectively.

The next lemma states that the algorithm indeed produces a solution of bounded size that satisfies both conditions.

Lemma 1. *If $A_t(c) \neq \infty$, then the algorithm outputs a set $H \subseteq \mathcal{H} \cap V(G_t)$, with $|H| \leq A_t(c)$, that satisfies* (C1) *and* (C2).

Let t_0 be the root of the tree decomposition and $c_\emptyset$ be the empty coloring. Since the bag corresponding to the root node is empty, we have $S_{t_0}(c_\emptyset) = \emptyset$ and thus $D_{t_0}(c_\emptyset) = \mathcal{D}$. Therefore, if $A_{t_0}(c_\emptyset) \leq k$, Lemma 1 implies that the set of hubs H computed by the algorithm is a feasible solution that satisfies each demand with cost at most $(1 + \epsilon)2r$. In the next section, we bound the size of H by the size of the global optimal solution H^*.

5 Analysis

For each node t of the tree decomposition, we want to show that the number of hubs computed by the algorithm for some coloring c of X_t is not larger than the number of hubs of H^* contained in G_t, that is, we would like to show that $A_t(c) \leq |H^* \cap V(G_t)|$, for some c. If the distances from each vertex $u \in X_t$ to its closest hub in H^* were stored exactly, then the partial solution corresponding to H^* would induce one such coloring c_t^*, and we could show the inequality for this particular coloring. More precisely, for each $u \in V(G)$, let $h^*(u)$ be a hub of H^* such that $d(u, h^*(u))$ is minimum and $P^*(u)$ be a corresponding shortest path. Assume that each $P^*(u)$ is obtained from a shortest path tree to $h^*(u)$ and that it has the minimum number of edges among the shortest paths. The *signature* of H^* corresponding to a partial solution in G_t is a function c_t^* on X_t such that

$$c_t^*(u) = \begin{cases} \downarrow d(u, h^*(u)) & \text{if } V(P^*(u)) \subseteq V(G_t), \\ \uparrow d(u, h^*(u)) & \text{otherwise.} \end{cases}$$

Since distances are stored approximately as integer powers of $(1 + \delta)$, the function c_t^* might not be a valid coloring. Instead, we show that the algorithm considers a coloring $\bar{c}_t$ with roughly the same values of c_t^* and that its values are computed by approximate addition trees. We say that an addition tree and an approximate addition tree are *corresponding* if they are isomorphic and have the same input values. Also, recall that a coloring c of X_t is *discarded* by the algorithm if the set $I_t(c)$, $F_t(c)$ or $J_t(c)$ corresponding to t is empty.

Lemma 2. *Let ℓ_{t_0} be the height of the tree decomposition. There exists a coloring $\bar{c}_t$ that is not discarded by the algorithm and such that, for every $u \in X_t$, the values $c_t^*(u)$ and $\bar{c}_t(u)$ are computed, respectively, by an addition tree and a corresponding approximate addition tree of height at most $2\ell_{t_0}$.*

By setting $\delta = \epsilon/(2\ell_{t_0} + 1)$, Theorem 1 implies the next lemma.

Lemma 3. *For every $u \in X_t$, if $c_t^*(u) \in \{\uparrow i, \downarrow i\}$ and $\bar{c}_t(u) \in \{\uparrow j, \downarrow j\}$, then $j \leq (1 + \epsilon)i$.*

Recall that H^* is a fixed global optimal solution that satisfies each demand with cost r. Our goal is to bound $A_t(\bar{c}_t) \leq |H^* \cap V(G_t)|$ for every node t, thus we would like to determine the subset of demands D_t^* that are necessarily satisfied by hubs $H^* \cap V(G_t)$ in the subproblem definition. This is made precise in the following.

Definition 5. $D_t^* = \{(a, b) \in \mathcal{D} : \min_{h \in H^* \setminus V(G_t)} d(a, h) + d(h, b) > r\}$.

Since the algorithm cannot determine D_t^*, we show that, for each node t, it outputs a solution H for the subproblem corresponding to $A_t(\bar{c}_t)$ that satisfies every demand in $D_t(\bar{c}_t)$. In Lemma 4, we show that every demand in $D_t(\bar{c}_t)$ is also in D_t^*, as, otherwise, there could be no solution with size bounded by $|H^* \cap V(G_t)|$. Conversely, we show in Lemma 5 that a demand in D_t^* that is not in $D_t(\bar{c}_t)$ must be in $S_t(\bar{c}_t)$, thus all demands are satisfied.

Lemma 4. $D_t(\bar{c}_t) \subseteq D_t^*$.

Proof. Let $(a, b) \in D_t(\bar{c}_t)$ and consider an arbitrary hub $h^* \in H^*$ that satisfies (a, b) with cost r. We will show that $h^* \in V(G_t)$, and thus $(a, b) \in D_t^*$. For the sake of contradiction, assume that $h^* \in V(G) \setminus V(G_t)$.

First we claim that $d(h^*, V(G_{ab}) \cap X_t) > r/2$. If not, then let $u \in V(G_{ab}) \cap X_t$ be a vertex with $\bar{c}_t(u) \in \{\uparrow i, \downarrow i\}$ such that $d(u, h^*) \leq r/2$. Because the closest hub to u has distance at least $i/(1 + \epsilon)$, we have $i \leq (1 + \epsilon)d(u, h^*) \leq (1 + \epsilon)r/2$, but since $u \in V(G_{ab})$, this implies that $(a, b) \in S_t(\bar{c}_t)$, and thus $(a, b) \notin D_t(\bar{c}_t)$. Then, it follows that indeed $d(h^*, V(G_{ab}) \cap X_t) > r/2$.

Now we show that it cannot be the case that $a, b \in V(G_t)$. Suppose that $a, b \in V(G_t)$. Consider the shortest path from a to h^*, and let u be the last vertex of this path that is in $V(G_t)$. Since X_t separates $V(G_t) \setminus X_t$ from $V(G) \setminus V(G_t)$, it follows that $u \in X_t$. From the previous claim, $d(h^*, u) > r/2$, and thus $d(h^*, a) > r/2$. Analogously, $d(h^*, b) > r/2$, but then $d(a, h^*) + d(h^*, b) > r$, which contradicts the fact that h^* satisfies (a, b) with cost r. This contradiction comes from supposing that $a, b \in V(G_t)$. Thus, either a or b is not in $V(G_t)$.

Assume without loss of generality that $a \in V(G_t)$ and $b \notin V(G_t)$. From the definition of $D_t(\bar{c}_t)$, we know that there exists $h \in V(G_{ab}) \cap V(G_t)$ such that $d(h, V(G_{ab}) \cap X_t) > r/2$. Let P be a path from a to b crossing h^* with length at most r. Similarly, since $h \in V(G_{ab})$, there exists a path Q from a to b crossing h with length at most r. Let u be the last vertex of P with $u \in X_t$, and let v be

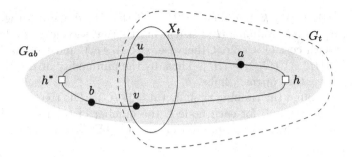

Fig. 1. Closed walk formed by P and Q.

the last vertex of Q with $v \in X_t$ (see Fig. 1). Concatenating P and Q leads to a closed walk of length at most $2r$. This walk crosses u, h^*, v and h, and thus

$$
\begin{aligned}
2r &\geq d(a, h^*) + d(h^*, b) + d(a, h) + d(h, b) \\
&= d(u, h^*) + d(h^*, v) + d(v, h) + d(h, u) \\
&> 2r,
\end{aligned}
\tag{1}
$$

where we used the fact that each term in (1) is greater than $r/2$. This is a contradiction, so $h^* \in V(G_t)$ and then $(a, b) \in D_t^*$.   □

Lemma 5. $D_t^* \subseteq D_t(\bar{c}_t) \cup S_t(\bar{c}_t)$.

Lemma 2 states that the coloring $\bar{c}_t$ is not discarded by the algorithm for each node t. Moreover, Lemmas 4 and 5 imply that the constraints of the recurrence are satisfied for this coloring. Thus, using induction, we can show that the algorithm does not open too many hubs.

Lemma 6. $A_t(\bar{c}_t) \leq |H^* \cap V(G_t)|$.

Theorem 5. *For every $\epsilon > 0$, there is a parameterized $(2 + \epsilon)$-approximation algorithm for MAkHC running in time $\mathcal{O}^*((\mathrm{tw}/\epsilon)^{\mathcal{O}(\mathrm{tw})})$.*

Proof. Consider a preprocessed instance $(G, \mathcal{C}, \mathcal{H}, \mathcal{D}, k)$ of MAkHC, in which the optimal value OPT is an integer bounded by $\mathcal{O}(\frac{1}{\epsilon}|V(G)|)$. We run the dynamic programming algorithm for each $r = 1, 2, \ldots$, and output the first solution with no more than k hubs. Next, we show that the dynamic programming algorithm either correctly decides that there is no solution of cost r that opens k hubs, or finds a solution of cost $(1 + \epsilon)2r$ that opens k hubs. Thus, when the main algorithm stops, $r \leq$ OPT, and the output is a $(2 + \epsilon')$-approximation, for a suitable ϵ'.

Assume H^* is a solution that satisfies each demand with cost r with minimum size. Recall t_0 is the root of the tree decomposition and $c_\emptyset$ is the coloring of an empty bag. If $A_{t_0}(c_\emptyset) \leq k$, then Lemma 1 states that the dynamic programming algorithm outputs a set of hubs H of size at most k that satisfies each demand in $D_{t_0}(c_\emptyset) = \mathcal{D}$ with cost $(1 + \epsilon)2r$. Otherwise, $k < A_{t_0}(c_\emptyset)$, and Lemma 6 implies $k < A_{t_0}(c_\emptyset) \leq |H^* \cap V(G_{t_0})| = |H^*|$. Thus, by the minimality of H^*, there is no solution of cost r that opens k hubs.

Finally, we bound the running time. Let $n = |V(G)|$. The tree decomposition has $\mathcal{O}(\text{tw} \cdot n)$ nodes and, for each node t, the number of colorings is $|\Sigma|^{\mathcal{O}(\text{tw})}$. Also, each recurrence can be computed in time $\mathcal{O}^*(|\Sigma|^{\mathcal{O}(\text{tw})})$. Since $r = \mathcal{O}(\frac{1}{\epsilon}n)$ and $\delta = \Theta\left(\frac{\epsilon}{\text{tw} \cdot \log n}\right)$, the size of Σ is

$$|\Sigma| = \mathcal{O}\left(\log_{1+\delta} r\right) = \mathcal{O}\left(\frac{\log r}{\log(1 + \delta)}\right) = \mathcal{O}\left(\frac{\log n + \log(1/\epsilon)}{\delta}\right)$$
$$= \mathcal{O}\left((\text{tw}/\epsilon)\left(\log^2 n + \log n \log(1/\epsilon)\right)\right) = \mathcal{O}\left((\text{tw}/\epsilon)^2 \log^2 n\right).$$

Notice that $\mathcal{O}(\log^{\mathcal{O}(\text{tw})} n) = \mathcal{O}^*(2^{\mathcal{O}(\text{tw})})$, thus the total running time is bounded by $\mathcal{O}^*\left(|\Sigma|^{\mathcal{O}(\text{tw})}\right) = \mathcal{O}^*\left((\text{tw}/\epsilon)^{\mathcal{O}(\text{tw})}\right)$. □

The algorithm for unweighted planar graphs is built upon the bidimensionality framework and some ideas applied to k-CENTER [15], where we use the fact that a graph with no large grid has treewidth at most a function of k and r.

Theorem 6. *For every $\epsilon > 0$, there is a parameterized $(2 + \epsilon)$-approximation algorithm for MAkHC when the parameters are k and r, and the input graph is unweighted and planar.*

6   Final Remarks

Our results are analogous to k-CENTER, which has a 2-approximation lower bound and does not admit an FPT algorithm. Unlike k-CENTER, however, we left open whether MAkHC admits an EPAS when parameterized by treewidth. The challenge seems to be the non-locality of the paths serving the demands, thus established techniques are not sufficient to tackle this issue. In this paper, we show how to compute a special subset of demands that must be served locally for each subproblem. We hope this technique may be of further interest. A possible direction of research is to consider the single allocation variant in the two-stop model, which is a well-studied generalization of MAkHC [3,16].

References

1. Alumur, S., Kara, B.Y.: Network hub location problems: the state of the art. Eur. J. Oper. Res. **190**(1), 1–21 (2008)
2. Ando, R., Matsui, T.: Algorithm for single allocation problem on hub-and-spoke networks in 2-dimensional plane. In: Algorithms and Computation, pp. 474–483 (2011)
3. Benedito, M.P.L., Pedrosa, L.L.C.: Approximation algorithms for median hub location problems. J. Comb. Optim. **38**(2), 375–401 (2019)

4. Benedito, M.P.L., Pedrosa, L.L.C.: An efficient parameterized approximation scheme for the star k-hub center. Procedia Comput. Sci. **195**, 49–58 (2021)
5. Blum, J.: W[1]-hardness of the k-center problem parameterized by the skeleton dimension. J. Comb. Optim. **44**, 2762–2781 (2022). https://doi.org/10.1007/s10878-021-00792-4
6. Bodlaender, H.L., Hagerup, T.: Parallel algorithms with optimal speedup for bounded treewidth. SIAM J. Comput. **27**(6), 1725–1746 (1998)
7. Bordini, C.F., Vignatti, A.L.: An approximation algorithm for the p-hub median problem. Electron. Notes Disc. Math. **62**, 183–188 (2017)
8. Campbell, J., Ernst, A., Krishnamoorthy, M.: Hub location problems. Facility location: application and theory (2002)
9. Campbell, J.F.: Hub location and the p-Hub median problem. Oper. Res. **44**(6), 923–935 (1996)
10. Charikar, M., Guha, S., Tardos, É., Shmoys, D.B.: A constant-factor approximation algorithm for the k-median problem. J. Comput. Syst. Sci. **65**(1), 129–149 (2002)
11. Chen, L.H., Cheng, D.W., Hsieh, S.Y., Hung, L.J., Lee, C.W., Wu, B.Y.: Approximation Algorithms for the Star k-Hub Center Problem in Metric Graphs, pp. 222–234. Springer International Publishing, Cham (2016). https://doi.org/10.1007/978-3-319-42634-1_18
12. Cornuéjols, G., Nemhauser, G., Wolsey, L.: The uncapacitated facility location problem. Cornell University Operations Research and Industrial Engineering, Technical report (1983)
13. Cygan, M., et al.: Parameterized Algorithms. Springer (2015). https://doi.org/10.1007/978-3-319-21275-3
14. Demaine, E.D., Fomin, F.V., Hajiaghayi, M.T., Thilikos, D.M.: Fixed-parameter algorithms for (k, r)-center in planar graphs and map graphs. ACM Trans. Algorithms **1**(1), 33–47 (2005)
15. Demaine, E.D., Fomin, F.V., Hajiaghayi, M.T., Thilikos, D.M.: Fixed-parameter algorithms for the (k, r)-center in planar graphs and map graphs. In: International Colloquium on Automata, Languages, and Programming, pp. 829–844. Springer (2003). https://doi.org/10.1007/3-540-45061-0_65
16. Farahani, R.Z., Hekmatfar, M., Arabani, A.B., Nikbakhsh, E.: Hub location problems: a review of models, classification, solution techniques, and applications. Comput. Ind. Eng. **64**(4), 1096–1109 (2013)
17. Feldmann, A.E.: Fixed-parameter approximations for k-center problems in low highway dimension graphs. Algorithmica **81**(3), 1031–1052 (2019)
18. Feldmann, A.E., Karthik, C., Lee, E., Manurangsi, P.: A survey on approximation in parameterized complexity: hardness and algorithms. Algorithms **13**(6), 146 (2020)
19. Feldmann, A.E., Marx, D.: The parameterized hardness of the k-center problem in transportation networks. Algorithmica **82**(7), 1989–2005 (2020)
20. Fomin, F.V., Golovach, P.A., Lokshtanov, D., Saurabh, S.: Algorithmic lower bounds for problems parameterized by clique-width. In: Proceedings of the Twenty-First Annual ACM-SIAM Symposium on Discrete Algorithms, pp. 493–502. SIAM (2010)
21. Ge, D., He, S., Ye, Y., Zhang, J.: Geometric rounding: a dependent randomized rounding scheme. J. Comb. Optim. **22**(4), 699–725 (2010)
22. Gelareh, S., Pisinger, D.: Fleet deployment, network design and hub location of liner shipping companies. Transp. Res. Part E: Logist. Transp. Rev. **47**(6), 947–964 (2011)

23. Gonzalez, T.F.: Clustering to minimize the maximum intercluster distance. Theoret. Comput. Sci. **38**, 293–306 (1985)
24. Goyal, D., Jaiswal, R.: Tight FPT approximation for constrained k-center and k-supplier. arXiv preprint arXiv:2110.14242 (2021)
25. Hochbaum, D.S.: Approximation algorithms for the set covering and vertex cover problems. SIAM J. Comput. **11**(3), 555–556 (1982)
26. Hochbaum, D.S., Shmoys, D.B.: A best possible heuristic for the k-center problem. Math. Oper. Res. **10**(2), 180–184 (1985)
27. Hochbaum, D.S., Shmoys, D.B.: A unified approach to approximation algorithms for Bottleneck problems. J. ACM **33**(3), 533–550 (1986)
28. Iwasa, M., Saito, H., Matsui, T.: Approximation algorithms for the single allocation problem in hub-and-spoke networks and related metric labeling problems. Discret. Appl. Math. **157**(9), 2078–2088 (2009)
29. Jaillet, P., Song, G., Yu, G.: Airline network design and hub location problems. Locat. Sci. **4**(3), 195–212 (1996)
30. Kara, B.Y., Tansel, B.: On the single-assignment p-hub center problem. Eur. J. Oper. Res. **125**(3), 648–655 (2000)
31. Karimi, H., Bashiri, M.: Hub covering location problems with different coverage types. Scientia Iranica **18**(6), 1571–1578 (2011)
32. Katsikarelis, I., Lampis, M., Paschos, V.T.: Structural parameters, tight bounds, and approximation for (k, r)-center. Discret. Appl. Math. **264**, 90–117 (2019)
33. Kloks, T.: Treewidth: Computations and Approximations, vol. 842. Springer Science & Business Media (1994). https://doi.org/10.1007/BFb0045388
34. Lampis, M.: Parameterized approximation schemes using graph widths. In: International Colloquium on Automata, Languages, and Programming, pp. 775–786. Springer (2014). https://doi.org/10.1007/978-3-662-43948-7_64
35. Liang, H.: The hardness and approximation of the star p-hub center problem. Oper. Res. Lett. **41**(2), 138–141 (2013)
36. Marx, D.: Parameterized complexity and approximation algorithms. Comput. J. **51**(1), 60–78 (2008)
37. O'Kelly, M.E.: The location of interacting hub facilities. Transp. Sci. **20**(2), 92–106 (1986)
38. O'Kelly, M.E.: A quadratic integer program for the location of interacting hub facilities. Eur. J. Oper. Res. **32**(3), 393–404 (1987)
39. Pedrosa, L.L.C., dos Santos, V.F., Schouery, R.C.S.: Uma aproximação para o problema de alocação de terminais. In: Anais do CSBC, ETC (2016)
40. Williamson, D.P., Shmoys, D.B.: The Design of Approximation Algorithms. Cambridge University Press, Cambridge (2011)
41. Yaman, H., Elloumi, S.: Star p-hub center problem and star p-hub median problem with bounded path lengths. Comput. Oper. Res. **39**(11), 2725–2732 (2012)
42. Yang, T.H.: Stochastic air freight hub location and flight routes planning. Appl. Math. Model. **33**(12), 4424–4430 (2009)

Theoretical Analysis of git bisect

Julien Courtiel, Paul Dorbec, and Romain Lecoq[(✉)]

Normandie University, UNICAEN, ENSICAEN, CNRS, GREYC, Caen, France
romain.lecoq@unicaen.fr

Abstract. In this paper, we consider the problem of finding a regression in a version control system (VCS), such as `git`. The set of versions is modelled by a Directed Acyclic Graph (DAG) where vertices represent versions of the software, and arcs are the changes between different versions. We assume that somewhere in the DAG, a bug was introduced, which persists in all of its subsequent versions. It is possible to query a vertex to check whether the corresponding version carries the bug. Given a DAG and a bugged vertex, the Regression Search Problem consists in finding the first vertex containing the bug in a minimum number of queries in the worst-case scenario. This problem is known to be NP-hard. We study the algorithm used in `git` to address this problem, known as `git bisect`. We prove that in a general setting, `git bisect` can use an exponentially larger number of queries than an optimal algorithm. We also consider the restriction where all vertices have indegree at most 2 (i.e. where merges are made between at most two branches at a time in the VCS), and prove that in this case, `git bisect` is a $\frac{1}{\log_2(3/2)}$-approximation algorithm, and that this bound is tight. We also provide a better approximation algorithm for this case.

1 Introduction

In the context of software development, it is essential to resort to Version Control Systems (VCS, in short), like `git` or `mercurial`. VCS enable many developers to work concurrently on the same system of files. Notably, all the *versions* of the project (that is to say the different states of the project over time) are saved by the VCS, as well as the different changes between versions.

Furthermore, many VCS offer the possibility of creating *branches* (i.e. parallel lines of development) and *merging* them, so that individuals can work on their own part of the project, with no risk of interfering with other developers work. Thereby the overall structure can be seen as a Directed Acyclic Graph (DAG), where the vertices are the versions, also named in this context *commits*, and the arcs model the changes between two versions.

The current paper deals with a problem often occurring in projects of large size: searching the origin of a so-called *regression*. Even with intensive testing techniques, it seems unavoidable to find out long-standing bugs which have been lying undetected for some time. Conveniently, one tries to fix this bug by finding the commit in which the bug appeared for the first time. The idea is that there

© Springer Nature Switzerland AG 2022
A. Castañeda and F. Rodríguez-Henríquez (Eds.): LATIN 2022, LNCS 13568, pp. 157–171, 2022.
https://doi.org/10.1007/978-3-031-20624-5_10

should be few differences between the code source of the commit that introduced the bug and the one from a previous bug-free commit, which makes it easier to find and fix the bug.

The identification of the faulty commit is possible by performing *queries* on existing commits. A query allows to figure out the status of the commit: whether it is *bugged* or it is *clean*. A single query can be very time-consuming: it may require running tests, manual checks, or the compilation of an entire source code. In some large projects, performing a query on a single commit can take up to a full day (for example, the Linux kernel project [8]). This is why it is essential to find the commit that introduced the bug with as few queries as possible.

The problem of finding an optimal solution in terms of number of queries, known as the Regression Search Problem, was proved to be NP-complete by Carmo, Donadelli, Kohayakawa and Laber in [6]. However, whenever the DAG is a tree (oriented from the leaves to the root), the computational complexity of the Regression Search Problem is polynomial [3,14], and even linear [13].

To our knowledge, very few papers in the literature deal with the Regression Search Problem in the worst-case scenario, as such. The Decision Tree problem, which is known to be NP-complete [11] as well as its approximation version [12], somehow generalises the Regression Search Problem, with this difference that the Decision Tree problem aims to minimise the average number of queries instead of the worst-case number of queries.

Many variations of the Regression Search problem exist:

- the costs of the queries may vary [9,10];
- the queries return the wrong result (say it is clean while the vertex is bugged or the converse) with a certain probability [10];
- one can just try to find a bugged vertex with at least one clean parent [4].

The most popular VCS today, namely `git`, proposes a tool for this problem: an algorithm named `git bisect`. It is a heuristic inspired by binary search that narrows down at each query the range of the possible faulty commits. This algorithm is widely used and shows excellent experimental results, though to our knowledge, no mathematical study of its performance have been carried out up to now.

In this paper, we fill this gap by providing a careful analysis on the number of queries that `git bisect` uses compared to an optimal strategy. This paper does not aim to find new approaches for the Regression Search Problem.

First, we show in Sect. 2 that in the general case, `git bisect` may be as bad as possible, testing about half the commits where an optimal logarithmic number of commits can be used to identify exactly the faulty vertex. But in all the cases where such bad performance occurs, there are large merges between more than two branches,[1] also named *octopus merges*. However, such merges are highly uncommon and inadvisable, so we carry out the study of `git bisect`

---

[1] According to https://www.destroyallsoftware.com/blog/2017/the-biggest-and-weirdest-commits-in-linux-kernel-git-history, a merge of 66 branches happened in the Linux kernel repository.

performances with the assumption that the DAG does not contain any octopus merge, that is every vertex has indegree at most two. Under such an assumption, we are able to prove in Sect. 3 that `git bisect` is an approximation algorithm for the problem, never using more than $\frac{1}{\log_2(3/2)} \approx 1.71$ times the optimal number of queries for large enough repositories. We also provide a family of DAGs for which the number of queries used by `git bisect` tends to $\frac{1}{\log_2(3/2)}$ times the optimal number of queries.

This paper also describes in Sect. 4 a new algorithm, which is a refinement of `git bisect`. This new algorithm, which we call `golden bisect`, offers a mathematical guaranteed ratio of $\frac{1}{\log_2(\phi)} \approx 1.44$ for DAGs with indegree at most 2 where $\phi = \frac{1+\sqrt{5}}{2}$ is the golden ratio. The search of new efficient algorithms for the Regression Search Problem seems to be crucial in software engineering (as evidenced by [4]); `golden bisect` is an example of progress in this direction.

1.1   Formal Definitions

Throughout the paper, we refer to VCS repositories as graphs, and more precisely as *Directed Acyclic Graphs* (DAG), i.e. directed graphs with no cycle. The set V of vertices corresponds to the versions of the software. An arc goes from a vertex p to another vertex v if v is obtained by a modification from p. We then say that p is a *parent* of v. A vertex may have multiple parents in the case of a merge. An *ancestor* of v is v itself or an ancestor of a parent of v.[2] Equivalently, a vertex is an ancestor of v if and only if it is co-accessible from v (i.e. there exists a path from this vertex to v).

We use the convention to write vertices in bold (for example v), and the number of ancestors of a vertex with the number letter between two vertical bars (for example $|v|$).

In our DAGs, we consider that a bug has been introduced at some vertex, named the *faulty commit*. This vertex is unique, and its position is unknown. The faulty commit is supposed to transmit the bug to each of its descendants (that is its children, its grand-children, and so on). Thus, vertices have two possible statuses: *bugged* or *clean*. A vertex is bugged if and only if it has the faulty commit as an ancestor. Other vertices are clean. This is illustrated by Fig. 1.

We consider the problem of identifying the faulty commit in a DAG D, where a bugged vertex b is identified. Usually, since the faulty commit is necessarily an ancestor of b, only the induced subgraph on b's ancestors is considered, and thus b is a *sink* (i.e. a vertex with no outgoing edge) accessible from all vertices in the DAG. When the bugged vertex is not specified, it is assumed to be the only unique sink of the DAG.

The problem is addressed by performing *queries* on vertices of the graph. Each query states whether the vertex is bugged or clean, and thus whether or not the faulty commit belongs to its ancestors or not. Once we find a bugged vertex whose parents are all clean, it is the faulty commit.

---

[2] Usually, v is not considered an ancestor of itself. Though for simplifying the terminology, we use this special convention here.

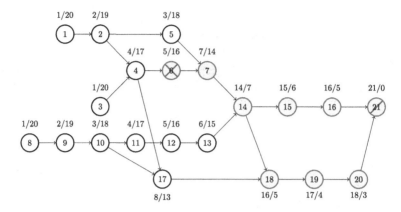

Fig. 1. An example of a DAG. The bugged vertices are colored. The crossed vertex (**6**) is the faulty commit. The notation a/b along each vertex indicates that a is the number of ancestors of the vertex, and b is the number of non-ancestors. The score (see Definition 2) is displayed in black.

The aim of the Regression Search Problem is to design a strategy for finding the faulty commit in a minimal number of queries.

Formally, a *strategy* (see for example [7]) for a DAG D is a binary tree S where the nodes are labelled by the vertices of D. Inner nodes of S represent queries. The root of S is the first performed query. If the queried vertex is bugged, then the following strategy is given by the left subtree. If it is clean, the strategy continues on the right subtree. At each query, there are fewer candidates for the faulty commit. Whenever a single candidate remains, the subtree is reduced to a leaf whose label is the only possible faulty commit.

For example, Fig. 2 shows a strategy tree for a directed path of size 5. Suppose that the faulty commit is **4**. In this strategy, we query in first **2**. Since it is clean, we query next **4**, which appears to be bugged. We finally query **3**: since it is clean, we infer that the faulty commit is **4**. We have found the faulty commit with 3 queries. Remark that if the faulty commit was **1**, **2** or **5**, the strategy would use only 2 queries.

The Regression Search Problem is formally defined as follows.

Definition 1. *Regression Search Problem.*
Input. *A DAG D with a marked vertex b, known to be bugged.*
Output. *A strategy which uses the least number of queries in the worst-case scenario.*

In terms of binary trees, the least number of queries in the worst-case scenario of a strategy corresponds to the height of the tree. For example, if the input DAG is a directed path of size n, we know that there exists a strategy with $\lceil \log_2(n) \rceil$ queries in the worst-case scenario. Indeed, a simple binary search enables to remove half of the vertices at each query.

A second interesting example is what we refer to as an *octopus*. In this digraph, there is a single sink and all other vertices are parent of the sink (see

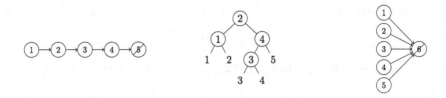

Fig. 2. *Left.* A directed path on 5 vertices. *Center.* A possible strategy for the Regression Search Problem on the path on 5 vertices. *Right.* An octopus of size 6.

Fig. 2). When the faulty commit is the sink, we must query all other vertices to make sure that the sink is faulty, regardless of the strategy. Thus, every strategy is equivalent for the Regression Search Problem on the n-vertices octopus, and uses $n - 1$ queries in the worst case.

These two examples actually constitute extreme cases for the Regression Search Problem, as shown by the following proposition.

Proposition 1. *For any DAG D with n vertices, any strategy that finds the faulty commit uses at least $\lceil \log_2(n) \rceil$ queries, and at most $n - 1$ queries.*

Proof. Remember that a strategy is a binary tree with at least n leaves, and the number of queries in the worst-case scenario corresponds to the height of the tree. But the height of such a binary tree is necessarily at least $\lceil \log_2(n) \rceil$, which proves the lower bound.

As for the upper bound, it is quite obvious because one can query at most $n - 1$ vertices in the Regression Search Problem.

From a complexity point of view, the Regression Search Problem is hard: Carmo, Donadelli, Kohayakawa and Laber proved in [6] that computing the least number of queries for the Regression Search Problem is NP-complete.[3]

1.2 Description of git bisect

As said in the introduction, some VCS provide a tool for the Regression Search Problem. The most known one is git bisect, but it has its equivalent in mercurial (hg bisect [5]).

The algorithm git bisect is a greedy algorithm based on the classical binary search. The local optimal choice consists in querying the vertices that split the digraph in the most balanced way. To be more precise, let us define the notion of score.

---

[3] In reality, the problem they studied has an extra restriction: a query cannot be performed on a vertex which was eliminated from the set of candidates for the faulty commit (which occurs for example when an ancestor is known to be bugged). However, the widget they used in the proof of NP-completeness also works for our problem where we do not necessarily forbid such queries.

Definition 2 (Score). *Given a DAG with n vertices, the* score *of a vertex* x *is*

$$\min(|x|, n - |x|),$$

where $|x|$ *is the number of ancestors of* x *(recall that* x *is an ancestor of itself).*

If vertex x is queried and appears to be bugged, then there remain $|x|$ candidates for the faulty commit: the ancestors of x. If the query of x reveals on the contrary that it is clean, then the number of candidates for the faulty commit is $n - |x|$, which is the number of non-ancestors. This is why the score of x can be interpreted as the least number of vertices to be eliminated from the set of possible candidates for the faulty commit, when x is queried. For a DAG, each vertex has a score and the maximum score is the score with the maximum value among all.

For example, let us refer to Fig. 1: vertex **6** has 5 ancestors (**1**, **2**, **3**, **4** and **6**). Its score is so $\min(5, 21 - 5) = 5$.

We give now a detailed description of `git bisect`.

Algorithm 1 (git bisect)
Input. *A DAG D and a bugged vertex* ***b.***
Output. *The faulty commit of D.*
Steps:

1. *Remove from D all non-ancestors of* ***b.***
2. *If D has only one vertex, return this vertex.*
3. *Compute the score for each vertex of D.*
4. *Query the vertex with the maximum score. If there are several vertices which have the maximum score, select any one then query it.*
5. *If the queried vertex is bugged, remove from D all non-ancestors of the queried vertex. Otherwise, remove from D all ancestors of the queried vertex.*
6. *Go to Step 2.*

Take for example the DAG from Fig. 1. Vertex **17** has the maximum score (8) so constitutes the first vertex to be queried. If we assume that the faulty commit is **6**, then the query reveals that **17** is clean. So all ancestors of **17** are removed (that are $1, 2, 3, 4, 8, 9, 10, 17$). Vertex **14** is then queried because it has the new maximum score 6, and so on.

The whole `git bisect` strategy tree is shown in Fig. 3. Notice that for this DAG, the `git bisect` algorithm is optimal since in the worst-case scenario it uses 5 queries and by Proposition 1, we know that any strategy uses at least $\lceil \log_2(21) \rceil = 5$ queries.

The greedy idea behind `git bisect` (choosing the query which partitions the commits as evenly as possible) is quite widespread in the literature. For example, it was used to find a $(\log(n) + 1)$-approximation for the Decision Tree Problem [1], in particular within the framework of geometric models [2].

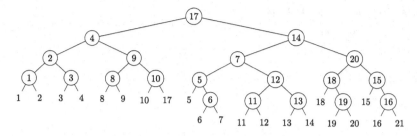

Fig. 3. The `git bisect` strategy corresponding to the graph of Fig. 1. In case of score equality, the convention we choose consists in querying the vertex with the smallest label.

2 Worst-Case Number of Queries

This section addresses the complexity analysis of `git bisect` in the worst-case scenario.

2.1 The Comb Construction

We describe in this subsection a way to enhance any DAG in such a way the Regression Search Problem can always be solved in a logarithmic number of queries.

Definition 3 (Comb addition). *Let D be a Directed Acyclic Graph with n vertices. Let $v_1 < v_2 < \ldots < v_n$ be a topological ordering of D, that is a linear ordering of the vertices such that if $v_i v_j$ is an arc, then $v_i < v_j$.*

We say that we add a comb to D if we add to D:

- *n new vertices $u_1, \ldots, u_n$;*
- *the arcs $v_i u_i$ for $i \in \{1, \ldots, n\}$;*
- *the arcs $u_i u_{i+1}$ for $i \in \{1, \ldots, n-1\}$.*

The resulting graph is denoted $comb(D)$. The new identified bugged vertex of $comb(D)$ is u_n.

Examples of comb addition are shown by Fig. 4 and Fig. 5.

The comb addition depends on the initial topological ordering, but the latter will not have any impact on the following results. This is why we take the liberty of writing $comb(D)$ without any mention to the topological ordering.

Theorem 2. *Let D be a Directed Acyclic Graph with n vertices and such that the number of queries used by the `git bisect` algorithm is x. If we add a comb to D, then the resulting DAG $comb(D)$ is such that:*

- *the optimal strategy uses only $\lceil log_2(2n) \rceil$ queries;*
- *when n is odd, the `git bisect` algorithm uses $x + 1$ queries.*

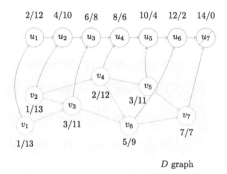

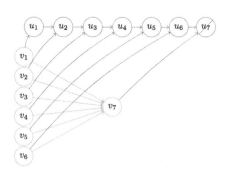

Fig. 4. Illustration of the comb addition. The initial digraph is highlighted in pink. (Color figure online)

Fig. 5. $Comb(D)$ graph where D is an octopus of size 7.

Proof. (Idea) On one hand, the optimal strategy for $comb(D)$ can be naturally achieved with a dichotomic search on the u_i vertices. On the other hand, with the assumption that n is odd, `git bisect` will necessarily query v_n first since its score is n, and the scores of the u_i vertices are all even. This explains why the `git bisect` algorithm uses $x + 1$ queries.

If the initial number of vertices n is even, there is no guarantee that `git bisect` will perform $x+1$ queries on $comb(D)$ – it will depend on whether the first queried vertex is v_n or $u_{n/2}$. However a referee of this paper rightly mentioned that the odd hypothesis could be (almost) removed by tweaking the comb construction whenever n is even. Indeed, by deleting the edge between $u_{n/2}$ and $v_{n/2}$, `git bisect` is forced to use $x+1$ queries in the worst-case scenario while a dichotomy strategy uses $\lceil \log_2(2n) + 1 \rceil$ queries.

2.2   A Pathological Example for `git bisect`

The following corollary shows the existence of digraphs for which the `git bisect` algorithm totally fails. The optimal number of queries is linear, while the `git bisect` algorithm effectively uses an exponential number of queries.

Theorem 3. *For any integer $k > 2$, there exists a DAG such that the optimal number of queries is k, while the `git bisect` algorithm always uses $2^{k-1} - 1$ queries.*

Proof. Choose D as an octopus with $2^{k-1} - 1$ vertices. The number of `git bisect` queries is $2^{k-1} - 2$ (like every other strategy). The wanted digraph is then $comb(D)$ (see Fig. 5 for an illustration). Indeed, by Theorem 2, the `git bisect` algorithm uses $2^{k-1} - 1$ `git bisect` queries to find the faulty commit in $comb(D)$, while an optimal strategy uses $\lceil \log_2 (2^k - 2) \rceil = k$ queries.

This also shows that the `git bisect` algorithm is not a C-approximation algorithm for the Regression Search Problem, for any constant C.

3   Approximation Ratio for Binary DAGs

The pathological input for the `git bisect` algorithm has a very particular shape (see Fig. 5): it involves a vertex with a gigantic indegree. However, in the context of VCS, this structure is quite rare. It means that many branches have been merged at the same time (the famous *octopus merge*). Such an operation is strongly discouraged, in addition to the fact that we just showed that `git bisect` becomes inefficient in this situation.

This motivates to define a new family of DAGs, closer to reality:

Definition 4 (Binary digraph). *A digraph is* binary *if each vertex has indegree (that is the number of ingoing edges) at most equal to 2.*

Fig. 6. *Left.* A binary DAG. *Right.* A non-binary DAG.

Figure 6 illustrates this definition. If we restrict the DAG to be binary, `git bisect` proves to be efficient.

We, the authors, reckon binary DAGs to be more natural in practice. Thus we have made the choice to only present the binary case in the main part of this paper, even if the following results can be easily generalised to DAGs whose indegree is bounded by an arbitrary integer Δ.

Theorem 4. *On any binary DAG with n vertices, the number of queries of the* git bisect *algorithm is at most* $\frac{\log_2(n)}{\log_2(\frac{3}{2})}$.

Corollary 1. *The algorithm* git bisect *is a* $\frac{1}{log_2(3/2)} \approx 1.71$ *approximation algorithm on binary DAGs.*

The key ingredient of the proof lies in the next lemma, which exhibits a core property of the binary DAGs. It states that if the DAG is binary, there must be a vertex with a "good" score.

Lemma 1. *In every binary DAG with n vertices, there exists a vertex v such that $|v|$, its number of ancestors, satisfies $\frac{n}{3} \leq |v| \leq \frac{2n+1}{3}$.*

By this lemma, we infer that `git bisect` removes at least approximately one third of the remaining vertices at each query. The overall number of queries is then equal to $\log_{3/2}(n)$.

The idea of the proof, in a few words, is to consider the vertex whose number of ancestors exceeds the number of non-ancestors by the narrowest of margins.

If this vertex does *not* satisfy Lemma 1, then it must have two parents, and at least one of them satisfies the lemma.

The upper bound of Theorem 4 is asymptotically sharp, as stated by the following proposition.

Proposition 2. *For any integer k, there exists a binary DAG J_k such that*

- *the number of git $bisect$ queries on $comb(J_k)$ is $k + \lceil \log_2(k) \rceil + 3$;*
- *an optimal strategy for $comb(J_k)$ uses at most $\log_2(\frac{3}{2}) k + \log_2(3k + 6) + 2$ queries.*

(Remember that the comb operation is described by Definition 3.)

Figure 7 shows what J_k looks like for $k = 3$. The number of queries for git $bisect$ in the worst-case scenario is 7 (which occurs for example when c is bugged).

By Proposition 2, we cannot find a better approximation ratio than $1/\log_2(3/2)$ for git $bisect$.

Corollary 2. *For any $\varepsilon > 0$, the git $bisect$ algorithm is not a $\left(\frac{1}{\log_2(3/2)} - \varepsilon \right)$ approximation algorithm for binary DAGs.*

4   A New Algorithm with a Better Approximation Ratio for Binary DAGs

In this section, we describe a new algorithm improving the number of queries in the worst-case scenario compared to git $bisect$ – theoretically at least.

4.1   Description of golden bisect

We design a new algorithm for the Regression Search Problem, which we name *golden bisect*, which is a slight modification of git $bisect$. It is so called because it is based on the *golden ratio*, which is defined as $\phi = \frac{1+\sqrt{5}}{2}$.

The difference of golden bisect with respect to git $bisect$ is that it may not query a vertex with the maximum score if the maximum score is too "low".

Let us give some preliminary definitions.

Definition 5. (Subsets $B^{\geq}$ and $B^{<}$). *Let D be a DAG. We define $V^{\geq}$ as the set of vertices which have more ancestors than non-ancestors. Let $B^{\geq}$ (for "Best" or "Boundary") denote the subset of vertices v of $V^{\geq}$ such that no parent of v belongs to $V^{\geq}$ and $B^{<}$ be the set of parents of vertices of $B^{\geq}$.*

The reader can look at Fig. 8 for an illustrative example.

Now, let us describe the golden bisect algorithm.

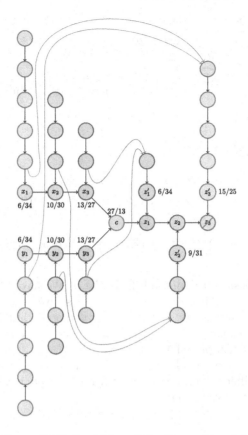

Fig. 7. Binary DAG J_3 which satisfies Proposition 2 for $k = 3$.

Algorithm 5 (golden bisect)
Input. A DAG D and a bugged vertex $\mathbf{b}$.
Output. The faulty commit of D.
Steps:

1. *Remove from D all non-ancestors of $\mathbf{b}$.*
2. *If D has only one vertex, return this vertex.*
3. *Compute the score for each vertex of D.*
4. ***If the maximum score is at least $\frac{n}{\phi^2} \approx 38.2\% \times n$ (where $\phi = \frac{1+\sqrt{5}}{2}$), query a vertex with the maximum score.***
5. ***Otherwise, query a vertex of $B^{\geq} \cup B^{<}$ which has the maximum score among vertices of $B^{\geq} \cup B^{<}$, even though it may not be the overall maximum score.***
6. *If the queried vertex is bugged, remove from D all non-ancestors of the queried vertex. Otherwise, remove from D all ancestors of the queried vertex.*
7. *Go to Step 2.*

(The differences with `git bisect` *are displayed in bold.)*

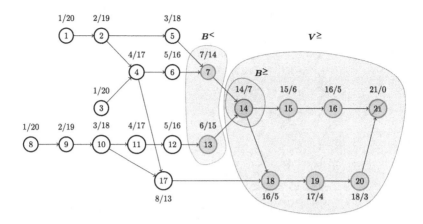

Fig. 8. A binary DAG with the 3 sets of vertices $V^{\geq}$, $B^{\geq}$ and $B^{<}$.

For example, consider the digraph from Fig. 8. We have $21/\phi^2 \approx 8.02$. The maximum score 8 is smaller than this number, so we run Step 5 instead of Step 4. Thus as its first query, `golden bisect` chooses indifferently **7** or **14**, which respectively belong to $B^{<}$ and $B^{\geq}$, and which have score 7. It diverges from `git bisect`, which picked **17** (score 8) instead.

For a full example, the reader can refer to the strategy tree in Fig. 9. Note that, even if it is different from `git bisect`, the `golden bisect` strategy uses 5 queries in the worst-case scenario.

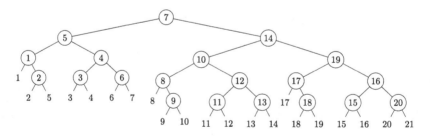

Fig. 9. The `golden bisect` strategy tree for the digraph of Fig. 8. In case of equality of score, the vertex with the smallest label is chosen.

4.2   Results for `golden bisect` on Binary DAGs

This subsection lists the main results about the complexity analysis of `golden bisect`. First, note that Theorem 3 also holds for `golden bisect`, so the general case (i.e. whenever the DAGs are not necessarily binary) is as bad as `git bisect`.

As for binary DAGs, we establish that the `golden bisect` algorithm has a better upper bound for the number of queries, in comparison with `git bisect`.

Theorem 6. *On any binary DAG with n vertices, the number of* `golden` `bisect` *queries is at most* $\log_\phi(n) + 1 = \frac{\log_2(n)}{\log_2(\phi)} + 1$, *where ϕ is the golden ratio.*

Proof. (Idea) The `golden bisect` algorithm has the remarkable following property: starting from a graph with n vertices, either the subgraph remaining after one query is of size at most $\frac{n}{\phi}$, or the subgraph obtained after two queries is of size at most $\frac{n}{\phi^2}$. If we admit this point, the proof of Theorem 6 has no difficulty.

The reason why we have such a guarantee on the size of the remaining graph after one or two queries comes from the choices of the sets $B^\geq$ and $B^<$. If `golden` `bisect` first queries a bugged vertex of $B^\geq$ with a "bad" score, then the parents of this vertex must have a "really good" score in the new resulting graph.

Let us take a critical example: `golden bisect` queries a bugged vertex of $B^\geq$, let us say b, with a score of $(n-1)/3$—which is the worst possible score for such vertices, by Lemma 1. In this case, each of the two parents of b will have the really good score of $(n-1)/3$ in the new graph, which is approximately half of its size. So, even if the first query has just removed one third of the vertices, the size of the graph after two queries is more or less $n/3$ (which is smaller than n/ϕ^2).

The ratio $1/\phi$ appears in fact whenever we try to balance what could go wrong after one query and what could go wrong after two queries.

Similar arguments hold whenever the first query concerns a vertex of $B^<$.

As first corollary, since no power of ϕ is an integer, the number of `golden` `bisect` queries for a binary DAG of size n is also at most $\lceil \log_\phi(n) \rceil = \left\lceil \frac{\log_2(n)}{\log_2(\phi)} \right\rceil$. We can also deduce that it is a better approximation algorithm than `git bisect` (in the binary case):

Corollary 3. *For every* $\varepsilon > 0$, `golden bisect` *is a* $\left(\frac{1}{\log_2(\phi)} + \varepsilon \right)$-*approximation algorithm on binary DAGs with a sufficiently large size.*

Moreover, we can find a family of graphs whose ratio "number of `golden` `bisect` queries"/"optimal number of queries" tends to $1/\log_2(\phi)$. This point will be more detailed in a longer version of this article.

Finally, this also gives an upper bound for the optimal number of queries in the worst-case scenario, given a binary DAG of size n.

Corollary 4. *For any binary DAG D with n vertices, the optimal number opt of queries for the Regression Search Problem satisfies*

$$\lceil \log_2(n) \rceil \leq opt \leq \lceil \log_\phi(n) \rceil.$$

Note that the latter corollary is an analogue of Proposition 1, but for binary DAGs. The lower bound is satisfied for a large variety of DAGs, the most obvious ones being the directed paths. The upper bound is also reached for some families of graphs.

5   Conclusion

In summary, this paper has established that `git bisect` can be very inefficient on very particular digraphs, but under the reasonable hypothesis that merges must not concern more than 2 branches each, it is proved to be a good approximation algorithm. This study has also developed a new algorithm, `golden bisect`, which displays better theoretical results than `git bisect`.

The natural next step will be to conduct experimental studies. The authors are currently implementing `git bisect` and `golden bisect`, and are going to put them to the test on benchmarks.

Notably, some open questions remain, and hopefully answers will be found through the experiments. Here is a list of such open questions:

- Even if `golden bisect` is a better approximation algorithm than the `git bisect` algorithm, it does not mean that `golden bisect` is overall better than `git bisect`. Does there exist some instance of binary DAG for which `golden bisect` is worst than the `git bisect` algorithm?
- In `git bisect` and in `golden bisect`, one never queries vertices which were eliminated from the set of candidates for the faulty commit. However, we could speed up the procedure by never removing any vertex after queries. For example, consider the DAG from Fig. 5. If we choose v_7 as first query and it is bugged, then we remove all u_i (the non-ancestors of v_7). However, querying the vertices u_i in the comb would be more efficient. Could we improve `git bisect` by authorising such queries?
- When we restrict the DAGs to be binary, is the Regression Search Problem still NP-complete?
- If we restrict the DAGs to be trees (oriented from the leaves to the root), is `git bisect` a good approximation algorithm? We conjecture that `git bisect` is a $2-$approximation algorithm for trees. (We have found examples where the ratio is 2.)

Finally we envisage studying the number of queries in the worst-case scenario, but whenever the input DAG is taken at random. Indeed, most of the examples described in this paper are not very likely to exist in reality. The notion of randomness for a digraph emanating from a VCS is therefore quite interesting and deserves to be developed. We could for example define a theoretical probabilistic model based on existing workflows. It will be also quite useful to use random samplers for VCS repositories in order to constitute benchmarks on demand.

References

1. Adler, M., Heeringa, B.: Approximating optimal binary decision trees. Algorithmica **62**(3–4), 1112–1121 (2012). https://doi.org/10.1007/s00453-011-9510-9
2. Arkin, E.M., Meijer, H., Mitchell, J.S.B., Rappaport, D., Skiena, S.S.: Decision trees for geometric models. Internat. J. Comput. Geom. Appl. **8**(3), 343–363 (1998). https://doi.org/10.1142/S0218195998000175

3. Ben-Asher, Y., Farchi, E., Newman, I.: Optimal search in trees (1999). https://doi.org/10.1137/S009753979731858X

4. Bendík, J., Benes, N., Cerna, I.: Finding regressions in projects under version control systems. CoRR (2017). arXiv:1708.06623

5. Boissinot, B.: hg bisect mercurial manpage. https://www.mercurial-scm.org/wiki/BisectExtension

6. Carmo, R., Donadelli, J., Kohayakawa, Y., Laber, E.: Searching in random partially ordered sets. Theoret. Comput. Sci. **321**(1), 41–57 (2004). https://doi.org/10.1016/j.tcs.2003.06.001

7. Cicalese, F., Jacobs, T., Laber, E.S., Molinaro, M.: On the complexity of searching in trees and partially ordered structures. Theoret. Comput. Sci. **412**, 6879–6896 (2011)

8. Couder, C.: Fighting regressions with git bisect (2009). https://git-scm.com/docs/git-bisect-lk2009

9. Dereniowski, D., Kosowski, A., Uznański, P., Zou, M.: Approximation strategies for generalized binary search in weighted trees. In: 44th International Colloquium on Automata, Languages, and Programming, volume 80 of LIPIcs. Leibniz Int. Proc. Inform., pages Art. No. 84, 14. Schloss Dagstuhl. Leibniz-Zent. Inform., Wadern (2017)

10. Emamjomeh-Zadeh, E., Kempe, D., Singhal, V.: Deterministic and probabilistic binary search in graphs. In: STOC 2016–Proceedings of the 48th Annual ACM SIGACT Symposium on Theory of Computing, pp. 519–532. ACM, New York (2016). https://doi.org/10.1145/2897518.2897656

11. Hyafil, L., Rivest, R.L.: Constructing optimal binary decision trees is NP-complete. Inf. Process. Lett. **5**(1), 15–17, 1976/77. https://doi.org/10.1016/0020-0190(76)90095-8

12. Laber, E.S., Nogueira, L.T.: On the hardness of the minimum height decision tree problem. Discrete Appl. Math. **144**(1–2), 209–212 (2004). https://doi.org/10.1016/j.dam.2004.06.002

13. Mozes, S., Onak, K., Weimann, O.: Finding an optimal tree searching strategy in linear time. In: Proceedings of the Nineteenth Annual ACM-SIAM Symposium on Discrete Algorithms, pp. 1096–1105. ACM, New York (2008)

14. Onak, K., Parys, P.: Generalization of binary search: searching in trees and forest-like partial orders. In: 47th Annual IEEE Symposium on Foundations of Computer Science (FOCS 2006), 21–24 October 2006, Berkeley, California, USA, Proceedings, pp. 379–388. IEEE Computer Society (2006). https://doi.org/10.1109/FOCS.2006.3

Pathlength of Outerplanar Graphs

Thomas Dissaux[(⊠)] and Nicolas Nisse

Université Côte d'Azur, Inria, CNRS, I3S, Nice, France
`thomas.dissaux@inria.fr`

Abstract. A *path-decomposition* of a graph $G = (V, E)$ is a sequence of subsets of V, called *bags*, that satisfy some connectivity properties. The *length* of a path-decomposition of a graph G is the greatest distance between two vertices that belong to a same bag and the *pathlength*, denoted by $p\ell(G)$, of G is the smallest length of its path-decompositions. This parameter has been studied for its algorithmic applications for several classical metric problems like the minimum eccentricity shortest path problem, the line-distortion problem, *etc.* However, deciding if the pathlength of a graph G is at most 2 is NP-complete, and the best known approximation algorithm has a ratio 2 (there is no c-approximation with $c < \frac{3}{2}$ unless $P = NP$). In this work, we focus on the study of the pathlength of simple sub-classes of planar graphs. We start by designing a linear-time algorithm that computes the pathlength of trees. Then, we show that the pathlength of cycles with n vertices is equal to $\lfloor \frac{n}{2} \rfloor$. Finally, our main result is a (+1)-approximation algorithm for the pathlength of outerplanar graphs. This algorithm is based on a characterization of almost optimal (of length at most $p\ell(G) + 1$) path-decompositions of outerplanar graphs.

Keywords: Path-decomposition · Pathlength · Outerplanar graph · Dual

1 Introduction

Path-decompositions of graphs have been extensively studied since their introduction in the Graph Minor theory of Robertson and Seymour, for their various algorithmic applications. A *path-decomposition* of a graph $G = (V, E)$ is a sequence $(X_1, \ldots, X_p)$ of subsets (called *bags*) of V such that (1) $\bigcup_{i \le p} X_i = V$, (2) for all edges $\{u, v\} \in E$, there exists $1 \le i \le p$ such that $u, v \in X_i$, and (3) for all $1 \le i \le z \le j \le p$, $X_i \cap X_j \subseteq X_z$. These constraints imply the following fundamental property (widely used in the proofs): for all $1 \le i < p$, $S = X_i \cap X_{i+1}$ *separates* $A = \bigcup_{j \le i} X_j \setminus S$ and $B = \bigcup_{j > i} X_j \setminus S$ (i.e. every path between A and B goes through S).

This work is partially founded by projects UCA JEDI (ANR-15-IDEX-01), the ANR project Digraphs (ANR-19-CE48-001) and EUR DS4H (ANR-17-EURE-004) Investments in the Future. Due to lack of space, some proofs are omitted or sketched. Full proofs can be found in [5].

A. Castañeda and F. Rodríguez-Henríquez (Eds.): LATIN 2022, LNCS 13568, pp. 172–187, 2022.
https://doi.org/10.1007/978-3-031-20624-5_11

The most classical measure of path-decompositions is their *width* corresponding to the maximum size of the bags (minus one). The *pathwidth* of a graph G is the minimum width of its path-decompositions. Typically, the famous theorem of Courcelle implies that numerous NP-hard problems can be solved in polynomial time in graphs of bounded pathwidth [2].

We focus on another measure of path-decompositions which, while less studied, has also numerous algorithmic applications. This measure, *the length* $\ell(D)$ of a path-decomposition D, is the maximum diameter of the bags of D, where the diameter $\ell(X)$ of a bag X is the largest distance (in G) between two vertices of X. The *pathlength* $p\ell(G)$ of a graph G, is the minimum length among all its path-decomposition [6]. In particular, this measure captures several metric properties of graphs. For example, the line distortion problem can be approximated (up to a constant factor) when the pathlength is bounded by a constant [7], which has many applications in computer vision [16], computational chemistry and biology [12], in network design and distributed protocol [10], *etc.* Moreover, since the pathlength is an upper bound of the *treelength*: the Traveling Salesman Problem admits a FPTAS in bounded pathlength graphs [14]; efficient compact routing schemes and sparse additive spanners can be built in the class of graphs with bounded pathlength [13]; computing the *metric dimension* is FPT in the pathlength plus the maximum degree [1], *etc.*

Unfortunately, deciding if the pathlength of a graph is at most 2 is NP-complete and there does not exist a c-approximation for any $c < \frac{3}{2}$ (unless $P = NP$) [9]. On the other hand, there exists a 2-approximation in general graphs [7]. While computing the pathwidth of planar graphs is known to be NP-complete [15], the case of pathlength has not been studied yet. In this paper, we initiate this study by considering outerplanar graphs. Note that the pathwidth of outerplanar graphs is known to be polynomial-time solvable, but the best known algorithm to compute the pathwidth of outerplanar n-node graphs has complexity $O(n^{11})$ [3]. Moreover, there exist 2-approximation algorithms for this problem, with time complexity $O(n log(n))$, that deal with the problem by relating the pathwidth of an outerplanar graph with the one of its weak dual [3,4].

Our Contributions. In Sect. 3, we first present a linear-time algorithm that computes the pathlength of trees, and prove that $p\ell(C_n) = \lfloor \frac{n}{2} \rfloor$ for any cycle C_n with n vertices. Section 4 is devoted to our main contribution. We design an algorithm that computes, in time $O(n^3(n + p\ell(G)^2))$, a path-decomposition of length at most $p\ell(G) + 1$ of any outerplanar n-node graph G. This algorithm is based on a structural characterization of almost optimal (of length at most $p\ell(G) + 1$) path-decompositions of outerplanar graphs.

2   Preliminaries

Let $G = (V, E)$ be any graph. When it will not be specified below, n will always be the number $|V|$ of vertices. In what follows, any edge $\{x, y\}$ is also considered as the set of two vertices x and y. In particular, we say that $X \subseteq V$ contains an edge e if $e \subseteq X$. Given a vertex $v \in V$, let $N(v) = \{u \in V \mid \{v, u\} \in E\}$ be the neighbourhood of v and let $N[v] = N(v) \cup \{v\}$ be its *closed neighbourhood*. Given $S \subseteq V$, let $N(S) = \{v \in V \setminus S \mid \exists u \in S, \{u, v\} \in E\}$ and let $G[S] = (S, E \cap (S \times S))$ be the subgraph of G induced by the vertices of S. The *distance* $dist_G(u, v)$ (or $dist(u, v)$ if there is no ambiguity) between $u \in V$ and $v \in V$ is the minimum length (number of edges) of a path between u and v in G. A subgraph H of G is *isometric* if $dist_H(u, v) = dist_G(u, v)$ for all $u, v \in V(H)$. In what follows, we will use the following result.

Lemma 1. *[6] For every isometric subgraph H of G, $p\ell(H) \le p\ell(G)$.*

Let us say that a path-decomposition of G is *optimal* if it has length $p\ell(G)$. Let us say that a path-decomposition is *reduced* if no bag is contained in another one. It is easy to check that any graph admits an optimal reduced path-decomposition.

Given two sequences $D = (X_1, \cdots, X_p)$ and D' of subsets of V and $S \subseteq V$, let $D \cup S = (X_1 \cup S, \cdots, X_p \cup S)$ (If $S = \{v\}$, we write $D \cup v$ instead of $D \cup \{v\}$). Let $D \cap S$ and $D \setminus S$ be defined in a similar way (in these cases, the empty bags that may be created are removed). Finally, let $D \odot D'$ be the sequence obtained by concatenation of D and D'. The following proposition is straightforward (and well known).

Proposition 1. *Let D be a path-decomposition of $G = (V, E)$ and $S \subseteq V$:*

1. *Then, $D' = D \cap S$ (resp., $D' = D \setminus S$) is a path-decomposition of $G[V \cap S]$ (resp., of $G \setminus S$). Moreover, if $G[V \cap S]$ (resp., $G \setminus S$) is an isometric subgraph of G, then $\ell(D') \le \ell(D)$.*
2. *Let D' be a path-decomposition of $G[V \setminus S]$. Then, $D' \cup S$ is a path-decomposition of G.*
3. *Let $V = A \cup B$ with $A \cap B = S$ and S separating $A \setminus S$ and $B \setminus S$ (there does not exist any edge $\{u, v\} \in E$ with $u \in A \setminus S$ and $v \in B \setminus S$), let D_1 be a path-decomposition of $G[A]$ with last bag containing S, and let D_2 be a path-decomposition of $G[B]$ with the first bag containing S. Then, $D_1 \odot D_2$ is a path-decomposition of G of length at most $\max\{\ell(D_1), \ell(D_2)\}$.*

3   Pathlength of Trees and Cycles

We first begin by the trees, the easiest sub-class of planar graphs. Note that, intuitively, the algorithm presented in Sect. 4, that computes the pathlength of outerplanar graphs follows similar ideas as the one we present for trees.

Theorem 1. *The pathlength of any tree T and an optimal path-decomposition can be computed in linear time.*

Sketch of Proof. Let $P = (v_0, \ldots, v_r)$ be a diameter of T, i.e., a longest path in T. For any $0 \le i < r$, let $e_i = \{v_i, v_{i+1}\}$ and T_i be the connected component of $T \setminus \{e_i, e_{i+1}\}$ containing v_i. Let D be a path-decomposition built as follows. The first bag of D is e_0, then sequentially (from $1 \le i < r$), we "add" the bags containing a path between v_i and a leaf of T_i (for each leaf of T_i ordered by any

DFS ordering of T_i starting from v_i), and then the bag e_{i+1} (see an illustration in Fig. 1). Clearly, P and such a decomposition can be computed in linear time using two BFS and one DFS. Moreover, this decomposition is optimal. Indeed, let X be a bag with maximum diameter $k = \ell(D)$, then it contains a leaf at distance k from the path P. By maximality of P, T contains an isometric spider S_k (three paths of length k joint by one of their end). Therefore, by Lemma 1, $p\ell(T) \geq k$ [6]. The full proof can be found in [5]. ◇

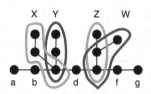

Fig. 1. Example of a tree T (with $p\ell(T) = 2$) and a path-decomposition $(\{ab\}, \{bc\}, X, Y, \{cd\}, \{de\}, Z, W, \{ef\}, \{fg\})$ obtained as in the proof of Theorem 1.

Remark 1. The above proof actually shows that, in any tree T, $p\ell(T)$ equals its minimum eccentricity shortest-path. Note that it was already known that the minimum eccentricity shortest-path of trees can be computed in linear time [8].

We will focus now on cycles. The full proof can be found in [5].

Theorem 2. *Let C_n be a cycle of length n. We have that $p\ell(C_n) = \lfloor \frac{n}{2} \rfloor$.*

Sketch of Proof. Note first that, a decomposition with only one bag containing all the vertices of C_n is a path-decomposition of length $\lfloor \frac{n}{2} \rfloor$. It remains to prove that this decomposition is optimal. Let u and $v \in V$ be at distance $\lfloor \frac{n}{2} \rfloor$. We suppose that there is a path-decomposition $D = (X_1, \ldots, X_p)$ such that $\ell(D) < \lfloor \frac{n}{2} \rfloor$. It implies that there exists $1 \leq i < j \leq p$ such that X_i is the last bag containing u and X_j is the first bag containing v. Moreover, for any $i < k \leq j$ $X_k \cap X_{k-1}$ separates u and v. We show that there exists $i < k \leq j$ such that X_k contains u' and v' such that $d(u', v') = \lfloor \frac{n}{2} \rfloor$, a contradiction. ◇

4 Outerplanar Graphs

This section is devoted to our main result: a polynomial-time algorithm for computing a path-decomposition of any simple outerplanar graph G with length at most $p\ell(G) + 1$. Due to lack of space and for simplicity, we only present the proof for 2-connected simple outerplanar graphs. The proof of the result in the case when the graph is not 2-connected can be found in [5].

A graph $G = (V, E)$ is *outerplanar* if it can be embedded in the plane without crossing edges and such that all vertices lie on the outer face (the unbounded face). An edge of an outerplanar graph is called an *internal* edge if it does not lie on the outer face. Note that, since we only consider simple graphs, the fact that an edge is internal or not does not depend on the outerplanar embedding. Let $E_{int} \subseteq E$ be the set of internal edges and $E_{out} = E \setminus E_{int}$ be the set of *outer* edges. Note that any internal edge $e \in E_{int}$ of an outerplanar graph $G = (V, E)$ is a separator (i.e., $G \setminus V(e)$ has several connected components). Moreover, since we are considering 2-connected outerplanar graphs, for every $e = \{u, v\} \in E_{int}$, $G \setminus \{u, v\}$ has exactly two connected components.

Path-Decompositions with Fixed First and Last Elements: Let $G = (V, E)$ be a connected graph and let $e \in E$. A path-decompostion $D = (X_1, \cdots, X_p)$ of G *starts* from e if $e \subseteq X_1$. Similarly, D *finishes* with e if $e \subseteq X_p$. Let $x, y \in E$, a path-decomposition of G starting from x and finishing with y is called a $\{x, y\}$-*path-decomposition*. Let $p\ell(G, x, y)$ be the minimum length among all $\{x, y\}$-path-decompositions of G. A $\{x, y\}$-path-decomposition is an *optimal* $\{x, y\}$-path-decomposition if its length is $p\ell(G, x, y)$. Clearly, any $\{x, y\}$-path-decomposition $(X_1, \cdots, X_p)$ corresponds to a $\{y, x\}$-path-decomposition $(X_p, \cdots, X_1)$ of same length, and so:

Claim. For any connected graph $G = (V, E)$ and $x, y \in E$, $p\ell(G, x, y) = p\ell(G, y, x)$.

The following claim directly holds by definition and because there always exists a reduced optimal path-decomposition (note that, for any reduced path-decomposition D of a connected graph, there exist $x, y \in E$ such that D starts from x and finishes with y).

Claim. For any connected graph $G = (V, E)$, $p\ell(G) = \min_{x,y \in E} p\ell(G, x, y)$.

For our purpose, we need to refine the above claim as follows. The proof can be found in [5].

Lemma 2. *For any 2-connected outerplanar graph G, $p\ell(G) = \min_{x,y \in E_{out}} p\ell(G, x, y)$.*

By Lemma 2, the computation of $p\ell(G)$ can be restricted to the $O(n^2)$ computations of $p\ell(G, x, y)$ for all fixed $x, y \in E_{out}$ (since G is planar, $|E| = O(n)$). Most of what follows is devoted to this task. Therefore, in Sects. 4.1 to 4.2, G will always be a 2-connected outerplanar graph and $x, y \in E_{out}$ will be fixed. Section 4.1 is devoted to the "easy" cases. If x and y are separated by an internal edge e, then it is possible to reduce the problem to the two connected components of $G \setminus e$, and if $x = y$, we present a greedy algorithm that computes an optimal $\{x, y\}$-path-decomposition of G. Sect. 4.2 is devoted to the remaining case: roughly, when x and y are distinct and belong to a same internal (bounded) face of G. In this latter case, we show that we can restrict our attention to particular $\{x, y\}$-path-decompositions of G and that such an optimal decomposition can be computed in polynomial time by dynamic programming. Finally, Sect. 4.3 formally states our main result and describes our algorithm to compute a path-decomposition of a 2-connected outerplanar graph with length at most $p\ell(G) + 1$.

4.1   Cases When Recursion or Greedy Algorithm are Possible

We say that x and y are *separated* in G if there exists $z \in E$ such that $x' = x \setminus z \neq \emptyset$, $y' = y \setminus z \neq \emptyset$ and all paths from x' to y' intersect z (recall that an edge is seen as a set of two vertices). Note also that, since any outer edge does not separate G in several connected components (because G is 2-connected), $z \in E_{int}$. Note that the following lemma directly holds by Proposition 1. The proof can be found in [5].

Lemma 3. *Let $G = (V, E)$ be a 2-connected simple outerplanar graph and $x, y \in E_{out}$ such that x and y are separated by $z \in E_{int}$. Let C^x and C^y be the connected component of $G \setminus z$ (the graph obtained from G by removing both ends of z) containing (or intersecting) respectively x and y. Let D_1 be an optimal $\{x, z\}$-path-decomposition of $G[V(C^x) \cup z]$ and D_2 be an optimal $\{z, y\}$-path-decomposition of $G[V(C^y) \cup z]$. Then, $D_1 \odot D_2$ is an optimal $\{x, y\}$-path-decomposition of G.*

Now, if $x = y$, then a *greedy path-decomposition P of G based on x* is any $\{x, x\}$-path-decomposition of G that can be obtained by the following recursive algorithm called *Greedy*.

– If G is a cycle $(v_1, \ldots, v_n)$ (w.l.o.g., $x = \{v_1, v_n\}$), $D = (X_1, \ldots, X_{n-1})$ with, for every $1 \leq i \leq n - 1$, $X_i = x \cup \{v_i, v_{i+1}\}$.
– Else, let $(v_1, \ldots, v_q)$ be the unique internal face containing x (the face is unique since x is an outer edge) such that $x = \{v_1, v_q\}$. Since G is not a cycle, there exists $1 \leq j < q$ such that $f = \{v_j, v_{j+1}\} \in E_{int}$. Let C and C' be the two connected components of $G \setminus f$ and, w.l.o.g., C intersects x. Let D_1 be a greedy path-decomposition of $G[V(C') \cup f]$ based on f and let $D_2 = (X_1, \cdots, X_p)$ be a greedy path-decomposition of $G[V(C) \cup f]$ based on x and let $1 \leq h \leq p$ be any integer such $f \subseteq X_h$ and $v_{j-1}, v_{j+2} \notin X_h$ (if $j = 1$, then $v_{j+2} \notin X_h$ and if $j = q - 1$, then $v_{j-1} \notin X_h$) (such an index h exists by induction). Then, $D = (X_1, \cdots, X_h) \odot (D_1 \cup (X_h \cap X_{h+1})) \odot (X_{h+1}, \cdots, X_p)$ is a greedy path-decomposition of G based on x.

It is easy to show by induction on $|V(G)|$ (see [5]) that:

Proposition 2. *Any sequence $D = (X_1, \cdots, X_p)$ returned by Algorithm Greedy is a $\{x, x\}$-path-decomposition of G. Moreover, Algorithm Greedy proceeds in linear time. Moreover, for all $1 < i \leq p$, $|X_i \setminus X_{i-1}| \leq 1$ and, if $X_i \setminus X_{i-1} = \{u\}$, then X_i consists of u, one of its neighbors u' such that $\{u, u'\} \in E_{out}$, of x and of all the vertices of each internal edge that separates u from x.*

Finally, from this description of the bags, we can prove that these greedy path-decompositions are optimal. The proof can be found in [5].

Theorem 3. *Let $G = (V, E)$ be a 2-connected simple outerplanar graph and $x \in E_{out}$. An optimal $\{x, x\}$-path-decomposition of G can be computed in linear time (in $O(|E|)$).*

4.2 Case When x and y Belong to a Same Face

In this section, we consider the last remaining case which is much more technical than the previous ones. Namely, $x \neq y$ and x and y lie on a same internal face F of G (i.e., they are not "separated").

In this setting, we first show that there always exists an almost optimal $\{x, y\}$-path-decomposition satisfying specific properties (first, *contiguous*, then *g-contiguous* and finally *LtR g-contiguous*, see formal definitions below). Then, a dynamic programming algorithm to compute a $\{x, y\}$-path-decomposition with minimum length (among such decompositions) is presented. We first need further notation.

Let $x = \{x_1, x_2\}$ and $y = \{y_1, y_2\}$. Let F (the internal face containing x and y) consist of two internally disjoint paths P_{up} between x_1 and y_1 and P_{down} between x_2 and y_2 (x and y may share one vertex, in which case, we assume that $x_2 = y_2$ and P_{down} is reduced to x_2). Let $\mathcal{C}$ be the set of connected components of $G \setminus F$. Let $\mathcal{C}_{up}$ (resp., $\mathcal{C}_{down}$) be the set of connected components C of $G \setminus F$ such that $N(C) \subseteq V(P_{up})$ (resp., $N(C) \subseteq V(P_{down})$). For every $C \in \mathcal{C}_{up} \cup \mathcal{C}_{down}$, let $\bar{C} = C \cup N(C)$ and let $s_C = N(C)$ (since G is outerplanar and 2-connected, s_C is an edge of F). Before describing the contiguous property, we need the following simple property which can be easily obtained thanks to Proposition 1 and to the third property of path-decompositions. The proof can be found in [5].

Lemma 4. *Let $G = (V, E)$ be a 2-connected simple outerplanar graph and $x, y \in E_{out}$ such that $x \neq y$ and x and y lie on the same internal face F of G.*
If $p\ell(G, x, y) \leq k$, then there exists a $\{x, y\}$-path-decomposition $(X_1, \cdots, X_p)$ of G with length at most k such that, for every $C \in \mathcal{C}$, if $X_i \cap C \neq \emptyset$, then $s_C \subseteq X_i$.

Let C be a component of $G \setminus F$. Let $s_C = \{l_C, r_C\}$ and let $d_C = \max_{v \in C \cup s_C}$ $\max\{dist(v, r_C), dist(v, l_C)\}$ and let $\mathcal{M}_C = \{v \in C \cup s_C \mid \max\{dist(v, r_C), dist(v, l_C)\} = d_C\}$. Note that, for every $v \in C \cup s_C$, $dist(v, l_C) - 1 \leq dist(v, r_C) \leq dist(v, l_C) + 1$. If there exists $v \in \mathcal{M}_C$ such that $dist(v, r_C) = dist(v, l_C) = d_C$, then let h_C^* be such a vertex. Otherwise, let h_C^* be any vertex of $\mathcal{M}_C$.

If there exists a vertex $v \in C \cup s_C$ with $dist(v, r_C) = dist(v, l_C) = d_C$, we say that C is a *convenient* component. The following claim is straightforward.

Proposition 3. *Let C be a connected component of $G \setminus F$ and let $v \in C$ and $u \in G \setminus C$. If C is convenient or $v \notin \mathcal{M}_C$, then $dist(v, u) \leq dist(h_C^*, u)$. Otherwise, $dist(v, u) \leq dist(h_C^*, u) + 1$.*

Toward g-Contiguous Decompositions. Let $D = (X_1, \cdots, X_p)$ be any $\{x, y\}$-path-decomposition of G and let $C \in \mathcal{C}$. The component C is said to be *contiguous (with respect to D)* if there exist $1 \leq a_C \leq b_C \leq p$ such that (1) $C \cap X_i = \emptyset$ if and only if $i \notin \{a_C, \cdots, b_C\}$, and $s_C \subseteq X_j$ for all $a_C \leq j \leq b_C$, and (2) there exists $R_C \subseteq V(F)$ such that $X_i \setminus C = R_C$ for every $a_C \leq i \leq b_C$. Intuitively, C is contiguous w.r.t. D if, once a vertex of C has been introduced in D, no vertex of $G \setminus C$ can be introduced in D before all vertices of C have been introduced.

A path-decomposition D is *contiguous* if every component of $G \setminus F$ is contiguous w.r.t. D.

In what follows, we show that there always exists an optimal (or almost optimal) $\{x, y\}$-path-decomposition of G which is contiguous. Note that if every component of $\mathcal{C}$ is convenient, then there exists a contiguous $\{x, y\}$-path-decomposition of length $p\ell(G, x, y)$. Unfortunately, if there is at least one component in $\mathcal{C}$ that is not convenient, then every contiguous $\{x, y\}$-path-decomposition of G might have length $p\ell(G, x, y) + 1$. Later, we show that there exist 2-connected simple outerplanar graphs for which the increase cannot be avoided.

Theorem 4. *Let $G = (V, E)$ be a connected simple outerplanar graph and $x, y \in E_{out}$ such that $x \neq y$ and x and y lie on the same internal face F of G.*

If $p\ell(G, x, y) \leq k$, then there exists a contiguous $\{x, y\}$-path-decomposition D' of G with length at most $k + 1$.

Proof. Let us say that a $\{x, y\}$-path-decomposition $D = (X_1, \cdots, X_p)$ of G satisfies Property $(*)$ if: $\ell(D) \leq k + 1$; for every $1 \leq i \leq p$, every $C \in \mathcal{C}$ and every $u, v \in C \cap X_i$, $dist(u, v) \leq k$; and, if $C \cap X_i \neq \emptyset$, then $s_C \subseteq X_i$.

Given a path-decomposition D, let $\mathcal{Q}(D) \subseteq \mathcal{C}$ be a set of components C of $\mathcal{C}$ such that there exist $1 \leq a_C \leq b_C \leq p$ such that (1) $C \cap X_i \neq \emptyset$ if and only if $a_C \leq i \leq b_C$, and $s_C \in X_j$ for all $a_C \leq j \leq b_C$, and (2) there exists $R_C \subseteq F \cup \bigcup_{C' \notin \mathcal{Q}(D)} C'$ such that, for every $a_C \leq i \leq b_C$, $X_i \setminus C = R_C$, and (3) for every $1 \leq j \leq p$ such that there exists no $C \in \mathcal{Q}(D)$ with $a_C \leq j \leq b_C$, then $\ell(X_j) \leq k$. Moreover, if there exist $C, C' \in \mathcal{Q}(D)$ with $b_C = j = a_{C'} - 1$, then $\ell(X_j \cap X_{j+1}) \leq k$.

Let D be a $\{x, y\}$-path-decomposition of G with length k. By Lemma 4, we may assume that, for every $C \in \mathcal{C}$, if $X_i \cap C \neq \emptyset$, then $s_C \subseteq X_i$. Hence, note that D satisfies Property $(*)$. Then, such a set $\mathcal{Q}(D)$ is well defined (possibly, $\mathcal{Q}(D)$ is empty).

Let us consider a $\{x, y\}$-path-decomposition $D' = (X_1, \cdots, X_p)$ of G with length at most $k + 1$ and satisfying Property $(*)$, and that maximizes $|\mathcal{Q}(D')|$. If $\mathcal{Q}(D') = \mathcal{C}$, then D' is the desired path-decomposition. For purpose of contradiction, let us assume that $\mathcal{C} \setminus \mathcal{Q}(D') \neq \emptyset$. Let $C \in \mathcal{C} \setminus \mathcal{Q}(D')$ and let $1 \leq i \leq p$ be the smallest integer such that $h_C^* \in X_i$. Note that, $s_C \in X_i$. Note also that, there is no $C' \in \mathcal{Q}(D')$ such that $a_{C'} < i \leq b_{C'}$ by definition of $\mathcal{Q}(D')$.

Let $Y = (D' \cap C) \cup (s_C \cup (X_{i-1} \cap X_i) \setminus C)$ (Recall the definition of $D \cap X$ and $D \cup X$, when D is a path-decomposition of G and $X \subseteq V(G)$, as defined in Sect. 2. Recall also that an edge is a set of two vertices). By Property $(*)$ and Proposition 3 and third item above, $\ell(Y) \leq k + 1$ (if C is convenient, we even have $\ell(Y) \leq k$). Therefore, $D'' = (X_1 \setminus C, \cdots, X_{i-1} \setminus C) \odot Y \odot (X_i \setminus C, \cdots, X_p \setminus C)$ is a $\{x, y\}$-path-decomposition of G, with length at most $k + 1$ and satisfying Prop. $(*)$, and such that $\mathcal{Q}(D') \cup \{C\} \subseteq \mathcal{Q}(D'')$, contradicting the maximality of $|\mathcal{Q}(D')|$. $\qquad \square$

Unfortunately, the previous theorem cannot be improved:

Lemma 5. *There exists 2-connected outerplanar graphs G and $x, y \in E_{out}$ such that every contiguous $\{x, y\}$-path-decomposition of G has length at least $p\ell(G, x, y) + 1$.*

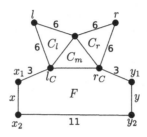

Fig. 2. Example of a 2-connected simple outerplanar graph G and $x, y \in E_{out}$ such that every contiguous $\{x, y\}$-path-decomposition of G has length at least $p\ell(G, x, y) + 1$. An edge with label q represents a path of length q.

Sketch of Proof. Let G be the graph depicted in Fig. 2. It can be proved that $p\ell(G, x, y) = 10$, but in every $\{x, y\}$-path-decomposition of G of length 10, any bag containing l contains only x_2 in P_{down}, while any bag containing r contains only y_2 in P_{down}. Hence, the component $C = G \setminus F$ cannot be added in a contiguous way: to preserve the minimum length, some vertices of P_{down} have to be added in such decompositions after l is added and before r is added in the decomposition. The full proof can be found in [5].   ◇

A contiguous $\{x, y\}$-path-decomposition $D = (X_1, \cdots, X_p)$ of G is said to be *g-contiguous* if, for every $C \in \mathcal{C}$, $(D_{a_C} \cap (\bar{C}), \cdots, D_{b_C} \cap (\bar{C}))$ is an optimal greedy path-decomposition of $G[\bar{C}] = G[C \cup s_C]$ based on s_C.

Note that transforming a contiguous path-decomposition $D = (X_1, \cdots, X_p)$ into a g-contiguous one can be done easily without increasing the length. Indeed, for each component $C \in \mathcal{C}$, there exists $R_C \subseteq V(F)$ such that $(X_{a_C} \setminus R_C, \cdots, X_{b_C} \setminus R_C)$ is a path-decomposition of $\bar{C}$ where each bag contains s_C. It is sufficient to replace, in D, this subsequence by the union of R_C and a greedy path-decomposition of $\bar{C}$ based on s_C. More precisely, the proof of the following theorem can be found in [5].

Theorem 5. *Let $G = (V, E)$ be a connected simple outerplanar graph and $x, y \in E_{out}$ such that $x \neq y$ and x and y lie on the same internal face F of G.*

If $p\ell(G, x, y) \leq k$, then there exists a g-contiguous $\{x, y\}$-path-decomposition $D = (X_1, \cdots, X_p)$ of G with length at most $k + 1$. Moreover, for every $C \in \mathcal{C}$ and $a_C \leq i \leq b_C$ and $u, v \in X_i$, $dist(u, v) = k + 1$ only if $u \in \mathcal{M}_C$ and $v \in F \setminus \bar{C}$.

Toward Left-to-Right g-Contiguous Decompositions: Recall that we are considering a n-node 2-connected simple outerplanar graph $G = (V, E)$ and $x, y \in E_{out}$ such that $x \neq y$ and x and y lie on the same internal face F of G.

Let $x = \{x_1, x_2\}$ and let $y = \{y_1, y_2\}$. Let $P_{up} = (x_1 = u_1, \cdots, u_t = y_1)$ and $P_{down} = (x_2 = d_1, \cdots, d_s = y_2)$ be the two internally disjoint paths that consist of all vertices of the face F. Note that, it may be possible that $x_1 = y_1$ or $x_2 = y_2$ but not both (in what follows, we assume that $x_1 \neq y_1$).

For every $C \in \mathcal{C}_{up}$ (i.e., such that $s_C \subseteq V(P_{up})$), let l_C (resp., r_C) be the vertex of s_C that is closest (resp., furthest) to x_1 in P_{up}. Similarly, for every $C \in \mathcal{C}_{down}$ (i.e., such that $s_C \subseteq V(P_{down})$), let l_C (resp., r_C) be the vertex of s_C that is closest (resp., furthest) to x_2 in P_{down}.

An edge $e \in E(F) \setminus \{x, y\}$ is said to be *trivial* if there does not exist $C \in \mathcal{C}$ such that $e = s_C$. While trivial edges are not related to any component of $G \setminus F$, we need to include them in the analysis that follows. To unify the notation, let $\bar{\mathcal{C}} = \{\bar{C} = C \cup s_C \mid C \in \mathcal{C}\} \cup \{e \in E(F) \setminus \{x, y\} \mid e \text{ is a trivial edge}\}$. Intuitively, every trivial edge $e \in E(F) \setminus \{x, y\}$ may be seen as $e = s_C$ for some dummy empty component $C = \emptyset$. Similarly, let $\bar{\mathcal{C}}_{up} = \{\bar{C} = C \cup s_C \mid C \in \mathcal{C}_{up}\} \cup \{e \in E(P_{up}) \mid e \text{ is a trivial edge}\}$ and $\bar{\mathcal{C}}_{down} = \{\bar{C} = C \cup s_C \mid C \in \mathcal{C}_{down}\} \cup \{e \in E(P_{down}) \mid e \text{ is a trivial edge}\}$. Note that $\bar{\mathcal{C}} = \bar{\mathcal{C}}_{up} \cup \bar{\mathcal{C}}_{down}$.

Let $\mathcal{O}_{up} = (C_1^u, \cdots, C_{s'}^u)$ be the ordering of $\bar{\mathcal{C}}_{up}$ such that, if $l_{C_i^u}$ is strictly closer to x_1 in P_{up} than $l_{C_j^u}$ (and so $r_{C_i^u}$ is strictly closer to x_1 in P_{up} than $r_{C_j^u}$), then $i < j$. Similarly, let $\mathcal{O}_{down} = (C_1^d, \cdots, C_{s'}^d)$ be the ordering of $\bar{\mathcal{C}}_{down}$ such that, if $l_{C_i^u}$ is strictly closer to x_2 in P_{down} than $l_{C_j^u}$ (and so if $r_{C_i^u}$ is strictly closer to x_1 in P_{up} than $r_{C_j^u}$), then $i < j$. Intuitively, we order the components of $\mathcal{C}_{up}$ and the trivial edges of P_{up} from x_1 to y_1 (resp., of $\mathcal{C}_{down}$ and the trivial edges of P_{down} from x_2 to y_2), i.e., from "left to right" ("from x to y").

In this section, we only consider g-contiguous $\{x, y\}$-path-decompositions $D = (X_1, \cdots, X_p)$ of G. That is, for every $C \in \mathcal{C}$, there exist $1 \leq a_C \leq b_C \leq p$, and an interval $I_C = [a_C, b_C]$ such that $X_i \cap C \neq \emptyset$ if and only if $i \in I_C$, $s_C \in X_i$ for all $i \in I_C$ and $X_i \setminus C = R_C \subseteq V(F)$ for all $i \in I_C$. In particular, $I_C \cap I_{C'} = \emptyset$ for all distinct $C, C' \in \mathcal{C}$ (since C, C' are distinct components of $G \setminus F$). We say that C appears in D in the bag X_{a_C}. Moreover, $(X_{a_C} \cap \bar{C}, \cdots, X_{b_C} \cap \bar{C})$ is a greedy path-decomposition of $G[\bar{C}]$ based on s_C. Recall also that we may assume that the property of the last statement in Theorem 5 holds.

By definition, D induces a total order $\mathcal{O}_D = (\bar{C}_1, \cdots, \bar{C}_b)$ on $\bar{C}$ such that, for any $1 \leq i < j \leq b$, $\bar{C}_i$ appears in D before $\bar{C}_j$ (i.e., $b_{C_i} < a_{C_j}$). We aim at considering such g-contiguous $\{x, y\}$-path-decompositions D such that the total orders $\mathcal{O}_D$ they induce satisfy some extra property defined below.

Let $H = H^1 \cup H^2$ be any set with $H^1 \cap H^2 = \emptyset$. Let $\mathcal{O} = (H_1, \cdots, H_q)$ be a total ordering on H, and let $\mathcal{O}^i$ be a total ordering of H^i for $i \in \{1, 2\}$. A prefix $\mathcal{P} = (H_1, \cdots, H_{q'})$ $(q' \leq q)$ of H is *compatible* with $\mathcal{O}^i$ if $\mathcal{P} \cap H^i$ is a prefix of $\mathcal{O}^i$ for $i \in \{1, 2\}$. If $q' = q$, then $\mathcal{O}$ is said to be *compatible* with $\mathcal{O}^1$ and $\mathcal{O}^2$.

Roughly, a contiguous $\{x, y\}$-path-decomposition D of G is said to be *LtR* (*letf to right*) if $\mathcal{O}_D$ is compatible with $\mathcal{O}_{up}$ and $\mathcal{O}_{down}$. More precisely,

Definition 1. *A contiguous $\{x, y\}$-path-decompositions $D = (X_1, \cdots, X_p)$ of G is LtR if and only if (1), for every $C_i^u, C_j^u \in \mathcal{O}_{up}$ (resp., $C_i^d, C_j^d \in \mathcal{O}_{down}$) with $i < j$, $b_{C_i^u} < a_{C_j^u}$ (resp., with $b_{C_i^d} < a_{C_j^d}$) and, moreover, (2), for every $C \in \bar{\mathcal{C}}_{up}$*

(resp., $\bar{\mathcal{C}}_{down}$), and $i \in I_C$, $X_i \cap F = \{l_C, r_C, f_C\}$ where f_C is one vertex of $V(P_{down})$ (resp., of $V(P_{up})$).

In what follows, we will iteratively transform a given g-contiguous $\{x,y\}$-path-decomposition of G into different path-decompositions. During these transformations, the obtained path-decomposition will always remain a g-contiguous $\{x,y\}$-path-decomposition of G, but its length may be increased temporarily. To deal with this difficulty, let us define the *weak length*, denoted by $w\ell(D)$, of an $\{x,y\}$-path-decomposition $D = (X_1, \cdots, X_p)$ of an outerplanar graph G (where x and y are outer edges of a same face of G). The *weak length*, denoted by $w\ell(X_i)$, of a bag X_i ($1 \le i \le p$) is $\max_{u \in X_i, v \in X_i \cap Y} dist(u,v)$ where $Y = V(P_{up})$ (resp., $Y = V(P_{down})$) if $a_C \le i \le b_C$ for a component $\bar{C} \in \bar{\mathcal{C}}_{down}$ (resp., $C \in \bar{\mathcal{C}}_{up}$). Then, $w\ell(D) = \max_{i \le p} w\ell(X_i)$. By definition of LtR and of the weak length:

Lemma 6. *[5] Let $D = (X_1, \cdots, X_p)$ be a LtR g-contiguous $\{x,y\}$-path-decomposition of G of weak length k such that, for every $C \in \mathcal{C}$ and $a_C \le i \le b_C$ and $u,v \in X_i$, $dist(u,v) > k$ only if $u \in \mathcal{M}_C$ and $v \in F \setminus \bar{C}$. Then, $\ell(D) \le k$.*

The next theorem roughly says that, from a g-contiguous $\{x,y\}$-path-decomposition, we can add the property that it is LtR without increasing the length.

Theorem 6. *Let $G = (V, E)$ be a connected simple outerplanar graph and $x, y \in E_{out}$ such that $x \ne y$ and x and y lie on the same internal face F of G.*

Let us assume that there exists a g-contiguous $\{x,y\}$-path-decomposition $D = (X_1, \cdots, X_p)$ of G with length k and such that, for every $C \in \mathcal{C}$ and $a_C \le i \le b_C$ and $u,v \in X_i$, $dist(u,v) = k$ only if $u \in \mathcal{M}_C$ and $v \in F \setminus \bar{C}$.

Then, there exists a LtR g-contiguous $\{x,y\}$-path-decomposition of G with weak length at most k and such that, for every $C \in \mathcal{C}$ and $a_C \le i \le b_C$ and $u,v \in X_i$, $dist(u,v) = k$ only if $u \in \mathcal{M}_C$ and $v \in F \setminus \bar{C}$.

Proof. Let $D = (X_1, \cdots, X_p)$ be a g-contiguous $\{x,y\}$-path-decomposition of G with weak length k. We say that D satisfies Property $(*)$ if, for every $C \in \mathcal{C}$ and $a_C \le i \le b_C$ and $u,v \in X_i$, $dist(u,v) = k$ only if $u \in \mathcal{M}_C$ and $v \in F \setminus \bar{C}$.

Recall that $\mathcal{O}_{up}$ (resp., $\mathcal{O}_{down}$) are uniquely defined since for any $1 \le i < t$, there is at most one component C such that $s_C = \{u_i, u_{i+1}\}$.

Let $D = (X_1, \cdots, X_p)$ be g-contiguous $\{x,y\}$-path-decomposition of G with weak length at most k satisfying Property $(*)$ that maximize $1 \le h \le p$ such that $(X_1, \cdots, X_h)$ is compatible with $\mathcal{O}_{up}$ and $\mathcal{O}_{down}$. Note that, if $h = p$, then D is the desired path-decomposition. Hence, for purpose of contradiction, let us assume that $h < p$.

Note that, because D is a contiguous $\{x,y\}$-path-decomposition and because $(X_1, \cdots, X_h)$ is compatible with $\mathcal{O}_{up}$ and $\mathcal{O}_{down}$, there exist $1 \le i \le s$ and $1 \le j \le t$ (recall that t and s are the number of vertices of P_{up} and P_{down} respectively) such that $X_h \cap X_{h+1} = \{u_i, d_j\}$.

Let $\mathcal{O}_D = \mathcal{O} \odot (C_1, \cdots, C_q)$ where $\mathcal{O}$ is the prefix of $\mathcal{O}_D$ that corresponds to the components appearing in $(X_1, \cdots, X_h)$. W.l.o.g., let us assume that $C_1 \in \bar{\mathcal{C}}_{up}$. Let $\mathcal{O}_{up} = \mathcal{O}' \odot (C_1', \cdots, C_{q'}')$ where $\mathcal{O}' = \mathcal{O} \cap \bar{\mathcal{C}}_{up}$. By maximality of h,

$C_1 \neq C_1'$. More precisely, $C_1 = C_z'$ for some $1 < z \leq q'$. There are two cases to be considered.

- First, let us assume that, for every $1 \leq \alpha < z$, for all $h_{C_\alpha'}^* \in \mathcal{M}_{C_\alpha'}$, $dist(h_{C_\alpha'}^*, d_j) \leq k$. Let

$$D' = (X_1, \cdots, X_h) \odot ((X_{a_{C_1'}}, \cdots, X_{b_{C_1'}}) \cap \bar{C}_1') \cup \{d_j\}$$

$$\odot ((X_{a_{C_2'}}, \cdots, X_{b_{C_2'}}) \cap \bar{C}_2') \cup \{d_j\} \odot \cdots \odot ((X_{a_{C_{z-1}'}}, \cdots, X_{b_{C_{z-1}'}}) \cap \bar{C}_{z-1}') \cup \{d_j\}$$

$$\odot ((X_{h+1}, \cdots, X_p) \setminus ((\bigcup_{1 \leq \alpha < z} \bar{C}'_\alpha) \setminus \{l_{C_z'}\})).$$

Intuitively, all components that are between (in $\mathcal{O}_{up}$) the last component of $\mathcal{O}$ and C_1 are "moved" just before C_1 in the decomposition (in D all these components were appearing after C_1).

Because D is g-contiguous and satisfies Property $(*)$, then D' is a g-contiguous $\{x, y\}$-path-decomposition of G satisfying Property $(*)$.

Moreover, its weak length is at most k. In particular, for every bag B of $((X_{a_{C_\alpha'}}, \cdots, X_{b_{C_\alpha'}}) \cap \bar{C}_\alpha') \cup \{d_j\}$ for $1 \leq \alpha < z$, $wl(B) \leq k$ because D satisfies Property $(*)$ and because, by assumption, $dist(h_{C_\alpha'}^*, d_j) \leq k$.

To conclude this case, D' is g-contiguous $\{x, y\}$-path-decomposition of G with $wl(D') \leq k$ satisfying Property $(*)$ and with a larger prefix than D that is compatible with $\mathcal{O}_{up}$ and $\mathcal{O}_{down}$, contradicting the maximality of h.

- Else, for every decomposition D defined as above and maximizing h (where $z := z(D)$, defined as above, depends on D), there exists an integer $1 \leq \alpha < z(D)$ and a vertex $h_{C_\alpha'}^* \in \mathcal{M}_{C_\alpha'}$ such that $dist(h_{C_\alpha'}^*, d_j) > k$ (otherwise, we are back to the previous case). Let $\alpha(D)$ be the smallest such integer α for the decomposition D.

Let $1 < \alpha^*(D) \leq q$ be such that $C_{\alpha(D)}' = C_{\alpha^*(D)}$.

Consider such a decomposition D (still maximizing h) that minimizes $\alpha^*(D)$. From now on, we denote the integer $\alpha(D)$ (for this particular decomposition D) by α and $\alpha^*(D)$ is denoted by α^*.

Let $\alpha < \beta \leq z \leq \gamma \leq q'$ be defined such that $[\beta, \gamma]$ is the inclusion-maximal interval (containing z) such that every component C_m' with $m \in [\beta, \gamma]$ appears before C_{α^*} in $\mathcal{O}_D$ (i.e., for every $m \in [\beta, \gamma]$, setting m' such that $C_m' = C_{m'}$, then $1 \leq m' < \alpha^*$). Note that, since the interval $[\beta, \gamma]$ is inclusion-maximal, it implies that the component $C_{\beta-1}'$ is either C_α' or some component that appears after C_α', and that the component $C_\gamma + 1$ does not exists (i.e., $\gamma = t - 1$) or C_γ appears after C_α in D. In all cases, every vertices from $(l_{C_\beta}, r_{C_\gamma})$-path can be remove from the bags before $X_{a_{C_{\alpha^*}}}$ without breaking any property of path-decomposition. Hence, let

$$D' = (X_1, \cdots, X_h) \odot ((X_{h+1}, \cdots, X_{a_{C_{\alpha^*}}-1}) \setminus (\bigcup_{\beta \leq m \leq \gamma} \bar{C}_m')) \odot$$

$$(X_{a_{C_{\alpha^*}}}, \cdots, X_{b_{C_{\alpha^*}}}) \odot ((X_{h+1} \cdots, X_{a_{C_{\alpha^*}}-1}) \cap \bigcup_{\beta \le m \le \gamma} \bar{C}'_m) \cup$$

$$(X_{b_{C_{\alpha^*}}} \cap V(P_{down})) \odot (X_{b_{C_{\alpha^*}}+1}, \cdots, X_p).$$

Intuitively, all components $C'_\beta, \cdots, C'_\gamma$ (and in particular, $C_1 = C'_z$) that were appearing before C_{α^*} in D (but that are greater than C_{α^*} in $\mathcal{O}_{up}$) are "moved" after C_{α^*} in D.

Because D is g-contiguous and satisfies Property $(*)$, then D' is a g-contiguous $\{x, y\}$-path-decomposition of G satisfying Property $(*)$. In particular, each edge of G belongs to some bag because we have ensured that all components intersecting $\bigcup_{\beta \le m \le \gamma} \bar{C}'_m$ do not appear in $(X_{h+1}, \cdots, X_{a_{C_{\alpha^*}}-1})$ (otherwise, the interval is not inclusion-maximal). It remains to prove that its weak length is at most k.

We will prove that every bag in $((X_{h+1} \cdots, X_q) \cap \bigcup_{\beta \le m \le \gamma} \bar{C}'_m) \cup (X_{b_{C_{\alpha^*}}} \cap V(P_{down}))$ has weak length at most k (that are the only bags where some vertices may be added compared with the bags of D). Since we are considering the weak length, we actually need to prove that, for every $v \in \bigcup_{\beta \le m \le \gamma} \bar{C}'_m$ and every $w \in X_{b_{C_{\alpha^*}}} \cap V(P_{down})$, $dist(v, w) \le k$. Actually, we will show that $dist(v, w) \le dist(h^*_{C'_\alpha}, w)$ (note that $dist(h^*_{C'_\alpha}, w) \le k$ since w belongs to every bag in $X_{a_{C'_\alpha}}, \cdots, X_{b_{C'_\alpha}}$).

Note that, because D is a path-decomposition, $X_h \cap X_{h+1} \cap V(P_{down}) = \{d_j\}$ and $y_2 \in X_p$, then w is between d_j and y_2 in P_{down}. Moreover, since $dist(h^*_{C'_\alpha}, d_j) > k$ and $dist(h^*_{C'_\alpha}, w) \le k$, then the shortest path between w and $h^*_{C'_\alpha}$ goes through y_2.

Let $\beta \le m \le \gamma$ such that $v \in \bar{C}'_m$ and let $1 \le \delta < a_{C'_\alpha}$ be such that $v \in X_\delta$ (in D). Because D is a path-decomposition and $d_j \in X_h$ and $w \in X_{a_{C'_\alpha}}$, there must be a vertex $w' \in X_\delta$ which is between d_j and w in P_{down}, and so $dist(v, w') \le k$ since D has weak length at most k. If the shortest path between v and w' goes through y_2, we get that $dist(v, w) \le dist(v, w') \le k$ and we are done. Otherwise, w is strictly between w' and y_2 (in particular $w \ne w'$) (because the shortest path between $h^*_{C'_\alpha}$ and w goes through y_2, the one between v and w' goes through x_2 and $s_{C'_\alpha}$ is closer to x_2 than $s_{C'_m}$). Note also that w' belongs to every bag in $X_{a_{C'_m}}, \cdots, X_{b_{C'_m}}$ and, in particular, one of these bags contains a vertex $h^*_{C'_m}$, and so $dist(h^*_{C'_m}, w') \le k$.

To sum up, $dist(w', l_{C'_m}) + dist(l_{C'_m}, h^*_{C'_m}) = dist(h^*_{C'_m}, w') \le k < dist(h^*_{C'_\alpha}, d_j) \le dist(w', l_{C'_\alpha}) + dist(l_{C'_\alpha}, h^*_{C'_\alpha})$. Because $l_{C'_\alpha}$ is between x_2 and $l_{C'_m}$ in P_{up} and the shortest path between $l_{C'_\alpha}$ and w' goes through x_2, we get that $dist(w', l_{C'_m}) \ge dist(w', l_{C'_\alpha})$ and so $dist(l_{C'_m}, h^*_{C'_m}) < dist(l_{C'_\alpha}, h^*_{C'_\alpha})$. Finally, $k \ge dist(h^*_{C'_\alpha}, w) = dist(h^*_{C'_\alpha}, r_{C'_\alpha}) + dist(r_{C'_\alpha}, w) \ge dist(r_{C'_m}, h^*_{C'_m}) + dist(r_{C'_m}, w) = dist(h^*_{C'_m}, w)$. The last inequality comes from the fact that $r_{C'_m}$ is between y_1 and $r_{C'_\alpha}$ in P_{up} and the shortest path between $l_{C'_\alpha}$ and w goes through y_1 and y_2. Hence, we get that $dist(h^*_{C'_m}, w) \le k$ and so, $dist(v, w) \le k$ by Property $(*)$.

It remains to show that D' contradicts the minimality of α^*. Let C be the component that appears in D' just after X_h.

- First, let us assume that $C \in \mathcal{C}_{up}$.

  If $C \leq C'_\alpha$ in $\mathcal{O}_{up}$, then by definition of α, D' corresponds to the first case of the proof of the Theorem, i.e., C and all components smaller than C in $\mathcal{O}_{up}$ can be moved just after X_h in D' (recall that all these components can be added with d_j by definition of α).

  Otherwise, by definition of α (and of d_j), we have that $\alpha(D') = \alpha(D) = \alpha$ and, then $C_{\alpha^*(D')} = C_{\alpha^*(D)}$. Since $C_{\alpha^*(D)}$ is "closer" from X_h in D' than in D (since D' is obtained from D by moving at least C_1 after $C_{\alpha^*(D)}$), we get that $\alpha^*(D') < \alpha^*(D)$, contradicting the minimality of $\alpha^*(D)$.

- If $C \in \mathcal{C}_{down}$, then we repeat the process. Either we fall in the first case, which contradicts the maximality of h, or we have to repeat the transformation of the second case. This leads to a new decomposition D'' with same prefix $(X_1, \cdots, X_h)$ and a component C' that appears just after this prefix in D''. If $C' \in \mathcal{C}_{down}$, then applying the above paragraph with D' and D'' instead of D and D' leads to a contradiction. Otherwise, since the prefix (and so d_j) is the same, then applying the paragraph above for C' also contradicts the minimality of $\alpha^*(D)$ (i.e., $\alpha^*(D'') < \alpha^*(D)$). □

Computation of Optimal LtR g-Contiguous Decompositions. Finally, we show how to compute LtR g-contiguous $\{x, y\}$-path-decompositions.

Theorem 7. *Let $G = (V, E)$ be a 2-connected simple outerplanar n-node graph and $x, y \in E_{out}$ such that $x \neq y$ and x and y lie on the same internal face F.*

Then, an LtR g-contiguous $\{x, y\}$-path-decomposition of G with length at most $p\ell(G, x, y) + 1$ can be computed in time $O(n + p\ell(G)^2)$.

Sketch of Proof. Note first that a greedy path-decompositions D_C based on s_C can be computed in linear time for each component $C \in \mathcal{C}$. Then, we only need to compute optimal orderings of the components in $\mathcal{C}_{up} = (C_1^u, \ldots, C_q^u)$ and in $\mathcal{C}_{down} = (C_1^d, \ldots, C_{q'}^d)$. For every $0 \leq i \leq q$, $0 \leq j \leq q'$, let $D[i, j]$ be a part of an optimal LtR g-contiguous $\{x, y\}$-path-decomposition of G, i.e. the first bag of $D[i, j]$ contains x, the last bag of $D[i, j]$ consist exactly in $\{r_{C_i^{up}}, r_{C_j^{down}}\}$. Moreover, $D[i, j]$ can be computed in constant time, i.e., $D[i, j]$ is either $D[i - 1, j] \odot D_{C_i^u} \cup \{r_{C_j^d}\}$ or $D[i, j - 1] \odot D_{C_j^d} \cup \{r_{C_i^u}\}$ depending on which one has the smallest length. Hence, this ordering can be computed in time $O(|F|^2) = O(p\ell(G)^2)$ (because F is an isometric cycle) by dynamic programming.

By Theorem 5, there exists a g-contiguous $\{x, y\}$-path-decomposition of length at most $p\ell(G, x, y) + 1$, satisfying the properties of Theorem 5. Hence, by Theorem 6, there exists a LtR g-contiguous $\{x, y\}$-path-decomposition of G satisfying the properties of Theorem 5 and with weak length at most $p\ell(G, x, y) + 1$. Finally, by Lemma 6, it follows that there exists a LtR g-contiguous $\{x, y\}$-path-decomposition of G with length at most $p\ell(G, x, y) + 1$. This guaranties the length of the computed decomposition since our algorithm computes a decomposition with minimum length among the LtR g-contiguous $\{x, y\}$-path-decompositions of G. The full proof can be found in [5]. ◇

4.3 Polynomial-Time +1 Approximation

We are finally ready to prove our main theorem.

Theorem 8. *There exists an algorithm that, for every n-node connected simple outerplanar graph $G = (V, E)$, computes in time $O(n^3(n + p\ell(G)^2))$ a path-decomposition of G with length at most $p\ell(G) + 1$.*

Proof. Due to lack of space, we assume that G is 2-connected, the full proof can be found in [5]. For every $x, y \in E_{out}$ (possibly $x = y$), the algorithm computes a $\{x, y\}$-path-decomposition of G length at most $p\ell(G, x, y) + 1$. By Lemma 2, such a decomposition with minimum length will be a path-decomposition of G with length at most $p\ell(G) + 1$.

Let us fix $x, y \in E_{out}$, we design an algorithm that computes a $\{x, y\}$-path-decomposition of G length at most $p\ell(G, x, y) + 1$ in time $O(n(n + p\ell(G)^2))$. If $x = y$, the Algorithm *Greedy* computes an optimal $\{x, y\}$-path-decomposition of G in linear time by Theorem 3 and we are done. So let us assume that $x \neq y$.

First, in linear time, the internal edges separating x and y are computed (this can be done, e.g., using SPQR trees [11]). Let $\{e_0 = x, e_1, \cdots, e_{q-1}, e_q = y\}$ where $e_i \in E_{int}$ for every $0 < i < q$ be the set of those separators in order they are met when going from x to y. For every $0 \leq i < q$, $e_i \neq e_{i+1}$ (they may intersect) and e_i and e_{i+1} share a same internal face F_i. Note that this decomposition into several connected components is done once for all at the beginning of the execution of the algorithm. It is not done anymore in the recursive calls described below and therefore it counts only for a linear time in the time complexity.

Assume first that $e_1 \neq y$. Let C' be the connected component of $G \setminus e_1$ containing (or intersecting if $e_1 \cap x \neq \emptyset$) x. Let $G_y = G[V \setminus C']$ and let $G_x = G[C' \cup e_1]$. Our algorithm first recursively computes a $\{e_1, y\}$-path-decomposition D_y of G_y with length at most $p\ell(G_y, e_1, y) + 1$ in time $O(|V(G_y)|(|V(G_y)| + p\ell(G)^2))$. Then, note that $x = e_0$ and e_1 are outer edges of G_x, $x \neq e_1$, and x and e_1 share a same internal face F_0. Hence, the condition of Theorem 7 are fulfilled and a $\{x, e_1\}$-path-decomposition D_x of G_x with length at most $p\ell(G_x, x, e_1) + 1$ can be computed in time $O(|V(G_x)| + p\ell(G)^2)$. Finally, from Lemma 3, the desired $\{x, y\}$-path-decomposition of G with length at most $p\ell(G, x, y) + 1$ is obtained from D_x and D_y. So in total, in time $O(|V(G_y)|(|V(G_y)| + k^2)) + O(|V(G_x)| + k^2) = O(n(n + p\ell(G)^2))$. In the case where $e_1 = y$, note that $x \neq y$ and x and y share an internal face F_0. Hence, the conditions of Theorem 7 are fulfilled and a $\{x, e_1\}$-path-decomposition D_x of G_x with length at most $p\ell(G, x, e_1) + 1$ can be computed in time $O(|V(G_x)| + p\ell(G)^2)$. □

5 Further Work

The next step would be to design a polynomial time exact algorithm (if it exists) to compute the pathlength of outerplanar graphs. Note that the increase of the length (+1) in our approximation algorithm comes from the contiguous property. The example of Fig. 2 shows that we cannot avoid this increase if we keep the contiguous property. Moreover, the *LtR* property has been proved from a contiguous path-decomposition. Therefore, this proof needs also to be adapted for the exact case. Another question would be to know whether our algorithm for trees can be adapted to chordal graphs. Moreover, the complexity of computing the pathlength (or treelength) of planar graphs is still open.

References

1. Belmonte, R., Fomin, F.V., Golovach, P.A., Ramanujan, M.S.: Metric dimension of bounded tree-length graphs. SIAM J. Discrete Math. **31**(2), 1217–1243 (2017)
2. Blumensath, A., Courcelle, B.: Monadic second-order definable graph orderings. Log. Methods Comput. Sci. **10**(1) (2014)
3. Bodlaender, H.L., Fomin, F.V.: Approximation of pathwidth of outerplanar graphs. J. Algorithms **43**(2), 190–200 (2002)
4. Coudert, D., Huc, F., Sereni, J.-S.: Pathwidth of outerplanar graphs. J. Graph Theory **55**(1), 27–41 (2007)
5. Dissaux, T., Nisse, N.: Pathlength of Outerplanar graphs. Research report, Inria & Université Nice Sophia Antipolis, CNRS, I3S, Sophia Antipolis, France, April 2022. https://hal.archives-ouvertes.fr/hal-03655637
6. Dourisboure, Y., Gavoille, C.: Tree-decompositions with bags of small diameter. Discrete Math. **307**(16), 2008–2029 (2007)
7. Dragan, F.F., Köhler, E., Leitert, A.: Line-distortion, bandwidth and path-length of a graph. Algorithmica **77**(3), 686–713 (2017)
8. Dragan, F.F., Leitert, A.: Minimum eccentricity shortest paths in some structured graph classes. J. Graph Algorithms Appl. **20**(2), 299–322 (2016)
9. Ducoffe, G., Legay, S., Nisse, N.: On the complexity of computing treebreadth. Algorithmica **82**(6), 1574–1600 (2020)
10. Herlihy, M., Kuhn, F., Tirthapura, S., Wattenhofer, R.: Dynamic analysis of the arrow distributed protocol. Theory Comput. Syst. **39**(6), 875–901 (2006)
11. Hopcroft, J.E., Tarjan, R.E.: Dividing a graph into triconnected components. SIAM J. Comput. **2**(3), 135–158 (1973)
12. Indyk, P.: Algorithmic applications of low-distortion geometric embeddings. In: 42nd Annual Symposium on Foundations of Computer Science, FOCS, pp. 10–33. IEEE (2001)
13. Kosowski, A., Li, B., Nisse, N., Suchan, K.: *k*-chordal graphs: From cops and robber to compact routing via treewidth. Algorithmica **72**(3), 758–777 (2015)
14. Krauthgamer, R., Lee, J.R.: Algorithms on negatively curved spaces. In: 47th Annual IEEE Symposium on Foundations of Computer Science (FOCS 2006), pp. 119–132 (2006)
15. Monien, B., Sudborough, I.H.: Min cut is NP-complete for edge weighted trees. Theor. Comput. Sci. **58**, 209–229 (1988)
16. Tenenbaum, J.B., de Silva, V., Langford, J.C. : A global geometric framework for nonlinear dimensionality reduction. Science **290**(5500), 2319–2323 (2000)

Approximations for the Steiner Multicycle Problem

Cristina G. Fernandes[1], Carla N. Lintzmayer[2(✉)],
and Phablo F. S. Moura[3]

[1] Department of Computer Science, University of São Paulo, São Paulo, Brazil
cris@ime.usp.br
[2] Center for Mathematics, Computing and Cognition, Federal University of ABC,
Santo André, São Paulo, Brazil
carla.negri@ufabc.edu.br
[3] Computer Science Department, Federal University of Minas Gerais, Belo Horizonte,
Minas Gerais, Brazil
phablo@dcc.ufmg.br

Abstract. The STEINER MULTICYCLE problem consists in, given a complete graph G, a weight function $w \colon E(G) \to \mathbb{Q}_+$, and a partition of $V(G)$ into terminal sets, finding a minimum-weight collection of disjoint cycles in G such that, for every terminal set T, all vertices of T are in a same cycle of the collection. This problem, which is motivated by applications on routing problems with pickup and delivery locations, generalizes the TRAVELING SALESMAN problem (TSP) and therefore is hard to approximate in general. Using an algorithm for the SURVIVABLE NETWORK DESIGN problem and T-joins, we obtain a 3-approximation for its metric case, improving on the previous best 4-approximation. Furthermore, inspired by a result by Papadimitriou and Yannakakis for the $\{1, 2\}$-TSP, we present an $(11/9)$-approximation for the particular case of the STEINER MULTICYCLE in which each edge weight is 1 or 2. This algorithm can be adapted into a $(7/6)$-approximation when every terminal set contains at least 4 vertices.

Keywords: Approximation algorithms · Steiner problems · Steiner multicycle · Traveling salesman problem

1 Introduction

In the STEINER MULTICYCLE problem, one is given a complete graph G, a weight function $w \colon E(G) \to \mathbb{Q}_+$, and a collection $\mathcal{T} \subseteq \mathcal{P}(V(G))$ of pairwise disjoint non-unitary sets of vertices, called *terminal sets*. We say that a cycle C *respects* $\mathcal{T}$ if, for all $T \in \mathcal{T}$, either every vertex of T is in C or no vertex of T is in C, and a set $\mathcal{C}$ of vertex-disjoint cycles *respects* $\mathcal{T}$ if all cycles in $\mathcal{C}$ respect $\mathcal{T}$ and every vertex in a terminal set is in some cycle of $\mathcal{C}$. The cost of such set $\mathcal{C}$ is the sum of the edge weights over all cycles in $\mathcal{C}$, a value naturally denoted by $w(\mathcal{C})$. The goal of the STEINER MULTICYCLE problem is to find a set of vertex-disjoint

© Springer Nature Switzerland AG 2022
A. Castañeda and F. Rodríguez-Henríquez (Eds.): LATIN 2022, LNCS 13568, pp. 188–203, 2022.
https://doi.org/10.1007/978-3-031-20624-5_12

cycles of minimum cost that respects $\mathcal{T}$. We denote by $\mathrm{opt}(G, w, \mathcal{T})$ the cost of such a minimum cost set. Note that the number of cycles in a solution might be smaller than $|\mathcal{T}|$, that is, it might be cheaper to join some terminal sets in the same cycle.

We consider that, in a graph G, a cycle is a non-empty connected subgraph of G all of whose vertices have degree two. Consequently, such cycles have at least three vertices. Here, as it is possible for a set $T \in \mathcal{T}$ to have only two vertices, we would like to consider a single edge as a cycle, of length two, whose cost is twice the weight of the edge, so that the problem also includes solutions that choose to connected some set from $\mathcal{T}$ with two vertices through such a length-2 cycle. So, for each set $T \in \mathcal{T}$ with $|T| = 2$, we duplicate in G the edge linking the vertices in T, and allow the solution to contain length-2 cycles.

The STEINER MULTICYCLE problem is a generalization of the TRAVELING SALESMAN problem (TSP), thus it is NP-hard and its general form admits the same inapproximability results as the TSP. It was proposed by Pereira *et al.* [16] as a generalization of the so called STEINER CYCLE problem (see Salazar-González [18]), with the assumption that the graph is complete and the weight function satisfies the triangle inequality. In this case, we may assume that the terminal sets partition the vertex set. Indeed, because the graph is complete and the weight function is metric, any solution containing non-terminal vertices does not have its cost increased by shortcutting these vertices (that is, removing them and adding the edge linking their neighbors in the cycle). We refer to such an instance of the STEINER MULTICYCLE problem as *metric*, and the problem restricted to such instances as the metric STEINER MULTICYCLE problem.

Pereira *et al.* [16] presented a 4-approximation algorithm for the metric STEINER MULTICYCLE problem, designed Refinement Search and GRASP based heuristics, and proposed an integer linear programming formulation for the problem. Lintzmayer *et al.* [13] then considered the version restricted to the Euclidean plane and presented a randomized approximation scheme for it, which combines some techniques for the EUCLIDEAN TSP [2] and for the EUCLIDEAN STEINER FOREST [5].

On the practical side, the STEINER MULTICYCLE problem models a collaborative less-than-truckload problem with pickup and delivery locations. In this scenario, several companies, operating in the same geographic regions, need to periodically transport products between different locations. To reduce the costs of transporting their goods, these companies can collaborate in order to create routes for shared cargo vehicles that visit the places defined by them for the collection and delivery of their products (see Ergun *et al.* [9,10]).

In this paper, we address the metric case of the problem as well as the so called $\{1, 2\}$-STEINER MULTICYCLE problem, in which the weight of each edge is either 1 or 2. Note that the latter is a particular case of the metric one and it is a generalization of the $\{1, 2\}$-TSP, therefore it is also APX-hard [15]. In some applications, there might be little information on the actual cost of the connections between points, but there might be at least some distinction between cheap connections and expensive ones. These situations could be modeled as instances of the $\{1, 2\}$-STEINER MULTICYCLE. For the metric case, we present a 3-approximation, improving on the previous best known. The proposed algorithm uses an approximate solution S for a derived instance of the SURVIVABLE NETWORK DESIGN problem and a minimum weight T-join in S, where T is the set of odd degree vertices in S. Considering the $\{1,2\}$-STEINER MULTICYCLE, we design an $\frac{11}{9}$-approximation following the strategy for the $\{1,2\}$-TSP proposed by Papadimitrou and Yannakakis [15].

The 3-approximation for the metric STEINER MULTICYCLE is presented in Sect. 2, together with a discussion involving the previous 4-approximation and the use of perfect matchings on the set of odd degree vertices of intermediate structures. The $\{1,2\}$-STEINER MULTICYCLE problem is addressed in Sect. 3, and we make some final considerations in Sect. 4.

2   Metric Steiner Multicycle Problem

An instance for the STEINER MULTICYCLE is also an instance for the well-known STEINER FOREST problem [21, Chapter 22], but the goal in the latter is to find a minimum weight forest in the graph that connects vertices in the same terminal set, that is, every terminal set is in some connected component of the forest. The optimum value of the STEINER FOREST is a lower bound on the optimum for the STEINER MULTICYCLE: one can produce a feasible solution for the STEINER FOREST from an optimal solution for the STEINER MULTICYCLE by throwing away one edge in each cycle, without increasing its cost.

The existing 4-approximation [16] for the metric STEINER MULTICYCLE problem is inspired in the famous 2-approximation for the metric TSP [17], and consists in doubling the edges in a Steiner forest for the terminal sets, and shortcutting an Eulerian tour in each of its components to a cycle. As there are 2-approximations for the STEINER FOREST problem, this leads to a 4-approximation.

It is tempting to try to use a perfect matching on the odd degree vertices of the approximate Steiner forest solution, as Christofides' algorithm [6] does to achieve a better ratio for the metric TSP. However, the best upper bound we can prove so far on such a matching is the weight of the approximate Steiner forest solution, which implies that such a matching has weight at most twice the optimum. With this bound, we also derive a ratio of at most 4.

Another problem that can be used with this approach is known as the SURVIVABLE NETWORK DESIGN problem [21, Chapter 23]. An instance for this problem consists of the following: a graph G, a weight function $w \colon E(G) \to \mathbb{Q}_+$, and a non-negative integer r_{ij} for each pair of vertices i, j with $i \neq j$, representing a connectivity requirement. The goal is to find a minimum weight subgraph G' of G such that, for every pair of vertices $i, j \in V(G)$ with $i \neq j$, there are at least r_{ij} edge-disjoint paths between i and j in G'.

From an instance of the STEINER MULTICYCLE problem, we can naturally define an instance of the SURVIVABLE NETWORK DESIGN problem: set $r_{ij} = 2$ for every two vertices i, j in the same terminal set, and set $r_{ij} = 0$ otherwise. As all vertices are terminals, all connectivity requirements are defined in this way. The optimum value of the SURVIVABLE NETWORK DESIGN problem is also a lower bound on the optimum for the STEINER MULTICYCLE problem: indeed an optimal solution for the STEINER MULTICYCLE problem is a feasible solution for the SURVIVABLE NETWORK DESIGN problem with the same cost.

There also exists a 2-approximation for the SURVIVABLE NETWORK DESIGN problem [12]. By applying the same approach of the 2-approximation for the metric TSP, of doubling edges and shortcutting, we achieve again a ratio of 4 for the metric STEINER MULTICYCLE. However, next we will show that one can obtain a 3-approximation for the metric STEINER MULTICYCLE problem, from a 2-approximate solution for the SURVIVABLE NETWORK DESIGN problem, using not a perfect matching on the odd degree vertices of such solution, but the related concept of T-joins.

2.1   A 3-Approximation Algorithm

Let T be a set of vertices of even size in a graph G. A set J of edges in G is a T-join if the collection of vertices of G that are incident to an odd number of edges in J is exactly T. Any perfect matching on the vertices of T is a T-join, so T-joins are, in some sense, a generalization of perfect matching on a set T. It is known that a T-join exists in G if and only if the number of vertices from T in each component of G is even. Moreover, there are polynomial-time algorithms that, given a connected graph G, a weight function $w\colon E(G) \to \mathbb{Q}_+$, and an even set T of vertices of G, find a minimum weight T-join in G. For these and more results on T-joins, we refer the reader to the book by Schrijver [19, Chapter 29].

The idea of our 3-approximation is similar to Christofides [6]. It is presented in Algorithm 1. Let $(G, w, \mathcal{T})$ be a metric instance of the STEINER MULTICYCLE problem. The first step is to build the corresponding SURVIVABLE NETWORK DESIGN problem instance, and to obtain a 2-approximate solution G' for this instance. The procedure 2APPROXSND represents the algorithm by Jain [12] for the SURVIVABLE NETWORK DESIGN. The second step considers the set T of the vertices in G' of odd degree and finds a minimum weight T-join J in G'. The procedure MINIMUMTJOIN represents the algorithm by Edmonds and Johnson [8] for this task. Finally, the Eulerian graph H obtained from G' by doubling the edges in J is built and, by shortcutting an Eulerian tour for each component of H, one obtains a collection $\mathcal{C}$ of cycles in G that is the output of the algorithm. The procedure SHORTCUTTING represents this part in Algorithm 1.

Because the number of vertices of odd degree in any connected graph is even, the number of vertices with odd degree in each component of G' is even. Therefore there is a T-join in G'. Moreover, the collection $\mathcal{C}$ produced by Algorithm 1 is indeed a feasible solution for the STEINER MULTICYCLE.

Next we prove that the proposed algorithm is a 3-approximation.

Theorem 1. *Algorithm 1 is a 3-approximation for the metric* STEINER MULTI-CYCLE *problem.*

Proof. Let us first observe that it suffices to prove that $w(J) \leq \frac{1}{2}w(G')$. Indeed, because G' is a 2-approximate solution for the SURVIVABLE NETWORK DESIGN

Algorithm 1. STEINERMULTICYCLEAPPROX_GENERAL(G, w, $\mathcal{T}$)

Input: a complete graph G, a weight function $w\colon E(G) \to \mathbb{Q}_+$ satisfying the triangle
    inequality, and a partition $\mathcal{T} = \{T_1, \ldots, T_k\}$ of $V(G)$
Output: a collection $\mathcal{C}$ of cycles that respects $\mathcal{T}$
1: $r_{ij} \leftarrow 2$ for every $i, j \in T_a$ for some $1 \leq a \leq k$
2: $r_{ij} \leftarrow 0$ for every $i \in T_a$ and $j \in T_b$ for $1 \leq a < b \leq k$
3: $G' \leftarrow 2\text{APPROXSND}(G, w, r)$
4: Let T be the set of odd degree vertices in G'
5: Let w' be the restriction of w to the edges in G'
6: $J \leftarrow \text{MINIMUMTJOIN}(G', w', T)$
7: $H \leftarrow G' + J$
8: $\mathcal{C} \leftarrow \text{SHORTCUTTING}(H)$
9: **return** $\mathcal{C}$

---

problem, and the optimum for this problem is a lower bound on $\text{opt}(G, w, \mathcal{T})$, we
have that $w(G') \leq 2\,\text{opt}(G, w, \mathcal{T})$. Hence we deduce that $w(J) \leq \text{opt}(G, w, \mathcal{T})$,
and therefore that $w(\mathcal{C}) \leq w(G') + w(J) \leq 3\,\text{opt}(G, w, \mathcal{T})$. We now proceed to
show that inequality $w(J) \leq \frac{1}{2}w(G')$ holds.

A *bridge* is an edge uv in a graph whose removal leaves u and v in different
components of the resulting graph. First, observe that we can delete from G'
any bridges, and the remaining graph, which we still call G', remains a solution
for the SURVIVABLE NETWORK DESIGN problem instance. Indeed a bridge is
not enough to assure the connectivity requirement between two vertices in the
same terminal set, so it will not separate any such pair of vertices and hence it
can be removed. In other words, we may assume that each component of G' is
2-edge-connected.

Edmonds and Johnson [8] gave an exact description of a polyhedra related
to T-joins. This description will help us to prove the claim. For a set S of edges
in a graph (V, E), let $v(S)$ denote the corresponding $|E|$-dimensional incidence
vector (with 1 in the i-th coordinate if edge i lies in S and 0 otherwise). For a
set X of vertices, let $\delta(X)$ denote the set of edges with one endpoint in X and
the other in $V \setminus X$. An *upper T-join* is any superset of a T-join. Let $P(G, T)$ be
the convex hull of all vectors $v(J)$ corresponding to the incidence vector of upper
T-joins J of a graph $G = (V, E)$. The set $P(G, T)$ is called the *up-polyhedra of
T-joins*, and it is described by

$$\sum_{e \in \delta(W)} x(e) \geq 1 \quad \text{for every } W \subseteq V \text{ such that } |W \cap T| \text{ is odd}, \qquad (1)$$

$$0 \leq x(e) \leq 1 \quad \text{for every edge } e \in E. \qquad (2)$$

(For more on this, see [19, Chapter 29].)

So, as observed in [3], any feasible solution x to the system of inequalities
above can be written as a convex combination of upper T-joins, that is, $x = \sum \alpha_i\, v(J_i)$, where $0 \leq \alpha_i \leq 1$ and $\sum_i \alpha_i = 1$, leading to the following.

Corollary 1 (Corollary 1 in [3]). *If all the weights $w(e)$ are non-negative, then, given any feasible assignment $x(e)$ satisfying the inequalities above, there exists a T-join with weight at most $\sum_{e \in E} w(e)x(e)$.*

Recall that, for each component C of G', $|V(C) \cap T|$ is even. Hence, for every $W \subseteq V(G')$ such that $|W \cap T|$ is odd, there must exist a component C of G' with $V(C) \cap W \neq \emptyset$, and $V(C) \setminus W \neq \emptyset$. As a consequence, it holds that $|\delta(W)| \geq 2$ because every component of G' is 2-edge-connected. Consider now the $|E(G')|$-dimensional vector $\bar{x}$ which assigns value $1/2$ to each edge of G'. From the discussion above, it is clear that $\bar{x}$ satisfies inequalities (1) and (2) for G' and T. Then Corollary 1 guarantees that there is a T-join J in G' such that $w(J) \leq \frac{1}{2} w(G')$. This completes the proof of the theorem.   □

2.2   Matchings, T-Joins, and Steiner Forests

Because G is complete and w is metric, the proof of Theorem 1 in fact implies that a minimum weight perfect matching in the graph $G[T]$ has weight at most $w(G')/2$, and therefore at most $\mathrm{opt}(G, w, T)$. However, we have no direct proof for this fact; only this argument that goes through a minimum weight T-join. But this fact means that one can exchange line 6 to compute, instead, a minimum weight perfect matching J in $G[T]$.

We investigated the possibility that one could achieve a ratio of 3 using a Steiner forest instead of a survivable network design solution. However the idea of using a T-join does not work so well with the Steiner forest, once its components are not 2-edge-connected. Indeed, if T is the set of odd degree vertices in a Steiner forest F, a bound as in the proof of Theorem 1 on a minimum weight T-join in F would not hold in general: there are examples for which such a T-join in F has weight $w(F)$.

In this paragraph, let $\mathrm{opt}_{\mathrm{SND}}$ denote the optimum value for the SURVIV-ABLE NETWORK DESIGN instance used in Algorithm 1, and $\mathrm{opt}_{\mathrm{SF}}$ denote the optimum value for the STEINER FOREST instance used in the 4-approximation from the literature [16]. Let $\mathrm{opt}_{\mathrm{SMC}}$ be the STEINER MULTICYCLE optimum value. Note that $\mathrm{opt}_{\mathrm{SF}} \leq \mathrm{opt}_{\mathrm{SND}} \leq \mathrm{opt}_{\mathrm{SMC}} \leq 2\,\mathrm{opt}_{\mathrm{SF}}$, where the last inequality holds because a duplicated Steiner forest solution leads to a cheaper feasible solution for the SURVIVABLE NETWORK DESIGN and the STEINER MULTICYCLE instances. Let G' and J be the subgraph and the T-join used in Algorithm 1, respectively, and let M be a minimum weight perfect matching in $G[T]$. Then $w(M) \leq w(J) \leq \frac{1}{2}w(G') \leq \mathrm{opt}_{\mathrm{SND}} \leq w(G')$. (For the first inequality, recall that J is a T-join in G' while M is a minimum weight perfect matching in $G[T]$.) If T' is the set of odd degree vertices in an optimal Steiner forest and M' is a minimum weight perfect matching in $G[T']$, then $w(M') \leq 2\mathrm{opt}_{\mathrm{SF}}$, and there are instances for which this upper bound is tight. So, as far as we know, there might be an instance where $w(M') > \mathrm{opt}_{\mathrm{SMC}}$. Even if this is not the case, in fact, what we can compute in polynomial time is a minimum weight perfect matching M'' for the set of odd degree vertices in a 2-approximate Steiner forest solution, so it would still be possible that $w(M'') > \mathrm{opt}_{\mathrm{SMC}}$ for some instances. We tried to find an instance where this is the case, but we have not succeeded so far.

3   {1, 2}-Steiner Multicycle Problem

In this section, we will address the particular case of the metric STEINER MUL-
TICYCLE problem that allows only edge weights 1 or 2.

A *2-factor* in a graph is a set of vertex-disjoint cycles that span all vertices of
the graph. It is a well-known result that there exists a polynomial-time algorithm
for finding a 2-factor of minimum weight in weighted graphs [14,20].

The algorithm for this case of the STEINER MULTICYCLE problem starts from
a minimum weight 2-factor of the given weighted graph, and then repeatedly
joins two cycles until a feasible solution is obtained. The key to guarantee a
good approximation ratio is a clever choice of the cycles to join at each step. To
proceed with the details, we need the following definitions.

Let $(G, w, \mathcal{T})$ be an instance of the STEINER MULTICYCLE problem with
$w \colon E(G) \to \{1, 2\}$. Recall that $\bigcup_{T \in \mathcal{T}} T = V(G)$, and that, for each set $T \in \mathcal{T}$
with $|T| = 2$, we duplicated in G the edge linking the vertices in T, to allow
the solution to contain length-2 cycles. We say an edge $e \in E(G)$ is an *i-edge* if
$w(e) = i$, for $i \in \{1, 2\}$. A cycle containing only 1-edges is called *pure*; otherwise,
it is called *nonpure*.

All steps of the procedure are summarized in Algorithm 2 below. The auxiliary procedures shall be explained later on.

Algorithm 2. STEINERMULTICYCLEAPPROX_12WEIGHTS$(G, w, \mathcal{T})$

Input: a complete graph G, a weight function $w \colon E(G) \to \{1, 2\}$, and a partition
    $\mathcal{T} = \{T_1, \ldots, T_k\}$ of $V(G)$
Output: a collection $\mathcal{C}$ of cycles that respects $\mathcal{T}$
 1: $F \leftarrow$ SPECIAL2FACTOR(G, w)
 2: $B \leftarrow$ BUILDBIPARTITEGRAPH$(G,w,\mathcal{T},F)$
 3: $M \leftarrow$ MAXIMUMMATCHING(B)
 4: Let D be a digraph such that $V(D) = F$ and there is an arc $(C, C') \in E(D)$ if C
    is matched by M to a vertex of C'
 5: $D' \leftarrow$ SPECIALSPANNINGGRAPH(D)
 6: $\mathcal{C}' \leftarrow$ JOINCOMPONENTCYCLES(F, D') (see Section 3.1)
 7: $\mathcal{C} \leftarrow$ JOINDISRESPECTINGCYCLES$(\mathcal{C}', D', \mathcal{T})$ (see Section 3.1)
 8: **return** $\mathcal{C}$

Procedure SPECIAL2FACTOR finds a minimum weight 2-factor F of (G, w)
with the two following properties:

(i) F contains at most one nonpure cycle; and
(ii) if F contains a nonpure cycle, no 1-edge in G connects an endpoint of a
     2-edge in the nonpure cycle to a pure cycle in F.

Given any minimum weight 2-factor F', one can construct in polynomial time a 2-
factor F from F' having properties (i) and (ii) as follows. To ensure property (i),
recall that the graph is complete, so we repeatedly join two nonpure cycles by

removing one 2-edge from each and adding two appropriate edges that turn them into one cycle. This clearly does not increase the weight of the 2-factor and reduces the number of cycles. To ensure property (ii), while there is a 1-edge yz in G connecting a 2-edge xy of the nonpure cycle to a 1-edge wz of a pure cycle, we remove xy and wz and add yz and xw, reducing the number of cycles without increasing the weight of the 2-factor. The resulting 2-factor is returned by SPECIAL2FACTOR.

In order to modify F into a collection of cycles that respect $\mathcal{T}$, without increasing too much its weight, Algorithm 2 builds some auxiliary structures that capture how the cycles in F attach to each other.

The second step of Algorithm 2 is to build a bipartite graph B (line 2) as follows. Let $V(B) = V(G) \cup \{C \in F \colon C$ is a pure cycle$\}$ and there is an edge vC in $E(B)$ if (i) $v \notin V(C)$ and C does not respect $\mathcal{T}$, and (ii) there is a vertex $u \in V(C)$ such that uv is a 1-edge. Note that the only length-2 cycles in G, and thus in F, are those that connecting a terminal set of size 2. So such cycles respect $\mathcal{T}$ and, hence, if they are in B (that is, if they are pure), they are isolated vertices in B. Procedure MAXIMUMMATCHING in line 3 computes in polynomial time a maximum matching M in B (e.g., using Edmonds' algorithm [7]).

Algorithm 2 then proceeds by building a digraph D where $V(D) = F$ and there is an arc $(C, C') \in E(D)$ if C is matched by M to a vertex of C'. Note that the vertices of D have outdegree 0 or 1, and the cycles in B unmatched by M have outdegree 0 in D. In particular, all pure length-2 cycles in F have outdegree 0 in D, because they are isolated in B, and therefore unmatched. If there is a nonpure cycle in F, it also has outdegree 0 in D. Therefore, any length-2 cycle in F, pure or nonpure, has outdegree 0 in D. These vertices with outdegree 0 in D might however have indegree different from 0. Next, Algorithm 2 applies procedure SPECIALSPANNINGGRAPH(D) to find a spanning digraph D' of D whose components are in-trees of depth 1, length-2 paths, or trivial components that correspond to isolated vertices of D. This takes linear time, and consists of a procedure described by Papadimitrou and Yannakakis [15], applied to each nontrivial component of D. See Fig. 1 for an example of these constructions.

At last, Algorithm 2 joins some cycles of F in order to obtain a collection of cycles that respect $\mathcal{T}$. This will happen in two phases. In the first phase, we join cycles that belong to the same component of D'. In the second (and last) phase, we repeatedly join cycles if they have vertices from the same set in $\mathcal{T}$, to obtain a feasible solution to the problem. This final step prioritizes joining cycles that have at least one 2-edge.

Details of these two phases, done by procedures JOINCOMPONENTCYCLES and JOINDISRESPECTINGCYCLES, as well as the analysis of the cost of joining cycles are given in Sect. 3.1. For now, observe that all cycles at the end of this process respect $\mathcal{T}$. Also note that length-2 cycles exist in the final solution only if they initially existed in F and connected terminals of some set $T \in \mathcal{T}$ with $|T| = 2$. The analysis of the approximation ratio of the algorithm is discussed in Sect. 3.2.

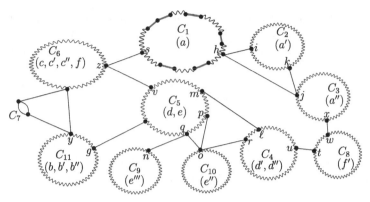

(a) Original graph G and the 2-factor (depicting only 1-edges in black and straight lines and some 2-edges in bold and red lines; squiggly lines correspond to one or more 1-edges). Inside each C_i, in parenthesis, we list some of the terminal vertices it contains.

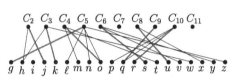

(b) Bipartite graph B and a matching M highlighted in red. Note that there is no edge incident to C_{11} because it already respects $\mathcal{T}$.

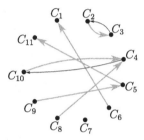

(c) Digraph D and corresponding subgraph D' highlighted in green.

Fig. 1. Auxiliar graphs and structures built by Algorithm 2. (Color figure online)

3.1   Joining Cycles

In the first phase, we join cycles in F if they belong to the same component of D', which can be either an in-tree of depth 1 or a length-2 path.

An in-tree of depth 1 of D' consists of a root C and some other cycles $\{C_j\}_{j=1}^{t}$, with $t \geq 1$. Note that each arc (C_j, C) can be associated with a 1-edge from G such that no two edges incide on the same vertex in C, because they came from the matching M. Also note that, if the nonpure cycle or a length-2 cycle appears in some in-tree, it could only be the root C. Let v_j be the endpoint in C of the edge associated with arc (C_j, C), for every $j \in \{1, \ldots, t\}$. Rename the cycles $\{C_j\}_{j=1}^{t}$ so that, if we go through the vertices of C in order, starting from v_1, these vertices appear in the order $v_1, \ldots, v_t$. We join all cycles in this in-tree into one single cycle in the following manner. For each v_i in C, if v_{i+1} is adjacent to v_i in C, then we join C_i and C_{i+1} with C as in Fig. 2a. Otherwise, we join C and C_i as in Fig. 2b. We shall consider that the new cycle contains at least one 2-edge.

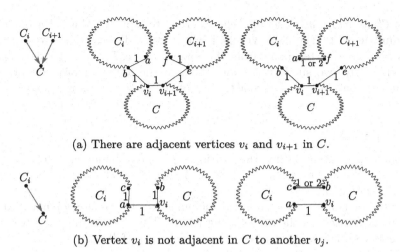

(a) There are adjacent vertices v_i and v_{i+1} in C.

(b) Vertex v_i is not adjacent in C to another v_j.

Fig. 2. Joining cycles that belong to in-trees of D' into a unique cycle. The bold red edges will be considered as 2-edges even if they are 1-edges. (Color figure online)

As for a component of D' which is a length-2 path, let C_i, C_j, and C_k be the three cycles that compose it, being C_i the beginning of the path and C_k its end. Note that if the nonpure cycle appears in some length-2 path, it could only be C_k. The arcs (C_i, C_j) and (C_j, C_k) are also associated with 1-edges of G, but now it may be the case that such edges share their endpoint in C_j. If that is not the case, then we join these three cycles as shown in Fig. 3a. Otherwise, we join the three cycles as shown in Fig. 3b. We shall also consider that the new cycle contains at least one 2-edge.

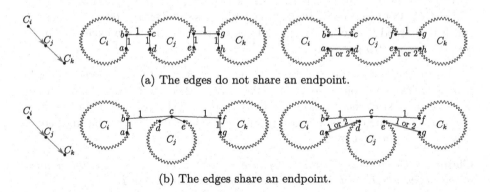

(a) The edges do not share an endpoint.

(b) The edges share an endpoint.

Fig. 3. Joining cycles that belong to length-2 paths of D' into a unique cycle. The bold red edges will be considered as 2-edges even if they are 1-edges. (Color figure online)

Let $\mathcal{C}'$ be the resulting set of cycles after the first phase. This is the output of procedure JOINCOMPONENTCYCLES. It may still be the case that two separated cycles in $\mathcal{C}'$ contain terminals from the same set $T \in \mathcal{T}$. So, in the last phase, while there are two such cycles, join them in the following order of priority: both cycles contain a 2-edge, exactly one of the cycles contains a 2-edge, none of the cycles contain a 2-edge. The resulting set of cycles of this phase, denoted by $\mathcal{C}$, is computed by JOINDISRESPECTINGCYCLES and is the one returned by Algorithm 2.

Now we proceed analyzing the cost increase caused by joining cycles in these two phases. Note that $w(\mathcal{C})$ is equal to $w(F)$ plus some value due to the increases caused by joining cycles.

For the first phase, we charge the increment of the cost for joining cycles to some of the vertices in the cycles being joined. This is done in such a way that each vertex is charged at most once according to the following.

Claim 2. *Each vertex not incident to a 2-edge of F is charged at most $2/9$ during the first phase, and no other vertex is charged.*

Proof. Consider an in-tree of depth 1 with root C and cycles $C_1, \ldots, C_t$ with $t \geq 1$. When we join cycles C_i and C_{i+1} with C, as in Fig. 2a, note that the increase on the cost is at most 1. We charge this cost to the vertices in C_i and C_{i+1}, which are at least 6 (3 per cycle), thus costing at most $1/6$ per vertex. When we only join a cycle C_i with C, as in Fig. 2b, the increase is also at most 1. We charge this cost to the vertices in C_i and also to the two vertices involved in C. Since there are at least 3 vertices in C_i, each of these vertices is charged at most $1/5$. Note that indeed each vertex is charged at most once. Moreover, if C is the nonpure cycle, then, by property (ii), the edges in C incident to v_i and to the next vertex in C must be 1-edges.

Consider now a length-2 path with vertices C_i, C_j, and C_k. The cycles C_i, C_j, and C_k were joined, as in Figs. 3a and 3b, so the extra cost is at most 2, which is charged to the at least 9 vertices that belong to these cycles, giving a cost of at most $2/9$ per vertex. □

As for the last phase, the increase in the cost will be considered for each pair of cycles being joined. If both cycles contain 2-edges, joining them will not increase the cost of the solution. If only one of the cycles contains a 2-edge, then the increase in the cost is at most 1. Joining cycles that do not contain 2-edges may increase the cost by 2.

Claim 3. *The increase in the last phase is at most c_p, where c_p is the number of pure cycles in F that do not respect $\mathcal{T}$ and are isolated in D'.*

Proof. In the last phase, note that cycles that were generated in the first phase will always contain a 2-edge. Therefore, the only possible increases in cost come from joining one of these c_p cycles. The increase is at most 2 if two such cycles are joined, and at most 1 if one such cycle is joined to some cycle other than these c_p ones. So the increase in this phase is at most c_p. □

3.2   Approximation Ratio

Theorem 4 shows how Algorithm 2 guarantees an 11/9 approximation ratio while Corollary 2 shows a case in which Algorithm 2 can be adapted to guarantee a 7/6 approximation ratio.

Theorem 4. *Algorithm 2 is an $\frac{11}{9}$-approximation for the $\{1,2\}$-STEINER MULTICYCLE problem.*

Proof. Let $(G, w, \mathcal{T})$ be an instance of the $\{1,2\}$-STEINER MULTICYCLE problem. Let $n = |V(G)|$ and denote by $e_2(X)$ the total amount of 2-edges in a collection X of cycles.

We start with two lower bounds on $\mathrm{opt}(G, w, \mathcal{T})$. Let F be the 2-factor used in Algorithm 2 when applied to $(G, w, \mathcal{T})$. The first one is $w(F)$, because any solution for STEINER MULTICYCLE problem is a 2-factor in G. Thus

$$\mathrm{opt}(G, w, \mathcal{T}) \geq w(F) = n + e_2(F). \tag{3}$$

The other one is related to pure cycles in F. Consider an optimal solution $\mathcal{C}^*$ for instance $(G, w, \mathcal{T})$. Thus $\mathrm{opt}(G, w, \mathcal{T}) = n + e_2(\mathcal{C}^*)$. Let $C_1^*, \ldots, C_r^*$ be the cycles of $\mathcal{C}^*$, where $C_i^* = (v_{i0}, \ldots, v_{i|C_i^*|})$ for each $i \in \{1, \ldots, r\}$, with $v_{i0} = v_{i|C_i^*|}$. Let $U = \{v_{ij} : i \in \{1, \ldots, r\}, j \in \{0, \ldots, |C_i^*| - 1\}$, and $v_{ij}v_{i\,j+1}$ is a 2-edge$\}$ and note that $|U| = e_2(\mathcal{C}^*)$. Let ℓ be the number of pure cycles in the 2-factor F that contain vertices in U. Clearly $e_2(\mathcal{C}^*) \geq \ell$, which gives us

$$\mathrm{opt}(G, w, \mathcal{T}) \geq n + \ell. \tag{4}$$

Now let $\mathcal{C}$ be the collection of cycles produced by Algorithm 2 for input $(G, w, \mathcal{T})$. Let us show an upper bound on the cost of $\mathcal{C}$. Solution $\mathcal{C}$ has cost $w(F)$ plus the increase in the cost made in the first phase, and then in the final phase of joining cycles. Let us start bounding the total cost increase in the first phase. Let c_p be as in Claim 3. Recall that these c_p cycles are not matched by M. Let $n(c_p)$ be the number of vertices in these c_p cycles, and note that $n(c_p) \geq 3c_p$, because each such cycle does not respect $\mathcal{T}$ and hence has at least three vertices. By Claim 2, the vertices incident to 2-edges of F are never charged. So there are at least $e_2(F)$ vertices of the nonpure cycle of F not charged during the first phase. Thus, at most $n - n(c_p) - e_2(F) \leq n - 3c_p - e_2(F)$ vertices were charged in the first phase. Also, by Claim 2, each such vertex was charged at most 2/9.

By Claim 3, the increase in this phase is at most c_p. Thus we have

$$
\begin{aligned}
w(\mathcal{C}) &\leq w(F) + \frac{2}{9}(n - 3c_p - e_2(F)) + c_p \\
&= n + e_2(F) + \frac{2}{9}(n - 3c_p - e_2(F)) + c_p \\
&= \frac{11}{9}n + \frac{7}{9}e_2(F) + \frac{1}{3}c_p \\
&\leq \frac{11}{9}n + \frac{7}{9}e_2(F) + \frac{1}{3}\ell \\
&\leq \frac{7}{9}(n + e_2(F)) + \frac{4}{9}(n + \ell) \\
&\leq \frac{7}{9}\operatorname{opt}(G, w, \mathcal{T}) + \frac{4}{9}\operatorname{opt}(G, w, \mathcal{T}) = \frac{11}{9}\operatorname{opt}(G, w, \mathcal{T}),
\end{aligned}
$$

$\qquad$ (5)

$\qquad$ (6)

where (5) holds by Claim 5, and (6) holds by (3) and (4). It remains to prove the following.

Claim 5. $c_p \leq \ell$.

Proof. Recall that c_p is the number of pure cycles in F that are isolated in D' and do not respect $\mathcal{T}$, and observe that $\ell \leq |U|$.

We will describe a matching in the bipartite graph B with at most ℓ unmatched cycles. From this, because M is a maximum matching in B and there are at least c_p cycles not matched by M, we conclude that $c_p \leq \ell$.

For each $i \in \{1, \ldots, r\}$, go through the vertices of C_i^* from $j = 0$ to $|C_i^*| - 1$ and if, for the first time, we find a vertex $v_{ij} \notin U$ that belongs to a pure cycle C (which does not respect $\mathcal{T}$) such that $v_{i\,j+1}$ is not in C, we match C to $v_{i\,j+1}$ in B. Note that, as $v_{ij} \notin U$, the edge between C and $v_{i\,j+1}$ is indeed in B. Every pure cycle that does not respect $\mathcal{T}$ will be matched by this procedure, except for at most ℓ. $\diamond$ $\qquad\qquad$ $\square$

This analysis is tight. Consider the instance depicted in Fig. 4a, with 9 vertices and $\mathcal{T} = \{\{a_1, a_2, a_3\}, \{b_1, b_2, c_1, c_2, d_1, d_2\}\}$. There is a Hamiltonian cycle in the graph with only 1-edges, so the optimum costs 9. However, there is also a 2-factor of cost 9 consisting of the three length-3 cycles C_1, C_2 and C_3, as in Fig. 4a. The matching in the graph B might correspond to the 1-edge between C_2 and C_1, and the 1-edge between C_3 and C_2, as in Fig. 4b. This leads to a length-2 path in D', as in Fig. 4c. The process of joining these cycles, as the algorithm does, might lead to an increase of 2 in the cost, resulting in the solution of cost 11 depicted in Fig. 4e, which achieves a ratio of exactly $11/9$. This example can be generalized to have $n = 9k$ vertices, for any positive integer k.

Similarly to what Papadimitrou and Yannakakis [15] achieve for the $\{1, 2\}$-TSP, we also derive the following.

Corollary 2. *Algorithm 2 is a $\frac{7}{6}$-approximation for the $\{1, 2\}$-STEINER MULTI-CYCLE problem when $|T| \geq 4$ for all $T \in \mathcal{T}$.*

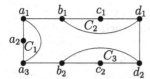

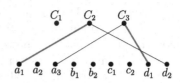

(a) Initial graph G and 2-factor F. All depicted edges are 1-edges while the missing ones are 2-edges.

(b) Bipartite graph built from F and matching M highlighted.

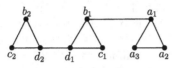

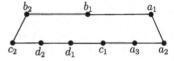

(c) Digraph D which coincides with D'.

(d) Joining cycles of the length-2 path in D'.

(e) Final solution C, with two 2-edges, and cost 11.

Fig. 4. Tight example for Algorithm 2. (Color figure online)

Proof. For weights 1 and 2, there is a polynomial-time algorithm that computes a minimum-weight 2-factor that contains no triangle [11, Section 3, Chapter 3]. Using this algorithm within SPECIAL2FACTOR in Algorithm 2, we can guarantee that there are at least 4 vertices per cycle in the produced collection C. The charging argument presented in Claim 2 can use the fact that the cycles have length at least 4, which increases the number of vertices to distribute the cost increase. For instance, when we join a cycle C_i with C, as in Fig. 2b, the increase is at most 1, and we charge this cost to the vertices in C_i and also to the two vertices involved in C. Now there are at least 4 vertices in C_i, so each of these vertices is charged at most 1/6. The other case in which the charged cost was more than 1/6 was when three cycles were joined, as in Figs. 3a and 3b. In this case, the extra cost is at most 2, which is now charged to the at least 12 vertices that belong to these cycles, giving a cost of at most 1/6 per vertex. So the value charged per vertex is at most 1/6 in all cases, and the result follows.   □

4   Final Remarks

When there is only one terminal set, the STEINER MULTICYCLE turns into the TSP. There is a $\frac{3}{2}$-approximation for the metric TSP, so the first natural question is whether there is also a $\frac{3}{2}$-approximation for the metric STEINER MULTICYCLE, or at least some approximation with a ratio better than 3.

The difficulty in the Steiner forest is also a major difficulty in the STEINER MULTICYCLE problem: how to find out what is the right way to cluster the terminal sets. Indeed, if the number k of terminal sets is bounded by a constant, then one can use brute force to guess the way an optimal solution clusters the terminal sets, and then, in the case of the STEINER MULTICYCLE, apply any approximation for the TSP to each instance induced by one of the clusters. This leads to a $\frac{3}{2}$-approximation for any metric instance with bounded number of terminal sets. It also leads to better approximations for hereditary classes of instances for which there are better approximations for the TSP.

It would be nice to find out whether or not the cost of a minimum weight perfect matching on the set of odd vertices of a minimum weight Steiner forest is at most the optimum value for the STEINER MULTICYCLE.

Observe that, for the $\{1, 2\}$-STEINER MULTICYCLE, we can achieve the same approximation ratio than the modified algorithm for the $\{1, 2\}$-TSP, but for the more general metric case, our ratio is twice the best ratio for the metric TSP. This comes from the fact that the backbone structure used in the solution for the metric TSP (the MST and the minimum weight 2-factor) can be computed in polynomial time. For the $\{1, 2\}$-STEINER MULTICYCLEwe can still use the 2-factor, but the two adaptations of the MST for the metric STEINER MULTICYCLE (the Steiner forest and the survivable network design) are hard problems, for which we only have 2-approximations, not exact algorithms.

In fact, for the $\{1, 2\}$-TSP, better approximation algorithms are known: there is an $\frac{8}{7}$-approximation by Berman and Karpinski [4], and a $\frac{7}{6}$-approximation and a faster $\frac{8}{7}$-approximation by Adamaszek et al. [1]. The latter algorithms rely on some tools that we were not able to extend to the $\{1, 2\}$-STEINER MULTICYCLE. On the other hand, the $\frac{8}{7}$-approximation due to Berman and Karpinski seems to be more amenable to an adaptation.

Acknowledgement. C. G. Fernandes was partially supported by the National Council for Scientific and Technological Development – CNPq (Proc. 310979/2020-0 and 423833/2018-9). C. N. Lintzmayer was partially supported by CNPq (Proc. 312026/2021-8 and 428385/2018-4). P. F. S. Moura was partially supported by the Fundação de Amparo à Pesquisa do Estado de Minas Gerais – FAPEMIG (APQ-01040-21). This study was financed in part by the Coordenação de Aperfeiçoamento de Pessoal de Nível Superior - Brasil (CAPES) - Finance Code 001, and by Grant #2019/13364-7, São Paulo Research Foundation (FAPESP).

References

1. Adamaszek, A., Mnich, M., Paluch, K.: New approximation algorithms for $(1, 2)$-TSP. In: Chatzigiannakis, I., Kaklamanis, C., Marx, D., Sannella, D. (eds.) 45th International Colloquium on Automata, Languages, and Programming (ICALP 2018). Leibniz International Proceedings in Informatics (LIPIcs), vol. 107, pp. 9:1–9:14. Schloss Dagstuhl-Leibniz-Zentrum fuer Informatik, Dagstuhl (2018). https://doi.org/10.4230/LIPIcs.ICALP.2018.9
2. Arora, S.: Polynomial time approximation schemes for Euclidean Traveling Salesman and other geometric problems. J. ACM **45**(5), 753–782 (1998). https://doi.org/10.1145/290179.290180
3. Bansal, N., Bravyi, S., Terhal, B.M.: Classical approximation schemes for the ground-state energy of quantum and classical Ising spin Hamiltonians on planar graphs. Quant. Inf. Comput. **9**(7), 701–720 (2009)

4. Berman, P., Karpinski, M.: 8/7-approximation algorithm for $(1, 2)$-TSP. In: Proceedings of the 17th Annual ACM-SIAM Symposium on Discrete Algorithm (SODA), pp. 641–648 (2006)
5. Borradaile, G., Klein, P.N., Mathieu, C.: A polynomial-time approximation scheme for Euclidean Steiner forest. ACM Trans. Algorithms **11**(3), 19:1–19:20 (2015). https://doi.org/10.1145/2629654
6. Christofides, N.: Worst-case analysis of a new heuristic for the traveling salesman problem. Technical report 388, Carnegie Mellon University (1976)
7. Edmonds, J.: Paths, trees, and flowers. Can. J. Math. **17**, 449–467 (1965). https://doi.org/10.4153/CJM-1965-045-4
8. Edmonds, J., Johnson, E.L.: Matchings, Euler tours and the Chinese postman problem. Math. Program. **5**, 88–124 (1973)
9. Ergun, O., Kuyzu, G., Savelsbergh, M.: Reducing truckload transportation costs through collaboration. Transp. Sci. **41**(2), 206–221 (2007). https://doi.org/10.1287/trsc.1060.0169
10. Ergun, O., Kuyzu, G., Savelsbergh, M.: Shipper collaboration. Compute. Oper. Res. **34**(6), 1551–1560 (2007). https://doi.org/10.1016/j.cor.2005.07.026. Part Special Issue: Odysseus 2003 Second International Workshop on Freight Transportation Logistics
11. Hartvigsen, D.: An extension of matching theory. Ph.D. thesis, Department of Mathematics, Carnegie Mellon University, Pittsburgh, PA, USA (1984). https://david-hartvigsen.net/?page_id=33
12. Jain, K.: A factor 2 approximation algorithm for the generalized Steiner network problem. Combinatorica **21**(1), 39–60 (2001). Preliminary version in FOCS 1998
13. Lintzmayer, C.N., Miyazawa, F.K., Moura, P.F.S., Xavier, E.C.: Randomized approximation scheme for Steiner Multi Cycle in the Euclidean plane. Theor. Comput. Sci. **835**, 134–155 (2020). https://doi.org/10.1016/j.tcs.2020.06.022
14. Lovász, L., Plummer, M.D.: Matching Theory. North-Holland Mathematics Studies, vol. 121. Elsevier, Amsterdam (1986)
15. Papadimitriou, C.H., Yannakakis, M.: The Traveling Salesman Problem with distances one and two. Math. Oper. Res. **18**(1), 1–11 (1993). https://doi.org/10.1287/moor.18.1.1
16. Pereira, V.N.G., Felice, M.C.S., Hokama, P.H.D.B., Xavier, E.C.: The Steiner Multi Cycle Problem with applications to a collaborative truckload problem. In: 17th International Symposium on Experimental Algorithms (SEA 2018), pp. 26:1–26:13 (2018). https://doi.org/10.4230/LIPIcs.SEA.2018.26
17. Rosenkrantz, D.J., Stearns, R.E., Lewis, P.M.: An analysis of several heuristics for the traveling salesman problem. SIAM J. Comput. **6**, 563–581 (1977)
18. Salazar-González, J.J.: The Steiner cycle polytope. Eur. J. Oper. Res. **147**(3), 671–679 (2003). https://doi.org/10.1016/S0377-2217(02)00359-4
19. Schrijver, A.: Combinatorial Optimization: Polyhedra and Efficiency. Springer, Heidelberg (2003)
20. Tutte, W.T.: A short proof of the factor theorem for finite graphs. Can. J. Math. **6**, 347–352 (1954)
21. Vazirani, V.V.: Approximation Algorithms. Springer, Heidelberg (2002). https://doi.org/10.1007/978-3-662-04565-7

Approximation Schemes for Packing Problems with ℓ_p-norm Diversity Constraints

Waldo Gálvez$^{(\boxtimes)}$ and Víctor Verdugo

Institute of Engineering Sciences, Universidad de O'Higgins, Rancagua, Chile
{waldo.galvez,victor.verdugo}@uoh.cl

Abstract. We consider the problem of packing a set of items, each of them from a specific category, to obtain a solution set of high total profit, respecting the capacities, and exhibiting a good balance in terms of the categories represented by the chosen solution. Formally, this diversity constraint is captured by including a general family of ℓ_p-norm constraints. These constraints make the problem considerably harder, and, in particular, the relaxation of the feasible region of the optimization problem we get is no longer convex. We show first that approximating this family of problems up to any extent is hard, and then we design two types of approximation schemes for them, depending on whether we are willing to violate the capacity or the ℓ_p-norm constraints by a negligible amount. As a corollary, we get approximation schemes for Packing problems with constraints on the Hill diversity of the solution, which is a classical measure to quantify the diversity of a set of categorized elements.

Keywords: Approximation algorithms · Packing problems · Diversity

1 Introduction

We consider the following classical packing problem:

$$\max \left\{ p^\top x : Wx \le c \text{ and } x \in \{0,1\}^n \right\}, \tag{1}$$

where $p \in \mathbb{Z}_+^n$ is a non-negative profit vector, $W \in \mathbb{Z}_+^{k \times n}$ is a non-negative weight matrix and $c \in \mathbb{Z}_+^k$ is a non-negative capacity vector, which together induce a set of k *capacity constraints*. This problem can be interpreted as selecting a subset of items that satisfies certain capacity constraints given by the matrix W and vector c so as to maximize its total profit, and it has been extensively studied in the literature; in particular, it is known that this problem is NP-hard even if $k = 1$ [31], but admits a PTAS when k is constant [20].

Following a current trend of recent results [1,5,39,41], we will also assume that the set of items is partitioned into $R \in [n]$ categories, with the goal in mind not only to maximize the total obtained profit but also to control how

© Springer Nature Switzerland AG 2022
A. Castañeda and F. Rodríguez-Henríquez (Eds.): LATIN 2022, LNCS 13568, pp. 204–221, 2022.
https://doi.org/10.1007/978-3-031-20624-5_13

each category is represented in the computed solution. In order to achieve such a goal, we incorporate the following type of constraints on the norm of the characteristic vector induced by the chosen items, denoted as *norm constraints*:

$$\max \left\{ p^\top x : Wx \le c, \ell_j \le \|\chi(x)\|_{q_j} \le u_j \text{ for each } j \in [t] \text{ and } x \in \{0,1\}^n \right\} \quad (2)$$

In the previous formulation, $\chi(x) \in \mathbb{Z}_+^R$ denotes the characteristic vector of x, which contains, in each coordinate $r \in [R]$, the sum of the values x_i such that the corresponding item i belongs to category r. For each such extra constraint $j \in [t]$, we are given a lower bound $\ell_j \ge 0$ and an upper bound $u_j \ge \ell_j$, plus a parameter q_j to specify the norm of the constraint. Recall that, for each $q \ge 1$ and $y = (y_1, \ldots, y_m) \in \mathbb{R}^m$, the ℓ_q-norm of y is defined as $\|y\|_q = (\sum_{i=1}^m |y_i|^q)^{1/q}$. In particular, the ℓ_1-norm of y is defined as $\|y\|_1 = \sum_{i=1}^m |y_i|$. We denote the problem defined by formulation (2) as *norm-constrained packing problem*, and in what follows we will assume that both k and t are constant.

The inclusion of ℓ_p-norm based constraints for Resource Allocation problems has been done before in the literature (for instance, see [4,7,14] in the context of load balancing), as it seems to accurately merge the objectives of optimizing the cost while ensuring the assignment to be "fair". In our context, and following the same train of thought, the main motivation behind considering the afore-mentioned norm-constrained packing problem is encoding a family of diversity measures introduced by Hill, known as *Hill numbers* [26].

Definition 1 (Hill [26]). *Given $q \in (1, \infty)$, $R \in \mathbb{N}$ and $y = (y_1, \ldots, y_R) \in \mathbb{Z}_+^R$, the Hill number of order q of y is defined as*

$$D_q(y) = \left(\sum_{i=1}^R \frac{y_i^q}{\|y\|_1^q} \right)^{1/(1-q)}. \quad (3)$$

For any $q > 1$, $D_q(y)$ ranges from 1 to R, being 1 when there exists $i \in [R]$ such that $y_i > 0$ and $y_j = 0$ for any $j \ne i$, and being R when $y_1 = y_2 = \cdots = y_R$. Hill numbers define diversity indices based solely on abundance, aiming to gauge the variety or heterogeneity of a community without focusing on the specific attributes of each individual. In particular, there is a longstanding consensus in ecology on recommending the usage of Hill numbers [11,18,25,26] as they satisfy key mathematical axioms and possess other desired properties [18]. This class of indices includes the well-known Simpson dominance index [44], which corresponds to the Hill number of order 2.

1.1 Our Results

We study the norm-constrained packing problem through the lens of polynomial time approximation algorithms. More in detail, our results are the following:

1. We show first that, for any $k \ge 2$ and $t \ge 1$, the problem does not admit any polynomial time approximation algorithm unless P = NP, as checking feasibility of an instance is NP-hard (see Lemma 1).

2. We provide polynomial time algorithms that, for any $\varepsilon > 0$, compute a $(1+\varepsilon)$-approximate solution by either slightly violating the capacity constraints (see Theorem 2) or the norm constraints (see Theorem 3). For the special case of $k = 1$, we provide a PTAS (see Theorem 4).
3. We show how to apply the norm-constrained packing problem to incorporate diversity constraints based on Hill numbers, and discuss some further applications to search-diversification problems (see Sect. 4).

In order to achieve our main results, we first provide a useful subroutine that, given a category $r \in [R]$ and an integer $s \in [n]$, computes the best possible solution to the classical packing problem defined by (1) of cardinality exactly s, restricted to items of category r (see Definition 2). Similarly to Lemma 1, this problem cannot be approximated in polynomial time, but it can be solved exactly in pseudo-polynomial time. Via rounding techniques, it is possible to adapt this subroutine so as to work in polynomial time by either increasing the capacities or by discarding a negligible amount of items; this procedure is then used to compute a family of *candidate sets* of items for each category, that are then used to create a *multiple-choice packing* instance (see Definition 3), for which there are known algorithms with good approximation guarantees. The multiple-choice packing instance ensures that at most one candidate set per category is chosen, while respecting the corresponding capacity and norm constraints.

1.2 Related Work

The classical packing problem defined by (1) has been extensively studied through the years. The case of $k = 1$ is known as the knapsack problem and is one of Karp's 21 NP-complete problems [31], but admits a FPTAS [28,42]; for the case of $k \geq 2$, it is proved by Magazine et al. [35] that achieving a FPTAS is NP-hard, but a PTAS is known for constant k [20]. The landscape when k is not a part of the input is much different, as formulation (1) encodes problems with strong inapproximability results such as the independent set problem [24,46] (see [6] for some positive results). Many different variants of this problem have been studied, such as multiple knapsack [13,29], geometric knapsack [22,23], online packing [2,37], among many others. We refer the reader to [8,17,32] for a comprehensive treatment of packing problems.

Diversity is a complex and multidimensional concept that is typically used to summarize the structure of a community. Incorporating diversity constraints in decision-making tasks and socio-technical systems at large has become a goal to achieve fairness and equity. This has motivated the algorithms, machine learning and artificial intelligence communities to study the notion of diversity and the related concepts of inclusion and representation [12,15,19,36]. Diversity has become essential in many areas of algorithmic design and learning, including fairness in data summarization [9,10,27] and fair clustering [3,16,21,30].

Close to our setting is the recent work by Patel, Khan and Louis [39], where the authors design algorithms for the knapsack problem under group fairness constraints. In this framework, each item belongs to a certain category, and the goal is to select a subset of items so that each category is represented in the solution according to pre-specified ranges of total weight, total profit or total number of items. Also recently, Pérez-Salazar, Torrico and Verdugo [41] incorporated Hill diversity constraints to a partitioning problem. Their approach, and consequently ours, is mainly motivated from the ecology perspective, where there is a spectrum of viewpoints to address diversity; on one side, rare species are a main focus, but on the other side, communities are essential and only measuring common species matters. In his influential work [44], Simpson introduced a sample-driven metric of diversity based on the abundance and richness of a population, which was later generalized to the Hill numbers [26], and more generally, it has been derived as special case of entropy indices [33].

1.3   Organization of the Article

In Sect. 2, we describe more in detail our general algorithmic approach and review some useful known results. Then, in Sect. 3, we describe in detail our main technical results, namely the approximation schemes for norm-constrained packing problems. Finally, in Sect. 4, we describe how to use our results to incorporate Hill diversity constraints to packing problems and some applications. Due to space constraints, some proofs are deferred to the full version of this article.

2   Preliminaries

We will first provide some useful notation. For a given matrix $A \in \mathbb{Z}_+^{m \times n}$, we denote by A_j the j-th row of A and by $A_{.,i}$ the i-th column of A. Given two vectors $y, y' \in \mathbb{Z}_+^n$, we say that $y \leq y'$ if, for each coordinate $i \in [n]$, it holds that $y_i \leq y_i'$. We also define $\|y\|_\infty = \max_{i \in [n]} y_i$, i.e. the maximum entry of vector y. For a given $x \in \{0,1\}^n$, we define as $I(x) \subseteq [n]$ the set of indices where x contains a 1; on the other hand, for a given set $S \subseteq [n]$, we define a vector $\phi(S) \in \{0,1\}^n$ containing a value 1 in each index $i \in S$ and a value 0 otherwise.

In the following, we provide some results regarding the complexity of the norm-constrained packing problem, and we introduce the algorithmic approach that we use to devise approximation schemes in Sect. 3. First, we can show that deciding the feasibility of a norm-constrained packing instance is NP-complete.

Lemma 1. *The problem of deciding the existence of a feasible solution for a norm-constrained packing instance is* NP-*complete.*

Proof (sketch). We can reduce the problem of checking feasibility for a norm-constrained packing problem with $k \geq 2$ and $t \geq 1$ from Equipartition, where we are given $2m$ integer numbers and an objective value B, and the goal is to decide whether the set can be partitioned into two sets of equal size, each one having total sum B. Indeed, for each number in the Equipartition instance we create an item of profit equal to its value, and can use the weights and capacity constraints to ensure that the total sum of each side of the partition is exactly B, while we can use the norm constraint to ensure that the size of each side of the partition is exactly m.   □

2.1   Algorithmic Approach: Efficient Search of Candidate Solutions

We define the following auxiliary problem, which will be key in the design of our algorithms.

Definition 2. *Given an instance of the packing problem defined by (1) and a number $s \in [n]$, the s-packing problem consists of finding the most profitable feasible solution for the problem that contains exactly s items. That is,*

$$\max \left\{ p^\top x : Wx \leq c, \sum_{i \in [n]} x_i = s \text{ and } x \in \{0,1\}^n \right\}. \tag{4}$$

Our general algorithmic approach consists of solving a family of well defined s-packing instances so as to reduce the search space to a polynomially bounded family of candidate item sets, which we can then feed to an algorithm with the goal to decide which candidate set from each category will be selected for the final solution. For the last part of the algorithm, we use techniques devised for the *multiple-choice* packing problem (see Definition 3).

Analogously to Lemma 1 regarding the norm-constrained packing problem, computing a feasible solution for the s-packing problem in general is hard.

Proposition 1. *The problem of deciding the existence of a feasible solution for an s-packing problem is* NP*-complete.*

Despite of the previous fact, the s-packing problem can be solved in pseudo-polynomial time using dynamic programming. This is a classical result in the field, whose proof can be found in the full version of this article.

Lemma 2. *There exists an algorithm that solves the s-packing problem exactly in time $O(ns \prod_{j=1}^{k} c_j)$.*

The following problem will be helpful to compute our final solution once we have defined a sensible family of candidate sets for each category.

Definition 3. *In the multiple-choice packing problem, we are given an instance (p, W, c) of the packing problem defined by (1), and furthermore the items are divided into m categories. The goal is to find a feasible solution that contains at most one item per category and maximizes the total profit.*

Multiple-choice constraints have been studied in many different contexts [43, 45]. For this particular case, Patt-Shamir and Rawitz developed a PTAS [40] when k is a constant, which will come useful for our purposes. For the special case of $k = 1$, a faster approximation scheme is known [34]. We summarize these two results in the following theorem.

Theorem 1 ([34,40]). *For every $\varepsilon > 0$, there is a $(1 + \varepsilon)$-approximation algorithm for the multiple-choice packing problem with running time $O((nm)^{\lceil k/\varepsilon \rceil})$. For the case of $k = 1$, there is a $(1 + \varepsilon)$-approximation algorithm with running time $O(n \log n + mn/\varepsilon)$.*

3   Approximation Schemes

Our main results are $(1+\varepsilon)$-approximation algorithms for the norm-constrained packing problem under slight resource augmentation assumptions, in the sense that our algorithms will always return a solution of almost optimal profit, but may violate either the capacity constraints or the norm constraints by a ε-fraction.

More specifically, we develop a first algorithm that computes an integral solution x_A of almost optimal profit, satisfying the norm constraints and satisfying that $Wx_A \leq (1+\varepsilon)c$ (see Theorem 2), and a second algorithm computing an integral solution x_B of almost optimal profit that satisfies the capacity constraints and, for each $j \in \{1,\dots,t\}$, satisfies that $(1-\varepsilon)\ell_j \leq \|\chi(x_B)\|_{q_j} \leq u_j$ (see Theorem 3). In the special case where the lower bound vector ℓ is zero, our algorithm is indeed a PTAS, but in general the returned solution might not be feasible for the original instance. However, we remark that no algorithm can consistently return feasible solutions when $k \geq 2$ and $t \geq 1$, unless P = NP, as showed in Lemma 1.

3.1   Approximation Scheme with Small Capacity Violation

As mentioned before, our algorithm consists of two main steps: First, we show that there exists an algorithm that solves almost optimally the s-packing problem if we are allowed to increase the capacities by a factor $(1+\varepsilon)$. In order to do that, we show that it is possible to modify the instance so that the total weight of any feasible solution now belongs to a set of candidates of polynomial size; this modification requires to increase the capacities, but we show that capacities increase by at most a $(1+\varepsilon)$-factor. Then, with respect to the modified instance (in particular, using the polynomially bounded set of possible total weights), we define a family of profitable candidate sets of items that may be included in our solution, and then feed them into a multiple-choice packing instance so as to ensure that only one such set is chosen per category, while satisfying the norm constraints.

The following proposition will imply that, under a slight resource augmentation assumption on the given capacity constraints, the s-packing problem can be solved almost optimally in polynomial time.

Proposition 2. *Given an instance (p, W, c) of the packing problem defined by (1), and given $\varepsilon > 0$, there exists an algorithm that computes another instance (p, W', c') such that:*

(a) *Every feasible solution for the original instance (p, W, c) is feasible for the modified instance (p, W', c');*

(b) *For every feasible solution y of the instance (p, W', c'), we have $Wy \leq (1 + 2\varepsilon)c$; and*

(c) *$\|c'\|_\infty \leq (1 + \varepsilon)n/\varepsilon$.*

Proof. For each $j \in [k]$, let $\mu_j = \varepsilon c_j / n$; we define the modified weight matrix W' by setting $W'_{j,i} = \lfloor W_{j,i}/\mu_j \rfloor = \lfloor nW_{j,i}/\varepsilon c_j \rfloor$ for each $j \in [k]$ and $i \in [n]$, and we define the modified capacity c' by its coordinates $c'_j = \lfloor (1+\varepsilon)c_j/\mu_j \rfloor$ for each $j \in [k]$. Notice that $c'_j = \lfloor (1+\varepsilon)n/\varepsilon \rfloor$, and computing W' and c' takes time $O(nk)$. Let x be a feasible solution for the original instance (p, W, c). It holds that, for any $j \in [k]$,

$$
\begin{aligned}
W'_j x &\leq \left\lfloor \sum_{i \in I(x)} \left(\frac{W_{j,i}}{\mu_j} + 1 \right) \right\rfloor \\
&= \left\lfloor \sum_{i \in I(x)} \frac{W_{j,i}}{\mu_j} + |I(x)| \right\rfloor \\
&\leq \left\lfloor \frac{1}{\mu_j}(W_j x + n\mu_j) \right\rfloor \leq \left\lfloor \frac{c_j + n\mu_j}{\mu_j} \right\rfloor = \left\lfloor \frac{(1+\varepsilon)c_j}{\mu_j} \right\rfloor = c'_j.
\end{aligned}
$$

Let y be a feasible solution for the instance (p, W', c'). If we consider the same solution but for the original instance (p, W, c), then, for any $j \in [k]$, we have that:

$$
\begin{aligned}
W_j y &\leq \sum_{i \in I(y)} \mu_j(W'_{j,i} + 1) \\
&\leq \mu_j(W'_j y + |I(y)|) \leq \mu_j c'_j + n\mu_j \leq (1+2\varepsilon)c_j,
\end{aligned}
$$

which concludes the proof of the proposition.   □

Thanks to Proposition 2, the algorithm from Lemma 2 has polynomial running time in the modified instance, and hence it can be used to compute good candidate sets of s items for each category and each possible $s \in [n]$, requiring resource augmentation on the capacity constraints. With this tool we can prove next our first main result. In what follows, x_{OPT} denotes a fixed optimal solution for a given instance of the norm-constrained packing problem, and OPT denotes its total profit.

Theorem 2. *For every $\varepsilon > 0$, there exists an algorithm for the norm-constrained packing problem that computes a solution $x \in \{0, 1\}^n$ such that:*

(i) $p^\top x \geq (1-\varepsilon)\mathsf{OPT}$,
(ii) $\ell_j \leq \|\chi(x)\|_{q_j} \leq u_j$ for each $j \in [t]$, and
(iii) $Wx \leq (1+\varepsilon)c$.

The running time of this algorithm is $(n/\varepsilon)^{O((k+t)^2/\varepsilon)}$.

Proof. We start by applying Proposition 2 with parameter $\varepsilon' = \varepsilon/2$ to the instance (p, W, c), to get a modified instance (p, W', c'). As a consequence, the algorithm from Lemma 2 takes time $O(n^{k+2}((1+\varepsilon)/\varepsilon)^k)$ when applied to any s-packing instance derived from this modified instance (i.e., defined by a subset of the items and possibly smaller capacities), which is polynomial. Furthermore, since the profit is preserved through the modification performed by Proposition 2, the profit of the obtained solutions cannot decrease.

For each possible category $r \in [R]$, each possible capacity vector $c'' \leq c'$ and each possible $s \in [n]$, we apply Lemma 2 to an instance restricted to items of category r, so as to obtain a solution $x_{r,c'',s}$ for the corresponding s-packing problem such that its total (modified) weight is at most c'' (coordinate-wise), it has s items of category r and its profit is at least the optimal profit for the corresponding instance. These sets will act as candidates for the final solution to our problem. If we denote by K_c the number of candidates, we have that $K_c \in O\left(Rn^{k+1}((1+\varepsilon)/\varepsilon)^k\right)$. This step will then take time $O(Rn^{2k+3}((1+\varepsilon)/\varepsilon)^{2k})$.

We now define a multiple-choice packing instance to obtain the desired solution as follows: For each $i \in [K_c]$, we define an item representing each possible candidate set $I(x_{r,c'',s})$ computed before, and these items are partitioned into R groups according to the specified categories; the matrix $\widetilde{W}$ of dimensions $(k+2t) \times K_c$ is defined so that, for each item i representing the candidate set $I(x_{r,c'',s})$, we have that:

(i) $\widetilde{W}_{j,i} = W'_j \cdot x_{r,c'',s}$ for every $j \in [k]$,
(ii) $\widetilde{W}_{k+j,i} = |I(x_{r,c'',s})|^{q_j}$ for each $j \in [t]$, and
(iii) $\widetilde{W}_{k+t+j,i} = n^{q_j} - |I(x_{r,c'',s})|^{q_j}$ for each $j \in [t]$.

The vector $\tilde{c} \in \mathbb{Z}^{k+2t}$ is given by $\tilde{c}_j = c'_j$ for every $j \in [k]$, $\tilde{c}_{k+j} = u_j^{q_j}$ and $\tilde{c}_{k+t+j} = Rn^{q_j} - \ell_j^{q_j}$ for each $j \in [t]$. On this instance, we apply Theorem 1 and finally return the solution x_{ALG} induced by the selected candidate sets. The time required to compute this solution can thus be bounded by $(n/\varepsilon)^{O((k+t)^2/\varepsilon)}$ thanks to the guarantees of Theorem 1.

We claim that any feasible solution x for the original problem defines a feasible solution y for the multiple choice instance if we consider $I(x)$ partitioned according to the categories as the selected candidate sets. Indeed, in terms of capacity constraints, this holds thanks to Proposition 2. Regarding the norm constraints, notice that, for each $j \in [t]$, we have that

$$\widetilde{W}_{k+j}y = \|\chi(x)\|_{q_j}^{q_j} \quad \text{and} \quad \widetilde{W}_{k+t+j}y = Rn^{q_j} - \|\chi(x)\|_{q_j}^{q_j},$$

and hence the corresponding constraints imply that $\|\chi(x_{\mathsf{ALG}})\|_{q_j} \leq u_j$ and $\ell_j \leq \|\chi(x_{\mathsf{ALG}})\|_{q_j}$ respectively. Thanks to the approximation guarantees of Theorem 1, the retrieved solution x_{ALG} has profit at least $(1 - \varepsilon)\mathsf{OPT}$, as in particular the optimal solution for the original problem induces a feasible family of candidate sets for the multiple-choice packing instance. Furthermore, it satisfies the norm constraints, and we have that $Wx_{\mathsf{ALG}} \leq (1 + 2\varepsilon')c = (1 + \varepsilon)c$ thanks to the guarantees of Proposition 2 for each selected candidate. $\square$

3.2 Approximation Scheme with Small Norm Violation

Now we develop a complementary algorithm to the one from Theorem 2, in the sense that it computes a solution satisfying the capacity constraints but that might violate the norm constraints. Our main tool will be an approximation scheme for the s-packing problem that returns an almost optimal solution in terms of profit, but potentially using slightly less than s items. This result is achieved, roughly speaking, by using the algorithm from Lemma 2 and then removing items of negligible total profit in order to satisfy the capacity constraints. This removal might violate the lower bounds on the ℓ_p-norm constraints. However, we choose the removed items in such a way that at most an ε-fraction from each category is removed, and hence in the end the norm constraints will be violated by a factor $(1-\varepsilon)$ at most. With this tool, we can compute a polynomial family of candidate sets and write a multiple-choice packing instance, where we can use Theorem 1 and retrieve our final solution. We first prove the following couple of technical results.

Proposition 3. *Consider a feasible solution x for an instance (p, W, c) of the packing problem defined by (1), and let $\varepsilon > 0$. Suppose that $|I(x)| \geq k^2/\varepsilon^4$. Then, there exist three integral vectors $x_L, x_S, x_D \in \{0,1\}^n$ such that the following holds:*

(a) $x = x_L + x_S + x_D$.
(b) $p^\top(x_L + x_S) \geq (1 - \varepsilon^2)p^\top x$.
(c) $|I(x_D)| \leq k/\varepsilon^2$.
(d) $|I(x_L)| \leq k^2/\varepsilon^4$ and $W_{\cdot,i} \leq \varepsilon^2(c - Wx_L)$ for each $i \in I(x_S)$.

Proof. Recall that, given a set of items $S \subseteq [n]$, we denote by $\phi(S) \in \{0,1\}^n$ the vector having value 1 for each $i \in S$ and zero otherwise. Consider a fixed $j \in [k]$. Let L_1 be the set of items $i \in I(x)$ satisfying that $W_{j,i} > \varepsilon^2 c_j$. If L_1 has total profit at most $\varepsilon^2 p^\top x/k$, then we can define the vectors in the following way: $x_L = 0$, $x_D = \phi(L_1)$ and $x_S = x - x_L - x_D$. In particular, $I(x_L) = \emptyset$ and $I(x_D) = L_1$, and thus all the items $i \in I(x) \setminus L_1$ satisfy that $W_{j,i} \leq \varepsilon^2(c - Wx_L)$. Otherwise, consider the set L_2 of items $i \in I(x) \setminus L_1$ satisfying that

$$W_{j,i} > \varepsilon^2\left(c_j - \sum_{\ell \in L_1} W_{j,\ell}\right),$$

i.e., the items whose weight on dimension j is relatively large with respect to the residual instance defined by L_1. If L_2 has total profit at most $\varepsilon^2 p^\top x/k$, then we can define the vectors in the following way: $x_L = \phi(L_1)$, $x_D = \phi(L_2)$ and $x_S = x - x_L - x_D$. In particular, we have $I(x_L) = L_1$ and $I(x_D) = L_2$, and then all the items in $I(x_S)$ have weight relatively small with respect to the residual capacity as desired (notice that $|L_1| \leq 1/\varepsilon^2$).

We continue with this procedure, defining at each iteration $h \in \mathbb{N}$ the set L_h of items $i \in I(x) \setminus \cup_{e=1}^{h-1} L_e$ such that

$$W_{j,i} > \varepsilon\left(c_j - \sum_{\ell \in \cup_{i=1}^{h-1} L_i} W_{j,\ell}\right),$$

and checking whether the total profit of L_h is at most $\varepsilon^2 p^\top x/k$.

Notice that, by construction, the sets L_h are disjoint and have size at most $1/\varepsilon^2$. Thus, a set L_h of total profit at most $\varepsilon^2 p^\top x/k$ must be found after at most k/ε^2 iterations; if this does not occur, then we have a vector $\phi(\cup_{i=1}^{k/\varepsilon^2} L_i) \leq x$ whose total profit is strictly larger than $p^\top x$, which is not possible. If we apply the same procedure for each possible dimension $j = 1, \ldots, k$ iteratively, we then obtain the final vectors which, by slightly abusing notation, we denote as x_L and x_D. Notice that $|I(x_\mathsf{L})| \leq k \cdot k/\varepsilon^2 \cdot 1/\varepsilon^2 \leq k^2/\varepsilon^4$, $|I(x_\mathsf{D})| \leq k/\varepsilon^2$ and also $p^\top x_\mathsf{D} \leq \varepsilon^2 p^\top x$ as claimed.   □

Proposition 4. *Consider a feasible solution x for an instance (p, W, c) of the packing problem defined by (1), and let $\varepsilon > 0$ such that $1/\varepsilon$ is integral. Suppose that $W_{j,i} \leq \varepsilon^2 c_j$ for every $i \in I(x)$ and every $j \in [k]$. Then, it is possible to compute an integral vector x_D' such that:*

(a) $p^\top x_\mathsf{D}' \leq \varepsilon p^\top x$ and $x_\mathsf{D}' \leq x$.
(b) $|I(x_\mathsf{D}')| \leq \varepsilon |I(x)| + k$.
(c) $W(x - x_\mathsf{D}') \leq \left(1 - \frac{\varepsilon}{k}(1 - (k+1)\varepsilon)\right) c$.

Proof. Consider a fixed $j \in [k]$. If we have that $\sum_{i=1}^n W_{j,i} x_i \leq (1 - \varepsilon) c_j$, then the claims are satisfied in dimension j if we define $x_\mathsf{D}' = 0$. Otherwise, we partition the set of items into k/ε groups of relatively balanced total weight and cardinality, so that the least profitable of them provides the desired set x_D'.

We sort the items $i \in I(x)$ non-increasingly by their weight $W_{j,i}$ (breaking ties arbitrarily), and we assign them into sets $S_1, \ldots, S_{k/\varepsilon}$ in a round-robin fashion; that is, each set S_i receives the items whose position in the previous sorted list is equal to $i \bmod k/\varepsilon$, until the total weight of some set becomes at least $\varepsilon(1 - \varepsilon) c_j/k$. Notice that this must happen since the total weight of the items is at least $(1 - \varepsilon) c_j$; furthermore, the last set where an item is assigned must be S_1 since

$$\sum_{\ell \in S_i} W_{j,\ell} \geq \sum_{\ell \in S_{i+1}} W_{j,\ell}$$

for every $i \in \{1, 2, \ldots, k/\varepsilon - 1\}$.

Observe that $|S_i| \leq \varepsilon |I(x)|/k + 1$ for every $i \in \{1, \ldots, k/\varepsilon\}$. Moreover, the total profit of each set is at most $\varepsilon p^\top x/k$, and it also holds that the total weight of each set S_i on dimension j is at least $\varepsilon(1 - \varepsilon) c_j/k - \varepsilon^2 c_j$. In fact, due to the way items are assigned to the sets, the second item assigned to S_1 has weight smaller than the first item assigned to $S_{k/\varepsilon}$, the third item assigned to S_1 has weight smaller than the second item assigned to $S_{k/\varepsilon}$ and so on, and also we have that $|S_1| = |S_{k/\varepsilon}| + 1$. Hence, it holds that

$$\sum_{\ell \in S_{k/\varepsilon}} W_{j,\ell} \geq \sum_{\ell \in S_1} W_{j,\ell} - \overline{W} \geq \frac{\varepsilon}{k}(1 - \varepsilon) c_j - \varepsilon^2 c_j,$$

where $\overline{W}$ is the weight of the first item assigned to set S_1 (see Fig. 1). In conclusion, if we define x_D' so that $I(x_\mathsf{D}')$ corresponds to the least profitable of these sets (whose total profit would be at most $\varepsilon p^\top x/k$), then the claims of the lemma follow for dimension j. If we apply this procedure iteratively for each dimension, we obtain analogous properties for all the dimensions simultaneously as claimed. □

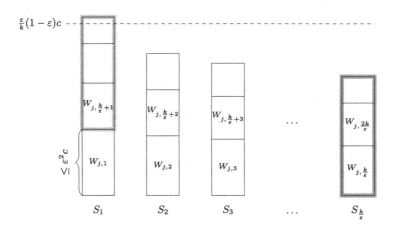

Fig. 1. Depiction of how items are partitioned into k/ε sets of balanced cardinality and lower bounded total weight as in the proof of Lemma 4. The total weight of the sets is non-increasing from left to right, but the highlighted set in $S_{k/\varepsilon}$ has total weight of at least the weight of the highlighted set in S_1.

We can now use the previous two results to devise a procedure to compute profitable candidates while slightly violating the cardinality constraints of the corresponding s-packing instance.

Lemma 3. *For every $0 < \varepsilon \le 2/(3k + 3)$, there exists an algorithm that computes, for every s-packing instance (p, W, c), a solution $x \in \{0,1\}^n$ such that:*

(i) $(1 - \varepsilon)s \le |I(x)| \le s$;
(ii) $p^\top x \ge (1-\varepsilon)\mathsf{OPT}$, where OPT is the optimal value of the s-packing instance; and
(iii) $Wx \le c$.

The running time of this algorithm is $(kn/\varepsilon)^{O(k^2/\varepsilon^4)}$.

Proof. Let $x_{P,s}$ be an optimal solution for the s-packing instance. We assume that $s > 16k^2/\varepsilon^4$, as otherwise we can simply guess the optimal solution. Let $x_\mathsf{L}, x_\mathsf{D}$ and x_S be the three integral vectors whose existence is ensured thanks to Proposition 3, applied with parameter $\varepsilon/2$.

Let $\varepsilon' = \varepsilon/(6k)$. As a first step, we guess x_L and x_D, which can be done efficiently since there are $O(n^{16k^2/\varepsilon^4})$ such possible vectors. Then, we remove from $[n] \setminus I(x_\mathsf{L} + x_\mathsf{D})$ every item i satisfying that there exists $j \in [k]$ such that $W_{j,i} > \varepsilon'^2(c_j - \sum_{i \in I(x_\mathsf{L})} W_{j,i})$. That is, the items which are not relatively small with respect to the residual capacity defined by x_L. From here we get a new instance defined by $(p, W, c - Wx_L)$. We apply Proposition 2 with parameter ε' to the instance $(p, W, c - Wx_L)$, obtaining a modified residual instance (p, W', c'), and then we run the algorithm from Lemma 2 with cardinality parameter $s - |I(x_L + x_D)|$ on this instance, which takes time $O(ns (k/\varepsilon + 1)^k)$. Finally, we apply Proposition 4 with parameter $\varepsilon/2$ on the previous solution for the residual instance (p, W', c'), and return the outcome $\tilde{x}$ plus x_L as the final solution x_{ALG}.

Notice that, since x_S defines a feasible solution for the residual instance (p, W', c'), the profit of the solution returned by Lemma 2 must be at least $p^\top x_S$, and accordingly, the profit of the final solution returned by our algorithm will be at least

$$(1 - \frac{\varepsilon}{2})p^\top x_S + p^\top x_L \geq (1 - \varepsilon)p^\top x_{P,s}.$$

On the other hand, thanks to Proposition 2, the total weight of the solution returned by Lemma 2 is at most $(1 + \varepsilon/(3k))(c - Wx_L)$, and hence the total weight of the solution after applying Proposition 4 is at most

$$\left(1 - \frac{\varepsilon}{2k}\left(1 - \frac{k+1}{2}\varepsilon\right)\right)\left(1 + \frac{\varepsilon}{3k}\right)(c - Wx_L) \leq c - Wx_L,$$

where we used the fact that $\varepsilon \leq 2/(3k+3)$; this implies that $Wx_{ALG} \leq c$. Finally, regarding the cardinality, we notice that $|I(x_{ALG})| \leq s$ as we only removed items from the solution computed by Lemma 2, and also that

$$|I(x_{ALG})| \geq |I(x_L)| + \left(1 - \frac{\varepsilon}{2}\right)(s - |I(x_L + x_D)|) - k$$

$$\geq \left(1 - \frac{\varepsilon}{2}\right)s - |I(x_D)| - k \geq (1 - \varepsilon)s,$$

where we used the fact that $|I(x_D)| \leq 4k/\varepsilon^2$ and $s \geq 16k^2/\varepsilon^4$. This concludes the proof. □

The previous lemma can be then used to compute profitable candidate sets for our final solution. In order to reduce the amount of candidate sets that are required to be computed, we provide first the following classical rounding result whose proof can be found in the full version of this article.

Proposition 5. *For any instance* (p, W, c) *of the packing problem defined by (1) and* $\varepsilon > 0$, *there exists a set* $\mathcal{C} \subseteq \mathbb{Z}_+^k$ *such that the following holds:*

(a) *For every feasible solution* x *of instance* (p, W, c), *there exists another solution* x' *and a vector* $\tilde{c} \in \mathcal{C}$ *such that* $\tilde{c} \leq c$, $x' \leq x$, $|I(x')| \geq (1 - \varepsilon)|I(x)|$, $p^\top x' \geq (1 - \varepsilon)p^\top x$ *and* $Wx' \leq \tilde{c}$.
(b) $|\mathcal{C}| \in O(n^{(k/\varepsilon+k)^2+2})$ *and it can be computed in time* $n^{O((k/\varepsilon)^2)}$.

Now we can prove our second main result.

Theorem 3. *For any* $\varepsilon > 0$, *there exists an algorithm for the norm-constrained packing problem that computes a solution* x *such that:*

(i) $p^\top x \geq (1 - \varepsilon)\mathsf{OPT}$.

(ii) $Wx \leq c$.

(iii) $(1 - 2\varepsilon)\ell_j \leq \|\chi(x)\|_{q_j} \leq u_j$ for each $j \in [t]$.

The running time of the algorithm is $(n/\varepsilon)^{O((k+2t)^5/\varepsilon^6)}$.

Proof. Let x_{OPT} be the optimal solution for the instance (p, W, c). Let $\mathcal{C}$ be the set guaranteed to exist from Proposition 5 with parameter ε. For each possible category $r \in [R]$, each $\tilde{c} \in \mathcal{C}$ and each $s \in [n]$, we compute a solution $x_{r,\tilde{c},s}$ to the instance of the s-packing problem defined by $(p, W, \tilde{c})$ restricted to items of category r, using Lemma 3 with parameter ε. The running time of this step is $O(Rn^{32k^2/\varepsilon^4+7}(k/\varepsilon + 1)^{2k})$.

We now construct a multiple-choice packing instance as follows: for each previous candidate set we define an item, and these items are partitioned into R groups according to their specified categories; the matrix $\widetilde{W}$ of $(k + 2t)$ rows is defined so that, for each item i representing the candidate set $x_{r,c'',s}$, we have:

1. $\widetilde{W}_{j,i} = W_j \cdot x_{r,\tilde{c},s}$ for every $j \in [k]$,
2. $\widetilde{W}_{k+j,i} = |I(x_{r,c'',s})|^{q_j}$ for each $j \in [t]$, and
3. $\widetilde{W}_{k+t+j,i} = n^{q_j} - |I(x_{r,c'',s})|^{q_j}$ for each $j \in [t]$.

Finally, the capacity vector $\tilde{c} \in \mathbb{Z}^{k+2t}$ is defined as $\tilde{c}_j = c_j$ for every $j \in [k]$, $\tilde{c}_{k+j} = u_j^{q_j}$ for each $j \in [t]$, and $\tilde{c}_{k+t+j} = Rn^{q_j} - (1 - \varepsilon)\ell_j^{q_j}$ for each $j \in [t]$. On this instance we apply Theorem 1 and obtain a solution induced by the selected candidate sets. The time required to compute this solution can be bounded by $(n/\varepsilon)^{O((k+2t)^5/\varepsilon^6)}$ thanks to the guarantees of Theorem 1.

For each $r \in [R]$, let $x_{\mathsf{OPT},r}$ be the optimal solution restricted to items of category r. Thanks to the guarantees of Lemma 3 and Proposition 5, there exists an integer value s such that $(1 - 2\varepsilon)|I(x_{\mathsf{OPT},r})| \leq s \leq |I(x_{\mathsf{OPT},r})|$ and a vector $\tilde{c} \in \mathcal{C}$ such that the solution $x_{r,\tilde{c},s}$ has profit at least $(1 - 2\varepsilon)p^\top x_{\mathsf{OPT},r}$, and hence their union defines a feasible solution of total profit at least $(1 - 2\varepsilon)p^\top x_{\mathsf{OPT}}$ and total weight at most c. Furthermore, regarding the norm constraints, we have that, for each $j \in [t]$, it holds that

$$\|\chi(x_{\mathrm{ALG}})\|_{q_j} \geq \|(1 - 2\varepsilon)\chi(x_{\mathsf{OPT}})\|_{q_j} \geq (1 - 2\varepsilon)\|\chi(x_{\mathsf{OPT}})\|_{q_j} \geq (1 - 2\varepsilon)\ell_j,$$

where x_{ALG} is the computed solution. Applying Theorem 1 will thus return a solution satisfying the claimed guarantees, hence proving the theorem. □

3.3   Improved Algorithm for the Case of $k = 1$

In this section, we provide a PTAS for the norm-constrained packing problem in the special case of $k = 1$. Our algorithm computes a feasible solution having arbitrarily close to optimal profit, as opposed to the previous algorithms where one of the family of restrictions is violated, which is unavoidable due to the restrictions imposed by Lemma 1.

We first round down the profits of the items so that they belong to a set of polynomial size while losing negligible profit. More specifically, for a given item $i \in [n]$, we define its rounded profit as $\tilde{p}_i = \left\lfloor \frac{p_i n}{\varepsilon \|p\|_\infty} \right\rfloor \frac{\varepsilon \|p\|_\infty}{n}$. In other words, we round the profits down to the closest multiple of $\varepsilon \|p\|_\infty / n$. We lose a factor of at most $(1 + \varepsilon)$ in the approximation factor from this rounding, and hence from now on we assume that the profits are rounded. The number of possible total rounded profits for any feasible solution is at most n^2/ε.

Following our algorithmic approach, we use the following result due to Patel et al. [39] to compute candidate sets for each possible category, so as to feed them as input for a multiple-choice packing instance. We restate this result adapted to our purposes, and its proof can be found in the full version of this article.

Lemma 4 (Patel et al. [39]). *Let $k = 1$. There exists an algorithm that, given a profit $p \in \mathbb{Z}_+$, computes the solution for the s-packing problem of total profit p having smallest total weight. The running time of this algorithm is $O\left(n^3 s/\varepsilon\right)$.*

With these tools, we can prove our main result for this setting. The proof follows our algorithmic approach in an analogous way to the previous cases, so we defer its proof to the full version of this article.

Theorem 4. *There exists a PTAS for the norm-constrained packing problem when $k = 1$ that runs in time $(tn/\varepsilon)^{O(1)}$.*

4   Packing Problems Under Hill Diversity Constraints

In this section, we discuss some consequences of our results in the context of incorporating diversity considerations on how categories frequencies appear in the computed solution. More in detail, we consider the following *packing problem with Hill diversity constraints* (see Definition 1 for details of the function $D_q(\cdot)$):

Definition 4. *In the packing problem with Hill diversity constraints, we are given an instance (p, W, c) of the classical packing problem defined by (1) and two values, $q, \delta \geq 1$. The goal is to compute a solution that maximizes the total profit among the feasible solutions x satisfying $D_q(\chi(x)) \geq \delta$.*

As mentioned before, in this problem we aim to balance the quality of the solution in terms of profit and its diversity. The main advantage of this formulation is that the decision-maker only fixes a threshold and the kind of measure to be used, and then the diversity of the solutions is specified solely based on the abundance of items from each category. In opposition, other frameworks require the decision-maker to specify for each category what is a fair representation (for instance, in the setting of group fairness studied by Patel et al. [39], it must be decided a priori how much profit, weight or how many items from each category are allowed in a feasible solution). Using Theorem 3, it is possible to obtain an efficient algorithm for this packing problem with Hill diversity constraints. Please refer to full version of this article.

Theorem 5. *Let $\varepsilon > 0$. There exists a polynomial-time algorithm that, given an instance of the packing problem with Hill diversity constraints defined by (p, W, c), q and δ, computes a solution x satisfying the following:*

(i) x is feasible for the packing problem defined by (1) with instance (p, W, c).
(ii) $p^\top x \geq (1 - \varepsilon)\mathsf{OPT}$, where OPT is the optimal value for the instance of the packing problem with Hill diversity constraints.
(iii) $D_q(\chi(x)) \geq (1 - \varepsilon)\delta$.

Notice that, if we use Theorem 2 instead of Theorem 3, we can obtain a solution that satisfies the Hill diversity constraints but violates the capacity constraints by a small multiplicative factor. This result finds applications in the context of *search-diversification*, where we receive a list of n categorized elements with non-negative valuations and a parameter $k \in [n]$, and the goal is to return a set of k elements of high total profit that satisfies diversity constraints. This problem models desired properties for search queries in databases, where we want to retrieve the most relevant elements but to avoid unfair or biased responses (see [1] for a detailed description and applications). In the context of our work, search-diversification with Hill diversity constraints can be formulated as

$$\max \left\{ p^\top x : \delta \leq D_q(\chi(x)), \|\chi(x)\|_1 = k \text{ and } x \in \{0,1\}^n \right\}, \tag{5}$$

where p corresponds to the valuations of the elements, and q and δ are given. Since there are no capacity constraints, by using our results we can directly obtain a PTAS for the problem.

Corollary 1. *There exists a PTAS for the search-diversification problem under Hill diversity constraints.*

Concluding Remarks

In this work, we have studied the norm-constrained packing problem and provided approximation schemes for it under slight violations of the constraints. We mention some interesting open questions:

1. Our results assume that the number of capacity and norm constraints is constant. As already mentioned, if the number of such constraints is not necessarily constant, the problem is much more challenging, but still, algorithms with different kinds of approximation guarantees could be designed (following the line of Bansal et al. [6]).
2. We study the inclusion of Hill diversity constraints in a fundamental problem such as the packing problem, but other classical problems can also incorporate these constraints, such as bin packing, independent set, clustering, among many others. We believe our techniques can be helpful in tackling this task in the context of resource allocation problems.
3. It would be interesting to consider other diversity measures such as the Gini coefficient [38] and entropy indices [33], or to incorporate more general norm constraints such as weighted ℓ_p-norms [7].

References

1. Abboud, A., Cohen-Addad, V., Lee, E., Manurangsi, P.: Improved approximation algorithms and lower bounds for search-diversification problems. In: 49th International Colloquium on Automata, Languages, and Programming (ICALP), vol. 229, pp. 7:1–7:18 (2022). https://doi.org/10.4230/LIPIcs.ICALP.2022.7
2. Albers, S., Khan, A., Ladewig, L.: Improved online algorithms for knapsack and GAP in the random order model. Algorithmica **83**(6), 1750–1785 (2021)
3. Anegg, G., Angelidakis, H., Kurpisz, A., Zenklusen, R.: A technique for obtaining true approximations for k-center with covering constraints. In: Bienstock, D., Zambelli, G. (eds.) Integer Programming and Combinatorial Optimization, pp. 52–65 (2020)
4. Awerbuch, B., Azar, Y., Grove, E.F., Kao, M., Krishnan, P., Vitter, J.S.: Load balancing in the l_p norm. In: 36th Annual Symposium on Foundations of Computer Science (FOCS), pp. 383–391. IEEE Computer Society (1995)
5. Bandyapadhyay, S., Inamdar, T., Pai, S., Varadarajan, K.R.: A constant approximation for colorful k-center. In: 27th Annual European Symposium on Algorithms (ESA), vol. 144, pp. 12:1–12:14 (2019)
6. Bansal, N., Korula, N., Nagarajan, V., Srinivasan, A.: Solving packing integer programs via randomized rounding with alterations. Theory Comput. **8**(1), 533–565 (2012)
7. Bansal, N., Pruhs, K.: Server scheduling in the weighted ℓ_p Norm. In: Farach-Colton, M. (ed.) LATIN 2004. LNCS, vol. 2976, pp. 434–443. Springer, Heidelberg (2004). https://doi.org/10.1007/978-3-540-24698-5_47
8. Cacchiani, V., Iori, M., Locatelli, A., Martello, S.: Knapsack problems - an overview of recent advances, Part II: multiple, multidimensional, and quadratic knapsack problems. Comput. Oper. Res. **143**, 105693 (2022)
9. Celis, E., Keswani, V., Straszak, D., Deshpande, A., Kathuria, T., Vishnoi, N.: Fair and diverse DPP-based data summarization. In: Dy, J., Krause, A. (eds.) Proceedings of the 35th International Conference on Machine Learning. Proceedings of Machine Learning Research, vol. 80, pp. 716–725. PMLR, 10–15 July 2018
10. Celis, L.E., Huang, L., Keswani, V., Vishnoi, N.K.: Fair classification with noisy protected attributes: a framework with provable guarantees (2021)
11. Chao, A., Chiu, C.H., Jost, L.: Unifying species diversity, phylogenetic diversity, functional diversity, and related similarity and differentiation measures through hill numbers. Annu. Rev. Ecol. Evol. Syst. **45**(1), 297–324 (2014)
12. Chasalow, K., Levy, K.: Representativeness in statistics, politics, and machine learning. In: Proceedings of the 2021 ACM Conference on Fairness, Accountability, and Transparency, FAccT 2021, pp. 77–89. Association for Computing Machinery, New York (2021)
13. Chekuri, C., Khanna, S.: A polynomial time approximation scheme for the multiple knapsack problem. SIAM J. Comput. **35**(3), 713–728 (2005)
14. Chen, L., Tao, L., Verschae, J.: Tight running times for minimum ℓ_p-norm load balancing: beyond exponential dependencies on $1/\varepsilon$. In: ACM-SIAM Symposium on Discrete Algorithms (SODA), pp. 275–315. SIAM (2022)
15. Chi, N., Lurie, E., Mulligan, D.K.: Reconfiguring diversity and inclusion for AI ethics. In: Proceedings of the 2021 AAAI/ACM Conference on AI, Ethics, and Society, pp. 447–457. Association for Computing Machinery, New York (2021)

16. Chierichetti, F., Kumar, R., Lattanzi, S., Vassilvitskii, S.: Fair clustering through fairlets. In: Proceedings of the 31st International Conference on Neural Information Processing Systems, NIPS 2017, pp. 5036–5044. Curran Associates Inc., Red Hook (2017)

17. Christensen, H.I., Khan, A., Pokutta, S., Tetali, P.: Approximation and online algorithms for multidimensional bin packing: a survey. Comput. Sci. Rev. **24**, 63–79 (2017)

18. Daly, A.J., Baetens, J.M., De Baets, B.: Ecological diversity: measuring the unmeasurable. Mathematics 6(7) (2018)

19. Drosou, M., Jagadish, H., Pitoura, E., Stoyanovich, J.: Diversity in big data: a review. Big Data **5**(2), 73–84 (2017)

20. Frieze, A.M., Clarke, M.R.B.: Approximation algorithms for the m-dimensional 0–1 knapsack problem: worst-case and probabilistic analyses. Eur. J. Oper. Res. **15**(1), 100–109 (1984)

21. Ghadiri, M., Samadi, S., Vempala, S.: Fair k-means clustering (2020). arXiv

22. Gálvez, W., Grandoni, F., Heydrich, S., Ingala, S., Khan, A., Wiese, A.: Approximating geometric knapsack via l-packings. ACM Trans. Algorithms 17(4), 33:1–33:67 (2021)

23. Gálvez, W., Grandoni, F., Khan, A., Ramírez-Romero, D., Wiese, A.: Improved approximation algorithms for 2-dimensional knapsack: Packing into multiple l-shapes, spirals, and more. In: 37th International Symposium on Computational Geometry (SoCG), vol. 189, pp. 39:1–39:17 (2021)

24. Håstad, J.: Clique is hard to approximate within n^ε. Electron. Colloq. Comput. Complex. (38) (1997)

25. Heip, C.H., Herman, P.M., Soetaert, K.: Indices of diversity and evenness. Oceanis **24**(4), 61–88 (1998)

26. Hill, M.O.: Diversity and evenness: a unifying notation and its consequences. Ecology **54**(2), 427–432 (1973)

27. Huang, L., Vishnoi, N.: Stable and fair classification. In: Chaudhuri, K., Salakhutdinov, R. (eds.) Proceedings of the 36th International Conference on Machine Learning. Proceedings of Machine Learning Research, vol. 97, pp. 2879–2890. PMLR, 09–15 June 2019

28. Ibarra, O.H., Kim, C.E.: Fast approximation algorithms for the knapsack and sum of subset problems. J. ACM **22**(4), 463–468 (1975)

29. Jansen, K.: Parameterized approximation scheme for the multiple knapsack problem. SIAM J. Comput. **39**(4), 1392–1412 (2009)

30. Jia, X., Sheth, K., Svensson, O.: Fair colorful k-center clustering. In: Bienstock, D., Zambelli, G. (eds.) Integer Programming and Combinatorial Optimization, pp. 209–222 (2020)

31. Karp, R.: Reducibility among combinatorial problems. In: Complexity of Computer Computations, pp. 85–103. Plenum Press (1972)

32. Kellerer, H., Pferschy, U., Pisinger, D.: Knapsack problems. Springer, Heidelberg (2004). https://doi.org/10.1007/978-3-540-24777-7

33. Keylock, C.J.: Simpson diversity and the Shannon-Wiener index as special cases of a generalized entropy. Oikos **109**(1), 203–207 (2005)

34. Lawler, E.L.: Fast approximation algorithms for knapsack problems. Math. Oper. Res. 4(4), 339–356 (1979). https://doi.org/10.1287/moor.4.4.339

35. Magazine, M.J., Chern, M.: A note on approximation schemes for multidimensional knapsack problems. Math. Oper. Res. **9**(2), 244–247 (1984)

36. Mitchell, M., et al.: Diversity and inclusion metrics in subset selection. In: Proceedings of the AAAI/ACM Conference on AI, Ethics, and Society, AIES 2020, pp. 117–123. Association for Computing Machinery, New York (2020)
37. Naori, D., Raz, D.: Online multidimensional packing problems in the random-order model. In: 30th International Symposium on Algorithms and Computation (ISAAC), vol. 149, pp. 10:1–10:15 (2019)
38. Papachristou, M., Kleinberg, J.M.: Allocating stimulus checks in times of crisis. In: The ACM Web Conference (WWW), pp. 16–26. ACM (2022)
39. Patel, D., Khan, A., Louis, A.: Group fairness for knapsack problems. In: 20th International Conference on Autonomous Agents and Multiagent Systems (AAMAS), pp. 1001–1009. ACM (2021)
40. Patt-Shamir, B., Rawitz, D.: Vector bin packing with multiple-choice. Discrete Appl. Math. **160**(10–11), 1591–1600 (2012). https://doi.org/10.1016/j.dam.2012.02.020
41. Perez-Salazar, S., Torrico, A., Verdugo, V.: Preserving diversity when partitioning: a geometric approach. In: ACM Conference on Equity and Access in Algorithms, Mechanisms, and Optimization (EAAMO), pp. 1–11 (2021)
42. Sahni, S.: Approximate algorithms for the 0/1 knapsack problem. J. ACM **22**(1), 115–124 (1975)
43. Sharma, E.: Harmonic algorithms for packing D-dimensional cuboids into bins. In: 41st IARCS Annual Conference on Foundations of Software Technology and Theoretical Computer Science (FSTTCS), vol. 213, pp. 32:1–32:22 (2021)
44. Simpson, E.H.: Measurement of diversity. Nature **163**(4148), 688 (1949)
45. Zhang, H., Jansen, K.: Scheduling malleable tasks. In: Handbook of Approximation Algorithms and Metaheuristics (2007). https://doi.org/10.1201/9781420010749.ch45
46. Zuckerman, D.: Linear degree extractors and the inapproximability of max clique and chromatic number. Theory Comput. **3**(1), 103–128 (2007)

Obtaining Approximately Optimal and Diverse Solutions via Dispersion

Jie Gao[1] , Mayank Goswami[2(✉)] , C. S. Karthik[1], Meng-Tsung Tsai[3] ,
Shih-Yu Tsai[4], and Hao-Tsung Yang[5]

[1] Department of Computer Science, Rutgers University, New Brunswick, USA
[2] Department of Computer Science, Queens College, City University of New York,
New York, USA
mayank.goswami@qc.cuny.edu
[3] Institute of Information Science, Academia Sinica, Taipei, Taiwan
[4] Department of Computer Science, Stony Brook University, Stony Brook, USA
[5] Department of Computer Science and Information Engineering, National Central
University, Taoyuan, Taiwan

Abstract. There has been a long-standing interest in computing diverse
solutions to optimization problems. In 1995 J. Krarup [28] posed the
problem of finding k-edge disjoint Hamiltonian Circuits of minimum
total weight, called the peripatetic salesman problem (PSP). Since then
researchers have investigated the complexity of finding diverse solutions
to spanning trees, paths, vertex covers, matchings, and more. Unlike the
PSP that has a constraint on the total weight of the solutions, recent
work has involved finding diverse solutions that are all optimal.

However, sometimes the space of exact solutions may be too small
to achieve sufficient diversity. Motivated by this, we initiate the study
of obtaining sufficiently-diverse, yet approximately-optimal solutions to
optimization problems. Formally, given an integer k, an approximation
factor c, and an instance I of an optimization problem, we aim to obtain
a set of k solutions to I that a) are all c approximately-optimal for I
and b) maximize the diversity of the k solutions. Finding such solutions,
therefore, requires a better understanding of the global landscape of the
optimization function.

Given a metric on the space of solutions, and the diversity measure as
the sum of pairwise distances between solutions, we first provide a general reduction to an associated budget-constrained optimization (BCO)
problem, where one objective function is to optimized subject to a bound
on the second objective function. We then prove that bi-approximations
to the BCO can be used to give bi-approximations to the diverse approximately optimal solutions problem.

As applications of our result, we present polynomial time approximation algorithms for several problems such as diverse c-approximate
*maximum matchings, $s - t$ shortest paths, global min-cut, and minimum
weight bases of a matroid.* The last result gives us *diverse c-approximate
minimum spanning trees,* advancing a step towards achieving diverse c-
approximate TSP tours.

© Springer Nature Switzerland AG 2022
A. Castañeda and F. Rodríguez-Henríquez (Eds.): LATIN 2022, LNCS 13568, pp. 222–239, 2022.
https://doi.org/10.1007/978-3-031-20624-5_14

We also explore the connection to the field of multiobjective optimization and show that the class of problems to which our result applies includes those for which the associated DUALRESTRICT problem defined by Papadimitriou and Yannakakis [35], and recently explored by Herzel et al. [26] can be solved in polynomial time.

Keywords: Diversity · Minimum spanning tree · Maximum matching · Shortest path · Travelling salesman problem · Dispersion problem

1   Introduction

Techniques for optimization problems focus on obtaining optimal solutions to an objective function and have widespread applications ranging from machine learning, operations research, computational biology, networks, to geophysics, economics, and finance. However, in many scenarios, the optimal solution is not only computationally difficult to obtain, but can also render the system built upon its utilization vulnerable to adversarial attacks. Consider a patrolling agent tasked with monitoring n sites in the plane. The most efficient solution (i.e., maximizing the frequency of visiting each of the n sites) would naturally be to patrol along the tour of shortest length[1] (the solution to TSP - the Traveling Salesman Problem). However, an adversary who wants to avoid the patroller can also compute the shortest TSP tour and can design its actions strategically [39]. Similarly, applications utilizing the minimum spanning tree (MST) on a communication network may be affected if an adversary gains knowledge of the network [13]; systems using solutions to a linear program (LP) would be vulnerable if an adversary gains knowledge of the program's function and constraints.

One way to address the vulnerability is to use a set of approximately optimal solutions and randomize among them. However, this may not help much to mitigate the problem, if these approximate solutions are combinatorially too "similar" to the optimal solution. For example, all points in a sufficiently small neighborhood of the optimal solution on the LP polytope will be approximately optimal, but these solutions are not too much different and the adversaries can still effectively carry out their attacks. Similarly one may use another tree instead of the MST, but if the new tree shares many edges with the MST the same vulnerability persists. Thus k-best enumeration algorithms [18,24,30,31,33] developed for a variety of problems fall short in this regard.

One of the oldest known formulations is the Peripatetic Salesman problem (PSP) by Krarup [28], which asks for k-edge disjoint Hamiltonian circuits of minimum total weight in a network. Since then, several researchers have tried to compute diverse solutions for several optimization problems [4,5,16,23]. Most of these works are on graph problems, and diversity usually corresponds to the size

---

[1] We assume without loss of generality that the optimal TSP is combinatorially unique by a slight perturbation of the distances.

of the symmetric difference of the edge sets in the solutions. Crucially, almost all of the aforementioned work demands either every solution individually be optimal, or the set of solutions in totality (as in the case of the PSP) be optimal. Nevertheless, **the space of optimal solutions may be too small to achieve sufficient diversity,** and it may just be singular (unique solution). In addition, for NP-complete problems finding just one optimal solution is already difficult. While there is some research that takes the route of developing FPT algorithms for this setting [5,17], to us it seems practical to also consider the relaxation to approximately-optimal solutions.

This motivates the problem of finding a set of diverse and *approximately optimal* solutions, which is the problem considered in this article. The number of solutions k and the desired approximation factor $c > 1$ is provided by the user as input. Working in the larger class gives one more hope of finding diverse solutions, yet every solution has a guarantee on its quality.

1.1 Our Contributions

We develop approximation algorithms for finding k solutions to the given optimization problem: for every solution, the quality is bounded by a user-given approximation ratio $c > 1$ to the optimal solution and the diversity of these k solutions is maximized. Given a metric on the space of solutions to the problem, we consider the diversity measure given by the sum (or average) of pairwise distances between the k solutions. Combining ideas from the well-studied problem on dispersion (which we describe next), we reduce the above problem to a budget constrained optimization (BCO) program.

1.2 Dispersion

Generally speaking, if the optimization problem itself is $\mathcal{NP}$-hard, finding diverse solutions for that problem is also $\mathcal{NP}$-hard (see Proposition 1 for more detail). On the other hand, interestingly, even if the original problem is not $\mathcal{NP}$-hard, finding diverse and approximately optimal solutions can still be $\mathcal{NP}$-hard. This is due to the connection of the diversity maximization objective with the general family of problems that consider selecting k elements from the given input set with maximum "dispersion", defined as max-min distance, max-average distance, and so on.

The dispersion problem has a long history, with many variants both in the metric setting and the geometric setting [15,29,38]. For example, finding a subset of size k from an input set of n points in a metric space that maximizes the distance between closest pairs or the sum of distances of the k selected points are both $\mathcal{NP}$-hard [1,37]. For the max-sum dispersion problem, the best known approximation factor is 2 for general metrics [7,25], although PTAS are available for Euclidean metrics or more generally, metrics of negative type, even with matroid constraints [10,11].

Dispersion in Exponentially-Sized Space. We make use of the general framework of the 2-approximation algorithm [8,37] to the max-sum k-dispersion problem, a greedy algorithm where the $i + 1$th solution is chosen to be the most distant/diverse one from the first i solutions. Notice that in our setting, there is an important additional challenge to understand the space within which the approximate solutions stay. In all of the problems we study, the total number of solutions can be *exponential in the input size*. Thus we need to have a non-trivial way of navigating within this large space and carry furthest insertion without considering all points in the space. This is where our reduction to budget constrained problem comes in.

Self Avoiding Dispersion. Furthermore, even after implicitly defining the $i + 1$th furthest point insertion via some optimization problem, one needs to take care that the (farthest, in terms of sum of distances) solution does not turn out to equal one of the previously found i solutions, as this is a requirement for the furthest point insertion algorithm. This is an issue one faces because of the implicit nature of the furthest point procedure in the exponential-sized space of solutions: in the metric k-dispersion problem, it was easy to guarantee distinctness as one only considered the $n - i$ points not yet selected.

1.3   Reduction to Budget Constrained Optimization

Combining with dispersion, we reduce the diversity computational problem to a budget constrained optimization (BCO) problem where the budget is an upper (resp. lower) bound if the quality of solution is described by a minimization (resp. maximization) problem. Intuitively the budget guarantees the quality of the solution, and the objective function maximizes diversity. Recall that the number of solutions k and the approximation factor c is input by the user; a larger c allows for more diversity.

We show how using an (a, b) bi-approximation algorithm for the BCO problem provides a set of $O(a)$-diverse, bc approximately-optimal solutions to the diversity computational problem (the hidden constant is at most 4). This main reduction is described in Theorem 1.

The main challenge in transferring the bi-approximation results because of a technicality that we describe next. Let $\mathcal{S}(c)$ be the space of c approximate solutions. A $(*, b)$ bi-approximation algorithm to the BCO relaxes the budget constraint by a factor b, and hence only promises to return a faraway point in the larger space $\mathcal{S}(b \cdot c)$. Thus bi-approximation of BCO do not simply give a farthest point insertion in the space of solutions, and instead *return a point in a larger space*. Nevertheless, we prove that in most cases, one loses a factor of at most 4 in the approximation factor for the diversity.

Once the reduction to BCOs is complete, for diverse approximate matchings, spanning trees and shortest paths we exploit the special characteristics of the corresponding BCO to solve it optimally ($a = b = 1$). For other problems such as global min-cut, diverse approximate minimum weight spanning trees, and the more general minimum weight bases of a matroid, we utilize known bi-approximations to the BCO to obtain bi-approximations for the diversity problem. For all problems except diverse (unweighted) spanning trees[2], our algorithms are the first polynomial time bi-approximations for these problems.

We also connect to the wide literature on multicriteria optimization and show that our result applies to the entire class of problems for which the associated DUALRESTRICT problem (defined by Papadimitriou and Yannakakis [35], and recently studied by Herzel et al. [26]) has a polynomial time solution. We discuss this in more detail after presenting our reduction.

Layout: The rest of this paper is organized as follows: we survey related work in Sect. 2, and formulate the problem in Sect. 3. In Sect. 4 we mention the connection to dispersion and describe the reduction to the budget constrained optimization problem (Theorem 1). Sections 5, 6, 7 and 8 describe four applications of our technique to various problems such as diverse approximate matchings, global min-cuts, shortest paths, minimum spanning trees, and minimum weight bases of a matroid. We remark that this list is by no means exhaustive, and we leave finding other interesting optimization problems which are amenable to our approach for future research. **Due to space constraints, all proofs can be found in the publicly available full version of this paper at [19].**

2 Related Work

Recently there has been a surge of interest in the tractability of finding diverse solutions for a number of combinatorial optimization problems, such as spanning trees, minimum spanning trees, k-paths, shortest paths, k-matchings, etc. [16,17,21–23]. Most of the existing work focuses on finding diverse optimal solutions. In cases when finding the optimal solution is NP-complete, several works have focused on developing FPT algorithms [5,17]. Nevertheless, as pointed out in [22], it would be more practical to consider finding a set of diverse "short" paths rather than one set of diverse shortest paths. They show that finding a set of approximately shortest paths with the maximum diversity is NP-hard, but leave the question of developing approximation algorithms open, a question that we answer in our paper for several problems. Similarly the problem of finding diverse maximum matchings was proved to be NP-hard in [16]. We remark that the main difference between our result and previous work is that our algorithms can find a diverse set of c-approximate solutions in polynomial time. If the attained diversity is not sufficient for the application, the user can input a larger c, in hopes of increasing it.

Multicriteria Optimization: In this domain, several optimization functions are given on a space of solutions. Clearly, there may not be a single solution that is the best for all objective functions, and researchers have focused on obtaining Pareto-optimal solutions, which are solutions that are non-dominated by other solutions. Put differently, a solution is Pareto-optimal if no other solution

---

[2] While an exact algorithm for diverse unweighted spanning trees is known [23], we give a faster (by a factor $\Omega(n^{1.5}k^{1.5}/\alpha(n,m))$ where $\alpha(\cdot)$ denotes the inverse of the Ackermann function), 2-approximation here.

can have a better cost for all criteria. Since exact solutions are hard to find, research has focused on finding ϵ Pareto-optimal solutions, which are a $1 + \epsilon$ factor approximations of Pareto-optimal solutions. Papadimitriou and Yannakakis [35] showed that under pretty mild conditions, any mutlicriteria optimization problem admits an ϵ Pareto-optimal set of fully polynomial cardinality. In terms of being able to *find* such an ϵ Pareto-optimal set, they show that a (FPTAS) PTAS exists for the problem if and only if an associated GAP problem can be solved in (fully) polynomial time. Very recently, Herzel et al. [26] study the class of problems for which an FPTAS or PTAS exists for finding ϵ Pareto-optimal solutions that are *exact* in one of the criteria. Such problems are a subset of the ones characterized by GAP. Herzel et al. [26] characterize the condition similarly: an FPTAS (PTAS) exists if and only if an associated DUALRESTRICT problem can be solved in (fully) polynomial time. For more details we refer the reader to the survey by Herzel at al. [27].

3   Diversity Computational Problem (DCP)

First, we define some notations. We use the definition of optimization problems given in [3] with additional formalism as introduced in [20].

Definition 1 (Optimization Problem). *An **optimization problem** Π is characterized by the following quadruple of objects $(I_\Pi, \mathsf{Sol}_\Pi, \Delta_\Pi, \mathsf{goal}_\Pi)$, where:*

- *I_Π is the set of instances of Π. In particular for every $n \in \mathbb{N}$, $I_\Pi(n)$ is the set of instances of Π of input size at most n (bits);*
- *Sol_Π is a function that associates to any input instance $x \in I_\Pi$ the set of feasible solutions of x;*
- *Δ_Π is the measure function[3], defined for pairs (x, y) such that $x \in I_\Pi$ and $y \in \mathsf{Sol}_\Pi(x)$. For every such pair (x, y), $\Delta_\Pi(x, y)$ provides a non-negative integer which is the value of the feasible solution y;*
- *$\mathsf{goal}_\Pi \in \{\min, \max\}$ specifies whether Π is a maximization or minimization problem.*

We would like to identify a subset of our solution space which are (approximately) optimal with respect to our measure function. To this effect, we define a notion of approximately optimal feasible solution.

Definition 2 (Approximately   Optimal   Feasible   Solution). *Let $\Pi(I_\Pi, \mathsf{Sol}_\Pi, \Delta_\Pi, \mathsf{goal}_\Pi)$ be an optimization problem and let $c \geq 1$. For every $x \in I_\Pi$ and $y \in \mathsf{Sol}_\Pi(x)$ we say that y is a c-**approximate optimal solution** of x if for every $y' \in \mathsf{Sol}_\Pi(x)$ we have $\Delta_\Pi(x, y) \cdot c \geq \Delta_\Pi(x, y')$ if $\mathsf{goal}_\Pi = \max$ and $\Delta_\Pi(x, y) \leq \Delta_\Pi(x, y') \cdot c$ if $\mathsf{goal}_\Pi = \min$.*

---

[3] We define the measure function only for feasible solutions of an instance. Indeed if an algorithm solving the optimization problem outputs a non-feasible solution, then the measure just evaluates to -1 in case of maximization problems and ∞ in case of minimization problems.

Definition 3 (Computational Problem). *Let $\Pi(I_\Pi, \mathsf{Sol}_\Pi, \Delta_\Pi, \mathsf{goal}_\Pi)$ be an optimization problem and let $\lambda : \mathbb{N} \to \mathbb{N}$. The **computational problem associated with** (Π, λ) is given as input an instance $x \in I_\Pi(n)$ (for some $n \in \mathbb{N}$) and real $c := \lambda(n) \geq 1$ find a c-approximate optimal feasible solution of x.*

Definition 4 (DCP - Diversity Computational Problem). *Let $\Pi(I_\Pi, \mathsf{Sol}_\Pi, \Delta_\Pi, \mathsf{goal}_\Pi)$ be an optimization problem and let $\lambda : \mathbb{N} \to \mathbb{N}$. Let $\sigma_{\Pi,t}$ be a diversity measure that maps every t feasible solutions of an instance of I_Π to a non-negative real number. The **diversity computational problem associated with** $(\Pi, \sigma_{\Pi,t}, k, \lambda)$ is given as input an instance $x \in I_\Pi(n)$ (for some $n \in \mathbb{N}$), an integer $k := k(n)$, and real $c := \lambda(n) \geq 1$, find k-many c-approximate solutions $y_1, \ldots, y_k$ to x which maximize the value of $\sigma_{\Pi,k}(x, y_1, \ldots, y_k)$.*

Proposition 1. *Let $\Pi(I_\Pi, \mathsf{Sol}_\Pi, \Delta_\Pi, \mathsf{goal}_\Pi)$ be an optimization problem and let $\lambda : \mathbb{N} \to \mathbb{N}$. Let $\sigma_{\Pi,t}$ be a diversity measure that maps every t feasible solutions of an instance of I_Π to a non-negative real number. If the computational problem associated with (Π, λ) is $\mathcal{NP}$-hard, then the diversity computational problem associated with $(\Pi, \sigma_{\Pi,t}, \lambda)$ also is $\mathcal{NP}$-hard.*

Therefore the interesting questions arise when we compute problems associated with (Π, λ) which are in $\mathcal{P}$, or even more when, $(\Pi, \mathbb{1})$ is in $\mathcal{P}$ where $\mathbb{1}$ is the constant function which maps every element of the domain to 1. For the remainder of this paper, we will consider $\lambda(n)$ to be the constant function, and will simply refer to the constant as c.

Finally, we define bicriteria approximations for the diversity computational problem:

Definition 5 ((α, β) Bi-approximation for the Diversity Computational Problem). *Consider the diversity computational problem associated with $(\Pi, \sigma_{\Pi,t}, k, c)$, and a given instance $x \in I_\Pi(n)$ (for some $n \in \mathbb{N}$). An algorithm is called an (α, β) bi-approximation for the diversity computational problem if it outputs k feasible solutions $y_1, \ldots, y_k$ such that a) y_i is a $\beta \cdot c$-approximate optimal feasible solution to x for all $1 \leq i \leq k$, and b) for any set $y_1', \ldots, y_k'$ of k-many c-approximate optimal feasible solutions, $\sigma_{\Pi,k}(y_1, \cdots, y_k) \cdot \alpha \geq \sigma_{\Pi,k}(y_1', \cdots, y_k')$. Furthermore, such an algorithm is said to run in polynomial time if the running time is polynomial in n and k.*

4   The Reduction: Enter Dispersion and Biobjective Optimization

As stated in the introduction, our problems are related to the classical dispersion problem in a metric space. Here we state the dispersion problem and use dispersion to reduce the problem of finding diverse, approximately optimal solutions to solving an associated budget constrained optimization problem.

4.1 Dispersion Problem

Definition 6 (*k*-Dispersion, Total Distance). *Given a finite set of points* P *whose pairwise distances satisfy the triangle inequality and an integer* $k \geq 2$, *find a set* $S \subseteq P$ *of cardinality* k *so that* $W(S)$ *is maximized, where* $W(S)$ *is the sum of the pairwise distances between points in* S.

The main previous work on the k-dispersion problem relevant to us is [37], where the problem was named as Maximum-Average Facility Dispersion problem with triangle inequality (MAFD-TI). The problems are equivalent as maximizing the average distance between the points also maximizes the sum of pairwise distances between them and vice-versa.

The k-dispersion problem is $\mathcal{NP}$-hard, but one can find a set S whose $W(S)$ is at least a constant factor of the maximum possible in polynomial time by a greedy procedure [37]. We call the greedy procedure *furthest insertion*. It works as follows. Initially, let S be a singleton set that contains an arbitrary point from the given set. While $|S| < k$, add to S a point $x \notin S$ so that $W(S \cup \{x\}) \geq W(S \cup \{y\})$ for any $y \notin S$. Repeat the greedy addition until S has size k. The final S is a desired solution, which is shown to be a 4-approximation in [37]. It is worth noting that the furthest insertion in [37] initializes S as a furthest pair of points in the given set, and the above change does not worsen the approximation factor. In a later paper [8], the greedy algorithm of choosing an arbitrary initial point is shown to be a 2-approximation, which is a tight bound for this algorithm [7].

Lemma 1 (Furthest Insertion in [8,37]). *The* k-*dispersion problem can be* 2-*approximated by the furthest insertion algorithm.*

The running time of the furthest insertion algorithm is polynomial in $|S|$ (the size of S), as it performs k iterations, each performing at most $O(k|S|)$ distance computations/lookups. Note that in our case S is the collection of objects of a certain type (matchings, paths, trees, etc.). Hence the size of our metric space is typically exponential in $|V|$ and $|E|$. This adds a new dimension of complexity to the traditional dispersion problems studied.

4.2 Reduction to Budget Constrained Optimization

Recall the definitions of the Diversity Computational Problem (Definition 4) and (a, b) bi-approximations (Definition 5). As the input instance $x \in I_\Pi$ will be clear from context, we drop the dependence on x, and assume a fixed input instance to a computational problem. Thus Sol_Π will denote the set of feasible solutions, and $\Delta_\Pi(y)$ the measure of the feasible solution y.

Diversity and Similarity Measures from Metrics. Let $d : \mathsf{Sol}_\Pi \times \mathsf{Sol}_\Pi \to \mathbb{R}^+$ be a metric on the space of feasible solutions. When such a metric is available, we will consider the diversity function $\sigma_{\Pi,t} : \mathsf{Sol}_\Pi \times \cdots \times \mathsf{Sol}_\Pi \to \mathbb{R}^+$ that assigns the diversity measure $\sum_{i,j} d(y_i, y_j)$ to a t-tuple of feasible solutions $(y_1, \cdots, y_t)$. Also, given such a metric d, define D to be the diameter of Sol_Π under d, i.e.,

$D = \max_{y,y' \in \mathsf{Sol}_\Pi} d(y, y')$. In many cases, we will be interested in the *similarity* measure $s_{\Pi,t}$ defined by $s_{\Pi,t}(y_1, \cdots, y_t) = \sum_{i,j} (D - d(y_i, y_j))$. The examples the reader should keep in mind are graph objects such as spanning trees, matchings, shortest paths, Hamiltonian circuits, etc., such that $d(y, y')$ denotes the Hamming distance, a.k.a. size of the symmetric difference of the edge sets of y and y', and s denotes the size of their intersection.

In the remainder of the paper we consider the above total distance (resp. similarity) diversity measures $\sigma_{\Pi,t}$ arising from the metric d (resp. similarity measure s), and we will parameterize the problem by d (resp. s) instead.

Definition 7 (Budget Constrained Optimization). *Given an instance of a computational problem Π, a constant $c \geq 1$, and a set $\{y_1, \ldots, y_i\}$ of feasible solutions in Sol_Π, define the **metric budget constrained optimization problem** $BCO(\Pi, (y_1, \ldots, y_i), c, d)$ as follows:*

- *If $\mathsf{goal}_\pi = \min$, define $\Delta^* := \min_{y \in \mathsf{Sol}_\Pi} \Delta_\Pi(y)$. Then $BCO(\Pi, (y_1, \ldots, y_i), c, d)$ is the problem*

$$
\begin{aligned}
\max \quad & f_d(y) := \sum_{j=1}^{i} d(y, y_j) \\
\text{s.t.} \quad & \Delta_\Pi(y) \leq c \cdot \Delta^* \\
& y \in \mathsf{Sol}_\Pi \setminus \{y_1, \ldots, y_i\}
\end{aligned}
\tag{1}
$$

- *If $\mathsf{goal}_\pi = \max$, define $\Delta^* := \max_{y \in \mathsf{Sol}_\Pi} \Delta_\Pi(y)$. Then $BCO(\Pi, (y_1, \ldots, y_i), c, d)$ is the problem*

$$
\begin{aligned}
\max \quad & f_d(y) := \sum_{j=1}^{i} d(y, y_j) \\
\text{s.t.} \quad & \Delta_\Pi(y) \cdot c \geq \Delta^* \\
& y \in \mathsf{Sol}_\Pi \setminus \{y_1, \ldots, y_i\}
\end{aligned}
\tag{2}
$$

- *Given a similarity measure s, define the **similarity budget constrained optimization problem** $BCO(\Pi, (y_1, \ldots, y_i), c, s)$ with the same constraint set as above (depending on goal_π), but with the objective function changed to $g_s(y) := \min \sum_{j=1}^{i} s(y, y_j)$ instead of $\max \sum_{j=1}^{i} d(y, y_j)$.*

Definition 8 (Bi-approximation to BCO). *An algorithm for an associated BCO is called an (a, b) bi-approximation algorithm if for any $1 \leq i \leq k$, it outputs a solution y such that the following holds.*

- *If $\mathsf{goal}_\Pi = \min$ and the associated BCO is $BCO(\Pi, (y_1, \ldots, y_i), c, d)$, then a) $y \in \mathsf{Sol}_\Pi \setminus \{y_1, \cdots, y_i\}$, b) $\Delta_\Pi(y) \leq b \cdot c \cdot \Delta^*$, and c) for all y' satisfying the constraints of $BCO(\Pi, (y_1, \ldots, y_i), c, d)$, $f_d(y) \cdot a \geq f_d(y')$.*
- *If $\mathsf{goal}_\Pi = \max$ and the associated BCO is $BCO(\Pi, (y_1, \ldots, y_i), c, d)$, then a) $y \in \mathsf{Sol}_\Pi \setminus \{y_1, \cdots, y_i\}$, b) $\Delta_\Pi(y) \cdot b \cdot c \geq \Delta^*$, and c) for all y' satisfying the constraints of $BCO(\Pi, (y_1, \ldots, y_i), c, d)$, $f_d(y) \cdot a \geq f_d(y')$.*

- *If* $\text{goal}_\Pi = \min$ *and the associated BCO is* $BCO(\Pi, (y_1, \ldots, y_i), c, s)$, *then a)* $y \in \text{Sol}_\Pi \backslash \{y_1, \cdots, y_i\}$, *b)* $\Delta_\Pi(y) \leq b \cdot c \cdot \Delta^*$, *and c) for all* y' *satisfying the constraints of* $BCO(\Pi, (y_1, \ldots, y_i), c, s)$, $g_s(y) \leq g_s(y') \cdot a$.
- *If* $\text{goal}_\Pi = \max$ *and the associated BCO is* $BCO(\Pi, (y_1, \ldots, y_i), c, s)$, *then a)* $y \in \text{Sol}_\Pi \backslash \{y_1, \cdots, y_i\}$, *b)* $\Delta_\Pi(y) \cdot b \cdot c \geq \Delta^*$, *and c) for all* y' *satisfying the constraints of* $BCO(\Pi, (y_1, \ldots, y_i), c, s)$, $g_s(y) \leq g_s(y') \cdot a$.

Remark: Minimization and maximization are essentially equivalent (by changing the sign), and so optimally solving one solves the other. The reason why we continue to treat them separately is because obtaining *an approximation* to minimizing total similarity $g_s(y) := \sum_{j=1}^{i} s(y, y_i)$ is not equivalent to an approximation to maximizing total distance $f_d(y) := \sum_{j=1}^{i} d(y, y_i)-$ in fact, these functions are the "opposite" of each other, as $f_d(y) = Di - g_s(y)$.

We are now ready to state our main theorem.

Theorem 1 (Reduction of DCP to BCO). *Consider an input* (Π, k, d, c) *to the diversity computational problem (DCP).*

- *For metric BCO,*
  1. *An* $(a, 1)$ *bi-approximation to* $BCO(\Pi, (y_1, \ldots, y_i), c, d)$ *can be used to give a* $(2a, 1)$ *bi-approximation to the DCP, and*
  2. *An* (a, b) *bi-approximation to* $BCO(\Pi, (y_1, \ldots, y_i), c, d)$ *can be used to give a* $(4a, b)$ *bi-approximation to the DCP.*
- *For similarity BCO,*
  3. *A* $(1, 1)$ *bi-approximation to* $BCO(\Pi, (y_1, \ldots, y_i), c, s)$ *can be used to give a* $(2, 1)$ *bi-approximation to the DCP,*
  4. *A* $(1, b)$ *bi-approximation to* $BCO(\Pi, (y_1, \ldots, y_i), c, s)$ *can be used to give* $(4, b)$ *bi-approximation to the DCP,*
  5. *A* $(1 + \epsilon, 1)$ *bi-approximation to* $BCO(\Pi, (y_1, \ldots, y_i), c, s)$ *can be used to give* $(4, 1)$ *bi-approximation to the DCP, under the condition that the average pairwise distance in the optimal solution to the DCP is at least* $D\frac{4\epsilon}{1+2\epsilon}$.

In all of the above, the overhead for obtaining a bi-approximation for the DCP, given a bi-approximation for BCO problem, is $O(k)$.

A few remarks are in order:

- The above theorem provides a recipe for solving the diversity computational problem for any given optimization problem. As long as *either* the metric or the similarity budget constrained optimization problems can be solved or approximated in polynomial time, one has an analogous result for the DCP.
- In the remainder of this paper we will see several applications that follow from the above 5 "types" of bi-approximations available. These include DCP for Maximum Matching and Global Min-Cut (Type 1), DCP for shortest path (Type 3), DCP for minimum weight bases of a matroid, minimum spanning trees (Types 4 and 5).

– Whenever either a or b (or both) is set to be $1 + \epsilon$, we call a bi-approximation for the BCO problem an FPTAS if the running time is polynomial in $1/\epsilon$ in addition to being polynomial in d and k. Otherwise we call it a PTAS.

Relation to Multicriteria Optimization: Observe that for similarity BCOs, we need either a or b to be 1. This class of biobjective problems that have a PTAS that is exact in one of the criteria is a special case of the multicriteria problems that have a PTAS that is exact in one of the criteria. Herzel et al. [26] showed that this class is exactly the class of problems for which the DUALRESTRICT version of the problem, posed by Diakonikolas and Yannakakis [14]), can be solved in polynomial time. These are also the class of problems having a polynomial-time computable approximate ϵ-Pareto set that is exact in one objective. This equivalence means that our theorem is applicable to this entire class of problems.

4.3   Relaxed BCOs and Self-avoidance

Before we delve into our applications, we describe another challenge in directly applying results from multicriteria optimization literature. For a BCO, the second constraint demands that $y \in \mathsf{Sol}_\Pi \backslash \{y_1, \cdots, y_i\}$. Intuitively y is the farthest point to the set of already discovered solutions $\{y_1, \cdots, y_i\}$, and because it is defined implicitly, without the second constraint y may equal one of the y_j ($1 \leq j \leq i$). Consider an alternate BCO, which we call BCO^r where the constraint is relaxed to $y \in \mathsf{Sol}_\Pi$. For many graph problems, solving BCO^r combined with the approach by Lawler [30] gives a solution to the original BCO. This is extremely useful because most of the literature on multicriteria optimization concerns optimization of the relaxed type of problems BCO^r, and one can borrow results derived before without worrying about the second constraint. We remark that for other problems, k-best enumeration algorithms (see [18,24,30,31,33] for examples) may be useful to switch from the BCO to its relaxed version. *Thus any algorithm for BCO^r can be used, modulo the self-avoiding constraint (to be handled using Lawler's approach), to give a polynomial time algorithm for the Diversity Computational Problem with the same guarantees as in Theorem 1.* We provide examples of the approach by Lawler in subsequent sections where we consider specific problems.

5   Application 1: Diverse Spanning Trees

In this section, we discuss the diverse spanning trees problem, which is the diversity computational problem for spanning trees with Hamming distance function as the diversity measure. Let $G = (V, E)$ be an n-node m-edge undirected graph. The problem aims to output a set S of k spanning trees $T_1, \cdots, T_k$ of G such that the sum of the pairwise distances $\sum_{i,j \in S} d(T_i, T_j)$ is maximized, where d is the Hamming distance between the edge sets of the trees. While this problem actually has an exact algorithm running in time $O((kn)^{2.5} \ m)$ [23], we get a faster approximation algorithm.

Theorem 2. *Given an n-node m-edge undirected graph $G = (V, E)$, there exists an $O(knm \cdot \alpha(n, m))$-time algorithm, where $\alpha(\cdot)$ is the inverse of the Ackermann function, that generates k spanning trees $T_1, \cdots, T_k$, such that the sum of all pairwise Hamming distances is at least half of an optimal set of k diverse spanning trees.*

We prove the above theorem by developing an exact $(1,1)$ polynomial time subroutine for the associated BCO problem. The proof can be found in the full version.

6 Application 2: Diverse Approximate Shortest Paths

Given a graph $G = (V, E)$, non-negative edge weights $w(e)$, two vertices s and t, and a factor $c > 1$, the diversity computational problem asks to output k many st paths, such that the weight of each path is within a factor c of the weight of the shortest st path, and subject to this constraint, the total pairwise distance between the paths is maximized. Here the distance between two paths is again the Hamming distance, or size of symmetric difference of their edge sets.

In [22], it is shown that finding k shortest paths with the maximum diversity (i.e. the average Hamming distance between solutions) can be solved in polynomial time, but finding k "short" paths with the maximum diversity is NP-hard. In contrast, in what follows, we will show that finding k "short" paths with constant approximate diversity is polynomial-time solvable.

We will show that the associated budget constrained optimization problem for this is of Type 3 in Theorem 1. In other words, we will show that the BCO can be solved exactly. This will result in a $(2,1)$ approximation algorithm for the diversity computational problem.

Hence, we need an algorithm that implements: given a set S of c-approximate shortest st-paths, find a c-approximate shortest st-path $P \notin S$ so that $W(S \cup \{P\})$ is maximum among all $W(S \cup \{P'\})$ for c-approximate shortest st-path $P' \notin S$. Here, $W(S')$ is the sum of all pairwise Hamming distances between two elements in S'. This is a special case of the bicriteria shortest paths, for which there is an algorithm in [34]. In our case, **one of the two weight functions is an integral function with range bounded in $[0, k]$**. Hence, it can be solved more efficiently than the solution in [34], which can be summarized as following.

Lemma 2 (Exact solution to the relaxed BCO^r problem). *Given a real $c \geq 1$ and a directed simple graph $G = (V \cup \{s, t\}, E)$ associated with two weight functions on edges $\omega : E \to \mathbb{R}^+$ and $f : E \to \{0, 1, \ldots, r\}$, there is an $O(r|V|^3)$-time algorithm to output an st-path P^* so that $\sum_{e \in E(P^*)} f(e)$ is minimized while retaining $\sum_{e \in E(P^*)} \omega(e) \leq c \sum_{e \in E(P)} \omega(e)$ for all st-paths P.*

Self-avoiding Constraint. We now turn to solving the associated (non-relaxed) BCO problem, by generalizing the above lemma to Corollary 1. Thus Corollary 1 will help us avoid the situation that a furthest insertion returns a path that is already picked by some previous furthest insertion.

Corollary 1 (Exact solution to the *BCO* problem). *Given a real $c \geq 1$, a directed simple graph $G = (V \cup \{s,t\}, E)$ associated with two weight functions on edges $\omega : E \to \mathbb{R}^+$, $f : E \to \{0, 1, \ldots, r\}$, and two disjoint subsets of edges $E_{in}, E_{ex} \subseteq E$ so that all edges in E_{in} together form a directed simple path P_{prefix} starting from node s, there exists an $O(r|V|^3)$-time algorithm to output an c-approximate shortest st-path P^* under ω so that $\sum_{e \in E(P^*)} f(e)$ is minimum among all the c-approximate shortest st-paths P that contain P_{prefix} as a prefix and contain no edges from E_{ex}, if such an c-approximate shortest st-path exists.*

We are ready to state our main result for the diverse c-approximate shortest st-paths.

Theorem 3 ($(2,1)$ Bi-approximation to the Diversity Problem on Shortest Paths). *For any directed simple graph $G = (V \cup \{s,t\}, E)$, given a constant $c > 1$ and an integer $k \in \mathbb{N}$, there exists an $O(k^3|V|^4)$-time algorithm that, if G contains at least k distinct c-approximate shortest st-paths, computes a set S of k distinct c-approximate shortest st-paths so that the sum of all pairwise Hamming distances between two paths in S is at least one half of the maximum possible; otherwise, reports "Non-existent".*

7 Application 3: Diverse Approximate Maximum Matchings, and Global Min-Cut

Consider the diversity computational problem for computing k many c-approximate maximum matchings for undirected graphs. In [16], the authors present an algorithm, among others, to find a pair of *maximum* matchings for *bipartite* graphs whose Hamming distance is maximized. In contrast, our result can be used to find $k \geq 2$ *approximate maximum* matchings for *any graph* whose diversity (i.e. the average Hamming distance) approximates the largest possible by a factor of 2.

We show that this problem can be restated into the budgeted matching problem [6]. As noted in [6], though the budgeted matching is in general $\mathcal{NP}$-hard, if both the weight and cost functions are integral and have a range bounded by a polynomial in $|V|$, then it can be solved in polynomial time with a good probability by a reduction to the exact perfect matching problem [9,32]. The exact running time for such a case is not stated explicitly in [6]. We combine the algorithm in [6] and the approach by Lawler [30] to prove:

Theorem 4. *There exists a $O(k^4|V|^7 \log^3 k|V|)$ time, $(2,1)$ bi-approximation to the diversity computational problem for c-approximate maximum matchings, with failure probability $1/|V|^{\Omega(1)}$.*

DCP for Global Min-Cuts: Next, consider the diversity computational problem for computing k many c-approximate global min-cuts: given a graph G and a positive weight function w on its edges, a c-approximate min-cut is a cut C whose cut-edge set $E(C)$ satisfies $\sum_{e \in E(C)} w(e) \leq c \sum_{e \in E(C')} w(e)$ for any other cut C'. Given i cuts, we define the (integral) cost of an edge as the number of cuts in which it appears as a cut edge. Consider the BCO with cost minimization in the objective function (as the cost of a cut is now inversely proportional to its sum of distances from the found cuts) and constraint with upper bound (the weight of the cut should be at most c times that of a global min weight cut). In [2] the authors provide a polynomial-time algorithm for this problem, implying that the BCO can be solved exactly in polynomial time. This gives us a $(2, 1)$ bi-approximation to the diversity computational problem for c-approximate global minimum cuts. We remark that one may be able to exploit integrality of our cost function to obtain a faster algorithm than that in [2].

8   Application 4: Diverse Minimum Weight Matroid Bases and Minimum Spanning Trees

One of the original ways to attack the peripatetic salesman problem (Krarup [28]) was to study the k edge-disjoint spanning trees problem [12]. Note that the existence of such trees is not guaranteed, and one can use our results in Sect. 5 to maximize diversity of the k trees found.

However, for an application to the TSP problem, cost conditions must be taken into account. Here we study the diverse computational problem (DCP) on minimum spanning trees: Given a weighted undirected graph $G = (V, E)$ with nonnegative weights $w(e)$, $c > 1$ and a $k \in \mathbb{N}$, return k spanning trees of G such that each spanning tree is a c-approximate minimum spanning tree, and subject to this, the diversity of the k trees is maximized. Here again the diversity of a set of trees is the sum of pairwise distances between them, and the distance between two trees is the size of their symmetric difference.

Our results in this section generalize to the problem of finding k diverse bases of a matroid such that every basis in the solution set is a c approximate minimum-weight basis. The DCP on MSTs is a special case of this problem. However, in order to not introduce extra notation and definitions here, we will describe our method for minimum spanning trees. We will then briefly sketch how to extend the algorithm to the general matroid case.

Starting with $T_1 = MST(G)$ (a minimum spanning tree on G, computable in polynomial time), assume we have obtained i trees $T_1, \cdots, T_i$, all of which are c-approximate minimum spanning trees. Assign to each edge a length $\ell(e)$ which equals $|\{j : 1 \leq j \leq i, e \in T_j\}|$.

Lemma 3. *Given $T_1, \cdots, T_i$, finding T_{i+1} that maximizes $\sum_{j=1}^{i} d(T, T_j)$ is equivalent to finding T that minimizes $\sum_{e \in T} \ell(e)$.*

Proof. An explicit calculation reveals that $\sum_{e \in T} \ell(e) = (n-1)i - \sum_{j=1}^{i} d(T, T_j)$.

Consider now the associated *similarity budget constrained optimization prob-lem*

$$\min \quad \sum_{e \in T} \ell(e)$$
$$\text{s.t.} \quad w(T) \le c \cdot w(MST(G)) \tag{3}$$
$$T \in \mathsf{Sol}_\Pi \backslash \{T_1, \ldots, T_i\}$$

Here Sol_Π is just the set of spanning trees on G. We will handle the self-avoiding constraints in a similar fashion as in Sect. 5. For the moment consider the relaxed BCO^r where the last constraint is simply $T \in \mathsf{Sol}_\Pi$. This is a budget constrained MST with two weights. This problem has been considered by Ravi and Goemans [36], who termed it the CMST problem. They provide a $(1, 2)$ bi-approximation that runs in near-linear time, and a $(1, 1 + \epsilon)$ bi-approximation that runs in polynomial time[4]. Also, they show that the $(1, 1 + \epsilon)$ bi-approximation can be used as a subroutine to compute a $(1 + \epsilon, 1)$ bi-approximation in pseudopolynomial time.

Applying their results and observing that we are in cases 4 and 5 of Theorem 1, we get

Theorem 5 (DCP for Mininum Spanning Trees). *There exists a*

- *polynomial (in n, m and k) time algorithm that outputs a $(4, 2)$ bi-approximation to the DCP problem for MSTs.*
- *polynomial (in n, m and k) and exponential in $1/\epsilon$ time algorithm that outputs a $(4, 1 + \epsilon)$ bi-approximation to the DCP problem for MSTs.*
- *pseudopolynomial time algorithm that outputs a $(4, 1)$ bi-approximation to the DCP problem for MSTs, as long as the average distance between the trees in the optimal solution to the k DCP on c-approximate minimum spanning trees does not exceed $\frac{4\epsilon(n-1)}{1+2\epsilon}$.*

Extension to Matroids: It is stated in the paper by Ravi and Goemans [36] that the same result holds if one replaces the set of spanning trees by the bases of any matroid. It is straightforward to show that the analog of Lemma 3 hold in the matroid setting too. With a bit of work, one can also generalize the approach of Lawler [30] to avoid self-intersection (the bases found so far), and thus all the techniques generalize to the matroid setting. In all of this, we assume an independence oracle for the matroid, as is standard. In [17], it is shown that, given integers k, d, finding k *perfect* matchings so that every pair of the found matchings have Hamming distance at least d is NP-hard. This hardness result also applies to finding weighted diverse bases and weighted diverse common independent sets.

Acknowledgements. Mayank Goswami would like to acknowledge support from the National Science Foundation grants CRII-1755791 and CCF-1910873. Jie Gao would like to acknowledge support OAC-1939459, CCF-2118953 and CCF-2208663. Karthik C. S. would like to acknowledge support from Rutgers University's Research Council

---

[4] The latter is a PTAS, not an FPTAS.

Individual Fulcrum Award (#AWD00010234) and a grant from the Simons Foundation, Grant Number 825876, Awardee Thu D. Nguyen. Meng-Tsung Tsai would like to acknowledge support from the Ministry of Science and Technology of Taiwan grant 109-2221-E-001-025-MY3.

References

1. Abbar, S., Amer-Yahia, S., Indyk, P., Mahabadi, S., Varadarajan, K.R.: Diverse near neighbor problem. In: Symposium on Computational Geometry (SoCG), pp. 207–214. ACM (2013)
2. Armon, A., Zwick, U.: Multicriteria global minimum cuts. Algorithmica **46**(1), 15–26 (2006)
3. Ausiello, G., Marchetti-Spaccamela, A., Crescenzi, P., Gambosi, G., Protasi, M., Kann, V.: Complexity and Approximation: Combinatorial Optimization Problems and Their Approximability Properties. Springer, Heidelberg (1999). https://doi.org/10.5555/554706
4. Baste, J., et al.: Diversity of solutions: an exploration through the lens of fixed-parameter tractability theory. Artif. Intell. **303**, 103644 (2022)
5. Baste, J., Jaffke, L., Masařík, T., Philip, G., Rote, G.: FPT algorithms for diverse collections of hitting sets. Algorithms **12**(12), 254 (2019)
6. Berger, A., Bonifaci, V., Grandoni, F., Schäfer, G.: Budgeted matching and budgeted matroid intersection via the gasoline puzzle. Math. Program. **128**(1–2), 355–372 (2011)
7. Birnbaum, B.E., Goldman, K.J.: An improved analysis for a greedy remote-clique algorithm using factor-revealing LPs. Algorithmica **55**(1), 42–59 (2009)
8. Borodin, A., Jain, A., Lee, H.C., Ye, Y.: Max-sum diversification, monotone submodular functions, and dynamic updates. ACM Trans. Algorithms **13**(3), 41:1–41:25 (2017)
9. Camerini, P.M., Galbiati, G., Maffioli, F.: Random pseudo-polynomial algorithms for exact matroid problems. J. Algorithms **13**(2), 258–273 (1992)
10. Cevallos, A., Eisenbrand, F., Zenklusen, R.: Max-sum diversity via convex programming. In: Fekete, S.P., Lubiw, A. (eds.) 32nd International Symposium on Computational Geometry, SoCG 2016, 14–18 June 2016, Boston, MA, USA. LIPIcs, vol. 51, pp. 26:1–26:14. Schloss Dagstuhl - Leibniz-Zentrum für Informatik (2016)
11. Cevallos, A., Eisenbrand, F., Zenklusen, R.: Local search for max-sum diversification. In: Proceedings of the Twenty-Eighth Annual ACM-SIAM Symposium on Discrete Algorithms, pp. 130–142. SIAM (2017)
12. Clausen, J., Hansen, L.A.: Finding k edge-disjoint spanning trees of minimum total weight in a network: an application of matroid theory. In: Rayward-Smith, V.J. (ed.) Combinatorial Optimization II. Mathematical Programming Studies, vol. 13, pp. 88–101. Springer, Heidelberg (1980). https://doi.org/10.1007/BFb0120910
13. Commander, C.W., Pardalos, P.M., Ryabchenko, V., Uryasev, S., Zrazhevsky, G.: The wireless network jamming problem. J. Comb. Optim. **14**(4), 481–498 (2007)
14. Diakonikolas, I., Yannakakis, M.: Small approximate pareto sets for biobjective shortest paths and other problems. SIAM J. Comput. **39**(4), 1340–1371 (2010)
15. Erkut, E.: The discrete P-dispersion problem. Eur. J. Oper. Res. **46**(1), 48–60 (1990). https://doi.org/10.1016/0377-2217(90)90297-O, https://www.sciencedirect.com/science/article/pii/037722179090297O

16. Fomin, F.V., Golovach, P.A., Jaffke, L., Philip, G., Sagunov, D.: Diverse pairs of matchings. In: 31st International Symposium on Algorithms and Computation (ISAAC). LIPIcs, vol. 181, pp. 26:1–26:12. Schloss Dagstuhl - Leibniz-Zentrum für Informatik (2020)

17. Fomin, F.V., Golovach, P.A., Panolan, F., Philip, G., Saurabh, S.: Diverse collections in matroids and graphs. In: Bläser, M., Monmege, B. (eds.) 38th International Symposium on Theoretical Aspects of Computer Science, STACS 2021, 16–19 March 2021, Saarbrücken, Germany (Virtual Conference). LIPIcs, vol. 187, pp. 31:1–31:14. Schloss Dagstuhl - Leibniz-Zentrum für Informatik (2021)

18. Gabow, H.N.: Two algorithms for generating weighted spanning trees in order. SIAM J. Comput. **6**(1), 139–150 (1977)

19. Gao, J., Goswami, M., Karthik, C.S., Tsai, M.T., Tsai, S.Y., Yang, H.T.: Obtaining approximately optimal and diverse solutions via dispersion (2022). https://doi.org/10.48550/ARXIV.2202.10028, https://arxiv.org/abs/2202.10028

20. Goldenberg, E., Karthik C. S.: Hardness amplification of optimization problems. In: Vidick, T. (ed.) 11th Innovations in Theoretical Computer Science Conference, ITCS 2020, 12–14 January 2020, Seattle, Washington, USA. LIPIcs, vol. 151, pp. 1:1–1:13. Schloss Dagstuhl - Leibniz-Zentrum für Informatik (2020). https://doi.org/10.4230/LIPIcs.ITCS.2020.1

21. Hanaka, T., Kiyomi, M., Kobayashi, Y., Kobayashi, Y., Kurita, K., Otachi, Y.: A framework to design approximation algorithms for finding diverse solutions in combinatorial problems. CoRR abs/2201.08940 (2022)

22. Hanaka, T., Kobayashi, Y., Kurita, K., Lee, S.W., Otachi, Y.: Computing diverse shortest paths efficiently: a theoretical and experimental study. CoRR abs/2112.05403 (2021)

23. Hanaka, T., Kobayashi, Y., Kurita, K., Otachi, Y.: Finding diverse trees, paths, and more. In: Thirty-Fifth AAAI Conference on Artificial Intelligence (AAAI), pp. 3778–3786. AAAI Press (2021)

24. Hara, S., Maehara, T.: Enumerate lasso solutions for feature selection. In: Singh, S.P., Markovitch, S. (eds.) Proceedings of the Thirty-First AAAI Conference on Artificial Intelligence (AAAI), pp. 1985–1991. AAAI Press (2017)

25. Hassin, R., Rubinstein, S., Tamir, A.: Approximation algorithms for maximum dispersion. Oper. Res. Lett. **21**(3), 133–137 (1997)

26. Herzel, A., Bazgan, C., Ruzika, S., Thielen, C., Vanderpooten, D.: One-exact approximate pareto sets. J. Global Optim. **80**(1), 87–115 (2021)

27. Herzel, A., Ruzika, S., Thielen, C.: Approximation methods for multiobjective optimization problems: a survey. INFORMS J. Comput. **33**(4), 1284–1299 (2021)

28. Krarup, J.: The peripatetic salesman and some related unsolved problems. In: Roy, B. (ed.) Combinatorial Programming: Methods and Applications. NATO Advanced Study Institutes Series, vol. 19, pp. 173–178. Springer, Dordrecht (1995). https://doi.org/10.1007/978-94-011-7557-9_8

29. Kuby, M.: Programming models for facility dispersion: the P-dispersion and maxisum dispersion problems. Geogr. Anal. **19**(4), 315–329 (1987). https://doi.org/10.1111/j.1538-4632.1987.tb00133.x

30. Lawler, E.L.: A procedure for computing the k best solutions to discrete optimization problems and its application to the shortest path problem. Manage. Sci. **18**(7), 401–405 (1972)

31. Lindgren, E.M., Dimakis, A.G., Klivans, A.: Exact map inference by avoiding fractional vertices. In: International Conference on Machine Learning, pp. 2120–2129. PMLR (2017)

32. Mulmuley, K., Vazirani, U.V., Vazirani, V.V.: Matching is as easy as matrix inversion. Comb. **7**(1), 105–113 (1987)
33. Murty, K.G.: An algorithm for ranking all the assignments in order of increasing cost. Oper. Res. **16**(3), 682–687 (1968)
34. Namorado Climaco, J.C., Queirós Vieira Martins, E.: A bicriterion shortest path algorithm. Eur. J. Oper. Res. **11**(4), 399–404 (1982)
35. Papadimitriou, C.H., Yannakakis, M.: On the approximability of trade-offs and optimal access of web sources. In: Proceedings 41st Annual Symposium on Foundations of Computer Science, pp. 86–92. IEEE (2000)
36. Ravi, R., Goemans, M.X.: The constrained minimum spanning tree problem. In: Karlsson, R., Lingas, A. (eds.) SWAT 1996. LNCS, vol. 1097, pp. 66–75. Springer, Heidelberg (1996). https://doi.org/10.1007/3-540-61422-2_121
37. Ravi, S.S., Rosenkrantz, D.J., Tayi, G.K.: Heuristic and special case algorithms for dispersion problems. Oper. Res. **42**(2), 299–310 (1994)
38. Wang, D., Kuo, Y.S.: A study on two geometric location problems. Inf. Process. Lett. **28**(6), 281–286 (1988) https://doi.org/10.1016/0020-0190(88)90174-3, https://www.sciencedirect.com/science/article/pii/0020019088901743
39. Yang, H.T., Tsai, S.Y., Liu, K.S., Lin, S., Gao, J.: Patrol scheduling against adversaries with varying attack durations. In: Proceedings of the 18th International Conference on Autonomous Agents and MultiAgent Systems, pp. 1179–1188 (2019)

Cryptography

On APN Functions Whose Graphs are Maximal Sidon Sets

Claude Carlet[1,2](✉) [ID]

[1] University of Bergen, Bergen, Norway
claude.carlet@gmail.com
[2] LAGA, Universities of Paris 8 and Paris 13 and CNRS, Saint-Denis, France

Abstract. The graphs $\mathcal{G}_F = \{(x, F(x)); x \in \mathbb{F}_2^n\}$ of those (n, n)-functions $F : \mathbb{F}_2^n \mapsto \mathbb{F}_2^n$ that are almost perfect nonlinear (in brief, APN; an important notion in symmetric cryptography) are, equivalently to their original definition by K. Nyberg, those Sidon sets (an important notion in combinatorics) S in $(\mathbb{F}_2^n \times \mathbb{F}_2^n, +)$ such that, for every $x \in \mathbb{F}_2^n$, there exists a unique $y \in \mathbb{F}_2^n$ such that $(x, y) \in S$. Any subset of a Sidon set being a Sidon set, an important question is to determine which Sidon sets are maximal relatively to the order of inclusion. In this paper, we study whether the graphs of APN functions are maximal (that is, optimal) Sidon sets. We show that this question is related to the problem of the existence/non-existence of pairs of APN functions lying at distance 1 from each others, and to the related problem of the existence of APN functions of algebraic degree n. We revisit the conjectures that have been made on these latter problems.

Keywords: Almost perfect nonlinear function · Sidon set in an Abelian group · Symmetric cryptography

1 Introduction

Almost perfect nonlinear (APN) functions, that is, vectorial functions $F : \mathbb{F}_2^n \mapsto \mathbb{F}_2^n$ whose derivatives $D_a F(x) = F(x) + F(x + a); a \neq 0$, are 2-to-1, play an important role in symmetric cryptography (see for instance the book [9]), since they allow an optimal resistance against the differential cryptanalysis of the block ciphers that use them as substitution boxes. Their mathematical study is an important domain of research, whose results (and in particular those by K. Nyberg in the early nineties) made possible the invention of the Advanced Encryption Standard (AES), chosen as a standard by the U.S. National Institute of Standards and Technology (NIST) in 2001, and today used worldwide as a cryptosystem dedicated to civilian uses. APN functions also play an important role in coding theory (see [11]).

Sidon sets, which are subsets S in Abelian groups such that all pairwise sums $x + y$ (with $\{x, y\} \subset S$, $x \neq y$), are different, are an important notion in combinatorics [1], whose name refers to the Hungarian mathematician Simon Sidon, who introduced the concept in relation to Fourier series.

© Springer Nature Switzerland AG 2022
A. Castañeda and F. Rodríguez-Henríquez (Eds.): LATIN 2022, LNCS 13568, pp. 243–254, 2022.
https://doi.org/10.1007/978-3-031-20624-5_15

These two notions are related: by definition, a vectorial function $F : \mathbb{F}_2^n \mapsto \mathbb{F}_2^n$ is APN if and only if its graph $\mathcal{G}_F = \{(x, F(x)); x \in \mathbb{F}_2^n\}$ is a Sidon set in $(\mathbb{F}_2^n \times \mathbb{F}_2^n, +)$. Since, given a Sidon set S, every subset of S is also a Sidon set, it is useful to study optimal Sidon sets (that is, Sidon sets that are maximal with respect to inclusion). In the present paper, we study the optimality of the graphs of APN functions as Sidon sets. We characterize such optimality in different ways (by the set $\mathcal{G}_F + \mathcal{G}_F + \mathcal{G}_F$ and by the Walsh transform of F) and we relate it to the two problems of the existence/non-existence of pairs of APN functions at Hamming distance 1 from each others, and of APN functions of algebraic degree n. We revisit the conjectures that have been made on these two problems. We address the case of the so-called plateaued APN functions by exploiting further a trick that Dillon used for showing that, for every APN function and every $c \neq 0$, there exist x, y, z such that $F(x) + F(y) + F(z) + F(x + y + z) = c$. The situation is more demanding in our case, but thanks to previous results on plateaued functions, we find a way to reduce the difficulty and this provides a much simpler proof that a plateaued APN function modified at one point cannot be APN, implying that its graph is an optimal Sidon set. We leave open the case of non-plateaued functions and list the known APN functions whose graphs could possibly be non-optimal Sidon sets (for values of n out of reach by computers).

2   Preliminaries

We call (n, m)-function any function F from $\mathbb{F}_2^n$ to $\mathbb{F}_2^m$ (we shall sometimes write that F is "in n variables"). It can be represented uniquely by its algebraic normal form (ANF) $F(x) = \sum_{I \subseteq \{1,\dots,n\}} a_I \prod_{i \in I} x_i$, where $a_I \in \mathbb{F}_2^m$. The algebraic degree of an (n, m)-function equals the global degree of its ANF. Function F is affine if and only if its algebraic degree is at most 1; it is called quadratic if its algebraic degree is at most 2; and it has algebraic degree n if and only if $\sum_{x \in \mathbb{F}_2^n} F(x) \neq 0$. In particular, if F is Boolean (that is, valued in $\mathbb{F}_2$, with $m = 1$) then its algebraic degree is n if and only if it has odd Hamming weight $w_H(F) = |\{x \in \mathbb{F}_2^n; F(x) \neq 0\}|$.

The vector space $\mathbb{F}_2^n$ can be identified with the field $\mathbb{F}_{2^n}$, since this field is an n-dimensional vector space over $\mathbb{F}_2$. If F is an (n, n)-function viewed over $\mathbb{F}_{2^n}$, then it can be represented by its (also unique) univariate representation $F(x) = \sum_{i=0}^{2^n - 1} a_i x^i$, $a_i \in \mathbb{F}_{2^n}$. Its algebraic degree equals then the maximum Hamming weight of (the binary expansion of) those exponents i in its univariate representation whose coefficients a_i are nonzero.

An (n, n)-function is called *almost perfect nonlinear* (APN) [2,16,17] if, for every nonzero $a \in \mathbb{F}_2^n$ and every $b \in \mathbb{F}_2^n$, the equation $D_a F(x) := F(x) + F(x + a) = b$ has at most two solutions. Equivalently, the system of equations

$$\begin{cases} x + y + z + t = 0 \\ F(x) + F(y) + F(z) + F(t) = 0 \end{cases}$$

has for only solutions quadruples (x, y, z, t) whose elements are not all distinct (i.e. are pairwise equal). The notion is preserved by extended affine (EA) equivalence (in other words, if F is APN then any function obtained by composing it on the left and on the right by affine permutations $x \to x \times M + u$, where M is a nonsingular $n \times n$ matrix over $\mathbb{F}_2$ and $u \in \mathbb{F}_2^n$, and adding an affine function to the resulting function is APN). It is also preserved by the more general CCZ-equivalence (two functions F and G are called CCZ-equivalent if their graphs $\mathcal{G}_F = \{(x, F(x)); x \in \mathbb{F}_2^n\}$ and $\mathcal{G}_G = \{(x, G(x)); x \in \mathbb{F}_2^n\}$ are the image of each other by an affine permutation of $((\mathbb{F}_2^n)^2, +)$, see more in [9]).

APN functions have been characterized by their Walsh transform [14]. Let us recall that the value at $u \in \mathbb{F}_2^n$ of the *Fourier-Hadamard transform* of a real-valued function φ over $\mathbb{F}_2^n$ is defined as $\widehat{\varphi}(u) = \sum_{x \in \mathbb{F}_2^n} \varphi(x)(-1)^{u \cdot x}$, (where "$\cdot$" denotes an inner product in $\mathbb{F}_2^n$). The Fourier-Hadamard transform is bijective. The value of the *Walsh transform* of F at $(u, v) \in \mathbb{F}_2^n \times \mathbb{F}_2^n$ equals the value at u of the Fourier-Hadamard transform of the function $(-1)^{v \cdot F(x)}$, that is, $W_F(u, v) = \sum_{x \in \mathbb{F}_2^n} (-1)^{v \cdot F(x) + u \cdot x}$. In other words, the Walsh transform of F equals the Fourier-Hadamard transform of the indicator function of its graph (which takes value 1 at the input (x, y) if and only if $y = F(x)$). Then F is APN if and only if $\sum_{u, v \in \mathbb{F}_2^n} W_F^4(u, v) = 3 \cdot 2^{4n} - 2^{3n+1}$. This is a direct consequence of the easily shown equality: $\sum_{u, v \in \mathbb{F}_2^n} W_F^4(u, v) = 2^{2n} |\{(x, y, z); F(x) + F(y) + F(z) + F(x + y + z) = 0\}|$.

The nonlinearity[1] of F equals the minimum Hamming distance between the component functions $v \cdot F$, $v \neq 0$, and the affine Boolean functions $u \cdot x + \begin{cases} 0 \\ 1 \end{cases}$. It equals $nl(F) = 2^{n-1} - \frac{1}{2} \max_{\substack{u, v \in \mathbb{F}_2^n \\ v \neq 0}} |W_F(u, v)|$.

A large part of known APN functions is made of functions EA-equivalent to power functions, that is, to functions of the form $F(x) = x^d$, after identification of $\mathbb{F}_2^n$ with the field $\mathbb{F}_{2^n}$ (which is possible since this field is an n-dimensional vector space over $\mathbb{F}_2$). The known APN power functions are all those whose exponents d are the conjugates $2^i d \pmod{2^n - 1}$ of those d given in Table 1 below, or of their inverses when they are invertible in $\mathbb{Z}/(2^n - 1)\mathbb{Z}$.

A subset of an elementary 2-group is called a *Sidon set* if it does not contain four distinct elements x, y, z, t such that $x + y + z + t = 0$. The notion is preserved by affine equivalence: if S is a Sidon set and A is an affine permutation, then $A(S)$ is a Sidon set.

By definition, an (n, n)-function F is then APN if and only if its graph $\mathcal{G}_F$ is a Sidon set in the elementary 2-group $((\mathbb{F}_2^n)^2, +)$.

Any set included in a Sidon set being a Sidon set, the most important for the study of Sidon sets in a given group is to determine those which are maximal (that is, which are not contained in larger Sidon sets); the knowledge of all maximal Sidon sets allows knowing all Sidon sets. A particular case of maximal set is when the set has maximal size, but the maximal size of Sidon sets is unknown. As far as we know, only an upper bound is known: the size $|S|$ of any Sidon set of $((\mathbb{F}_2^n)^2, +)$ satisfies $\binom{|S|}{2} = \frac{|S|(|S|-1)}{2} \leq 2^{2n} - 1$, that is (see e.g. [13]),

---

[1] The relationship between nonlinearity and almost perfect nonlinearity is not clear. The question whether all APN functions have a rather large nonlinearity is open.

$|S| \leq \left\lfloor \frac{1+\sqrt{2^{2n+3}-7}}{2} \right\rfloor \approx 2^{n+\frac{1}{2}}$. And an obvious lower bound on the maximal size of Sidon sets in $((\mathbb{F}_2^n)^2, +)$ is of course $|S| \geq 2^n$ since there exist APN functions whatever is the parity of n.

Table 1. Known APN exponents on $\mathbb{F}_{2^n}$ up to equivalence and to inversion.

Functions	Exponents d	Conditions
Gold	$2^i + 1$	$\gcd(i, n) = 1$
Kasami	$2^{2i} - 2^i + 1$	$\gcd(i, n) = 1$
Welch	$2^t + 3$	$n = 2t + 1$
Niho	$2^t + 2^{\frac{t}{2}} - 1$, t even $2^t + 2^{\frac{3t+1}{2}} - 1$, t odd	$n = 2t + 1$
Inverse	$2^{2t} - 1$ or $2^n - 2$	$n = 2t + 1$
Dobbertin	$2^{4t} + 2^{3t} + 2^{2t} + 2^t - 1$	$n = 5t$

We shall see that there are many cases of APN functions whose graphs are maximal Sidon set. The size 2^n of the graph is roughly $\sqrt{2}$ times smaller than what gives the upper bound on the size of Sidon sets, and there seems to be room for the existence of APN functions whose graphs are non-maximal Sidon sets. However, there is no known case where the graph is non-maximal. We relate the question of such existence to a known conjecture on APN functions, and this may lead to conjecturing that no APN function exists whose graph is non-maximal as a Sidon set (however, many conjectures made in the past on APN functions have subsequently been disproved; it may then be risky to state explicitly such conjecture).

3   Characterizations

Note that the property that $\mathcal{G}_F$ is an optimal Sidon set is preserved by CCZ equivalence.

The graph of an APN function F is a non-optimal Sidon set if and only if there exists an ordered pair (a, b) such that $b \neq F(a)$ and such that $\mathcal{G}_F \cup \{a, b\}$ is a Sidon set. It is easily seen that $\mathcal{G}_F \cup \{a, b\}$ is a Sidon set if and only if the system of equations

$$\begin{cases} x + y + z + a = 0 \\ F(x) + F(y) + F(z) + b = 0 \end{cases} \tag{1}$$

has no solution. Indeed, if this system has a solution (x, y, z) then x, y, z are necessarily distinct, because $b \neq F(a)$, and then, $\mathcal{G}_F \cup \{a, b\}$ is not a Sidon set, since the four points $(x, F(x)), (y, F(y)), (z, F(z))$, and (a, b) are pairwise distinct (because (a, b) by hypothesis cannot equal one of the other points) and sum to $(0, 0)$. Conversely, if the system (1) has no solution, then $\mathcal{G}_F \cup \{a, b\}$ is a Sidon set because, F being APN, four distinct points in $\mathcal{G}_F$ cannot sum to 0 and three points in $\mathcal{G}_F$ cannot sum to (a, b) either. Hence, the graph of an APN function is an optimal Sidon set if and only if, for every ordered pair (a, b) such that $b \neq F(a)$, the system (1) has a solution, that is, since $(x + y + z, F(x) + F(y) + F(z))$ lives outside $\mathcal{G}_F$ when x, y, z are distinct because F is APN[2], $\{(x + y + z, F(x) + F(y) + F(z)); x, y, z \in \mathbb{F}_2^n\}$ covers the whole set $(\mathbb{F}_2^n)^2 \setminus \mathcal{G}_F$. And since, for every (n, n)-function F, $(x + y + z, F(x) + F(y) + F(z))$ covers $\mathcal{G}_F$ when x, y, z are not distinct in $\mathbb{F}_2^n$, we have:

Proposition 1. *The graph of an APN (n, n)-function F is an optimal Sidon set in $((\mathbb{F}_2^n)^2, +)$ if and only if the set*

$$\mathcal{G}_F + \mathcal{G}_F + \mathcal{G}_F = \{(x + y + z, F(x) + F(y) + F(z)); x, y, z \in \mathbb{F}_2^n\}$$

covers the whole space $(\mathbb{F}_2^n)^2$.

Remark. For a vectorial function, APNness implies a behavior as different as possible from that of affine functions from the viewpoint of derivatives, since for F APN, $D_a F(x) = F(x) + F(x + a)$ covers a set of (maximal) size 2^{n-1} for every nonzero a, while for an affine function, this set has (minimal) size 1. Having a graph that is an optimal Sidon set also implies a behavior as different as possible from affine functions, from the viewpoint of $\mathcal{G}_F + \mathcal{G}_F + \mathcal{G}_F$, since if F is affine, then $(x + y + z, F(x) + F(y) + F(z)) = (x + y + z, F(x + y + z))$ covers a set of size 2^n, which is minimal.   ◇

Remark. J. Dillon (private communication) observed that, for every nonzero $c \in \mathbb{F}_{2^n}$, the equation $F(x) + F(y) + F(z) + F(x + y + z) = c$ must have a solution. In other words, there exists a in $\mathbb{F}_2^n$ such that the system in (1) with $b = F(a) + c$ has a solution.

Dillon's proof is given in [9] (after Proposition 161). Let us revisit this proof and say more: let v and c be nonzero elements of $\mathbb{F}_2^n$ and let $G(x) = F(x) + (v \cdot F(x)) c$. Then we have $G(x) + G(y) + G(z) + G(x + y + z) = F(x) + F(y) + F(z) + F(x + y + z) + (v \cdot (F(x) + F(y) + F(z) + F(x + y + z))) c$ and $G(x) + G(y) + G(z) + G(x + y + z) = 0$ if and only if $t := F(x) + F(y) + F(z) + F(x + y + z)$ satisfies $t = (v \cdot t) c$. If $v \cdot c = 1$, then this is equivalent to $t \in \{0, c\}$. Hence, we have $|\{(x, y, z); G(x) + G(y) + G(z) + G(x + y + z) = 0\}| = |\{(x, y, z); F(x) + F(y) + F(z) + F(x + y + z) \in \{0, c\}\}|$. The common size of these two sets is strictly larger than the number of triples (x, y, z) such that x, y, z are not distinct (that is, $3 \cdot 2^{4n} - 2^{3n + 1}$) since G having zero nonlinearity because $v \cdot G = 0$ (still assuming that $v \cdot c = 1$), it cannot be APN, as proved in [7]. This proves that $|\{(x, y, z); F(x) + F(y) + F(z) + F(x + y + z) = c\}| > 0$, since F being APN, we have $|\{(x, y, z); F(x) + F(y) + F(z) + F(x + y + z) = 0\}|$ if and only if x, y, z are not distinct.

Dillon's result shows (as we already observed) that for every nonzero c, there exists a in $\mathbb{F}_2^n$ such that the system in (1) with $b = F(a) + c$ has a solution, while in Proposition 1, we want that this same system has a solution for every a and every nonzero c in $\mathbb{F}_2^n$.

---

[2] This is a necessary and sufficient condition for APNness.

In the case of a quadratic function F, since the derivative $D_a F(x) = F(x) + F(x + a)$ is affine, its image set $Im(D_a F) = \{D_a F(x); x \in \mathbb{F}_2^n\}$ when $a \neq 0$ is an affine hyperplane, say equals $u_a + H_a$ where u_a is an element of $\mathbb{F}_2^n$ and H_a is a linear hyperplane of $\mathbb{F}_2^n$, say $H_a = \{0, v_a\}^\perp$, where $v_a \neq 0$. Since $F(x) + F(y) + F(z) + F(x + y + z)$ equals $D_a F(x) + D_a F(z)$ with $a = x + y$, and since $Im(D_a F) + Im(D_a F) = H_a + H_a = H_a$, Dillon's result means then in this particular case that $\bigcup_{\substack{a \in \mathbb{F}_2^n \\ a \neq 0}} H_a$ equals $\mathbb{F}_2^n$.   ◇

Let us now translate Proposition 1 in terms of the Walsh transform (by a routine method):

Corollary 1. *The graph of an APN (n, n)-function F is an optimal Sidon set in $((\mathbb{F}_2^n)^2, +)$ if and only if:*

$$\forall (a, b) \in (\mathbb{F}_2^n)^2, \quad \sum_{(u,v) \in (\mathbb{F}_2^n)^2} (-1)^{v \cdot b + u \cdot a} \, W_F^3(u, v) \neq 0. \tag{2}$$

Indeed we have:

$$\sum_{(u,v) \in (\mathbb{F}_2^n)^2} (-1)^{v \cdot b + u \cdot a} \, W_F^3(u, v)$$

$$= \sum_{x,y,z \in \mathbb{F}_2^n} \sum_{(u,v) \in (\mathbb{F}_2^n)^2} (-1)^{v \cdot (F(x) + F(y) + F(z) + b) + u \cdot (x + y + z + a)}$$

$$= 2^{2n} |\{(x, y, z) \in (\mathbb{F}_2^n)^3; (x + y + z, F(x) + F(y) + F(z)) = (a, b)\}|.$$

Remark. An APN function F has then a graph that is non-maximal as a Sidon set if and only if, making the product of all the expressions in Corollary 1 for (a, b) ranging over $(\mathbb{F}_2^n)^2$, we obtain 0:

$$\sum_{\substack{\mathcal{U} = (u_{a,b}, v_{a,b})_{(a,b) \in (\mathbb{F}_2^n)^2} \\ \in ((\mathbb{F}_2^n)^2)^{((\mathbb{F}_2^n)^2)}}} (-1)^{\sum_{(a,b) \in (\mathbb{F}_2^n)^2} (v_{a,b} \cdot b + u_{a,b} \cdot a)} \prod_{(a,b) \in (\mathbb{F}_2^n)^2} W_F^3(u_{a,b}, v_{a,b}) = 0.$$

◇

Remark. Without loss of generality (by changing $F(x)$ into $F(x) + F(0)$), let $F(0) = 0$. Then, since F is APN, we know that $\sum_{(u,v) \in (\mathbb{F}_2^n)^2} W_F^3(u, v) = 3 \cdot 2^{3n} - 2^{2n+1}$ (this can be easily calculated since $\sum_{(u,v) \in (\mathbb{F}_2^n)^2} W_F^3(u, v) = 2^{2n} |\{(x, y, z) \in (\mathbb{F}_2^n)^3; \ x + y + z = F(x) + F(y) + F(z) = 0\}| = 2^{2n} |\{(x, y, z) \in (\mathbb{F}_2^n)^3; \ x = 0 \text{ and } y = z \text{ or } y = 0 \text{ and } x = z \text{ or } z = 0 \text{ and } x = y\}|$). Hence, Inequality (2) is, under the condition $F(0) = 0$, equivalent to:

$$\forall (a, b) \in (\mathbb{F}_2^n)^2, \quad \sum_{\substack{(u,v) \in (\mathbb{F}_2^n)^2 \\ v \cdot b + u \cdot a = 0}} W_F^3(u, v) \neq 3 \cdot 2^{3n-1} - 2^{2n}.$$

Searching for APN (n, n)-functions whose graphs are non-maximal Sidon sets corresponds then to searching for APN (n, n)-functions F and linear hyperplanes H of $(\mathbb{F}_2^n)^2$ such that $\sum_{(u,v) \in H} W_F^3(u, v) = 3 \cdot 2^{3n-1} - 2^{2n}$.   ◇

4   Relation with the Problem of the (Non)existence of Pairs of APN Functions at Distance 1 from Each Others

The question of the existence of pairs of APN functions lying at Hamming distance 1 from each others, and the related question of the existence of APN functions of algebraic degree n have been studied in [4]. The question of the possible distance between APN functions has been studied further in [3]. The following proposition will show the close relationship between these two questions and the maximality of the graphs of APN functions as Sidon sets.

Proposition 2. *Let n be any positive integer and F any APN (n, n)-function. The graph of F is non-maximal as a Sidon set if and only if there exists an APN (n, n)-function G which can be obtained from F by changing its value at one single point (i.e. such that G lies at Hamming distance 1 from F).*

Proof. Assume first that the graph of F is non-maximal as a Sidon set. Then there exists $(a, b) \in (\mathbb{F}_2^n)^2$ such that $b \neq F(a)$ and $\mathcal{G}_F \cup \{(a, b)\}$ is a Sidon set. Then the set $(\mathcal{G}_F \cup \{(a, b)\}) \backslash \{(a, F(a))\}$ being automatically a Sidon set, the function G such that $G(x) = \begin{cases} F(x) \text{ if } x \neq a \\ b \text{ if } x = a \end{cases}$ is also APN.

Conversely, if a pair (F, G) of APN functions at distance 1 from each other exists, then there exists a unique $a \in \mathbb{F}_2^n$ such that $F(a) \neq G(a)$ (and $F(x) = G(x)$ for any $x \neq a$). Let us show that the set equal to the union of the graphs of F and G is then a Sidon set (and the graphs of F and G are then non-optimal as Sidon sets): otherwise, let X, Y, Z, T be distinct ordered pairs in this union and such that $X + Y + Z + T = 0$. Since F and G are APN, two elements among X, Y, Z, T have necessarily a for left term. Without loss of generality, we can assume that $Z = (a, F(a))$ and $T = (a, G(a))$. But then we have $X = (x, F(x))$ and $Y = (y, F(y))$ for some x, y and since $X + Y + Z + T = 0$ we must then have $x = y$ and therefore $X = Y$, a contradiction.   □

Note that if F and G are defined as in Proposition 2 and F has algebraic degree smaller than n, then G has algebraic degree n, since $\sum_{x \in \mathbb{F}_2^n} G(x) = \sum_{x \in \mathbb{F}_2^n} F(x) + b + F(a) = b + F(a) \neq 0$. Hence one function at least among F and G has algebraic degree n.

The following conjecture was stated in [4] (we number it as in this paper):

Conjecture 2: any function obtained from an APN function F by changing one value is not APN.

In other words, there do not exist two APN functions at Hamming distance 1 from each other. Another conjecture was even stated as follows (we number it as in [4] as well):

Conjecture 1: there does not exist any APN function of algebraic degree n for $n \geq 3$.

This conjecture is stronger than Conjecture 2 since if it is true then a pair (F, G) of APN functions at distance 1 from each other would need to be made with functions of degrees less than n, and this is impossible according to Proposition 2 and to the observation below it.

According to Proposition 2, Conjecture 2 is equivalent to:

Conjecture 3: the graphs of all APN functions are maximal Sidon sets.

Conjectures 1 and 2–3 are still completely open. Ref. [3] has studied further the Hamming distance between APN functions, but no progress was made on Conjectures 1 and 2.

5   The Case of Plateaued APN Functions

Let C be a class of (n, n)-functions that is globally preserved by any translation applied to the input of the functions or to their output. For proving that the graphs of all the APN functions in C are optimal Sidon sets by using Proposition 1, it is enough, thanks to a translation of the input by a and of the output by $F(a)$, to prove that, for any APN function F in this class, the system (1) with $a = F(0) = 0$ (and $b = c$) has a solution[3]. Moreover, according to what we have seen in the remark recalling Dillon's observation, if we define $G(x) = F(x) + (v \cdot F(x)) c$, where $v \cdot c = 1$, if G also belongs to C for every F in C, it is enough to show that, for every function $G \in C$ such that $G(0) = 0$ and having zero nonlinearity, the equation $G(x) + G(y) + G(x + y) = 0$ has solutions (x, y) where x and y are linearly independent over $\mathbb{F}_2$. Indeed, we have $|\{(x, y); G(x) + G(y) + G(x+y) = 0\}| = |\{(x, y); F(x) + F(y) + F(x+y) \in \{0, c\}\}|$, and since F is APN such that $F(0) = 0$, the equality $F(x) + F(y) + F(x + y) = 0$ requires that x and y are linearly dependent. Hence, the equation $F(x) + F(y) + F(x + y) = c$ has solutions if and only if the equation $G(x) + G(y) + G(x+y) = 0$ has solutions (x, y) where x and y are linearly independent.

Recall that an (n, n)-function is called *plateaued* (see e.g. [9]) if, for every $v \in \mathbb{F}_2^n$, there exists a number $\lambda_v \geq 0$ (which is necessarily a power of 2) such that $W_F(u, v) \in \{0, \pm\lambda_v\}$ for every $u \in \mathbb{F}_2^n$. All quadratic APN functions (and more generally all generalized crooked functions, that is, all functions F such that for every $a \neq 0$, the image set H_a of $D_a F$ is an affine hyperplane[4] are plateaued and some other non-quadratic functions are plateaued as well (e.g. all Kasami APN functions, see [19], and all AB functions).

The class of plateaued functions is preserved by translations of the input and by translations of the output; moreover, if F is plateaued then $G(x) = F(x) + (v \cdot F(x)) c$, where $v \cdot c = 1$, is plateaued (since the component functions of G are also component functions of F) and is non-APN since it has zero nonlinearity. We know from [8, Proposition 7] that, when G is plateaued, the condition "the equation $G(x) + G(y) + G(x + y) = 0$ has linearly independent solutions x, y" is equivalent to non-APNness. This provides a much simpler proof

---

[3] Note that we could also reduce ourselves to $a = b = 0$ but we could not reduce ourselves to $a = F(0) = b = 0$ without loss of generality. This is why we consider G in the sequel.

[4] See more in [9], where is recalled that no non-quadratic crooked function is known.

of the next proposition, which has been initially proved in [4, Theorem 3], but the proof was long, globally, and technical for n even.

Corollary 2. *Given any plateaued APN (n,n)-function F, changing F at one input gives a function which is not APN. Hence, the graphs of plateaued APN (n,n)-functions are all optimal Sidon sets.*

The proof is straightforward thanks to the observations above and to Proposition 2.

According to Proposition 1, we have then that, for every plateaued APN function, $\mathcal{G}_F + \mathcal{G}_F + \mathcal{G}_F$ covers the whole space $(\mathbb{F}_2^n)^2$, and according to Corollary 1, that $\forall (a,b) \in (\mathbb{F}_2^n)^2$, $\sum_{(u,v) \in (\mathbb{F}_2^n)^2} (-1)^{v \cdot b + u \cdot a} W_F^3(u,v) \neq 0$.

Among plateaued APN functions are almost bent functions. A vectorial Boolean function $F : \mathbb{F}_2^n \to \mathbb{F}_2^n$ is called *almost bent* (AB) [14] if its nonlinearity achieves the best possible value $2^{n-1} - 2^{\frac{n-1}{2}}$ (with n necessarily odd), that is, if all of the component functions $v \cdot F$, $v \neq 0$, satisfy $W_F(u,v) \in \{0, \pm 2^{\frac{n+1}{2}}\}$. All AB functions are APN. The converse is not true in general, even when n is odd, but it is true for n odd in the case of plateaued functions (and more generally in the case of functions whose Walsh transform values are all divisible by $2^{\frac{n+1}{2}}$). In Table 1, the AB functions are all Gold and Kasami functions for n odd and Welch and Niho functions.

Remark. The fact that the graphs of AB functions are optimal Sidon sets can also be directly shown by using the van Dam and Fon-Der-Flaass characterization of AB functions [18]: any (n,n)-function is AB if and only if the system
$$\begin{cases} x + y + z = a \\ F(x) + F(y) + F(z) = b \end{cases}$$
admits $3 \cdot 2^n - 2$ solutions if $b = F(a)$ (i.e. F is APN) and $2^n - 2$ solutions otherwise. It can also be deduced from Corollary 1; F being AB, we have:

$$\sum_{(u,v) \in (\mathbb{F}_2^n)^2} (-1)^{v \cdot b + u \cdot a} W_F^3(u,v)$$

$$= 2^{3n} + 2^{n+1} \left(\sum_{(u,v) \in (\mathbb{F}_2^n)^2} (-1)^{v \cdot b + u \cdot a} W_F(u,v) - 2^n \right)$$

and this equals $2^{3n} + 2^{3n+1} - 2^{2n+1} \neq 0$ if $b = F(a)$ and $2^{3n} - 2^{2n+1} \neq 0$ otherwise.   ◇

For n even there also exist plateaued APN functions: all Gold and all Kasami functions.

Quadratic functions (i.e. functions of algebraic degree at most 2) are plateaued, as well as the APN function in 6 variables that is now commonly called the Brinkmann-Leander-Edel-Pott function [15]:

$$x^3 + \alpha^{17}(x^{17} + x^{18} + x^{20} + x^{24}) + \alpha^{14}[\alpha^{18}x^9 + \alpha^{36}x^{18} + \alpha^9 x^{36} + x^{21} + x^{42}]$$
$$+ \alpha^{14} Tr_1^6(\alpha^{52}x^3 + \alpha^6 x^5 + \alpha^{19}x^7 + \alpha^{28}x^{11} + \alpha^2 x^{13}),$$

where α is primitive (see [9] for the history of this function).

5.1   An Interesting Particular Case

Some APN functions have all their component functions $v \cdot F$ unbalanced (i.e. of Hamming weight different from 2^{n-1}, that is, such that $W_F(0, v) \neq 0$); this is the case for instance of all APN power functions in even number n of variables. A simpler characterization than by Proposition 1 (and Corollary 1) is possible in such case, providing an interesting property of such functions:

Corollary 3. *For every APN plateaued (n, n)-function whose component functions are all unbalanced, the set $ImF + ImF = \{F(x) + F(y); (x, y) \in (\mathbb{F}_2^n)^2\}$ (where ImF is the image set of F) covers the whole space $\mathbb{F}_2^n$.*

Proof. Since we have $W_F(0, v) \neq 0$ for every v, we have then $W_F^3(u, v) = W_F^2(0, v) W_F(u, v)$, for every u, v and therefore:

$$
\sum_{(u,v) \in (\mathbb{F}_2^n)^2} (-1)^{v \cdot b + u \cdot a} W_F^3(u, v) = \sum_{(u,v) \in (\mathbb{F}_2^n)^2} (-1)^{v \cdot b + u \cdot a} W_F(u, v) W_F^2(0, v)
$$

$$
= \sum_{v \in \mathbb{F}_2^n} (-1)^{v \cdot b} W_F^2(0, v) \Big(\sum_{u \in \mathbb{F}_2^n} (-1)^{u \cdot a} W_F(u, v) \Big)
$$

$$
= 2^n \sum_{v \in \mathbb{F}_2^n} (-1)^{v \cdot b} W_F^2(0, v)(-1)^{v \cdot F(a)}
$$

$$
= 2^n \sum_{v, x, y \in \mathbb{F}_2^n} (-1)^{v \cdot (b + F(x) + F(y) + F(a))}
$$

$$
= 2^{2n} |\{(x, y) \in (\mathbb{F}_2^n)^2; F(x) + F(y) + F(a) = b\}|.
$$

Hence, since the graph of F is an optimal Sidon set, for every (a, b), the set $\{(x, y) \in (\mathbb{F}_2^n)^2; F(x) + F(y) + F(a) = b\}$ is not empty, that is, we have $ImF + ImF = \mathbb{F}_2^n$.   □

This can also be deduced from [9, Theorem 19] and Dillon's result recalled above.

We have then $ImF + ImF = \mathbb{F}_2^n$ in particular for every APN power function in even dimension n. Of course, this is also true for n odd, since APN power functions are in this case bijective.

Note the difference between the condition in Proposition 1, "$(x + y + z, F(x) + F(y) + F(z))$ covers the whole space $(\mathbb{F}_2^n)^2$" which lives in $(\mathbb{F}_2^n)^2$ and deals with three elements x, y, z, and that in Corollary 3, "$F(x) + F(y)$ covers the whole space $\mathbb{F}_2^n$", which lives in $\mathbb{F}_2^n$, involves two elements x, y and is simpler.

Remark. We know from [10,12] that the size of the image set of any APN (n, n)-function is at least $\frac{2^n + 1}{3}$ when n is odd and $\frac{2^n + 2}{3}$ when n is even. Since both numbers are considerably larger than $2^{\frac{n}{2}}$, the size of ImF is plenty sufficient for allowing the condition of Corollary 3 to be satisfied. Of course, the fact that ImF has size much larger than $2^{\frac{n}{2}}$ is not sufficient and the question whether some APN functions may have graphs that are not optimal as Sidon sets remains open.   ◇

6   Candidate APN Functions for Having Non-optimal Graphs as Sidon Sets

Plateaued functions are a large part of all known APN functions but they are most probably a tiny part of all APN functions. The only known APN functions that are not plateaued are power functions (the inverse and Dobbertin functions in Table 1) and some APN functions found in [6]. In [4] is proved that if F is an APN power function, then given $u \neq 0$, $ux^{2^n-1} + F$ is not APN (or equivalently $u\delta_0 + F$ is not APN, where δ_0 is the indicator of $\{0\}$) when either $u = 1$ or F is a permutation. But, for covering all cases of change at one point of an APN power function, we would need to address $ux^{2^n-1} + F$ for n even and F not plateaued, and $u(x+1)^{2^n-1} + F$ for every n and F not plateaued. This was done with the multiplicative inverse function x^{2^n-2} for n odd (which is APN): changing it at one point (any one) gives a function that is not APN. According to Proposition 2, the graph of the APN multiplicative inverse function is then a maximal Sidon set. But there is some uncertainty about general APN power functions (however, it was checked with a computer that for $n \leq 15$, changing any APN power function at one point makes it non-APN).

Given the APN functions covered in [4], the only possibility of finding known APN functions with a graph that is not maximal as a Sidon set is with:

- functions EA equivalent to Dobbertin functions in a number of variables divisible by 10, at least 20,
- the functions obtained in [6] as CCZ equivalent to Gold functions in even numbers of variables (because in odd numbers of variables, they are AB, since ABness is preserved by CCZ equivalence), that is: $x^{2^i+1} + (x^{2^i} + x + 1)tr(x^{2^i+1})$, $n \geq 4$ even, $\gcd(i,n) = 1$, and $[x + Tr_3^n(x^{2(2^i+1)} + x^{4(2^i+1)})) + tr(x)Tr_3^n(x^{2^i+1} + x^{2^{2i}(2^i+1)})]^{2^i+1}$, where $6|n$ and $\gcd(i,n) = 1$, and $Tr_3^n(x) = x + x^8 + x^{8^2} + \cdots + x^{8^{\frac{n}{3}-1}}$,
- and the following functions found in [5]: $x^3 + tr(x^9) + (x^2 + x + 1)tr((x^3))$, where $n \geq 4$ is even and $\gcd(i,n) = 1$, and $\left(x + Tr_3^n(x^6 + x^{12}) + tr(x)Tr_3^n(x^3 + x^{12})\right)^3 + tr\left((x + Tr_3^n(x^6 + x^{12}) + tr(x)Tr_3^n(x^3 + x^{12}))^9\right)$, where $6|n$ and $\gcd(i,n) = 1$.

But the investigation made in [4] did not find any example and it seems difficult to push it to larger values of n.

Acknowledgement. We thank Lilya Budaghyan and Nian Li for useful information. The research of the author is partly supported by the Trond Mohn Foundation and Norwegian Research Council.

References

1. Babai, L., Sós, V.T.: Sidon sets in groups and induced subgraphs of Cayley graphs. Eur. J. Combin. **6**(2), 101–114 (1985)
2. Beth, T., Ding, C.: On almost perfect nonlinear permutations. In: Helleseth, T. (ed.) EUROCRYPT 1993. LNCS, vol. 765, pp. 65–76. Springer, Heidelberg (1994). https://doi.org/10.1007/3-540-48285-7_7
3. Budaghyan, L., Carlet, C., Helleseth, T., Kaleyski, N.: On the distance between APN functions. IEEE Trans. Inf. Theory **66**(9), 5742–5753 (2020)
4. Budaghyan, L., Carlet, C., Helleseth, T., Li, N., Sun, B.: On upper bounds for algebraic degrees of APN functions. IEEE Trans. Inf. Theory **64**(6), 4399–4411 (2017)
5. Budaghyan, L., Carlet, C., Leander, G.: Constructing new APN functions from known ones. Finite Fields Appl. **15**(2), 150–159 (2009)
6. Budaghyan, L., Carlet, C., Pott, A.: New classes of almost bent and almost perfect nonlinear polynomials. IEEE Trans. Inf. Theory **52**(3), 1141–1152 (2006)
7. Carlet, C.: Boolean Models and Methods in Mathematics, Computer Science, and Engineering, chapter Vectorial Boolean Functions for Cryptography. Cambridge University Press (2010)
8. Carlet, C.: Boolean and vectorial plateaued functions and APN functions. IEEE Trans. Inf. Theory **61**(11), 6272–6289 (2015)
9. Carlet, C.: Boolean Functions for Cryptography and Coding Theory. Cambridge University Press, Cambridge (2021)
10. Carlet, C.: Bounds on the nonlinearity of differentially uniform functions by means of their image set size, and on their distance to affine functions. IEEE Trans. Inf. Theory **67**(12), 8325–8334 (2021)
11. Carlet, C., Charpin, P., Zinoviev, V.A.: Codes, bent functions and permutations suitable for DES-like cryptosystems. Des. Codes Cryptogr. **15**(2), 125–156 (1998)
12. Carlet, C., Heuser, A., Picek, S.: Trade-offs for S-boxes: cryptographic properties and side-channel resilience. In: Gollmann, D., Miyaji, A., Kikuchi, H. (eds.) ACNS 2017. LNCS, vol. 10355, pp. 393–414. Springer, Cham (2017). https://doi.org/10.1007/978-3-319-61204-1_20
13. Carlet, C., Mesnager, S.: On those multiplicative subgroups of $\mathbb{F}_{2^n}^*$ which are Sidon sets and/or sum-free sets. J. Algebraic Comb. **55**(1), 43–59 (2022). https://doi.org/10.1007/s10801-020-00988-7
14. Chabaud, F., Vaudenay, S.: Links between differential and linear cryptanalysis. In: De Santis, A. (ed.) EUROCRYPT 1994. LNCS, vol. 950, pp. 356–365. Springer, Heidelberg (1995). https://doi.org/10.1007/BFb0053450
15. Edel, Y., Pott, A.: A new almost perfect nonlinear function which is not quadratic. Adv. Math. Commun. **3**(1), 59–81 (2009)
16. Nyberg, K.: Differentially uniform mappings for cryptography. In: Helleseth, T. (ed.) EUROCRYPT 1993. LNCS, vol. 765, pp. 55–64. Springer, Heidelberg (1994). https://doi.org/10.1007/3-540-48285-7_6
17. Nyberg, K., Knudsen, L.R.: Provable security against differential cryptanalysis. In: Brickell, E.F. (ed.) CRYPTO 1992. LNCS, vol. 740, pp. 566–574. Springer, Heidelberg (1993). https://doi.org/10.1007/3-540-48071-4_41
18. van Dam, E.R., Fon-Der-Flaass, D.: Codes, graphs, and schemes from nonlinear functions. Eur. J. Combin. **24**(1), 85–98 (2003)
19. Yoshiara, S.: Plateaudness of Kasami APN functions. Finite Fields Appl. **47**, 11–32 (2017)

On the Subfield Codes of a Subclass of Optimal Cyclic Codes and Their Covering Structures

Félix Hernández[1]([✉])[iD] and Gerardo Vega[2][iD]

[1] Posgrado en Ciencia e Ingeniería de la Computación, Universidad Nacional
Autónoma de México, 04510 Ciudad de México, Mexico
`felixhdz@ciencias.unam.mx`
[2] Dirección General de Cómputo y de Tecnologías de Información y Comunicación,
Universidad Nacional Autónoma de México, 04510 Ciudad de México, Mexico
`gerardov@unam.mx`

Abstract. A class of optimal three-weight $[q^k - 1, k + 1, q^{k-1}(q-1) - 1]$
cyclic codes over $\mathbb{F}_q$, with $k \geq 2$, achieving the Griesmer lower bound,
was presented by Heng and Yue [IEEE Trans. Inf. Theory, 62(8) (2016)
4501–4513]. In this paper we study some of the subfield codes of this
class of optimal cyclic codes when $k = 2$. The weight distributions of
the subfield codes are settled. It turns out that some of these codes are
optimal and others have the best known parameters. The duals of the
subfield codes are also investigated and found to be almost optimal with
respect to the sphere-packing bound. In addition, the covering structure
for the studied subfield codes is determined. Some of these codes are
found to have the important property that any nonzero codeword is
minimal. This is a desirable property which is useful in the design of
a secret sharing scheme based on a linear code. Moreover, we present a
specific example of a secret sharing scheme based on one of these subfield
codes.

Keywords: Subfield codes · Optimal cyclic codes · Secret sharing
schemes · Covering structure · Sphere-packing bound

1 Introduction

Let $\mathbb{F}_q$ be the finite field with q elements. An $[n, l, d]$ linear code, $\mathcal{C}$, over $\mathbb{F}_q$ is
a l-dimensional subspace of $\mathbb{F}_q^n$ with minimum Hamming distance d. It is called
optimal if there is no $[n, l, d']$ code with $d' > d$, and *cyclic* if $(c_0, c_1, \ldots, c_{n-1}) \in \mathcal{C}$
implies $(c_{n-1}, c_0, \ldots, c_{n-2}) \in \mathcal{C}$.

Recently, a class of optimal three-weight $[q^k - 1, k + 1, q^{k-1}(q-1) - 1]$ cyclic
codes over $\mathbb{F}_q$ achieving the Griesmer lower bound was presented in [11], which
generalizes a result in [20] from $k = 2$ to arbitrary positive integer $k \geq 2$. Further,
the q_0-ary subfield codes of two families of q-ary optimal linear codes were studied

---

F. Hernández - PhD student, manuscript partially supported by CONACyT, México.

A. Castañeda and F. Rodríguez-Henríquez (Eds.): LATIN 2022, LNCS 13568, pp. 255–270, 2022.
https://doi.org/10.1007/978-3-031-20624-5_16

in [10], with q_0 being a power of a prime such that q is in turn a power of q_0 (that is, $\mathbb{F}_{q_0}$ is a proper subfield of $\mathbb{F}_q$). Also, some basic results on subfield codes were derived and the subfield codes of ovoid codes were determined in [5]. In addition, the subfield codes of several families of linear codes were obtained in [4], and the subfield codes of hyperoval and conic codes were studied in [9]. The basic idea in these last four references is to consider the subfield code of an optimal, or almost optimal, linear code over $\mathbb{F}_q$ and expect the subfield code over $\mathbb{F}_{q_0}$ to have good parameters. In all cases, subfield codes with very attractive parameters were found. Thus, the first objective of this paper is to study the q_0-ary subfield codes for a subclass of the optimal three-weight cyclic codes reported in [11] and determine their weight distributions. It turns out that the studied subfield codes also have three nonzero weights, which is of interest as cyclic codes with few weights have a wide range of applications in many research fields such as authentication codes [6], secret sharing schemes [3,13,16,17,21], association schemes [2], strongly walk regular graphs [19], and design of frequency hopping sequences [7]. As we will see, some of the subfield codes are optimal and others have the best known parameters. The duals of the subfield codes are also investigated and found to be almost optimal with respect to the sphere-packing bound. The second objective is to determine the covering structure for the studied subfield codes. By means of the Ashikhmin-Barg Lemma (see [1]) we show that some of these codes have the important property that all their nonzero codewords are minimal. This is a desirable property which is useful in the design of a secret sharing scheme based on a linear code. Moreover, we present a specific example of a secret sharing scheme based on one of these subfield codes.

This work is organized as follows: In Sect. 2, we fix some notation and recall some definitions and some known results to be used in subsequent sections. Section 3 is devoted to presenting preliminary results. In Sect. 4 we determine the subfield codes of a subclass of already known optimal three-weight cyclic codes. In Sect. 5, we investigate the covering structure for the studied subfield codes and present a specific example of a secret sharing scheme based on one of these codes. Finally, Sect. 6 is devoted to conclusions.

2   Notation, Definitions and Known Results

Unless otherwise specified, throughout this work we will use the following:

Notation. Let $q_0 = p^t$, where t is a positive integer and p is a prime number. For an integer $r > 1$ we are going to fix $q = q_0^r = p^{tr}$. For an integer $k > 1$, let $\mathbb{F}_{q^k}$ be the finite extension of degree k of the finite field $\mathbb{F}_q$ and let γ be a primitive element of $\mathbb{F}_{q^k}$. Let F be a finite field of characteristic p and E a finite extension of F. Then we will denote by "$\mathrm{Tr}_{E/F}$" the *trace mapping* from E to F, while "Tr" will denote the *absolute trace mapping* from E to the prime field $\mathbb{F}_p$.

The *weight enumerator* of a linear code $\mathcal{C}$ of length n is defined as $1 + A_1 z + \cdots + A_n z^n$, while the sequence $\{1, A_1, \ldots, A_n\}$ is called its *weight distribution*, where A_i $(1 \leq i \leq n)$ denote the number of codewords in $\mathcal{C}$ with Hamming weight i. If $\sharp\{1 \leq i \leq n : A_i \neq 0\} = M$, then $\mathcal{C}$ is called an *M-weight* code.

$\mathcal{C}^{\perp}$ will denote the *dual code* of $\mathcal{C}$ and we recall that if $\mathcal{C}$ is an $[n, l]$ linear code, then its dual code is an $[n, n - l]$ linear code. The sequence $\{1, A_1^{\perp}, \ldots, A_n^{\perp}\}$ will denote the weight distribution of the dual code $\mathcal{C}^{\perp}$. Suppose that the minimum weight of $\mathcal{C}^{\perp}$ is at least 2 (that is, $A_1^{\perp} = 0$) and fix $m = n(q - 1)$. Then, the first four Pless power moments (see [12, pp. 259–260]) for $\mathcal{C}$ are:

$$\sum_{i=1}^{n} A_i = q^l - 1,$$

$$\sum_{i=1}^{n} i A_i = q^{l-1} m,$$

$$\sum_{i=1}^{n} i^2 A_i = q^{l-2} [m(m + 1) + 2A_2^{\perp}],$$

$$\sum_{i=1}^{n} i^3 A_i = q^{l-3} [m(m(m + 3) - q + 2) + 6(m - q + 2)A_2^{\perp} - 6A_3^{\perp}]. \tag{1}$$

The *canonical additive character* of $\mathbb{F}_q$ is defined as follows

$$\chi(x) := e^{2\pi\sqrt{-1}\operatorname{Tr}(x)/p}, \qquad \text{for all } x \in \mathbb{F}_q.$$

Let $a \in \mathbb{F}_q$. The orthogonality relation for the canonical additive character χ of $\mathbb{F}_q$ is given by (see for example [14, Chapter 5]):

$$\sum_{x \in \mathbb{F}_q} \chi(ax) = \begin{cases} q & \text{if } a = 0, \\ 0 & \text{otherwise.} \end{cases}$$

This property plays an important role in numerous applications of finite fields. Among them, this property is useful for determining the Hamming weight of a given vector over a finite field; for example, if $\mathbf{w}(\cdot)$ stands for the usual Hamming weight function and if $V = (a_1, a_2, \ldots, a_n) \in \mathbb{F}_q^n$, then

$$\mathbf{w}(V) = n - \frac{1}{q} \sum_{i=1}^{n} \sum_{x \in \mathbb{F}_q} \chi(a_i x). \tag{2}$$

We now recall the class of optimal three-weight cyclic codes for which we are interested in obtaining their subfield codes.

Theorem 1 [11, Theorem 11]. *Let e_1 and e_2 be integers and let $\mathcal{C}_{(q,k,e_1,e_2)}$ be the cyclic code of length $q^k - 1$ over $\mathbb{F}_q$ given by*

$$\mathcal{C}_{(q,k,e_1,e_2)} := \left\{ \left(a\gamma^{\frac{q^k-1}{q-1}e_1 j} + \operatorname{Tr}_{\mathbb{F}_{q^k}/\mathbb{F}_q} \left(b\gamma^{e_2 j} \right) \right)_{j=0}^{q^k-2} : a \in \mathbb{F}_q, b \in \mathbb{F}_{q^k} \right\}. \tag{3}$$

If $\gcd(\frac{q^k-1}{q-1}, e_2) = 1$ *and* $\gcd(q-1, ke_1 - e_2) = 1$, *then* $\mathcal{C}_{(q,k,e_1,e_2)}$ *is an optimal three-weight* $[q^k - 1, k + 1, q^{k-1}(q-1) - 1]$ *cyclic code with weight enumerator*

$$1 + (q-1)(q^k - 1)z^{q^{k-1}(q-1)-1} + (q^k - 1)z^{q^{k-1}(q-1)} + (q-1)z^{q^k-1}.$$

In addition, if $q > 2$, *its dual code is a* $[q^k - 1, q^k - k - 2, 3]$ *cyclic code.*

Let $\mathcal{C}$ be an $[n, l]$ linear code over $\mathbb{F}_q$. The following describes a way to construct a new $[n, l']$ linear code, $\mathcal{C}^{(q_0)}$, over $\mathbb{F}_{q_0}$ (see [5]). Let G be a generator matrix of $\mathcal{C}$. Take a basis of $\mathbb{F}_q = \mathbb{F}_{q_0^r}$ over $\mathbb{F}_{q_0}$ and represent each entry of G as an $r \times 1$ column vector of $\mathbb{F}_{q_0}^r$ with respect to this basis. Replace each entry of G with the corresponding $r \times 1$ column vector of $\mathbb{F}_{q_0}^r$. With this method, G is modified into an $lr \times n$ matrix over $\mathbb{F}_{q_0}$ generating a new linear code, $\mathcal{C}^{(q_0)}$, over $\mathbb{F}_{q_0}$ of length n, called *subfield code*. It is known that the subfield code $\mathcal{C}^{(q_0)}$ is independent of both the choice of the basis of $\mathbb{F}_q$ over $\mathbb{F}_{q_0}$ and the choice of the generator matrix G of $\mathcal{C}$ (see Theorems 2.1 and 2.6 in [5]). Also, it is clear that the dimension l' of $\mathcal{C}^{(q_0)}$ satisfies $l' \leq lr$.

Remark 1. We recall that the *subfield subcode* of a linear code, $\mathcal{C}$, over $\mathbb{F}_q$ is the subset of codewords in $\mathcal{C}$ whose components are all in $\mathbb{F}_{q_0}$ (see for example [12, p. 116]). In consequence, observe that a subfield code and a subfield subcode are different codes in general. In addition, note that the subfield codes defined here are also different from the subfield codes in [18, Subsect. 4.1] defined as one-weight irreducible cyclic codes (see Proposition 4.1 therein).

For what follows, we are interested in obtaining the weight distributions for the subfield codes of a subclass of the optimal three-weight cyclic codes in Theorem 1. To that end, the following is a useful result that will allow us to represent a q_0-ary subfield code, $\mathcal{C}^{(q_0)}$, in terms of the trace function.

Lemma 1 [5, *Theorem 2.5*]. *Let* $\mathcal{C}$ *be an* $[n, l]$ *linear code over* $\mathbb{F}_q$. *Let* $G = [g_{ij}]_{1 \leq i \leq l, 1 \leq j \leq n}$ *be a generator matrix of* $\mathcal{C}$. *Then, the trace representation of the subfield code* $\mathcal{C}^{(q_0)}$ *is given by the following set*

$$\left\{ \left(\mathrm{Tr}_{\mathbb{F}_q/\mathbb{F}_{q_0}} \left(\sum_{i=1}^{l} a_i g_{i1} \right), \ldots, \mathrm{Tr}_{\mathbb{F}_q/\mathbb{F}_{q_0}} \left(\sum_{i=1}^{l} a_i g_{in} \right) \right) : a_1, \ldots, a_l \in \mathbb{F}_q \right\}.$$

3   Preliminary Results

Throughout this and the next section, we are interested in obtaining the weight distributions for the subfield codes of a subclass of the optimal cyclic codes in Theorem 1 when $k = 2$. Thus, from now on, we fix $k = 2$. That is, $\frac{q^2-1}{q-1} = q + 1$ and $\langle \gamma \rangle = \mathbb{F}_{q^2}^*$.

Note that if $\mathcal{C}_{(q,2,e_1,e_2)}$ is an optimal cyclic code in Theorem 1 then, in accordance with Lemma 1, its subfield code, $\mathcal{C}^{(q_0)}_{(q,2,e_1,e_2)}$, is given by (recall that $q = q_0^r$):

$$\mathcal{C}^{(q_0)}_{(q,2,e_1,e_2)} = \left\{ \mathbf{c}(a,b)^{(q_0)} : a \in \mathbb{F}_q, b \in \mathbb{F}_{q^2} \right\}, \tag{4}$$

where

$$\mathbf{c}(a,b)^{(q_0)} := \left(\mathrm{Tr}_{\mathbb{F}_q/\mathbb{F}_{q_0}} \left(a\gamma^{(q+1)e_1 j} \right) + \mathrm{Tr}_{\mathbb{F}_{q^2}/\mathbb{F}_{q_0}} \left(b\gamma^{e_2 j} \right) \right)_{j=0}^{q^2-2}. \tag{5}$$

Remark 2. Like $\mathcal{C}_{(q,2,e_1,e_2)}$, $\mathcal{C}^{(q_0)}_{(q,2,e_1,e_2)}$ is also a cyclic code of length $q^2 - 1$. Furthermore, if $h_a(x) \in \mathbb{F}_{q_0}[x]$ is the *minimal polynomial* of γ^{-a} (see [15, Ch. 4]) and if d is the smallest positive integer such that $aq_0^d \equiv a \pmod{q^2 - 1}$, then observe that $\deg(h_a(x)) = d$. Therefore, $h_{(q+1)e_1}(x) \neq h_{e_2}(x)$, $h_{(q+1)e_1}(x)h_{e_2}(x)$ is the *parity-check polynomial* of $\mathcal{C}^{(q_0)}_{(q,2,e_1,e_2)}$ (see [15, Ch. 7]), and if l' is its dimension, then $l' = d_1 + d_2$, where d_1 and d_2 are the smallest positive integers such that $(q+1)e_1 q_0^{d_1} \equiv (q+1)e_1 \pmod{q^2 - 1}$ and $e_2 q_0^{d_2} \equiv e_2 \pmod{q^2 - 1}$, respectively (see [14, Part (v) of Theorem 3.33]).

In order to obtain the weight distributions of the subfield codes of the form $\mathcal{C}^{(q_0)}_{(q,2,e_1,e_2)}$, we will need the following preliminary result.

Lemma 2. *Let χ and χ' be the canonical additive characters of $\mathbb{F}_q$ and $\mathbb{F}_{q^2}$, respectively. For $a \in \mathbb{F}_q$ and $b \in \mathbb{F}_{q^2}$, consider the exponential sum*

$$Z(a,b) := \sum_{y \in \mathbb{F}_{q_0}^*} \sum_{x \in \mathbb{F}_{q^2}^*} \chi(yax^{q+1})\chi'(ybx).$$

Then

$$Z(a,b) = \begin{cases} (q_0 - 1)(q_0^{2r} - 1) & \text{if } a = b = 0, \\ -(q_0 - 1)(q_0^r + 1) & \text{if } a \neq 0 \text{ and } b = 0, \\ -(q_0 - 1) & \text{if } a = 0 \text{ and } b \neq 0, \\ -(q_0 - 1)(q_0^r + 1) & \text{if } (a,b) \neq (0,0) \text{ and } \mathrm{Tr}_{\mathbb{F}_q/\mathbb{F}_{q_0}} \left(\frac{b^{q+1}}{a} \right) = 0, \\ q_0(q_0^{r-1} - 1) + 1 & \text{if } (a,b) \neq (0,0) \text{ and } \mathrm{Tr}_{\mathbb{F}_q/\mathbb{F}_{q_0}} \left(\frac{b^{q+1}}{a} \right) \neq 0. \end{cases}$$

Proof. Clearly, $Z(0,0) = (q_0 - 1)(q_0^{2r} - 1)$. If $a \neq 0$ and $b = 0$, then

$$Z(a,0) = \sum_{y \in \mathbb{F}_{q_0}^*} \sum_{x \in \mathbb{F}_{q^2}^*} \chi(yax^{q+1}) = (q+1) \sum_{y \in \mathbb{F}_{q_0}^*} \sum_{x \in \mathbb{F}_q^*} \chi(yax),$$

$$= (q+1) \sum_{y \in \mathbb{F}_{q_0}^*} (-1) = -(q+1)(q_0 - 1) = -(q_0 - 1)(q_0^r + 1).$$

Further, if $a = 0$ and $b \neq 0$,

$$Z(0,b) = \sum_{y\in\mathbb{F}_{q_0}^*}\sum_{x\in\mathbb{F}_{q^2}^*}\chi'(ybx) = \sum_{y\in\mathbb{F}_{q_0}^*}(-1) = -(q_0-1).$$

Now, let φ be the canonical additive character of $\mathbb{F}_{q_0}$ and suppose that $(a,b)\neq (0,0)$. By the transitivity and linearity of the trace function, we have

$$Z(a,b) = \sum_{y\in\mathbb{F}_{q_0}^*}\sum_{x\in\mathbb{F}_{q^2}^*}\varphi\left(y\mathrm{Tr}_{\mathbb{F}_q/\mathbb{F}_{q_0}}\left(ax^{q+1}\right)\right)\varphi\left(y\mathrm{Tr}_{\mathbb{F}_{q^2}/\mathbb{F}_{q_0}}(bx)\right),$$

$$= \sum_{y\in\mathbb{F}_{q_0}^*}\sum_{x\in\mathbb{F}_{q^2}^*}\varphi\left(y\left(\mathrm{Tr}_{\mathbb{F}_q/\mathbb{F}_{q_0}}\left(ax^{q+1}+\mathrm{Tr}_{\mathbb{F}_{q^2}/\mathbb{F}_q}(bx)\right)\right)\right),$$

$$= \sum_{y\in\mathbb{F}_{q_0}^*}\sum_{x\in\mathbb{F}_{q^2}^*}\varphi\left(y\left(\mathrm{Tr}_{\mathbb{F}_q/\mathbb{F}_{q_0}}\left(ax^q\left(x+\frac{b^q}{a}\right)+bx\right)\right)\right),$$

where the last equality holds because $\mathrm{Tr}_{\mathbb{F}_{q^2}/\mathbb{F}_q}(bx) = bx+b^q x^q$. Let $B = \mathbb{F}_{q^2}\backslash\{\frac{b^q}{a}\}$. Thus, after applying the variable substitution $x\mapsto w-\frac{b^q}{a}$, we obtain

$$Z(a,b) = \sum_{y\in\mathbb{F}_{q_0}^*}\sum_{w\in B}\varphi\left(y\left(\mathrm{Tr}_{\mathbb{F}_q/\mathbb{F}_{q_0}}\left(a\left(w^q-\frac{b^{q^2}}{a^q}\right)w+b\left(w-\frac{b^q}{a}\right)\right)\right)\right).$$

However, $b^{q^2} = b$ and $a^q = a$. Thus, since $B = \mathbb{F}_{q^2}\backslash\{\frac{b^q}{a}\}$,

$$Z(a,b) = \sum_{y\in\mathbb{F}_{q_0}^*}\sum_{w\in B}\varphi\left(y\left(\mathrm{Tr}_{\mathbb{F}_q/\mathbb{F}_{q_0}}\left(aw^{q+1}-\frac{b^{q+1}}{a}\right)\right)\right),$$

$$= -\sum_{y\in\mathbb{F}_{q_0}^*}\varphi(0)+\sum_{y\in\mathbb{F}_{q_0}^*}\varphi\left(-y\mathrm{Tr}_{\mathbb{F}_q/\mathbb{F}_{q_0}}\left(\frac{b^{q+1}}{a}\right)\right)$$

$$\times\sum_{w\in\mathbb{F}_{q^2}}\varphi\left(y\left(\mathrm{Tr}_{\mathbb{F}_q/\mathbb{F}_{q_0}}\left(aw^{q+1}\right)\right)\right),$$

$$= -(q_0-1)+\sum_{y\in\mathbb{F}_{q_0}^*}\varphi\left(-y\mathrm{Tr}_{\mathbb{F}_q/\mathbb{F}_{q_0}}\left(\frac{b^{q+1}}{a}\right)\right)\sum_{w\in\mathbb{F}_{q^2}}\chi\left(yaw^{q+1}\right),$$

where χ is the canonical additive character of $\mathbb{F}_q$ (note that $w^{q+1}\in\mathbb{F}_q$). But, since $a,y\neq 0$, we have

$$\sum_{w\in\mathbb{F}_{q^2}}\chi\left(yaw^{q+1}\right) = 1+\sum_{w\in\mathbb{F}_{q^2}^*}\chi\left(yaw^{q+1}\right),$$

$$= 1+(q+1)\sum_{w\in\mathbb{F}_q^*}\chi\left(yaw\right) = -q.$$

Therefore, finally, we obtain

$$Z(a,b) = -(q_0 - 1) - q \sum_{y \in \mathbb{F}_{q_0}^*} \varphi \left(-y \mathrm{Tr}_{\mathbb{F}_q / \mathbb{F}_{q_0}} \left(\frac{b^{q+1}}{a} \right) \right),$$

$$= \begin{cases} -(q_0 - 1)(q_0^r + 1) & \text{if } \mathrm{Tr}_{\mathbb{F}_q / \mathbb{F}_{q_0}} \left(\frac{b^{q+1}}{a} \right) = 0, \\ q_0(q_0^{r-1} - 1) + 1 & \text{otherwise.} \end{cases}$$

$\square$

4   The Subfield Codes of a Subclass of Optimal Cyclic Codes

By means of the following result we now determine the subfield codes, along with their weight distributions, for a subclass of the optimal three-weight cyclic codes in Theorem 1.

Theorem 2. *Let $r > 1$, e_1 and e_2 be integers and let $\mathcal{C}_{(q,2,e_1,e_2)}^{(q_0)}$ be the subfield code of length $q_0^{2r} - 1$, over $\mathbb{F}_{q_0}$, given by (4). Assume that $\gcd(q^2 - 1, e_2) = 1$ and $\gcd(q - 1, 2e_1 - e_2) = 1$. Then the following assertions hold true:*

(A) *If $(q-1) | (q_0 - 1)e_1$, then $\mathcal{C}_{(q,2,e_1,e_2)}^{(q_0)}$ is an optimal three-weight cyclic code of length $q_0^{2r} - 1$ and dimension $2r + 1$, over $\mathbb{F}_{q_0}$, that belongs to the class of optimal three-weight cyclic codes in Theorem 1 (therein $k = 2r$ and $q = q_0$).*
(B) *Let $\mathcal{I}$ be an integer such that $\mathcal{I}e_2 \equiv 1 \pmod{q^2 - 1}$. If $(q-1) \nmid (q_0 - 1)e_1$ and $\mathcal{I}e_1 \equiv 1 \pmod{q - 1}$, then $\mathcal{C}_{(q,2,e_1,e_2)}^{(q_0)}$ is a three-weight cyclic code of length $q_0^{2r} - 1$ and dimension $3r$, over $\mathbb{F}_{q_0}$, whose weight enumerator is*

$$1 + q_0^{r-1}(q_0^{2r} - 1)(q_0 - 1)z^{q_0^{r-1}(q_0^{r+1} - q_0^r - 1)} + (q_0^{2r} - 1)z^{q_0^{2r-1}(q_0 - 1)}$$
$$+ q_0^{r-1}(q_0^r - 1)(q_0^r - q_0 + 1)z^{q_0^{r-1}(q_0 - 1)(q_0^r + 1)}. \qquad (6)$$

In addition, $A_1^{\perp} = A_2^{\perp} = 0$, and

$$A_3^{\perp} = \frac{(q_0^{r+2} - 3q_0^{r+1} + q_0^2 + 3q_0^r - 6q_0 + 6)(q_0^{2r} - 1)(q_0 - 1)}{6}.$$

That is, the dual code, $\mathcal{C}_{(q,2,e_1,e_2)}^{(q_0)\perp}$, of $\mathcal{C}_{(q,2,e_1,e_2)}^{(q_0)}$ is a $[q_0^{2r} - 1, q_0^{2r} - 3r - 1, 3]$ cyclic code and is almost optimal with respect to the sphere-packing bound.

Proof. First of all, since $\gcd(\frac{q^2-1}{q-1}, e_2) \leq \gcd(q^2 - 1, e_2) = 1$ and $\gcd(q - 1, 2e_1 - e_2) = 1$, observe that $\mathcal{C}_{(q,2,e_1,e_2)}$ indeed belongs to the class of optimal three-weight cyclic codes in Theorem 1 (therein $k = 2$).

Part (A): Let $e_1' = \frac{(q_0-1)e_1}{q-1}$. Clearly $(q + 1)e_1 = \frac{q^2-1}{q_0-1}e_1'$. Let $h_{(q+1)e_1}(x) = h_{\frac{q^2-1}{q_0-1}e_1'}(x)$, $h_{e_2}(x) \in \mathbb{F}_{q_0}[x]$ be the minimal polynomials of $\gamma^{-\frac{q^2-1}{q_0-1}e_1'}$ and γ^{-e_2}, respectively. Hence, in accordance with Remark 2, note that $\deg(h_{(q+1)e_1}(x)) = 1$, because $\frac{q^2-1}{q_0-1}e'_1 q_0 \equiv \frac{q^2-1}{q_0-1}e'_1 \pmod{q^2-1}$. Also, as $\langle \gamma \rangle = \langle \gamma^{-e_2} \rangle = \mathbb{F}^*_{q^2} = \mathbb{F}^*_{q_0^{2r}}$, $\deg(h_{e_2}(x)) = 2r$. In consequence, $\mathcal{C}^{(q_0)}_{(q,2,e_1,e_2)}$ has dimension $2r+1$. In fact, since $\gamma^{(q+1)e_1} = \gamma^{\frac{q_0^r-1}{q_0-1}e'_1} \in \mathbb{F}^*_{q_0}$, note that the code $\mathcal{C}^{(q_0)}_{(q,2,e_1,e_2)}$ is given by the set (see (4))

$$\left\{ \left(\gamma^{(q+1)e_1 j} \operatorname{Tr}_{\mathbb{F}_q/\mathbb{F}_{q_0}}(a) + \operatorname{Tr}_{\mathbb{F}_{q^2}/\mathbb{F}_{q_0}} \left(b\gamma^{e_2 j} \right) \right)_{j=0}^{q^2-2} : a \in \mathbb{F}_q, b \in \mathbb{F}_{q^2} \right\}$$

$$= \left\{ \left(a_0 \gamma^{\frac{q_0^{2r}-1}{q_0-1}e'_1 j} + \operatorname{Tr}_{\mathbb{F}_{q_0^{2r}}/\mathbb{F}_{q_0}} \left(b\gamma^{e_2 j} \right) \right)_{j=0}^{q_0^{2r}-2} : a_0 \in \mathbb{F}_{q_0}, b \in \mathbb{F}_{q_0^{2r}} \right\}. \quad (7)$$

Clearly $(q_0-1)|(q_0^l-1)$, for every non-negative integer l (that is, $q_0^l \equiv 1 \pmod{q_0-1}$). Thus, since $\frac{q_0^r-1}{q_0-1} = q_0^{r-1} + q_0^{r-2} + \cdots + q_0 + 1$, $(q_0-1)|(\frac{q_0^r-1}{q_0-1} - r)$. Therefore, as $e'_1 = \frac{(q_0-1)e_1}{q_0^r-1}$ and $q-1 = \frac{q_0^r-1}{q_0-1}(q_0-1)$, we have

$$\gcd(q_0-1, 2re'_1 - e_2) = \gcd(q_0-1, 2re'_1 - e_2 + 2(\frac{q_0^r-1}{q_0-1} - r)e'_1),$$

$$= \gcd(q_0-1, 2\frac{q_0^r-1}{q_0-1}e'_1 - e_2),$$

$$= \gcd(q_0-1, 2e_1 - e_2) \le \gcd(q-1, 2e_1 - e_2) = 1.$$

That is, $\gcd(q_0-1, 2re'_1 - e_2) = 1$. Moreover, since $\gcd(q^2-1, e_2) = 1$, we also have $\gcd(\frac{q_0^{2r}-1}{q_0-1}, e_2) = 1$. This means, in consequence and in agreement with Theorem 1, that $\mathcal{C}^{(q_0)}_{(q,2,e_1,e_2)}$ is an optimal three-weight cyclic code of length $q_0^{2r}-1$ and dimension $2r+1$ that belongs to such a theorem. In fact, from (7) and (3), note that

$$\mathcal{C}^{(q_0)}_{(q,2,e_1,e_2)} = \mathcal{C}_{(q_0,2r,e'_1,e_2)},$$

where $e'_1 = \frac{(q_0-1)e_1}{q-1}$.

Part (B): Note that, by Remark 2, $\mathcal{C}^{(q_0)}_{(q,2,e_1,e_2)}$ is cyclic. Now, let $h_{(q+1)e_1}(x)$, $h_{e_2}(x) \in \mathbb{F}_{q_0}[x]$ be as before. Since $(q-1) \nmid (q_0-1)e_1$, observe that r is the smallest positive integer such that $(q+1)e_1 q_0^r = (q_0^r+1)e_1 q_0^r \equiv (q+1)e_1 \pmod{q_0^{2r}-1}$. Thus, $\deg(h_{(q+1)e_1}(x)) = r$, and since $\deg(h_{e_2}(x)) = 2r$, the dimension of $\mathcal{C}^{(q_0)}_{(q,2,e_1,e_2)}$ is $3r$.

Let φ, χ and χ' be the canonical additive characters of $\mathbb{F}_{q_0}$, $\mathbb{F}_q$ and $\mathbb{F}_{q^2}$, respectively. Let $a \in \mathbb{F}_q$, $b \in \mathbb{F}_{q^2}$, and $\mathbf{c}(a,b)^{(q_0)} \in \mathcal{C}^{(q_0)}_{(q,2,e_1,e_2)}$. Hence, from (5) and by the orthogonality relation for the character φ (see (2)), the Hamming weight of the codeword $\mathbf{c}(a,b)^{(q_0)}$, $\mathbf{w}(\mathbf{c}(a,b)^{(q_0)})$, is equal to

$$q^2 - 1 - \frac{1}{q_0} \sum_{y \in \mathbb{F}_{q_0}} \sum_{w \in \mathbb{F}_{q^2}^*} \varphi \left(y \left(\mathrm{Tr}_{\mathbb{F}_q/\mathbb{F}_{q_0}} (aw^{(q+1)e_1}) + \mathrm{Tr}_{\mathbb{F}_{q^2}/\mathbb{F}_{q_0}} (bw^{e_2}) \right) \right),$$

$$= q^2 - 1 - \frac{1}{q_0} \sum_{y \in \mathbb{F}_{q_0}} \sum_{w \in \mathbb{F}_{q^2}^*} \chi(yaw^{(q+1)e_1}) \chi'(ybw^{e_2}),$$

$$= \frac{(q_0 - 1)(q_0^{2r} - 1)}{q_0} - \frac{1}{q_0} \sum_{y \in \mathbb{F}_{q_0}^*} \sum_{w \in \mathbb{F}_{q^2}^*} \chi(yaw^{(q+1)e_1}) \chi'(ybw^{e_2}).$$

But $\mathcal{I}e_2 \equiv 1 \pmod{q^2 - 1}$ and $\mathcal{I}e_1 \equiv 1 \pmod{q - 1}$. Thus, after applying the variable substitution $w \mapsto x^{\mathcal{I}}$, we get

$$\sum_{y \in \mathbb{F}_{q_0}^*} \sum_{w \in \mathbb{F}_{q^2}^*} \chi(yaw^{(q+1)e_1}) \chi'(ybw^{e_2}) = \sum_{y \in \mathbb{F}_{q_0}^*} \sum_{x \in \mathbb{F}_{q^2}^*} \chi(yax^{(q+1)\mathcal{I}e_1}) \chi'(ybx^{\mathcal{I}e_2}),$$

$$= \sum_{y \in \mathbb{F}_{q_0}^*} \sum_{x \in \mathbb{F}_{q^2}^*} \chi(yax^{(q+1)}) \chi'(ybx),$$

$$= Z(a, b),$$

where $Z(a, b)$ is as in Lemma 2. In fact, due to this lemma, we have that $\mathbf{w}(\mathbf{c}(a, b)^{(q_0)})$ is equal to

$$\begin{cases} 0 & \text{if } a = b = 0, \\ q_0^{r-1}(q_0 - 1)(q_0^r + 1) & \text{if } a \neq 0 \text{ and } b = 0, \\ q_0^{2r-1}(q_0 - 1) & \text{if } a = 0 \text{ and } b \neq 0, \\ q_0^{r-1}(q_0 - 1)(q_0^r + 1) & \text{if } (a, b) \neq (0, 0) \text{ and } \mathrm{Tr}_{\mathbb{F}_q/\mathbb{F}_{q_0}} \left(\frac{b^{q+1}}{a} \right) = 0, \\ q_0^{r-1}(q_0^{r+1} - q_0^r - 1) & \text{if } (a, b) \neq (0, 0) \text{ and } \mathrm{Tr}_{\mathbb{F}_q/\mathbb{F}_{q_0}} \left(\frac{b^{q+1}}{a} \right) \neq 0, \end{cases}$$

which is in accordance with (6). Now observe that

$$A_{q_0^{r-1}(q_0-1)(q_0^r+1)} = \#\{a \in \mathbb{F}_q^*\} + \#\{(a, b) \in \mathbb{F}_q^* \times \mathbb{F}_{q^2}^* : \mathrm{Tr}_{\mathbb{F}_q/\mathbb{F}_{q_0}} \left(\frac{b^{q+1}}{a} \right) = 0\},$$

$$= (q - 1) + (q - 1)(q + 1)(\frac{q}{q_0} - 1),$$

$$= q_0^{r-1}(q_0^r - 1)(q_0^r - q_0 + 1).$$

Similarly, the frequencies of the other weights of $C_{(q,2,e_1,e_2)}^{(q_0)}$ can be computed and we omit the details here. Then the weight enumerator of $C_{(q,2,e_1,e_2)}^{(q_0)}$ follows.

Finally, note that $A_1^\perp = 0$, since otherwise $C_{(q,2,e_1,e_2)}^{(q_0)}$ would be the null code $\{0\}$. Thus, a direct application of the last two identities in (1) shows that $A_2^\perp = 0$ and that the value of $A_3^\perp$ is the announced one. Lastly, by the sphere-packing bound (see for example [12, Theorem 1.12.1]), it is not difficult to verify that for a code of length $q_0^{2r} - 1$ and dimension $q_0^{2r} - 3r - 1$, its minimum Hamming distance can be at most 4. Therefore, the code $\mathcal{C}_{(q,2,e_1,e_2)}^{(q_0)\perp}$ is almost optimal since its minimum Hamming distance is 3.   □

Example 1. The following are some examples of Theorem 2.

(a) Let $(q_0, r, e_1, e_2) = (3, 2, 4, 1)$. Then $q = 9$ and clearly $(q-1)|(q_0-1)e_1$. Thus, owing to Part (A) of Theorem 2, the subfield code $\mathcal{C}_{(9,2,4,1)}^{(3)} = \mathcal{C}_{(3,4,1,1)}$ is an optimal three-weight cyclic code of length 80 and dimension 5, over $\mathbb{F}_3$, whose weight enumerator is

$$1 + 160z^{53} + 80z^{54} + 2z^{80}.$$

(b) Let $(q_0, r, e_1, e_2) = (2, 2, 1, 1)$. Then $q = 4$, $\mathcal{I} = 1$, and clearly $(q-1) \nmid (q_0-1)e_1$. Thus, owing to Part (B) of Theorem 2, the subfield code $\mathcal{C}_{(4,2,1,1)}^{(2)}$ is a binary three-weight $[15, 6, 6]$ cyclic code with weight enumerator

$$1 + 30z^6 + 15z^8 + 18z^{10},$$

while its dual code is an almost optimal $[15, 9, 3]$ cyclic code with respect to the sphere-packing bound, with $A_1^\perp = A_2^\perp = 0$ and $A_3^\perp = 5$.

(c) Let $(q_0, r, e_1, e_2) = (3, 2, 1, 1)$. Then $q = 9$, $\mathcal{I} = 1$, and clearly $(q-1) \nmid (q_0-1)e_1$. Thus, owing to Part (B) of Theorem 2, the subfield code $\mathcal{C}_{(9,2,1,1)}^{(3)}$ is a three-weight $[80, 6, 51]$ cyclic code over $\mathbb{F}_3$ with weight enumerator

$$1 + 480z^{51} + 80z^{54} + 168z^{60},$$

while its dual code is an almost optimal $[80, 74, 3]$ cyclic code with respect to the sphere-packing bound, with $A_1^\perp = A_2^\perp = 0$ and $A_3^\perp = 640$.

(d) Let $(q_0, r, e_1, e_2) = (2, 4, 2, 2)$. Then $q = 16$ and $\mathcal{I} = 128$. Clearly $(q-1) \nmid (q_0-1)e_1$ and $\mathcal{I}e_1 \equiv 1 \pmod{q-1}$. Thus, owing to Part (B) of Theorem 2, the subfield code $\mathcal{C}_{(16,2,2,2)}^{(2)}$ is a binary three-weight $[255, 12, 120]$ cyclic code with weight enumerator

$$1 + 2040z^{120} + 255z^{128} + 1800z^{136},$$

while its dual code is an almost optimal $[255, 243, 3]$ cyclic code with respect to the sphere-packing bound, with $A_1^\perp = A_2^\perp = 0$ and $A_3^\perp = 595$.

Remark 3. According to the code tables at [8], note that the $[15, 6, 6]$ code in (b) is optimal, while the $[80, 6, 51]$ code in (c) and its dual code are optimal. Finally, the $[255, 12, 120]$ code in (d) has the best known parameters.

By fixing $k = 2$, it is important to observe that the condition on the integer e_2 is more restrictive in Theorem 2 ($\gcd(q^2 - 1, e_2) = 1$) than in Theorem 1 ($\gcd(q+1, e_2) = 1$). This implies, of course, that Theorem 2 can only determine the subfield codes for a subclass of the three-weight cyclic codes in Theorem

1. Specifically, this means that there are optimal three-weight cyclic codes in Theorem 1, whose subfield codes cannot be described through Theorem 2. For example, with the help of a computer, it is not difficult to verify that the subfield code $\mathcal{C}^{(2)}_{(4,2,1,3)}$ is a four-weight binary cyclic code with weight enumerator $1 + 25z^6 + 30z^8 + 3z^{10} + 5z^{12}$ (for this example note that $\gcd(q+1, e_2) = 1$, but $\gcd(q^2 - 1, e_2) \neq 1$). This subfield code, like the subfield code in (b) Example 1, is optimal. However, unlike the dual of $\mathcal{C}^{(2)}_{(4,2,1,1)}$, the dual of $\mathcal{C}^{(2)}_{(4,2,1,3)}$ is a binary optimal cyclic code with parameters $[15, 9, 4]$. This example let us know that, beyond Theorem 2, there are still other optimal three-weight cyclic codes whose subfield codes have good parameters.

5   The Covering Structure of the Subfield Codes

For any $\mathbf{c} = (c_0, c_1, \ldots, c_{n-1}) \in \mathbb{F}_{q_0}^n$, the *support* of $\mathbf{c}$ is defined by the set $\{i \mid 0 \leq i \leq n-1, c_i \neq 0\}$. Furthermore, for any two vectors $\mathbf{c}, \mathbf{c}' \in \mathbb{F}_{q_0}^n$, $\mathbf{c}$ is said to *cover* $\mathbf{c}'$ if the support of $\mathbf{c}$ contains that of $\mathbf{c}'$. A nonzero codeword is called a *minimal codeword* if it covers only its multiples in a linear code. The set of all minimal codewords in a linear code is called the *covering structure* of the code.

Determining the covering structure of a linear code is in general a difficult but at the same time interesting problem as it is closely related to the construction of secret sharing schemes (see for example [3,13,16,17,21]). In this section we determine the covering structure of the subfield codes in Theorem 2. As we will see, some of these codes have the important property that any nonzero codeword is minimal. These codes are suitable for constructing secret sharing schemes with nice access structures. Moreover, we present a specific example of a secret sharing scheme based on one of these subfield codes.

There are several ways to construct secret sharing schemes by using linear codes. One of them was proposed by Massey in [16,17] and is presented below (see [13,21]).

Let $\mathcal{C}$ be an $[n, l]$ linear code over $\mathbb{F}_{q_0}$. In the *secret sharing scheme based on a linear code* $\mathcal{C}$, the secret $\mathbf{s}$ is an element of $\mathbb{F}_{q_0}$, which is called the secret space. There is a dealer P_0 and $n-1$ parties $P_1, P_2, \ldots, P_{n-1}$ involved in the secret sharing scheme, the dealer being a trusted person. Let $G^\perp = (\mathbf{g}_0^\perp, \mathbf{g}_1^\perp, \ldots, \mathbf{g}_{n-1}^\perp)$ be a generator matrix of the dual code, $\mathcal{C}^\perp$, of $\mathcal{C}$ such that $\mathbf{g}_i^\perp$ is the i-th column vector of $G^\perp$ and $\mathbf{g}_i^\perp \neq 0$ for $0 \leq i \leq n-1$. Then, the secret sharing scheme based on $\mathcal{C}$ is described as follows:

Step 1) In order to compute the shares with respect to a secret $\mathbf{s}$, the dealer P_0 chooses randomly a vector $\mathbf{u} = (u_0, u_1, \ldots, u_{n-l-1}) \in \mathbb{F}_{q_0}^{n-l}$ such that $\mathbf{s} = \mathbf{u}\mathbf{g}_0^\perp$. There are altogether q_0^{n-l-1} such vectors $\mathbf{u} \in \mathbb{F}_{q_0}^{n-l}$.

Step 2) The dealer P_0 treats $\mathbf{u}$ as an information vector and computes the corresponding codeword $\mathbf{t} = \mathbf{u}G^\perp = (t_0, t_1, \ldots, t_{n-1})$ in $\mathcal{C}^\perp$. Then he sends t_i to party P_i as the share for every i ($1 \leq i \leq n-1$).

Step 3) The secret $\mathbf{s}$ is recovered as follows: since $t_0 = \mathbf{u}\mathbf{g}_0^\perp = \mathbf{s}$, a set of shares $\{t_{i_1}, t_{i_2}, \ldots, t_{i_m}\}$ can determine the secret $\mathbf{s}$ iff $\mathbf{g}_0^\perp$ is a linear combination of $\{\boldsymbol{g}_{i_1}^{\perp}, \boldsymbol{g}_{i_2}^{\perp}, \ldots, \boldsymbol{g}_{i_m}^{\perp}\}$, where $1 \leq i_1 < i_2 < \cdots < i_m \leq n-1$.

Clearly, if a group of participants $\mathcal{D}$ can recover the secret by combining their shares, then any group of participants containing $\mathcal{D}$ can also recover the secret. The set $\{i_1, i_2, \ldots, i_m\}$ is said to be a *minimal access set* if it can recover the secret **s** but none of its proper subsets can do so. The *access structure* of the secret sharing scheme refers to the set of all minimal access sets.

For a linear code $\mathcal{C}$, the following lemma from [16] presents a one-to-one correspondence between the set of minimal access sets of the secret sharing scheme based on $\mathcal{C}$ and the set of minimal codewords in $\mathcal{C}$ whose first coordinate is 1.

Lemma 3. *Let $\mathcal{C}$ be an $[n, l]$ linear code over $\mathbb{F}_{q_0}$. Then, the set $\{i_1, i_2, \ldots, i_m\} \subseteq \{1, 2, \ldots, n-1\}$ with $i_1 < i_2 < \cdots < i_m$ is a minimal access set in the secret sharing scheme based on $\mathcal{C}$ iff there is a minimal codeword $\mathbf{c} = \{c_0, c_1, \ldots, c_{n-1}\}$ in $\mathcal{C}$ such that the support of $\mathbf{c}$ is $\{0, i_1, i_2, \ldots, i_m\}$ and $c_0 = 1$.*

If **c** is a nonzero codeword whose first coordinate is 1 and the support of the codeword **c** is $\{0, i_1, i_2, \ldots, i_m\}$ such that $1 \leq i_1 < i_2 < \cdots < i_m \leq n-1$, we call the set $\{i_1, i_2, \ldots, i_m\}$ the *access support* of the codeword **c**.

From the discussion above, determining the access structure of the secret sharing scheme based on a linear code $\mathcal{C}$ is equivalent to determining the set of access supports of the minimal codewords in $\mathcal{C}$ whose first coordinate is 1. Thus, in the following we determine the covering structure of the subfield codes in Theorem 2. To that end, the next results found in [1] will be useful.

Lemma 4. *Let $\mathcal{C}$ be a linear code over $\mathbb{F}_{q_0}$ with minimum Hamming distance d. Then, every codeword whose weight is less than or equal to $\frac{dq_0 - q_0 + 1}{q_0 - 1}$ must be a minimal codeword.*

The following lemma states that if the weights of a linear code are close enough to each other, then all nonzero codewords of the code are minimal.

Lemma 5 *(Ashikhmin-Barg Lemma). Let $\mathcal{C}$ be an $[n, l]$ linear code over $\mathbb{F}_{q_0}$, and let $\mathbf{w}_{\min}$ and $\mathbf{w}_{\max}$ be the minimum and maximum nonzero weights of $\mathcal{C}$, respectively. If*

$$\frac{\mathbf{w}_{\min}}{\mathbf{w}_{\max}} > \frac{q_0 - 1}{q_0},$$

then all nonzero codewords of $\mathcal{C}$ are minimal.

We remark that the condition in the previous lemma is only a sufficient condition. There are codes such that all their nonzero codewords are minimal without satisfying this condition (see for example [21]).

Now, we are able to give the covering structure of the subfield codes in Theorem 2:

Theorem 3. *Assume the same notation as in Theorem 2. Then the covering structure of a subfield code of the form $C_{(q,2,e_1,e_2)}^{(q_0)}$ is as follows:*

(a) *If $C_{(q,2,e_1,e_2)}^{(q_0)}$ belongs to Part (A) of Theorem 2, then all its nonzero codewords with weight $q_0^{2r} - 1$ are not minimal, while the other nonzero codewords are minimal.*

(b) *If $C_{(q,2,e_1,e_2)}^{(q_0)}$ belongs to Part (B) of Theorem 2, then all its nonzero codewords are minimal.*

Proof. Part (a): Clearly, all nonzero codewords with weight $q_0^{2r} - 1$ are not minimal as the length of $C_{(q,2,e_1,e_2)}^{(q_0)}$ is $q_0^{2r} - 1$ (see Theorem 2). Now, since $2r > 2$, it is not difficult to verify that

$$ q_0^{2r-1}(q_0 - 1) \leq \frac{q_0^{2r+1} - q_0^{2r} - 2q_0 + 1}{q_0 - 1} . $$

Thus, since $q_0^{2r-1}(q_0 - 1) - 1 < q_0^{2r-1}(q_0 - 1)$, it follows from Lemma 4 that the assertion in Part (a) holds.

Part (b): Let $\mathbf{w}_{\min}$ and $\mathbf{w}_{\max}$ be as in Lemma 5. Thus, by (6), $\mathbf{w}_{\min} = q_0^{r-1}(q_0^{r+1} - q_0^r - 1)$ and $\mathbf{w}_{\max} = q_0^{r-1}(q_0 - 1)(q_0^r + 1)$. The result now follows directly from Lemma 5.   □

Linear codes whose nonzero codewords are all minimal have an interesting access structure as described in the following:

Proposition 1 [21, *Proposition 2*]. *Let C be an $[n,l]$ linear code over $\mathbb{F}_{q_0}$ and let $G = (\mathbf{g}_0, \mathbf{g}_1, \ldots, \mathbf{g}_{n-1})$ be a generator matrix of C such that $\mathbf{g}_i$ is the i-th column vector of G and $\mathbf{g}_i \neq 0$ for $0 \leq i \leq n - 1$. If each nonzero codeword of C is minimal, then the access structure of the secret sharing scheme based on C is composed of q_0^{l-1} minimal access sets, which is equal to the set of access supports of the nonzero codewords in C with first coordinate 1. In addition, we have the following:*

(a) *If $\mathbf{g}_i$ is a scalar multiple of $\mathbf{g}_0$, $1 \leq i \leq n - 1$, then participant P_i must be in every minimal access set. Such a participant is called a dictatorial participant.*

(b) *If $\mathbf{g}_i$ is not a scalar multiple of $\mathbf{g}_0$, $1 \leq i \leq n - 1$, then participant P_i must be in $(q_0 - 1)q_0^{l-2}$ out of q_0^{l-1} minimal access sets.*

We end this section by presenting a specific example of the secret sharing scheme described above. Thus, let $(q_0, r, e_1, e_2) = (2, 2, 1, 1)$. Then $q = 4$ and by (b) Example 1 we know that the subfield code $C_{(4,2,1,1)}^{(2)}$ is a binary three-weight $[15, 6, 6]$ cyclic code with weight enumerator $1 + 30z^6 + 15z^8 + 18z^{10}$. We take $\mathbb{F}_{16} = \mathbb{F}_2(\gamma)$ with $\gamma^4 + \gamma + 1 = 0$. With this choice, and by using the notation in Remark 2, $h_5(x) = x^2 + x + 1$ and $h_1(x) = x^4 + x^3 + 1$ (see [15, p. 99]). Therefore, $(x^{15} - 1)/h_5(x)h_1(x) = x^9 + x^6 + x^5 + x^4 + x + 1$ and $h_5(x)h_1(x) = x^6 + x^3 + x^2 + x + 1$ are the generator and parity-check polynomials of $C_{(4,2,1,1)}^{(2)}$, respectively. In consequence, the generator matrices, G and $G^\perp$, for $C_{(4,2,1,1)}^{(2)}$ and its dual are:

$$G = \begin{bmatrix} 1\,1\,0\,0\,1\,1\,1\,0\,0\,1\,0\,0\,0\,0\,0 \\ 0\,1\,1\,0\,0\,1\,1\,1\,0\,0\,1\,0\,0\,0\,0 \\ 0\,0\,1\,1\,0\,0\,1\,1\,1\,0\,0\,1\,0\,0\,0 \\ 0\,0\,0\,1\,1\,0\,0\,1\,1\,1\,0\,0\,1\,0\,0 \\ 0\,0\,0\,0\,1\,1\,0\,0\,1\,1\,1\,0\,0\,1\,0 \\ 0\,0\,0\,0\,0\,1\,1\,0\,0\,1\,1\,1\,0\,0\,1 \end{bmatrix}, \quad G^{\perp} = \begin{bmatrix} 0\,0\,0\,0\,0\,0\,0\,1\,0\,0\,1\,1\,1\,1 \\ 0\,0\,0\,0\,0\,0\,1\,0\,0\,1\,1\,1\,1\,0 \\ 0\,0\,0\,0\,0\,1\,0\,0\,1\,1\,1\,1\,0\,0 \\ 0\,0\,0\,0\,1\,0\,0\,1\,1\,1\,1\,0\,0\,0 \\ 0\,0\,0\,1\,0\,0\,1\,1\,1\,1\,0\,0\,0\,0 \\ 0\,0\,1\,0\,0\,1\,1\,1\,1\,0\,0\,0\,0\,0 \\ 0\,0\,1\,0\,0\,1\,1\,1\,0\,0\,0\,0\,0\,0 \\ 0\,1\,0\,0\,1\,1\,1\,1\,0\,0\,0\,0\,0\,0 \\ 1\,0\,0\,1\,1\,1\,1\,0\,0\,0\,0\,0\,0\,0 \end{bmatrix}.$$

Thus, in the secret sharing scheme based on $\mathcal{C}_{(4,2,1,1)}^{(2)}$, 14 participants and a dealer are involved. Owing to Lemma 3, Part (b) of Theorem 3, and Proposition 1, there are altogether $q_0^{l-1} = 2^5 = 32$ minimal access sets:

$$\{4,5,6,7,8,9,11,12,13\} \ \{1,2,3,4,6,7,8,10,14\} \ \{1,4,10,11,14\} \ \{2,5,7,8,13\}$$
$$\{2,3,4,6,10,11,12,13,14\} \ \{1,2,3,7,10,11,13\} \ \{5,7,9,12,14\} \ \{6,7,10,11,12\}$$
$$\{1,2,4,5,6,8,12,13,14\} \ \{1,2,4,8,9,10,11,12,13\} \ \{1,2,5,11,12\} \ \{1,4,5,6,9\}$$
$$\{1,3,4,5,7,11,12,13,14\} \ \{1,3,4,6,7,9,10,12,13\} \ \{2,4,7,9,10\} \ \{1,6,8,10,13\}$$
$$\{1,3,7,8,9,10,11,12,14\} \ \{1,2,3,5,6,7,9,13,14\} \ \{2,3,8,10,12\} \ \{3,9,10,13,14\}$$
$$\{2,3,5,6,8,9,11,12,14\} \ \{2,6,7,8,9,10,11,13,14\} \ \{3,4,5,8,14\} \ \{3,5,6,11,13\}$$
$$\{1,2,3,4,5,7,8,9,11\} \ \{1,3,5,6,7,8,12\} \ \{3,4,6,8,9,10,11\} \ \{2,4,5,6,7,11,14\}$$
$$\{4,7,8,10,12,13,14\} \ \{1,2,6,9,10,12,14\} \ \{1,5,8,9,11,13,14\} \ \{2,3,4,5,9,12,13\}.$$

Moreover, in accordance with Part (b) of Proposition 1, note that any participant P_i ($1 \le i \le 14$) appears in $(q_0 - 1)q_0^{l-2} = 16$ out of $q_0^{l-1} = 32$ minimal access sets. In order to appreciate the use of the previous minimal access sets, suppose that we wish to "split" a 4-bit secret, $\mathbf{s}$, into 4-bit shares for fourteen parties $P_1, P_2, \ldots, P_{14}$. Following [16], $\mathbf{s} \in \mathrm{GF}(2^4) = \mathbb{F}_{16} := \{0, 1, 2, \ldots, 9, a, b, c, d, e, f\}$ and suppose $\mathbf{s} = b = [1011]$. The dealer randomly chooses four codewords $\mathbf{c}_1, \mathbf{c}_2,$ $\mathbf{c}_3$ and $\mathbf{c}_4$, in the dual code of $\mathcal{C}_{(4,2,1,1)}^{(2)}$, with the condition that each bit in the secret $\mathbf{s}$ matches the first component of one of these four codewords. Suppose that the dealer's choice is:

$$\mathbf{c}_1 = [100101110111111],$$
$$\mathbf{c}_2 = [000000000000000],$$
$$\mathbf{c}_3 = [100010111101011],$$
$$\mathbf{c}_4 = [110100100100010].$$

By means of these codewords the dealer now proceeds to generate the 4-bit shares for the fourteen parties:

1	0	0	1	0	1	1	1	0	1	1	1	1	1	1
0	0	0	0	0	0	0	0	0	0	0	0	0	0	0
1	0	0	0	1	0	1	1	1	1	0	1	0	1	1
1	1	0	1	0	0	1	0	0	1	0	0	0	1	0
↓	↓	↓	↓	↓	↓	↓	↓	↓	↓	↓	↓	↓	↓	↓
b	1	0	9	2	8	b	a	2	b	8	a	8	b	a
↓	↓	↓	↓	↓	↓	↓	↓	↓	↓	↓	↓	↓	↓	↓
$\mathbf{s}$	P_1	P_2	P_3	P_4	P_5	P_6	P_7	P_8	P_9	P_{10}	P_{11}	P_{12}	P_{13}	P_{14}

In this way, the share for party P_1 is 1, the share for party P_2 is 0, and so on. Finally, note that any of the above minimal access sets can recover the secret $\mathbf{s} = b$. For example, by using the shares for the minimal access set $\{1, 4, 5, 6, 9\}$, we get $1 + 2 + 8 + b + b = b$.

6 Conclusions

In this paper we studied the q_0-ary subfield codes of a subclass of optimal three-weight cyclic codes of length $q^2 - 1$ and dimension 3 that belongs to the class of codes in Theorem 1. We proved that some of these subfield codes are optimal three-weight cyclic codes of length $q^2 - 1$ and dimension $2r + 1$ (where $q = q_0^r$) that belong, again, to the class of optimal three-weight cyclic codes in Theorem 1 (Part (A) of Theorem 2). For the other subfield codes studied here we showed that they are three-weight cyclic codes of length $q^2 - 1$ whose dimension is now $3r$ (Part (B) of Theorem 2). For the latter subfield codes, we also determined the minimum Hamming distance for their duals, and with this, we concluded that these duals are almost optimal with respect to the sphere-packing bound. Furthermore, it was shown that some subfield codes in Part (B) of Theorem 2 are optimal and others have the best known parameters according to the code tables at [8] (Example 1 and Remark 3). However, as pointed out at the end of Sect. 4, there is evidence of the existence of other subfield codes with good parameters. Therefore, as further work, it would be interesting to study those other subfield codes.

Finally, as an application of linear codes with few weights, the covering structure of the subfield codes in Theorem 2 was determined (Theorem 3) and used to present a specific example of a secret sharing scheme based on one of these subfield codes at the end of Sect. 5.

Acknowledgements. The authors would like to thank the anonymous reviewers for their valuable comments and suggestions that helped to improve the quality of the paper.

References

1. Ashikhmin, A., Barg, A.: Minimal vectors in linear codes. IEEE Trans. Inf. Theory **44**(5), 2010–2017 (1998). https://doi.org/10.1109/18.705584
2. Calderbank, A.R., Goethals, J.M.: Three-weight codes and association schemes. Philips J. Res. **39**(4–5), 143–152 (1984)
3. Carlet, C., Ding, C., Yuan, J.: Linear codes from perfect nonlinear mappings and their secret sharing schemes. IEEE Trans. Inf. Theory **51**(6), 2089–2102 (2005). https://doi.org/10.1109/TIT.2005.847722
4. Carlet, C., Charpin, P., Zinoviev, V.: Codes, bent functions and permutations suitable for des-like cryptosystems. Des. Codes Cryptogr. **15**, 125–156 (1998). https://doi.org/10.1023/A:1008344232130
5. Ding, C., Heng, Z.: The subfield codes of ovoid codes. IEEE Trans. Inf. Theory **65**(8), 4715–4729 (2019). https://doi.org/10.1109/TIT.2019.2907276

6. Ding, C., Wang, X.: A coding theory construction of new systematic authentication codes. Theor. Comput. Sci. **330**(1), 81–99 (2005). https://doi.org/10.1016/j.tcs.2004.09.011
7. Ding, C., Yin, J.: Sets of optimal frequency-hopping sequences. IEEE Trans. Inf. Theory **54**(8), 3741–3745 (2008). https://doi.org/10.1109/TIT.2008.926410
8. Grassl, M.: Bounds on the minimum distance of linear codes and quantum codes. http://www.codetables.de. Accessed 4 Aug 2022
9. Heng, Z., Ding, C.: The subfield codes of hyperoval and conic codes. Finite Fields Appl. **56**, 308–331 (2019). https://doi.org/10.1016/j.ffa.2018.12.006
10. Heng, Z., Wang, Q., Ding, C.: Two families of optimal linear codes and their subfield codes. IEEE Trans. Inf. Theory **66**(11), 6872–6883 (2020). https://doi.org/10.1109/TIT.2020.3006846
11. Heng, Z., Yue, Q.: Several classes of cyclic codes with either optimal three weights or a few weights. IEEE Trans. Inf. Theory **62**(8), 4501–4513 (2016). https://doi.org/10.1109/TIT.2016.2550029
12. Huffman, W.C., Pless, V.: Fundamentals of Error-Correcting Codes. Cambridge University Press, Cambridge (2003)
13. Li, C., Qu, L., Ling, S.: On the covering structures of two classes of linear codes from perfect nonlinear functions. IEEE Trans. Inf. Theory **55**(1), 70–82 (2009). https://doi.org/10.1109/TIT.2008.2008145
14. Lidl, R., Niederreiter, H.: Finite Fields. Cambridge University Press, Cambridge (1983)
15. MacWilliams, F.J., Sloane, N.J.A.: The Theory Of Error-Correcting Codes. North-Holland, Amsterdam, The Netherlands (1977)
16. Massey, J.L.: Minimal codewords and secret sharing. In: Proceedings of the 6th Joint Swedish-Russian International Workshop on Information Theory, pp. 276–279 (1993)
17. Massey, J.L.: Some applications of coding theory in cryptography. In: Codes and Ciphers: Cryptography and Coding IV, pp. 33–47 (1995)
18. Schmidt, B., White, C.: All two-weight irreducible cyclic codes? Finite Fields Their Appl. **8**(1), 1–17 (2002). https://doi.org/10.1006/ffta.2000.0293
19. Shi, M., Solé, P.: Three-weight codes, triple sum sets, and strongly walk regular graphs. Des. Codes Crypt. **87**(10), 2395–2404 (2019). https://doi.org/10.1007/s10623-019-00628-7
20. Vega, G.: A characterization of a class of optimal three-weight cyclic codes of dimension 3 over any finite field. Finite Fields Their Appl. **42**, 23–38 (2016). https://doi.org/10.1016/j.ffa.2016.07.001
21. Yuan, J., Ding, C.: Secret sharing schemes from three classes of linear codes. IEEE Trans. Inf. Theory **52**(1), 206–212 (2006). https://doi.org/10.1109/TIT.2005.860412

Social Choice Theory

Multidimensional Manhattan Preferences

Jiehua Chen$^{(\boxtimes)}$, Martin Nöllenburg, Sofia Simola, Anaïs Villedieu,
and Markus Wallinger

TU Wien, Vienna, Austria
{jiehua.chen,martin.noellenburg,sofia.simola,anais.villedieu,
markus.wallinger}@tuwien.ac.at

Abstract. A preference profile (i.e., a collection of linear preference orders of the voters over a set of alternatives) with m alternatives and n voters is d-Manhattan (resp. d-Euclidean) if both the alternatives and the voters can be placed into a d-dimensional space such that between each pair of alternatives, every voter prefers the one which has a shorter Manhattan (resp. Euclidean) distance to the voter.

We initiate the study of how d-Manhattan preference profiles depend on the values m and n. First, we provide explicit constructions to show that each preference profile with m alternatives and n voters is d-Manhattan whenever $d \geq \min(n, m-1)$. Second, for $d = 2$, we show that the smallest non d-Manhattan preference profile has either 3 voters and 6 alternatives, or 4 voters and 5 alternatives, or 5 voters and 4 alternatives. This is more complex than the case with d-Euclidean preferences (see [Bogomolnaia and Laslier, 2007] and [Bulteau and Chen, 2022]).

1 Introduction

Modelling voters' linear preferences (aka. rankings) over a set of alternatives as geometric distances is an approach popular in many research fields such as economics [11,13,17], political and social sciences [2,15,22,25], and psychology [3,9]. The idea is to consider the alternatives and voters as points in a d-dimensional space such that

$$\text{for each two alternatives, each voter prefers the one that is } \textit{closer} \text{ to her. } (*)$$

If the proximity is measured via the Euclidean distance, then *preference profiles* (i.e., a collection of distinct linear preference orders specifying voters' preferences) obeying $(*)$ are called d-Euclidean. While the d-Euclidean model seems to be canonical, in real life the shortest path between two points may be Manhattan rather than Euclidean. For instance, in urban geography, the alternatives (e.g., a shop or a supermarket) and the voters (e.g., individuals) are often located on grid-like streets. That is, the distance between an alternative and a voter is more likely to be measured according to the Manhattan distance (aka. Taxicab

---

J. Chen and S. Simola, and M. Wallinger are supported by the Vienna Science and Technology Fund (WWTF) under grant VRG18-012 and ICT19-035, respectively. Villedieu is supported by the Austrian Science Fund (FWF) grant P31119.

A. Castañeda and F. Rodríguez-Henríquez (Eds.): LATIN 2022, LNCS 13568, pp. 273–289, 2022.
https://doi.org/10.1007/978-3-031-20624-5_17

distance or ℓ_1-norm-distance), i.e., the sum of the absolute differences of the coordinates of the alternative and the voter. Similarly to the Euclidean preference notion, we call a preference profile d-*Manhattan* if there exists an embedding for the voters and the alternatives which satisfies condition ($*$) under the Manhattan distance. Indeed, Manhattan preferences have been studied for a wide range of applications such as facility location [19,26], group decision making [24], and voting and committee elections [12]. Many voting advice applications, such as the German Wahl-O-Mat [28] and Finnish Ylen Vaalikone [27] use Manhattan distances to measure the distance between a voter and an alternative, indicating that such distances may be perceived as more natural in human decision making.

Despite their practical relevance, Manhattan preferences have attracted far less attention than their close relative Euclidean preferences. Bogomolnaia and Laslier [2] studied how restrictive the assumption of Euclidean preferences is. They showed that every preference profile with n voters and m alternatives is d-Euclidean if $d \geq \min(n, m - 1)$. When indifference between alternatives is allowed, the only-if statement holds as well. For $d = 1$, their smallest non 1-Euclidean preference profile consists of either 3 voters and 3 alternatives or 2 voters and 4 alternatives. For $d = 2$, their smallest non 2-Euclidean profile consists of either 4 voters and 4 alternatives or 3 voters and 8 alternatives, which is also tight by Bulteau and Chen [5]. To the best of our knowledge, no such kind of characterization result on the d-Manhattan preferences exists. For maximally Manhattan preferences, however, Escoffier et al. [16] show that a 2-Manhattan preference profile for four alternatives can contain up to 19 distinct preference orders. From the computational point of view, it is known that for $d = 1$, deciding whether a given preference profile is Euclidean (and hence Manhattan) can be done in polynomial time [10,14,18]. For any fixed $d \geq 2$, however, testing Euclidean preferences is complete for the complexity class *existential theory of the reals* $\exists\mathbb{R}$ [23], while it is straightforward to see that the problem for the Manhattan case is contained in NP [21]; note that NP $\subseteq \exists\mathbb{R}$. Nothing about the complexity lower bound is known for Manhattan preferences.

Our Contribution. In this paper, we aim to close the gap and study how to find a d-Manhattan embedding for a given preference profile and what is the smallest dimension for such an embedding. First, we prove that, similarly to the Euclidean case, every preference profiles with m alternatives and n voters is d-Manhattan if $d \geq \min(m - 1, n)$ (Theorems 1 and 2). Then, focusing on the two-dimensional case, we seek to determine tight bounds on the smallest number of either alternatives or voters of a non d-Manhattan profile. We show that an arbitrary preference profile with n voters and m alternatives is 2-Manhattan if and only if either $m \leq 3$

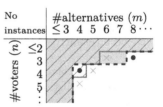

Fig. 1. Boundaries of non 2-Euclidean (resp. non 2-Manhattan) profiles with a given number of voters and alternatives. Each blue bullet (resp. red cross) represents the existence of such a non Euclidean (resp. non Manhattan) profile.

(Theorems 2 and 5), or $n \leq 2$ (Theorem 1), or $n \leq 3$ and $m \leq 5$ (Theorem 3 and Proposition 2), or $n \leq 4$ and $m \leq 4$ (Theorem 4 and Proposition 2). Note that this is considerably different than the Euclidean case: There exists a non Euclidean preference profile with $n = 4$ and $m = 4$, while every preference profile with $n \leq 3$ and $m \leq 7$ is Euclidean. The proof for the "only if" part is via presenting forbidden subprofiles (see Definitions 3 to 5), which may be of independent interests for determining 2-Manhattan preferences. The proof for the "if" part is computer-aided. See Fig. 1 for a summary for $d = 2$.

The paper is organized as follows: Sect. 2 introduces necessary definitions and notations. In Sects. 3 and 4, we present the first positive result and the negative findings, respectively. In Sect. 5, we show the remaining positive results by describing a computer program that finds a Manhattan embedding for every possible preference profile with three voters and five alternatives, or four voters and four alternatives. We conclude with a few future research directions in Sect. 6. Due to space constraints, proofs of results marked with ($\star$) are available in the full version [7].

2 Preliminaries

Given a non-negative integer t, we use $[t]$ to denote the set $\{1, \ldots, t\}$. Let $\boldsymbol{x}$ denote a vector of length d or a point in a d-dimensional space, and let i denote an index $i \in [d]$. We use $\boldsymbol{x}[i]$ to refer to the ith value in $\boldsymbol{x}$.

Let $\mathcal{A} := [m]$ be a set of alternatives. A *preference order* $\succ$ of $\mathcal{A}$ is a linear order (a.k.a. permutation or ranking) of $\mathcal{A}$; a linear order is a binary relation which is total, irreflexive, and transitive. For two distinct alternatives a and b, the relation $a \succ b$ means that a is preferred to (or in other words, ranked higher than) b in $\succ$. An alternative c is the *most-preferred* alternative in $\succ$ if for any alternative $b \in \mathcal{A} \setminus \{c\}$ it holds that $c \succ b$. Let $\succ$ be a preference order over $\mathcal{A}$. For a subset $B \subseteq \mathcal{A}$ of alternatives and an alternative c not in B, we use $B \succ c$ (resp. $c \succ B$) to denote that for each $b \in B$ it holds that $b \succ c$ (resp. $c \succ b$). A *preference profile* (or *profile* in short) $\mathcal{P}$ specifies the preference orders of a number of voters over a set of alternatives. Formally, $\mathcal{P} := (\mathcal{A}, \mathcal{V}, \mathcal{R})$, where $\mathcal{A}$ denotes the set of m alternatives, $\mathcal{V}$ denotes the set of n voters, and $\mathcal{R} := (\succ_1, \ldots, \succ_n)$ is a collection of n preference orders such that each voter $v_i \in \mathcal{V}$ ranks the alternatives according to the preference order $\succ_i$ on $\mathcal{A}$. We will omit the subscript i from $\succ_i$ if it is clear from the context. Throughout the paper, if not explicitly stated otherwise, we assume $\mathcal{P}$ is a preference profile of the form $(\mathcal{A}, \mathcal{V}, \mathcal{R})$. For notational convenience, for each alternative $a \in \mathcal{A}$ and each voter $v_i \in \mathcal{V}$, let $\mathsf{rk}_i(a)$ denote the rank of alternative a in the preference order $\succ_i$, which is the number of alternatives which are preferred to a by voter v_i, i.e., $\mathsf{rk}_i(a) = |\{b \in \mathcal{A} \mid b \succ_i a\}|$. For instance, if voter v_i has preference order $2 \succ_i 3 \succ_i 1 \succ_i 4$, then $\mathsf{rk}_i(3) = 1$.

Given a d-dimensional vector $\boldsymbol{x} \in \mathbb{R}^d$ and an ℓ_p-norm with $p \geq 1$, let $\|\boldsymbol{x}\|_p$ denote the ℓ_p-norm of $\boldsymbol{x}$, i.e., $\|\boldsymbol{x}\|_p = (|\boldsymbol{x}[1]|^p + \cdots + |\boldsymbol{x}[d]|^p)^{1/p}$, and let $\|\boldsymbol{x}\|_\infty$ denote the ℓ_∞-norm of $\boldsymbol{x}$, i.e., $\|\boldsymbol{x}\|_p = \max\{\boldsymbol{x}[i]\}_{i \in [d]}$. Given two points $\boldsymbol{u}, \boldsymbol{w}$

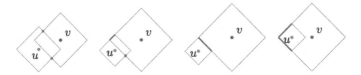

Fig. 2. The intersection (in red) of two circles under the Manhattan distance in $\mathbb{R}^2$ can be two points, one point and one line segment, one line segment, or two line segments. (Color figure online)

in $\mathbb{R}^d$ and $p \in \{1, 2, \infty\}$, we use the ℓ_p-norm of $\boldsymbol{u} - \boldsymbol{w}$, i.e., $\|\boldsymbol{u} - \boldsymbol{w}\|_p$, to denote the ℓ_p-distance of $\boldsymbol{u}$ and $\boldsymbol{w}$. By convention, we use *Manhattan*, *Euclidean*, and *Max* distances to refer to ℓ_1-, ℓ_2-, and ℓ_∞-distances, respectively.

For $d = 2$, the Manhattan distance of two points is equal to the length of a shortest path between them on a rectilinear grid. Hence, under Manhattan distances, a *circle* is a square rotated at a $45°$ angle from the coordinate axes. The intersection of two Manhattan-circles can range from two points to two segments as depicted in Fig. 2.

Basic Geometric Notation. Throughout this paper, we use lower case letters in boldface to denote points in a space. Given two points $\boldsymbol{q}$ and $\boldsymbol{r}$, we introduce the following notions: Let $\mathsf{BB}(\boldsymbol{q}, \boldsymbol{r})$ denote the set of points which are contained in the (smallest) rectilinear bounding box of points $\boldsymbol{q}$ and $\boldsymbol{r}$, i.e., $\mathsf{BB}(\boldsymbol{q}, \boldsymbol{r}) := \{\boldsymbol{x} \in \mathbb{R}^d \mid \min\{\boldsymbol{q}[i], \boldsymbol{r}[i]\} \leq \boldsymbol{x}[i] \leq \max\{\boldsymbol{q}[i], \boldsymbol{r}[i]\}$ for all $i \in [d]\}$. The *perpendicular bisector* (bisector in short) between two points $\boldsymbol{q}$ and $\boldsymbol{r}$ with respect to a norm ℓ_p is a set $\mathsf{H}_p(\boldsymbol{q}, \boldsymbol{r})$ of points which each have the same distance to both $\boldsymbol{q}$ and $\boldsymbol{r}$. Formally, $\mathsf{H}_p(\boldsymbol{q}, \boldsymbol{r}) := \{\boldsymbol{x} \in \mathbb{R}^d \mid \|\boldsymbol{x} - \boldsymbol{q}\|_p = \|\boldsymbol{x} - \boldsymbol{r}\|_p\}$. In a d-dimensional space, a bisector of two points under the Manhattan distance (i.e., ℓ_1-norm) can itself be a d-dimensional object, while a bisector under Euclidean distances is always $(d-1)$-dimensional; see e.g., Fig. 3 (right).

The Two-Dimensional Case. In a two-dimensional space, the vertical line and the horizontal line crossing any point divide the space into four non-disjoint quadrants: the north-east, south-east, north-west, and south-west quadrants. Given a point $\boldsymbol{q}$, we use $\mathsf{NE}(\boldsymbol{q})$, $\mathsf{SE}(\boldsymbol{q})$, $\mathsf{NW}(\boldsymbol{q})$, and $\mathsf{SW}(\boldsymbol{q})$ to denote these four quadrants. Formally, $\mathsf{NE}(\boldsymbol{q}) := \{\boldsymbol{z} \in \mathbb{R}^2 \mid \boldsymbol{z}[1] \geq \boldsymbol{q}[1] \wedge \boldsymbol{z}[2] \geq \boldsymbol{q}[2]\}$, $\mathsf{SE}(\boldsymbol{q}) := \{\boldsymbol{z} \in \mathbb{R}^2 \mid \boldsymbol{z}[1] \geq \boldsymbol{q}[1] \wedge \boldsymbol{z}[2] \leq \boldsymbol{q}[2]\}$, $\mathsf{NW}(\boldsymbol{q}) := \{\boldsymbol{z} \in \mathbb{R}^2 \mid \boldsymbol{z}[1] \leq \boldsymbol{q}[1] \wedge \boldsymbol{z}[2] \geq \boldsymbol{q}[2]\}$, and $\mathsf{SW}(\boldsymbol{q}) := \{\boldsymbol{z} \in \mathbb{R}^2 \mid \boldsymbol{z}[1] \leq \boldsymbol{q}[1] \wedge \boldsymbol{x}[2] \leq \boldsymbol{q}[2]\}$.

Embeddings. The Euclidean (resp. Manhattan) representation models the preferences of the voters over the alternatives using the Euclidean (resp. Manhattan) distance. A shorter distance indicates a stronger preference. For technical reason, we also introduce the ℓ-max preferences which are based on the ℓ_∞-distances.

Definition 1 (d-Manhattan, d-Euclidean, and d-Max embeddings). *Let* $\mathcal{P} := (\mathcal{A}, \mathcal{V} := \{v_1, \ldots, v_n\}, \mathcal{R} := (\succ_1, \ldots, \succ_n))$ *be a profile. Let* $E\colon \mathcal{A} \cup \mathcal{V} \to$

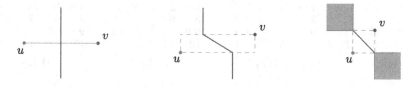

Fig. 3. The bisector (in green) between points u and v under the Manhattan distance. The green lines and areas extend to infinity. (Color figure online)

$\mathbb{R}^d$ be an embedding of the alternatives and the voters. For each $(\Lambda, p) \in \{(d\text{-}$ Manhattan, $1)$, $(d\text{-}Euclidean$, $2)$, $(d\text{-}Max$, $\infty)\}$, a voter $v_i \in V$ is called Λ with respect to embedding E if for each two alternatives $a, b \in \mathcal{A}$ it holds that

$$a \succ_i b \text{ if and only if } \|E(a) - E(v_i)\|_p < \|E(b) - E(v_i)\|_p.$$

Embedding E is called a d-Manhattan (resp. d-Euclidean, d-Max) embedding of profile $\mathcal{P}$ if each voter in V is d-Manhattan (resp. d-Euclidean, d-Max) wrt. E. A preference profile is d-Manhattan (resp. d-Euclidean, d-Max) if it admits a d-Manhattan (resp. d-Euclidean, d-Max) embedding.

To characterize necessary conditions for 2-Manhattan profiles, we need to define several notions which describe the relative orders of the points in each axis.

Definition 2 (BE- and EX-properties). *Let $\mathcal{P}$ be a profile containing at least 3 voters called u, v, w and let E be an embedding for $\mathcal{P}$. Then, E satisfies*

- *the (v, u, w)-BE-property if $E(v) \in \mathsf{BB}(E(u), E(w))$ (see the illustration in the first row of Fig. 4) and*
- *the (v, u, w)-EX-property if there exists (i, j) with $\{i, j\} = \{1, 2\}$ such that*

$$\min\{E(v)[i], E(w)[i]\} \le E(u)[i] \le \max\{E(v)[i], E(w)[i]\} \quad and$$
$$\min\{E(u)[j], E(v)[j]\} \le E(w)[j] \le \max\{E(u)[j], E(v)[j]\}$$

(see the illustrations in the last two rows of Fig. 4).

If E does not satisfy (v, u, w)-BE-property (-EX-property) we say it violates (v, u, w)-BE-property (resp. -EX-property). For brevity's sake, by symmetry, we omit voters u and w and just speak of the v-BE-property (resp. v-EX-property) if u, v, w are the only voters contained in $\mathcal{P}$.

Note that any embedding for three voters u, v, w must satisfy u-, v- or w-EX-property or u-, v- or w-BE-property, although it may satisfy more than one of these (consider for example three voters at the same point). However, each of these embeddings satisfying the (v, u, w)-BE-property (resp. (v, u, w)-EX-property) forbids certain types of preference profiles. The following two configurations describe preferences whose existence precludes an embedding from satisfying either BE-property or EX-property for some voters, as we will show in Lemmas 3 and 4.

```
1    2    3ʷ 4        1ᵘ 2    3    4        1    2    3ᵘ 4        1ʷ 2    3    4
5    6ᵛ 7    8        5    6ᵛ 7    8        5    6ᵛ 7    8        5    6ᵛ 7    8
9ᵘ 10   11   12       9    10   11ʷ12       9ʷ 10   11   12       9    10   11ᵘ12
13   14   15   16      13   14   15   16      13   14   15   16      13   14   15   16
      (BE)                  (BE)                  (BE)                  (BE)

1    2ʷ 3    4        1    2    3ᵛ 4        1    2ᵘ 3    4        1    2    3ᵛ 4
5ᵘ 6    7    8        5ᵘ 6    7    8        5ʷ 6    7    8        5ʷ 6    7    8
9    10   11ᵛ12       9    10ʷ11   12       9    10   11ᵛ12       9    10ᵘ11   12
13   14   15   16      13   14   15   16      13   14   15   16      13   14   15   16
      (EX)                  (EX)                  (EX)                  (EX)

1    2ʷ 3    4        1ᵛ 2    3    4        1    2ᵘ 3    4        1ᵛ 2    3    4
5    6    7ᵘ 8        5    6    7ᵘ 8        5    6    7ʷ 8        5    6    7ʷ 8
9ᵛ 10   11   12       9    10ʷ11   12       9ᵛ 10   11   12       9    10ᵘ11   12
13   14   15   16      13   14   15   16      13   14   15   16      13   14   15   16
      (EX)                  (EX)                  (EX)                  (EX)
```

Fig. 4. Two possible embeddings illustrating the properties in Definition 2 (the numbering will be used in some proofs). (BE) means "between" while (EX) "external".

Definition 3 (BE-configurations). *A profile $\mathcal{P}$ with 3 voters u, v, w and 3 alternatives a, b, x is a (v, u, w)-BE-configuration if the following holds:*

$$u\colon b \succ_u x \succ_u a, \qquad v\colon a \succ_v x \succ_v b, \qquad w\colon b \succ_w x \succ_w a.$$

Definition 4 (EX-configurations). *A profile $\mathcal{P}$ with 3 voters u, v, w and 6 alternatives x, a, b, c, d, e (c, d, e not necessarily distinct) is a (v, u, w)-EX-configuration if the following holds:*

$$\begin{aligned}
u\colon & \quad a \succ_u x \succ_u b, & c \succ_u x, & \quad d \succ_u x \\
v\colon & \quad \{a, b\} \succ_v x, & & \quad x \succ_v \{d, e\}, \\
w\colon & \quad b \succ_w x \succ_w a, & c \succ_w x, & \quad e \succ_w x.
\end{aligned}$$

Example 1. Consider two profiles $\mathcal{Q}_1$ and $\mathcal{Q}_2$ which satisfy the following:

$$\mathcal{Q}_1\colon v_1\colon 1 \succ_1 2 \succ_1 3, \qquad v_2\colon 3 \succ_2 2 \succ_2 1, \qquad v_3\colon 3 \succ_3 2 \succ_3 1$$
$$\mathcal{Q}_2\colon v_1\colon \{1, 2\} \succ_1 3 \succ_1 4, \qquad v_2\colon \{1, 4\} \succ_2 3 \succ_2 2, \qquad v_3\colon \{2, 4\} \succ_3 3 \succ_3 1.$$

Clearly, $\mathcal{Q}_1$ is a (v_1, v_2, v_3)-BE-configuration. Further, one can verify that $\mathcal{Q}_2$ contains a (v_1, v_2, v_3)-, (v_2, v_1, v_3)-, and (v_3, v_1, v_2)-EX-configuration, by setting $(a, b, x, c, d, e) := (1, 2, 3, 4, 4, 4)$, $(a, b, x, c, d, e) := (1, 4, 3, 2, 2, 2)$, and $(a, b, x, c, d, e) := (2, 4, 3, 1, 1, 1)$, respectively.

The next configuration is a restriction of the worst-diverse configuration. The latter is used to characterize the so-called single-peaked preferences [1].

Definition 5 (All-triples worst-diverse configuration). *A profile $\mathcal{P}$ is an all-triples worst-diverse configuration if for every triple of alternatives $\{x, y, z\} \subseteq \mathcal{A}$ there are three voters $u, v, w \in \mathcal{V}$ which form a worst-diverse configuration, i.e., their preferences satisfy $\{x, y\} \succ_u z$, $\{x, z\} \succ_v y$, and $\{y, z\} \succ_w x$.*

3   Manhattan Preferences: Positive Results

In this section, we show that for sufficiently high dimension d, i.e., $d \geq \min(n, m-1)$, a profile with n voters and m alternatives is always d-Manhattan. The same result holds for d-Euclidean profiles [2]. The idea of the proof for n voters is similar to the proof for d-Euclidean preferences in [2]. The proof for $m + 1$ voters is more different from d-Euclidean-case. Whilst the proof for d-Euclidean-case relies on abstract geometric properties, it is relatively straightforward to give a full concrete construction on the d-Manhattan case.

Theorem 1. *Every profile with n voters is n-Manhattan.*

Proof. Let $\mathcal{P} = (\mathcal{A}, \mathcal{V}, (\succ_i)_{i \in [n]})$ be a profile with m alternatives and n voters $\mathcal{V}$ such that $\mathcal{A} = \{1, \dots, m\}$. The idea is to first embed the voters from $\mathcal{V}$ onto n selected vertices of an n-dimensional hypercube, and then embed the alternatives such that each coordinate of an alternative reflects the preferences of a specific voter. More precisely, define an embedding $E \colon \mathcal{A} \cup \mathcal{V} \to \mathbb{Z}$ such that for each voter $v_i \in \mathcal{V}$ and each coordinate $z \in [n]$, we have $E(v_i)[z] := -m$ if $z = i$, and $E(v_i)[z] := 0$ otherwise.

It remains to specify the embedding of the alternatives. To ease notation, for each alternative $j \in \mathcal{A}$, let mk_j denote the maximum rank of the voters towards j, i.e., $\mathsf{mk}_j := \max_{v_i \in \mathcal{V}} \mathsf{rk}_i(j)$. Further, let $\hat{n}_j$ denote the index of the voter who has maximum rank over j; if there are two or more such voters, then we fix an arbitrary one. That is, $v_{\hat{n}_j} := \arg\max_{v_i \in \mathcal{V}} \mathsf{rk}_i(j)$. Then, the embedding of each alternative $j \in \mathcal{A}$ is defined as follows:

$$\forall z \in [n] \colon E(j)[z] := \begin{cases} \mathsf{rk}_z(j) - \mathsf{mk}_j, & \text{if } z \neq \hat{n}_j, \\ M + 2\mathsf{rk}_z(j) + \sum_{k \in [n]} (\mathsf{rk}_k(j) - \mathsf{mk}_j), & \text{otherwise.} \end{cases}$$

Herein, M is set to a large but fixed value such that the second term in the above definition is non-negative. For instance, we can set $M := n \cdot m$. Notice that by definition, the following holds for each alternative $j \in \mathcal{A}$.

$$-m \leq \mathsf{rk}_z(j) - \mathsf{mk}_j \leq 0, \quad \text{and} \tag{1}$$

$$M + 2\mathsf{rk}_z(j) + \sum_{k \in [n]} (\mathsf{rk}_k(j) - \mathsf{mk}_j) \geq M - n \cdot m \geq 0. \tag{2}$$

Thus, for each $i \in [n]$ and $j \in [m]$, it holds that

$$|E(j)[i] - E(v_i)[i]| \overset{(1),(2)}{=} E(j)[i] + m, \tag{3}$$

$$\|E(j)\|_1 = \sum_{z \in [n] \setminus \{\hat{n}\}} |\mathsf{rk}_z(j) - \mathsf{mk}_j| + |M + 2\mathsf{rk}_{\hat{n}}(j) + \sum_{k \in [n]} (\mathsf{rk}_k(j) - \mathsf{mk}_j)|$$

$$\overset{(1),(2)}{=} \sum_{z \in [n] \setminus \{\hat{n}_j\}} (-\mathrm{rk}_z[j] + \mathrm{mk}_j) + M + 2\mathrm{rk}_{\hat{n}_j}(j) + \sum_{k \in [n]} (\mathrm{rk}_k[j] - \mathrm{mk}_j)$$

$$\overset{\mathrm{mk}_j = \mathrm{rk}_{\hat{n}_j}(j)}{=} M + 2\mathrm{rk}_{\hat{n}}(j) \tag{4}$$

Now, in order to prove that this embedding is n-Manhattan we show that the Manhattan distance between an arbitrary voter v_i and an arbitrary alternative j is linear in the rank value $\mathrm{rk}_i(j)$. By definition, this distance is:

$$\|E(v_i) - E(j)\|_1 = \sum_{k \in [n]} |E(j)[k] - E(v_i)[k]| = |E(j)[i] - E(v_i)[i]| + \sum_{k \in [n] \setminus \{i\}} |E(j)[k]|$$

$$\overset{(3)}{=} E(j)[i] + m + \|E(j)\|_1 - |E(j)[i]|. \tag{5}$$

We distinguish between two cases.

Case 1: $i \neq \hat{n}_j$. Then, it follows that $\|E(v_i) - E(j)\|_1 \overset{(5)}{=} m + E(j)[i] + \|E(j)\|_1 - |E(j)[i]| \overset{(1),(4)}{=}$ $m + 2(\mathrm{rk}_i(j) - \mathrm{mk}_j) + M + 2\mathrm{rk}_{\hat{n}_j}(j) \overset{\mathrm{rk}_{\hat{n}}(j) = \mathrm{mk}_j}{=} m + M + 2\mathrm{rk}_i(j)$.

Case 2: $i = \hat{n}_j$. Then, by definition, it follows that
$\|E(v_i) - E(j)\|_1 \overset{(5)}{=} m + E(j)[i] + \|E(j)\|_1 - |E(j)[i]| \overset{(2)}{=} m + \|E(j)\|_1 \overset{(4)}{=} m + M + 2\mathrm{rk}_{\hat{n}_j}(j)$.
   In both cases, we obtain that $\|E(v_i) - E(j)\|_1 = m + M + 2\mathrm{rk}_i(j)$, which is linear in the ranks, as desired.   □

By Theorem 1, we obtain that any profile with two voters is 2-Manhattan. The following example provides an illustration.

Example 2. Consider profile $\mathcal{P}_1$ with 2 voters and 5 alternatives: $v_1: 1 \succ 2 \succ 3 \succ 4 \succ 5$ and $v_2: 5 \succ 4 \succ 3 \succ 1 \succ 2$. By the proof of Theorem 1, the maximum ranks and the voters with maximum rank, and the embedding of the voters and alternatives is as follows, where $M := n \cdot m = 10$; see Fig. 5a for an illustration.

j	1 2 3 4 5	$x \in \mathcal{V} \cup \mathcal{A}$	v_1	v_2	1	2	3	4	5
mk_j	3 4 2 3 4	$E(x)[1]$	-5	0	-3	-3	14	14	14
$\hat{n}_j$	2 2 1 1 1	$E(x)[2]$	0	-5	13	15	0	-2	-4

Theorem 2. *Every profile with $m + 1$ alternatives is m-Manhattan.*

Proof. Let $\mathcal{P} = (\mathcal{A}, \mathcal{V}, (\succ_i)_{i \in [n]})$ be a profile with $m + 1$ alternatives and n voters $\mathcal{V}$ such that $\mathcal{A} = \{1, \ldots, m + 1\}$. The idea is to first embed the alternatives from $\mathcal{A}$ onto $m + 1$ selected vertices of an m-dimensional hypercube, and then embed the voters such that the m-Manhattan distances from each voter to the alternatives increase as the preferences decrease. More precisely, define an embedding $E: \mathcal{A} \cup \mathcal{V} \to \mathbb{N}_0$ such that alternative $m + 1$ is embedded in the origin coordinate, i.e., $E(m + 1) = (0)_{z \in [m]}$. For each alternative $j \in [m]$ and each coordinate $z \in [m]$, we have $E(j)[z] := 2m$ if $z = j$, and $E(j)[z] := 0$ otherwise.

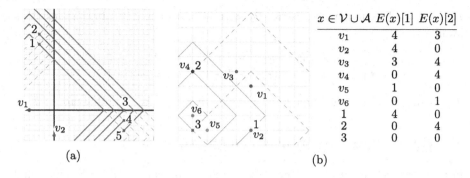

Fig. 5. (a): Illustration for Example 2. (b): Illustration for Example 3; the circles are with respect to v_2 and v_6, respectively.

Then, the embedding of each voter $v_i \in \mathcal{V}$ is defined as follows: $\forall j \in [m]$:

$$
E(v_i)[j] := \begin{cases} 2m - \mathsf{rk}_i(j), & \text{if } \mathsf{rk}_i(j) < \mathsf{rk}_i(m+1), \\ m - \mathsf{rk}_i(j), & \text{if } \mathsf{rk}_i(j) > \mathsf{rk}_i(m+1). \end{cases}
$$

Observe that $0 \le E(v_i)[j] \le 2m$. Before we show that E is 2-Manhattan for $\mathcal{P}$, let us establish a simple formula for the distance between a voter and an alternative.

Claim 1 ($\star$). *For each voter $v_i \in \mathcal{V}$ and each alternative $j \in \mathcal{A}$, we have*

$$
\|E(v_i) - E(j)\|_1 = \begin{cases} \|E(v_i)\|_1 + 2(m - E(v_i)[j]), & \text{if } j \neq m+1, \\ \|E(v_i)\|_1, & \text{otherwise.} \end{cases}
$$

Now, we proceed with the proof. Consider an arbitrary voter $v_i \in \mathcal{V}$ and let $j, k \in [m+1]$ be two consecutive alternatives in the preference order $\succ_i$ such that $\mathsf{rk}_i(j) = \mathsf{rk}_i(k) - 1$. We distinguish between three cases.

Case 1: $\mathsf{rk}_i(k) < \mathsf{rk}_i(m+1)$ or $\mathsf{rk}_i(j) > \mathsf{rk}_i(m+1)$. Then, by Claim 1 and by definition, it follows that $\|E(v_i) - E(j)\|_1 - \|E(v_i) - E(k)\|_1 = 2(E(v_i)[k] - E(v_i)[j]) = 2(\mathsf{rk}_i(j) - \mathsf{rk}_i(k)) < 0$.

Case 2: $\mathsf{rk}_i(k) = \mathsf{rk}_i(m+1)$, i.e., $k = m+1$ and $E(v_i)[j] = 2m - \mathsf{rk}_i(j)$. Then, by Claim 1 and by definition, it follows that $\|E(v_i) - E(j)\|_1 - \|E(v_i) - E(k)\|_1 = 2(m - E(v_i)[j]) = 2\mathsf{rk}_i(j) - 2m < 0$. Note that the last inequality holds since $\mathsf{rk}_i(j) = \mathsf{rk}_i(k) - 1 < m$.

Case 3: $\mathsf{rk}_i(j) = \mathsf{rk}_i(m+1)$, i.e., $j = m+1$ and $E(v_i)[k] = m - \mathsf{rk}_i(k)$. Then, by Claim 1 and by definition, it follows that $\|E(v_i) - E(j)\|_1 - \|E(v_i) - E(k)\|_1 = -2(m - E(v_i)[k]) = -2\mathsf{rk}_i(k) < 0$. Note that the last inequality holds since $\mathsf{rk}_i(k) = \mathsf{rk}_i(j) + 1 > 0$.

Since in all cases, we show that $\|E(v_i) - E(j)\|_1 - \|E(v_i) - E(k)\|_1 < 0$, E is indeed m-Manhattan for $\mathcal{P}$. $\square$

Theorem 2 implies that any profile for 3 alternatives is 2-Manhattan. The following example illustrates the corresponding Manhattan embedding.

Example 3. The following profile $\mathcal{P}_2$ with 6 voters and 3 alternatives is 2-Manhattan.

$$v_1: 1 \succ 2 \succ 3, \quad v_2: 1 \succ 3 \succ 2, \quad v_3: 2 \succ 1 \succ 3,$$
$$v_4: 2 \succ 3 \succ 1, \quad v_5: 3 \succ 1 \succ 2, \quad v_6: 3 \succ 2 \succ 1.$$

One can check that the embedding E given in Fig. 5b is 2-Manhattan for $\mathcal{P}_3$.

4   Manhattan Preferences: Negative Results

In this section, we consider minimally non 2-Manhattan profiles. We show that for $n \in \{3, 4, 5\}$ voters, the smallest non 2-Manhattan profile has $9 - n$ alternatives (Theorems 3 to 5). Before we show this, we first go through some technical but useful statements for 2-Manhattan preference profiles in Sect. 4.1. Then, we show the proofs of the main results in Sects. 4.2 to 4.4. For brevity's sake, given an embedding E and a voter $v \in \mathcal{V}$ (resp. an alternative $a \in \mathcal{A}$), we use boldface $\boldsymbol{v}$ (resp. $\boldsymbol{a}$) to denote the embedding $E(v)$ (resp. $E(a)$).

4.1   Technical Results

Lemma 1. *Let $\mathcal{P}$ be a 2-Manhattan profile and let E be a 2-Manhattan embedding for $\mathcal{P}$. For any two voters r, s and two alternatives x, y the following holds: (i) If $r, s: y \succ x$, then $\boldsymbol{x} \notin \mathsf{BB}(\boldsymbol{r}, \boldsymbol{s})$. (ii) If $r: x \succ y$ and $s: y \succ x$, then $\boldsymbol{s} \notin \mathsf{BB}(\boldsymbol{r}, \boldsymbol{x})$.*

Proof. Let $\mathcal{P}$, E, r, s, and x, y be as defined. Both statements follow from using simple calculations and the triangle inequality of Manhattan distances.

For Statement (i), suppose, towards a contradiction, that $r, s: y \succ x$ and $\boldsymbol{x} \in \mathsf{BB}(\boldsymbol{r}, \boldsymbol{s})$. By the definition of Manhattan distances, this implies that $\|\boldsymbol{r} - \boldsymbol{x}\|_1 + \|\boldsymbol{x} - \boldsymbol{s}\|_1 = \|\boldsymbol{r} - \boldsymbol{s}\|_1$. By the preferences of voters r and s we infer that $\|\boldsymbol{s} - \boldsymbol{y}\|_1 + \|\boldsymbol{r} - \boldsymbol{y}\|_1 < \|\boldsymbol{s} - \boldsymbol{x}\|_1 + \|\boldsymbol{r} - \boldsymbol{x}\|_1 = \|\boldsymbol{r} - \boldsymbol{s}\|_1$, a contradiction to the triangle inequality of $\|\cdot\|_1$.

For Statement (ii), suppose, towards a contradiction, that $r: x \succ y$ and $s: y \succ x$ and $\boldsymbol{s} \in \mathsf{BB}(\boldsymbol{r}, \boldsymbol{x})$. By the definition of Manhattan distances, this implies that $\|\boldsymbol{r} - \boldsymbol{x}\|_1 = \|\boldsymbol{r} - \boldsymbol{s}\|_1 + \|\boldsymbol{s} - \boldsymbol{x}\|_1$. By the preferences of voters r and s we infer that $\|\boldsymbol{r} - \boldsymbol{s}\|_1 + \|\boldsymbol{s} - \boldsymbol{y}\|_1 < \|\boldsymbol{r} - \boldsymbol{s}\|_1 + \|\boldsymbol{s} - \boldsymbol{x}\|_1 = \|\boldsymbol{r} - \boldsymbol{x}\|_1 < \|\boldsymbol{r} - \boldsymbol{y}\|_1$, a contradiction to the triangle inequality of $\|\cdot\|_1$.   □

The following is a summary of coordinate differences wrt. the preferences.

Observation 1 ($\star$)**.** *Let profile $\mathcal{P}$ admit a 2-Manhattan embedding E. For each voter s and each two alternatives x, y with $s: x \succ y$, the following holds:*

(i) *If $\boldsymbol{y} \in \mathsf{NE}(\boldsymbol{s})$, then $\boldsymbol{y}[1] + \boldsymbol{y}[2] > \boldsymbol{x}[1] + \boldsymbol{x}[2]$.*
(ii) *If $\boldsymbol{y} \in \mathsf{NW}(\boldsymbol{s})$, then $-\boldsymbol{y}[1] + \boldsymbol{y}[2] > -\boldsymbol{x}[1] + \boldsymbol{x}[2]$.*
(iii) *If $\boldsymbol{y} \in \mathsf{SE}(\boldsymbol{s})$, then $\boldsymbol{y}[1] - \boldsymbol{y}[2] > \boldsymbol{x}[1] - \boldsymbol{x}[2]$.*
(iv) *If $\boldsymbol{y} \in \mathsf{SW}(\boldsymbol{s})$, then $-\boldsymbol{y}[1] - \boldsymbol{y}[2] > -\boldsymbol{x}[1] - \boldsymbol{x}[2]$.*

The next technical lemma excludes two alternatives to be put in the same quadrant region of some voters.

Lemma 2. *Let profile $\mathcal{P}$ admit a 2-Manhattan embedding E. Let r, s, t and x, y be 3 voters and 2 alternatives in $\mathcal{P}$, respectively. The following holds.*

(i) *If $r\colon x \succ y$ and $s\colon y \succ x$, then for each $\Pi \in \{\mathsf{NE}, \mathsf{NW}, \mathsf{SE}, \mathsf{SW}\}$ it holds that if $\boldsymbol{x} \in \Pi(\boldsymbol{s})$, then $\boldsymbol{y} \notin \Pi(\boldsymbol{r})$.*

(ii) *If $r, t\colon x \succ y$, $s\colon y \succ x$, $\boldsymbol{r} \in \mathsf{SW}(\boldsymbol{s})$, and $\boldsymbol{t} \in \mathsf{NE}(\boldsymbol{s})$, then for each $\Pi \in \{\mathsf{NW}, \mathsf{SE}\}$ it holds that if $\boldsymbol{x} \in \Pi(\boldsymbol{s})$, then $\boldsymbol{y} \notin \Pi(\boldsymbol{s})$.*

Proof. Let $\mathcal{P}, E, r, s, t, x, y$ be as defined. The first statement follows directly from applying Observation 1. Hence, we only consider the case with $\Pi = \mathsf{NW}$. For a contradiction, suppose that $\boldsymbol{x} \in \mathsf{NW}(\boldsymbol{s})$ and $\boldsymbol{y} \in \mathsf{NW}(\boldsymbol{r})$. Since $r\colon x \succ y$ and $s\colon y \succ x$, by Observation 1(ii), we have $\boldsymbol{y}[2] - \boldsymbol{y}[1] > \boldsymbol{x}[2] - \boldsymbol{x}[1] > \boldsymbol{y}[2] - \boldsymbol{y}[1]$, a contradiction.

Statement (ii): We only show the case with $\Pi = \mathsf{NW}$ as the other case is symmetric. For a contradiction, suppose that $\boldsymbol{x}, \boldsymbol{y} \in \mathsf{NW}(\boldsymbol{s})$. Since $r, t\colon x \succ y$, $s\colon y \succ x$, $\boldsymbol{x} \in \mathsf{NW}(\boldsymbol{s})$, by the first statement, we have $\boldsymbol{y} \notin \mathsf{NW}(\boldsymbol{r}) \cup \mathsf{NW}(\boldsymbol{t})$. However, since $\boldsymbol{y} \in \mathsf{NW}(\boldsymbol{s})$, it follows that $\boldsymbol{y} \in \mathsf{BB}(\boldsymbol{r}, \boldsymbol{t})$, a contradiction to Lemma 1(i). □

The next two lemmas specify the relation between a BE-configuration and the BE-property, and between a EX-configuration and the EX-property, respectively.

Lemma 3 ($\star$). *If a profile contains a (v, u, w)-BE-configuration, then no 2-Manhattan embedding satisfies the (v, u, w)-BE-property.*

Lemma 4 ($\star$). *If a profile contains a (v, u, w)-EX-configuration, then no 2-Manhattan embedding satisfies the (v, u, w)-EX-property.*

4.2 The Case with 3 Voters and 6 Alternatives

Using Lemmas 3 and 4, we prove Theorem 3 with the help of Example 4.

Example 4. The following profile $\mathcal{P}_3$ with 3 voters and 6 alternatives is not 2-Manhattan.
$$v_1 : 1 \succ 2 \succ 3 \succ 4 \succ 5 \succ 6, \quad v_2 : 1 \succ 4 \succ 6 \succ 3 \succ 5 \succ 2, \quad v_3 : 6 \succ 5 \succ 2 \succ 3 \succ 1 \succ 4.$$

Theorem 3. *There exists a non 2-Manhattan profile with 3 voters and 6 alternatives.*

Proof. Consider profile $\mathcal{P}_3$ given in Example 4. Suppose, towards a contradiction, that E is a 2-Manhattan embedding for $\mathcal{P}_3$. Since each embedding for 3 voters must satisfy one of the two properties in Definition 2, we distinguish between two cases: there exists a voter who is embedded inside the bounding box of the other two, or there is no such voter.

Case 1: There exists a voter v_i, $i \in [3]$, such that E satisfies the v_i-BE-property. Since $\mathcal{P}$ contains a (v_1, v_2, v_3)-BE-configuration wrt. $(a, b, x) = (2, 6, 5)$, by Lemma 3 it follows that E violates the v_1-BE-property. Analogously, since $\mathcal{P}$ contains a (v_2, v_1, v_3)-BE-configuration regarding $a = 4, b = 2, x = 3$, and (v_3, v_1, v_2)-BE-configuration with $a = 5, b = 1, x = 3$, neither does E satisfy the v_2-BE-property or the v_3-BE-property.

Case 2: There exists a voter v_i, $i \in [3]$, such that E satisfies the v_i-EX-property. Now, consider the subprofile $\mathcal{P}'$ restricted to the alternatives $1, 2, 3, 6$. We claim that this subprofile contains an EX-configuration, which by Lemma 4 precludes the existence of such a voter v_i with the v_i-EX-property: First, since $\mathcal{P}'$ contains a (v_3, v_1, v_2)-EX-configuration (setting $(u, v, w) := (v_1, v_3, v_2)$ and $(x, a, b, c, d, e) = (3, 2, 6, 1, 1, 1)$), by Lemma 4, it follows that E violates the v_3-EX-property. In fact, $\mathcal{P}'$ also contains a v_2-EX-configuration (setting $(u, v, w) := (v_1, v_2, v_3)$ and $(x, a, b, c, d, e) = (3, 1, 6, 2, 2, 2)$) and a v_1-EX-configuration (setting $(u, v, w) := (v_2, v_1, v_3)$ and $(x, a, b, c, d, e) = (3, 1, 2, 6, 6, 6)$). By Lemma 4, it follows that E violates the v_2-EX-property and the v_1-EX-property.

Summarizing, we obtain a contradiction for E.   □

4.3   The Case with 4 Voters and 5 Alternatives

We prove Theorem 4 for 2-Max instead of for 2-Manhattan preferences since the reasoning for 2-Max is simpler and more intuitive. It is, however, possible to follow similar steps for 2-Manhattan preferences and obtain an analogous proof. The following proposition allows us to extend any result we obtain of the (non-) existence of 2-Manhattan embeddings to 2-Max embeddings and vice versa. The same claim has been made by Escoffier et al. [16, Proposition 2].

Proposition 1 [20]. *There is a natural isometry between $\mathbb{R}^2$ under ℓ_1-norm and $\mathbb{R}^2$ under ℓ_∞-norm.*

We first prove that any profile with at least 5 alternatives which contains an all-triples worst-diverse configuration is not 2-Max. Then we proceed to show that the example below with 4 voters and 5 alternatives is such a profile.

Example 5. The following profile $\mathcal{P}_4$ with 5 alternatives contains an all-triples worst-diverse configuration and will be shown to be not 2-Max.

$$v_1: 1 \succ 2 \succ 3 \succ 4 \succ 5, \qquad v_2: 1 \succ 2 \succ 3 \succ 5 \succ 4,$$
$$v_3: 1 \succ 4 \succ 5 \succ 3 \succ 2, \qquad v_4: 2 \succ 4 \succ 5 \succ 3 \succ 1.$$

To do this, we utilize the following two lemmas:

Lemma 5 (⋆). *Let $\mathcal{P}$ admit a d-Max embedding E. If $z \in \mathsf{BB}(x, y)$, then there is no voter v satisfying $\{x, y\} \succ_v z$.*

Lemma 6 (⋆). *For any set $\mathcal{S}$ of 5 points in $\mathbb{R}^2$, there must exist three distinct points $x, y, z \in \mathcal{S}$ such that $z \in \mathsf{BB}(x, y)$.*

Now, we are ready to show our second main result.

Theorem 4. *There exists a non 2-Manhattan profile with 4 voters and 5 alternatives.*

Proof. Suppose, towards a contradiction, that we have a profile $\mathcal{P}$ with at least 5 alternatives $\{a, b, c, d, e\}$ which contains an all-triples worst-diverse configuration and is 2-Max with a 2-Max embedding E.

As we have 5 alternatives, by Lemma 6 there must be a triple $\{x, y, z\} \subset \{a, b, c, d, e\}$ such that $z \in \mathsf{BB}(x, y)$. This together with Lemma 5 implies that no voter v can satisfy $\{x, y\} \succ_v z$. However, this is a contradiction to our assumption that $\mathcal{P}$ contains an all-triples worst-diverse configuration. Therefore we cannot have a profile $\mathcal{P}$ with at least 5 alternatives which contains an all-triples worst-diverse configuration and has a 2-Max embedding E.

One can verify that profile $\mathcal{P}_4$ given in Example 5 with 5 alternatives and 4 voters contains an all-triples worst-diverse configuration, and is not 2-Max: The alternatives $1, 2, 4$, and 5 are ranked last by voters v_4, v_3, v_2, and v_1, respectively. Therefore we can pick the corresponding voters for every triple involving only the alternatives $1, 2, 4$ and 5. It is straightforward to verify that there is a worst-diverse configuration for every triple of alternatives involving 3 as well. Thus we have shown that there is a profile with 4 voters and 5 alternatives that is not 2-Max. By Proposition 1 it is also not 2-Manhattan. □

4.4 The Case with 5 Voters and 4 Alternatives

The proof of Theorem 5 will be based on the following example.

Example 6. Any profile $\mathcal{P}_5$ satisfying the following is not 2-Manhattan.

$$v_1: 1 \succ 2 \succ 3 \succ 4, \quad v_2: 1 \succ 4 \succ 3 \succ 2, \quad v_3: \{2, 4\} \succ 3 \succ 1,$$
$$v_4: 3 \succ 2 \succ 1 \succ 4, \quad v_5: 3 \succ 4 \succ 1 \succ 2.$$

Before we proceed with the proof, we show a technical but useful lemma.

Lemma 7 ($\star$)**.** *Let $\mathcal{P}$ be a profile with 4 voters u, v, w, r and 4 alternatives a, b, c, d satisfying the following:*

$$u: \{a, b\} \succ c \succ d, \ v: \{b, d\} \succ c \succ a, \ w: \{a, d\} \succ c \succ b, \ r: c \succ \{a, b\} \succ d.$$

If E is a 2-Manhattan embedding for $\mathcal{P}$ with $v \in \mathsf{BB}(u, w)$, then $v \in \mathsf{BB}(r, w)$.

Theorem 5 ($\star$)**.** *There exists a non 2-Manhattan profile with 5 voters and 4 alternatives.*

Proof sketch. We show that profile $\mathcal{P}_5$ given in Example 6 is not 2-Manhattan. Suppose, towards a contradiction, that $\mathcal{P}_5$ admits a 2-Manhattan embedding E. First, we observe that one of v_1, v_2, v_3 is embedded inside the bounding box defined by the other two since the subprofile of $\mathcal{P}_5$ restricted to voters v_1, v_2 and

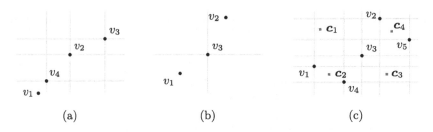

Fig. 6. Illustrations of possible embeddings for the proof of Theorem 5: Left: $v_2 \in$ BB(v_4, v_3). (6a): $v_2 \in$ BB(v_1, v_3). (6b): $v_3 \in$ BB$(v_1, v_2) \cup$ BB$(v_4, v_2) \cup$ BB(v_5, v_1). (6c): $v_1, v_4, c_2 \in$ SW(v_3), $v_2, v_5, c_4 \in$ NE(v_3) such that $c_2 \in$ SE(v_1), $c_2 \in$ NW(v_4), $c_4 \in$ SE(v_2), $c_4 \in$ NW(v_5).

v_3 is equivalent to profile $\mathcal{Q}_2$ which, by Lemma 4, violates the EX-property (for each of v_1, v_2, and v_3, respectively). Hence, we distinguish between two cases.

Case 1: $v_2 \in$ BB(v_1, v_3) or $v_1 \in$ BB(v_2, v_3). Note that these two subcases are equivalent in the sense that if we exchange the roles of alternatives 2 and 4, i.e., $1 \mapsto 1$, $3 \mapsto 3$, $2 \mapsto 4$, and $4 \mapsto 2$, we obtain an equivalent (in terms of the Manhattan property) profile where the roles of voters v_1 and v_2 (resp. v_4 and v_5) are exchanged. Hence, it suffices to consider the case of $v_2 \in$ BB(v_1, v_3). W.l.o.g., assume that $v_1[1] \leq v_2[1] \leq v_3[1]$ and $v_1[2] \leq v_2[2] \leq v_3[2]$ (see Fig. 6a). Then, by Lemma 7 (setting $(u, v, w, r) := (v_1, v_2, v_3, v_4)$), we obtain that $v_2 \in$ BB(v_4, v_3). This implies that $v_4[1] \leq v_2[1]$ and $v_4[2] \leq v_2[2]$.

By the preferences of v_4, v_2, v_3 regarding alternatives 2 and 1, and by Lemma 1, it follows that $c_2 \in$ NW$(v_2) \cup$ SE(v_2) and $c_1 \notin$ BB$(v_3, v_4) \cup$ NE$(v_3) \cup$ SW(v_4). Similarly, regarding the preferences over 3 and 1, it follows that $c_3 \in$ NW$(v_2) \cup$ SE(v_2). By Lemma 2(ii) (considering the preferences of v_1, v_2 and v_3 regarding alternatives 2 and 3), we further infer that either $c_2 \in$ NW(v_2) and $c_3 \in$ SE(v_2) or $c_2 \in$ SE(v_2) and $c_3 \in$ NW(v_2). Without loss of generality, assume that $c_2 \in$ NW(v_2) and $c_3 \in$ SE(v_2).

By the preferences of v_3 and v_2 (resp. v_4 and v_2) regarding 1 and 3 and by Lemma 2(i), it follows that $c_1 \notin$ SE(v_3) (resp. $c_1 \notin$ SE(v_4)). By prior reasoning, we have that $c_1 \in$ NW$(v_3) \cup$ NW(v_4). However, this is a contradiction due to the preferences of v_4 and v_2 (resp. v_3 and v_2) regarding 1 and 2: By Lemma 2(ii), it follows that $c_1 \notin$ NW$(v_3) \cup$ NW(v_4).

Case 2: $v_3 \in$ BB(v_1, v_2). Without loss of generality, assume that $v_1[1] \leq v_3[1] \leq v_2[1]$ and $v_1[2] \leq v_3[2] \leq v_2[2]$; see Fig. 6b. Then, by Lemma 7 (setting $(u, v, w, r) := (v_1, v_3, v_2, v_4)$ and $(u, v, w, r) := (v_2, v_3, v_1, v_5)$, respectively), we obtain that $v_3 \in$ BB(v_4, v_2) and $v_3 \in$ BB(v_5, v_1). This implies that

$$v_4[1] \leq v_3[1] \text{ and } v_4[2] \leq v_3[2], \text{ and } v_5[1] \geq v_3[1] \text{ and } v_5[2] \geq v_3[2]. \quad (6)$$

By applying Lemmas 1 and 2 repeatedly, we will infer that alternatives $1, 2, 3, 4$ are embedded to the northwest, southwest, southeast, and northeast of v_3, respectively. Moreover, the embeddings of voters v_1, v_2, v_3, and v_4 are as specified in Fig. 6c. Now, since v_2 and v_4 (which are on the opposite "diagonal" of v_3) both have $1 \succ 4$ and $3 \succ 2$, while v_3: $4 \succ_3 1$ and $2 \succ_3 3$, the bisector between alternatives 1 and 4 and the one between alternatives 2 and 3 *must* "cross" twice. Similarly, due to v_1 and v_5, and v_3, the bisector between alternatives 1 and 2 and the one between alternatives 3 and 4 "cross" twice. This is, however, impossible. The details of the remaining proof can be found in the full version [7].

5 2-Manhattan Embeddings

In this section, we identified the following positive result through exhaustive embedding all the possible preference profiles in the two-dimensional space.

Proposition 2. *If $(n, m) = (3, 5)$ or $(n, m) = (4, 4)$, then each profile with at most n voters and at most m alternatives is 2-Manhattan.*

Proof. Since the Manhattan property is monotone, to show the statement, we only need to look at profiles which have either 3 voters and 5 alternatives, or 4 voters and 4 alternatives. We achieve this by using a computer program employing the CPLEX solver that exhaustively searches for all possible profiles with either 3 voters and 5 alternatives, or 4 voters and 4 alternatives, and provide a 2-Manhattan embedding for each of them. Since the CPLEX solver accepts constraints on the absolute value of the difference between any two variables, our computer program is a simple one-to-one translation of the d-Manhattan constraints given in Definition 1, without any integer variables. Peters [21] has noted this formulation for d-Manhattan embeddings. The same program can also be used to show that the preference profiles from the Examples 4, 5 and 6 do not admit a 2-Manhattan embedding.

Following a similar line as in the work of Chen and Grottke [6], we did some optimization to significantly shrink the search space on all profiles: We only consider profiles with distinct preference orders and we assume that one of the preference orders is $1 \succ \ldots \succ m$. Hence, the number of relevant profiles with n voters and m alternatives is $\binom{m!-1}{n-1}$. For $(n, m) = (3, 5)$ and $(n, m) = (4, 4)$, we need to iterate through 7021 and 1771 profiles, respectively. We implemented a program which, for each of these produced profiles, uses the IBM ILOG CPLEX optimization software package to check and find a 2-Manhattan embedding. The verification is done by going through each voter's preference order and checking the condition given in Definition 1. All generated profiles, together with their 2-Manhattan embeddings and the distances used for the verification, are available at https://owncloud.tuwien.ac.at/index.php/s/s6t1vymDOx4EfU9.   □

6 Conclusion

Motivated by the questions of how restricted d-Manhattan preferences are, we initiated the study of the smallest dimension sufficient for a profile to be d-Manhattan. We provided algorithms for larger dimension d and forbidden sub-profiles for $d = 2$.

This work opens up several future research directions. One future research direction concerns the characterization of d-Manhattan profiles through forbidden subprofiles. Such work has been done for other restricted preference domains such as single-peakedness [1], single-crossingness [4], and 1-Euclideanness [8]. Another research direction is to establish the computational complexity of determining whether a given profile is d-Manhattan. To this end, let us mention that 1-Euclidean profiles cannot be characterized via finitely many finite forbidden subprofiles [8], but they can be recognized in polynomial time [10,14,18]. As for $d \geq 2$, recognizing d-Euclidean profiles becomes notoriously hard (beyond NP) [21]. This stands in stark contrast to recognizing d-Manhattan preferences, which is in NP. For showing NP-hardness, our forbidden subprofiles may be useful for constructing suitable gadgets. Finally, it would be interesting to see whether assuming d-Manhattan preferences can lower the complexity of some computationally hard social choice problems.

References

1. Ballester, M.Á., Haeringer, G.: A characterization of the single-peaked domain. Soc. Choice Welfare **36**(2), 305–322 (2011)
2. Bogomolnaia, A., Laslier, J.-F.: Euclidean preferences. J. Math. Econ. **43**(2), 87–98 (2007)
3. Borg, I., Groenen, P.J.F., Mair, P.: Applied Multidimensional Scaling and Unfolding. Springer, Cham (2018). https://doi.org/10.1007/978-3-319-73471-2
4. Bredereck, R., Chen, J., Woeginger, G.J.: A characterization of the single-crossing domain. Soc. Choice Welfare **41**(4), 989–998 (2013)
5. Bulteau, L., Chen, J.: 2-dimensional Euclidean preferences. Technical report. arXiv:2205.14687 (2022)
6. Chen, J., Grottke, S.: Small one-dimensional Euclidean preference profiles. Soc. Choice Welfare **57**(1), 117–144 (2021). https://doi.org/10.1007/s00355-020-01301-y
7. Chen, J., Nöllenburg, M., Simola, S., Villedieu, A., Wallinger, M.: Multidimensional Manhattan preferences. Technical report. arXiv:2201.09691 (2022)
8. Chen, J., Pruhs, K., Woeginger, G.J.: The one-dimensional Euclidean domain: finitely many obstructions are not enough. Soc. Choice Welfare **48**(2), 409–432 (2017)
9. Coombs, C.H.: A Theory of Data. Wiley, Hoboken (1964)
10. Doignon, J.-P., Falmagne, J.-C.: A polynomial time algorithm for unidimensional unfolding representations. J. Algorithms **16**(2), 218–233 (1994)
11. Downs, A.: An Economic Theory of Democracy. Harper and Row, New York (1957)
12. Eckert, D., Klamler, C.: An equity-efficiency trade-off in a geometric approach to committee selection. Eur. J. Polit. Econ. **26**(3), 386–391 (2010)
13. Eguia, J.X.: Foundations of spatial preferences. J. Math. Econ. **47**(2), 200–205 (2011)
14. Elkind, E., Faliszewski, P.: Recognizing 1-Euclidean preferences: an alternative approach. In: Lavi, R. (ed.) SAGT 2014. LNCS, vol. 8768, pp. 146–157. Springer, Heidelberg (2014). https://doi.org/10.1007/978-3-662-44803-8_13
15. Enelow, J.M., Hinich, M.J.: Advances in the Spatial Theory of Voting. Cambridge University Press, Cambridge (2008)

16. Escoffier, B., Spanjaard, O., Tydrichová, M.: Euclidean preferences in the plane under ℓ_1, ℓ_2 and ℓ_∞ norms. Technical report. arXiv:2202.03185 (2021)
17. Hotelling, H.: Stability in competition. Econ. J. **39**(153), 41–57 (1929)
18. Knoblauch, V.: Recognizing one-dimensional Euclidean preference profiles. J. Math. Econ. **46**(1), 1–5 (2010)
19. Larson, R.C., Sadiq, G.: Facility locations with the Manhattan metric in the presence of barriers to travel. Oper. Res. **31**(4), 652–669 (1983)
20. Lee, D.-T., Wong, C.K.: Voronoi diagrams in $l_1(l_\infty)$ metrics with 2-dimensional storage applications. SIAM J. Comput. **9**(1), 200–211 (1980)
21. Peters, D.: Recognising multidimensional Euclidean preferences. In: Proceedings of the 31st AAAI Conference on Artificial Intelligence (AAAI 2017), pp. 642–648 (2017)
22. Poole, K.T.: Spatial Models of Parliamentary Voting. Cambridge University Press, Cambridge (1989)
23. Schaefer, M.: Complexity of some geometric and topological problems. In: Eppstein, D., Gansner, E.R. (eds.) GD 2009. LNCS, vol. 5849, pp. 334–344. Springer, Heidelberg (2010). https://doi.org/10.1007/978-3-642-11805-0_32
24. Shiha, H.-S., Shyur, H.-J., Lee, S.: An extension of TOPSIS for group decision making. Math. Comput. Model. **45**(7–8), 801–813 (2007)
25. Stokes, D.E.: Spatial models of party competition. Am. Polit. Sci. Rev. **57**(2), 368–377 (1963)
26. Sui, X., Boutilier, C.: Optimal group manipulation in facility location problems. In: Walsh, T. (ed.) ADT 2015. LNCS (LNAI), vol. 9346, pp. 505–520. Springer, Cham (2015). https://doi.org/10.1007/978-3-319-23114-3_30
27. The Finnish Ylen Vaalikone Website. https://vaalikone.yle.fi
28. The German Wahl-O-Mat Website. https://www.bpb.de/themen/wahl-o-mat

Theoretical Machine Learning

Exact Learning of Multitrees
and Almost-Trees Using Path Queries

Ramtin Afshar$^{(\boxtimes)}$ ⓘ and Michael T. Goodrich ⓘ

Department of Computer Science, University of California, Irvine, USA
{afsharr,goodrich}@uci.edu

Abstract. Given a directed graph, $G = (V, E)$, a **path** query, $\mathsf{path}(u, v)$, returns whether there is a directed path from u to v in G, for $u, v \in V$. Given only V, exactly learning all the edges in G using path queries is often impossible, since path queries cannot detect transitive edges. In this paper, we study the query complexity of exact learning for cases when learning G is possible using path queries. In particular, we provide efficient learning algorithms, as well as lower bounds, for multitrees and almost-trees, including butterfly networks.

Keywords: Graph reconstruction · Exact learning · Directed acyclic graphs

1 Introduction

The exact learning of a graph, which is also known as **graph reconstruction**, is the process of learning how a graph is connected using a set of queries, each involving a subset of vertices of the graph, to an all-knowing oracle. In this paper, we focus on learning a directed acyclic graph (DAG) using path queries. In particular, for a DAG, $G = (V, E)$, we are given the vertex set, V, but the edge set, E, is unknown and learning it through a set of path queries is our goal. A **path** query, $\mathsf{path}(u, v)$, takes two vertices, u and v in V, and returns whether there is a directed path from u to v in G.

This work is motivated by applications in various disciplines of science, such as biology [34,37,47,48], computer science [11,13,18–20,22,31,39], economics [26,27], psychology [38], and sociology [24]. For instance, it can be useful for learning phylogenetic networks from path queries. Phylogenetic networks capture ancestry relationships between a group of objects of the same type. For example, in a digital phylogenetic network, an object may be a multimedia file (a video or an image) [13,18–20], a text document [35,44], or a computer virus [22,39]. In such a network, each node represents an object, and directed edges show how an object has been manipulated or edited from other objects [5]. In a digital phylogenetic network, objects are usually archived and we can issue path queries between a pair of objects (see, e.g., [18]).

The full version of this paper is available in [3].

© Springer Nature Switzerland AG 2022
A. Castañeda and F. Rodríguez-Henríquez (Eds.): LATIN 2022, LNCS 13568, pp. 293–311, 2022.
https://doi.org/10.1007/978-3-031-20624-5_18

Learning a phylogenetic network has several applications. For instance, learning a multimedia phylogeny can be helpful in different areas such as security, forensics, and copyright enforcement [18]. Afshar *et al.* [5] studied learning phylogenetic trees (rooted trees) using path queries, where each object is the result of a modification of a single parent. Our work extends this scenario to applications where objects can be formed by merging two or more objects into one, such as image components. In addition, our work also has applications in biological scenarios that involve hybridization processes in phylogenetic networks [10].

Another application of our work is to learn the directed acyclic graph (DAG) structure of a causal Bayesian network (CBN). It is well-known that observational data (collected from an undisturbed system) is not sufficient for exact learning of the structure, and therefore interventional data is often used, by forcing some independent variables to take some specific values through experiments. An interventional path query requires a small number of experiments, since, $\mathsf{path}(i, j)$, intervenes the only variable correspondent to i. Therefore, applying our learning methods (similar to the method by Bello and Honorio, see [11]) can avoid an exponential number of experiments [33], and it can improve the results of Bello and Honorio [11] for the types of DAGs that we study.

We measure the efficiency of our methods in terms of the number of vertices, $n = |V|$, using these two complexities:

- Query complexity, $Q(n)$: This is the total number of queries that we perform. This parameter comes from the learning theory [2,14,21,46] and complexity theory [12,51].
- Round complexity, $R(n)$: This is the number of rounds that we perform our queries. The queries performed in a round are in a batch and they may not depend on the answer of the queries in the same round (but they may depend on the queries issued in the previous rounds).

Related Work. The problem of exact learning of a graph using a set of queries has been extensively studied [1,4–7,25,29,30,32,36,41–43,50]. With regard to previous work on learning directed graphs using path queries, Wang and Honorio [50] present a sequential randomized algorithm that takes $Q(n) \in O(n \log^2 n)$ path queries in expectation to learn rooted trees of maximum degree, d. Their divide and conquer approach is based on the notion of an even-separator, an edge that divides the tree into two subtrees of size at least n/d. Afshar *et al.* [5] show that learning a degree-d rooted tree with n nodes requires $\Omega(nd + n \log n)$ path queries [5] and they provide a randomized parallel algorithm for the same problem using $Q(n) \in O(n \log n)$ queries in $R(n) \in O(\log n)$ rounds with high probability (w.h.p.)[1], which instead relies on finding a near-separator, an edge that separates the tree into two subtrees of size at least $n/(d + 2)$, through a "noisy" process that requires noisy estimation of the number of descendants of a node by sampling. Their method, however, relies on the fact the ancestor set

---

[1] We say that an event happens with high probability if it occurs with probability at least $1 - \frac{1}{n^c}$, for some constant $c \geq 1$.

of a vertex in a rooted tree forms a total order. In Sect. 4, we extend their work to learn a rooted spanning tree for a DAG.

Regarding the reconstruction of trees with a specific height, Jagadish and Anindya [29] present a sequential deterministic algorithm to learn undirected fixed-degree trees of height h using $Q(n) \in O(nh \log n)$ separator queries, where a separator query given three vertices a, b, and c, it returns "true" if and only if b is on the path from a to c. Janardhanan and Reyzin [30] study the problem of learning an almost-tree of height h (a directed rooted tree with an additional cross-edge), and they present a randomized sequential algorithm using $Q(n) \in O(n \log^3 n + nh)$ queries.

Our Contributions. In Sect. 3, we present our learning algorithms for multitrees—a DAG with at most one directed path for any two vertices. We begin, however, by first presenting a deterministic result for learning directed rooted trees using path queries, giving a sequential deterministic approach to learn fixed-degree trees of height h, with $O(nh)$ queries, which provides an improvement over results by Jagadish and Anindya [29]. We then show how to use a tree-learning method to design an efficient learning method for a multitree with a roots using $Q(n) \in O(an \log n)$ queries and $R(n) \in O(a \log n)$ rounds w.h.p. We finally show how to use our tree learning method to design an algorithm with $Q(n) \in O(n^{3/2} \cdot \log^2 n)$ queries to learn butterfly networks w.h.p.

In Sect. 4, we introduce a separator theorem for DAGs, which is useful in learning a spanning-tree of a rooted DAG. Next, we present a parallel algorithm to learn almost-trees of height h, using $O(n \log n + nh)$ path queries in $O(\log n)$ parallel rounds w.h.p. We also provide a lower bound of $\Omega(n \log n + nh)$ for the worst case query complexity of a deterministic algorithm or an expected query complexity of a randomized algorithm for learning fixed-degree almost-trees proving that our algorithm is optimal. Moreover, this asymptotically-optimal query complexity bound, improves the sequential query complexity for this problem, since the best known results by Janardhanan and Reyzin [30] achieved a query complexity of $O(n \log^3 n + nh)$ in expectation.

2   Preliminaries

For a DAG, $G = (V, E)$, we represent the in-degree and out-degree of vertex $v \in V$ with $d_i(v)$ and $d_o(v)$ respectively. Throughout this paper, we assume that an input graph has maximum degree, d, i.e., for every $v \in V$, $d_i(v) + d_o(v) \leq d$. A vertex, v, is a root of the DAG if $d_i(v) = 0$. A DAG may have several roots, but we call a DAG rooted if it has only one root. Note that in a rooted DAG with root r, there is at least one directed path from r to every $v \in V$.

Definition 1 (arborescence). *An arborescence is a rooted DAG with root r that has exactly one path from r to each vertex $v \in V$. It is also referred to as a spanning directed tree at root r of a directed graph.*

We next introduce multitree, which is a family of DAGs useful in distributed computing [16,28] that we study in Sect. 3.

Definition 2 (multitree). *A multitree is a DAG in which the subgraph reachable from any vertex induces a tree, that is, it is a DAG with at most one directed path for any pair of vertices.*

We next review the definition of a butterfly network, which is a multitree used in high speed distributed computing [17,23,40] for which we provide efficient learning method in Sect. 3.

Definition 3 (Butterfly network). *A butterfly network, also known as depth-k Fast Fourier Transform (FFT) graph is a DAG recursively defined as $F^k = (V, E)$ as follows:*

- *For $k = 0$: F^0 is a single vertex, i.e. $V = \{v\}$ and $E = \{\}$.*
- *Otherwise, suppose $F_A^{k-1} = (V_A, E_A)$ and $F_B^{k-1} = (V_B, E_B)$ each having m sources and m targets $(t_0, ..., t_{m-1}) \in V_A$ and $(t_m, ..., t_{2m-1}) \in V_B$. Let $V_C = (v_0, v_1, ..., v_{2m-1})$ be $2m$ additional vertices. We have $F^k = (V, E)$, where $V = V_A \cup V_B \cup V_C$ and $E = E_A \cup E_B \bigcup_{0 \leq i \leq m-1}(t_i, v_i) \cup (t_i, v_{i+m}) \cup (t_{i+m}, v_i) \cup (t_{i+m}, v_{i+m})$ (See Fig. 1 for illustration).*

Definition 4 (ancestory). *Given a directed acyclic graph, $G = (V, E)$, we say u is a **parent** of a vertex v (v is a **child** of u), if there exists a directed edge $(u, v) \in E$. The **ancestor** relationship is a transitive closure of the parent relationship, and **descendant** relationship is a transitive closure of child relationship. We denote the descendant (resp. ancestor) set of vertex v, with $D(v)$, (resp. $A(v)$). Also, let $C(v)$ denote children of v.*

Definition 5. *A **path query** in a directed graph, $G = (V, E)$, is a function that takes two vertices u and v, and returns 1, if there is a directed path from u to v, and returns 0 otherwise. We also let $count(s, X) = \Sigma_{x \in X} path(s, x)$.*

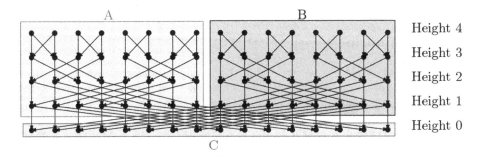

Fig. 1. An example of a butterfly network with height 4 (Depth 4), F^4, as a composition of two F^3 (A and B) and 2^4 additional vertices, C, in Height 0.

As Wang and Honorio observed [50], transitive edges in a directed graph are not learnable by path queries. Thus, it is not possible using path queries to learn all the edges for a number of directed graph types, including strongly connected graphs and DAGs that are not equal to their transitive reductions (i.e., graphs that have at least one transitive edge). Fortunately, transitive edges are not likely in phylogenetic networks due to their derivative nature, so, we focus on learning DAGs without transitive edges.

Definition 6. *In a directed graph, $G = (V, E)$, an edge $(u, v) \in E$ is called a* **transitive edge** *if there is a directed path from u to v of length greater than 1.*

Definition 7 (almost-tree). *An almost-tree is a rooted DAG resulting from the union of an arborescence and an additional cross edge. The* **height** *of an almost-tree is the length of its longest directed path.*

Note: some researchers define almost-trees to have a constant number of cross edges (see, e.g., [8,9]. But allowing more than one cross edge can cause transitive edges; hence, almost-trees with more than one cross edge are not all learnable using path queries, which is why we follow Janardhanan and Reyzin [30] to limit almost-trees to have one cross edge. We next introduce even-separator, which will be used in Sect. 4.

Definition 8 (even-separator). *Let $G = (V, E)$ be a rooted degree-d DAG. We say that vertex $v \in V$ is an even-separator if $\frac{|V|}{d} \leq count(v, V) \leq \frac{|V|(d-1)}{d}$.*

3 Learning Multitrees

In this section, we begin by presenting a deterministic algorithm to learn a rooted tree (a multitree with a single root) of height h, using $O(nh)$ path queries. This forms the building blocks for the main results of this section, which are an efficient algorithm to learn a multitree of arbitrary height with a number of roots and an efficient algorithm to learn a butterfly network.

Rooted Trees. Let $T = (V, E, r)$ be a directed tree rooted at r with maximum degree that is a constant, d, with vertices, V, and edges, E. At the beginning of any exactly learning algorithm, we only know V, and $n = |V|$, and our goal is to learn r, and E by issuing path queries.

To begin with, learning the root of the tree can be deterministically done using $O(n)$ path queries as suggested by Afshar *et al.* [5, Corollary 10]. Their approach is to first pick an arbitrary vertex v, (ii) learning its ancestor set and establishing a total order on them, and (iii) finally applying a maximum-finding algorithm [15,45,49] by simulating comparisons using path queries.

Next, we show how to learn the edges, E. Jagadish and Anindya [29] propose an algorithm to reconstruct fixed-degree trees of height h using $O(nh \log n)$ queries. Their approach is to find an edge-separator—an edge that splits the tree into two subtrees each having at least n/d vertices—and then to recursively build the two subtrees. In order to find such an edge, (i) they pick an arbitrary vertex, v, and learn an arbitrary neighbor of it such as, u, (ii) if (u, v) is not an edge-separator, they move to the neighboring edge that lies on the direction of maximum vertex set size. Hence, at each step after performing $O(n)$ queries, they get one step closer to the edge-separator. Therefore, they learn the edge-separator using $O(nh)$ queries, and they incur an extra $O(\log n)$ factor to build the tree recursively due to their edge-separator based recursive approach.

We show that finding an edge-separator for a deterministic algorithm is unnecessary, however. We instead propose a vertex-separator based learning algorithm. Our learn-short-tree(V, r) method takes as an input, the vertex set, V, and root vertex, r, and returns edges of the tree, E. Let $\{r_1, \ldots, r_d\}$ be a tentative set of children for vertex r initially set to $Null$, and for $1 \leq i \leq d$, let V_i represents the vertex set of the subtree rooted at r_i. For $1 \leq i \leq d$, we can find child r_i, by starting with an arbitrary vertex r_i, and looping over $v \in V$ to update r_i if for $v \neq r$, $\mathsf{path}(v, r_i) = 1$. Since, in a rooted tree, an ancestor relationship for ancestor set of any vertex is a total order, r_i will be a child of root r. Once we learn r_i, its descendants are the set of nodes $v \in V$ such that $\mathsf{path}(r_i, v) = 1$. We then remove V_i from the set of vertices of V to learn another child of r in the next iteration. It finally returns the union of edges (r, r_i) and edges returned by the recursive calls learn-short-tree(V_i, r_i), for $1 \leq i \leq d$. The full pseudo-code of function learn-short-tree(V, r) is provided in the full version of the paper [3].

The query complexity, $Q(n)$, for learning the tree is as following:

$$Q(n) = \Sigma_{i=1}^{d} Q(|V_i|) + O(n) \qquad (1)$$

Since the height of the tree is reduced by at least 1 for each recursive call, $Q(n) \in O(nh)$. Hence, we have the following theorem.

Theorem 1. *One can deterministically learn a fixed-degree height-h directed rooted tree using $O(nh)$ path queries.*

This provides an improvement upon the results of Jagadish and Anindya [29] (see the full version of this paper [3]).

Multitrees of Arbitrary Height. We next provide a parallel algorithm to learn a multitree of arbitrary height with a number of roots. Remind that Wang and Honorio [50, Theorem 8] prove that learning a multitree with $\Omega(n)$ roots requires $\Omega(n^2)$ queries. Suppose that $G = (V, E)$ is a multitree with a roots. We show that we can learn G using $Q(n) \in O(an \log n)$ queries in $R(n) \in O(a \log n)$ parallel rounds w.h.p.

Let us first explain how to learn a root. Our learn-root method learns a root using $Q(n) \in O(n)$ queries in $R(n) \in O(1)$ rounds w.h.p. Note that in a multitree with more than one root, the ancestor set of an arbitrary vertex does not necessarily form a total order, so, we may not directly apply a parallel maximum finding algorithm on the ancestor set to learn a root.

Our learn-root method takes as input vertex set V, and returns a root of the DAG. It first learns in parallel, Y, the ancestor set of v (the nodes $u \in V$ such that $path(u, v) = 1$). While $|Y| > m$, where $m = C_1 * \sqrt{|V|}$ for some constant C_1 fixed in the analysis, it takes a sample, S, of expected size of m from Y uniformly at random. Then, it performs path queries for every pair $(a, b) \in S \times S$ in parallel to learn a partial order of S, that is, we say $a < b$ if and only if $path(a, b) = 1$. Hence, a root of the DAG should be an ancestor of a minimal element in S. Using this fact, we keep narrowing down Y until $|Y| \leq m$, when we can afford to generate a partial order of Y in Line 7, and return a minimal element of Y (see Algorithm 1).

Algorithm 1: Our algorithm to find a root in V

Function learn-root(V):

1   $m = C_1 * \sqrt{|V|}$ Pick an arbitrary vertex $v \in V$ **for** *each* $u \in V$ **do in parallel**

2   Perform query $path(u, v)$ to find ancestor set Y

3   **while** $|Y| > m$ **do**

4       $S \leftarrow$ a random sample of expected size m from Y **for** $(a, b) \in S \times S$ **do in parallel**
            Perform query $path(a, b)$

5   Pick a vertex $y \in S$ such that for all $a \in S$: $path(a, y) == 0$ **for** $a \in Y$ **do in parallel**
            Perform query $path(a, y)$ to find ancestors of y, Y'

6       $Y \leftarrow Y'$

7   **for** $(a, b) \in Y \times Y$ **do in parallel**
        Perform query $path(a, b)$

8   $y \leftarrow$ a vertex in Y such that for all $a \in Y$: $path(a, y) == 0$ **return** y

---

Before providing the anlaysis of our efficient learn-root method, let us present Lemma 1, which is an important lemma throughout our analysis, as it extends Afshar *et al.* [5, Lemma 14] to directed acyclic graphs.

Lemma 1. *Let $G = (V, E)$ be a DAG, and let Y be the set of vertices formed by the union of at most c directed (not necessarily disjoint) paths, where $c \leq |V|$ and $|Y| > m = C_1 \sqrt{|V|}$. If we take a sample, S, of m elements from Y, then with probability $1 - \frac{1}{|V|^2}$, for each of these c paths such as P, every two consecutive nodes of S in the sorted order of P are within distance $O(|Y| \log |V| / \sqrt{|V|})$ from each other in P.*

Proof. Since we pick our sample S independently and uniformly at random, some nodes of Y may be picked more than once, and each vertex will be picked with probability $p = \frac{m}{|Y|} = \frac{C_1 \cdot \sqrt{|V|}}{|Y|}$. Let P be the set of vertices of an arbitrary path among these c paths. Divide P into consecutive sections of size, $s = \frac{|Y| \log |V|}{\sqrt{|V|}}$. The last section on P can have any length from 1 to $\frac{|Y| \log |V|}{\sqrt{|V|}}$. Let R be the set of vertices of an arbitrary section of path P (any section except the last one). We have that expected size of $|R \cap S|$, $E[|R \cap S|] = s \cdot p = C_1 \log |V|$. Since we pick our sample independently, using standard Chernoff bound for any constant

$C_1 > 8 \ln 2$, we have that $Pr[|R \cap S| = 0] < 1/|V|^4$. Using a union bound, with probability at least $1 - c/|V|^3$, our sample S will pick at least one node from all sections except the last section of all paths. Therefore, if $c \leq |V|$, with probability at least $1 - \frac{1}{|V|^2}$, the distance between any two consecutive nodes on a path in our sample is at most 2 s.

Lemma 2. *Let $G = (V, E)$ be a DAG, and suppose that roots have at most $c \in O(n^{1/2-\epsilon})$ for constant $0 < \epsilon < 1/2$ paths (not necessarily disjoint) in total to vertex v, then, learn-root(V) outputs a root with probability at least $1 - \frac{1}{|V|}$, with $Q(n) \in O(n)$ and $R(n) \in O(1)$.*

Proof. The correctness of the learn-root method relies on the fact that if Y is a set of ancestors of vertex v, then for vertex r, a root of the network, and for all $y \in Y$, we have: $path(y, r) = 0$. Using Lemma 1 and a union bound, after at most $1/\epsilon$ iterations of the **While** loop, with probability at least $1 - \frac{1/\epsilon}{|V|^2}$, the size of $|Y|$ will be $O(m)$. Hence, we will be able to find a root using the queries performed in Line 7. Note that this Las Vegas algorithm always returns a root correctly. We can simply derive a Monte Carlo algorithm by replacing the **while** loop with a **for** loop of two iterations.

Therefore, the query complexity of the algorithm is as follows w.h.p:

- We have $O(|V|)$ queries in 1 round to find ancestors of v.
- Then, we have $1/\epsilon$ iterations of the **while** loop, each having $O(m^2) + O(|Y|) \in O(|V|)$ queries in $1/\epsilon$ rounds.
- Finally, we have $O(m^2)$ in 1 round in Line 7.

Overall, this amounts to $Q(n) \in O(n)$, $R(n) \in O(1)$ w.h.p. $\qed$

Since in a multitree with $a \in O(n^{1/2-\epsilon})$ roots (for $0 < \epsilon < 1/2$), each root has at most one path to a given vertex v, we have at most $a \in O(n^{1/2-\epsilon})$ directed paths in total from roots to an arbitrary vertex v. Therefore, we can apply Lemma 2 to learn a root w.h.p. Note that if $a \notin O(n^{1/2-\epsilon})$, as an alternative, we can learn a root w.h.p. using $O(n \log n)$ queries with $R(n) \in O(\log n)$ rounds by (i) picking an arbitrary vertex $v \in V$ and learning its ancestors, $A(v) \cap V$ in parallel (ii) replacing path queries with inverse-path queries (inverse-path$(u, v) = 1$ if and only if v has a directed path to u), (ii) and applying the rooted tree learning method by Afshar *et al.* [5, Algorithm 2] to learn the tree with inverse direction to v. Note that any of the leaves of the inverse tree rooted at v is a root of the multitree.

Our multitree learning algorithm works by repetitively learning a root, r, from the set of candidate roots, R ($R = V$ at the beginning). Then, it learns a tree rooted at R by calling the rooted tree learning method by Afshar *et al.* [5, Algorithm 2]. Finally, it removes the set of vertices of the tree from R to perform another iteration of the algorithm so long as $|R| > 0$. We give the details of the algorithm below.

1. Let R be the set of candidate roots for the multitree initialized with V.
2. Let $r \leftarrow$ learn-root(R).
3. Issue queries in parallel, $path(r, v)$ for all $v \in V$ to learn descendants, $D(r)$.
4. Learn the tree rooted at r by calling learn-rooted-tree($r, D(r)$).
5. Let $R = R \setminus D(r)$, and if $|R| > 0$ go to step 2..

Theorem 2 analyzes the complexity of our multitree learning algorithm.

Theorem 2. *One can learn a multitree with a roots using $Q(n) \in O(an \log n)$ path queries in $R(n) \in O(a \log n)$ parallel rounds w.h.p.*

Proof. The query complexity and the round complexity of our multitree learning method is dominated by the calls to the learn-rooted-tree by Afshar *et al.* [5, Algorithm 2] which takes $Q(n) \in O(n \log n)$ queries in $R(n) \in O(\log n)$ parallel rounds w.h.p. Hence, using a union bound and by adjusting the sampling constants for learn-rooted-tree by Afshar *et al.* [5, Algorithm 2] we can establish the high probability bounds.

Butterfly Networks. Next, we provide an algorithm to learn a butterfly network. Suppose that $F^h = (V, E)$ is a butterfly network with height h (i.e., a depth-h FFT graph, see definition 3). We show that we can learn F^h using $Q(n) \in O(2^{3h/2}h^2)$ path queries with high probability. Note that in a butterfly networks of height h, the number of nodes will be $n = 2^h \cdot (h + 1)$. Also, note that the graph has a symmetry property, that is, all leaves are reachable from the root, and all roots are reachable from the leaves if we reverse the directions of the edges, and that each node but the leaves has exactly two children, and each node but the roots have exactly two parents, and so on. Due to this symmetry property, we can apply learn-short-tree but with inverse path query (inverse-path(u, v) = 1 if and only if v has a directed path to u) to find the tree with inverse direction to a leaf.

Our algorithm first learns all the roots and all the leaves of the graph. We first perform a sequential search to find an arbitrary root of the network, r. Note that we can learn r by picking an arbitrary vertex x and looping over all the vertices and updating x to y if path(y, x) = 1. After learning its descendants, $D(r)$, we make a call to our learn-short-tree method to build the tree rooted at r, which enables us to learn all the leaves, L. Then, we pick an arbitrary leaf, $l \in L$, and after learning its ancestors, $A(l)$, we call the learn-short-tree method (with inverse path query) to learn the tree with inverse direction to l, which enables us to learn all the roots, R. We then take two sample subsets, S, and T, of expected size $O(2^{h/2}h)$ from R, and L respectively, and uniformly at random. We will show that the union of the edges of trees rooted at r for all $r \in S$ and the inverse trees rooted at l for all $l \in T$ includes all the edges of the network w.h.p. We give the details of our algorithm below.

1. Learn a root, r, using a sequential search.
2. Perform path queries to learn descendant set, $D(r)$, of r.

3. Call learn-short-tree$(r, D(r))$ method to learn the leaves of the network, L.
4. Let $l \in L$ be an arbitrary leaf in the network, then perform path queries to learn the ancestors of l, $A(l)$.
5. Call learn-short-tree$(l, A(l))$ with inverse path query definition to learn the roots of the network, R.
6. Pick a sample S of size $c \cdot 2^{h/2}h$ from R, and a sample T of size $c \cdot 2^{h/2}h$ from L uniformly at random for a constant $c > 0$.
7. Perform queries to learn descendant set, $D(s)$, for every $s \in S$, and to learn ancestor set $A(t)$, for every $t \in T$.
8. Call learn-short-tree$(s, D(s))$ to learn the tree rooted at s for all $s \in S$.
9. Call learn-short-tree$(t, A(t))$ using inverse reverse path query to learn the tree rooted at t for all $t \in T$.
10. Return the union of all the edges learned.

Theorem 3. *One can learn a butterfly network of height, h, using $Q(n) \in O(2^{3h/2}h^2)$ path queries with high probability.*

Proof. The query complexity of the algorithm is dominated by $O(2^{h/2}h)$ times the running time of our learn-short-tree method, which takes $O(2^h h)$ queries for each tree. Consider a directed edge from vertex x at height k to vertex y at height $k - 1$ in the network. If $k \leq h/2$, then x has at least $2^{\lfloor h/2 \rfloor}$ ancestors in the root, that is, $|A(x) \cap R| \geq 2^{\lfloor h/2 \rfloor}$. Since our sample, S, has an expected size of $2^{h/2} \cdot ch$, the expected size of $|S \cap A(x) \cap R| \geq ch/2$. Using a standard Chernoff bound, the probability, $Pr[|S \cap A(x) \cap R| = 0] \leq e^{-ch/4}$. Hence, for large enough c, this probability is less than $1/2^{2h}$. Therefore, we will be able to learn edge (x, y) through a tree rooted at $s \in S$. Similarly, we can show that if $k > h/2$, then y has at least $2^{\lfloor h/2 \rfloor}$ descendants in the leaves, that is, $|D(y) \cap L| \geq 2^{\lfloor h/2 \rfloor}$. Since, our sample T, has an expected size of $2^{h/2} \cdot ch$, the expected size of $|T \cap D(y) \cap L| \geq ch/2$. Using a standard Chernoff bound, the probability, $Pr[|T \cap D(y) \cap L| = 0] \leq e^{-ch/4}$. Hence, for large enough c, this probability is less than $1/2^{2h}$. Therefore, we will be able to learn edge (x, y) through a tree inversely rooted at $t \in T$ in this case. A union bound establishes the high probability.

4   Parallel Learning of Almost-Trees

Let $G = (V, E)$ be an almost-tree of height h. We learn G with $Q(n) \in O(n \log n + nh)$ path queries in $R(n) \in O(\log n)$ rounds w.h.p. Note that we can learn the root of an almost-tree by Algorithm 1, and given that the root has at most 2 paths to any vertex, it will take $Q(n) \in O(n)$ queries and $R(n) \in O(1)$ w.h.p. by Lemma 2. We then learn a spanning rooted tree for it, and finally we learn the cross-edge. We will also prove that our algorithm is optimal by showing that any randomized algorithm needs an expected number of $\Omega(n \log n + nh)$.

Learning an Arborescence in a DAG. Our parallel algorithm learns an arborescence, a spanning directed rooted tree, of the graph with a divide and conquer approach based on our separator theorem, which is an extension of Afshar *et al.* [5, Lemma 5] for DAGs.

Theorem 4. *Every degree-d rooted DAG, $G = (V, E)$, has an even-separator (see Definition 8).*

Proof. We prove through a iterative process that there exists a vertex v such that $\frac{|V|}{d} \leq |D(v)| \leq \frac{|V| \cdot (d-1)}{d}$. Let r be the root of the DAG. We have that $|D(r)| = |V|$. Since r has at most d children and each $v \in V$ is a descendent of at least one of the children of r, r has a child x, such that $D(x) \geq |V|/d$. If $D(x) \leq \frac{|V| \cdot (d-1)}{d}$, x is an even-separator. Otherwise, since $d_o(x) \leq d - 1$, x has a child, y, such that $|D(y)| \geq |V|/d$. If $|D(y)| \leq \frac{|V| \cdot (d-1)}{d}$, y is an even-separator. Otherwise, we can repeat this iterative procedure with a child of y having maximum number of descendants. Since, $|D(y)| < |D(x)|$, and a directed path in a DAG ends at vertices of out-degree 0 (with no descendants), this iterative procedure will return an even-separator at some point.

Next, we introduce Lemma 3 which shows that for fixed-degree rooted DAGs, if we pick a vertex v uniformly at random, there is an even separator in $A(v)$, ancestor set of v, with probability depending on d.

Lemma 3. *Let $G = (V, E)$ be a degree-d DAG with root r, and let v be a vertex chosen uniformly at random from v. Let Y be the ancestor set for v in V. Then, with probability at least $\frac{1}{d}$, there is an even-separator in Y.*

Proof. By Theorem 4, G has an even-separator, e. Since $|D(e)| \geq \frac{|V|}{d}$, with probability at least $\frac{1}{d}$, v will be one of the descendants of e.

Although a degree-d rooted DAG has an even-separator, checking if a vertex is an even-separator requires a lot of queries for exact calculation of the number of descendants. Thus, we use a more relaxed version of the separator, which we call **near-separator**, for our divide and conquer algorithm.

Definition 9. *Let $G = (V, E)$ be a rooted degree-d DAG. We say that vertex $v \in V$ is a near-separator if $\frac{|V|}{d+2} \leq |D(v)| \leq \frac{|V|(d+1)}{d+2}$.*

Note that every even-separator is also a near-separator. We show if an even-separator exists among $A(v)$ for an arbitrary vertex v, then we can locate a near-separator among $A(v)$ w.h.p. Incidentally, Afshar *et al.* [5] used a similar divide and conquer approach to learn directed rooted trees, but their approach relied on the fact that there is exactly one path from root to every vertex of the tree. We will show how to meet the challenge of having multiple paths to a vertex from the root in learning an arborescence for a rooted DAG.

Our learn-spanning-tree method takes as input vertex set, V, of a DAG rooted at r, and returns the edges, E, of an arborescence of it. In particular, it enters a repeating **while** loop to learn a near-separator by (i) picking a random vertex $v \in V$, (ii) learning its ancestors, $Y = A(v) \cap V$, (iii) and checking if Y has a near-separator, w, by calling learn-separator method, which we describe next. Once learn-separator returns a vertex, w, we split V into $V_1 = D(w) \cap V$ and $V_2 = V \setminus V_1$ given that $path(w, z) = 1$ if and only if $z \in V_1$. If $\frac{|V|}{d} \leq |V_1| \leq \frac{|V|(d-1)}{d}$, we verify w is a near-separator. If w is a near separator, then it calls learn-parent method, to learn a parent, u, for w. Finally, it makes two recursive calls to learn a spanning tree rooted at w for vertex set V_1, and a spanning tree rooted at r with vertex set V_2 (see full version of the paper [3] for a full pseudo-code of the algorithm). Note that our learn-parent(v, V) method is similar to our learn-root(V) method except that it passes closest nodes to v to the next iteration rather than the farthest nodes (please refer to the full version of the paper [3] for details).

Next, we show how to adapt an algorithm to learn a near-separator for DAGs by extending the work of Afshar et al.[5, Algorithm 3]. Our learn-separator method takes as input vertex v, its ancestors, Y, vertex set V of a DAG rooted at r, and returns w.h.p. a near-separator among vertices of Y provided that there is an even-separator in Y. If $|Y|$ is too large ($|Y| > |V|/K$), then it enters **Phase 1**. The goal of this phase is to remove the nodes that are unlikely to be a separator in order to pass a smaller set of candidate separator to **Phase 2**. It chooses a random sample, S, of expected size $m = C_1\sqrt{|V|}$, where $C_1 > 0$ is a fixed constant. It adds $\{v, r\}$ to the sample S. It then estimates $|D(s) \cap V|$ for each $s \in S$, using a random sample, X_s, of size $K = O(\log |V|)$ from V by issuing path queries. If all of the estimates, $count(s, X_s)$, are smaller than $\frac{K}{d+1}$, we return $Null$, as we argue that in this case the nodes in Y do not have enough descendants to act as a separator. Similarly, If all of the estimates, are greater than $\frac{Kd}{d+1}$, we return $Null$, as we show that in this case the nodes in Y have too many descendants to act as a separator. If one of these estimates for a vertex s lies in the range of $[\frac{K}{d+1}, \frac{Kd}{d+1}]$, we return it as a near-separator. Otherwise, we filter the set of Y by removing the nodes that are unlikely to be a separator through a call to filter-separator method, which we present next. Then, we enter **Phase 2**, where for every $s \in Y$, we take a random sample X_s of expected size of $O(log|V|)$ from V to estimate $|D(s) \cap V|$. If one of these estimates for a vertex s lies in the range of $[\frac{K}{d+1}, \frac{Kd}{d+1}]$, we return it as a near-separator. We will show later that the output is a near-separator w.h.p (please refer to the full version of the paper [3] for a pseudo-code description of learn-separator method).

Next, let us explain our filter-separator method, whose purpose is to remove some of the vertices in Y that are unlikely to be a separator to shrink the size of Y. We first establish a partial order on elements of S by issuing path queries in parallel. Since there are at most $c = 2$ directed paths from root to vertex v, for path $1 \leq i \leq c$, let $l_i \in S$ be the oldest node on path i having $count(l_i, X_{l_i}) < \frac{K}{d+1}$ (resp. $g_i \in S$ be the youngest node on path i having $count(g_i, X_{g_i}) > \frac{Kd}{d+1}$). We then perform queries to remove ancestors of g_i, and descendants of l_i from Y. We will prove later that this filter reduces $|Y|$ considerably without filtering an even-separator. We will give the details of this method in Algorithm 2.

Lemma 4 shows that our filter-separator method efficiently in parallel eliminates the nodes that are unlikely to act as a separator.

Lemma 4. *Let $G = (V, E)$ be a DAG rooted at r, with at most c directed (not necessarily disjoint) paths from r to vertex v, and let $Y = A(v) \cap V$, and let*

Algorithm 2: Filter out the vertices unlikely to be a separator

---

Function filter-separator(S, Y, V):

1   **for** *each* $\{a, b\} \in S$ **do in parallel**

2   perform query $path(a, b)$

3   Let $P_1, P_2, \ldots, P_c$ be the c paths from r to v. For $1 \leq i \leq c$: let $l_i \in (S \cap P_i)$ such that $count(l_i, X_{l_i}) < \frac{K}{d+1}$, and there exists no $b \in (S \cap A(l_i))$ where $count(b, X_b) < \frac{K}{d+1}$. For $1 \leq i \leq c$: let $g_i \in (S \cap P_i)$ such that $count(g_i, X_{g_i}) > \frac{K \cdot d}{d+1}$, and there exists no $b \in (S \cap D(g_i))$ where $count(b, X_b) > \frac{K \cdot d}{d+1}$. **for** $1 \leq i \leq c$ *and* $v \in V$ **do in parallel**

4   perform query $path(v, g_i)$ to find $(A(g_i) \cap V)$. Remove $(A(g_i) \cap V)$ from Y. perform query $path(l_i, v)$ to find $(D(l_i) \cap V)$. Remove $(D(l_i) \cap V)$ from Y.

5 **return** Y

S *be a random sample of expected size* m *that includes* v, *and* r *as well. The call to* **filter-separator**(S, Y, V) *in our* **learn-separator** *method returns a set of size* $O(c \cdot |Y| \log |V| / \sqrt{|V|})$, *and If* Y *has an even-separator, the returned set includes an even-separator with probability at least* $1 - \frac{|S|+1}{|V|^2}$.

Proof. The proof idea is to first employ Lemma 1 to show that with very high probability the size of the returned set is at most $c \cdot O(|Y| \log |V| / \sqrt{|V|})$. Then, it follows by arguing that if e is an even-separator it is unlikely for e to be an ancestor of g_i or a descendant of d_i in Lines 4, 4 of filter-separator method. Please refer to the full version of the paper [3] for details.

Lemma 5 establishes the fact that our learn-separator finds w.h.p. a near-separator among ancestors $A(v) \cap V$, if there is an even-separator in $A(v) \cap V$.

Lemma 5. *Let* $G = (V, E)$ *be a DAG rooted at* r, *with at most* c *directed (not necessarily disjoint) paths from* r *to vertex* v, *and let* $Y = A(v) \cap V$. *If* Y *has an even-separator, then our* **learn-separator** *method returns a near-separator w.h.p.*

Proof. See full version of the paper [3]. ∎

Lemma 6. *Let* $G = (V, E)$ *be a DAG rooted at* r, *with at most* c *directed (not necessarily disjoint) paths from* r *to vertex* v. *Then, our* **learn-separator**(v, Y, V, r) *method, takes* $Q(n) \in O(c|V|)$ *queries in* $R(n) \in O(1)$ *rounds.*

Proof. – In **phase 1**, it takes $O(mK) \in o(|V|)$ queries in 1 round to estimate the number of descendants for sample S.

– The call to filter-separator in **phase 1** takes m^2 queries in one round to derive a partial order for S, and since there are at most c paths from r to v, it takes $O(c \cdot |V|)$ in one round to remove nodes from Y.

– In **Phase 2**, it takes $O(|Y|K) \in O(|V|)$ queries in 1 round to estimate the number of descendants for all nodes of Y.

Algorithm 3: lean a cross-edge for an almost tree

Function learn-cross-edge(V, E):

1   **for** $v \in V$ **do**

2   **for** $c \in C(v)$ **do**

3   **for** $t \in (D(V) \setminus D(c))$ **do in parallel**

4   Perform query $path(c, t)$

5   Let c be the only node and let t be the node with maximum height having $path(c, t) = 1$ for $s \in D(c)$ **do in parallel**

6   Perform query $path(s, t)$

7   Let s be the node with minimum height having $path(s, t) = 1$. **return** (s, t)

Theorem 5. *Suppose $G = (V, E)$ is a rooted DAG with $|V| = n$, and maximum constant degree, d, with at most constant, c directed (not necessarily disjoint) paths from root, r, to each vertex. Our* **learn-spanning-tree** *algorithm learns an arborescence of G using $Q(n) \in O(n \log n)$ and $R(n) \in O(\log n)$ w.h.p.*

Proof. See full version of the paper [3].   ∎

__Learning a Cross-Edge.__ Next, we will show that a cross-edge can be learnt using $O(nh)$ queries in just 2 parallel rounds for an almost-tree of height h. Our learn-cross-edge algorithm takes as input vertices V and edges E of an arborescence of a almost-tree, and returns the cross-edge from the source vertex, s, to the destination vertex, t. In this algorithm, we refer to $D(v)$ for a vertex v as the set of descendants of v according to E (the only edges learned by the arborescence). We will show later that there exists a vertex, c, whose parent is vertex, v, such that the cross-edge has to be from a source vertex $s \in D(c)$ to a destination vertex $t \in (D(v) \setminus D(c))$. In particular, this algorithm first learns t and c with $O(nh)$ queries in 1 parallel round. Note that $t \in (D(v) \setminus D(c))$ is a node with maximum height having $path(c, t) = 1$. Once it learns t and c, then it learns source s, where $s \in D(c)$ is the node with minimum height satisfying $path(s, t) = 1$, using $O(n)$ queries in 1 round. We give the details in Algorithm 3.

The following lemma shows that Algorithm 3 correctly learns the cross-edge using $O(nh)$ queries in just 2 rounds.

Lemma 7. *Given an arborescence with vertex set V, and edge set, E, of an almost-tree, Algorithm 3 learns the cross-edge using $O(nh)$ queries in 2 rounds.*

Proof. Suppose that the cross-edge is from a vertex s to to a vertex t. Let v be the least common ancestor of s and t in the arborescence, and let c be a child of v on the path from v to s. Since $t \in (D(v) \setminus D(c))$, we have that $path(c, t) = 1$ in Line 4. Note that since there is only one cross-edge, there will be exactly one node such as c satisfying $path(c, t) = 1$. Note that in Line 4 we can also learn t, which is the node with maximum height satisfying $path(c, t) = 1$. Finally, we just do a parallel search in the descendant set of c to learn s in Line 6.

We charge each $path(c, t)$ query in Line 4 to the vertex v. Since each vertex has at most d children the number of queries associated with vertex v will be at most $O(|D(v)| \cdot d)$. Hence, using a double counting argument and the fact that each vertex is a descendant of $O(h)$ vertices, the sum of the queries performed Line 4 will be, $\Sigma_{v \in V} O(|D(v)| \cdot d) = O(nh)$. Finally, we need $O(n)$ queries 1 round to learn s in Line 6.

Theorem 6. *Given vertices, V, of an almost-tree, we can learn root, r, and the edges, E, using $Q(n) \in O(n \log n + nh)$ path queries, and $R(n) \in O(\log n)$ w.h.p.*

Proof. Note that in almost-trees there are at most $c = 2$ paths from root r to each vertex. Therefore, by Lemma 2, we can learn root of the graph using $O(n)$ queries in $O(1)$ rounds with probability at least $1 - \frac{1}{|V|}$. Then, by Theorem 5, we can learn a spanning tree of the graph using $O(n \log n)$ queries in $O(\log n)$ rounds with probability at least $1 - \frac{1}{|V|}$. Finally, by Lemma 7 we can deterministically learn a cross-edge using $O(nh)$ queries in just 2 rounds.

Lower Bound. The following lower bound improves the one by Janardhanan and Reyzin [30] and proves that our algorithm to learn almost-trees in optimal.

Theorem 7. *Let G be a a degree-d almost-tree of height h with n vertices. Learning G takes $\Omega(n \log n + nh)$ queries. This lower bound holds for both worst case of a deterministic algorithm and for an expected cost of a randomized algorithm.*

Proof. We use the same graph as the one used by Janardhanan and Reyzin [30], but we improve their bound using an information-theoretic argument. Consider a caterpillar graph with height h, and a complete d-ary tree with $\Omega(n)$ leaves attached to the last level of it. If there is a cross-edge from one of the leaves of the caterpillar to one of the leaves of the d-ary tree, it takes $\Omega(nh)$ queries involving a leaf of the caterpillar and a leaf of the d-ary tree. Suppose that a querier, Bob, knows the internal nodes of the d-ary, and he wants to know that for each leaf l of the d-ary, what is the parent of l in the d-ary tree. If there are m leaves for the caterpillar, the number of possible d-ary trees will be at least $\frac{m!}{(d!)^{m/d}}$. Therofore, using an information-theoretic lower bound, we need $\Omega \left(\log \left(\frac{m!}{(d!)^{m/d}} \right) \right)$ bit of information to be able to learn the parent of the leaves of d-ary tree. Since the queries involving a leaf of the caterpillar and a leaf of the d-ary tree do not provide any information about how the d-ary tree is built, it takes $\Omega(n \log n)$ queries to learn the d-ary tree.

References

1. Abrahamsen, M., Bodwin, G., Rotenberg, E., Stöckel, M.: Graph reconstruction with a betweenness oracle. In: Ollinger, N., Vollmer, H. (eds.) 33rd Symposium on Theoretical Aspects of Computer Science, STACS 2016, 17–20 February 2016, Orléans, France. LIPIcs, vol. 47, pp. 5:1–5:14. Schloss Dagstuhl - Leibniz-Zentrum für Informatik (2016). https://doi.org/10.4230/LIPIcs.STACS.2016.5
2. Afshani, P., Agrawal, M., Doerr, B., Doerr, C., Larsen, K.G., Mehlhorn, K.: The query complexity of finding a hidden permutation. In: Brodnik, A., López-Ortiz, A., Raman, V., Viola, A. (eds.) Space-Efficient Data Structures, Streams, and Algorithms. LNCS, vol. 8066, pp. 1–11. Springer, Heidelberg (2013). https://doi.org/10.1007/978-3-642-40273-9_1

3. Afshar, R., Goodrich, M.T.: Exact learning of multitrees and almost-trees using path queries. 10.48550/ARXIV.2208.04216, https://arxiv.org/abs/2208.04216

4. Afshar, R., Goodrich, M.T., Matias, P., Osegueda, M.C.: Reconstructing binary trees in parallel. In: Scheideler, C., Spear, M. (eds.) SPAA 2020: 32nd ACM Symposium on Parallelism in Algorithms and Architectures, Virtual Event, USA, 15–17 July 2020, pp. 491–492. ACM (2020). https://doi.org/10.1145/3350755.3400229

5. Afshar, R., Goodrich, M.T., Matias, P., Osegueda, M.C.: Reconstructing biological and digital phylogenetic trees in parallel. In: Grandoni, F., Herman, G., Sanders, P. (eds.) 28th Annual European Symposium on Algorithms, ESA 2020, 7–9 September 2020, Pisa, Italy (Virtual Conference). LIPIcs, vol. 173, pp. 3:1–3:24. Schloss Dagstuhl - Leibniz-Zentrum für Informatik (2020). https://doi.org/10.4230/LIPIcs.ESA.2020.3

6. Afshar, R., Goodrich, M.T., Matias, P., Osegueda, M.C.: Parallel network mapping algorithms. In: Agrawal, K., Azar, Y. (eds.) SPAA 2021: 33rd ACM Symposium on Parallelism in Algorithms and Architectures, Virtual Event, USA, 6–8 July 2021, pp. 410–413. ACM (2021). https://doi.org/10.1145/3409964.3461822

7. Afshar, R., Goodrich, M.T., Matias, P., Osegueda, M.C.: Mapping networks via parallel kth-hop traceroute queries. In: Berenbrink, P., Monmege, B. (eds.) 39th International Symposium on Theoretical Aspects of Computer Science, STACS 2022, 15–18 March 2022, Marseille, France (Virtual Conference). LIPIcs, vol. 219, pp. 4:1–4:21. Schloss Dagstuhl - Leibniz-Zentrum für Informatik (2022). https://doi.org/10.4230/LIPIcs.STACS.2022.4

8. Akutsu, T.: A polynomial time algorithm for finding a largest common subgraph of almost trees of bounded degree. IEICE Trans. Fundam. Electron. Commun. Comput. Sci. **76**(9), 1488–1493 (1993)

9. Bannister, M.J., Eppstein, D., Simons, J.A.: Fixed parameter tractability of crossing minimization of almost-trees. In: Wismath, S., Wolff, A. (eds.) GD 2013. LNCS, vol. 8242, pp. 340–351. Springer, Cham (2013). https://doi.org/10.1007/978-3-319-03841-4_30

10. Barton, N.H.: The role of hybridization in evolution. Mol. Ecol. **10**(3), 551–568 (2001)

11. Bello, K., Honorio, J.: Computationally and statistically efficient learning of causal Bayes nets using path queries. In: Bengio, S., Wallach, H.M., Larochelle, H., Grauman, K., Cesa-Bianchi, N., Garnett, R. (eds.) Advances in Neural Information Processing Systems 31: Annual Conference on Neural Information Processing Systems 2018, NeurIPS 2018(December), pp. 3–8, 2018. Montréal, Canada, pp. 10954–10964 (2018). https://proceedings.neurips.cc/paper/2018/hash/a0b45d1bb84fe1bedbb8449764c4d5d5-Abstract.html

12. Bernasconi, A., Damm, C., Shparlinski, I.E.: Circuit and decision tree complexity of some number theoretic problems. Inf. Comput. **168**(2), 113–124 (2001). https://doi.org/10.1006/inco.2000.3017

13. Bestagini, P., Tagliasacchi, M., Tubaro, S.: Image phylogeny tree reconstruction based on region selection. In: 2016 IEEE International Conference on Acoustics, Speech and Signal Processing, ICASSP 2016, Shanghai, China, 20–25 March 2016, pp. 2059–2063. IEEE (2016). https://doi.org/10.1109/ICASSP.2016.7472039

14. Choi, S., Kim, J.H.: Optimal query complexity bounds for finding graphs. Artif. Intell. **174**(9–10), 551–569 (2010). https://doi.org/10.1016/j.artint.2010.02.003

15. Cole, R., Vishkin, U.: Deterministic coin tossing and accelerating cascades: micro and macro techniques for designing parallel algorithms. In: Hartmanis, J. (ed.) Proceedings of the 18th Annual ACM Symposium on Theory of Computing, 28–30 May 1986, Berkeley, California, USA, pp. 206–219. ACM (1986). https://doi.org/10.1145/12130.12151

16. Colombo, C., Lepage, F., Kopp, R., Gnaedinger, E.: Two SDN multi-tree approaches for constrained seamless multicast. In: Pop, F., Negru, C., González-Vélez, H., Rak, J. (eds.) 2018 IEEE International Conference on Computational Science and Engineering, CSE 2018, Bucharest, Romania, 29–31 October 2018, pp. 77–84. IEEE Computer Society (2018). https://doi.org/10.1109/CSE.2018.00017

17. Comellas, F., Fiol, M.A., Gimbert, J., Mitjana, M.: The spectra of wrapped butterfly digraphs. Networks **42**(1), 15–19 (2003). https://doi.org/10.1002/net.10085

18. Dias, Z., Goldenstein, S., Rocha, A.: Exploring heuristic and optimum branching algorithms for image phylogeny. J. Vis. Commun. Image Represent. **24**(7), 1124–1134 (2013). https://doi.org/10.1016/j.jvcir.2013.07.011

19. Dias, Z., Goldenstein, S., Rocha, A.: Large-scale image phylogeny: tracing image ancestral relationships. IEEE Multim. **20**(3), 58–70 (2013). https://doi.org/10.1109/MMUL.2013.17

20. Dias, Z., Rocha, A., Goldenstein, S.: Image phylogeny by minimal spanning trees. IEEE Trans. Inf. Forensics Secur. **7**(2), 774–788 (2012). https://doi.org/10.1109/TIFS.2011.2169959

21. Dobzinski, S., Vondrák, J.: From query complexity to computational complexity. In: Karloff, H.J., Pitassi, T. (eds.) Proceedings of the 44th Symposium on Theory of Computing Conference, STOC 2012, New York, NY, USA, 19–22 May 2012, pp. 1107–1116. ACM (2012). https://doi.org/10.1145/2213977.2214076

22. Goldberg, L.A., Goldberg, P.W., Phillips, C.A., Sorkin, G.B.: Constructing computer virus phylogenies. J. Algorithms **26**(1), 188–208 (1998). https://doi.org/10.1006/jagm.1997.0897

23. Goodrich, M.T., Jacob, R., Sitchinava, N.: Atomic power in forks: a super-logarithmic lower bound for implementing butterfly networks in the nonatomic binary fork-join model. In: Marx, D. (ed.) Proceedings of the 2021 ACM-SIAM Symposium on Discrete Algorithms, SODA 2021, Virtual Conference, 10–13 January 2021, pp. 2141–2153. SIAM (2021). https://doi.org/10.1137/1.9781611976465.128

24. Heckerman, D., Meek, C., Cooper, G.: A Bayesian approach to causal discovery. In: Holmes, D.E., Jain, L.C. (eds.) Innovations in Machine Learning, vol. 194, pp. 1–28. Springer, Heidelberg (2006). https://doi.org/10.1007/3-540-33486-6_1

25. Hein, J.J.: An optimal algorithm to reconstruct trees from additive distance data. Bull. Math. Biol. **51**(5), 597–603 (1989)

26. Hünermund, P., Bareinboim, E.: Causal inference and data fusion in econometrics. arXiv preprint arXiv:1912.09104 (2019)

27. Imbens, G.W.: Potential outcome and directed acyclic graph approaches to causality: relevance for empirical practice in economics. J. Econ. Lit. **58**(4), 1129–79 (2020)

28. Itai, A., Rodeh, M.: The multi-tree approach to reliability in distributed networks. Inf. Comput. **79**(1), 43–59 (1988). https://doi.org/10.1016/0890-5401(88)90016-8

29. Jagadish, M., Sen, A.: Learning a bounded-degree tree using separator queries. In: Jain, S., Munos, R., Stephan, F., Zeugmann, T. (eds.) ALT 2013. LNCS (LNAI), vol. 8139, pp. 188–202. Springer, Heidelberg (2013). https://doi.org/10.1007/978-3-642-40935-6_14

30. Janardhanan, M.V., Reyzin, L.: On learning a hidden directed graph with path queries. CoRR abs/2002.11541 (2020). https://arxiv.org/abs/2002.11541
31. Ji, J.H., Park, S.H., Woo, G., Cho, H.G.: Generating pylogenetic tree of homogeneous source code in a plagiarism detection system. Int. J. Control Autom. Syst. **6**(6), 809–817 (2008)
32. King, V., Zhang, L., Zhou, Y.: On the complexity of distance-based evolutionary tree reconstruction. In: Proceedings of the Fourteenth Annual ACM-SIAM Symposium on Discrete Algorithms, 12–14 January 2003, Baltimore, Maryland, USA, pp. 444–453. ACM/SIAM (2003). http://dl.acm.org/citation.cfm?id=644108.644179
33. Kocaoglu, M., Shanmugam, K., Bareinboim, E.: Experimental design for learning causal graphs with latent variables. In: Guyon, I., von Luxburg, U., Bengio, S., Wallach, H.M., Fergus, R., Vishwanathan, S.V.N., Garnett, R. (eds.) Advances in Neural Information Processing Systems 30: Annual Conference on Neural Information Processing Systems 2017(December), pp. 4–9, 2017. Long Beach, CA, USA, pp. 7018–7028 (2017), https://proceedings.neurips.cc/paper/2017/hash/291d43c696d8c3704cdbe0a72ade5f6c-Abstract.html
34. Lagani, V., Triantafillou, S., Ball, G., Tegnér, J., Tsamardinos, I.: Probabilistic computational causal discovery for systems biology. In: Geris, L., Gomez-Cabrero, D. (eds.) Uncertainty in Biology. SMTEB, vol. 17, pp. 33–73. Springer, Cham (2016). https://doi.org/10.1007/978-3-319-21296-8_3
35. Marmerola, G.D., Oikawa, M.A., Dias, Z., Goldenstein, S., Rocha, A.: On the reconstruction of text phylogeny trees: evaluation and analysis of textual relationships. PLoS ONE **11**(12), e0167822 (2016)
36. Mathieu, C., Zhou, H.: A simple algorithm for graph reconstruction. In: Mutzel, P., Pagh, R., Herman, G. (eds.) 29th Annual European Symposium on Algorithms, ESA 2021, 6–8 September 2021, Lisbon, Portugal (Virtual Conference). LIPIcs, vol. 204, pp. 68:1–68:18. Schloss Dagstuhl - Leibniz-Zentrum für Informatik (2021). https://doi.org/10.4230/LIPIcs.ESA.2021.68
37. Meinshausen, N., Hauser, A., Mooij, J.M., Peters, J., Versteeg, P., Bühlmann, P.: Methods for causal inference from gene perturbation experiments and validation. Proc. Natl. Acad. Sci. **113**(27), 7361–7368 (2016)
38. Moffa, G.: Using directed acyclic graphs in epidemiological research in psychosis: an analysis of the role of bullying in psychosis. Schizophr. Bull. **43**(6), 1273–1279 (2017)
39. Pfeffer, A., et al.: Malware analysis and attribution using genetic information. In: 2012 7th International Conference on Malicious and Unwanted Software, pp. 39–45. IEEE (2012)
40. Ranade, A.G.: Optimal speedup for backtrack search on a butterfly network. In: Leighton, T. (ed.) Proceedings of the 3rd Annual ACM Symposium on Parallel Algorithms and Architectures, SPAA '91, Hilton Head, South Carolina, USA, 21–24 July 1991, pp. 40–48. ACM (1991). https://doi.org/10.1145/113379.113383
41. Reyzin, L., Srivastava, N.: On the longest path algorithm for reconstructing trees from distance matrices. Inf. Process. Lett. **101**(3), 98–100 (2007). https://doi.org/10.1016/j.ipl.2006.08.013
42. Rong, G., Li, W., Yang, Y., Wang, J.: Reconstruction and verification of chordal graphs with a distance oracle. Theor. Comput. Sci. **859**, 48–56 (2021). https://doi.org/10.1016/j.tcs.2021.01.006
43. Rong, G., Yang, Y., Li, W., Wang, J.: A divide-and-conquer approach for reconstruction of $\{c_{\geq 5}\}$-free graphs via betweenness queries. Theor. Comput. Sci. **917**, 1–11 (2022). https://doi.org/10.1016/j.tcs.2022.03.008

44. Shen, B., Forstall, C.W., de Rezende Rocha, A., Scheirer, W.J.: Practical text phylogeny for real-world settings. IEEE Access **6**, 41002–41012 (2018). https://doi.org/10.1109/ACCESS.2018.2856865

45. Shiloach, Y., Vishkin, U.: Finding the maximum, merging and sorting in a parallel computation model. In: Brauer, W., et al. (eds.) CONPAR 1981. LNCS, vol. 111, pp. 314–327. Springer, Heidelberg (1981). https://doi.org/10.1007/BFb0105127

46. Tardos, G.: Query complexity, or why is it difficult to seperate NP ^{a} cap co np^a from p^a by random oracles a? Comb. **9**(4), 385–392 (1989). https://doi.org/10.1007/BF02125350

47. Tennant, P.W., et al.: Use of directed acyclic graphs (DAGS) to identify confounders in applied health research: review and recommendations. Int. J. Epidemiol. **50**(2), 620–632 (2021)

48. Triantafillou, S., Lagani, V., Heinze-Deml, C., Schmidt, A., Tegner, J., Tsamardinos, I.: Predicting causal relationships from biological data: applying automated causal discovery on mass cytometry data of human immune cells. Sci. Rep. **7**(1), 1–11 (2017)

49. Valiant, L.G.: Parallelism in comparison problems. SIAM J. Comput. **4**(3), 348–355 (1975). https://doi.org/10.1137/0204030

50. Wang, Z., Honorio, J.: Reconstructing a bounded-degree directed tree using path queries. In: 57th Annual Allerton Conference on Communication, Control, and Computing, Allerton 2019, Monticello, IL, USA, 24–27 September 2019, pp. 506–513. IEEE (2019). https://doi.org/10.1109/ALLERTON.2019.8919924

51. Yao, A.C.: Decision tree complexity and betti numbers. J. Comput. Syst. Sci. **55**(1), 36–43 (1997). https://doi.org/10.1006/jcss.1997.1495

Almost Optimal Proper Learning and Testing Polynomials

Nader H. Bshouty$^{(\boxtimes)}$

Department of Computer Science, Technion, Haifa, Israel
bshouty@cs.technion.ac.il

Abstract. We give the first almost optimal polynomial-time proper learning algorithm of Boolean sparse multivariate polynomial under the uniform distribution. For s-sparse polynomial over n variables and $\epsilon = 1/s^\beta$, $\beta > 1$, our algorithm makes

$$q_U = \left(\frac{s}{\epsilon}\right)^{\frac{\log \beta}{\beta} + O\left(\frac{1}{\beta}\right)} + \tilde{O}\left(s\right)\left(\log\frac{1}{\epsilon}\right)\log n$$

queries. Notice that our query complexity is sublinear in $1/\epsilon$ and almost linear in s. All previous algorithms have query complexity at least quadratic in s and linear in $1/\epsilon$.

We then prove the almost tight lower bound

$$q_L = \left(\frac{s}{\epsilon}\right)^{\frac{\log \beta}{\beta} + \Omega\left(\frac{1}{\beta}\right)} + \Omega\left(s\right)\left(\log\frac{1}{\epsilon}\right)\log n,$$

Applying the reduction in [9] with the above algorithm, we give the first almost optimal polynomial-time tester for s-sparse polynomial. Our tester, for $\beta > 3.404$, makes

$$\tilde{O}\left(\frac{s}{\epsilon}\right)$$

queries.

Keywords: Proper learning · Property testing · Polynomial

1 Introduction

In this paper, we study the learnability and testability of the class of sparse (multivariate) polynomials over $GF(2)$. A polynomial over $GF(2)$ is the sum in $GF(2)$ of monomials, where a monomial is a product of variables. It is well known that every Boolean function has a unique representation as a (multilinear) polynomial over $GF(2)$. A Boolean function is called s-sparse polynomial if its unique polynomial expression contains at most s monomials.

Israel—Center for Theoretical Sciences, Guangdong Technion, (GTIIT), China.

A. Castañeda and F. Rodríguez-Henríquez (Eds.): LATIN 2022, LNCS 13568, pp. 312–327, 2022.
https://doi.org/10.1007/978-3-031-20624-5_19

1.1 Learning

In the learning model [1,27], the learning algorithm has access to a black-box query oracle to a function f that is s-sparse polynomial. The goal is to run in $poly(n, s, 1/\epsilon)$ time, make $poly(n, s, 1/\epsilon)$ black-box queries and, with probability at least $2/3$, learn a Boolean function h that is ϵ-close to f under the uniform distribution, i.e., $\mathbf{Pr}_x[f(x) \neq h(x)] \leq \epsilon$. The learning algorithm is called *proper learning* if it outputs an s-sparse polynomial. The learning algorithm is called *exact learning algorithm* if $\epsilon = 0$.

Proper and non-proper learning algorithms of s-sparse polynomials that run in polynomial-time and make a polynomial number of queries have been studied by many authors [2–4,6,8,9,11,13,16,17,20,23,26].

For learning s-sparse polynomial without black-box queries (PAC-learning without black-box queries, [27]) and for exact learning ($\epsilon = 0$), the following results are known. In [20], Hellerstein and Servedio gave a non-proper learning algorithm that learns only from random examples under any distribution that runs in time $n^{O(n \log s)^{1/2}}$. Roth and Benedek, [23], show that for any $s \geq 2$, polynomial-time proper PAC-learning without black-box queries of s-sparse polynomials implies RP=NP. They gave a proper exact learning ($\epsilon = 0$) algorithm that makes $(n/\log s)^{\log s}$ black-box queries. They also show that to exactly learn s-sparse polynomial, we need at least $(n/\log s)^{\log s}$ black-box queries. Some generalizations of the above algorithms for any field and Ring are studied in [13,17].

For polynomial-time non-proper and proper learning s-sparse polynomial with black-box queries under the uniform distribution, all the algorithms in the literature [2–4,6,8,9,11,23,26] have query complexities that are at least quadratic in s and linear in $1/\epsilon$. In [2], Beimel et al. give a non-proper algorithm that returns a Multiplicity Automaton equivalent to the target. Their algorithm asks $O(s^2 \log n + s/\epsilon)$ queries. See also [3,4]. In [23], it is shown that there is a *deterministic* learning algorithm that makes $O(sn/\epsilon)$ queries. The learning algorithms in [8,9,11] are based on collecting enough small monomials that their sum approximates the target. The query complexities they achieve are $O((s/\epsilon)^c \log n)$, $c > 16$, $O((s/\epsilon)^{16} \log n)$, and $O((s^2/\epsilon) \log n)$, respectively. In [26], Schapire and Sellie gave an exact learning algorithm from $O(ns^3)$ membership queries and $O(ns)$ equivalent queries. Using Angluin's reduction in [1], their algorithm can be changed to a learning algorithm (under the uniform distribution) that makes[1] $O(ns^3 + n(s/\epsilon))$ queries.

In this paper, we prove

Theorem 1. *Let $\epsilon = 1/s^\beta$. There is a proper learning algorithm for s-sparse polynomial that runs in polynomial-time and makes*

$$q_U = \left(\frac{s}{\epsilon}\right)^{\frac{\log \beta}{\beta} + O(\frac{1}{\beta})} + \tilde{O}(s)\left(\log \frac{1}{\epsilon}\right) \log n$$

queries.

---

[1] It is not clear from their algorithm how to use the technique in this paper to improve the query complexity. But even if there is a way, the query complexity will be at least the number of membership queries $O(ns^3)$.

To the best of our knowledge, this is the first learning algorithm whose query complexity is sublinear in $1/\epsilon$ and almost linear in s. See Theorems 4 for the exact query complexity.

We then give the following lower bound that shows that our query complexity is almost optimal.

Theorem 2. *Let $\epsilon = 1/s^\beta$. Any learning algorithm for s-sparse polynomial must make at least*

$$q_L = \left(\frac{s}{\epsilon}\right)^{\frac{\log \beta}{\beta} + \Omega\left(\frac{1}{\beta}\right)} + \Omega\left(s\left(\log \frac{1}{\epsilon}\right)\log n\right)$$

queries.

1.2 Property Testing

A problem closely related to learning polynomial is the problem of property testing polynomial: Given black-box query access to a Boolean function f. Distinguish, with high probability, the case that f is s-sparse polynomial versus the case that f is ϵ-far from every s-sparse polynomial. Property testing of Boolean function was first considered in the seminal works of Blum, Luby and Rubinfeld [7] and Rubinfeld and Sudan [24] and has recently become a very active research area. See the surveys and books [18,19,21,22].

In the uniform distribution framework, where the distance between two functions is measured with respect to the uniform distribution, the first testing algorithm for s-sparse polynomial runs in exponential time [14] and makes $\tilde{O}(s^4/\epsilon^2)$ queries. Chakraborty et al. [12], gave another exponential time algorithm that makes $\tilde{O}(s/\epsilon^2)$ queries. Diakonikolas et al. gave in [15] the first polynomial-time testing algorithm that makes $poly(s, 1/\epsilon) > s^{10}/\epsilon^3$ queries. In [9], Bshouty gave a polynomial-time algorithm that makes $\tilde{O}(s^2/\epsilon)$ queries. As for the lower bound for the query complexity, the lower bound $\Omega(1/\epsilon)$ follows from Bshouty and Goldriech lower bound in [10]. Blais et al. [5], and Saglam, [25], gave the lower bound $\Omega(s \log s)$.

Applying the reduction in [9] (see the discussion after Theorem 51 in Sect. 6.3 in [9]), with Theorem 1, we get

Theorem 3. *For any $\epsilon = 1/s^\beta$ there is an algorithm for ϵ-testing s-sparse polynomial that makes*

$$Q = \left(\frac{s}{\epsilon}\right)^{\frac{\log \beta}{\beta} + O\left(\frac{1}{\beta}\right)} + \tilde{O}\left(\frac{s}{\epsilon}\right)$$

queries.

In particular, for $\beta > 3.404$,

$$Q = \tilde{O}\left(\frac{s}{\epsilon}\right).$$

Notice that the query complexity of the tester in Theorem 3 is $\tilde{O}((1/\epsilon)^{1+1/\beta})$. This is within a factor of $(1/\epsilon)^{1/\beta}$ of the lower bound $\Omega(1/\epsilon)$. Therefore, the query complexity in Theorem 3 is almost optimal.

2   Techniques

In this section, we give a brief overview of the techniques used for the main results, Theorems 1, 2, and 3.

2.1   Upper Bound

This section gives a brief overview of the proof of Theorem 1.

Our algorithm first reduces the learning of s-sparse polynomial to exact learning s-sparse polynomials with monomials of size at most $d = O(\log(s/\epsilon))$, i.e., degree-d s-sparse polynomials. Given an s-sparse polynomial f, we project each variable to 0 with probability $O(\log s/\log(1/\epsilon))$. In this projection, monomials of size greater than $\Omega(d)$ vanish, with high probability. Then we learn the projected function. We take enough random zero projections of f so that, with high probability, for every monomial M of f of size at most $\log(s/\epsilon)$, there is a projection q such that M does not vanish under q. Collecting all the monomials of degree at most $\log(s/\epsilon)$ in all the projections gives a hypothesis that is ϵ-close to the target function f.

Now to exactly learn the degree-d s-sparse polynomials, where $d = O(\log(s/\epsilon))$, we first give an algorithm that finds a monomial of a degree-d s-sparse polynomial that makes

$$Q = 2^{dH_2\left(\frac{\log s}{d}\right)(1-o_s(1))} \log n = s^{\log(\log(s/\epsilon)/\log s)+O(1)} \log n \tag{1}$$

queries where H_2 is the binary entropy. The best-known algorithm for this problem has query complexity $Q' = 2^d \log n \approx poly(s/\epsilon) \log n$, [9,11]. For small enough ϵ, $Q' \gg Q$. The previous algorithm in [9] chooses uniformly at random assignments until it finds a positive assignment a, i.e., $f(a) = 1$. Then recursively do the same for $f(a * x)$, where $a * x = (a_1 x_1, a_2 x_2, \ldots, a_n x_n)$, until no more a with smaller Hamming weight can be found. Then $f(a*x) = \prod_{a_i=1} x_i$ is a monomial of f. To find a positive assignment in a degree d polynomial from uniformly at random assignments, we need to make, on average, 2^d queries. The number of nonzero entries in $a * x$ is on average $n/2$. Therefore, this algorithm makes $O(2^d \log n)$ queries. In this paper, we study the probability $\mathbf{Pr}_{\mathcal{D}_p}[f(a) = 1]$ when a is chosen according to the product distribution $\mathcal{D}_p$, where each a_i is equal to 1 with probability p and is 0 with probability $1 - p$. We show that to maximize this probability, we need to choose $p = 1 - (\log s)/d$. Replacing the uniform distribution with the distribution $\mathcal{D}_p$ in the above algorithm gives the query complexity in (1).

Now, let f be a degree-d s-sparse polynomial, and suppose we have learned some monomials $M_1, \ldots, M_t$ of f. To learn a new monomial of f, we learn a monomial of $f + h$ where $h = M_1 + M_2 + \cdots + M_t$. This gives an algorithm that makes,

$$q = \left(\frac{s}{\epsilon}\right)^{\frac{\log \beta}{\beta}+O(\frac{1}{\beta})} \log n \tag{2}$$

queries where $\epsilon = 1/s^\beta$. All previous algorithms have query complexity that are at least quadratic in s and linear in $1/\epsilon$.

Now, notice that the query complexity in (2) is not the query complexity that is stated in Theorem 1. To get the query complexity in the theorem, we use another reduction. This reduction is from exact learning degree-d s-sparse polynomials over n variables to exact learning degree-d s-sparse polynomials over $m = O(d^2 s^2)$ variables. Given a degree-d s-sparse polynomials f over n variables. We choose uniformly at random a projection $\phi : [n] \to [m]$ and learn the polynomial $F(x_1, \ldots, x_m) = f(x_{\phi(1)}, \ldots, x_{\phi(n)})$ over m variables. This is equivalent to distributing the n variables, uniformly at random, into m boxes, assigning different variables for different boxes, and then learning the function with the new variables. We choose $m = O(d^2 s^2)$ so that different variables in f fall into different boxes. By (2), the query complexity of learning F is

$$q' = \left(\frac{s}{\epsilon}\right)^{\frac{\log \beta}{\beta} + O(\frac{1}{\beta})} \log m = \left(\frac{s}{\epsilon}\right)^{\frac{\log \beta}{\beta} + O(\frac{1}{\beta})}. \tag{3}$$

After we learn F, we find the relevant variables of F, i.e., the variables that F depends on. Then, for each relevant variable of F, we search for the relevant variable of f that corresponds to this variable. Each search makes $O(\log n)$ queries. The number of relevant variables of f is at most ds and here $d = O(\log(s/\epsilon))$, which adds

$$\tilde{O}(s) \left(\log \frac{1}{\epsilon} \right) (\log n)$$

to the query complexity in (3). This gives the query complexity in Theorem 1. We also show that all the above can be done in time $O(qn)$ where q is the query complexity.

See more details in Sect. 4.3.

2.2   Lower Bound

This section gives a brief overview of Theorem 2.

In this paper, we give two lower bounds. One that proves the right summand of the lower bound

$$\Omega \left(s \left(\log \frac{1}{\epsilon} \right) \log n \right), \tag{4}$$

and the second proves the left summand

$$\left(\frac{s}{\epsilon}\right)^{\frac{\log \beta}{\beta} + \Omega(\frac{1}{\beta})}. \tag{5}$$

To prove (4), we consider the class of $\log(1/(2\epsilon))$-degree s-sparse polynomials. We show that any learning algorithm for this class can be modified to an exact learning algorithm. Then, using Yao's minimax principle, the query complexity of exactly learning this class is at least log of the class size. This gives the first lower bound in (4).

To prove (5), we consider the class

$$C = \left\{ \prod_{i \in I} x_i \prod_{j \in J} (1 + x_j) \; \middle| \; I, J \subseteq [n], |J| \leq \log s, |I| \leq \log(1/\epsilon) - \log s - 1 \right\}.$$

It is easy to see that every polynomial in C is a s-sparse polynomial.

Again, we show that any learning algorithm for this class can be modified to an exact learning algorithm. We then use Yao's minimax principle to show that, to exactly learn C, we need at least $\Omega(|C|)$ queries. This gives the lower bound in (5).

2.3   Upper Bound for Testing

For the result in testing, we use the reduction in [9]. In [9] it is shown that given a learning algorithm for s-sparse polynomial that makes $q(s, n)$ queries, one can construct a testing algorithm for s-sparse polynomial that makes

$$q(s, \tilde{O}(s)) + \tilde{O}\left(\frac{s}{\epsilon}\right)$$

queries. Using Theorem 1 we get a testing algorithm with query complexity (recall that $\epsilon = 1/s^\beta$)

$$\left(\frac{s}{\epsilon}\right)^{\frac{\log \beta}{\beta} + O(\frac{1}{\beta})} + \tilde{O}\left(\frac{s}{\epsilon}\right).$$

We then show that for $\beta \geq 6.219$, this query complexity is $\tilde{O}(s/\epsilon)$. In the full paper, we give another learning algorithm that has query complexity better than Theorem 1 for $\beta < 6.219$. Using this algorithm, we get a tester that has query complexity $\tilde{O}(s/\epsilon)$ for $\beta \geq 3.404$.

3   Definitions and Preliminary Results

In this section, we give some definitions and preliminary results.

We will denote by $P_{n,s}$ the class of s-sparse polynomials over $GF(2) = \{0, 1\}$ with the Boolean variables $(x_1, \ldots, x_n)$ and $P_{n,d,s} \subset P_{n,s}$, the class of degree-$d$ s-sparse polynomials over $GF(2)$. Formally, let $\mathcal{S}_{n,\leq d} = \cup_{i \leq d} \mathcal{S}_{n,i}$, where $\mathcal{S}_{n,i} = \binom{[n]}{i}$ is the set of all i-subsets of $[n] = \{1, 2, \ldots, n\}$. The class $P_{n,d,s}$ is the class of all the polynomials over $GF(2)$ of the form $\sum_{I \in S} \prod_{i \in I} x_i$ where $S \subseteq \mathcal{S}_{n,\leq d}$ and $|S| \leq s$. The class $P_{n,s}$ is $P_{n,n,s}$.

Let B_n be the uniform distribution over $\{0, 1\}^n$. The following result is well known. See for example [4].

Lemma 1. *For any $f \in P_{n,d,s}$ we have*

$$\mathbf{Pr}_{x \in B_n}[f(x) \neq 0] = \mathbf{Pr}_{x \in B_n}[f(x) = 1] \geq 2^{-d}.$$

We will now extend Lemma 1 to other distributions.

Let $W_{n,s}$ be the set of all the assignments in $\{0,1\}^n$ of Hamming weight at least $n - \lfloor \log s \rfloor$. We prove the following in the full paper for completeness [13, 23].

Lemma 2. *[13, 23]. For any $0 \neq f \in P_{n,s}$ there is an assignment $a \in W_{n,s}$ such that $f(a) = 1$.*

The *p-product distribution* $\mathcal{D}_{n,p}$ is a distribution over $\{0,1\}^n$ where $\mathcal{D}_{n,p}(a) = p^{\text{wt}(a)}(1-p)^{n-\text{wt}(a)}$ where $\text{wt}(a)$ is the Hamming weight of a. Let $H_2(x) = -x \log_2 x - (1-x) \log_2(1-x)$ be the binary entropy function.

We prove

Lemma 3. *Let $p \geq 1/2$. For every $f \in P_{n,d,s}$, $f \neq 0$, we have*

$$\mathbf{Pr}_{x \in \mathcal{D}_{n,p}}[f(x) = 1] \geq \begin{cases} p^{d-\lfloor \log s \rfloor}(1-p)^{\lfloor \log s \rfloor} & d \geq \lfloor \log s \rfloor \\ (1-p)^d & d < \lfloor \log s \rfloor \end{cases}.$$

In particular, if $d \geq 2\lfloor \log s \rfloor$, then for $p' = (d - \lfloor \log s \rfloor)/d$

$$\max_{p \geq 1/2} \mathbf{Pr}_{x \in \mathcal{D}_{n,p}}[f(x) = 1] = \mathbf{Pr}_{x \in \mathcal{D}_{n,p'}}[f(x) = 1] \geq 2^{-H_2\left(\frac{\lfloor \log s \rfloor}{d}\right)d}$$

and if $d < 2\lfloor \log s \rfloor$, then for $p' = 1/2$

$$\max_{p \geq 1/2} \mathbf{Pr}_{x \in \mathcal{D}_{n,p}}[f(x) = 1] = \mathbf{Pr}_{x \in \mathcal{D}_{n,p'}}[f(x) = 1] \geq 2^{-d}.$$

Proof. We first consider the case $n = d$. Let $0 \neq f(x) \in P_{d,d,s}$. When $d \geq \lfloor \log s \rfloor$, by Lemma 2, there is $a \in W_{d,s}$ such that $f(a) = 1$. Therefore

$$\mathbf{Pr}_{x \in \mathcal{D}_{d,p}}[f(x) = 1] \geq \mathcal{D}_{d,p}(a) = p^{\text{wt}(a)}(1-p)^{d-\text{wt}(a)}$$
$$\geq p^{d-\lfloor \log s \rfloor}(1-p)^{\lfloor \log s \rfloor}. \tag{6}$$

When $d < \lfloor \log s \rfloor$, we have $P_{d,d,s} = P_{d,d,2^d}$. This is because $2^d < s$ and any polynomial in d variables of degree d has at most 2^d monomials. Therefore, by (6), we have

$$\mathbf{Pr}_{x \in \mathcal{D}_{d,p}}[f(x) = 1] \geq p^{d-\lfloor \log 2^d \rfloor}(1-p)^{\lfloor \log 2^d \rfloor} = (1-p)^d.$$

Thus, the result follows for nonzero functions with d variables.

Let $0 \neq f(x) \in P_{n,d,s}$. Let M be a monomial of f of maximal degree $d' \leq d$. Assume wlog that $M = x_1 x_2 \cdots x_{d'}$. First notice that for any $(a_{d+1}, \ldots, a_n) \in \{0,1\}^{n-d}$ we have $g(x_1, \ldots, x_d) := f(x_1, \ldots, x_d, a_{d+1}, \ldots, a_n) \neq 0$ and $g \in P_{d,d,s}$. Consider the indicator random variable $X(x)$ that is equal to 1 if $f(x) = 1$. Then

$$\mathbf{Pr}_{x \in \mathcal{D}_{n,p}}[f(x) = 1] = \mathbf{E}[X]$$
$$= \mathbf{E}_{(a_{d+1}, \ldots, a_n) \in \mathcal{D}_{n-d,p}}[\mathbf{E}_{(x_1, x_2, \ldots, x_d) \in \mathcal{D}_{d,p}}[X(x_1, \ldots, x_d, a_{d+1}, \ldots, a_n)]]$$
$$= \mathbf{E}_{(a_{d+1}, \ldots, a_n) \in \mathcal{D}_{n-d,p}}[\mathbf{Pr}_{(x_1, x_2, \ldots, x_d) \in \mathcal{D}_{d,p}}[f(x_1, \ldots, x_d, a_{d+1}, \ldots, a_n) = 1]]$$
$$\geq \begin{cases} p^{d-\lfloor \log s \rfloor}(1-p)^{\lfloor \log s \rfloor} & d \geq \lfloor \log s \rfloor \\ (1-p)^d & d < \lfloor \log s \rfloor \end{cases}.$$

$\square$

In particular, since $f(x) \neq g(x)$ is equivalent to $f(x) + g(x) = 1$ and $f + g \in P_{n,2s,d}$, we have the following result for $\mathbf{Pr}_{x \in \mathcal{D}_{n,p}}[f(x) \neq g(x)]$.

Lemma 4. *Let $p \geq 1/2$. For every $f, g \in P_{n,d,s}$, $f \neq g$, we have*

$$\mathbf{Pr}_{x \in \mathcal{D}_{n,p}}[f(x) \neq g(x)] \geq \begin{cases} p^{d - \lfloor \log s \rfloor - 1}(1 - p)^{\lfloor \log s \rfloor + 1} & d \geq \lfloor \log s \rfloor + 1 \\ (1 - p)^d & d < \lfloor \log s \rfloor + 1 \end{cases}.$$

In particular, if $d \geq 2\lfloor \log s \rfloor + 2$, then for $p' = (d - \lfloor \log s \rfloor - 1)/d$

$$\max_{p \geq 1/2} \mathbf{Pr}_{x \in \mathcal{D}_{n,p}}[f(x) \neq g(x)] = \mathbf{Pr}_{x \in \mathcal{D}_{n,p'}}[f(x) \neq g(x)] \geq 2^{-H_2\left(\frac{\lfloor \log s \rfloor + 1}{d}\right)d}$$

and if $d < 2\lfloor \log s \rfloor + 2$, then for $p' = 1/2$

$$\max_{p \geq 1/2} \mathbf{Pr}_{x \in \mathcal{D}_{n,p}}[f(x) \neq g(x)] = \mathbf{Pr}_{x \in \mathcal{D}_{n,p'}}[f(x) \neq g(x)] \geq 2^{-d}.$$

In particular, for $p' = \max((d - \lfloor \log s \rfloor - 1)/d, 1/2)$,

$$\max_{p \geq 1/2} \mathbf{Pr}_{x \in \mathcal{D}_{n,p}}[f(x) \neq g(x)] = \mathbf{Pr}_{x \in \mathcal{D}_{n,p'}}[f(x) \neq g(x)] \geq 2^{-H_2\left(\min\left(\frac{1}{2}, \frac{\lfloor \log s \rfloor + 1}{d}\right)\right)d}.$$

Consider the algorithm **Test** in Fig. 1. We now prove

Lemma 5. *The algorithm* **Test**(f, g, δ) *for $f, g \in P_{n,d,s}$ given as black-boxes, makes*

$$q = 2^{H_2\left(\min\left(\frac{\lfloor \log s \rfloor + 1}{d}, \frac{1}{2}\right)\right)d} \ln \frac{1}{\delta}$$

queries, runs in time $O(qn)$, and if $f \neq g$, with probability at least $1 - \delta$, returns an assignment a such that $f(a) \neq g(a)$. If $f = g$ then with probability 1 returns "$f = g$".

Proof. If $f \neq g$ then, by Lemma 4 and since $1 - x \leq e^{-x}$, the probability that $f(a) = g(a)$ for all a is at most

$$\left(1 - 2^{-H_2\left(\min\left(\frac{\lfloor \log s \rfloor + 1}{d}, \frac{1}{2}\right)\right)d}\right)^{H_2\left(\min\left(\frac{\lfloor \log s \rfloor + 1}{d}, \frac{1}{2}\right)\right)d \ln(1/\delta)} \leq \delta.$$

$\square$

4   The Learning Algorithm

In this section, we give the learning algorithm for $P_{n,s}$.

Test(f, g, δ)
Input: Black-box access to $f, g \in P_{m,d,s}$.
Output: If $f \neq g$ then find a assignment a such that $f(a) \neq g(a)$.

1. Let $p = \max\left(1 - \frac{\lfloor \log s \rfloor + 1}{d}, \frac{1}{2}\right)$.
2. Repeat $2^{H_2\left(\min\left(\frac{\lfloor \log s \rfloor + 1}{d}, \frac{1}{2}\right)\right)d} \ln \frac{1}{\delta}$ times
3.   Draw $a \in \mathcal{D}_{m,p}$
4.   If $f(a) \neq g(a)$ then return a
5. Return "$f = g$"

Fig. 1. For $f, g \in P_{m,d,s}$, if $f \neq g$ then, with probability at least $1 - \delta$, returns an assignment a such that $f(a) \neq g(a)$.

4.1   The Reduction Algorithms

In this subsection, we give the reductions we use for the learning. In the introduction, we gave a brief explanation of the proof. The full proof can be found in the full paper.

Let $f(x_1, \ldots, x_n)$ be any Boolean function. A *p-zero projection* of f is a random function, $f(z) = f(z_1, \ldots, z_n)$ where each z_i is equal to x_i with probability p and is equal to 0 with probability $1 - p$.

We now give the first reduction,

Lemma 6. $(P_{n,s} \rightarrow P_{n,d,s})$. *Let* $0 < p < 1$ *and* $w = (s/\epsilon)^{\log(1/p)} \ln(16s)$. *Suppose there is a proper learning algorithm that exactly learns* $P_{n,d,s}$ *with* $Q(d, \delta)$ *queries in time* $T(d, \delta)$ *and probability of success at least* $1 - \delta$. *Then there is a proper learning algorithm that learns* $P_{n,s}$ *with* $O(w \cdot Q(D, 1/(16w)) \log(1/\delta))$ *queries where*

$$D = \log \frac{s}{\epsilon} + \frac{\log s + \log \log s + 6}{\log(1/p)},$$

in time $w \cdot T(D, 1/(16w)) \log(1/\delta)$, *probability of success at least* $1 - \delta$ *and accuracy* $1 - \epsilon$.

We now give the second reduction.

Lemma 7. $(P_{n,d,s} \rightarrow P_{(2ds)^2, d, s})$. *Suppose there is a proper learning algorithm that exactly learns* $P_{(2ds)^2, d, s}$ *with* $Q(d, \delta)$ *queries in time* $T(d, \delta)$ *and probability of success at least* $1 - \delta$. *Then there is a proper learning algorithm that exactly learns* $P_{n,d,s}$ *with* $q = (Q(d, 1/16) + ds \log n) \log(1/\delta)$ *queries in time* $(T(d, 1/16) + dsn \log n) \log(1/\delta)$ *and probability of success at least* $1 - \delta$.

4.2   The Algorithm for $P_{m,d,s}$

In this section, we give a learning algorithm that exactly learns $P_{m,d,s}$.

Consider the algorithm **FindMonomial** in Fig. 2. For two assignments a and b in $\{0, 1\}^m$ we define $a * b = (a_1 b_1, \ldots, a_m b_m)$. We prove

Lemma 8. *For* $f \in P_{m,d,s}$, $f \neq 0$, **FindMonomial**(f, d, δ) *makes at most*

$$Q = O\left(2^{dH_2\left(\min\left(\frac{\lfloor \log s \rfloor + 1}{d}, \frac{1}{2}\right)\right)} d \log(m/\delta)\right)$$

queries, runs in time $O(Qn)$, *and with probability at least* $1 - \delta$ *returns a monomial of* f.

Proof. Let $t = 8d \ln(m/\delta)$. Let $a^{(1)}, \ldots, a^{(t)}$ be the assignments generated in the "Repeat" loop of **FindMonomial**(f, d, δ). Define the random variable $X_i = wt(a^{(i)}) - d_i$, $i \in [t]$, where d_i is the degree of the minimal degree monomial of $f(a^{(i)} * x)$. First notice that every monomial of $f(a^{(i)} * x)$ is a monomial of $f(a^{(i-1)} * x)$ and of f. Therefore, $d \geq d_{i+1} \geq d_i$. Also, if $X_i = 0$ then $f(a^{(i)} * x) = \prod_{a_j^{(i)}=1} x_j$ is a monomial of f.

Given $a^{(i)}$ such that $g(x) = f(a^{(i)} * x) \neq 0$. Notice that $g \in P_{m,d,s}$ and if $f(b * a^{(i)} * x) = 0$ then $a^{(i+1)} = a^{(i)}$, $d_{i+1} = d_i$ and $X_{i+1} = X_i$. Let M be any monomial of g of degree d_i and suppose, wlog, $M = x_1 x_2 \cdots x_{d_i}$. For $b \in \mathcal{D}_{m,p}$ where $p = 2^{-1/d}$, with probability $\eta := (2^{-1/d})^{d_i} \geq 1/2$, $b_1 = b_2 = \cdots = b_{d_i} = 1$. If $b_1 = b_2 = \cdots = b_{d_i} = 1$ then $f(b * a^{(i)} * x) \neq 0$. This is because M remains a monomial of $f(b * a^{(i)} * x)$. Therefore, $\eta' := \mathbf{Pr}[f(b * a^{(i)} * x) \neq 0] \geq \eta$. The expected weight of $(b_{d_i+1} a_{d_i+1}^{(i)}, \ldots, b_m a_m^{(i)})$ is $2^{-1/d} X_i$. Also, if $f(b * a^{(i)} * x) \neq 0$ then, with probability at least $1/2$, **Test** succeed to detect that $f(b * a^{(i)} * x) \neq 0$. Therefore,

$$\mathbf{E}[X_{i+1}|X_i] \leq \frac{1}{2}\eta'(2^{-1/d}X_i) + \left(1 - \frac{\eta'}{2}\right)X_i$$

$$\leq \frac{1}{2}\eta'\left(1 - \frac{1}{2d}\right)X_i + \left(1 - \frac{\eta'}{2}\right)X_i$$

$$= \left(1 - \frac{\eta'}{4d}\right)X_i \leq \left(1 - \frac{1}{8d}\right)X_i.$$

Now, $X_0 \leq m$ and therefore $\mathbf{E}[X_t] \leq m\left(1 - \frac{1}{8d}\right)^t \leq \delta$. Thus, by Markov's bound, the probability that $f(a^{(t)} * x)$ is not a monomial is

$$\mathbf{Pr}[X_t \neq 0] = \mathbf{Pr}[X_t \geq 1] \leq \delta.$$

Now, by Lemma 5, the query complexity in the lemma follows. □

We now prove

Lemma 9. *There is a proper learning algorithm that exactly learns* $P_{m,d,s}$, *makes*

$$Q(m, d, \delta) = O\left(s 2^{H_2\left(\min\left(\frac{\lfloor \log s \rfloor + 1}{d}, \frac{1}{2}\right)\right)d} d \log(ms/\delta)\right)$$

queries, and runs in time $O(Q(m, d, \delta)n)$.

Proof. In the first iteration of the algorithm, we run **FindMonomial**$(f, d, \delta/s)$ to find one monomial. Suppose at iteration t the algorithm has t monomials $M_1, \ldots, M_t$ of f. In the $t + 1$ iteration, we run **FindMonomial**$(f + \sum_{i=1}^{t} M_i, d, \delta/s)$ to find a new monomial of f.

The correctness and query complexity follows from Lemma 8. □

FindMonomial(f, d, δ)
Input: Black-box access to $f \in P_{m,d,s}$
Output: Find a monomial of f.
Procedure **Test**(f, g, δ) in Figure 1 tests, with confidence $1 - \delta$, if $f = g$ using Lemma 5.
If $f \neq g$ then it returns an assignment u such that $f(u) \neq g(u)$.

1. $a = (1, 1, 1, \ldots, 1) \in \{0, 1\}^m$.
2. Repeat $t = 8d \ln \frac{m}{\delta}$ times
3.   Draw $b \in \mathcal{D}_{m,p}$ where $p = 2^{-1/d}$.
4.   **Test**$(f(b * a * x), 0, 1/2)$
5.   If $f(b * a * x) \neq 0$ then $a \leftarrow a * b$.
6. Return $\prod_{a_i=1} x_i$

Fig. 2. For $f \in P_{m,d,s}$ returns a monomial of f.

4.3   The Algorithm

In this section, we give the algorithm for $P_{n,s}$. We prove

Theorem 4. *Let $\epsilon = 1/s^\beta$, $\beta \geq 1$. There is a proper learning algorithm for s-sparse polynomial with probability of success at least $2/3$ that makes*

$$q_U = \left(\frac{s}{\epsilon} \right)^{\gamma(\beta) + o_s(1)} + O\left(s \left(\log \frac{1}{\epsilon} \right) \log n \right) \tag{7}$$

queries and runs in time $O(q_U \cdot n)$ where

$$\gamma(\beta) = \min_{0 \leq \eta \leq 1} \frac{\eta + 1}{\beta + 1} + (1 + 1/\eta) H_2 \left(\frac{1}{(1 + 1/\eta)(\beta + 1)} \right).$$

In particular

1. $\gamma(\beta) = \frac{\log \beta}{\beta} + \frac{4.413}{\beta} + \Theta\left(\frac{1}{\beta^2} \right)$.
2. $\gamma(\beta) < 1$ *for $\beta > 6.219$. That is, the query complexity is sublinear in $1/\epsilon$ and almost linear in s when $\beta > 6.219$.*
3. *For $\beta > 4.923$ the query complexity is better than the best known query complexity (which is $s^2/\epsilon = (s/\epsilon)^{(2+\beta)/(1+\beta)}$).*
4. $\gamma(\beta) \leq 4$ *for all β, and γ is a monotone decreasing function in β.*

We note here that in the full paper, we give another algorithm that improves the bounds in items 2–4. In particular, the query complexity of the algorithm with the above algorithm is better than the best-known query complexity for $\beta > 1$. The above algorithm also works for $\beta < 1$, but the one in full paper has a better query complexity.

We now give the proof of the Theorem

Proof. Since $\log(1/\epsilon)/(\log s) = \beta \geq 1$, we have $\log s < \log(1/\epsilon)$. Let $p > 1/2$ and

$$D = \log\frac{s}{\epsilon} + \frac{\log s + \log\log s + 6}{\log(1/p)} > 2\log s + 2.$$

We will choose p later such that $\log(1/p) = \Theta((\log s)/(\log(1/\epsilon)))$ and therefore $D = O(\log(s/\epsilon)) = O(\log(1/\epsilon))$.

We start from the algorithm in Lemma 9 that exactly learns $P_{(2Ds)^2,D,s}$ with

$$Q_1(D,\delta) = O\left(s2^{D\cdot H_2\left(\frac{\lfloor\log s\rfloor+1}{D}\right)}D\log((2Ds)^2s/\delta)\right)$$

$$= \tilde{O}\left(s2^{D\cdot H_2\left(\frac{\lfloor\log s\rfloor+1}{D}\right)}\right)\log(1/\delta)$$

queries, time $O(Q_1 n)$ and probability of success at least $1 - \delta$.

By the second reduction, Lemma 7, there is a proper learning algorithm that exactly learns $P_{n,D,s}$ with

$$Q_2(D,\delta) = (Q_1(D,1/16) + Ds\log n)\log(1/\delta)$$

$$= \left(\tilde{O}\left(s2^{D\cdot H_2\left(\frac{\lfloor\log s\rfloor+1}{D}\right)}\right) + O\left(s\left(\log\frac{1}{\epsilon}\right)\log n\right)\right)\log(1/\delta).$$

queries in time $O(Q_2(D,\delta)n)$ and probability of success at least $1 - \delta$.

If we now use the first reduction in Lemma 6 as is we get the first summand in the query complexity in (7), but the second summand, $O(s\log(1/\epsilon)\log(n))$, becomes $(s^{1+\log(1/p)}/\epsilon^{\log(1/p)})(\log(1/\epsilon))\log(n)$, which is not what we stated in the Theorem. Instead, we use the first reduction with the following changes.

Notice that the $\log n$ in the summand $ds\log(n)$ in the first reduction resulted from searching for the relevant variable in the set $\{x_u\}_{\phi(u)=t}$ for some $t \in [m]$. See the proof of Lemma 7. Suppose the algorithm knows a priori w relevant variables of the function f and is required to run the second reduction. Then the term $ds\log n$ can be replaced by $(ds - w)\log n$. This is because the reduction needs to search only for the other at most $ds - w$ relevant variables of f. Now, if we use the first reduction in Lemma 6, when we find the relevant variables of f in the p-zero projections $f(z^{(1)}),\ldots,f(z^{(i)})$, we do not need to search for them again in the following p-zero projections $f(z^{(i+1)}),\ldots,f(z^{(w)})$. Therefore, the query complexity of the search of all the variables remains $O(s\log(1/\epsilon)\log(n))$.

Therefore, after using the first reduction with the above modification, we get a proper learning algorithm that learns $P_{n,s}$ that makes

$$Q_3(d,\delta,\epsilon) = \left(\tilde{O}\left(s\left(\frac{s}{\epsilon}\right)^{\log(1/p)}2^{D\cdot H_2\left(\frac{\lfloor\log s\rfloor+1}{D}\right)}\right) + O\left(s\left(\log\frac{1}{\epsilon}\right)\log n\right)\right)\log\frac{1}{\delta}$$

queries in time $O(n\cdot Q_3)$, probability of success at least $1-\delta$ and accuracy $1-\epsilon$.

Now recall that $\epsilon = 1/s^\beta$ and choose $p = 2^{-\eta/(\beta+1)}$ for a constant $0 < \eta < 1$. Then $p > 1/2$, $\log(1/p) = \eta/(\beta + 1) = \Theta((\log s)/\log(1/\epsilon))$,

$$D = (\beta + 1)\left(1 + \frac{1}{\eta}\right)\log s + \Theta(\beta\log\log s) = \left(1 + \frac{1}{\eta}\right)\log\frac{s}{\epsilon} + \Theta(\beta\log\log s)$$

and

$$s\left(\frac{s}{\epsilon}\right)^{\log(1/p)} \quad 2^{D\cdot H_2\left(\frac{\lfloor \log s\rfloor+1}{D}\right)}$$

$$= \left(\frac{s}{\epsilon}\right)^{\frac{\eta+1}{\beta+1}}\left(\frac{s}{\epsilon}\right)^{\left((1+1/\eta)+\Theta\left(\frac{\log\log s}{\log s}\right)\right)H_2\left(\frac{1}{(1+1/\eta)(\beta+1)}\left(1-\Theta\left(\frac{\log\log s}{\beta\log s}\right)\right)\right)}$$

$$= \left(\frac{s}{\epsilon}\right)^{\frac{\eta+1}{\beta+1}+(1+1/\eta)H_2\left(\frac{1}{(1+1/\eta)(\beta+1)}\right)(1-o_s(1))}.$$

This completes the proof.   □

5   Lower Bounds

In this section, we prove the following lower bound for learning sparse polynomials.

Theorem 5. *Let $\epsilon = 1/s^\beta$. Any learning algorithm for s-sparse polynomial with a confidence probability of at least 2/3 must make at least*

$$\tilde{\Omega}\left(\left(\frac{s}{\epsilon}\right)^{\frac{\beta\cdot H_2(\min(1/\beta,1/2))}{\beta+1}}\right) + \Omega\left(s\left(\log\frac{1}{\epsilon}\right)\log n\right)$$

$$= \left(\frac{s}{\epsilon}\right)^{\frac{\log\beta}{\beta}+\frac{1}{(\ln 2)\beta}+\Omega\left(\frac{\log\beta}{\beta^2}\right)} + \Omega\left(s\left(\log\frac{1}{\epsilon}\right)\log n\right)$$

queries.

We first give the following lower bound that proves the second summand in the lower bound

Lemma 10. *Any learning algorithm for $P_{n,s}$ with a confidence probability of at least 2/3 must make at least $\Omega(s(\log(1/\epsilon))\log n)$ queries.*

Proof. Consider the class $C = P_{n,\log(1/(2\epsilon)),s}$. Consider a (randomized) learning algorithm A_R for $P_{n,s}$ with a confidence probability of at least 2/3 and accuracy ϵ. Then A_R is also a (randomized) learning algorithm for C. Since by Lemma 1, any two distinct functions in C have distance 2ϵ, A_R exactly learns C with a confidence probability of at least 2/3. This is because, after learning an ϵ-close formula h, since any two distinct functions in C have distance 2ϵ, the closest function in C to h is the target function. By Yao's minimax principle, there is a deterministic non-adaptive exact learning algorithm A_D with the same query complexity as A_R that learns at least $(2/3)|C|$ functions in C. By the standard information-theoretic lower bound, the query complexity of A_D is at least $\log((2/3)|C|)$. Since

$$\log|C| = \log\left(\binom{n}{\log(1/(2\epsilon))}_s\right) = \Omega\left(\left(\log\frac{1}{\epsilon}\right)s\log n\right)$$

the result follows.   □

We now give the following lower bound that proves the second summand in the lower bound

Lemma 11. *Let* $\epsilon = 1/s^\beta$. *Any learning algorithm for* $P_{n,s}$ *with a confidence probability of at least* 2/3 *must make at least*

$$\Omega\left(\left(\frac{s}{\epsilon}\right)^{\frac{\beta \cdot H_2(\min(1/\beta, 1/2))}{\beta+1}}\right)$$

queries.

Proof. We first prove the lower bound for $\beta > 1$. Let $t = \log(1/\epsilon) - \log s - 1$ and $r = \log s$. Let W be the set of all pairs (I, J) where I and J are disjoint sets, $I \cup J = [t+r]$, $|I| \geq t$ and $|J| = t+r-|I| \leq r$. For every $(I, J) \in W$ define $f_{I,J} = \prod_{i \in I} x_i \prod_{j \in J} (1+x_j)$. Obviously, for two distinct $(I_1, J_1), (I_2, J_2)$, $f_{I_1,J_1} \cdot f_{I_2,J_2} = 0$. Consider the set $C = \{f_{I,J} | (I, J) \in W\}$. First notice that $C \subset P_{n,t+r,s} \subseteq P_{n,s}$ and, by Lemma 1, $\Pr[f_{I,J} = 1] \geq 2^{-(t+r)} = 2^{-\log(1/\epsilon)+1} = 2\epsilon$. Furthermore, since for $(I_1, J_1) \neq (I_2, J_2)$ the degree of $f_{I_1,J_1} + f_{I_2,J_2}$ is $\log(1/\epsilon) - 1$, we also have

$$\Pr[f_{I_1,J_1} \neq f_{I_2,J_2}] \geq 2\epsilon. \tag{8}$$

Therefore, any learning algorithm for $P_{n,s}$ (with accuracy ϵ and confidence 2/3) is a learning algorithm for C and thus is an exact learning algorithm for C. This is because, after learning an ϵ-close formula h, by (8), the closest function in C to h is the target function.

Consider now a (randomized) non-adaptive exact learning algorithm A_R for C with probability of success at least 2/3 and accuracy ϵ. By Yao's minimax principle, there is a deterministic non-adaptive exact learning algorithm A_D such that, for uniformly at random $f \in C$, with a probability at least 2/3, A_D returns f. We will show that A_D must make more than $q = (1/10)|C|$ queries. Now since,

$$|C| = \sum_{i=0}^{\log s} \binom{\log \frac{1}{\epsilon} - 2}{i} \geq \tilde{\Omega}\left(2^{H_2\left(\min\left(\frac{\log s}{\log(1/\epsilon)}, \frac{1}{2}\right)\right)\log(1/\epsilon)}\right)$$

$$= \tilde{\Omega}\left(\left(\frac{1}{\epsilon}\right)^{H_2(\min(1/\beta, 1/2))}\right) = \tilde{\Omega}\left(\left(\frac{s}{\epsilon}\right)^{\frac{\beta \cdot H_2(\min(1/\beta, 1/2))}{\beta+1}}\right)$$

the result follows.

To this end, suppose for the contrary, A_D makes q queries. Let $S = \{a^{(1)}, \ldots, a^{(q)}\}$ be the queries that A_D makes. For every $(I, J) \in W$ let $S_{I,J} = \{a \in S | f_{I,J}(a) = 1\}$. Since for any two distinct $(I_1, J_1), (I_2, J_2) \in W$ we have $f_{I_1,J_1} \cdot f_{I_2,J_2} = 0$, the sets $\{S_{I,J}\}_{(I,J) \in W}$ are disjoint sets.

Let $f = f_{I',J'}$ be uniformly at random function in C. We will show that, with probability at least 4/5, A_D fails to learn f, which gives a contradiction. Since

$$E_{(I,J) \in W}[|S_{I,J}|] = \frac{\sum_{(I,J) \in W} |S_{I,J}|}{|W|} = \frac{q}{w} = \frac{1}{10},$$

at least $(9/10)|W|$ of the $S_{I,J}$ are empty sets. Therefore, with probability at least $9/10$, $S_{I',J'}$ is an empty set. In other words, with probability at least $9/10$, the answers to the all the queries are 0. If the answers to all the queries are zero, then with probability at most $1/10$, the algorithm can guess I', J', and therefore, the failure probability of the algorithm is at least $4/5$. This proves the case $\beta > 1$.

Now we prove the result for $0 < \beta \leq 1$. By Lemma 10, we get the lower bound $\Omega(s)$. Since $s = \left(\frac{s}{\epsilon}\right)^{1/(\beta+1)}$ and for $0 < \beta \leq 1$

$$\frac{1}{\beta+1} \geq \frac{\beta \cdot H_2(\min(1/\beta, 1/2))}{\beta+1}$$

the result follows. $\square$

References

1. Angluin, D.: Queries and concept learning. Mach. Learn. **2**(4), 319–342 (1987)
2. Beimel, A., Bergadano, F., Bshouty, N.H., Kushilevitz, E., Varricchio, S.: Learning functions represented as multiplicity automata. J. ACM, **47**(3), 506–530, 2000. https://doi.org/10.1145/337244.337257
3. Bergadano, F., Bshouty, N.H., Varricchio, S.: Learning multivariate polynomials from substitution and equivalence queries. Electron. Colloquium Comput. Complex. **8** 1996. https://eccc.weizmann.ac.il/eccc-reports/1996/TR96-008/index.html
4. Bisht, L., Bshouty, N.H., Mazzawi, H.: On optimal learning algorithms for multiplicity automata. In: Lugosi, G., Simon, H.U. (eds.) COLT 2006. LNCS (LNAI), vol. 4005, pp. 184–198. Springer, Heidelberg (2006). https://doi.org/10.1007/11776420_16
5. Blais, E., Brody, J., Matulef, K.: Property testing lower bounds via communication complexity. In Proceedings of the 26th Annual IEEE Conference on Computational Complexity, CCC 2011, San Jose, California, USA, 8–10 June 2011, pp. 210–220, 2011. https://doi.org/10.1109/CCC.2011.31
6. Blum, A., Singh, M.: Learning functions of PHK terms. In Fulk, M.A., Case, J. (eds.) Proceedings of the Third Annual Workshop on Computational Learning Theory, COLT 1990, University of Rochester, Rochester, NY, USA, 6–8 August 1990, pp. 144–153. Morgan Kaufmann, 1990. http://dl.acm.org/citation.cfm?id=92620
7. Blum, M., Luby, M., Rubinfeld, R.: Self-testing/correcting with applications to numerical problems. J. Comput. Syst. Sci. **47**(3), 549–595 (1993). https://doi.org/10.1016/0022-0000(93)90044-W
8. Bshouty, N.H.: On learning multivariate polynomials under the uniform distribution. Inf. Process. Lett. **61**(6), 303–30 (1997). https://doi.org/10.1016/S0020-0190(97)00021-5
9. Bshouty, N.H.: Almost optimal testers for concise representations. Electron. Colloquium Comput. Complex. (ECCC) **26**, 156 (2019). https://eccc.weizmann.ac.il/report/2019/156
10. Bshouty, N.H., Goldreich, O.: On properties that are non-trivial to test. Electron. Colloquium Comput. Complex. (ECCC) **13** (2022). https://eccc.weizmann.ac.il/report/2022/013/

11. Bshouty, N.H., Mansour, Y.: Simple learning algorithms for decision trees and multivariate polynomials. SIAM J. Comput. **31**(6), 1909–192 (2002). https://doi.org/10.1137/S009753979732058X
12. Chakraborty, S., García-Soriano, D., Matsliah, A.: Efficient sample extractors for juntas with applications. In Automata, Languages and Programming - 38th International Colloquium, ICALP 2011, Zurich, Switzerland, 4–8 July 2011, Proceedings, Part I, pp. 545–556, 2011. https://doi.org/10.1007/978-3-642-22006-7_46
13. Clausen, M., Dress, A.W.M., Grabmeier, J., Karpinski, M.: On zero-testing and interpolation of k-sparse multivariate polynomials over finite fields. Theor. Comput. Sci. **84**(2), 151–164 (1991). https://doi.org/10.1016/0304-3975(91)90157-W
14. Diakonikolas, I., et al.: Testing for concise representations. In: 48th Annual IEEE Symposium on Foundations of Computer Science (FOCS 2007), 20–23 October 2007, Providence, RI, USA, Proceedings, pp. 549–558 (2007). https://doi.org/10.1109/FOCS.2007.32
15. Diakonikolas, I., Lee, H.K., Matulef, K., Servedio, R.A., Wan, A.: Efficiently testing sparse phGF(2) polynomials. Algorithmica **61**(3), 580–605 (2011). https://doi.org/10.1007/s00453-010-9426-9
16. Dür, A., Grabmeier, J.: Applying coding theory to sparse interpolation. SIAM J. Comput. **22**(4), 695–704 (1993).https://doi.org/10.1137/0222046
17. Fischer, P., Simon, H.U.: On learning ring-sum-expansions. SIAM J. Comput. **21**(1), 181–192 (1992). https://doi.org/10.1137/0221014
18. Goldreich, O. (ed.): Property Testing. LNCS, vol. 6390. Springer, Heidelberg (2010). https://doi.org/10.1007/978-3-642-16367-8
19. Goldreich, O.: Introduction to Property Testing. Cambridge University Press (2017). http://www.cambridge.org/us/catalogue/catalogue.asp?isbn=9781107194052
20. Hellerstein, L., Servedio, R.A.: On PAC learning algorithms for rich Boolean function classes. Theor. Comput. Sci. **384**(1), 66–76 (2007). https://doi.org/10.1016/j.tcs.2007.05.018
21. Ron, D.: Property testing: a learning theory perspective. Found. Trends Mach. Learn. **1**(3), 307–402 (2008). https://doi.org/10.1561/2200000004
22. Ron, D.: Algorithmic and analysis techniques in property testing. Found. Trends Theor. Comput. Sci. **5**(2), 73–20 (2009). https://doi.org/10.1561/0400000029
23. Roth, R.M., Benedek, G.M.: Interpolation and approximation of sparse multivariate polynomials over GF(2). SIAM J. Comput. **20**(2), 291–314 (1991). https://doi.org/10.1137/0220019
24. Rubinfeld, R., Sudan, M.: Robust characterizations of polynomials with applications to program testing. SIAM J. Comput. **25**(2), 252–271 (1996). https://doi.org/10.1137/S0097539793255151
25. Saglam, M.: Near log-convexity of measured heat in (discrete) time and consequences. In: 59th IEEE Annual Symposium on Foundations of Computer Science, FOCS 2018, Paris, France, 7–9 October 2018, pp. 967–978 (2018). https://doi.org/10.1109/FOCS.2018.00095
26. Schapire, R.E., Sellie, L.: Learning sparse multivariate polynomials over a field with queries and counterexamples. J. Comput. Syst. Sci. **52**(2), 201–213 (1996). https://doi.org/10.1006/jcss.1996.0017
27. Valiant, L.G.: A theory of the learnable. Commun. ACM **27**(11), 1134–1142 (1984). https://doi.org/10.1145/1968.1972

Estimating the Clustering Coefficient Using Sample Complexity Analysis

Alane M. de Lima$^{(\boxtimes)}$, Murilo V. G. da Silva, and André L. Vignatti

Department of Computer Science, Federal University of Paraná, Curitiba, Brazil
{amlima,murilo,vignatti}@inf.ufpr.br

Abstract. In this work we present a sampling algorithm for estimating the local clustering of each vertex of a graph. Let G be a graph with n vertices, m edges, and maximum degree Δ. We present an algorithm that, given G and fixed constants $0 < \varepsilon, \delta, p < 1$, outputs the values for the local clustering coefficient within ε error with probability $1 - \delta$, for every vertex v of G, provided that the (exact) local clustering of v is not "too small." We use VC dimension theory to give a bound for the number of edges required to be sampled by the algorithm. We show that the algorithm runs in time $\mathcal{O}(\Delta \lg \Delta + m)$. We also show that the running time drops to, possibly, sublinear time if we restrict G to belong to some well-known graph classes. In particular, for planar graphs the algorithm runs in time $\mathcal{O}(\Delta)$. In the case of bounded-degree graphs the running time is $\mathcal{O}(1)$ if a bound for the value of Δ is given as a part of the input, and $\mathcal{O}(n)$ otherwise.

Keywords: Clustering coefficient · Approximation algorithm · Sampling · VC dimension

1 Introduction

The occurrence of *clusters* in networks is a central field of investigation in the area of network theory [9,31]. The existence of such phenomena motivated the creation of a variety of measures in order to quantify its prevalence; the *clustering coefficient* [2,6,24] is one of the most popular of these measures.

There are global and local versions of the clustering coefficient. Given a graph, its *global clustering coefficient* is a value that quantifies the overall clustering of the graph in terms of the number of existing triangles. If the objective, however, is to analyze features of complex networks such as modularity, community structure, assortativity, and hierarchical structure, then the concept of *local clustering coefficient* is a better fit. This measure quantifies the degree in which a vertex is a part of a cluster in a graph. Simply speaking, the measure is related to the ratio of the number of triangles existing in the neighborhood of the target vertex to the total number of pair of nodes in the neighborhood. A precise definition for this measure is provided in Definition 1.

© Springer Nature Switzerland AG 2022
A. Castañeda and F. Rodríguez-Henríquez (Eds.): LATIN 2022, LNCS 13568, pp. 328–341, 2022.
https://doi.org/10.1007/978-3-031-20624-5_20

An exact algorithm for computing the local clustering coefficient of each vertex of a graph typically runs in cubic time. When dealing with large scale graphs, however, this is inefficient in practice, and high-quality approximations obtained with high confidence are usually sufficient. More specifically, given an error parameter ε, a confidence parameter δ, and an adjustable lower bound parameter p (treated as constants), the idea is to sample a subset of edges in the graph such that the values for the local clustering coefficient can be estimated within ε error from the exact value with probability $1 - \delta$, for each vertex that respects a certain function of the parameter p. At the core of our strategy we use VC-dimension theory to give a bound on the size of the sample in order to meet the desired quality guarantees.

1.1   Related Work

The local clustering coefficient was originally proposed by Watts and Strogatz (1998) [31] in order to determine if a graph has the property of being *small-world*. Intuitively, this coefficient measures how close the neighborhood of a vertex is to being a clique. Over the years, many variants of this measure have been proposed, making it somewhat difficult to provide a unified comparison between all these approaches under the light of algorithmic complexity.

One of these variations is the study of Soffer and Vázquez (2005) [30] on the influence of the degree of a vertex on the local clustering computation, a modification on the original measure where the *degree-correlation* is filtered out. The work of Li et al. (2018) [16] provides a measure combining the local clustering coefficient and the local-degree sum of a vertex, but focused on a specific application of influence spreading. Other extensions of the measure and their applications in particular scenarios include link prediction [8,33] and community detection [11,20,23,25,36]. In the theoretical front, working on random graphs, Kartun–Gilles and Bianconi (2019) [12] gives a statistical analysis of the topology of nodes in networks from different application scenarios. There are also many recent bounds for the average clustering of *power-law* graphs [3,7,10,14], a graph model that represents many social and natural phenomena.

The algorithmic complexity of the exact computation of the local clustering coefficient of each vertex of a graph typically runs in cubic time. In our work, however, we are interested in faster approximation algorithms for obtaining good quality estimations. Let $n = |V|$ and $m = |E|$ in a graph G. In the work of Kutzkov and Pagh (2013) [15], the authors show an ε-approximation streaming algorithm for the local clustering coefficient of each vertex of degree at least d in expected time $\mathcal{O}(\frac{m}{\alpha\varepsilon^2} \log \frac{1}{\varepsilon} \log \frac{n}{\delta})$, where α is the local clustering coefficient of such vertex. This holds with probability at least $1 - \delta$. In the work of Zhang et al. (2017) [34], the authors propose an ε-approximation *MapReduce*-based algorithm for the problem, and empirically compare its performance with other approximation algorithms designed using this type of approach [13,28].

Results for the computation of the top k vertices with the highest local clustering coefficient were also proposed [4,17,35]. In particular, Zhang et al. (2015) [35] use VC dimension and the ϵ-sample theorem on their algorithm analysis, but in a different sample space than the one that we are facing here, and for a scenario which is not exactly the one that we are tackling. In fact, sample complexity analysis has been shown to be an effective tool in the design of some graph algorithms, e.g. the computation of betweenness [26,27] and percolation centralities [5,19].

1.2 Our Results

In this paper we present an algorithm that samples edges from an input graph G and, for fixed constants $0 < \varepsilon, \delta, p < 1$, outputs an estimate $\tilde{l}(v)$ for the exact value $l(v)$ of the local clustering coefficient of each vertex $v \in V$, such that $|l(v) - \tilde{l}(v)| \leq \varepsilon l(v)$, with probability at least $1 - \delta$ whenever $l(v)$ is at least $pm/\binom{\delta_v}{2}$, where δ_v is the degree of v. The main theme in our work is that, by using Vapnik–Chervonenkis (VC) dimension theory, we can obtain an upper bound for the sample size that is tighter than the ones given by standard Hoeffding and union-bound sampling techniques. In particular, we show that the sample size does not depend of the size of G, but on a specific property of it, more precisely, its maximum degree Δ.

In Sect. 3.1 we give a definition for the VC dimension of a graph and show in Theorem 2 that, for any graph, the VC dimension is at most $\lfloor \lg{(\Delta - 1)} \rfloor + 1$. The sample size used in the algorithm depends, roughly speaking, on this value. In Corollary 1, we show that our analysis is tight by presenting an explicit construction of a class of graphs for which the VC dimension reaches this upper bound. Even so, we also provide a tighter analysis for the case in which the input graph belongs to certain graph classes. In the class of *bounded-degree graphs* the VC dimension is bounded by a constant. In the case of *planar graphs*, we show, in Corollary 2, that the VC dimension is at most 2.

In Sect. 3.2, we show that the running time for the general case of our algorithm is $\mathcal{O}(\Delta \lg \Delta + m)$. In Corollaries 3 and 4 we present an analysis for planar graphs and for bounded-degree graphs, cases where the running time drops to, possibly, sublinear time. In the case of planar graphs, our algorithm has running time $\mathcal{O}(\Delta)$. In the case of bounded-degree graphs the running time is $\mathcal{O}(1)$ if a bound for the value of Δ is given as a part of the input, and $\mathcal{O}(n)$, otherwise.

2   Preliminaries

In this section, we present the definitions, notation, and results that are the groundwork of our proposed algorithm. In all results of this paper, we assume w.l.o.g. that the input graph is connected, since otherwise the algorithm can be applied separately to each of its connected components.

2.1   Graphs and Local Clustering Coefficient

Let $G = (V, E)$ be a graph where V is the set of vertices and E the set of edges. We use the convention that $n = |V|$ and $m = |E|$. For each vertex $v \in V$, let δ_v

be the degree of v, and $\Delta_G = \max_{v \in V}\{\delta_v\}$ the maximum degree of the graph G. When the context is clear, we simply use Δ instead of Δ_G. We refer to a triangle as being a complete graph with three vertices. Given $v \in V$, we let T_v be the number of triangles that contain v.

Definition 1 *(Local Clustering Coefficient). Given a graph $G = (V, E)$, the local clustering coefficient of a vertex $v \in V$ is*

$$l(v) = \frac{2T_v}{\delta_v(\delta_v - 1)}.$$

2.2 Sample Complexity and VC Dimension

In sampling algorithms, we typically want to estimate a certain quantity observing some parameters of quality and confidence. The sample complexity analysis relates the minimum size of a random sample required to obtain results that are consistent with the desired parameters. An upper bound to the Vapnik–Chervonenkis Dimension (VC dimension) of a class of binary functions, a central concept in sample complexity theory, is especially defined in order to model the particular problem that we are facing. An upper bound to the VC dimension is also an upper bound to the sample size that respects the desired quality and confidence parameters.

Generally speaking, the VC dimension measures the expressiveness of a class of subsets defined on a set of points [26]. An in-depth exposition of the definitions and results presented below can be found in the books of Anthony and Bartlett (2009) [1], Mohri *et al.* (2012) [22], Shalev-Shwartz and Ben-David (2014) [29], and Mitzenmacher and Upfal (2017) [21].

Definition 2 (Range space). *A* range space *is a pair $\mathcal{R} = (U, \mathcal{I})$, where U is a* domain *(finite or infinite) and $\mathcal{I}$ is a collection of subsets of U, called* ranges.

For a given $S \subseteq U$, the *projection* of $\mathcal{I}$ on S is the set $\mathcal{I}_S = \{S \cap I : I \in \mathcal{I}\}$. If $|\mathcal{I}_S| = 2^{|S|}$ then we say S is *shattered* by $\mathcal{I}$. The VC dimension of a range space is the size of the largest subset S that can be shattered by $\mathcal{I}$, i.e.,

Definition 3 (VC dimension). *Given a range space $\mathcal{R} = (U, \mathcal{I})$, the VC dimension of $\mathcal{R}$, denoted VCDim($\mathcal{R}$), is*

$$VCDim(\mathcal{R}) = \max\{d : \exists S \subseteq U \text{ such that } |S| = d \text{ and } |\mathcal{I}_S| = 2^d\}.$$

The following combinatorial object, called a *relative (p, ε)-approximation*, is useful in the context when one wants to find a sample $S \subseteq U$ that estimates the size of ranges in $\mathcal{I}$, with respect to an adjustable parameter p, within relative error ε, for $0 < \varepsilon, p < 1$. This holds with probability at least $1 - \delta$, for $0 < \delta < 1$, where π is a distribution on U and $\Pr_\pi(I)$ is the probability of a sample from π belongs to I.

Definition 4 (Relative (p, ε)-Approximation, see [26], Def. 5). *Given the parameters $0 < p, \varepsilon < 1$, a set S is called a (p, ε)-approximation w.r.t. a range space $\mathcal{R} = (U, \mathcal{I})$ and a distribution π on U if for all $I \in \mathcal{I}$,*

(i) $\left| \mathrm{Pr}_\pi(I) - \frac{|S \cap I|}{|S|} \right| \leq \varepsilon \, \mathrm{Pr}_\pi(I), \quad$ *if* $\mathrm{Pr}_\pi(I) \geq p$,

(ii) $\frac{|S \cap I|}{|S|} \leq (1 + \varepsilon)p$, *otherwise.*

An upper bound to the VC dimension of a range space allows to build a sample S that is a (p, ε)-approximation set.

Theorem 1 (see [18], Theorem 5). *Given $0 < \varepsilon, \delta, p < 1$, let $\mathcal{R} = (U, \mathcal{I})$ be a range space with $VCDim(\mathcal{R}) \leq d$, let π be a given distribution on U, and let c' be an absolute positive constant. A collection of elements $S \subseteq U$ sampled w.r.t. π with*

$$|S| \geq \frac{c'}{\varepsilon^2 p} \left(d \log \frac{1}{p} + \log \frac{1}{\delta} \right)$$

is a relative (p, ε)-approximation with probability at least $1 - \delta$.

3   Estimation for the Local Clustering Coefficient

We first define the range space associated to the local clustering coefficient of a graph G and its corresponding VC dimension, and then we describe the proposed approximation algorithm.

3.1   Range Space and VC Dimension Results

Let $G = (V, E)$ be a graph. The range space $\mathcal{R} = (U, \mathcal{I})$ associated with G is defined as follows. The universe U is defined to be the set of edges E. We define a range τ_v, for each $v \in V$, as $\tau_v = \{e \in E : \text{both endpoints of } e \text{ are neighbors of } v \text{ in } G\}$, and the range set corresponds to $\mathcal{I} = \{\tau_v : v \in V\}$. For the sake of simplicity, we often use $VCDim(G)$ (instead of $VCDim(\mathcal{R})$) to denote the VC dimension of the range space $\mathcal{R}$ associated with G.

Theorem 2 shows an upper bound for $VCDim(G)$.

Theorem 2. $VCDim(G) \leq \lfloor \lg (\Delta - 1) \rfloor + 1$.

Proof. By definition, an edge $e \in E$ belongs to a range τ_v if both endpoints of e, say, a and b, are neighbors of v. That is, the number of ranges that contain e corresponds to the common neighbors of a and b. Let N be the set of such common neighbors. The maximum number of common neighbors a pair of vertices may have is Δ. Therefore, e is contained in at most $\Delta - 1$ ranges. Assuming that $VCDim(\mathcal{R}) = d$, then from Definition 3, the edge e must appear in 2^{d-1} ranges. We have

$$2^{d-1} \leq \Delta - 1 \Longrightarrow d - 1 \leq \lg (\Delta - 1) \Longrightarrow d \leq \lfloor \lg(\Delta - 1) \rfloor + 1.$$

$\square$

One may ask when the bound given in Theorem 2 is tight. We now present an explicit construction of a family of graphs $\mathcal{G} = (G_d)_{d \geq 3}$ in order to show that this bound is tight with relation to Δ. A graph G_d, for $d \geq 3$, of this family is constructed as follows. Initially, we create d disjoint edges $e_1, \ldots, e_d$. The endpoints of these edges are called *non-indexed vertices*. For every non-empty subset of k edges $e_{i_1}, \ldots, e_{i_k}$, for $1 \leq i_1 < i_2 < \ldots < i_k \leq d$, we create a vertex $v_{(i_1, i_2, \ldots, i_k)}$ and connect it to both endpoints of each edge in the subset. These vertices are called *indexed vertices*. Figure 1 illustrates G_3 and G_4.

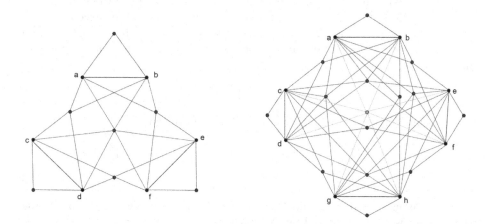

Fig. 1. The first two graphs of the construction of the family $\mathcal{G}$. In the case of G_3 (left), the edges of S are $e_1 = \{a, b\}$, $e_2 = \{c, d\}$, and $e_3 = \{e, f\}$. In the case of G_4 (right), the edges of S are $e_1 = \{a, b\}$. $e_2 = \{c, d\}$, $e_3 = \{e, f\}$, and $e_4 = \{(g, h\}$. Non-indexed vertices are labeled and depicted in black. We depict the indexed vertices in different colors, depending on the size of its neighborhood in S.

Claim. $\Delta_{G_d} = 2^{d-1} + 1$.

Proof. A vertex v in a graph $\mathcal{G}$ can be either indexed or non-indexed. We analyze each case separately.

Let v be a non-indexed vertex that is an endpoint of an edge e_j. W.l.o.g., we may assume that $j = 1$. The vertex v is adjacent to every indexed vertex with indices of the form $(1, i_1, \ldots, i_k)$. The first index is fixed, so there are 2^{d-1} indices of this form. So v is adjacent to 2^{d-1} indexed vertices. Also, v is adjacent to the other endpoint of e_1. Therefore, the degree of any non-indexed vertex is $2^{d-1} + 1$.

The degree of an indexed vertex cannot be larger than $2d$, since such vertex is adjacent to, at most, both endpoints of each edge $e_1, \ldots, e_d$. Since $2^{d-1} + 1 \geq 2d$, the result follows.   □

Theorem 3. *For every $d \geq 3$, $VCDim(G_d) \geq \lfloor \lg(\Delta_{G_d} - 1) \rfloor + 1$.*

Proof. Remember, $\mathcal{R} = (U, \mathcal{I})$, where $U = E$ and $\mathcal{I} = \{\tau_v : v \in V\}$ where $\tau_v = \{e \in E : \text{the endpoints of } e \text{ are neighbors of } v \text{ in } G\}$. First, we present a sample $S \subseteq U$, $|S| = d$, which is shattered, i.e. $|\mathcal{I}_S| = 2^d$, concluding that the VC dimension is at least d. After that, we show that $d = \lfloor \log(\Delta_{G_d} - 1) \rfloor + 1$, which proves the theorem.

Let $S = \{e_1, \ldots, e_d\}$. Consider an indexed vertex $v' = v_{(i_1, i_2, \ldots, i_k)}$. By the construction of the graph, we have that $S \cap \tau_{v'} = \{e_{i_1}, \ldots, e_{i_k}\}$, for all $\tau_{v'}$, i.e., there is a one-to-one mapping of each v' to each $S \cap \tau_{v'}$. Since there are $2^d - 1$ indexed vertices v' (there is an indexed vertex for every subset except for the empty set), then there are $2^d - 1$ different intersections. Finally, the intersection that generates the empty set can be obtained by $S \cap \tau_{v''}$, where v'' is any non-indexed vertex. In other words,

$$|\{S \cap \tau_v \mid \tau_v \in \mathcal{I}\}| = |\mathcal{I}_S| = 2^d,$$

i.e., $VCDim(G_d) \geq d$. Now, using Claim 3.1, we have that

$$\lfloor \log(\Delta_{G_d} - 1) \rfloor + 1 = \lfloor \log(2^{d-1} + 1 - 1) \rfloor + 1 = \lfloor d - 1 \rfloor + 1 = d.$$

$\square$

Combining Theorems 2 and 3, we conclude that the VC dimension of the range space is tight, as stated by Corollary 1.

Corollary 1. *For every $d \geq 3$, there is a graph G such that*

$$VCDim(G) = d = \lfloor \lg(\Delta - 1) \rfloor + 1.$$

Next we define a more general property that holds for a graph G_d.

Property P. We say that a graph $G = (V, E)$ has the *Property P* if exists $S \subseteq E$, $|S| \geq 3$, such that:
(i) For each $e = \{u, v\} \in S$, $|e \cap \{S \setminus \{e\}\}| \leq 1$.
(ii) For each subset $S' \subseteq S$, there is at least one vertex $v_{S'}$ that is adjacent to both endpoints of each edge of S'.

For every $d \geq 3$, Theorem 4 gives conditions based on Property P that a graph must obey in order to have VC dimension at least d.

Theorem 4. *Let G be a graph. If $VCdim(G) \geq 3$, then G has Property P.*

Proof. We prove the contrapositive of the statement, i.e., we show that if G does not have Property P, then $VCdim(G) < 3$. Note that if we assume that G does not have Property P, then for all $S \subseteq E, |S| \geq 3$, we have that either condition (i) or condition (ii) is false.

If it is the case that (ii) is false, then for all $S \subseteq E, |S| \geq 3$, there is a set $S' \subseteq S$ such that there is no $v_{S'} \in V$ which is adjacent to both endpoints of each edge in S'. We have that the number of subsets of S is $2^{|S|}$, so G must have at least $2^{|S|}$ vertices so that $\mathcal{I}_S = 2^{|S|}$. From the definition of shattering, if $\mathcal{I}_S < 2^{|S|}$, then it is not possible that VCdim$(G) \geq |S|$. Since $|S| \geq 3$, it cannot be the case that VCdim$(G) \geq 3$.

Now consider the case where (i) is false. In this case, for all $S \subseteq E, |S| \geq 3$, there is an edge $e = \{u, v\} \in S$ where both u and v are endpoints of other edges in S (i.e., $|e \cap \{S \setminus \{e\}\}| = 2$). We name such edge $e_2 = \{b, c\}$. Suppose w.l.o.g. that e_2 shares its endpoints with the edges $e_1 = \{a, b\}$ and $e_3 = \{c, d\}$. Then every triangle containing e_1 and e_3 necessarily contains e_2. Denote by z the vertex which forms triangles with e_1 and e_2. Then z also forms a triangle with e_2, since it is adjacent to both b and c, which are the endpoints of e_2. Hence, the subset $\{e_1, e_3\}$ cannot be generated from the intersection of $\mathcal{I}$ with e_1, e_2, and e_3. Therefore it cannot be the case that VCdim$(G) \geq 3$.   □

Although Theorem 2 gives a tight bound for the VC dimension, if we have more information about the type of graph that we are working, we can prove better results. In Corollary 2, we show that if G is a graph from the class of planar graphs, then the VC dimension of G is at most 2. Another very common class of graphs where we can achieve a constant bound for the VC dimension is the class of *bounded-degree graphs*, i.e. graphs where Δ is bounded by a constant. For this class, the upper bound comes immediately from Theorem 2.

Note that, even though planar graphs and bounded-degree graphs are both classes of sparse graphs, such improved bounds for the VC dimension for these classes do not come directly from the sparsity of these graphs, since we can construct a (somewhat arbitrary) class of sparse graphs $\mathcal{G}'$ where the VC dimension is as high as the one given by Theorem 2. The idea is that $\mathcal{G}' = (G_d')_{d \geq 3}$, where each graph G_d' is the union of G_d with a sufficiently large sparse graph. In the other direction, one should note that dense graphs can have small VC dimension as well, since complete graphs have VC dimension at most 2. This comes from the fact that complete graphs do not have the Property P. In fact, for a K_q, $q \geq 4$, the VC dimension is exactly 2, since any set of two edges that have one endpoint in common can be shattered in this graph.

Corollary 2. *If G is a planar graph, then $VCDim(G) \leq 2$.*

Proof. We prove that the VC dimension of the range space of a planar graph is at most 2 by demonstrating the contrapositive statement. More precisely, from Theorem 4, we have that if VCDim$(G) \geq 3$, then G has *Property P*. In this case we show that G must contain a subdivision of a $K_{3,3}$, concluding that G cannot be planar, according to the Theorem of Kuratowski [32].

From Theorem 4, G has a subset of edges $\{e_1, e_2, e_3\}$ respecting conditions (i) and (ii) of *Property P*. Let $e_1 = \{a, b\}$, $e_2 = \{c, d\}$, and $e_3 = \{e, f\}$. Note that these three edges may have endpoints in common. By *condition (i)*, we may assume w.l.o.g. that $a \neq c \neq e$. By symmetry, w.l.o.g., there are three possibilities for the vertices b, d, and f: (1) they are all distinct vertices, (2) we have $d = f$, but $b \neq f$, and (3) they are the same vertex, i.e. $b = d = f$. In Fig. 2 we show three graphs, one for each of these three possible configurations for the arrangement of edges e_1, e_2, and e_3. By *condition (ii)* there are at least four vertices, say, u, v, w, and x respecting the following:

- u is adjacent to all vertices of $\{a, b, c, d, e, f\}$;
- v is adjacent to all vertices of $\{a, b, c, d\}$ and not adjacent to both e and f;
- w is adjacent to all vertices of $\{a, b, e, f\}$ and not adjacent to both c and d;
- x is adjacent to all vertices of $\{c, d, e, f\}$ and not adjacent to both a and b.

Note that, even though every edge depicted in Fig. 2 is mandatory in G, there may be other edges in G that are not shown in the picture.

Since all of $\{v, c\}, \{c, x\}$, and $\{x, e\}$ are edges in G, then there is a path from v to e in G. Let $E(P)$ be the edges of this path. Consider $A = \{a, b, e\}$ and $B = \{u, v, w\}$, and let X be the set of edges with one endpoint in A and one endpoint in B. We can obtain a subgraph H of G that contains a subdivision of a bipartite graph $K_{3,3}$ with bipartition (A, B) in the following way. The vertex set of H is $A \cup B \cup \{c, x\}$ and the edge set of H is $X \cup E(P)$. Therefore G cannot be a planar graph. □

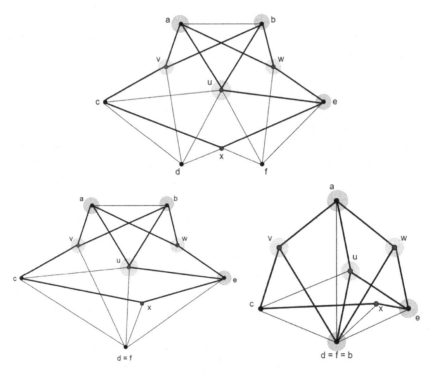

Fig. 2. Three possible arrangements for the edges $e_1 = \{a, b\}$, $e_2 = \{c, d\}$, and $e_3 = \{e, f\}$ from the proof of Corollary 2. The case where b, d, and f are distinct vertices is depicted above. The case where $d = f$, but $b \neq f$ is shown below in the left. Below in the right, we show the case where $b = d = f$. In all three cases, these edges are part of a subgraph H of G that contains a subdivision of a $K_{3,3}$.

3.2   Algorithm

The algorithm takes as input a graph $G = (V, E)$, the quality and confidence parameters $0 < \varepsilon, \delta < 1$, and a parameter $0 < p < 1$, all assumed to be constants. It outputs the estimation $\tilde{l}(v)$ for the exact value $l(v)$ of the local clustering coefficient for each vertex $v \in V$, such that

$$|l(v) - \tilde{l}(v)| \leq \varepsilon l(v), \text{ with prob. at least } 1 - \delta \text{ whenever } l(v) \geq \sigma_v(p),$$

where $\sigma_v(p) = pm/\binom{\delta_v}{2}$ is an adjustable function, depending on p. The idea, roughly speaking, is that $l(v) \geq \sigma_v(p)$ holds if the neighborhood of v is not too small.

Next we present Algorithm 1. At the beginning all $\tilde{T}_v$ are set to zero.

Algorithm 1: LOCALCLUSTERINGESTIMATION(G,ε,δ,p)

input : Graph $G = (V, E)$ with m edges, parameters $0 < \varepsilon, \delta, p < 1$.
output: Local clustering coefficient estimation $\tilde{l}(v)$, $\forall v \in V$ s.t. $l(v) \geq \sigma_v(p)$.

1   $r \leftarrow \left\lceil \frac{c'}{\varepsilon^2 p} \left((\lfloor \lg \Delta - 1 \rfloor + 1) \log \frac{1}{p} + \log \frac{1}{\delta} \right) \right\rceil$
2   **for** $i \leftarrow 1$ *to* r **do**
3   sample an edge $e = \{a, b\} \in E$ uniformly at random
4   **forall** the $v \in N_a$ **do**
5   **if** $v \in N_b$ **then**
6              $\tilde{T}_v \leftarrow \tilde{T}_v + \frac{m}{r}$
7   **return** $\tilde{l}(v) \leftarrow \frac{2\tilde{T}_v}{\delta_v(\delta_v-1)}$, for each $v \in V$.

Theorem 5. *Given a graph $G = (V, E)$, let $S \subset E$ be a sample of size*

$$r = \left\lceil \frac{c'}{\varepsilon^2 p} \left((\lfloor \lg \Delta - 1 \rfloor + 1) \log \frac{1}{p} + \log \frac{1}{\delta} \right) \right\rceil,$$

for given $0 < p, \varepsilon, \delta < 1$. Algorithm 1 returns with probability at least $1 - \delta$ an approximation $\tilde{l}(v)$ to $l(v)$ within ε relative error, for each $v \in V$ such that $l(v) \geq \sigma_v(p)$.

Proof. For each $v \in V$, let $\mathbb{1}_v(e)$ be the function that returns 1 if $e \in \tau_v$ (and 0 otherwise). Thus, $T_v = \sum_{e \in E} \mathbb{1}_v(e)$. The estimated value $\tilde{T}_v$, computed by Algorithm 1, is incremented by m/r whenever an edge $e \in S$ belongs to τ_v, i.e.,

$$\tilde{T}_v = \sum_{e \in S} \frac{m}{r} \mathbb{1}_v(e).$$

Note that

$$\tilde{T}_v = \sum_{e \in S} \frac{m}{r} \mathbb{1}_v(e) = \frac{m}{r} \sum_{e \in S} \mathbb{1}_v(e) = m \cdot \frac{|S \cap \tau_v|}{|S|}.$$

Thus, assuming that we have a relative (p, ε)-approximation (Definition 4),

$$\frac{|T_v - \tilde{T}_v|}{T_v} = \frac{\left| m \cdot \Pr_\pi(\tau_v) - m \cdot \frac{|S \cap \tau_v|}{|S|} \right|}{m \cdot \Pr_\pi(\tau_v)} = \frac{\left| \Pr_\pi(\tau_v) - \frac{|S \cap \tau_v|}{|S|} \right|}{\Pr_\pi(\tau_v)} \leq \varepsilon.$$

Or, simply put, $|T_v - \tilde{T}_v| \leq \varepsilon T_v$. Therefore,

$$|l(v) - \tilde{l}(v)| = \frac{2|T_v - \tilde{T}_v|}{\delta_v(\delta_v - 1)} \leq \frac{2\varepsilon T_v}{\delta_v(\delta_v - 1)} = \varepsilon l(v).$$

Combining this with Theorems 1 and 2, and using a sample S with size

$$r = \left\lceil \frac{c'}{\varepsilon^2 p} \left((\lfloor \lg \Delta - 1 \rfloor + 1) \log \frac{1}{p} + \log \frac{1}{\delta} \right) \right\rceil,$$

we have that Algorithm 1 provides an ε-error estimation for $l(v)$ with probability $1 - \delta$ for all $v \in V$ s.t. $\Pr(\tau_v) \geq p$. But $\Pr(\tau_v) \geq p$ if and only if $l(v) \geq \sigma_v(p)$ since

$$l(v) = \frac{T_v}{\binom{\delta_v}{2}} = \frac{m \Pr(\tau_v)}{\binom{\delta_v}{2}}.$$

$\square$

We remark that $\tilde{T}_v$ is an unbiased estimator for T_v, since

$$\mathbb{E}[\tilde{T}_v] = \mathbb{E}\left[\sum_{e \in S} \frac{m}{r} \mathbb{1}_v(e) \right] = \frac{m}{r} \sum_{e \in S} \Pr(e \in \tau_v) = \frac{m}{r} \sum_{e \in S} \frac{|\tau_v|}{m} = T_v.$$

Theorem 6. *Given a graph $G = (V, E)$ and a sample of size*

$$r = \left\lceil \frac{c'}{\varepsilon^2 p} \left((\lfloor \lg \Delta - 1 \rfloor + 1) \log \frac{1}{p} + \log \frac{1}{\delta} \right) \right\rceil,$$

Algorithm 1 has running time $\mathcal{O}(\Delta \lg \Delta + m)$.

Proof. In line 1, the value of Δ can be computed in time $\Theta(m)$. Given an edge $\{a, b\}$ we first store the neighbors of b in a directed address table. Then, lines 4, 5, and 6 take time $\mathcal{O}(\Delta)$ by checking, for each $v \in N_a$, if v is in the table. Hence, the total running time of Algorithm 1 is $\mathcal{O}(r \cdot \Delta + m) = \mathcal{O}(\Delta \lg \Delta + m)$. $\square$

As mentioned before, for specific graph classes, the running time proved in Theorem 6 can be reduced. We can achieve this either by proving that graphs in such classes have a smaller VC dimension, or by looking more carefully at the algorithm analysis for such classes. In Corollaries 3 and 4 we present results for two such classes.

Corollary 3. *If G is a planar graph, then Algorithm 1 has running time $\mathcal{O}(\Delta)$.*

Proof. By Corollary 2, VCDim$(G) \leq 2$. So, the sample size in the Algorithm 1 changes from a function of Δ to a constant. Note that, in particular, since we do not need to find the value of Δ, line 1 can be computed in time $\mathcal{O}(1)$. As with the proof of Theorem 6, lines 4, 5, and 6 still take time $\mathcal{O}(\Delta)$. Since r is constant, line 2 takes constant time. So, the total running time of Algorithm 1 is $\mathcal{O}(r \cdot \Delta) = \mathcal{O}(\Delta)$. □

Another case where we can provide a better running for the algorithm is the case for *bounded-degree graphs*, i.e. the case where the maximum degree of any graph in the class is bounded by a constant.

Corollary 4. *Let G be a bounded-degree graph, where d is such bound. Algorithm 1 has running time $\mathcal{O}(1)$ or $\mathcal{O}(n)$, respectively, depending on whether d is part of the input or not.*

Proof. If d is part of the input, then the number of samples r in line 1 can be computed in time $\mathcal{O}(1)$. Line 2 is executed $\mathcal{O}(1)$ times, and the remaining of the algorithm, in lines 4, 5, and 6, takes $\mathcal{O}(1)$ time, since the size of the neighborhood of every vertex is bounded by a constant.

On the other hand, if d is not part of the input, then Δ must be computed for the execution of line 1. In this case we check the degree of every vertex by traversing its adjacency list. All these adjacency lists have constant size. Performing this for all vertices takes time $\mathcal{O}(n)$. The other steps of the algorithm take constant time. □

4 Conclusion

We present a sampling algorithm for local clustering problem. In our analysis we define a range space associated to the input graph, and show how the sample size of the algorithm relates to the VC dimension of this range space. This kind of analysis takes into consideration the combinatorial structure of the graph, so the size of the sample of edges used by the algorithm depends on the maximum degree of the input graph.

Our algorithm executes in time $\mathcal{O}(\Delta \lg \Delta + m)$ in the general case and guarantees, for given parameters ε, δ, and p, that the approximation value has relative error ε with probability at least $1 - \delta$, for every node whose clustering coefficient is greater than a certain function adjusted by the parameter p. For planar graphs we show that the sample size can be bounded by a constant, an the running time in this case is $\mathcal{O}(\Delta)$. In the case of bounded-degree graphs, where there is also a constant bound on the sample size, the running time drops to $\mathcal{O}(1)$ or $\mathcal{O}(n)$, depending on whether the bound on the degree is part of the input or not.

References

1. Anthony, M., Bartlett, P.L.: Neural Network Learning: Theoretical Foundations, 1st edn. Cambridge University Press, New York (2009)
2. Barabási, A.L., Pósfai, M.: Network Science. Cambridge University Press, Cambridge (2016)
3. Bloznelis, M.: Degree and clustering coefficient in sparse random intersection graphs. Ann. Appl. Probab. **23**(3), 1254–1289 (2013)
4. Brautbar, M., Kearns, M.: Local algorithms for finding interesting individuals in large networks. In: Innovations in Computer Science (2010)
5. de Lima, A.M., da Silva, M.V., Vignatti, A.L.: Percolation centrality via rademacher complexity. Discret. Appl. Math. (2021)
6. Easley, D.A., Kleinberg, J.M.: Networks, Crowds, and Markets - Reasoning About a Highly Connected World. Cambridge University Press, NY (2010)
7. Fronczak, A., Fronczak, P., Hołyst, J.A.: Mean-field theory for clustering coefficients in Barabási-Albert networks. Phys. Rev. E **68**(4), 046126 (2003)
8. Gupta, A.K., Sardana, N.: Significance of clustering coefficient over Jaccard Index. In: 2015 Eighth International Conference on Contemporary Computing (IC3), pp. 463–466. IEEE (2015)
9. Holland, P.W., Leinhardt, S.: Transitivity in structural models of small groups. Comp. Group Stud. **2**(2), 107–124 (1971)
10. Iskhakov, L., Kamiński, B., Mironov, M., Prałat, P., Prokhorenkova, L.: Local clustering coefficient of spatial preferential attachment model. J. Complex Netw. **8**(1), cnz019 (2020)
11. Ji, Q., Li, D., Jin, Z.: Divisive algorithm based on node clustering coefficient for community detection. IEEE Access **8**, 142337–142347 (2020)
12. Kartun-Giles, A.P., Bianconi, G.: Beyond the clustering coefficient: a topological analysis of node neighbourhoods in complex networks. Chaos Solit. Fractals: X **1**, 100004 (2019)
13. Kolda, T.G., Pinar, A., Plantenga, T., Seshadhri, C., Task, C.: Counting triangles in massive graphs with MapReduce. SIAM J. Sci. Comput. **36**(5), S48–S77 (2014)
14. Krot, A., Ostroumova Prokhorenkova, L.: Local clustering coefficient in generalized preferential attachment models. In: Gleich, D.F., Komjáthy, J., Litvak, N. (eds.) WAW 2015. LNCS, vol. 9479, pp. 15–28. Springer, Cham (2015). https://doi.org/10.1007/978-3-319-26784-5_2
15. Kutzkov, K., Pagh, R.: On the streaming complexity of computing local clustering coefficients. In: Proceedings of the Sixth ACM International Conference on Web Search and Data Mining, pp. 677–686 (2013)
16. Li, M., Zhang, R., Hu, R., Yang, F., Yao, Y., Yuan, Y.: Identifying and ranking influential spreaders in complex networks by combining a local-degree sum and the clustering coefficient. Int. J. Mod. Phys. B **32**(06), 1850118 (2018)
17. Li, X., Chang, L., Zheng, K., Huang, Z., Zhou, X.: Ranking weighted clustering coefficient in large dynamic graphs. World Wide Web **20**(5), 855–883 (2017)
18. Li, Y., Long, P.M., Srinivasan, A.: Improved bounds on the sample complexity of learning. J. Comput. Syst. Sci. **62**(3), 516–527 (2001)
19. de Lima, A.M., da Silva, M.V., Vignatti, A.L.: Estimating the percolation centrality of large networks through pseudo-dimension theory. In: Proceedings of the 26th ACM SIGKDD International Conference on Knowledge Discovery & Data Mining, pp. 1839–1847 (2020)

20. Liu, S., Xia, Z.: A two-stage BFS local community detection algorithm based on node transfer similarity and local clustering coefficient. Phys. A **537**, 122717 (2020)
21. Mitzenmacher, M., Upfal, E.: Probability and Computing: Randomization and Probabilistic Techniques in Algorithms and Data Analysis, 2nd edn. Cambridge University Press, New York (2017)
22. Mohri, M., Rostamizadeh, A., Talwalkar, A.: Foundations of Machine Learning. The MIT Press, Cambridge (2012)
23. Nascimento, M.C.: Community detection in networks via a spectral heuristic based on the clustering coefficient. Discret. Appl. Math. **176**, 89–99 (2014)
24. Newman, M.E.J.: Networks: an introduction. Oxford University Press (2010)
25. Pan, X., Xu, G., Wang, B., Zhang, T.: A novel community detection algorithm based on local similarity of clustering coefficient in social networks. IEEE Access **7**, 121586–121598 (2019)
26. Riondato, M., Kornaropoulos, E.M.: Fast approximation of betweenness centrality through sampling. Data Min. Knowl. Disc. **30**(2), 438–475 (2016)
27. Riondato, M., Upfal, E.: ABRA: approximating betweenness centrality in static and dynamic graphs with rademacher averages. ACM Trans. Knowl. Discov. Data **12**(5), 61:1–61:38 (2018)
28. Seshadhri, C., Pinar, A., Kolda, T.G.: Fast triangle counting through wedge sampling. In: Proceedings of the SIAM Conference on Data Mining, vol. 4, p. 5. Citeseer (2013)
29. Shalev-Shwartz, S., Ben-David, S.: Understanding Machine Learning: From Theory to Algorithms. Cambridge University Press, New York (2014)
30. Soffer, S.N., Vazquez, A.: Network clustering coefficient without degree-correlation biases. Phys. Rev. E **71**(5), 057101 (2005)
31. Watts, D.J., Strogatz, S.H.: Collective dynamics of 'small-world' networks. Nature **393**(6684), 440–442 (1998)
32. West, D.B.: Introduction to Graph Theory, 2 edn. Prentice Hall (2000)
33. Wu, Z., Lin, Y., Wang, J., Gregory, S.: Link prediction with node clustering coefficient. Phys. A **452**, 1–8 (2016)
34. Zhang, H., Zhu, Y., Qin, L., Cheng, H., Yu, J.X.: Efficient local clustering coefficient estimation in massive graphs. In: Candan, S., Chen, L., Pedersen, T.B., Chang, L., Hua, W. (eds.) DASFAA 2017. LNCS, vol. 10178, pp. 371–386. Springer, Cham (2017). https://doi.org/10.1007/978-3-319-55699-4_23
35. Zhang, J., Tang, J., Ma, C., Tong, H., Jing, Y., Li, J.: Panther: fast top-k similarity search on large networks. In: Proceedings of the 21th ACM SIGKDD International Conference on Knowledge Discovery and Data Mining, pp. 1445–1454 (2015)
36. Zhang, R., Li, L., Bao, C., Zhou, L., Kong, B.: The community detection algorithm based on the node clustering coefficient and the edge clustering coefficient. In: Proceeding of the 11th World Congress on Intelligent Control and Automation, pp. 3240–3245. IEEE (2014)

Automata Theory and Formal Languages

Binary Completely Reachable Automata

David Casas⬭ and Mikhail V. Volkov$^{(\boxtimes)}$⬭

Institute of Natural Sciences and Mathematics, Ural Federal University, 620000 Ekaterinburg, Russia
m.v.volkov@urfu.ru

Abstract. We characterize complete deterministic finite automata with two input letters in which every non-empty set of states occurs as the image of the whole state set under the action of a suitable input word. The characterization leads to a polynomial-time algorithm for recognizing this class of automata.

Keywords: Deterministic finite automaton · Transition monoid · Complete reachability · Strongly connected digraph · Cayley digraph

1 Introduction

Completely reachable automata are complete deterministic finite automata in which every non-empty subset of the state set occurs as the image of the whole state set under the action of a suitable input word. Such automata appeared in the study of descriptional complexity of formal languages [2,10] and in relation to the Černý conjecture [4]. A systematic study of completely reachable automata was initiated in [2,3] and continued in [1]. In [1,3] completely reachable automata were characterized in terms of a certain finite sequence of directed graphs (digraphs): the automaton is completely reachable if and only if the final digraph in this sequence is strongly connected. In [1, Theorem 11] it was shown that given an automaton $\mathscr{A}$ with n states and m input letters, the k-th digraph in the sequence assigned to $\mathscr{A}$ can be constructed in $O(mn^{2k}\log n)$ time. However, this does not yet ensure a polynomial-time algorithm for recognizing complete reachability: a series of examples in [1] demonstrates that the length of the digraph sequence for an automaton with n states may reach $n - 1$.

Here we study completely reachable automata with two input letters; for brevity, we call automata with two input letters *binary*. Our main results provide a new characterization of binary completely reachable automata, and the characterization leads to a quasilinear time algorithm for recognizing complete reachability for binary automata.

Our prerequisites are minimal: we only assume the reader's acquaintance with basic properties of strongly connected digraphs, subgroups, and cosets.

The authors were supported by the Ministry of Science and Higher Education of the Russian Federation, project FEUZ-2020-0016.

A. Castañeda and F. Rodríguez-Henríquez (Eds.): LATIN 2022, LNCS 13568, pp. 345–358, 2022.
https://doi.org/10.1007/978-3-031-20624-5_21

2 Preliminaries

A *complete deterministic finite automaton* (DFA) is a triple $\mathscr{A} = \langle Q, \Sigma, \delta \rangle$ where Q and Σ are finite sets called the *state set* and, resp., the *input alphabet* of $\mathscr{A}$, and $\delta \colon Q \times \Sigma \to Q$ is a totally defined map called the *transition function* of $\mathscr{A}$.

The elements of Σ are called *input letters* and finite sequences of letters are called *words over Σ*. The empty sequence is also treated as a word, called the *empty word* and denoted ε. The collection of all words over Σ is denoted Σ^*.

The transition function δ extends to a function $Q \times \Sigma^* \to Q$ (still denoted by δ) via the following recursion: for every $q \in Q$, we set $\delta(q, \varepsilon) = q$ and $\delta(q, wa) = \delta(\delta(q, w), a)$ for all $w \in \Sigma^*$ and $a \in \Sigma$. Thus, every word $w \in \Sigma^*$ induces the transformation $q \mapsto \delta(q, w)$ of the set Q. The set $T(\mathscr{A})$ of all transformations induced this way is called the *transition monoid* of $\mathscr{A}$; this is the submonoid generated by the transformations $q \mapsto \delta(q, a)$, $a \in \Sigma$, in the monoid of all transformations of Q. A DFA $\mathscr{B} = \langle Q, \Theta, \zeta \rangle$ with the same state set as $\mathscr{A}$ is said to be *syntactically equivalent* to $\mathscr{A}$ if $T(\mathscr{B}) = T(\mathscr{A})$.

The function δ can be further extended to non-empty subsets of the set Q. Namely, for every non-empty subset $P \subseteq Q$ and every word $w \in \Sigma^*$, we let $\delta(P, w) = \{\delta(q, w) \mid q \in P\}$.

Whenever there is no risk of confusion, we tend to simplify our notation by suppressing the sign of the transition function; this means that we write $q.w$ for $\delta(q, w)$ and $P.w$ for $\delta(P, w)$ and specify a DFA as a pair $\langle Q, \Sigma \rangle$.

We say that a non-empty subset $P \subseteq Q$ is *reachable* in $\mathscr{A} = \langle Q, \Sigma \rangle$ if $P = Q.w$ for some word $w \in \Sigma^*$. A DFA is called *completely reachable* if every non-empty subset of its state set is reachable. Observe that complete reachability is actually a property of the transition monoid of $\mathscr{A}$; hence, if a DFA $\mathscr{A}$ is completely reachable, so is any DFA that is syntactically equivalent to $\mathscr{A}$.

Given a DFA $\mathscr{A} = \langle Q, \Sigma \rangle$ and a word $w \in \Sigma^*$, the *image* of w is the set $Q.w$ and the *excluded set* $\mathrm{excl}(w)$ of w is the complement $Q \backslash Q.w$ of the image. The number $|\mathrm{excl}(w)|$ is called the *defect* of w. If a word w has defect 1, its excluded set consists of a unique state called the *excluded state* for w. Further, for any $w \in \Sigma^*$, the set $\{p \in Q \mid p = q_1.w = q_2.w$ for some $q_1 \neq q_2\}$ is called the *duplicate set* of w and is denoted by $\mathrm{dupl}(w)$. If w has defect 1, its duplicate set consists of a unique state called the *duplicate state* for w. We identify singleton sets with their elements, and therefore, for a word w of defect 1, $\mathrm{excl}(w)$ and $\mathrm{dupl}(w)$ stand for its excluded and, resp., duplicate states.

For any $v \in \Sigma^*$, $q \in Q$, let $qv^{-1} = \{p \in Q \mid p.v = q\}$. Then for all $u, v \in \Sigma^*$,

$$\mathrm{excl}(uv) = \{q \in Q \mid qv^{-1} \subseteq \mathrm{excl}(u)\}, \tag{1}$$

$$\mathrm{dupl}(uv) = \{q \in Q \mid qv^{-1} \cap \mathrm{dupl}(u) \neq \varnothing \text{ or } |qv^{-1} \backslash \mathrm{excl}(u)| \geq 2\}. \tag{2}$$

The equalities (1) and (2) become clear as soon as the definitions of $\mathrm{excl}(\ )$ and $\mathrm{dupl}(\ )$ are deciphered.

Recall that DFAs with two input letters are called *binary*. The question of our study is: under which conditions is a binary DFA completely reachable? The rest of the section presents a series of reductions showing that to answer this question, it suffices to analyze DFAs of a specific form.

Let $\mathscr{A} = \langle Q, \{a,b\}\rangle$ be a binary DFA with $n > 1$ states. If neither a nor b has defect 1, no subset of size $n - 1$ is reachable in $\mathscr{A}$. Therefore, when looking for binary completely reachable automata, we must focus on DFAs possessing a letter of defect 1. We will always assume that a has defect 1.

The image of every non-empty word over $\{a,b\}$ is contained in either $Q.a$ or $Q.b$. If the defect of b is greater than or equal to 1, then at most two subsets of size $n - 1$ are reachable (namely, $Q.a$ and $Q.b$), whence $\mathscr{A}$ can only be completely reachable provided that $n = 2$. The automaton $\mathscr{A}$ is then nothing but the classical flip-flop, see Fig. 1.

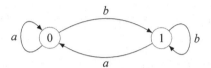

Fig. 1. The flip-flop. Here and below a DFA $\langle Q, \Sigma\rangle$ is depicted as a digraph with the vertex set Q and a labeled edge $q \xrightarrow{a} q'$ for each triple $(q,a,q') \in Q \times \Sigma \times Q$ such that $q.a = q'$.

Having isolated this exception, we assume from now on that $n \geq 2$ and the letter b has defect 0, which means that b acts as a permutation of Q. The following fact was first stated in [2]; for a proof, see, e.g., [1, Sect. 6].

Lemma 1. *If $\mathscr{A} = \langle Q, \{a,b\}\rangle$ is a completely reachable automaton in which the letter b acts as a permutation of Q, then b acts as a cyclic permutation.*

Taking Lemma 1 into account, we restrict our further considerations to DFAs with $n \geq 2$ states and two input letters a and b such that a has defect 1 and b acts a cyclic permutation. Without any loss, we will additionally assume that these DFAs have the set $\mathbb{Z}_n = \{0, 1, \ldots, n - 1\}$ of all residues modulo n as their state set and the action of b at any state merely adds 1 modulo n. Let us also agree that whenever we deal with elements of $\mathbb{Z}_n$, the signs $+$ and $-$ mean addition and subtraction modulo n, unless the contrary is explicitly specified.

Further, we will assume that $0 = \mathrm{excl}(a)$ as it does not matter from which origin the cyclic count of the states start.

Since b is a permutation, for each $k \in \mathbb{Z}_n$, the transformations $q \mapsto q.b^k a$ and $q \mapsto q.b$ generate the same submonoid in the monoid of all transformations of $\mathbb{Z}_n$ as do the transformations $q \mapsto q.a$ and $q \mapsto q.b$. This means that if one treats the word $b^k a$ as a new letter a_k, say, one gets the DFA $\mathscr{A}_k = \langle \mathbb{Z}_n, \{a_k, b\}\rangle$ that is syntactically equivalent to $\mathscr{A}$. Therefore, $\mathscr{A}$ is completely reachable if and only if so is $\mathscr{A}_k$ for some (and hence for all) k. Hence we may choose k as we wish and study the DFA $\mathscr{A}_k$ for the specified value of k instead of $\mathscr{A}$.

What can we achieve using this? From (1) we have $\mathrm{excl}(b^k a) = \mathrm{excl}(a) = 0$. Further, let $q_1 \neq q_2$ be such that $q_1.a = q_2.a = \mathrm{dupl}(a)$. Choosing $k = q_1$ (or $k = q_2$), we get $0.b^k a = \mathrm{dupl}(a)$. Thus, we will assume that $0.a = \mathrm{dupl}(a)$.

Summarizing, we will consider DFAs $\langle \mathbb{Z}_n, \{a,b\}\rangle$ such that:

– the letter a has defect 1, $\mathrm{excl}(a) = 0$, and $0.a = \mathrm{dupl}(a)$;
– $q.b = q + 1$ for each $q \in \mathbb{Z}_n$.

We call such DFAs *standardized*. For the purpose of complexity considerations at the end of Sect. 5, observe that given a binary DFA $\mathscr{A}$ in which one letter acts as a cyclic permutation while the other has defect 1, one can 'standardize' the automaton, that is, construct a standardized DFA syntactically equivalent to $\mathscr{A}$, in linear time with respect to the size of $\mathscr{A}$.

3   A Necessary Condition

Let $\langle \mathbb{Z}_n, \{a,b\} \rangle$ be a standardized DFA and $w \in \{a,b\}^*$. A subset $S \subseteq \mathbb{Z}_n$ is said to be *w-invariant* if $S.w \subseteq S$.

Proposition 1. *If $\langle \mathbb{Z}_n, \{a,b\} \rangle$ is a completely reachable standardized DFA, then no proper subgroup of $(\mathbb{Z}_n, +)$ is a-invariant.*

Proof. Arguing by contradiction, assume that $H \subsetneq \mathbb{Z}_n$ is a subgroup such that $H.a \subseteq H$. Let d stand for the index of the subgroup H in the group $(\mathbb{Z}_n, +)$. The set $\mathbb{Z}_n$ is then partitioned into the d cosets

$$H_0 = H, \; H_1 = H.b = H+1, \; \ldots, \; H_{d-1} = H.b^{d-1} = H+d-1.$$

For $i = 0, 1, \ldots, d-1$, let T_i be the complement of the coset H_i in $\mathbb{Z}_n$. Then we have $T_i = \cup_{j \neq i} H_j$ and $T_i.b = T_{i+1 \,(\mathrm{mod}\, d)}$ for each $i = 0, 1, \ldots, d-1$.

Since $\mathscr{A}$ is completely reachable, each subset T_i is reachable. Take a word w of minimum length among words with the image equal to one of the subsets $T_0, T_1, \ldots, T_{d-1}$. Write w as $w = w'c$ for some letter $c \in \{a,b\}$.

If $c = b$, then for some $i \in \{0, 1, \ldots, d-1\}$, we have

$$\mathbb{Z}_n.w'b = T_i = T_{i-1 \,(\mathrm{mod}\, d)}.b.$$

Since b^n acts as the identity mapping, applying the word b^{n-1} to this equality yields $\mathbb{Z}_n.w' = T_{i-1 \,(\mathrm{mod}\, d)}$ whence the image of w' is also equal to one of the subsets $T_0, T_1, \ldots, T_{d-1}$. This contradicts the choice of w.

Thus, $c = a$, whence the set $\mathbb{Z}_n.w$ is contained in $\mathbb{Z}_n.a$. The only T_i that is contained in $\mathbb{Z}_n.a$ is T_0 because each T_i with $i \neq 0$ contains H_0, and $H_0 = H$ contains 0, the excluded state of a. Hence, $\mathbb{Z}_n.w = T_0$, that is, $\mathbb{Z}_n.w'a = T_0$. For each state $q \in \mathbb{Z}_n.w'$, we have $q.a \in T_0$, and this implies $q \in T_0$ since H_0, the complement of T_0, is a-invariant. We see that $\mathbb{Z}_n.w' \subseteq T_0$ and the inclusion cannot be strict because T_0 cannot be the image of its proper subset. However, the equality $\mathbb{Z}_n.w' = T_0$ again contradicts the choice of w.   □

We will show that the condition of Proposition 1 is not only necessary but also sufficient for complete reachability of a standardized DFA. The proof of sufficiency requires a construction that we present in full in Sect. 5, after studying its simplest case in Sect. 4.

4   Rystsov's Graph of a Binary DFA

Recall a sufficient condition for complete reachability from [2]. Given a (not necessarily binary) DFA $\mathscr{A} = \langle Q, \Sigma \rangle$, let $W_1(\mathscr{A})$ stand for the set of all words in Σ^* that have defect 1 in $\mathscr{A}$. Consider a digraph with the vertex set Q and the edge set $E = \{(\mathrm{excl}(w), \mathrm{dupl}(w)) \mid w \in W_1(\mathscr{A})\}$. We denote this digraph by $\Gamma_1(\mathscr{A})$. The notation comes from [2], but much earlier, though in a less explicit form, the construction was used by Rystsov [11] for some special species of DFAs. Taking this into account, we refer to $\Gamma_1(\mathscr{A})$ as the *Rystsov graph* of $\mathscr{A}$.

Theorem 1 ([2, Theorem 1]).   *If a DFA $\mathscr{A} = \langle Q, \Sigma \rangle$ is such that the graph $\Gamma_1(\mathscr{A})$ is strongly connected, then $\mathscr{A}$ is completely reachable.*

It was shown in [2] that the condition of Theorem 1 is not necessary for complete reachability, but it was conjectured that the condition might characterize binary completely reachable automata. However, this conjecture has been refuted in [1, Example 2] by exhibiting a binary completely reachable automaton with 12 states whose Rystsov graph is not strongly connected. Here we include a similar example which we will use to illustrate some of our results.

Consider the standardized DFA $\mathscr{E}'_{12} = \langle \mathbb{Z}_{12}, \{a,b\} \rangle$ where the action of the letter a is specified as follows:

$$\begin{array}{c|cccccccccccc} q & 0 & 1 & 2 & 3 & 4 & 5 & 6 & 7 & 8 & 9 & 10 & 11 \\ \hline q.a & 10 & 1 & 2 & 8 & 4 & 5 & 10 & 9 & 3 & 7 & 6 & 11 \end{array}.$$

(The DFA $\mathscr{E}'_{12}$ only slightly differs from the DFA $\mathscr{E}_{12}$ used in [1, Example 2], hence the notation.) The DFA $\mathscr{E}'_{12}$ is shown in Fig. 2, in which we have replaced edges that should have been labeled a and b with solid and, resp., dashed edges.

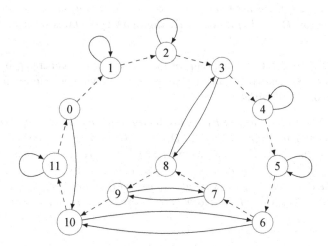

Fig. 2. The DFA $\mathscr{E}'_{12}$; solid and dashed edges show the action of a and, resp., b

We postpone the description of the digraph $\Gamma_1(\mathscr{E}'_{12})$ and the proof that the DFA $\mathscr{E}'_{12}$ is completely reachable until we develop suitable tools that make the description and the proof easy.

We start with a characterization of Rystsov's graphs of standardized DFAs. Let $\mathscr{A} = \langle \mathbb{Z}_n, \{a,b\} \rangle$ be such a DFA. It readily follows from (1) and (2) that $\mathrm{excl}(w) \cdot b = \mathrm{excl}(wb)$ and $\mathrm{dupl}(w) \cdot b = \mathrm{dupl}(wb)$ for every word $w \in W_1(\mathscr{A})$. Therefore, the edge set E of the digraph $\Gamma_1(\mathscr{A})$ is closed under the *translation* $(q,p) \mapsto (q \cdot b, p \cdot b) = (q+1, p+1)$. As a consequence, for any edge $(q,p) \in E$ and any k, the pair $(q+k, p+k)$ also constitutes an edge in E.

Denote by $D_1(\mathscr{A})$ the set of ends of edges of $\Gamma_1(\mathscr{A})$ that start at 0, that is, $D_1(\mathscr{A}) = \{p \in \mathbb{Z}_n \mid (0,p) \in E\}$. We call $D_1(\mathscr{A})$ the *difference set* of $\mathscr{A}$. Our first observation shows how to recover all edges of $\Gamma_1(\mathscr{A})$, knowing $D_1(\mathscr{A})$.

Lemma 2. *Let $\mathscr{A} = \langle \mathbb{Z}_n, \{a,b\} \rangle$ be a standardized DFA. A pair $(q,p) \in \mathbb{Z}_n \times \mathbb{Z}_n$ forms an edge in the digraph $\Gamma_1(\mathscr{A})$ if and only if $p - q \in D_1(\mathscr{A})$.*

Proof. If $p - q \in D_1(\mathscr{A})$, the pair $(0, p-q)$ is an edge in E, and therefore, so is the pair $(0+q, (p-q)+q) = (q,p)$. Conversely, if (q,p) is an edge in E, then so is $(q + (n-q), p + (n-q)) = (0, p-q)$, whence $p - q \in D_1(\mathscr{A})$. $\qquad \square$

By Lemma 2, the presence or absence of an edge in $\Gamma_1(\mathscr{A})$ depends only on the difference modulo n of two vertex numbers. This means that $\Gamma_1(\mathscr{A})$ is a *circulant* digraph, that is, the Cayley digraph of the cyclic group $(\mathbb{Z}_n, +)$ with respect to some subset of $\mathbb{Z}_n$. Recall that if D is a subset in a group G, the *Cayley digraph of G with respect to D*, denoted $\mathrm{Cay}(G,D)$, has G as its vertex set and $\{(g, gd) \mid g \in G, \, d \in D\}$ as its edge set. The following property of Cayley digraphs of finite groups is folklore[1].

Lemma 3. *Let G be a finite group, D a subset of G, and H the subgroup of G generated by D. The strongly connected components of the Cayley digraph $\mathrm{Cay}(G,D)$ have the right cosets Hg, $g \in G$, as their vertex sets, and each strongly connected component is isomorphic to $\mathrm{Cay}(H,D)$. In particular, the digraph $\mathrm{Cay}(G,D)$ is strongly connected if and only if G is generated by D.*

Let $H_1(\mathscr{A})$ stand for the subgroup of the group $(\mathbb{Z}_n, +)$ generated by the difference set $D_1(\mathscr{A})$. Specializing Lemma 3, we get the following description for Rystsov's graphs of standardized DFAs.

Proposition 2. *Let $\mathscr{A} = \langle \mathbb{Z}_n, \{a,b\} \rangle$ be a standardized DFA. The digraph $\Gamma_1(\mathscr{A})$ is isomorphic to the Cayley digraph $\mathrm{Cay}(\mathbb{Z}_n, D_1(\mathscr{A}))$. The strongly connected components of $\Gamma_1(\mathscr{A})$ have the cosets of the subgroup $H_1(\mathscr{A})$ as their vertex sets, and each strongly connected component is isomorphic to the Cayley digraph $\mathrm{Cay}(H_1(\mathscr{A}), D_1(\mathscr{A}))$. In particular, the digraph $\Gamma_1(\mathscr{A})$ is strongly connected if and only if the set $D_1(\mathscr{A})$ generates $(\mathbb{Z}_n, +)$ or, equivalently, if and only if the greatest common divisor of $D_1(\mathscr{A})$ is coprime to n.*

---

[1] In fact, our definition is the semigroup version of the notion of a Cayley digraph, but this makes no difference since in a finite group, every subsemigroup is a subgroup.

Proposition 2 shows that structure of the Rystsov graph of a standardized DFA $\mathscr{A}$ crucially depends on its difference set $D_1(\mathscr{A})$. The definition of the edge set of $\Gamma_1(\mathscr{A})$ describes $D_1(\mathscr{A})$ as the set of duplicate states for all words w of defect 1 whose excluded state is 0, that is, $D_1(\mathscr{A}) = \{\mathrm{dupl}(w) \mid \mathrm{excl}(w) = 0\}$. Thus, understanding of difference sets amounts to a classification of transformations caused by words of defect 1. It is such a classification that is behind the following handy description of difference sets.

Proposition 3. *Let* $\mathscr{A} = \langle \mathbb{Z}_n, \{a, b\} \rangle$ *be a standardized DFA. Let* $r \neq 0$ *be such that* $r \,.\, a = \mathrm{dupl}(a)$. *Then*

$$D_1(\mathscr{A}) = \{\mathrm{dupl}(a) \,.\, v \mid v \in \{a, b^r a\}^*\}. \tag{3}$$

Proof. Denote by N the image of the letter a, that is, $N = \mathbb{Z}_n \setminus \{0\}$. If $q \,.\, a = p$ for some $q \in \mathbb{Z}_n$ and $p \in N$, then, clearly, $(q - r) \,.\, b^r a = p$. Hence the only state in N that has a preimage of size 2 under the actions of both a and $b^r a$ is

$$\mathrm{dupl}(a) = \begin{cases} 0 \,.\, a = r \,.\, a, \\ (n - r) \,.\, b^r a = 0 \,.\, b^r a, \end{cases}$$

and in both cases 0 belongs to the preimage. Thus, the preimage of every $p \in N$ under both a and $b^r a$ contains a unique state in N, which means that both a and $b^r a$ act on the set N as permutations. Hence every word $v \in \{a, b^r a\}^*$ acts on N as a permutation. Then the word av has defect 1 and $\mathrm{excl}(av) = 0$. Applying the equality (2) with a in the role of u, we derive that $\mathrm{dupl}(av) = \mathrm{dupl}(a) \,.\, v$. Thus, denoting the right-hand side of (3) by D, we see that every state in D is the duplicate state of some word whose only excluded state is 0. This means that $D_1(\mathscr{A}) \supseteq D$.

To verify the converse inclusion, take an arbitrary state $p \in D_1(\mathscr{A})$ and let w be a word of defect 1 such that $\mathrm{excl}(w) = 0$ and $\mathrm{dupl}(w) = p$. Since $\mathrm{excl}(w) = 0$, the word w ends with the letter a. We prove that p lies in D by induction on the number of occurrences of a in w. If a occurs in w once, then $w = b^k a$ for some $k \in \mathbb{Z}_n$. We have $p = \mathrm{dupl}(w) = \mathrm{dupl}(b^k a) = \mathrm{dupl}(a) \in D$.

If a occurs in w at least twice, write $w = w' b^k a$ where w' ends with a. Then the word w' has defect 1 and $\mathrm{excl}(w') = 0$. As w' has fewer occurrences of a, the inductive assumption applies and yields $\mathrm{dupl}(w') \in D$. Denoting $\mathrm{dupl}(w')$ by p', we have $p = p' \,.\, b^k a$. If we prove that $k \in \{0, r\}$, we are done since the set D is both a-invariant and $b^r a$-invariant by its definition. Arguing by contradiction, assume $k \notin \{0, r\}$. Let $\ell = k \,.\, a$; then k is the only state in ℓa^{-1}. Hence $\ell a^{-1} = \mathrm{excl}(w' b^k)$, and the equality (1) (with $u = w' b^k$ and $v = a$) shows that $\ell \in \mathrm{excl}(w' b^k a) = \mathrm{excl}(w)$. Clearly, $\ell \neq 0$ as ℓ lies in the image of a. Therefore the conclusion $\ell \in \mathrm{excl}(w)$ contradicts the assumption $\mathrm{excl}(w) = 0$. $\square$

For an illustration, we apply (3) to compute the difference set for the DFA $\mathscr{E}'_{12}$ shown in Fig. 2. In $\mathscr{E}'_{12}$, we have $r = 6$ and $\mathrm{dupl}(a) = 10$. Acting by a and $b^6 a$ gives $10 \,.\, a = 6$ and $10 \,.\, b^6 a = (10 + 6) \,.\, a = 4 \,.\, a = 4$. Thus, $4, 6 \in D_1(\mathscr{E}'_{12})$. Acting by a or $b^6 a$ at 4 and 6 does not produce anything new: $4 \,.\, a = 4$ and $4 \,.\, b^6 a = (4 + 6) \,.\, a = 10 \,.\, a = 6$ while $6 \,.\, a = 10$ and $6 \,.\, b^6 a = (6 + 6) \,.\, a = 0 \,.\, a = 10$. We conclude that $D_1(\mathscr{E}'_{12}) = \{4, 6, 10\}$. Since 2, the greatest common divisor of $\{4, 6, 10\}$, divides 12, we see that the digraph $\Gamma_1(\mathscr{E}'_{12})$ is not strongly connected. The subgroup $H_1(\mathscr{E}'_{12})$ consists of even residues modulo 12 and has index 2. Hence the digraph $\Gamma_1(\mathscr{E}'_{12})$ has two strongly connected components whose vertex sets are $\{0, 2, 4, 6, 8, 10\}$ and $\{1, 3, 5, 7, 9, 11\}$, and for each $q \in \mathbb{Z}_{12}$, it has the edges $(q, q+4)$, $(q, q+6)$, and $(q, q+10)$.

In fact, formula (3) leads to a straightforward algorithm that computes the difference set of any standardized DFA $\mathscr{A}$ in time linear in n. This, together with Proposition 2, gives an efficient way to compute the Rystsov graph of $\mathscr{A}$.

Let $D_1^0(\mathscr{A}) = D_1(\mathscr{A}) \cup \{0\}$. It turns out that $D_1^0(\mathscr{A})$ is always a union of cosets of a nontrivial subgroup.

Proposition 4. *Let $\mathscr{A} = \langle \mathbb{Z}_n, \{a, b\} \rangle$ be a standardized DFA. Let $r \neq 0$ be such that $r \cdot a = \mathrm{dupl}(a)$. Then the set $D_1^0(\mathscr{A})$ is a union of cosets of the subgroup generated by r in the group $H_1(\mathscr{A})$.*

Proof. It is easy to see that the claim is equivalent to the following implication: if $d \in D_1^0(\mathscr{A})$, then $d + r \in D_1^0(\mathscr{A})$. This clearly holds if $d + r = 0$. Thus, assume that $d \in D_1^0(\mathscr{A})$ is such that $d + r \neq 0$. Then $(d + r) \cdot a \in D_1(\mathscr{A})$. Indeed, if $d = 0$, then $(d + r) \cdot a = r \cdot a = \mathrm{dupl}(a) \in D_1(\mathscr{A})$. If $d \neq 0$, then $d \in D_1(\mathscr{A})$, whence $(d + r) \cdot a = d \cdot b^r a \in D_1(\mathscr{A})$ as formula (3) ensures that the set $D_1(\mathscr{A})$ is closed under the action of the word $b^r a$.

We have observed in the first paragraph of the proof of Proposition 3 that a acts on the set $N = \mathbb{Z}_n \setminus \{0\}$ as a permutation. Hence for some k, the word a^k acts on N as the identity map. Then $d + r = (d + r) \cdot a^k = ((d + r) \cdot a) \cdot a^{k-1} \in D_1(\mathscr{A})$ since we have already shown that $(d + r) \cdot a \in D_1(\mathscr{A})$ and formula (3) ensures that the set $D_1(\mathscr{A})$ is a-invariant. □

In our running example $\mathscr{E}'_{12}$, $r = 6$ and the set $D_1^0(\mathscr{E}'_{12}) = \{0, 4, 6, 10\}$ is the union of the subgroup $\{0, 6\}$ with its coset $\{4, 10\}$ in the group $H_1(\mathscr{E}'_{12})$.

Let $\mathscr{A} = \langle \mathbb{Z}_n, \{a, b\} \rangle$ be a standardized DFA. Proposition 4 shows that then the set $D_1^0(\mathscr{A})$ is situated between the subgroup $H_1(\mathscr{A})$ and the subgroup R generated by $r \neq 0$ such that $r \cdot a = \mathrm{dupl}(a)$:

$$R \subseteq D_1^0(\mathscr{A}) \subseteq H_1(\mathscr{A}). \qquad (4)$$

Formula (3) implies that the difference set $D_1(\mathscr{A})$ is a-invariant, and so is the set $D_1^0(\mathscr{A})$ since $0 \cdot a = \mathrm{dupl}(a) \in D_1(\mathscr{A})$. By Proposition 1, if the automaton $\mathscr{A}$ is completely reachable, then either $H_1(\mathscr{A}) = \mathbb{Z}_n$ or $H_1(\mathscr{A})$ is a proper subgroup and both inclusions in (4) are strict. Recall that by Proposition 2 $H_1(\mathscr{A}) = \mathbb{Z}_n$ if and only if the digraph $\Gamma_1(\mathscr{A})$ is strongly connected. In the other case, n must be a product of at least three (not necessarily distinct) prime numbers. Indeed, the subgroups of $(\mathbb{Z}_n, +)$ ordered by inclusion are in a 1–1 correspondence to the divisors of n ordered by division, and no product of only two primes can have two different proper divisors d_1 and d_2 such that d_1 divides d_2. We thus arrive at the following conclusion.

Corollary 1. *A binary DFA $\mathscr{A}$ with n states where n is a product of two prime numbers is completely reachable if and only if one of its letters acts as a cyclic permutation of the state set, the other letter has defect 1, and the digraph $\Gamma_1(\mathscr{A})$ is strongly connected.*

Corollary 1 allows one to show that the number of states in a binary completely reachable automata whose Rystsov graph is not strongly connected is at least 12. (Thus, our examples of such automata ($\mathscr{E}_{12}$ from [1, Example 2] and $\mathscr{E}'_{12}$ from the present paper) are of minimum possible size.) Indeed, Corollary 1 excludes all sizes less than 12 except 8. If a standardized DFA $\mathscr{A}$ has 8 states and the digraph $\Gamma_1(\mathscr{A})$ is not strongly connected, then the group $H_1(\mathscr{A})$ has size at most 4 and its subgroup R generated by the non-zero state in $\mathrm{dupl}(a)a^{-1}$ has size at least 2. By Proposition 4 the set $D_1^0(\mathscr{A})$ is a union of cosets of the subgroup R in the group $H_1(\mathscr{A})$, whence either $D_0(\mathscr{A}) = R$ or $D_0(\mathscr{A}) = H_1(\mathscr{A})$. In either case, we get a proper a-invariant subgroup, and Proposition 1 implies that the DFA $\mathscr{A}$ is not completely reachable.

5   Subgroup Sequences for Standardized DFAs

In [1,3] Theorem 1 is generalized in the following way. A sequence of digraphs $\Gamma_1(\mathscr{A})$, $\Gamma_2(\mathscr{A})$, ..., $\Gamma_k(\mathscr{A})$, ... is assigned to an arbitrary (not necessarily binary) DFA $\mathscr{A}$, where $\Gamma_1(\mathscr{A})$ is the Rystsov graph of $\mathscr{A}$ while the 'higher level' digraphs $\Gamma_2(\mathscr{A})$, ..., $\Gamma_k(\mathscr{A})$, ... are defined via words that have defect 2, ..., k, ... in $\mathscr{A}$. (We refer the interested reader to [1,3] for the precise definitions; here we do not need them.) The length of the sequence is less than the number of states of $\mathscr{A}$, and $\mathscr{A}$ is completely reachable if and only if the final digraph in the sequence is strongly connected.

For the case when $\mathscr{A}$ is a standardized DFA, Proposition 2 shows that the Rystsov graph $\Gamma_1(\mathscr{A})$ is completely determined by the difference set $D_1(\mathscr{A})$ and the subgroup $H_1(\mathscr{A})$ that $D_1(\mathscr{A})$ generates. This suggests that for binary automata, one may substitute the 'higher level' digraphs of [1,3] by suitably chosen 'higher level' difference sets and their generated subgroups.

Take a standardized DFA $\mathscr{A} = \langle \mathbb{Z}_n, \{a,b\} \rangle$ and for each $k > 1$, inductively define the set $D_k(\mathscr{A})$ and the subgroup $H_k(\mathscr{A})$:

$$D_k(\mathscr{A}) = \{ p \in \mathbb{Z}_n \mid p \in \mathrm{dupl}(w) \text{ for some } w \in \{a,b\}^*$$
$$\text{such that } 0 \in \mathrm{excl}(w) \subseteq H_{k-1}(\mathscr{A}), \ |\mathrm{excl}(w)| \leq k \}, \tag{5}$$
$$H_k(\mathscr{A}) \text{ is the subgroup of } (\mathbb{Z}_n, +) \text{ generated by } D_k(\mathscr{A}).$$

Observe that if we let $H_0(\mathscr{A}) = \{0\}$, the definition (5) makes sense also for $k = 1$ and leads to exactly the same $D_1(\mathscr{A})$ and $H_1(\mathscr{A})$ as defined in Sect. 4.

Using the definition (5), it is easy to prove by induction that $D_k(\mathscr{A}) \subseteq D_{k+1}(\mathscr{A})$ and $H_k(\mathscr{A}) \subseteq H_{k+1}(\mathscr{A})$ for all k.

Proposition 5. *If $\mathscr{A} = \langle \mathbb{Z}_n, \{a,b\} \rangle$ is a standardized DFA and $H_\ell(\mathscr{A}) = \mathbb{Z}_n$ for some ℓ, then $\mathscr{A}$ is a completely reachable automaton.*

Proof. As $\mathscr{A}$ is fixed, we write D_k and H_k instead of $D_k(\mathscr{A})$ and, resp., $H_k(\mathscr{A})$.

Take any non-empty subset $S \subseteq \mathbb{Z}_n$. We prove that S is reachable in $\mathscr{A}$ by induction on $n - |S|$. If $n - |S| = 0$, there is nothing to prove as $S = \mathbb{Z}_n$ is reachable via the empty word. Now let S be a proper subset of $\mathbb{Z}_n$. We aim to find a subset $T \subseteq \mathbb{Z}_n$ such that $S = T \cdot v$ for some word $v \in \{a,b\}^*$ and $|T| > |S|$. Since $n - |T| < n - |S|$, the induction assumption applies to the subset T whence $T = \mathbb{Z}_n \cdot u$ for some word $u \in \{a,b\}^*$. Then $S = \mathbb{Z}_n \cdot uv$ is reachable as required.

Thus, fix a non-empty subset $S \subsetneq \mathbb{Z}_n$. Since cosets of the trivial subgroup H_0 are singletons, S is a union of cosets of H_0. On the other hand, since $H_\ell = \mathbb{Z}_n$, the only coset of H_ℓ strictly contains S, and so S is not a union of cosets of H_ℓ. Now choose $k \geq 1$ to be the maximal number for which S is a union of cosets of the subgroup H_{k-1}. The subgroup H_k already has a coset, say, $H_k + t$ being neither contained in S nor disjoint with S; in other words, $\varnothing \neq S \cap (H_k + t) \subsetneq H_k + t$.

By Lemma 3, the coset $H_k + t$ serves as the vertex set of a strongly connected component of the Cayley digraph $\mathrm{Cay}(\mathbb{Z}_n, D_k)$. Therefore, some edge of $\mathrm{Cay}(\mathbb{Z}_n, D_k)$ connects $(H_k + t) \setminus S$ with $S \cap (H_k + t)$ in this strongly connected component, that is, the head q of this edge lies in $(H_k + t) \setminus S$ while its tail p belongs to $S \cap (H_k + t)$. Let $p' = p - q$; then $p' \in D_k$ by the definition of the Cayley digraph. By (5) there exists a word $w \in \{a,b\}^*$ such that $p' \in \mathrm{dupl}(w)$ and $\mathrm{excl}(w) \subseteq H_{k-1}$. Then $p = p' + q = p' \cdot b^q \in \mathrm{dupl}(w) \cdot b^q = \mathrm{dupl}(wb^q)$ and $\mathrm{excl}(wb^q) = \mathrm{excl}(w) \cdot b^q = \mathrm{excl}(w) + q \subseteq H_{k-1} + q$. From $p \in \mathrm{dupl}(wb^q)$ we conclude that there exist $p_1, p_2 \in \mathbb{Z}_n$ such that $p = p_1 \cdot wb^q = p_2 \cdot wb^q$. Since S is a union of cosets of the subgroup H_{k-1}, the fact that $q \notin S$ implies that the whole coset $H_{k-1} + q$ is disjoint with S, and the inclusion $\mathrm{excl}(wb^q) \subseteq H_{k-1} + q$ ensures that S is disjoint with $\mathrm{excl}(wb^q)$. Therefore, for every $s \in S \setminus \{p\}$, there exists a state $s' \in \mathbb{Z}_n$ such that $s' \cdot wb^q = s$. Now letting $T = \{p_1, p_2\} \cup \{s' \mid s \in S \setminus \{p\}\}$, we conclude that $S = T \cdot wb^q$ and $|T| = |S| + 1$.   □

For an illustration, return one last time to the DFA $\mathscr{E}'_{12}$ shown in Fig. 2. We have seen that the subgroup $H_1(\mathscr{E}'_{12})$ consists of even residues modulo 12. Inspecting the word ab^3a gives $\mathrm{excl}(ab^3a) = \{0,8\} \subseteq H_1(\mathscr{E}'_{12})$ and $1 \in \mathrm{dupl}(ab^3a)$, whence $1 \in D_2(\mathscr{E}'_{12})$. Therefore the subgroup $H_2(\mathscr{E}'_{12})$ generated by $D_2(\mathscr{E}'_{12})$ is equal to $\mathbb{Z}_{12}$, and $\mathscr{E}'_{12}$ is a completely reachable automaton by Proposition 5.

To illustrate the next level of the construction (5), consider the standardized DFA $\mathscr{E}_{48} = \langle \mathbb{Z}_{48}, \{a,b\} \rangle$ shown in Fig. 3. We have replaced edges that should have been labeled a and b with solid and, resp., dashed edges and omitted all loops to lighten the picture. The action of a in $\mathscr{E}_{48}$ is defined by $0 \cdot a = 24 \cdot a = 18$, $13 \cdot a = 14$, $14 \cdot a = 13$, $18 \cdot a = 24$, $30 \cdot a = 32$, $32 \cdot a = 30$, and $k \cdot a = k$ for all other $k \in \mathbb{Z}_{48}$.

One can calculate that $D_1(\mathscr{E}_{48}) = \{18,24,42\}$ whence the subgroup $H_1(\mathscr{E}_{48})$ consists of all residues divisible by 6. Computing $D_2(\mathscr{E}_{48})$, one sees that this set consists of even residues and contains 2 (due to the word $ab^{32}a$ that has $\mathrm{excl}(ab^{32}a) = \{0,30\} \subseteq H_1(\mathscr{E}_{48})$ and $\mathrm{dupl}(ab^{32}a) = \{2,18\}$). Hence the subgroup $H_2(\mathscr{E}_{48})$ consists of all even residues. Finally, the word $ab^{24}ab^{12}ab^8$ has $\{0,8,20\} \subseteq H_1(\mathscr{E}_{48})$ as its excluded set while its duplicate set contains 13. Hence $13 \in D_3(\mathscr{E}_{48})$ and the subgroup $H_3(\mathscr{E}_{48})$ coincides with $\mathbb{Z}_{48}$. We conclude that the DFA $\mathscr{E}_{48}$ is completely reachable by Proposition 5.

As mentioned, the subgroups of $(\mathbb{Z}_n, +)$ ordered by inclusion correspond to the divisors of n ordered by division whence for any standardized DFA $\mathscr{A}$ with n states, the number of different subgroups of the form $H_k(\mathscr{A})$ is $O(\log n)$. Therefore, if the subgroup sequence $H_0(\mathscr{A}) \subseteq H_1(\mathscr{A}) \subseteq \cdots \subseteq H_k(\mathscr{A}) \subseteq \ldots$ strictly grows at each step, then it reaches $\mathbb{Z}_n$ after at most $O(\log n)$ steps, and by Proposition 5 $\mathscr{A}$ is a completely reachable automaton. What happens if the sequence stabilizes earlier? Our next result answers this question.

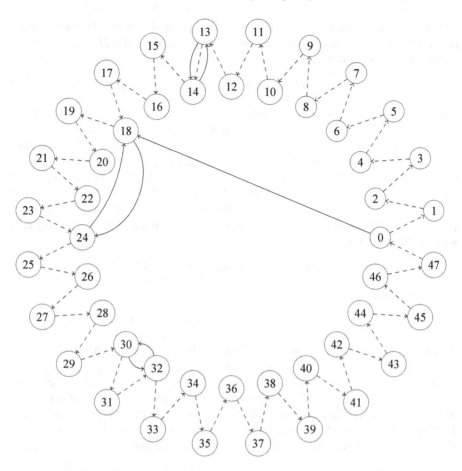

Fig. 3. The DFA $\mathscr{E}_{48} = \langle \mathbb{Z}_{48}, \{a,b\} \rangle$ with $H_2(\mathscr{E}_{48}) \neq \mathbb{Z}_{48}$. Solid and dashed edges show the action of a and, resp., b; loops are not shown

Proposition 6. *If for a standardized DFA* $\mathscr{A} = \langle \mathbb{Z}_n, \{a,b\} \rangle$, *there exists* ℓ *such that* $H_\ell(\mathscr{A}) = H_{\ell+1}(\mathscr{A}) \subsetneqq \mathbb{Z}_n$, *then* $\mathscr{A}$ *is not completely reachable.*

Proof. As in the proof of Proposition 5, we use D_k and H_k instead of $D_k(\mathscr{A})$ and, resp., $H_k(\mathscr{A})$ in our arguments.

It suffices to prove the following claim:

Claim: the equality $H_\ell = H_{\ell+1}$ implies that the subgroup H_ℓ is a-invariant.

Indeed, since $H_\ell \subsetneqq \mathbb{Z}_n$, we get a proper a-invariant subgroup, and Proposition 1 then shows that $\mathscr{A}$ is not completely reachable.

Technically, it is more convenient to show that if $H_\ell = H_{\ell+1}$, then $H_k . a \subseteq H_\ell$ for every $k = 0, 1, \ldots, \ell$. We induct on k. The base $k = 0$ is clear since $H_0 = \{0\}$ and $0 . a = \mathrm{dupl}(a) \in D_1 \subseteq H_1 \subseteq H_\ell$.

Let $k < \ell$ and assume $H_k.a \subseteq H_\ell$; we aim to verify that $p.a \in H_\ell$ for every $p \in H_{k+1}$. Since the subgroup H_{k+1} is generated by D_{k+1} and contains H_k, we may choose a representation of p as the sum

$$p = q + d_1 + \cdots + d_m, \quad q \in H_k, \ d_1, \ldots, d_m \in D_{k+1} \setminus H_k,$$

with the least number m of summands from $D_{k+1} \setminus H_k$. We show that $p.a \in H_\ell$ by induction on m. If $m = 0$, we have $p = q \in H_k$ and $p.a \in H_\ell$ since $H_k.a \subseteq H_\ell$.

If $m > 0$, we write p as $p = d_1 + s$ where $s = q + d_2 + \cdots + d_m$. By (5), there exists a word $w \in \{a,b\}^*$ such that $d_1 \in \mathrm{dupl}(w)$, $0 \in \mathrm{excl}(w) \subseteq H_k$ and $|\mathrm{excl}(w)| \leq k + 1$. Consider the word $wb^s a$. We have $p.a = (d_1 + s).a = d_1.b^s a$, and the equality (2) gives $p.a \in \mathrm{dupl}(wb^s a)$. From the equality (1), we get $\mathrm{excl}(wb^s a) = (\mathrm{excl}(w) + s).a \cup \{0\}$ if $\mathrm{dupl}(a)a^{-1}$ is either contained in or disjoint with $\mathrm{excl}(w) + s$, and $\mathrm{excl}(wb^s a) = ((\mathrm{excl}(w) + s) \setminus \mathrm{dupl}(a)a^{-1}).a \cup \{0\}$ if $|\mathrm{dupl}(a)a^{-1} \cap (\mathrm{excl}(w) + s)| = 1$. In any case, we have the inclusion

$$\mathrm{excl}(wb^s a) \subseteq (\mathrm{excl}(w) + s).a \cup \{0\} \tag{6}$$

and the inequality

$$|\mathrm{excl}(wb^s a)| \leq |(\mathrm{excl}(w) + s)).a| + 1 \leq |\mathrm{excl}(w))| + 1 \leq (k+1) + 1 \leq \ell + 1. \tag{7}$$

For any $t \in \mathrm{excl}(w) \subseteq H_k$, the number of summands from $D_{k+1} \setminus H_k$ in the sum $t + s = t + q + d_2 + \cdots + d_m$ is less than m. By the induction assumption, we have $(t+s).a \in H_\ell$. Hence, $(\mathrm{excl}(w) + s).a \subseteq H_\ell$, and since 0 also lies in the subgroup H_ℓ, we conclude from (6) that $\mathrm{excl}(wb^s a) \subseteq H_\ell$. From this and the inequality (7), we see that the word $wb^s a$ satisfies the conditions of the definition of $D_{\ell+1}$ (cf. (5)) whence every state in $\mathrm{dupl}(wb^s a)$ belongs to $D_{\ell+1}$. We have observed that $p.a \in \mathrm{dupl}(wb^s a)$. Hence $p.a \in D_{\ell+1} \subseteq H_{\ell+1}$. Since $H_\ell = H_{\ell+1}$, we have $p.a \in H_\ell$, as required. $\qquad\square$

Now we deduce a criterion for complete reachability of binary automata.

Theorem 2. *A binary DFA $\mathscr{A}$ with n states is completely reachable if and only if either $n = 2$ and $\mathscr{A}$ is the flip-flop or one of the letters of $\mathscr{A}$ acts as a cyclic permutation of the state set, the other letter has defect 1, and in the standardized DFA $\langle \mathbb{Z}_n, \{a,b\} \rangle$ syntactically equivalent to $\mathscr{A}$, no proper subgroup of $(\mathbb{Z}_n, +)$ is a-invariant.*

Proof. Necessity follows from the reductions in Sect. 2 and Proposition 1.

For sufficiency, we can assume that $\mathscr{A} = \langle \mathbb{Z}_n, \{a,b\} \rangle$ is standardized. If no proper subgroup of $(\mathbb{Z}_n, +)$ is a-invariant, then the claim from the proof of Proposition 6 implies that the sequence $H_0(\mathscr{A}) \subseteq H_1(\mathscr{A}) \subseteq \cdots \subseteq H_k(\mathscr{A}) \subseteq \ldots$ strictly grows as long as the subgroup $H_k(\mathscr{A})$ remains proper. Hence, $H_\ell(\mathscr{A}) = \mathbb{Z}_n$ for some ℓ and $\mathscr{A}$ is a completely reachable automaton by Proposition 5. $\qquad\square$

Remark 1. The proof of Theorem 2 shows that only subgroups that contain $H_1(\mathscr{A})$ matter. Therefore, one can combine Theorem 1, Proposition 2 and Theorem 2 as follows: *a standardized DFA $\mathscr{A} = \langle \mathbb{Z}_n, \{a,b\} \rangle$ is completely reachable if and only if either $H_1(\mathscr{A}) = \mathbb{Z}_n$ or no proper subgroup of $(\mathbb{Z}_n, +)$ containing the subgroup $H_1(\mathscr{A})$ is a-invariant.*

The condition of Theorem 2 can be verified in low polynomial time. We sketch the corresponding algorithm.

Given a binary DFA $\mathscr{A}$ with n states, we first check if $n = 2$ and $\mathscr{A}$ is the flip-flop. If **yes**, $\mathscr{A}$ is completely reachable. If **not**, we check whether one of the letters of $\mathscr{A}$ acts as a cyclic permutation of the state set while the other letter has defect 1. If **not**, $\mathscr{A}$ is not completely reachable. If **yes**, we pass to the standardized DFA $\langle \mathbb{Z}_n, \{a, b\} \rangle$ syntactically equivalent to $\mathscr{A}$. As a preprocessing, we compute and store the set $\{(k, k \cdot a) \mid k \in \mathbb{Z}_n\}$.

The rest of the algorithm can be stated in purely arithmetical terms. Call a positive integer d a *nontrivial divisor* of n if d divides n and $d \neq 1, n$. We compute all nontrivial divisors of n by checking through all integers $d = 2, \ldots, \lfloor \sqrt{n} \rfloor$: if such d divides n, we store d and $\frac{n}{d}$. If for some nontrivial divisor d of n, all numbers $(td) \cdot a$ with $t = 0, 1, \ldots, \frac{n}{d} - 1$ are divisible by d, then d generates a proper a-invariant subgroup in $(\mathbb{Z}_n, +)$ and $\mathscr{A}$ is not completely reachable. If for every nontrivial divisor d of n, there exists $t \in \{0, 1, \ldots, \frac{n}{d} - 1\}$ such that $(td) \cdot a$ is not divisible by d, then no proper subgroup of $(\mathbb{Z}_n, +)$ is a-invariant and $\mathscr{A}$ is completely reachable.

To estimate the time complexity of the described procedure, observe that one has to check at most $\frac{n}{d}$ numbers for each nontrivial divisor d of n. Clearly,

$$\sum_{\substack{1 < d < n \\ d \mid n}} \frac{n}{d} = \sum_{\substack{1 < d < n \\ d \mid n}} d = \sigma(n) - n - 1,$$

where $\sigma(n)$ stands for the sum of all divisors of n, a well-studied function in the theory of numbers; see, e.g., [8, Chapters XVI–XVIII]. It is known that $\limsup \frac{\sigma(n)}{n \log \log n} = e^\gamma$ where γ is the Euler–Mascheroni constant [8, Theorem 323]; this implies that the number of checks in our procedure is $O(n \log \log n)$. The total complexity depends on the time spent for verifying the divisibility condition. If one uses the transdichotomous model [7] (as suggested by one of the referees), assuming constant time for division, the whole procedure can be implemented in $O(n \log \log n)$ time.

One can speed up the above algorithm, using Remark 1, which implies that only the divisors $d > 1$ of the g.c.d. of n and $0 \cdot a$ have to be checked. However, the improvement only reduces the constant behind the $O(\ )$ notation.

6   Conclusion

We have characterized binary completely reachable automata; our characterization leads to an algorithm that given a binary DFA $\mathscr{A}$, decides whether or not $\mathscr{A}$ is completely reachable in quasilinear time with respect to the size of $\mathscr{A}$. Very recently, after the present paper was submitted, Ferens and Szykuła [6] have devised a polynomial-time algorithm for recognizing complete reachability of arbitrary DFAs, but the complexity of their algorithm is higher.

Our results heavily depend on the fact that apart from a single exception, binary completely reachable automata are *circular*, that is, have a letter acting as a cyclic permutation of the state set. In the literature, one can find several situations when a problem that remains open in general, admits quite a nontrivial solution when restricted to circular automata. Here we mention only Dubuc's result [5] on the Černý conjecture and the recent paper by Yong He *et al.* [9] on Trahtman's conjecture. It appears that circular automata may behave in a similar way with respect to complete reachability, and our follow-up work aims at extending the results of the present paper to arbitrary (not necessarily binary) circular automata. We also plan to study an 'orthogonal' extension, aiming to characterize completely reachable automata in which one letter has defect 1 while the other letters act as permutations and generate a group that transitively acts on the state set.

Acknowledgement. We thank the anonymous reviewers for their careful reading of our paper and their many useful comments and suggestions.

References

1. Bondar, E.A., Casas, D., Volkov, M.V.: Completely reachable automata: an interplay between automata, graphs, and trees. CoRR abs/2201.05075 (2022). https://arxiv.org/abs/2201.05075
2. Bondar, E.A., Volkov, M.V.: Completely reachable automata. In: Câmpeanu, C., Manea, F., Shallit, J. (eds.) DCFS 2016. LNCS, vol. 9777, pp. 1–17. Springer, Cham (2016). https://doi.org/10.1007/978-3-319-41114-9_1
3. Bondar, E.A., Volkov, M.V.: A characterization of completely reachable automata. In: Hoshi, M., Seki, S. (eds.) DLT 2018. LNCS, vol. 11088, pp. 145–155. Springer, Cham (2018). https://doi.org/10.1007/978-3-319-98654-8_12
4. Don, H.: The Černý conjecture and 1-contracting automata. Electr. J. Comb. **23**(3), 3–12 (2016). https://doi.org/10.37236/5616
5. Dubuc, L.: Sur les automates circulaires et la conjecture de Černý. RAIRO Inform. Théorique App. **32**, 21–34 (1998). in French. http://www.numdam.org/item/ITA_1998__32_1-3_21_0
6. Ferens, R., Szykuła, M.: Completely reachable automata: A polynomial solution and quadratic bounds for the subset reachability problem. CoRR abs/2208.05956 (2022). https://arxiv.org/abs/2208.05956
7. Fredman, M.L., Willard, D.E.: Surpassing the information theoretic bound with fusion trees. J. Comput. Syst. Sci. **47**(3), 424–436 (1993). https://doi.org/10.1016/0022-0000(93)90040-4
8. Hardy, G.H., Wright, E.M.: An Introduction to the Theory of Numbers, 6th edn. Oxford University Press, Oxford (2008)
9. He, Y., Chen, X., Li, G., Sun, S.: Extremal synchronizing circular automata. Inf. Comput. **281**, 104817 (2021). https://doi.org/10.1016/j.ic.2021.104817
10. Maslennikova, M.I.: Reset complexity of ideal languages. In: Bieliková, M., Friedrich, G., Gottlob, G., Katzenbeisser, S., Špánek, R., Turán, G. (eds.) SOFSEM 2012. vol. II, pp. 33–44. Institute of Computer Science Academy of Sciences of the Czech Republic (2012). http://arxiv.org/abs/1404.2816
11. Rystsov, I.K.: Estimation of the length of reset words for automata with simple idempotents. Cybern. Syst. Anal. **36**(3), 339–344 (2000). https://doi.org/10.1007/BF02732984

Conelikes and Ranker Comparisons

Viktor Henriksson[1] and Manfred Kufleitner[2]([✉])

[1] Loughborough University, Loughborough, UK
b.v.d.henriksson@lboro.ac.uk
[2] University of Stuttgart, FMI, Stuttgart, Germany
kufleitner@fmi.uni-stuttgart.de

Abstract. For every fixed class of regular languages, there is a natural hierarchy of increasingly more general problems: Firstly, the membership problem asks whether a given language belongs to the fixed class of languages. Secondly, the separation problem asks for two given languages whether they can be separated by a language from the fixed class. And thirdly, the covering problem is a generalization of separation problem to more than two given languages. Most instances of such problems were solved by the connection of regular languages and finite monoids. Both the membership problem and the separation problem were also extended to ordered monoids. The computation of pointlikes can be interpreted as the algebraic counterpart of the separation problem. In this paper, we consider the extension of computation of pointlikes to ordered monoids. This leads to the notion of conelikes for the corresponding algebraic framework.

We apply this framework to the Trotter-Weil hierarchy and both the full and the half levels of the FO^2 quantifier alternation hierarchy. As a consequence, we solve the covering problem for the resulting subvarieties of **DA**. An important combinatorial tool are uniform ranker characterizations for all subvarieties under consideration; these characterizations stem from order comparisons of ranker positions.

1 Introduction

For a given variety of regular languages, there is a hierarchy of decision problems: First, we can ask whether a given regular language is in the variety; this is known as the *membership problem*. Very often, the membership problem is solved by giving an effective characterization. Famous solutions to the membership problem includes Simon's characterization of the piecewise testable languages in terms of $\mathcal{J}$-trivial monoids [22], and Schützenberger's characterization of the star-free languages by aperiodic monoids [20]. Inspired by these results, Eilenberg showed that there exists a one to one correspondence between varieties of regular languages and varieties of finite monoids [6]. This correspondence leads to an important approach for deciding the membership problem: one verifies some equivalent algebraic property of the syntactic monoid. The challenge here, however, is to identify the algebraic property and to prove its equivalence.

© Springer Nature Switzerland AG 2022
A. Castañeda and F. Rodríguez-Henríquez (Eds.): LATIN 2022, LNCS 13568, pp. 359–375, 2022.
https://doi.org/10.1007/978-3-031-20624-5_22

A more general problem is the *separation problem*. Given two languages, it asks whether there exists a language in the fixed variety which contains the first language and is disjoint with the second language. By applying the separation problem to a language and its complement, we obtain an answer to the membership problem. Thus, the separation problem is more general than the membership problem. Moreover, the separation problem can be used as a tool to solve the membership problem for varieties where this was not previously known; see e.g. [17]. A further generalization is given by the *covering problem* [18]. This problem considers a finite set of languages and a distinguished language, and asks how well the finite set of languages can be separated by a cover of the distinguished language.

As noted by Almeida, the separation problem for regular languages can also be solved via algebra by deciding so-called pointlikes [1]. The problem of deciding pointlikes is well studied, and there are effective characterizations for many varieties, e.g. aperiodics [8], $\mathcal{R}$-trivial monoids [3], $\mathcal{J}$-trivial monoids [4,23] and finite groups [5].

A well studied fragment of first order logic is two-variable first-order logic FO^2. The languages definable in FO^2 form a variety, with the corresponding monoid variety **DA**. In the study of FO^2 and **DA**, two natural hierarchies have emerged: the Trotter-Weil hierarchy defined by a deep connection to the hierarchy of bands, and the quantifier alternation hierarchy. In stark contrast to the full FO quantifier alternation hierarchy, membership of the FO^2 quantifier alternation hierarchy is solved for all levels [7,11,14]. In particular, a tight connection between the Trotter-Weil and the quantifier alternation hierarchy has appeared; Weil and the second author showed that the join levels of the quantifier alternation hierarchy (i.e., the FO^2_m levels) correspond to the intersection levels of the Trotter-Weil hierarchy [14], and combining two results from [7,13] shows that the join levels of the Trotter-Weil hierarchy correspond to the intersection levels of the quantifier alternation hierarchy.

Rankers have emerged as an important tool in the study of FO^2. These were first introduced by Schwentick, Thérien, and Vollmer by the name of turtle programs [21]. Using comparisons of restricted sets of rankers, Weis and Immerman gave a combinatorial characterization of the full levels of the quantifier alternation hierarchy [25]. This approach was extended to the half-levels of the quantifier alternation hierarchy in the PhD-thesis of Lauser [16]. The corners of the Trotter-Weil hierarchy also admit ranker characterizations using the concept of condensed rankers [15].

This article solves the covering problem (and thus also the separation problem) for all levels of the Trotter-Weil hierarchy and quantifier alternation hierarchy inside FO^2. For this, we rely on two main tools, *conelikes* and *ranker comparisons*. Conelikes are introduced in Sect. 3. They extend pointlikes to ordered monoids, and are algebraic versions of the imprints used by Place and Zeitoun; see e.g. [18]. Thus, they have a strong connection to the covering problem; an algorithm for computing the conelikes with respect to a monoid variety can be used to solve the covering problem for the corresponding language variety and vice versa.

Sections 4 and 5 deals with ranker comparisons. In Sect. 4, we give a framework for ranker comparisons using general sets of ranker pairs. We show that any set of pairs of rankers which is closed under ranker subwords gives rise to a stable relation and thus defines a monoid.

In Sect. 5, we use this framework to give uniform characterizations for all levels of the Trotter-Weil and quantifier alternation hierarchy. In particular, we give a characterization of the corners of the Trotter-Weil hierarchy in terms of ranker comparisons. Together, these sets of ranker comparisons form a natural hierarchy, the *ranker comparison hierarchy* which encompasses both the quantifier alternation hierarchy and the Trotter-Weil hierarchy.

The rest of the article is devoted to the solution of the covering problem. In Sect. 6, we present sets of subsets of a monoid which can be computed effectively. Our main theorem states that these sets coincide with the conelikes (or the pointlikes for the unordered varieties). Our main theorem also provides optimal separators: relational morphisms such that the conelikes with respect to these morphisms are the same as the conelikes with respect to the corresponding variety. The co-domains in these morphisms are defined using ranker comparisons.[1]

2 Preliminaries

2.1 Words and Monoids

Let A be a collection of symbols, called an *alphabet*. The set of concatenations of symbols in A is A^*. In other words, A^* is the *free monoid* of A. An element $u \in A^*$ is a *word* and a subset $L \subseteq A^*$ a *language*. The empty word is ε. A *(scattered) subword* of u is a word $v = a_1 \ldots a_n$ such that $u = u_1 a_1 \ldots u_n a_n u_{n+1}$ for some (possibly empty) words u_i. Let $u = u_1 u_2 u_3$ for some (possibly empty) words u_1, u_2, u_3. Then u_1 is a *prefix* and u_2 is a *factor* of u.

If $u = u_1 \ldots u_n$ where each u_i is a word, then $u_1 \ldots u_n$ is a *factorization* of u. This concept extends to subsets of A^*; if $L \subseteq A^*$, a factorization of L is $U_1 \ldots U_n$ where each $u \in L$ can be factored as $u = u_1 \ldots u_n$ in such a way that $u_i \in U_i$.

For an alphabet A, let J_A denote the monoid whose elements are subsets of A and whose operation is the union operation. This monoid has a natural ordering defined by $U \leq V$ if $U \subseteq V$ for $U, V \subseteq A$. Let $\mathsf{alph}_A : A^* \to J_A$ be the extension of $\mathsf{alph}_A(a) = \{a\}$ for each $a \in A$. We drop the subscript when A is clear from context. Given a surjective homomorphism $\mu : A^* \to M$, a morphism $\alpha : M \to J_A$ is called a *content morphism* if $\mathsf{alph}_A = \alpha \circ \mu$.

If M is a monoid and $e \in M$ satisfies $ee = e$, then e is *idempotent*. Given a monoid M there exists a (smallest) number ω_M such that u^{ω_M} is idempotent for each $u \in M$. If M is clear from context, we write ω for this number. For sets $S, T \subseteq M$, we have

---

[1] The results in this paper appeared in the first author's PhD thesis [9].

$$ST = \{st \in M \mid s \in S, t \in T\}.$$

Note that 2^M is a monoid under this operation. This definition does not coincide with the factorizations of languages given above.[2] To resolve this ambiguity, we always consider concatenation to mean factorization when dealing with $L \subseteq A^*$ and monoid multiplication otherwise.

An important tool in monoid theory are the *Green's relations*, out of which we introduce the following three. Given a monoid M and $s, t \in M$, we define $s \leq_{\mathcal{R}} t$ if $sM \subseteq tM$, $s \leq_{\mathcal{L}} t$ if $Ms \subseteq Mt$, $s \leq_{\mathcal{J}} t$ if $MsM \subseteq MtM$. We define $s \mathcal{R} t$ if $s \leq_{\mathcal{R}} t$ and $t \leq_{\mathcal{R}} s$ and we define $s \mathcal{L} t$ and $s \mathcal{J} t$ correspondingly. We say that $s <_{\mathcal{R}} t$ if $s \leq_{\mathcal{R}} t$ but not $s \mathcal{R} t$ and equivalently for $\mathcal{L}$ and $\mathcal{J}$. Let $u \in A^*$ and $\mu : A^* \to M$. Then there is a unique factorization $u = u_1 a_1 \ldots u_{n-1} a_{n-1} u_n$ such that $\mu(u_1 a_1 \ldots a_i) \mathcal{R} \mu(u_1 a_1 \ldots a_i u_{i+1}) <_{\mathcal{R}} \mu(u_1 a_1 \ldots a_i u_{i+1} a_{i+1})$. This is the *$\mathcal{R}$-factorization of u with respect to μ*. The *$\mathcal{L}$-factorization of u with respect to μ* is defined symmetrically.

Given a monoid M with a binary relation $\preceq$, we say that $\preceq$ is *stable* if for all $s, t, x, y \in M$, $s \preceq t$ implies $xsy \preceq xty$. We say that a monoid is *ordered* if it is equipped with a stable order. A *congruence* is a stable equivalence relation. In particular, any stable preorder $\preceq$ induces the congruence given by $s \sim t$ if and only if $s \preceq t$ and $t \preceq s$. If M is ordered, and $s \in M$, then $\uparrow s = \{t \in M \mid s \leq t\}$. If M is a monoid, and $\preceq$ is a stable preorder, then $M/\preceq$ is the monoid whose elements are the equivalence classes of the induced congruence, the multiplication is that induced by the multiplication in M and where, for $s, t \in M$ with $[s], [t]$ the corresponding equivalence classes, we have $[s] \leq [t]$ if and only if $s \preceq t$. Given a language L, the *syntactic preorder* is the relation $u \leq_L v$ if and only if $xuy \in L \Rightarrow xvy \in L$ for all $x, y \in A^*$. Let $\mu : A^* \to A^*/\leq_L$ be the canonical projection, then π is the *syntactic morphism* and $A^*/\leq_L$ the *syntactic monoid* of L. A language is *regular* if and only if the syntactic monoid is finite.

Let M and N be (possibly ordered) monoids. A *relational morphism* is a relation $\tau : M \to N$ (or mapping $M \to 2^N$) which satisfies $1_N \in \tau(1_M)$, for all $s \in M$, $\tau(s) \neq \emptyset$, for all $s, t \in M$, $\tau(s)\tau(t) \subseteq \tau(st)$. If there is a relational morphism $\tau : M \to N$ such that $\tau(s) \cap \tau(s') \neq \emptyset$ implies $s = s'$ we say that M *divides* N. A *division of ordered monoids* is a division where we also assume $t \leq t'$ for some $t \in \tau(s)$, $t' \in \tau(s')$ implies $s \leq s'$. A *variety of monoids* is a collection of monoids closed under division and finite direct products. A collection of ordered monoids is a *positive variety* if it is closed under finite direct products and division of ordered monoids.

For a relational morphism $\tau : M \to N$, a set $S \subseteq M$ such that $\bigcap_{s \in S} \tau(s) \neq \emptyset$ is *pointlike* with respect to τ. If $t \in \bigcap_{s \in S} \tau(s)$, then t is a *witness* of S being pointlike. If $\mathbf{V}$ is a variety and S is pointlike for every relational morphism $\tau : M \to N$ where $N \in \mathbf{V}$, then S is pointlike with respect to $\mathbf{V}$. The set of all pointlikes in M with respect to τ is $\mathsf{PL}_\tau(M)$, and the set of all pointlikes with respect to $\mathbf{V}$ is $\mathsf{PL}_\mathbf{V}(M)$.

---

[2] Indeed, $\{aa, bb\}$ can be factored as $\{a, b\} \{a, b\}$, but $ab, ba \in \{a, b\} \{a, b\}$ if seen as a multiplication.

A useful way to define varieties is through the use of ω-*identities* and ω-*relations*. An ω-*term* is either x where x is taken from some (usually infinite) set of variables X, or tt' or t^ω where t and t' are ω-terms. An ω-identity is given by $t = t'$ or $t \leq t'$ where t and t' are ω-terms. Given a monoid M, an *interpretation* of ω-terms is any extension of a map $\chi : X \to M$ for which $\chi(tt') = \chi(t)\chi(t')$ and $\chi(t^\omega) = \chi(t)^{\omega_M}$. We say that a monoid M *satisfy* an ω-identity $t = t'$ if $\chi(t) = \chi(t')$ for all interpretations χ. It similarly satisfies $t \leq t'$ if $\chi(t) \leq \chi(t')$ for all interpretations. If $R_1, \ldots, R_n$ are ω-identities or -relations, then $[\![R_1, \ldots, R_n]\!]$ denotes the collection of all monoids which satisfy all R_i. Some varieties that are of importance in this text are

- **DA** $= [\![(xzy)^\omega = (xzy)^\omega z(xzy)^\omega]\!]$.
- **J** $= [\![(st)^\omega s(xy)^\omega = (st)^\omega y(xy)^\omega]\!]$, or equivalently all monoids whose $\mathcal{J}$-classes are trivial.
- **J**$_1$ $= [\![x^2 = x, xy = yx]\!]$,
- **J**$^+$ $= [\![1 \leq z]\!]$.

2.2 Logic

We consider FO[<], first order logic using the following syntax:

$$\varphi ::= \top \mid \bot \mid a(x) \mid x = y \mid x < y \mid \neg\varphi \mid \varphi \wedge \varphi \mid \varphi \vee \varphi \mid \exists x \varphi.$$

Here $a \in A$ for some fixed alphabet A, and $\varphi \in \text{FO}[<]$. We interpret formulae in FO[<] over words as follows. If $i, j \in \mathbb{N}$, then $u, i, j \vDash x < y$ if and only if $i < j$, and $u, i \vDash a(x)$ if and only if $u[i] = a$. The logical connectives and existential quantifier are interpreted as usual. We use the macro $x \leq y$ to mean $x < y \vee x = y$ and the macro $\forall x \varphi$ to mean $\neg \exists x \neg \varphi$. If φ is a formula without free variables over the alphabet A, we define $L(\varphi) = \{u \in A^* \mid u \vDash \varphi\}$. If $\mathscr{F}$ is a collection of formulae, we say that $L \subseteq A^*$ is *definable in* $\mathscr{F}$ if there exists $\varphi \in \mathscr{F}$ such that $L = L(\varphi)$.

In particular, we are interested in FO^2[<], i.e. the fragment of FO[<] where we only allow the use (and reuse) of two variable names. Thus

$$\exists x : a(x) \wedge (\exists y : y > x \wedge b(y) \wedge (\exists x : x > y \wedge c(x)))$$

is allowed in FO^2[<] whereas

$$\exists x : a(x) \wedge (\exists y : y > x \wedge b(y) \wedge (\exists z : z > x \wedge y > z \wedge c(z)))$$

is not. It is well known that FO^2[<] is a proper fragment of FO[<]. We are primarily interested in some fragments of FO^2[<]. Consider the syntax

$$\varphi_0 ::= \top \mid \bot \mid a(x) \mid x = y \mid x < y \mid \neg\varphi_0 \mid \varphi_0 \vee \varphi_0 \mid \varphi_0 \wedge \varphi_0$$

$$\varphi_m ::= \varphi_{m-1} \mid \neg\varphi_{m-1} \mid \varphi_m \vee \varphi_m \mid \varphi_m \wedge \varphi_m \mid \exists x \varphi_m$$

The collection of formulae $\varphi_m[<]$ is denoted by $\Sigma_m^2[<]$, the collection of negations of formulae in $\Sigma_m^2[<]$ is $\Pi_m^2[<]$ and the Boolean closure of $\Sigma_m^2[<]$ is FO$_m^2[<]$. In what follows, we drop the reference to the predicate symbol $<$, and assume it to be understood from context.

2.3 Ramsey Numbers

A *graph* is a pair $\mathcal{G} = (V, E)$ where V is a set of *vertices* and $E \subseteq \{S \subseteq 2^V \mid |S| = 2\}$ is a set of *edges*. An *edge-coloring* is a map $c : E \to C$ where C is some set of colors. A graph is *complete* if $E = \{S \subseteq 2^V \mid |S| = 2\}$, i.e. if there is an edge between any two elements. A set $F \subseteq E$ of edges is *monochrome* if $c(e) = c(e')$ for all $e \in F$. A *triangle* is a set of three distinct edges $e_1, e_2, e_3 \in E$ where $e_i \cap e_j \neq \emptyset$ for $1 \leq i, j \leq 3$. The following theorem is a special case of Ramsey's Theorem [19].

Theorem 1. *Let C be a finite set of colours. Then there exists a number R, called the* Ramsey number *of C such that any complete graph $\mathcal{G} = (V, E)$ with $R \leq |V|$ contains a monochrome triangle.*

2.4 Hierarchies Inside DA

A variety of special importance for this article is **DA**. This monoid variety has a natural correspondence to FO^2 since a language is definable in the latter if and only if its syntactic monoid is in **DA** [24].

We are interested in hierarchies of subvarieties of **DA**. One important such hierarchy is the *Trotter-Weil hierarchy*. Its original motivation comes from an intimate relation with the hierarchy of bands, but here we give a more explicit definition.

Definition 1. *Let M be a monoid, and let $s, t \in M$. Then*

- *$s \sim_{\mathbf{K}} t$ if for all idempotents $e \in M$, either $es, et <_{\mathcal{J}} e$ or $es = et$,*
- *$s \sim_{\mathbf{D}} t$ if for all idempotents $f \in M$, either $sf, tf <_{\mathcal{J}} f$ or $sf = tf$.*

The join of these relations is $\sim_{\mathbf{KD}}$.

It is straight-forward to check that these relations are congruences (see e.g. [12]). Let $\mathbf{R}_1 = \mathbf{L}_1 = \mathbf{J}_1$, and let $M \in \mathbf{R}_m$ if $M/\sim_{\mathbf{K}} \in \mathbf{L}_{m-1}$ and $M \in \mathbf{L}_m$ if $M/\sim_{\mathbf{D}} \in \mathbf{R}_{m-1}$. When defining $\mathbf{R}_m$ and $\mathbf{L}_m$ for $m \geq 2$, starting with $\mathbf{J}_1$ yields the same result as starting with $\mathbf{J}$. For our purposes, starting with $\mathbf{J}_1$ is more natural.

The varieties $\mathbf{R}_m$ and $\mathbf{L}_m$ are all contained in **DA**. Together with their joins and intersections they make up the Trotter-Weil hierarchy shown in Fig. 1.

There is an intimate connection between the quantifier alternation hierarchy, also shown in Fig. 1, and the Trotter-Weil hierarchy. Indeed, it was shown by Weil and the second author that the languages definable in FO_m^2 are exactly those whose syntactic monoid is in $\mathbf{R}_{m+1} \cap \mathbf{L}_{m+1}$ [14]. Furthermore, combining the results in [13] and [7] gives the following proposition.

Proposition 1. *A language is definable in both Σ_m^2 and Π_m^2 if and only if its syntactic monoid is in $\mathbf{R}_m \vee \mathbf{L}_m$.*

The *corners* of the quantifier alternation hierarchy, Σ_m^2 and Π_m^2, also have algebraic characterizations, given by Fleischer, Kufleitner and Lauser [7]. We define these recognizing varieties using the stable relation $\preceq_{\mathbf{KD}}$ introduced by the authors [10].

Definition 2. *Let M be a monoid, and let $s, t \in M$. We say that $s \preceq_{\mathbf{KD}} t$ if for all $x, y \in M$, the following holds:*

(i) If $x \mathrel{\mathcal{R}} xty$, then $x \mathrel{\mathcal{R}} xsy$,
(ii) If $xty \mathrel{\mathcal{L}} y$, then $xsy \mathrel{\mathcal{L}} y$,
(iii) If $x \mathrel{\mathcal{R}} xt$ and $ty \mathrel{\mathcal{L}} y$, then $xsy \leq xty$.

If $u \preceq_{\mathbf{KD}} v$ and $v \preceq_{\mathbf{KD}} u$, we say that $u \equiv_{\mathbf{KD}} v$.[3]

Let $\mathbf{Si}_1 = \mathbf{J}^+ = [\![1 \leq z]\!]$ and let $M \in \mathbf{Si}_m$ if $M/{\preceq_{\mathbf{KD}}} \in \mathbf{Si}_{m-1}$. For every m, the collection $\mathbf{Si}_m$ is a positive variety. A language is definable in Σ_m^2 if and only if its syntactic monoid is in $\mathbf{Si}_m$. We say that an ordered monoid M is in $\mathbf{Pi}_m$ if and only if M with the order reversed, is in $\mathbf{Si}_m$. It is clear that a language is definable in Π_m^2 if and only if its syntactic monoid is in $\mathbf{Pi}_m$.

3 Conelikes and the Covering Problem

In this section, we introduce the main problems of the article, the *separation problem* and the *covering problem*. Given a variety $\mathcal{V}$, the (asymmetric) separation problem is defined as follows:

Given $L, L' \subseteq A^*$, determine if there is a language $K \in \mathcal{V}$ such that $L \subseteq K$ and $L' \cap K = \emptyset$.

If there exists such a K, we say that L is $\mathcal{V}$-separable from L'. The symmetric separation problem is to determine whether both L is $\mathcal{V}$-separable from L' and L' is $\mathcal{V}$-separable from L. If $\mathcal{V}$ is a full variety, i.e. closed under complements, these two problems are equivalent (just choose $A^* \setminus K$ to separate L' from L).

There is a strong connection between the (symmetric) separation and the problem of deciding pointlikes [1]. In this section, we introduce the more general covering problem, together with a generalization of pointlikes which works well with the asymmetric setting. This generalization, which we call *conelikes*, is folklore. However, to the knowledge of the authors they have not been made precise in the algebraic setting.[4]

Let $\mathbf{K}$ be a set of languages, and $\mathbf{L}$ a finite set of languages. Then $\mathbf{K}$ is *separating* for $\mathbf{L}$ if for all $K \in \mathbf{K}$, there exists $L' \in \mathbf{L}$ such that $K \cap L' = \emptyset$. We

---

[3] The name $\preceq_{\mathbf{KD}}$ was originally inspired by the relation $\sim_{\mathbf{KD}}$ since they share some properties. However, it should be noted that the relation $\equiv_{\mathbf{KD}}$ is not the same as $\sim_{\mathbf{KD}}$. As a counter example, note the syntactic monoid of $a^+b^+cA^*da^+b^+$ where the equivalence class of ab is $\equiv_{\mathbf{KD}}$-related to all elements in the minimal $\mathcal{J}$-class, whereas it is not $\sim_{\mathbf{K}}$- or $\sim_{\mathbf{D}}$-related to anything.

[4] The imprints used by Place and Zeitoun in [18] yield corresponding objects in the language setting.

only consider situations when $\mathbf{K}$ is a *cover* of some language L, i.e. $\mathbf{K}$ is finite and $L \subseteq \bigcup \mathbf{K}$.

Definition 3. *Let* $\mathcal{V}$ *be a (positive) variety. The* covering problem *for* $\mathcal{V}$ *is defined as follows:*

Given $L \subseteq A^*$ *and* $\mathbf{L} \subseteq 2^{A^*}$ *where* $\mathbf{L}$ *is finite, determine if there is* $\mathbf{K} \subseteq \mathcal{V}$ *which covers* L *and is separating for* $\mathbf{L}$.

If such a $\mathbf{K}$ exists, we say that $(L, \mathbf{L})$ is $\mathcal{V}$-coverable. Separation is exactly the special case when $\mathbf{L}$ is a singleton. Whenever $\mathcal{V}$ is a full variety, it is equivalent to answer the covering problem for $(L, \mathbf{L})$ and $(A^*, \{L\} \cup \mathbf{L})$ [18], and for regular languages this is in turn equivalent to computing the $\mathbf{V}$-pointlikes of a finite monoid recognizing all languages of $\mathbf{L}$ [1].

We want to use algebraic methods to solve the covering problem. However, the covering problem is relevant for positive varieties, whereas pointlikes do not take orders into account. This motivates the following generalization of pointlikes.

Definition 4. *Let* M *be a monoid, and let* $\tau : M \to N$ *be a relational morphism. For* $s \in M$, $S \subseteq M$, *we say that* (s, S) *is* conelike *with respect to* τ *if there exists an element* $x \in \tau(s)$ *such that* $S \subseteq \tau^{-1}(\uparrow x)$. *We call* x *a* witness *of* (s, S) *being conelike. As with pointlikes we say that a pair* (s, S) *is conelike with respect to a variety* $\mathbf{V}$ *if it is conelike for any* $\tau : M \to N \in \mathbf{V}$. *We denote by* $\mathsf{Cone}_\tau(M)$ *the conelikes of* M *with respect to* τ, *and by* $\mathsf{Cone}_\mathbf{V}(M)$ *the conelikes of* M *with respect to* $\mathbf{V}$.

Note that if N is unordered, we can define an order $u \leq v$ if and only if $u = v$. In this case, a pair (s, S) is conelike if and only if $S \cup \{s\}$ is pointlike. In particular, this means that for non-positive varieties, calculating the pointlikes and the conelikes is the same problem.

The concept of pointlikes is in general not expressive enough to solve the covering problem. However, it is still possible to define pointlikes for a variety of ordered monoids, and such pointlikes are used throughout the article. Note that if $S \subseteq M$ is pointlike with respect to some variety of ordered monoids, then (s, S) is conelike for any $s \in S$.

Proposition 2. *Let* $\mathbf{L} = \{L_i\}$ *be a finite collection of regular languages over some alphabet* A, *and for each* L_i *let* $\mu_i : A^* \to M_i$ *be its syntactic monoid. Let* $\mu : A^* \to M_1 \times \cdots \times M_n$ *be defined by* $\mu(a) = (\mu_1(a), \ldots, \mu_n(a))$ *and let* $M = \mu(A^*)$. *Let* $\mathbf{V}$ *be a variety of monoids recognizing a variety* $\mathcal{V}$ *of languages, and let* $L \subseteq A^*$. *Then the following are equivalent*

(i) $(L, \mathbf{L})$ *is* **not** $\mathcal{V}$-coverable,
(ii) *there exists a conelike* (s, S) *with respect to* $\mathbf{V}$ *such that* $L \cap \mu^{-1}(s) \neq \emptyset$ *and for all* $L' \in \mathbf{L}$ *there exists* $s' \in S$ *such that* $L' \cap \mu^{-1}(s') \neq \emptyset$.

Before leaving the topic of conelikes, we introduce a common tool for determining pointlikes and solving the covering problem (see e.g. [18]). The idea is to construct sets of subsets of M which have closure properties analogous to those of pointlikes and conelikes.

Definition 5. *Given a monoid M, a subset of 2^M is closed if it contains the singletons and is closed under multiplication and subsets. Similarly, a set $\mathcal{C} \subseteq M \times 2^M$ is closed if it has the following closure properties:*

- *$(s, \{s\}) \in \mathcal{C}$ for all elements $s \in M$,*
- *$(s, S), (t, T) \in \mathcal{C}$ implies $(st, ST) \in \mathcal{C}$,*
- *$(s, S) \in \mathcal{C}$ implies $(s, S') \in \mathcal{C}$ for all subsets $S' \subseteq S$.*

We note that the set of conelikes (or pointlikes) with respect to some variety **V** or some relational morphism τ is closed.

4   A Framework for Ranker Comparisons

One of the main techniques in this paper is ranker comparisons. This concept has close connections to fragments of FO^2, a connection we explore in Sect. 5 (see. [16,25]). In this section, we introduce a general framework and give sufficient conditions for instances of this framework to define a monoid.

Definition 6. *Let A be some alphabet. A* ranker *over A is a nonempty word over $\{X_a, Y_a\}_{a \in A}$, which can be interpreted as a partial function from A^* to $\mathbb{N}$. The interpretation is defined inductively as follows:*

- *$X_a(u) = \inf \{n \in \mathbb{N} \mid u[n] = a\}$ if it exists and undefined otherwise,*
- *$Y_a(u) = \sup \{n \in \mathbb{N} \mid u[n] = a\}$ if it exists and undefined otherwise,*
- *$rX_a(u) = \inf \{n \in \mathbb{N} \mid n > r(u), u[n] = a\}$ if $r(u)$ is defined and the infimum is finite, and undefined otherwise,*
- *$rY_a(u) = \sup \{n \in \mathbb{N} \mid n < r(u), u[n] = a\}$ if $r(u)$ is defined and the supremum is finite, and undefined otherwise.*

Note that we read rankers from left to right (as opposed to function composition). Thus $X_a Y_b(bab) = 1$, whereas $Y_b X_a(bab)$ is undefined. If $p = a_1 \cdots a_n$, we define $X_p = X_{a_1} \cdots X_{a_n}$ and $Y_p = Y_{a_1} \cdots Y_{a_n}$ for compactness.

We define the following collections of rankers:

$$R_1^X = \{X_a\}_{a \in A}^+ , \qquad\qquad R_1^Y = \{Y_a\}_{a \in A}^+ ,$$
$$R_{m+1}^X = \{X_a\}_{a \in A}^* \, R_m^Y , \qquad\qquad R_{m+1}^Y = \{Y_a\}_{a \in A}^* \, R_m^X ,$$

Here the juxtaposition on the left denotes concatenation. We furthermore define $R_m = R_m^X \cup R_m^Y$, and $R = \bigcup_m R_m$. Note that these sets depend on the alphabet A although this dependence is not written out explicitly. We always assume that the alphabet is clear from context. Given a ranker r, the *alternation depth* of r is the smallest m such that $r \in R_m$. The *depth* of r is the length of r as a word.

Since rankers are $\mathbb{N}$-valued functions, there is a natural way of comparing them given a speficied word u. In other words, given $u \in A^*$ and rankers r, s, we are interested in whether $r(u) \leq s(u)$ and $r(u) < s(u)$ hold. The following definition introduces a general framework, inspired by the comparisons of Weis and Immerman [25] and Lauser [16].

Definition 7. *Let A be some alphabet, and let $\mathscr{C} \subseteq R \times R$ be some set of pairs of rankers over A. We define $[\mathscr{C}] = \bigcup_{(r,s) \in \mathscr{C}} \{r, s\}$, i.e. the set of rankers that occurs on some position in some pair of $\mathscr{C}$. For $u, v \in A^*$, we have $u \leq^{\mathscr{C}} v$ if:*

(i) The same set of rankers in $[\mathscr{C}]$ are defined on u and v,
(ii) For each $(r, s) \in \mathscr{C}$ such that r and s are defined on u and v, we have
   $r(u) \leq s(u) \Rightarrow r(v) \leq s(v)$, and $r(u) < s(u) \Rightarrow r(v) < s(v)$.

If $u \leq^{\mathscr{C}} v$ and $v \leq^{\mathscr{C}} u$, we say that $u \equiv^{\mathscr{C}} v$.

For rankers r, s, and words $u, v \in A^*$ we have $r(u) \leq s(u) \Rightarrow r(v) \leq s(v)$ if and only if $s(v) < r(v) \Rightarrow s(u) < r(u)$. In particular, this means that if $\mathscr{C}$ is symmetric, i.e. $(r, s) \in \mathscr{C} \Leftrightarrow (s, r) \in \mathscr{C}$, then $\leq^{\mathscr{C}}$ and $\equiv^{\mathscr{C}}$ are equal. Note that for a certain choice of symmetric $\mathscr{C}$ we get the relation introduced in [25]. We say that a language L is *definable* by $\mathscr{C}$ if L is an ideal under the relation $\leq^{\mathscr{C}}$. Furthermore, we say that a language is an $\mathscr{C}$-*set* if it is a subset of such an ideal.

For a language A and some sets of rankers $\mathscr{C} \subseteq R \times R$, we want to consider the monoid $A^*/\leq^{\mathscr{C}}$, which is a well defined monoid only when $\leq^{\mathscr{C}}$ is stable. For general $\mathscr{C}$, this is not the case. However, Proposition 3 provides a large class of sets for which it does hold.[5]

Proposition 3. *Let R be some collection of rankers and let $\mathscr{C} \subseteq R \times R$ be closed under subwords, i.e. be such that $(r, s) \in \mathscr{C}$ implies $(r', s') \in \mathscr{C}$ for any subwords r' of r and s' of s. Then the preorder $\leq^{\mathscr{C}}$ is stable.*

If $\mathscr{C}$ furthermore is finite, then $A^*/\leq^{\mathscr{C}}$ is a finite monoid. This monoid can be constructed explicitly; given representatives u and v for some elements, one can check which ranker comparisons uv satisfy.

5   The Ranker Comparison Hierarchy

Rankers and ranker comparisons have a long tradition in the study of fragments of FO^2. Indeed, rankers were first introduced by Schwentick, Thérien and Vollmer as a characterization of FO^2 itself [21]. Ranker comparisons were used by Weis and Immerman to give a characterization of the languages definable in FO_m^2 [25]; this was later expanded to the full alternation hierarchy in the PhD thesis of

---

[5] As an example on when it does not hold, consider the singleton $\{(\mathsf{X}_{aa}, \mathsf{Y}_{aa})\}$. We note that neither X_{aa} nor Y_{aa} are defined on ε nor on a. Thus $\varepsilon \leq^{\mathscr{C}} a$. However, $a \not\leq^{\mathscr{C}} aa$. The following proposition gives a condition on $\mathscr{C}$ which implies that $\leq^{\mathscr{C}}$ is stable.

Lauser [16]. A ranker characterization of the corners of the Trotter-Weil hierarchy is also known, using so called condensed rankers [15].

In this section, we place these results into our general framework. In particular, we rephrase the characterization of the Trotter-Weil corners in terms of ranker comparisons. This leads to a natural hierarchy containing both the Trotter-Weil and quantifier alternation hierarchies: the *ranker comparison hierarchy*, shown in Fig. 1.

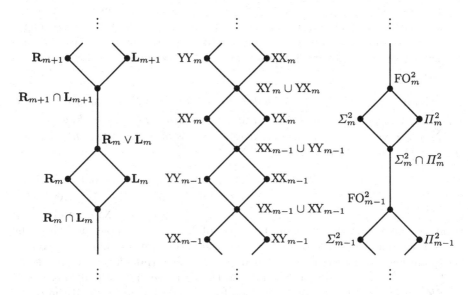

Fig. 1. The Ranker Comparison Hierarchy surrounded by the Trotter-Weil hierarchy (left) and the Quantifier Alternation Hierarchy (right)

The levels of the ranker comparison hierarchy are built using the following collections of ranker comparisons. We note that they are finite and closed under subwords, and thus define finite monoids by Proposition 3. For $m \geq 1$:

$$\mathrm{XX}_{m,n} = \left\{ (r,s) \in R_m^X \times R_m^X \mid |r|, |s| \leq n \right\},$$
$$\mathrm{YY}_{m,n} = \left\{ (r,s) \in R_m^Y \times R_m^Y \mid |r|, |s| \leq n \right\}$$

and for $m \geq 2$:

$$\mathrm{XY}_{m,n} = \left\{ (r,s) \in R_m^X \times R_m^Y \mid |r|, |s| \leq n \right\},$$
$$\mathrm{YX}_{m,n} = \left\{ (r,s) \in R_m^Y \times R_m^X \mid |r|, |s| \leq n \right\}.$$

We also consider unions of these sets. We use the notation $\leq_{m,n}^{\mathrm{XY}}$ instead of $\leq^{\mathrm{XY}_{m,n}}$ and similarly for the other sets. Note that we have $u \leq_{m,n}^{\mathrm{XY}} v$ if and only if $v \leq_{m,n}^{\text{YX}} u$ and $v \leq_{m,n}^{\text{XY} \cup \text{YX}} u$ if and only if both $v \leq_{m,n}^{\text{XY}} u$ and $v \leq_{m,n}^{\text{YX}} u$.[6] We also give the following names for the induced monoids.

Definition 8. *We define*

$$N_{m,n}^{\text{XX}} = A^*/\leq_{m,n}^{\text{XX}} \qquad N_{m,n}^{\text{YY}} = A^*/\leq_{m,n}^{\text{YY}} \qquad N_{m,n}^{\text{XX} \cup \text{YY}} = A^*/\leq_{m,n}^{\text{XX} \cup \text{YY}}$$
$$N_{m,n}^{\text{XY}} = A^*/\leq_{m,n}^{\text{XY}} \qquad N_{m,n}^{\text{YX}} = A^*/\leq_{m,n}^{\text{YX}} \qquad N_{m,n}^{\text{XY} \cup \text{YX}} = A^*/\leq_{m,n}^{\text{XY} \cup \text{YX}} .$$

The fragments $\Sigma_{1,n}^2$, $\Pi_{1,n}^2$ and $\text{FO}_{1,n}^2$ can be characterized using existence of subwords. Since our framework require the same subwords up to a certain length to be present in both words of interest (condition *(i)* in Definition 7), it can not properly handle the two former cases. Therefore, we make the following special definition.

Definition 9. *Let A be some alphabet, we say that $u \leq_{1,n}^{\text{XY}} v$ if any ranker $r \in R_1^X \cup R_1^Y$ with $|r| \leq 2n$ which is defined on u is also defined on v. Equivalently, $u \leq_{1,n}^{\text{XY}} v$ if every subword of length $2n$ which exists in u also exists in v. We say that $u \leq_{1,n}^{\text{YX}} v$ if $v \leq_{1,n}^{\text{XY}} u$ and $u \leq_{1,n}^{\text{XY} \cup \text{YX}} v$ if $u \equiv_{1,n}^{\text{XY}} v$.*

With some abuse of notation, we say that a language is definable by $\text{XY}_{1,n}$ (resp. $\text{YX}_{1,n}$ or $\text{XY}_{1,n} \cup \text{YX}_{1,n}$) if it is an ideal under the relation $\leq_{1,n}^{\text{XY}}$ (resp. $\leq_{1,n}^{\text{YX}}$ or $\leq_{1,n}^{\text{YX} \cup \text{YX}}$). However, we note that there are no actual sets $\text{XY}_{1,n}$ and $\text{YX}_{1,n}$; trying to interpret such sets in the sense of Definition 7 does not yield the desired outcome. This abuse of notation ensures that we can define all levels of the quantifier alternation hierarchy using consistent terminology. We also define $N_{1,n}^{\text{XY}}$ and $N_{1,n}^{\text{YX}}$ to be the monoids induced by the respective (stable) preorders.

We now restate the known correspondences between rankers and the quantifier alternation hierarchy in our framework. The following characterization of the FO_m^2 levels is due to Weis and Immerman (*(i)* and *(ii)* [25]) and Kufleitner and Weil (*(i)* and *(iii)* [14]).

Proposition 4. *Given a language L, the following are equivalent:*

(i) L *is definable in* FO_m^2,
(ii) L *is definable by* $\text{XY}_{m,n} \cup \text{YX}_{m,n}$ *for some n,*
(iii) *the syntactic morphism of L is in* $\mathbf{R}_{m+1} \cap \mathbf{L}_{m+1}$.

Similarly, we have the following characterization of the Σ_m^2 levels, due to Fleischer et al. (*(i)* and *(iii)* [7]) and Lauser (*(i)* and *(ii)* [16, Thm. 11.3]). One easily gets the symmetric characterization of the Π_m^2-levels.

Proposition 5. *Given a language L, the following are equivalent:*

---

[6] Note that although the relation $\leq_{m,n}^{\text{XY} \cup \text{YX}}$ is similar to the relation introduced by Weis and Immerman [25], there is a slight difference regarding the variable n. The length of rankers allowed by Weis and Immerman is made to ensure correspondence with the depth of formulae in FO^2. The results of this contribution does not consider depths of FO^2 formulae, and thus this difference is not important here.

(i) L is definable in Σ_m^2,
(ii) L is definable by $\mathrm{XY}_{m,n}$ for odd m and $\mathrm{YX}_{m,n}$ for even m for some n,[7]
(iii) the syntactic monoid of L is in $\mathbf{Si}_m$.

The corners of the Trotter-Weil hierarchy have a ranker characterization in terms of condensed rankers [15]. We reformulate this result in terms of ranker comparisons. The ranker comparison characterization of the join levels then follows directly. We also use Proposition 1 to relate the join levels to the intersection levels of the quantifier alternation hierarchy.

Proposition 6. Let $m \geq 1$. Then L is definable by $\mathrm{XX}_{m,n}$ (resp. $\mathrm{YY}_{m,n}$) for some n if and only if its syntactic monoid M is in $\mathbf{R}_{m+1}$ (resp. $\mathbf{L}_{m+1}$). Furthermore, the following are equivalent:

(i) L is definable by $\mathrm{XX}_{m,n} \cup \mathrm{YY}_{m,n}$ for some n,
(ii) the syntactic monoid of L is in $\mathbf{R}_{m+1} \vee \mathbf{L}_{m+1}$,
(iii) the syntactic monoid of L is in $\mathbf{Pi}_m \cap \mathbf{Si}_m$,
(iv) L is definable in Σ_m^2 and in Π_m^2.

Taken together, these three propositions gives us a new way of considering the Trotter-Weil hierarchy and the quantifier alternation hierarchy together, as a ranker comparison hierarchy; see Fig. 1.

6 Saturations for Fragments of FO^2

In this section, we present computable closed sets for all levels of the Trotter-Weil and quantifier alternation hierarchies, in other words for all levels of the ranker comparison hierarchy. We also state our main results: that these sets agree with the corresponding sets of pointlikes. The proof thereof is the subject of the subsequent sections.

The sets presented below relies on the monoids having content morphisms (intuitively on the monoid elements having a fixed alphabet). This is not true for all monoids; consider for example $M = \{1, a\}$ with $aa = 1$. However, it is always possible to *alphabetize* a monoid by explicitly distinguishing elements with different alphabets. If M is a monoid with a generating set A, then the submonoid of $M \times J_A$ generated by $(a, \{a\})_{a \in A}$ has a content morphism. It also has a surjective morphism onto M.

We now introduce the relevant closed sets. Note that for our purposes, $\mathbf{R}_1 = \mathbf{L}_1 = \mathbf{J}_1$. We first give the sets for the corners of the Trotter-Weil hierarchy. These are important building blocks for the other sets.

Definition 10. Let M be a monoid with a content morphism α. We define:

– $\mathsf{Sat}_{\mathbf{J}_1}(M) = \mathsf{Sat}_{\mathbf{R}_1}(M) = \mathsf{Sat}_{\mathbf{L}_1}(M) = \{S \subseteq M \mid \alpha(s) = \alpha(t) \text{ for all } s, t \in S\}$,

---

[7] The relation in [16] has only one-sided inclusion of definedness of rankers with the maximum number of alternations for all m as an explicit assumption. However, for $m \geq 2$, two sided inclusion follow implicitly.

– for $m \geq 2$, $\mathsf{Sat}_{\mathbf{R}_m}(M)$ is the smallest closed set of M such that if $Z \in \mathsf{Sat}_{\mathbf{L}_{m-1}}(M)$, $U \in \mathsf{Sat}_{\mathbf{R}_m}(M)$, $\alpha(Z) \leq \alpha(U)$ and U is idempotent in 2^M, then $UZ \in \mathsf{Sat}_{\mathbf{R}_m}(M)$

– for $m \geq 2$, $\mathsf{Sat}_{\mathbf{L}_m}(M)$ is the smallest closed set of M such that if $Z \in \mathsf{Sat}_{\mathbf{R}_{m-1}}(M)$, $V \in \mathsf{Sat}_{\mathbf{L}_m}(M)$, $\alpha(Z) \leq \alpha(V)$ and V is idempotent in 2^M, then $ZV \in \mathsf{Sat}_{\mathbf{L}_m}(M)$

The definition inductively ensures that for any W in any of the introduced sets, we have $\alpha(w) = \alpha(w')$ for all $w, w' \in W$. Thus, $\alpha(W)$ is a well defined element of J_A, making the comparisons $\alpha(Z) \leq \alpha(U)$ and $\alpha(Z) \leq \alpha(V)$ meaningful.

The other closed sets build on so-called $\mathbf{RL}_m$-factors. If one think of the elements of a monoid as the languages they represent, one can think of $\mathbf{RL}_m$-factors as collections of languages which can not be distinguished from any side using rankers of alternation depth at most m, while containing words of arbitrary length.

Definition 11. *Let M be a monoid with a content morphism α. Let $S, E \in \mathsf{Sat}_{\mathbf{R}_m}(M)$, $T, F \in \mathsf{Sat}_{\mathbf{L}_m}(M)$ with $\alpha(S), \alpha(T) \leq \alpha(E) = \alpha(F)$ and E, F idempotent in 2^M. Let W be such that $\alpha(w) \leq \alpha(E)$ for all $w \in W$. Then $SEWFT$ is an $\mathbf{RL}_m$-factor.*

Since $\mathsf{Sat}_{\mathbf{R}_m}(M)$ and $\mathsf{Sat}_{\mathbf{L}_m}(M)$ can be constructed for each m, the $\mathbf{RL}_m$-factors can also be effectively constructed. Note that the alphabet of an $\mathbf{RL}_m$-factor is well defined. Using these factors, we construct the following sets.

Definition 12. *Let M be a monoid. Then*

– $\mathsf{ConeSat}_{\mathbf{Si}_1}(M)$ *is the smallest closed set for which* $(1, S) \in M$ *for all* $S \subseteq M$.
– $\mathsf{ConeSat}_{\mathbf{Pi}_1}(M)$ *is the smallest closed set for which* $(s, \{1, s\}) \in M$ *for all* $s \in M$.

Suppose further that M has a content morphism α, then for $m \geq 2$:

– $\mathsf{Sat}_{\mathbf{J}}(M)$, *is the smallest closed set such that* $E \in \mathsf{Sat}_{\mathbf{J}}(M)$ *for all idempotent sets E satisfying $\alpha(s) = \alpha(t)$ for all $s, t \in E$.*
– $\mathsf{Sat}_{\mathbf{R}_{m+1} \cap \mathbf{L}_{m+1}}(M)$ *is the smallest closed set which for all n contain the product*

$$U_1 V_1 U_2 \ldots V_{n-1} U_n$$

where every U_i is an $\mathbf{RL}_m$-factor while $V_i \in \mathsf{Sat}_{\mathbf{R}_m \cap \mathbf{L}_m}(M)$ (or $V_i \in \mathsf{Sat}_{\mathbf{J}}(M)$ for $m = 2$), and $\alpha(v_i') \leq \alpha(U_i), \alpha(U_{i+1})$ for all $v_i' \in V_i$,
– $\mathsf{Sat}_{\mathbf{R}_m \vee \mathbf{L}_m}(M)$ *is the smallest closed set which for all n contain the product*

$$U_1 V_1 U_2 \ldots V_{n-1} U_n$$

where every U_i is an $\mathbf{RL}_m$-factor while $V_i \in \mathsf{Sat}_{\mathbf{R}_{m-1} \vee \mathbf{L}_{m-1}}(M)$ (or $V_i \in 2^M$ for $m = 2$), and $\alpha(v_i') \leq \alpha(U_i), \alpha(U_{i+1})$ for all $v_i' \in V_i$,

– $\mathsf{ConeSat}_{\mathbf{Si}_m}(M)$ *is the smallest closed set which for all n contain the product*

$$(u_1, U_1)(v_1, V_1)(u_2, U_2)\ldots(v_{n-1}, V_{n-1})(u_n, U_n)$$

where for all i, we have $u_i \in U_i$ *and* U_i *is an* $\mathbf{RL}_m$*-factor while* $(v_i, V_i) \in$ $\mathsf{ConeSat}_{\mathbf{Si}_{m-1}}(M)$, *and* $\alpha(v_i') \le \alpha(U_i), \alpha(U_{i+1})$ *for all* $v_i' \in V_i$,
– $\mathsf{ConeSat}_{\mathbf{Pi}_m}(M)$ *is the smallest closed set which for all n contain the product*

$$(u_1, U_1)(v_1, V_1)(u_2, U_2)\ldots(v_{n-1}, V_{n-1})(u_n, U_n)$$

where for all i, we have $u_i \in U_i$ *and* U_i *is an* $\mathbf{RL}_m$*-factor while* $(v_i, V_i) \in$ $\mathsf{ConeSat}_{\mathbf{Pi}_{m-1}}(M)$, *and* $\alpha(v_i') \le \alpha(U_i), \alpha(U_{i+1})$ *for all* $v_i' \in V_i$,

We now state our main theorem. Apart from giving the pointlikes of the different levels, it also provides *separators*. These are monoids with relational morphisms which are optimal in $\mathbf{V}$ for separating the elements of M. In other words, the relational morphisms τ satisfy $\mathsf{PL}_\tau(M) = \mathsf{PL}_{\mathbf{V}}(M)$. The theorem states only the monoids explicitly; the relational morphisms are the natural relational morphisms, obtained by mapping every element in M to their preimage in A^* and projecting onto the relevant monoids.

Theorem 2. *Let M be a finite monoid, and let* $n = \lceil R/2 \rceil - 1$ *where R is the Ramsey number of M. Then*

(i) $\mathsf{Cones}_{\mathbf{Si}_1}(M) = \mathsf{ConeSat}_{\mathbf{Si}_1}(M)$ *with separator* $N_{1,n}^{\mathrm{XY}}$,
(ii) $\mathsf{Cone}_{\mathbf{Pi}_1}(M) = \mathsf{ConeSat}_{\mathbf{Pi}_1}(M)$ *with separator* $N_{1,n}^{\mathrm{YX}}$,

Furthermore, suppose M has a content morphism $\alpha : M \twoheadrightarrow J_A$, *and let* $n = (m + |A|)(R - 1)$ *and* $n' = (m - 1 + 3|A|)(R - 1) + |A|$ *where R is the Ramsey number of* 2^M.

(iii) $\mathsf{PL}_{\mathbf{J}_1}(M) = \mathsf{Sat}_{\mathbf{J}_1}(M)$ *with the separator* J_A,
(iv) $\mathsf{PL}_{\mathbf{R}_m}(M) = \mathsf{Sat}_{\mathbf{R}_m}(M)$ *with the separator* $N_{m,n}^{\mathrm{XX}}$,
(v) $\mathsf{PL}_{\mathbf{L}_m}(M) = \mathsf{Sat}_{\mathbf{L}_m}(M)$ *with the separator* $N_{m,n}^{\mathrm{YY}}$,
(vi) $\mathsf{PL}_{\mathbf{J}}(M) = \mathsf{Sat}_{\mathbf{J}}(M)$ *with separator* $N_{1,|A|R+R-1}^{\mathrm{XY}\cup\mathrm{YX}}$,[8]
(vii) $\mathsf{PL}_{\mathbf{R}_{m+1}\cap\mathbf{L}_{m+1}}(M) = \mathsf{Sat}_{\mathbf{R}_{m+1}\cap\mathbf{L}_{m+1}}(M)$ *with separator* $N_{m,n'}^{\mathrm{XX}\cup\mathrm{YY}}$,
(viii) $\mathsf{PL}_{\mathbf{R}_m\vee\mathbf{L}_m}(M) = \mathsf{Sat}_{\mathbf{R}_m\vee\mathbf{L}_m}(M)$ *with separator* $N_{m,n'}^{\mathrm{XX}\cup\mathrm{YY}}$,
(ix) $\mathsf{Cones}_{\mathbf{Si}_m}(M) = \mathsf{ConeSat}_{\mathbf{Si}_m}(M)$ *with separator* $N_{m,n'}^{\mathrm{XY}}$ *for odd m and* $N_{m,n'}^{\mathrm{YX}}$ *for even m,*
(x) $\mathsf{Cone}_{\mathbf{Pi}_m}(M) = \mathsf{ConeSat}_{\mathbf{Pi}_m}(M)$ *with separator* $N_{m,n'}^{\mathrm{YX}}$ *for odd m and* $N_{m,n'}^{\mathrm{XY}}$ *for even m,*

The following is an immediate corollary, given Proposition 2.

Corollary 1. *The covering problem has a solution for all language varieties associated with the levels of the quantifier alternation hierarchy. In particular, this implies solutions to the separation problems for all of these varieties.*

---

[8] See [2]. We reprove it in order to get a separator which is defined using rankers.

References

1. Almeida, J.: Some algorithmic problems for pseudovarieties. Publ. Math. Debrecen **54**(1), 531–552 (1999)
2. Almeida, J., Costa, J.C., Zeitoun, M.: Pointlike sets with respect to **R** and **J**. J. Pure Appl. Algebra **212**(3), 486–499 (2008)
3. Almeida, J., Silva, P.V.: SC-hyperdecidability of **R**. Theoret. Comput. Sci. **255**(1–2), 569–591 (2001)
4. Almeida, J., Zeitoun, M.: The pseudovariety **J** is hyperdecidable. RAIRO Inform. Théor. Appl. **31**(5), 457–482 (1997)
5. Ash, C.J.: Inevitable graphs: a proof of the type II conjecture and some related decision procedures. Internat. J. Algebra Comput. **1**(1), 127–146 (1991)
6. Eilenberg, S.: Automata, Languages, and Machines, vol. 59B. Academic Press (1976)
7. Fleischer, L., Kufleitner, M., Lauser, A.: Block products and nesting negations in FO^2. In: Hirsch, E.A., Kuznetsov, S.O., Pin, J.É., Vereshchagin, N.K. (eds.) CSR 2014. LNCS, vol. 8476, pp. 176–189. Springer, Cham (2014). https://doi.org/10.1007/978-3-319-06686-8_14
8. Henckell, K.: Pointlike sets: the finest aperiodic cover of a finite semigroup. J. Pure Appl. Algebra **55**(1–2), 85–126 (1988)
9. Henriksson, V.: Membership and separation problems inside two-variable first order logic. Loughborough University, Thesis (2021)
10. Henriksson, V., Kufleitner, M.: Nesting negations in FO^2 over infinite words. CoRR, abs/2012.01309 (2020)
11. Krebs, A., Straubing, H.: An effective characterization of the alternation hierarchy in two-variable logic. ACM Trans. Comput. Log. **18**(4), 30:1–30:22 (2017)
12. Krohn, K., Rhodes, J.L., Tilson, B.: Homomorphisms and semilocal theory. In: Algebraic Theory of Machines, Languages, and Semigroups, chapter 8, pp. 191–231. Academic Press (1968)
13. Kufleitner, M., Lauser, A.: The join levels of the Trotter-Weil hierarchy are decidable. In: Rovan, B., Sassone, V., Widmayer, P. (eds.) MFCS 2012. LNCS, vol. 7464, pp. 603–614. Springer, Heidelberg (2012). https://doi.org/10.1007/978-3-642-32589-2_53
14. Kufleitner, M., Weil, P.: The FO^2 alternation hierarchy is decidable. In: Proceedings CSL 2012, LIPIcs, vol. 16, pp. 426–439. Dagstuhl Publishing (2012)
15. Kufleitner, M., Weil, P.: On logical hierarchies within FO^2-definable languages. Log. Methods Comput. Sci. **8**(3:11), 30 (2012)
16. Lauser, A.: Formal language theory of logic fragments. Ph.D. thesis, University of Stuttgart (2014)
17. Place, T., Zeitoun, M.: Going higher in the first-order quantifier alternation hierarchy on words. In: Esparza, J., Fraigniaud, P., Husfeldt, T., Koutsoupias, E. (eds.) ICALP 2014. LNCS, vol. 8573, pp. 342–353. Springer, Heidelberg (2014). https://doi.org/10.1007/978-3-662-43951-7_29
18. Place, T., Zeitoun, M.: The covering problem. Log. Methods Comput. Sci. **14**(3) (2018)
19. Ramsey, F.P.: On a problem of formal logic. Proc. London Math. Soc. **s(2)-30**(4), 264–286 (1929)
20. Schützenberger, M.-P.: Sur le produit de concaténation non ambigu. Semigroup Forum **13**(1), 47–75 (1976)

21. Schwentick, T., Thérien, D., Vollmer, H.: Partially-ordered two-way automata: a new characterization of DA. In: Kuich, W., Rozenberg, G., Salomaa, A. (eds.) DLT 2001. LNCS, vol. 2295, pp. 239–250. Springer, Heidelberg (2002). https://doi.org/10.1007/3-540-46011-X_20

22. Simon, I.: Piecewise testable events. In: Brakhage, H. (ed.) GI-Fachtagung 1975. LNCS, vol. 33, pp. 214–222. Springer, Heidelberg (1975). https://doi.org/10.1007/3-540-07407-4_23

23. Steinberg, B.: On pointlike sets and joins of pseudovarieties. Internat. J. Algebra Comput. 8(2), 203–234 (1998). With an addendum by the author

24. Thérien, D., Wilke, T.: Over words, two variables are as powerful as one quantifier alternation. In: Proceedings of the Thirtieth Annual ACM Symposium on Theory of Computing, pp. 234–240. ACM Press (1998)

25. Weis, P., Immerman, N.: Structure theorem and strict alternation hierarchy for FO^2 on words. Log. Methods Comput. Sci. 5(3:4), 23 (2009)

The Net Automaton of a Rational Expression

Sylvain Lombardy[1]([✉]) and Jacques Sakarovitch[2]

[1] LaBRI - UMR 5800 - Bordeaux INP - Bordeaux University - CNRS,
Bordeaux, France
`sylvain.lombardy@labri.fr`
[2] IRIF-CNRS/Paris Cité University and LTCI/Télécom Paris, IPP, Paris, France
`jacques.sakarovitch@telecom-paris.fr`

Abstract. In this paper, we present a new construction of a finite automaton associated with a rational (or regular) expression. It is very similar to the one of the so-called Thompson automaton, but it overcomes the failure of the extension of that construction to the case of *weighted* rational expressions. At the same time, it preserves all (or almost all) of the properties of the Thompson automaton. This construction has two supplementary outcomes. The first one is the reinterpretation in terms of automata of a data structure introduced by Champarnaud, Laugerotte, Ouardi, and Ziadi for the efficient computation of the *position* (or Glushkov) automaton of a rational expression, and which consists in a duplicated syntactic tree of the expression decorated with some additional links. The second one supposes that this construction devised for the case of weighted expressions is brought back to the domain of Boolean expressions. It allows then to describe, *in terms of automata*, the construction of the Star Normal Form of an expression that was defined by Brüggemann-Klein, and also with the purpose of an efficient computation of the position automaton.

Keywords: Rational expression · Thompson automaton · Weighted automaton

1 Introduction

This paper deals, once more, with the question of building a finite automaton that accepts the language, or — more important for us — the series, denoted by a rational (or regular) expression. This effective view of one direction of the fundamental Kleene Theorem has attracted much attention, works and publications as it corresponds to one of the building bricks of compilation and text retrieval. In the recently published Handbook of Automata Theory [20], we have given a survey on the many aspects of the transformation of an expression into an automaton (and vice-versa), together with a comprehensive bibliography [23].

Surprisingly enough, this long lasting problem — more than 60 years now — may still reveal new aspects, especially when one considers the case of *weighted*

© Springer Nature Switzerland AG 2022
A. Castañeda and F. Rodríguez-Henríquez (Eds.): LATIN 2022, LNCS 13568, pp. 376–392, 2022.
https://doi.org/10.1007/978-3-031-20624-5_23

expressions, which is precisely our concern. For instance, we have recently shown that the *derived term automaton* (aka partial derivatives automaton) may be defined without any reference to derivation or quotient — which in particular makes the construction valid for building transducers [18].

At the opposite of the compact — and rather confidential — derived term automaton, the Thompson automaton [26], famous and universally used via the `grep` command of the Unix system, is the largest (in terms of number of states) of the automata computed from an expression. It is so close to the expression it translates that it can be used to derive all other automata computed from the expression; this is what is explained in [1]. On the other hand, it is not difficult to extend the original construction of Ken Thompson to the case of weighted expressions — this is also done in [1].

The problem is that this extension is *illegitimate* in the sense that via this construction a valid expression may produce an automaton which is not valid. This is known for a long time already (see [17] for instance). The purpose of this paper is the definition of *a Thompson-like construction which is consistent with weights*, that is, which yields a valid automaton when applied to a valid expression. We call this automaton the *nailed expression tree automaton*, or *net automaton* for short.

Let us be more specific and explain the notion of validity which is central to this work. The classical models of formal languages and automata refer to a 'Boolean universe' where a word belongs, or not, to a language, or is accepted, or not, by an automaton. The extension, or generalization, of these models to quantitative and more versatile concepts where every word is given a *coefficient*, be it called probability, distance, cost, weight — we use weight — goes back to the very beginning of automata theory ([21, 25]). It is to be acknowledged that in these early times the coefficients themselves were used mostly to decide whether a word is accepted or not (being non zero, or above a given threshold). But this part of the theory has known a renewed interest in the recent years, in works where the value of the coefficients itself is taken into consideration and enriches the model.

This extension does not come for free and the definition of the star operator in such a framework raises a real problem. An axiomatic approach has been developped in many works (*e.g.* [5, 6, 15]) in order to address it. It allows to define different classes of semirings, depending on which axioms are satisfied. But it does not cover natural weight sets such as integers or rational numbers. In our previous works on the subject, [16, 17, 22] for instance, we have chosen the intuitive extension of the star of an element as the sum of its powers. This implies that we are able to define infinite sums in the semiring of weights and, to this end, that the semiring be endowed with a topology. All usual semirings are in this case. The star of an element is then partially defined.

An expression is valid when in the inductive process of evaluation of the series it denotes, all stars are defined. The validity of an automaton may be slightly trickier to define, as we have shown in [17], but in any case it boils down to the definition of infinite sums of elements in the semiring. It is of course essential that the validity be preserved in the correspondence between expressions and automata realised by the constructions that establish Kleene's Theorem. This is what fails with the 'classical' weighted extension of Thompson construction and that is achieved with the net automaton.

As we shall see, the net automaton of an expression shares all, or almost all, properties of the Thompson automaton of the expression. Both automata look very much the same, like chiefly two hammocks of ε-transitions stretched between their unique initial and final states, their states with in- and out-degrees at most 2. Both automata give the *position automaton* of the expression when the ε-transitions are (adequatly) removed. The main differences between the two constructions being first that the net automaton of an invalid expression is not defined whereas the Thompson automaton of any expression is defined even if the expression is not valid and second that the initial state of the net automaton may be final, with a weight equal to the constant term of the expression, whereas the Thompson automaton has only one final state, with weight 1.

The first property of the net automaton, besides the fact it fulfils the purpose it is meant to, is that it can be given a global and direct description in addition to the inductive construction used for its definition, a description from which its name is taken.

Let E be a rational expression. We take two copies of the syntactic tree of E and transform them into weighted directed graphs: in one copy the edges are directed downward, upward in the other, and they are labeled with the empty word ε. The corresponding leaves in the two copies are connected, from the downward to the upward copies, with transitions labeled by the letter labeling the leaf. The root of the downward copy is made initial, the root of the upward copy is made final. Few rules are added that put weights on some of the ε-transitions and connect some nodes in the upward copy to nodes in the downward copy. The result is the net automaton of E.

This construction enlights the fact that the net automaton is indeed very close to a data structure that the 'Rouen school' has introduced in order to compute efficiently first the *partial derivative automaton* of an expression [10] and second the coefficient of a word in the series denoted by a weighted expression [9]. If they had transformed this data structure into an automaton, they would have probably built something very close to the net automaton.

The second property of the net automaton we report on is more surprising as it occurs in the Boolean case, a domain where one could think it has nothing to add to the Thompson automaton. If E is a Boolean regular expression, it so happens that the net automaton of E contains some ε-transitions that can be characterized as, and coined, 'superfluous'. When these superfluous transitions are deleted, not only the remaining automaton of course accepts the same language (otherwise the deleted transitions would not have been called superfluous) but this automaton is 'almost' isomorphic to the net automaton of the *Star Normal Form* of E.

This means that the construction of the net automaton of a Boolean regular expression somehow simultaneously carries out the computation of the Star Normal Form of the expression. An unexpected outcome for a process that has been devised for solving a problem in the weighted case and which shows that the net automaton truly translates in the automaton world the deep structure of the expression.

The paper is organized as follow. In the preliminary section, we recall the definition and notation for the weighted version of automata theory and the construction of the position automaton of an expression. The next section recalls the definition of the Thompson automaton and gives the one of the net automaton to make clear their likeness and differences. The last two sections present the two results quoted above. The first one gives the direct construction of the net automaton from the syntactic tree of the expression. The second one shows how the net automaton of the Star Normal Form of a (Boolean) expression is built directly from the net automaton of the expression.

For want of space, not only most of the proofs of the statements but also the formal definitions of the constructions are not given. We hope that figures will give enough intuition of the notions we describe. As a result, the body of the paper is more an extended abstract than a regular one.

2 Definitions and Notation: The Quantitative Perspective

The definition of usual notions, such as words, languages, free monoids, expressions, automata, rational (or regular) sets, recognisable sets, etc. may be found in numerous textbooks (*e.g.* [14]). The corresponding notions of multiplicity (or weight) semirings, (formal power) series, weighted automata, etc. are probably less common knowledge but are still presented in quite a few books [3,4,11,24] to which we refer the reader. Let us be more explicit for the two notions we study: the *weighted rational expressions* and the *weighted finite automata*. Before, we recall the notions of semiring and series; at the end, the construction of the position (or Glushkov) automaton.

The set of words over an alphabet A is denoted by A^*, the empty word by ε. A semiring $\mathbb{K}$ is a set endowed with an associative and commutative addition and an associative multiplication with neutral elements respectively denoted by $0_\mathbb{K}$ and $1_\mathbb{K}$; the multiplication is distributive over the addition and $0_\mathbb{K}$ is an annihilator for the multiplication. Every semiring is supposed to be equipped with a topology. For instance, $\mathbb{N}$, $\mathbb{Z}$, $(\mathbb{Z}, \min, +)$ are equipped with the discrete topology, $\mathbb{Q}$, $\mathbb{R}$ with the topology defined by the distance.

If k is an element of a semiring $\mathbb{K}$, k^* is the sum of all powers of k: $k^* = \sum_{n \in \mathbb{N}} k^n$. This infinite sum may be defined — k is said to be *starrable* — or not defined — k is said to be *non starrable*. In any $\mathbb{K}$, $0_\mathbb{K}$ is starrable and $0_\mathbb{K}^* = 1_\mathbb{K}$.

The extension of languages to the weighted case are called *series*. Formally, a series over A^* with coefficients in $\mathbb{K}$ is a *map* from A^* into $\mathbb{K}$. The value of a series s at a word w of A^* is called the *coefficient* or rather here the *weight* of w in s and is denoted by $\langle s, w \rangle$. The set of series over A^* with weights in $\mathbb{K}$, equipped with the pointwise addition and the Cauchy product is a *semiring* denoted by $\mathbb{K}\langle\!\langle A^* \rangle\!\rangle$. It is endowed with the topology inherited from the one on $\mathbb{K}$ by the simple convergence topology.

The *constant term* $\mathsf{c}(s)$ of a series s is the weight of ε in s. A series is *proper* if its constant term is $0_\mathbb{K}$. The *proper part* s_p of a series s is the series obtained from s by zeroing the weight of ε and keeping all other weights unchanged.

A proper series is always starrable. This property is generalised by the following[1].

Theorem 1 ([16]).   *Let* $\mathbb{K}$ *be a strong semiring. A series* s *in* $\mathbb{K}\langle\langle A^*\rangle\rangle$ *is starrable if and only if* $\mathsf{c}(s)$ *is starrable in* $\mathbb{K}$ *and in this case* $s^* = (\mathsf{c}(s)^* . s_\mathsf{p})^* . \mathsf{c}(s)^*$.

2.1   Weighted Automata

A weighted automaton is a labeled weighted directed graph whose vertices (called *states*) are endowed with initial and final functions. More formally:

Definition 1. *Let* $\mathbb{K}$ *be a semiring and* A *a finite alphabet. A* $\mathbb{K}$-*automaton over* A *is denoted by a tuple* $\mathcal{A} = \langle Q, E, I, T \rangle$ *where* Q *is the finite set of* states, I *is the* initial function *from* Q *into* $\mathbb{K}$, T *is the* final function *from* Q *into* $\mathbb{K}$, *and* E *is the set of* transitions, *a subset of* $Q \times (A \cup \{\varepsilon\}) \times \mathbb{K} \times Q$.

The semantics of computations of an automaton requires that E be the graph of a (partial) map from $Q{\times}(A{\cup}\{\varepsilon\}){\times}Q$ into $\mathbb{K}\backslash\{0_\mathbb{K}\}$ and hence be finite. Likewise, a state is initial (*resp.* final) if and only if its initial (*resp.* final) value is not $0_\mathbb{K}$. A (classical) finite automaton is a weighted automaton whose weight semiring is the Boolean semiring $\mathbb{B}$: I and T are then subsets of Q and E is a set of triples.

A transition in E is a 4-tuple $t = (r, x, h, s)$: x is the *label*, h the *weight* of t. If $x = \varepsilon$, t is called an ε-*transition*. A path in $\mathcal{A}$ is defined as in graph theory and is a sequence of transitions. The label (*resp.* the weight) of a path is the concatenation of the labels (*resp.* the product of the weights) of its transitions. A computation is a path from an initial state p to a final state q; its label is the label of the path and its weight is the product $I(p) . k . T(q)$, where k is the weight of the path.

In an automaton with ε-transitions, a given word may be the label of an infinite number of computations. Hence the infinite summations implied, and the topology required, in the next definition.

Definition 2. *Let* $\mathcal{A}$ *be a* $\mathbb{K}$-*automaton over* A^*. *The* evaluation *of a word* w *in* $\mathcal{A}$ *is the sum, if it is defined, of the weights of all computations with label* w.

If the evaluation of every word in A^* *is defined, the automaton* $\mathcal{A}$ *is said to be* valid. *And the* behaviour *of* $\mathcal{A}$ *is the series denoted by* $|\mathcal{A}|$ *where the coefficient of every word* w *is the evaluation of* w *in* $\mathcal{A}$.

Lemma 1. *Let* $\mathcal{A}$ *be a* $\mathbb{K}$-*automaton. Every word is the label of a finite number of computations if and only if there is no circuit of* ε-*transitions in the trim part of* $\mathcal{A}$.

---

[1] A topological semiring is *strong* if the product of two summable families is a summable family. It is a sufficient condition in order to establish Theorem 1. Not all semirings are strong ([19]) but all usual semirings are. This precision is not of importance here but on the other hand Theorem 1 is essential for the constructions that follow and we wanted to have a correct statement.

Hence, every weighted automaton with no circuit of ε-transitions is valid.

Remark 1. In [17], we have given a more complex definition for the validity of automata in order to encompass the most general (topological) weight semirings and thus to ensure that ε-removal procedures are licit for any valid automaton. For $\mathbb{Z}$-automata — as is the counter-example we give below — this definition of validity boils down to the more intuitive one given in Definition 2.

2.2   Weighted Rational Expressions

Definition 3. *Let $\mathbb{K}$ be a semiring and A a finite alphabet. A $\mathbb{K}$-rational expression over A is an expression defined as follows.*

  (i) *The atoms are* 0, 1, *and* a *for every letter* a *in* A;
 (ii) *there are two binary operators* + *and* $\cdot$;
(iii) *there is a unary (postfix) operator* $*$;
 (iv) *for every element k of $\mathbb{K}$ there are two unary operators (one prefix and one postfix) denoted by concatenation: if* E *is a rational expression, so are* $k\,\mathsf{E}$ *and* $\mathsf{E}\,k$.

Remark 2. Trivial identities ($\mathsf{E}+0 = \mathsf{E}$, $\mathsf{E}\cdot 1 = \mathsf{E}$, $\mathsf{E}\cdot 0 = 0$, *etc.*) can be applied to rational expressions. In some constructions, like the computation of the derived term automaton of a rational expression [16], they are necessary. They are useless in the construction of the Thompson automaton or the net automaton, which is defined in this paper. Therefore, we consider rational expressions without any simplification.

For every rational expression E, the *size* $\sigma(\mathsf{E})$ is the number of atoms and operators in E, while the *litteral length* $\ell(\mathsf{E})$ is the number of letters in the expression. These values can be computed inductively.

Definition 4. *The constant term* $\mathsf{c}(\mathsf{E})$ *of a $\mathbb{K}$-rational expression E is an element of $\mathbb{K}$ inductively defined by:*

$$\mathsf{c}(0) = 0_{\mathbb{K}}, \quad \mathsf{c}(1) = 1_{\mathbb{K}}, \quad \forall a \in A,\ \mathsf{c}(a) = 0_{\mathbb{K}}, \quad \mathsf{c}(k\,\mathsf{E}) = k\,.\,\mathsf{c}(\mathsf{E}),$$

$$\mathsf{c}(\mathsf{E}\,k) = \mathsf{c}(\mathsf{E})\,.\,k, \qquad \mathsf{c}(\mathsf{F}+\mathsf{G}) = \mathsf{c}(\mathsf{F}) + \mathsf{c}(\mathsf{G}), \qquad \mathsf{c}(\mathsf{F}\cdot\mathsf{G}) = \mathsf{c}(\mathsf{F})\,.\,\mathsf{c}(\mathsf{G}),$$

$$\text{and} \qquad \mathsf{c}(\mathsf{F}^*) = \mathsf{c}(\mathsf{F})^* \quad \textit{if } \mathsf{c}(\mathsf{F}) \textit{ is starrable.}$$

Notice that the constant term is not defined for every expression. Actually, the star operator is involved in its definition and it may happen that, for an expression F^*, the star of $\mathsf{c}(\mathsf{F})$ is not defined.

Definition 5. *A $\mathbb{K}$-rational expression E is* valid *if $\mathsf{c}(\mathsf{E})$ is defined.*

Definition 6. *The interpretation of a valid $\mathbb{K}$-rational expression over A is the (rational) series inductively defined by:*

$$|0| = 0_{\mathbb{K}}, \quad |1| = 1_{\mathbb{K}}, \quad \forall a \in A,\ |a| = a, \quad |k\,\mathsf{E}| = k\,.\,|\mathsf{E}|, \quad |\mathsf{E}\,k| = |\mathsf{E}|\,.\,k,$$

$$|\mathsf{F}+\mathsf{G}| = |\mathsf{F}| + |\mathsf{G}|, \qquad |\mathsf{F}\cdot\mathsf{G}| = |\mathsf{F}|\,.\,|\mathsf{G}|,$$

$$|\mathsf{F}^*| = (\mathsf{c}(\mathsf{F}^*)\,.\,|\mathsf{F}|_{\mathsf{p}})^*\,.\,\mathsf{c}(\mathsf{F}^*) \quad \textit{if } \mathsf{c}(\mathsf{F}) \textit{ is starrable.}$$

By Theorem 1, in strong semirings, $|F^*| = |F|^*$ if it is defined, and it is straightforward that $c(E)$ is the constant term of $|E|$.

Proposition 1. *For every valid $\mathbb{K}$-rational expression* E, $c(E) = \langle |E|, \varepsilon \rangle$.

2.3   The Position Automaton of an Expression

There are several classical constructions of automata without any ε-transitions from valid rational expressions. Most of them lead actually to the same automaton, sometimes call Glushkov, standard, or position automaton, depending on the algorithm used for its construction [8, 13].

In this paper, we call it *position automaton* because we use this description to analyse its connections with the net automaton. The position automaton of a rational expression E has a unique initial state; every other state corresponds to the position of a letter in the expression. Every transition arriving in the state corresponding to the position p is labelled with the letter at position p in E. Let p and q be two positions. There is a transition from the initial state to the state q if the letter in position q appears as the first letter of some words in $|E|$; there is a transition from the state p to the state q if the letter in position p is followed by the the letter in position q in some words in $|E|$; likewise, the state p is final if the letter in position p is the last letter of some words in $|E|$. The position automaton of a rational expression E has no ε-transition, it has $\ell(E) + 1$ states, and one unique initial state on which no transition arrives.

3   The Nailed Expression Tree Automaton

This section recalls the classical construction of the Thompson automaton, points out the problem when it is generalised to weighted expressions, proposes a *new inductive construction* that solves the problem, and shows the similarity between the two resulting automata.

3.1   The Thompson Automaton

The so-called 'Thompson automaton' has been described by Ken Thompson as the core of the function `grep` implemented in Unix for regular expression search [26]. The extension of the construction to weighted expressions is quite straightforward (see [1] for instance). We denote by $\mathcal{T}(E)$ the Thompson automaton of an expression E. The inductive construction of the Boolean Thompson automaton is presented in every automata textbook; it is shown, together with the extension to the weighted case, in Fig. 1.

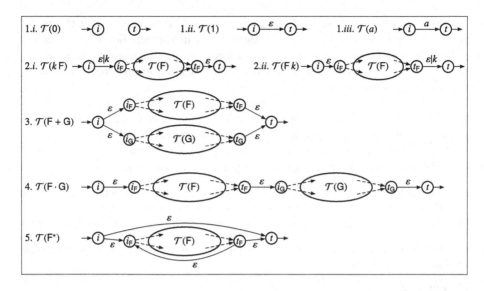

Fig. 1. The construction of the weighted Thompson automaton.

If E is a Boolean expression, $|\mathcal{T}(E)| = |E|$ of course holds. The extension of that fundamental equation proves to be indeed problematic in the weighted case, when not all expressions, nor automata, are valid.

First note that the very definition makes it possible to build $\mathcal{T}(E)$ for an expression E which is not valid. It is immediate to check that in this case $\mathcal{T}(E)$ is not valid as well: there exist some pairs of states which are connected by an infinite number of paths of ε-transitions such that the sum of the weights of these paths is not defined. The problem with the definition of $\mathcal{T}(E)$ is that the converse of this statement does not hold and that a *valid expression* E *may be associated with a Thompson automaton* $\mathcal{T}(E)$ *which is not valid*, as shown by Example 1.

Example 1. Consider the $\mathbb{Z}$-rational expression $E_1 = (a^* + (-1)b^*)^*$.
Since $c(a^* + (-1)b^*) = c(a^*) - c(b^*) = 0$ is starrable, E_1 is a valid expression. On the other hand, $\mathcal{T}(E_1)$ contains circuits of ε-transitions (with weight 1 or -1) and is therefore not valid. It is drawn in Fig. 3(a), postponed to allow comparison with the other new construction.

However, at least the following statement holds.

Proposition 2. *Let* E *be a* $\mathbb{K}$-*rational expression. If* $\mathcal{T}(E)$ *is valid, then* E *is valid and* $|\mathcal{T}(E)| = |E|$.

3.2 The Nailed Expression Tree Automata

The construction we are aiming at is inductive, quite similar and parallel to the original Thompson construction. The resulting automaton shares almost all

384   S. Lombardy and J. Sakarovitch

characteristics and properties of the Thompson automaton. We call it the *nailed expression tree automaton* — which we shorten in the handier *net automaton* — by reference to another method to constructing it, that consists in duplicating and 'decorating' the syntactic tree of the rational expression, and that is described in Sect. 4.

We denote by $\mathcal{N}(\mathsf{E})$ the net automaton of an expression E. It is defined inductively by the rules drawn in Fig. 2. Notice that by rule (5) the construction of the net automaton is possible if and only if the expression is valid.

The fundamental property of the construction is expressed by Proposition 3 which is easy to establish by induction on the expression.

Proposition 3. *Let* E *be a* $\mathbb{K}$-*rational expression. Then, in* $\mathcal{N}(\mathsf{E})$, *there is no circuit of* ε-*transitions nor any path of* ε-*transitions from the initial state to the final state.*

The soundness of the construction of the net automaton is then expressed by the following statement.

Theorem 2.
Let E *be a valid* $\mathbb{K}$-*rational expression. Then,* $\mathcal{N}(\mathsf{E})$ *is valid and* $|\mathcal{N}(\mathsf{E})| = |\mathsf{E}|$.

Example 2. The net automaton $\mathcal{N}(\mathsf{E}_1)$ is drawn in Fig. 3(b).

Example 3. Let E_2 be the $\mathbb{Q}$-rational expression $\mathsf{E}_2 = (a^* \cdot (\frac{-1}{2}b^*)^*)^*$. Then $\mathsf{c}((\frac{-1}{2}b^*)^*) = (\frac{-1}{2})^* = \frac{2}{3}$. Hence $\mathsf{c}(\mathsf{E}_2) = (\frac{2}{3})^* = 3$. The net automaton $\mathcal{N}(\mathsf{E}_2)$ is drawn in Fig. 4 left.

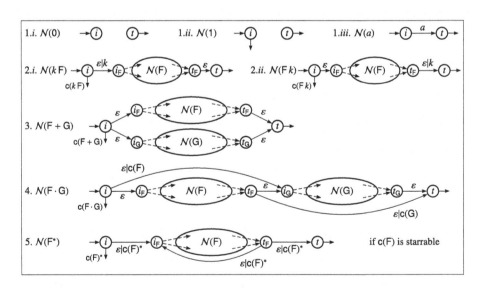

Fig. 2. The inductive construction of the net automaton of an expression

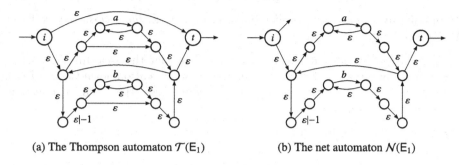

(a) The Thompson automaton $\mathcal{T}(\mathsf{E}_1)$ (b) The net automaton $\mathcal{N}(\mathsf{E}_1)$

Fig. 3. Two automata for E_1.

3.3   The Common Properties of the Thompson and the Net Automata

There are of course (and hopefully) differences between the Thompson and the net automata of a given rational expression. But there are also many similarities that are listed in the following statement, obvious by induction on the expression (recall that $\sigma(\mathsf{E})$ is the size of the expression E, *cf.* Sect. 2.2).

Proposition 4. *Let* E *be a valid* $\mathbb{K}$-*rational expression. Then,* $\mathcal{T}(\mathsf{E})$ *and* $\mathcal{N}(\mathsf{E})$ *have the following properties in common:*

(i) *they contain exactly* $2\,\sigma(\mathsf{E})$ *states and less than* $4\,\sigma(\mathsf{E})$ *transitions;*

(ii) *there are at most* two *transitions outgoing from (resp. ingoing to) each state; if there are two transitions, these are* ε-*transitions;*

(iii) *they have a single initial state to which there is no incoming transition and a single final state from which there is no outgoing transition. In the case of* $\mathcal{N}(\mathsf{E})$ *however, the initial state is also final, with weight* $\mathsf{c}(\mathsf{E})$.

A further common property between Thompson and net automata requires the definition of the ε-*closure* of an automaton. Let $\mathcal{A} = \langle Q, E, I, T \rangle$ be an automaton over the alphabet A and with ε-transitions. Let R be the subset of Q of states q with at least one ingoing transition labeled by a letter in A — that is, states whose ingoing transitions are not all ε-transitions. And let $P = I \cup R$. For every p in P and every q in R, an ε/a-path (from p to q) is the concatenation of an ε-path from p to a state s with a transition with label a from s to q.

The (backwards) ε-*closure* of $\mathcal{A}$ is the automaton $\langle P, F, I, U \rangle$, where:

- for every p in P, $U(p)$ is equal to the sum for all t in T of the sum of the weights of all ε-paths from p to t multiplied by $T(t)$;
- for every p in P, every q in R, and every letter a, there is a transition from p to q with label a with a weight equal to the sum of weights of all ε/a-paths from p to q.

In general, this construction requires that the automaton is valid [17]. As the net automaton has no circuit of ε-transitions, it is always valid.

Proposition 5. *For every valid rational expression* E, *the ε-closure of* N(E) *is the position automaton of* E.

In the Boolean case, it is common knowledge that the ε-closure of $\mathcal{T}(\mathsf{E})$ is the position automaton of E. In the weighted case, it is also known that, if $\mathcal{T}(\mathsf{E})$ is valid, its ε-closure is the position automaton of E (see [1], even if the question of validity is not raised there).

Example 4. The position automaton of E_2 is drawn in Fig. 4 right.

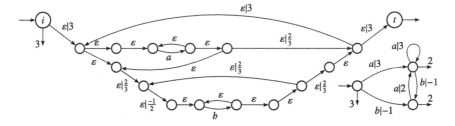

Fig. 4. The net automaton and the position automaton of E_2.

4   From the Expression Syntactic Tree to the Net Automaton

We now build the net automaton directly from an expression, or rather from its syntactic tree, and not by induction on the formation of the expression.

Let E be a *valid* $\mathbb{K}$-rational expression and t_E its syntactic tree. Every node of t_E is labelled by a *rational operator* or an *atom* to which it corresponds in the expression and on top of this labelling there is a 1–1 correspondence between the *subexpressions* of E and the *nodes* of t_E. From this tree, we build an automaton in *three steps* that we describe now. For further reference, we call this construction 'the procedure K'.

1. The states and a first set of transitions; initial and final states

1.a We make two copies of t_E, D and U. For every node n in t_E, n_D is the image of n in D, and n_U is the image of n in U.

1.b In D, every edge is labelled with ε and directed downward. The root of D is an initial state, with initial weight $1_\mathbb{K}$.

1.c In U, every edge is labelled with ε and directed upward. The root of U is a final state, with final weight $1_\mathbb{K}$.

1.d The initial state is final, with final weight c(E).

Because of rule 1.d, the expression E has to be valid.

2. The connection of atoms yields a new set of transitions.

2.a For every leaf ℓ in t_E labelled with a letter a, a transition from ℓ_D to ℓ_U is added, with label a.

3. Weighting the transitions and a last set of bridging transitions.

The transitions are weighted according to the operator that labels the adjacent states and new transitions from U to D are set up.

3.a *The 'exterior multiplication' nodes.* For every node n labelled with k. in $\mathsf{t_E}$, let c be its child. The transition from n_D to c_D is weighted by k. For every node n labelled with $.k$ in $\mathsf{t_E}$, let c be its child. The transition from c_U to n_U is weighted by k.

3.b *The 'star' nodes.* Let n be a node in $\mathsf{t_E}$ labelled with $*$ and c its child; let F be the subexpression of E rooted in c. The transition from n_D to c_D and the transition from c_U to n_U are both weighted by $\mathsf{c(F^*)}$. A new ε-transition with weight $\mathsf{c(F^*)}$ is set up from c_U in U to c_D in D.

3.c *The 'product' nodes.* Let n be a node in $\mathsf{t_E}$ labelled with $\cdot$ and l and r respectively its left and right child. Let F (resp. G) be the subexpression of E rooted in l (*resp.* in r). The transition that goes from n_D to r_D in D is weighted by $\mathsf{c(F)}$. The transition that goes from l_U to n_U in U is weighted by $\mathsf{c(G)}$. A new ε-transition with weight $1_{\mathbb{K}}$ is set up from l_U in U to r_D in D.

Example 5. Let $\mathsf{E_3}$ be the $\mathbb{Q}$-rational expression $\mathsf{E_3} = \left(\left(\frac{1}{2}a^*\right)^* \cdot (b + (-1).1)\right).3$. In Fig. 5, the downward graph D and the upward graph U are clearly identified. Transitions connecting leaves (step 2.a) are drawn with double lines. Finally, the ε-transitions which are added in steps 3.b and 3.c are drawn with dashed and dotted lines respectively.

The Result

Proposition 6. *Let E be a $\mathbb{K}$-rational expression and $\mathsf{t_E}$ its syntactic tree. The automaton built from $\mathsf{t_E}$ by the procedure K is equal to $\mathcal{N}(\mathsf{E})$.*

The proof is straightforward by induction on the formation of the rational expression.

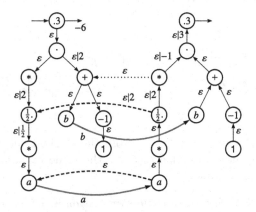

Fig. 5. The net automaton of $\mathsf{E_3}$.

Relation with Previous Work

Procedure K gives a new perspective on the so-called *ZPC structure* defined in [9] and used in several works of its authors (see [9] for instance). Seen from that point of view, the net automaton of an expression E and the ZPC structure associated with E bear a general ressemblance and many similarities. Let us point out their differences.

- The main difference is that the ZPC structure *is not an automaton*, but a directed weighted graph. There is no initial nor final states in the ZPC structure and, even more important, there is no connection from the leaves of the top-down graph to those of the bottom-up graph.
- There is another slight difference in the ε-transitions which we add from the upward graph to the downward graph and which correspond to the 'follow' links in the ZPC structure: for every star node n in t_E and its child c which is the root of a subexpression F, in the ZPC structure there is a link from c_U to n_D (with weight $1_\mathbb{K}$), while in the net automaton the link is from c_U to c_D with weight $c(F)^*$.

5   An Automaton Interpretation of the Star Normal Form

We focus now on the net automaton of unweighted rational expressions that we call regular expressions. This automaton gives a new point of view on the star normal form.

5.1   The Star Normal Form

In [7], Brüggeman-Klein defined the Star Normal Form (SNF) of a regular expression. The main property of a regular expression E in SNF is that it contains no subexpression F^* such that $|F|$ contains the empty word.

Definition 7. *The star normal form of a regular expression* E *is the expression* $E^\blacksquare$ *which is computed by a double induction as follows.*

$$0^\blacksquare = 0 \ , \quad 1^\blacksquare = 1 \ , \quad a^\blacksquare = a \ , \qquad 0^\square = 0 \ , \quad 1^\square = 0 \ , \quad a^\square = a \ ,$$

$$(F+G)^\blacksquare = F^\blacksquare + G^\blacksquare \ , \qquad\qquad (F+G)^\square = F^\square + G^\square \ ,$$

$$(F \cdot G)^\blacksquare = F^\blacksquare \cdot G^\blacksquare \ , \qquad\qquad (F \cdot G)^\square = \begin{cases} F^\square + G^\square & \text{if } c(F) = c(G) = 1 \\ F^\blacksquare \cdot G^\blacksquare & \text{otherwise} \end{cases} \ ,$$

$$(F^*)^\blacksquare = (F^\square)^* \ , \qquad\qquad (F^*)^\square = F^\square .$$

This definition is slightly different from the original one given in [7]. It is proven in [2] that these formulae give yet the same expression and they are better fit to induction proofs.

Proposition 7 ([7]). *For every regular expression* E, $E^\blacksquare$ *is equivalent to* E.

Proposition 8 ([7]). *For every regular expression* E, *the position automaton of* E *is isomorphic to the position automaton of* E$^{\blacksquare}$.

The procedure to convert a regular expression into SNF is linear; moreover, the classical construction of the position automaton from some regular expression is cubic in the size of the expression, while it is quadratic if the regular expression is in SNF. Using SNF is therefore a way to efficiently convert a regular expression into its position automaton.

5.2   The SNF Automaton

Definition 8. *Let us call* SNF automaton *of a regular expression* E *the automaton* $\mathcal{N}(E^{\blacksquare})$, *that is, the net automaton of the SNF of* E.

In order to describe the relationship between the net automaton of a regular expression E and its SNF automaton, we define the notion of *superfluous transitions*. In an automaton $\mathcal{A}$ with ε-transitions, an ε-transition (p, ε, q) is *superfluous* if there exists in $\mathcal{A}$ a path of ε-transitions from p to q that does not contain (p, ε, q). A superfluous transition (p, ε, q) can then be removed from $\mathcal{A}$ without changing the language accepted by $\mathcal{A}$.

Let us denote by $\mathbf{S}(\mathcal{A})$ the automaton obtained from $\mathcal{A}$ by deleting all superfluous transitions.

Lemma 2. *Let* $\mathcal{A}$ *be an automaton with* ε-transitions. *If* $\mathcal{A}$ *has no circuit of* ε-transitions, *then* $\mathbf{S}(\mathcal{A})$ *is equivalent to* $\mathcal{A}$.

The lemma expresses that when $\mathcal{A}$ has no circuit of ε-transitions, all superfluous transitions of $\mathcal{A}$ are independant: each of them remains superfluous even if the other ones are removed. Since a net automaton $\mathcal{N}(E)$ has no circuit of ε-transitions, the automaton $\mathbf{S}(\mathcal{N}(E))$ is equivalent to $\mathcal{N}(E)$.

Since the sizes of an expression and its SNF may be different, and since deleting the superfluous transitions does not change the number of states of an automaton, Proposition 4 (i) already implies that it is not true that $\mathbf{S}(\mathcal{N}(E))$ is equal to $\mathcal{N}(E^{\blacksquare})$. However, we shall see that $\mathcal{N}(E^{\blacksquare})$ is 'essentially' equal to $\mathbf{S}(\mathcal{N}(E))$ and that the computation of $\mathcal{N}(E)$ contains also the computation of the SNF of E since removing the superfluous transitions from $\mathcal{N}(E)$ yields an automaton with the same structure as $\mathcal{N}(E^{\blacksquare})$.

To give evidences of this fact, we take a new look at the definition of the net automaton as given in Fig. 2 and see it as the definition of *operations on automata*, provided they have a unique initial state with no incoming transition that may be final, and another unique final state with no outgoing transition. If $\mathcal{A}$ and $\mathcal{B}$ are two such automata, $\mathcal{A} \oplus \mathcal{B}$, $\mathcal{A} \odot \mathcal{B}$ and $\mathcal{A}^{\circledast}$ are the automata obtained by the application of the rules described respectively in Figs. 2.3, 2.4, and 2.5. Thus, $\mathcal{N}(F + G) = \mathcal{N}(F) \oplus \mathcal{N}(G)$, $\mathcal{N}(F \cdot G) = \mathcal{N}(F) \odot \mathcal{N}(G)$, and $\mathcal{N}(F^{*}) = \mathcal{N}(F)^{\circledast}$.

We give two inductive definitions of automata, $\mathcal{N}_{\blacksquare}(E)$ and $\mathcal{N}_{\blacklozenge}(E)$, in Fig. 6.

1.i. $\mathcal{N}_\blacksquare(0) = \mathcal{N}(0)$ 1'.i. $\mathcal{N}_\square(0) = \mathcal{N}(0)$
           1.ii. $\mathcal{N}_\blacksquare(1) = \mathcal{N}(1)$ 1'.ii. $\mathcal{N}_\square(1) = \mathcal{N}(0)$
                      1.iii. $\mathcal{N}_\blacksquare(a) = \mathcal{N}(a)$ 1'.iii. $\mathcal{N}_\square(a) = \mathcal{N}(a)$

3. $\mathcal{N}_\blacksquare(F + G) = \mathcal{N}_\blacksquare(F) \oplus \mathcal{N}_\blacksquare(G)$ 3'. $\mathcal{N}_\square(F + G) = \mathcal{N}_\square(F) \oplus \mathcal{N}_\square(G)$

4. $\mathcal{N}_\blacksquare(F \cdot G) = \mathcal{N}_\blacksquare(F) \odot \mathcal{N}_\blacksquare(G)$ 4'. $\mathcal{N}_\square(F \cdot G) = \begin{cases} \mathcal{N}_\square(F) \oplus \mathcal{N}_\square(G) & \text{if } c(F) = c(G) = 1 \\ \mathcal{N}_\blacksquare(F) \odot \mathcal{N}_\blacksquare(G) & \text{otherwise} \end{cases}$

5. $\mathcal{N}_\blacksquare(F^*) = (\mathcal{N}_\blacksquare(F))^\circledast$ 5'. $\mathcal{N}_\square(F^*) = \mathcal{N}_\square(F) =$

1.i. $\mathcal{N}_\blacklozenge(0) = \mathcal{N}(0)$ 1'.i. $\mathcal{N}_\lozenge(0) = \mathcal{N}(0)$
           1.ii. $\mathcal{N}_\blacklozenge(1) = \mathcal{N}(1)$ 1'.ii. $\mathcal{N}_\lozenge(1) = \mathcal{N}(0)$
                      1.iii. $\mathcal{N}_\blacklozenge(a) = \mathcal{N}(a)$ 1'.iii. $\mathcal{N}_\lozenge(a) = \mathcal{N}(a)$

3. $\mathcal{N}_\blacklozenge(F + G) = \mathcal{N}_\blacklozenge(F) \oplus \mathcal{N}_\blacklozenge(G)$ 3'. $\mathcal{N}_\lozenge(F + G) = \mathcal{N}_\lozenge(F) \oplus \mathcal{N}_\lozenge(G)$

4. $\mathcal{N}_\blacklozenge(F \cdot G) = \mathcal{N}_\blacklozenge(F) \odot \mathcal{N}_\blacklozenge(G)$ 4'. $\mathcal{N}_\lozenge(F \cdot G) = \begin{cases} \mathcal{N}_\lozenge(F) \oplus \mathcal{N}_\lozenge(G) & \text{if } c(F) = c(G) = 1 \\ \mathcal{N}_\blacklozenge(F) \odot \mathcal{N}_\blacklozenge(G) & \text{otherwise} \end{cases}$

5. $\mathcal{N}_\blacklozenge(F^*) = (\mathcal{N}_\blacklozenge(F))^\circledast$ 5'. $\mathcal{N}_\lozenge(F^*) =$

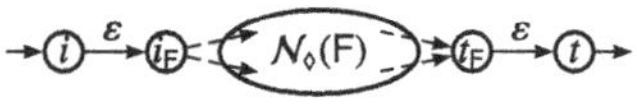

Fig. 6. The inductive definitions of $\mathcal{N}_\blacksquare(E)$ (top) and of $\mathcal{N}_\blacklozenge(E)$ (bottom)

As the definition of $\mathcal{N}_\blacksquare(E)$ exactly follows the definition of the SNF of E, the following proposition is straightforward.

Proposition 9. *For every regular expression* E, $\mathcal{N}_\blacksquare(E) = \mathcal{N}(E^\blacksquare)$.

Notice that the definition of $\mathcal{N}_\blacklozenge(E)$ differs from the one of $\mathcal{N}_\blacksquare(E)$ by the rule 5' only. The following result establishes the similarity between $\mathbf{S}(\mathcal{N}(E))$ and $\mathcal{N}(E^\blacksquare)$.

Theorem 3. *For every regular expression* E, $\mathbf{S}(\mathcal{N}(E)) = \mathcal{N}_\blacklozenge(E)$ *are equal.*

The proof requires to study when superfluous transitions appear in the definition of the net automaton.

Example 6. Let $E_3 = ((a^*)^*)^*$. The star normal form of E_3 is $E_3^\blacksquare = a^*$. Figure 7 (a) shows the net automaton $\mathcal{N}(E_3)$; its superfluous transitions are drawn with dashed lines. Their deletion leads to $\mathcal{N}_\blacklozenge(E_3)$ in Fig. 7(b). The ε-transitions drawn with double lines can be *contracted* to obtain $\mathcal{N}_\blacksquare(E_3)$ in Fig. 7(c).

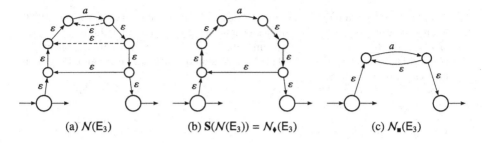

(a) $\mathcal{N}(\mathsf{E}_3)$ (b) $\mathsf{S}(\mathcal{N}(\mathsf{E}_3)) = \mathcal{N}_\blacklozenge(\mathsf{E}_3)$ (c) $\mathcal{N}_\blacksquare(\mathsf{E}_3)$

Fig. 7. Three automata for E_3.

Then, Proposition 8 appears as a corollary of Theorem 3. Actually, for every regular expression E, the ε-closure of $\mathcal{N}(\mathsf{E})$ and $\mathcal{N}_\blacklozenge(\mathsf{E})$ are equal since the transitive closure of ε-transitions is the same with or without superfluous transitions. Moreover, the ε-closure of $\mathcal{N}_\blacklozenge(\mathsf{E})$ and $\mathcal{N}_\blacksquare(\mathsf{E})$ are equal since they only differ by the contraction of some ε-transitions.

6 Conclusion

We well know it is unlikely that the Thompson construction or automaton be replaced — whether in textbooks or software — by the net automaton that we have just described. All the more that the Thompson automaton is totally correct, and remarkably efficient, in the domain for which it has been designed: the search of (Boolean) regular expressions.

We think however that it was important and necessary to have a well-founded construction in the domain of quantitative models, expressions and automata, a correct and safe one, independently of the nature of weights.

In the domain of Boolean automata and expressions, that is, in the domain where Thompson construction plays its role and where ours could be considered as useless and superfluous, the fact that this new automaton appears to contain also the a priori unrelated notion of Star Normal Form is an evidence of its significance for the expression it translates. This is another instance of the philosophy expressed by Eilenberg in his treatise [12] that the consideration of multiplicity helps in understanding the true nature of concepts developped in the Boolean setting.

References

1. Allauzen, C., Mohri, M.: A unified construction of the Glushkov, follow, and Antimirov automata. In: Královič, R., Urzyczyn, P. (eds.) MFCS 2006. LNCS, vol. 4162, pp. 110–121. Springer, Heidelberg (2006). https://doi.org/10.1007/11821069_10
2. Angrand, P.Y., Lombardy, S., Sakarovitch, J.: On the number of broken derived terms of a rational expression. J. Automata Lang. Comb. **15**, 27–51 (2010)

3. Berstel, J., Reutenauer, C.: Rational Series and Their Languages. Springer (1988). Translation of Les séries rationnelles et leurs langages Masson (1984)
4. Berstel, J., Reutenauer, C.: Noncommutative Rational Series with Applications. Cambridge University Press (2011). New version of Rational Series and Their Languages. Springer (1988)
5. Bloom, S.L., Ésik, Z.: Matrix and matricial iteration theories. I. J. Comput. Syst. Sci. **46**, 381–408 (1993)
6. Bloom, S.L., Ésik, Z., Kuich, W.: Partial Conway and iteration semirings. Fundam. Inform. **86**, 19–40 (2008)
7. Brügemann-Klein, A.: Regular expressions into finite automata. Theoret. Comput. Sci. **120**, 197–213 (1993)
8. Caron, P., Flouret, M.: Glushkov construction for series: the non commutative case. Int. J. Comput. Math. **80**(4), 457–472 (2003)
9. Champarnaud, J.M., Laugerotte, É., Ouardi, F., Ziadi, D.: From regular weighted expressions to finite automata. Int. J. Found. Comput. Sci. **15**(5), 687–700 (2004)
10. Champarnaud, J.M., Ponty, J.L., Ziadi, D.: From regular expressions to finite automata. Int. J. Comput. Math. **72**(4), 415–431 (1999)
11. Droste, M., Kuich, W., Vogler, H.: (ed.): Handbook of Weighted Automata, Springer(2009). https://doi.org/10.1007/978-3-642-01492-5
12. Eilenberg, S.: Automata, Languages and Machines, vol. A. Academic Press (1974)
13. Glushkov, V.M.: The abstract theory of automata. Russ. Math. Surv. **16**, 1–53 (1961)
14. Hopcroft, J.E., Motwani, R., Ullman, J.D.: Introduction to Automata Theory, Languages and Computation. Addison-Wesley, 3rd edn. (2006)
15. Kuich, W.: Automata and languages generalized to ω-continuous semirings. Theoret. Comput. Sci. **79**, 137–150 (1991)
16. Lombardy, S., Sakarovitch, J.: Derivatives of rational expressions with multiplicity. Theoret. Comput. Sci. **332**, 141–177 (2005)
17. Lombardy, S., Sakarovitch, J.: The validity of weighted automata. Int. J. Algebra Comput. **23**(4), 863–914 (2013)
18. Lombardy, S., Sakarovitch, J.: Derived terms without derivation. J. Comput. Sci. Cybern. **37**(3), 201–221 (2021). arXiv: arxiv.org/abs/2110.09181
19. Madore, D., Sakarovitch, J.: An example of a non strong Banach algebra (in preparation)
20. Pin, J.É.: Handbook of Automata Theory, vol. I and II. European Mathematical Society Press (2021)
21. Rabin, M.O.: Probabilistic automata. Inform. Control **6**, 230–245 (1963)
22. Sakarovitch, J.: Rational and recognisable power series. In: Droste, M., Kuich, W., Vogler, H. (eds.) Handbook of Weighted Automata, pp. 105–174. Springer (2009). https://doi.org/10.1007/978-3-642-01492-5_4
23. Sakarovitch, J.: Automata and expressions. In: Pin, J.É. (ed.) Handbook of Automata Theory, vol. I, pp. 39–78. European Mathematical Society Press (2021)
24. Salomaa, A., Soittola, M.: Automata-Theoretic Aspects of Formal Power Series. Springer (1977). https://doi.org/10.1007/978-1-4612-6264-0
25. Schützenberger, M.P.: On the definition of a family of automata. Inform. Control **4**, 245–270 (1961)
26. Thompson, K.: Regular expression search algorithm. Comm. Assoc. Comput. Mach. **11**, 419–422 (1968)

Embedding Arbitrary Boolean Circuits into Fungal Automata

Augusto Modanese$^{(\boxtimes)}$ and Thomas Worsch

Karlsruhe Institute of Technology (KIT), Karlsruhe, Germany
{modanese,worsch}@kit.edu

Abstract. Fungal automata are a variation of the two-dimensional sandpile automaton of Bak, Tang and Wiesenfeld (Phys. Rev. Lett., 1987). In each step toppling cells emit grains only to *some* of their neighbors chosen according to a specific update sequence. We show how to embed any Boolean circuit into the initial configuration of a fungal automaton with update sequence HV. In particular we give a constructor that, given the description B of a circuit, computes the states of all cells in the finite support of the embedding configuration in $O(\log |B|)$ space. As a consequence the prediction problem for fungal automata with update sequence HV is P-complete. This solves an open problem of Goles et al. (Phys. Lett. A, 2021).

1 Introduction

The two-dimensional sandpile automaton by Bak, Tang, and Wiesenfeld [1] has been investigated from different points of view. Because of the simple local rule, it is easily generalized to the d-dimensional case for any integer $d \geq 1$.

Several *prediction problems* for these cellular automata (CA) have been considered in the literature. Their difficulty varies with the dimensionality. The recent survey by Formenti and Perrot [3] gives a good overview. For one-dimensional sandpile CA the problems are known to be easy (see, e.g., [7]). For d-dimensional sandpile CA where $d \geq 3$, they are known to be P-complete [9]. In the two-dimensional case the situation is unclear; analogous results are not known.

Fungal automata (FA) as introduced by Goles et al. [6] are a variation of the two-dimensional sandpile automaton where a toppling cell (i.e., a cell with state ≥ 4) emits 2 excess grains of sand either to its two horizontal ("H") or to its two vertical neighbors ("V"). These two modes of operation may alternate depending on an *update sequence* specifying in which steps grains are moved horizontally and in which steps vertically.

The construction in [6] shows that some natural prediction problem is P-complete for two-dimensional fungal automata with update sequence H^4V^4 (i.e., grains are first transferred horizontally for 4 steps and then vertically for 4 steps, alternatingly). The paper leaves open whether the same holds for shorter update

---

An extended version may be found at https://arxiv.org/abs/2208.08779 [8].

A. Castañeda and F. Rodríguez-Henríquez (Eds.): LATIN 2022, LNCS 13568, pp. 393–408, 2022.
https://doi.org/10.1007/978-3-031-20624-5_24

sequences. The shortest non-trivial sequence is HV (and its complement VH); at the same time this appears to be the most difficult to use. By a reduction from the well-known *circuit value problem* (CVP), which is P-complete, we will show:

Theorem 1. *The following prediction problem is P-complete for FA with update sequence HV:*

    Given *as inputs initial states for a finite rectangle R of cells, a cell index y (encoded in binary), and an upper bound T (encoded in unary) on the number of steps of the FA,*

    decide *whether cell x is in a state $\neq 0$ or not at some time $t \leq T$ when the FA is started with R surrounded by cells all in state 0.*

We assume readers are familiar with cellular automata (see Sect. 2 for the definition). We also assume knowledge of basic facts about Boolean circuits and complexity theory, some of which we recall next.

1.1   Boolean Circuits and the CVP

A Boolean circuit is a directed acyclic graph of *gates*: NOT gates (with one input), AND and OR gates with two inputs, $n \geq 1$ INPUT gates and one OUTPUT gate. The output of a gate may be used by an arbitrary number of other gates. Since a circuit is a dag and each gate obtains its inputs from gates in previous layers, ultimately the output of each gate can be computed from a subset of the input gates in a straightforward way.

    It is straightforward to realize NOT, AND, and OR gates in terms of NAND gates with two inputs (with an only constant overhead in the number of gates). To simplify the construction later on, we assume that circuits consist exclusively of NAND gates.

    Each gate of a circuit is described by a 4-tuple (g, t, g_1, g_2) where g is the number of the gate, t describes the type of the gate, and g_1 and g_2 are the numbers of the gates (called sources of g) which produce the inputs for gate g; all numbers are represented in binary. If gate g has only one input, then $g_2 = g_1$ by convention. Without loss of generality the INPUT gates have numbers 1 to n and since their predecessors g_1 and g_2 will never be used, assume they are set to 0. All other gates have subsequent numbers starting at $n + 1$ such that the inputs for gate g are coming from gates with strictly smaller numbers. Following Ruzzo [10] the description B of a complete circuit is the concatenation of the descriptions of all of its gates, sorted by increasing gate numbers.

    Problem instances of the *circuit value problem* (CVP) consist of the description B of a Boolean circuit C with n inputs and a list x of n input bits. The task is to decide whether $C(x) = 1$ holds or not. It is well known that the CVP is P-complete.

1.2   Challenges

A standard strategy for showing P-completeness of a problem Π in some computational model $\mathcal{M}$ (and also the one employed by Goles et al. in [6]) is by a reduction from the CVP to Π, which entails describing how to "embed" circuits in $\mathcal{M}$.

In our setting of fungal automata with update sequence HV, while realizing wires and signals as in [6] is possible, there is no obvious implementation for negation nor for a reliable wire crossing. Hence, it seems one can only directly construct circuits that are *both* planar and monotone. Although it is known that the CVP is P-complete for *either* planar or monotone circuits [5], it is unlikely that one can achieve the same under both constraints. This is because the CVP for circuits that are both monotone and planar lies in NC^2 (and is thus certainly not P-complete unless $P \subseteq NC^2$) [2].

We are able to overcome this barrier by exploiting features that are present in fungal automata but not in general circuits: *time* and *space*. Namely, we deliberately *retard* signals in the circuits we implement by extending the length of the wires that carry them. We show how this allows us to realize a primitive form of transistor. From this, in turn, we are able to construct a NAND gate, thus allowing both wire crossings and negations to be implemented.

Our construction is not subject to the limitations that apply to the two-dimensional case that were previously shown by Gajardo and Goles in [4] since the FA starting configuration is not a fixed point. The resulting construction is also significantly more complex than that of [6].

1.3   Overview of the Construction

In the rest of the paper we describe how to embed any Boolean circuit with description B and an assignment of values to the inputs into a configuration c of a fungal automaton in such a way that the following holds:

- "Running" the FA for a sufficient number of steps results in the "evaluation" of all simulated gates. In particular, after reaching a stable configuration, a specific cell of the FA is in state 1 or 0 if and only if the output of the circuit is 1 or 0, respectively.
- The initial configuration F of the FA is simple in the sense that, given the description of a circuit and an input to it, we can produce its embedding F using $O(\log n + \log |B|)$ space. Thus we have a log-space reduction from the CVP to the prediction problem for FA.

The construction consists of several layers:

Layer 0: The underlying model of fungal automata.

Layer 1: As a first abstraction we subdivide the space into "*blocks*" of 2×2 cells and always think of update "*cycles*" consisting of 4 steps of the CA, using the update sequence $(HV)^2$.

Layer 2: On top of that we will implement "*polarized circuits*" processing "*polarized signals*" that run along "*wires*".

Layer 3: Polarized circuitry is then used to implement "*Boolean circuits with delay*": "*bits*" are processed by "*gates*" connected by "*cables*".[1]

---

[1] Here we slightly deviate from the standard terminology of Boolean circuits and reserve the term "*wire*" for the more primitive wires defined in layer 2.

Layer 4: Finally a given Boolean circuit (without delay) can be embedded in a
fungal automaton (as a circuit with delay) in a systematic fashion that needs
only logarithmic space to construct.

The rest of this paper has a simple organization: Each layer i will be described
separately in section $i + 2$.

2   Layer 0: The Fungal Automaton

Let $\mathbb{N}_+$ denote the set of positive integers and $\mathbb{Z}$ that of all integers. For $d \in \mathbb{N}_+$,
a "*d-dimensional CA*" is a tuple (S, N, δ) where:

- S is a finite set of states
- N is a finite subset of $\mathbb{Z}^d$, called the "*neighborhood*"
- $\delta \colon S^N \to S$ is the "*local transition function*"

In the context of CA, the elements of $\mathbb{Z}^d$ are referred to as *cells*. The function
δ induces a "*global transition function*" $\Delta \colon S^{\mathbb{Z}^d} \to S^{\mathbb{Z}^d}$ by applying δ to each
cell simultaneously. In the following, we will be interested in the case $d = 2$ and
the so-called *von Neumann neighborhood* $N = \{(a, b) \in \mathbb{Z}^2 \mid |a| + |b| \leq 1\}$ of
radius 1.

Except for the updating of cells the fungal automaton is just a two-
dimensional CA with the von Neumann neighborhood of radius 1 and $S =
\{0, 1, \dots, 7\}$ as the set of states.[2] A "*configuration*" is thus a mapping $c \colon \mathbb{Z}^2 \to S$.

Depending on the their states cells will be depicted as follows in diagrams:
– state 0 as ☐   – state 1 as ⊡   – state $i \in S \setminus \{0, 1\}$ as $\boxed{i}$

We will use colored background for cells in states 2, 3, and 4 since their
presence determines the behavior of the polarized circuit. The state 1 is only
a "side effect" of an empty cell receiving a grain of sand from some neighbors;
hence it is represented as a dot. Cells which are not included in a figure are
always assumed to be in state 0.

For a logical predicate P denote by $[P]$ the value 1 if P is true and the value
0 if P is false. For $i \in \mathbb{Z}^2$ denote by $h(i)$ the two horizontal neighbors of cell i
and by $v(i)$ its two vertical neighbors. Cells are updated according to 2 functions
H and V mapping from $S^{\mathbb{Z}^2}$ to $S^{\mathbb{Z}^2}$ where for each $i \in \mathbb{Z}^2$ the following holds:

$$H(c)(i) = c(i) - 2 \cdot [c(i) \geq 4] + \sum_{j \in h(i)} [c(j) \geq 4];$$

$$V(c)(i) = c(i) - 2 \cdot [c(i) \geq 4] + \sum_{j \in v(i)} [c(j) \geq 4].$$

The updates are similar to the sandpile model by Bak, Tang, and Wiesenfeld
[1], but toppling cells only emit grains of sand either to their horizontal or
their vertical neighbors. Therefore whenever a cell is non-zero, it stays non-zero
forever.

---

[2] We use states as in [1]; however, the states 6 and 7 never occur in our construction.

The composition of these functions applying first H and then V is denoted HV. For the transitions of a fungal automaton with update sequence HV these functions are applied alternatingly, resulting in a computation c, $H(c)$, $V(H(c))$, $H(V(H(c)))$, $V(H(V(H(c))))$, and so on. In examples we will often skip three intermediate configurations and only show c, $HVHV(c)$, etc. Figure 1 shows a simple first example.

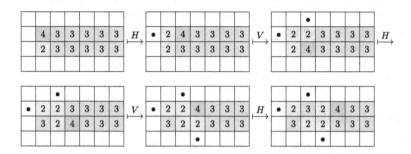

Fig. 1. Five transitions according to $HVHVH$

3   Layer 1: Coarse Graining Space and Time

As a first abstraction from now on one should always think of the space as subdivided into "*blocks*" of 2×2 cells. Furthermore we will look at update "*cycles*" consisting of 4 steps of the CA, thus using the update sequence $HVHV$ which we will abbreviate to Z. As an example Fig. 2 shows the same cycle as Fig. 1 and the following cycle in a compact way. Block boundaries are indicated by thicker lines.

Fig. 2. compact representation of two cycles

Cells outside the depicted area of a figure are assumed to be $\boxed{0}$ initially and they will never become critical and topple during the shown computation.

4   Layer 2: Polarized Components

We turn to the second lowest level of abstraction. Here we work with two types of signals, which we refer to as *positive* (denoted ⊞) and *negative* (denoted ⊟). Both types will have several representations as a block in the FA.

– All representations of a ⊞ signal have in common that the **upper** *left corner* of the block is a ④ and the other cells are ② or ③.
– All representations of a ⊟ signal have in common that the **lower** *left corner* of the block is a ④ and the other cells are ② or ③.

Not all representations will be appropriate in all situations as will be discussed in the next subsection.

The rules of fungal automata allow us to perform a few basic operations on these *polarized* signals (e.g., duplicating, merging, or crossing them under certain assumptions). The highlight here is that we can implement a (delay-sensitive) form of transistor that works with polarized signals, which we refer to as a *switch*.

As a convention, in the figures in this section, we write x and y for the inputs of a component and z, z_1, and z_2 for the outputs.

4.1   Polarized Signals and Wires

Representations of ⊞ and ⊟ signals are shown in Fig. 3. We will refer to a block initially containing a ⊞ or ⊟ signal as a ⊞ or ⊟ *source*, respectively. (This will be used, for instance, to set the inputs to the embedded CVP instance.)

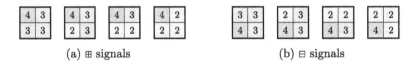

(a) ⊞ signals   (b) ⊟ signals

Fig. 3. Representations of ⊞ and ⊟ signals

A comparison of Fig. 2 and Fig. 3a shows that in the former a ⊞ signal is "moving from left to right". In general we will use *wires* to propagate signals. Wires extending horizontally or vertically can be constructed by juxtaposing *wire blocks* consisting of 2 × 2 blocks of cells in state ③.

While one can use the same wire blocks for both types of signals, each block is destroyed upon use and thus can only be used once. In particular, this means a wire will either be used by a ⊞ or a ⊟ signal. We refer to the respective wires as ⊞ and ⊟ *wires*, accordingly.

4.2   Diodes

Note that ⊞ and ⊟ signals do not encode any form of direction in them (regarding their propagation along a wire). In fact, a signal propagates in any direction a wire is placed in. In order for our components to operate correctly, it will be necessary to ensure a signal is propagated in a single direction. To realize this, we use *diodes*.

A diode is an element on a horizontal wire that only allows a signal to flow from left to right. A signal coming from right to left is not allowed through.

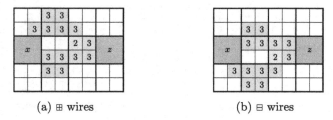

(a) ⊞ wires   (b) ⊟ wires

Fig. 4. Diode implementations

As the other components, the diode is intended to be used only once. For the implementation, refer to Fig. 4. (Recall that x denotes the component's input and z its output.)

For all the remaining elements described in this section, we implicitly add diodes to their inputs and outputs. This ensures that the signals can only flow from left to right (as intended). This is probably not necessary for all elements, but doing so makes the construction simpler while the overhead is only a constant factor blowup in the size of the elements.

4.3 Duplicating, Merging, and Crossing Wires

Wires of the same polarity can be *duplicated* or *merged*. By duplicating a wire we mean we create two wires z_1 and z_2 from a single wire x in such a way that, if any signal arrives from x, then this signal is duplicated and propagated on both z_1 and z_2. (Equivalently, one might imagine that $x = z_1$ and z_2 is a wire copy of x.) In turn, a wire merge realizes in some sense the reverse operation: We have two wires x and y *of the same polarity* and create a wire z such that, if a signal arrives from x or y (or both), then a signal of the same polarity will emerge at z. (Hence one could say the wire merge realizes a polarized OR gate.) See Fig. 5 for the implementations.

As discussed in the introduction, there is no straightforward realization of a wire crossing in fungal automata in the traditional sense. Nevertheless, it turns out we *can* cross wires under the following constraints:

1. The two wires being crossed are a ⊞ and a ⊟ wire.
2. The crossing is used only once and by a single input wire; that is, once a signal from either wire passes through the crossing, it is destroyed. (If two signals arrive from both wires at the same time, then the crossing is destroyed without allowing any signal to pass through.)

To elicit these limitations, we refer to such crossings as *semicrossings*.

We actually need two types of semicrossings, one for each choice of polarities for the two input wires. The semicrossings are named according to the polarity of the top input wire: A ⊞ semicrossing has a ⊞ wire as its top input (and a ⊟ wire as its bottom one) whereas a ⊟ semicrossing has a ⊟ wire at the top (and a ⊞ wire at the bottom). For the implementations, see Fig. 6.

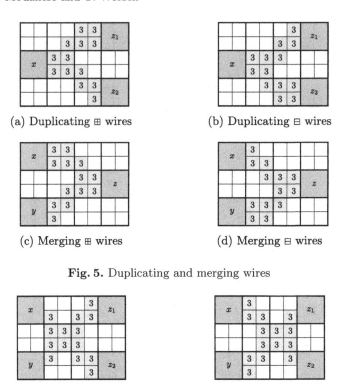

(a) Duplicating ⊞ wires

(b) Duplicating ⊟ wires

(c) Merging ⊞ wires

(d) Merging ⊟ wires

Fig. 5. Duplicating and merging wires

(a) ⊞ semicrossing

(b) ⊟ semicrossing

Fig. 6. Semicrossing implementations

4.4 Switches

A *switch* is a rudimentary form of transistor. It has two inputs and one output. Adopting the terminology of field-effect transistors (FETs), we will refer to the two inputs as the *source* and *gate* and the output as the *drain*. In its initial state, the switch is *open* and does not allow source signals to pass through. If a signal arrives from the gate, then it turns the switch *closed*. A subsequent signal arriving from the source will then be propagated on to the drain. This means that switches are *delay-sensitive*: A signal arriving at the source only continues on to the drain if the gate signal has arrived beforehand (or simultaneously to the source).

Similar to semicrossings, our switches come in two flavors. In both cases the top input is a ⊞ wire and the bottom one a ⊟. The difference is that, in a ⊞ switch, the source (and thus also the drain) is the ⊞ input and the gate is the ⊟ input. Conversely, in a ⊟ switch the source and drain are ⊟ wires and the gate is a ⊞ wire. Refer to Fig. 7 for the implementation of the two types of switches.

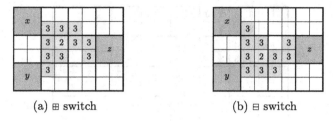

(a) ⊞ switch   (b) ⊟ switch

Fig. 7. Switch implementations

4.5 Delays and Retarders

As mentioned in the introduction, the circuits we construct are sensitive to the time it takes for a signal to flow from one point to the other. To render this notion precise, we define for every component a *delay* which results from the time taken for a signal to pass through the component. This is defined as follows:

- The delay of a source is zero.
- The delay of a wire (including bends) at some block B is the delay of the wire's source S plus the length (in blocks) of the shortest contiguous path along the wire that leads from S to B according to the von Neumann neighborhood. We will refer to this length as the *wire distance* between S and B. For example, the wire distance between the inputs and outputs in all of Figs. 4, 5, 6 and 7 is 4; similarly, the distance between x and z in Fig. 8 (see below) is 15.
- The delay of a gate (i.e., a diode, wire duplication, wire merge, or semicrossing) is the maximum over the delays of its inputs plus the gate width (in blocks).

Notice our definition of wire distance may grossly estimate the actual number of steps a signal requires to propagate from S to B. This is fine for our purposes since we only need to reason about upper bounds later in Sect. 6.3.

Finally we will also need a *retarder* element, which is responsible for adding a variable amount of delay to a wire. Refer to Fig. 8 for their realization.

Retarders can have different dimensions. Evidently, one can ensure a delay of t with a retarder that is $O(\sqrt{t}) \times O(\sqrt{t})$ large. We are going to use retarders of delay at most D, where D depends on the CVP instance and is set later in Sect. 6.3. Hence, it is safe to assume all retarders in the same configuration are of the same size horizontally and vertically, but realize different delays. This allows one to use retarders of a *single* size for any fixed circuit, which simplifies the layout significantly (see also Sects. 6.3 and 6.4).

5 Layer 3: Working with Bits

We will now use the elements from Sect. 4 (represented as in Fig. 9) to construct planar delay-sensitive Boolean circuits. Our circuits will use NAND gates as their basis. We discuss how to overcome the planarity restriction in Sect. 5.4.

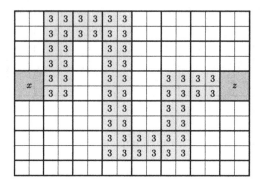

Fig. 8. Implementation of a basic retarder (for both ⊞ and ⊟ signals) that ensures a delay of ≥ 12 at z (relative to x). Retarders for greater delays can be realized by increasing (i) the height of the meanders, (ii) the number of up-down meanders, and (iii) the positions of the input and output.

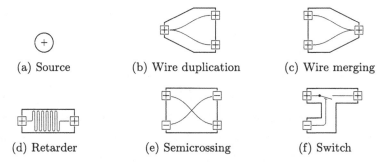

(a) Source   (b) Wire duplication   (c) Wire merging

(d) Retarder   (e) Semicrossing   (f) Switch

Fig. 9. Representations of the elements from abstraction layer 2 as used in layer 3. The polarities indicate whether the ⊞ or ⊟ version of the component is used.

5.1 Representation of Bits

For the representation of a bit, we use a pair consisting of a polarized ⊞ wire and a polarized ⊟ wire. Such a pair of polarized wires is called a *cable*. As mentioned earlier, most of the time signals will travel from left to right. It is straightforward to generalize the notion of wire distance (see Sect. 4.5) to cables simply by setting it to the maximum of the respective wire distances.

A signal on a cable's ⊞ wire represents a binary 1, and a signal on the ⊟ wire represents a binary 0. By convention we will always draw the ⊞ wire "above" the ⊟ wire of the same cable. When referring to a gate's inputs and outputs, we indicate the ⊞ and ⊟ components of a cable with subscripts. For instance, for an input cable x, we write x_+ for its ⊞ and x_- for its ⊟ component.

5.2   Bit Duplication

To duplicate a cable, we use the *Boolean branch* depicted in Fig. 10. The circuit consists of two wire duplications (one of each polarity) and a crossing.

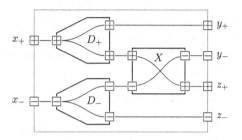

Fig. 10. Boolean branch

5.3   NAND Gates

As a matter of fact the NAND gate is inspired by the implementation of such a gate in CMOS technology[3]. Refer to Fig. 11 for the implementation.

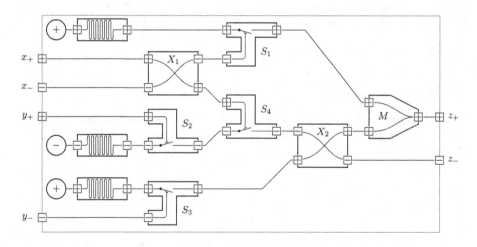

Fig. 11. NAND gate

Notice the usage of switches means these gates are *delay-sensitive*; that is, the gate only operates correctly (i.e., computing the NAND function) if the retarders have *strictly greater delays* than the inputs x and y. In fact, for our construction we will need to instantiate this same construction using *varying* values for

---

[3] E.g. https://en.wikipedia.org/wiki/NAND_gate#/media/File:CMOS_NAND.svg.

the retarders' delays (but not their size as mentioned in Sect. 4.5). This seems necessary in order to chain NAND gates in succession (since each gate in a chain incurs a certain delay which must be compensated for in the next gate down the chain).

In addition, notice that in principle NAND gates have *variable* size as their dimension depends on that of the three retarders. As is the case for retarders, in the same embedding we insist on having all NAND gates be of the same size. We defer setting their dimensions to Sect. 6.3; for now, it suffices to keep in mind that NAND gates (and retarder elements) in the same embedding only vary in their delay (and not their size).

Claim. Assuming the retarders have larger delay than the input cables x and y, the circuit on Fig. 11 realizes a NAND gate.

Proof. Consider first the case where both x_+ and y_+ are set. Since x_- is not set, X_1 is consumed by x_+, turning S_4 on. In addition, since y_+ is set, S_2 is also turned on. Hence, using the assumption on the delay of the inputs, the negative source flows through S_2, S_4, and X_2 on to z_-. Since both the switches S_1 and S_3 remain open, the z_+ output is never set. Notice the crossings X_1 and X_2 are each used exactly once.

Let now x_- or y_- (or both) be set. Then either S_2 or S_4 is open, which means z_- is never set. As a result, X_2 is used at most once (namely in case y_- is set). If x_- is set, then S_1 is opened, thus allowing the positive source to flow on to M. The same holds if y_- is set, in which case M receives the positive source arriving from S_3. Hence, at least one positive signal will flow to the M gate, causing z_+ to be set eventually. $\qquad\square$

5.4 Cable Crossings

There is a more or less well-known idea to cross to bits using three XOR gates which can for example be found in the paper by Goldschlager [5]. Figure 12 shows the idea.

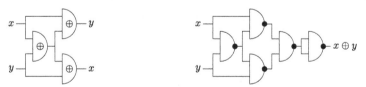

(a) Crossing two cables using XOR gates   (b) Implementing XOR with NAND gates

Fig. 12. Implementing cable crossings as in [5]

This construction can be used in FA. Because of the delays, there is not *the* crossing gate, but a whole family of them. Depending on the position in the whole circuit layout, each crossing needs NAND gates with specific builtin delays (which will be set in Sect. 6.3).

6   Layer 4: Layout of a Whole Circuit

Finally we describe one possibility to construct a finite rectangle of cells F of a FA containing the realization of a complete circuit, given its description B. The important point here is that, in order to produce F from B, the constructor only needs logarithmic space. (Therefore the simplicity of the layout has precedence over any form of "optimization".)

6.1   Arranging the Circuit in Tiles

Let C be the circuit that is to be embedded as an FA configuration F. Letting n be the length of inputs to C and m its number of gates, notice we have an upper bound of m on the circuit depth of C. Without restriction, we may assume $m \geq n$, which also implies an upper bound of $m + n = O(m)$ on the number of cables of C (since C has bounded fan-in). The logical gates of C are denoted by $G_1, \ldots, G_m$ and we assume that G_i has number $n + i$ in description B of C (recall Sect. 1.1).

In the configuration F we have cables $x_1, \ldots, x_n$ originating from the input gates as well as cables $g_1, \ldots, g_m$ coming from (the embedding of) the gates of C. The x_i and g_i flow in and out of equal-sized *tiles* $T_1, \ldots, T_m$, where in the i-th tile T_i we implement the i-th gate G_i of C. The inputs to T_i are $I_i = \{x_1, \ldots, x_n, g_1, \ldots, g_{i-1}\}$ and its outputs $O_i = I_i \cup \{g_i\}$; hence $I_{i+1} = O_i$.

Recall that, unlike standard circuits, the behavior of our layer 3 circuits is subject to spatial considerations, that is, to both gate placement and wire length. For the sake of simplicity, each tile is shaped as a square and all tiles are of the same size. In addition, the tiles are placed in ascending order from left to right and with no space in-between. The only objects in F that lie outside the tiles are the inputs and output of C itself. The inputs are placed immediately next to corresponding cables that go into T_1 whereas the output is placed next to its corresponding wire g_m at the outgoing end of T_m.

6.2   Layout for Tile i

As depicted in Fig. 13, each tile is subdivided into two areas. The *upper part* contains the wires that pass through it, while the *lower part* implements the gate G_i proper.

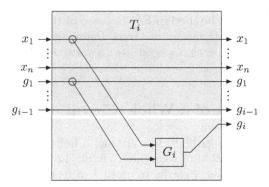

Fig. 13. Overview of the tile T_i. The *upper part* of the tile has green background, the *lower part* has blue background. (Color figure online)

We give a broad overview of the process for constructing T_i. First determine the numbers y_1 and y_2 of the inputs to G_i. Then duplicate the bits on cables y_1 and y_2 (as in Sect. 5.2) and cross the copies over to the lower part of the tile. These crossings require setting adequate delays, which will be adressed in the next section. (In case $y_1 = y_2$, duplicate the cable twice and proceed as otherwise described.) Next instantiate G_i with a proper amount of delay (again, see the next section) and plug in y_1 and y_2 as inputs into G_i. Finally connect all inputs in I_i as well as the output wire g_i of G_i to their respective outputs. Notice the tile contains $O(m)$ crossings and thus also $O(m)$ NAND gates in total.

6.3   Choosing Suitable Delays for All Gates

The two details that remain are setting the dimensions and the delays for the retarders in all NAND gates. This requires certain care since we may otherwise end up running into a chicken-and-egg problem: The retarders' dimensions are determined by the required delays (in order to have enough space to realize them); in turn, the delays depend on the aforementioned dimensions (since the input wires in the NAND gates must be laid so as to "go around" the retarders).

The solution is to assume we already have an upper bound D on the maximum delay in F. This allows us to fix the size of the components as follows:

- The retarders and NAND gates have side length $O(\sqrt{D})$.
- Each tile has side length $O(m\sqrt{D})$.
- The support of F fits into a square with side length $O(m^2\sqrt{D})$.

With this in place, we determine upper bounds on the delays of the upper gates in a tile (i.e., the gates in the upper part of the tile), then of the lower gates G_i, then of the tiles themselves, and finally of the entire embedding of C. In the end we obtain an upper bound for the maximum possible delay in F. Simply setting D to be at least as large concludes the construction.

Upper Gates. In order to set the delays of a NAND gate G in a tile T_i, we first need an upper bound d_{input} on the delays of the two inputs to G. Suppose the origins O_1 and O_2 of these inputs (i.e., either a NAND gate output or an input to T_i) have delay at most d_{origin}. Then certainly we have $d_{\text{input}} \leq d_{\text{origin}} + d_{\text{cable}}$, where d_{cable} is the maximum of the cable distances between either one of O_1 and O_2 and the switches they are connected to inside G. Due to the layout of a tile and since a NAND gate has $O(\sqrt{D})$ side length, we know d_{cable} is at most $O(\sqrt{D})$. Hence, if G is in the j-th layer of T_i, then we may safely upper-bound its delay by $d_i + (j+1)d_{\text{cable}}$, where d_i is the maximum over the delays of the inputs to T_i.

Lower Gates. Since there are $O(m)$ cables inside a tile, there are $O(m)$ cable crossings and thus $O(m)$ NAND gates realizing these crossings. Hence the inputs to the gate G_i in the lower part of T_i have delay at most $d_i + O(m) \cdot d_{\text{cable}} + O(m\sqrt{D})$, where the last factor is due to the side length of T (i.e., the maximum cable length needed to connect the last of the upper gates with G_i).

Tiles. Clearly the greatest delay amongst the output cables of T_i is that of g_i (since every other cable originates from a straight path across T_i). As we have determined in the last paragraph, at its output g_i has delay $d_{i+1} \leq d_i + O(m\sqrt{D})$. Since the side length of a tile is $O(m\sqrt{D})$, we may upper-bound the delays of the inputs of T_i by $i \cdot O(m\sqrt{D})$.

Support of F. Since there are m tiles in total, it suffices to choose a maximum delay D that satisfies $D \geq cm^2\sqrt{D}$ for some adequate constant c (that results from the considerations above). In particular, this means we may set $D = \Theta(m^4)$ independently of C.

6.4   Constructor

In this final section we describe how to realize a logspace constructor R which, given a CVP instance consisting of the description of a circuit C and an input x to it, reduces it to an instance as in Theorem 1. Due to the structure of F, this is relatively straightforward.

The constructor R outputs the description of F column for column. (Computing the coordinates of an element or wire is clearly feasible in logspace.) In the first few columns R sets the inputs to the embedded circuit according to x. Next R constructs F tile for tile. To construct tile T_i, R determines which cables are the inputs to G_i and constructs crossings accordingly. To estimate the delays of each wire, R uses the upper bounds we have determined in Sect. 6.3, which clearly are all computable in logspace (since the maximum delay D is polynomial in m).

Finally R also needs to produce y and T as in the statement of Theorem 1. Let c_i be the cable of T_m that corresponds to the output of the embedded circuit C. Then we let y be the index of the cell next to the ⊞ wire of c_i at the output of T_m. (Hence y assumes a non-zero state if and only if c_i contains a 1, that is, $C(x) = 1$.) As for T, certainly setting it to the number of cells in F suffices (since a signal needs to visit every cell in F at most once).

7 Summary

We have shown that, for fungal automata with update sequence HV, the prediction problem is P-complete, solving an open problem of Goles et al. [6].

References

1. Bak, P., Tang, C., Wiesenfeld, K.: Self-organized criticality: an explanation of the $1/f$ noise. Phys. Rev. Lett. **59**(4), 381–384 (1987). https://doi.org/10.1103/PhysRevLett.59.381
2. Dymond, P.W., Cook, S.A.: Hardware complexity and parallel computation (preliminary version). In: 21st Annual Symposium on Foundations of Computer Science, Syracuse, New York, USA, 13–15 October 1980, pp. 360–372. IEEE Computer Society (1980). https://doi.org/10.1109/SFCS.1980.22
3. Formenti, E., Perrot, K.: How hard is it to predict sandpiles on lattices? A survey. Fundam. Informaticae **171**(1–4), 189–219 (2020). https://doi.org/10.3233/FI-2020-1879
4. Gajardo, A., Goles, E.: Crossing information in two-dimensional sandpiles. Theor. Comput. Sci. **369**(1–3), 463–469 (2006). https://doi.org/10.1016/j.tcs.2006.09.022
5. Goldschlager, L.M.: The monotone and planar circuit value problems are log space complete for P. SIGACT News **9**(2), 25–29 (1977). https://doi.org/10.1145/1008354.1008356
6. Goles, E., Tsompanas, M.A.I., Adamatzky, A., Tegelaar, M., Wosten, H.A.B., Martínez, G.J.: Computational universality of fungal sandpile automata. Phys. Lett. A **384**(22), 126541:1–126541:8 (2020). https://doi.org/10.1016/j.physleta.2020.126541
7. Miltersen, P.B.: The computational complexity of one-dimensional sandpiles. Theory Comput. Syst. **41**(1), 119–125 (2007). https://doi.org/10.1007/s00224-006-1341-8
8. Modanese, A., Worsch, T.: Embedding arbitrary boolean circuits into fungal automata. CoRR abs/2208.08779 (2022). http://arxiv.org/abs/2208.08779
9. Moore, C., Nilsson, M.: The computational complexity of sandpiles. J. Stat. Phys. **96**(1), 205–224 (1999). https://doi.org/10.1023/A:1004524500416
10. Ruzzo, W.L.: On uniform circuit complexity. J. Comput. Syst. Sci. **22**(3), 365–383 (1981). https://doi.org/10.1016/0022-0000(81)90038-6

How Many Times Do You Need to Go Back to the Future in Unary Temporal Logic?

Thomas Place[1,2] and Marc Zeitoun[1(✉)]

[1] CNRS, Bordeaux INP, LaBRI, Univerasity of Bordeaux, UMR 5800,
33400 Talence, France
{thomas.place,marc.zeitoun}@labri.fr
[2] Institut Universitaire de France, Paris, France

Abstract. Unary temporal logic (UTL) can express properties on finite words with the temporal modalities "sometimes in the future/past". The languages definable in UTL are well-understood. In particular, they correspond to *unambiguous languages*, which are built by applying successively three standard operators to the trivial class of languages (consisting of the empty language and the universal one): polynomial, Boolean, and finally unambiguous polynomial closures. Moreover, it is known that one can decide whether a given regular language is expressible in UTL.

We extend these results in two ways. First, we use *generalized* temporal modalities "sometimes in the future/past", which depend on a class "$\mathcal{C}$" of languages. Second, we investigate a hierarchy inside such a variant of UTL: its *future/past hierarchy*. Each level in this hierarchy consists of all languages definable with a *bounded* number of *alternations* between the "sometimes in the future" and "sometimes in the past" modalities.

We show that if $\mathcal{C}$ is a class of *group languages* with mild properties, there is a correspondence between levels of such a $\mathcal{C}$-specified hierarchy and classes of languages obtained from $\mathcal{C}$ by applying standard operators: the polynomial, the Boolean, and the left/right deterministic closures.

We also show that if $\mathcal{C}$ has decidable "separation problem", then one can decide membership of a regular language within any level of the corresponding future/past hierarchy. Finally, these results extend to the case where we allow "tomorrow" and "yesterday" temporal modalities.

1 Introduction

The goal of this paper is to understand the fine-grained structure of fragments of a standard temporal logic, and in particular, their expressive power. Temporal logics are common formalisms in computer science, whose purpose is to specify properties of finite or infinite structures, such as words or trees. Their success stems from the good balance between their ease of use and their expressiveness.

Funded by the DeLTA project (ANR-16-CE40-0007).

A. Castañeda and F. Rodríguez-Henríquez (Eds.): LATIN 2022, LNCS 13568, pp. 409–425, 2022.
https://doi.org/10.1007/978-3-031-20624-5_25

For instance, it is well known [6] that on words, linear time temporal logic is exactly as powerful as first-order logic. Another standard logic is *unary temporal logic* (UTL, or TL for short), which shares ties with navigational logics for trees. Here, we focus on fragments of TL on finite words over some alphabet A.

The logic TL has two temporal operators: "sometimes in the future" and "sometimes in the past". Each TL formula defines a regular language on A^*. Therefore, TL defines a class of regular languages, which turns out to be one of the most robust ones [26]: first, Etessami, Vardi and Wilke [3,4] proved that TL has the same expressiveness as FO^2, the restriction of first-order logic to two variables. Second, Thérien and Wilke [27] designed a decidable characterization of FO^2, *i.e.*, a *membership* algorithm that decides whether a given regular language is definable in this logic. Obtaining such an algorithm is important, as this requires a solid understanding on the investigated languages. Third, Schützenberger [24], Pin, Straubing and Thérien [13,15] described this class in terms of languages built from $\emptyset$ and A^* by applying standard operators: the Boolean, polynomial and unambiguous polynomial closures. Here, the Boolean closure $Bool(\mathcal{C})$ of a class of languages $\mathcal{C}$ is the smallest Boolean algebra containing $\mathcal{C}$. The polynomial closure $Pol(\mathcal{C})$ of $\mathcal{C}$ is the smallest class containing $\mathcal{C}$ closed under union and (marked) language concatenation. Finally, the unambiguous polynomial closure $UPol(\mathcal{C})$ of $\mathcal{C}$ is a subclass of $Pol(\mathcal{C})$, defined by semantic restrictions on the allowed unions and marked products. The results of [15] show that languages definable in TL are exactly those of $UPol(Bool(Pol(\{\emptyset, A^*\})))$.

This smoothly generalizes to $TL(\mathcal{C})$, a version of TL parameterized by a "base class" $\mathcal{C}$ of languages [19,23] (TL corresponds to $TL(\{\emptyset, A^*\})$). On the logical side, this boils down to enriching TL with temporal operators (or FO^2 with predicates) built from $\mathcal{C}$ in a natural way. Moreover, if $\mathcal{C}$ is a class of *group languages* satisfying mild properties, $TL(\mathcal{C})$ is exactly $UPol(Bool(Pol(\mathcal{C})))$. In this case, [19,23] provide yet another definition of the languages in $TL(\mathcal{C})$. They are built from $Bool(Pol(\mathcal{C}))$ by applying two other closure operators in alternation: the left (resp. right) deterministic closure $LPol$ (resp. $RPol$). Finally, membership remains decidable for $TL(\mathcal{C})$, provided that $\mathcal{C}$ satisfies some properties.

Contributions. These multiple equivalent definitions of $TL(\mathcal{C})$ lead to natural hierarchies: the quantifier alternation hierarchy of FO^2, the $LPol/RPol$ alternation hierarchy, called the *deterministic hierarchy*, and the future-past hierarchy, which counts the number of alternations between future and past operators. While the first has been already investigated [5,8,11] for a particular base class, this is not the case for the future-past hierarchy. Our first contribution connects the future-past hierarchy with the deterministic hierarchy: we show that the future-past hierarchy inside $TL(\mathcal{C})$ coincides with the deterministic hierarchy of base class $Bool(Pol(\mathcal{C}))$. This holds when $\mathcal{C}$ is a class of group languages satisfying mild properties, and for extensions of such classes capturing the variants of TL allowing the "tomorrow" and "yesterday" operators. In practice, this makes it possible to cope with temporal operators controlling the words that may be used along future/past jumps (typical properties are modulo tests on the length of such words, or on the number of occurrences of a specific letter). The second result is that membership is decidable for all levels of this hierarchy. For some of them, we use the language-theoretic characterization together with generic results of [18, 19, 23]. For others (the so-called *join levels*), this requires specific work. Altogether, these results generalize work by Kufleitner and Lauser [8, 9].

Other Hierarchies. Two alternative hierarchies inside TL are known, albeit specific to $TL(\{\emptyset, A^*\})$. First, a hierarchy based on the notion of "ranker" is considered in [10]. Another hierarchy is investigated in [8]. It is based on unambiguous interval temporal logic (this is another logic equivalent to TL introduced in [12]). These hierarchies are independent from our work.

Organization. We present terminology in Sect. 2 and future-past hierarchies in Sect. 3. Their characterization by deterministic hierarchies is proved in Sect. 4. Finally, Sect. 5 is devoted to membership for the *join levels*. Due to space limitations, several proofs are postponed to the full version of the paper.

2   Preliminaries

We fix a finite alphabet A for the whole paper. As usual, A^* is the set of all finite words over A, including the empty word ε. We let $A^+ = A^* \setminus \{\varepsilon\}$. For $u, v \in A^*$, we let uv be the word obtained by concatenating u and v. Given $w \in A^*$, we write $|w| \in \mathbb{N}$ for the length of w. We also consider *positions*. A word $w = a_1 \cdots a_{|w|} \in A^*$ is viewed as an *ordered set* $\mathsf{Pos}(w) = \{0, 1, \ldots, |w|, |w| + 1\}$ *of* $|w| + 2$ *positions*. Each position i such that $1 \leq i \leq |w|$ carries the label $a_i \in A$. Positions 0 and $|w| + 1$ are *artificial* leftmost and rightmost positions, which carry *no label*. For every $i \in \mathsf{Pos}(w)$, we define an element $w[i] \in A \cup \{min, max\}$ (where "*min*" and "*max*" do not belong to A). We let $w[0] = min$, $w[i] = a_i$ if $1 \leq i \leq |w|$ and $w[|w| + 1] = max$. Finally, given $i, j \in \mathsf{Pos}(w)$ such that $i < j$, we write $w(i, j) = a_{i+1} \cdots a_{j-1} \in A^*$ (*i.e.*, we keep the letters carried by all positions that are *strictly* between i and j). Note that $w(0, |w| + 1) = w$.

A language is a subset of A^*. We look at *regular* languages, *i.e.*, that can be equivalently defined by a regular expression, an automaton or a morphism into a finite monoid. We work with the latter definition. A *monoid* is a set M equipped with a multiplication $s, t \mapsto st$, which is associative and has an identity element written "1_M". An element $e \in M$ is *idempotent* if it satisfies $ee = e$. For all $S \subseteq M$, we write $E(S)$ for the set of all idempotents in S. It is standard that when M is *finite*, there exists $\omega(M) \in \mathbb{N}$ (written ω when M is understood) such that s^ω is idempotent for every $s \in M$. Clearly, A^* equipped with concatenation is a monoid (ε is the identity). Hence, we may consider morphisms $\alpha : A^* \to M$ into a monoid M. We say that $L \subseteq A^*$ is *recognized* by such a morphism α when there exists $F \subseteq M$ such that $L = \alpha^{-1}(F)$. It is well known that a language is regular if and only if it can be recognized by a morphism into a *finite* monoid.

Remark 1. The only infinite monoid that we consider is A^*. From now on, we implicitly assume that every other monoid $M, N, \ldots$ in this paper is finite.

Classes of Languages. A *class of languages* $\mathcal{C}$ is a set of languages. Such a class forms a *lattice* if it is closed under both union and intersection, and contains the languages $\emptyset$ and A^*. It is a *Boolean algebra* if it is additionally closed under complement. Finally, a class $\mathcal{C}$ is *quotient-closed* when for all $L \in \mathcal{C}$ and $u, v \in A^*$, the language $\{w \in A^* \mid uwv \in L\}$ belongs to $\mathcal{C}$ as well. A class $\mathcal{C}$ is a *prevariety* when it is a quotient-closed Boolean algebra and contains *only regular languages*.

We use a decision problem called *membership* as a means to investigate classes. For a class $\mathcal{C}$, the $\mathcal{C}$-membership problem takes as input a regular language L and asks if $L \in \mathcal{C}$. Intuitively, obtaining a procedure for $\mathcal{C}$-membership requires a solid understanding of $\mathcal{C}$. We also look at a more involved problem, called separation. For a class $\mathcal{C}$ and two languages L_0 and L_1, we say that L_0 is $\mathcal{C}$ *-separable* from L_1 when there exists $K \in \mathcal{C}$ such that $L_0 \subseteq K$ and $L_1 \cap K = \emptyset$. The $\mathcal{C}$-separation problem takes two regular languages L_0 and L_1 as input and asks whether L_0 is $\mathcal{C}$-separable from L_1. We do *not* present separation algorithms in this paper: we only use them as an intermediary to investigate *membership*.

We turn to a key tool. Let $\mathcal{C}$ be a prevariety. A $\mathcal{C}$ *-morphism* is a *surjective* morphism $\eta : A^* \to N$ such that $\eta^{-1}(F) \in \mathcal{C}$ for all $F \subseteq N$. We use this notion to handle membership. It is well known that for every regular language L, there exists a canonical morphism recognizing L. We briefly recall its definition. We associate with L an equivalence $\equiv_L$ on A^*: given $u, v \in A^*$, we let $u \equiv_L v$ when $xuy \in L \Leftrightarrow xvy \in L$ for all $x, y \in A^*$. One may verify that $\equiv_L$ is a congruence and that, since L is regular, it has finite index. Thus, the quotient set $M_L = A^*/\equiv_L$ is a finite monoid. The *syntactic morphism* of L is the morphism $\alpha_L : A^* \to M_L$ which maps a word to its $\equiv_L$-class. It can be computed from any representation of L. We have the following standard property (see, e.g., [17, Proposition 3]).

Proposition 1. *Let $\mathcal{C}$ be a prevariety. A regular language L belongs to $\mathcal{C}$ if and only if its syntactic morphism $\alpha_L : A^* \to M_L$ is a $\mathcal{C}$-morphism.*

In view of Proposition 1, getting an algorithm for $\mathcal{C}$-membership boils down to finding a procedure that decides whether an input morphism $\alpha : A^* \to M$ is a $\mathcal{C}$-morphism. This is how we approach the question in the paper. We shall also use $\mathcal{C}$-morphisms as mathematical tools in proof arguments. In this context, we shall need the following simple corollary of Proposition 1.

Proposition 2. *Let $\mathcal{C}$ be a prevariety, $k \geq 1$ and $L_1, \ldots, L_k \in \mathcal{C}$. There exists a $\mathcal{C}$-morphism $\eta : A^* \to N$ such that $L_1, \ldots, L_k$ are recognized by η.*

Group Languages. A group is a monoid G such that every element $g \in G$ has an inverse $g^{-1} \in G$, *i.e.*, such that $gg^{-1} = g^{-1}g = 1_G$. We call *group language* a language recognized by a morphism into a *finite group*. We shall consider classes $\mathcal{G}$ that are group prevarieties (*i.e.*, containing group languages only).

We let GR be the class of *all* group languages. Another example is the class AMT of *alphabet modulo testable languages*. For all $w \in A^*$ and $a \in A$, we let $\#_a(w) \in \mathbb{N}$ be the number of occurrences of "a" in w. The class AMT consists of all finite Boolean combinations of languages $\{w \in A^* \mid \#_a(w) \equiv k \bmod m\}$ where $a \in A$ and $k, m \in \mathbb{N}$ are such that $k < m$ (these are exactly the languages recognized by commutative groups). We also look at MOD, which consists of all finite Boolean combinations of languages $\{w \in A^* \mid |w| \equiv k \bmod m\}$ with $k, m \in \mathbb{N}$ such that $k < m$. Finally, we write ST for the trivial class ST $= \{\emptyset, A^*\}$. One may verify that GR, AMT, MOD and ST are all group prevarieties.

By definition, $\{\varepsilon\}$ and A^+ are *not* group languages. This motivates the next definition: the *well-suited extension of a class* $\mathcal{C}$, written $\mathcal{C}^+$, consists of all languages $L \cap A^+$ and $L \cup \{\varepsilon\}$ for $L \in \mathcal{C}$ (hence, $\mathcal{C} \subseteq \mathcal{C}^+$). The following lemma is easy.

Lemma 1. *We have* $\{\varepsilon\}, A^+ \in \mathcal{C}^+$. *In addition, if* $\mathcal{C}$ *is a prevariety, so is* $\mathcal{C}^+$.

We conclude with a lemma concerning $\mathcal{G}$-morphisms and $\mathcal{G}^+$-morphisms.

Lemma 2. *Let* $\mathcal{G}$ *be a group prevariety and* $\eta : A^* \to M$ *be a morphism. If* η *is a* $\mathcal{G}$-morphism, then M is a group. If η is a $\mathcal{G}^+$-morphism, then $\eta(A^+)$ is a group.*

3 Future/Past Hierarchies of Unary Temporal Logic

We define unary temporal logic and its future/past hierarchies. We work with a generalized definition of unary temporal logic introduced in [23]: with every class $\mathcal{C}$, we associate a logic $\mathrm{TL}(\mathcal{C})$ and its future/past hierarchy.

3.1 Definition

Syntax. We first recall the definition of the full logic used in [23]. A TL formula is built from atomic formulas using Boolean connectives and temporal operators. The atomic formulas are $\top$, $\bot$, *min*, *max* and a for every letter $a \in A$. All Boolean connectives are allowed: if ψ_1 and ψ_2 are TL formulas, then so are $(\psi_1 \vee \psi_2)$, $(\psi_1 \wedge \psi_2)$ and $(\neg \psi_1)$. We also associate *two temporal operators* with every language $L \subseteq A^*$ which we write F_L and P_L: if ψ is a TL formula, then so are $(\mathrm{F}_L \ \psi)$ and $(\mathrm{P}_L \ \psi)$. We omit parentheses when there is no ambiguity.

We now classify the TL formulas by counting the alternations between future and past operators in their parse tree. With all $n \in \mathbb{N}$, we associate three sets of TL formulas: FL^n, PL^n and BL^n (where F, P and B stand for future, past and Boolean combinations respectively). The first two are defined by induction on $n \in \mathbb{N}$. The FL^0 and PL^0 formulas are the Boolean combinations of atomic formulas (*i.e.*, they do not contain temporal operators). Assume now that $n \geq 1$.

- FL^n is the least set containing the PL^{n-1} formulas and closed under Boolean connectives and *future* operators: if $\psi \in \mathrm{FL}^n$ and $L \subseteq A^*$, then $\mathrm{F}_L \ \psi \in \mathrm{FL}^n$.
- PL^n is the least set containing the FL^{n-1} formulas and closed under Boolean connectives and *past* operators: if $\psi \in \mathrm{PL}^n$ and $L \subseteq A^*$, then $\mathrm{P}_L \ \psi \in \mathrm{PL}^n$.

Finally, for each $n \in \mathbb{N}$, we define the BL^n formulas as the Boolean combinations of FL^n and PL^n formulas.

Semantics. Evaluating a TL formula φ requires a word $w \in A^*$ and a position $i \in \mathsf{Pos}(w)$. We use structural induction on φ to define what it means for (w, i) *to satisfy* φ. We denote this property by $w, i \models \varphi$:

- $w, i \models \top$ always holds and $w, i \models \bot$ never holds.
- for $\ell \in A \cup \{min, max\}$, $w, i \models \ell$ holds if $w[i] = \ell$.
- $w, i \models \psi_1 \vee \psi_2$ if $w, i \models \psi_1$ or $w, i \models \psi_2$.
- $w, i \models \psi_1 \wedge \psi_2$ if $w, i \models \psi_1$ and $w, i \models \psi_2$.
- $w, i \models \neg\psi$ if $w, i \models \psi$ does not hold.
- $w, i \models F_L \psi$ if there is $j \in \mathsf{Pos}(w)$ such that $i < j$, $w(i, j) \in L$ and $w, j \models \psi$.
- $w, i \models P_L \psi$ if there is $j \in \mathsf{Pos}(w)$ such that $j < i$, $w(j, i) \in L$ and $w, j \models \psi$.

When no distinguished position is specified, we evaluate formulas at *the two unlabeled positions* 0 *and* $|w|+1$ *simultaneously*. Given a TL formula φ and $w \in A^*$, we write $w \models \varphi$ and say that w *satisfies* φ when $w, 0 \models \varphi$ *and* $w, |w|+1 \models \varphi$. The *language defined by* φ is $L(\varphi) = \{w \in A^* \mid w \models \varphi\}$.

Each language L is defined by "$(min \wedge F_L\ max) \vee max$". Thus, we restrict the available formulas using a class $\mathcal{C}$. Let $n \in \mathbb{N}$ and $Z \in \{\mathrm{TL}, \mathrm{FL}^n, \mathrm{PL}^n, \mathrm{BL}^n\}$. We write $Z[\mathcal{C}]$ for the set of all Z formulas φ such that every operator F_L or P_L occurring in φ satisfies $L \in \mathcal{C}$. Finally, we write $Z(\mathcal{C})$ for the class of all languages that can be defined by a $Z[\mathcal{C}]$ formula. We are mainly interested in the case when $\mathcal{C}$ is a group prevariety $\mathcal{G}$ or its well-suited extension $\mathcal{G}^+$.

Example 1. Let $\mathrm{ST} = \{\emptyset, A^*\}$. Then, $\mathrm{TL}(\mathrm{ST})$ corresponds to a classic variant of unary temporal logic: F_{A^*} or P_{A^*} are the standard operators "sometimes in the future" (F) and "sometimes in the past" (P). Let $A = \{a, b, c\}$. The language $L = a^* b A^* c a^*$ is defined by $(min \wedge F (c \wedge \neg F b)) \vee (max \wedge P (b \wedge \neg P c))$, which is a $\mathrm{BL}^1[\mathrm{ST}]$ formula. Note that it is important here that formulas be evaluated at *both* unlabeled positions simultaneously: the formula states that $F (c \wedge \neg F b)$ holds at "0" and $P (b \wedge \neg P c))$ holds at "$|w| + 1$". We get $L \in \mathrm{BL}^1(\mathrm{ST})$.

The logic $\mathrm{TL}(\mathrm{MOD})$ is also interesting. Let $K = A^* b (aa)^*$ which is defined by the $\mathrm{FL}^1[\mathrm{MOD}]$ formula $(min \wedge F (b \wedge (\neg F (b \vee c)) \wedge F_{(AA)^*}\ max)) \vee max$ (clearly, $A^*, (AA)^* \in \mathrm{MOD}$). We get $K \in \mathrm{FL}^1(\mathrm{MOD})$. Finally, it is also natural to consider $\mathrm{TL}(\mathrm{AMT})$: when using a temporal operator, one may then count the number of occurrences of each letter between two positions, modulo some integer.

Remark 2. Well-suited extensions are natural inputs as well. It is shown in [23] that for all prevarieties $\mathcal{C}$, we have $\mathrm{TL}(\mathcal{C}^+) = \mathrm{TLX}(\mathcal{C})$ where TLX is a stronger variant which allows additional operators "tomorrow" (X) and "yesterday" (Y). Roughly, the idea is that since $\{\varepsilon\} \in \mathcal{C}^+$ by Lemma 1, one may use the operators $F_{\{\varepsilon\}}$ and $P_{\{\varepsilon\}}$ in $\mathrm{TL}[\mathcal{C}^+]$ formulas. They are clearly equivalent to X and Y. Hence, $\mathrm{TL}(\mathrm{ST}^+) = \mathrm{TLX}(\mathrm{ST})$ is the standard variant of unary temporal logic (which allows F, P, X and Y). While we do not detail this point due to space limitations, the result of [23] extends to future/past hierarchies (the proof is identical). For example, $\mathrm{BL}^n(\mathcal{C}^+) = \mathrm{BLX}^n(\mathcal{C})$ for all $n \in \mathbb{N}$.

3.2   Tools

We define equivalence relations that we shall use as tools. For each TL formula φ, we define the *rank* of φ as the length of the longest sequence of *nested temporal operators* within its parse tree: the rank of an atomic formula is 0, the rank of $(\psi_1 \vee \psi_2)$ and $(\psi_1 \wedge \psi_2)$ is the maximum between the ranks of ψ_1 and ψ_2, the rank of $(\neg\psi)$ is the rank of ψ and the rank of $(\mathrm{F}_L\ \psi)$ and $(\mathrm{P}_L\ \psi)$ is the rank of ψ plus one. Moreover, given $n \in \mathbb{N}$ and a morphism $\eta : A^* \to N$, an $\mathrm{FL}^n[\eta]$ (resp. $\mathrm{PL}^n[\eta]$) formula is an FL^n (resp. PL^n) formula φ such that for all operators F_L or P_L occurring in φ, the language $L \subseteq A^*$ is recognized by η.

We are ready to define our equivalences. Let $\eta : A^* \to N$ be a morphism and let $k, n \in \mathbb{N}$. We define two relations on pairs (w, i), where $w \in A^*$ and $i \in \mathsf{Pos}(w)$. Let $w, w' \in A^*$, $i \in \mathsf{Pos}(w)$ and $i' \in \mathsf{Pos}(w')$. We write $w, i \blacktriangleright_{n,\eta,k} w', i'$ (resp. $w, i \blacktriangleleft_{n,\eta,k} w', i'$) to mean that for every $\mathrm{FL}^n[\eta]$ (resp. $\mathrm{PL}^n[\eta]$) formula φ of rank at most k, we have $w, i \models \varphi \Leftrightarrow w', i' \models \varphi$. Note that despite the notation, $w, i \blacktriangleright_{n,\eta,k} w', i'$ does not entail $w', i' \blacktriangleleft_{n,\eta,k} w, i$ in general. By definition, $\blacktriangleright_{n,\eta,k}$ and $\blacktriangleleft_{n,\eta,k}$ are *equivalences* of finite index (one may verify that there are finitely many non-equivalent $\mathrm{FL}^n[\eta]$ (resp. $\mathrm{PL}^n[\eta]$) formulas of rank at most k).

We adapt these relations to words. Let $\cong \in \{\blacktriangleright_{n,\eta,k}, \blacktriangleleft_{n,\eta,k}\}$ and $w, w' \in A^*$. We write $w \cong w'$ when $w, 0 \cong w', 0$ *and* $w, |w| + 1 \cong w', |w'| + 1$. This defines equivalences of finite index on A^*. We use them to characterize the two classes $\mathrm{FL}^n(\mathcal{C})$ and $\mathrm{PL}^n(\mathcal{C})$ for a given prevariety $\mathcal{C}$ (the proof is presented in the full version of the paper).

Lemma 3. *Let $\mathcal{C}$ be a prevariety, $n \geq 1$ and $L \subseteq A^*$. Then, $L \in \mathrm{FL}^n(\mathcal{C})$ (resp. $L \in \mathrm{PL}^n(\mathcal{C})$) if and only if there exists a $\mathcal{C}$-morphism $\eta : A^* \to N$ and $k \in \mathbb{N}$ such that L is a union of $\blacktriangleright_{n,\eta,k}$-classes (resp. of $\blacktriangleleft_{n,\eta,k}$-classes).*

We complete the definition with properties. The proofs rely on arguments similar to Ehrenfeucht-Fraïssé games (strictly speaking, we use alternative inductive definitions of $\blacktriangleright_{n,\eta,k}$ and $\blacktriangleleft_{n,\eta,k}$ rather than a "game"). They are presented in the full version of the paper. First, these equivalences are congruences.

Lemma 4. *Let $\eta : A^* \to N$ be a morphism, and let $n \geq 1$, $k \in \mathbb{N}$ and $\cong \in \{\blacktriangleright_{n,\eta,k}, \blacktriangleleft_{n,\eta,k}\}$. For every $u, v, u', v' \in A^*$, if $u \cong v$ and $u' \cong v'$, then $uu' \cong vv'$.*

We now consider the special case of morphisms $\eta : A^* \to N$ such that $\eta(A^+)$ is a *group* (this is mandatory in the next two results). In view of Lemma 2, this corresponds to the case $\mathcal{C} \in \{\mathcal{G}, \mathcal{G}^+\}$ for a group prevariety $\mathcal{G}$. We present two properties for the cases $n = 1$ and $n > 1$. We shall use them in Sect. 4 to establish the language theoretic characterization of future/past hierarchies.

We start with the case $n = 1$. Let $\eta : A^* \to N$ be a morphism and $k \in \mathbb{N}$. We define an equivalence $\sim_{\eta,k}$ on A^*, in *two* steps. First, given $w, w' \in A^*$, $i \in \mathsf{Pos}(w)$ and $i' \in \mathsf{Pos}(w')$, we write $w, i \equiv_{\eta,k} w', i'$ if the following conditions hold:

1. We have $w[i] = w'[i']$.
2. If $k \geq 1$, then $\eta(wi) = \eta(wi')$ and $\eta(wi) = \eta(w'i')$.
3. If $\eta^{-1}(1_N) = \{\varepsilon\}$, then for every $h \in \mathbb{Z}$ such that $-k \leq h \leq k$, we have $i + h \in \mathsf{Pos}(w) \Leftrightarrow i' + h \in \mathsf{Pos}(w')$ and in that case, $w[i + h] = w'[i' + h]$.

When $\eta^{-1}(1_N) \cap A^+ \neq \emptyset$, Condition 3 is trivial: it can be discarded (roughly, this hypothesis distinguishes the case $\mathcal{C} = \mathcal{G}$ from $\mathcal{C} = \mathcal{G}^+$). Finally, given $w, w' \in A^*$, we let $w \sim_{\eta,k} w'$ if for all $i \in \mathsf{Pos}(w)$ (resp. $i' \in \mathsf{Pos}(w')$) there exists $i' \in \mathsf{Pos}(w')$ (resp. $i \in \mathsf{Pos}(w)$) such that $w, i \equiv_{\eta,k} w', i'$. We may now state our first property.

Lemma 5. *Let $\eta : A^* \to N$ be a morphism such that $\eta(A^+)$ is a group and let $k \in \mathbb{N}$. Let $p = \omega(N)$. For all $u, v \in A^*$, if $u \sim_{\eta,k} v$, then $uv^{2kp} \blacktriangleright_{1,\eta,k} v^{2kp+1}$ and $v^{2kp}u \blacktriangleleft_{1,\eta,k} v^{2kp+1}$.*

We now present a second property, which we shall use in order to handle the case when $n > 1$.

Proposition 3. *Let $\eta : A^* \to N$ be a morphism such that $\eta(A^+)$ is a group and $k \in \mathbb{N}$. Let $p = \omega(N)$. The following properties hold for all $n \geq 1$ and $u, v \in A^*$,*

- *If n is even and $u \blacktriangleright_{n,\eta,k^2} v$, then $uv^{2kp} \blacktriangleright_{n+1,\eta,k} v^{2kp+1}$.*
- *If n is odd and $u \blacktriangleleft_{n,\eta,k^2} v$, then $uv^{2kp} \blacktriangleleft_{n+1,\eta,k} v^{2kp+1}$.*
- *If n is odd and $u \blacktriangleright_{n,\eta,k^2} v$, then $v^{2kp}u \blacktriangleright_{n+1,\eta,k} v^{2kp+1}$.*
- *If n is even and $u \blacktriangleleft_{n,\eta,k^2} v$, then $v^{2kp}u \blacktriangleleft_{n+1,\eta,k} v^{2kp+1}$.*

Remark 3. It might be surprising that there are *four* cases in Proposition 3. This is because the property does not only depend on the outermost kind of temporal operator in formulas (*i.e.*, future for $\mathrm{FL}^n[\eta]$ and past for $\mathrm{PL}^n[\eta]$) but also on the *innermost* kind (which depends on whether n is odd or even).

4   Characterization By Deterministic Hierarchies

We present a generic language theoretic characterization of future/past hierarchies associated with *group* prevarieties. It generalizes the characterization of the full logic presented in [23]. It is based on variants of *polynomial closure*.

4.1   Polynomial Closure

Given finitely many languages $L_0, \ldots, L_n \subseteq A^*$, a *marked product* of $L_0, \ldots, L_n$ is a product of the form $L_0 a_1 L_1 \cdots a_n L_n$ where $a_1, \ldots, a_n \in A$. In particular, a single language L_0 is a marked product (this is the case $n = 0$).

The *polynomial closure* of a class $\mathcal{C}$, denoted by $Pol(\mathcal{C})$, consists of all *finite unions* of marked products $L_0 a_1 L_1 \cdots a_n L_n$ such that $L_0, \ldots, L_n \in \mathcal{C}$. If $\mathcal{C}$ is a prevariety, then $Pol(\mathcal{C})$ is a quotient-closed lattice (this is due to Arfi [2], see also [14,20] for recent proofs). Yet, $Pol(\mathcal{C})$ need not be closed under complement. Hence, it is often combined with another operator. The Boolean closure of a class $\mathcal{D}$, denoted by $Bool(\mathcal{D})$, is the smallest Boolean algebra containing $\mathcal{D}$. Finally, we write $BPol(\mathcal{C})$ for $Bool(Pol(\mathcal{C}))$. The following result is standard (see [20], for example).

Proposition 4. *If $\mathcal{C}$ is a prevariety, then so is $BPol(\mathcal{C})$.*

Remark 4. The classes $Pol(\mathcal{C})$ and $BPol(\mathcal{C})$ are quite prominent. For example, $Pol(\mathrm{ST})$ contains exactly the finite unions of languages $A^* a_1 A^* \cdots a_n A^*$ where $n \in \mathbb{N}$ and $a_1, \cdots, a_n \in A$ are letters. Moreover, $BPol(\mathrm{ST})$ consists of all finite Boolean combinations of such languages: this is the famous class of *piecewise testable languages* [25]. In the literature, such classes are more often associated with classical logic rather than with temporal logic. Indeed, it is well known [20, 28] that $BPol$ corresponds to the quantifier alternation free fragment of first-order logic $(\mathcal{B}\Sigma_1)$. For each prevariety $\mathcal{C}$, there exists a set of first-order predicates $\mathbb{I}_\mathcal{C}$ such that $BPol(\mathcal{C})$ contains exactly the languages that can be defined by a $\mathcal{B}\Sigma_1$ sentence using only predicates in $\mathbb{I}_\mathcal{C}$. On the other hand, no characterization of $BPol$ based on temporal logic is known. In order to establish a connection with unary temporal logic, we have to apply additional operators on top of $BPol$.

4.2   Deterministic Variants

We consider four restrictions of *Pol*: *UPol*, *LPol*, *RPol* and *MPol*. The first three are standard (see for example [13,15,24]). On the other hand, *MPol* was introduced recently in [18]. We restrict the marked products to those satisfying specific semantic conditions and the unions to *disjoint* ones. Consider a marked product $L_0 a_1 L_1 \cdots a_n L_n$. For $1 \leq i \leq n$, we let $L'_i = L_0 a_1 L_1 \cdots a_{i-1} L_{i-1}$ and $L''_i = L_i a_{i+1} \cdots L_{n-1} a_n L_n$. In particular, $L'_1 = L_0$ and $L''_n = L_n$. We say that,

- $L_0 a_1 L_1 \cdots a_n L_n$ is *left deterministic* when $L'_i \cap L'_i a_i A^* = \emptyset$ for $1 \leq i \leq n$.
- $L_0 a_1 L_1 \cdots a_n L_n$ is *right deterministic* when $L''_i \cap A^* a_i L''_i = \emptyset$ for $1 \leq i \leq n$.
- $L_0 a_1 L_1 \cdots a_n L_n$ is *mixed deterministic* when either $L'_i \cap L'_i a_i A^* = \emptyset$, or $L''_i \cap A^* a_i L''_i = \emptyset$ for $1 \leq i \leq n$.
- $L_0 a_1 L_1 \cdots a_n L_n$ is *unambiguous* when for all $w \in L_0 a_1 L_1 \cdots a_n L_n$, there is a *unique* decomposition $w = w_0 a_1 w_1 \cdots a_n w_n$ where $w_i \in L_i$ for $1 \leq i \leq n$.

By definition, a left or right deterministic marked product is also mixed deterministic. It is also simple to verify that mixed deterministic marked products are unambiguous. Note that these four notions depend on the product itself and not only on the resulting language. For example, $A^* a A^*$ (which is not unambiguous) and $(A \setminus \{a\})^* a A^*$ (which is left deterministic) evaluate to the same language.

Remark 5. A mixed deterministic product need not be left or right deterministic. For example, let $L_1 = (ab)^+$, $L_2 = c^+$ and $L_3 = (ba)^+$. The product $L_1 c L_2 c L_3$ is mixed deterministic since $L_1 \cap L_1 c A^* = \emptyset$ and $L_3 \cap A^* c L_3 = \emptyset$. However, it is neither left deterministic nor right deterministic. Similarly, a unambiguous product need not be mixed deterministic. If $L_4 = (ca)^+$, the product $L_1 a L_4$ is unambiguous but it neither left nor right deterministic.

The *left polynomial closure* of a class $\mathcal{C}$, written $LPol(\mathcal{C})$, consists of all *finite disjoint unions* of *left deterministic marked products* $L_0 a_1 L_1 \cdots a_n L_n$ such that $L_0, \ldots, L_n \in \mathcal{C}$ (by "disjoint" we mean that the languages in the union must be pairwise disjoint). The *right polynomial closure* of $\mathcal{C}$ ($RPol(\mathcal{C})$), the *mixed polynomial closure* of $\mathcal{C}$ ($MPol(\mathcal{C})$) and the *unambiguous polynomial closure* of $\mathcal{C}$ ($UPol(\mathcal{C})$) are defined analogously by replacing the requirement to be "left deterministic" for marked products by the appropriate one.

We introduce a key property of these operators, which is not apparent on the definition: when applied to a prevariety, they also yield a prevariety. Moreover, in that case, the four operators preserve the decidability of *membership*. This is proved in [19,23] for $UPol$ and in [18] for $LPol$, $RPol$ and $MPol$. From this, we shall obtain decidability of membership for classes built with FL^n, PL^n and BL^n.

Theorem 1. *([18,23]).* *Let* $X \in \{UPol, LPol, RPol, MPol\}$. *For every prevariety* $\mathcal{C}$, *the class* $X(\mathcal{C})$ *is a prevariety as well. Moreover, if* $\mathcal{C}$ *has decidable membership, then so does* $X(\mathcal{C})$.

For each operator, Theorem 1 is based on a generic algebraic characterization of the classes that it builds. In Theorem 2 below, we recall the symmetric characterizations of $LPol$ and $RPol$, as we shall need them in order to establish the correspondence with future/past hierarchies. This requires two notions.

Let $\mathcal{C}$ be a prevariety and let $\alpha : A^* \to M$ be a morphism. We define *two* relations on M (both depending on α). Given $s, t \in M$, we say that (s, t) is a $\mathcal{C}$-*pair* if $\alpha^{-1}(s)$ is *not* $\mathcal{C}$-separable from $\alpha^{-1}(t)$. This relation is not very robust: it is reflexive (if α is surjective) and symmetric (this is tied to $\mathcal{C}$ being closed under complement). The second relation is an equivalence "$\sim_{\mathcal{C}}$" on M. For $s, t \in M$, we write $s \sim_{\mathcal{C}} t$ when $s \in F \Leftrightarrow t \in F$ for every $F \subseteq M$ such that $\alpha^{-1}(F) \in \mathcal{C}$. By definition, $\sim_{\mathcal{C}}$ is an equivalence relation. In fact, it is shown in [20,23] that it is the reflexive transitive closure of the "$\mathcal{C}$-pair" relation (we do not use this property). We now present the characterizations of $LPol$ and $RPol$ taken from [18]. They are crucial for proving Theorem 3 below, which expresses $\mathrm{FL}^n(\mathcal{C}), \mathrm{PL}^n(\mathcal{C}), \mathrm{BL}^n(\mathcal{C})$ in terms of the operators $LPol$, $RPol$ and $BPol$.

Theorem 2. *([18]).* *Let* $\mathcal{C}$ *be a prevariety and let* $\alpha : A^* \to M$ *be a surjective morphism. The three following properties are equivalent:*

1. *The morphism* α *is an* $LPol(\mathcal{C})$-*morphism (resp. an* $RPol(\mathcal{C})$-*morphism).*
2. *For all* $\mathcal{C}$-*pairs* $(s, t) \in M^2$, *we have* $s^{\omega+1} = s^{\omega}t$ *(resp.* $s^{\omega+1} = ts^{\omega}$*).*
3. *For all* $s, t \in M$ *such that* $s \sim_{\mathcal{C}} t$, *we have* $s^{\omega+1} = s^{\omega}t$ *(resp.* $s^{\omega+1} = ts^{\omega}$*).*

This implies the statement on membership in Theorem 1 for $LPol$ and $RPol$. Let us explain how on $LPol$, for instance. By Proposition 1, deciding $LPol(\mathcal{C})$-membership boils down to deciding if an input morphism $\alpha : A^* \to M$ is an $LPol(\mathcal{C})$-morphism. By the third assertion in Theorem 2, this is possible if one can compute the equivalence $\sim_{\mathcal{C}}$ on M. By definition, this boils down to $\mathcal{C}$-membership (it suffices to compute all subsets $F \subseteq M$ such that $\alpha^{-1}(F) \in \mathcal{C}$).

4.3   Characterization of Future/Past Hierarchies

It is well known that there is a correspondence between full unary temporal logic and $UPol$. This was first proved for the standard variants in [4,16,27], and was then generalized to our extended definition in [23]. More precisely, for each *group* prevariety $\mathcal{G}$, we have $\mathrm{TL}(\mathcal{G}) = UPol(BPol(\mathcal{G}))$ and $\mathrm{TL}(\mathcal{G}^+) = UPol(BPol(\mathcal{G}^+))$. Here, we generalize these results to future/past hierarchies.

We use $LPol$ and $RPol$ to define hierarchies (the definition is taken from [18]). It is shown in [19,23] that for all prevarieties $\mathcal{C}$, the class $UPol(\mathcal{C})$ is the least one containing $\mathcal{C}$ and closed under left and right deterministic marked products as well as disjoint union. Thus, applying $LPol$ and $RPol$ in alternation builds a classification of $UPol(\mathcal{C})$: the *deterministic hierarchy of basis* $\mathcal{C}$. For all $n \in \mathbb{N}$, we define two levels $LPol_n(\mathcal{C})$ and $RPol_n(\mathcal{C})$. We let $LPol_0(\mathcal{C}) = RPol_0(\mathcal{C}) = \mathcal{C}$. For $n \geq 1$, $LPol_n(\mathcal{C}) = LPol(RPol_{n-1}(\mathcal{C}))$ and $RPol_n(\mathcal{C}) = RPol(LPol_{n-1}(\mathcal{C}))$. The union of all levels is exactly $UPol(\mathcal{C})$. These are strict hierarchies and the levels $LPol_n(\mathcal{C})$ and $RPol_n(\mathcal{C})$ are incomparable for all $n \geq 1$, in general. This motivates additional intermediary levels "combining" $LPol_n(\mathcal{C})$ and $RPol_n(\mathcal{C})$: for all $n \geq 1$, we let $LPol_n(\mathcal{C}) \vee RPol_n(\mathcal{C})$ be the least Boolean algebra containing both $LPol_n(\mathcal{C})$ and $RPol_n(\mathcal{C})$.

For every prevariety $\mathcal{C}$, we connect the future/past hierarchy of $\mathrm{TL}(\mathcal{C})$ with the deterministic hierarchy of basis $BPol(\mathcal{C})$. In the general case, we only prove that the latter is included in the former. This inclusion is *strict* in general: an example of prevariety $\mathcal{C}$ such that $UPol(BPol(\mathcal{C}))$ is *strictly* included in $\mathrm{TL}(\mathcal{C})$ is provided in [23] (strictness follows from results of [7]).

Proposition 5. *Let $\mathcal{C}$ be prevariety. The following properties hold for all $n \geq 1$:*

1. *If n is odd, $RPol_n(BPol(\mathcal{C})) \subseteq \mathrm{FL}^n(\mathcal{C})$ and $LPol_n(BPol(\mathcal{C})) \subseteq \mathrm{PL}^n(\mathcal{C})$.*
2. *If n is even, $LPol_n(BPol(\mathcal{C})) \subseteq \mathrm{FL}^n(\mathcal{C})$ and $RPol_n(BPol(\mathcal{C})) \subseteq \mathrm{PL}^n(\mathcal{C})$.*

Remark 6. There are *four* cases in Proposition 5. This is because for every level $RPol_n(BPol(\mathcal{C}))$ or $LPol_n(BPol(\mathcal{C}))$, the notation highlights the last operator used in its construction from $BPol(\mathcal{C})$. However, the logic corresponding to this level is determined by the *first* operator in the construction. For example, we have $RPol(BPol(\mathcal{C})) \subseteq \mathrm{FL}^1(\mathcal{C})$ and all classes which are built from $RPol(BPol(\mathcal{C}))$ by applying $LPol$ and $RPol$ in alternation are included in a level $\mathrm{FL}^n(\mathcal{C})$.

Proof. (of Proposition 5). We use induction on n. There are four cases. We prove that if n is odd, then $RPol_n(BPol(\mathcal{C})) \subseteq \mathrm{FL}^n(\mathcal{C})$ (the other cases are symmetric). Let $\mathcal{D} = LPol_{n-1}(BPol(\mathcal{C}))$ and fix $L \in RPol_n(BPol(\mathcal{C})) = RPol(\mathcal{D})$. We prove that $L \in \mathrm{FL}^n(\mathcal{C})$. We need the next easy lemma.

Lemma 6. *Let $K \in BPol(\mathcal{C})$. There exists an $\mathrm{FL}^1[\mathcal{C}]$ formula ξ_K such that for all $w \in A^*$ and all $i \in \mathsf{Pos}(w)$, we have*

$$w, i \models \xi_K \iff i \leq |w| \text{ and } wi \in K.$$

Proof. By definition, $K \in BPol(\mathcal{C})$ is a Boolean combination of languages of the form $K_0 a_1 K_1 \cdots a_n K_n$ with $a_1, \ldots, a_n \in A$ and $K_0, \ldots, K_n \in \mathcal{C}$. Since we may use Boolean connectives freely in $\mathrm{FL}^1[\mathcal{C}]$, we may assume without loss of generality that K itself is of the form $K_0 a_1 K_1 \cdots a_n K_n$. It now suffices to verify that the following formula ξ_K satisfies the desired property:

$$\xi_K = \mathrm{F}_{K_0} \left(a_1 \wedge \mathrm{F}_{K_1} \left(a_2 \wedge \mathrm{F}_{K_2} \left(\cdots a_n \wedge \mathrm{F}_{K_n} \, max \right) \right) \right).$$

□

Since $L \in RPol(\mathcal{D})$ with $\mathcal{D} = LPol_{n-1}(BPol(\mathcal{C}))$, it is shown in [18, Proposition 5.3] that L is a finite union of products $L_0 a_1 L_1 \cdots a_m L_m$ satisfying the two following conditions: (1) $L_h \in \mathcal{D}$ for every $h \leq m$, and (2) there exists a right deterministic marked product $K_0 a_1 K_1 \cdots a_m K_m$ such that $K_h \in BPol(\mathcal{C})$ and $L_h \subseteq K_h$ for every $h \leq m$. Hence, by closure under union, it suffices to prove that every such product $L_0 a_1 L_1 \cdots a_m L_m$ belongs to $FL^n(\mathcal{C})$. We use a subinduction on h to prove that $L_0 a_1 L_1 \cdots a_h L_h \in FL^n(\mathcal{C})$ for $0 \leq h \leq m$.

When $h = 0$, there are two cases. If $n = 1$, then $\mathcal{D} = BPol(\mathcal{C})$ which means that $L_0 \in BPol(\mathcal{C})$ and it is defined by the $FL^1[\mathcal{C}]$ formula $(min \wedge \xi_{L_0}) \vee max$, where ξ_{L_0} is given by Lemma 6. Otherwise, $n > 1$ and $\mathcal{D} = LPol_{n-1}(BPol(\mathcal{C}))$ with $n - 1 \geq 1$. Consequently, since $(n - 1)$ is even, the main induction on n yields $L_0 \in \mathcal{D} \subseteq FL^{n-1}(\mathcal{C}) \subseteq FL^n(\mathcal{C})$, which completes this case.

We now assume that $h \geq 1$. Let $R = L_0 a_1 L_1 \cdots a_{h-1} L_{h-1}$. We prove that $R a_h L_h \in FL^n(\mathcal{C})$. Induction on h yields $FL^n[\mathcal{C}]$ formulas φ_R and φ_{L_h} defining R and L_h. We also use the $FL^1[\mathcal{C}]$ formula ξ_{K_h} associated with $K_h \in BPol(\mathcal{C})$ by Lemma 6. We write ψ for the $FL^1[\mathcal{C}]$ formula $a_h \wedge \xi_{K_h}$. Since $K_0 a_1 K_1 \cdots a_m K_m$ is right deterministic, one may verify that $A^* a_h K_h$ is unambiguous. Hence, by definition of ψ, for every $w \in A^*$, there exists *at most one* position $i \in Pos(w)$ such that $w, i \models \psi$. Therefore, since $L_h \subseteq K_h$, it follows that for every $w \in A^*$, we have $w \in R a_h L_h$ if and only if w satisfies the three following conditions:

1. There exists $i \in Pos(w)$ (which must be unique) such that $w, i \models \psi$.
2. The prefix wi belongs to R (*i.e.*, we have $wi \models \varphi_R$).
3. The suffix wi belongs to L_h (*i.e.*, we have $wi \models \varphi_{L_h}$).

It remains to prove that these properties can be expressed in $FL^n[\mathcal{C}]$. Condition 1 is expressed by the $FL^1[\mathcal{C}]$ formula $(min \wedge F \psi) \vee max$. We turn to Condition 2. We modify φ_R into a new formula φ'_R expressing the desired property. For every word $w \in A^*$, we restrict the evaluation of φ_R to the positions $j \in Pos(w)$ such that either $w, j \models F \psi$ or $j = |w| + 1$. We build φ'_R by applying the two following modifications to φ_R. First, we recursively replace every subformula $P_U \zeta$ by $(max \wedge P (\psi \wedge P_U \zeta)) \vee (F \psi \wedge P_U \zeta)$. Second, we replace every subformula $F_U \zeta$ by $(F_U (\zeta \wedge F \psi)) \vee (F_U (\psi \wedge F (max \wedge \zeta)))$. Since φ_R is an $FL^n[\mathcal{C}]$ formula and n is *odd*, one may verify that φ'_R is also an $FL^n[\mathcal{C}]$ formula. The key point is that since n is odd, the $FL^n[\mathcal{C}]$ formulas are defined inductively from the $FL^1[\mathcal{C}]$ formulas and inserting the $FL^1[\mathcal{C}]$ formula ψ in an $FL^n[\mathcal{C}]$ formula yields a new $FL^n[\mathcal{C}]$ formula. It can be verified that φ'_R expresses Condition 2.

We turn to Condition 3. We modify φ_{L_h} into another formula φ'_{L_h} expressing the desired property. For all $w \in A^*$, we restrict the evaluation of φ_{L_h} to the positions $j \in Pos(w)$ such that $j = 0$ or $w, j \models P \psi$. We build φ'_{L_h} by applying the two following modifications to φ_{L_h}. We recursively replace every subformula $P_U \zeta$ by $(P_U (\psi \wedge P (min \wedge \zeta))) \vee (P_U (\zeta \wedge P \psi))$. Moreover, we replace every subformula $F_U \zeta$ by $(min \wedge F (\psi \wedge F_U \zeta)) \vee (P \psi \wedge F_U \zeta)$. As in the previous case, since n is odd, one may verify that φ'_{L_h} remains an $FL^n[\mathcal{C}]$ formula and that it expresses Condition 3. Finally, the language $R a_h L_h$ is now defined by the following $FL^n[\mathcal{C}]$ formula: $((min \wedge F \psi) \vee max) \wedge \varphi'_R \wedge \varphi'_{L_h}$.   □

With Proposition 5 in hand, we may now consider the case when $\mathcal{C} \in \{\mathcal{G}, \mathcal{G}^+\}$ for a group prevariety $\mathcal{G}$. In this case, there is an exact correspondence.

Theorem 3. *Let $\mathcal{G}$ be a group prevariety and let $\mathcal{C} \in \{\mathcal{G}, \mathcal{G}^+\}$. The three following properties hold for every $n \geq 1$:*

1. *If n is odd, $\mathrm{FL}^n(\mathcal{C}) = RPol_n(BPol(\mathcal{C}))$ and $\mathrm{PL}^n(\mathcal{C}) = LPol_n(BPol(\mathcal{C}))$.*
2. *If n is even, $\mathrm{FL}^n(\mathcal{C}) = LPol_n(BPol(\mathcal{C}))$ and $\mathrm{PL}^n(\mathcal{C}) = RPol_n(BPol(\mathcal{C}))$.*
3. *We have $\mathrm{BL}^n(\mathcal{C}) = LPol_n(BPol(\mathcal{C})) \vee RPol_n(BPol(\mathcal{C}))$.*

Before we prove Theorem 3, let us discuss its consequences. An important application is membership for future/past hierarchies. Let $\mathcal{G}$ be a group prevariety and $\mathcal{C} \in \{\mathcal{G}, \mathcal{G}^+\}$. In view of Theorem 3 and Theorem 1, it is immediate that membership is decidable for all levels $\mathrm{FL}^n(\mathcal{C})$ and $\mathrm{PL}^n(\mathcal{C})$ as soon as this problem is decidable for $BPol(\mathcal{C})$. Since $\mathcal{C} \in \{\mathcal{G}, \mathcal{G}^+\}$, it follows from results of [22] that $BPol(\mathcal{C})$-membership boils down to $\mathcal{G}$-separation (this is based on independent techniques). Hence, we obtain the following corollary.

Corollary 1. *Consider a group prevariety $\mathcal{G}$ with decidable separation and let $\mathcal{C} \in \{\mathcal{G}, \mathcal{G}^+\}$. For every $n \geq 1$, membership is decidable for $\mathrm{FL}^n(\mathcal{C})$ and $\mathrm{PL}^n(\mathcal{C})$.*

Remark 7. We do not mention the levels $\mathrm{BL}^n(\mathcal{C})$ yet as this requires more work. This is the topic of Sect. 5: we prove that Corollary 1 also holds for them.

Proof. (of Theorem 3). By definition, the third assertion is an immediate consequence of the others. Hence, we focus on the first two. We use induction on n. By symmetry, we only treat the case when n is odd and show that $\mathrm{FL}^n(\mathcal{C}) = RPol_n(BPol(\mathcal{C}))$. Proposition 5 yields the right to left inclusion. We fix $L \in \mathrm{FL}^n(\mathcal{C})$ and prove that $L \in RPol_n(BPol(\mathcal{C}))$. Let $\mathcal{D} = LPol_{n-1}(BPol(\mathcal{C}))$: we prove that $L \in RPol(\mathcal{D})$. Let $\alpha_L : A^* \to M_L$ be the syntactic morphism of L. By Proposition 1 and Theorem 2, it suffices to prove that for every $\mathcal{D}$-pair $(s,t) \in M_L^2$, we have $s^{\omega+1} = ts^\omega$. Since $L \in \mathrm{FL}^n(\mathcal{C})$, Lemma 3 yields a $\mathcal{C}$-morphism $\eta : A^* \to N$ and $k \in \mathbb{N}$ such that L is a union of $\blacktriangleright_{n,\eta,k}$-classes. Note that since $\mathcal{C} \in \{\mathcal{G}, \mathcal{G}^+\}$, we know that $\eta(A^+)$ is a group by Lemma 2. Let $p = \omega(N)$. We claim that since (s,t) is $\mathcal{D}$-pair, there exist $u,v \in A^*$ such that $\alpha_L(u) = t$, $\alpha_L(v) = s$ and $v^{2kp+1} \blacktriangleright_{n,\eta,k} uv^{2kp}$. Let us first explain why this completes the proof. Since $\blacktriangleright_{n,\eta,k}$ is a congruence by Lemma 4, we get $xv^{2kp+1}y \blacktriangleright_{n,\eta,k} xuv^{2kp}y$ for every $x,y \in A^*$. Since L is a union of $\blacktriangleright_{n,\eta,k}$-classes, this yields $xv^{2kp+1}y \in L \Leftrightarrow xuv^{2kp}y \in L$ for all $x,y \in A^*$, i.e., $v^{2kp+1} \equiv_L uv^{2kp}$. Hence, $\alpha_L(v^{2kp+1}) = \alpha_L(uv^{2kp})$, i.e., $s^{2kp+1} = ts^{2kp}$. We now multiply by enough copies of s on the right to get $s^{\omega+1} = ts^\omega$, as desired.

We now build $u,v \in A^*$. There are two cases depending on n. If $n = 1$, then $\mathcal{D} = BPol(\mathcal{C})$. We use the equivalence $\sim_{\eta,k}$ defined in Sect. 3. Observe that the $\sim_{\eta,k}$-classes belong to $BPol(\mathcal{C})$. There are two subcases depending on η.

– If $\eta^{-1}(1_N) \cap A^+ \neq \emptyset$, then Condition 3 in the definition of $\sim_{\eta,k}$ is trivial. Hence, one may verify that the $\sim_{\eta,k}$-classes are Boolean combinations of languages of the form $\eta^{-1}(g)$ and $\eta^{-1}(g_1)a\eta^{-1}(g_2)$ for $a \in A$ and $g, g_1, g_2 \in N$, i.e., languages in $BPol(\mathcal{C})$, since η is a $\mathcal{C}$-morphism.

– Otherwise, $\eta^{-1}(1_N) = \{\varepsilon\}$ and one may verify that the $\sim_{\eta,k}$-classes are Boolean combinations of languages $w\eta^{-1}(g)$, $\eta^{-1}(g)w$ and $\eta^{-1}(g_1)w\eta^{-1}(g_2)$ for $w \in A^*$ and $g, g_1, g_2 \in N$. These are languages in $BPol(\mathcal{C})$ since η is a $\mathcal{C}$-morphism (which implies that $\{\varepsilon\} \in \mathcal{C}$, since it is recognized by η).

Since (s,t) is a $BPol(\mathcal{C})$-pair, it follows that there exist $u, v \in A^*$ such that $\alpha_L(u) = t$, $\alpha_L(v) = s$ and $u \sim_{\eta,k} v$ (otherwise, $\alpha_L^{-1}(s)$ can be separated from $\alpha_L^{-1}(t)$ by a union of $\sim_{\eta,k}$-classes, i.e., by a language in $BPol(\mathcal{C})$). Thus, since $\alpha(A^+)$ is a group, Lemma 5 yields $v^{2kp+1} \blacktriangleright_{1,\eta,k} uv^{2kp}$, completing the case $n = 1$.

We now assume that $n > 1$. Lemma 3 implies that the $\blacktriangleright_{n-1,\eta,k^2}$-classes belong to $FL^{n-1}(\mathcal{C})$. Hence, since $n-1$ is even (n is odd by hypothesis) induction yields that the $\blacktriangleright_{n-1,\eta,k^2}$-classes belong to $\mathcal{D} = LPol_{n-1}(BPol(\mathcal{C}))$. Since (s,t) is a $\mathcal{D}$-pair, this yields $u, v \in A^*$ such that $\alpha(u) = t$, $\alpha(v) = s$ and $u \blacktriangleright_{n-1,\eta,k^2} v$ (otherwise, $\alpha^{-1}(s)$ can be separated from $\alpha^{-1}(t)$ by a union of $\blacktriangleright_{n-1,\eta,k^2}$-classes, i.e., by a language in $\mathcal{D}$). Since $n-1$ is even and $\alpha(A^+)$ is a group, it then follows from Proposition 3 that $uv^{2kp} \blacktriangleright_{n,\eta,k} v^{2kp+1}$, completing the proof.   □

5   Intermediary Levels

We now consider the levels $BL^n(\mathcal{C})$ in future/past hierarchies. We prove that when $\mathcal{C} \in \{\mathcal{G}, \mathcal{G}^+\}$ where $\mathcal{G}$ is a group prevariety with decidable separation, membership is decidable for all levels $BL^n(\mathcal{C})$. By Theorem 3, we know that $BL^n(\mathcal{C}) = LPol_n(BPol(\mathcal{C})) \vee RPol_n(BPol(\mathcal{C}))$ for every $n \geq 1$. A key ingredient in our approach is a result of [18] based on mixed polynomial closure ($MPol$).

Theorem 4. *([18]). Let $\mathcal{D}$ be a prevariety. For every number $n \geq 1$, we have $LPol_{n+1}(\mathcal{D}) \vee RPol_{n+1}(\mathcal{D}) = MPol(LPol_n(\mathcal{D}) \vee RPol_n(\mathcal{D}))$.*

Combining Theorem 4 with Theorem 3 yields $BL^{n+1}(\mathcal{C}) = MPol(BL^n(\mathcal{C}))$. Now, recall that by Theorem 1, $MPol$ preserves the decidability of membership (this is shown in [18]). Therefore, an immediate induction reduces membership for $BL^n(\mathcal{C})$ to membership for $BL^1(\mathcal{C})$. Thus, we concentrate on this case: we prove that for every group prevariety $\mathcal{G}$ with decidable separation, if $\mathcal{C} \in \{\mathcal{G}, \mathcal{G}^+\}$, then membership is decidable for $BL^1(\mathcal{C}) = LPol(BPol(\mathcal{C})) \vee RPol(BPol(\mathcal{C}))$.

We present algebraic characterizations of $LPol(BPol(\mathcal{C})) \vee RPol(BPol(\mathcal{C}))$. There are two statements depending on whether $\mathcal{C} = \mathcal{G}$ or $\mathcal{C} = \mathcal{G}^+$. We use the $\mathcal{G}$-pair relation defined in Sect. 4 (this is how the statement depends on $\mathcal{G}$-separation). We first characterize the classes $LPol(BPol(\mathcal{G})) \vee RPol(BPol(\mathcal{G}))$.

Theorem 5. *Let $\mathcal{G}$ be a group prevariety and let $\alpha : A^* \to M$ be a surjective morphism. Then, α is an $LPol(BPol(\mathcal{G})) \vee RPol(BPol(\mathcal{G}))$-morphism if and only if it satisfies the following property:*

$$(sq(tq')^\omega)^\omega s((r't)^\omega rs)^\omega = (sq(tq')^\omega)^\omega t((r't)^\omega rs)^\omega \quad (1)$$
$$\text{for all } q, q', r, r' \in M \text{ and all } \mathcal{G}\text{-pairs } (s,t) \in M^2.$$

We complete Theorem 5 with a second statement, which applies to the classes $LPol(BPol(\mathcal{G}^+)) \vee RPol(BPol(\mathcal{G}^+))$.

Theorem 6. *Let $\mathcal{G}$ be a group prevariety and let $\alpha : A^* \to M$ be a surjective morphism. Then, α is an $LPol(BPol(\mathcal{G}^+)) \vee RPol(BPol(\mathcal{G}^+))$-morphism if and only if it satisfies the following property:*

$$(esfq(etfq')^\omega)^\omega esf((r'etf)^\omega resf)^\omega = (esfq(etfq')^\omega)^\omega etf((r'etf)^\omega resf)^\omega \tag{2}$$
$$\text{for all } q, q', r, r' \in M, \text{ all } e, f \in E(\alpha(A^+)) \text{ and all } \mathcal{G}\text{-pairs } (s,t) \in M^2.$$

Recall that the $\mathcal{G}$-pairs associated with a morphism can be computed provided that $\mathcal{G}$-separation is decidable (by definition, (s, t) is a $\mathcal{G}$-pair if and only if $\alpha^{-1}(s)$ is not $\mathcal{G}$-separable from $\alpha^{-1}(t)$). Hence, Theorem 5 and Theorem 6 imply that if $\mathcal{G}$ is a group prevariety *with decidable separation* and $\mathcal{C} \in \{\mathcal{G}, \mathcal{G}^+\}$, then membership is decidable for the class $LPol(BPol(\mathcal{C})) \vee RPol(BPol(\mathcal{C}))$. Using Theorem 1 and Theorem 4, one may then lift decidability to *all* levels $LPol_n(BPol(\mathcal{C})) \vee RPol_n(BPol(\mathcal{C}))$ for $n \geq 1$. Finally, Theorem 3 yields the following corollary.

Corollary 2. *Let $\mathcal{G}$ be a group prevariety with decidable separation. For every $n \geq 1$, membership is decidable for $\mathrm{BL}^n(\mathcal{G})$ and $\mathrm{BL}^n(\mathcal{G}^+)$.*

Remark 8. Theorem 5 generalizes a known result in the special case when $\mathcal{G}$ is the trivial class $\mathrm{ST} = \{\emptyset, A^*\}$. In this case, it is known [1,9] that a surjective morphism $\alpha : A^* \to M$ is an $LPol(BPol(\mathrm{ST})) \vee RPol(BPol(\mathrm{ST}))$-morphism if and only if M satisfies the equation $(sq)^\omega s(rs)^\omega = (sq)^\omega(rs)^\omega$ for all $q, r, s \in M$. This equation is equivalent to (1) in this case. Indeed, since ST is trivial, *every* pair in M^2 is an ST-pair. In particular, if $q, r, s \in M$, then $(s, 1_M)$ is an ST-pair. Hence, the above equation is the special case of (1) when $t = q' = r' = 1_M$. Conversely, if the above equation holds, then given elements $s, t, q, q', r, r' \in M$, we may apply the equation twice to get $(sq(tq')^\omega)^\omega s((r't)^\omega rs)^\omega = (sq(tq')^\omega)^\omega((r't)^\omega rs)^\omega$ and $(tq')^\omega t(r't)^\omega = (tq')^\omega(r't)^\omega$. When combined, the two imply that (1) holds.

The proofs of Theorem 5 and Theorem 6 are presented in the full version of the paper. The two arguments are similar (though the one of Theorem 6 is technically more involved). Let us point out that these proofs are nontrivial. As for most algebraic characterizations of this kind, the challenging direction consists in proving that if some morphism $\alpha : A^* \to M$ satisfies (1) (resp. (2)), then it must be an $LPol(BPol(\mathcal{G})) \vee RPol(BPol(\mathcal{G}))$-morphism (resp. an $LPol(BPol(\mathcal{G}^+)) \vee RPol(BPol(\mathcal{G}^+))$-morphism). In particular, this part of the proof relies heavily on properties of the operators $LPol$, $RPol$ and $UPol$ established in [18] and [19,23].

6   Conclusion

For all group prevarieties $\mathcal{G}$, we characterized the future/past hierarchies within the variants $\mathrm{TL}(\mathcal{G})$ and $\mathrm{TL}(\mathcal{G}^+)$ of unary temporal logic with the deterministic hierarchies of bases $BPol(\mathcal{G})$ and $BPol(\mathcal{G}^+)$. We used these results to prove that if $\mathcal{G}$-separation is decidable, then membership is also decidable for *all* levels $FL^n(\mathcal{G})$, $PL^n(\mathcal{G})$, $BL^n(\mathcal{G})$, $FL^n(\mathcal{G}^+)$, $PL^n(\mathcal{G}^+)$ and $BL^n(\mathcal{G}^+)$ in such hierarchies.

A natural question is whether decidability can be pushed to more general problems than membership, *e.g.*, separation. When $\mathcal{G}$ is the trivial class ST $= \{\emptyset, A^*\}$, it is known that separation is decidable for all levels $FL^n(\text{ST})$ and $PL^n(\text{ST})$ (this is shown for their counterparts in deterministic hierarchies [18]). Moreover, it is also known that if $\mathcal{G}$ is a group prevariety with decidable separation, then $BPol(\mathcal{G})$- and $BPol(\mathcal{G}^+)$-separation are also decidable [21]. In view of our characterizations, this suggests that similar results may hold for the whole future/past hierarchies of $TL(\mathcal{G})$ and $TL(\mathcal{G}^+)$.

References

1. Almeida, J., Azevedo, A.: The join of the pseudovarieties of $\mathcal{R}$-trivial and $\mathcal{L}$-trivial monoids. J. Pure Appl. Algebra **60**(2), 129–137 (1989)
2. Arfi, M.: Polynomial operations on rational languages. In: Brandenburg, F.J., Vidal-Naquet, G., Wirsing, M. (eds.) STACS 1987. LNCS, vol. 247, pp. 198–206. Springer, Heidelberg (1987). https://doi.org/10.1007/BFb0039607
3. Etessami, K., Vardi, M.Y., Wilke, T.: First-order logic with two variables and unary temporal logic. In: Proceedings of the 12th Annual IEEE Symposium on Logic in Computer Science, pp. 228–235. LICS 1997 (1997)
4. Etessami, K., Vardi, M.Y., Wilke, T.: First-order logic with two variables and unary temporal logic. Inf. Comput. **179**(2), 279–295 (2002)
5. Fleischer, L., Kufleitner, M., Lauser, A.: The half-levels of the FO^2 alternation hierarchy. Theory Comput. Syst. **61**(2), 352–370 (2017)
6. Kamp, H.W.: Tense Logic and the Theory of Linear Order. Phd thesis, Computer Science Department, University of California at Los Angeles, USA (1968)
7. Krebs, A., Lodaya, K., Pandya, P.K., Straubing, H.: Two-variable logics with some betweenness relations: expressiveness, satisfiability and membership. Log. Methods Comput. Sci. **16**(3) (2020)
8. Kufleitner, M., Lauser, A.: The join levels of the Trotter-Weil hierarchy are decidable. In: Proceedings of the 37th International Symposium on Mathematical Foundations of Computer Science. MFCS 2012, vol. 7464, pp. 603–614 (2012)
9. Kufleitner, M., Lauser, A.: The join of $\mathcal{R}$-trivial and $\mathcal{L}$-trivial monoids via combinatorics on words. Discrete Math. Theor. Comput. Sci. **14**(1), 141–146 (2012)
10. Kufleitner, M., Weil, P.: On logical hierarchies within FO2-definable languages. Logical Methods Comput. Sci. **8**(3:11), 1–30 (2012)
11. Kufleitner, M., Weil, P.: The FO^2 alternation hierarchy is decidable. In: Proceedings of the 21st International Conference on Computer Science Logic, pp. 426–439. CSL 2012 (2012)
12. Lodaya, K, Pandya, PK., Shah, SS.: Marking the chops: an unambiguous temporal logic. In: Ausiello, G., Karhumäki, J., Mauri, G., Ong, L. (eds.) TCS 2008. IIFIP, vol. 273, pp. 461–476. Springer, Boston, MA (2008). https://doi.org/10.1007/978-0-387-09680-3_31
13. Pin, J.E.: Propriétés syntactiques du produit non ambigu. In: Proceedings of the 7th International Colloquium on Automata, Languages and Programming, pp. 483–499. ICALP 1980 (1980)

14. Pin, J.: An Explicit Formula for the Intersection of Two Polynomials of Regular Languages. In: Béal, M.-P., Carton, O. (eds.) DLT 2013. LNCS, vol. 7907, pp. 31–45. Springer, Heidelberg (2013). https://doi.org/10.1007/978-3-642-38771-5_5

15. Pin, J.E., Straubing, H., Thérien, D.: Locally trivial categories and unambiguous concatenation. J. Pure Appl. Algebra **52**(3), 297–311 (1988)

16. Pin, J.E., Weil, P.: Polynomial closure and unambiguous product. Theory Comput. Syst. **30**(4), 383–422 (1997)

17. Place, T.: Deciding classes of regular languages: the covering approach. In: Leporati, A., Martín-Vide, C., Shapira, D., Zandron, C. (eds.) LATA 2020. LNCS, vol. 12038, pp. 89–112. Springer, Cham (2020). https://doi.org/10.1007/978-3-030-40608-0_6

18. Place, T.: The amazing mixed polynomial closure and its applications to two-variable first-order logic. In: Proceedings of the 37th Annual ACM/IEEE Symposium on Logic in Computer Science. LICS 2022 (2022)

19. Place, T., Zeitoun, M.: Separating without any ambiguity. In: Proceedings of the 45th International Colloquium on Automata, Languages, and Programming, pp. 137:1–137:14. ICALP 2018 (2018)

20. Place, T., Zeitoun, M.: Generic results for concatenation hierarchies. Theory Comput. Syst. **63**(4), 849–901 (2019). (selected papers from CSR 2017)

21. Place, T., Zeitoun, M.: Separation and covering for group based concatenation hierarchies. In: Proceedings of the 34th Annual ACM/IEEE Symposium on Logic in Computer Science, pp. 1–13. LICS 2019 (2019)

22. Place, T., Zeitoun, M.: Characterizing level one in group-based concatenation hierarchies. In: Proceeding of the 17th International Computer Science Symposium in Russia. CSR 2022 (2022)

23. Place, T., Zeitoun, M.: Unambiguous polynomial closure explained (2022). 10.48550/ARXIV.2205.12703, arxiv.org/abs/2205.12703

24. Schützenberger, M.P.: Sur le produit de concaténation non ambigu. Semigroup Forum **13**, 47–75 (1976)

25. Simon, I.: Piecewise testable events. In: Brakhage, H.. (ed.) GI-Fachtagung 1975. LNCS, vol. 33, pp. 214–222. Springer, Heidelberg (1975). https://doi.org/10.1007/3-540-07407-4_23

26. Tesson, P., Therien, D.: Diamonds are forever: The variety DA. In: Semigroups, Algorithms, Automata and Languages, Coimbra (Portugal) 2001, pp. 475–500. World Scientific (2002)

27. Thérien, D., Wilke, T.: Over words, two variables are as powerful as one quantifier alternation. In: Proceedings of the 30th Annual ACM Symposium on Theory of Computing. pp. 234–240. STOC 1998, ACM, New York, NY, USA (1998)

28. Thomas, W.: Classifying regular events in symbolic logic. J. Comput. Syst. Sci. **25**(3), 360–376 (1982)

String Attractors and Infinite Words

Antonio Restivo⬭, Giuseppe Romana⬭, and Marinella Sciortino$^{(\boxtimes)}$⬭

Dipartimento di Matematica e Informatica, Università di Palermo, Palermo, Italy
{antonio.restivo,giuseppe.romana01,marinella.sciortino}@unipa.it

Abstract. The notion of *string attractor* has been introduced by Kempa
and Prezza (STOC 2018) in the context of Data Compression and it represents a set of positions of a finite word in which all of its factors can be
"attracted". The smallest size γ^* of a string attractor for a finite word is a
lower bound for several repetitiveness measures associated with the most
common compression schemes, including BWT-based and LZ-based com-
pressors. The combinatorial properties of the measure γ^* have been studied in [Mantaci et al., TCS 2021]. Very recently, a complexity measure,
called *string attractor profile function*, has been introduced for infinite
words, by evaluating γ^* on each prefix. Such a measure has been studied
for automatic sequences and linearly recurrent infinite words in [Scha-
effer and Shallit, arXiv 2021]. In this paper, we study the relationship
between such a complexity measure and other well-known combinatorial
notions related to repetitiveness in the context of infinite words, such
as the factor complexity and the recurrence. Furthermore, we introduce
new string attractor-based complexity measures, in which the structure
and the distribution of positions in a string attractor of the prefixes of
infinite words are considered. We show that such measures provide a
finer classification of some infinite families of words.

Keywords: String attractor · Sturmian word · Recurrent word ·
Morphism · Repetitiveness measure · Factor complexity

1 Introduction

Compressibility and repetitiveness are two fundamental aspects in processing
huge text collections [24]. In many application domains, massive and highly
repetitive data needs to be stored, analysed and queried. The main purpose
of compressed indexing data structures is to store the texts and the structures
needed to handle them by requiring space close to the size of the compressed data
[22]. In such a context, finding good measures able to capture the repetitiveness
of texts is strictly related to having effective parameters to evaluate the performance, both in terms of time and space, of such compressed data structures. For
this reason, some of the most widely used repetitiveness measures are associated with effective compression schemes. For instance, we recall the number z of
phrases in the LZ77 parsing and the number r of equal-letter runs produced by

© Springer Nature Switzerland AG 2022
A. Castañeda and F. Rodríguez-Henríquez (Eds.): LATIN 2022, LNCS 13568, pp. 426–442, 2022.
https://doi.org/10.1007/978-3-031-20624-5_26

the Burrows-Wheeler Transform [23]. In such a framework, Kempa and Prezza proposed in [14] a repetitiveness measure that, instead of being associated with a specific compressor, is related to some combinatorial properties of the text with the aim of unifying existing compressor-based measures. A *string attractor* Γ for a text w is a set of positions in w such each factor of w must have an occurrence crossing some position in Γ. Intuitively, the more repetitive the text, the lower the number of positions of its attractor. The measure $\gamma^*(w)$ is the size of a string attractor of smallest size for w. On the one hand, it has been proved that γ^* is a lower bound to all other compressor-based repetitiveness measure, on the other it is NP-complete to find the smallest attractor size γ^* for a given text w. Combinatorial properties of the measure γ^* for finite words have been explored in [21].

In Combinatorics on words, the notion of repetitiveness has been declined in several ways and under a variety of aspects. For instance, the *factor complexity function* p_x of an infinite word x is a function that counts, for any $n > 0$, the number of distinct factors of length n. Intuitively, the lower the factor complexity, the more repetitive an infinite word is. That is, the most repetitive words one can think of are those obtained by repeating the same factor infinitely many times, i.e. *periodic words*, for which factor complexity takes on a constant value definitively. Among aperiodic words, *Sturmian words* are the infinite words with minimal factor complexity. An infinite word x is *recurrent* if each factor of x occurs infinitely often. The *recurrence function* R_x for an infinite word x, gives for each n, the size of the smallest window containing each factor of x of length n, whatever such a window is located in x. Intuitively, it is strictly related to the maximum gap between successive occurrences of any factor, when all factors of length n are considered. If such a gap is finite for each n, then the word is called *uniformly recurrent*. For the *linearly recurrent words* such a gap grows at most linearly with n.

Very recently, a bridge between these two different approaches has been presented in [26], where the *string attractor profile function* s_x has been introduced. It measures, for each n, the size of a string attractor of smallest size for the prefix of length n of an infinite word x. The behaviour of s_x has been studied when x is linearly recurrent word or an automatic sequence, whose symbols can be defined through a finite automaton [1].

In this paper, we explore the relationship between the string attractor profile function of an infinite word x and the other combinatorial notions of repetitiveness. In particular, we prove that the values that s_x takes for infinitely many n give an upper bound to the factor complexity. On the other hand, we face the problem of searching for the necessary conditions, in terms of repetitiveness combinatorial properties, for the string attractor profile function to take values bounded by a constant. Moreover, we study the behavior of the string attractor profile function for infinite words that are fixed point of a morphism, which represent a mathematical mechanism to generate repetitive words.

Another contribution of this paper is to introduce two new complexity measures based on the notion of string attractor, which allow to obtain a finer classification of some infinite families of words. More in detail, we define the *string attractor span complexity* (denoted by span_x) and the *string attractor leftmost complexity* (denoted by lm_x) of an infinite word x, which are related for each $n > 0$ to the distribution of the positions within a string attractor of the prefix of x of length n. These measures make it possible to distinguish infinite words that are indistinguishable under the action of the string profile function. In addition to exploring the relations between such measures and the periodicity and recurrence properties of an infinite word, we consider the class of infinite words for which the span complexity takes on a constant value infinitely many times. This allows us to obtain a new characterization of Sturmian words that are the infinite words with span complexity function equal to 1 for infinitely many n. More in general, we prove that if the span complexity span_x takes a constant value for each $n > 0$, the aperiodic infinite word x is a quasi-Sturmian word. Quasi-Sturmian words represent the simplest generalization of Sturmian words in terms of factor complexity.

2   Preliminaries

Let $\Sigma = \{a_1, a_2, \ldots, a_\sigma\}$ be a finite alphabet. We denote by Σ^* the set of finite words over Σ. An infinite word $x = x_1 x_2 \ldots$ is an infinite sequence of characters in Σ. Given a finite word $w = w_1 w_2 \cdots w_n$, we denote with $|w| = n$ the length of the word. The *empty-word* is denoted by ε. The *reverse* of a word w is the word read from right to left, that is $w^R = w_n w_{n-1} \cdots w_1$. A finite word v is called *factor* of a word x (finite or infinite) if there exist $i, j > 0$ such that $j - i + 1 = |v|$ and $x[i, j] = x_i x_{i+1} \cdots x_j = v$. We assume that $x[i, j] = \varepsilon$ if $j < i$. We denote by $F(x)$ the set of all factors of x. The word u is a *prefix* (resp. *suffix*) of x if $x = uy$ (resp. $x = yu$) for some word y. A factor u of x is *right special* if there exist $a, b \in \Sigma$ with $a \neq b$ such that both ua and ub are factors of x.

String Attractor of a Finite Word. A string attractor for a word w is a set of positions in w such that all distinct factors of w have an occurrence *crossing* at least one of the attractor's elements. More formally, a *string attractor* of a finite word $w \in \Sigma^n$ is a set of γ positions $\Gamma = \{j_1, \ldots, j_\gamma\}$ such that every factor $w[i, j]$ of w has an occurrence $w[i', j'] = w[i, j]$ with $j_k \in \{i', i' + 1, \ldots, j'\}$, for some $j_k \in \Gamma$. We denote by $\gamma^*(w)$ the size of a smallest string attractor for w. We denote by $\mathsf{alph}(w)$ the set of the characters of Σ appearing in w, i.e. $F(w) \cap \Sigma$. It is easy to see that $\gamma^*(w) \geq |\mathsf{alph}(w)|$.

Example 1. Let $w = adcbaadcbadc$ be a word on the alphabet $\Sigma = \{a, b, c, d\}$. The set $\Gamma = \{1, 4, 6, 8, 11\}$ is a string attractor for w. Note that the set $\Gamma' = \{4, 6, 8, 11\}$ obtained from Γ by removing the position 1 is still a string attractor for w, since all the factors that cross position 1 have a different occurrence that crosses a different position in Γ. The positions of Γ' are underlined in

$$w = adc\underline{b}a\underline{a}d\underline{c}ba\underline{d}c.$$

Γ' is also a smallest string attractor since $|\Gamma'| = |\Sigma|$. Then $\gamma^*(w) = 4$. Remark that the sets $\{3, 4, 5, 11\}$ and $\{3, 4, 6, 7, 11\}$ are also string attractors for w. It is easy to verify that the set $\Delta = \{1, 2, 3, 4\}$ is not a string attractor since, for instance, the factor aa does not intersect any position in Δ.

Factor Complexity. Let x be an infinite word. The *factor complexity function* p_x of x counts, for any positive integer n, all the distinct factors of x of length n, i.e. $p_x(n) = |F(x) \cap \Sigma^n|$.

Periodicity. Given a word x, a natural number $p > 0$ is called *period* of x if $x_i = x_j$ when $i \equiv j \mod p$. An infinite word x is called *ultimately periodic* if there exist $u \in \Sigma^*$ and $v \in \Sigma^+$ such that $x = uv^\omega$, i.e. x is the concatenation of u followed by infinite copies of a non-empty word v. If $u = \varepsilon$, then x is called *periodic*. An infinite word is *aperiodic* if it is not ultimately periodic.

Recurrence and Appearance Functions. An infinite word x is said to be *recurrent* if every factor that occurs in x occurs infinitely often in x. The *recurrence function* $R_x(n)$ gives, for each n, the least integer m (or ∞ if no such m exists) such that every block of m consecutive symbols in x contains at least an occurrence of each factor of x of length n. An infinite word x is *uniformly recurrent* if $R_x(n) < \infty$ for each $n > 0$. If $R_x(n)$ is linear, x is called *linearly recurrent*. It is easy to see that an ultimately periodic word $x = uv^\omega$ with $u \neq \varepsilon$ is not recurrent. On the other hand, if x is periodic (the case $u = \varepsilon$) then x is linearly recurrent. Therefore, a recurrent word is either aperiodic or periodic. Given an infinite word x, $A_x(n)$ denotes the length of the shortest prefix containing all the factors of x of length n. The function $A_x(n)$ is called *appearance function* of x.

Remark 1. It is known that $A_x(n) \leq R_x(n)$ (see [1]). Moreover, for any infinite word x and for each $n > 0$, since $|\Sigma|$ is finite, $A_x(n)$ is always defined and $A_x(n) \geq p_x(n) + n - 1 = \Omega(n)$.

Power Freeness. An infinite word x is said k-*power free*, for some $k > 1$, if for every factor w of x, w^k is not a factor of x. If for every factor w of x there exists k such that w^k is not a factor of x, then x is called ω-*power free*.

Morphisms. They represent a very interesting way to generate an infinite family of words. Let Σ and Σ' be alphabets. A *morphism* is a map φ from Σ^* to Σ'^* that obeys the identity $\varphi(uv) = \varphi(u)\varphi(v)$ for all words $u, v \in \Sigma^*$. A morphism φ is called *prolongable* on a letter $a \in \Sigma$ if $\varphi(a) = au$ with $u \in \Sigma^+$. If for all $a \in \Sigma$ holds that $\varphi(a) \neq \varepsilon$, then the morphism φ is called *non-erasing*. Given a non-erasing morphism φ prolongable on some $a \in \Sigma$, the infinite family of finite words $\{a, \varphi(a), \ldots, \varphi^i(a), \ldots\}$ are prefixes of a unique infinite word $\varphi^\infty(a) = \lim_{i \to \infty} \varphi^i(a)$, that is called *purely morphic word* or *fixed point of* φ. A morphism φ is called *primitive* if exists $t > 0$ such that $b \in F(\varphi^t(a))$, for every pair of symbols $a, b \in \Sigma$. If exists k such that $|\varphi(a)| = k$ for every $a \in \Sigma$, then the morphism is called k-*uniform*.

String Attractor Profile Function. Let x be an infinite word. For any $n > 0$, we denote by $s_x(n)$ the size of a smallest string attractor for the prefix of x of length n. The function s_x is called *string attractor profile function* of x. This notion has been introduced in [26].

Example 2. Let us consider the Thue-Morse word

$$t = 0110100110010110 \cdots,$$

that is the fixed point of the morphism $0 \mapsto 01$, $1 \mapsto 10$. It has been proved in [26] (cf. also [17]) that $s_t(n) \leq 4$ for any $n > 0$. Moreover, it is known that the functions $p_t(n)$, $R_t(n)$ and $A_t(n)$ are $\Theta(n)$. See [1] for details.

3   String Attractor Profile Function, Factor Complexity and Recurrence

In this section we explore the relationships among different functions that aim to measure the repetitiveness of factors within infinite sequences of symbols.

The following theorem extends to infinite words a result proved in [8, Lemma 5.6] that states the measure δ, defined using the factor complexity on finite words, is a lower bound for the measure γ^*. Here, we establish a relationship among appearance, factor complexity and string attractor profile functions. In particular, in the following theorem we show that upper bounds on s_x can induce upper bounds on p_x.

Theorem 1. *Let x be an infinite word. For all $n > 0$, one has*

$$p_x(n) \leq n \cdot s_x(A_x(n)).$$

Proof. Let us consider the value $A_x(n)$ representing the length of the smallest prefix of x containing all the factors of x of length n. Since the alphabet is finite, the value $A_x(n)$ is finite. By definition $s_x(A_x(n))$ is the size of the smallest string attractor of the prefix of length $A_x(n)$. Therefore, each factor of x of length n crosses at least one element of the string attractor. Since each element of the string attractor is crossed by at most n distinct factors of x of length n, one has $p_x(n) \leq n \cdot s_x(A_x(n))$.   □

From previous theorem, the following corollary can be deduced.

Corollary 1. *Let x be an infinite word. If there exists $k > 0$ such that $s_x(n) < k$ for each $n > 0$, then $p_x(n) \leq n \cdot k$.*

In other words, Corollary 1 states that if an infinite word has the string attractor profile function bounded by some constant value, then it has at most linear factor complexity. We know that, in general, the converse of Corollary 1 is not true. In fact, there are infinite words x such that the factor complexity is linear and $s_x(n)$ is not bounded. For instance, in Example 3 we consider the characteristic sequence c of the powers of 2.

Example 3. Let us consider the characteristic sequence c of the powers of 2, i.e. $c_i = 1$ if $i = 2^j$ for some $j \geq 0$, 0 otherwise.

$$c = 1101000100000001 \cdots .$$

It is easy to see that c is aperiodic and not recurrent because the factor 11 has just one occurrence. It is known that $p_c(n)$ and $A_c(n)$ are $\Theta(n)$ ([1]). One can prove that $s_c(n) = \Theta(\log n)$ ([16,21,26]).

We raise the following:

Question 1. Let x be an uniformly recurrent word such that p_x is linear. Is $s_x(n)$ bounded by a constant value?

Remark that, by assuming a stronger hypothesis on the recurrence of x, a positive answer to Question 1 can be given, as stated in the following theorem proved in [26]. Such a result can be applied to describe the behaviour of the string profile function $s_t(n)$ for the Thue-Morse word t, as shown in Example 2.

Theorem 2 ([26]). *Let x be an infinite word. If x is linearly recurrent (i.e. $R_x(n) = \Theta(n)$), then $s_x(n) = \Theta(1)$.*

The following proposition shows that also in case of ultimately periodic words, the string attractor profile function is bounded by a constant value.

Proposition 1. *Let x be an infinite word. If x is ultimately periodic, then $s_x(n) = \Theta(1)$.*

Proof. Let $u \in \Sigma^*$ and $v \in \Sigma^+$ such that $x = uv^\omega$. Since every periodic word is linearly recurrent, if $u = \varepsilon$ by Theorem 2 the thesis holds. If $u \neq \varepsilon$, then for every $n > |u|$ we can use a bound on the size γ^* with respect to the concatenation provided in [21], which says that for any $u, v \in \Sigma^+$, it holds that $\gamma^*(uv) \leq \gamma^*(u) + \gamma^*(v) + 1$. Therefore, $s_x(n) \leq \gamma^*(u) + s_{v^\omega}(n - |u|) + 1 \leq |u| + k' + 1$. On the other hand, for all the prefixes of length $n \leq |u|$ it holds that $s_x(n) \leq |u| < |u| + k' + 1$. Since $|u|$ and k' are constant, we can choose $k = |u| + k' + 1$ and the thesis follows. $\square$

An interesting upper bound on the function s_x can be obtained by assuming that the appearance function is linear, as shown in [26] and reported in the following theorem.

Theorem 3 ([26]). *Let x be an infinite word. If $A_x(n) = \Theta(n)$, then $s_x(n) = O(\log n)$.*

On the other hand, if the function s_x is bounded by some constant value, the property of power freeness can be deduced, as proved in the following proposition.

Proposition 2. *Let x be an infinite word. If $s_x(n) = \Theta(1)$, then x is either ultimately periodic or ω-power free.*

Proof. If x is ultimately periodic, then by Proposition 1 $s_x(n) = \Theta(1)$. So, let us assume x is aperiodic. By contraposition, suppose x is not ω-power free. Then there exists a factor w of x such that, for every $q > 0$, w^q is factor of x. Moreover, $x \neq uw^\omega$ for every $u \in \Sigma^*$, otherwise x would be ultimately periodic. It follows that we can write $x = v_0 \cdot \prod_{i=1}^{\infty} w^{q_i} v_i$, with $v_0 \in \Sigma^*$, and, for every $i \geq 1$, $q_i > 0$ and $v_i \in \Sigma^+$ such that v_i does not have w neither as prefix nor as suffix. Observe that there exist infinitely many distinct factors of the form $v_j w^{q_j} v_{j+1}$ for some $j \geq 0$ and for each of these distinct factors we have at least one position in the string attractor. Thus, for every $k > 0$ exists $n > 0$ such that $s_x(n) > k$ and the thesis follows.   ☐

On the other hand, the converse of Proposition 2 is not true for ω-power free words. Such a result leads to the formulation of the following Question 2. Note that a positive answer to Question 2 implies a positive answer to Question 1:

Question 2. Let x be ω-power free word such that p_x is linear. Is $s_x(n)$ bounded by a constant value?

The following examples show that for many infinite words known in literature the string attractor profile function is not bounded by a constant. So, it could be interesting to study its behaviour. In particular, Example 4 shows that there exist recurrent (not uniformly) infinite words x such that the function s_x is unbounded. However, one can find a uniformly recurrent infinite word t such that s_t is unbounded, as shown in Example 5 .

Example 4. Let $n_0, n_1, n_2, n_3, \ldots$ be an increasing sequence of positive integers. Let us define the following sequence of finite words: $v_0 = 1$, $v_{i+1} = v_i 0^{n_i} v_i$, for $i > 0$. Let us consider $v = \lim_{i \to \infty} v_i$. It is possible to verify that v is recurrent, but not uniformly, and s_v is unbounded.

Example 5. Toeplitz words are infinite words constructed by an iterative process, specified by a Toeplitz pattern, which is a finite word over the alphabet $\Sigma \cup \{?\}$, where ? is a distinguished symbol not belonging to Σ [4]. Let us consider the alphabet $\Sigma = \{1, 2\}$ and the pattern $p = 12???$. The Toeplitz word z (also called $(5, 3)$-Toeplitz word) is generated by the pattern p by starting from the infinite word p^ω, obtained by repeating p infinitely. Next, each ? is replaced by a single symbol from p^ω, and so forth. So,

$$z = 121211221112221121121222112121121211222212112\cdots.$$

It is known that all Toeplitz words are uniformly recurrent and, as shown in [4], $p_z(n) = \Theta(n^r)$ with $r = (\log 5)/(\log 5 - \log 3) \approx 3.15066$. By applying Corollary 1, we can deduce that s_z is unbounded.

On the other hand, in support of the fact that $s_x(n)$ can be bounded by a constant value by using weaker assumptions than those of Theorem 2, we can show there exist uniformly (and not linearly) recurrent words for which $s_x(n)$ is bounded. A large class of examples is represented by some Sturmian words, as shown in Sect. 6.

All the infinite words considered in the paper, with information on string attractor profile function, factor complexity and recurrence properties, are summarized in Fig. 1.

Infinite word x	$p_x(n)$	Recurrence	$s_x(n)$
Period-doubling word p (Ex. 7)	$\Theta(n)$	Linearly recurrent	2
Thue-Morse word t (Ex. 2)	$\Theta(n)$	Linearly recurrent	4
Charact. Sturmian word s (Thm. 7)	$\Theta(n)$	Uniformly recurrent	2
Power of 2 charact. sequ. c (Ex. 3)	$\Theta(n)$	Not recurrent	$\Theta(\log n)$
$(5,3)$-Toeplitz word z (Ex. 5)	$\Theta(n^{\frac{\log 5}{\log 5 - \log 3}})$	Uniformly recurrent	Not constant

Fig. 1. Factor complexity function p_x, recurrence, and string attractor profile function s_x for some infinite words.

Finally, we pose the problem of what values the string attractor profile function can assume, and in particular, whether an upper bound exists for these values. We therefore prove the following proposition.

Proposition 3. *Let x be an infinite word. Then $s_x(n) = O(\frac{n}{\log n})$.*

Proof. The proposition can be proved by combining results from [14] and [18]. In fact, in [14] it has been proved that, for a given finite word, there exists a string attractor of size equal to the number z of phrases of its LZ77 parsing. In [18] it has been proved that an upper bound on z for a word of length n is $\frac{n}{(1-\epsilon_n)\log_\sigma n}$, where $\epsilon_n = 2\frac{1+\log_\sigma(\log_\sigma(\sigma n))}{\log_\sigma n}$ and σ is the size of the finite alphabet. $\square$

We wonder if the bound of Proposition 3 is tight, i.e. if there exists an infinite word x such that $s_x = \Theta(\frac{n}{\log n})$ for each $n \geq n_0$, for some positive n_0. Certainly, it is possible to construct an infinite word x for which there exists a sub-sequence of positive integers n_i, for $i > 0$, such that $s_x(n_i) = \Theta(\frac{n_i}{\log n_i})$. For instance, such a word x can be constructed by using a suitable sequence of de Brujin words. However, having information about the values of the string attractor profile function on a sub-sequence n_i does not allow us to determine its behavior for the remaining values of n.

4 String Attractor Profile Function on Purely Morphic Words

In this section, we consider the behavior of string attractor profile function for an infinite word x, when it is a fixed point of a morphism. Note that morphisms represents an interesting mechanism to generate infinite families of repetitive sequences, which has many mathematical properties ([1,2,11]). Some repetitiveness measures have been explored when applied to words x generated by morphisms. In [12] the number r of BWT equal-letter runs has been studied for all prefixes obtained by iterating a morphism. In [9] the measure $z_x(n)$ that gives the number z of phrases in the LZ77 parsing of the prefix $x[1,n]$ has been studied. It has been proved that both z and r are upper bound for the measure γ^*, when they are applied to finite words. The bounds on the measure z proved in [9] can be used to prove the following theorem.

Theorem 4. Let $x = \varphi^\infty(a)$ be the fixed point of a morphism φ prolongable on $a \in \Sigma$. Then, $s_x(n) = O(i)$, where i is such that $|\varphi^i(a)| \le n < |\varphi^{i+1}(a)|$.

In the following, we provide a finer result in the case of binary purely morphic word.

Theorem 5. Let $x = \mu^\infty(a)$ be the binary fixed-point of a morphism $\mu : \{a,b\}^* \mapsto \{a,b\}^*$ prolongable on a. Then, either $s_x(n) = \Theta(1)$ or $s_x(n) = \Theta(\log n)$, and it is decidable when the first or the latter occurs.

Proof. If x is ultimately periodic, then by Proposition 1 follows that $s_x(n) = \Theta(1)$. Suppose now x is aperiodic. For morphisms defined on a binary alphabet, it holds that if $x = \mu^\infty(a)$ is aperiodic, then $|\mu^i(a)|$ grows exponentially with respect to i (see [12]). Moreover, if μ is primitive, then by [10, Theorem 1] and [1, Theorem 10.9.4] x is linearly recurrent, and by Theorem 2 we have that $s_x(n) = \Theta(1)$. If μ is not primitive, as summed up in [12], then only one of the following cases occurs: (1) there exist a coding $\tau : \Sigma \mapsto \{a,b\}^+$ and a primitive morphism $\varphi : \Sigma^* \mapsto \Sigma^*$ such that $x = \mu^\infty(a) = \tau(\varphi^\infty(a))$ [25]; (2) x contains arbitrarly large factors on $\{b\}^*$. For case (1), since τ preserves the recurrence of a word and that $\varphi^\infty(a)$ is linearly recurrent, then x is linearly recurrent as well, and by Theorem 2 $s_x(n) = \Theta(1)$. For case (2), one can notice that x is not ω-power free, and by Proposition 2 for every $k > 0$ exists n' such that $s_x(n) > k$, for every $n \ge n'$. More in detail, the number of distinct maximal runs of b's grows logarithmically with respect to the length of the prefixes of x [12], i.e. $s_x(n) = \Omega(\log n)$. On the other hand, by Theorem 4 we know that $s_x(n) = O(i)$, where $i > 0$ is such that $|\mu^i(a)| \le n < |\mu^{i+1}(a)|$. Since $i = \Theta(\log n)$, we can further deduce an upper bound for the string attractor profile function and it follows that $s_x(n) = \Theta(\log n)$. Finally, from a classification in [12] we can decide, only from μ, if either $s_x(n) = \Theta(1)$ or $s_x(n) = \Theta(\log n)$.   □

Note that the result of Theorem 5 does not contradict a possible positive answer to the Questions 1 and 2, because the infinite words x with linear factor complexity and such that $s_x(n) = \Theta(\log n)$ are not ω-power free. Moreover, the same bounds of Theorem 5 have been obtained for a related class of words, i.e. the automatic sequences, as reported in the following theorem. In short, an infinite word x is k-*automatic* if and only if there exists a coding $\tau : \Sigma \mapsto \Sigma$ and a k-uniform morphism μ_k such that $x = \tau(\mu_k^\infty(a))$, for some $a \in \Sigma$ ([1]).

Theorem 6 ([26]). *Let x be a k-automatic infinite word. Then, either $s_x(n) = \Theta(1)$ or $s_x(n) = \Theta(\log n)$, and it is decidable when the first or the latter occurs.*

5   New String Attractor-Based Complexities

In this section we introduce two new string attractor-based complexity measures, called *span complexity* and *leftmost complexity*, that allow us to obtain a finer classification for infinite families of words that takes into account the distribution of positions in a string attractor of each prefix of an infinite word. Examples 7 and 8 show two infinite words, Period-Doubling word and Fibonacci word, which are not distinguishable if we consider their respective string attractor profile function. In fact, they are point by point equal to 2, definitively. The situation is very different if we look at how the positions within a string attractor are arranged.

Span and Leftmost String Attractor of a Finite Word. Let w be a a finite word and let $\mathcal{G}$ be set of all string attractors $\Gamma = \{\delta_1, \delta_2, \ldots, \delta_\gamma\}$ for w, with $\delta_1 < \delta_2 < \ldots < \delta_\gamma$ and $1 \le \gamma \le |w|$. We define *span* of a finite word the value $\mathsf{span}(w) = \min_{\Gamma \in \mathcal{G}}\{\delta_\gamma - \delta_1\}$. In other words, $\mathsf{span}(w)$ gives the minimum distance between the rightmost and the leftmost positions of any string attractor for w. Moreover, given two string attractors Γ_1 and Γ_2, we say that Γ_1 is more to the left of Γ_2 if the rightmost position of Γ_1 is smaller than the rightmost position of Γ_2. Then, we define $\mathsf{lm}(w) = \min_{\Gamma \in \mathcal{G}}\{\delta_\gamma \in \Gamma\}$. Any $\Gamma \in \mathcal{G}$ such that $\delta_\gamma = \mathsf{lm}(w)$ is called *leftmost string attractor* for w.

Example 6. Let us consider the word $w = \overline{abc}\overline{c}\underline{abc}$. One can see that the sets $\Gamma_1 = \{4, 5, 6\}$ (underlined positions) and $\Gamma_2 = \{1, 2, 4\}$ (overlined positions) are two suitable string attractors for w. Even if both string attractors are of smallest size ($|\Gamma_1| = |\Gamma_2| = |\Sigma|$), only the set Γ_1 is of minimum span, since all of its positions are consecutive, and therefore $\mathsf{span}(w) = 6 - 4 = 2$. On the other hand, one can see that $\max\{\Gamma_2\} < \max\{\Gamma_1\}$. Moreover, one can notice that the set $\Delta = \{1, 2, 3\}$ is not a string attractor for w, and therefore $\mathsf{lm}(w) = \max\{\Gamma_2\} = 4$.

Example 6 shows that for a finite word w, these two measures can be obtained by distinct string attractors. In fact, the set $\{2, 3, 4\}$ is not a string attractor for $w = abccabc$, hence it does not exists $\Gamma'(w) = \{\delta_1, \delta_2, \ldots, \delta_{\gamma'}\} \in \mathcal{G}$ such that $\delta_\gamma = 4$ and $\delta_{\gamma'} - \delta_1 = 2$.

The value $\mathsf{span}(w)$ can be used to derive an upper-bound on the number of distinct factors of w, as shown in the following lemma. Such a result will be used to find upper bounds on the factor complexity of an infinite word.

Lemma 1. *Let w be a finite word. Then, for all $0 < n \le |w|$, it holds that $|F(w) \cap \Sigma^n| \le n + span(w)$.*

Proof. Let $\Gamma = \{\delta_1, \delta_2, \ldots, \delta_\gamma\}$ be a string attractor for w such that $\delta_\gamma - \delta_1 = span(w)$. Then, the superset $X = \{i \in \mathbb{N} \mid \delta_1 \le i \le \delta_\gamma\}$ of Γ is a string attractor for w as well. Since every factor has an occurrence crossing a position in X, it is possible to find all factors in $F(w) \cap \Sigma^n$ by considering a sliding window of length n, starting at position $\max\{\delta_1 - n + 1, 1\}$ and ending at $\min\{\delta_\gamma, |w| - n + 1\}$. One can see that this interval is of size at most $\delta_\gamma - (\delta_1 - n + 1) + 1 = \delta_\gamma - \delta_1 + n = n + span(w)$ and the thesis follows. $\square$

The following proposition shows upper bounds for the measures γ^*, span and Im, when a morphism is applied to a finite word w.

Proposition 4. *Let $\varphi : \Sigma^* \mapsto \Sigma'^*$ be a morphism. Then, there exists $K > 0$ which depends only from φ such that, for every $w \in \Sigma^+$, it holds that:*

1. $\gamma^*(\varphi(w)) \le 2\gamma^*(w) + K$;
2. $span(\varphi(w)) \le K \cdot span(w)$;
3. $Im(\varphi(w)) \le K \cdot Im(w)$.

Span Complexity and Leftmost Complexity. The following measures take into account the distribution of the positions within a string attractor for each prefix of an infinite word x. More in detail, we define the *string attractor span complexity* (or simply *span complexity*) of an infinite word x as $span_x(n) = span(x[1, n])$. We also introduce the *string attractor leftmost complexity* (or simply *leftmost complexity*) of an infinite word x, defined as $Im_x(n) = Im(x[1, n])$. Example 7 shows the behaviour of such measures when the period-doubling word is considered. Proposition 5 shows the relationship between the measures s_x, $span_x$ and Im_x.

Example 7. Let us consider the period-doubling sequence

$$pd = 101110101011 \cdots,$$

that is the fixed point of the morphism $1 \mapsto 10$, $0 \mapsto 11$. It has been proved in [26] that $s_{pd}(n) = 2$ for any $n > 1$, while $span_x(n) = 1$ when $1 < n \le 5$, and $span_x(n) = 2^i$ when $3 \cdot 2^i \le n < 3 \cdot 2^{i+1}$ and $i \ge 1$.

Proposition 5. *Let x be an infinite word. Then,*

$$s_x(n) - 1 \le span_x(n) \le Im_x(n).$$

Proof. Let $\Gamma = \{\delta_1, \delta_2, \ldots, \delta_\gamma\}$ be a leftmost string attractor, i.e. $\delta_\gamma = Im_x(n)$. It is possible to check that $Im_x(n) = \delta_\gamma \ge \delta_\gamma - \delta_1 \ge span_x(n)$. Let $\mathcal{G}$ be the set of all string attractors for $x[1..n]$ and let $\Gamma' = \{\lambda_1, \ldots, \lambda_{\gamma'}\} \in \mathcal{G}$ be a string attractor such that $\lambda_{\gamma'} - \lambda_1 = span_x(n)$. Recall that, for every string attractor $\Gamma' \in \mathcal{G}$ and a set X such that $\Gamma' \subseteq X$, it holds that $X \in \mathcal{G}$ as well. Then, the set $X = \{i \in \mathbb{N} \mid \lambda_1 \le i \le \lambda_{\gamma'}\}$ is a string attractor for $x[1..n]$. Finally, $span_x(n) = \lambda_{\gamma'} - \lambda_1 = |X| - 1 \ge s_x(n) - 1$ and the thesis follows. $\square$

The following two propositions show that the boundedness of the two new complexity measures here introduced can be related to some properties of repetitiveness for infinite words, such as periodicity and recurrence.

Proposition 6. *Let x be an infinite word. x is ultimately periodic if and only if there exists $k > 0$ such that $\mathsf{lm}_x(n) \leq k$, for infinitely many $n > 0$.*

Proof. First we prove the first implication. If x is ultimately periodic, then there exist $u \in \Sigma^*$ and $v \in \Sigma^+$ such that $x = uv^\omega$. Observe that, for any $n \geq |uv|$, the set $\Gamma = \{i \in \mathbb{N} \mid 1 \leq i \leq |uv|\}$ is a string attractor for $x[1..n]$, since every factor that starts in uv is clearly covered, and every factor that lies within two or more consecutive v's has another occurrence starting in the first v. It follows that we can pick $k = |uv|$ such that $\mathsf{lm}_x(n) \leq k$ for every $n > 0$.

We now show the other direction of the implication. By hypothesis, for all $n > 0$ there exists $n' \geq n$ and a set Γ' such that $\Gamma' = \{\delta_1, \delta_2, \ldots, \delta_{\gamma'}\}$ is a string attractor for $x[1..n']$, with $\delta_1 < \delta_2 < \ldots < \delta_{\gamma'} \leq k$. Hence, also the superset $\Gamma'' = \{i \in \mathbb{N} \mid 1 \leq i \leq \min\{n', k\}\}$ is a string attractor for $x[1, n']$. One can notice that, for each $n' > k$, Γ'' can capture at most k distinct factors of length n, i.e. one factor starting at each position of Γ''. Therefore, for all $n' > k$ we have that $p_x(n') \leq k = \Theta(1)$. One can observe that for each $n > 0$ there exists $n' \geq n$ such that $p_x(n') = \Theta(1)$, and from the monotonicity of the factor complexity we have $p_x(n) \leq p_x(n') = \Theta(1)$, and the thesis follows. ☐

Proposition 7. *Let x be an infinite word. If there exists $k > 0$ such that $\mathsf{span}_x(n) \leq k$ for infinitely many n, then x is ultimately periodic or recurrent.*

Proof. Let x be an ultimately periodic word. By Propositions 6 and 5, there exists $k > 0$ such that $k \geq \mathsf{lm}_x(n) \geq \mathsf{span}_x(n)$, for every $n > 0$. So, let us suppose that x is aperiodic, and by contraposition assume that x is not recurrent. Then, for a sufficiently large value n', there exists a factor u of $x[1, n']$ that occurs exactly once in x. It follows that in order to cover the factor u, any suitable string attractor $\Gamma(x[1, n])$ with $n > n'$ must have its first position $\delta_1 \leq n'$. Let us consider then all the prefixes of length $n > n'$ (ignoring a finite set does not affect the correctness of the proof). From Proposition 6 one can observe that x being aperiodic implies that, for each $k > 0$, we can find only a finite number of $n > 0$ such that $k > \mathsf{lm}_x(n)$. In other terms, any string attractor of a prefix of x ultimately has the first position bounded above by the constant value n', while $\mathsf{lm}_x(n)$ must grow after the concatenation of a finite number of symbols and the thesis follows. ☐

6   Words with Constant Span Complexity

In this section, we consider infinite words for which the span complexity measure takes a constant value for infinite points. By using Proposition 7, we know that, under this assumption, an infinite word is either ultimately periodic or recurrent. In this section we focus our attention on aperiodic words by showing that different constant values for the span complexity characterize different infinite families of words.

Sturmian Words. They are very well-known combinatorial objects having a large number of mathematical properties and characterizations. Sturmian words have also a geometric description as digitized straight lines [19, Chapt.2]. Among aperiodic binary infinite words, they are those with minimal factor complexity, i.e. an infinite word x is a *Sturmian word* if $p_x(n) = n + 1$, for $n \geq 0$. Moreover, Sturmian words are uniformly recurrent. An important class of Sturmian words is that of *Characteristic Sturmian words*. A Sturmian word x is *characteristic* if both $0x$ and $1x$ are Sturmian words. An important property of characteristic Sturmian words is that they can be constructed by using finite words, called *standard Sturmian words*, defined recursively as follows. Given an infinite sequence of integers $(d_0, d_1, d_2, \ldots)$, with $d_0 \geq 0, d_i > 0$ for all $i > 0$, called a *directive sequence*, $x_0 = b, x_1 = a, x_{i+1} = x_i^{d_i-1} x_{i-1}$, for $i \geq 1$. A characteristic Sturmian word is the limit of a infinite sequence of standard Sturmian words, i.e. $x = \lim_{i \to \infty} x_i$. Standard Sturmian words are finite words with many interesting combinatorial properties and appear as extremal case for several algorithms and data structures [5–7, 15, 20, 27].

The following theorem shows that each prefix of a characteristic Sturmian word has a smallest string attractor of span 1, i.e. consisting of two consecutive positions.

Theorem 7. *Let x be a characteristic Sturmian word and let $x_0, x_1, \ldots, x_k, \ldots$ be the sequence of standard Sturmian words such that $x = \lim_{k \to \infty} x_k$. Let $\overline{n}$ be the smallest positive integer such that $\mathsf{alph}(x[1..\overline{n}]) = 2$. Then, $s_x(n) = 2$ and $\mathsf{span}_x(n) = 1$ for $n \geq \overline{n}$. In particular, a string attractor for $x[1, n]$ is given by*

$$
\Gamma_n = \begin{cases} \{1\}, & \text{if } n < \overline{n}; \\ \{|x_{k'-1}| - 1, |x_{k'-1}|\}, & \text{if } |x_{k'}| \leq n \leq |x_{k'}| + |x_{k'-1}| - 2; \\ \{|x_{k'}| - 1, |x_{k'}|\}, & \text{if } |x_{k'}| + |x_{k'-1}| - 1 \leq n < |x_{k'+1}| \end{cases}
$$

where k' is the greatest integer $k \geq 2$ such that x_k (with $|x_k| \geq \overline{n}$) is prefix of $x[1, n]$. Moreover, Γ_n is the leftmost string attractor of $x[1, n]$.

Example 8. Consider the infinite Fibonacci word $x = abaababaabaababaababa\ldots$ that is a characteristic Sturmian word with directive sequence $1, 1, \ldots, 1, \ldots$.

In Fig. 2 are shown the first prefixes of x of length n and their respective leftmost string attractor Γ_n, with $n \geq 2$.

The following proposition shows that there is a one-to-one correspondence between each characteristic Sturmian word and the sequence of the leftmost string attractors of its prefixes.

Proposition 8. *Let x be a characteristic Sturmian word and, for each $n \geq 1$, let Γ_n be the string attractor of the prefix $x[1, n]$ defined in Theorem 7. Let y be*

$$\begin{array}{ll}
\mathbf{x[1]} = \underline{a} & \varGamma_1 = \{1\} \\
\mathbf{x[1,2]} = \underline{ab} & \varGamma_2 = \{1,2\} \\
\mathbf{x[1,3]} = \underline{ab}a & \varGamma_3 = \{1,2\} \\
x[1,4] = ab\underline{a}a & \varGamma_4 = \{2,3\} \\
\mathbf{x[1,5]} = ab\underline{aa}b & \varGamma_5 = \{2,3\} \\
x[1,6] = ab\underline{aa}ba & \varGamma_6 = \{2,3\} \\
x[1,7] = abaa\underline{ab}ab & \varGamma_7 = \{4,5\} \\
\mathbf{x[1,8]} = abaa\underline{ab}aba & \varGamma_8 = \{4,5\}
\end{array}$$

Fig. 2. Prefixes of the Fibonacci word x of length up to 8 and their leftmost string attractor $\varGamma_n$. For Fibonacci words we have $\bar{n} = 2$. The underlined positions in $x[1,n]$ correspond to those in $\varGamma_n$, while the prefixes in bold are standard Sturmian words.

a characteristic Sturmian word such that $\varGamma_n$ is the leftmost string attractor for $y[1,n]$ for any $n \geq 1$. Then, $x = y$ (up to exchanging a and b).

Remark 2. There are non-characteristic Sturmian words such that some of their prefixes do not admit any string attractor of span 1. For instance, let $x = aaaaaabaaaaaabaaaaaaab\ldots$ be the characteristic Sturmian word obtained by the directive sequence $(6, 2, \ldots)$. Consider the non-characteristic Sturmian word x' such that $x = aaaa \cdot x'$, hence $x' = aabaaaaaabaaaaaaab\ldots$. Let us consider the prefix $x'[0,13] = aabaaaaaaabaaaa$. Since the b's occur only at positions 2 and 9 and the factor $aaaaaa$ only in $x'[3,8]$, the candidates as string attractor with two consecutive positions are $\Delta_1 = \{2,3\}$ and $\Delta_2 = \{8,9\}$. However, one can check that the factors $aaab$ and $baaaaa$ do not cross any position in Δ_1 and Δ_2 respectively. Nonetheless, there exists a string attractor of size 2 that does not contain two consecutive positions, that is $\varGamma = \{3,9\}$.

The following theorem shows that a new characterization of Sturmian words can be obtained in terms of span of the prefixes.

Theorem 8. *Let x be an infinite aperiodic word. Then, x is Sturmian if and only if $\mathsf{span}_x(n) = 1$ for infinitely many $n > 0$.*

Proof. Observe that every Sturmian word x has an infinite number of right special factors as prefixes, as for every aperiodic and uniformly recurrent word. Moreover, for every right special factor v of a Sturmian word, there is a characteristic Sturmian word s with v^R as prefix [19, Proposition 2.1.23]. Since for every string $v \in \Sigma^*$ and every string attractor $\varGamma(v)$ of v it holds that the set $\varGamma(v^R) = \{n - i - 1 \mid i \in \varGamma(v)\}$ is a suitable string attractor of v^R [21], and from Theorem 7 we know that $\mathsf{span}_s(n) = 1$ for every prefix of every characteristic Sturmian word s, it follows that exist infinite prefixes v of x such that $\mathsf{span}_x(|v|) = 1$, that is our thesis.

For the other direction of the implication, recall that an infinite word x is aperiodic if and only if $p_x(k) \geq k + 1$ for all $k > 0$. Moreover, by hypothesis for every $n > 0$ exists $n' > n$ such that $\mathsf{span}_x(n') = 1$. It follows that $|F(x[1,n]) \cap \Sigma^k| \leq |F(x[1,n']) \cap \Sigma^k| \leq n + \mathsf{span}_x(n') = n + 1$ for every $n > 0$, and therefore x is Sturmian. □

Quasi-Sturmian Words. Let us consider now the *quasi-Sturmian* words, defined in [3] as follows: a word x is *quasi-Sturmian* if there exist integers d and n_0 such that $p_x(n) = n + d$, for each $n \geq n_0$. The infinite words having factor complexity $n + d$ have been also studied in [13] where they are called "words with minimal block growth". Quasi-Sturmian words can be considered the simplest generalizations of Sturmian words in terms of factor complexity. In [3] the following characterization of quasi-Sturmian words has been given.

Theorem 9 ([3]). *An infinite word x over the alphabet Σ is quasi-Sturmian if and only if it can be written as $x = w\varphi(y)$, where w is a finite word and y is a Sturmian word on the alphabet $\{a, b\}$, and φ is a morphism from $\{a, b\}^*$ to Σ^* such that $\varphi(ab) \neq \varphi(ba)$.*

The following proposition shows that constant values for the span complexity at infinitely many points imply quasi-Sturmian words, i.e., the most repetitive infinite aperiodic words after the Sturmian words.

Proposition 9. *Let x be an aperiodic infinite word. If there exists $k > 0$ such that $\mathsf{span}_x(n) \leq k$ for infinitely many $n > 0$, then x is quasi-Sturmian.*

Proof. By hypothesis, for all $n > 0$ exists $n' \geq n$ such that $\mathsf{span}_x(n') \leq k$, for some $k > 0$. Then, for every finite n-length prefix of x and every $m > 0$, by using Lemma 1 it holds that $|F(x[1,n]) \cap \Sigma^m| \leq |F(x[1,n']) \cap \Sigma^m| \leq m + sp_x(n') \leq m + k$. Moreover, it is known that for every aperiodic word it holds that $p_x(n) \geq n + 1$ and $p_x(n+1) > p_x(n)$, for every $n \geq 0$. Hence, there exist $k' \leq k$ and $n_0 \geq 0$ such that $p_x(n) = n + k'$, for every $n \geq n_0$ and the thesis follows. □

Remark 3. Note that, in general, the converse of Proposition 9 is not true. In fact, let w be a finite word, y a Sturmian word and φ a non-periodic morphism. Then, $x = w\varphi(y)$ is quasi-Sturmian. We can choose as finite prefix w a symbol $c \notin \mathsf{alph}(\varphi(y))$. One can notice that in this case x is not recurrent, and by Proposition 7 the function span_x is not bounded by constant. Instead, if $w = \varepsilon$, then the converse of Proposition 9 is true. It can be derived from Proposition 4 and Theorem 8.

7 Conclusions

In this paper, we have shown that the notion of string attractor introduced in the context of Data Compression can be useful in the study of combinatorial properties of infinite words. The string attractor- based span complexity measure has indeed been used to characterize some infinite word families. The problem of characterizing words with bounded string attractor profile function remains open. On the other hand, the two new complexity measures here introduced could be useful to represent, in a more succinct way, information on infinite sequences of words. Finally, it might be interesting to explore how the span and lm measures are related to the compressor-based measures.

References

1. Allouche, J.P., Shallit, J.: Automatic Sequences: Theory, Applications. Cambridge University Press, Generalizations (2003)
2. Béal, M., Perrin, D., Restivo, A.: Decidable problems in substitution shifts. CoRR abs/2112.14499 (2021)
3. Cassaigne, J.: Sequences with grouped factors. In: Developments in Language Theory, pp. 211–222. Aristotle University of Thessaloniki (1997)
4. Cassaigne, J., Karhumäki, J.: Toeplitz words, generalized periodicity and periodically iterated morphisms. Eur. J. Comb. **18**(5), 497–510 (1997)
5. Castiglione, G., Restivo, A., Sciortino, M.: Circular Sturmian words and Hopcroft's algorithm. Theor. Comput. Sci. **410**(43), 4372–4381 (2009)
6. Castiglione, G., Restivo, A., Sciortino, M.: On extremal cases of Hopcroft's algorithm. Theor. Comput. Sci. **411**(38–39), 3414–3422 (2010)
7. Castiglione, G., Restivo, A., Sciortino, M.: Hopcroft's algorithm and cyclic automata. In: Martín-Vide, C., Otto, F., Fernau, H. (eds.) LATA 2008. LNCS, vol. 5196, pp. 172–183. Springer, Heidelberg (2008). https://doi.org/10.1007/978-3-540-88282-4_17
8. Christiansen, A.R., Ettienne, M.B., Kociumaka, T., Navarro, G., Prezza, N.: Optimal-time dictionary-compressed indexes. ACM Trans. Algorithms **17**(1), 8:1-8:39 (2021)
9. Constantinescu, S., Ilie, L.: The Lempel-Ziv complexity of fixed points of morphisms. SIAM J. Discret. Math. **21**(2), 466–481 (2007)
10. Damanik, D., Lenz, D.: Substitution dynamical systems: characterization of linear repetitivity and applications. J. Math. Anal. Appl. **321**(2), 766–780 (2006)
11. Durand, F., Perrin, D.: Dimension Groups and Dynamical Systems: Substitutions, Bratteli Diagrams and Cantor Systems. Cambridge Studies in Advanced Mathematics, Cambridge University Press (2022)
12. Frosini, A., Mancini, I., Rinaldi, S., Romana, G., Sciortino, M.: Logarithmic equal-letter runs for BWT of purely morphic words. In: Diekert, V., Volkov, M. (eds.) Developments in Language Theory. Lecture Notes in Computer Science, vol. 13257, pp. 139–151. Springer, Cham (2022). https://doi.org/10.1007/978-3-031-05578-2_11
13. Heinis, A.: Languages under substitutions and balanced words. Journal de Théorie des Nombres de Bordeaux **16**, 151–172 (2004)
14. Kempa, D., Prezza, N.: At the roots of dictionary compression: string attractors. In: STOC 2018, pp. 827–840. ACM (2018)
15. Knuth, D., Morris, J., Pratt, V.: Fast pattern matching in strings. SIAM J. Comput. **6**(2), 323–350 (1977)
16. Kociumaka, T., Navarro, G., Prezza, N.: Towards a definitive measure of repetitiveness. In: Kohayakawa, Y., Miyazawa, F.K. (eds.) LATIN 2021. LNCS, vol. 12118, pp. 207–219. Springer, Cham (2020). https://doi.org/10.1007/978-3-030-61792-9_17

17. Kutsukake, K., Matsumoto, T., Nakashima, Y., Inenaga, S., Bannai, H., Takeda, M.: On repetitiveness measures of Thue-Morse words. In: Boucher, C., Thankachan, S.V. (eds.) SPIRE 2020. LNCS, vol. 12303, pp. 213–220. Springer, Cham (2020). https://doi.org/10.1007/978-3-030-59212-7_15
18. Lempel, A., Ziv, J.: On the complexity of finite sequences. IEEE T. Inform. Theory **22**(1), 75–81 (1976)
19. Lothaire, M.: Algebraic Combinatorics on Words. Cambridge University Press, Cambridge (2002)
20. Mantaci, S., Restivo, A., Sciortino, M.: Burrows-Wheeler transform and Sturmian words. Inform. Process. Lett. **86**, 241–246 (2003)
21. Mantaci, S., Restivo, A., Romana, G., Rosone, G., Sciortino, M.: A combinatorial view on string attractors. Theor. Comput. Sci. **850**, 236–248 (2021)
22. Navarro, G.: Indexing highly repetitive string collections, part I: repetitiveness measures. ACM Comput. Surv. **54**(2), 29:1-29:31 (2021)
23. Navarro, G.: Indexing highly repetitive string collections, part II: compressed indexes. ACM Comput. Surv. **54**(2), 26:1-26:32 (2021)
24. Navarro, G.: The compression power of the BWT: technical perspective. Commun. ACM **65**(6), 90 (2022)
25. Pansiot, J.-J.: Complexité des facteurs des mots infinis engendrés par morphismes itérés. In: Paredaens, J. (ed.) ICALP 1984. LNCS, vol. 172, pp. 380–389. Springer, Heidelberg (1984). https://doi.org/10.1007/3-540-13345-3_34
26. Schaeffer, L., Shallit, J.: String attractors for automatic sequences. CoRR abs/2012.06840 (2021)
27. Sciortino, M., Zamboni, L.Q.: Suffix automata and standard Sturmian words. In: Harju, T., Karhumäki, J., Lepistö, A. (eds.) DLT 2007. LNCS, vol. 4588, pp. 382–398. Springer, Heidelberg (2007). https://doi.org/10.1007/978-3-540-73208-2_36

Combinatorics and Graph Theory

On the Zero-Sum Ramsey Problem over $\mathbb{Z}_2^d$

José D. Alvarado[1]([⊠])[ID], Lucas Colucci[1][ID], Roberto Parente[2][ID],
and Victor Souza[3][ID]

[1] Instituto de Matemática e Estatística, Universidade de São Paulo, São Paulo, Brazil
`josealvarado.mat17@gmail.com, lcolucci@ime.usp.br`
[2] Instituto de Computação, Universidade Federal da Bahia, Salvador, Brazil
`roberto.parente@ufba.br`
[3] Department of Pure Mathematics and Mathematical Statistics,
University of Cambridge, Cambridge, UK
`vss28@cam.ac.uk`

Abstract. Let Γ be a finite abelian group and let G be a graph. The zero-sum Ramsey number $R(G, \Gamma)$ is the least integer N (if it exists) such that, for every edge-colouring $E(K_N) \mapsto \Gamma$, one can find a copy $G \subseteq K_N$ where the sum of the colours of the edges of G is zero.

A large body of research on this problem has emerged in the past few decades, paying special attention to the case where Γ is cyclic. In this work, we start a systematic study of $R(G, \Gamma)$ for groups Γ of small exponent, in particular, $\Gamma = \mathbb{Z}_2^d$. For the Klein group $\mathbb{Z}_2^2$, the smallest non-cyclic abelian group, we compute $R(G, \mathbb{Z}_2^2)$ for all odd forests G and show that $R(G, \mathbb{Z}_2^2) \leq n + 2$ for all forests G on at least 6 vertices. We also show that $R(C_4, \mathbb{Z}_2^d) = 2^d + 1$ for any $d \geq 2$, and determine the order of magnitude of $R(K_{t,r}, \mathbb{Z}_2^2)$ as $d \to \infty$ for all t, r.

We also consider the related setting where the ambient graph K_N is substituted by the complete bipartite graph $K_{N,N}$. Denote the analogue bipartite zero-sum Ramsey number by $B(G, \Gamma)$. We show that $B(C_4, \mathbb{Z}_2^d) = 2^d + 1$ for all $d \geq 1$ and $B(\{C_4, C_6\}, \mathbb{Z}_2^d) = 2^{d/2} + 1$ for all even $d \geq 2$. Additionally, we show that $B(K_{t,r}, \mathbb{Z}_2^d)$ and $R(K_{t,r}, \mathbb{Z}_2^d)$ have the same asymptotic behaviour as $d \to \infty$, for all t, r. Finally, we conjecture the value of $B(\{C_4, \ldots, C_{2m}\}, \mathbb{Z}_2^d)$ and provide the corresponding lower bound.

Keywords: Zero-sum Ramsey theory · Finite abelian groups · Klein group · Forests · Bicliques · Cycle graphs

1 Introduction

The celebrated theorem of Erdős-Ginzburg-Ziv [8] asserts that if $a_1, \ldots, a_{2m-1}$ is a sequence of $2m - 1$ elements from $\mathbb{Z}_m$, then there is $S \subseteq \{1, \ldots, 2m - 1\}$ with $|S| = m$ such that $\sum_{i \in S} a_i = 0 \pmod{m}$. This result can be viewed as a generalisation of the pigeonhole principle, obtained when $m = 2$. Motivated

© Springer Nature Switzerland AG 2022
A. Castañeda and F. Rodríguez-Henríquez (Eds.): LATIN 2022, LNCS 13568, pp. 445–459, 2022.
https://doi.org/10.1007/978-3-031-20624-5_27

by this observation, Bialostocki and Dierker [2–4] connected this result with Ramsey theory. Indeed, we consider the colours as elements in an abelian group and instead of monochromatic substructures, we find zero-sum substructures.

Let Γ be a finite abelian group and G be a graph. The *zero-sum Ramsey number* $R(H, G)$ is the least integer N such that for every edge-colouring $c\colon E(K_N) \to G$, one can find a copy of $G \subseteq K_N$ such that $\sum_{e \in E(G)} c(e) = 0$ in Γ. We say that such copy of G has zero sum. Recall that the order $\mathrm{ord}(x)$ of $x \in \Gamma$ is the minimum integer $m > 0$ such that $mx = 0$. The exponent $\exp(\Gamma)$ of a finite abelian group is the maximum order of its elements. If $\exp(\Gamma)$ does not divide the number of edges $e(G)$ of G, then $R(G, \Gamma) = \infty$. Indeed, the constant colouring $c(e) = a$, where $\mathrm{ord}(a) = \exp(\Gamma)$, has no zero-sum copy of G. If $\exp(\Gamma)$ divides $e(G)$, then $R(G, \Gamma)$ is indeed finite. Recall that the r-coloured Ramsey number $R_r(G)$ is minimum N such that every r-colouring of the edges of K_N has a monochromatic copy of G. So in that case, $R(G, \Gamma) \leq R_{|\Gamma|}(G) < \infty$, as a copy of G for which all edges are assigned the same group element has zero sum.

Like in the classical Ramsey theory, we can define *zero-sum bipartite Ramsey number* $B(G, \Gamma)$ as the least integer N such that for every edge-colouring $c\colon E(K_{N,N}) \to \Gamma$ one can find a copy of $G \subseteq K_{N,N}$ with zero sum. As before, if $\exp(\Gamma) \nmid e(G)$, $B(G, \Gamma) = \infty$, and if $\exp(\Gamma) \mid e(G)$, we have $B(G, \Gamma) \leq B_{|\Gamma|}(G)$, the $|\Gamma|$-coloured bipartite Ramsey number. Additionally, for a collection of graphs $G_1, \ldots, G_m$, we define $R(\{G_1, \ldots, G_m\}, \Gamma)$ to be the least integer N such that every colouring $c\colon E(K_N) \to \Gamma$ has a zero-sum copy of G_i, for some $1 \leq i \leq m$. $B(\{G_1, \ldots, G_m\}, \Gamma)$ is defined analogously.

For a comprehensive review of zero-sum Ramsey problems going back to the early 90's, we recommend the survey of Caro [6]. The case $\Gamma = \mathbb{Z}_2$ is now very well understood. Indeed, due to the work of Caro [5], extending on earlier work of Alon and Caro [1], $R(G, \mathbb{Z}_2)$ was determined completely. We say that a graph is *odd* if all its vertices have odd degree. The result of Caro [5] is then the following.

Theorem 1. *For any graph G with n vertices and an even number of edges,*

$$R(G, \mathbb{Z}_2) = \begin{cases} n+2 & \text{if } G = K_n \text{ and } n = 0, 1 \pmod 4, \\ n+1 & \text{if } G = K_p \cup K_q \text{ and } \binom{p}{2} + \binom{q}{2} = 0 \pmod 4, \text{ or } G \text{ is odd}, \\ n & \text{otherwise}, \end{cases}$$

where $K_p \cup K_q$ is the disjoint union of a copy of K_p and a copy of K_q, with $p + q = n$.

In the bipartite setting, Caro and Yuster [7] determined $B(G, \mathbb{Z}_2)$ for all bipartite graphs G with an even number of edges.

In this work, we consider $\Gamma = \mathbb{Z}_2^d$ with $d \geq 2$. Naturally, we are restricted to graphs with an even number of edges. Taking $d = 2$, we obtain the Klein group $\mathbb{Z}_2^2$, the smallest non-cyclic abelian group. In our first result, we determine precisely the value of $R(F, \mathbb{Z}_2^2)$ for odd forests F and give some general bounds for a general one.

Theorem 2. *If F is a forest on $n \geq 6$ vertices and an even number of edges, then $R(F, \mathbb{Z}_2^2) \leq n + 2$, with equality if F is an odd forest. Moreover, if F has only one vertex of even degree, then $R(F, \mathbb{Z}_2^2) \geq n + 1$.*

We remark that the equality $R(F, \mathbb{Z}_2^2) = n+2$ cannot hold for all odd forests. Indeed, we show in Sect. 5 that $R(K_2 \cup K_2, \mathbb{Z}_2^2) = 7$.

We turn now to graphs that contain a cycle and any $d \geq 2$. Indeed, we start with cycles themselves, the smallest with an even number of edges being C_4. In this case, we obtain a precise result.

Theorem 3. *For $d \geq 2$, we have $R(C_4, \mathbb{Z}_2^d) = B(C_4, \mathbb{Z}_2^d) = 2^d + 1$.*

We note here that the equality between $R(C_4, \mathbb{Z}_2^d)$ and $B(C_4, \mathbb{Z}_2^d)$ seems to be coincidental. The lower bound in each case follows from a similar construction. However, determining $B(C_4, \mathbb{Z}_2^d)$ is considerably easier, and employing the same strategy to the complete graph gives $R(C_4, \mathbb{Z}_2^d) \leq 2^d + 3$. Reducing this bound to $2^d + 1$ is the main difficulty in Theorem 3. Moreover, with the characterisations for $\Gamma = \mathbb{Z}_2$ discussed above, we have $R(C_4, \mathbb{Z}_2) = 4$ and $B(C_4, \mathbb{Z}_2) = 3$.

Generalising the construction for the lower bound in Theorem 3, we also provide a family of constructions that are suitable to avoid longer cycles.

Theorem 4. *For any $m \geq 2$ and $d \geq m - 1$, we have*

$$B(\{C_{2k} : 2 \leq k \leq m\}, \mathbb{Z}_2^d) \geq 2^{\lfloor d/(m-1) \rfloor} + 1. \tag{1}$$

Furthermore, equality holds if $m = 2$ or if $m = 3$ and d is even.

Even though we only showed that this bound is tight when $m = 3$ and d is even, we can determine $B(\{C_4, C_6\}, \mathbb{Z}_2^d)$ asymptotically for $d \to \infty$. However, this construction does not adapt well to the complete graph case, and we provide no analogous result to $R(\{C_{2k} : 2 \leq k \leq m\}, \mathbb{Z}_2^d)$. We conjecture in Sect. 5 that Theorem 4 is asymptotically tight for $m \geq 4$.

Another class of graphs we consider are the bicliques $K_{t,r}$ where $1 \leq t \leq r$ are not both odd. We completely solve this case asymptotically.

Theorem 5. *Fix $1 \leq t \leq r$, not both odd. As $d \to \infty$, we have*

$$R(K_{t,r}, \mathbb{Z}_2^d) = \begin{cases} \Theta(4^{d/r}) & \text{if } r \text{ is even,} \\ \Theta(4^{d/t}) & \text{if } r \text{ is odd and } t \text{ is even,} \end{cases}$$

where the implicit constants depend on r and t. The same holds for $B(K_{t,r}, \mathbb{Z}_2^d)$.

We provide a connection between the zero-sum Ramsey numbers of $K_{t,r}$ and the size of the longest sequences in a group Γ without any zero-sum subsequence of a prescribed length. We rely on a result of Sidorenko [11] for the upper bound in Theorem 5, and we modify his construction to obtain the lower bound.

A small interesting graph that does not fit any of the above categories is K_4. We only provide the weak bounds that $10 \leq R(K_4, \mathbb{Z}_2^2) \leq 91$, and leave the exact value as a challenge to the reader.

2 Forests

In this section, we write $\mathbb{Z}_2^2 = \{0, \alpha, \beta, \gamma\}$, where $x + x = 0$ for every $x \in \mathbb{Z}_2^2$, and $\alpha + \beta + \gamma = 0$. The following easy lemma gives a useful condition to check whether a given graph edge-coloured with $\mathbb{Z}_2^2$ has zero-sum.

Lemma 6. *Let G be a graph with an even number of edges. Then, an edge-colouring $c\colon E(G) \to \mathbb{Z}_2^2$ is such that G is zero-sum if, and only if, $|c^{-1}(0)| = |c^{-1}(\alpha)| = |c^{-1}(\beta)| = |c^{-1}(\gamma)|$ (mod 2).*

The goal of this section is to prove Theorem 2. The upper and lower bounds will be treated separately in the subsections below.

2.1 Lower Bounds

Clearly, $R(F, \Gamma) \geq n$ for every graph F on n vertices and any group Γ. We will provide some general results that improve on this lower bound for $\Gamma = \mathbb{Z}_2^2$, depending on the parity of the degrees of the vertices of F.

Proposition 7. *Let F be an odd graph on n vertices. Then, $R(F, \mathbb{Z}_2^2) \geq n + 2$.*

Proof. Let $V(K_{n+1}) = \{v_1, \ldots, v_{n+1}\}$ and consider an edge-colouring c of K_{n+1} as defined below. Let $c(v_1 v_i) = \alpha$ for every $2 \leq i \leq n + 1$, $c(v_2 v_j) = \beta$ for every $3 \leq j \leq n + 1$, $c(e) = \gamma$ for all other edges e in K_{n+1}. We claim that this colouring has no zero-sum copy of F. Indeed, by Lemma 6, a copy of F in K_{n+1} is zero-sum if and only if every colour appears in an even number of edges of F, as the colour 0 does not appear in c.

But no vertex of F can be mapped to v_1, otherwise the colour α is used an odd number of times. Similarly, the vertex v_2 cannot be used, otherwise the colour β is used an odd number of times. But F has n vertices, so it cannot avoid both v_1 and v_2 simultaneously. Thus, no copy of F has zero sum. $\square$

Proposition 8. *Let F be a graph on n vertices wherein the degrees of all but one vertex are odd. Then, $R(F, \mathbb{Z}_2^2) \geq n + 1$.*

Proof. Let $V(K_n) = \{v_1, \ldots, v_n\}$ and consider an edge-colouring c if K_n as defined below. Let $c(v_1 v_2) = \alpha$, $c(v_1 v_i) = \beta$ for every $3 \leq i \leq n$, $c(v_2 v_j) = \gamma$ for $3 \leq j \leq n$ and $c(e) = 0$ for all other edges e in K_n. We claim that this colouring has no zero-sum copy of F. By Lemma 6, a copy of F in K_n with zero sum if and only if every colour appears with the same parity in the edges of F.

As F has n vertices, every vertex in K_n is the image of a vertex of F. If the edge $v_1 v_2$ is used in F, every colour must be used an odd number of times. As F has only one vertex with even degree, v_1 or v_2 must be the image of an odd vertex of F. Thus, either β or γ is used an even number of times in F. Now assume that $v_1 v_2$ is not used in F, so every colour must be used an even number of times. But that is again not possible, either v_1 or v_2 must be the image of an odd vertex of F. In any case, c has no zero-sum copy of F. $\square$

Propositions 7 and 8 clearly imply the lower bound of Theorem 2.

2.2 Upper Bound

To finish the proof of Theorem 2, we deal with the upper bound in the result below. The case of stars will be missing, but that is covered in Sect. 2.3.

Theorem 9. *If F is a forest on $n \geq 6$ vertices and even number of edges which is not a star, then $R(F, \mathbb{Z}_2^2) \leq n + 2$.*

Proof. The idea of the proof of the upper bound is the following: suppose that a given colouring c of the K_{n+2} has two vertex disjoint paths on three vertices (we will call such graphs *cherries*) $x_1 u x_2$ and $y_1 v y_2$ such that $c(u\,x_1) + c(u\,x_2) \neq c(v\,y_1) + c(v\,y_2)$, and these two sums are not zero. If F is not a star (since the case of stars will be treated separately in Theorem 10), and if we let ℓ_1 and ℓ_2 be leaves of F such that their neighbours p_1 and p_2 are distinct, we can map $F \setminus \{\ell_1, \ell_2\}$ into $K_{n+2} \setminus \{x_1, x_2, y_1, y_2\}$ in a way that p_1 is mapped to u and p_2 is mapped to v. In this case, we have four choices to complete the embedding of F in K_{n+2}, and it is immediate to check that this four sums are distinct elements of $\mathbb{Z}_2^2$. In particular, one of those four copies of F is zero-sum.

Let F be a forest on $n \geq 6$ vertices and an even number of edges, and suppose first that F has a vertex p with two leaves ℓ_1 and ℓ_2 as neighbours. Then, if K_{n+2} is coloured so that a vertex u have three neighbours v_1, v_2, v_3 such that the colours of the edges uv_1, uv_2 and uv_3 are distinct, say, x, y and z, we can take a vertex v_4 and mapping $F \setminus \{\ell_1, \ell_2\}$ to $K_{n+2} \setminus \{v_1, v_2, v_3, v_4\}$ in a way that p is mapped to u, we have six choices (all the 2-subsets of $\{v_1, v_2, v_3, v_4\}$) for embedding $\{\ell_1, \ell_2\}$ in K_{n+2}. One can check that, regardless of the colour of uv_4, these six choices cover $\mathbb{Z}_2^2$, and hence one of these choices yields a zero-sum copy of F.

Otherwise, for every vertex $u \in V(K_{n+2})$, at most two colours appear in the edges incident to u. For $x \neq y \in \mathbb{Z}_2^2$, let V_{xy} be the set of vertices u for which the colours x and y appear in the edges incident to u. Note that if $\{x, y\} \cap \{z, w\} = \emptyset$, then either V_{xy} or V_{zw} is empty. Hence, we have that the set of nonempty V_{xy} are contained in either $\{ab, ac, ad\}$ or in $\{ab, bc, ac\}$ (where a, b, c and d are distinct colours).

In the first case, assume without loss of generality that the possibly nonempty V_{xy} are $V_{0\alpha}$, $V_{0\beta}$ and $V_{0\gamma}$. In this case, if at least two of these sets contain edges of colour distinct from zero, then it is simple to see that they have two vertex-disjoint cherries that have distinct nonzero sum. By the argument above, we can find a zero-sum copy of F. Otherwise, the colouring uses at most two colours, and the result follows from Theorem 1 (by identifying the colours with 0 and 1).

In the second case, without loss of generality let us suppose that the possibly nonempty V_{xy} are $V_{\alpha\beta}$, $V_{\beta\gamma}$, $V_{\gamma\alpha}$. In this case, it is also possible to find two vertex-disjoint cherries that have distinct nonzero sum, unless at most two colours are used in the colouring. By the same argument as above, we can find a zero-sum copy of F.

Finally, we have to deal with the case where F does not have a vertex p with two leaves ℓ_1 and ℓ_2 as neighbours. In this case, it is possible to prove (considering a longest path in F) that F is not odd or a matching.

First, note that the result is trivial if every cherry in K_{n+2} has sum zero, as it implies that the whole graph is monochromatic. Otherwise, there is a cherry uxv with nonzero sum, say, α. Consider the graph $K_{n+2} \setminus \{u, x, v\}$. If it has any cherries of sum β or γ, we can find a zero-sum copy of F as before. Otherwise, this subgraph contains only cherries of sum 0 and α.

Then, it is clear that, on the edges incident to any given vertex, either only the colours 0 and α, or β and γ may appear. It follows that this whole subgraph is coloured with only either colours 0 and α or β and γ. Assume the former without loss of generality.

First, suppose that all cherries inside $K_{n+2} \setminus \{u, x, v\}$ have sum zero, i.e., $K_{n+2} \setminus \{u, x, v\}$ is monochromatic with colour c. In this case, note that if any edge joining the vertices $\{u, x, v\}$ to $K_{n+2} \setminus \{u, x, v\}$ has colour c, say, xy, we may embed F in K_{n+2} without using vertices u and v, and in a way that a leaf of F is mapped into x. In this way, all edges of F get colour c, so F is zero-sum. Otherwise, suppose every edge joining the vertices $\{u, x, v\}$ to $K_{n+2} \setminus \{u, x, v\}$ has a colour distinct from c. We claim that all these edges have the same colour. Indeed, if two independent edges joining the vertices $\{u, x, v\}$ to $K_{n+2} \setminus \{u, x, v\}$ have distinct colour, together with edges in $K_{n+2} \setminus \{u, x, v\}$, they form two vertex-disjoint cherries of nonzero distinct sums. From this, we can conclude that every edge joining these two sets have the same colour, say, d. Now, we embed F in $K_{n+2} \setminus \{v\}$ in a way that two leaves of F are mapped into x and u. It follows that this copy of F is zero-sum.

Finally, suppose that there is a cherry in $K_{n+2} \setminus \{u, x, v\}$ with sum α. If any edge joining the vertices $\{u, x, v\}$ to $K_{n+2} \setminus \{u, x, v\}$ has colour β or γ, it is immediate that we can find two vertex-disjoint cherries that have distinct nonzero sum (one using the edge coloured with β or γ with an edge from $K_{n+2} \setminus \{u, x, v\}$ and the other with sum α inside $K_{n+2} \setminus \{u, x, v\}$). Hence, all the edges joining $\{u, x, v\}$ to $K_{n+2} \setminus \{u, x, v\}$ have colour 0 or α. In particular, $K_{n+2} \setminus \{x\}$ is a graph on $n+1$ vertices wherein all edges are coloured with 0 or α. The result then follows once more from Theorem 1. □

2.3 Stars

To finish the proof of Theorem 2, we need to determine $R(F, \mathbb{Z}_2^2)$ for stars.

Theorem 10. *Let $S_n = K_{1,n-1}$ be the star on n vertices, where $n \geq 3$ is odd. Then, $R(S_n, \mathbb{Z}_2^2) = n + 2$ for all $n \geq 3$.*

Proof. For the lower bound, let $V(K_{n+1}) = \{v_1, \ldots, v_{n+1}\}$ and consider the colouring $c \colon E(K_{n+1}) \to \mathbb{Z}_2^2 = \{0, \alpha, \beta, \gamma\}$ defined as follows. Fix a cycle $C = v_1 v_2 \cdots v_{n+1}$, set $c(v_i v_{i+1}) := \alpha$ for each i odd, $c(v_i v_{i+1}) := \beta$ for each i even (where $v_{n+2} = v_1$), and finally, set $c(uw) := \gamma$ for every edge $uw \in E(K_{n+1}) \setminus E(C)$. We claim that this colouring has no zero-sum copy of S_n. By contradiction, suppose that one can find a copy S of S_n with zero-sum. Without loss of generality, we can assume that the star is centred at v_1. Since our colouring does not use colour 0, by Lemma 6, we deduce that a, b, c are even, where a, b, c is the number of edges of S using colour α, β and γ, respectively. This means that vertices v_{n+1} and v_2 are not neighbourhoods to v_1 in S. Therefore, the degree of v_1 in S has cardinality at most $n + 1 - 3 = n - 2$, a contradiction.

Now, we show the upper bound. We leave the special case $n = 3$ to the end of the proof. Thus, assume that $n \geq 5$ is odd and let $c\colon E(K_{n+2}) \to \mathbb{Z}_2^2$ be a colouring. Our goal is to show that we can find a copy of S_n with zero-sum. We divide the proof into two cases.

First, suppose that some vertex v has the property that at most two colours appear in the incident edges of the vertex. If just one colour is used, we have a monochromatic star S_n centred in v, and we are happy. If we use exactly two colours, say a and b, then $V(K_{n+2}) \setminus \{v\} = A \sqcup B$ where A (resp. B) is the set of vertices u in $V(K_{n+2}) \setminus \{v\}$ such that uv has colour a (resp. b). Since n is odd, we deduce that either $|A|$ and $|B|$ are even or $|A|$ and $|B|$ are odd. In the even case, we remove exactly two vertices of either A or B (whichever contains at least two vertices). In the odd case, we remove exactly one vertex of each set A and B. In both cases, the remaining graph between v and $A \sqcup B$ is a star with n vertices and zero-sum.

Secondly, suppose that any vertex has the property that at least three colours appear in its incident edges. Fix any vertex v. Since $(n + 1)/4 \geq 6/4 > 1$, by the pigeonhole principle, some colour x appears in two edges incident with v. Let y and z be another two colours which appear in the incident edges of x. Let u_1, u_2, u_3 and u_4 be neighbourhoods of x such that $c(v\,u_1) = x$, $c(v\,u_2) = x$, $c(v\,u_3) = y$ and $c(v\,u_4) = z$. Note that $\{x + x, x + y, x + z, y + z\} = \mathbb{Z}_2^2$. Hence, if we consider the spanning star in K_{n+2} centred in v, say S_v, with colour sum α, say, we can write α of the form $x + x$, $x + y$, $x + z$ or $y + z$. In any case, we can remove the two vertices of the set $\{u_1, u_2, u_3, u_4\}$ which corresponds to the summands of α expressed as in the previous form, i.e., u_1 and u_2 if $\alpha = x + x$, etc. Of course, the star S that arises from removing such two vertices from S_v has n vertices and colour sum equal to $c(S) = c(S_v) - \alpha = 0$.

To finish, we consider the special case $n = 3$. Note that if incident edges have the same colour, then we are happy (since we are finding a monochromatic S_3). Thus, assume that our colouring c is proper (i.e. two incident edges have different colours). Since the degree of any vertex is exactly four, the set of colours of the incident edges of each vertex is precisely $\mathbb{Z}_2^2$. Hence, we are shown that each fixed colour induces a perfect matching, which is a contradiction in K_5. □

3   Cycles

In this section, we aim to find some exact zero-sum Ramsey numbers involving short cycles, culminating in the proofs of Theorem 3 and Theorem 4. We start by describing a construction that will be useful not only for cycles, but for bicliques later in Sect. 4.

Proposition 11 (Product colouring). *Let $\mathbb{F}$ be a finite field with additive group Γ. If there is a length N sequence A of elements of Γ such that there are* *no zero-sum subsequence of lengths r or t, then*

$$B(K_{t,r}, \Gamma) \geq N + 1,$$
$$R(K_{t,r}, \Gamma) \geq N + 1.$$

Proof. We consider the bipartite graph $K_{N,N}$ with vertex set $A \sqcup A$. Consider the colouring $c \colon E(K_{N,N}) \to \Gamma$ defined as $c(x\,y) = xy$, where the product is taken on $\mathbb{F}$. Suppose that there is a $K_{t,r} \subseteq K_{N,N}$ with zero sum. Then there are subsequences $a_1, \ldots, a_t$ and $a_1', \ldots, a_r'$ in A such that

$$\sum_{i \in [t]} \sum_{j \in [r]} a_i a_j' = \left(\sum_{i \in [t]} a_i \right) \left(\sum_{j \in [r]} a_j' \right) = 0.$$

As this holds in $\mathbb{F}$, one of the sums has to be equal to zero, a contradiction.

For the clique, consider K_N with vertex set A and define, as before, the colouring $c \colon E(K_N) \to \Gamma$ as $c(x\,y) = xy$. The same argument applies, and this colouring has no $K_{t,r}$ with zero sum. $\qquad\square$

We now proceed to exact results about C_4. We split Theorem 3 into two results: Theorem 12 and Theorem 13. We start with the easier one.

Theorem 12. *For any $d \geq 1$, we have $B(C_4, \mathbb{Z}_2^d) = 2^d + 1$.*

Proof. Let $N = 2^d + 1$ and $c \colon E(K_{N,N}) \to \mathbb{Z}_2^d$ be an arbitrary colouring. Let $A \sqcup B = V(K_{N,N})$ be a bipartition and pick two distinct vertices $a_1, a_2 \in A$. Define the sequence $\{x_b\}_{b \in B}$ as

$$x_b := c(a_1\, b) + c(a_2\, b) \in \mathbb{Z}_2^d.$$

Since the length of B is $N > |\mathbb{Z}_2^d|$, by pigeonhole principle, one can find two distinct elements of B, say b_1 and b_2, such that $x_{b_1} = x_{b_2}$. We conclude that the 4-cycle induced by a_1, a_2, b_1, b_2 has zero-sum. Indeed, we have

$$c(a_1\, b_1) + c(b_1\, a_2) + c(a_2\, b_2) + c(b_2\, a_1) = x_{b_1} + x_{b_2} = 0.$$

Hence, $B(C_4, \mathbb{Z}_2^d) \leq 2^d + 1$. For the lower bound, we use the product colouring of Proposition 11 with $A = \mathbb{Z}_2^d$, $N = 2^d$. It is easy to check that A has no zero-sum subsequence of length 2, thus

$$B(C_4, \mathbb{Z}_2^d) = B(K_{2,2}, \mathbb{Z}_2^d) \geq N + 1 = 2^d + 1,$$

as we wanted. $\qquad\square$

We now turn to the problem of determining $R(C_4, \mathbb{Z}_2^d)$. As $K_{N,N} \subseteq K_{2N}$, Theorem 12 implies that $R(C_4, \mathbb{Z}_2^d) \leq 2(2^d + 1) = 2^{d+1} + 2$. However, in the proof of Theorem 12, we have never used more than two vertices in one of the parts. With this observation, we improve the upper bound to $2^d + 3$. Surprisingly, we are able to show that $R(C_4, \mathbb{Z}_2^d) = 2^d + 1$, the same value as $B(C_4, \mathbb{Z}_2^d)$.

Theorem 13. *For any $d \geq 2$, we have $R(C_4, \mathbb{Z}_2^d) = 2^d + 1$.*

Proof. For the lower bound, we apply Proposition 11 with $A = \mathbb{Z}_2^d$, which as argued before, has no zero-sum subsequence of length 2. This gives

$$R(C_4, \mathbb{Z}_2^d) = R(K_{2,2}, \mathbb{Z}_2^d) \geq |A| + 1 = 2^d + 1.$$

For the upper bound, we divide the proof into two steps. First, we claim that $R(C_4, \mathbb{Z}_2^d) \in \{2^d + 1, 2^d + 2\}$. We then proceed by contradiction and show that if there was a colouring of K_{2^d+1} without zero-sum C_4, such colouring could be extended to a colouring of K_{2^d+2} with the same property. This is enough to conclude that $R(C_4, \mathbb{Z}_2^d) = 2^d + 1$.

Claim. $R(C_4, \mathbb{Z}_2^d) \in \{2^d + 1, 2^d + 2\}$.

Proof of Claim. We already have the lower bound, so we just need to show that $R(C_4, \mathbb{Z}_2^d) \leq 2^d + 2$. Observe that a zero-sum C_4 in K_N is equivalent, in a group of exponent 2, to a pair of triangles sharing a single edge with the same colour sum. Fortunately, if $N = 2^d + 2$, we can find this pair of triangles in the collection of triangles with the 'most popular colour', as we show below.

Indeed, consider any colouring $c\colon E(K_N) \to \mathbb{Z}_2^d$. There are at least

$$\frac{\binom{N}{3}}{|\mathbb{Z}_2^d|} = \frac{N(N-1)(N-2)}{6(N-2)} = \frac{N(N-1)}{6}$$

triangles sharing the same colour sum $\alpha \in \mathbb{Z}_2^d$.

We claim that two of such triangles have a common edge. Let G^α be the graph which is the union of such triangles (i.e., with colour sum equal to α). Indeed, if none of these triangles share an edge, then $e(G^\alpha) \geq 3N(N-1)/6 = e(K_N)$. In particular, $G^\alpha = K_N$. However, each triangle contributes with 0 or 2 to the degree of a vertex. This is a contradiction, as $N-1$ is odd. $\qquad\square$

With this claim in hand, we proceed by contradiction. Let $N = 2^d + 1$ and suppose that there exists a *zero-sum-C_4-free* colouring $c\colon E(K_N) \to \mathbb{Z}_2^d$, that is, a colouring for which each copy of $C_4 \subseteq K_N$ has a nonzero sum. Our goal is to extend c to a colouring $\hat{c}\colon E(K_{N+1}) \to \mathbb{Z}_2^d$ that is also zero-sum-C_4-free. This contradicts our claim and finishes the proof.

Our construction is based on the following observation: a 4-cycle has zero-sum if and only if the pairs of triangles created by adding any of the two non-edge diagonals have the same colour-sum.

For convenience, assume that $V(K_N) = V$ and $V(K_{N+1}) = V \sqcup \{w\}$. Since we construct $\hat{c}$ as an extension of c, it is enough to show that the new copies of C_4, namely those containing w, have non-zero sum.

Given three distinct vertices x, y, z, we denote by $\hat{c}(x\,y\,z) := \hat{c}(x\,y) + \hat{c}(y\,z) + \hat{c}(z\,x)$, the colour-sum of the triangle spanned by x, y and z. By the previous observation, the condition $\hat{c}$ needs to satisfy is that

$$\hat{c}(u\,v\,w) \notin \left\{ \hat{c}(u\,v\,x) \;:\; x \in V \setminus \{u, v\} \right\},$$

for every $u, v \in V$. The observation also implies that $\hat{c}(u\,v\,x)$ have to be all distinct, as $x \in V \setminus \{u, v\}$. This gives $N - 2 = 2^d - 1$ obstructions to $\hat{c}(u\,v\,w)$. Therefore, there is a unique candidate for the value of $\hat{c}(u\,v\,w)$, which we denote by $c^*(u\,v)$. Recall that the total sum of all elements of $\mathbb{Z}_2^d$ is equal to zero for $d \geq 2$, so we have

$$c^*(u\,v) = \sum_{z \in V \setminus \{u,v\}} c(u\,v\,z).$$

Now, note that the condition that $\hat{c}(u\,v\,w) = c^*(u\,v)$ is equivalent to

$$\hat{c}(u\,w) + \hat{c}(v\,w) = c(u\,v) + c^*(u\,v). \tag{2}$$

This can be seen as an over-determined linear system of $\binom{N}{2}$ equations on N variables, namely $\hat{c}(u\,w)$ for $u \in V$. Nonetheless, this system has a solution. We exhibit a solution by fixing an element $x \in V$ and setting $\hat{c}(x\,w) = 0$. This immediately forces all other values of $\hat{c}(y\,w)$ as (2) implies that

$$\hat{c}(y\,w) = c(x\,y) + c^*(x\,y) + \hat{c}(x\,w) = c(x\,y) + c^*(x\,y),$$

which we take as a definition for $\hat{c}(y\,w)$, for all $y \in V \setminus \{x\}$.

We check that this is indeed a solution to the system of equations, namely, that (2) holds for all $u, v \in V$. If $u = x$ or $v = x$, then this follows directly by the way we defined $\hat{c}$. For distinct $u, v \in V \setminus \{x\}$, we have,

$$\hat{c}(u\,w) + \hat{c}(v\,w) = c(x\,u) + c^*(x\,u) + c(x\,v) + c^*(x\,v)$$
$$= c(x\,u) + c(x\,v) + \sum_{z \in V \setminus \{x,u\}} c(x\,u\,z) + \sum_{z' \in V \setminus \{x,v\}} c(x\,v\,z')$$
$$= c(x\,u) + c(x\,v) + \sum_{z \in V \setminus \{x,u,v\}} \Big(c(x\,u\,z) + c(x\,v\,z) \Big)$$
$$= c(x\,u) + c(x\,v) + \sum_{z \in V \setminus \{x,u,v\}} \Big(c(x\,u) + c(u\,z) + c(x\,v) + c(v\,z) \Big)$$
$$= c(x\,u) + c(x\,v) + \sum_{z \in V \setminus \{x,u,v\}} \Big(c(u\,z) + c(v\,z) \Big)$$
$$= \sum_{z \in V \setminus \{u,v\}} \Big(c(u\,z) + c(v\,z) \Big) = c(u\,v) + c^*(u\,v),$$

as we wanted. Therefore, c can indeed be extended to a zero-sum-C_4-free colouring $\hat{c}$ on K_{N+1}, a contradiction. □

Finally, we note that the product colouring of Proposition 11 can be modified to yield a lower bound for a family of short cycles, leading to Theorem 4.

Proof of Theorem 4. Let $d = (m-1)k + h$, where $0 \leq h < m - 1$. Since $d \geq m$, we have $k \geq 1$. As $\mathbb{Z}_2^k$ is the additive group of $\mathbb{F}_{2^k}$, we can represent any element of $\mathbb{Z}_2^d = \mathbb{Z}_2^{(m-1)k+h}$ as a vector $(x_1, \ldots, x_{m-1}, y)$, where $x_i \in \mathbb{F}_{2^k}$ and $y \in \mathbb{Z}_2^h$. Let A and B be two disjoint copies of $\mathbb{F}_{2^k}$ and consider the complete bipartite graph $K_{N,N}$ with bipartition $A \sqcup B$. Define the colouring $c \colon E(K_{N,N}) \to \mathbb{Z}_2^d$ as

$$c(a\,b) := (ab, a^2 b, \ldots, a^{m-1} b, 0).$$

We claim that this colouring has no zero-sum cycle of length $2k$, for $2 \leq k \leq m$. Indeed, suppose it has a zero-sum cycle of length 2ℓ, $1 \leq \ell \leq m$. Then there are distinct $a_1, \ldots, a_\ell \in A$ and $b_1, \ldots, b_\ell \in B$ such that $\sum_{i=1}^{\ell} c(a_i\, b_i) + c(a_i\, b_{i+1}) = 0$, where we index a_i and b_i with $i \pmod{\ell}$. Therefore, we have

$$\sum_{i=1}^{\ell} a_i^k (b_i + b_{i+1}) = 0, \tag{3}$$

for every $1 \leq k \leq m - 1$. We also have (3) with $k = 0$ as $\mathbb{Z}_2^d$ has exponent two. Therefore, if $M = (M_{ij})$ is the matrix with $M_{ij} = a_j^{i-1}$, then $Mu = 0$, where u is the vector with $u_j = b_j + b_{j+1}$. As $u \neq 0$, M is singular and $\det M = 0$. However, M is a Vandermonde matrix, so $\det M = \prod_{1 \leq i < j \leq \ell} (a_i - a_j) = 0$, so $a_i = a_j$ for some $i < j$, which is a contradiction. Therefore, (1) holds.

Now, note that Theorem 12 gives the equality for $m = 2$. We show here equality for $m = 3$. Consider a colouring $c \colon E(K_{N,N}) \to \mathbb{Z}_2^d$ with $N = 2^{\lfloor d/2 \rfloor} + 1$. We prove that if there is no zero-sum C_4 in c, then there must be a zero-sum C_6 in c. Indeed, consider a bipartition of $V(K_{N,N}) = A \sqcup B$ and fix $a \in A$, $b \in B$. Label $A' = A \setminus \{a\} = \{a_1, \ldots, a_{N-1}\}$ and $B' = B \setminus \{b\} = \{a_1, \ldots, b_{N-1}\}$. Define an auxiliary colouring of the bipartite graph induced by A' and B' as follows,

$$\chi(a_i\, b_j) = c(a\, b_j) + c(a_i\, b_j) + c(a_i\, b).$$

Now, note that if $\chi(a_i\, b_j) = c(a\, b)$ for some i, j, then $\{a, b, a_i, b_j\}$ form a zero-sum C_4. As we assumed there is no such C_4, there are only $2^d - 1$ available colours for each of $\chi(a_i\, b_j)$. Since d is even, $(N - 1)^2 > 2^d - 1$, so there are two edges with the same value of χ, say $\chi(a_i\, b_j) = \chi(a_u\, b_v)$. If $a_i = a_u$, then $b_j \neq b_v$ and $\{a, b_j, a_i, b_v\}$ forms a zero-sum C_4. Indeed, $c(a\, b_j) + c(a_i\, b_j) + c(a_i\, b_v) + c(a\, b_v) = \chi(a_i\, b_j) + \chi(a_u\, b_v) = 0$. Analogously, if $b_j = b_v$, then $a_i \neq a_u$ and $\{a_i, b, a_v, b_j\}$ is a zero-sum C_4. Therefore, we may assume that $a_i \neq a_u$ and $b_j \neq b_v$. This implies, that $\{a, b_j, a_i, b, a_u, b_v\}$ is a zero-sum C_6. In fact, $c(a\, b_j) + c(a_i\, b_j) + c(a_1\, b) + c(b\, a_u) + c(a_u\, b_v) + c(a\, b_v) = \chi(a_i\, b_j) + \chi(a_u\, b_v) = 0$, and we are done. $\square$

Remark 14. When $m = 3$, we have not shown equality when d is odd. However, if we choose N to the smallest integer with $(N - 1)^2 > 2^d - 1$, then we would have found a zero-sum C_4 or C_6 following the same proof. This translates to the asymptotic bound $B(\{C_4, C_6\}, \mathbb{Z}_2^d) = \Theta(2^{d/2})$.

In Sect. 5, we conjecture that Theorem 4 is asymptotically sharp.

4   Bicliques

In this section, we prove Theorem 5. We do so by connecting the zero-sum Ramsey problems $R(K_{t,r}, \Gamma)$ and $B(K_{t,r}, \Gamma)$ into a problem about sequences in Γ without zero-sum subsequences of prescribed length.

If Γ is an abelian group and $\exp(\Gamma) \mid r$, then define $s_r(\Gamma)$ to be the minimum number k such that any sequence of k elements in Γ has a subsequence of length r with zero sum. This parameter introduced by Gao [9] generalises the Erdős-Ginzburg-Ziv constant of a group $s(\Gamma) := s_{\exp(\Gamma)}(\Gamma)$. See the survey of Gao and Geroldinger [10] for more on this and related constants.

Proposition 15. *Let Γ be an abelian group and r be such that $\exp(\Gamma) \mid r$. Then for any $t \geq 1$, we have*

$$B(K_{t,r}, \Gamma) \leq \max\{s_r(\Gamma), t\}, \tag{4}$$

$$R(K_{t,r}, \Gamma) \leq s_r(\Gamma) + t. \tag{5}$$

Proof. Let $N = \max\{s_r(\Gamma), t\}$ and consider a colouring $c \colon E(K_{N,N}) \to \Gamma$. Let $A \sqcup B = V(K_{N,N})$ be a bipartition and fix distinct elements $a_1, \ldots, a_t \in B$. For each $a \in A$, consider the element $x_b = c(a_1 b) + \cdots + c(a_t b)$ in Γ. The sequence $\{x_b\}_{b \in B}$ has length $|B| \geq s_r(\Gamma)$, therefore there are r of them with zero-sum, let's say $x_{b_1} + \cdots + x_{b_r} = 0$. Expanding the definition of x_a, we see that the $K_{t,r}$ spanned by $\{a_1, \ldots, a_t\}$ and $\{b_1, \ldots, b_r\}$ has zero-sum.

Finally, the proof of (5) is almost identical to (4). If $N = s_r(\Gamma) + t$, firstly fix t elements $a_1, \ldots, a_t \in V(K_N)$ and finish as before, taking the sequence x_b with $b \notin \{a_1, \ldots, a_t\}$. $\qquad\square$

For the group $\Gamma = \mathbb{Z}_2^d$, Sidorenko [11] showed that, as $d \to \infty$,

$$s_{2r}(\mathbb{Z}_2^d) \leq \Theta(2^{d/r}). \tag{6}$$

Together with Proposition 15, this provides an upper bound sufficient for Theorem 5. In view of Proposition 11, (6) gives us $R(K_{2r,2r}, \mathbb{Z}_2^d) = \Theta(2^{d/r})$. As it is, (6) does not imply a lower bound other blicliques. Nonetheless, Sidorenko's construction also avoids zero-sum subsequences of other lengths, as he already noted. We reproduce his construction below to show that it indeed has this stronger property, allowing us to deal with all blicliques in Theorem 5.

Proposition 16. *For any $m, d \geq 1$, there is a subset $A \subseteq \mathbb{Z}_2^d$ of size $|A| = 2^{\lfloor d/m \rfloor}$ with no zero-sum subset of size ℓ, for all $2 \leq \ell \leq 2m$.*

Proof. Let $d = mk + h$, where $0 \leq h < m$. Since $\mathbb{Z}_2^k$ is the additive group of $\mathbb{F}_{2^k}$, we can represent the elements in $\mathbb{Z}_2^d = \mathbb{Z}_2^{mk+h}$ as vectors $(x_1, \ldots, x_m, y)$ with $x_i \in \mathbb{F}_{2^k}$ and $y \in \mathbb{Z}_2^h$. Define the set A as

$$A := \left\{ (x, x^3, \ldots, x^{2m-1}, 0) \in \mathbb{Z}_2^d : x \in \mathbb{F}_{2^k} \right\}.$$

Clearly $|A| = 2^k = 2^{\lfloor d/m \rfloor}$. Suppose that there are distinct $a_1, \ldots, a_\ell \in A$ with $2 \leq \ell \leq 2m$ and $a_1 + \cdots + a_\ell = 0$. Write $a_i = (x_i, x_i^3, \ldots, x_i^{2m-1}, 0)$, for $1 \leq i \leq \ell$, where the x_i are all distinct. We can assume that $x_i \neq 0$, as if $x_j = 0$, then $\ell \geq 3$ and we can remove a_j from the sum.

The zero-sum condition translates to $\sum_{i=1}^{\ell} x_i^{2j-1} = 0$ for $1 \le j \le m$. For even exponents, we have $\left(\sum_{i=1}^{\ell} x_i^j\right)^2 = \sum_{i=1}^{\ell} x_i^{2j} = 0$. Therefore, $\sum_{i=1}^{\ell} x_i^j = 0$ for each $1 \le j \le 2\ell$.

Consider the matrix $M = (M_{ij})$, where $M_{ij} = x_j^i$. We have $Mu = 0$ where u is all 1's vector, so M is singular and $\det M = 0$. However $\det M = x_1 \cdots x_\ell \prod_{1 \le i < j \le \ell}(x_i - x_j)$ via the Vandermonde determinant. Since $x_i \ne 0$ for all i, we must have $x_i = x_j$ for some $i < j$, a contradiction.   □

Remark 17. When exhibiting his lower bound for (6), Sidorenko [11] constructs a sequence A' from A by picking a single element $a \in A$ and making it appear exactly $2r-1$ times in A', other elements unchanged. As no subsequence of A' of length $2 \le \ell \le 2m$ has zero sum, he concludes that $s_{2m}(\mathbb{Z}_2^d) \ge 2^{\lfloor d/m \rfloor} + 2m - 1$.

We are now ready for the main proof of this section.

Proof of Theorem 5. Let $1 \le t \le r$ be not both odd numbers. Assume initially that $r = 2m$ is even. From Proposition 16, we have a set $A \subseteq \mathbb{Z}_2^d$ with no zero-sum subset of any size $2 \le \ell \le 2m = r$. Hence it has no zero-sum subset of size t, where we may have to remove 0 from A if $t = 1$. Using the product colouring, Proposition 11, we have $B(K_{t,r}, \mathbb{Z}_2^d) \ge 2^{\lfloor d/m \rfloor} - 1 = \Omega(4^{d/r})$. On the other hand, Proposition 15 gives $B(K_{t,r}, \mathbb{Z}_2^d) \le \max\{s_{2m}(\mathbb{Z}_2^d), t\}$. Together with Sidorenko's bound (6), this yields $B(K_{t,r}, \mathbb{Z}_2^d) = O(2^{d/m}) = O(4^{d/r})$. Similarly, we obtain $2^{\lfloor d/m \rfloor} - 1 \le R(K_{t,r}, \mathbb{Z}_2^d) \le s_{2m}(\mathbb{Z}_2^d) + t = O(2^{d/m})$.

Now suppose that r is odd and $t = 2m$ is even. From Proposition 16, we have a set $A \subseteq \mathbb{Z}_2^{d-1}$ with no zero-sum subset of any size $2 \le \ell \le 2m = t$. Consider now the set $A' = \{(x,1) \in \mathbb{Z}_2^d : x \in A\}$. Clearly A' still has no zero-sum subset of size t. However, the addition of the coordinate with 1 implies that there is also no zero-sum subset of odd size, in particular, of size r. By Proposition 11, we have $B(K_{t,r}, \mathbb{Z}_2^d) \ge 2^{\lfloor d-1/m \rfloor} = \Omega(4^{d/t})$. As before, Proposition 15 and (6) gives $B(K_{t,r}, \mathbb{Z}_2^d) \le \max\{s_{2m}(\mathbb{Z}_2^d), t\} = O(2^{d/m}) = O(4^{d/t})$. Finally, the same argument also shows that $2^{\lfloor d-1/m \rfloor} \le R(K_{t,r}, \mathbb{Z}_2^d) \le s_{2m}(\mathbb{Z}_2^d) + t = O(2^{d/m})$.   □

5   Further Questions

We conjecture that Theorem 4 is asymptotically tight for all $m \ge 2$.

Conjecture 18. For every $m \ge 2$, the following holds as $d \to \infty$,

$$B(\{C_{2k} : k = 2, \ldots, m\}, \mathbb{Z}_2^d) = \Theta(2^{d/(m-1)}).$$

In Theorem 4, we show that this holds for $m = 2, 3$. It would not be a surprise if the lower bound (1) in Theorem 4 is attained for infinitely many d, given any fixed $m \ge 4$, but that seems too strong to conjecture with the current evidence.

For $\mathbb{Z}_2^2$, the smallest open case is K_4. We made some progress showing that,

$$10 \le R(K_4, \mathbb{Z}_2^2) \le 91$$

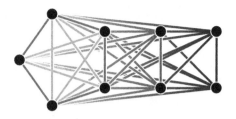

Fig. 1. A colouring of the edges of K_9 by elements of $\mathbb{Z}_2^2$ without zero-sum K_4. Different colours are different elements of $\mathbb{Z}_2^2$. Any assignment works. (Color figure online)

The lower bound is in Fig. 1. As any copy of K_4 use different colours with different parities, Lemma 6 implies it has non-zero sum.

The upper bound is a repeated pigeonhole argument. Indeed, consider an edge colouring c of K_{91} with colours $\mathbb{Z}_2^2 = \{0, \alpha, \beta, \gamma\}$. Fix one edge xy and partition the vertices $z \in V(K_{91}) \setminus \{x, y\}$ into four classes V_0, V_α, V_β and V_γ, according to the value of $c(x\,z) + c(y\,z)$. At least one the classes has 23 vertices, let's say V_f, for some $f \in \mathbb{Z}_2^2$.

Inside V_f, if there is an edge with colour $c(x\,y)$ we form a zero-sum K_4 together with x and y, so we may assume only three colours appear. Let zw be an edge inside V_f and partition the remaining 21 vertices into four classes U_0, U_α, U_β and U_γ according to the value of $c(z\,s) + c(w\,s)$, where $s \in V_f \setminus \{z, w\}$. Note that even though three colours are possible, the sum can be any of the four values. Again, one of the classes have at least 6 vertices. But now, the colour $c(z\,w)$ is also forbidden, so in this six vertices, only two colours are allowed. In $\mathbb{Z}_2^2$, any two elements are isomorphic to $\mathbb{Z}_2$, so we can find a zero-sum copy of K_4 as $R(K_4, \mathbb{Z}_2) = 6$ by Theorem 1.

Question 19. What is $R(K_4, \mathbb{Z}_2^2)$?

For a general graph G with an even number of edges, Theorem 1 implies that $R(G, \mathbb{Z}_2) \leq n + 2$. We believe that there is a reasonably small constant C such that $R(G, \mathbb{Z}_2^2) \leq n + C$ for all G.

Question 20. Is there a constant C such that $R(G, \mathbb{Z}_2^2) \leq n + C$ for all graphs G with an even number of edges? If so, what is the smallest such C?

If such C exists, Theorem 2 implies that $C \geq 2$. We can further show that $C \geq 3$. Indeed, the colouring in Fig. 2 shows that $R(K_2 \sqcup K_2, \mathbb{Z}_2^2) \geq 7$, just observe that no two disjoint edges can have the same colour, so their sum is not zero in $\mathbb{Z}_2^2$.

In fact, we have $R(K_2 \cup K_2, \mathbb{Z}_2^2) = 7$. Indeed, suppose that an edge colouring of K_7 by elements of $\mathbb{Z}_2^2$ has no zero-sum $K_2 \cup K_2$. In particular, no parallel edges have the same colour. As K_7 has 21 edges, one colour is used 6 times. As edges of this colour cannot be parallel, they have to form a spanning star. Removing this star, we have now a colouring of K_6 with precisely three colours. But K_6 has 15 edges, so one colour appear in 5 edges. Again, they have to form

Fig. 2. A colouring of the edges of K_6 by elements of $\mathbb{Z}_2^2$ without zero-sum $K_2 \cup K_2$. Different colours are different elements of $\mathbb{Z}_2^2$. Any assignment works. (Color figure online)

a spanning star and we can reduce to a K_5, coloured in two colours. But a two coloured K_5 must have a monochromatic $K_2 \cup K_2$, and we are done.

Acknowledgments. J.D. Alvarado is supported by FAPESP (2020/10796-0). L. Colucci is supported by FAPESP (2020/08252-2). R. Parente was supported by CNPq (406248/2021-4) and (152074/2020-1) while a postdoctoral researcher at Universidade de São Paulo. V. Souza would like to thank his PhD supervisor Professor Béla Bollobás for his support and encouragement.

References

1. Alon, N., Caro, Y.: On three zero-sum Ramsey-type problems. J. Graph Theory **17**(2), 177–192 (1993). https://doi.org/10.1002/jgt.3190170207
2. Bialostocki, A., Dierker, P.: On zero sum Ramsey numbers: small graphs. In: Ars Combinatoria, vol. 29(A), pp. 193–198. Charles Babbage Research Centre (1990)
3. Bialostocki, A., Dierker, P.: Zero sum Ramsey theorems. In: Congressus Numerantium, vol. 70, pp. 119–130. Utilitas Math. (1990)
4. Bialostocki, A., Dierker, P.: On the Erdős-Ginzburg-Ziv theorem and the Ramsey numbers for stars and matchings. Discret. Math. **110**(1–3), 1–8 (1992). https://doi.org/10.1016/0012-365x(92)90695-c
5. Caro, Y.: A complete characterization of the zero-sum (mod 2) Ramsey numbers. J. Comb. Theory Ser. A **68**(1), 205–211 (1994). https://doi.org/10.1016/0097-3165(94)90098-1
6. Caro, Y.: Zero-sum problems - a survey. Discret. Math. **152**(1), 93–113 (1996). https://doi.org/10.1016/0012-365X(94)00308-6
7. Caro, Y., Yuster, R.: The characterization of zero-sum (mod 2) bipartite Ramsey numbers. J. Graph Theory **29**(3), 151–166 (1998). https://doi.org/10.1002/(SICI)1097-0118(199811)29:3⟨151::AID-JGT3⟩3.0.CO;2-P
8. Erdős, P., Ginzburg, A., Ziv, A.: Theorem in the additive number theory. Bull. Res. Council Israel Sect. F **10F**(1), 41–43 (1961)
9. Gao, W.: On zero-sum subsequences of restricted size II. Discret. Math. **271**(1), 51–59 (2003). https://doi.org/10.1016/S0012-365X(03)00038-4
10. Gao, W., Geroldinger, A.: Zero-sum problems in finite abelian groups: a survey. Expo. Math. **24**(4), 337–369 (2006). https://doi.org/10.1016/j.exmath.2006.07.002
11. Sidorenko, A.: Extremal problems on the hypercube and the codegree Turán density of complete r-Graphs. SIAM J. Discret. Math. **32**(4), 2667–2674 (2018). https://doi.org/10.1137/17M1151171

On χ-Diperfect Digraphs with Stability Number Two

Caroline Aparecida de Paula Silva[1]([✉]) [iD], Cândida Nunes da Silva[2] [iD], and Orlando Lee[1] [iD]

[1] Institute of Computing, University of Campinas, Campinas, SP, Brazil
{caroline.silva,lee}@ic.unicamp.br
[2] Department of Computing, Federal University of São Carlos, Sorocaba, SP, Brazil
candida@ufscar.br

Abstract. Let D be a digraph. A proper coloring $\mathcal{C}$ and a path P of D are *orthogonal* if P contains exactly one vertex of each color class in $\mathcal{C}$. In 1982, Berge defined the class of χ-diperfect digraphs. A digraph D is χ-diperfect if for every minimum coloring $\mathcal{C}$ of D, there exists a path P orthogonal to $\mathcal{C}$ and this property holds for every induced subdigraph of D. Berge showed that some super-orientations of an odd cycle of length at least five and of its complement are not χ-diperfect. In 2022, de Paula Silva, Nunes da Silva and Lee characterized which super-orientations of such graphs are χ-diperfect. In this paper, we show that there are other minimal non-χ-diperfect digraphs with stability number two and three. In particular, the underlying graph of these digraphs with stability number two that we have found are subgraphs of the complement of an odd cycle with at least seven vertices. Motivated by this fact, we introduce a class of graphs, called *nice* graphs, which consist of all 2-connected graphs in which every odd cycle has length exactly five. We characterize which super-orientations of the complement of a nice graph are χ-diperfect.

Keywords: χ-diperfect digraphs · Coloring · Rainbow paths

1 Introduction

Let $G = (V(G), E(G))$ be a graph. We use the concepts of path and cycle as defined in [2]. We may think of a path or cycle as a subgraph of G. The *length* of a path (respectively, cycle) is its number of edges. The *order of a path P*, denoted by $|P|$, defined as its number of vertices, that is, $|P| = |V(P)|$. Similarly, the *order of a cycle* is its number of vertices. Let C_k denote the graph isomorphic to a cycle of length $k \geq 3$ and let $\overline{G}$ denote the *complement of G*.

We also use the concepts of stable set and clique as defined in [2]. The *stability number* of G is the cardinality of a maximum stable set, denoted by $\alpha(G)$. The cardinality of a maximum clique is denoted by $\omega(G)$.

Supported by CAPES - Finance Code 001, FAPESP Proc. 2020/06116-4 and 2015/11937-9, and CNPq Proc. 303766/2018-2 and Proc 425340/2016-3.

A. Castañeda and F. Rodríguez-Henríquez (Eds.): LATIN 2022, LNCS 13568, pp. 460–475, 2022.
https://doi.org/10.1007/978-3-031-20624-5_28

A *(proper) coloring* $\mathcal{C} = \{C_1, C_2, ..., C_m\}$ of a graph G is a partition of $V(G)$ into stable sets, also called *color classes*. A coloring $\mathcal{C}$ of G is *minimum* if $\mathcal{C}$ has the smallest possible number of color classes. The cardinality of a minimum coloring, denoted by $\chi(G)$, is the *chromatic number* of G.

For every concept for graphs, we may have an analogue for digraphs. Let $D = (V(D), A(D))$ be a digraph. The *underlying graph* of D, denoted by $U(D)$, is the simple graph with vertex set $V(D)$ such that u and v are adjacent in $U(D)$ if and only if either $(u, v) \in A(D)$ or $(v, u) \in A(D)$ or both. We borrow terminology from undirected graphs when dealing with a digraph D by considering its underlying graph $U(D)$. In particular, a *coloring* of a digraph D is a coloring of its underlying graph $U(D)$. Similarly, we may obtain a directed graph D from a graph G by replacing each edge uv of G by an arc (u, v), or an arc (v, u), or both; such directed graph D is called a *super-orientation* of G. A super-orientation which does not contain a *digon* (a directed cycle of length two) is an *orientation*. A digraph D is *symmetric* if D is a super-orientation of a graph G in which every edge uv of G is replaced by both arcs (u, v) and (v, u).

If (u, v) is an arc of D, then we say that u *dominates* v and v is *dominated* by u. If v is not dominated by any of its neighbors, then we say that v is a *source*. Similarly, if v does not dominate any of its neighbors in D, then we say that v is a *sink*. A *directed path* or *directed cycle* is an orientation of a path or cycle, respectively, in which each vertex dominates its successor in the sequence.

Henceforth, when we say path of a digraph, we mean directed path but we will not use the same convention for cycles. When we say a cycle of a digraph, we mean either a super-orientation of an undirected cycle with length at least three or a digon. We denote by $\lambda(G)$ ($\lambda(D)$) the cardinality of a maximum path in a graph (digraph). A path in a graph or digraph is *hamiltonian* if it contains all of its vertices. In 1934, Rédei [10] proved the following classical result.

Theorem 1 (Rédei [10]). *Every super-orientation of a complete graph has a hamiltonian path.*

A graph G is *perfect* if $\chi(H) = \omega(H)$ for every induced subgraph H of G. It is easy to show that if G is perfect, then G cannot contain either an odd cycle of order at least five or its complement as an induced subgraph. Berge [3] conjectured that the converse was true as well. In 2006, Chudnovsky, Robertson, Seymour and Thomas [3] proved this long standing open conjecture and it became known as the Strong Perfect Graph Theorem:

Theorem 2 (Chudnovsky et al. [3]). *A graph G is perfect if and only if G does not contain an odd cycle with five or more vertices or its complement as an induced subgraph.*

In 1982, Berge [1] introduced the concept of χ-diperfection of digraphs. Analogously to Theorem 2, he was interested in obtaining a characterization of such digraphs in terms of forbidden subdigraphs. Let $\mathcal{C}$ be a coloring and let P be a path of D. We say that $\mathcal{C}$ and P are *orthogonal* if $|V(P) \cap C| = 1$ for every $C \in \mathcal{C}$. We also say that $\mathcal{C}$ is orthogonal to P and vice versa. A digraph D is

χ-*diperfect* if every induced subdigraph H of D has the following property: for every minimum coloring $\mathcal{C}$ of H, there exists a path P of H such that $\mathcal{C}$ and P are orthogonal. A digraph D is *diperfect* if $U(D)$ is perfect. Berge [1] showed that diperfect digraphs and symmetric digraphs are χ-diperfect. For ease of further references, let us state the following.

Theorem 3 (Berge [1]). *Let D be a diperfect digraph. Then, D is χ-diperfect.*

Berge also showed that for a cycle of length five and for the complement of an odd cycle with at least five vertices, there are orientations which are not χ-diperfect. In [9], de Paula Silva, Nunes da Silva and Lee present a characterization of super-orientations of odd cycles (Theorem 4) and a characterization of super-orientations of complements of odd cycles that are χ-diperfect (Theorem 5). Let D be a super-orientation of an odd cycle $C = (x_1, x_2, \ldots, x_{2\ell+1}, x_1)$ of order at least five. Let $P = (x_i, \ldots, x_j)$ be a subpath of C. We say that the subdigraph $D[V(P)]$ is a *sector* if each of x_i and x_j is a source or a sink in D; we say that the sector is *odd* if P has odd length, otherwise it is *even*. We also use $(x_i, \ldots, x_j)$ or $x_i C x_j$ to denote the corresponding sector in D. We say that D is a *conflicting odd cycle* if it contains at least two arc-disjoint odd sectors.

Theorem 4 (de Paula Silva et al. [9]). *Let D be a super-orientation of an odd cycle with order at least five. Then, D is χ-diperfect if and only if D is not a conflicting odd cycle.*

Theorem 5 (de Paula Silva et al. [9]). *Let D be a super-orientation of the complement of an odd cycle with order at least five. Then, D is χ-diperfect if and only if every vertex of D belongs to a path of order $\chi(D)$.*

We say that a digraph D is an *obstruction* if D a minimal non-χ-diperfect digraph, i.e., D is non-χ-diperfect but every proper induced subdigraph of D is χ-diperfect. Conflicting odd cycles and non-χ-diperfect super-orientations of complements of odd cycles are examples of obstructions. Given the Strong Perfect Graph Theorem and Theorems 4 and 5, it may be tempting to conjecture that the set of obstructions is exactly the set of non-χ-diperfect super-orientations of odd cycles and their complements. Investigating this question we found new obstructions with stability number two and three. In this paper, we focus on those with stability number two. Curiously, we noted that the underlying graph of such obstructions that we have found so far are spanning $(k + 1)$-chromatic subgraphs of a $\overline{C_{2k+1}}$ with $k \geq 3$. Later we found out that we may build an obstruction by deleting an arc from some non-χ-diperfect super-orientation of a $\overline{C_{2k+1}}$ with $k \geq 3$.

Motivated by these observations, we decided to investigate digraphs with stability number two whose underlying graph does not contain spanning $(k+1)$-chromatic subgraphs of a $\overline{C_{2k+1}}$ with $k \geq 3$. In other words, we are interested in a family $\mathcal{H}$ of digraphs in which $D \in \mathcal{H}$ if and only if $\alpha(D) = 2$ and, for every induced subdigraph D' of D, it follows that $U(D')$ is not a spanning $(k + 1)$-chromatic subgraph of a $\overline{C_{2k+1}}$ with $k \geq 3$. One may note that this is equivalent to saying that a digraph $D \in \mathcal{H}$ if and only if every odd cycle of $\overline{U(D)}$ has length five.

2   Related Results

In this section we present some results related to the problem we study in this paper. The first theorem we present was proved independently by Roy in 1967 [11] and Gallai in 1968 [5].

Theorem 6 (Gallai-Roy [5,11]**).** *Let D be a digraph. For every maximum path P of D, there is a coloring $\mathcal{C}$ of D such that P and $\mathcal{C}$ are orthogonal. In particular, $\chi(D) \leq \lambda(D)$.*

We may compare this with the definition of χ-diperfection. In a χ-diperfect digraph, we require that for every minimum coloring $\mathcal{C}$, there exists a path P such that $\mathcal{C}$ and P are orthogonal. It is known that this property does not hold for every digraph [1].

A *path partition* of D is a collection of vertex-disjoint paths of D that cover $V(D)$. Let $\pi(D)$ denote the cardinality of a smallest path partition of D. We use the terms *initial vertex* and *terminal vertex* for paths to indicate the first and the last vertex in the sequence of a given path. We denote by $ter(P)$ (respectively, $ini(P)$) the terminal (respectively, initial) vertex of a path P. Similarly, if $\mathcal{P}$ is a collection of paths, we denote by $ter(\mathcal{P})$ $(ini(\mathcal{P}))$ the set of terminal (respectively, initial) vertices of each path in $\mathcal{P}$. A stable set S and a path partition $\mathcal{P}$ are *orthogonal* if $|S \cap P| = 1$ for every $P \in \mathcal{P}$.

Theorem 6 has a *dual* version in which we exchange the roles of stable sets and paths. This result is the celebrated Gallai-Milgram's Theorem which follows from the next lemma.

Lemma 1 (Gallai-Milgram [6]**).** *Let D be a digraph and let $\mathcal{P}$ be a path partition of D. Then,*

(i) *there is a path partition $\mathcal{Q}$ of D such that $ini(\mathcal{Q}) \subset ini(\mathcal{P})$, $ter(\mathcal{Q}) \subset ter(\mathcal{P})$ and $|\mathcal{Q}| = |\mathcal{P}| - 1$, or*

(ii) *there is a stable set S which is orthogonal to $\mathcal{P}$.*

Theorem 7 (Gallai and Milgram [6]**).** *Let D be a digraph. For every minimum path partition $\mathcal{P}$ of D, there is a stable set S such that $\mathcal{P}$ and S are orthogonal. In particular, $\pi(D) \leq \alpha(D)$.*

For the sake of conciseness, we refer the interested reader to Berge's paper [1] and Sambinelli's PhD thesis [12] for a survey of the known results on this subject.

3   Properties of Obstructions

Let D be a digraph with a fixed minimum coloring $\mathcal{S}$. We say that a subdigraph H of D is *rainbow* if no two vertices in H are in the same color class of $\mathcal{S}$. Similarly, a path P (respectively, cycle) in D is a *rainbow path* (respectively, *rainbow cycle*) if no two vertices in P are in the same color class of $\mathcal{S}$; moreover, if $|P| = k$, then we may say that P is a *k-rainbow path*. Let D_1 and D_2 be two disjoint rainbow subdigraphs of D. We say that D_1 and D_2 are *color-compatible* if $D_1 \cup D_2$ contains exactly one vertex of each color class of $\mathcal{S}$, i.e., $D_1 \cup D_2$ is a rainbow subdigraph of D with exactly $\chi(D)$ vertices. We also say that D_1 is *color-compatible with* D_2 and vice versa. A graph G is *(vertex) color-critical* if $\chi(G - X) < \chi(G)$ for every non-empty $X \subseteq V(G)$. Equivalently, color-critical graphs may be characterized in the following way.

Theorem 8. *A graph G is color-critical if and only if for every $v \in V(G)$ there is a minimum coloring of G in which v belongs to a singleton color class.* ∎

Now we present the relation between color-critical graphs and χ-diperfection.

Lemma 2. *If G is the underlying graph of an obstruction, then G is color-critical.*

Proof. Towards a contradiction, suppose that G is not color-critical. Let D be a minimal non-χ-diperfect super-orientation of G. Let $\mathcal{S}$ be a minimum coloring of D that does not admit a $\chi(D)$-rainbow path. Let D' be a proper subdigraph of D such that $\chi(D) = \chi(D')$. Clearly, $\mathcal{S}$ restricted to D' is a $\chi(D')$-coloring of D'. Since D is a minimal non-χ-diperfect digraph, there is a $\chi(D')$-rainbow path in D'. However, a $\chi(D')$-rainbow path of D' is also a $\chi(D)$-rainbow path of D, a contradiction. ∎

Let D be a digraph and let $G = U(D)$. We now look at $\overline{G}$. If such graph is disconnected, it is easy to see that G can be partitioned into two disjoint subgraphs, say G_1 and G_2, such that every vertex of G_1 is adjacent to every vertex of G_2. So in any coloring of $\mathcal{S}$ of G, no color class of $\mathcal{S}$ contains vertices of both G_1 and G_2. Based on this fact, we show in Lemma 4 that $\overline{G}$ must be 2-connected when D is an obstruction. Before we state such result, we need to present a lemma that is a straightforward application of Lemma 1 but it is also useful in other places of the text.

Lemma 3. *Let D be a digraph and let P_1 and P_2 be two disjoint paths of D. If every vertex of P_1 is adjacent to every vertex of P_2, then D has a path P such that (i) $V(P) = V(P_1) \cup V(P_2)$, (ii) $ini(P) \in \{ini(P_1), ini(P_2)\}$ and (iii) $ter(P) \in \{ter(P_1), ter(P_2)\}$.*

Proof. Let $D' = D[V(P_1) \cup V(P_2)]$ and let $\mathcal{P} = \{P_1, P_2\}$ be a path partition of D'. Since every vertex of P_1 is adjacent to every vertex of P_2, there is no stable set orthogonal to $\mathcal{P}$. By Lemma 1, D' has a path partition $\mathcal{P}'$ such that $|\mathcal{P}'| = |\mathcal{P}| - 1 = 1$, $ini(\mathcal{P}') \subset ini(\mathcal{P})$ and $ter(\mathcal{P}') \subset ter(\mathcal{P})$. Hence, the path P of $\mathcal{P}'$ satisfies properties (i)–(iii). ∎

Lemma 4. *If G is the underlying graph of an obstruction, then $\overline{G}$ is 2-connected.*

Proof. Let D be an obstruction and let $\mathcal{S}$ be a minimum coloring of D that does not admit a $\chi(D)$-rainbow path. Let $F = \overline{G}$ where G is the underlying graph of D. Towards a contradiction, suppose that F is not 2-connected. Suppose first that F is disconnected. Let X be the vertex set of a component of F and let $Y = V(F) - X$. Note that, in D, every vertex of X is adjacent to every vertex of Y. Thus, no color class of $\mathcal{S}$ has vertices in both X and Y. Moreover, $\mathcal{S}$ restricted to X and $\mathcal{S}$ restricted to Y are minimum colorings of $D[X]$ and $D[Y]$, respectively. Since D is a minimal non-χ-diperfect digraph, there is a $\chi(D[X])$-rainbow path P_1 in $D[X]$ and a $\chi(D[Y])$-rainbow path P_2 in $D[Y]$. Since P_1 and P_2 are color-compatible paths and every vertex of P_1 is adjacent to every vertex of P_2, we may apply Lemma 3 to P_1 and P_2 and obtain a $\chi(D)$-rainbow path of D, a contradiction.

So we may assume that F is connected but has a cut vertex v. Let X be the vertex set of a component of $F - v$ and let $Y = V(F) \setminus (X \cup \{v\})$. Similarly to the previous case, in D, every vertex of X is adjacent to every vertex of Y. Without loss of generality, we may assume that, if there are other vertices in the same color class of v, those vertices belong to X. Hence, no color class of $\mathcal{S}$ has vertices in both $X \cup \{v\}$ and Y. Moreover, $\mathcal{S}$ restricted to $X \cup \{v\}$ and $\mathcal{S}$ restricted to Y are minimum colorings of $D[X \cup \{v\}]$ and $D[Y]$, respectively. Let $k = \chi(D[X \cup \{v\}])$ and $\ell = \chi(D[Y])$ (so, $\chi(D) = k + \ell$). Since D is a minimal non-χ-diperfect digraph, there is a k-rainbow path in $D[X \cup \{v\}]$ and an ℓ-rainbow path P_2 in $D[Y]$. First assume that there is a k-rainbow path P_1 in $D[X \cup \{v\}]$ such that $v \notin V(P_1)$. Similarly to the previous case, P_1 and P_2 are color-compatible paths and every vertex of P_1 is adjacent to every vertex of P_2. Thus, we may apply Lemma 3 to P_1 and P_2 and obtain a $\chi(D)$-rainbow path of D, a contradiction.

Thus, we may assume that every k-rainbow path P_1 of $D[X \cup \{v\}]$ contains v. Let $\mathcal{S}'$ be $\mathcal{S}$ restricted to $Y \cup \{v\}$. Suppose that $\mathcal{S}'$ is not a minimum coloring of $D[Y \cup \{v\}]$, i.e., $\chi(D[Y \cup \{v\}]) = \ell$. Note that this implies that v does not belong to a singleton color class of $\mathcal{S}$. Hence, $\mathcal{S}$ restricted to X must be a minimum k-coloring of $D[X]$. However there is a k-rainbow path in $D[X]$ that does not contain v, a contradiction to our assumption. So we may assume that $\mathcal{S}'$ is a minimum coloring of $D[Y \cup \{v\}]$ and hence $\chi(D[Y \cup \{v\}]) = \ell + 1$. Since, by our assumption, no vertex in Y belongs to the same color class of v in $\mathcal{S}$, it follows that v must belong to a singleton color class in $\mathcal{S}'$. Thus, there is an $(\ell + 1)$-rainbow path P_3 in $D[Y \cup \{v\}]$ that contains v. Hence, assume that $P_1 = (x_1, \ldots, x_i = v, x_{i+1}, \ldots, x_k)$ and $P_3 = (y_1, \ldots, y_j = v, y_{j+1}, \ldots, y_{\ell+1})$. Moreover, every vertex in $\{x_1, \ldots, x_{i-1}, x_{i+1}, \ldots, x_k\}$ is adjacent to every vertex in $\{y_1, \ldots, y_{j-1}, y_{j+1}, \ldots, y_{\ell+1}\}$. By Lemma 3 there is a path R_1 such that $V(R_1) = \{x_1, \ldots, x_{i-1}, y_1, \ldots, y_{j-1}\}$ and $ter(R_1) \in \{x_{i-1}, y_{j-1}\}$. Similarly, there is a path R_2 such that $V(R_2) = \{x_{i+1}, \ldots, x_k, y_{j+1}, \ldots, y_{\ell+1}\}$ and $ini(R_2) \in \{x_{i+1}, y_{j+1}\}$. Since v is dominated by both x_{i-1} and y_{j-1} and v dominates both x_{i+1} and y_{j+1}, it follows that $R_1 v R_2$ is a $\chi(D)$-rainbow path of D, a contradiction. ∎

We may now restrict our attention to color-critical graphs whose complement is connected. We present below some results that provide us information on the number of vertices and on the properties of minimum colorings in such graphs (and so, in minimal non-χ-diperfect digraphs).

466   C. A. de Paula Silva et al.

In 1963, Gallai [4] showed a lower bound on the number of vertices of a graph that is color-critical and whose complement is connected. In 2002, Stehlík [13] proved a slightly stronger result that implies Gallai's lower bound.

Theorem 9 (Gallai [4]). *Let G be a color-critical graph. If $\overline{G}$ is connected, then G has at least $2\chi(G) - 1$ vertices.*

Theorem 10 (Stehlík [13]). *Let G be a color-critical graph and let $v \in V(G)$. If $\overline{G}$ is connected, then $G - v$ has a $(\chi(G) - 1)$-coloring in which every color class has at least two vertices.*

The following results are specific for digraphs with stability number two.

Lemma 5. *Let G be the underlying graph of an obstruction. If G has stability number two, then G has exactly $2\chi(G) - 1$ vertices.*

Proof. By Lemmas 2 and 4, G is a color-critical and $\overline{G}$ is connected. Let $n = |V(G)|$. By Theorem 9, it follows that $n \geq 2\chi(G) - 1$. Towards a contradiction, suppose that $n > 2\chi(G) - 1$. Let $v \in V(G)$ and let $G' = G - v$. Note that G' has at least $2\chi(G) - 1$ vertices and $\alpha(G') \leq 2$. Thus, $\chi(G') \geq \left\lceil \frac{2\chi(G)-1}{2} \right\rceil \geq \chi(G)$. However, this is a contradiction since G is color-critical and, hence, $\chi(G') = \chi(G) - 1$. ∎

Corollary 1. *Let G be the underlying graph of an obstruction. If G has stability number two, then in every minimal coloring of G there exists exactly one singleton color class and every other color class has size two.* ∎

We use the concepts of (perfect) matching as defined in [2]. A graph F is *factor-critical* if $F - v$ has a perfect matching, for any $v \in V(F)$.

Corollary 2. *Let G be the underlying graph of an obstruction. If G has stability number two, then $\overline{G}$ is factor-critical. Moreover, every minimum coloring of G corresponds to a maximum matching of $\overline{G}$ and vice versa.*

Proof. Let $u \in V(G)$ and let $\mathcal{S}$ be a minimum coloring in which u is a singleton color class (such coloring exists by Theorem 8). By Corollary 1, it follows that $\{u\}$ must be the only singleton color class of $\mathcal{S}$ and every other color class of $\mathcal{S}$ has size two. Let $F = \overline{G}$. We may build a perfect matching M of $F - u$ by converting each color class $\{v_1, v_2\}$ of $\mathcal{S} \setminus \{u\}$ into an edge $v_1 v_2$ of M. ∎

4   New Obstructions

We were able to find a few more examples of obstructions that were not yet known. All these digraphs have stability number two or three and they can be found in de Paula Silva's master's dissertation [8]. At this moment, we do not know if there are obstructions with stability number greater than three distinct from the conflicting odd cycles.

In this paper, we particularly focus on obstructions with stability number two. Two examples of such digraphs are depicted in Fig. 1. One may verify by inspection that neither of these digraphs contain conflicting odd cycles or non-χ-diperfect super-orientations of $\overline{C_{2k+1}}$, with $k \geq 2$. Note that the underlying graph of both digraphs are spanning 4-chromatic subgraphs of a $\overline{C_7}$. We observe that the underlying graph of all the obstructions with stability number two that we have found are spanning $(k+1)$-chromatic subgraphs of a $\overline{C_{2k+1}}$ with $k \geq 3$. In fact, we may show that we can build an obstruction from some non-χ-diperfect super-orientation of a $\overline{C_{2k+1}}$ with $k \geq 3$, as we state in Lemma 6. Due to space limitation, we are not able to present the proof of this result here, but it can also be found in [8].

Lemma 6 (de Paula Silva, Nunes da Silva and Lee [8]). *For every $k \geq 3$, there is an obstruction that is obtained by deleting an arc from some non-χ-diperfect super-orientation of a $\overline{C_{2k+1}}$.* ■

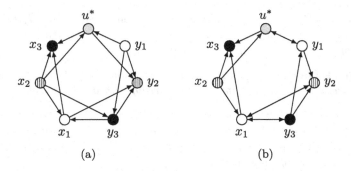

Fig. 1. Obstructions with stability number two.

In view of such observations, we decided to investigate digraphs with stability number two whose underlying graph does not contain a spanning $(k+1)$-chromatic subgraph of a $\overline{C_{2k+1}}$ with $k \geq 3$. In fact, we were able to characterize which of these digraphs are χ-diperfect and we present our result in next section.

5   Characterization of a Special Class

Recall that $\mathcal{H}$ is the family of digraphs such that $D \in \mathcal{H}$ if and only if $\alpha(D) = 2$ and for every induced subdigraph D' of D it follows that $U(D)$ is not a spanning $(k+1)$-chromatic subgraph of $\overline{C_{2k+1}}$ with $k \geq 3$. Note that this is equivalent to saying that every (not necessarily induced) odd cycle of $\overline{U(D)}$ has length five.

Let G be the underlying graph of an obstruction. By Lemma 4, $\overline{G}$ is 2-connected. Moreover, Corollary 2 states that if $\alpha(G) = 2$, then $\overline{G}$ is factor-critical and every maximum matching of $\overline{G}$ corresponds to a minimum coloring of G and vice versa. We present now some auxiliary results on 2-connected factor-critical graphs that are helpful in understanding the structure of these digraphs.

Lemma 7. *Let F be a factor-critical graph and let $u^* \in V(F)$. Let M be a perfect matching of $F - u^*$. Then, there is an odd cycle C in F such that $u^* \in V(C)$ and M restricted to $F - V(C)$ is a perfect matching.*

Proof. Let v be a vertex that is adjacent to u^*. Let M' be a perfect matching of $F - v$. Then, in $M \triangle M'$ there is an even path P from u^* to v whose edges alternate between M and M'. So, $C = P + u^*v$ is an odd cycle. Since the only vertex not covered by M restricted to C is u^*, M restricted to $M - C$ is a perfect matching. ∎

Let G be a graph and let G' be a subgraph of G. A path $P = (v_1, v_2 \ldots, v_\ell)$ is an *ear* of G' if $v_1, v_\ell \in V(G)$ and $v_2, \ldots, v_{\ell-1} \in V(G) \setminus V(G')$. In other words, the extremes of P belong to G' but the internal vertices do not. An *ear decomposition* of G is a sequence $(G_1, G_2, \ldots, G_\ell)$ of subgraphs of G such that

- G_1 is a cycle,
- $G_{i+1} = G_i \cup P_i$, where P_i is an ear of G_i with $1 \leq i < \ell$, and
- $G_\ell = G$.

If every ear in $\{P_1, \ldots, P_{\ell-1}\}$ has odd length, then we say that $(G_1, G_2, \ldots, G_\ell)$ is an *odd-ear decomposition* of G. In 1972, Lovász [7] proved the following characterization of 2-connected factor-critical graphs.

Theorem 11 (Lovász [7]). *A 2-connected graph F is factor-critical if and only if F has an odd-ear decomposition starting with an odd cycle.*

Let F be a graph with $2k + 1$ vertices, for $k \geq 2$ and with at least one induced cycle C of length five. We say that F is *nice* if the vertices of C can be labelled as $u_1, \ldots, u_5$ and the vertices of $F - V(C)$ can be labelled as $x_1, \ldots, x_{k-2}, y_1, \ldots, y_{k-2}$ so that

- for $i \in \{1, \ldots, k-2\}$, the neighbors of x_i are y_i and u_1, and
- for $i \in \{1, \ldots, k-2\}$, the neighbors of y_i are x_i and u_3.

Thus, for every $i \in \{1, \ldots, k-2\}$, it follows that $(u_1, x_i, y_i, u_3, u_2, u_1)$ is an induced C_5 (see Fig. 2 for an example with $k = 4$).

We may characterize nice graphs in terms of odd-ear decompositions. We state such characterization below for ease of further reference.

Proposition 1. *A graph F is nice if and only if there is an odd-ear decomposition $(F_1, F_2, \ldots, F_\ell)$ of F such that:*

(a) F_1 is an odd cycle of length five,
(b) $F_{i+1} = F_i \cup P_i$, where P_i is an ear of length three of F_1 with $1 \leq i < \ell$, and
(c) all the ears $P_1, \ldots, P_{\ell-1}$ have as extremes the same pair of non-adjacent vertices of F_1. ∎

We can easily check that every odd cycle of a nice graph must have length five. Moreover, by Theorem 11, a nice graph is 2-connected and factor-critical. In fact, we prove that every 2-connected factor-critical graph in which every odd cycle has length five must be isomorphic to a nice graph. Before we present such result, we need an auxiliary lemma.

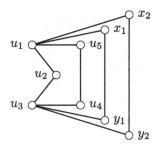

Fig. 2. Nice graph with nine vertices.

Lemma 8. *Let F be a graph in which every odd cycle has length five and let C be an odd cycle of F. Let P be an ear of C. If the extremes of P are non-adjacent in C, then the length of P is two or three. Otherwise, the length of P is four.*

Proof. Let $C = (u_1, \ldots, u_5)$ be an odd cycle of F and let $P = (v_1, \ldots, v_\ell)$ be an ear of C. Suppose first that the extremes of P are non-adjacent, say $v_1 = u_1$ and $v_\ell = u_3$. Since F has no cycle of length three, the length of P is at least two. Towards a contradiction, assume that the length of P is greater than three (so $\ell \geq 5$). Then, either $(v_1 = u_1, v_2, \ldots, v_\ell = u_3, u_2, u_1)$ or $(v_1 = u_1, v_2, \ldots, v_\ell = u_3, u_4, u_5, u_1)$ is an odd cycle of length greater than five, a contradiction. So suppose that the extremes of P are adjacent, say $v_1 = u_1$ and $v_\ell = u_2$. Towards a contradiction, suppose that the length of P is distinct from four (so $\ell \neq 5$). Then either $(v_1 = u_1, v_2, \ldots, v_\ell = u_2, u_1)$ or $(v_1 = u_1, v_2, \ldots, v_\ell = u_2, u_3, u_4, u_5, u_1)$ is an odd cycle of length distinct from five, a contradiction. ∎

Lemma 9. *Let F be a 2-connected factor-critical graph. If every odd cycle of F has length five, then F is isomorphic to a nice graph.*

Proof. By Theorem 11, F has an odd-ear decomposition $(F_1, \ldots, F_\ell)$ in which F_1 is an odd cycle. Let $\{P_1, \ldots, P_{\ell-1}\}$ be the ears in such ear-decomposition. We show by induction on ℓ that conditions (a) to (c) from Proposition 1 hold for $(F_1, \ldots, F_\ell)$. Since every odd cycle of F has length five, condition (a) immediately holds. Let $C = F_1 = (u_1, u_2, u_3, u_4, u_5, u_1)$. Clearly, if $\ell = 1$ then $F = C$ is a nice graph. Suppose now that $\ell = 2$. Since P_1 is an ear of C (of odd length), by Lemma 8, its length is exactly three and its extremes are non-adjacent vertices of C. So property (b) is satisfied and property (c) trivially holds. So we may assume that $\ell \geq 3$. By Theorem 11, $F_{\ell-1}$ is a 2-connected factor-critical graph and, clearly, every odd cycle of $F_{\ell-1}$ must have length five. Thus, by induction hypothesis, $F_{\ell-1}$ is a nice graph. By properties (b) and (c) from Proposition 1, we may assume that $P_i = (u_1, x_i, y_i, u_3)$ for every $i \in \{1, \ldots, \ell-2\}$. Now, it suffices to show that $P_{\ell-1}$ is an ear of C whose extremes are u_1 and u_3.

   First, we show that $P_{\ell-1} = (z_1, \ldots, z_t)$ is an ear of C. Towards a contradiction, suppose that at least one extreme of $P_{\ell-1}$ does not belong to C. We may assume without loss of generality that $z_1 = x_1$. Suppose first that $z_t \notin V(C)$. If there is j such that $z_t = y_j$ then $(u_1, z_1 = x_1, \ldots, z_t = y_j, u_3)$ is an ear that contradicts Lemma 8 (see Fig. 3a for the case where $j = 1$ and see Fig. 3b for the case where $j > 1$). So we may assume that there is j such that $z_t = x_j$. Then $(u_3, y_1, x_1 = z_1, \ldots, z_t = x_j, u_1)$ is an ear of C that contradicts Lemma 8 (see Fig. 3c). Hence, we may assume that $z_t \in V(C)$.

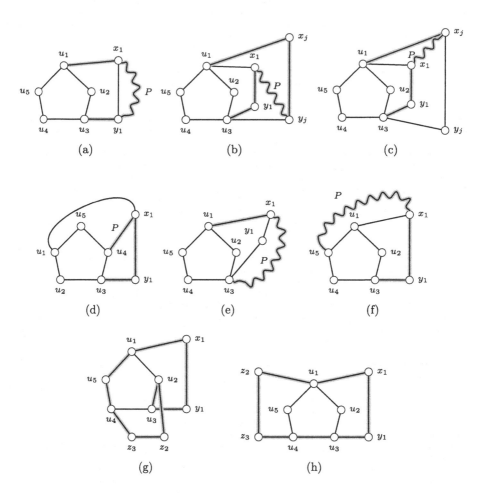

Fig. 3. Auxiliary illustration for the proof of Lemma 9.

Since F has no C_3, it follows that $z_1 = x_1$ is non-adjacent to vertices u_2, u_3 and u_5. So, if the length of $P_{\ell-1}$ is one, then $P_{\ell-1} = (x_1, u_4)$ and (u_3, y_1, x_1, u_4) is an ear of C that contradicts Lemma 8 (see Fig. 3d). Thus, we may assume that the length of $P_{\ell-1}$ is at least three (so $t \geq 4$). If $z_t = u_3$, then $(u_1, z_1 = x_1, \ldots, z_t = u_3)$ is an ear of C that contradicts Lemma 8 (see Fig. 3e). Otherwise, $(u_3, y_1, x_1 = z_1, \ldots, z_t)$ is an ear of C of length greater than four, a contradiction to Lemma 8 (see Fig. 3f for an example with $z_t = u_5$). Since all cases lead us to a contradiction, it follows that $P_{\ell-1}$ must be an ear of C. Also, recall that

$P_{\ell-1}$ has odd length by definition. By Lemma 8, it must have length three, i.e., $P_{\ell-1} = (z_1, z_2, z_3, z_4)$ and, furthermore, z_1 and z_4 are non-adjacent. Thus, property (b) is satisfied.

Now we show that property (c) of Proposition 1 holds for $P_{\ell-1}$, i.e., we show that $z_1 = u_1$ and $z_t = u_3$. Towards a contradiction, suppose that $z_1 = u_2$ and $z_4 = u_4$. Then, $(u_1, x_1, y_1, u_3, z_1 = u_2, z_2, z_3, z_4 = u_4, u_5, u_1)$ is a C_9, a contradiction (see Fig. 3g). The argument is analogous to show that the extremes of $P_{\ell-1}$ cannot be u_2 and u_5. So suppose that $z_1 = u_1$ and $z_4 = u_4$. Then, $(z_1 = u_1, x_1, y_1, u_3, u_4 = z_4, z_3, z_2, z_1 = u_1)$ is a C_7, a contradiction (see Fig. 3h). The argument is analogous to show that the extremes of $P_{\ell-1}$ cannot be u_3 and u_5. Hence, the extremes of $P_{\ell-1}$ must be u_1 and u_3.   ∎

The next proposition is an immediate consequence of Lemma 7 and it can be easily verified recalling that, by Corollary 2, the complement of a color-critical graph G with $\alpha(G) = 2$ is factor-critical and every maximum matching of $\overline{G}$ corresponds to a minimum coloring of G.

Proposition 2. *Let G be the complement of a nice graph and let S be a minimum coloring of G. Then, there is an induced cycle $C = C_5$ of G such that the vertices of C can be labelled as $(v_1, \ldots, v_5, v_1)$ and the vertices in $V(G) - V(C)$ can be labelled as $x_1, \ldots, x_{k-2}, y_1, \ldots, y_{k-2}$ so that:*

(a) *the vertex u^* in the singleton color class of S belongs to $V(C)$,*
(b) *$\{x_i, y_i\}$ is a color class of S for $i \in \{1, \ldots, k-2\}$,*
(c) *for $i \in \{1, \ldots, k-2\}$, the non-neighbors of x_i are y_i and v_1,*
(d) *for $i \in \{1, \ldots, k-2\}$, the non-neighbors of y_i are x_i and v_2, and*
(e) *$(v_1, y_i, v_4, x_i, v_2, v_1)$ is an induced odd cycle of length five.*

    ∎

Henceforth, we may assume that every complement of a nice graph has a fixed minimum coloring S and its vertices are labelled as described in Proposition 2. Note that the cycle $(v_1, \ldots, v_5, v_1)$ of G is the complement of the cycle $(u_1, \ldots, u_5, u_1)$ mentioned on the definition of a nice graph and on Lemma 9 (see Fig. 4). We use the notation $X \mapsto Y$ to denote that every vertex of X dominates every vertex of Y in D and no vertex of Y dominates a vertex of X in D. If $X = \{u\}$ (respectively, $Y = \{v\}$), we may write directly $u \mapsto Y$ (respectively, $X \mapsto v$).

Lemma 10. *Let D be a super-orientation of the complement of a nice graph with $2k + 1$ vertices, for $k \geq 2$. If D contains no conflicting odd cycle as an induced subdigraph, then D has a $(k + 1)$-rainbow path.*

Proof. By hypothesis, D contains no conflicting odd cycle. Then, by Theorem 4, C has a 3-rainbow path P. Let $X = \{x_1, \ldots, x_{k-2}\}$ and let $Y = \{y_1, \ldots, y_{k-2}\}$. Note that $D[X]$ and $D[Y]$ induce semicomplete digraphs and they are both color-compatible with P. Suppose first that $v_1 \notin V(P)$. In this case, every vertex in X is adjacent to every vertex of P. By Theorem 1, $D[X]$ has a hamiltonian path

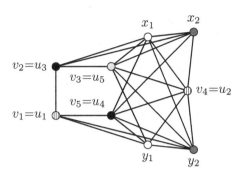

x_1 x_2

$v_2 = u_3$

$v_3 = u_5$

$v_5 = u_4$ $v_4 = u_2$

$v_1 = u_1$

y_1 y_2

Fig. 4. Example of labelling of a complement of a nice graph with nine vertices.

P'. We may apply Lemma 3 to P' and P, and obtain a $(k+1)$-rainbow of D. The argument is symmetric in the case where $v_2 \notin V(P)$. Hence, we may assume that every 3-rainbow path P of C contains both v_1 and v_2. Note that, since v_4 is non-adjacent to both v_1 and v_2, we know that $v_4 \neq u^*$. Also, we may assume without loss of generality, that $u^* = v_1$ or $u^* = v_5$. In both cases, note that v_1 and v_4 belong to distinct color classes. We consider the following two cases.

Case 1. *There is no digon between v_1 and v_2.*

By the Principle of Directional Duality, we may assume that $v_1 \mapsto v_2$. Let $i \in \{1, \dots, k-2\}$. Suppose that $v_1 \mapsto y_i$ and $x_i \mapsto v_2$. Since $C' = (v_4, y_i, v_1, v_2, x_i, v_4)$ is an induced C_5 by definition, it follows that v_4 must be dominated by y_i. Otherwise, (v_1, y_i) and (v_1, v_2) would be two arc-disjoint odd sectors of C'. Hence, C' is a conflicting odd cycle, a contradiction (see Fig. 5a). Thus, $P' = (v_1, y_i, v_4)$ is a 3-rainbow path of C'. Let $Z = (Y \setminus \{y_i\}) \cup \{v_5\}$. Similarly to what happened before, $D[Z]$ induces a semicomplete digraph that is color-compatible with P'. Moreover, every vertex of Z is adjacent to every vertex of P'. Since, by Theorem 1, there is a hamiltonian path R in $D[Z]$, we may obtain a $(k+1)$-rainbow path of D by applying Lemma 3 to P' and R.

Hence, we may assume that, there is no $i \in \{1, \dots, k-2\}$ such that $v_1 \mapsto y_i$ and $x_i \mapsto v_2$. Let Y^- be the subset of vertices of Y that dominate v_1 and let X^+ be the subset of vertices of X such that $x_i \in X$ if and only if $y_i \notin Y$. Note that, by our assumption, v_2 must dominate every vertex in X^+.

Recall that every 3-rainbow path P of C contains both v_1 and v_2 and $v_1 \mapsto v_2$. Hence, $P = (v_5, v_1, v_2)$ or $P = (v_1, v_2, v_3)$. Thus, $\{v_1, v_2, v_5\} \cup Y^- \cup X^+$ or $\{v_1, v_2, v_3\} \cup Y^- \cup X^+$ contains exactly one vertex of each color class of $\mathcal{S}$. If $P = (v_5, v_1, v_2)$, let $T^- = Y^- \cup \{v_5\}$ and let $T^+ = X^+$; otherwise let $T^- = Y^-$ and let $T^+ = X^+ \cup \{v_3\}$. By Theorem 1, $D[T^-]$ has a hamiltonian path P_1 and $D[T^+]$ has a hamiltonian path P_2. Since every vertex of T^- dominates v_1 and v_2 dominates every vertex of T^+, it follows that $P_1 v_1 v_2 P_2$ is a $(k+1)$-rainbow path of D (see Fig. 5b).

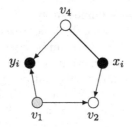

(a) the subpaths (y_i, v_1) and (v_1, v_2) are two arc-disjoint odd sectors.

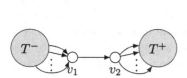

(b) the sets $T^- \cup \{v_1\}$ and $T^+ \cup \{v_2\}$ induce semicomplete and color-compatible digraphs in D.

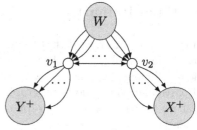

(c) the sets W, X^+ and Y^+ induce semicomplete digraphs in D. Moreover, $D[W \cup \{v_1, v_2\}]$ is color-compatible with $D[X^+]$ and $D[Y^+]$.

Fig. 5. Auxiliary illustration for the proof of Lemma 10

Case 2. *There is a digon between v_1 and v_2.*

By the Principle of Directional Duality, we may assume that v_5 dominates v_1. Let Y^- be the subset of vertices of Y that dominate v_1. Let X^- be the subset of vertices of X such that $x_i \in X^-$ if and only if $y_i \notin Y^-$ and x_i dominates v_2. Let $W = X^- \cup Y^- \cup \{v_5\}$.

So, if there is a color class $\{x_i, y_i\}$ such that $x_i \notin W$ and $y_i \notin W$, then v_1 dominates y_i and v_2 dominates x_i. Let X^+ be the subset of vertices of X such that $x_i \notin W$ and $y_i \notin W$ and let Y^+ be the subset of vertices of Y such that $x_i \notin W$ and $y_i \notin W$. Note that $x_i \in X^+$ if and only if $y_i \in Y^+$ and, hence, $|X^+| = |Y^+|$. Moreover, note that $D[W \cup \{v_1, v_2\}]$ is color-compatible with $D[X^+]$ and $D[Y^+]$, and $D[W], D[X^+], D[Y^+]$ induce semicomplete digraphs (see Fig. 5c). Let P', P_1 and P_2 be hamiltonian paths of $D[W], D[X^+], D[Y^+]$, respectively (such paths exist by Theorem 1). If $ter(P')$ dominates v_1, then $P'v_1v_2P_2$ is a $(k+1)$-rainbow path of D. Otherwise, $ter(P')$ dominates v_2 and $P'v_2v_1P_1$ is a $(k+1)$-rainbow path of D. ∎

Theorem 12. *Let D be a digraph in which every odd cycle of $\overline{U(D)}$ has length five. Then, D is χ-diperfect if and only if D does not contain a conflicting odd cycle as an induced subdigraph.*

Proof. (Necessity) The necessity immediately follows by Theorem 4.

(Sufficiency) To show that D is χ-diperfect, it suffices to prove that, for any minimum coloring of D, there is a $\chi(D)$-rainbow path in D. So towards a contradiction, suppose that D is an obstruction i.e. there is a minimum coloring of D that does not admit a $\chi(D)$-rainbow path but every proper induced subdigraph is χ-diperfect. Let $G = U(D)$. By Lemmas 2 and 4, we may assume that G is color-critical and that $\overline{G}$ is 2-connected. Since, by hypotheses, every odd cycle of $\overline{G}$ has length five, it follows that $\alpha(G) = 2$. By Corollary 2, $\overline{G}$ is a factor-critical graph. Hence, by Lemma 9, $\overline{G}$ is isomorphic to a nice graph. Thus, by Lemma 10, D has a $\chi(D)$-rainbow path, a contradiction. ■

6   Final Remarks

In this paper we showed that there are minimal non-χ-diperfect digraphs whose underlying graphs are neither an odd cycle of length at least five nor its complement; all these obstructions we have found have stability number two or three. In particular, the underlying graph of an obstruction with stability number two that we have found is a subgraph of some complement of an odd cycle of length at least seven.

Motivated by this fact we investigated a class of digraphs whose underlying graphs have stability number two such that every odd cycle of their complement has length exactly five. We proved that a digraph in this class is χ-diperfect if and only if it does not contain an induced conflicting odd cycle. The proof we presented is not straightforward, which suggests that figuring out the set of obstructions may be difficult. It is still open whether there is some obstruction with stability number at least four that is not a conflicting odd cycle.

References

1. Berge, C.: Diperfect graphs. Combinatorica **2**(3), 213–222 (1982)
2. Bondy, J.A., Murty, U.S.R.: Graph Theory. Springer, Heidelberg (2008)
3. Chudnovsky, M., Robertson, N., Seymour, P., Thomas, R.: The strong perfect graph theorem. Ann. Math. **164**, 51–229 (2006)
4. Gallai, T.: Kritische graphen ii. Magyar Tud. Akad. Mat. Kutato Int. Kozl. **8**, 373–395 (1963)
5. Gallai, T.: On directed paths and circuits. In: Theory of Graphs, pp. 115–118 (1968)
6. Gallai, T., Milgram, A.N.: Verallgemeinerung eines graphentheoretischen Satzes von Rédei. Acta Sci. Math. **21**, 181–186 (1960)
7. Lovász, L.: A note on factor-critical graphs. Studia Sci. Math. Hungar **7**(279–280), 11 (1972)
8. de Paula Silva, C.A.: χ-diperfect digraphs. Master's thesis, State University of Campinas - UNICAMP (2022)
9. de Paula Silva, C.A., da Silva, C.N., Lee, O.: χ-diperfect digraphs. Discret. Math. **345**(9), 112941 (2022)
10. Rédei, L.: Ein kombinatorischer satz. Acta Litt. Szeged **7**(39–43), 97 (1934)

11. Roy, B.: Nombre chromatique et plus longs chemins d'un graphe. ESAIM: Math. Model. Numer. Anal.-Modélisation Mathématique et Analyse Numérique **1**(5), 129–132 (1967)
12. Sambinelli, M.: Partition problems in graphs and digraphs. Ph.D. thesis, State University of Campinas - UNICAMP (2018)
13. Stehlík, M.: Critical graphs with connected complements. J. Comb. Theory Ser. B **89**(2), 189–194 (2003)

Percolation and Epidemic Processes in One-Dimensional Small-World Networks
(Extended Abstract)

Luca Becchetti[1], Andrea Clementi[2], Riccardo Denni[1], Francesco Pasquale[2],
Luca Trevisan[3], and Isabella Ziccardi[4(✉)]

[1] Sapienza Università di Roma, Rome, Italy
{becchetti,denni}@dis.uniroma1.it
[2] Università di Roma Tor Vergata, Rome, Italy
{clementi,pasquale}@mat.uniroma2.it
[3] Bocconi University, Milan, Italy
l.trevisan@unibocconi.it
[4] Università dell'Aquila, L'Aquila, Italy
isabella.ziccardi@graduate.univaq.it

Abstract. We obtain tight thresholds for bond percolation on one-dimensional small-world graphs, and apply such results to obtain tight thresholds for the *Independent Cascade* process and the *Reed-Frost* process in such graphs.

Although one-dimensional small-world graphs are an idealized and unrealistic network model, a number of realistic qualitative epidemiological phenomena emerge from our analysis, including the epidemic spread through a sequence of local outbreaks, the danger posed by random connections, and the effect of super-spreader events.

Keywords: Percolation · Epidemic models · Small-world graphs

1 Introduction and Related Works

Given a graph $G = (V, E)$ and a bond percolation probability p, the bond percolation process is to subsample a random graph $G_p = (V, E_p)$ by independently choosing each edge of G to be included in E_p with probability p and to be omitted with probability $1 - p$. We will call G_p the *percolation graph* of G. The main

LT's work on this project has received funding from the European Research Council (ERC) under the European Union's Horizon 2020 research and innovation programme (grant agreement No. 834861). LB's work on this project was partially supported by the ERC Advanced Grant 788893 AMDROMA, the EC H2020RIA project "SoBigData++" (871042), the MIUR PRIN project ALGADIMAR. AC's and FP's work on this project was partially supported by the University of Rome "Tor Vergata" under research program "Beyond Borders" project ALBLOTECH (grant no. E89C20000620005).

© Springer Nature Switzerland AG 2022
A. Castañeda and F. Rodríguez-Henríquez (Eds.): LATIN 2022, LNCS 13568, pp. 476–492, 2022.
https://doi.org/10.1007/978-3-031-20624-5_29

questions that are studied about this process are whether G_p is likely to contain a large connected component, and what are the typical distances of reachable nodes in G_p.

The study of percolation originates in mathematical physics, where it has often been studied in the setting of infinite graphs, for example infinite lattices and infinite trees [24,32,35]. The study of percolation on finite graphs is of interest in computer science, because of its relation, or even equivalence, to a number of fundamental problems in network analysis [1,17,23,27] and in distributed and parallel computing [14,22]. For example, the percolation process arises in the study of *network reliability* in the presence of independent link failures [22,25]; in this case one is typically interested in *inverse problems*, such as designing networks that have a high probability of having a large connected component for a given edge failure probability $1 - p$.

This paper is motivated by the equivalence of the percolation process with the *Independent Cascade* process, which models the spread of information in networks [17,23], and with the *Reed-Frost* process of *Susceptible-Infectious-Recovered (SIR)* epidemic spreading [12,34].

In a SIR epidemiological process, every person, at any given time, is in one of three possible states: either susceptible (S) to the infection, or actively infectious and able to spread the infection (I), or recovered (R) from the illness, and immune to it. In a network SIR model, we represent people as nodes of a graph, and contacts between people as edges, and we have a probability p that each contact between an infectious person and a susceptible one transmits the infection. The Reed-Frost process, which is the simplest SIR network model, proceeds through synchronous time steps, the infectious state lasts for only one time step, and the graph does not change with time.

The Information Cascade process is meant to model information spreading in a social network, but it is essentially equivalent to the Reed-Frost process.[1] If we run the Reed-Frost process on a graph $G = (V, E)$ with an initial set I_0 and with a probability p that each contact between an infectious and a susceptible person leads to transmission, then the resulting process is equivalent to percolation on the graph G with parameter p in the following sense: the set of vertices reachable from I_0 in the percolation graph G_p has the same distribution as the set of nodes that are recovered at the end of the Reed-Frost process in G with I_0 as the initial set of infected nodes. Furthermore, the set of nodes infected in the first t steps (that is, the union of infectious and recovered nodes at time t) has the same distribution as the set of nodes reachable in the percolation graph G_p from I_0 in at most t steps.

Information Cascade and Reed-Frost processes on networks are able to capture a number of features of real-world epidemics, such as the fact that people

---

[1] The main difference is that Information Cascade allows the probability of "transmission" along an edge (u, v) to be a quantity $p_{(u,v)}$, but this generalization would also make sense and be well defined in the Reed-Frost model and in the percolation process. The case in which all the probabilities are equal is called the *homogenous* case.

typically have a small set of close contacts with whom they interact frequently, and more rare interactions with people outside this group, that different groups of people have different social habits that lead to different patterns of transmissions, that outbreaks start in a localized way and then spread out, and so on. Complex models that capture all these features typically have a large number of tunable parameters, that have to be carefully estimated, and have a behavior that defies rigorous analysis and that can be studied only via simulations.

In this work we are interested in finding the simplest model, having few parameters and defining a simple process, in which we could see the emergence of complex phenomena.

One-Dimensional Small-World Graphs. We choose to analyze the Reed-Frost process on *one-dimensional small-world* graphs, which is a fundamental generative model of networks in which there is a distinction between local connection (corresponding to close contacts such as family and coworkers) and long-range connections (corresponding to occasional contacts such as being seated next to each other in a restaurant or a train). Small-world graphs are a class of probabilistic generative models for graphs introduced by Watts and Strogatz [36], which are obtained by overlaying a low-dimensional lattice with additional random edges. A one-dimensional small-world graph is a cycle overlayed with additional random edges.

Because of our interest in studying the most basic models, with the fewest number of parameters, in which we can observe complex emergent behavior, we consider the following simplified generative model which was introduced in [30] and often adopted in different network applications [18,31,33]: the distribution of *one-dimensional small-world graphs* with parameter q on n vertices is just the union of a cycle with n vertices with an Erdős-Rényi random graph $\mathcal{G}_{n,q}$, in which edges are sampled independently and each pair of nodes has probability q of being an edge. We will focus on the sparse case in which $q = c/n$, with c constant, so that the overall graph has average degree $c + 2$ and maximum degree that is, with high probability, $O(\log n / \log \log n)$. As we will see, we are able to determine, for every value of c, an exact threshold for the critical probability of transmission and to establish that, above the threshold, the epidemic spreads with a realistic pattern of a number of localized outbreaks that progressively become more numerous.

We are also interested in modeling, again with the simplest possible model and with the fewest parameters, the phenomenon of *superspreading*, encountered both in practice and in simulations of more complex models. This is the phenomenon by which the spread of an epidemic is disproportionately affected by rare events in which an infectious person contacts a large number of susceptible ones. To this end, we also consider a generative model of small-world one-dimensional graphs obtained as the union of a cycle with a random perfect matching. This generative model has several statistical properties in common with the $c = 1$ instantiation of the above generative model: the marginal distribution of each edge is the same, and edges are independent in the first case and have low correlation in the random matching model. The only difference is the degree distribution, which is somewhat irregular (but with a rapidly decreasing exponential tail) in the first case and essentially 3-regular in the second case.

Before proceeding with a statement of our results, we highlight for future reference the definitions of our generative models we consider.

Definition 1 (1-D Small-World Graphs - $\mathcal{SWG}(n, q)$). *For every $n \geq 3$ and $0 \leq q \leq 1$, the distribution $\mathcal{SWG}(n, q)$ is sampled by generating a one-dimensional small-world graph $G = (V, E)$, where $|V| = n$, $E = E_1 \cup E_2$, (V, E_1) is a cycle, and E_2 is the set of random edges, called* bridges, *of an Erdős-Rényi random graph $\mathcal{G}_{n,q}$.*

Definition 2 (3-regular 1-D Small-World Graphs - 3-$\mathcal{SWG}(n)$). *For every even $n \geq 4$, the distribution 3-$\mathcal{SWG}(n)$ is sampled by generating a one-dimensional small-world graph $G = (V, E)$, where $|V| = n$, $E = E_1 \cup E_2$, (V, E_1) is a cycle, and E_2 is the set of edges, called* bridges, *of a uniformly chosen perfect matching on V.*

In the definition of 3-$\mathcal{SWG}(n)$, we allow edges of the perfect matching to belong to E_1. If this happens, only edges in $E_2 - E_1$ are called bridges. The graphs sampled from 3-$\mathcal{SWG}(n)$ have maximum degree 3, and every node has degree 3 or 2. On average, only $\mathcal{O}(1)$ nodes have degree 2. This is why, with a slight abuse of terminology, we refer to these graphs as being "3-regular"[2].

2 Our Contribution

2.1 Tight Thresholds for Bond Percolation

Our main technical results are to establish sharp bounds for the critical percolation probability p in the $\mathcal{SWG}(n, q)$ model. In particular, we are interested in fully rigorous analysis that hold in high concentration (i.e., with high probability), avoiding mean-field approximations or approximations that treat certain correlated events as independent, which are common in the analysis of complex networks in the physics literature. While such approximations are necessary when dealing with otherwise intractable problems, they can fail to capture subtle differences between models. For example, for $q = 1/n$, the marginal distributions of bridge edges are the same in the two models above, while correlations between edges are non-existing in the $\mathcal{SWG}(n, q)$ model and very small in the 3-$\mathcal{SWG}(n)$ model. Yet, though the two models have similar expected behaviors and are good approximations of each other, our rigorous analysis shows that the two models exhibit notably different thresholds.

As for the $\mathcal{SWG}(n, q)$ model, we show the following threshold behaviour of the bond-percolation process.

---

[2] We recall that the 3-$\mathcal{SWG}(n)$ model and random 3-regular graphs are *contiguous*, i.e. each property that holds with probability $1 - o(1)$ on one of the two models, holds with probability $1 - o(1)$ also in the other one [21].

Theorem 1 (Percolation on the $\mathcal{SWG}(n,q)$ model). *Let $p > 0$ be a bond percolation probability. For any constant $c > 0$, sample a graph $G = (V, E_1 \cup E_2)$ from the $\mathcal{SWG}(n, c/n)$ distribution, and consider the percolation graph G_p. If we define*

$$p_c = \frac{\sqrt{c^2 + 6c + 1} - c - 1}{2c}$$

we have that, for any constant $\varepsilon > 0$:

1. *If $p > p_c + \varepsilon$, w.h.p.[3] a subset of nodes of size $\Omega_\varepsilon(n)$ exists that induces a subgraph of G_p having diameter $\mathcal{O}_\varepsilon(\log n)$;*
2. *If $p < p_c - \varepsilon$, w.h.p. all the connected components of G_p have size $\mathcal{O}_\varepsilon(\log n)$.*

Some remarks are in order. In the theorem above, probabilities are taken both over the randomness in the generation of the graph G and over the randomness of the percolation process. We highlight the sharp result on the $\mathcal{SWG}(n, c/n)$ model for the case $c = 1$: similarly to the regular 3-$\mathcal{SWG}(n)$ model, each node here has one bridge edge in average, and the obtained critical value for the percolation probability p turns out to be $\sqrt{2} - 1$.

Sharp bounds on the percolation threshold for the 3-$\mathcal{SWG}(n)$ model have been obtained by Goerdt in [19]. His analysis shows that such bounds hold with probability $1 - o(1)$ (converging to 1 when the number of nodes in the graph tends to infinity), while the analysis approach we introduce to obtain Theorem 1 also provides an alternative proof that achieves better concentration probability.

Theorem 2 (Percolation on the 3-$\mathcal{SWG}(n)$ model). *Let $p > 0$ be a bond percolation probability. Sample a graph $G = (V, E_1 \cup E_2)$ from the 3-$\mathcal{SWG}(n)$ distribution, and consider the percolation graph G_p. For any constant $\varepsilon > 0$:*

1. *If $p > 1/2 + \varepsilon$, w.h.p. a subset of nodes of size $\Omega_\varepsilon(n)$ exists that induces a connected subgraph (i.e. a giant connected component) of G_p;*
2. *If $p < 1/2 - \varepsilon$, w.h.p. all the connected components of G_p have size $\mathcal{O}_\varepsilon(\log n)$.*

A detailed comparison of the two models is provided in Subsect. 2.2, after Theorem 2. An overall view of our analysis, leading to all the theorems above, is provided in Sect. 4, while in the next subsection, we describe the main consequences of our analysis for the Independent-Cascade protocol on the considered small-world models.

2.2   Applications to Epidemic Processes

As remarked in Sect. 1, bond percolation with percolation probability p is equivalent to the Reed-Frost process (for short, RF process) with transmission probability p. Informally speaking, the nodes at hop-distance t in the percolation graph G_p, from any fixed source subset, are distributed exactly as those that will be informed (and activated) at time t, according to the RF process.

---

[3] As usual, we say that an event E holds with high probability (for short, w.h.p.) if a constant $\gamma > 0$ exists such that $\mathbf{Pr}\,(E) > 1 - n^{-\gamma}$.

In this setting, our analysis and results, we described in Subsect. 2.1, have the following important consequences.

Theorem 3 (The RF process on the $\mathcal{SWG}(n,q)$ model). *Let $I_0 \subseteq V$ be a set of source nodes, and $p > 0$ a constant probability. For any constant $c > 0$, sample a graph $G = (V, E_1 \cup E_2)$ from the $\mathcal{SWG}(n, c/n)$ distribution, and run the RF process with transmission probability p over G from I_0. If we define*

$$p_c = \frac{\sqrt{c^2 + 6c + 1} - c - 1}{2c},$$

for every $\varepsilon > 0$, we have the following:

1. *If $p > p_c + \varepsilon$, with probability $\Omega_\varepsilon(1)$ a subset of $\Omega_\varepsilon(n)$ nodes will be infectious within time $\mathcal{O}_\varepsilon(\log n)$, even if $|I_0| = 1$. Moreover, if $|I_0| \geq \beta \log n$ for a sufficiently large constant $\beta = \beta(\varepsilon)$, then the above event occurs w.h.p.;*
2. *If $p < p_c - \varepsilon$, w.h.p. the process will stop within $\mathcal{O}_\varepsilon(\log n)$ time steps, and the number of recovered nodes at the end of the process will be $\mathcal{O}_\varepsilon(|I_0| \log n)$.*

As for the 3-$\mathcal{SWG}(n)$ model, we get the following results for the Reed-Frost process.

Theorem 4 (The RF process on the 3-$\mathcal{SWG}(n)$ model). *Let V be a set of n vertices, $I_0 \subseteq V$ be a set of source nodes, and $p > 0$ be a bond percolation probability. Sample a graph $G = (V, E_1 \cup E_2)$ from the 3-$\mathcal{SWG}(n)$ distribution, and run the RF protocol with transmission probability p over G from I_0. For every $\varepsilon > 0$, we have the following:*

1. *If $p > 1/2 + \varepsilon$, with probability $\Omega_\varepsilon(1)$, a subset of $\Omega_\varepsilon(n)$ nodes will be infectious within time $\mathcal{O}_\varepsilon(n)$, even if $|I_0| = 1$. Moreover, if $|I_0| \geq \beta \log n$ for a sufficiently large constant $\beta = \beta(\varepsilon)$, then the above event occurs w.h.p.;*
2. *If $p < 1/2 - \varepsilon$, then, w.h.p., the process will stop within $O_\varepsilon(\log n)$ time steps, and the number of recovered nodes at the end of the process will be $O_\varepsilon(|I_0| \log n)$.*

We notice that the first claim of each of the above two theorems, concerning the multi-source case, i.e. the case $|I_0| \geq \beta \log n$), are not direct consequences of (the corresponding first claims of) Theorems 1 and 2: although each element of I_0 has constant probability of belonging to the "giant component" of the graph G_p, these events are not independent, and so it is not immediate that, when $|I_0|$ is of the order of $\log n$, at least an element of I_0 belongs to the giant component with high probability. Such claims instead are non-trivial consequences of our technical analysis.

On the other hand, the second claims of the above two theorems are simple consequences of the corresponding claims of Theorems 1 and 2.

From a topological point of view, because of a mix of local and random edges, epidemic spreading in the above models proceeds as a sequence of outbreaks, a process that is made explicit in our rigorous analysis, where we see the emergence of two qualitative phenomena that are present in real-world epidemic spreading.

One is that the presence of long-distance random connections has a stronger effect on epidemic spreading than local connections, that, in epidemic scenarios, might motivate lockdown measures that shut down long-distance connections. This can be seen, quantitatively, in the fact that the critical probability in a cycle is $p = 1$, corresponding to a critical *basic reproduction number*[4] R_0 equal to 2. On the other hand, the presence of random matching edges or random $\mathcal{G}_{n,c/n}$ edges in the setting $c = 1$ defines networks in which the critical R_0 is, respectively, 1.5 and $3 \cdot (\sqrt{2} - 1) \approx 1.24$, meaning that notably fewer local infections can lead to large-scale contagion on a global scale.

The other phenomenon is that the irregular networks of the $\mathcal{SWG}(n, c/n)$ model in the case $c = 1$ show a significantly lower critical probability, i.e. $\sqrt{2}-1 \approx .41$, than the critical value .5 of the nearly regular networks of the 3-$\mathcal{SWG}(n)$ model, though they have the same number of edges (up to lower order terms) and very similar distributions. As a further evidence of this phenomenon, we remark the scenario yielded by the random irregular networks sampled from the $\mathcal{SWG}(n, c/n)$ distribution with c even smaller than 1: for instance, the setting $c = .7$, though yielding a much sparser topology than the 3-$\mathcal{SWG}(n)$ networks, has a critical probability which is still smaller than .5. Moreover, this significant difference between the $\mathcal{SWG}(n, c/n)$ model and the regular 3-$\mathcal{SWG}(n)$ one holds even for more dense regimes. In detail, the almost-regular version of $\mathcal{SWG}$ in which c independent random matchings are added to the ring of n nodes has a critical probability $1/(c + 1)$[5]. Then, simple calculus shows that the critical probability given by Theorem 3 for the $\mathcal{SWG}(n, c/n)$ model is smaller than $1/(c + 1)$, for any choice of the density parameter c.

The most significant difference between the two distributions above is the presence of a small number of high-degree vertices in $\mathcal{SWG}(n, c/n)$, suggesting that even a small number of "super-spreader" nodes can have major global consequences.

2.3 Extensions of Our Results for Epidemic Models

Non-homogenous Transmission Probability. Our techniques allow extensions of our results to a natural non-homogenous bond-percolation process on small-world graphs, in which local edges percolate with probability p_1, while bridges percolates with probability p_2: our analysis in fact keeps the role of the two type of connections above well separated from each other. We are inspired,

---

[4] The quantity R_0 in a SIR process is the expected number of people that an infectious person transmits the infection to, if all the contacts of that person are susceptible. In the percolation view of the process, it is the average degree of the percolation graph G_p.

[5] This result follows from the following two facts: the $(c+2)$-regular version of the $\mathcal{SWG}$ model is contiguous to the random $(c + 2)$-regular model [21], and the threshold for the latter obtained in [19].

for instance, by epidemic scenarios in which the chances for any node to get infected/informed by a local tie are significantly higher than those from sporadic, long ties.

In this non-homogenous setting, for the $\mathcal{SWG}(n,q)$ model with $q = c/n$ for some absolute constant $c > 0$, a direct consequence of our results is that, w.h.p., the Independent-Cascade protocol reaches $\Omega(n)$ nodes within $\mathcal{O}(\log n)$ time iff the following condition on the three parameters of the process is satisfied

$$p_1 + c \cdot p_1 p_2 + c \cdot p_2 > 1 \, .$$

Some comments are in order. In the case $c = 1$, the formula above shows a perfect symmetry in the role of the two bond probabilities p_1 and p_2. In a graph sampled from $\mathcal{SWG}(n, 1/n)$, however, the overall number of local ties (i.e. ring edges) is n, while the number of bridges is highly concentrated on $n/2$ (it is w.h.p. $\leqslant n/2 + \sqrt{n \log n}$). This means that a public-health intervention aimed at reducing transmission has to suppress twice as much local transmissions in order to obtain the same effect of reducing by a certain amount the number of long-range transmissions. If we consider the case $c = 2$, in which the number of bridges is about equal to the number of local edges, we see that the impact of a change in p_2 weighs roughly twice as much as a corresponding change p_1.

So, even in the fairly unrealistic one-dimensional small-world model, it is possible to recover analytical evidences for the effectiveness of public-health measures that block or limit long-range mobility and super-events (such as football matches, international concerts, etc.). The generalization to non-homogenous transmission probabilities is provided in the full version of the paper [4].

Longer Node Activity and Incubation. Natural generalizations of the setting considered in this work include models in which i) the interval of time during which a node is active (i.e., the *activity period*) follows some (possibly node-dependent) distribution and/or ii) once infected, a node only becomes active after an *incubation* period, whose duration again follows some distribution. While the introduction of activity periods following general distributions may considerably complicate the analysis, our approach rather straightforwardly extends to two interesting cases, in which the incubation period of each node is a random variable (as long as incubation periods are independent) and/or the activity period of a node consists of k consecutive units of time, with k a fixed constant. This generalized model with random, node-dependent incubation periods corresponds to a discrete, synchronous version of the SEIR model,[6] which was recently considered as a model of the COVID-19 outbreak in Wuhan [28]. These extensions are shown in the full version of the paper [4].

Roadmap. Due to the page limit, the paper is organized as follows. In Sect. 3 further related work is summarized, while the most related, important previous contributions have been already mentioned in the previous sections. Section 4 gives an overall description of the main ideas and technical results behind our

---

[6] With respect to SIR, for each node we have a fourth, *Exposed* state, corresponding to the incubation period of a node.

analysis of bond-percolation in one-dimensional small-world graphs. The full proofs and the generalizations can be found in the full version of the paper [4].

3 Further Related Work

The *fully-mixed* SIR model [34] is the simplest SIR epidemiological model, and it treats the number of people in each of the three possible states as continuous quantities that evolve in time in accordance with certain differential equations. In this setup, the evolution of the process is governed by the expected number R_0 of people that each infectious person would infect, if all the contacts of that person were susceptible. If $R_0 < 1$, the process quickly ends, reaching a state with zero infectious people and a small number of recovered ones. If $R_0 > 1$, the process goes through an initial phase in which the number of infectious people grows exponentially with time, until the number of recovered people becomes a $1 - 1/R_0$ fraction of the population (the *herd immunity threshold*); the number of infectious people decreases after that, and eventually the process ends with a constant fraction of the population in the recovered state.

If we consider the Reed-Frost process on a graph G that is a clique on n vertices, then the percolation graph G_p is an Erdős-Rényi random graph with edge probability sampled from $\mathcal{G}_{n,p}$. Classical results from the analysis of random graphs give us that if $pn < 1$ then, with high probability, all the connected components of the graph have size $\mathcal{O}(\log n)$, and so the set of vertices that is reachable from I_0 has cardinality at most $\mathcal{O}(|I_0| \cdot \log n)$ and if $pn > 1$ then there is a connected component of cardinality $\Omega(n)$, and, except with probability exponentially small in I_0, at least one vertex of I_0 belongs to the giant component and is able to reach $\Omega(n)$ vertices. The parameter R_0 of the fully mixed continuous model corresponds to the average degree of G_p, which is pn if G_p is distributed as $\mathcal{G}_{n,p}$, so we see that the fully mixed continuous model agrees with the Reed-Frost process on a clique.

A number of techniques have been developed to study percolation in graphs other than the clique, and there is a vast body of work devoted to the study of models of bond percolation and epidemic spreading, as surveyed in [34,37]. Below, we review analytical studies of such processes on finite graphs. As far as we know, our results are the first rigorous ones to establish threshold phenomena in small-world graphs for the bond-percolation process (and, thus, for the Reed-Frost process).

There has been some previous work on studying sufficient conditions for the RF process to reach a sublinear number of vertices.

In [15], for a symmetric, connected graph $G = (V, E)$, Draief et al. prove a general lower bound on the critical point for the IC process in terms of spectral properties. Further versions of such bounds for special cases have been subsequently derived in [26,27]. Specifically, if one lets P be the matrix such that $P(u,v) = p(u,v)$ is the percolation probability of the edge $\{u,v\}$, and $P(u,v) = 0$ if $\{u,v\} \notin E$, and if one call λ the largest eigenvalue of P, then $\lambda < 1 - \epsilon$ implies that for a random start vertex s we have that the expected number of vertices to which s spreads the infection is $o_\epsilon(n)$.

In the RF process, in which all probabilities are the same, $P = p \cdot A$, where A is the adjacency matrix of G, and so the condition is asking for $p < (1-\epsilon)/\lambda_{\max}(A)$.

This condition is typically not tight, and it is never tight in the "small-worlds" graphs we consider:

- In the $3\text{-}\mathcal{SWG}(n)$ model, the largest eigenvalue of the adjacency matrix is $3 - o(1)$, but the critical probability is $1/2$ and not $1/3$;
- In the $\mathcal{SWG}(n, 1/n)$ model of a cycle plus Erdős-Rényi edges, the largest eigenvalue of the adjacency matrix is typically $\Omega(\sqrt{\log n / \log \log n})$ because we expect to see vertices of degree $\Omega(\log n / \log \log n)$ and the largest eigenvalue of the adjacency matrix of a graph is at least the square root of its maximal degree. The spectral bound would only tell us that the infection dies out if $p = \mathcal{O}(\sqrt{\log \log n / \log n})$, which goes to zero with n. A better way to use the spectral approach is to model the randomness of the small-world graph and the randomness of the percolation together; in this case, we have matrix $P(u, v)$ such that $P(u, v) = p$ for edges of the cycle and $P(u, v) = p/n$ for the other edges. This matrix has the largest eigenvalue $3p - o(1)$, so the spectral method would give a probability of $1/3$, while we can locate the threshold at $\sqrt{2} - 1 \approx .41$.

We are not aware of previous rigorous results that provide sufficient conditions for the IC process to reach $\Omega(n)$ nodes (either on average or with high probability) in general graphs, or for the equivalent question of proving that the percolation graph of a given graph has a connected component with $\Omega(n)$ vertices.

A fundamental and rigorous study of bond percolation in random graphs has been proposed by Bollobás et al. in [9]. They establish a coupling between the bond percolation process and a suitably defined branching process. In the general class of inhomogenous Erdős-Rényi random graphs, they derived the critical point (threshold) of the phase transition and the size of the giant component above the transition. The class of inhomogeneous random graphs to which their analysis applies includes generative models that have been studied in the complex network literature. For instance, a version of the Dubin's model [16] can be expressed in this way, and so can the *mean-field scale-free model* [10], which is, in turn, related to the Barabási-Albert model [3], having the same individual edge probabilities, but with edges present independently. Finally, we observe that the popular CHKNS model introduced by Callaway et al. [11] can be analyzed using an edge-independent version of this model. Indeed, they consider a random graph-formation process where, after adding each node, a Poisson number of edges is added to the graph, again choosing the endpoints of these edges uniformly at random. For all such important classes of random graph models, they show tight bounds for the critical points and the relative size of the giant component beyond the phase transition.

In our setting, if we sample a graph from $\mathcal{SWG}(n, q)$ and then consider the percolation graph G_p, the distribution of G_p is that of an inhomogenous Erdős-Rényi graph in which the cycle edges have probability p and the remaining edges have probability pq (the 3-$\mathcal{SWG}(n)$ model, however, cannot be expressed as an inhomogenous Erdős-Rényi graph). Unfortunately, if we try to apply the results of [9] to the inhomogeneous random graph equivalent to percolation with parameter p in the $\mathcal{SWG}(n,q)$ model, we do not obtain tractable conditions on the critical value p for which the corresponding graph has a large connected component of small diameter, which is the kind of result that we are interested in proving.

Bond percolation and the IC process on the class of one-dimensional small-world networks have been studied in [30]: using numerical approximations on the moment generating function, non-rigorous bounds on the critical threshold have been derived while analytical results are given neither for the expected size of the number of informed nodes above the transition phase of the process nor for its completion time. Further non-rigorous results on the critical points of several classes of complex networks have been derived in [26,27] (for good surveys see [34,37]).

In [6–8], different versions of the bond percolation process has been studied in small-world structures formed by a d-dimensional grid augmented by random edges that follow a power-law distribution: a bridge between points x and y is selected with probability $\sim 1/\text{dist}(x,y)^\alpha$, where $\text{dist}(x,y)$ is the grid distance between x and y and α is a fixed power-law parameter. Besides other aspects, each version is characterized by: (1) whether the grid is infinite or finite, and (2) whether the grid edges (local ties) do percolate with probability p or not. Research in this setting has focused on the emergence of a large connected component and on its diameter as functions of the parameters d and α, while, to the best of our knowledge, no rigorous threshold bounds are known for the bond percolation probability p. In [5], the authors study the bond percolation process when $d = 1$ as α changes: they identify three different behaviors depending on α, and for each of these they give an approximation of the percolation threshold.

In the computer science community, to the best of our knowledge, Kempe et al. [23] were the first to investigate the IC process from an optimization perspective, in the context of viral marketing and opinion diffusion. In particular, they introduced the *Influence Maximization* problem, where the goal is to find a source subset of k nodes of an underlying graph to inform at time $t = 0$, so as to maximize the expected number of informed nodes at the end of the IC process. They prove this is an NP-hard problem and show a polynomial time algorithm achieving constant approximation. Further approximation results on a version of *Influence Maximization* in which the completion time of the process is considered can be found in [13,29].

4   Overview of Our Analysis

A standard technique in bond percolation, applied for example to percolation in infinite trees and in random graphs, is to analyze the process of running a BFS in the percolation graph, delaying decisions about the percolation of edges from a node w to unvisited vertices until the time w is taken out of the BFS queue.

In random graphs and infinite trees, the distribution of unvisited neighbors of w in the percolation graph remains simple, even conditioned on previous history, and one can model the size of the BFS queue as a Galton-Watson process, thus reducing the percolation analysis to standard results about branching processes. Let us briefly recall the definition of the Galton-Watson branching process.

Definition 3 (Galton-Watson Branching Process). *Let W be a nonnegative integer valued random variable, and let $\{W_t\}_{t\geq 1}$ be an infinite sequence of i.i.d. copies of W. The Galton-Watson Branching Process generated by W is the process $\{B_t\}_{t\geq 0}$ defined by the recursion*

$$B_t = \begin{cases} 1 & \text{if } t = 0 \\ B_{t-1} + W_t - 1 & \text{if } t > 0 \text{ and } B_{t-1} > 0 \\ 0 & \text{if } t > 0 \text{ and } B_{t-1} = 0. \end{cases} \quad (1)$$

We define $\sigma = \min\{t > 0 : B_t = 0\}$ (if no such t exists we set $\sigma = +\infty$).

A standard result about branching process is summarized in the following theorem.

Theorem 5 ([2]). *Let $\{B_t\}_{t\geq 0}$ be a Galton-Watson Branching Process generated by the integer valued random variable W. Then*

1. *If $\mathbf{E}[W] < 1$, the process dies with probability 1, i.e. $\mathbf{Pr}(\sigma < +\infty) = 1$;*
2. *If $\mathbf{E}[W] > 1$, there exists a constant $c > 0$ s.t. $\mathbf{Pr}(\sigma = +\infty) > c$.*

Basically, the above theorem implies that, if the number of vertices that we add at each step to the queue is less than one on average, the visit will reach on average a constant number of vertices, and if it is more than one and the graph is infinite the visit will reach on average an infinite number of vertices.

4.1 Analysis of Bond Percolation in the $\mathcal{SWG}(n,q)$ Model

In this section, we describe the key ingredients of our analysis of the $\mathcal{SWG}(n,q)$ model proving Theorems 1 and 3.

It would be very difficult to analyze a BFS exploration of the percolation graph of $\mathcal{SWG}(n,q)$, since the distribution of unvisited neighbors of a vertex w in the percolation graph is highly dependent on the previous history of the BFS (in particular, it matters whether none, one, or both of the neighbors of w along the cycle are already visited). Instead, and this is one of the technical innovations of our work, we define a modified BFS visit whose process is more tractable to analyze.

The main idea of our modified BFS is that in one step we do the following: after we pull a node w from the queue, we first look at the neighbors x of w that are reachable through bridge edges in the percolation graphs; then, for each "bridge neighbor" x of w, we visit the "local cluster" of x, that is, we explore the vertices reachable from x along paths that only consist of edges of the cycle

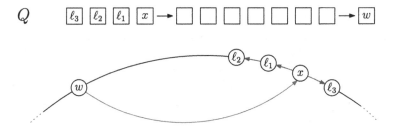

Fig. 1. The figure shows an example in which the visit first proceeds from a node w extracted from the queue to a new node x over a bridge of the percolation graph and then reaches further nodes, starting from x and proceeding along ring edges of the percolation graph.

that are in the percolation graph (we indicate the local cluster of x with $\mathrm{LC}(x)$); finally, we add to the queue all non-visited vertices in the local clusters of the bridge neighbors of w. These steps are exemplified in Fig. 1.

In fact, if we delay decisions about the random choice of the bridge edges and the random choices of the percolation, then we have a good understanding of the following two key random variables:

1. the number of bridge neighbors x of w along percolated bridge edges, which are, on average pqn' if the graph comes from $\mathcal{SWG}(n,q)$, p is the percolation probability, and n' is the number of unvisited vertices at that point in time;
2. the size of the "local cluster" of each such vertex x, that is of the vertices reachable from x along percolated cycle edges, which has expectation

$$\mathbf{E}\left[\mathrm{LC}(x)\right] = (1+p)/(1-p)\,. \tag{2}$$

Intuitively, we would hope to argue that in our modified visit of a graph sampled from $\mathcal{SWG}(n,q)$ to which we apply percolation with probability p, the following happens in one step: we remove one node from the queue, and we add on average

$$N = pqn' \cdot (1+p)/(1-p) \tag{3}$$

new nodes. As long as $n' = n - o(n)$ we can approximate n' with n, so the number $n - n'$ of visited vertices is $\Omega(n)$. This way, we would have modeled the size of the queue with a Galton-Watson process and we would be done. The threshold behavior would occur at a p such that $pqn \cdot (1+p)/(1-p) = 1$. A smaller value of p would imply that we remove one node at every step and, on average, add less than one node to the queue, leading the process to die out quickly. A larger value of p would imply that we remove one node at every step and, on average, add more than one node to the queue, leading the process to blow up until we reach $\Omega(n)$ vertices.

We are indeed able to prove this threshold behavior, at least for $q = c/n$ for constant c. However, we encounter significant difficulty in making this idea rigorous: if we simply proceeded as described above, we would be double-counting vertices, because in general, the "local cluster" of a node added to the queue at a certain point may collide with the local cluster of another node added at a later point. This may be fine as long as we are trying to upper bound the number of reachable vertices, but it is definitely a problem if we are trying to establish a lower bound.

To remedy this difficulty, we truncate the exploration of each local cluster at a properly chosen constant size L (we denote as $\mathrm{LC}^L(x)$ the truncated local cluster of a node x). In our visit, we consider only unvisited neighbors x of w that are sufficiently far along the cycle from all previously visited vertices so that there is always "enough space" to grow a truncated local cluster around x without hitting already visited vertices. In more detail, we introduce the notion of "*free* node" used in the algorithm and its analysis.

Definition 4 (*free* node). *Let $G_{SW} = (V, E_1 \cup E_2)$ be a small-world graph and let $L \in \mathbb{N}$. We say that a node $x \in V$ is free for a subset of nodes $X \subseteq V$ if x is at distance at least $L+1$ from any node in X in the subgraph (V, E_1) induced by the edges of the ring.*

Thanks to the above definition, we can now formalize our modified BFS.

Algorithm 1. SEQUENTIAL L-VISIT

Input: A small-world graph $G_{SW} = (V, E_{SW})$; a subgraph H of G_{SW}; a set of initiators $I_0 \subseteq V$;

1: $Q = I_0$
2: $R = \emptyset$
3: **while** $Q \neq \emptyset$ **do**
4:     $w = \mathrm{dequeue}(Q)$
5:     $R = R \cup \{w\}$
6:     **for** each bridge neighbor x of w in H **do**
7:        **if** x is *free* for $R \cup Q$ in G_{SW} **then**
8:           **for** each node y in the L-truncated local cluster $\mathrm{LC}^L(x)$ **do**
9:              $\mathrm{enqueue}(y, Q)$

To sum up, the L-truncation negligibly affects the average size of local clusters, the restriction to a subset of unvisited vertices negligibly affects the distribution of unvisited neighbors, and the analysis carries through with the same parameters and without the "collision of local clusters" problem.

In more detail, thanks to the arguments we described above, from (2) and (3), we can prove that, if p is above the critical threshold $p_c = \frac{\sqrt{c^2+6c+1}-c-1}{2c}$, then, with probability $\Omega(1)$, the connected components of G_p containing the initiator subset have overall size $\Omega(n)$. In terms of our BFS visit in Algorithm 1, we in fact derive the following result[7] (its full proof is in the full version of the paper).

---

[7] We state the result for the case $|I_0| = 1$.

Lemma 1. *Assume that we are under the hypothesis of Theorem 1. Let $s \in V$ an initiator node. If $p > p_c + \varepsilon$, there are positive parameters $L = L(c, \varepsilon)$, $k = k(c, \varepsilon)$, $t_0 = t_0(c, \varepsilon)$, $\varepsilon' = \varepsilon'(c, \varepsilon)$, and $\gamma = \gamma(c, \varepsilon)$ such that the following holds. Run the* SEQUENTIAL L-VISIT *procedure in Algorithm 1 on input (G, G_p, s): if n is sufficiently large, for every t larger than t_0, at the end of the t-th iteration of the while loop we have*

$$\mathbf{Pr}\left(|R \cup Q| \geq n/k \text{ OR } |Q| \geq \varepsilon' t\right) \geq \gamma.$$

The truncation is such that our modified BFS does not discover all vertices reachable from I_0 in the percolation graph, but only a subset. However, this is sufficient to prove lower bounds to the number of reachable vertices when p is above the threshold. Proving upper bounds, when p is under the threshold (i.e., the second claims of Theorems 1 and 3) is easier because, as mentioned, we can allow double-counting of reachable vertices.

The above line of reasoning is our key idea, when p is above the threshold, to get $\Omega_\varepsilon(1)$ confidence probability for: (i) the existence of a linear-size, induced connected subgraph in G_p (i.e., a "weaker" version of Claim 1 of Theorem 1), and (ii) the existence of a large epidemic outbreak, starting from an arbitrary source subset I_0 (i.e., Claim 1 of Theorem 3). In the full description of this analysis, we also describe the further technical steps to achieve *high-probability* for event (i) and, also, for event (ii) when the size of the source subset is $|I_0| = \Omega(\log n)$.

Bounding the Number of Hops: Parallelization of the BFS Visit. To get bounds on the number of the BFS levels, we study the BFS-visit in Algorithm 1 only up to the point where there are $\Omega(\log n)$ nodes in the queue (this first phase is not needed if I_0 already has size $\Omega(\log n)$), and then we study a *"parallel"* visit in which we add at once all nodes reachable through an L-truncated local cluster and through the bridges from the nodes currently in the queue, skipping those that would create problems with our invariants: to this aim, we need a stronger version of the notion of *free* node.

Here we can argue that, as long as the number of visited vertices is $o(n)$, the number of nodes in the queue grows by a constant factor in each iteration, and so we reach $\Omega(n)$ nodes in $\mathcal{O}(\log n)$ number of iterations that corresponds to $\mathcal{O}(\log n)$ distance from the source subset in the percolation graph G_p.

A technical issue that we need to address in the analysis of our parallel visit is that the random variables that count the contribution of each L-truncated local cluster, added during one iteration of the visit, are not mutually independent. To prove concentration results for this exponential growth, we thus need to show that such a mutual correlation satisfies a certain *local* property and then apply suitable bounds for partly-dependent random variables [20]

References

1. Alon, N., Benjamini, I., Stacey, A.: Percolation on finite graphs and isoperimetric inequalities. Ann. Probab. **32**(3), 1727–1745 (2004)

2. Alon, N., Spencer, J.H.: The Probabilistic Method. Wiley Publishing, 2nd edn. (2000)
3. Barabási, A.L., Albert, R.: Emergence of scaling in random networks. Science **286**(5439), 509–512 (1999). https://doi.org/10.1126/science.286.5439.509
4. Becchetti, L., Clementi, A., Denni, R., Pasquale, F., Trevisan, L., Ziccardi, I.: Percolation and epidemic processes in one-dimensional small-world networks. arXiv e-prints pp. arXiv-2103 (2021)
5. Becchetti, L., Clementi, A., Pasquale, F., Trevisan, L., Ziccardi, I.: Bond percolation in small-world graphs with power-law distribution. arXiv preprint arXiv:2205.08774 (2022)
6. Benjamini, I., Berger, N.: The diameter of long-range percolation clusters on finite cycles. Random Struct. Algorithms **19**(2), 102–111 (2001). https://doi.org/10. 1002/rsa.1022
7. Biskup, M.: On the scaling of the chemical distance in long-range percolation models. Ann. Probability **32**(4), 2938–2977 (2004). https://doi.org/10.1214/ 009117904000000577
8. Biskup, M.: Graph diameter in long-range percolation. Random Struct. Algorithms **39**(2), 210–227 (2011). https://doi.org/10.1002/rsa.20349
9. Bollobás, B., Janson, S., Riordan, O.: The phase transition in inhomogeneous random graphs. Random Struct. Algorithms **31**(1), 3–122 (2007). https://doi.org/10. 1002/rsa.20168. https://onlinelibrary.wiley.com/doi/abs/10.1002/rsa.20168
10. Bollobás, B., Riordan, O.: The diameter of a scale-free random graph. Comb. **24**(1), 5–34 (2004). https://doi.org/10.1007/s00493-004-0002-2
11. Callaway, D.S., Hopcroft, J.E., Kleinberg, J.M., Newman, M.E.J., Strogatz, S.H.: Are randomly grown graphs really random? Phys. Rev. E **64**, 041902 (2001). https://doi.org/10.1103/PhysRevE.64.041902
12. Chen, W., Lakshmanan, L., Castillo, C.: Information and influence propagation in social networks. Synthesis Lectures Data Manage. **5**, 1–177 (10 2013). https://doi. org/10.2200/S00527ED1V01Y201308DTM037
13. Chen, W., Lu, W., Zhang, N.: Time-critical influence maximization in social networks with time-delayed diffusion process. In: Proceedings of the Twenty-Sixth AAAI Conference on Artificial Intelligence, AAAI 2012, pp. 592–598. AAAI Press (2012)
14. Choi, H., Pant, M., Guha, S., Englund, D.: Percolation-based architecture for cluster state creation using photon-mediated entanglement between atomic memories. NPJ Quantum Information **5**(1), 104 (2019)
15. Draief, M., Ganesh, A., Massoulié, L.: Thresholds for virus spread on networks. In: Proceedings of the 1st International Conference on Performance Evaluation Methodolgies and Tools, p. 51-es. valuetools 2006. Association for Computing Machinery, New York (2006). https://doi.org/10.1145/1190095.1190160
16. Durrett, R., Kesten, H.: The critical parameter for connectedness of some random graphs. A Tribute to P. Erdos, pp. 161–176 (1990)
17. Easley, D., Kleinberg, J.: Networks, Crowds, and Markets: Reasoning About a Highly Connected World. Cambridge University Press, Cambridge (2010)
18. Garetto, M., Gong, W., Towsley, D.: Modeling malware spreading dynamics. In: IEEE INFOCOM 2003. Twenty-second Annual Joint Conference of the IEEE Computer and Communications Societies (IEEE Cat. No.03CH37428), vol. 3, pp. 1869–1879 (2003). https://doi.org/10.1109/INFCOM.2003.1209209
19. Goerdt, A.: The giant component threshold for random regular graphs with edge faults. Theor. Comput. Sci. **259**(1-2), 307–321 (2001). https://doi.org/10.1016/ S0304-3975(00)00015-3

20. Janson, S.: Large deviations for sums of partly dependent random variables. Random Struct. Algorithms **24**, May 2004. https://doi.org/10.1002/rsa.20008
21. Janson, S., Rucinski, A., Luczak, T.: Random Graphs. Wiley, New York (2011)
22. Karlin, A.R., Nelson, G., Tamaki, H.: On the fault tolerance of the butterfly. In: Proceedings of the Twenty-Sixth Annual ACM Symposium on Theory of Computing, pp. 125–133 (1994)
23. Kempe, D., Kleinberg, J., Tardos, E.: Maximizing the spread of influence through a social network. Theory Comput. **11**(4), 105–147 (2015). https://doi.org/10.4086/toc.2015.v011a004, (An extended abstract appeared in Proc. of 9th ACM KDD '03)
24. Kesten, H.: The critical probability of bond percolation on the square lattice equals 1/2. Commun. Math. Phys. **74**(1), 41–59 (1980)
25. Kott, A., Linkov, I. (eds.): Cyber Resilience of Systems and Networks. RSD, Springer, Cham (2019). https://doi.org/10.1007/978-3-319-77492-3
26. Lee, E.J., Kamath, S., Abbe, E., Kulkarni, S.R.: Spectral bounds for independent cascade model with sensitive edges. In: 2016 Annual Conference on Information Science and Systems, CISS 2016, Princeton, NJ, USA, 16–18 March 2016, pp. 649–653. IEEE (2016). https://doi.org/10.1109/CISS.2016.7460579
27. Lemonnier, R., Seaman, K., Vayatis, N.: Tight bounds for influence in diffusion networks and application to bond percolation and epidemiology. In: Proceedings of the 27th International Conference on Neural Information Processing Systems - Volume 1, NIPS 2014, pp. 846–854. MIT Press, Cambridge (2014)
28. Lin, Q., et al.: A conceptual model for the coronavirus disease 2019 (COVID-19) outbreak in Wuhan, China with individual reaction and governmental action. Int. J. Infect. Dis. **93**, 211–216 (2020)
29. Liu, B., Cong, G., Xu, D., Zeng, Y.: Time constrained influence maximization in social networks. In: Proceedings of the IEEE International Conference on Data Mining, ICDM, pp. 439–448, December 2012. https://doi.org/10.1109/ICDM.2012.158
30. Moore, C., Newman, M.E.: Exact solution of site and bond percolation on small-world networks. Phys. Rev. E **62**, 7059–7064 (2000). https://doi.org/10.1103/PhysRevE.62.7059
31. Moore, C., Newman, M.E.: Epidemics and percolation in small-world networks. Phys. Rev. E **61**, 5678–5682 (2000). https://doi.org/10.1103/PhysRevE.61.5678
32. Newman, C.M., Schulman, L.S.: One dimensional $1/|j-i|^s$ percolation models: the existence of a transition for $s \leq 2$. Commun. Math. Phys. **104**(4), 547–571 (1986)
33. Newman, M.E., Watts, D.J.: Scaling and percolation in the small-world network model. Phys. Rev. E **60**(6), 7332 (1999)
34. Pastor-Satorras, R., Castellano, C., Van Mieghem, P., Vespignani, A.: Epidemic processes in complex networks. Rev. Mod. Phys. **87**, 925–979 (2015). https://doi.org/10.1103/RevModPhys.87.925
35. Shante, V.K., Kirkpatrick, S.: An introduction to percolation theory. Adv. Phys. **20**(85), 325–357 (1971). https://doi.org/10.1080/00018737100101261
36. Watts, D.J., Strogatz, S.H.: Collective dynamics of 'small-world' networks. Nature **393**(6684), 440–442 (1998)
37. Wei, W., Ming, T., Eugene, S., Braunstein, L.A.: Unification of theoretical approaches for epidemic spreading on complex networks. Reports Progress Phys. **80**(3), 036603 (2017). https://doi.org/10.1088/1361-6633/aa5398

A Combinatorial Link Between Labelled Graphs and Increasingly Labelled Schröder Trees

Olivier Bodini[1], Antoine Genitrini[2], and Mehdi Naima[2(✉)]

[1] Université Sorbonne Paris Nord – LIPN – CNRS, UMR 7030, Villetaneuse, France
Olivier.Bodini@lipn.univ-paris13.fr
[2] Sorbonne Université – LIP6 – CNRS UMR 7606, Paris, France
{Antoine.Genitrini,Mehdi.Naima}@lip6.fr

Abstract. In this paper we study a model of Schröder trees whose labelling is increasing along the branches. Such tree family is useful in the context of phylogenetic. The tree nodes are of arbitrary arity (i.e. out-degree) and the node labels can be repeated throughout different branches of the tree. Once a formal construction of the trees is formalized, we then turn to the enumeration of the trees inspired by a renormalisation due to Stanley on acyclic orientations of graphs. We thus exhibit links between our tree model and labelled graphs and prove a one-to-one correspondence between a subclass of our trees and labelled graphs. As a by-product we obtain a new natural combinatorial interpretation of Stanley's renormalising factor. We then study different combinatorial characteristics of our tree model and finally, we design an efficient uniform random sampler for our tree model based on the classical recursive generation method.

Keywords: Evolution process · Schröder trees · Increasing trees · Monotonic trees · Labelled graphs · Combinatorics · Uniform sampling

1 Introduction

Increasing trees are ubiquitous in combinatorics especially because they aim at modelling various classical phenomena: phylogenetics, the frequencies of family names or the graph of the Internet [24] for example. Meir and Moon [19] studied the distance between nodes in their now classical model of recursive trees. Bergeron *et al.* [1] studied several families of increasingly-labelled trees for a wide range of models embedded in the simple families of trees. We also refer to [7] where recent results on various families of increasing trees and the methods to study them, from a quantitative point of view, are surveyed.

Increasing trees can often be described as the result of a dynamical construction: the nodes are added one by one at successive integer-times in the tree (their labels being the time when they are added). This dynamical process sometimes

© Springer Nature Switzerland AG 2022
A. Castañeda and F. Rodríguez-Henríquez (Eds.): LATIN 2022, LNCS 13568, pp. 493–509, 2022.
https://doi.org/10.1007/978-3-031-20624-5_30

allows us to apply probabilistic methods to show results about different characteristics on the trees and often gives an efficient way to uniformly sample large
trees using simple, iterative and local rules.

In the recent years, many links were found between evolution processes in
the form of increasing trees and classical combinatorial structures, for instance
permutations are known to be in bijection with increasing binary trees [10, p.
143], increasing even trees and alternating permutations are put in bijection
in [5,16], plane recursive trees are related to Stirling permutations [15] and more
recently increasing Schröder trees have been proved in one-to-one correspondence
with even permutations and with weak orderings on sets of n elements (counted
by ordered Bell numbers) in [2,3]. By adding some constraint in the increasing
labelling of the latter model, Lin *et al.* [17] exhibited closed relationships between
various families of polynomials (especially Eulerian, Narayana and Savage and
Schuster polynomials).

The theory of analytic combinatorics developed in [10] gives a framework to
study many classes of discrete structures by applying principles based on the now
classical *symbolic method*. In various situations we get direct answers to questions
concerning the count of the number of objects, the study of typical shapes and
the development of methods for the uniform sampling of objects. Using this
approach we explore links between labelled directed graphs and an evolution
process that generates increasing trees seen as enriched Schröder trees. Schröder
in [23] studied trees with possible multifurcations to model phylogenetic. The
trees he studied were counted by their number of leaves which represent the
number of species. We pursue enriching Schröder trees in the same vein as [2,3]
but with a more general model.

Our evolution process can be reinterpreted as a builder for phylogenetic tree
that represents the evolutionary relationship among species. At each branching
node of the tree, the descendant species from distinct branches have distinguished
themselves in some manner and are no more dependent: the past is shared but
the futures are independent. For more information on the phylogenetic links the
reader may refer to the thesis [20].

The study of this evolution process leads to unexpected links between our
trees and labelled graphs: we then prove a bijection between both families of
structures. The links we find also give a new combinatorial interpretation of the
renormalisation factor that Stanley used in [25] based on ideas of [6] and more
recently for *graphic generating functions* e.g. in [13,22].

Our Main Contributions: A study of an evolution process that produces
increasing trees with label repetitions. The study of this evolution process using
tools of analytic combinatorics produces functional equations for generating
functions that are divergent. Next, using a renormalisation we provide a quantitative study for the enumeration problem and the asymptotic analysis of several
parameters. After that, we introduce a one-to-one correspondence between a
sub-family of our increasing trees and labelled graphs. Finally, we design a very
efficient unranking method for the sampling for such increasing trees which easily
translates to a uniform random sampler for the trees.

This work is part of a long term project, in which we aim at relaxing the classical rules of increasing labelling (described in, e.g., [1]), by allowing label repetitions in the tree.

In Table 1 we provide the main statistics of our enriched Schröder tree model that we will call *strict monotonic general tree model*. Due to its relationship with Schröder trees the size of a tree is given by its number of leaves, independently of its number of internal nodes.

Table 1. Main analytic results for the characteristics of a large typical tree. n stands for the size of the trees and the results are asymptotic when $n \to +\infty$.

	Number of trees	Average number of distinct labels	Average number of internal nodes	Average height
Strict monotonic general trees	$c\,(n-1)! \\ \cdot\, 2^{(n-1)(n-2)/2}$	$\Theta(n)$	$\Theta(n^2)$	$\Theta(n)$

Plan of the Paper: The paper is organized as follows. First, in Sect. 2 we present our evolution process and then extract from it a general recursive formula to count the number of trees of a given size. We end this section by giving the statement of Theorem 2 on the asymptotic enumeration of the trees. We postpone the proof to next section. Next, in Sect. 3 we count the trees according to their sizes, we also study the distribution of the number of iteration steps to construct the trees of a given size. As a result, we can simply prove Theorem 2 and give detailed characteristics of the shape of large trees. Based on the previous results, in Sect. 4 we make quantitative studies of several tree parameters: the number of internal labels and the height of the trees. We then turn to show the relationship between our trees and labelled graphs. We exhibit a bijection in Sect. 5. Finally, in Sect. 6 we design an unranking method for the sampling of strict monotonic general trees.

2   The Model and Its Enumeration

Definition 1. *A* strict monotonic general tree *is a labelled tree constructed by the following evolution process:*

- *Start with a single (unlabelled) leaf.*
- *At every step $\ell \geq 1$, select a non-empty subset of leaves, replace all of them by internal nodes labelled by ℓ, attach to at least one of them a sequence of two leaves or more, and attach to all others a unique leaf.*

The two trees in Fig. 1 are sampled uniformly among all strict monotonic general trees of respective sizes (i.e. number of leaves) 15 and 500. The left-hand-side tree has 14 distinct node-labels, i.e. it can be built in 14 steps using Definition 1. The right-hand-side tree is represented as a circular tree with stretched edges:

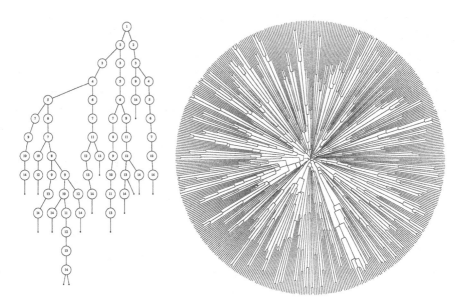

Fig. 1. Two strict monotonic general trees, with respective sizes 15 and 500

the length of an edge is proportional to the label difference of the two nodes it connects. Here the tree contains 500 leaves built with 499 iterations of the growth process. Thus the maximal arity is 2. This tree contains 62494 internal nodes almost all of them (except 499) being unary nodes.

We can specify strict monotonic general trees using the symbolic method [10]; the internal node labelling is transparent and does not appear in the specification in consequence, we use ordinary generating functions. We denote by $F(z)$ the generating function of strict monotonic general trees and by $\mathcal{F}_n$ the set of all strict monotonic general trees of size n; from Definition 1, we get

$$F(z) = z + F\left(z + \frac{z}{1-z}\right) - F(2z). \tag{1}$$

The combinatorial meaning of this specification is the following: A tree is either a single leaf, or it is obtained by taking an already constructed tree, and then each leaf is either replaced by a leaf (i.e. there is no change) or by an internal node attached to a sequence of at least one leaf. Furthermore we omit the case where no leaf is replaced by an internal node with at least two children (this is encoded in the subtracting $F(2z)$).

From this equation we extract the recurrence for the number f_n of strict monotonic general trees with n leaves. In fact we get

$$f_n = [z^n]F(z) = [z^n]\left(z + F\left(z + \frac{z}{1-z}\right) - F(2z)\right)$$

$$= \delta_{n,1} - 2^n f_n + [z^n]\sum_{\ell \geq 1} f_\ell \left(2z + \frac{z^2}{1-z}\right)^\ell$$

$$= \delta_{n,1} - 2^n f_n + \sum_{\ell \geq 1} f_\ell \ [z^{n-\ell}]\sum_{i=0}^{\ell}\binom{\ell}{i}2^{\ell-i}\left(\frac{z}{1-z}\right)^i,$$

which implies that

$$f_n = \begin{cases} 1 & \text{if } n = 1, \\ \displaystyle\sum_{\ell=1}^{n-1}\sum_{i=1}^{\min(n-\ell,\ell)}\binom{\ell}{i}\binom{n-\ell-1}{i-1}2^{\ell-i}f_\ell & \text{for all } n \geq 2. \end{cases} \quad (2)$$

The inner sum can be explained combinatorially: starting with a tree of size ℓ we reach a tree of size n in one iteration by adding $n - \ell$ leaves. The index i in the inner sum stands for the number of leaves that are replaced by internal nodes or arity at least 2, by definition of the model we have $1 \leq i \leq \min(n - \ell, \ell)$. There are $\binom{\ell}{i}$ possible choices for the i leaves that are replaced by nodes of arity at least 2. Each of the remaining $\ell - i$ leaves is either kept unchanged or replaced by a unary node, which gives $2^{\ell-i}$ possible choices. And finally, there are $\binom{n-\ell-1}{i-1}$ possible ways to distribute the (indistinguishable) $n - \ell$ additional leaves among the i new internal nodes so that each of the i nodes is given at least one additional leaf (it already has one leaf, which is the leaf that was replaced by an internal node). The first terms of the sequence are the following:

$$(f_n)_{n\geq 0} = (0, 1, 1, 5, 66, 2209, 180549, 35024830, 15769748262, \dots).$$

Theorem 2. *There exists a constant c such that the number f_n of strict monotonic general trees of size n satisfies, asymptotically when n tends to infinity,*

$$f_n \underset{n\to\infty}{\sim} c\ (n-1)!\ 2^{\frac{(n-1)(n-2)}{2}},$$

with $1.4991 < c < 8.9758$.

Through several experimentations we note that c is smaller than $3/2$ but it is close to it. For instance when $n = 1000$, we get $c \approx 1.49913911$. We postpone the proof to the next section to make use of the number of iteration steps.

3   Iteration Steps and Asymptotic Enumeration

In this section, we look at the number of distinct internal-node labels that occur in a typical strict monotonic general tree, i.e. the number of iterations needed to build it.

Proposition 1. *Let $f_{n,k}$ denotes the number of strict monotonic general trees of size n with k distinct node-labels, then, for all $n \geq 1$,*

$$f_{n,n-1} = (n-1)! \, 2^{\frac{(n-1)(n-2)}{2}}.$$

Note that the first terms are

$$(f_{n,n-1})_{n \geq 0} = (0, 1, 1, 4, 48, 1536, 122880, 23592960, 10569646080, \dots).$$

This is a shifted version of the sequence OEIS A011266 used by Stanley in [25] that is in relation with acyclic orientations of graphs. In particular $f_{n+1,n}$ is the renormalisation he used in the generating function context.

Proof. We use a new variable u to mark the number of iterations (i.e. the number of distinct node-labels) in the iterative Eq. (1). We get

$$F(z, u) = z + u \, F\left(z + \frac{z}{1-z}, u\right) - u \, F(2z, u). \tag{3}$$

Using either Eq. (3) or a direct combinatorial argument, we get that, for all $k \geq n$, $f_{n,k} = 0$ and

$$f_{n,k} = \begin{cases} 1 & \text{if } n = 1 \text{ and } k = 0, \\ \sum_{\ell=k}^{n-1} \sum_{i=1}^{\min(n-\ell,\ell)} \binom{\ell}{i} 2^{\ell-i} \binom{n-\ell-1}{i-1} f_{\ell,k-1} & \text{if } 1 \leq k < n. \end{cases}$$

In particular, for $k = n - 1$, we get $f_{n,n-1} = (n-1) \, 2^{n-2} \, f_{n-1,n-2}$. Solving the recurrence we get

$$f_{n,n-1} = f_{1,0} \prod_{j=1}^{n-1} j \, 2^{j-1} = (n-1)! \, 2^{\sum_{j=0}^{n-2} j} = (n-1)! \, 2^{\frac{(n-1)(n-2)}{2}},$$

because $f_{1,0} = 1$. This concludes the proof.   □

Note that alternatively the recurrence of $f_{n,n-1}$ can be obtained by extracting the coefficient $[z^n]$ in the following functional equation $T(z) = z + z^2 \, T'(2z)$.

Lemma 1. *Both sequences (f_n) and $(f_{n,n-1})$ have the same asymptotic behaviour up to a multiplicative constant.*

Proof. Let us start with the definition of a new sequence

$$g_n = \begin{cases} 1 & \text{if } n = 1, \\ f_n / f_{n,n-1} & \text{otherwise.} \end{cases}$$

This sequence g_n satisfies the following recurrence:

$$g_n = \begin{cases} 1 & \text{if } n = 1, \\ \sum_{\ell=1}^{n-1} \sum_{i=1}^{\min(n-\ell,\ell)} \binom{\ell}{i} 2^{\ell-i} \binom{n-\ell-1}{i-1} g_\ell \frac{(\ell-1)! \, 2^{(\ell-1)(\ell-2)/2}}{(n-1)! \, 2^{(n-1)(n-2)/2}} & \text{otherwise.} \end{cases}$$

When $n > 1$, extracting the term g_{n-1} from the sum we get

$$g_n = g_{n-1} + \sum_{\ell=1}^{n-2} \sum_{i=1}^{\min(n-\ell,\ell)} \binom{\ell}{i} 2^{\ell-i} \binom{n-\ell-1}{i-1} g_\ell \frac{(\ell-1)! \, 2^{(\ell-1)(\ell-2)/2}}{(n-1)! \, 2^{(n-1)(n-2)/2}}.$$

Since all summands are non-negative, this implies that $g_n \geq g_{n-1}$, and thus that this sequence is non-decreasing. To prove that this sequence converges, it only remains to prove that it is (upper-)bounded.

Equation (2) implies that, for $n \geq 2$,

$$f_n \leq \sum_{\ell=1}^{n-1} 2^{\ell-1} \sum_{i=1}^{\min(n-\ell,\ell)} \binom{\ell}{i} \binom{n-\ell-1}{i-1} f_\ell.$$

Chu-Vandermonde's identity states that, for all $\ell \leq n$,

$$\sum_{i=1}^{\min(n-\ell,\ell)} \binom{\ell}{i} \binom{n-\ell-1}{i-1} = \binom{n-1}{\ell-1}.$$

This implies the following upper-bound for f_n:

$$f_n \leq \sum_{\ell=1}^{n-1} 2^{\ell-1} \binom{n-1}{\ell-1} f_\ell = \sum_{\ell=1}^{n-1} 2^{n-\ell-1} \binom{n-1}{\ell} f_{n-\ell}.$$

Using the same argument for g_n we get

$$g_n \leq g_{n-1} + \sum_{\ell=2}^{n-1} \frac{2^{(\ell-1)(\ell-2n+2)/2}}{\ell!} g_{n-\ell}.$$

We look at the exponent of 2 in the sum: For all $\ell \geq 2$ (as in the sum), we have $2\ell \geq \ell + 2$, and thus $2n - \ell - 2 \geq 2(n - \ell)$. This implies that for all $\ell \geq 2$, $(\ell-1)(\ell - 2n + 2)/2 \leq -(n - \ell)$, and thus that

$$g_n \leq g_{n-1} + \sum_{\ell=2}^{n-1} \frac{1}{\ell! \, 2^{n-\ell}} g_{n-\ell}.$$

Since the sequence $(g_n)_n$ is non-decreasing, we obtain

$$g_n \leq g_{n-1} + \frac{g_{n-1}}{2^n} \sum_{\ell=2}^{n-1} \frac{2^\ell}{\ell!} \leq g_{n-1} + g_{n-1} \frac{e^2 - 3}{2^n}.$$

We set $\alpha = e^2 - 3$. Iterating the last inequality, we get that

$$g_n \leq g_{n-1} \left(1 + \frac{\alpha}{2^n}\right) \leq g_1 \prod_{i=2}^{n} \left(1 + \frac{\alpha}{2^i}\right) = \exp\left(\sum_{i=2}^{n} \ln\left(1 + \frac{\alpha}{2^i}\right)\right),$$

because $g_1 = 1$. Note that, when $i \to +\infty$, we have $\ln(1+\alpha2^{-i}) \le \alpha2^{-i}$ (because $\ln(1+x) \le x$ for all $x \ge 0$). This implies that, for all $n \ge 1$,

$$g_n \le \exp\left(\alpha \sum_{i=2}^{\infty} 2^{-i}\right) = \exp(\alpha/2) \approx 8.975763927.$$

In other words, the sequence $(g_n)_n$ is bounded. Since it is also non-decreasing, it converges to a constant c, which is also non-zero since $g_n \ge g_1 \ne 0$ for all $n \ge 1$. This is equivalent to $f_n \sim c f_{n,n-1}$ when $n \to +\infty$ as claimed. To get a lower bound on c, note that, for all $n \ge 1$, $c \ge g_n \ge g_{1000} = f_{1000}/f_{1000,999} \approx 1.49913911$.   □

The proof of Lemma 1 gives also the one of Theorem 2. This result means that asymptotically a constant fraction of the strict monotonic general trees of size n are built in $(n-1)$ steps. For these trees, at each step of construction only one single leaf expands into a binary node. All other leaves either become a unary node or stay unchanged, meaning that on average half of the leaves will expand into unary node with one leaf expanding into a binary node. The number of internal nodes of these trees then grows like $n^2/4$.

4   Analysis of Typical Parameters

In this section we are interested in typical parameters describing the structure of large monotonic general trees. The proofs of next three Theorems have the same structure: an upper bound derived from strict bounds on all trees and a lower bound following mean analysis strict monotonic general trees.

4.1   Quantitative Analysis of the Number of Internal Nodes

Theorem 3. *Let $I_n^{\mathcal{F}}$ be the number of internal nodes in a tree taken uniformly at random among all strict monotonic general trees of size n. Then for all $n \ge 1$, we have*

$$\frac{(n-1)(n+2)}{2c} \le \mathbb{E}[I_n^{\mathcal{F}}] \le \frac{(n-1)n}{2}.$$

To prove this theorem, we use the following proposition.

Proposition 2. *Let us denote by $s_{n,k}$ the number of strict monotonic general trees of size n that have $n-1$ distinct node-labels and k internal nodes. For all $n \ge 1$ and $k \ge 0$,*

$$s_{n,k} = (n-1)!\binom{(n-1)(n-2)/2}{k-(n-1)},$$

and thus, if $I_n^{\mathcal{S}}$ is the number of internal nodes in a tree taken uniformly at random among all strict monotonic general trees of size n that have $n-1$ distinct label nodes, then, for all $n \ge 1$,

$$\mathbb{E}[I_n^{\mathcal{S}}] = \frac{(n-1)(n+2)}{4}.$$

We are now ready to prove the main theorem of this section.

Proof (of Theorem 3). Note that the number of internal nodes of a strict monotonic general tree of size n belongs to $\{1, \ldots, n(n-1)/2\}$. The upper bound follows directly from an induction. A tree of size 1 contains 0 internal node. Suppose for all $i < n$ that the maximal number of internal node for a tree of size i to be $i(i-1)/2$. We take a size-n tree and we are interested in the maximal number of internal nodes it contains. Let us denote by ℓ the maximal internal node label. We remove all leaves of internal nodes labelled by ℓ and this nodes are becoming leaves. The resulting tree has size $1 \leq i < n$ and by induction contains at most $i(i-1)/2$ internal nodes. The maximal number of internal nodes that were labelled by ℓ is i, thus the initial tree contained at most $i(i-1)/2 + i = i(i+1)/2 \leq n(n-1)/2$. internal nodes.

For the lower bound, we denote by $\mathcal{S}_n$ the set of strict monotonic general trees of size n that have $n-1$ distinct node-labels. Moreover, we denote by t_n a tree taken uniformly at random in $\mathcal{F}_n$, and by $I_n^{\mathcal{F}}$ its number of internal nodes. We have, for all $n \geq 1$,

$$\mathbb{E}[I_n^{\mathcal{F}}] = \mathbb{E}[I_n^{\mathcal{F}} \mid t_n \in \mathcal{S}_n] \cdot \mathbb{P}(t_n \in \mathcal{S}_n) + \mathbb{E}[I_n^{\mathcal{F}} \mid t_n \notin \mathcal{S}_n] \cdot \mathbb{P}(t_n \notin \mathcal{S}_n)$$

$$\geq \mathbb{E}[I_n^{\mathcal{F}} \mid t_n \in \mathcal{S}_n] \cdot \mathbb{P}(t_n \in \mathcal{S}_n) = \mathbb{E}[I_n^{\mathcal{S}}] \cdot \frac{f_{n,n-1}}{f_n},$$

where we have used conditional expectations and the fact that conditionally on being in $\mathcal{S}_n$, t_n is uniformly distributed in this set, and, in particular, $\mathbb{E}[I_n^{\mathcal{F}} \mid t_n \in \mathcal{S}_n] = \mathbb{E}I_n^{\mathcal{S}}$. Using Proposition 2 and the upper bound of Lemma 1, we thus get

$$\mathbb{E}[I_n^{\mathcal{F}}] \geq \frac{1}{c} \frac{(n-1)(n+2)}{4},$$

which concludes the proof. $\square$

4.2   Quantitative Analysis of the Number of Distinct Labels

Theorem 4. *Let $X_n^{\mathcal{F}}$ denote the number of distinct internal-node labels (or construction steps) in a tree taken uniformly at random among all strict monotonic general trees of size n, then for all $n \geq 1$,*

$$\frac{n-1}{c} \leq \mathbb{E}[X_n^{\mathcal{F}}] \leq n - 1.$$

Proof. First note that since at every construction step in Definition 1 we add at least one leaf in the tree, then after ℓ construction steps, there are exactly ℓ distinct labels and at least $\ell + 1$ leaves in the tree. Therefore, $n \geq X_n^{\mathcal{F}} + 1$ for all $n \geq 1$, which implies in particular that $\mathbb{E}[X_n] \leq n - 1$, as claimed.

For the lower bound, we reason as in the proof of Lemma 1, and using the same notations:

$$\mathbb{E}[X_n^{\mathcal{F}}] \geq \mathbb{E}[X_n^{\mathcal{F}} \mid t_n \in \mathcal{S}_n] \cdot \mathbb{P}(t_n \in \mathcal{S}_n) = (n-1) \frac{f_{n,n-1}}{f_n},$$

because $\mathbb{E}[X_n^{\mathcal{F}} \mid t_n \in \mathcal{S}_n] = n - 1$ by definition of $\mathcal{S}_n$ (being the set of all strict monotonic general trees of size n that have $n - 1$ distinct node-labels). Using the upper bound of Lemma 1 gives that $\mathbb{E}[X_n^{\mathcal{F}}] \geq (n-1)/c$, which concludes the proof.   □

4.3   Quantitative Analysis of the Height of the Trees

Theorem 5. *Let $H_n^{\mathcal{F}}$ denotes the height of a tree taken uniformly at random in $\mathcal{F}_n$, the set of all strict monotonic general trees of size n. Then we have, for all $n \geq 0$,*

$$\frac{n}{2c} \leq \mathbb{E}[H_n^{\mathcal{F}}] \leq n - 1.$$

To prove this theorem, we first prove the following:

Proposition 3. *Let us denote by $H_n^{\mathcal{S}}$ the height of a tree taken uniformly at random in $\mathcal{S}_n$, the set of all strict monotonic general trees of size n that have $n - 1$ distinct labels. Then we have, for all $n \geq 0$,*

$$\frac{n}{2} \leq \mathbb{E}[H_n^{\mathcal{S}}] \leq n - 1.$$

Proof. Define the sequence of random trees $(t_n)_{n \geq 0}$ recursively as: t_1 is a single leaf; and given t_{n-1}, we define t_n as the tree obtained by choosing a leaf uniformly at random among all leaves of t_{n-1}, replacing it by an internal nodes to which two leaves are attached, and, for each of the other leaves of t_{n-1}, choose with probability $1/2$ (independently from the rest) whether to leave it unchanged or to replace it by a unary node to which one leaf is attached.

One can prove by induction on n that for all $n \geq 1$, t_n is uniformly distributed in $\mathcal{S}_n$. We denote by H_n the height of t_n and therefore $H_n \overset{d}{=} H_n^{\mathcal{S}}$. Since the height of t_n is at most the height of t_{n-1} plus 1 for all $n \geq 2$, we get that $H_n \leq n - 1$.

For the lower bound, we note that, for the height of t_n to be larger than the height of t_{n-1}, we need to have replaced at least one of the maximal-height leaves in t_{n-1}. There is at least one leaf of t_{n-1} which is at height H_{n-1} and this leaf is replaced by an internal node with probability

$$\frac{1}{2}\left(1 - \frac{1}{n-1}\right) + \frac{1}{n-1} \geq \frac{1}{2}.$$

Therefore, for all $n \geq 1$, we have

$$\mathbb{P}(H_n = H_{n-1} + 1) \geq \frac{1}{2},$$

which implies, since $H_n \in \{H_{n-1}, H_{n-1} + 1\}$ almost surely,

$$\mathbb{E}[H_n] = \mathbb{E}[H_{n-1}] + \mathbb{P}(H_n = H_{n-1} + 1) \geq \mathbb{E}[H_{n-1}] + \frac{1}{2}.$$

Therefore, for all $n \geq 1$, we have $\mathbb{E}[H_n] \geq \mathbb{E}[H_0] + n/2 = n/2$, as claimed.   □

Proof (of Theorem 5). By Definition 1, it is straightforward to see that the height of a tree built in ℓ steps is at most ℓ since the height increases by at most one per construction step. Since a tree of size n is built in at most $n - 1$ steps, we get that $H_n^{\mathcal{F}} \leq n - 1$, which implies, in particular, that $\mathbb{E}[H_n^{\mathcal{F}}] \leq n - 1$.

For the lower bound, note that, if t_n is a tree taken uniformly at random in $\mathcal{F}_n$ and $H_n^{\mathcal{F}}$ is its height, then

$$\mathbb{E}[H_n^{\mathcal{F}}] \geq \mathbb{E}[H_n^{\mathcal{F}} | t_n \in \mathcal{S}_n] \cdot \mathbb{P}(X \in \mathcal{S}_n) \geq \frac{1}{c} \mathbb{E}[H_n^{\mathcal{S}}],$$

where we have used Proposition 1 and the fact that t_n conditioned on being in $\mathcal{S}_n$ is uniformly distributed in this set and thus $E[H_n^{\mathcal{F}} | t_n \in \mathcal{S}_n] = \mathbb{E}H_n^{\mathcal{S}}$. By Proposition 3, we thus get $\mathbb{E}[H_n^{\mathcal{F}}] \geq n/(2c)$, as claimed.   □

4.4   Quantitative Analysis of the Depth of the Leftmost Leaf

Theorem 6. *Let us denote by $D_n^{\mathcal{F}}$ the depth of the leftmost leaf of a tree taken uniformly at random in $\mathcal{F}_n$, the set of all strict monotonic general trees of size n. Then we have, for all $n \geq 0$,*

$$\frac{n}{2c} \leq \mathbb{E}[H_n^{\mathcal{F}}] \leq n - 1.$$

Proposition 4. *Let us denote by $D_n^{\mathcal{S}}$ the depth of the leftmost leaf of a tree taken uniformly at random in $\mathcal{S}_n$, the set of all strict monotonic general trees of size n that have $n - 1$ distinct labels. Then we have, for all $n \geq 0$,*

$$\frac{n}{2} \leq \mathbb{E}[D_n^{\mathcal{S}}] \leq n - 1.$$

Proof. Given the uniform process of trees t_n presented in Proposition 3. The depth of the leftmost leaf is always smaller than $n - 1$. Let X_n be a Bernoulli variable taking value 1 if the leftmost leaf of t_n has been expanded at iteration n and the value 0 otherwise. Then for $n \geq 1$,

$$\mathbb{P}(X_n = 1) = \frac{1}{n} + \frac{(n-1)}{n}\frac{1}{2} = \frac{n+1}{2n} \geq \frac{1}{2}.$$

Since at each iteration step either the leftmost leaf expand to make a binary node which gives $\frac{1}{n}$ or it has not created a binary and then it has $\frac{1}{2}$ probability to make a unary node. The depth of the leftmost leaf is $D_n^{\mathcal{S}} = \sum_{k=1}^{n} X_k$. Therefore for $n \geq 1$,

$$\mathbb{E}[D_n^{\mathcal{S}}] \geq \frac{n}{2}.$$

Which concludes the proof.   □

Proof (of Theorem 6). By the same arguments than in Theorem 5 the result follows directly since we have the same bounds on the depth of leftmost leaf as we had in the height of the tree.   □

5   Correspondence with Labelled Graphs

In Sect. 3 we defined $f_{n,k}$ the number of strict monotonic general trees of size n with exactly k distinct node-labels. Then we have shown that, for all $n \geq 1$,

$$f_{n,n-1} = (n-1)! \, 2^{\frac{(n-1)(n-2)}{2}}.$$

The factor $2^{(n-1)(n-2)/2} = 2^{\binom{n-1}{2}}$ in the context of graphs with $n-1$ vertices counts the different combinations of undirected edges between vertices. The factor $(n-1)!$ accounts for all possible permutations of vertices. We will denote $\mathcal{S}_n$ to be the trees that $f_{n,n-1}$ counts and exhibit a bijection between strict monotonic general trees of $\mathcal{S} = \cup_{n \geq 1} \mathcal{S}_n$ with a class of labelled graphs with $n-1$ vertices defined in the following.

For all $n \geq 1$, we denote by $\mathcal{G}_n$ the set of all labelled graphs (V, ℓ, E) such that $V = \{1, \ldots, n\}$, $E \subseteq \{\{i, j\}: i \neq j \in V\}$ and $\ell = (\ell_1, \ldots, \ell_n)$ is a permutation of V (see Fig. 2 for an example). We set $\mathcal{G} = \cup_{n=0}^{\infty} \mathcal{G}_n$. Choosing a graph in $\mathcal{G}_n$ is equivalent to (1) choosing ℓ (there are $n!$ choices) and (2) for each of the $\binom{n}{2}$ possible edges, choose whether it belongs to E or not (there are $2^{\binom{n}{2}}$ choices in total). In total, we thus get that $|\mathcal{G}_n| = n! \, 2^{\binom{n}{2}}$.

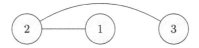

Fig. 2. The graph $\mathcal{G}_3$. In this representation, the vertices $V = \{1, 2, 3\}$, $E = \{\{1, 2\}, \{2, 3\}\}$ are drawn from left to right (node 1 is the leftmost one, node 3 is the rightmost one), and their label is their image by ℓ: in this example $\ell = (2, 1, 3)$.

A size-n permutation σ is denoted by $(\sigma_1, \ldots, \sigma_n)$, and σ_i is its i-th element (the image of i), while $\sigma^{-1}(k)$ is the preimage of k (the position of k in the permutation).

We define $\mathcal{M} : \mathcal{S} \to \mathcal{G}$ recursively on the size of the tree it takes as an input: first, if t is the tree of size 1 (which contains only one leaf) then we set $\mathcal{M}(t)$ to be the empty graph $(\emptyset, \epsilon, \emptyset)$, where ϵ is the empty permutation. Now assume we have defined $\mathcal{M}$ on $\cup_{\ell=1}^{n-1} \mathcal{S}_\ell$, and consider a tree $t \in \mathcal{S}_n$. By Definition 1 and since $t \in \mathcal{S}_n$, there exists a unique binary node in t labelled by $n-1$, and this node is attached to two leaves. Consider $\hat{t}$ the tree obtained when removing all internal nodes labelled by $n-1$ (and all the leaves attached to them) from t and replacing them by leaves. Denote by v_n the position (in, e.g., depth-first order) of the leaf of $\hat{t}$ that previously contained the binary node labelled by $n-1$ in t. Denote by $u_1, \ldots, u_m$ the positions of the leaves of $\hat{t}$ that previously contained unary nodes labelled by $n-1$ in t. We set $\mathcal{M}(\hat{t}) = (\{1, \ldots, n-1\}, \hat{\ell}, \hat{E})$ and define $\mathcal{M}(t) = (\{1, \ldots, n\}, \ell, E)$ where

$$\ell_i = \begin{cases} v_n & \text{if } i = n \\ \hat{\ell}_i & \text{if } \hat{\ell}_i < v_n \\ \hat{\ell}_i + 1 & \text{if } \hat{\ell}_i \geq v_n, \end{cases}$$

$E = \hat{E} \cup \{\{\ell^{-1}(u_j), n\} : 1 \leq j \leq m\}$. From v_n we know that $\ell^{-1}(v_n) = n$ and since any $u_i \neq v_n$, then $\ell^{-1}(u_i) \in \{1, \ldots, n-1\}$, we never create any loop in the resulting graph. An example of the bijection is depicted in Fig. 3.

Theorem 7. *The mapping $\mathcal{M}$ is bijective, and $\mathcal{M}(\mathcal{S}_n) = \mathcal{G}_{n-1}$ for $n \geq 1$.*

Proof. From the definition, it is clear that two different trees have two distinct images by $\mathcal{M}$, thus implying that $\mathcal{M}$ is injective; this is enough to conclude since $|\mathcal{G}_{n-1}| = |\mathcal{S}_n|$ (see Theorem 1 for the cardinality of $\mathcal{S}_n$).

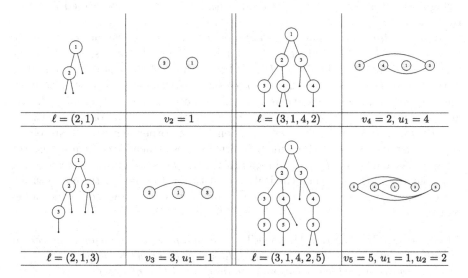

Fig. 3. Bijection between an evolving tree in $\mathcal{S}$ from size 3 to 6 and its corresponding graph in $\mathcal{G}$

6 Uniform Random Sampling

In this section we exhibit a very efficient way for the uniform sampling of strict monotonic general trees using the described evolution process. We finally explain how this same uniform sampler can be used to generate Erdős-Rényi graphs.

The global approach for our algorithmic framework deals with the *recursive generation method* adapted to the analytic combinatorics point of view in [11].

But in our context we note that we can obtain for free (from a complexity view) an *unranking algorithm*[1]. This fact is sufficiently rare to mention it: usually unranking algorithm are less efficient than recursive generation ones. Unranking algorithmic has been developed in the 70's by Nijenhuis and Wilf [21] and then has been introduced to the context of analytic combinatorics by Martínez and Molinero [18]. We use the same method as the one described in [3].

In our recurrence when r grows, the sequence $(f_{n-r})_r$ decreases extremely fast. Thus for the uniform random sampling, it will appear more efficient to read Eq. (2) in the following way:

$$f_n = \binom{n-1}{1} 2^{n-2} f_{n-1} + \sum_{i=1}^{2} \binom{n-2}{i} 2^{n-2-i} f_{n-2}$$

$$+ \sum_{i=1}^{3} \binom{n-3}{i} 2^{n-3-i} \binom{2}{i-1} f_{n-3} + \cdots + f_1. \tag{4}$$

Using the latter decomposition the algorithm can now be described as Algorithm 1.

In Algorithm 1 note that the While loop allows us to determine the values for ℓ, i and r (see Eq. (2) to identify the variables). Then the recursive call is done using the adequate rank $r \mod f_{n-\ell}$. The last lines of the algorithm (for 21 to 27) are necessary to modify the tree T of size $n - \ell$ that has just been built. In line 22 we determine which leaves of T will be substituted by internal nodes (of arity at most 2) with new leaves. It is based on the unranking of combinations, see [12] for a survey in this context. Then for the other leaves that are either kept as they are of replaced by unary internal nodes attached to a leaf we use the integer F seen as a $(n-\ell-i)$-bit integer: if the bit $\#s$ is 0 then the corresponding leaf is kept, and if it is 1 then the leaf is substituted. And finally the composition unranking allows us to determine how many leaves are attached to the nodes selected with B.

Theorem 8. *The function* UNRANKTREE *is an unranking algorithm and calling it with the parameters n and a uniformly-sampled integer s in $\{0, \ldots, f_n - 1\}$ gives as output a uniform strict monotonic general tree of size n.*

The correctness of the algorithm follows directly from the total order over the trees deduced from the decomposition Eq. (4).

Theorem 9. *Once the pre-computations have been done, the function* UNRANK-TREE *needs in average $\Theta(n)$ arithmetic operations to construct a tree of size n.*

Proof. The proof for this theorem is analogous to the one for Theorem 3.6.5 in [3] after showing that both UNRANKBINOMIAL and UNRANKCOMPOSITION run in $\Theta(n)$ in the number of arithmetic operations.

---

[1] Unranking algorithms are based on a total order on the combinatorial objects under consideration and aim at building an object directly using its rank in the total order.

Algorithm 1. Strict Monotonic General Tree Unranking

```
 1: function UNRANKTREE(n, s)
 2:     if n = 1 then
 3:         return the tree reduced to a single leaf
 4:     ℓ := 1
 5:     r := s
 6:     i := 1
 7:     while r >= 0 do
```
8:     $t := \binom{n-\ell}{i} 2^{n-\ell-i} \binom{\ell-1}{i-1}$

9:     $r := r - t \cdot f_{n-\ell}$

10:     $i := i + 1$

11:     **if** $i > min(\ell, n - \ell)$ **then**

12:         $i := 1$

13:         $\ell := \ell + 1$

14:     **if** $i > 1$ **then**

15:         $i := i - 1$

16:     **else**

17:         $\ell := \ell - 1$

18:         $i := min(\ell, n - \ell)$

19:     $r := r + t \cdot f_{n-\ell}$

20:     $T := $ UNRANKTREE$(n - \ell, r \mod f_{n-\ell})$

21:     $r := r \mathbin{//} f_{n-\ell}$ ▷ // stands for the integer division

22:     $B := $ UNRANKBINOMIAL$(n - \ell, i, r \mathbin{//} \binom{n-\ell}{i})$ ▷ see Algorithm 2 in [3]

23:     $r := r \mod \binom{n-\ell}{i}$

24:     $F := r \mathbin{//} \binom{\ell-1}{i-1}$

25:     $C := $ UNRANKCOMPOSITION$(\ell, i, r \mod \binom{\ell-1}{i-1})$ ▷ see Algorithm 2 in [3]

26:     Using F, substitute in T, using any traversal, the leaves selected with B with

27:     internal nodes and new leaves according to C; the other leaves are changed as

28:     unary nodes with a leaf or not

29:     **return** the tree T

The sequences $(f_\ell)_{\ell \le n}$ and $(\ell!)_{\ell \in \{1,\dots,n\}}$ have been pre-computed and stored.

Note that the uniform sampling of trees from $\mathcal{S}_n$ corresponds through the bijection to sampling a random graph $\mathcal{G}_n(1/2) = (V, E)$ defined as follows: $V = \{1, \dots, n\}$ and each edge belong to E with probability $1/2$, independently from the other edges. This model, also called the Erdős-Rényi random graph was originally introduced by Erdős and Rényi [9], and simultaneously by Gilbert [14], and has been since then extensively studied in the probability and combinatorics literature (see, for example, the books [4] and [8] for introductory surveys). Algorithm 1 samples uniform trees from $\mathcal{S}_n$ with constant time rejection and therefore can be used to sample Erdős-Rényi graphs.

Acknowledgement. The authors thank Cécile Mailler for fruitful discussions to relate this model of Schröder trees to the ones presented in [3].

The authors thank the anonymous referees for their comments and suggested improvements. All these remarks have highly increased the quality of the paper.

References

1. Bergeron, F., Flajolet, P., Salvy, B.: Varieties of increasing trees. In: CAAP, pp. 24–48 (1992)
2. Bodini, O., Genitrini, A., Naima, M.: Ranked Schröder trees. In: Proceedings of the Sixteenth Workshop on Analytic Algorithmics and Combinatorics, ANALCO 2019, pp. 13–26 (2019)
3. Bodini, O., Genitrini, A., Mailler, C., Naima, M.: Strict monotonic trees arising from evolutionary processes: combinatorial and probabilistic study. Adv. Appl. Math. **133**, 102284 (2022)
4. Bollobás, B.: Random Graphs. Cambridge Studies in Advanced Mathematics, 2nd edn. Cambridge University Press, Cambridge (2001)
5. Donaghey, R.: Alternating permutations and binary increasing trees. J. Comb. Theory Ser. A **18**(2), 141–148 (1975)
6. Doubilet, P., Rota, G.C., Stanley, R.: On the foundations of combinatorial theory. VI. the idea of generating function. In: Proceedings of the Sixth Berkeley Symposium on Mathematical Statistics and Probability, Volume 2: Probability Theory, pp. 267–318. University of California Press, Berkeley (1972)
7. Drmota, M.: Random Trees. Springer, Vienna-New York (2009). https://doi.org/10.1007/978-3-211-75357-6
8. Durrett, R.: Random Graph Dynamics. Cambridge Series in Statistical and Probabilistic Mathematics. Cambridge University Press, Cambridge (2006)
9. Erdős, P., Rényi, A.: On random graphs. Publicationes Mathematicae **6**(26), 290–297 (1959)
10. Flajolet, P., Sedgewick, R.: Analytic Combinatorics. Cambridge University Press, Cambridge (2009)
11. Flajolet, P., Zimmermann, P., Van Cutsem, B.: A calculus for the random generation of labelled combinatorial structures. Theoret. Comput. Sci. **132**(1–2), 1–35 (1994)
12. Genitrini, A., Pépin, M.: Lexicographic unranking of combinations revisited. Algorithms **14**(3) (2021)
13. Gessel, I.M.: Enumerative applications of a decomposition for graphs and digraphs. Discret. Math. **139**(1–3), 257–271 (1995)
14. Gilbert, E.N.: Random graphs. Ann. Math. Stat. **30**(4), 1141–1144 (1959)
15. Janson, S.: Plane recursive trees, Stirling permutations and an urn model. Discrete Math. Theor. Comput. Sci. (2008)
16. Kuznetsov, A.G., Pak, I.M., Postnikov, A.E.: Increasing trees and alternating permutations. Russ. Math. Surv. **49**(6), 79 (1994)
17. Lin, Z., Ma, J., Ma, S.M., Zhou, Y.: Weakly increasing trees on a multiset. Adv. Appl. Math. **129**, 102206 (2021)
18. Martínez, C., Molinero, X.: Generic algorithms for the generation of combinatorial objects. In: Rovan, B., Vojtáš, P. (eds.) MFCS 2003. LNCS, vol. 2747, pp. 572–581. Springer, Heidelberg (2003). https://doi.org/10.1007/978-3-540-45138-9_51
19. Meir, A., Moon, J.W.: On the altitude of nodes in random trees. Canad. J. Math. **30**(5), 997–1015 (1978)
20. Naima, M.: Combinatorics of trees under increasing labellings: asymptotics, bijections and algorithms. Ph.D. thesis, Université Sorbonne Paris Nord (2020)
21. Nijenhuis, A., Wilf, H.S.: Combinatorial Algorithms. Computer Science and Applied Mathematics, Academic Press, Cambridge (1975)

22. de Panafieu, E., Dovgal, S.: Symbolic method and directed graph enumeration. Acta Math. Universitatis Comenianae **88**(3) (2019)
23. Schröder, E.: Vier Combinatorische Probleme. Z. Math. Phys. **15**, 361–376 (1870)
24. Sornette, D.: Critical Phenomena in Natural Sciences: Chaos, Fractals, Selforganization and Disorder: Concepts and Tools. Springer, Heidelberg (2006). https://doi.org/10.1007/3-540-33182-4
25. Stanley, R.P.: Acyclic orientations of graphs. Discret. Math. **5**(2), 171–178 (1973)

Min Orderings and List Homomorphism Dichotomies for Signed and Unsigned Graphs

Jan Bok[1]([✉])(ID), Richard C. Brewster[2](ID), Pavol Hell[3](ID), Nikola Jedličková[4](ID), and Arash Rafiey[5]

[1] Computer Science Institute, Faculty of Mathematics and Physics, Charles University, Prague, Czech Republic
`bok@iuuk.mff.cuni.cz`
[2] Department of Mathematics and Statistics, Thompson Rivers University, Kamloops, Canada
`rbrewster@tru.ca`
[3] School of Computing Science, Simon Fraser University, Burnaby, Canada
`pavol@cs.sfu.ca`
[4] Department of Applied Mathematics, Faculty of Mathematics and Physics, Charles University, Prague, Czech Republic
`jedlickova@kam.mff.cuni.cz`
[5] Mathematics and Computer Science, Indiana State University, Terre Haute, IN, USA
`arash.rafiey@indstate.edu`

Abstract. The CSP dichotomy conjecture has been recently established, but a number of other dichotomy questions remain open, including the dichotomy classification of list homomorphism problems for signed graphs. Signed graphs arise naturally in many contexts, including for instance nowhere-zero flows for graphs embedded in non-orientable surfaces. For a fixed signed graph $\widehat{H}$, the list homomorphism problem asks whether an input signed graph $\widehat{G}$ with lists $L(v) \subseteq V(\widehat{H}), v \in V(\widehat{G})$, admits a homomorphism f to $\widehat{H}$ with all $f(v) \in L(v), v \in V(\widehat{G})$.

Usually, a dichotomy classification is easier to obtain for list homomorphisms than for homomorphisms, but in the context of signed graphs a structural classification of the complexity of list homomorphism problems has not even been conjectured, even though the classification of the complexity of homomorphism problems is known.

Kim and Siggers have conjectured a structural classification in the special case of "weakly balanced" signed graphs. We confirm their conjecture for reflexive and irreflexive signed graphs; this generalizes previous results on weakly balanced signed trees, and weakly balanced separable signed graphs [1–3]. In the reflexive case, the result was first presented in [19], where the proof relies on a result in this paper. The irreflexive result is new, and its proof depends on first deriving a theorem on extensions of min orderings of (unsigned) bipartite graphs, which is interesting on its own.

© Springer Nature Switzerland AG 2022
A. Castañeda and F. Rodríguez-Henríquez (Eds.): LATIN 2022, LNCS 13568, pp. 510–526, 2022.
https://doi.org/10.1007/978-3-031-20624-5_31

1 Introduction

The CSP Dichotomy Theorem [8,24] guarantees that each homomorphism problem for a fixed template relational structure **H** ("does a corresponding input relational structure **G** admit a homomorphism to **H**?") is either polynomial-time solvable or NP-complete, the distinction being whether or not the structure **H** admits a certain symmetry. In the context of graphs **H** = H, there is a more natural structural distinction, namely the tractable problems correspond to the graphs H that have a loop, or are bipartite [16]. For list homomorphisms (when each vertex $v \in V(G)$ has a list $L(v) \subseteq V(H)$), the distinction turns out to be whether or not H is a "bi-arc graph", a notion related to interval graphs [10]. In the special case of bipartite graphs H, the distinction is whether or not H has a min ordering. A *min ordering* of a bipartite graph with parts A, B is a pair of linear orders $<_A, <_B$ of A and B respectively, such that if there are edges $ab, a'b'$ with $a \in A, a' \in A, a < a'$ and $b \in B, b' \in B, b' < b$, then there is also the edge ab'. If a bipartite graph H has a min ordering, then the list homomorphism problem to H is polynomial-time solvable; otherwise, it is NP-complete [9,15].

An *invertible pair* in a bipartite graph with parts A, B is a pair of vertices $x, x' \in A$ (or $x, x' \in B$), with a pair of walks $x = v_1, v_2, \ldots, v_k = x'$ and $x' = v'_1, v'_2, \ldots, v'_k = x$ of equal length, and another pair of walks $x' = w_1, w_2, \ldots, w_m = x$ and $x = w'_1, w'_2, \ldots, w'_m = x'$ of equal length, such that each v_i is non-adjacent to v'_{i+1} for all $i = 1, 2, \ldots, k - 1$ and each w_j is non-adjacent to w'_{j+1}, for all $j = 1, 2, \ldots, m-1$. It is easy to see that if an invertible pair exists, then there can be no min ordering, and the converse is proved in [15]. Specifically, it is proved there (by a reduction to the reflexive case proved in [11]) that if a bipartite graph has no invertible pairs, then it has a min ordering. However, for reflexive graphs, the direct proof in [11] implies a stronger result—namely, if a set of ordered pairs of vertices does not violate transitivity, then it can be extended to a min ordering if and only if there is no invertible pair. (Invertible pairs and min orderings in reflexive graphs are defined analogously to bipartite graphs, see [11]. A set of ordered pairs is said to violate transitivity if it contains some pairs $(t_0, t_1), (t_1, t_2), (t_2, t_3), \ldots, (t_{k-1}, t_k), (t_k, t_0)$ with $t_0 < t_1 < \cdots < t_k < t_0$). Such a result was not known for bipartite graphs.

In this paper, we fill the gap and prove an analogous extension version of the min ordering characterization for bipartite graphs, see Corollary 1. This result is then used in the following section to prove the bipartite case of the conjecture of Kim and Siggers.

Since the reflexive case already had an extension result in [11], we can apply a similar method to prove the conjecture for reflexive graphs[1].

A *signed graph* $\hat{H}$ is a graph H together with an assignment of *signs* $+, -$ to the edges of H. Edges may be assigned both signs, or equivalently, there may be two parallel edges with opposite signs between the same two vertices. (There may be edges that are loops, and there may also be two parallel loops of opposite signs at the same vertex). Edges with a $+$ sign are called *positive*, or

---

[1] The details of this can be found in our arXiv paper [4].

blue, edges with a − sign are called *negative*, or *red*. Edges with both signs are called *bicoloured*, while purely red or purely blue edges are called *unicoloured*. Two signed graphs are called *switch-equivalent* if one can be obtained from the other by a sequence of vertex switchings, where a *switching* at a vertex v flips the signs of all edges incident with v. (A bicoloured edge remains bicoloured). Signed graphs arise in many contexts in mathematics and in applications. This includes knot theory, qualitative matrix theory, gain graphs, psychosociology, chemistry, and statistical physics [23]. In graph theory, they are of particular interest in nowhere-zero flows for graphs embedded in non-orientable surfaces [18].

A *homomorphism* of a signed graph $\widehat{G}$ to a signed graph $\widehat{H}$ is a vertex mapping f which is a sign-preserving homomorphism of $\widehat{G'}$ to $\widehat{H}$ for some signed graph $\widehat{G'}$ switch-equivalent to $\widehat{G}$. Equivalently, a homomorphism of a signed graph $\widehat{G}$ to a signed graph $\widehat{H}$ is a homomorphism f of the underlying graph G of $\widehat{G}$ to the underlying graph H of $\widehat{H}$, which maps bicoloured edges of $\widehat{G}$ to bicoloured edges of $\widehat{H}$, and for which any unicoloured closed walk W in $\widehat{G}$ with unicoloured image $f(W)$ in $\widehat{H}$ has the same product of the signs of its edges. (In other words, closed walks with only unicoloured edges map to closed walks that either contain a bicoloured edge or have the same parity of the number of negative edges). We will use this definition in the last section, as it does not require switching in the input graph before mapping it. The equivalence of the two definitions follows from the theorem of Zaslavsky [22], and the actual switching required for $\widehat{G}$ before the mapping if one exists, as well as the two violating closed walks if such a mapping does not exist, can be found in polynomial time [20].

The study of homomorphisms of signed graphs was pioneered by Guenin [14] and introduced more systematically by Naserasr, Rollová and Sopena, see the survey [20].

The *homomorphism problem* for the signed graph $\widehat{H}$ asks whether an input signed graph $\widehat{G}$ admits a homomorphism to $\widehat{H}$. The *s-core* of a signed graph $\widehat{H}$ is the smallest homomorphic image of $\widehat{H}$ that is a subgraph of $\widehat{H}$. (The s-core is unique up to isomorphism [6]). It was conjectured in [6] that the homomorphism problem for $\widehat{H}$ is polynomial if the s-core of $\widehat{H}$ has at most two edges (a bicoloured edge counts as two edges), and is NP-complete otherwise. The conjecture was verified in [6] for all signed graphs that do not simultaneously contain a bicoloured edge and a unicoloured loop of each colour. Finally, the full conjecture was established in [7].

The *list homomorphism problem* for a signed graph $\widehat{H}$ asks whether an input signed graph $\widehat{G}$ with lists $L(v) \subseteq V(\widehat{H}), v \in V(\widehat{G})$, admits a homomorphism f to $\widehat{H}$ with all $f(v) \in L(v), v \in V(\widehat{G})$. The complexity classification for these list homomorphism problems appears to be difficult, and no structural classification conjecture has arisen. (Even though these are not directly CSP problems, the fact that dichotomy holds can be derived from the CSP Dichotomy Theorem). Some special cases have been treated [1,2,5,19], including a full classification for signed trees [3].

In [19], H. Kim and M.H. Siggers focus on a special class of signed graphs: we say that a signed graph $\widehat{H}$ is *weakly balanced* if any closed walk of uni-coloured edges has an even number of negative edges. Equivalently, there is a switch-equivalent signed graph $\widehat{H'}$ in which there are no purely red edges [3]. (Our terminology comes from [3], in [19] these signed graphs correspond to the so-called *pr-graphs*. We also remark that while balanced signed graphs are some-times called bipartite [13,21], weakly balanced signed graphs have no relation to weakly bipartite signed graphs as defined therein).

Kim and Siggers [19] conjectured a classification of the complexity of the list homomorphism problems for weakly balanced signed graphs $\widehat{H}$, and announced it holds in the special case of signed graphs that are *reflexive* (each vertex has at least one loop). In the last version of [19] they use a result from this paper for a proof (see the footnote on page 4 of [19], version v4). Their paper also highlights the importance of irreflexive signed graphs, by reducing parts of the problem for general signed graphs to their bipartite translations. Their conjec-ture is particularly elegant when stated for irreflexive signed graphs. (We note that non-bipartite irreflexive signed graphs are not relevant because their list homomorphism problems are NP-complete by [16]; it is also easy to see that they always contain an invertible pair).

To be specific, we assume that $\widehat{H}$ is a bipartite signed graph without purely red edges, and define a *special min ordering* of $\widehat{H}$ to be a min ordering of the underlying graph H of $\widehat{H}$, such that at each vertex its bicoloured neighbours precede its unicoloured neighbours. The conjectured classification for weakly balanced signed graphs states that the list homomorphism problem for $\widehat{H}$ is polynomial-time solvable if $\widehat{H}$ has a special min ordering, and is NP-complete otherwise.

This implies that there are two natural obstructions to $\widehat{H}$ having a polynomial-time solvable list homomorphism problem – namely invertible pairs, which obstruct the existence of a min ordering, and chains, which obstruct a min ordering from being made special. *Invertible pairs* are defined above for unsigned bipartite graphs, and for signed bipartite graphs they are just invertible pairs in the underlying unsigned graph. A *chain* in a signed graph $\widehat{H}$ consists of two walks of equal length, a walk U with vertices $u = u_0, u_1, \ldots, u_k = v$ and a walk D, with vertices $u = d_0, d_1, \ldots, d_k = v$ such that the edges $u u_1, d_{k-1} v$ are uni-coloured, and the edges $u d_1, u_{k-1} v$ are bicoloured, and for each $i, 1 \leq i \leq k-2$, we have both $u_i u_{i+1}$ and $d_i d_{i+1}$ edges of H while $d_i u_{i+1}$ is not an edge of H, or both $u_i u_{i+1}$ and $d_i d_{i+1}$ bicoloured edges of H while $d_i u_{i+1}$ is not a bicoloured edge of H. See Fig. 1 for an example.

Kim and Siggers also conjectured that a weakly balanced signed graph $\widehat{H}$ has a special min ordering if and only if it has no invertible pairs and no chains. We prove both conjectures (cf. Theorem 3 below), in the case of irreflexive and reflexive signed graphs. In both cases, the result for signed graphs is derived using the extension results for unsigned graphs. While for reflexive graphs the extension result is obvious from the proof of [11], we provide a new proof in the irreflexive case in the next section.

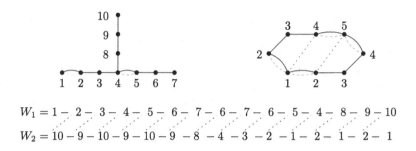

$$W_1 = 1 - 2 - 3 - 4 - 5 - 6 - 7 - 6 - 7 - 6 - 5 - 4 - 8 - 9 - 10$$
$$W_2 = 10 - 9 - 10 - 9 - 10 - 9 - 8 - 4 - 3 - 2 - 1 - 2 - 1 - 2 - 1$$

Fig. 1. An example of a signed graph (on the left) with a chain (on the right) and an invertible pair $(1, 10)$ certified by the pair of walks W_1, W_2 and the pair consisting of the reverse of both walks.

2   Min Orderings of Bipartite Graphs

In this section we only deal with unsigned bipartite graphs H, with a fixed bipartition A, B. The *pair digraph* H^+ has as vertices all ordered pairs of distinct equicoloured vertices of H, i.e., $V(H^+) = \{(a, a') : a, a' \in A, a \neq a'\} \cup \{(b, b') : b, b' \in B, b \neq b'\}$. There is in H^+ an arc from (a, a') to (b, b') if and only if $ab, a'b'$ are edges of H while ab' is not an edge of H. In that case we also say that (a, a') *dominates* (b, b'). We note that (a, a') dominates (b, b') if and only if (b', b) dominates (a', a), a property we call *skew symmetry* of H^+. We also note that (a, a') is an invertible pair if and only if (a, a') and (a', a) are in the same strong component of H^+.

Theorem 1. *The following statements are equivalent for a bipartite graph H:*

1. *H has a min ordering.*
2. *H has no invertible pairs.*
3. *The vertices of H^+ can be partitioned into sets D, D' such that*
   (a) $(x, y) \in D$ if and only if $(y, x) \in D'$,
   (b) $(x, y) \in D$ and (x, y) dominates (x', y') in H^+ implies $(x', y') \in D$,
   (c) $(x, y), (y, z) \in D$ implies $(x, z) \in D$.

Proof. We may assume that H is connected, in particular it has no isolated vertices.

It is straightforward to see that 1 implies 2, and 3 implies 1 (by defining $x < y$ if $(x, y) \in D$). Thus it remains to show that 2 implies 3.

Therefore, we assume that H has no invertible pairs. Note that for each strong component C of H^+, there is a corresponding reversed strong component C' whose pairs are precisely the reversed pairs of the pairs in C; we shall say that C, C' are *coupled* strong components. Note that a strong component C may be coupled with itself - it is easy to check that all pairs in a self-coupled component are invertible.

The partition of $V(H^+)$ into D, D' will correspond to separating each pair of coupled strong components C, C' of H^+. The vertices of one strong component will be placed in the set D, their reversed pairs will go to D'. We wish to make these choices without having a *circuit* in D, i.e., a sequence of pairs $(x_0, x_1), (x_1, x_2), \ldots, (x_n, x_0) \in D$. Note that the vertices in any circuit are either all in A or all in B. We will build these sets D, D' iteratively, making sure they satisfy the following properties.

(i) There is no circuit in D;
(ii) each strong component of H^+ belongs entirely to D, D', or to $V(H^+) - D - D'$;
(iii) the pairs in D' are precisely the reversed pairs of the pairs in D;
(iv) there is no arc of H^+ from D to a vertex outside of D.

Initially, we can choose D, D' to be any sets satisfying these properties, and at each iterative step we add one strong component to D and its coupled component to D'. This will terminate when each strong component is in D or D', i.e., when each pair (x, y) with $x \neq y$ belongs either to D or to D'. Since the final D has no circuit, it satisfies the transitivity property from statement 3(c), and because of (iii), it satisfies 3(a). Moreover, (iv) implies it satisfies 3(b).

Let C be a strong component of H^+. We say that C is *trivial* if it consists of just one pair. We say that C is *ripe* if it has no arc *to* another strong component in $H^+ - D - D'$. Note that whether C is ripe or not depends on what is currently in D, and hence strong components become ripe as D gets larger. We say that a pair (a, b) is a *sink pair* if $N(a)$ contains $N(b)$. Note that there are no arcs in H^+ from a sink pair (a, b), and so in particular it forms a trivial strong component which is ripe for all sets D.

In the general step, our algorithm shall choose a strong component C that is currently ripe, add all of its pairs to D, and add all pairs of C' to D'. Note that this process is guaranteed to maintain the validity of (ii), (iii), (iv), but it may fail (i), creating a circuit in D. However, we will prove that there always exists a ripe strong component C that can be added without creating a circuit in D. In fact, the first failed choice will identify a subsequent choice that will be guaranteed to succeed.

Thus assume we chose C to be an arbitrary ripe strong component of the graph on $V(H^+) - D - D'$. If $C \cup D$ has no circuit, we can add the pairs in C to D and the pairs in C' to D'. Otherwise, suppose $(x_0, x_1), (x_1, x_2), \ldots, (x_n, x_0)$ is a shortest circuit in $C \cup D$. (Subscripts in the vertices in the circuit will be treated modulo $n + 1$).

Since there are no invertible pairs, and since we never place both a pair and its reverse in D, we must have $n \geq 2$. We may assume without loss of generality that $(x_n, x_0) \in C$; note that other pairs of the circuit could also be in C.

Observation 1. *If a pair (x_i, x_{i+1}) in the circuit lies in a non-trivial strong component and the next pair (x_{i+1}, x_{i+2}) lies in a trivial strong component, then (x_{i+1}, x_{i+2}) is a sink pair.*

Since (x_i, x_{i+1}) lies in a non-trivial strong component, it is easy to see that there are two edges $x_i p, x_{i+1} q$ in H such that $x_i q, x_{i+1} p$ are non-edges. (If (x_i, x_{i+1}) dominates (a, b) and is dominated by (u, v), then set $p = u, q = b$). We say that $x_i p, x_{i+1} q$ are *independent edges* in H. If (x_{i+1}, x_{i+2}) is not a sink pair it dominates some (q, r). (If it dominates some (q', r), it also dominates (q, r)). Therefore, $p x_{i+2}$ is not an edge of H, else (p, q) would dominate (x_{i+2}, x_{i+1}), putting it in D (since (p, q) is in D), contradicting the minimality of our circuit. It follows that $q x_{i+2}$ is also not an edge of H, else (p, q) would dominate (x_i, x_{i+2}) which must then lie in D, again contradicting the minimality of the circuit. Therefore in this case (x_{i+1}, x_{i+2}) lies in two independent edges $q x_{i+1}, r x_{i+2}$, and hence belongs to a non-trivial component as claimed. Note that it follows from this that *if $x_i p, x_{i+1} q$ are independent edges, then $p x_{i+2}, q x_{i+2}$ are non-edges.*

Case 1. Assume each pair in the circuit belongs to a trivial component. We claim that in this case some (x_i, x_{i+2}) lies in a trivial ripe component C^* which can be added to D without creating a circuit. Note that it suffices to prove that (x_i, x_{i+2}) forms a trivial ripe component, say X. Indeed, X could not be part of D by the assumed minimality of the circuit $(x_0, x_1), (x_1, x_2), \ldots, (x_n, x_0)$ in $C \cup D$. Moreover, adding X to D cannot creat a circuit; if such a circuit was, say (x_i, x_{i+2}), (x_{i+2}, t_1), $(t_1, t_2), \ldots, (t_k, x_i)$, then there was already a circuit in D, namely (x_i, x_{i+1}), (x_{i+1}, x_{i+2}), (x_{i+2}, t_1), $(t_1, t_2), \ldots, (t_k, x_i)$, contrary to assumption. (Note that since (x_i, x_{i+2}) is a trivial component, adding component X amounts to adding only that pair).

If all (x_i, x_{i+1}) are sink pairs, then (x_0, x_2) is also a sink pair, because $N(x_2) \subseteq N(x_1) \subseteq N(x_0)$. Otherwise, some (x_i, x_{i+1}) is not a sink pair, say (x_i, x_{i+1}) dominates (p, q). Then $x_i q$ is not an edge; on the other hand, $x_{i+1} p$ must be an edge, since (x_i, x_{i+1}) lies in a trivial component, and this is true for any neighbour p of x_i. To prove the claim, we shall show that any (p, r) dominated by (x_i, x_{i+2}) must lie in D, whence (x_i, x_{i+2}) lies in a ripe component. If $x_{i+1} r$ is an edge, then (x_i, x_{i+1}) dominates (p, r) and hence $(p, r) \in D$. If $x_{i+1} r$ is not an edge, then (x_{i+1}, x_{i+2}) dominates (p, r) and hence $(p, r) \in D$. Moreover, $x_{i+2} p$ must be an edge (for any $p \in N(x_i)$), else (p, r) dominates (x_i, x_{i+2}) which is then also in D, contradicting the minimality of our circuit $(x_0, x_1), (x_1, x_2), \ldots, (x_n, x_0)$.

Case 2. Assume each pair in the circuit belongs to a non-trivial component.

We first claim that there exists a set of mutually independent edges $x_0 y_0$, $x_1 y_1, \ldots, x_n y_n$. We have already seen (see the comment right after Observation 1) that there exist independent edges $x_0 y_0, x_1 y_1$, so let $x_0 y_0, x_1 y_1, \ldots, x_k y_k$ be independent edges and $k < n$. We note that $y_0 x_{k+1}$ cannot be an edge, otherwise (y_0, y_1) (which is in D because it is dominated by $(x_0, x_1) \in D$) dominates (x_{k+1}, x_1), completing a shorter circuit in $C \cup D$. Similarly, $y_1 x_{k+1}$ cannot be an edge, otherwise (y_1, y_2) (which is in D) dominates (x_{k+1}, x_2), also completing a shorter circuit. Continuing this way, we conclude $y_k x_{k+1}$ cannot be an edge, else (y_{k-1}, y_k) (which is in D) dominates (x_{k-1}, x_{k+1}), also yielding a shorter circuit. Since x_{k+1} is not adjacent to any of $y_0, \ldots, y_k$, and there are no isolated vertices, there exists a vertex different from $y_0, \ldots, y_k$ that is adjacent to x_{k+1}; let that vertex be y_{k+1}. Analogously, y_{k+1} is not adjacent to $x_0, x_1, \ldots, x_k$, so that $x_0 y_0, x_1 y_1, \ldots, x_k y_k, x_{k+1} y_{k+1}$ is also an independent set of edges, and by induction on k we obtain an independent set of edges $x_0 y_0, x_1 y_1, \ldots, x_n y_n$.

Observation 2. *Any vertex p adjacent to at least two of the vertices $x_0, x_1, \ldots, x_n$ is adjacent to all of them, and any vertex q adjacent to at least two of the vertices $y_0, y_1, \ldots, y_n$ is adjacent to all of them.*

Otherwise, there is an index j such that p is not adjacent to x_j but is adjacent to x_{j+1}, and an index $k \neq j + 1$ such that p is adjacent to x_k. Then the pair (y_j, p) is dominated by the pair (x_j, x_{j+1}) (which is in D), and dominates the pair (x_j, x_k). Note that we have $j \neq k - 1$ and by the definition of D and D' also $j \neq k + 1$. Thus there are two possible non-trivial cases ($n > 2$): either $k + 1 < j \leq n$, or $0 \leq j < k - 1$. In both cases we obtain a shorter circuit and thus a contradiction. (The proof for q is analogous).

For future reference, we note that there are in $C \cup D$ other circuits similar to (and of the same length as) $(x_0, x_1), (x_1, x_2), \ldots, (x_n, x_0)$: in particular

(1) the circuit $(y_0, y_1), (y_1, y_2), \ldots, (y_n, y_0)$,
(2) any circuit $(y_0, y_1), \ldots, (y_{i-1}, y_i'), (y_i', y_{i+1}), \ldots, (y_n, y_0)$ where y_i' is adjacent to x_i but not to x_{i-1},
(3) and any circuit $(x_0, x_1), \ldots, (x_{i-1}, x_i'), (x_i', x_{i+1}), \ldots, (x_n, x_0)$ where x_i' is adjacent to y_i but not to y_{i-1}.

In each of these cases it can be easily checked that all pairs are in $C \cup D$.

Since each of these circuits is also minimal, Observation 2 applies to any of these alternate circuits. For ease of the explanations, we will assume that the vertices x_i are *white* in the bipartition of the graph, and the vertices y_j are *black*.

Let K denote the set of (black) vertices of H adjacent to all $x_i, i = 1, \ldots, n$ and K' the set of (white) vertices adjacent to all y_i. Each of the remaining vertices (of either colour) has at most one neighbour amongst $x_0, x_1, \ldots, x_n$ and at most one neighbour amongst $y_0, y_1, \ldots, y_n$.

Observation 3. *The graph $H \setminus (K \cup K')$ has components $S_0, S_1, \ldots, S_m$ where, for $i = 1, \ldots, m$, the vertices x_i and y_i are in S_i, and if $p \in K$ is adjacent to a (white) vertex of S_i, then it is adjacent to all white vertices of S_i, and if $q \in K'$ is adjacent to a (black) vertex of S_i, then it is adjacent to all black vertices of S_i.*

Moreover, if $x_0', x_1', \ldots, x_n'$ are any white vertices with $x_i' \in S_i$, then (x_0', x_1'), $(x_1', x_2'), \ldots, (x_n', x_0')$ is also a circuit in $C \cup D$; and similarly, if $y_0', y_1', \ldots, y_n'$ are any black vertices with $y_i' \in S_i$, then $(y_0', y_1'), (y_1', y_2'), \ldots, (y_n', y_0')$ is also a circuit in $C \cup D$.

First, we show that any path joining two different vertices x_i, x_j must contain a vertex of $K \cup K'$. Let $x_i, b_1, a_2, \ldots, a_t, b_t, x_j$ be such a path. If $x_r b_1 \in E(H)$ for some $r \neq i$, then by Observation 2, b_1 is adjacent to all $x_0, x_1, \ldots, x_n$, implying that $b_1 \in K$. Thus, suppose that b_1 is adjacent to only x_i. Now as in (2), we have the circuit $(y_0, y_1), (y_1, y_2), \ldots, (y_{i-1}, b_1), (b_1, y_{i+1}), \ldots,$

(y_n, y_0) in $C \cup D$. (We say we "replaced" y_i by b_1). Note that (y_{i-1}, b_1) and (y_{i-1}, y_i) are in the same strong component of H^+ and similarly for (y_i, y_{i+1}) and (b_1, y_{i+1}). As before, if $a_2 y_r \in E(H)$, $r \neq i$, then a_2 is adjacent to all $y_0, y_1, \dots, y_n$, and hence, $a_2 \in K'$. Thus assume a_2 is adjacent to only b_1. As in (3), we can replace x_i by a_2, and obtain the circuit $(x_0, x_1), \dots, (x_{i-1}, a_2), (a_2, x_{i+1}), \dots, (x_n, x_0)$. By continuing this way, we eventually obtain the circuit $(x_0, x_1), \dots, (x_{i-1}, a_t), (a_t, x_{i+1}), \dots, (x_n, x_0)$ in $C \cup D$. Noting that $a_t b_t$ and $b_t x_j$ are in $E(H)$, we conclude that b_t is adjacent to all $x_0, x_1, \dots, x_n$, implying that $b_t \in K$.

This means that $H \backslash (K \cup K')$ has components S_i with $x_i, y_i \in S_i$ for $i = 1, 2, \dots, n$, as well as possibly other components $S_j, j > n$. The observations above now imply that each x_i can be replaced in the circuit by any of y_i's neighbours $x_i' \in S_i$, and by repeating the argument, by any x_i' in the component S_i. Thus any $x_i' \in S_i$ lies in a suitable circuit and $p \in K$ is adjacent to each black vertex of $S_i, i \leq n$.

We note that $K \cup K'$ induces a biclique. Otherwise, suppose $a \in K$ and $b \in K'$ where $ab \notin E(H)$. Now $(x_0, x_1), (a, y_1), (x_1, b), (y_1, y_0), (x_1, x_0)$ is a directed path in H^+ and $(x_1, x_0), (a, y_0), (x_0, b), (y_0, y_1), (x_0, x_1)$ is also a directed path in H^+, implying that x_0, x_1 is an invertible pair, a contradiction.

Observation 4. *Let C_i be the component of H^+ containing (x_i, x_{i+1}), and let W be any directed path in H^+, starting at $(u, v) \in C_i$. Then for every $(p, q) \in W$, either $(p, q) \in C_i$ with $p \in S_i$ and $q \in S_{i+1}$, or (p, q) is the last vertex of W, and (p, q) is a sink pair with $p \in K \cup K'$ and $q \in S_{i+1}$.*

Let (p, q) be the second vertex of W, following (u, v). Since uq is not an edge, $q \notin K \cup K'$ and $u \notin K \cup K'$. Since $u \in S_i, v \in S_{i+1}$, we have $q \in S_{i+1}$ and $p \in S_i$ or $p \in K \cup K'$. In the former case, (p, q) is in C_i; in the latter case, when $p \in K \cup K'$, we have $N(q) \subseteq N(p)$, implying that (p, q) is a sink pair.

From Observations 3 and 4 we conclude that all pairs (x_i, x_{i+1}) lie in different strong components of H^+, and in particular that *in the circuit $(x_0, x_1), (x_1, x_2), \dots, (x_n, x_0)$ only the pair (x_n, x_0) lies in C*, i.e., that all the other pairs (x_i, x_{i+1}), $i < n$, belong to D.

By a similar logic, we can deduce that the strong component C' containing (x_0, x_n) is also ripe. Indeed, consider the strong component C' (coupled with C), containing (x_0, x_n), and a pair $(p, q) \notin C'$ dominated by some $(u, v) \in C'$. This means that up, vq are edges, and uq is not an edge, of H. Note that $(v, u) \in C$ belongs to the component C_n from Observation 4, thus $v \in S_n, u \in S_0$. It follows that p is in S_0 or in $K \cup K'$ and $q \in S_n$ (because the absence of the edge uq means it is not in $K \cup K'$). In the former case we would have (p, q) is in C', contrary to assumption. In the latter case, $p \in K \cup K'$ and $q \in S_n$. Now (p, q) is also dominated by any pair $(w, v) \in C_{n-1}$, with w in S_{n-1}. Since all pairs of C_{n-1} are in D, we have $(p, q) \in D$ and so C' is ripe.

In conclusion if adding to D the strong component C containing (x_n, x_0) created a circuit $(x_0, x_1), (x_1, x_2), \dots, (x_n, x_0)$, then adding its coupled component C' containing (x_0, x_n) cannot create a circuit. Indeed, if such circuit (x_n, z_1),

$(z_1, z_2), \ldots, (z_m, x_0)$, (x_0, x_n) existed, then the circuit

$$(x_0, x_1), (x_1, x_2), \ldots, (x_{n-1}, x_n), (x_n, z_1), (z_1, z_2), \ldots, (z_m, x_0)$$

was already present in D, contrary to our assumption.

It remains to consider the following case.

Case 3. Assume some pair (x_i, x_{i+1}) in the circuit lies in a non-trivial strong component and the next pair (x_{i+1}, x_{i+2}) lies in a trivial strong component. By Observation 1 (x_{i+1}, x_{i+2}) is a sink pair, and there are two independent edges px_i, qx_{i+1}. Since there are no isolated vertices, we have an edge rx_{i+2}, and by the discussion following Observation 1 again, px_{i+2} and qx_{i+2} are non-edges giving $r \neq p, r \neq q$. Moreover, rx_{i+1} is an edge (else rx_{i+2} and qx_{i+1} would be independent edges putting (x_{i+1}, x_{i+2}) into a non-trivial strong component). Finally, we observe that rx_i must be an edge, otherwise (p, r) is dominated by (x_i, x_{i+1}) and dominates (x_i, x_{i+2}) which would mean (x_i, x_{i+2}) is in D, contradicting the minimality of our circuit. Since this is true for any $r \in N(x_{i+2})$, this means that (x_i, x_{i+2}) is also a sink pair and hence lies in a trivial ripe strong component. It cannot be in D (by the circuit minimality), so we can add it to D instead of the original strong component C. This cannot create a circuit, as there would have been a circuit in D already. (The proof is similar to the proof in Case 1).

Thus in each case we have identified a ripe strong component that can be added to D maintaining the validity of our four conditions (i–iv).   □

From the proof of Theorem 1 we derive the following corollary, that will be used in the next section.

Corollary 1. *Suppose D is a set of pairs of vertices of a bipartite graph H, such that*

1. *if $(x, y) \in D$ and (x, y) dominates (x', y') in H^+, then $(x', y') \in D$, and*
2. *D has no circuit.*

Then there exists a bipartite min ordering $<$ of H such that $x < y$ for each $(x, y) \in D$ if and only if H has no invertible pair.

3 Min Orderings of Weakly Balanced Bipartite Signed Graphs

Suppose $\widehat{H}$ is a weakly balanced signed graph, represented by a signed graph without purely red edges. The underlying graph of $\widehat{H}$ is denoted by H. Define D_0 to consist of all pairs (x, y) in H^+ such that for some vertex z there is a bicoloured edge zx and a blue edge zy. Let D be the reachability closure of D_0, i.e., the set of all vertices reachable from D_0 by directed paths in H^+. It is easy to see that a min ordering of H is a special min ordering of $\widehat{H}$ if and only if it extends D (in the sense that each pair $(x, y) \in D$ has $x < y$). Note that in bipartite graphs, for any $(x, y) \in D$ the vertices x and y are on the same side of any bipartition.

Theorem 2. *If $\widehat{H}$ has no chain, then the set D can be extended to a special min ordering.*

Clearly, the set D by its definition satisfies condition 1 of Corollary 1. It remains to verify that it also satisfies condition 2.

Define a *petal* in $\widehat{H}$ to be two walks $x, l_1, l_2, \ldots, l_k$ and $x, u_1, u_2, \ldots, u_k$ where xl_1 is bicoloured, xu_1 is unicoloured, and $l_i u_{i+1}$ is not an edge for $i = 1, 2, \ldots, k-1$. We denote the petal by $x, (l_1, u_1), (l_2, u_2), \ldots, (l_k, u_k)$. We say the petal has length k. The petal has *terminals*, or *terminal pair*, (l_k, u_k). We call l_k the *lower terminal*, and u_k the *upper terminal*. In a special min ordering $l_i < u_i$ for $i = 1, 2, \ldots, k$ (Fig. 2).

A *flower* is a collection of petals $P_1, P_2, \ldots, P_n$ with the following structure. If (l_k, u_k) is the terminal pair of P_i and $(l'_{k'}, u'_{k'})$ is the terminal pair of P_{i+1}, then $u_k = l'_{k'}$. (The petal indices are treated modulo n so that the lower terminal of P_1 equals the upper terminal of P_n). Suppose $P_1, P_2, \ldots, P_n$ is a flower with terminal pairs $(l^{(1)}, u^{(1)}), (l^{(2)}, u^{(2)}), \ldots, (l^{(n)}, u^{(n)})$. Then the following circular implication shows $\widehat{H}$ does not admit a special min ordering:

$$l^{(1)} < u^{(1)} = l^{(2)} < \cdots < l^{(n)} < u^{(n)} = l^{(1)}.$$

It is clear that a flower yields a circuit in the set D (of H^+) defined at the start of this section, and conversely, each such circuit arises from a flower. Thus, it remains to prove that if $\widehat{H}$ contains a flower, then it also contains a chain.

We present two observations that allow us to extend the length of a petal, or modify the terminal pair.

Observation 5. *Suppose $x, (l_1, u_1), \ldots, (l_k, u_k)$ is a petal. Let v be a vertex such that $u_k v$ is an edge and $l_k v$ is not an edge. Then $x, (l_1, u_1), \ldots, (l_k, u_k), (w, v)$ is a petal of length $k+1$ for any neighbour w of l_k.*

Observation 6. *Suppose $x, (l_1, u_1), \ldots, (l_k, u_k)$ is a petal. Then $x, (l_1, u_1), \ldots, (w, u_k)$ is a petal of length k for any neighbour w of l_{k-1} (where $l_0 = x$ in the case $k = 1$ in which case xw must be bicoloured).*

Each petal in $\widehat{H}$ enforces an order on the pairs (l_i, u_i). Our aim is to prove that if (l_i, u_i) belongs to several petals, then all petals in $\widehat{H}$ enforce the same ordering, or we discover a chain in $\widehat{H}$.

We are now ready to prove the lemma needed.

Lemma 1. *Suppose $P_1, P_2, \ldots, P_n$ is a flower in $\widehat{H}$. Then $\widehat{H}$ contains a chain.*

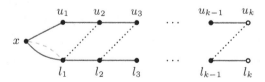

Fig. 2. A petal of length k with terminals (l_k, u_k). Dotted edges are missing.

Proof. We proceed by induction on n. The statement is clearly true if $n = 2$ as the flower is precisely a chain.

Thus assume $n \geq 3$. Without loss of generality suppose the length of P_2 is minimal over all petals. We begin by proving we may reduce the length of P_2 to one. Thus, assume P_2 has length at least two. Suppose the terminal pairs and their predecessors are labelled as in Fig. 3 on the left.

We first observe that if as is an edge, then by Observation 6 we can change the terminal pair of P_2 to be (s, e). Now $P_2, P_3, \ldots, P_n$ is a flower with $n-1$ petals and by induction $\widehat{H}$ has a chain. Hence, assume as is not an edge. By Observation 5 we can extend P_1 to $x, \ldots, (t, c), (s, b), (r, a)$. Using similar reasoning, we see that eu is not an edge and P_3 can be extended so its terminal pair is (d, u). Thus we remove the terminal pair from P_2 so that its terminal pair is (a, d). At this point, P_1, P_2, P_3 are the first three petals of a flower where the length of P_2 has been reduced by one from its initial length.

If n is even, then we similarly reduce the length of $P_4, P_6, \ldots, P_n$ (recall P_2 has minimal length over all petals) and extend the length of all petals with odd indices by one. We obtain a flower where all petals with even subscripts have their length reduced by one from their initial length. Continuing in this manner we can reduce P_2 to length one.

On the other hand, if n is odd, then we modify the extension of P_1 so its terminal pair is (a, t). Namely at the first step we extend P_1 to $x, \ldots, (t, c), (s, b), (t, a)$. As n is odd, the petal P_n will be extended. The extension of P_n will extend its upper terminal from s to t. Thus, the upper terminal of P_n equals the lower terminal of P_1 and the result is again a flower. As in the previous case, we can reduce P_2 to have length one. (As an aid to the reader, we note that when $n = 3$, $s = v$ and $t = u$. The extensions of P_1 and P_3 are such that P_1 will terminate in (a, t) and P_3 will terminate in (d, t)).

Thus, we may assume we have a flower where P_2 has length one as shown in Fig. 3 on the right. If as is a unicoloured edge, then we modify the terminal pair of P_2 to be (b, s). Hence, P_1, P_2 is a flower with two petals and thus a chain. If as is a bicoloured edge, then we modify P_2 to have terminal pair (s, e). Now $P_2, P_3, \ldots, P_n$ is a flower with $n-1$ petals, and by induction $\widehat{H}$ contains a chain. Therefore, as is not an edge.

If et is an edge, then we can modify P_1 to have terminal pair (e, b) by Observation 6. Thus, P_1, P_2 is a flower with two petals, i.e., a chain. Hence, et is not an edge, and we can now extend P_1 by Observation 5 to be $x, \ldots, (t, c), (s, b)$, $(t, a), (s, e)$ incorporating P_2 into P_1. Now we have a flower $P_1, P_3, \ldots, P_n$ and by induction $\widehat{H}$ has a chain.   □

Thus if a weakly balanced bipartite signed graph has no invertible pair and no chain, it has no flowers by Theorem 2, and hence by Corollary 1 it has a special min ordering.

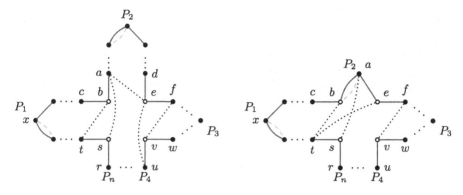

Fig. 3. The labellings used in Lemma 1. On the left is the case when P_2 has length greater than 1 and on the right when P_2 has length 1. Dotted edges are missing.

Finally, we remark that the proofs are algorithmic, allowing us to construct the desired min ordering (if there is no invertible pair) or special min ordering (if there is no invertible pair and no chain).

We have proved our main theorem, which was conjectured by Kim and Siggers.

Theorem 3. *A weakly balanced bipartite signed graph $\widehat{H}$ has a special min ordering if and only if it has no chain and no invertible pair. If $\widehat{H}$ has a special min ordering, then the list homomorphism problem for $\widehat{H}$ can be solved in polynomial time. Otherwise $\widehat{H}$ has a chain or an invertible pair and the list homomorphism problem for $\widehat{H}$ is NP-complete.*

The NP-completeness results are known, and the polynomial time algorithm is presented in the next section.

4   A Polynomial Time Algorithm for the Bipartite Case

Kim and Siggers have proved that the list homomorphism problem for weakly balanced bipartite or reflexive signed graphs with a special min ordering is polynomial time solvable. Their proof however depends on the dichotomy theorem [8,24], and is algebraic in nature. For the bipartite case, we offer a simple low-degree algorithm that effectively uses the special min ordering.

We begin by a review of the usual polynomial time algorithm to solve the list homomorphism problem to a bipartite graph H with a min ordering [12], cf. [17]. Recall that we assume H has a bipartition A, B.

Given an input graph G with lists $L(v) \subseteq V(H), v \in V(G)$, we may assume G is bipartite (else there is no homomorphism at all), with a bipartition U, V, where lists of vertices in U are subsets of A, and lists of vertices in V are subsets of B. We first perform a consistency test, which reduces the lists $L(v)$ to $L'(v)$ by repeatedly removing from $L(v)$ any vertex x such that for some edge $vw \in E(G)$

no $y \in L(w)$ has $xy \in E(H)$. If at the end of the consistency check some list is empty, there is no list homomorphism. Otherwise, the mapping $f(v) = \min L(v)$ (in the min ordering) ensures f is a homomorphism (because of the min ordering property [17]).

We will apply the same logic to a weakly balanced bipartite signed graph $\widehat{H}$; we assume that $\widehat{H}$ has been switched to have no purely red edges. If the input signed graph $\widehat{G}$ is not bipartite, we may again conclude that no homomorphism exists, regardless of lists. Otherwise, we refer to the alternate definition of a homomorphism of signed graphs, and seek a list homomorphism f of the underlying graph of $\widehat{G}$ to the underlying graph of $\widehat{H}$, that:

- maps bicoloured edges of $\widehat{G}$ to bicoloured edges of $\widehat{H}$, and
- maps unicoloured closed walks in $\widehat{G}$ that have an odd number of red edges to closed walks in $\widehat{H}$ that include bicoloured edges.

Indeed, as observed in the first section, this is equivalent to having a list homomorphism of $\widehat{G}$ to $\widehat{H}$, since $\widehat{H}$ does not have unicoloured closed walks with any purely red (i.e., negative) edges.

The above basic algorithm can now be applied to the underlying graphs; if it finds there is no list homomorphism, we conclude there is no list homomorphism of the signed graphs either. However, if the algorithm finds a list homomorphism of the underlying graphs which takes a closed walk R with odd number of red edges to a closed walk M with only purely blue edges, we need to adjust it. (As noted in the introduction, Zaslavsky's algorithm will identify such a closed walk if one exists). Since the algorithm assigns to each vertex the smallest possible image (in the min ordering), we will remove all vertices of M from the lists of all vertices of R, and repeat the algorithm. The following result ensures that vertices of M are not needed for the images of vertices of R.

Theorem 4. *Let $\widehat{H}$ be a weakly balanced bipartite signed graph with a special min ordering $\leq$.*

Suppose C is a closed walk in $\widehat{G}$ and f, f' are two homomorphisms of $\widehat{G}$ to $\widehat{H}$ such that $f(v) \leq f'(v)$ for all vertices v of $\widehat{G}$, and such that $f(C)$ contains only blue edges but $f'(C)$ contains a bicoloured edge.

Then the homomorphic images $f(C)$ and $f'(C)$ are disjoint.

Proof. We begin with three simple observations.

Observation 7. *There exists a blue edge $ab \in f(C)$ and a bicoloured edge $uv \in f'(C)$ such that $a < u, b < v$.*

Indeed, let u be the smallest vertex in A incident to a bicoloured edge in $f'(C)$, and let v be the smallest vertex in B joined to u by a bicoloured edge in $f'(C)$. Let xy be an edge of C for which $f'(x) = u, f'(y) = v$, and let $a = f(x), b = f(y)$. By assumption, $a = f(x) \leq f'(x) = u$ and $b = f(y) \leq f'(y) = v$. Moreover, $a \neq u$ and $b \neq v$ by the special property of min ordering.

Observation 8. *For every $r \in f'(C)$, there exists an $s \in f(C)$ with $s \leq r$.*

This follows from the fact that some x in $\widehat{G}$ has $s = f(x) \leq f'(x) = r$.

Observation 9. *There do not exist edges ab, bc, de with $a < d < c$ and $b < e$, such that ab is blue and de is bicoloured.*

Since $<$ is a min ordering, the existence of such edges would require db to be an edge and the special property of $<$ at d would require this edge to be bicoloured, contradicting the special property at b.

The following observation enhances Observation 9.

Observation 10. *There does not exist a walk $a_0 b_0, b_0 a_1, a_1 b_1, \ldots, b_k c$ of blue edges, and a bicoloured edge de such that $a_0 < d < c$ and $b_0 < e$.*

This is proved by induction on the (even) length k. Observation 9 applies if $k = 0$. For $k > 0$, Observation 9 still applies if $a_0 < d < a_1$ (using the blue walk $a_0 b_0, b_0 a_1$ and the bicoloured edge de). If $d > a_1$, we can apply the induction hypothesis to $a_1 < d < c$ and de as long as $b_1 < e$. The special property of $<$ ensures that $b_1 \neq e$. Finally, if $e < b_1$, then Observation 9 applies to the edges $b_0 a_1, a_1 b_1, ed$.

Having these observations, we can now prove the conclusion. Indeed, suppose that $f(C)$ and $f'(C)$ have a common vertex g. Let us take the largest vertex g, and by symmetry assume it is in A, like a, u, where a, b, u, v are the vertices from Observation 7. Recall that we have chosen u to be the smallest vertex in A incident with a bicoloured edge of $f'(C)$, and v is smallest vertex in B adjacent to u by a bicoloured edge in $f'(C)$.

Suppose first that $g > u$. In $f(C)$ there is a path with edges $ab, ba_1, \ldots, hg$ which has $a < u < g$ and $b < v$, contradicting Observation 10.

If $g = u$ then the path with edges $ba, ab_1, b_1 a_1, \ldots, a_k h, hg$ in $f(C)$ has all edges blue, and thus $h > v$ as $<$ is special. Therefore $b < v < h$ and $a < g$, also contradicting Observation 10.

Finally, suppose that $g < u$. Here we use the path in $f'(C)$ with edges gv_1, $v_1 u_1, u_1 v_2, \ldots, u_{k-1} v_k, v_k u, uv$. A small complication arises if $v_1 > v$, so extend the path to also include ab by preceding it with the path in $f(C)$ with edges $ab, ba_1, a_1 b_1, b_1 a_2, \ldots, b_t g$. Of course the result is now a walk W, not necessarily a path. Note that the first edges of W are blue (being in $f(C)$), but the last edge uv is bicoloured.

If uv is the first bicoloured edge, then $v < v_k$ by the special property, and we have $b < v < v_k$ and $a < u$, a contradiction with Observation 10. Otherwise, the first bicoloured edge on the walk must be some $u_j v_{j+1}$ (where $v_j u_j$ is unicoloured and $u_j \neq u$), or some $v_j u_j$ (where $u_{j-1} v_j$ is unicoloured).

In the first case, where $u_j v_{j+1}$ is the first bicoloured edge, $u_j > u$ by the definition of u. Then $a < u < u_j$ and $b < v$, implying again a contradiction with Observation 10. In the second case, where $v_j u_j$ is the first bicoloured edge, we have again $a < u \leq u_j < u_{j-1}$ (using the special property at v_j), and therefore we have $a < u < u_{j-1}$ and $b < v$ contrary to Observation 10. □

We observe that each phase removes at least one vertex from at least one list, and since $\widehat{H}$ is fixed, the algorithm consists of $O(n)$ phases of arc consistency, where n is the number of vertices (and m number of edges) of $\widehat{G}$. Since arc consistency admits an $O(m + n)$ time algorithm, our overall algorithm has complexity $O(n(m + n))$.

Acknowledgements. This project has received funding from the European Research Council (ERC) under the European Union's Horizon 2020 research and innovation programme (grant agreement No 810115 - DYNASNET). J. Bok and N. Jedličková were also partially supported by GAUK 370122 and SVV-2020-260578. R. Brewster gratefully acknowledges support from the NSERC Canada Discovery Grant programme. P. Hell gratefully acknowledges support from the NSERC Canada Discovery Grant programme. A. Rafiey gratefully acknowledges support from the grant NSF1751765.

We thank Reza Naserasr and Mark Siggers for helpful discussions.

References

1. Bok, J., Brewster, R., Feder, T., Hell, P., Jedličková, N.: List homomorphisms to separable signed graphs. In: Balachandran, N., Inkulu, R. (eds.) CALDAM 2022. LNCS, vol. 13179, pp. 22–35. Springer, Cham (2022). https://doi.org/10.1007/978-3-030-95018-7_3

2. Bok, J., Brewster, R.C., Feder, T., Hell, P., Jedličková, N.: List homomorphism problems for signed graphs. In: Esparza, J., Král, D. (eds.) 45th International Symposium on Mathematical Foundations of Computer Science (MFCS 2020). Leibniz International Proceedings in Informatics (LIPIcs), vol. 170, pp. 20:1–20:14. Schloss Dagstuhl-Leibniz-Zentrum für Informatik, Dagstuhl (2020). https://drops.dagstuhl.de/opus/volltexte/2020/12688

3. Bok, J., Brewster, R.C., Feder, T., Hell, P., Jedličková, N.: List homomorphism problems for signed graphs (2021, submitted). https://arxiv.org/abs/2005.05547

4. Bok, J., Brewster, R.C., Hell, P., Jedličková, N., Rafiey, A.: Min orderings and list homomorphism dichotomies for signed and unsigned graphs. CoRR abs/2206.01068 (2022). https://arxiv.org/abs/2206.01068

5. Bok, J., Brewster, R.C., Hell, P., Jedličková, N.: List homomorphisms of signed graphs. In: Bordeaux Graph Workshop, pp. 81–84 (2019)

6. Brewster, R.C., Foucaud, F., Hell, P., Naserasr, R.: The complexity of signed graph and edge-coloured graph homomorphisms. Discret. Math. **340**(2), 223–235 (2017)

7. Brewster, R.C., Siggers, M.: A complexity dichotomy for signed H-colouring. Discret. Math. **341**(10), 2768–2773 (2018)

8. Bulatov, A.A.: A dichotomy theorem for nonuniform CSPs. In: 2017 IEEE 58th Annual Symposium on Foundations of Computer Science (FOCS), pp. 319–330. IEEE (2017)

9. Feder, T., Hell, P., Huang, J.: List homomorphisms and circular arc graphs. Combinatorica **19**(4), 487–505 (1999)

10. Feder, T., Hell, P., Huang, J.: Bi-arc graphs and the complexity of list homomorphisms. J. Graph Theory **42**(1), 61–80 (2003)

11. Feder, T., Hell, P., Huang, J., Rafiey, A.: Interval graphs, adjusted interval digraphs, and reflexive list homomorphisms. Discret. Appl. Math. **160**(6), 697–707 (2012)

12. Feder, T., Vardi, M.Y.: The computational structure of monotone monadic SNP and constraint satisfaction: a study through Datalog and group theory. In: STOC, pp. 612–622 (1993)

13. Guenin, B.: A characterization of weakly bipartite graphs. J. Combin. Theory Ser. B **83**(1), 112–168 (2001)
14. Guenin, B.: Packing odd circuit covers: a conjecture (2005). Manuscript
15. Hell, P., Mastrolilli, M., Nevisi, M.M., Rafiey, A.: Approximation of minimum cost homomorphisms. In: Epstein, L., Ferragina, P. (eds.) ESA 2012. LNCS, vol. 7501, pp. 587–598. Springer, Heidelberg (2012). https://doi.org/10.1007/978-3-642-33090-2_51
16. Hell, P., Nešetřil, J.: On the complexity of H-coloring. J. Combin. Theory Ser. B **48**(1), 92–110 (1990)
17. Hell, P., Nešetřil, J.: Graphs and Homomorphisms. Oxford Lecture Series in Mathematics and its Applications, vol. 28. Oxford University Press, Oxford (2004)
18. Kaiser, T., Lukoťka, R., Rollová, E.: Nowhere-zero flows in signed graphs: a survey. In: Selected Topics in Graph Theory and Its Applications. Lecture Notes of Seminario Interdisciplinare di Matematica, vol. 14, pp. 85–104. Seminario Interdisciplinare di Matematica (S.I.M.), Potenza (2017)
19. Kim, H., Siggers, M.: Towards a dichotomy for the switch list homomorphism problem for signed graphs (2021). https://arxiv.org/abs/2104.07764
20. Naserasr, R., Sopena, E., Zaslavsky, T.: Homomorphisms of signed graphs: an update. European J. Combin. **91**, Paper No. 103222, 20 (2021). https://doi.org/10.1016/j.ejc.2020.103222
21. Schrijver, A.: A short proof of Guenin's characterization of weakly bipartite graphs. J. Combin. Theory Ser. B **85**(2), 255–260 (2002)
22. Zaslavsky, T.: Signed graph coloring. Discrete Math. **39**(2), 215–228 (1982)
23. Zaslavsky, T.: A mathematical bibliography of signed and gain graphs and allied areas. Electron. J. Combin. **5**, Dynamic Surveys 8, 124 (1998). Manuscript prepared with Marge Pratt
24. Zhuk, D.: A proof of CSP dichotomy conjecture. In: 58th Annual IEEE Symposium on Foundations of Computer Science—FOCS 2017, pp. 331–342. IEEE Computer Society, Los Alamitos (2017)

On the Zombie Number of Various Graph Classes

Prosenjit Bose[1], Jean-Lou De Carufel[2(✉)], and Thomas Shermer[3]

[1] School of Computer Science, Carleton University, Ottawa, Canada
jit@scs.carleton.ca
[2] School of Electrical Engineering and Computer Science, University of Ottawa, Ottawa, Canada
jdecaruf@uottawa.ca
[3] School of Computing Science, Simon Fraser University, Burnaby, Canada
shermer@sfu.ca

Abstract. The game of *zombies and survivor* on a graph is a pursuit-evasion game that is a variant of the game of cops and robber. The game proceeds in rounds where first the zombies move then the survivor moves. Zombies must move to an adjacent vertex that is on a shortest path to the survivor's location. The survivor can move to any vertex in the closed neighborhood of its current location. The *zombie number* of a graph G is the smallest number of zombies required to catch the survivor on G. The graph G is said to be *k-zombie-win* if its zombie number is $k \geq 1$.

We first examine bounds on the zombie number of the Cartesian and strong products of various graphs. We also introduce graph classes which we call *capped products* and provide some bounds on zombie number of these as well. Fitzpatrick et al. (Discrete Applied Math., 2016) provided a sufficient condition for a graph to be 1-zombie-win. Using a capped product, we show that their condition is not necessary. We also provide bounds for two variants called *lazy zombies* and *tepid zombies* on some graph products. A lazy zombie is a zombie that does not need to move from its current location on its turn. A tepid zombie is a lazy zombie that can move to a vertex whose distance to the survivor does not increase. Finally, we design an algorithm (polynomial in n for constant k) that can decide, given a graph G, whether or not it is k-zombie-win for all the above variants of zombies.

1 Introduction

Zombies and Survivor is a type of pursuit-evasion game played on a graph[1]. It is a variant of the game of Cops and Robber (see Bonato and Nowakowski for an excellent introduction to this topic [3]). The deterministic version of Zombies and Survivor was introduced by Fitzpatrick et al. [5]. This game is played on a

---

[1] We use standard graph theoretic notation, terminology and definitions (see [4]).

Supported by the Natural Sciences and Engineering Research Council of Canada (NSERC).

A. Castañeda and F. Rodríguez-Henríquez (Eds.): LATIN 2022, LNCS 13568, pp. 527–543, 2022.
https://doi.org/10.1007/978-3-031-20624-5_32

connected graph G and the game proceeds in rounds. Initially, the k *zombies* with $k \geq 1$ are either placed strategically on $V(G)$ or an adversary places them on $V(G)$. Then the survivor selects an initial position $s \in V(G)$. During a round, first each zombie moves from its current position to an adjacent vertex that belongs to some shortest path to the survivor's position. Then, the survivor can move to any vertex in $N[s]$ (the closed neighborhood of s in G). Here is where one can see the distinction between this game and the game of Cops and Robber. A cop on vertex v can move to any vertex in $N[v]$ like a survivor, however, a zombie must always move from its current position and is restricted to only follow an edge of a shortest path to the survivor. The goal of the zombies is to have at least one zombie land on the current position of the survivor on its turn. If it does so, then G is considered k-*zombie-win*. The goal of the survivor is to perpetually evade all k zombies. A graph that is not k-zombie-win is considered *survivor-win with respect to k zombies*. When $k = 1$, we simply say zombie-win or survivor-win.

In contrast to the game of Cops and Robber, the starting configuration of the zombies plays an important role in determining whether a survivor is caught. For example, if several zombies are placed on the same vertex in a cycle with more than 4 vertices, then they will never catch a survivor. However, 2 zombies can be strategically placed in order to always catch a survivor. As such, we define two types of zombie number, one where the zombies are strategically placed and the other where an adversary determines the initial position of the zombies, i.e. in a worst-case starting position.

The *zombie number* $z(G)$ of a graph G is defined as the minimum number of zombies required to catch the survivor on G, and the *universal zombie number* $u(G)$ is defined as the minimum number of zombies required to catch the survivor when the starting configuration of the zombies is determined by an adversary. The above example shows that $z(G) = 2$ but $u(G) = \infty$ when G is a cycle on more than 4 vertices. We denote the cop number of a graph as $c(G)$. Since a cop has more power than a zombie, we have $c(G) \leq z(G) \leq u(G)$. From this observation, we get that zombie-win graphs are also cop-win graphs.

In their paper, Fitzpatrick et al. [5] establish the first results on the deterministic version of the game of Zombies and Survivor. They provide an example showing that if a graph is cop-win, then it is not necessarily zombie-win. They provide a sufficient condition for a graph to be zombie-win. They also establish several results about the zombie number of the Cartesian product of graphs. The main aspect that makes different variants of these pursuit-evasion problems quite challenging is the fact that the cop number and zombie number is not a monotonic property with respect to subgraphs. For example, both the cop number and zombie number of a clique is 1 but the cop number and zombie number of a cycle on more than 3 vertices is 2.

In this paper, we consider bounds on the zombie number as we progressively give more power to the zombies. In addition to normal zombies, we study *lazy zombies* and *tepid zombies*. A lazy zombie is a zombie that has the ability to remain on its current location on its turn. A tepid zombie is a lazy zombie that can move to a vertex whose distance to the survivor does not increase. The notation for the corresponding zombie numbers for these zombies are $z_L(G), u_L(G), z_T(G), u_T(G)$ for the lazy zombie number, universal lazy zombie number, tepid zombie number and universal tepid zombie number of a graph, respectively. Note that universality includes the case of all pursuers starting on the same vertex. This means that many graphs have infinite $u(G)$; for instance, $u(C_n) = \infty$ for $n \geq 5$.

1.1 Contributions

We first focus on the relationship between the zombie numbers of some graphs and the zombie numbers of different products of those graphs. We examine bounds on Cartesian products of graphs and strong products of graphs. We also introduce a variant which we call a capped product and provide some upper and lower bounds on these products as well. Using the capped product, we provide an example showing that the sufficient condition for a graph to be zombie-win established by Fitzpatrick et al. [5] is not a necessary condition. Our bounds on Cartesian and strong products are summarized in Table 1. We then design an algorithm that can decide whether or not a graph G is k-zombie win for all the above variants of zombies. Unfortunately, due to space constraints, some proofs are missing.

Table 1. Summary of our bounds for Cartesian product and strong product.

	Cartesian product	Strong product
Zombies	$z(G \square H) \leq$ $z(G) + z(H)$ [1,6]	$z(G \boxtimes H) \leq z(G) \cdot z(H)$
Universal zombies	$u(G \square H) \leq$ $u(G) + u(H)$	$u(G \boxtimes H) \leq u(G) + u(H) - 1$
Lazy zombies	$z_L(G \square H) \quad\leq$ $z_L(G) + z_L(H)$ $z_L(P_2 \square C_n) =$ $2, n \geq 3$	$z_L(G \boxtimes H) \leq z_L(G) \cdot z_L(H)$
Universal lazy zombies	$u_L(G \square H) \quad\leq$ $u_L(G) + u_L(H)$ $u_L(P_2 \square C_n) =$ $2, n \geq 3$	$u_L(G \boxtimes H) \leq u_L(G) + u_L(H) - 1$
Tepid zombies	$z_T(G \square H) \leq$ $z_T(G) + z_T(H)$	$z_T(G \boxtimes H) \leq z_T(G) \cdot z_T(H)$
Universal tepid zombies	$u_T(G \square H) \leq$ $u_T(G) + u_T(H)$	$u_T(G \boxtimes H) \leq u_T(G) + u_T(H) - 1$

2   Graph Products

Given connected graphs $G_1 = (V_1, E_1)$ and $G_2 = (V_2, E_2)$, we denote their Cartesian product as $G_1 \square G_2$ and their strong product as $G_1 \boxtimes G_2$. Given a path $P = (a_1, b_1), \ldots, (a_i, b_j)$ in $G_1 \square G_2$ (resp. $G_1 \boxtimes G_2$), we define the *shadow* of P onto G_1, $\Psi_{G_1}(P)$, as the walk $a_1, \ldots, a_i$. Similarly, $\Psi_{G_2}(P)$ is the walk $b_1, \ldots, b_j$ in G_2.

In the following, we number the vertices of a path P_j as $0, 1, \ldots, j - 1$ (in order) and the vertices of a cycle C_j similarly. When it is clear from context, we drop the fixed parts of the Cartesian product pair notation, writing simply ab rather than (a, b).

2.1   Cartesian Products

In this subsection, we study the zombie number of the Cartesian products of graph. Fitzpatrick *et al.* ([5], Question 10) asked whether $z(G \square H) \leq z(G) + z(H)$, which was answered in the affirmative independently by Bartier et al. [1] and Keramatipour and Bahrak [6]. Both proofs are similar and hinge on the following observation that is embedded in their proofs.

Observation 1. *[1,6] Let $\pi_{G_1}(a_1, a_i) = a_1, \ldots, a_i$ and $\pi_{G_2}(b_1, b_j) = b_1, \ldots, b_j$ (with $i, j \geq 2$) be shortest paths in G_1 and G_2, respectively. There is a shortest path from $a_1 b_1$ to $a_i b_j$ in $G_1 \square G_2$ where either $a_2 b_1$ or $a_1 b_2$ is the second vertex in the path.*

Observe that the strategy presented by Bartier et al. [1] and Keramatipour and Bahrak [6] can be used both by lazy zombies and tepid zombies on $G \square H$, which allows us to conclude the following:

Corollary 1. *For all graphs G and H, $z_L(G \square H) \leq z_L(G) + z_L(H)$ and $z_T(G \square H) \leq z_T(G) + z_T(H)$.*

We show how their bound can also be generalized for universal zombies.

Lemma 1. $u(G_1 \square G_2) \leq u(G_1) + u(G_2)$.

Proof. The adversary places $u(G_1) + u(G_2)$ zombies in $G_1 \square G_2$. Partition these zombies into two sets U_1 and U_2 where $|U_1| = u(G_1)$ and $|U_2| = u(G_2)$. Let the survivor's current position be denoted as (s_1, s_2). Let (a, b) (resp. (c, d)) be the coordinates of an arbitrary zombie in U_2 (resp. U_1). If $a = s_1$ (resp. $d = s_2$) then the U_2-zombies (U_1-zombies) move in one turn of a winning strategy for G_2 (resp. G_1) on the second (resp. first) coordinate. This is a valid zombie move by Observation 1. If $a \neq s_1$ (resp. $d \neq s_2$) then the U_2-zombies (U_1-zombies) move toward s_1 (resp. s_2) on the first (resp. second) coordinate. Again, this is a valid move due to Observation 1.

Let $S_{G_1}(U_1)$ (resp. $S_{G_2}(U_2)$) be an upper bound on the number of moves for $u(G_1)$ (resp. $u(G_2)$) zombies to catch a survivor in G_1 (resp. G_2). Let $M =$

$S_{G_1}(U_1) + S_{G_2}(U_2) + |V(G_1)| + |V(G_2)|$. Notice that if the survivor is stationary on G_2 (resp. G_1) for $|V(G_2)| + S_{G_1}(U_1)$ (resp. $|V(G_1)| + S_{G_2}(U_2)$) turns, then a U_1-zombie (resp. U_2-zombie) catches the survivor. On its turn, by definition of the Cartesian product, the survivor must be stationary on one coordinate. In other words, if it moves in G_1, then it is stationary in G_2 and vice versa. Therefore, after M turns, the survivor must have been stationary for at least $|V(G_2)| + S_{G_1}(U_1)$ turns on G_2 or at least $|V(G_1)| + S_{G_2}(U_2)$ turns on G_1. Without loss of generality, assume that the survivor was stationary on the second coordinate for $|V(G_2)| + S_{G_1}(U_1)$ turns. Then, a U_1-zombie must have caught the survivor. After $|V(G_2)|$ turns where the survivor is stationary on G_2, the U_1-zombies catch the shadow on the second coordinate. Then, after another $S_{G_1}(U_1)$ turns where the survivor is stationary on G_2, a U_1-zombie catches the survivor since the U_1-zombies are playing the winning strategy on G_1.   □

The strategy presented in the proof of Lemma 1 can be used both by lazy zombies and tepid zombies on $G\square H$, which allows us to conclude the following:

Corollary 2. *For all graphs G and H, $u_L(G\square H) \leq u_L(G) + u_L(H)$ and $u_T(G\square H) \leq u_T(G) + u_T(H)$.*

Fitzpatrick *et al.* [5] showed that $z(P_2\square C_n) > 2$ for odd $n \geq 5$ which is optimal since $z(G\square H) \leq z(G) + z(H)$. Although the general bounds of Corollaries 1 and 2 imply an upper bound of 3 for lazy and tepid zombies, we prove that these general upper bounds are not tight by showing optimal bounds.

Theorem 1. *For $n \geq 3$, $z_L(P_2\square C_n) = 2$.*

Proof. That one zombie cannot catch a survivor on this graph is trivial. Start with z_1 on $(0,0)$ and z_2 on $(1, \lceil \frac{n}{2} \rceil)$. Without loss of generality, the survivor starts on (s_x, s_y) with $0 \leq s_y \leq \lceil \frac{n}{2} \rceil$. On the zombies' turn, the survivor is at (s'_x, s'_y) while z_1 is at $(0, a)$ and z_2 is at $(1, b)$; $a \leq s'_y \leq b$. If survivor is adjacent to a zombie, the zombie captures. Otherwise, if $s'_x = 0$, move z_1 to $(0, a + 1)$; if $s'_x = 1$, move z_2 to $(1, b - 1)$. Only one zombie moves and the other stays still. This strategy keeps the survivor between the zombies (i.e. $a \leq s'_y \leq b$) but reduces $b - a$ by one each turn, so the survivor is eventually caught.   □

Fitzpatrick *et al.* [5] showed that for $n \geq 5$ odd, $z(P_2\square C_n) = 3 > 2 = c(P_2\square C_n)$. In the following theorem, we show that $u_L(P_2\square C_n) = c(P_2\square C_n)$.

Theorem 2. *For $n \geq 3$, $u_L(P_2\square C_n) = 2$.*

Proof. (Sketch) Two lazy zombies eventually force a situation where the zombies can play the strategy from the proof of Theorem 1.   □

2.2   Strong Products

We now turn our attention to the relationship between strong products and zombie numbers. Although the strong product and Cartesian product of two graphs are related, the strong product introduces its own set of complications. We begin by introducing the concept of *lifting*. Given two graphs G_1 and G_2, let $W_1 = a_1, \ldots, a_i$ be a walk in G_1 and $W_2 = b_1, \ldots, b_j$ be a walk in G_2. We assume that $i < j$. We define a walk in $G_1 \boxtimes G_2$ as the *lifting* of W_1 and W_2, denoted $\Lambda(W_1, W_2)$, as the path $(a_1, b_1), (a_2, b_2), \ldots, (a_i, b_i), (a_i, b_{i+1}), (a_i, b_{i+2}), \ldots, (a_i, b_j)$. Similarly, when $i \geq j$, $\Lambda(W_1, W_2)$ is the path $(a_1, b_1), (a_2, b_2), \ldots, (a_j, b_j)$, $(a_{j+1}, b_j), (a_{j+2}, b_j), \ldots, (a_i, b_j)$. We first establish some properties of shortest paths.

Lemma 2. *Let P_1 be a shortest path from a_1 to a_i in G_1 and let P_2 be a shortest path from b_1 to b_j in G_2. The path $\Lambda(P_1, P_2)$ is a shortest path in $G_1 \boxtimes G_2$ from (a_1, b_1) to (a_i, b_j) of length $\max\{i, j\}$.*

Lemma 3. *If $a_1 a_2$ is an edge on a shortest path P_1 from a_1 to a_i in G_1 and $b_1 b_2$ is an edge on a shortest path P_2 from b_1 to b_j in G_2, then the edge from (a_1, b_1) to (a_2, b_2) is on a shortest path from (a_1, b_1) to (a_i, b_j) in $G_1 \boxtimes G_2$.*

Lemma 4. *Let P be a shortest path between two vertices in $G_1 \boxtimes G_2$. $\Psi_{G_1}(P)$ is a shortest path in G_1 or $\Psi_{G_2}(P)$ is a shortest path in G_2.*

Using the above lemmas, we are able to give an upper bound on $z(G_1 \boxtimes G_2)$.

Lemma 5. $z(G_1 \boxtimes G_2) \leq z(G_1) \cdot z(G_2)$.

Proof. Without loss of generality, assume that $z(G_1) \leq z(G_2)$. Let $i = z(G_1)$ and $j = z(G_2)$. Let a winning starting configuration of zombies on G_1 be a placement of the zombies on vertices $a_1, \ldots, a_i$ and a winning starting configuration of zombies on G_2 be a placement of zombies on vertices $b_1, \ldots, b_j$. On $G_1 \boxtimes G_2$, the initial placement of the ij zombies is on the vertices (a_x, b_y) for $x \in \{1, \ldots, i\}$ and $y \in \{1, \ldots, j\}$. Since $z(G_1) = i$, there is a winning strategy where one zombie ultimately catches the survivor on G_1 after a finite number of moves. Similarly, there is a winning strategy where 1 zombie ultimately catches the survivor on G_2 after a finite number of moves. By Lemma 3, this strategy is played simultaneously on $G_1 \boxtimes G_2$. Suppose that a zombie on vertex x on G_1, on its turn, moves to vertex y. And on the same turn, a zombie on vertex v in G_2 moves to vertex w on its turn. Then the zombie on (x, v) in $G_1 \boxtimes G_2$ moves to (y, w), which is a valid move by Lemma 2. Now, suppose that the zombie catches the survivor on G_1 at vertex s. Let $b_1', \ldots, b_j'$ be the current positions of the zombies in G_2 after having started on an initial winning configuration. By construction, we have a zombie on every vertex of the form (s, b_y') where $y \in \{1, \ldots, j\}$. Therefore, when the pursuit continues, the zombie catches the survivor on G_2 which is guaranteed since $z(G_2) = j$. At this point, the zombie will have caught the survivor on $G_1 \boxtimes G_2$. □

Notice that the zombie number of $z(G \boxtimes H)$ is at most $z(G) \cdot z(H)$ as opposed to $z(G) + z(H)$ as is the case of the Cartesian product. The main reason is that in the Cartesian product, on its turn, the survivor had to remain stationary on one coordinate whereas in the strong product, the survivor can move on both coordinates. We are able to overcome this difficulty in the case of universal zombies. Given $k \geq 1$, the k-variant of the game ends when k zombies are at the survivor's location.

Lemma 6. *With $u(G) + k - 1$ zombies in G, for $k \geq 1$, there exists a strategy for the k-variant of the game where k zombies catch the survivor after a finite number of steps.*

Lemma 7. $u(G_1 \boxtimes G_2) \leq u(G_1) + u(G_2) - 1$.

Proof. Let $i = u(G_1)$ and $j = u(G_2)$. Place $i + j - 1$ zombies anywhere in G. By Lemma 6, j zombies can catch survivor in G_1 when $i + j - 1$ zombies are initially placed anywhere in G_1. Similarly, i zombies can catch the survivor on G_2 when $i + j - 1$ zombies are initially placed anywhere in G_2. By Lemma 3, both the strategies on G_1 and G_2 can be played simultaneously on $G_1 \boxtimes G_2$ until there are j zombies with first coordinate s_1 and i zombies with second coordinate s_2, where (s_1, s_2) is the survivor's position. By the pigeonhole principle, since there are only $i + j - 1$ zombies total in $G_1 \boxtimes G_2$, there must be at least 1 zombie on (s_1, s_2), which implies that the survivor has been caught.   □

3   Capped Products and Cop-Win Spanning Trees

Taking inspiration from examples in Fitzpatrick et al., we define the *capped product*. Imagine some graph $G \boxtimes P_j$ or $G \square P_j$. In these products, there are two special copies G_0 and G_{j-1} of G, called the *end copies* of G, corresponding to each of the ends of the path P_j. To *cap* this product, we take a graph H with $V(G) \subseteq V(H)$ and identify the vertices common to G_0 and H. We denote such a graph as $G \boxtimes P_j \pitchfork H$ or $G \square P_j \pitchfork H$. (Generally the identification of vertices in G and H is understood and not specified.) We call any vertex of H that is not placed in correspondence with a vertex of G_0 an *extra* vertex.

If G has n vertices, the specific caps H that seem useful are K_n, where any 1-to-1 correspondence with the vertices of G is the same; K_{n+1}, where one vertex of the cap is extra but the rest are in 1-to-1 correspondence; and $K_{1,n}$, where the leaves of $K_{1,n}$ are placed in 1-to-1 correspondence with G, and the internal vertex is extra. However, we write our results in more general terms. For instance, both K_{n+1} and $K_{1,n}$ can be characterized as caps that have an extra universal vertex. Figure 1 shows (a) $C_6 \boxtimes P_3$, (b) $C_6 \boxtimes P_3 \pitchfork K_6$, (c) $C_6 \boxtimes P_3 \pitchfork K_7$, and (d) $C_6 \boxtimes P_3 \pitchfork K_{1,6}$. In general, we label the vertices of C_j from 0 to $j - 1$ in counterclockwise order. Fitzpatrick et al. [5] use this type of construction for their Figure 1 ($C_5 \boxtimes P_3 \pitchfork K_5$) and Figure 5 ($C_5 \boxtimes P_3 \pitchfork K_6$).

For rooted trees T, we also extend the notation to $G \square T \pitchfork H$ and $G \boxtimes T \pitchfork H$, where the copy of G corresponding to the root is used as the "end copy" to identify vertices of H with. We consider the path P_k on k vertices as a tree rooted at the vertex 0.

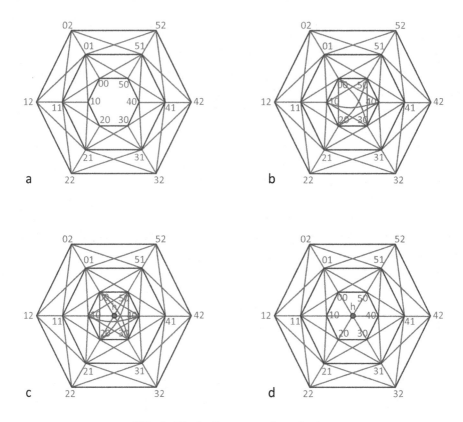

Fig. 1. Illustrating capped products.

We want to discuss Theorem 6 by Fitzpatrick et al. [5], but we first need to define what a *cop-win spanning tree* is. In a graph G, we denote the *neighborhood* of a vertex u as $N(u)$. It corresponds to the set of all neighbors of u in G. The *closed neighborhood* of u is $N[u] = N(u) \cup \{u\}$. Given two different vertices $u, v \in V(G)$, we say that u is a *corner* of v (or v is a *cover* of u) whenever $N[u] \subseteq N[v]$. Let $u \in V(G)$ be a corner of $v \in V(G)$. Let G_1 be the graph obtained from G by removing u and all its incident edges. We call this operation a (one-point) *retraction* and we denote it by $u \to v$. Note that this differs from the standard definition of one-point retraction [3] in that a self-loop is not required or created on v. The graph G_1 is called a *retract* of G. We know that $c(G_1) \leq c(G)$ (refer to [2]). A *dismantling* is a finite sequence of retractions, which produces a finite sequence of graphs $G_0 = G, G_1, G_2, ..., G_k$, where G_{i+1} is a retract of G_i ($0 \leq i \leq k-1$). G is cop-win if and only if there exists a dismantling of length n (where $n = |V(G)|$) such that G_n is the trivial graph on one vertex [7,8]. Given a cop-win graph G, a *cop-win spanning tree* is a tree T with vertex set $V(G)$ that is associated to a dismantling of G. In T, u is a child of v if and only if

$u \to v$ is a retraction in the associated dismantling. In their paper, Fitzpatrick et al. [5] prove the following theorem.

Theorem 3. *If there exists a breadth-first search of a graph such that the associated spanning tree is also a cop-win spanning tree, then G is zombie-win.*

They then go on to show that this is a stronger statement (in the sense that more graphs satisfy the hypothesis) than saying that *bridged*[2] graphs are zombie-win. However, they do not say whether they believe this theorem is a characterization; i.e. whether it captures all graphs that are zombie-win. Here we answer that question in the negative, providing an example of a zombie-win graph that does not meet the hypothesis of Theorem 3.

Lemma 8. *The graph $C_6 \boxtimes P_3 \pitchfork K_{1,6}$ (Fig. 1d) is zombie-win but does not have a breadth-first-search spanning tree that is a cop-win spanning tree.*

We note that $C_6 \boxtimes P_3 \pitchfork K_{1,6}$ has the symmetric cop-win spanning tree consisting of all of the radial edges (that is, all edges $(h, x0)$, $(x0, x1)$, and $(x1, x2)$). This corresponds to the zombie strategy in the proof of the lemma. However, we note that this spanning tree, while not a breadth-first-search spanning tree, is still a *shortest-path* spanning tree. We therefore investigated whether shortest-path cop-win spanning trees imply a zombie-win graph. The answer is no; the subgraph of $C_6 \boxtimes P_3 \pitchfork K_{1,6}$ shown in Fig. 2a, which we call the *katydid*, provides the counterexample.

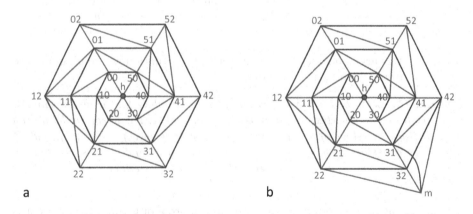

Fig. 2. The katydid and katydid plus graphs

Lemma 9. *The katydid (Fig. 2a) has a shortest-path cop-win spanning tree but is survivor-win.*

---

[2] A graph is *bridged* if it contains no isometric cycles of length greater than 3.

However, we did find that the presence of a shortest-path cop-win spanning tree implies that a tepid zombie can win.

To prove this, we first need a few technical lemmas. Let $level(w)$ denote the level of vertex w (distance to root) in some rooted tree. First, note that for any edge uv of G in a shortest-path tree of G, $|level(u) - level(v)| \leq 1$.

Lemma 10. *Let T be a cop-win spanning tree of graph G that is also a shortest path tree from some vertex r, and uv be some edge of G. If u' and v' are distinct ancestors of u and v respectively, with $level(u') = level(v')$, then the edge $u'v'$ exists.*

Proof. Let u^* be the parent of u and v^* be the parent of v in T. Since $u' \neq v'$, then $u^* \neq v^*$. Assume without loss of generality that $level(u) \leq level(v)$ in T. If $level(u) < level(v)$, then $level(u) = level(v) - 1$ by the previous note. So we consider two cases: (1) $level(u) = level(v)$, or (2) $level(u) = level(v) - 1$.

1. Without loss of generality, assume that u gets dismantled before v. Since u is a corner of u^* (when we dismantle u), there must be an edge u^*v in G. Hence, u^* and v are in Case 2, where $level(u^*) < level(u)$. So we apply Case 2 with $u := u^*$.

2. Assume $level(u) = level(v) - 1$. Observe that v must be dismantled before u. Otherwise, u would be a corner of u^* (when we dismantle u), and hence, there would be an edge u^*v in G. This would break the shortest-path tree T since $level(u^*) = level(u) - 1 = level(v) - 2$. Consider the dismantling of v. Since v is a corner of v^* (when we dismantle v), then there is an edge uv^* in G. Hence, u and v^* are in Case 1, where $level(v^*) < level(v)$. So we apply Case 1 with $v := v^*$.

Repeating this argument, we eventually get to the situation where u' and v' are in Case 1, from which there is an edge $u'v'$ in G.   □

Lemma 11. *Let T be a cop-win spanning tree of graph G that is also a shortest path tree from some vertex r, and uv be some edge of G. Furthermore, let u' and v' be distinct ancestors of u and v respectively, with $level(u') = level(v')$, and u'' and v'' be ancestors of u and v that are children of u' and v', if such children exist. Then, if one or more of u'' and v'' exist, there is either an edge $u'v''$ or an edge $v'u''$.*

Proof. Repeating the argument presented in the proof of Lemma 10, we eventually get to the situation where u' and v'' are in Case 2, or u'' and v' are in Case 2, from which there is an edge $u'v''$ or $u''v'$ in G.   □

Theorem 4. *Let G be a graph with a cop-win spanning tree T that is also a shortest path tree from some vertex r. Then one tepid zombie can capture a survivor on G.*

Proof. (Sketch) The tepid zombie starts at r, the root of T. On its turn, it will always move to an ancestor of the survivor's position, possibly maintaining the same distance to the survivor. In addition, it will move down the tree if it can. Within n turns, the survivor must repeat its location, and one can show that in this loop there is always one zombie position where it can move down the tree. Capture follows when the zombie reaches the same depth as the survivor.   □

A version of the Katydid graph, which we call the Katydid plus, shows that the previous theorem is not a characterization of tepid zombie-win graphs.

Lemma 12. *The Katydid plus (Fig. 2b) is tepid zombie-win but does not have a shortest-path spanning tree that is a cop-win spanning tree.*

3.1   General Results on Capped Products

In this section, we state some general results on zombie numbers of capped products, but first, we examine two good strategies–one for a zombie and one for the survivor.

Stay-On-Top Strategy. This is a zombie strategy for a graph $G \square T$ or $G \boxtimes T$, where T is a tree. Here, if the survivor is at (a_s, b_s), the zombie tries to move to (a_s, b_z) where b_z is *above* b_s—either at b_s or an ancestor of b_s in T. If it is able to achieve this, then it moves to the b_z, which is closest to b_s, that satisfies the constraints.

In $G \boxtimes T$, once a zombie is able to get to (a_s, b_z) where b_z is above b_s, then the survivor will eventually be caught: A survivor move from (a_s, b_s) to (a_s', b_s') is responded to by a zombie move from (a_s, b_z) to (a_s', b_z') where b_z' is one level lower than b_z in T. Eventually b_z' will equal b_s', catching the survivor.

In $G \square T$, once a zombie is able to get to (a_s, b_z) where b_z is above b_s, then the survivor is constrained to making a finite number of moves that change its second coordinate. A survivor move from (a_s, b_s) to (a_s, b_s') is met with a zombie move from (a_s, b_z) to (a_s, b_z') where b_z' is one level lower than b_z. On the other hand, a survivor move from (a_s, b_s) to (a_s', b_s) is followed by a zombie move from (a_s, b_z) to (a_s', b_z), which does not move the zombie down the tree. So the survivor can move as much as it likes in G and never be caught by this zombie.

Plus-Two Cycle Strategy. This is a survivor strategy for a graph $C_k \boxtimes G$, for $k \geq 4$. Suppose the zombie is at a location (a_i, b_i) and the survivor can get to a location (a_i', b_i') where $|a_i' - a_i| \pmod k \geq 2$. If the survivor is not adjacent to the zombie in a-coordinates, then the survivor waits where it is. When the zombie gets to an adjacent a-coordinate, the survivor moves away from it in a-coordinate, attaining a distance of two in a-coordinate after its move. If the zombie then increases the distance in a-coordinate (while at the same time decreasing the distance in b-coordinate), the survivor mimics the zombie's move, moving towards the zombie in a-coordinate so as to reattain a distance of two in a.

Theorem 5. *Let G be any n-vertex graph, and H be any graph on more than n vertices with a universal vertex. Then, $z(G \boxtimes P_j \cap H) = 1$.*

Proof. Let h be the universal vertex of H. The zombie chooses h as its starting position. If the survivor chooses any vertex of H (including the vertices of G_0) as its starting position, the zombie captures them in the first turn. So the survivor chooses some vertex (u, v) where $u \in G$ and $v \geq 1$. The zombie, on its first turn, can move to $(u, 0)$ as this vertex is on a shortest path from h to (u, v). The survivor starts their k-th turn at (u_k, v_k) with the zombie at (u_k, z_k) where $z_k < v_k$. They move to (u_{k+1}, v_{k+1}) where $u_k u_{k+1} \in G$ and $|v_k - v_{k+1}| \leq 1$. The zombie, on its next turn, moves along a shortest path to (u_{k+1}, z_{k+1}) where z_{k+1} is either v_{k+1}, in which case they capture the survivor, or is $z_k + 1 < v_{k+1}$. Since the zombie moves one step down in the P_j in each turn, and the survivor cannot get past the zombie, the survivor will be captured after at most j moves.   □

We can prove the following by a similar method to the previous theorem.

Theorem 6. *Let G be any n-vertex graph, T be any tree, and H be any graph on more than n vertices with an extra universal vertex. Then, $z(G \boxtimes T \pitchfork H) = 1$.*

Lemma 13. *The graph $C_j \boxtimes P_k \pitchfork K_{1,j}$ is*

(1) universal zombie win if $k = 1$ or $j = 3$, or
(2) zombie-win from h, and survivor-win from other vertices if $k \geq 2$ and $j \geq 4$.

Proof.

1. Assume $k = 1$. C_j with the cap $K_{1,j}$ glued on is the *wheel* on $j + 1$ vertices; it is universal zombie-win because it has a universal vertex.

   Assume $j = 3$. Let $z = (z_a, t)$ and $x = (s_a, t')$. If $t < t'$, then the zombie plays the stay-on-top strategy in $C_3 \boxtimes P_k$, eventually cornering the survivor. Otherwise, the zombie keeps pursuing the survivor. Observe that the survivor can never be at the same level as the zombie otherwise it gets caught. Therefore, the zombie catches the survivor in at most k rounds.

2. Assume now $k \geq 2$ and $j \geq 4$.

   Suppose we play the game with the zombie starting at vertex h. The survivor picks a location (a_0, b_0) to start. The zombie moves on its first move to $(a_0, 0)$, and thereafter plays the stay-on-top strategy in $C_j \boxtimes P_k$, eventually cornering the survivor.

   Suppose now that we play the game with the zombie starting at vertex (a_0, b_0). The survivor can then choose $((a_0 + 2) \pmod{j}, k - 1)$ as its starting location, and follow the plus-two cycle strategy on $C_j \boxtimes P_k$. This strategy works, and the survivor survives, if the zombie cannot move to h.

   We now show that the zombie cannot move to h. Suppose the zombie is adjacent to h at $(a_n, 0)$ after its n-th move. The zombie's distance to the survivor is now $\max(2, k - 1)$. If the zombie moved to h, its distance to the survivor would become k. However, it has a move to $((a_0 + 1) \pmod{j}, 1)$ which has distance $\max(1, k - 2)$ to the survivor, which is less than k. Therefore h is not on a shortest path to the survivor, and the survivor's strategy works.

                                                                            □

Lemma 14. *The graph $C_j \boxtimes P_k \pitchfork K_j$ is universal zombie-win if $k = 1$ or $k = 2$ or $j = 3$. It is survivor-win against 1 zombie, whenever $k \geq 3$ and $j \geq 4$.*

Proof.

1. The zombie always wins on the first turn if the graph is K_j. If the graph is $C_j \boxtimes P_2 \pitchfork K_j$, then the zombie either wins on the first turn or can move along a shortest to-survivor path to some vertex $(a_1, 0)$. The survivor then moves to (a'_1, b'_1) and then the zombie moves to $(a'_1, 0)$, either capturing the zombie (if $b'_1 = 0$) or capturing it on the next turn (if $b'_1 = 1$).

   Assume $j = 3$. Let $z = (z_a, t)$ and $s = (s_a, t')$. If $t < t'$, then the zombie plays the stay-on-top strategy in $C_3 \boxtimes P_k$, eventually cornering the survivor. Otherwise, the zombie keeps pursuing the survivor. Observe that the survivor can never be at the same level as the zombie otherwise it gets caught. Therefore, the zombie catches the survivor in at most $k - 1$ rounds.

2. Assume now $k \geq 3$ and $j \geq 4$.

   Suppose we play the game with the zombie starting at some vertex (a_0, b_0). The survivor will start at $((a_0 + 2) \pmod{j}, k - 1)$, and play the plus-two cycle strategy on $C_j \boxtimes P_k$. This strategy works if the zombie can never move from $(a_i, 0)$ to some $(a_{i+1}, 0)$, which effectively limits the play to $C_j \boxtimes P_k$.

   Suppose therefore that the zombie is at $(a_i, 0)$; the survivor will be at $((a_i + 2) \pmod{j}, k - 1)$. The distance from the zombie to the survivor in $C_j \boxtimes P_k \pitchfork K_j$ is $\max(2, k - 1)$. Since $k \geq 3$ this max is equal to $k - 1$. The distance from $(a_{i+1}, 0)$ to the survivor in $C_j \boxtimes P_k \pitchfork K_j$ is $\max(x, k - 1)$, where x is the distance, in C_j, from a_{i+1} to $(a_i + 2) \pmod{j}$. Regardless of what x is, the to-survivor distance is at least $k - 1$, meaning that $(a_{i+1}, 0)$ is no improvement over $(a_i, 0)$ and therefore cannot be moved to by the zombie. $\quad\square$

Lemma 15. *Let $G = C_j \square P_2 \pitchfork K_j$. Then $z(G) \geq \frac{j}{5} = \frac{n}{10}$.*

Proof. The result holds for $j \leq 5$ since all graphs have zombie number at least 1. Hence, assume that $j \geq 6$. Suppose we have fewer than $\frac{j}{5}$ zombies, and they have been placed in the initial round of the game. Consider the shadow of these zombies on C_j. There are some five consecutive vertices of C_j without a shadow; without loss of generality we may assume these are the vertices $1, 2, 3, 4$, and 5 where vertex 0 has a shadow. We initially place the survivor at $(2, 1)$. If there is a zombie or zombies at $(0, 1)$ and/or $(j - 1, 1)$, these zombies have distance at most 3 to the survivor and may move to $(1, 1)$ and/or $(0, 1)$. Other zombies at $(x, 0)$ have a shortest path to the survivor of length 2 going through $(2, 0)$, so they move to $(2, 0)$. Other zombies at $(x, 1)$ have a shortest path to the survivor of length 3, going through $(x, 0)$ and $(2, 0)$; each of these zombies moves to its corresponding $(x, 0)$.

On the survivor's first turn, there are potentially zombies at $(0, 1), (1, 1)$, $(2, 0)$, and various $(x, 0)$'s. The survivor moves to $(3, 1)$. The zombies near the survivor follow but stay one behind the survivor, and the zombies at $(x, 0)$ move to $(3,0)$, also one behind the survivor. The survivor can continue through the whole cycle C_j (repeatedly) by always increasing its first coordinate.

Thus, fewer than $\frac{j}{5}$ zombies fail to catch a survivor.   □

For fixed rooted trees other than P_2, we still require a linear number of zombies.

Lemma 16. *Let T be a tree having depth $d \geq 2$. If $G = C_j \Box T \text{ ⋒ } K_j$ or $C_j \Box T \text{ ⋒ } K_{j+1}$, then $z(G) \geq \frac{j}{d+4}$. If $G = C_j \Box T \text{ ⋒ } K_{1,j}$, then $z(G) \geq \frac{j}{d+5}$.*

Proof. (proof for $G = C_j \Box T \text{ ⋒ } K_j$; other cases are similar.) The result holds for $j \leq d + 4$ since all graphs have zombie number at least 1. Hence, assume that $j \geq d + 5$. Again we examine shadows of zombies and place a largest gap in the shadows at locations $1 \ldots d + 4$ with a zombie shadow at location 0. We start with the survivor at $(2, a)$, where a is a vertex of T adjacent to the root.

As in the previous proof, the survivor simply walks forward on C_j, never changing its coordinate in T. A few zombies may tag along behind the survivor, but most will approach it by going through the root level of the tree, which is a clique.   □

Lemma 17. *Let $G = C_j \Box P_2 \text{ ⋒ } K_j$. Then $u_L(G) \leq 2$.*

Proof. (Sketch) We proceed in two phases. In the first, we get one of the zombies to have second coordinate 0. In the second, this zombie will be stationary, and prohibits the survivor from any vertex having second coordinate 0, and also one vertex of second coordinate 1. Effectively, this traps the survivor on a path of size $j - 1$; the second zombie pursues them along this path, eventually capturing.   □

Lemma 18. *Let T be a tree rooted at r, H and J be connected graphs such that $J \subseteq H$, and $G = J \Box T \text{ ⋒ } H$. Let $z = (z_j, z_t)$ and $s = (s_j, s_t)$ be two vertices of G with z_t an ancestor of s_t in T. Then, there is a shortest path from z to s with one of the following two forms: either a sequence of edges going down in T to (z_j, s_t) followed by a sequence of edges in J or a sequence of edges going up in T to (z_j, r) followed by a sequence of edges in H to (s_j, r), followed by a sequence of edges going down in T to (s_j, s_t).*

Proof. If there is a shortest path that does not use edges from H, then by Observation 1, it has the first form. If there is a shortest path that does use edge(s) from H, then there is one that goes directly to the root. Indeed, if such a shortest path uses any edge in J, then we can as well use it at the root level.   □

Lemma 19. *Suppose we have the same situation as Lemma 18, with a zombie at $z = (z_j, z_t)$ and survivor at s, and a shortest path from z to $s = (s_j, s_t)$ which has the second form. Then, regardless of the survivor moves, the zombie may make a sequence of moves going up in T until it reaches (z_j, r).*

Proof. Let d be the distance along a shortest path from z to s of form 2. After k turns, the zombie is at z', k levels above z_t, and the survivor is at s', at most k steps away from (s_j, s_t). Thus there is a path of form 2 having length $d' \leq d$ from the zombie to the survivor; following the remainder of the original path of form 2 to s and then the survivor's moves from s to s' gives a path of exactly $(d - k) + k = d$ so the shortest one can be no larger.

Suppose that z' is not at the root in T and there is a shortest path of length $d'' < d'$ from the zombie z' to the survivor s'; this path must not reach the root in T as otherwise d' would be smaller. This path, followed by the reverse of the survivor's path from s to s', gives a path of length at most $d'' + k$ from z' to s. Since this path does not use any root vertices, there must be an equal-length path of form 1. The first k edges of this path go directly down from z' to z; removing these edges from the path gives us a path from z to s of length at most $(d'' + k) - k = d''$. This is a contradiction, as $d'' < d$ and the (shortest) path from z to s was of length d. So our supposition of a shorter path is false and there is an upward shortest path from z' which the zombie can follow.   □

Lemma 20. *Let $G = C_j \square T \pitchfork K_j$, where T is a tree. Then $u_L(G) \leq 3$.*

Proof. (Sketch) Let z_1, z_2, and z_3 be three lazy zombies starting anywhere in G. We proceed in three phases: the objective of the first and second phase is to get a zombie in place to play the stay-on-top strategy. The objective of the third phase is to repeatedly force the survivor to move in the second coordinate; since this happens repeatedly, the survivor is eventually captured by the stay-on-top zombie.   □

By computer, we have verified that $C_{13} \square P_3 \pitchfork K_{13}$ is a graph that requires at least three lazy zombies, so the lemma is tight.

4 Decision Algorithm

We present an algorithm to determine whether a graph $G = (V, E)$ is survivor-win, zombie-win or universal zombie-win. G is zombie-win provided that there exists a vertex $v \in V$ such that if the zombie's starting position is v, then it wins regardless of the survivor's starting position. G is universal zombie-win if G is zombie-win from any vertex. If G is neither zombie-win nor universal zombie-win, it is survivor-win. We then generalize this approach to determine whether a graph is k-zombie win or universal k-zombie win for $k \geq 1$. Our approach easily generalizes to lazy and tepid zombies. Our approach is similar to the approach taken by Berarducci and Intrigila [2] to determine whether a graph is k-cop win.

4.1 Algorithm

The setting for the decision algorithm is the following. Given the graph $G = (V, E)$ as input, first construct an auxiliary graph G' from G in the following way. The graph G' is a directed bipartite graph. The vertex set $V(G') = V \times$

$V \times \{t_z, t_s\}$, where t_z and t_s are flags to represent whether it is the zombie's turn to move or the survivor's turn to move. The edge set $E(G')$ is defined as follows. Let a, b, c be vertices of G. There is a directed edge from vertex (a, b, t_z) to (c, b, t_s) in G' if ac is an edge on a shortest path from a to b in G. There is a directed edge from (a, b, t_s) to (a, c, t_z) in G' if $b = c$ or if bc is an edge in G. The edges model the allowable moves by the zombie and the survivor. An edge from (a, b, t_z) to (c, b, t_s) signifies that the zombie is currently on vertex a, the survivor is on vertex b, the t_z means it is the zombie's turn to move and the zombie can move to c since ac is on a shortest path from a to b. As such, the auxiliary graph encodes all the possible moves by the zombie and the survivor. The algorithm then *marks* the nodes that represent states from which the zombie wins. As such, if all the nodes in the auxiliary graph are marked, then G is universal zombie win. If at the end of the algorithm, $\exists a \in V$, such that $\forall b \in V$, (a, b, t_z) is marked, then G is zombie win provided that the zombie's initial position is vertex a.

Theorem 7. *ZombieDecision(G) correctly determines in $n^{O(1)}$ time whether G is universal 1-zombie win, G is 1-zombie win provided the initial zombie position is on a vertex $a \in G$, or G is survivor win.*

This theorem follows by induction. We can generalize the above algorithm to work for lazy or tepid zombies, and also for k zombies rather than 1 in time $n^{O(k)}$.

5 Conclusion

In this paper, we proved some upper and lower bounds on the zombie number of the Cartesian, strong and capped product of various graphs in terms of the zombie number of the individual graphs that make up the product. Using capped products, we were able to disprove some conjectures about zombie numbers. We also studied these bounds for lazy and tepid zombies. Finally, we design an algorithm $n^{O(k)}$ that can decide given a graph G, whether or not it is k-zombie-win for all the above variants of zombies. Most of the upper and lower bounds on the products of graphs are not tight. We leave as an open problem to find tight upper and lower bounds.

References

1. Bartier, V., Bénéteau, L., Bonamy, M., La, H., Narboni, J.: A note on deterministic zombies. Discret. Appl. Math. **301**, 65–68 (2021)
2. Berarducci, A., Intrigila, B.: On the cop number of a graph. Adv. Appl. Math. **14**, 389–403 (1993)
3. Bonato, A., Nowakowski, R.: The Game of Cops and Robbers on Graphs. AMS (2011)
4. Diestel, R.: Graph Theory, 4th Edition, volume 173 of Graduate texts in mathematics. Springer (2012)

5. Fitzpatrick, S., Howell, J., Messinger, M.-E., Pike, D.: A deterministic version of the game of zombies and survivors on graphs. Discret. Appl. Math. **213**, 1–12 (2016)
6. Keramatipour, A., Bahrak, B.: Zombie number of the cartesian product of graphs. Disc. Appl. Math **289**, 545–549 (2021)
7. Nowakowski, R., Winkler, P.: Vertex-to-vertex pursuit in a graph. Discret. Math. **43**(2–3), 235–239 (1983)
8. Quilliot, A.: Jeux et pointes fixes sur les graphes. PhD thesis, Université de Paris VI, Paris, France (1978)

Patterns in Ordered (random) Matchings

Andrzej Dudek[1][(✉)], Jarosław Grytczuk[2], and Andrzej Ruciński[3]

[1] Department of Mathematics, Western Michigan University, Kalamazoo, MI, USA
`andrzej.dudek@wmich.edu`
[2] Faculty of Mathematics and Information Science, Warsaw University of
Technology, Warsaw, Poland
`j.grytczuk@mini.pw.edu.pl`
[3] Department of Discrete Mathematics, Adam Mickiewicz University, Poznań, Poland
`rucinski@amu.edu.pl`

Abstract. An *ordered matching* is an ordered graph which consists of
vertex-disjoint edges (and have no isolated vertices). In this paper we
focus on unavoidable patterns in such matchings. First, we investigate
the size of canonical substructures in ordered matchings and generalize
the Erdős-Szekeres theorem about monotone sequences. We also estimate
the size of canonical substructures in a *random* ordered matching. Then
we study *twins*, that is, pairs of order-isomorphic, disjoint sub-matchings.
Among other results, we show that every ordered matching of size n
contains twins of length $\Omega(n^{3/5})$, but the length of the longest twins in
almost every ordered matching is $O(n^{2/3})$.

Keywords: Ordered matchings · Unavoidable patterns · Twins

1 Introduction

A graph G is said to be *ordered* if its vertex set is linearly ordered. Let G and H be
two ordered graphs with $V(G) = \{v_1 < \cdots < v_m\}$ and $V(H) = \{w_1 < \cdots < w_m\}$
for some integer $m \geq 1$. We say that G and H are *order-isomorphic* if for all
$1 \leq i < j \leq m$, $v_i v_j \in G$ if and only if $w_i w_j \in H$. Note that every pair of
order-isomorphic graphs is isomorphic, but not vice-versa. Also, if G and H are
distinct graphs on the same linearly ordered vertex set V, then G and H are
never order-isomorphic, and so all $2^{\binom{|V|}{2}}$ labeled graphs on V are pairwise non-
order-isomorphic. It shows that the notion of order-isomorphism makes sense
only for pairs of graphs on distinct vertex sets.

One context in which order-isomorphism makes quite a difference is that of
subgraph containment. If G is an ordered graph, then any subgraph G' of G
can be also treated as an ordered graph with the ordering of $V(G')$ inherited

---

The first author was supported in part by Simons Foundation Grant #522400.
The second author was supported in part by Narodowe Centrum Nauki, grant
2020/37/B/ST1/03298. The third author was supported in part by Narodowe Centrum Nauki, grant 2018/29/B/ST1/00426.

A. Castañeda and F. Rodríguez-Henríquez (Eds.): LATIN 2022, LNCS 13568, pp. 544–556, 2022.
https://doi.org/10.1007/978-3-031-20624-5_33

from the ordering of $V(G)$. Given two ordered graphs, (a "large" one) G and (a "small" one) H, we say that a subgraph $G' \subset G$ is *an ordered copy of H in G* if G' and H are order-isomorphic.

All kinds of questions concerning subgraphs in ordinary graphs can be posed for ordered graphs as well (see, e.g., [11]). For example, in [3], the authors studied Turán and Ramsey type problems for ordered graphs. In particular, they showed that there exists an ordered matching on n vertices for which the (ordered) Ramsey number is super-polynomial in n, a sharp contrast with the linearity of the Ramsey number for ordinary (i.e. unordered) matchings. This shows that it makes sense to study even such seemingly simple structures as ordered matchings. In fact, Jelínek [7] counted the number of matchings avoiding (i.e. not containing) a given small ordered matching.

In this paper we focus exclusively on *ordered matchings*, that is, ordered graphs which consist of vertex-disjoint edges (and have no isolated vertices). For example, in Fig. 1, we depict two ordered matchings, $M = \{\{1,3\}, \{2,4\}, \{5,6\}\}$ and $N = \{\{1,5\}, \{2,3\}, \{4,6\}\}$ on vertex set $[6] = \{1, 2, \ldots, 6\}$ with the natural linear ordering.

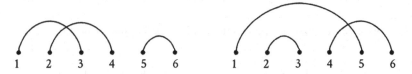

Fig. 1. Exemplary matchings, M and N, of size 3.

A convenient representation of ordered matchings can be obtained in terms of *double occurrence words* over an n-letter alphabet, in which every letter occurs exactly twice as the label of the ends of the corresponding edge in the matching. For instance, our two exemplary matchings can be written as $M = ABABCC$ and $N = ABBCAC$ (see Fig. 2).

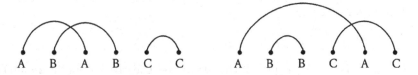

Fig. 2. Exemplary matchings M and N.

Unlike in [7], we study what sub-structures are *unavoidable* in ordered matchings. A frequent theme in both fields, the theory of ordered graphs as well as enumerative combinatorics, are unavoidable sub-structures, that is, *patterns* that appear in every member of a prescribed family of structures. A good example providing everlasting inspiration is the famous theorem of Erdős and Szekeres [5] on monotone subsequences. It states that any sequence $x_1, x_2, \ldots, x_n$ of distinct real numbers contains an increasing or decreasing subsequence of length at least $\sqrt{n}$.

And, indeed, our first goal is to prove its analog for ordered matchings. The reason why the original Erdős-Szekeres Theorem lists only two types of subsequences is, obviously, that for any two elements x_i and x_j with $i < j$ there are just two possible relations: $x_i < x_j$ or $x_i > x_j$. For matchings, however, for every two edges $\{x, y\}$ and $\{u, w\}$ with $x < y$, $u < w$, and $x < u$, there are three possibilities: $y < u$, $w < y$, or $u < y < w$ (see Fig. 3). In other words, every two edges form either *an alignment*, *a nesting*, or *a crossing* (the first term introduced by Kasraoui and Zeng in [8], the last two terms coined by Stanley [10]). These three possibilities give rise, respectively, to three "unavoidable" ordered *canonical* sub-matchings (*lines*, *stacks*, and *waves*) which play an analogous role to the monotone subsequences in the classical Erdős-Szekeres Theorem.

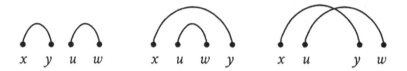

Fig. 3. An alignment, a nesting, and a crossing of a pair of edges.

Informally, lines, stacks, and waves are defined by demanding that every pair of edges in a sub-matching forms, respectively, an alignment, a nesting, or a crossing (see Fig. 5). Here we generalize the Erdős-Szekeres Theorem as follows.

Theorem 1. *Let ℓ, s, w be arbitrary positive integers and let $n = \ell s w + 1$. Then, every ordered matching M on $2n$ vertices contains a line of size $\ell + 1$, or a stack of size $s + 1$, or a wave of size $w + 1$.*

It is not hard to see that the above result is optimal. For example, consider the case $\ell = 5$, $s = 3$, $w = 4$. Take 3 copies of the wave of size $w = 4$: $ABCDABCD$, $PQRSPQRS$, $XYZTXYZT$. Arrange them into a stack-like structure (see Fig. 4):

$$ABCDPQRSXYZTXYZTPQRSABCD.$$

Now, concatenate $\ell = 5$ copies of this structure. Clearly, we obtain a matching of size $\ell s w = 5 \cdot 3 \cdot 4$ with no line of size 6, no stack of size 4, and no wave of size 5.

Also observe that the symmetric case of Theorem 1 implies that M always contains a canonical structure of size at least $n^{1/3}$.

Finally, notice that forbidding an alignment yields a so called *permutational matching* (for definition see the paragraph after Theorem 4). Permutational matchings are in a one-to-one correspondence with permutations of order n. Moreover, under this bijection waves and stacks in a permutational matching M become, respectively, increasing and decreasing subsequences of the permutation

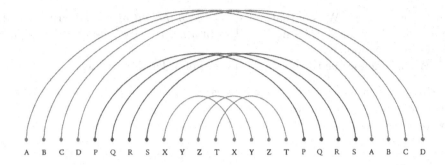

A B C D P Q R S X Y Z T X Y Z T P Q R S A B C D

Fig. 4. A stack of waves.

which is the image of M. Thus, we recover the original Erdős-Szekeres Theorem as a special case of Theorem 1.

We also examine the question of unavoidable sub-matchings for *random* matchings. A random (ordered) matching $\mathbb{RM}_n$ is selected uniformly at random from all $(2n)!/(n!2^n)$ matchings on vertex set $[2n]$. It follows from a result of Stanley (Theorem 17 in [10]) that $a.a.s.$[1] the size of the largest stack and wave in $\mathbb{RM}_n$ is $(1 + o(1))\sqrt{2n}$. In Sect. 2 we complement his result and prove that the maximum size of lines is also about $\sqrt{n}$.

Theorem 2.

(i) A.a.s. the random matching $\mathbb{RM}_n$ contains no lines of size at least $e\sqrt{n}$.
(ii) A.a.s. the random matching $\mathbb{RM}_n$ contains lines of size at least $\sqrt{n}/8$.

Our second goal is to estimate the size of the largest (ordered) twins in ordered matchings. The problem of twins has been widely studied for other combinatorial structures, including words, permutations, and graphs (see, e.g., [1,9]). We say that two edge-disjoint (ordered) subgraphs G_1 and G_2 of an (ordered) graph G form *(ordered) twins in G* if they are (order-)isomorphic. The size of the (ordered) twins is defined as $|E(G_1)| = |E(G_2)|$. For ordinary matchings, the notion of twins becomes trivial, as every matching of size n contains twins of size $\lfloor n/2 \rfloor$ – just split the matching into two as equal as possible parts. But for ordered matchings the problem becomes interesting. The above mentioned generalization of Erdős-Szekeres Theorem immediately (again by splitting into two equal parts) yields ordered twins of length $\lfloor n^{1/3}/2 \rfloor$. In Sect. 3 we provide much better estimates on the size of largest twins in ordered matchings which, not so surprisingly, are of the same order of magnitude as those for twins in permutations (see [2] and [4]).

2   Unavoidable Sub-matchings

Let us start with formal definitions. Let M be an ordered matching on the vertex set $[2n]$, with edges denoted as $e_i = \{a_i, b_i\}$ so that $a_i < b_i$, for all $i = 1, 2, \ldots, n$,

---

[1] *Asymptotically almost surely.*

and $a_1 < \cdots < a_n$. We say that an edge e_i is *to the left of* e_j and write $e_i < e_j$ if $a_i < a_j$. That is, in ordering the edges of a matching we ignore the positions of the right endpoints.

We now define the three canonical types of ordered matchings:

- *Line*: $a_1 < b_1 < a_2 < b_2 < \cdots < a_n < b_n$,
- *Stack*: $a_1 < a_2 < \cdots < a_n < b_n < b_{n-1} < \cdots < b_1$,
- *Wave*: $a_1 < a_2 < \cdots < a_n < b_1 < b_2 < \cdots < b_n$.

Assigning letter A_i to edge $\{a_i, b_i\}$, their corresponding double occurrence words look as follows:

- Line: $A_1 A_1 A_2 A_2 \cdots A_n A_n$,
- Stack: $A_1 A_2 \cdots A_n A_n A_{n-1} \cdots A_1$,
- Wave: $A_1 A_2 \cdots A_n A_1 A_2 \cdots A_n$.

Each of these three types of ordered matchings can be equivalently characterized as follows. Let us consider all possible ordered matchings with just two edges. In the double occurrence word notation these are $AABB$ (an alignment), $ABBA$ (a nesting), and $ABAB$ (a crossing). Now a line, a stack, and a wave is an ordered matching in which *every* pair of edges forms, respectively, an alignment, a nesting, and a crossing (see Fig. 5).

Fig. 5. A line, a stack, and a wave of size three.

Consider a sub-matching M' of M and an edge $e \in M \setminus M'$, whose left endpoint is to the left of the left-most edge f of M'. Note that if M' is a line and e and f form an alignment, then $M' \cup \{e\}$ is a line too. Similarly, if M' is a stack and $\{e, f\}$ form a nesting, then $M' \cup \{e\}$ is a stack too. However, an analogous statement fails to be true for waves, as e, though crossing f, may not necessarily cross all other edges of the wave M'. Due to this observation, in the proof of our first result we will need another type of ordered matchings combining lines and waves. We call a matching $M = \{\{a_i, b_i\} : i = 1, \ldots, n\}$ with $a_i < b_i$, for all $i = 1, 2, \ldots, n$, and $a_1 < \cdots < a_n$, a *landscape* if $b_1 < b_2 < \cdots < b_n$, that is, the right-ends of the edges of M are also linearly ordered (a first-come-first-serve pattern). Notice that there are no nestings in a landscape. In the double occurrence word notation, a landscape is just a word obtained by a *shuffle* of the two copies of the word $A_1 A_2 \cdots A_n$. Examples of landscapes for $n = 4$ are, among others, $ABCDABCD$, $AABCBCDD$, $ABCABDCD$ (see Fig. 6). Now we are ready to prove Theorem 1.

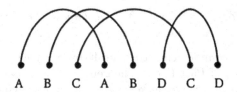

Fig. 6. A landscape of size four.

Proof of Theorem 1. Let M be any ordered matching with edges $\{a_i, b_i\}$, $i = 1, 2, \ldots, n$. Let s_i denote the size of a largest stack whose left-most edge is $\{a_i, b_i\}$. Similarly, let L_i be the largest size of a landscape whose left-most edge is $\{a_i, b_i\}$. Consider the sequence of pairs (s_i, L_i), $i = 1, 2, \ldots, n$. We argue that no two pairs of this sequence may be equal. Indeed, let $i < j$ and consider the two edges $\{a_i, b_i\}$ and $\{a_j, b_j\}$. These two edges may form a nesting, an alignment, or a crossing. In the first case we get $s_i > s_j$, since the edge $\{a_i, b_i\}$ enlarges the largest stack starting at $\{a_j, b_j\}$. In the two other cases, we have $L_i > L_j$ by the same argument. Since the number of pairs (s_i, L_i) is $n > s \cdot \ell w$, it follows that either $s_i > s$ for some i, or $L_j > \ell w$ for some j. In the first case we are done, as there is a stack of size $s + 1$ in M.

In the second case, assume that L is a landscape in M of size at least $p = \ell w + 1$. Let us order the edges of L as $e_1 < e_2 < \cdots < e_p$, accordingly to the linear order of their left ends. Decompose L into edge-disjoint waves, $W_1, W_2, \ldots, W_k$, in the following way. For the first wave W_1, pick e_1 and all edges whose left ends are between the two ends of e_1, say, $W_1 = \{e_1 < e_2 < \ldots < e_{i_1}\}$, for some $i_1 \geqslant 1$. Clearly, W_1 is a true wave since there are no nesting pairs in L. Notice also that the edges e_1 and e_{i_1+1} are non-crossing since otherwise the latter edge would be included in W_1. Now, we may remove the wave W_1 from L and repeat this step for $L - W_1$ to get the next wave $W_2 = \{e_{i_1+1} < e_{i_1+2} < \ldots < e_{i_2}\}$, for some $i_2 \geqslant i_1 + 1$. And so on, until exhausting all edges of L, while forming the last wave $W_k = \{e_{i_{k-1}+1} < e_{i_{k-1}+2} < \ldots < e_{i_k}\}$, with $i_k \geqslant i_{k-1} + 1$. Clearly, the sequence $e_1 < e_{i_1+1} < \ldots < e_{i_{k-1}+1}$ of the leftmost edges of the waves W_i must form a line. So, if $k \geqslant \ell + 1$, we are done. Otherwise, we have $k \leqslant \ell$, and because $p = \ell w + 1$, some wave W_i must have at least $w + 1$ edges. This completes the proof. □

It is not hard to see that the above result is optimal.

Now we change gears a little bit and investigate the size of unavoidable structures in *random* ordered matchings. Let $\mathbb{RM}_n$ be a random (ordered) matching of size n, that is, a matching picked uniformly at random out of the set of all

$$\alpha_n := \frac{(2n)!}{n! 2^n}$$

matchings on the set $[2n]$.

Stanley determined very precisely the maximum size of two of our three canonical patterns, stacks and waves, contained in a random ordered matching.

Theorem 3 (Theorem 17 in [10]). *The largest stack and the largest wave contained in* $\mathbb{RM}_n$ *are each a.a.s. of size* $(1 + o(1))\sqrt{2n}$.

Our Theorem 2 complements this result by estimating the maximum size of lines. In the proof of Part (ii) of Theorem 2 we will make use of the following lemma that can be easily checked by a standard application of the second moment method, and, therefore, its proof is omitted here. Define the *length* of an edge $\{i, j\}$ in a matching on $[2n]$ as $|j - i|$.

Lemma 1. *Let a sequence* $k = k(n)$ *be such that* $1 \ll k \ll n$. *Then, a.a.s. the number of edges of length at most* k *in* $\mathbb{RM}_n$ *is* $k(1 + o(1))$.

Proof of Theorem 2. Part (i) is an easy application of the first moment method. Let X_k be a random variable counting the number of ordered copies of lines of size k in $\mathbb{RM}_n$. Then,

$$\mathbb{E}X_k = \binom{2n}{2k} \cdot 1 \cdot \frac{\alpha_{n-k}}{\alpha_n} = \frac{2^k}{(2k)!} \frac{n!}{(n-k)!} \le \frac{2^k}{(2k)!} \cdot n^k \le \frac{2^k}{(2k/e)^{2k}} \cdot n^k = \left(\frac{e^2 n}{2k^2}\right)^k.$$

Thus, for $k = e\sqrt{n}$ we have

$$\Pr(X_k > 0) \le \mathbb{E}X_k \le \left(\frac{e^2 n}{2k^2}\right)^k = 2^{-e\sqrt{n}} = o(1).$$

It remains to prove Part (ii). Let $k = \sqrt{n}/2$. Due to Lemma 1, a.a.s. the number of edges of length at most k in $\mathbb{RM}_n$ is at least $\sqrt{n}/4$. We will show that among the edges of length at most k, there are a.a.s. at most $\sqrt{n}/8$ crossings or nestings. After removing one edge from each crossing and nesting we obtain a line of size at least $\sqrt{n}/4 - \sqrt{n}/8 = \sqrt{n}/8$.

For a 4-element subset $S = \{u_1, u_2, v_1, v_2\} \subset [2n]$ with $u_1 < v_1 < u_2 < v_2$, let X_S be an indicator random variable equal to 1 if both $\{u_1, u_2\} \in \mathbb{RM}_n$ and $\{v_1, v_2\} \in \mathbb{RM}_n$, that is, if S spans a crossing in $\mathbb{RM}_n$. Clearly,

$$\Pr(X_S = 1) = \frac{1}{(2n-1)(2n-3)}.$$

Let $X = \sum X_S$, where the summation is taken over all sets S as above and such that $u_2 - u_1 \le k$ and $v_2 - v_1 \le k$. Note that this implies that $v_1 - u_1 \le k-1$. Let $f(n, k)$ denote the number of terms in this sum. We have

$$f(n, k) \le \left(2n(k-1) - \binom{k}{2}\right)\binom{k}{2} \le \left(nk - \frac{1}{2}\binom{k}{2}\right)k^2,$$

as we have at most $2n(k-1) - \binom{k}{2}$ choices for u_1 and v_1 and, once u_1, v_1 have been selected, at most $\binom{k}{2}$ choices of u_2 and v_2. It is easy to see that $f(n, k) = \Omega(nk^3)$. Hence, $\mathbb{E}X = \Omega(k^3/n) \to \infty$, while

$$\mathbb{E}X = \sum_S \mathbb{E}X_S = \frac{f(n, k)}{(2n-1)(2n-3)} \le k^3/4n = \frac{1}{32}\sqrt{n}.$$

To apply Chebyshev's inequality, we need to estimate $\mathbb{E}(X(X-1))$, which can be written as

$$\mathbb{E}(X(X-1)) = \sum_{S,S'} \Pr(\{\{u_1, u_2\}, \{v_1, v_2\}, \{u_1', u_2'\}, \{v_1', v_2'\}\} \subset \mathrm{RM}_n),$$

where the summation is taken over all (ordered) pairs of sets $S = \{u_1, u_2, v_1, v_2\} \subset [2n]$ with $u_1 < v_1 < u_2 < v_2$ and $S' = \{u_1', u_2', v_1', v_2'\} \subset [2n]$ with $u_1' < v_1' < u_2' < v_2'$ such that $u_2 - u_1 \leq k$, $v_2 - v_1 \leq k$, $u_2' - u_1' \leq k$, and $v_2' - v_1' \leq k$. We split the above sum into two sub-sums Σ_1 and Σ_2 according to whether $S \cap S' = \emptyset$ or $|S \cap S'| = 2$ (for all other options the above probability is zero). In the former case,

$$\Sigma_1 \leq \frac{f(n,k)^2}{(2n-1)(2n-3)(2n-5)(2n-7)} = (\mathbb{E}X)^2(1 + O(1/n)).$$

In the latter case, the number of such pairs (S, S') is at most $f(n,k) \cdot 4k^2$, as given S, there are four ways to select the common pair and at most k^2 ways to select the remaining two vertices of S'. Thus,

$$\Sigma_2 \leq \frac{f(n,k) \cdot 4k^2}{(2n-1)(2n-3)(2n-5)} = O(k^5/n^2) = O(\sqrt{n})$$

and, altogether,

$$\mathbb{E}(X(X-1)) \leq (\mathbb{E}X)^2(1 + O(1/n)) + O(\sqrt{n}) = (\mathbb{E}X)^2 + O(\sqrt{n}).$$

By Chebyshev's inequality,

$$\Pr(|X - \mathbb{E}X| \geq \mathbb{E}X) \leq \frac{\mathbb{E}(X(X-1)) + \mathbb{E}X - (\mathbb{E}X)^2}{(\mathbb{E}X)^2}$$

$$\leq 1 + O(1/\sqrt{n}) + \frac{1}{\mathbb{E}X} - 1 = O\left(\frac{1}{\sqrt{n}}\right) \to 0.$$

Thus, a.a.s. $X \leq 2\mathbb{E}X \leq \sqrt{n}/16$.

We deal with nestings in a similar way. For a 4-element subset $S = \{u_1, u_2, v_1, v_2\} \subset [2n]$ with $u_1 < v_1 < v_2 < u_2$, let Y_S be an indicator random variable equal to 1 if both $\{u_1, u_2\} \in \mathrm{RM}_n$ and $\{v_1, v_2\} \in \mathrm{RM}_n$, that is, if S spans a nesting in RM_n. Further, let $Y = \sum Y_S$, where the summation is taken over all sets S as above and such that $u_2 - u_1 \leq k$ and so $v_2 - v_1 \leq k$ (in fact, $k - 2$). It is crucial to observe that, again, $\mathbb{E}Y \leq k^3/n = \sqrt{n}/32$. Indeed, this time there are at most $2nk - \binom{k+1}{2}$ choices for u_1 and u_2 and, once u_1, u_2 have been selected, at most $\binom{k}{2}$ choices of v_1, and v_2, while the probability of both pairs appearing in RM_n remains the same as before. The remainder of the proof goes mutatis mutandis.

We conclude that a.a.s. the number of crossings and nestings of length at most k in RM_n is at most $\sqrt{n}/8$ as was required. $\qquad\square$

3 Twins

Recall that by *twins* in an ordered matching M we mean any pair of disjoint, order-isomorphic sub-matchings M_1 and M_2 and that their *size* is defined as the number of edges in just one of them. For instance, the matching $M = AABCDDEBCFGHIHEGFI$ contains twins $M_1 = BCDDBC$ and $M_2 = EFHHEF$ of size three (see Fig. 7).

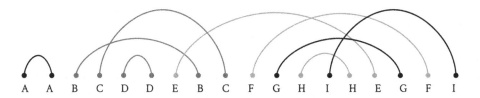

$$A \quad A \quad B \quad C \quad D \quad D \quad E \quad B \quad C \quad F \quad G \quad H \quad I \quad H \quad E \quad G \quad F \quad I$$

Fig. 7. Twins of size 3 with pattern $XYZZXY$.

Let $t(M)$ denote the maximum size of twins in a matching M and $t^{\mathrm{match}}(n)$ – the minimum of $t(M)$ over all matchings on $[2n]$.

We first point to a direct correspondence between twins in permutations and ordered twins in a certain kind of matchings. By a *permutation* we mean any finite sequence of pairwise distinct positive integers. We say that two permutations $(x_1, \ldots, x_k)$ and $(y_1, \ldots, y_k)$ are *similar* if their entries preserve the same relative order, that is, $x_i < x_j$ if and only if $y_i < y_j$ for all $1 \leqslant i < j \leqslant k$. Any two similar and disjoint sub-permutations of a permutation π are called *twins*. For example, in permutation

$$(6, 1, 4, 7, 3, 9, 8, 2, 5),$$

the red and blue subsequences form a pair of twins of length 3, both similar to permutation $(1, 3, 2)$.

For a permutation π, let $t(\pi)$ denote the maximum integer k such that π contains twins of length k each. Further, let $t^{\mathrm{perm}}(n)$ be the minimum of $t(\pi)$ over all permutations of $[n]$, called also *n-permutations*. By the first moment method Gawron [6] proved that $t^{\mathrm{perm}}(n) \leqslant cn^{2/3}$ for some constant $c > 0$.

As for a lower bound, notice that by the Erdős-Szekeres Theorem, we have $t^{\mathrm{perm}}(n) \geqslant \lfloor \frac{1}{2} n^{1/2} \rfloor$. This bound was substantially improved by Bukh and Rudenko [2]

Theorem 4 (Bukh and Rudenko [2]). *For all n, $t^{\mathrm{perm}}(n) \geqslant \frac{1}{8} n^{3/5}$.*

We call an ordered matching M on the set $[2n]$ *permutational* if the left end of each edge of M lies in the set $[n]$. In the double occurrence word notation such a matching can be written as $M = A_1 A_2 \ldots A_n A_{i_1} A_{i_2} \ldots A_{i_n}$, where $\pi_M = (i_1, i_2, \ldots, i_n)$ is a permutation of $[n]$ (see Fig. 8).

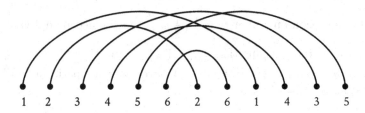

Fig. 8. The permutational matching that corresponds to the permutation $(2, 6, 1, 4, 3, 5)$.

Clearly there are only $n!$ permutational matchings, nevertheless the connection to permutations turned out to be quite fruitful. Indeed, it is not hard to see that ordered twins in a permutational matching M are in one-to-one correspondence with twins in π_M. Thus, we have $t(M) = t(\pi_M)$ and, consequently, $t^{\mathrm{match}}(n) \leq t^{\mathrm{perm}}(n)$. In particular, by the above mentioned result of Gawron, it follows that $t^{\mathrm{match}}(n) = O(n^{2/3})$.

More subtle is the opposite relation.

Proposition 1. *For all $1 \leq m \leq n$, where $n - m$ is even,*

$$t^{\mathrm{match}}(n) \geq \min\left\{ t^{\mathrm{perm}}(m),\ 2t^{\mathrm{match}}\left(\frac{n - m + 2}{2}\right)\right\}.$$

Proof. Let M be a matching on $[2n]$. Split the set of vertices of M into two halves, $A = [n]$ and $B = [n + 1, 2n]$ and let $M(A, B)$ denote the set of edges of M with one end in A and the other end in B. Note that $M' := M(A, B)$ is a permutational matching. We distinguish two cases. If $|M'| \geqslant m$, then

$$t(M) \geq t(M') = t(\pi_{M'}) \geq t^{\mathrm{perm}}(|M'|) \geq t^{\mathrm{perm}}(m).$$

If, on the other hand, $e(A, B) \leq m - 2$, then we have sub-matchings M_A and M_B of M of size at least $(n - m + 2)/2$ in sets, respectively, A and B. Thus, in this case, by concatenation,

$$t(M) \geq t(M_A) + t(M_B) \geq 2t^{\mathrm{match}}\left(\frac{n - m + 2}{2}\right).$$

$\square$

Proposition 1 allows, under some mild conditions, to „carry over" any lower bound on $t^{\mathrm{perm}}(n)$ to one on $t^{\mathrm{match}}(n)$.

Lemma 2. *If for some $0 < \alpha, \beta < 1$, we have $t^{\mathrm{perm}}(n) \geq \beta n^\alpha$ for all $n \geq 1$, then $t^{\mathrm{match}}(n) \geq \beta(\gamma n)^\alpha$ for any $0 < \gamma \leq \min\{1 - 2^{1-1/\alpha}, 1/4\}$ and all $n \geq 1$.*

Proof. Assume that for some $0 < \alpha, \beta < 1$, we have $t_r^{\mathrm{perm}}(n) \geq \beta n^\alpha$ for all $n \geq 1$ and let $0 < \gamma \leq \min\{1 - 2^{1-1/\alpha}, 1/4\}$ be given. We will prove that $t^{\mathrm{match}}(n) \geq$

$\beta(\gamma n)^{\alpha}$ by induction on n. For $n \leqslant \frac{1}{\gamma}\left(\frac{1}{\beta}\right)^{1/\alpha}$ the claimed bound is at most 1, so it is trivially true. Assume that $n \geq \frac{1}{\gamma}\left(\frac{1}{\beta}\right)^{1/\alpha}$ and that $t^{\mathrm{match}}(n') \geq \beta(\gamma n)^{\alpha}$ for all $n' < n$. Let $n_{\gamma} \in \{\lceil \gamma n \rceil, \lceil \gamma n \rceil + 1\}$ have the same parity as n. Then, by Proposition 1 with $m = n_{\gamma}$,

$$t^{\mathrm{match}}(n) \geq \min \left\{ t^{\mathrm{perm}}(n_{\gamma}), 2t^{\mathrm{match}}\left(\frac{n - n_{\gamma} + 2}{2}\right) \right\}.$$

By the assumption of the lemma, $t^{\mathrm{perm}}(n_{\gamma}) \geq \beta n_{\gamma}^{\alpha} \geq \beta(\gamma n)^{\alpha}$. Since $\gamma \leq 1/4$ and so, $n \geq 4$, we have $(n - n_{\gamma} + 2)/2 \leq n - 1$. Hence, by the induction assumption, also

$$2t^{\mathrm{match}}\left(\frac{n - n_{\gamma} + 2}{2}\right) \geq 2\beta\left(\gamma \frac{n - n_{\gamma} + 2}{2}\right)^{\alpha} \geq 2\beta\left(\gamma n \frac{1 - \gamma}{2}\right)^{\alpha} \geq \beta(\gamma n)^{\alpha}$$

where the last inequality follows by the assumption on γ.   □

In particular, Theorem 4 and Lemma 2 with $\beta = 1/8$, $\alpha = 3/5$, and $\gamma = 1/4$ imply immediately the following result.

Corollary 1. *For every n, $t^{\mathrm{match}}(n) \geq \frac{1}{8}\left(\frac{n}{4}\right)^{3/5}$.*

Moreover, any future improvement of the bound in Theorem 4 would automatically yield a corresponding improvement of the lower bound on $t^{\mathrm{match}}(n)$.

As for an upper bound, we already mentioned that $t^{\mathrm{match}}(n) = O\left(n^{2/3}\right)$. This means that for each n there is a matching M of size n with $t(M) \leq cn^{2/3}$, where $c > 0$ is a fixed constant. In fact, this holds for *almost all M*.

Proposition 2. *A.a.s. $t(\mathbb{RM}_n) = O\left(n^{2/3}\right)$.*

Proof. Consider a random (ordered) matching $\mathbb{RM}_n$. The expected number of twins of size k in $\mathbb{RM}_n$ is

$$\frac{1}{2}\binom{2n}{2k, 2k, 2n - 4k}\frac{\alpha_k \cdot 1 \cdot \alpha_{n-2k}}{\alpha_n} = \frac{2^k n!}{2(2k)! k! (n - 2k)!} < \left(\frac{e^3 n^2}{2k^3}\right)^k,$$

which tends to 0 with $n \to \infty$ if $k \geq cn^{2/3}$, for any $c > e2^{-1/3}$. This implies that a.a.s. there are no twins of size at least $cn^{2/3}$ in $\mathbb{RM}_n$.   □

4   Final Remarks

Proposition 2 asserts that a.a.s. $t(\mathbb{RM}_n) = O\left(n^{2/3}\right)$. In the journal version of this extended abstract we intend to prove the matching lower bound: a.a.s. $t(\mathbb{RM}_n) = \Omega\left(n^{2/3}\right)$. The real challenge, however, would be to prove (or disprove) that the bound holds for *all* matchings of size n.

Conjecture 1. For each n there is a matching M of size n with $t(M) \geq cn^{2/3}$, where $c > 0$ is a fixed constant. Consequently, $t^{\mathrm{match}}(n) = \Theta\left(n^{2/3}\right)$.

The same statement is conjectured for twins in permutations (see [4]). By our results from Sect. 3, we know that both conjectures are actually equivalent.

In a similar way twins may be defined and studied in general ordered graphs.

Problem 1. How large twins must occur in every ordered graph with n edges?

For *unordered* graphs there is a result of Lee, Loh, and Sudakov [9] giving an asymptotically exact answer of order $\Theta(n \log n)^{2/3}$. It would be nice to have an analogue of this result for ordered graphs.

Finally, it seems natural to look for Erdős-Szekeres type results like Theorem 1 for more general structures. One possible direction to pursue is to consider, for some fixed $k \geqslant 3$, ordered k-*uniform* matchings. In full analogy with graph ordered matchings $(k = 2)$, these structures correspond to k-*occurrence words*, in which every letter appears exactly k times. For instance, for $k = 3$ there are exactly $\frac{1}{2}\binom{6}{3} = 10$ ways two triples AAA and BBB can intertwine which, somewhat surprisingly, give rise to 9 canonical structures, analogous to lines, stacks, and waves in the graph case. In fact, they correspond to different pairs of the three graph structures. Using this correspondence, in the journal version we intend to prove that every 3-occurrence word of length $3n$ contains one of these 9 structures of size $\Omega\left(n^{1/9}\right)$. We suspect that similar phenomena hold for each $k \geq 4$ or even for words in which the occurrences of particular letters may vary.

Acknowledgements. We would like to thank all four anonymous referees for a careful reading of the manuscript and suggesting a number of editorial improvements.

References

1. Axenovich, M., Person, Y., Puzynina, S.: A regularity lemma and twins in words. J. Combin. Theor. Ser. A **120**(4), 733–743 (2013)
2. Bukh, B., Rudenko, O.: Order-isomorphic twins in permutations. SIAM J. Discret. Math. **34**(3), 1620–1622 (2020)
3. Conlon, D., Fox, J., Lee, C., Sudakov, B.: Ordered Ramsey numbers. J. Combin. Theor. Ser. B **122**, 353–383 (2017)
4. Dudek, A., Grytczuk, J., Ruciński, A.: Variations on twins in permutations. Electron. J. Combin. 28(3), Paper No. 3.19, 18 (2021)
5. Erdős, P., Szekeres, G.: A combinatorial problem in geometry. Compos. Math. **2**, 463–470 (1935)
6. Gawron, M.: Izomorficzne podstruktury w słowach i permutacjach, Master Thesis (2014)
7. Jelínek, V.: Dyck paths and pattern-avoiding matchings. Eur. J. Combin. **28**(1), 202–213 (2007)
8. Kasraoui, A., Zeng, J.: Distribution of crossings, nestings and alignments of two edges in matchings and partitions. Electron. J. Combin. 13(1), Research Paper 33, 12 (2006)

9. Lee, C., Loh, P.S., Sudakov, B.: Self-similarity of graphs. SIAM J. Discret. Math. **27**(2), 959–972 (2013)

10. Stanley, R.P.: Increasing and decreasing subsequences and their variants. In: International Congress of Mathematicians, Vol. I, pp. 545–579. European Mathematical Society, Zürich (2007)

11. Tardos, G.: Extremal theory of ordered graphs. In: Proceedings of the International Congress of Mathematicians-Rio de Janeiro 2018, Vol. IV, invited lectures, pp. 3235–3243. World Scientific Publishing, Hackensack, NJ (2018)

Tree 3-Spanners on Generalized Prisms
of Graphs

Renzo Gómez[1][(✉)] ⓘ, Flávio K. Miyazawa[1] ⓘ, and Yoshiko Wakabayashi[2] ⓘ

[1] Institute of Computing, University of Campinas, Campinas, Brazil
{rgomez,fkm}@ic.unicamp.br
[2] Institute of Mathematics and Statistics, University of São Paulo, São Paulo, Brazil
yw@ime.usp.br

Abstract. Let $t \geq 1$ be a rational constant. A t-spanner of a graph G is a spanning subgraph of G in which the distance between any pair of vertices is at most t times their distance in G. This concept was introduced by Peleg & Ullman in 1989, in the study of optimal synchronizers for the hypercube. Since then, spanners have been used in multiple applications, especially in communication networks, motion planning and distributed systems. The problem of finding a t-spanner with the minimum number of edges is NP-hard for every $t \geq 2$. Cai & Corneil, in 1995, introduced the TREE t-SPANNER PROBLEM (TREES$_t$), that asks whether a given graph admits a tree t-spanner (a t-spanner that is a tree). They showed that TREES$_t$ can be solved in linear time when $t = 2$, and is NP-complete when $t \geq 4$. The case $t = 3$ has not been settled yet, being a challenging problem. The prism of a graph G is the graph obtained by considering two copies of G, and by linking its corresponding vertices by an edge (also defined as the Cartesian product $G \times K_2$). Couto & Cunha (2021) showed that TREES$_t$ is NP-complete even on this class of graphs, when $t \geq 5$. We investigate TREES$_3$ on prisms of graphs, and characterize those that admit a tree 3-spanner. As a result, we obtain a linear-time algorithm for TREES$_3$ (and the corresponding search problem) on this class of graphs. We also study a partition of the edges of a graph related to the distance condition imposed by a t-spanner, and derive a necessary condition —checkable in polynomial time— for the existence of a tree t-spanner on an arbitrary graph. As a consequence, we show that TREES$_3$ can be solved in polynomial time on the class of generalized prisms of trees.

Keywords: Tree spanner · 3-spanner · Prisms of graphs · Generalized prisms

1 Introduction

Let G be a connected graph. The *distance* between a pair of vertices u and v, in G, is the minimum length of a path between them, and it is denoted by $d_G(u,v)$.

© Springer Nature Switzerland AG 2022
A. Castañeda and F. Rodríguez-Henríquez (Eds.): LATIN 2022, LNCS 13568, pp. 557–573, 2022.
https://doi.org/10.1007/978-3-031-20624-5_34

Let $t \geq 1$ be a rational constant. We say that a spanning subgraph H of G is a *t-spanner* of G if the following condition holds.

$$d_H(u, v) \leq t \cdot d_G(u, v), \text{ for all } u, v \in V(G).$$

Peleg & Ullman [26] introduced this concept in 1989, and showed how to use a t-spanner to construct a synchronizer (an algorithm that can be applied to a synchronous algorithm to produce an asynchronous one). In this context, the quality of the spanner is measured by the constant t, which is known as the *stretch factor*. Since then, spanners have appeared in many practical applications such as motion planning [30], routing tables in communication networks [1], distance oracles [3,28], etc. In these cases, besides looking for a spanner of small stretch factor, it is also desirable that it has few number of edges. Motivated by the fact that trees are the sparsest connected graphs, Cai & Corneil [6] introduced the TREE t-SPANNER PROBLEM (TREES$_t$); where one asks whether a given graph admits a t-spanner that is a tree. They showed a linear-time algorithm for TREES$_t$ when $t \leq 2$, and also showed that the problem becomes NP-complete when $t \geq 4$. The complexity of the case $t = 3$ is still open since then. In the search for an answer to this question, TREES$_3$ has been investigated on several classes of graphs. It is known that it admits a polynomial-time algorithm on planar graphs [16], convex graphs [29], split graphs [29], line graphs [11], etc. In all results mentioned above, we have considered only cases in which the stretch factor t is an integer. Indeed, results for this case carry over to the case t is a rational number. (This is valid only when the input graph is unweighted.) We summarize the complexity status of TREES$_t$ on some classes of graphs in Table 1.

Table 1. Computational complexity of TREES$_t$ on some classes of graphs. When $t \leq 2$, the problem can be solved in linear. The shaded cells indicate the results obtained here.

Graph class	$t = 3$	$t = 4$	$t \geq 5$
chordal	open	NP-complete [14]	NP-complete [14]
strongly chordal	open	P [4]	P [4]
interval	P [23]	P [23]	P [23]
planar	P [16]	NP-complete [13]	NP-complete [13]
bounded-degree	P [17]	P [17]	P [17]
prisms of graphs	P	open	NP-complete [9]
gen. prisms of trees	P	open	open
bipartite	open	open	NP-complete [5]
chordal bipartite	open	open	NP-complete [5]
ATE-free	P [5]	P [5]	P [5]
convex bipartite	P [29]	P [29]	P [29]

An optimization problem related to TREES$_t$ is the MINIMUM MAX-STRETCH SPANNING TREE PROBLEM (MMST). In this problem, we seek a tree t-spanner of a graph such that t is minimized. The NP-completeness of TREES$_t$ implies that MMST is NP-hard. Due to this fact, this problem has been studied from the perspective of exact and approximation algorithms. Álvarez-Miranda & Sinnl [24] proposed a mixed integer linear programming formulation for this problem. Also, Couto et. al. [10] proposed some heuristics for MMST. In terms of approximation, Emek & Peleg [15] proposed an $\mathcal{O}(\log n)$-approximation algorithm for MMST on graphs of order n. On the other hand, Galbiati [18] proved that MMST has no $(2 - \epsilon)$-approximation algorithm, for any $\epsilon > 0$, unless P $=$ NP. More recently, MMST has also been approached from a theoretical point-of-view. In this case, the aim is to search for upper and lower bounds for its optimal value in different classes of graphs. Let $\sigma_T(G)$ denote the optimal value of MMST when the input graph is G. Lin & Lin [22] showed a tight bound for $\sigma_T(G)$ when G is the result of the Cartesian product of paths, complete graphs or cycles. Couto & Cunha [9] studied $\sigma_T(G)$ on two classes of graphs defined as follows. First, the *prism* of a graph G is the graph obtained by considering the union of two copies of G, and by linking its corresponding vertices. In case we consider the union of G and $\overline{G}$ (the complement of G), and link its corresponding vertices, the resulting graph is called the *complementary prism* of G. Couto & Cunha [9] characterized $\sigma_T(G)$ when G is a complementary prism. Moreover, they showed that TREES$_t$ is NP-complete on prisms of graphs, when $t \geq 5$. Our main result is a characterization of prisms of graphs that admit a tree 3-spanner. As a result, we show a linear-time algorithm that solves TREES$_3$ on this class of graphs. Furthermore, we show a necessary condition for the admissibility of a tree t-spanner on arbitrary graphs. As a consequence, we extend our previous result on a particular case of generalized prisms.

This work is organized as follows. In Sect. 2, we present some definitions and the terminology that we use in the text. We also state some important results that will be used in the subsequent sections. In Sect. 3, we study the tree 3-spanner problem on prisms of graphs. We obtain a characterization of such graphs that admit a tree 3-spanner, showing a linear-time algorithm for TREES$_3$ on this class. After that, in Sect. 4, we study a partition of the edges of a graph related to the distance condition imposed by a t-spanner. As a result, we derive a necessary condition for the existence of a tree t-spanner of a graph, which can be tested in polynomial time. As a consequence, we generalize the result obtained in Sect. 3 for a class of graphs that contains the prisms of trees. Owing to space limitation, most of the proofs are sketched and in some cases they are omitted.

2   Preliminaries

In this section, we present some results, definitions and the terminology that will be used in this text. We always consider that the input graph (for the problems under consideration) is connected, even if this is not stated explicitly. The *length* of a path P (resp. cycle C) is its number of edges, and it is denoted by $|P|$ (resp.

$|C|$). In the study of t-spanners, paths and cycles of length at most t play an important role. We call a path (resp. cycle) of length k, a k-*path* (resp. k-*cycle*). The *girth* of a graph G, denoted as $g(G)$, is the minimum length of a cycle in G. If G is a tree, we consider its girth to be infinite. The following result gives an equivalent definition for a t-spanner. It states that, to verify whether a subgraph is a t-spanner of a graph G, we only need to check the distance condition on pairs of adjacent vertices in G.

Proposition 1 (Peleg & Schäffer, 1989). *Let H be a spanning subgraph of a graph $G = (V, E)$. Then, H is a t-spanner of G if and only if for every edge $uv \in E$, $d_H(u, v) \leq t$.*

We define now the classes of graphs that are the objects of our study. Let G be a graph. The *prism* of G, denoted by $\mathcal{P}(G)$, is the graph defined as follows. First, consider two copies of G, say G_1 and G_2, such that each vertex $v_i \in V(G_i)$, $i = 1, 2$, corresponds to the vertex v in $V(G)$. Then,

$$V(\mathcal{P}(G)) = V(G_1) \cup V(G_2),$$
$$E(\mathcal{P}(G)) = E(G_1) \cup E(G_2) \cup \{v_1 v_2 : v \in V(G)\}.$$

That is, $\mathcal{P}(G)$ is obtained from the union of G_1 and G_2, by linking the corresponding vertices. We observe that the prism of G is also known as the Cartesian product $\mathcal{P}(G) = G \times K_2$. We refer to the perfect matching that links G_1 to G_2 as the *linkage* of $\mathcal{P}(G)$. We show an example of this construction in Fig. 1, where the linkage of $\mathcal{P}(G)$ is depicted by double edges. This class of graphs first appeared in the study of a conjecture regarding the hamiltonicity of 3-connected planar graphs [2,25]. It has also been studied on problems regarding domination [19] and vertex-coloring [8]. Throughout this text, whenever we are dealing with a prism $\mathcal{P}(G)$, we will denote by G_1 and G_2 the copies of G in $\mathcal{P}(G)$. Furthermore, v_1 and v_2 will denote the copies of the vertex v in G_1 and G_2, respectively.

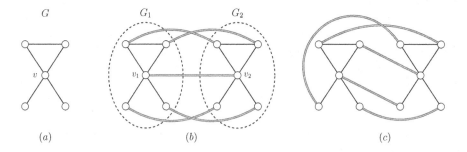

Fig. 1. (a) a graph G; (b) $\mathcal{P}(G)$; and (c) a generalized prism of G.

Note that the linkage of $\mathcal{P}(G)$ can be seen as a function $f : V(G) \to V(G)$, where we interpret an edge $a_1 b_2$, $a_1 \in V(G_1)$ and $b_2 \in V(G_2)$ as $f(a) = b$. So, in the case of $\mathcal{P}(G)$, f is the identity function. On the other hand, if we consider any bijection, say $g : V(G) \to V(G)$, and define the graph G' as

$$V(G') = V(G_1) \cup V(G_2),$$
$$E(G') = E(G_1) \cup E(G_2) \cup \{u_1 v_2 : u, v \in V(G), g(u) = v\},$$

we say that G' is a *generalized prism* of G. Moreover, we call g the *function induced* by the linkage of G'. We show an example of a generalized prism in Fig. 1(c).

We conclude this section presenting some results concerning the connectivity of graphs. Let k be a positive integer. We say that a graph G is *k-connected* if (a) it has at least $k + 1$ vertices; and if (b) the removal of any set of $k - 1$ vertices from G does not disconnect it. In our results regarding tree t-spanner admissibility, we frequently analyse the 2-connected subgraphs of a graph. The following result gives a useful property of 2-connected graphs (see Diestel [12]).

Proposition 2. *If G is a 2-connected graph, then for any pair of distinct vertices or edges of G, there exists a cycle in G containing them.*

Let G be a graph. A *block* of G is a maximal connected subgraph of G that does not contain a cut-vertex. Thus, every block of G is either a 2-connected graph or an edge (that disconnects G). The latter is called a *trivial* block. We denote by $\mathcal{B}(G)$ the set of blocks of G. An important structure to be used in what follows concerns the *block graph* of G. Let $A \subseteq V(G)$ be the set of cut-vertices of G, and let B be a set of vertices that represents the blocks of G. Then, the (A, B)-bipartite graph H such that $ab \in E(H)$ if and only if a belongs to the block represented by b, is the block graph of G. As H is a tree (see Diestel [12]), it is also referred to as the *block tree* of G.

3 Tree 3-spanner on Prisms of Graphs

This section is devoted to characterizing prisms of graphs that admit a tree 3-spanner. In order to understand better the structure of such spanners, we start by looking at the prisms of some well-known classes of graphs. For example, if we consider trees, their prisms always admit a tree 3-spanner. To show this, consider a tree T, and let M_T be the linkage of $\mathcal{P}(T)$. Then, the subgraph T' induced by $T_1 \cup M_T$ is a tree 3-spanner of $\mathcal{P}(T)$. This follows from the fact that, for each edge $u_2 v_2$ in T_2, the path $\langle u_2, u_1, v_1, v_2 \rangle$ links u_2 and v_2 in T'. We show an example of such construction in Fig. 2(a) (the edges in the spanner are indicated by double edges).

On the other hand, if we consider cycles, this is not always the case. Lin & Lin [22] showed that the prism of a cycle C admits a tree 3-spanner if and only if $|C| \leq 3$. When $|C| \geq 4$, a minimum 3-spanner of $\mathcal{P}(C)$ has $|V(\mathcal{P}(C))|$ edges, see Fig. 2(b) (and see also Gómez et. al [20]).

In what follows, we show that some edges in the linkage of a prism $\mathcal{P}(G)$ always belong to a tree 3-spanner of $\mathcal{P}(G)$, if it exists. For instance, let us

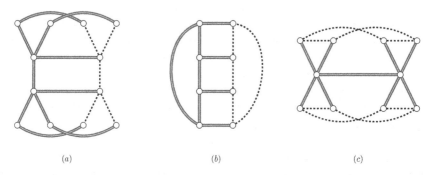

Fig. 2. Minimum 3-spanners of (a) a prism of a tree; (b) a prism of a 4-cycle; and (c) a prism of two triangles with a common vertex.

consider a star $G = K_{1,m}$, $m \geq 2$. Let u be the center of G. Since G is acyclic, for each edge $uv \in E(G)$, the unique 4-cycle that contains the edges u_1v_1 or u_2v_2, in $\mathcal{P}(G)$, also contains u_1u_2. This observation implies that, if S is a 3-spanner of G that does not contain u_1u_2, then S contains a cycle (otherwise it violates the 3-spanner condition). Indeed, this reasoning can be extended to prisms of trees, when we consider the cut-vertices of a tree. Next, we show a generalization of this observation for every graph G (We omit this proof owing to space limitation).

Lemma 1. *Let $\mathcal{P}(G)$ be the prism of a graph G. If $\mathcal{P}(G)$ admits a tree 3-spanner, say T. Then, the edge v_1v_2 belongs to T, for every cut-vertex $v \in V(G)$.*

The previous result tells us that the two copies of a cut-vertex of a graph G are adjacent in any tree 3-spanner of $\mathcal{P}(G)$, if it exists. In what follows, we focus on the blocks of G, specifically on the structure of a tree 3-spanner inside the copies of a block of G.

We say that a vertex u of a graph G is *universal* if u is adjacent to each vertex in $V(G) \setminus \{u\}$. Let G be the graph shown in Fig. 2(c). In this case, $\mathcal{P}(G)$ admits a tree 3-spanner. In particular, each of the blocks of G contains a vertex that is universal inside the block (the cut-vertex). If we restrict our analysis to the subgraph induced by a block, we obtain K_2 or a 2-connected graph, say G'. In this case, the existence of a universal vertex is sufficient for $\mathcal{P}(G')$ to admit a tree 3-spanner. On the other hand, as we stated in the beginning of this section, a prism of a k-cycle, for $k \geq 4$, does not admit a tree 3-spanner. Since these cycles do not contain a universal vertex, that may suggest that the existence of a universal vertex is also necessary. This observation is the crux of our characterization, and we show it in what follows.

Theorem 1. *Let G be a 2-connected graph. Then, $\mathcal{P}(G)$ admits a tree 3-spanner if and only if G has a universal vertex.*

Proof (sketch). First, if G has a universal vertex, say u, then the bistar induced by the set of edges

$$F = \{u_1 w_1 : w \in V(G) \setminus \{u\}\} \cup \{u_2 w_2 : w \in V(G) \setminus \{u\}\} \cup \{u_1 u_2\}$$

is a tree 3-spanner of $\mathcal{P}(G)$.

Next, suppose that $\mathcal{P}(G)$ admits a tree 3-spanner, say T. By contradiction, suppose that G does not have a universal vertex. As T is a spanning tree of $\mathcal{P}(G)$, there exists an edge $u_1 u_2 \in E(T)$. Let $S = V(G) \setminus N_G[u]$, where $N_G[u]$ denotes the closed neighborhood of u (i.e. $N_G(u) \cup \{u\}$). Observe that

$$
\begin{aligned}
d_T(u_1, v_1) \geq d_G(u, v) \geq 2, &\qquad d_T(u_2, v_1) \geq d_G(u, v) + 1 \geq 3, \\
d_T(u_1, v_2) \geq d_G(u, v) + 1 \geq 3, &\qquad d_T(u_2, v_2) \geq d_G(u, v) \geq 2,
\end{aligned}
\tag{1}
$$

for each $v \in S$. Let T_1 and T_2 be the components of $T - u_1 u_2$ that contain u_1 and u_2, respectively. By (1), the path between v_1 and v_2 in T does not contain u_1. That is,

$$v_1, v_2 \text{ belong to the same component of } T - u_1 u_2 \text{ for each } v \in S. \tag{2}$$

Let C be a shortest cycle in G that contains u and a vertex in S (such a cycle exists by Proposition 2). Let $C := \langle u, w^0, \ldots, w^k \rangle$. Note that, the minimality of C implies that $w^1, \ldots, w^{k-1}$ belongs to S. Thus, by (2), the vertices w_1^1 and w_2^1 belong either to T_1 or T_2. Without loss of generality, suppose that $w_1^1, w_2^1 \in V(T_1)$ (the proof of the other case is symmetric). The following claim shows how the copies of the vertices in $V(C) \cap S$ are connected in T. We leave its proof to the reader.

Claim 1. *The vertices w_1^i and w_2^i belong to T_1, for $i = 1, \ldots, k - 1$.*

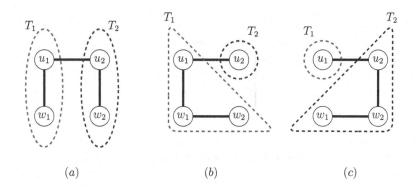

(a) (b) (c)

Fig. 3. Vertices of type $a)$, $b)$ and $c)$ in T are shown in (a), (b) and (c). The black edges represent the edges in T (a tree 3-spanner).

Next, we focus on the vertices w^0 and w^k in C. Given a vertex $w \in N_G(u)$, we consider the following three cases in T (depicted in Fig. 3).

a) the edges $w_1 u_1$ and $w_2 u_2$ belong to T;
b) the edges $w_1 u_1$ and $w_1 w_2$ belong to T;
c) the edges $w_2 u_2$ and $w_1 w_2$ belong to T.

In particular, we say that a vertex $w \in N_G(u)$ is of type a), b) or c) if its copies w_1 and w_2 satisfy the corresponding condition above. The following claim shows that every vertex adjacent to u in G is either of type a), b) or c). Its proof is omitted owing to space limitation.

Claim 2. *Let $w \in N_G(u)$. Then, w is of type a), b) or c).*

To conclude the proof of the theorem, we show that there exists an edge in C whose copy in G_1 or G_2 violates the 3-spanner condition in T. Since $w^0 \in N_G(u)$, we distinguish three cases.
Case 1: w^0 is of type a)

By Claim 1, the vertex $w_2^1 \in V(T_1)$. Since $w^1 \in S$, by (1) we have that $d_T(w_2^1, u_1) \geq 3$. Then,

$$d_T(w_2^1, w_2^0) = d_T(w_2^1, u_1) + d_T(u_1, u_2) + d_T(u_2, w_2^0) \geq 3 + 1 + 1 = 5,$$

which contradicts the fact that T is a 3-spanner of $\mathcal{P}(G)$.
Case 2: w^0 is of type b)

In this case, the edge $w_1^0 u_1 \in E(T)$. Let T' be the subtree of $T - w_1^0 u_1$ that contains w_1^0. We depict the subtrees T_1, T_2 and T' in Fig. 4. The black edges represent edges of T. The following claim holds. We leave its proof to the reader.

Claim 3. *The vertices w_1^i and w_2^i belong to T', for $i = 0, 1, \ldots, k$.*

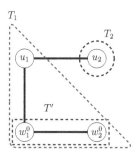

Fig. 4. The subtrees of T in Case 2.

The previous claim implies that

$$d_T(w_2^k, u_2) = d_T(w_2^k, w_1^0) + d_T(w_1^0, u_2) \geq 2 + 2 = 4,$$

a contradiction.
Case 3: w_0 is of type c)

In this case, $d_T(w_1^1, u_1) \geq 2$. Thus, we have that

$$d_T(w_1^1, w_1^0) = d_T(w_1^1, u_1) + d_T(u_1, u_2) + d_T(u_2, w_1^0) \geq 2 + 1 + 2 = 5,$$

a contradiction. $\square$

In the proof of the previous result, we have used the following two facts:

- G is 2-connected (required to obtain the cycle C),
- the edge $u_1 u_2$ belongs to a tree 3-spanner T of $\mathcal{P}(G)$.

Observe that both conditions are satisfied by a nontrivial block and a cut-vertex inside this block. Indeed, by analogous arguments, we obtain the following result.

Corollary 1. *Let $\mathcal{P}(G)$ be the prism of a graph G. If $\mathcal{P}(G)$ admits a tree 3-spanner, then every cut-vertex u in G is a universal vertex inside the blocks that contain u.*

To show our characterization, we just need the following last observation. (Its proof is omitted owing to space limitation.)

Lemma 2. *Let $\mathcal{P}(G)$ be the prism of a graph G. If $\mathcal{P}(G)$ admits a tree 3-spanner, then every nontrivial block of G has at most one cut-vertex.*

Now, let us consider the block tree $\mathcal{T}(G)$ of a graph G. Lemma 2 implies that the nontrivial blocks of G correspond to vertices of degree at most one in $\mathcal{T}(G)$. Thus, either they correspond to leaves or to a vertex of degree zero (G is 2-connected).

Theorem 2. *Let G be a graph. Then $\mathcal{P}(G)$ admits a tree 3-spanner if and only if each nontrivial block B of G satisfies the following two conditions:*
c_1) *B contains a universal vertex in B; and*
c_2) *B contains at most one cut-vertex of G.*

Proof (sketch). The previous results imply that if G satisfies c_1) and c_2), then $\mathcal{P}(G)$ admits a tree 3-spanner. Now, suppose that each nontrivial block B of G satisfies c_1) and c_2). Without loss of generality, suppose that G is not 2-connected. The following procedure constructs a tree 3-spanner of $\mathcal{P}(G)$.

1: $T \leftarrow \emptyset$
2: **for** each block $B \in \mathcal{B}(G)$ **do**
3:     Let u be a cut-vertex in B
4:     $T \leftarrow T \cup \{u_1 w_1 : w \in V(B) \setminus \{u\}\}$
5:     **if** B is nontrivial **then**
6:         $T \leftarrow T \cup \{u_2 w_2 : w \in V(B) \setminus \{u\}\}$
7:     **end if**
8: **end for**
9: $T \leftarrow T \cup \{u_1 u_2 : u$ is a cut-vertex of $G\}$

We show an example of this construction in Fig. 5. $\square$

Based on the previous result, we describe below an algorithm, called Algorithm 1, that decides, given a graph G, whether $\mathcal{P}(G)$ admits a tree 3-spanner, and in the affirmative case outputs one such spanner.

Theorem 3. *Algorithm 1 solves* TREES$_3$ *on prisms of graphs in linear time.*

Proof. The correctness of the algorithm follows from Theorem 2. Regarding its time complexity, Tarjan [27] showed a linear-time algorithm that finds the cut-vertices and blocks of a graph. Moreover, each of the steps in the loop at lines 3–20 takes constant (amortized) time. Therefore, Algorithm 1 runs in linear time on the size of G. We observe that, even if we are given $\mathcal{P}(G)$ as input (instead of G), Algorithm 1 runs in linear time. We refer to Imrich & Petering [21] for an algorithm to recognize Cartesian products of graphs in linear time.   □

Algorithm 1. TREE3S-PRISM(G)

Input: A connected graph G
Output: A set of edges that induces a tree 3-spanner of $\mathcal{P}(G)$ if it exists
1: Find the blocks and cut-vertices of G
2: $T \leftarrow \emptyset$ ▷ set of edges that induces a tree 3-spanner of $\mathcal{P}(G)$
3: **for** $B \in \mathcal{B}(G)$ **do**
4:     **if** B is a trivial block **then**
5:         Let $B = \{uv\}$
6:         $T \leftarrow T \cup \{u_1 v_1\}$
7:     **else**
8:         **if** B has no cut-vertex **then**   ▷ G is 2-connected
9:             Let u be a vertex of maximum degree in B
10:         **else if** B contains more than one cut-vertex of G **then**
11:             **return** $\emptyset$
12:         **else**
13:             Let u be the cut-vertex of B
14:         **end if**
15:         **if** u is not a universal vertex in B **then**
16:             **return** $\emptyset$
17:         **end if**
18:         $T \leftarrow T \cup \{u_1 w_1, u_2 w_2 : w \in V(B) \setminus \{u\}\}$
19:     **end if**
20: **end for**
21: $T \leftarrow T \cup \{u_1 u_2 : u \text{ is a cut-vertex of } G\}$
22: **return** T

Fekete & Kremer [16] designed an algorithm to decide whether a planar graph admits a tree 3-spanner (and find one if it exists). Given a planar graph G, their algorithm transforms G into a planar graph G' where the boundary of each face has length at most four. Finally, they solve TREES$_3$ on G', and if a solution is obtained, they show how to obtain a solution for G. We observe that Algorithm 1

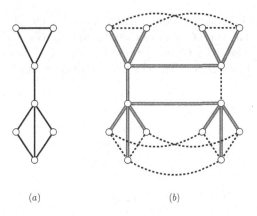

Fig. 5. (a) A graph G; and (b) a tree 3-spanner of $\mathcal{P}(G)$.

can be seen as an alternative solution for TREES$_3$ on a subclass of planar graphs. Namely, the prisms of outerplanar graphs (which are also planar).

To conclude, we observe that 2-connected graphs H with $g(H) \geq 4$ do not contain a universal vertex. Moreover, a graph G has $g(G) \geq 4$ if and only if $g(B) \geq 4$ for each nontrivial block of G. Thus, by Theorem 2, we have to exclude graphs G whose nontrivial blocks have girth at least 4. This means that if G is bipartite, then G has to be a tree. Therefore, we obtain the following result.

Corollary 2. *Let G be a bipartite graph. Then, $\mathcal{P}(G)$ admits a tree 3-spanner if and only if G is a tree.*

4   Generalized Prisms

In this section, we study a decomposition of a graph that is related to t-spanners. This technique was used on the class of bounded-degree graphs by Cai & Keil [7] and by Gómez et. al. [20] in the study of 2-spanners and 3-spanners, respectively. We will show that this decomposition gives a necessary condition for the existence of a tree t-spanner on an arbitrary graph.

Let $G = (V, E)$ be a graph, and let H be a t-spanner of G. Observe that if an edge $e = uv \in E$ does not belong to H, then there exists a path P between u and v, in H, such that $|P| \leq t$. Thus, the edge e belongs to a cycle of size at most $t + 1$ in G. This fact motivates the definition of the following auxiliary graph. Let $\mathcal{L}_t(G)$ be the graph defined from G as follows.

$$V(\mathcal{L}_t(G)) = \{v_e : e \in E\},$$
$$E(\mathcal{L}_t(G)) = \{v_e v_f : e, f \in E \text{ belong to a cycle } C \text{ in } G, |C| \leq t+1\}.$$

We denote by $\mathcal{C}_t(G)$ the partition of E induced by the connected components of $\mathcal{L}_t(G)$. That is, two edges $e, f \in E$ belong to the same class in $\mathcal{C}_t(G)$ if and only if v_e and v_f belong to the same connected component of $\mathcal{L}_t(G)$. In Fig. 6, we show an example of a graph G, its associated graph $\mathcal{L}_2(G)$, and its decomposition in $\mathcal{C}_2(G)$. The vertices of $\mathcal{L}_2(G)$ are represented by full squares, and its edges are depicted by (full) black edges. The classes in $\mathcal{C}_2(G)$ are represented by different types of edges.

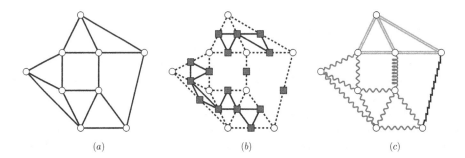

(a)   (b)   (c)

Fig. 6. (a) a graph G; (b) the graph $\mathcal{L}_2(G)$ (its vertices are shaded rectangles, and its edges are depicted by full black edges); (c) the four classes in $\mathcal{C}_2(G)$ (represented by different types of edges).

The importance of $C_t(G)$ comes from the following result. It shows that any t-spanner of G is composed of t-spanners for each graph in $\mathcal{C}_t(G)$.

Proposition 3 (Gómez et. al. [20], 2022). *A subgraph S of a graph G is a t-spanner if and only if $S \cap H$ is a t-spanner of H, for every $H \in \mathcal{C}_t(G)$.*

In the case of tree t-spanners, Cai & Corneil [6] also noted the following analogous result (see Observation 1.4. in [6]).

Proposition 4 (Cai & Corneil, 1995). *Let T be a spanning tree of a graph G. Then, T is a tree t-spanner if and only if $T \cap H$ is a tree t-spanner of H, for each $H \in \mathcal{B}(G)$.*

The following lemma relates the subgraphs in $\mathcal{C}_t(G)$ to the blocks of G.

Lemma 3. *Let G be a graph, and let $H \in \mathcal{C}_t(G)$ such that $H \not\cong K_2$. Then, H is 2-connected.*

Proof. For this, we show that if a graph H has a cut-vertex, then $\mathcal{C}_t(H) \neq \{H\}$. Let u be a cut-vertex of H, and let C be a component of $H - \{u\}$. Let H_1 be the graph induced by $C \cup \{u\}$, and let $H_2 := H - C$. Since any path between H_1 and H_2 contains u, there is no cycle that contains an edge in H_1 and another edge in H_2. Therefore, $\mathcal{C}_t(H) \neq \{H\}$.   □

Since the decomposition of a graph G into blocks is unique, we have the following result.

Corollary 3. *Let G be a graph. Then, for each $H \in \mathcal{C}_t(G)$, there exists a unique block $B_H \in \mathcal{B}(G)$ such that H is a subgraph of B_H.*

When using a decomposition technique, it is frequently preferable that the subproblems we obtain are as small as possible. In this respect, Corollary 3 says that the decomposition $\mathcal{C}_t(G)$ is at least as good as the block decomposition when dealing with t-spanner problems. Now, we show the main result of this section. It says that, if G admits a tree t-spanner, then the subgraphs in $\mathcal{C}_t(G)$ are precisely the blocks of G.

Theorem 4. *Let G be a graph. If G admits a tree t-spanner, then $\mathcal{C}_t(G) = \mathcal{B}(G)$.*

Proof (sketch). Suppose that $\mathcal{C}_t(G) \neq \mathcal{B}(G)$. We will prove that G does not admit a tree t-spanner. As $\mathcal{C}_t(G) \neq \mathcal{B}(G)$, there exists $H \in \mathcal{C}_t(G)$ and $B_H \in \mathcal{B}(G)$ such that $H \subseteq B_H$ and $H \neq B_H$. Let $H_1, \ldots, H_k$ be the subgraphs in $\mathcal{C}_t(G)$ such that

(i) $H_1 = H$,
(ii) $H_i \subseteq B_H$, and
(iii) $E(B_H) = \bigcup_{i=1}^{k} E(H_i)$.

In what follows, we consider each edge set $E(H_i)$ as a color class. We say that a cycle C, in B_H, is *colorful* if it contains edges of at least two colors. Since $H_1 = H \subseteq B_H$ and $H_1 \neq B_H$, we have that $k \geq 2$. Furthermore, if we consider an edge $e \in E(H_1)$ and an edge $f \in E(H_2)$, Proposition 2 (see Sect. 2) implies that there exists a colorful cycle in B_H.

Let C^* be a colorful cycle in B_H, and let S be a minimum t-spanner of G. Without loss of generality, suppose that $E(C^*) \cap E(H_i) \neq \emptyset$, for $i = 1, \ldots, k$; otherwise we do not consider the subgraph H_i. From C^*, we will construct a closed trail in S, concluding that S contains a cycle which is a contradiction. For this, let u and v be the ends of a path in $E(C^*) \cap E(H_1)$. Starting from the vertex u, we traverse C^* in the direction of v, and label its vertices as

$$C^* := \langle u = w_1, w_2, \ldots, w_\ell \rangle.$$

Let $S_i := S \cap H_i$. We construct a closed trail C' in S as follows.

1: $i \leftarrow 1$
2: **while** $i \leq \ell$ **do**
3:     Let c be the color of the edge $w_i w_{i+1}$ (i.e. $w_i w_{i+1} \in E(H_c)$)
4:     $j \leftarrow \max\{j' : i+1 \leq j' \leq \ell,$ the edge $w_{j'-1} w_{j'}$ has color $c\}$
5:     Replace path $\langle w_i, \ldots, w_j \rangle$ in C^* with a path between w_i and w_j in S_c
6:     $i \leftarrow j$
7: **end while**

We show an example of this construction in Fig. 7. We show the cycle C^* in Fig. 7 (a) (depicted by full edges). In Fig. 7 (b), we show a minimum 2-spanner of G (depicted by double edges). Finally, the closed trail C' is depicted by wavy edges in Fig. 7 (c). □

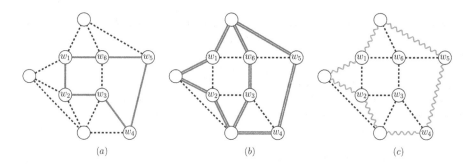

Fig. 7. (a) The cycle C^*; (b) a minimum 2-spanner of G; and (c) the closed trail C'.

On the one hand, Theorem 4 tells us that, regarding tree t-spanners, the decomposition $\mathcal{C}_t(G)$ is not better than the block decomposition of the graph. On the other hand, it gives a simple criterion to test whether a graph does not admit a tree t-spanner. In what follows, we use the previous result to characterize the generalized prisms of trees that admit a tree 3-spanner. We omit its proof owing to space limitation.

Theorem 5. *Let T be a tree, and let T' be a generalized prism of T. Then, T' admits a tree 3-spanner if and only if the linkage of T' induces an automorphism on T.*

The previous theorem tells us that a generalized prism of a tree T admits a tree 3-spanner if and only if it is isomorphic to $\mathcal{P}(T)$. In the previous proof, the fact that there is no 4-cycle inside the copies of T implies that the linkage induces an automorphism on T, otherwise we violate the condition given in Theorem 4. We observe that the same condition is satisfied by a graph with large girth.

Corollary 4. *Let G be a graph such that $g(G) \geq 5$, and let G' be a generalized prism of G. Then, G' admits a tree 3-spanner if and only if*

a) G is a tree, and
b) the linkage of G' induces an automorphism on G.

Proof. If G' satisfies a) and b), then Theorem 5 implies that G' admits a tree 3-spanner. On the other hand, suppose that G' admits a tree 3-spanner, say T'. Let f be the function induced by the linkage of G'. Since $g(G) \geq 5$, we have that $f(u)f(v) \in E(G')$, for each edge $uv \in E(G)$, otherwise $\mathcal{C}_3(G') \neq \mathcal{B}(G')$. Thus, f is an automorphism on G. Finally, suppose that G is not a tree, and let C be a cycle in G. Next, consider the copy of the cycle C in G_1, say C_1. Since $g(G_1) \geq 5$, for each edge $u_1 v_1 \in E(C_1) \setminus E(T)$, the path $\langle u_1, u_2, v_2, v_1 \rangle$ exists in T. Therefore, T contains a closed trail, a contradiction. □

The previous result implies that, excluding prisms of trees, no generalized prism of a graph G with $g(G) \geq 5$ admits a tree 3-spanner. Let G' be a generalized prism of a graph G. The previous result may suggest that a necessary

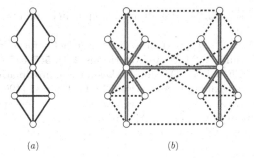

Fig. 8. (a) a graph G; and (b) a generalized prism of G that admits a tree 3-spanner.

condition for G' to admit a tree 3-spanner is to be isomorphic to $\mathcal{P}(G)$. However, this is not true, as shown in Fig. 8.

5   Concluding Remarks and Future Work

In this work, we focused on the problem TREES$_3$ whose complexity is still unknown since 1995. We studied this problem on the class of prisms of graphs. We characterized which of these graphs admit a tree 3-spanner, and designed a linear-time algorithm that solves TREES$_3$ on this class. Moreover, we obtained a necessary condition for the existence of a tree t-spanner on a general graph, that is checkable in polynomial time. As a byproduct, we characterized the generalized prisms of trees that admit a tree 3-spanner. Currently, we are working on the extension of our results to generalized prisms of arbitrary graphs.

Other directions for further research involve extending our characterization for graphs of the form $G \times H$, where G is a general graph, and H is a path, cycle or complete graph. Finally, we observe that a prism of a graph is a special case of a graph that contains a perfect matching whose removal disconnects it. In that sense, it would be an interesting and challenging problem to derive sufficient conditions for the nonexistence of a tree 3-spanner in terms of separating cuts. This kind of result may possibly help understanding better the properties of the graphs that admit a tree 3-spanner.

Acknowledgements. This research has been partially supported by FAPESP - São Paulo Research Foundation (Proc. 2015/11937-9). R. Gómez is supported by FAPESP (Proc. 2019/14471-1); F. K. Miyazawa is supported by FAPESP (Proc. 2016/01860-1) and CNPq (Proc. 314366/2018-0); Y. Wakabayashi is supported by CNPq (Proc. 311892/2021-3 and 423833/2018-9).

References

1. Awerbuch, B., Bar-Noy, A., Linial, N., Peleg, D.: Improved routing strategies with succinct tables. J. Algorithms **11**(3), 307–341 (1990)

2. Barnette, D., Rosenfeld, M.: Hamiltonian circuits in certain prisms. Discret. Math. **5**, 389–394 (1973)

3. Baswana, S., Sen, S.: Approximate distance oracles for unweighted graphs in expected $O(n^2)$ time. ACM Trans. Algorithms **2**(4), 557–577 (2006)

4. Brandstädt, A., Chepoi, V., Dragan, F.: Distance approximating trees for chordal and dually chordal graphs. J. Algorithms **30**(1), 166–184 (1999)

5. Brandstädt, A., Dragan, F., Le, H., Le, V., Uehara, R.: Tree spanners for bipartite graphs and probe interval graphs. Algorithmica **47**(1), 27–51 (2007)

6. Cai, L., Corneil, D.: Tree spanners. SIAM J. Discret. Math. **8**(3), 359–387 (1995)

7. Cai, L., Keil, M.: Spanners in graphs of bounded degree. Networks **24**(4), 233–249 (1994)

8. Chudá, K., Škoviera, M.: $L(2,1)$-labelling of generalized prisms. Discret. Appl. Math. **160**(6), 755–763 (2012)

9. Couto, F., Cunha, L.: Hardness and efficiency on t-admissibility for graph operations. Discret. Appl. Math. **304**, 342–348 (2021)

10. Couto, F., Cunha, L., Juventude, D., Santiago, L.: Strategies for generating tree spanners: algorithms, heuristics and optimal graph classes. Inform. Process. Lett. 177, Paper No. 106265, 10 (2022)

11. Couto, F., Cunha, L., Posner, D.: Edge tree spanners. In: Gentile, C., Stecca, G., Ventura, P. (eds.) Graphs and Combinatorial Optimization: from Theory to Applications. ASS, vol. 5, pp. 195–207. Springer, Cham (2021). https://doi.org/10.1007/978-3-030-63072-0_16

12. Diestel, R.: Graph Theory. GTM, vol. 173. Springer, Heidelberg (2017). https://doi.org/10.1007/978-3-662-53622-3

13. Dragan, F., Fomin, F., Golovach, P.: Spanners in sparse graphs. J. Comput. Syst. Sci. **77**(6), 1108–1119 (2011)

14. Dragan, F., Köhler, E.: An approximation algorithm for the tree t-spanner problem on unweighted graphs via generalized chordal graphs. Algorithmica **69**(4), 884–905 (2014)

15. Emek, Y., Peleg, D.: Approximating minimum max-stretch spanning trees on unweighted graphs. SIAM J. Comput. **38**(5), 1761–1781 (2008)

16. Fekete, S., Kremer, J.: Tree spanners in planar graphs. Discret. Appl. Math. **108**(1–2), 85–103 (2001)

17. Fomin, F., Golovach, P., van Leeuwen, E.: Spanners of bounded degree graphs. Inform. Process. Lett. **111**(3), 142–144 (2011)

18. Galbiati, G.: On finding cycle bases and fundamental cycle bases with a shortest maximal cycle. Inform. Process. Lett. **88**(4), 155–159 (2003)

19. Goddard, W., Henning, M.: A note on domination and total domination in prisms. J. Comb. Optim. **35**(1), 14–20 (2018)

20. Gómez, R., Miyazawa, F., Wakabayashi, Y.: Minimum t-spanners on subcubic graphs. In: WALCOM: Algorithms and Computation, pp. 365–380. Lecture Notes in Computer Science, Springer, Cham (2022). https://doi.org/10.1007/978-3-030-96731-4_30

21. Imrich, W., Peterin, I.: Recognizing Cartesian products in linear time. Discret. Math. **307**(3–5), 472–483 (2007)

22. Lin, L., Lin, Y.: Optimality computation of the minimum stretch spanning tree problem. Appl. Math. Comput. **386**, 125502 (2020)

23. Madanlal, M., Venkatesan, G., Pandu Rangan, C.: Tree 3-spanners on interval, permutation and regular bipartite graphs. Inform. Process. Lett. **59**(2), 97–102 (1996)

24. Álvarez Miranda, E., Sinnl, M.: Mixed-integer programming approaches for the tree t^*-spanner problem. Optim. Lett. **13**(7), 1693–1709 (2019)
25. Paulraja, P.: A characterization of Hamiltonian prisms. J. Graph Theor. **17**(2), 161–171 (1993)
26. Peleg, D., Ullman, J.: An optimal synchronizer for the hypercube. SIAM J. Comput. **18**(4), 740–747 (1989)
27. Tarjan, R.: Depth-first search and linear graph algorithms. SIAM J. Comput. **1**(2), 146–160 (1972)
28. Thorup, M., Zwick, U.: Approximate distance oracles. J. ACM **52**(1), 1–24 (2005)
29. Venkatesan, G., Rotics, U., Madanlal, M., Makowsky, J., Pandu Rangan, C.: Restrictions of minimum spanner problems. Inform. Comput. **136**(2), 143–164 (1997)
30. Wang, W., Balkcom, D., Chakrabarti, A.: A fast online spanner for roadmap construction. Int. J. Rob. Res. **34**(11), 1418–1432 (2015)

A General Approach to Ammann Bars for Aperiodic Tilings

Carole Porrier[1,2]([⊠])[iD] and Thomas Fernique[1,3][iD]

[1] Université Sorbonne Paris Nord, Villetaneuse, France
carole.porrier@gmail.com
[2] Université du Québec à Montréal, Montreal, Canada
[3] CNRS, Paris, France

Abstract. Ammann bars are formed by segments (decorations) on the tiles of a tiling such that forming straight lines with them while tiling forces non-periodicity. Only a few cases are known, starting with Robert Ammann's observations on Penrose tiles, but there is no general explanation or construction. In this article we propose a general method for cut and project tilings based on the notion of *subperiods* and we illustrate it with an aperiodic set of 36 decorated prototiles related to what we called *Cyrenaic tilings*.

Keywords: Aperiodic tilings · Ammann bars · Cut and project tilings

1 Introduction

Shortly after the famous Penrose tilings were introduced by Roger Penrose in 1974 [13] and popularized by Martin Gardner in 1977 [9], amateur mathematician Robert Ammann [17] found particularly interesting decorations of the tiles (Fig. 1): if one draws segments in the same way on all congruent tiles then on any valid tiling all those segments compose straight lines, going in five different directions. Conversely if one follows the assembly rule consisting of prolonging every segment on the tiles into a straight line then the obtained tiling is indeed a Penrose tiling. Those lines are called *Ammann bars* and the corresponding matching rule is locally equivalent to the ones given by Penrose using arrows on the sides or alternative decorations [14].

Penrose tilings have many interesting properties and can be generated in several ways. The *cut and project* method[1] follows their algebraic study by de Bruijn in 1981 [7]. Beenker soon proposed a whole family of tilings based on it [5], including the Ammann-Beenker tilings that Ammann found independently. A cut and project tiling can be seen as a digitization of a two-dimensional plane in a n-dimensional Euclidean space ($n > 2$), and we will talk about $n \to 2$ *tilings* in that sense. When the *slope* of the plane does not contain any rational line, the

---

[1] Terms in italic which are not defined in the introduction are defined formally in further sections. The introduction is meant to give a general idea of the article.

© Springer Nature Switzerland AG 2022
A. Castañeda and F. Rodríguez-Henríquez (Eds.): LATIN 2022, LNCS 13568, pp. 574–589, 2022.
https://doi.org/10.1007/978-3-031-20624-5_35

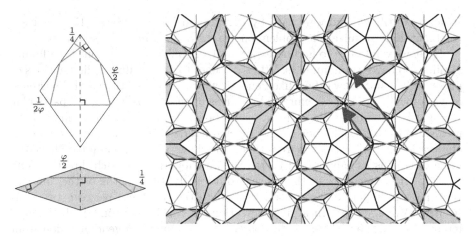

Fig. 1. Left: Penrose tiles with Ammann segments (in orange). On each rhombus the dashed line is an axis of symmetry and the sides have length $\varphi = (1 + \sqrt{5})/2$. Right: Ammann bars on a valid pattern of Penrose tiles, where each segment is correctly prolonged on adjacent tiles. The red vectors are "integer versions" of one subperiod. (Color figure online)

tiling is non-periodic. This is the case for Penrose tilings for instance, so the set of tiles defining them is *aperiodic*: one can tile the plane with its tiles but only non-periodically. The first aperiodic tileset was found by Berger, thus proving the undecidability of the *Domino Problem* [6] and relating tilings to logic. Since then, relatively few others were exhibited: many non-periodic tilings exist (even infinitely many using the cut and project method), but we usually do not have a corresponding aperiodic tileset.

Links were made between such tilings and quasicrystals [16,19], that is crystals whose diffraction pattern is not periodic but still ordered, with rotational symmetries. The study of *local rules*, i.e. constraints on the way tiles can fit together in finite patterns, can help modeling the long range aperiodic order of quasicrystals. For instance, Penrose tilings are defined by their 1-*atlas*, which is a small number of small patterns: any and all tilings containing only those patterns (of the given size) are Penrose tilings. Alternately, they can also be defined by their *Ammann local rules*, as stated in the first paragraph. On the contrary, it was proven [8] that Ammann-Beenker tilings, also known as 8-fold tilings, do not have *weak local rules*, i.e. no finite set of patterns is enough to characterize them. Socolar found sort of Ammann bars for them [18], but they extend outside the boundary of the tiles, thus do not fit the framework considered here.

Grünbaum and Shephard [10] detail the properties of Ammann bars in the case of Penrose tilings and their close relation to the *Fibonacci word*. They also present two tilesets by Ammann with Ammann bars (A2 and A3) but these are substitutive and not cut and project tilings. Generally speaking, we do not know much about Ammann bars and for now each family of aperiodic tilings has to be treated on a case-by-case basis. Yet they can reveal quite useful to study the structure of tilings, and were used by Porrier and Blondin Massé [15] to solve a combinatorial optimization problem on graphs defined by Penrose tilings.

Here, we would like to find necessary and/or sufficient conditions for a family of tilings to have Ammann bars. When it comes to $4 \to 2$ tilings (digitizations of planes in $\mathbb{R}^4$) and a few others like Penrose, which are $5 \to 2$ tilings, the existence of weak local rules can be expressed in terms of *subperiods*, which are particular vectors of the slope [2,3]. As mentioned above, Ammann-Beenker tilings have no local rules and their slope cannot be characterized by its subperiods. Careful observation of Penrose tilings from this angle shows that Ammann bars have the same directions as subperiods: there are two subperiods in each direction, one being φ times longer than the other. Additionally, the lengths of the "integer versions" of subperiods are closely related to the distances between two consecutive Ammann bars in a given direction, as can be seen in Fig. 1. Though interesting, this special case is too particular to hope for a generalization from it alone. Nonetheless, we think that Ammann bars are related to subperiods.

Since subperiods are simpler for $4 \to 2$ tilings, for which we also have a stronger result regarding weak local rules, we focus on those. Namely, Bédaride and Fernique [3] showed that a $4 \to 2$ tiling has weak local rules if and only if its slope is characterized by its subperiods. It seems some conditions of alignment play a part in the existence of Ammann bars. This led us to introduce the notion of *good projection* (Definition 1 p. 9) on a slope. We propose a constructive method to find Ammann bars for $4 \to 2$ tilings which are characterized by subperiods and for which we can find a good projection. We prove the following result:

Proposition 1. *The tileset obtained with our method is always finite.*

We found several examples of $4 \to 2$ tilings characterized by their subperiods and admitting a good projection. For each of them, we have been able to show that the finite tileset given by our method is aperiodic. We conjecture that this actually always holds but we have not yet been able to prove that. Here, we detail one of these examples, namely $4 \to 2$ tilings with a slope based on the irrationality of $\sqrt{3}$ that we called *Cyrenaic tilings* in reference to Theodorus of Cyrene who proved $\sqrt{3}$ to be irrational. They have "short" subperiods, which facilitates observations on drawings. In this case, our method yields the set of decorated tiles depicted in Fig. 2. Those tiles give Ammann bars to Cyrenaic tilings and we were able to prove the following:

Theorem 1. *The tileset $\mathcal{C}$ in Fig. 2 is aperiodic.*

The case of Penrose indicates that our construction could (and should) be adapted in order to work for $5 \to 2$ tilings, or general cut and project ($n \to d$) tilings. In particular, for Penrose the lines are shifted and the number of lines is reduced compared with our method, so that only two decorated tiles are needed. Besides, in each direction the distance between two consecutive lines can take only two values, and the sequence of intervals is substitutive. In the case of Cyrenaic tilings, the bi-infinite word defined by each sequence of intervals

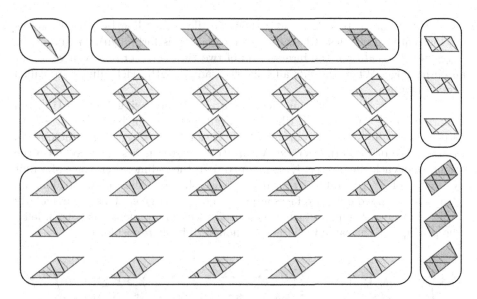

Fig. 2. Set $\mathcal{C}$ of 36 decorated prototiles obtained from Cyrenaic tilings. Any tiling by these tiles where segments extend to lines is non-periodic (Theorem 1).

between Ammann bars seems to be substitutive so maybe we could compose them after finding the substitution. Lines could also be shifted as it is the case for Penrose tilings, instead of passing through vertices. An optimal shift (reducing the number of lines or tiles) would then have to be determined. Our SageMath code as well as some more technical explanations are given in the following repository:

https://github.com/cporrier/Cyrenaic

The paper is organized as follows. Section 2 introduces the settings, providing the necessary formal definitions, in particular local rules and subperiods. In Sect. 3 we present our method to construct a set of decorated prototiles yielding Ammann bars. We rely on subperiods characterizing a slope as well as a good projection, and prove Proposition 1. Finally, in Sect. 4 we show that Ammann bars of the set $\mathcal{C}$ force any tiling with its tiles to have the same subperiods as Cyrenaic tilings, thus proving Theorem 1.

2 Settings

2.1 Canonical Cut and Project Tilings

A **tiling** of the plane is a covering by **tiles**, i.e. compact subsets of the space, whose interiors are pairwise disjoint. In this article we focus on **tilings by parallelograms**: let $v_0, ..., v_{n-1}$ ($n \geq 3$) be pairwise non-collinear vectors of the Euclidean plane, they define $\binom{n}{2}$ parallelogram *prototiles* which are the sets

$T_{ij} := \{\lambda v_i + \mu v_j \mid 0 \leq \lambda, \mu \leq 1\}$; then the tiles of a tiling by parallelograms are translated prototiles (tile rotation or reflection is forbidden), satisfying the edge-to-edge condition: the intersection of two tiles is either empty, a vertex or an entire edge. When the v_i's all have the same length, such tilings are called *rhombus tilings*.

Let $e_0, ..., e_{n-1}$ be the canonical basis of $\mathbb{R}^n$. Following Levitov [12] and Bédaride and Fernique [2], a tiling by parallelograms can be **lifted** in $\mathbb{R}^n$, to correspond to a "stepped" surface of dimension 2 in $\mathbb{R}^n$, which is unique up to the choice of an initial vertex. An arbitrary vertex is first mapped onto the origin, then each tile of type T_{ij} is mapped onto the 2-dimensional face of a unit hypercube of $\mathbb{Z}^n$ generated by e_i and e_j, such that two tiles adjacent along an edge v_i are mapped onto two faces adjacent along an edge e_i. This is particularly intuitive for $3 \to 2$ tilings which are naturally seen in 3 dimensions (Fig. 3, left). The principle is the same for larger n, though difficult to visualize.

Fig. 3. Examples. Left: Rauzy tiling from which you can visualize the lift in $\mathbb{R}^3$. Center: Ammann-Beenker tiling. Right: Penrose tiling.

If a tiling by parallelograms can be lifted into a tube $E + [0, t]^n$ where $E \subset \mathbb{R}^n$ is a plane and $t \geq 1$, then this tiling is said to be **planar**. In that case, **thickness** of the tiling is the smallest suitable t, and the corresponding (unique up to translation) E is called the **slope** of the tiling. A planar tiling by parallelograms can thus be seen as an approximation of its slope, which is as good as the thickness is small. Planarity is said **strong** if $t = 1$ and **weak** otherwise.

Strongly planar tilings by parallelograms can also be obtained by the so-called **(canonical) cut and project method**. For this, consider a d-dimensional affine plane $E \subset \mathbb{R}^n$ such that $E \cap \mathbb{Z}^n = \emptyset$, select ("cut") all the d-dimensional facets of $\mathbb{Z}^n$ which lie within the tube $E + [0, 1]^n$, then "project" them onto $\mathbb{R}^d$. If this projection π yields a tiling of $\mathbb{R}^d$ it is called **valid** (see Fig. 4), and the tiling is a strongly planar tiling by parallelograms with slope E. Such tilings

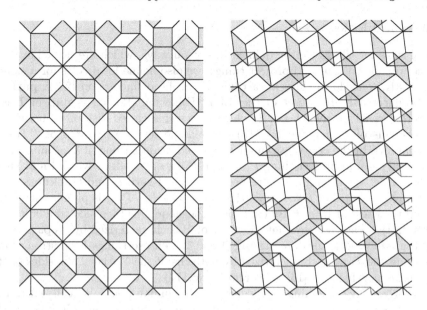

Fig. 4. Golden octagonal tiling with the usual valid projection (left) and a non-valid projection on the same slope (right). Colors of the tiles are the same with respect to the $\pi(e_i)$'s, with an opacity of 50% in both images. (Color figure online)

are called canonical cut and project tilings or simply $n \to d$ **tilings**. Not every projection is suitable, but the orthogonal projection onto E seen as $\mathbb{R}^d$ is known to be valid [11]. Here we only consider the case of a 2-dimensional slope E which is totally irrational, that is, which does not contain any rational line. This yields aperiodic tilings of the plane.

Figure 3 illustrates the above notions with three well-known examples. Rauzy tilings are $3 \to 2$ tilings whose slope E is generated by

$$\vec{u} = (\alpha - 1, -1, 0) \qquad \text{and} \qquad \vec{v} = (\alpha^2 - \alpha - 1, 0, -1),$$

where $\alpha \approx 1.89$ is the only real root of $x^3 - x^2 - x - 1$. Ammann-Beenker tilings, composed of tiles of the set A5 in the terminology of Grünbaum and Shephard [10], are the $4 \to 2$ tilings with slope E generated by

$$\vec{u} = (\sqrt{2}, 1, 0, -1) \qquad \text{and} \qquad \vec{v} = (0, 1, \sqrt{2}, 1).$$

Generalized Penrose tilings are the $5 \to 2$ tilings with slope E generated by

$$\vec{u} = (\varphi, 0, -\varphi, -1, 1) \qquad \text{and} \qquad \vec{v} = (-1, 1, \varphi, 0, -\varphi),$$

where $\varphi = (1 + \sqrt{5})/2$ is the golden ratio. The "strict" Penrose tilings as defined by Roger Penrose in [14] (set P3 in the terminology of [10]) correspond to the case when E contains a point whose coordinates sum to an integer.

2.2   Local Rules

Local rules for tilings can be defined in several ways, which are not equivalent. Since we focus on cut and project tilings, we also define local rules for a slope.

Firstly, weak local rules for a tiling T can be defined as in [2]. A **pattern** is a connected finite subset of tiles of T. Following [12], an **r-map** of T is a pattern formed by the tiles of T which intersect a closed disk of radius $r \geq 0$. The **r-atlas** of T, denoted by $T(r)$, is then the set of all r-maps of T (up to translation). In the case of a canonical cut and project tiling, it is a finite set. A canonical cut and project tiling $\mathcal{P}$ of slope E is said to admit **weak local rules** if there exist $r \geq 0$ and $t \geq 1$, respectively called **radius** and **thickness**, such that any $n \rightarrow d$ tiling T whose r-atlas is contained in $\mathcal{P}(r)$ is planar with slope E and thickness at most t. By extension, the slope E is then said to admit local rules. In that case, we say that the slope of $\mathcal{P}$ is characterized by its patterns of a given size. Local rules are **strong** if $t = 1$. Penrose tilings have strong local rules and the slope is characterized by patterns of the 1-atlas if the sides of the tiles have length 1 (see [16], Theorem 6.1, p.177).

Another way of defining local rules is with Ammann bars. We call **Ammann segments** decorations on tiles which are segments whose endpoints lie on the borders of tiles, such that when tiling with those tiles, each segment has to be continued on adjacent tiles to form a straight line. We say that a slope E admits **Ammann local rules** if there is a finite set of prototiles decorated with Ammann segments such that any tiling with those tiles is planar with slope E. In particular, no periodic tiling of the plane should be possible with those tiles if E is irrational. For instance, the marking of the Penrose tiles yielding Ammann bars is shown in Fig. 1, along with a valid pattern where each segment is correctly prolonged on adjacent tiles.

2.3   Subperiods

Adapted from Bédaride and Fernique [1], the $i_1, ..., i_{n-3}$-**shadow** of an $n \rightarrow 2$ tiling T is the orthogonal projection $\pi_{i_1,...,i_{n-3}}$ of its lift on the space generated by $\{e_j \mid 0 \leq j \leq n-1, j \neq i_1, ..., i_{n-3}\}$. This corresponds to reducing to zero the lengths of $\pi(e_{i_1}), ..., \pi(e_{i_{n-3}})$ in the tiling, so that the tiles defined by these vectors disappear. This is illustrated in Fig. 5. An $n \rightarrow 2$ tiling thus has $\binom{n}{3}$ shadows.

An $i_1, ..., i_{n-3}$-**subperiod** of an $n \rightarrow 2$ tiling T is a prime period of its $i_1, ..., i_{n-3}$-shadow, hence an integer vector in $\mathbb{R}^3$. By extension, we call subperiod of a slope E any vector of E which projects on a subperiod in a shadow of T. A subperiod is thus a vector of E with 3 integer coordinates: those in positions $j \notin \{i_1, ..., i_{n-3}\}$. We say that a slope is *determined* or *characterized* by its subperiods if only finitely many slopes have the same subperiods (in the shadows).

(a) Starting from an Ammann-Beenker tiling (on the left), progressively reduce the **length of one of the four vectors defining the tiles, until it is null (on the right). The** shadow thus obtained is periodic in one direction.

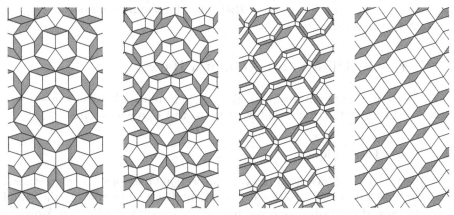

(b) Starting from a Penrose tiling (on the left), progressively reduce the lengths of two **of the five vectors defining the tiles, until they are null (on the right). The shadow thus** obtained is periodic in one direction.

Fig. 5. Shadows of Ammann-Beenker and Penrose tilings.

For instance, the slope of Ammann-Beenker tilings has four subperiods:

$$p_0 = (\sqrt{2}, 1, 0, -1),$$
$$p_1 = (1, \sqrt{2}, 1, 0),$$
$$p_2 = (0, 1, \sqrt{2}, 1),$$
$$p_3 = (-1, 0, 1, \sqrt{2}).$$

while that of Penrose tilings has ten, each with two non-integer coordinates.

This notion was first introduced by Levitov [12] as the *second intersection condition* and then developed by Bédaride and Fernique, who showed in [2] and

[3] that in the case of $4 \to 2$ tilings, a plane admits weak local rules if and only if it is determined by its subperiods. It was shown in [1] that this is not the case for Ammann-Beenker tilings: indeed, their subperiods are also subperiods of all Beenker tilings (introduced in [5]), that are the planar tilings with a slope generated, for any $s \in (0, \infty)$, by

$$u = (1, 2/s, 1, 0) \qquad \text{and} \qquad v = (0, 1, s, 1).$$

The Ammann-Beenker tilings correspond to the case $s = \sqrt{2}$ and do not admit local rules. On the other hand, generalized Penrose tilings have a slope characterized by its subperiods [2] and do admit local rules.

In this article, we focus on $4 \to 2$ tilings with irrational slope E characterized by four subperiods. In this case, each subperiod of E has exactly one non-integer coordinate. Since the vertices of the tiling are projected points of $\mathbb{Z}^4$, we define "integer versions" of subperiods: if $p_i = (x_0, x_1, x_2, x_3)$ is a subperiod, then its floor and ceil versions are respectively $\lfloor p_i \rfloor = (\lfloor x_0 \rfloor, \lfloor x_1 \rfloor, \lfloor x_2 \rfloor, \lfloor x_3 \rfloor)$ and $\lceil p_i \rceil = (\lceil x_0 \rceil, \lceil x_1 \rceil, \lceil x_2 \rceil, \lceil x_3 \rceil)$. Note that only the non-integer coordinate x_i is affected, and that $\lfloor p_i \rfloor, \lceil p_i \rceil \notin E$.

3   Cyrenaic Tilings and Ammann Bars

In this section, we present a construction to get Ammann bars for some $4 \to 2$ tilings and we give the example of what we named *Cyrenaic tilings*.

3.1   Good Projections

In Subsect. 2.1, we defined what is a valid projection for a slope E and mentioned the classical case of the orthogonal projection. There are however other valid projections, and this will play a key role here. We will indeed define Ammann bars as lines directed by subperiods and it will be convenient for the projected i-th subperiod $\pi(p_i)$ to be collinear with $\pi(e_i)$, so that the image of a line directed by p_i is still a line in the i-th shadow (Fig. 6). This leads us to introduce the following definition:

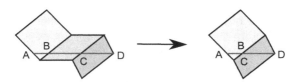

Fig. 6. Aligned segments in a pattern remain aligned in the shadow corresponding to the direction of the line.

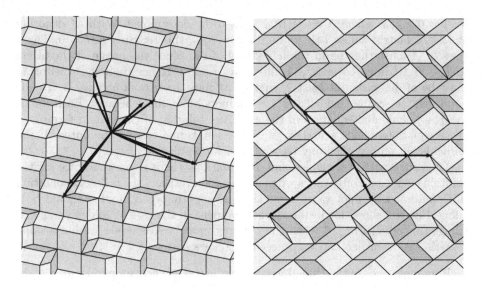

Fig. 7. Cyrenaic tiling with $\pi(\lfloor p_i \rfloor)$ and $\pi(\lceil p_i \rceil)$ for each subperiod p_i. On the left, we used the orthogonal projection which is valid but not *good*; on the right we used a good projection. Colors of the tiles are the same on both images with respect to the $\pi(e_i)$'s. Starting from the central pattern, one can see how one tiling is merely a deformation of the other. (Color figure online)

Definition 1. *A **good projection** for a 2-dimensional slope $E \subset \mathbb{R}^4$ is a valid projection $\pi : \mathbb{R}^4 \to \mathbb{R}^2$ such that for every $i \in \{0, 1, 2, 3\}$, $\pi(p_i)$ and $\pi(e_i)$ are collinear.*

Figure 7 illustrates the difference between two valid projections, one being good but not the other, on the slope of Cyrenaic tilings which we present in the next subsection. With the good projection, projected subperiods have the same directions as the sides of the tiles. This is why if segments on the tiles of a tiling T are directed by $\pi(p_i)$ then continuity of the lines in direction i is preserved in the i-shadow of T, for any $i \in \{0, 1, 2, 3\}$, as illustrated in Fig. 6. Indeed, consider a line L in direction i, then it is parallel to the sides of the tiles which disappear in the i-shadow of T. Now consider a tile t_0 which disappears in this shadow, containing a segment $[BC] \subset L$, and its neighbors t_{-1} and t_1 containing segments $[AB], [CD] \subset L$. Taking the i-shadow corresponds to translating remaining tiles in direction i, hence by such a translation the endpoint of an Ammann segment is mapped to a point on the same line (namely the image of the other endpoint of the same segment). As a result, the images of points B and C are on the same line, so that points A, B, C, D are still aligned.

3.2 Finding Good Projections

Given a slope E with subperiods $p_0, \dots, p_3$, we search for a good projection π as follows. We will define it by its 2×4 matrix A, which must satisfy $Ae_i = \lambda_i Ap_i$

for $i = 0, \ldots, 3$, where $\Lambda := (\lambda_i)_{i=0,\ldots,3}$ is to be determined. With M denoting the 4×4 matrix whose i-th column is $e_i - \lambda_i p_i$, this rewrites $AM = 0$. The 2 rows of A must thus be in the left kernel of M. Since the image of the facets in $E + [0,1]^4$ must cover $\mathbb{R}^2$, A must have rank 2. Hence the left kernel of M must be of dimension at least 2, that is, M must have rank at most 2. This is equivalent to saying that all the 3×3 minors of M must be zero. Each minor yields a polynomial equation in the λ_i's. Any solution of the system formed by these equations yields a matrix M whose left kernel can be computed. If the kernel is not empty, then any basis of it yields a suitable matrix A.

Of course with 4 variables and 16 equations there is no guarantee that a solution exists, and oftentimes when a projection respects the collinear condition in Definition 1 it is not valid: some tiles are superimposed in what should be a tiling. Figure 4 shows for instance what happens in the case of golden octagonal tilings (introduced in [2]) when the obtained matrix A is used. To find a slope E with a good projection, we proceed as follows:

1. Randomly choose the three integer coordinates of each subperiod p_i;
2. Check that only finitely many slopes admit these subperiods;
3. Use the above procedure to find a good projection (if any);
4. Repeat until a good projection is found.

We easily found several examples using this method. In particular, the following caught our attention because it has very short subperiods. Here are the integer coordinates of these:

$$p_0 = (*, 0, 1, 1),$$
$$p_1 = (1, *, -1, 1),$$
$$p_2 = (1, -1, *, 0),$$
$$p_3 = (2, 1, -1, *),$$

where $*$ stands for the non-integer coordinate. We checked that there are only two ways to choose these non-integer coordinates so that the subperiods indeed define a plane, namely:

$$p_0 = (a, 0, 1, 1),$$
$$p_1 = (1, a - 1, -1, 1),$$
$$p_2 = (1, -1, a + 1, 0),$$
$$p_3 = (2, 1, -1, a),$$

with $a = \pm\sqrt{3}$. Proceeding as explained at the beginning of this subsection yields

$$M = \frac{1}{6} \begin{pmatrix} 3 & -a & -a & -2a \\ 0 & a+3 & a & -a \\ -a & a & -a+3 & a \\ -a & -a & 0 & 3 \end{pmatrix},$$

whose left kernel is generated, for example, by the rows of the matrix

$$A := \frac{1}{2} \begin{pmatrix} 2 & 0 & a+1 & a-1 \\ 0 & 2 & -a-1 & a+1 \end{pmatrix}$$

Only $a = \sqrt{3}$ defines a valid projection, so we choose this value. We denote by E_c the slope generated by the p_i's and call **Cyrenaic tilings** the $4 \to 2$ tilings with slope E_c. Figure 7 illustrates this.

Fig. 8. A Cyrenaic tiling with all the lines in the directions of the subperiods, through every vertex of the tiling. Directions are shown separately to ease visualization, and lines are dashed so that one can see the edges of the tiling.

3.3 Defining the Prototiles

We describe here the method we used to obtain the tileset $\mathcal{C}$ depicted in Fig. 2. Let E be a 2-dimensional irrational plane in $\mathbb{R}^4$ characterized by its subperiods and which admits a good projection π. Consider a tiling with slope E obtained using the good projection π. Draw through each vertex of this tiling four lines directed by each of the projected subperiods $\pi(p_i)$'s. Figure 8 shows what we obtain for a Cyrenaic tiling. These lines decorate the tiles of the tiling with segments that can take four different directions. All these decorated tiles, considered up to translation, define the wanted tileset. Note that the tileset does not depend on the initially considered tiling, because the $4 \to 2$ tilings with a given irrational slope share the same finite patterns (this known fact is e.g. proven by Prop. 1 in [3]). We can now prove:

Proposition 1. *The tileset obtained by the above method is always finite.*

Proof. We prove that the number of different intervals (distances) between two consecutive lines in a given direction is finite. This yields finitely many ways to decorate a tile by parallel segments, hence finitely many different tiles.

Consider a subperiod p_i and the set $\mathcal{D}_i$ of all lines in E directed by $\pi(p_i)$ and passing through the vertices of the tiling, that is by all points $\pi(x)$ with $x \in \mathbb{Z}^4 \cap (E + [0,1]^4)$. Since the distance from a vertex to its neighbors is $\|\pi(e_k)\|$ for some k, the interval between two consecutive lines of $\mathcal{D}_i$ is at most $d_1 := \max_{j \neq i}\{\|\pi(e_j)\|\}$.

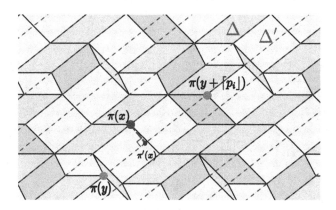

Fig. 9. Illustration of the proof of Proposition 1. $\lceil p_i \rfloor$ stands for $\lfloor p_i \rfloor$ or $\lceil p_i \rceil$.

Let $\Delta \in \mathcal{D}_i$, $x \in \mathbb{R}^4$ such that $\pi(x) \in \Delta$, and $\Delta' \in \mathcal{D}_i$ which is closest to Δ (Fig. 9). Then the distance from $\pi(x)$ to its orthogonal projection $\pi'(x)$ on Δ' is at most d_1. Besides, the distance between two vertices lying on Δ' is at most $d_2 := \max(\|\pi(\lfloor p_i \rfloor)\|, \|\pi(\lceil p_i \rceil)\|)$. Indeed, if $y \in \mathbb{Z}^4 \cap (E + [0,1]^4)$ then $y + p_i \in E + [0,1]^4$ and has three integer coordinates so that it lies on an edge of $\mathbb{Z}^4$ (seen as a grid in $\mathbb{R}^4$), between $y + \lfloor p_i \rfloor$ and $y + \lceil p_i \rceil$; now at least one of these two points is in $\mathbb{Z}^4 \cap (E + [0,1]^4)$, therefore its projection is also a vertex of the tiling, which lies on Δ' (since $\pi(p_i)$, $\pi(\lfloor p_i \rfloor)$ and $\pi(\lceil p_i \rceil)$ are collinear). Hence the distance between $\pi'(x)$ and the closest vertex $\pi(y)$ of the tiling which lies on Δ' is at most $d_2/2$. As a result, $\text{dist}(\pi(x), \pi(y)) \leq d := \sqrt{d_1^2 + d_2^2/4}$, i.e. at least one vertex on Δ' is in the ball $B(\pi(x), d)$. Consequently, measuring the intervals around a line Δ in the d-maps of the tiling is enough to list all possible intervals between two consecutive lines in the whole tiling. Since the d-atlas is finite, so is the number of intervals. □

Although the previous proof does not give an explicit bound on the number of tiles, it does give a constructive procedure to obtain these tiles. It is indeed sufficient to compute the constant d (which depends on the subperiods and the projection), then to enumerate the d-maps (for example by enumerating all patterns of size d and keeping only those which can be lifted in a tube $E + [0, 1]^4$ – in practice we used a more efficient algorithm based on the notion of region [4] which we do not detail here – and, for each d-map, to draw the lines and enumerate the new decorated tiles obtained. In the case of Cyrenaic tilings, it is sufficient to enumerate the tiles which appear in the 5-atlas in terms of graph distance[2]. We obtain 2 or 3 intervals in each direction, and the set $\mathcal{C}$ of 36 decorated prototiles in Fig. 2.

4   Tiling with the Tileset $\mathcal{C}$

By construction, the tileset $\mathcal{C}$ can be used to form all the Cyrenaic tilings (with the decorations by lines). However, nothing yet ensures that these tiles cannot be used to tile in other ways, and obtain for instance tilings which would be periodic or not planar. We shall here prove that this actually cannot happen.

Say we have a set S of tiles decorated with Ammann segments obtained from a given slope $E \subset \mathbb{R}^4$ characterized by subperiods $(p_i)_{i \in \{0,1,2,3\}}$ with a good projection π, and we want to show that any tiling with those tiles is planar with slope E. Let $\mathcal{T}$ be the set of all tilings that can be made with (only) tiles of S. By construction (assembly rules for the tiles in S), four sets of lines appear on any $T \in \mathcal{T}$ and the lines of each set are parallel to a projected subperiod $\pi(p_i)$ and to $\pi(e_i)$ for the same i. We can therefore talk about the i-shadow of T as the tiling obtained when reducing to zero the length of sides of tiles which are parallel to $\pi(p_i)$. Then as shown in Subsect. 3.1, for any $i \in \{0, 1, 2, 3\}$, continuity of the lines in direction i is preserved in the i-shadow of T.

Note that this is true for any set of tiles obtained with the method described above. We can then use the lines to show that a shadow is periodic and determine its prime period: starting from a vertex of the shadow, we follow the line in the chosen direction until we hit another vertex, for each valid configuration of the tiles. If the vector from the first vertex to the next is always the same, then it is a prime period of the shadow.

Proposition 2. *Every tiling composed with tiles of $\mathcal{C}$ has the same subperiods as Cyrenaic tilings.*

Proof. For the set $\mathcal{C}$, we observe that each i-shadow is periodic with period $q_i := \pi_i(p_i)$ where p_i is the i-subperiod of Cyrenaic tilings. This is shown in Fig. 10. In each shadow there are three original (non-decorated) tiles, each of

---

[2] To get the set $\mathcal{C}$ we used the 6-atlas as a precaution.

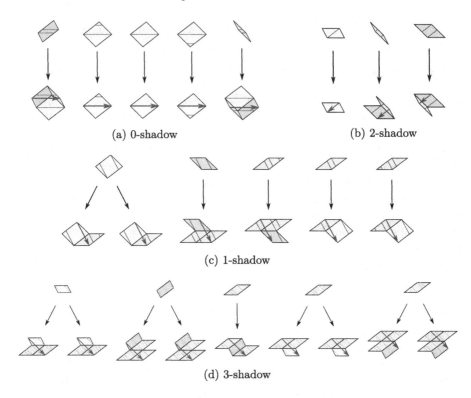

Fig. 10. Periods of the 4 shadows of tilings that can be realized with the set $\mathcal{C}$: starting at any vertex and following a line in direction i, depending on the first traversed tile, there are at most two possibilities until reaching another vertex, and the vector between both vertices is always the same.

which can appear in different versions when taking the decorations into account. For each i-shadow here we only look at the decorations in direction i, where we have the continuity of the lines (other decorations are irrelevant). All possible tiles are given on the top row, and following the arrows from each tile one can see all different possibilities[3] to place other tiles in order to continue the line directing the red vector. For each shadow, the vector is the same for all possible configurations, which means that the shadow is periodic, and we find exactly the subperiods of Cyrenaic tilings.   □

The main result in [2] thus yields the following:

Corollary 1. *Every tiling composed of tiles of $\mathcal{C}$ is planar with slope E_c.*

There is no guarantee that their thickness is always 1. Yet since the slope E_c is totally irrational, Theorem 1 follows.

---

[3] Remember that a line passes through every vertex, in each direction.

Acknowledgements. We would like to thank Alexandre Blondin Massé and the five reviewers for their careful proofreading of this article.

References

1. Bédaride, N., Fernique, T.: The Ammann-Beenker tilings revisited. In: Schmid, S., Withers, R.L., Lifshitz, R. (eds.) Aperiodic Crystals, pp. 59–65. Springer, Dordrecht (2013). https://doi.org/10.1007/978-94-007-6431-6_8

2. Bédaride, N., Fernique, T.: When periodicities enforce aperiodicity. Commun. Math. Phys. **335**(3), 1099–1120 (2015)

3. Bédaride, N., Fernique, T.: Weak local rules for planar octagonal tilings. Israel J. Math. **222**(1), 63–89 (2017)

4. Bédaride, N., Fernique, Th.: Canonical projection tilings defined by patterns. Geometriae Dedicata **208**(1), 157–175 (2020)

5. Beenker, F.P.M.: Algebraic theory of non periodic tilings of the plane by two simple building blocks: a square and a rhombus. Technical report. TH Report 82-WSK-04, Technische Hogeschool Eindhoven (1982)

6. Berger, R.: The Undecidability of the Domino Problem. American Mathematical Society (1966)

7. de Bruijn, N.G.: Algebraic theory of Penrose's non-periodic tilings of the plane. Math. Proc. **A84**, 39–66 (1981). Reprinted in [19]

8. Burkov, S.E.: Absence of weak local rules for the planar quasicrystalline tiling with the 8-fold rotational symmetry. Commun. Math. Phys. **119**(4), 667–675 (1988)

9. Gardner, M.: Mathematical games. Sci. Am. **236**(1), 110–121 (1977)

10. Grünbaum, B., Shephard, G.C.: Tilings and Patterns. W. H. Freeman and Company, New York (1987)

11. Harriss, E.O.: On canonical substitution tilings. Ph.D. thesis, University of London (2004)

12. Levitov, L.: Local rules for quasicrystals. Commun. Math. Phys. **119**, 627–666 (1988)

13. Penrose, R.: The Rôle of aesthetics in pure and applied mathematical research. Bull. Inst. Math. Appl. 266–271 (1974). Reprinted in [19]

14. Penrose, R.: Pentaplexity. Math. Intelligencer **2**(1), 32–37 (1978)

15. Porrier, C., Blondin Massé, A.: The Leaf function of graphs associated with Penrose tilings. Int. J. Graph Comput. **1**(1), 1–24 (2020)

16. Senechal, M.: Quasicrystals and Geometry. Cambridge University Press, Cambridge (1995)

17. Senechal, M.: The Mysterious Mr. Ammann. Math. Intelligencer **26**, 10–21 (2004)

18. Socolar, J.E.S.: Simple octagonal and dodecagonal quasicrystals. Phys. Rev. B **39**, 10519–10551 (1989)

19. Steinhardt, P.J., Ostlund, S.: The Physics of Quasicrystals. World Scientific (1987). Collection of reprints. ISBN 9971-50-226-7

Complexity Theory

List Homomorphism: Beyond the Known Boundaries

Sriram Bhyravarapu[1]([✉]), Satyabrata Jana[1], Fahad Panolan[3], Saket Saurabh[1,2], and Shaily Verma[1]

[1] The Institute of Mathematical Sciences, HBNI, Chennai, India
{sriramb,saket,shailyverma}@imsc.res.in, satyamtma@gmail.com
[2] University of Bergen, Bergen, Norway
[3] Indian Institute of Technology Hyderabad, Hyderabad, India
fahad@cse.iith.ac.in

Abstract. Given two graphs G and H, and a list $L(u) \subseteq V(H)$ associated with each $u \in V(G)$, a list homomorphism from G to H is a mapping $f : V(G) \to V(H)$ such that (i) for all $u \in V(G)$, $f(u) \in L(u)$, and (ii) for all $u, v \in V(G)$, if $uv \in E(G)$ then $f(u)f(v) \in E(H)$. The LIST HOMOMORPHISM problem asks whether there exists a list homomorphism from G to H. Enright, Stewart and Tardos [SIAM J. Discret. Math., 2014] showed that the LIST HOMOMORPHISM problem can be solved in $O(n^{k^2-3k+4})$ time on graphs where every connected induced subgraph of G admits "a multichain ordering" (see the introduction for the definition of multichain ordering of a graph), that includes permutation graphs, biconvex graphs, and interval graphs, where $n = |V(G)|$ and $k = |V(H)|$. We prove that LIST HOMOMORPHISM parameterized by k even when G is a bipartite permutation graph is W[1]-hard. In fact, our reduction implies that it is not solvable in time $n^{o(k)}$, unless the Exponential Time Hypothesis (ETH) fails. We complement this result with a matching upper bound and another positive result.

1. There is a $O(n^{8k+3})$ time algorithm for LIST HOMOMORPHISM on bipartite graphs that admit a multichain ordering that includes the class of bipartite permutation graphs and biconvex graphs.
2. For bipartite graph G that admits a multichain ordering, LIST HOMOMORPHISM is fixed parameter tractable when parameterized by k and the number of layers in the multichain ordering of G.

In addition, we study a variant of LIST HOMOMORPHISM called LIST LOCALLY SURJECTIVE HOMOMORPHISM. We prove that LIST LOCALLY SURJECTIVE HOMOMORPHISM parameterized by the number of vertices in H is W[1]-hard, even when G is a chordal graph and H is a split graph.

Keywords: List homomorphism · FPT · W[1]-hardness · Bipartite permutation graphs · Chordal graphs

© Springer Nature Switzerland AG 2022
A. Castañeda and F. Rodríguez-Henríquez (Eds.): LATIN 2022, LNCS 13568, pp. 593–609, 2022.
https://doi.org/10.1007/978-3-031-20624-5_36

1   Introduction

Given a graph G, a *proper coloring* is an assignment of colors to the vertices of G such that adjacent vertices are assigned different colors. Given a graph G and an integer k, the k-COLORING problem asks if there exists a proper coloring of G using k colors. The k-COLORING problem is known to be NP-complete even when $k = 3$ [17]. It is a very well-studied problem due to its practical applications. Many variants of coloring have been studied. In 1970's Vizing [27] and Erdős et al. [11] independently, introduced LIST k-COLORING which is a generalization of k-COLORING. Given a graph G and a list of admissible colors $L(v) \subseteq [k]$ for each vertex v in $V(G)$, the LIST k-COLORING problem asks whether there exists a proper coloring of G where each vertex is assigned a color from its list. Here, $[k] = \{1, 2, \ldots, k\}$. LIST k-COLORING has found practical applications in wireless networks, for example in frequency assignment problem [18,28].

Given two graphs G and H, a *graph homomorphism* from G to H is a mapping $f : V(G) \rightarrow V(H)$ such that if $uv \in E(G)$, then $f(u)f(v) \in E(H)$. Given two graphs G and H, and a list $L(v) \subseteq V(H)$ for each $v \in V(G)$, a *list homomorphism* from G to H is a graph homomorphism f from G to H such that $f(v) \in L(v)$ for each vertex v in $V(G)$. Given an instance (G, H, L), the LIST HOMOMORPHISM problem (LHOM for short) asks whether there exists a list homomorphism from G to H. Observe that LIST k-COLORING is a special case of LIST HOMOMORPHISM where H is a simple complete graph on k vertices.

LIST k-COLORING is NP-complete for $k \geq 3$ as it is an extension of k-COLORING problem. The problem remains NP-complete even for planar bipartite graphs [22]. On the positive side, for a fixed k, the problem is known to be polynomial time solvable on co-graphs [20], P_5-free graphs [19] and partial t-trees [20]. Considering the LIST HOMOMORPHISM problem, given a fixed integer $k = |V(H)|$, polynomial time algorithms are available for graphs of bounded tree-width [8], interval graphs, permutation graphs [10] and convex bipartite graphs [7]. Recently LIST HOMOMORPHISM on graphs with bounded tree-width has been studied in [23]. The list homomorphism has also been studied as *list H-coloring* in the literature and is a well studied problem [4,5,9,24]. Feder et al. [12–14] gave classifications of the complexity of LHOM based on the restrictions on graph H. Recently, LHOM has been studied for signed graphs [1,2,21].

Enright, Stewart and Tardos [10] showed that the LIST HOMOMORPHISM problem can be solved in $O(n^{k^2-3k+4})$ time on bipartite permutation graphs, interval graphs and biconvex graphs, where $n = |V(G)|$ and $k = |V(H)|$. It is natural to ask whether the running time can be improved or can we obtain a FPT algorithm when parameterized by k. Towards that we prove the following results.

Theorem 1. LHOM *can be solved in time* $O(n^{4k+3})$ *on bipartite permutation graphs.*

Theorem 2. LHOM *can be solved in* $O(n^{8k+3})$ *time on biconvex graphs.*

Theorem 3. LIST k-COLORING *parameterized by k is W[1]-hard on bipartite permutation graphs. Furthermore, there is no $f(k)n^{o(k)}$-time algorithm for* LIST k-COLORING, *for any computable function f unless* ETH *fails.*

Since LIST k-COLORING is a particular case of LHOM, similar hardness results hold for LHOM. However, we design fixed-parameter tractable (FPT) algorithms when parameterized by $|V(H)|$ and the diameter of the input graph G, where diameter of a graph is the maximum distance between any pair of vertices.

Theorem 4. LHOM *is FPT on bipartite permutation graphs and biconvex graphs, when parameterized by $|V(H)|$ and the diameter of the input graph G.*

We also a study a variant of LHOM called LIST LOCALLY SURJECTIVE HOMOMORPHISM. Given two graphs G and H, and a list $L(v) \subseteq V(H)$ for each $v \in V(G)$, a *list locally surjective homomorphism* from G to H is a list homomorphism $f : V(G) \to V(H)$ that is surjective in the neighborhood of each vertex in G. In other words, if $f(v) = v'$, then for every vertex $u' \in N_H(v')$, there is a vertex $u \in N_G(v)$, such that $f(u) = u'$. That is, for each connected component C of H if one vertex in C is "used" by the homomorphism, then all the vertices are used. Given as an input (G, H, L), the LIST LOCALLY SURJECTIVE HOMOMORPHISM problem (LLSHOM for short) asks whether there exists a list locally surjective homomorphism from G to H. We prove the following result about LLSHOM.

Theorem 5 ($\star^1$)**.** *Given an instance (G, H, L) such that G is a chordal graph, and H is a split graph, it is W[1]-hard to decide whether there is a list locally surjective homomorphism from G to H, when parameterized by $|H|$.*

Other Related Works. In 1999, Feder et al. [15] studied LIST M-PARTITION problem. The input to the problem is a graph $G = (V, E)$ and a $m \times m$ matrix M with entries $M(i,j) \in \{0, 1, *\}$. The goal is to check whether there exists a partition of $V(G)$ into m parts (called M-partition) such that for distinct vertices x and y of G placed in parts i and j respectively, we have that (i) if $M(i,j) = 0$, then $xy \notin E(G)$, (ii) if $M(i,j) = 1$, then $xy \in E(G)$, and (iii) if $M(i,j) = *$, then xy may or may not be an edge of G. By considering H as a graph on m vertices and M as a matrix obtained from the adjacency matrix of H by replacing each 1 with $*$, each homomorphism corresponds to a M-partition of G. Thus LIST M-PARTITION generalizes LIST k-COLORING and LHOM.

Valadkhan [25,26] gave polynomial time algorithms for LIST M-PARTITION for various graph classes. They gave $O(m^2 n^{4m+2})$ time algorithms for interval and permutation graphs, $O(m^2 n^{8m+2})$ time algorithms for interval bigraphs, interval containment bigraphs, and circular-arc graphs, $O(m^2 n^{4mt+2})$ time algorithm for comparability graphs with bounded clique-covering number t. The algorithm on interval graphs is an improvement over the algorithm by Enright,

---

[1] Due to paucity of space the proofs of results marked with $\star$ are omitted here.

Stewart and Tardos [10]. Feder et al. [16] showed that LIST M-PARTITION can be solved in $O(t^{t+1} \cdot n)$ time on graphs of treewidth at most t.

Our Methods. In this paper, we study LHOM on sub-classes of bipartite graphs by exploiting their structural properties. In particular, the sub-classes of bipartite graphs studied in this paper admit a "multichain" ordering (see Definition 3 in Preliminaries). Some of the graph classes that admit a multichain ordering include interval graphs, permutation graphs, bipartite permutation graphs, biconvex graphs, etc [10]. Towards proving Theorems 1 and 2, we prove that there is a list homomorphism such that if we know the labels of $O(k)$ vertices in a layer, in polynomial time we can extend that to a list homomorphism.

In Sect. 3, we present a $O(n^{8k+3})$ time algorithm for LHOM on bipartite graphs that admit a multichain ordering (Theorem 6). It is known that biconvex graphs and bipartite permutation graphs admit a multichain ordering. Hence Theorem 2 follows from Theorem 6. Since there are additional properties for bipartite permutation graphs, we provide an improved algorithm to bipartite permutation graphs that runs in $O(n^{4k+3})$ time (Theorem 1). These are improvements over the results from [10].

In Sect. 4, we show that LIST k-COLORING is W[1]-hard on bipartite permutation graphs (Theorem 3). We prove this result by giving a parameter preserving reduction from the MULTI-COLORED INDEPENDENT SET problem.

2   Preliminaries

Let $f : D \to R$ be a function from a set D to a set R. For a subset $A \subseteq D$, we use $f|_A : A \to R$ to denote the restriction of f to A. We will also use the words labelings and mappings for functions. A *partial labeling* on a set D is a function on a strict subset of D.

Let $G = (V, E)$ be a graph. We also use $V(G)$ and $E(G)$ to denote the vertex set and the edge set of the graph G, respectively. For a vertex $v \in V(G)$, the number of vertices adjacent with v is called the *degree* of v in G and it is denoted by $deg_G(v)$ (or simply $deg(v)$ if the graph G is clear from the context). The set of all the vertices adjacent with v is called as the neighborhood of v and it is denoted by $N_G(v)$ (or simply $N(v)$). The *distance* between two vertices $u, v \in V(G)$ is the length of a shortest path between u and v in G. Let X and Y be two disjoint subsets of $V(G)$, then $E(X, Y)$ denotes the set of edges with one endpoint in X and the other is in Y. A graph G is called a *split graph* if the vertices of G can be partitioned into two sets C and I such that $G[C]$ is a clique and $G[I]$ is an independent set. A graph is a *permutation graph* if there is some pair P_1, P_2 of permutations of the vertex set such that there is an edge between vertices x and y if and only if x precedes y in one of $\{P_1, P_2\}$, while y precedes x in the other. A graph is a *bipartite permutation graph* if it is both bipartite and a permutation graph.

Let (G, H, L) be an instance for LIST HOMOMORPHISM, where $V(H) = \{1, 2, \ldots, k\}$. First notice that if G is not connected, then (G, H, L) is a yes-instance if and only if for all connected components C of G, $(C, H, L|_{V(C)})$ is a yes-instance. Thus, throughout the paper, we assume that for an instance (G, H, L) of LIST HOMOMORPHISM, G is connected.

Definition 1 (Chain Graph [10]). *A bipartite graph* $G = (A \uplus B, E)$ *is a chain graph if and only if for any two vertices* $u, v \in A$, *either* $N(u) \subseteq N(v)$ *or* $N(v) \subseteq N(u)$. *It follows that, for any two vertices* $u, v \in B$, *either* $N(u) \subseteq N(v)$ *or* $N(v) \subseteq N(u)$.

Definition 2. *For a graph* G *and a vertex subset* U, *we say that an ordering* σ *of* U *is* increasing *in* G, *if for any* $x <_\sigma y$, $N_G(x) \subseteq N_G(y)$. *We say that an ordering* σ' *of* U *is* decreasing *in* G, *if for any* $x <_{\sigma'} y$, $N_G(y) \subseteq N_G(x)$.

For a chain graph $G = (A \uplus B, E)$, there is an ordering σ of A which is increasing in G and there is an ordering σ' of B which is decreasing in G. For a vertex $u \in A$, a vertex $v \in N(u)$ is called a *private neighbor* of u if for any vertex w such that $w <_\sigma u$, v is not a neighbor of w. In fact, for a chain graph $G = (A \uplus B, E)$, any ordering of A that is non-decreasing in its degrees increases in G. Also, any ordering of B that is non-increasing in its degrees decreases in G.

Definition 3 (Multichain ordering). *For a connected graph* G, *the* distance layers *of* G *from a vertex* v_0 *is a sequence* $L_0, L_1, \ldots, L_r$ *where* $L_0 = \{v_0\}$, L_i *is the set of vertices that are at distance* i *from* v_0 *for each* $i \in [r]$, *and* r *is the largest integer such that* $L_r \neq \emptyset$. *These layers form a* multichain ordering *of* G *if for every two consecutive layers* L_i *and* L_{i+1}, *the edges connecting these two layers form a chain graph. That is, the graph* $(L_i \cup L_{i+1}, E(L_i, L_{i+1}))$ *is a chain graph. We say that* G *admits a multichain ordering if there is a vertex* v_0 *such that the distance layers of* G *from* v_0 *forms a multichain ordering.*

It is known that all connected permutation graphs and connected interval graphs have multichain orderings [10]. Let G be a graph and let $L_0, L_1, \ldots, L_r$ be a multichain ordering of G. Then, for any $i \in [r]$, let G_i be the bipartite graph with vertex set $L_{i-1} \cup L_i$ and edge set $E(L_{i-1}, L_i)$. Then, we know that for each $i \in [r]$, G_i is a chain graph. Thus, for each $i \in [r-1]$, there are two orderings $\sigma_{i,1}$ and $\sigma_{i,2}$ of L_i such that $\sigma_{i,1}$ is decreasing in G_i and $\sigma_{i,2}$ is increasing in G_{i+1}. The following result implies that for connected bipartite permutation graphs there is a multichain ordering where for each layer L_i, $\sigma_{i,1}$ is same as $\sigma_{i,2}$.

Proposition 1 ([3]). *A connected graph* $G = (V, E)$ *is a bipartite permutation graph if and only if the vertex set* $V(G)$ *can be partitioned into independent sets* $V_0, V_1, \ldots, V_q$ *such that the following holds.*

1. *Any two vertices in non-consecutive sets are non-adjacent.*
2. *Any two consecutive sets* V_{i-1} *and* V_i, *induce a chain graph denoted by* G_i.
3. *For each* $j \in \{0, \ldots, q\}$, *there is an ordering* σ_j *of* V_j *with the following properties. For each* $i \in \{1, 2, \ldots, q-1\}$, σ_i *is decreasing in* G_i *and increasing in* G_{i+1}. *Moreover,* σ_0 *is increasing in* G_1 *and* σ_q *is decreasing in* G_q.
4. $|V_0| = 1$ *and* $V_0, V_1, \ldots, V_q$ *is the distance layers of* G *from the vertex in* V_0.

Observation 1. *Let* G *be a connected bipartite graph that admits a multichain ordering* $V_0, \ldots, V_q$. *Then for each* $i \in \{0, 1, \ldots, q\}$, V_i *is an independent set.*

3   XP Algorithms: Proofs of Theorems 1 and 2

In this section, we give an $O(n^{8k+3})$ time algorithm for LIST HOMOMORPHISM on bipartite graphs that admit a multichain ordering. We first discuss an algorithm for LHOM on bipartite permutation graphs that runs in $O(n^{4k+3})$ time (Theorem 1). We then extend this algorithm to bipartite graphs admitting a multichain ordering that includes biconvex graphs. Thereby, settling Theorem 2.

We first prove the following lemma, which is crucial to our algorithm.

Lemma 1. *Let (G, H, L) be an instance of LIST HOMOMORPHISM, where G is a connected bipartite permutation graph. Let $V_0, \ldots, V_q$ be a sequence of independent sets such that the properties mentioned in Proposition 1 hold. For each $i \in \{0, \ldots, q\}$, σ_i is the ordering of V_i and for each $j \in [q]$, G_j is the graph $G[V_{j-1} \cup V_j]$ mentioned in Proposition 1. If there exists a list homomorphism from G to H, then there exists a list homomorphism f from G to H such that for any $i \in \{0, 1, \ldots, q\}$, and any $w \in V_i$, at least one of the following is true.*

1. *w is the first vertex or the last vertex in σ_i that is assigned the label $f(w)$.*
2. *$f(w)$ is the least integer in $L(w)$ such that there exist $x, y \in V_i$ with $x <_{\sigma_i} w <_{\sigma_i} y$ and $f(x) = f(y) = f(w)$.*

Proof. Let f be a list homomorphism such that maximum number of vertices satisfy the stated properties (1) or (2). If all the vertices satisfy the stated properties, then f is our desired list homomorphism. Otherwise, let w be a vertex such that it does not satisfy (1) and (2). Let $w \in V_i$ for some $i \in \{0, 1, \ldots, q\}$. Since w does not satisfy (1), we know that there exist $v, x \in V_i$ such that $v <_{\sigma_i} w <_{\sigma_i} x$ and $f(v) = f(w) = f(x)$. Since w does not satisfy (2), there exists an integer $c \in L(w)$ and two vertices $x', y' \in V_i$ such that $c < f(w)$, $x' \leq_{\sigma_i} w \leq_{\sigma_i} y'$ and $f(x') = f(y') = c$. Without loss of generality, let c be the least integer with the above property. Now consider the following function $f' : V(G) \to V(H)$. For each $z \neq w$, $f'(z) = f(z)$ and $f'(w) = c$.

Now we claim that f' is a list homomorphism from G to H and the number of vertices in G that satisfies (1) or (2) with respect to f' is strictly more than the number of vertices in G that satisfies (1) or (2) with respect to f, which leads to a contradiction.

Claim. f' is a list homomorphism from G to H.

Proof. Since f is a list homomorphism and $c = f'(w) \in L(w)$, we have that for any vertex $u \in V(G)$, $f'(u) \in L(u)$. Recall that, $N(w) \cap V_i = \emptyset$ and all the neighbors of w are in $V_{i-1} \cup V_{i+1}$. Since $x' \leq_{\sigma_i} w \leq_{\sigma_i} y'$, we have $N(w) \subseteq N(x')$ in G_i and $N(w) \subseteq N(y')$ in G_{i+1}. Thus, any neighbor w' of w is adjacent to either x' or y'. This implies that $f(w')f'(w)$ is an edge in H. For any edge $zz' \in E(G)$ with $w \notin \{z, z'\}$, $f'(z)f'(z') = f(z)f(z')$ and hence $f'(z)f'(z')$ is an edge in H. Thus, we have proved that f' is a list homomorphism.   □

Claim. The number of vertices in G that satisfies (1) or (2) with respect to f' is strictly more than the number of vertices in G that satisfies (1) or (2) with respect to f.

Proof. Notice that w does not satisfy (1) and (2) with respect to f, but it satisfies (2) with respect to f'.

Now we want to prove that for other vertices if they were satisfying (1) or (2) in f, then they so do in f'. Let x be the first vertex and y be the last vertex in σ_i such that $f'(x) = f'(y) = f'(w)$. Since w satisfies (2) with respect to f', we have that $x <_{\sigma_i} w <_{\sigma_i} y$. Let x_1 be the first vertex and y_1 be the last vertex in σ_i such that $f(x_1) = f(y_1) = f(w)$. Since w does not satisfy (1) with respect to f, we have that $x_1 <_{\sigma_i} w <_{\sigma_i} y_1$.

Let u be a vertex in G such that $u \neq w$ and u satisfies (1) or (2) with respect to f. We prove that u satisfies (1) or (2) with respect to f' also. If $f'(u) \notin \{f(w), f'(w)\}$ or $u \notin V_i$, then clearly u satisfies (1) or (2) with respect to f'. So we assume that $u \in V_i$ and $f'(u) \in \{f(w), f'(w)\}$

Case 1: $f'(u) = f(w)$, and u satisfies (1) with respect to f. Then u is the first vertex or the last vertex that is assigned a label $f(u)$ by f. Since w is the only vertex such that $f'(w) \neq f(w)$ and $f'(u) = f(w)$, u is the first vertex or the last vertex that is assigned a label $f'(u) = f(u)$ by f'.

Case 2: $f'(u) = f(w)$, and u satisfies (2) with respect to f. Then, $f(u)$ (which is equal to $f(w)$ and $f'(u)$) is the least integer in $L(u)$ such that there exist $x_1, y_1 \in V_i$ with $x_1 <_{\sigma_i} u <_{\sigma_i} y_1$ and $f(x_1) = f(y_1) = f(u) = f(w)$. Thus, by the definition of x_1 and y_1, we have that $x_1 <_{\sigma_i} u <_{\sigma_i} y_1$ and $f(x_1) = f(y_1) = f(u)$. This implies that u and w appears between x_1 and y_1 in the ordering σ_i. We consider the case $x_1 <_{\sigma_i} u <_{\sigma_i} w <_{\sigma_i} y_1$, and we omit the case $x_1 <_{\sigma_i} w <_{\sigma_i} u <_{\sigma_i} y_1$ as the arguments are symmetric. If $f'(w) \notin L(u)$, then u satisfies (2) with respect to f'. Now, if $f'(w) \in L(u)$, then there will not be any vertex $z <_{\sigma_i} u$ such that $f(z) = f'(w)$. Otherwise, we get $f(z) = f'(w) = f(y)$, $f'(w) \in L(u)$, and $z <_{\sigma_i} u <_{\sigma_i} y$, and it contradicts the assumption that $f(u)$ is the least integer satisfying property (2) for u with respect to f. This implies that u satisfies (2) with respect to f'.

Case 3: $f'(u) = f'(w)$ and u satisfies (1) with respect to f. Suppose u is the first vertex in σ_i that is assigned a label $f(u)$ by f. We claim that $u <_{\sigma_i} w$. For the sake of contradiction, let $w <_{\sigma_i} u$. We know that $x <_{\sigma_i} w$ and $f(x) = f'(w) = f'(u)$. This contradicts the assumption that u is the first vertex in σ_i that is assigned a label $f(u)$ by f. Since $u <_{\sigma_i} w$, u is the first vertex in σ_i that is assigned a label $f'(u)$ (which is equal to $f(u)$) by f' and hence u satisfies (1) with respect to f'. The case when u is the last vertex in σ_i that is assigned a label $f(u)$ by f, is symmetric in arguments and hence is omitted.

Case 4: $f'(u) = f'(w)$ and u satisfies (2) with respect to f. Since u satisfies (2) with respect to f, and $f'(u) = f'(w)$, we have that $x <_{\sigma_i} u <_{\sigma_i} y$ because x is the first vertex and y is the last vertex in σ_i which are assigned the label $f(u) = f'(u)$ by f. This implies that u satisfies (2) with respect to f'.

Thus, all the vertices which satisfy (1) or (2) with respect to f also satisfy (1) or (2) with respect to f'. The vertex w does not satisfy (1) or (2) with respect to f, but satisfies (2) with respect to f'. This completes the proof of the claim. □

This completes the proof of the lemma.   □

Proof (Proof of Theorem 1). Let (G, H, L) be an instance of LHOM where $G = (V, E)$ is a bipartite permutation graph and $V(H) = \{1, 2, \ldots, k\}$. By Proposition 1, there exists a partition of V into $V_0, V_1, \ldots, V_q$ satisfying properties (1)-(4). Because of property (4), such a partition can be constructed in polynomial time.

We now discuss the overall idea of the algorithm. In each set V_i, $i \in \{0, 1, \ldots, q\}$, for each label $j \in [k]$, we guess whether the label j is assigned to 0, 1 or at least 2 vertices in V_i. For the latter case, when at least two vertices are assigned the label j, we guess two vertices with label j and extend the labeling to other vertices. Depending on the guess for the label j, we guess the first vertex and the last vertex (the first and the last vertices are the same when there is exactly one vertex assigned the label j) in σ_i that are assigned the label j, in a list homomorphism from G to H. Using the partial labeling obtained from each guess, we obtain a full labeling of V_i maintaining the property of list homomorphism using Lemma 1. That is, for each vertex that is not assigned a label, we choose a label satisfying property (2) of Lemma 1. Then we construct a directed graph G' using the labelings obtained at each V_i and solve the directed s-t path problem on G' to decide if a list homomorphism exists from G to H.

Now we explain the algorithm in detail. We process the vertices of G in the order $V_0, V_1, \ldots, V_q$. From (3) of Definition 1, there exists an ordering σ_i of V_i that is decreasing in G_i and increasing in G_{i+1}. At each V_i, $0 \le i \le q$, for each label $j \in [k]$, we guess the first and the last vertices in σ_i that are assigned the label j. That is, we guess a partial labeling $\widehat{c}$ of V_i such that at most $2k$ vertices are assigned labels. Then we extend $\widehat{c}$ to a full labeling $c : V_i \to \{1, 2, \ldots, k\}$. For each vertex u labeled under $\widehat{c}$, we set $c(u) = \widehat{c}(u)$. For each of the remaining vertices, we use Lemma 1 to assign a label. We say a labeling c of V_i is *feasible* if there exists a partial labeling $\widehat{c}$ of V_i that can be extended to c using Lemma 1. Let C_i denote the set of all feasible labelings of V_i. Hence $|C_i| \le n^{2k}$.

We now construct an auxiliary directed graph G' with $V(G') = \{s, t\} \cup C_0 \cup C_1 \cup \cdots \cup C_q$, where C_i contains a vertex corresponding to every feasible labeling of V_i, $0 \le i \le q$. We add edges between vertices of two consecutive sets C_i and C_{i+1}, for each $0 \le i \le q-1$, in the following manner. We add a directed edge from $c \in C_i$ to $c' \in C_{i+1}$ if the labeling $c \cup c'$ is a list homomorphism from $G[V_i \cup V_{i+1}]$ to H, where c and c' are feasible labelings of V_i and V_{i+1}, respectively. We add directed edges from s to all vertices in C_0. Similarly, we add directed edges from all vertices in C_q to t. If we find a $s - t$ path in G', then such a path indicates the existence of list homomorphism from G to H.

Next, we show that there exists a list homomorphism from G to H if and only if there is a directed path from s to t in G'. If there exists a directed path from s to t in G', say P, then the number of vertices in P is $q + 3$. Moreover, $|P \cap C_i| = 1$, for each $0 \le i \le q$. This is due to the fact that there are edges only between consecutive sets C_i and C_{i+1} and the directed edges are from vertices in C_i to C_{i+1}. In addition, the edge from $c \in C_i$ to $c' \in C_{i+1}$ indicates the existence of list homomorphism from $G[V_i \cup V_{i+1}]$ to H, where c and c' are feasible labelings of V_i and V_{i+1} respectively. Let c_i be the vertex at distance $i+1$ from s in P. The vertex c_i represents a feasible labeling of V_i. Thus the feasible labelings $c_1, \ldots, c_q$ assigned to $V_0, V_1, \ldots, V_q$, respectively, together obtain a list homomorphism from G to H.

For the forward direction, let f be a list homomorphism from G to H such that f satisfies the properties mentioned in Lemma 1. Then, there exists a vertex $c_i \in V(G')$ that captures the labeling of V_i with respect to the labeling f, for each $0 \leq i \leq q$. Since f is a list homomorphism, there exists an edge from c_i to c_{i+1}, for all $0 \leq i \leq q - 1$. This leads to a directed path from s to t.

Next, we do the runtime analysis. Because of the property (4) in Proposition 1, the partition $V_0, \ldots, V_q$ can be computed in $O(n^3)$ time. In our process, for each V_i, $i \in \{0, \ldots, q\}$, we guess whether a label is assigned to none of the vertices, one vertex, or more than one vertex in V_i. Since the number of labels is k, the above guessing takes $O(3^k)$ time. Then, we guess at most $2k$ vertices from each V_i that are "critical" (the first and the last vertices assigned a label in V_i) for the labeling resulting in $O(3^k n^{2k})$ partial labelings. We extend a partial labeling to a full labeling by assigning a label to an unlabelled vertex using (2) of Lemma 1, which takes $O(n^2)$ time. The number of edges between a pair of layers in G' is $O(3^{2k} n^{4k})$. Since there are q pairs of layers in G', the total number of edges is $O(q 3^{2k} n^{4k})$. Since $q \leq n$, and checking if an edge corresponds to a valid list homomorphism takes $O(n^2)$ time, we need $O(3^{2k} n^{4k+3})$ time to complete the construction of G'. The final step of the algorithm is to find a directed $s - t$ path in G' which can be done in $O(9^k n^{4k+3})$ time. Thus, the overall running time is $O(9^k n^{4k+3})$. □

The above algorithm can be extended when the input graph is a bipartite graph that admits a multichain ordering property. Theorem 2 is a corollary of Theorem 6.

Theorem 6 ($\star$). LIST HOMOMORPHISM *can be solved in* $O(n^{8k+3})$ *time on bipartite graphs that admit a multichain ordering property.*

4   Hardness: Proof of Theorem 3

In this section, we prove Theorem 3. To prove that, we use a specific type of chain graph. Let $G = (A, B, E)$ be a bipartite graph with $|A| = r$ and $|B| = s$ such that $(x_1, x_2, \ldots, x_r)$ and $(y_1, y_2, \ldots, y_s)$ be the chain orderings of A and B, respectively. That is $(x_1, x_2, \ldots, x_r)$ is increasing in G and $(y_1, y_2, \ldots, y_s)$ is decreasing in G. We call G, an *incremental* chain graph if $E(G) = \{x_i y_j : 1 \leq i \leq r, \ 1 \leq j \leq i, \ j \leq s\}$.

Towards proving the hardness, we give a polynomial-time parameter preserving reduction from the MULTICOLORED INDEPENDENT SET (McIS for short) problem to LIST k-COLORING. In McIS, the input is a graph G, a positive integer k, and a partition $(X_1, \ldots, X_k)$ of $V(G)$. The goal is to check if there exists a k-sized independent set $S \subseteq V(G)$ such that for all $i \in [k]$, $|S \cap X_i| = 1$. The problem is known to be W[1]-hard [6]. In fact it is known that McIS can not be solved in time $n^{o(k)}$ unless the Exponential Time Hypothesis fails. Let $(G, k, (X_1, \ldots, X_k))$ be an instance of McIS such that m be the number of edges in G and X_i be an independent set with cardinality n, for each $i \in [k]$ (without loss of generality we can assume this).

For our reduction, we require that m is a multiple of 2 and 3. Suppose m is not a multiple of 6. In this case, we can modify our instance (G, k) to $(G', k+1)$ such that the number of edges in G' is a multiple of 6. Let $m = b \mod 6$ where $b \in \{1, 2, \ldots 5\}$. We add one new set of vertices X_{k+1} of size n, and add b number of edges between some vertex of X_{k+1} to b vertices in X_k. Additionally, we update the parameter k to $k+1$. Observe that G' has a multicolored independent set of size $k+1$ if and only if G has a multicolored independent set of size k. Thus without loss of generality, we can assume that m is a multiple of 6 for the given instance $(G, k, (X_1, X_2, \ldots, X_k))$.

First of all, we fix an arbitrary ordering of the vertices in X_i, for each $i \in [k]$. Let $\sigma(V(G))$ be a vertex ordering of G such that the vertices of X_1 appear in the above mentioned fixed order in $\sigma(V(G))$ and then the vertices of X_2 and so on. Let $E(G) = \{e_1, e_2, \ldots, e_m\}$ be the set of edges in G. From this we construct an instance (G', k') of LIST k-COLORING.

Construction of a Block. First we explain a construction of a *block* and later we explain how to construct G' from the blocks.

1. Let $X_i' = X_i \cup \{x_i\}$ where x_i is a new element. We take one copy of each part X_i' of G and we call the union of these copies, a *layer*. We mention that later we add two or three more vertices to each layer. We take $2m$ copies of a layer, say $D_j = (X_{j1} \cup X_{j2} \cup \cdots \cup X_{jk})$, for $j \in [2m]$. We call this union of $2m$ layers, a *block*. For each D_j, we define an order σ_j on the vertices in D_j as follows. In the order $\sigma(V(G))$ insert x_i just before the first vertex of X_i for all $i \in [k]$.

2. For each edge e_ℓ, we add three new vertices in $D_{2\ell-1}$ and two new vertices in $D_{2\ell}$. Towards explaining this, let us fix an edge $e_\ell = uv \in E(G)$ such that $u \in X_i$ and $v \in X_j$ for some $i, j \in [k]$, where $i < j$. Let $u^{2\ell-1} \in X_{(2\ell-1)i}$ and $v^{2\ell-1} \in X_{(2\ell-1)j}$ be the copies of the vertices u and v in layer $D_{2\ell-1}$, respectively. We add two new vertices $\alpha(e_\ell)$ and $\beta(e_\ell)$ just before the vertices $u^{2\ell-1}$ and $v^{2\ell-1}$ in $\sigma_{2\ell-1}$, respectively. Similarly, we add two new vertices $\alpha'(e_\ell)$ and $\beta'(e_\ell)$ just before the vertices $u^{2\ell}$ and $v^{2\ell}$ in $\sigma_{2\ell}$, respectively. Also, we add one new vertex $\gamma(e_\ell)$ at the end of the ordering $\sigma_{2\ell-1}$ in layer $D_{2\ell-1}$ and this vertex we call an *edge vertex*. Let $Q = \{\alpha(e_\ell), \alpha'(e_\ell), \beta(e_\ell), \beta'(e_\ell), \gamma(e_\ell) : \ell \in [m]\}$. So, notice that now a layer D_j is a union of $X_{j1} \cup X_{j2} \cup \cdots \cup X_{jk}$ and two or three vertices from Q. This completes the description of the vertex set of a block B.

   Moreover, we use $\sigma_{2\ell-1}$ and $\sigma_{2\ell}$ to represent the order of vertices in $D_{2\ell-1}$ and $D_{2\ell}$, respectively, (including the new vertices) that is naturally derived from the old $\sigma_{2\ell-1}$ and $\sigma_{2\ell}$ as per the explanation of the new vertices added.

3. Next we explain the edges of G'. For each $i \in \{1, 3, \ldots, 2m-1\}$, we add edges between D_i and D_{i+1} such that the graph induced on $D_i \cup D_{i+1}$ is an incremental chain graph. Observe that except the edge vertex, every vertex in

D_i has a private neighbor in D_{i+1} now and the edge vertex in D_i is adjacent with every vertex in D_{i+1}. In the next step, we will add more edges.

4. For each $\ell \in [m]$, we add the following edges between $D_{2\ell-1}$ and $D_{2\ell}$. Let $e_\ell = uv$. We make $\alpha(e_\ell)$ adjacent to $u^{2\ell}$ and $\beta(e_\ell)$ adjacent to $v^{2\ell}$. Additionally, we add an edge between the vertex $\alpha'(e_\ell)$ and the vertex that appears just before the vertex $\alpha(e_\ell)$ in the ordering $\sigma_{2\ell-1}$ of layer $D_{2\ell-1}$. Similarly, we add an edge between the vertex $\beta'(e_\ell)$ and the vertex that appears just before the vertex $\beta(e_\ell)$ in the ordering $\sigma_{2\ell-1}$ of layer $D_{2\ell-1}$. See Fig. 1 for an illustration.

5. Next we explain the edges between D_i and D_{i+1}, where $i \in \{2, 4, \ldots, 2m\}$. Recall that $Q = \{\alpha(e_\ell), \alpha'(e_\ell), \beta(e_\ell), \beta'(e_\ell), \gamma(e_\ell) : \ell \in [m]\}$. Let us fix an $i \in \{2, 4, \ldots, 2m\}$. We add edges between D_i and D_{i+1} such that the graph induced on $(D_i \cup D_{i+1}) \setminus Q$ with bipartition $D_i \setminus Q$ and $D_{i+1} \setminus Q$ forms an incremental chain graph with respect to the orderings σ_i and σ_{i+1} restricted on $D_i \setminus Q$ and $D_{i+1} \setminus Q$, respectively. Let ℓ and ℓ' be the integers such that $2\ell = i$ and $2\ell' - 1 = i + 1$. Notice that $\ell' = \ell + 1$. Also, notice that $\alpha'(e_\ell), \beta'(e_\ell) \in D_i$ and $\alpha(e_{\ell'}), \beta(e_{\ell'}) \in D_{i+1}$. Add the minimum number of edges on $\alpha'(e_\ell), \beta'(e_\ell), \alpha(e_{\ell'})$ and $\beta(e_{\ell'})$, between D_i and D_{i+1} such that it forms a chain graph with respect to orders σ_i and σ_{i+1}. For example, let w be the endpoint of the edge e_ℓ in graph G with minimum index in the ordering σ of graph G'. Let w^i and w^{i+1} be the copies of w in D_i and D_{i+1}, respectively. Note that w^i be the vertex that appears just before $\alpha'(e_\ell)$. Then, add edges between $\alpha'(e_\ell)$ and all the vertices that appear before w^{i+1} in σ_{i+1}. We also add an edge between $\alpha'(e_\ell)$ and w^{i+1}. Similarly, let z be the vertex that appears just after the minimum index endpoint of the edge $e_{\ell'}$ in the ordering σ of graph G. Let z^{i+1} be the copy of vertex z in layer D_{i+1} and appears just after $\alpha(e_{\ell'})$. Then add edges between $\alpha(e_{\ell'})$ and vertices that appear after z^i in σ_i. We also add an edge between $\alpha(e_{\ell'})$ and z^i. The cases of $\beta'(e_\ell)$ and $\beta(e_{\ell'})$ are symmetric. This completes the description of the edge set of the block B.

6. We denote the first vertex and the last vertex of any set X_{ij} in layer D_i as $first(X_{ij})$ and $last(X_{ij})$, respectively. Notice that $first(X_{ij})$ is the copy of x_j in $X_{i,j}$ and it will not corresponds to a vertex in $V(G)$. Next, we define a list function $L: V(B) \to [3k] \cup \{c_1, c_2, c_3, c_4, \hat{c}_1, \hat{c}_2, \hat{c}_3, \hat{c}_4\}$. For each $i \in [2m]$, $j \in [k]$, and $v \in X_{ij}$

$$\text{If } i = 1 \mod 3 \text{ , then } L(v) = \begin{cases} \{3j-2\} & \text{if } v = first(X_{ij}) \\ \{3j-1\} & \text{if } v = last(X_{ij}) \\ \{3j-2, 3j-1\} & \text{if } first(X_{ij}) < v < last(X_{ij}) \end{cases}$$

$$\text{If } i = 2 \mod 3 \text{ , then } L(v) = \begin{cases} \{3j\} & \text{if } v = first(X_{ij}) \\ \{3j-2\} & \text{if } v = last(X_{ij}) \\ \{3j-2, 3j\} & \text{if } first(X_{ij}) < v < last(X_{ij}) \end{cases}$$

$$\text{If } i = 0 \mod 3 \text{ , then } L(v) = \begin{cases} \{3j-1\} & \text{if } v = first(X_{ij}) \\ \{3j\} & \text{if } v = last(X_{ij}) \\ \{3j-1, 3j\} & \text{if } first(X_{ij}) < v < last(X_{ij}) \end{cases}$$

7. For each $\ell \in [m]$, we explain the lists of $\alpha(e_\ell), \alpha'(e_\ell), \beta(e_\ell), \beta'(e_\ell)$, and $\gamma(e_\ell)$ as follows. Towards that let us fix $\ell \in [m]$. Let $e_\ell = uv$, where $u \in X_i$ and $v \in X_j$ for some $1 \le i < j \le k$. If ℓ is an odd number, then

$$L(\alpha(e_\ell)) = L(first(X_{(2\ell-1)i})) \cup \{c_1\}$$
$$L(\alpha'(e_\ell)) = L(first(X_{(2\ell-1)i})) \cup \{c_1, c_3\}$$
$$L(\beta(e_\ell)) = L(first(X_{(2\ell-1)j})) \cup \{c_2\}$$
$$L(\beta'(e_\ell)) = L(first(X_{(2\ell-1)j})) \cup \{c_2, c_4\}$$
$$L(\gamma(e_\ell)) = \{c_3, c_4\}$$

If ℓ is a even number, then replace each c_r with $\widehat{c}_r$ in the above equations, where $r \in \{1, 2, 3, 4\}$.

Construction of G'. We take $(nk + 1)$ copies of a block with the same list function, say $B_1, B_2, \ldots, B_{nk+1}$. For any two consecutive blocks B_i and B_{i+1}, where $i \in [nk]$, we add edges between the last layer of B_i and the first layer of B_{i+1} according to item (5) in the construction of a block. Observe that the color lists of the vertices of the last layer of B_i and the first layer of B_{i+1} are *compatible* as the number of layers in each block is a multiple of m, which is a multiple of 3 and by the definition of list function. This completes the construction of graph G' with a list function $L \colon V(G') \to \mathcal{C}$, where $\mathcal{C} = [3k] \cup \{c_1, c_2, c_3, c_4, \widehat{c}_1, \widehat{c}_2, \widehat{c}_3, \widehat{c}_4\}$.

It is easy to verify that the obtained graph G' is a bipartite permutation graph as the layers of each block partition the vertex set of G' and the edges connecting any two consecutive layers induce a chain graph. Moreover, every vertex v in G' has a list $L(v) \subseteq \mathcal{C}$.

Note that for every layer D_i, the first vertex and the last vertex in each part X_{ij} has exactly one color in its list, say c and c', respectively such that $c \neq c'$, where $i \in [2m]$, $j \in [k]$ and $c, c' \in \mathcal{C}$. All the other vertices in the same part X_{ij} contain c and c' in their lists. It follows that in any list coloring, the first vertex in X_{ij} gets the color c and the last vertex in X_{ij} gets the color c' and all the other vertices get the color either c or c'. Note that there exists a vertex $w \in X_{ij}$ such that w is the first vertex in the ordering of X_{ij} which gets the color c', we call such a vertex *switch*. Moreover, a switch corresponds to a vertex in $V(G)$. Observe that each X_{ij} ($i \in [2m]$, $j \in [k]$) contains at least one switch because of the list assignment of the vertices of X_{ij}. Additionally, each X_{ij} contains at most one switch by the definition of a switch. Therefore, in each layer D_i there are k switches, exactly one in each part X_{ij}, where $i \in [2m]$ and $j \in [k]$. We call a block B, a *consistent block* if for any pair of layers D_i and $D_{i'}$ and for any $j \in [k]$, the switches in X_{ij} and $X_{i'j}$ corresponds to the same vertex in G (i.e., these switches are copies of "a vertex" in $V(G)$).

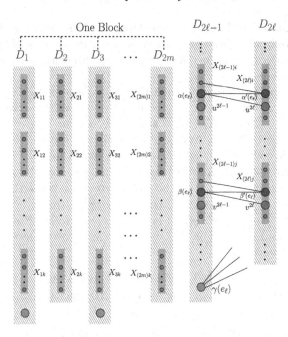

Fig. 1. A block on $2m$ layers in G' is illustrated on the left side of the figure, where each D_i represents a layer. The vertices and edges (apart from the edges of induced incremental chain graph) between two consecutive layers $D_{2\ell-1}$ and $D_{2\ell}$ are illustrated on the right side of the figure (the last two layers), where the green and red colored vertices are the newly added vertices corresponding to the edge $e_\ell = uv$ in G. (Color figure online)

Correctness Proof. Next, we show that G has a multicolored independent set of size k if and only if G' has a list k'-coloring, where $k' = 3k + 8$.

Lemma 2. *If G has a multicolored independent set of size k, then G' has a list k'-coloring, where $k' = 3k + 8$.*

Proof. Let $I = \{y_1, y_2, \ldots, y_k\}$ be an independent set in G such that $y_i \in X_i$. Our goal is to construct a list coloring $\phi : V(G') \to \mathcal{C}$. First, in each layer, we color the first and the last vertex of each set X_{ij} with the (only) color present in their lists. Next, we color all other vertices (except the vertices from Q) in every layer D_i ($i \in [2m]$) of each block such that the block gets consistent; that is, in every layer D_i of each block, switches corresponds to same vertices. Here, they correspond to the copies of $y_1, \ldots, y_k$. That is, for any vertex $y_j \in I$ and layer D_i, we make the vertex corresponding to y_j a switch. That means we color any vertex (except the vertices in Q) that appears before the vertex corresponding to y_j in the set X_{ij} with the color given to the first vertex of X_{ij} (which is a copy of x_j) and color all the other vertices (except the vertices in Q) in X_{ij} with the color given to the last vertex of the set X_{ij}, for $i \in [2m]$ and $j \in [k]$.

Observe that all the colored vertices maintain the proper coloring property by the chain ordering of each layer at this step. Also, every vertex gets color from its associated list. Thus, all the colored vertices maintain the list coloring property. The only uncolored vertices are the vertices in Q and the last vertices (called edge vertices, $\gamma(e_\ell)$ for all $\ell \in [m]$) in each layer. Next, we explain how to color those vertices.

Let $e_\ell = uv$ be an edge in G such that $u \in X_{i'}$ and $v \in X_{j'}$, where $i', j' \in [k]$ and $i' < j'$. Recall that corresponding to edge e_ℓ, we have three vertices $\alpha(e_\ell), \in X_{(2\ell-1)i'}$, $\beta(e_\ell) \in X_{(2\ell-1)j'}$, and $\gamma(e_\ell)$ in layer $D_{2\ell-1}$; and two vertices $\alpha'(e_\ell) \in X_{(2\ell)i'}$ and $\beta'(e_\ell) \in X_{(2\ell)j'}$ of layer $D_{2\ell}$. These are the only uncolored vertices so far in layer $D_{2\ell-1}$ and $D_{2\ell}$ of each block, for all $\ell \in [m]$. Note that according to our obtained (partial) list coloring, for any $j \in [k]$ and $i \in [2m]$, the switch in X_{ij} is y_j^i (i.e., the copy of y_j in the layer D_i). Now, there are three cases based on the position of the vertex $u^{2\ell-1}$ (or $v^{2\ell-1}$) with respect to the switches $y_{i'}^{2\ell-1}, y_{j'}^{2\ell-1}, y_{i'}^{2\ell}, y_{j'}^{2\ell}$ in layers $D_{2\ell-1}$ and $D_{2\ell}$. First we explain the colors of $\alpha(e_\ell)$ and $\alpha'(e_\ell)$. For this, we have three cases based the position of $u^{2\ell-1}$ compared with $y_{i'}^{2\ell-1}$.

Case 1: $u^{2\ell-1} <_{\sigma_{2\ell-1}} y_{i'}^{2\ell-1}$. In this case, we have $\alpha(e_\ell) <_{\sigma_{2\ell-1}} u^{2\ell-1} <_{\sigma_{2\ell-1}} y_{i'}^{2\ell-1}$ and $\alpha'(e_\ell) <_{\sigma_{2\ell}} u^{2\ell} <_{\sigma_{2\ell}} y_{i'}^{2\ell}$. Recall that the list of the vertex $\alpha(e_\ell)$ contains the color given to the first vertex $x_{i'}^{2\ell-1}$ of $X_{(2\ell-1)i'}$. Moreover, in our partial coloring, we colored $u^{2\ell-1}$ with the color of $x_{i'}^{2\ell-1}$. We color $\alpha(e_\ell)$ with the color of $x_{i'}$. Notice that the neighbours of $\alpha(e_\ell)$ in $D_{2\ell}$ is a subset of the neighbours of $u^{2\ell-1}$ in $D_{2\ell}$. Similarly, the neighbours of $\alpha(e_\ell)$ in $D_{2\ell-2}$ is a subset of the neighbours of $x_{i'}^{2\ell-1}$ in $D_{2\ell-2}$. So, as long as the colors on $x_{i'}^{2\ell-1}$ and $u^{2\ell-1}$ do not violate the proper coloring property, it holds on the vertex $\alpha(e_\ell)$. We color $\alpha'(e_\ell)$ with the unique color c_1' in $L(\alpha'(e_\ell)) \cap \{c_1, \widehat{c_1}\}$. Notice that this color is available only in the list of $\alpha(e_\ell)$ in $D_{2\ell-1}$ and we colored that vertex with a different color. Moreover, c_1' is not present in the list of any vertex in the layer $D_{2\ell+1}$.

Case 2: $y_{i'}^{2\ell-1} <_{\sigma_{2\ell-1}} u^{2\ell-1}$. In this case, we have $y_{i'}^{2\ell-1} <_{\sigma_{2\ell-1}} \alpha(e_\ell) <_{\sigma_{2\ell-1}} u^{2\ell-1}$ and $y_{i'}^{2\ell} <_{\sigma_{2\ell}} \alpha'(e_\ell) <_{\sigma_{2\ell}} u^{2\ell}$. Notice that $y_{i'}^{2\ell}$ and $u^{2\ell}$ are colored with the color q of $last(X_{(2\ell)i'})$ (which is same as the color of $x_{i'}^{2\ell-1}$). We color $\alpha'(e_\ell)$ with color q. Using arguments similar to that in the Case 1, one can argue that as long as the colors on $y_{i'}^{2\ell}$ and $u^{2\ell}$ do not violate the proper coloring property, it holds on the vertex $\alpha'(e_\ell)$. Now we color $\alpha(e_\ell)$ with the unique color c_1' in $L(\alpha(e_\ell)) \cap \{c_1, \widehat{c_1}\}$. Notice that this color is available only in the list of $\alpha'(e_\ell)$ in $D_{2\ell}$ and we colored that vertex with a different color. Moreover, c_1' is not present in the list of any vertex in the layer $D_{2\ell-2}$.

Case 3: $u^{2\ell-1} = y_{i'}^{2\ell-1}$. In this case, we have $u^{2\ell} = y_{i'}^{2\ell}$ and e_ℓ incident on $y_{i'}$ in G. Note that in this case, the vertices $\alpha(e_\ell)$ and $\alpha'(e_\ell)$ appear just before $y_{i'}^{2\ell-1}$ and $y_{i'}^{2\ell}$, respectively. Recall that the lists of both the vertices $\alpha(e_\ell)$ and $\alpha'(e_\ell)$ contain the color given to the first vertex $x_{i'}$ of $X_{(2\ell-1)i'}$. Observe that the vertex $u^{2\ell}$ is the switch in $X_{(2\ell)i'}$ and $\alpha(e_\ell)$ is adjacent to $u^{2\ell}$. Since the vertex $u^{2\ell}$ is the switch, $u^{2\ell}$ is colored with the color $\phi(last(X_{(2\ell)i'}))$, which is same as $\phi(x_{i'}^{2\ell-1})$. Therefore, we cannot color the vertex $\alpha(e_\ell)$ with the same color

$\phi(x_{i'}^{2\ell-1})$. In this case, we color the vertex $\alpha(e_\ell)$ with the unique color present in $L(\alpha(e_\ell)) \cap \{c_1, \widehat{c}_1\}$. Also, we color the vertex $\alpha'(e_\ell)$ with the unique color present in $L(\alpha(e_\ell)) \cap \{c_3, \widehat{c}_3\}$. It is easy to argue that the obtained partial coloring does not violate any constraint so far.

Similar to the Cases 1–3, we color the vertices $\beta(e_\ell)$ and $\beta'(e_\ell)$ based on one of the cases. Lastly, we color the edge vertex $\gamma(e_\ell)$ from its list. Recall that if ℓ is odd, $L(\gamma(e_\ell)) = \{c_3, c_4\}$ and if ℓ is even $L(\gamma(e_\ell)) = \{\widehat{c}_3, \widehat{c}_4\}$. We consider the case when ℓ is odd. The other case is symmetric in arguments and hence omitted. Notice that $L(\gamma(e_\ell)) = \{c_3, c_4\}$. Observe that we use a color from the set $\{c_3, c_4\}$ to color a vertex $\alpha'(e_\ell)$ or $\beta'(e_\ell)$, only in the Case 3. In order to (properly) color the vertex $\gamma(e_\ell)$ from its list, we need to prove that at most one of the vertices $\alpha'(e_\ell)$ and $\beta'(e_\ell)$ belong to Case 3 and use a color from the set $\{c_3, c_4\}$. Therefore, we prove the following claim.

Claim. For $e_\ell = uv \in E(G)$, exactly one of the vertices $\alpha'(e_\ell)$ or $\beta'(e_\ell)$ use the color from the set $\{c_3, c_4\}$.

Proof. Suppose that both the vertices $\alpha'(e_\ell)$ and $\beta'(e_\ell)$ use the color from the set $\{c_3, c_4\}$. It follows that both the vertices $\alpha'(e_\ell)$ and $\beta'(e_\ell)$ are colored by the Case 3. It implies, the vertices $u^{2\ell-1} = y_{i'}^{2\ell-1}$ and $v^{2\ell-1} = y_{j'}^{2\ell-1}$. Moreover, the endpoints of the edge e_ℓ are $y_{i'}$ and $y_{j'}$. This is a contradiction to the fact that $y_{i'}$ and $y_{j'}$ belong to the independent set I. □

Thus, when ℓ is odd, we can color the edge vertex $\gamma(e_\ell)$ with an available color from its list $\{c_3, c_4\}$ that is not used to color the vertices in Q that are corresponding to the edge e_ℓ. Also, notice that $\gamma(e_\ell)$ does not have any neighbours in the layer $D_{2\ell-2}$. To argue that the given coloring does not violate any edge constraints between two consecutive blocks, one can use the subset of the arguments used in Cases 1–3 and the fact that the last layer of a block is an even layer and m is a multiple of 3. Hence, we obtained a list k'-coloring of G'. □

Lemma 3 ($\star$). *If G' has a list k'-coloring, then G has a multicolored independent set of size k, where $k' = 3k + 8$.*

5 Conclusion

In this paper, we study LHOM on bipartite graphs that admit a multichain ordering and give efficient algorithms. However, we could not extend the algorithm to interval graphs because the graph induced by the vertices in a layer (in a multichain ordering) need not be an independent set. It is interesting to get faster algorithms for LHOM on interval graphs.

Acknowledgements. We would like to thank anonymous referees for their helpful comments. The first author acknowledges SERB-DST for supporting this research via grant PDF/2021/003452.

References

1. Bok, J., Brewster, R., Feder, T., Hell, P., Jedličková, N.: List homomorphism problems for signed graphs. arXiv preprint arXiv:2005.05547 (2020)
2. Bok, J., Brewster, R., Feder, T., Hell, P., Jedličková, N.: List homomorphisms to separable signed graphs. In: Balachandran, N., Inkulu, R. (eds.) CALDAM 2022. LNCS, vol. 13179, pp. 22–35. Springer, Cham (2022). https://doi.org/10.1007/978-3-030-95018-7_3
3. Brandstädt, A., Lozin, V.V.: On the linear structure and clique-width of bipartite permutation graphs. Ars Comb. **67** (2003)
4. Chen, H., Jansen, B.M.P., Okrasa, K., Pieterse, A., Rzazewski, P.: Sparsification lower bounds for list H-coloring. In: Cao, Y., Cheng, S.-W., Li, M. (eds.) 31st International Symposium on Algorithms and Computation, ISAAC 2020, 14–18 December 2020. LIPIcs, vol. 181, pp. 58:1–58:17 (2020)
5. Chitnis, R., Egri, L., Marx, D.: List h-coloring a graph by removing few vertices. Algorithmica **78**(1), 110–146 (2017)
6. Cygan, M., et al.: Parameterized Algorithms, vol. 5. Springer, Cham (2015). https://doi.org/10.1007/978-3-319-21275-3
7. Díaz, J., Diner, Ö.Y., Serna, M., Serra, O.: On list k-coloring convex bipartite graphs. In: Gentile, C., Stecca, G., Ventura, P. (eds.) Graphs and Combinatorial Optimization: from Theory to Applications. ASS, vol. 5, pp. 15–26. Springer, Cham (2021). https://doi.org/10.1007/978-3-030-63072-0_2
8. Díaz, J., Serna, M.J., Thilikos, D.M.: Counting h-colorings of partial k-trees. Theor. Comput. Sci. **281**(1–2), 291–309 (2002)
9. Egri, L., Krokhin, A., Larose, B., Tesson, P.: The complexity of the list homomorphism problem for graphs. Theory Comput. Syst. **51**(2), 143–178 (2012)
10. Enright, J.A., Stewart, L., Tardos, G.: On list coloring and list homomorphism of permutation and interval graphs. SIAM J. Discret. Math. **28**(4), 1675–1685 (2014)
11. Erdös, P., Rubin, A.L., Taylor, H.: Choosability in graphs. In: Proceedings West Coast Conference on Combinatorics, Graph Theory and Computing, Congressus Numerantium, vol. 26, pp. 125–157 (1979)
12. Feder, T., Hell, P.: List homomorphisms to reflexive graphs. J. Comb. Theory Ser. B **72**(2), 236–250 (1998)
13. Feder, T., Hell, P., Huang, J.: List homomorphisms and circular arc graphs. Combinatorica **19**(4), 487–505 (1999)
14. Feder, T., Hell, P., Huang, J.: Bi-arc graphs and the complexity of list homomorphisms. J. Graph Theory **42**(1), 61–80 (2003)
15. Feder, T., Hell, P., Klein, S., Motwani, R.: List partitions. SIAM J. Discret. Math. **16**(3), 449–478 (2003)
16. Feder, T., Hell, P., Klein, S., Nogueira, L.T., Protti, F.: List matrix partitions of chordal graphs. Theor. Comput. Sci. **349**(1), 52–66 (2005)
17. Garey, M.R., Johnson, D.S., Stockmeyer, L.: Some simplified NP-complete problems. In: Proceedings of the Sixth Annual ACM Symposium on Theory of Computing, pp. 47–63 (1974)
18. Garg, N., Papatriantafilou, M., Tsigas, P.: Distributed list coloring: how to dynamically allocate frequencies to mobile base stations. In: Proceedings of SPDP 1996: 8th IEEE Symposium on Parallel and Distributed Processing, pp. 18–25 (1996)
19. Hoàng, C.T., Kamiński, M., Lozin, V., Sawada, J., Shu, X.: Deciding k-colorability of P_5-free graphs in polynomial time. Algorithmica **57**(1), 74–81 (2010)

20. Jansen, K., Scheffler, P.: Generalized coloring for tree-like graphs. Discret. Appl. Math. **75**(2), 135–155 (1997)
21. Kim, H., Siggers, M.: Towards a dichotomy for the switch list homomorphism problem for signed graphs. arXiv preprint arXiv:2104.07764 (2021)
22. Kratochvil, J., Tuza, Z.: Algorithmic complexity of list colorings. Discret. Appl. Math. **50**(3), 297–302 (1994)
23. Okrasa, K., Piecyk, M., Rzazewski, P.: Full complexity classification of the list homomorphism problem for bounded-treewidth graphs. In: 28th Annual European Symposium on Algorithms, ESA 2020, Pisa, Italy, 7–9 September 2020 (Virtual Conference), pp. 74:1–74:24 (2020)
24. Okrasa, K., Rzazewski, P.: Complexity of the list homomorphism problem in hereditary graph classes. In: Bläser, M., Monmege, B. (eds.) 38th International Symposium on Theoretical Aspects of Computer Science, STACS 2021. LIPIcs, vol. 187, pp. 54:1–54:17 (2021)
25. Valadkhan, P.: List matrix partitions of special graphs. Ph.D. thesis, Applied Sciences: School of Computing Science (2013)
26. Valadkhan, P.: List matrix partitions of graphs representing geometric configurations. Discret. Appl. Math. **260**, 237–243 (2019)
27. Vizing, V.G.: Vertex colorings with given colors. Diskret. Analiz **29**, 3–10 (1976)
28. Wang, W., Liu, X.: List-coloring based channel allocation for open-spectrum wireless networks. In: VTC-2005-Fall. 2005 IEEE 62nd Vehicular Technology Conference, vol. 1, pp. 690–694. Citeseer (2005)

On the Closures of Monotone Algebraic Classes and Variants of the Determinant

Prasad Chaugule[1](✉) and Nutan Limaye[2]

[1] Indian Institute of Technology, Bombay, Mumbai, India
prasad@cse.iitb.ac.in
[2] IT University of Copenhagen, Copenhagen, Denmark

Abstract. In this paper we prove the following two results.
- We show that for any $C \in \{\mathsf{mVF}, \mathsf{mVP}, \mathsf{mVNP}\}$, $C = \overline{C}$. Here, $\mathsf{mVF}, \mathsf{mVP}$, and mVNP are monotone variants of VF, VP, and VNP, respectively. For an algebraic complexity class C, $\overline{C}$ denotes the closure of C. For mVBP a similar result was shown in [4]. Here we extend their result by adapting their proof.
- We define polynomial families $\{\mathcal{P}(k)_n\}_{n \geq 0}$, such that $\{\mathcal{P}(0)_n\}_{n \geq 0}$ equals the Determinant polynomial. We show that $\{\mathcal{P}(k)_n\}_{n \geq 0}$ is VBP complete for $k = 1$ and it becomes VNP complete when $k \geq 2$. In particular, $\mathcal{P}(k)_n$ is $\mathtt{Det}_n^{\neq k}(\mathtt{X})$, a polynomial obtained by summing over all signed cycle covers that avoid length k cycles. We show that $\mathtt{Det}_n^{\neq 1}(\mathtt{X})$ is complete for VBP and $\mathtt{Det}_n^{\neq k}(\mathtt{X})$ is complete for VNP for all $k \geq 2$ over any field $\mathbb{F}$.

1 Introduction

Valiant [17] initiated the study of the complexity of the algebraic computation. Given a polynomial, how efficiently can we compute it? In order to formalise the notion of efficiency, Valiant defined many natural models of computation. These include algebraic circuits, algebraic formulas, and algebraic branching programs.

An *algebraic circuit* is a directed acyclic graph. The nodes with in-degree zero are leaf nodes, which are labelled with variables $X = \{x_1, x_2, \ldots, x_n\}$ or field constants. The other nodes are either $+$ or $\times$ operators, which compute polynomials of their inputs. The $+$ node adds its inputs and $\times$ node multiplies its inputs. There is a designated output node which has out-degree 0. The output of the circuit is the polynomial computed by this node. An *algebraic formula* is a circuit in which the underlying DAG is a tree. The size of the circuit/formula is the number of nodes in it.

An *algebraic branching program* (ABP) A is a layered DAG with two special nodes s and t called the source node and the sink node, respectively. The edges are labelled with linear forms $\sum_{i=1}^{n} c_i x_i + c_0$, where $c_i \in \mathbb{F}$. For every directed path ρ from s to t, we associate a polynomial P_ρ which is formed by multiplying all the labels along the edges in path ρ. The polynomial computed by the ABP A is equal to $\sum_\rho P_\rho$ where the sum is over all $s - t$ paths in A. The size of the algebraic branching program A is the number of nodes in it. We assume that the length of every path from s to t is the same.

A polynomial family f_n is called a p-bounded family if both the number of variables and the degree of f_n are polynomially bounded in n. The class of p-bounded families computed by polynomial sized circuits is called VP. Similarly, the class of p-bounded polynomial families computed by formulas of polynomial size are called VF. Finally, the class of p-bounded polynomial families computed by ABPs of polynomial size is called VBP.

Closures of Monotone Complexity Classes. Instead of studying the complexity of the exact computation of any polynomial, we could also study the algebraic approximation of a polynomial. A polynomial Q over $\mathbb{F}(\varepsilon)[X]$ is said to be an algebraic approximation of a polynomial P over $\mathbb{F}[X]$ if there exist polynomials $Q_1, Q_2, \ldots$ in $\mathbb{F}[X]$ such that $Q \equiv P + \sum_{i \geq 1} \varepsilon^i Q_i$. It is possible that approximating a polynomial is computationally less expensive than computing it exactly. More formally, let $\mathcal{C}$ be any class of polynomials and $\overline{\mathcal{C}}$, the *closure of $\mathcal{C}$*, be the polynomials approximated by the polynomials in $\mathcal{C}$. One can always ask the *strict containment* question: Is $\mathcal{C} \subsetneq \overline{\mathcal{C}}$? One of the important and well-known questions is the VP vs. $\overline{\text{VP}}$ question. Here $\overline{\text{VP}}$ stands for a class of p-bounded polynomials that are approximated by polynomial sized circuits[1]. From the definition, it is clear that $\text{VP} \subseteq \overline{\text{VP}}$, but whether the containment is strict or not is an open question, which has a rich and long history [2,3,8,13].

In some cases, strict containment holds. Let $\sum^{[k]} \prod \sum$ denotes the class of polynomials which can be represented as the sum of product of affine forms, where k denotes the fan-in of the top-most addition gate in the circuit. Kumar [10] showed that the class of polynomials $\sum^{[3]} \prod \sum$ is strictly contained in $\overline{\sum^{[3]} \prod \sum}$. Let VP_k denotes the class of algebraic branching programs of width k. Bringmann, Ikenmeyer and Zuiddam [5] looked at the class of polynomials computed by *width-2 algebraic branching programs* and showed that for any constant k, $\text{VP}_k \subseteq \overline{\text{VP}_2}$. This along with a result of Allender and Wang [1] shows that $\text{VP}_2 \subsetneq \overline{\text{VP}_2}$.

In this work, however, we consider monotone algebraic complexity classes and for each class $\mathcal{C}$ that we consider, we show that $\mathcal{C} = \overline{\mathcal{C}}$. That is, we answer the strict containment question negatively for monotone algebraic complexity classes. Recently, Bläser, Ikenmeyer, Mahajan, Pandey and Saurabh [4] showed that mVBP equals its closure, where mVBP is a class of polynomials computable by *monotone algebraic branching program* (formal definition in Sect. 2). We extend their result to other monotone algebraic complexity classes. We show that $\text{mVP} = \overline{\text{mVP}}$, that is, anything that can be approximated by monotone circuits of polynomial size (computing polynomials of polynomial degree) can also be computed by them. We also show that mVF, a class of polynomials computed by monotone algebraic *formulas*, is equal to its closure and that mVNP,

---

[1] Formally, $\overline{\text{VP}}$ is defined using *topological approximations*. However, for reasonably well-behaved fields $\mathbb{F}$, the two notions of approximation are equivalent. We will focus on algebraic approximation in this note.

the monotone analogue of VNP^2 (formally defined in Sect. 3) equals its closure. Overall, our results imply that approximation does not add extra power to the monotone algebraic models of computation. Our proof is elementary and a simple generalisation of the proof of $\mathsf{mVBP} = \overline{\mathsf{mVBP}}$ from [4].

Variants of the Determinant. We recall the combinatorial definition of the determinant polynomial (See for instance [11]).

Definition 1. *Let G_n be a directed complete graph on n vertices, where the edge label of edge (i,j) is $x_{i,j}$ for $i,j \in [n]$. Let $\mathcal{C}$ denotes the set of cycle covers[3] of G_n. We define $\mathrm{Det}_n(X) = \sum_{C \in \mathcal{C}} s(C) \cdot m_C$, where $s(C) = (-1)^{n+t}$, where t is the number of cycles in the cycle cover C and m_C is the monomial formed by multiplying the labels on all the edges in C.*

We define a variant $\mathrm{Det}_n^{\neq k}(X)$ of the determinant, which sums over all signed cycle covers that avoid length k cycles. Formally,

Definition 2. *Let k be a fixed constant. Let G_n be a directed complete graph on n vertices, where the edge label of edge (i,j) is $x_{i,j}$ for $i,j \in [n]$. Let $\mathcal{C}^{\neq k}$ denote all those cycle covers of G_n in which all the cycles have length $\neq k$. We define*

$$\mathrm{Det}_n^{\neq k}(X) = \sum_{C \in \mathcal{C}^{\neq k}} s(C) \cdot m_C$$

where $s(C) = (-1)^{n+t}$, where t is the number of cycles in the cycle cover C and m_C is the monomial formed by multiplying the labels on all the edges in C.

We prove the following Lemma.

Lemma 1. *$\mathrm{Det}_n^{\neq 1}(X)$ is complete[4] for VBP and $\mathrm{Det}_n^{\neq k}(X)$ is complete for VNP for all $k \geq 2$ over any field $\mathbb{F}$.*

Using Valiants criteria [17], it is easy to observe that for $k \geq 0$, $\mathrm{Det}_n^{\neq k}(X)$ is in VNP. We prove VNP hardness in Sect. 4.

The family $\mathrm{Det}_n^{\neq k}(X)$ can be viewed as a parametrized family with k as the parameter such that $\mathrm{Det}_n^{\neq 1}(X)$ is VBP-complete and $\mathrm{Det}_n^{\neq k}(X)$ for all $k \geq 2$ is VNP-complete over all fields. In previous works by Durand et al. [7] and Chaugule

---

[2] Let $f_n(x_1, x_2, \ldots, x_{k(n)})$ be a p-bounded polynomial family. f_n is said to be in VNP if there exists a family $g_n \in$ VP such that $f_n = \sum_{y_1, y_2 \ldots y_{k'(n)} \in \{0,1\}} g_n(x_1, x_2, \ldots, x_{k(n)}, y_1, y_2, \ldots, y_{k'(n)})$, where $k'(n)$ is polynomially bounded in n.

[3] A cycle cover of a directed graph H is a set of vertex disjoint cycles which are subgraphs of graph H and contains all the vertices of H.

[4] A family $f_n(X)$ is said to be a p-projection of $g_n(Y)$ (denoted as $f_n(X) \leq_p g_n(Y)$) if there exists a polynomially bounded function $t : \mathbb{N} \longrightarrow \mathbb{N}$ such that $f_n(X)$ can be computed from $g_{t(n)}(Y)$ by setting its variables to one of the variables of $f_n(X)$ or to field constants. A p-bounded family f_n is said to be hard for a complexity class $\mathcal{C}$ if for any $g_n \in \mathcal{C}$, $g_n \leq_p f_n$. A p-bounded family f_n is complete for a class $\mathcal{C}$ if it is in $\mathcal{C}$ and is hard for $\mathcal{C}$.

et al. [6], many polynomials were proposed which characterised VP. These were defined using homomorphism polynomials. It would be interesting to come up with a parametrized polynomial family which characterizes not just VBP and VNP, but all natural complexity classes, namely VF, VBP, VP, and VNP.

2   Preliminaries

2.1   Border Complexity

Analogous to algebraic complexity classes VP, VF, VBP, we define the border complexity variants of these classes. We denote the corresponding complexity classes as $\overline{\mathrm{VP}}$, $\overline{\mathrm{VF}}$, and $\overline{\mathrm{VBP}}$ respectively.

Definition 3. *A polynomial family $h_n \in \mathbb{F}[X]$ is said to be in class $\overline{VP}$ ($\overline{VF}$, or $\overline{VBP}$ respectively) if and only if there exist polynomials $h_{n,i} \in \mathbb{F}[X]$ and a function $t : \mathbb{N} \longrightarrow \mathbb{N}$ such that the polynomial $g_n(x) = h_n(x) + \varepsilon h_{n,1}(x) + \varepsilon^2 h_{n,2}(x) + \ldots + \varepsilon^{t(n)} h_{n,t(n)}(x)$ can be computed by an algebraic circuit $C \in \mathbb{F}(\varepsilon)[X]$ (algebraic formula $F \in \mathbb{F}(\varepsilon)[X]$ or an algebraic branching program $A \in \mathbb{F}(\varepsilon)[X]$, respectively) of size polynomial in n, where ε is the new indeterminate.*

2.2   Monotone Complexity

Since, we are in the monotone setting, hereafter, we will work only with the field of real numbers or the field of rational numbers, for the sake of the presentation, we set, $\mathbb{F} = \mathbb{R}$.

An algebraic circuit (algebraic formula and algebraic branching programs, respectively) is said to be a monotone algebraic circuit (monotone algebraic formula and monotone algebraic branching programs, respectively) if all the input constants (from $\mathbb{R}$) are non-negative.

Definition 4. *The class of p-bounded families computed by polynomial sized monotone circuits (monotone formula/ABPs) is called mVP (mVF, mVBP, respectively).*

2.3   Monotone Border Complexity

Analagous to our complexity classes mVP, mVF, mVBP, we now define the border complexity variants of these classes. We denote the corresponding complexity classes as $\overline{\mathrm{mVP}}$, $\overline{\mathrm{mVF}}$, $\overline{\mathrm{mVBP}}$ respectively.

Let $\mathbb{R}_+[\varepsilon, \varepsilon^{-1}]$ denote the ring of Laurent polynomials that are non-negative for all sufficiently small $\varepsilon > 0$. In other words, the monotonicity condition means that for each $\alpha \in \mathbb{R}_+[\varepsilon, \varepsilon^{-1}]$, there is a $\beta > 0$ such that for all $\varepsilon \in (0, \beta]$, $\alpha(\varepsilon) \geq 0$.

Definition 5. *A polynomial family $h_n \in \mathbb{F}[X]$ is said to be in class $\overline{mVP}$ ($\overline{mVF}$, $\overline{mVBP}$) if and only if there exist polynomials $h_{n,i} \in \mathbb{F}[X]$ such that the polynomial $g_n(x) = h_n(x) + \varepsilon h_{n,1}(x) + \varepsilon^2 h_{n,2}(x) + \ldots + \varepsilon^t h_{n,t}(x)$ can be computed by an algebraic circuit C (algebraic formula, branching program, resp.) of size polynomial in n where the input are either labelled by variables or polynomials from $\mathbb{R}_+[\varepsilon, \varepsilon^{-1}]$.*

The definition of $\overline{\mathsf{mVNP}}$ is slightly involved. We define it in Sect. 3.3.

3 Closures of Monotone Algebraic Classes

The first result we prove in this section is about algebraic closure of mVP.

Theorem 1. $\mathsf{mVP} = \overline{\mathsf{mVP}}$

In [4], Bläser, Ikenmeyer, Mahajan, Pandey, and Saurabh proved that "$\mathsf{mVBP} = \overline{\mathsf{mVBP}}$". Our proof has an analogous proof outline. It consists of the following three steps.

1. Let $\mathcal{C}_n$ be a monotone alebraic circuit of size $s = \mathrm{poly}(n)$ computing the polynomial $Q = \varepsilon^0 f_0 + \sum_{i=1}^{k} \varepsilon^i f_i$. We convert the circuit $\mathcal{C}_n$ into another monotone circuit $\mathcal{C}'_n$ of size $\mathrm{poly}(s)$, such that the circuit $\mathcal{C}'_n$ computes the polynomial $\varepsilon^t Q$ for some positive integer t, and no input gate in $\mathcal{C}'_n$ has a label with negative exponent of ε as its input (Sect. 3.1).
2. We give a general procedure to convert a monotone circuit computing the polynomial $\varepsilon^t Q$ to $\varepsilon^{t-1} Q$ (with edge labels). Moreover, no input gate or edge in the converted circuit has a negative exponent of ε as its label (Sect. 3.2).
3. We repeatedly apply step 2 and construct a monotone circuit which computes the polynomial $\varepsilon^0 Q$. By finally substituting ε to 0, the result follows.

Although, the overall proof idea is very similar to the proof idea from [4], the details in Step 1 and Step 2 are slightly different. Informally, we use the structural properties of the universal circuit to get Step 1, whereas we exploit the monotonicity property of the given circuit to achieve Step 2.

3.1   Step 1 of Theorem 1

Before going into the details of the proof this step of Theorem 1, we recall the following definitions.

Definition 6 ([14, 16]). *A family of circuits $\{\mathcal{U}_n\}_{n \in \mathbb{N}}$ is called a universal circuit family if for every polynomial $f_n(x_1, x_2, \ldots, x_n)$ of degree $d(n)$ which is computed by a circuit of size $s(n)$, there exists another circuit Ψ which computes f_n such that the underlying Directed Acyclic Graph (DAG) of Ψ is the same as that of $\mathcal{U}_m$ for some $m \in \mathrm{poly}(n, s, d)$. We assume that both $s(n)$ and $d(n)$ are polynomially bounded in n.*

A universal circuit C_n is said to be in a normal form if it satisfies the following properties.

- *The circuit C_n is a semi-bounded circuit, i.e., the indegree of $\times$ gate is 2, whereas the indegree of $+$ gates is unbounded.*
- *The circuit C_n is a multiplicatively disjoint circuit.*
- *The circuit C_n is a layered circuit where layeres are alternately labelled with $+$ and $\times$ gates. We assume that the root node is a $\times$ gate. Also, the distance between the root node to every leaf node of circuit C_n is the same.*

- *The degree of the circuit C_n is n. The depth of the circuit $C_n = 2c\lceil \log n \rceil$, for some constant c.*
- *The number of variables $v(n)$ and the size $s(n)$ of the circuit C_n are both polynomially bounded in n.*

It is known that a universal circuit can be assumed to be in a normal form.

We recall the definition of a parse tree of an algebraic circuit.

Definition 7 ([12]). *The set of parse trees of a circuit C, $\mathcal{T}(C)$ is defined inductively on the size of circuit C*

- *If the size of C is 1, then the circuit C is its own parse tree.*
- *If the circuit size is greater than 1, then the output gate is either a multiplication gate or an addition gate*
    1. *Let the output gate t be an addition gate. Then the parse trees of circuit C are formed by taking a parse tree of any one of its children, say t' along with the edge (t', t).*
    2. *Let the output gate t be a multiplication gate. Let t_1 and t_2 be the children gates of t. Let C_{t_1} and C_{t_2} be the subcircuits rooted at gates t_1 and t_2 in circuit C. The parse trees of C are formed by taking a node disjoint copy of a parse tree of subcircuit C_{t_1} and a parse tree of subcircuit C_{t_2} along with the edges (t_1, t) and (t_2, t).*

Remark 1. Given an algebraic circuit C computing a polynomial f. Let $\mathcal{T}(C)$ denote the set of all parse trees of circuit C. Let $mon(T)$ be the monomial associated with $T \in \mathcal{T}(C)$, where $mon(T)$ is equal to the product of the labels of the leaves of T. It is known that f is equal to $\sum_{T \in \mathcal{T}(C))} mon(T)$ (see [12]).

Remark 2. Note that the shape of every parse tree T of the universal circuit (in normal form) C_n is the same. Moreover, the number of leaf nodes in any parse tree T of C_n is equal to $2^{c\lceil \log n \rceil} = n^c$.

The following Lemma gives the details for the first step.

Lemma 2. *Consider a monotone algebraic circuit C_n of size $s = \text{poly}(n)$ computing a polynomial $Q = \varepsilon^0 f_0 + \sum_{i=1}^{k} \varepsilon^i f_i$ (that is, circuit C_n computes the polynomial f_0 in the border sense), then there exists another monotone circuit C'_n (in normal form) of size $\text{poly}(s)$ such that the circuit C'_n computes the polynomial $\varepsilon^t Q$, for some positive integer t. Moreover, no input gate in circuit C'_n has a label with a negative exponent of ε as its input. And the underlying DAG structure of C'_n is the same as that of C_n.*

Proof. Consider a circuit C_n which computes the polynomial Q and say it has size s. We know that there exists a universal circuit in normal form $\mathcal{U}_m$, where $m = \text{poly}(s)$ such that the circuit $\mathcal{U}_m$ also computes Q. Moreover, it is known that the monotonicity of circuit C_n can be preserved in circuit $\mathcal{U}_m$. Let j be the largest negative exponent in any of the input gates of C_n. We multiply the label of every input gate of circuit $\mathcal{U}_m$ by ε^j. We call this new circuit C'_n. Since, the circuit C_n is in the normal form, the shape of every parse tree of circuit C_n is the same and the number of leaf nodes in any parse tree is equal to n^c (see Remark 1 and Remark 2). Therefore, the polynomial computed by circuit C'_n is equal to a scaled version of polynomial computed by circuit C_n, that is, the polynomial computed by C'_n is equal to $\varepsilon^t Q$, where $t = j \times n^c$. We did not change the underlying graph structure of C_n to obtain C'_n. $\square$

3.2   Step 2 of Theorem 1

We now present the details regarding the second step of our main proof.

Here we will use algebraic circuits with edge labels. A circuit C with edge labels over a field $\mathbb{F}$ and variable set $X = \{x_1, x_2, \ldots, x_n\}$ is exactly similar to the algebraic circuit we have been considering, except for the edge function $w : E \longrightarrow X \cup \mathbb{F}$. Here E denotes the edge set of C. The polynomial computed at any input gate u is equal to the label of u. Let P_u denote the polynomial computed at the node u in the circuit C. The polynomial computed at any computation gate u with operation op, with u_ℓ and u_r as its children nodes is equal to the $(w((u, u_\ell)) \times P_{u_\ell}) \text{op}(w((u, u_r)) \times P_{u_r})$. The polynomial computed by the circuit C is the polynomial computed by the output gate of circuit C. The size of the algebraic circuit C is the number of nodes in it.

Definition 8. *A node u in any monotone circuit over $\mathbb{R}_+[\varepsilon^{-1}, \varepsilon]$ is called a good node, if the polynomial f_u computed at node u is divisible by ε.*

Lemma 3. *Consider any monotone circuit $\mathcal{D}_n$ of size s with edge label function $w : E \longrightarrow \{\varepsilon^i | i \in \mathbb{Z}_{\geq 0}\}$, where E is the set of edges of $\mathcal{D}_n$. Suppose it computes a polynomial $\varepsilon^b Q$ for some $b \geq 1$ where $Q = f_0 + \sum_{i=1}^{b} \varepsilon^i f_i$ and the circuit $\mathcal{D}_n$ satisfies the following properties:*

- *No input gate or any edge in $\mathcal{D}_n$ has a label with a negative exponent of ε.*
- *The underlying graph structure of $\mathcal{D}_n$ is the same as the graph structure of a universal circuit as in Definition 6.*

Then there exists a circuit $\mathcal{D}'_n$ of size s such that $\mathcal{D}'_n$ computes $\varepsilon^{b-1} Q$. Moreover, no input gate (or an edge label) in $\mathcal{D}'_n$ is labelled with a negative exponent of ε. Also, the underlying graph structure of the circuit $\mathcal{D}'_n$ is same as that of $\mathcal{D}_n$, upto the relabelling of the input gates and edge labels of the circuit $\mathcal{D}_n$.

Let $\mathcal{G}$ denote the set of all good nodes of circuit $\mathcal{D}_n$. In order to prove Lemma 3, we need to describe the circuit $\mathcal{D}'_n$. The circuit $\mathcal{D}'_n$ is exactly similar to the circuit $\mathcal{D}_n$, with labels of some input nodes $\mathcal{L}' \subseteq \mathcal{G}$ scaled by $1/\epsilon$, and an updated edge function w'. Moreover, the new function w' also satisfies the property that no edge has a label with a negative exponent of ε.

Instead of $\mathcal{L}'$, we construct a larger set $\mathcal{S}$ such that $\mathcal{L}' \subseteq \mathcal{S} \subseteq \mathcal{G}$ and the root node $r \in \mathcal{S}$. Let f_u denote the polynomial computed at node u in circuit $\mathcal{D}_n$. Let $\widehat{f_u}$ denote the polynomial computed at node u in circuit $\mathcal{D}'_n$. It suffices to prove the following lemma.

Lemma 4. *In the circuit* $\mathcal{D}'_n$, *for every node* $u \in \mathcal{S}$, $\widehat{f}_u = \frac{1}{\varepsilon} f_u$.

Since, the root node $r \in \mathcal{S}$, the proof of Lemma 3 immediately follows (see Sect. 3.2 for proof details).

Identifying the Set $\mathcal{S}$ and the Edge Label Function $\widehat{w}$. The main idea is to mark some of the good nodes of circuit $\mathcal{D}_n$ which will form the set $\mathcal{S}$. First of all, we will mark the root node of circuit $\mathcal{D}_n$. The main goal is to scale the polynomial computed at the marked root node by $1/\epsilon$. We will propagate this effect from the root node layer to the layer immediately after it and so on till we reach to the layer of the leaf nodes. In this process, we may also change the labels of the edges thereby changing the old w function to the new updated w' function. Upon reaching the last layer, we scale all the marked leaf nodes of $\mathcal{D}_n$ by $1/\epsilon$ (by relabelling). We now discuss the procedure in detail.

Let us number the layers of $\mathcal{D}_n$ from 1 to m such that the layer containing the root node is numbered as 1 and so on till the layer containing the leaf nodes is numbered m.

Marking the Root Node: We mark the root node of circuit $\mathcal{D}_n$. Note that by our assumption of circuit $\mathcal{D}_n$, the root node is always a good node.

Marking the Nodes at Layer $i + 1$: Given the marking of nodes upto layer numbered i, we give a procedure to mark the nodes in layer numbered $i + 1$. We break this case into two parts depending on whether $i + 1$ is even or odd.

Case 1: $i + 1$ is Odd and $i + 1 \leq m$: We know that the layer numbered i consists of the $+$ gates.. Let $u_1, u_2, \ldots, u_t$ be the marked nodes of layer i. Inductively, we know that $u_1, u_2, \ldots, u_t$ are all good nodes. For each $j \in [t]$, let $u_{j,1}, u_{j,2}, \ldots, u_{j,f(j)}$ be children of node u_j. Note that, for all $j \in [t]$, the nodes $u_{j,1}, u_{j,2}, \ldots, u_{j,f(j)}$ are in layer numbered $i + 1$. For each $k \in [f(j)]$, the condition of the monotonicity of $\mathcal{D}_n$ (along with the property that u_j is a good node and is an addition gate) guarantees that

1a. either $u_{j,k}$ is a good node,
1b. or the edge $(u_{j,k}, u_j)$ is labelled by ε^β for some $\beta \geq 1$.

We now describe a procedure which essentially scales the polynomials computed via $u_{j,1}, u_{j,2}, \ldots, u_{j,f(j)}$ by $1/\varepsilon$, which in turn implies that the polynomial computed at u_j is also scaled by $1/\varepsilon$. This scaling is done either immediately by reducing the exponent of ε along the edge between the child node and the $+$ gate or is postponed to the next layer by marking the child node. We now state it in Algorithm 1 below.

Case 2: $i + 1$ is Even and $i + 1 \leq m$: We know that the layer numbered i consists of $\times$ gates. Let $u_1, u_2, \ldots, u_t$ be the marked nodes of layer i. Inductively, we know that $u_1, u_2, \ldots, u_t$ are all good nodes. For each $j \in [t]$, let $u_{j,\ell}$ and $u_{j,r}$ be the left child and the right child of node u_j, respectively. Note that, for all $j \in [t]$, the nodes $u_{j,\ell}$ and $u_{j,r}$ are in layer numbered $i + 1$. Since the circuit $\mathcal{D}_n$ is multiplicatively disjoint, for each $j \in [t]$, nodes $u_{j,\ell}$ and $u_{j,r}$ are always distinct.

---

for $j = 1$ *to* t **do**
    for $k = 1$ *upto* $f(j)$ **do**
        if *the node $u_{j,k}$ is already marked* **then**
            we do nothing;
        end
        else if *Case 1a holds* **then**
            we mark the node $u_{j,k}$;
        end
        else if *Case 1b holds* **then**
            we relabel the edge $(u_{j,k}, u_j)$ initially labelled by ε^β to $\varepsilon^{\beta-1}$;
        end
    end
end

Algorithm 1: Procedure to mark nodes on layer $i + 1$ when $i + 1$ is odd.

The property of monotonicity of $\mathcal{D}_n$ (along with the property that u_j is a good node and is a multiplication gate) guarantees that one of the following two cases holds.

2a. If both $u_{j,\ell}$ and $u_{j,r}$ are not good nodes then there exist at least one edge from $\{(u_{j,\ell}, u_j), (u_{j,r}, u_j)\}$ which is labelled by ε^γ for some $\gamma \geq 1$. Let us call that edge e'.

2b. Either $u_{j,\ell}$ or $u_{j,r}$ is a good node. (If both are good we arbitrarily fix one and call it as z)

The main idea of the following procedure is to scale either the polynomial fed via $u_{j,\ell}$ or the polynomial fed via $u_{j,r}$ by $1/\varepsilon$, which in turn scales the polynomial computed at u_j by $1/\varepsilon$. This scaling is done either immediately by reducing/increasing the exponent of ε appropriately along the edge between the child node and the $\times$ gate and/or is postponed to the next layer by marking the child node. We now state the procedure in Algorithm 2 below.

We know that $\mathcal{S}$ consists of all the marked nodes of $\mathcal{D}_n$. Let $\ell_1, \ell_2, \ldots, \ell_{t'}$ be the marked leaf nodes of $\mathcal{D}_n$. By the construction, we know that these nodes are good nodes. Therefore, their labels have the exponent of ε at least 1. We reduce the exponent of ε in each of these marked leaves by 1. We call the new circuit $\mathcal{D}'_n$ where $\widehat{w}$ denotes the edge label function of $\mathcal{D}'_n$.

Proof of Lemma 4 We prove by using induction on the layer numbers of the nodes in $\mathcal{S}$ in the circuit $\mathcal{D}'_n$.

Base Case: Consider the layer m of $\mathcal{D}'_n$. This layer consists of the leaf nodes of circuit $\mathcal{D}'_n$. By our construction, it is easy to note that the base case holds.

Inductive Case: We assume that the inductive hypothesis holds for layer numbered $i + 1$ and show that it holds for layer i. We break this case into two parts depending on whether i is even or odd.

Set $\mathcal{E} = \phi$
  for $j = 1$ *to* t **do**
    if *exactly one of the nodes from* $u_{j,\ell}$ *and* $u_{j,r}$ *is marked* **then**
    | we do nothing
    end
    else if *both the nodes* $u_{j,\ell}$ *and* $u_{j,r}$ *are marked* **then**
    | we arbitrarily pick one of the edges from $\{(u_{j,\ell}, u_j), (u_{j,r}, u_j)\}$. Let us
    | call that edge e. we increase the exponent of ε on edge e by 1.
    end
    else if *Case 2a holds* **then**
    | we reduce the exponent of ε on edge e' by 1, that is, we relabel the edge
    | e' by $\varepsilon^{\gamma-1}$.
    end
    else if *Case 2b holds* **then**
    | we mark the node z. Let $\mathcal{K} \subseteq \mathcal{E}$ such that each node $a \in \mathcal{K}$ is a parent of
    | z.
    | **for** *each* $a \in \mathcal{K}$ **do**
    | | the exponent of the ε on edge (z, a) is increased by 1.
    | **end**
    end
  | Update $\mathcal{E} = \mathcal{E} \cup \{u_j\}$.
end

Algorithm 2: Procedure to mark nodes on layer $i+1$ when $i+1$ is even.

i **is Even:** We know that the layer numbered i consists of the addition gates. Let $u_1, u_2, \ldots, u_t$ be the marked nodes of layer i. For each $j \in [t]$, let $u_{j,1}, u_{j,2}, \ldots, u_{j,f(j)}$ be children of node u_j. For each $j \in [t]$ we know that $f_{u_j} = \sum_{k=1}^{f(j)} w(u_{j,k}, u_j) \times f_{u_{j,k}}$ and therefore, $\widehat{f}_{u_j} = \sum_{k=1}^{f(j)} \widehat{w}(u_{j,k}, u_j) \times \widehat{f}_{u_{j,k}}$. By our construction, for all $k \in [f(j)]$, either $u_{j,k}$ is a marked node or the edge label function $\widehat{w}$ updates the weight on edge $(u_{j,k}, u_j)$ as follows : $\widehat{w}(u_{j,k}, u_j) = \frac{1}{\varepsilon} w(u_{j,k}, u_j)$. Let $\mathcal{M} \subseteq [f(j)]$ is such that for each $a \in \mathcal{M}$, $u_{j,a}$ is a marked node. Inductively, we know that $\widehat{f}_{u_{j,a}} = \frac{1}{\varepsilon} f_{u_{j,a}}$. Therefore,

$$\widehat{f}_{u_j} = \sum_{k=1}^{f(j)} \widehat{w}(u_{j,k}, u_j) \times \widehat{f}_{u_{j,k}} \tag{1}$$

$$= \sum_{a \in \mathcal{M}} \widehat{w}(u_{j,a}, u_j) \times \widehat{f}_{u_{j,a}} + \sum_{b \in \overline{\mathcal{M}}} \widehat{w}(u_{j,b}, u_j) \times \widehat{f}_{u_{j,b}} \tag{2}$$

$$= \sum_{a \in \mathcal{M}} w(u_{j,a}, u_a) \times \frac{1}{\varepsilon} f_{u_{j,a}} + \sum_{b \in \overline{\mathcal{M}}} \frac{1}{\varepsilon} w(u_{j,b}, u_j) \times f_{u_{j,b}} = \frac{1}{\varepsilon} f_{u_j} \tag{3}$$

i **is odd** : This case is similar to the previous one. We skip the other details of this case. $\square$

By repeatedly applying the step 2, we obtain a circuit $\mathcal{C}'_n$ which computes the polynomial $\varepsilon^0 Q$. This finishes the proof of Theorem 1.

3.3   mVNP and mVF

We state the definition of the monotone variant of VNP.

Definition 9. *Let $f_n(x_1, x_2, \ldots, x_{k(n)})$ be a p-bounded polynomial family. We say that f_n is in mVNP if there exists a family $g_n \in$ mVP such that $f_n = \sum_{y_1, y_2 \ldots, y_{k'(n)} \in \{0,1\}} g_n(x_1, x_2, \ldots, x_k, y_1, y_2, \ldots, y_{k'(n)})$, where $k'(n)$ is polynomially bounded in n.*

We now recall the definition of $\overline{\mathsf{VNP}}$ from [9].

Definition 10 ([9]). *Let $f_n(x_1, x_2, \ldots, x_{k(n)})$ be a p-bounded polynomial family. We say that f_n is in $\overline{\mathsf{VNP}}$ if there exists a p-bounded family $\hat{f}_n \in$ VNP over the field $\mathbb{F}(\varepsilon)$ and a function $t : \mathbb{N} \longrightarrow \mathbb{N}$ such that $\hat{f}_n(x) = f_n(x) + \varepsilon f_{n,1}(x) + \varepsilon f_{n,2}(x) + \ldots + \varepsilon^{t(n)} f_{n,t(n)}(x)$ for some $f_{n,1}, f_{n,2}, \ldots, f_{n,t(n)}$ defined over $\mathbb{F}$.*

We can now define the class $\overline{\mathsf{mVNP}}$.

Definition 11. *Let $f_n(x_1, x_2, \ldots, x_{k(n)})$ be a p-bounded polynomial family. We say that f_n is in $\overline{\mathsf{mVNP}}$ if there exist a p-bounded family $\hat{f}_n \in$ mVNP over $\mathbb{R}_+[\varepsilon, \varepsilon^{-1}]$ and a function $t : \mathbb{N} \longrightarrow \mathbb{N}$ such that $\hat{f}_n(x) = f_n(x) + \varepsilon f_{n,1}(x) + \varepsilon f_{n,2}(x) + \ldots + \varepsilon^{t(n)} f_{n,t(n)}(x)$ for some $f_{n,1}, f_{n,2}, \ldots, f_{n,t(n)}$ defined over $\mathbb{R}$.*

Lemma 5. *mVNP $= \overline{\mathsf{mVNP}}$*

Proof. Let $f_n(x_1, x_2, \ldots, x_{k(n)}) \in \overline{\mathsf{mVNP}}$. By the definition of $\overline{\mathsf{mVNP}}$, we know that there exists a p-family $\hat{f}_n \in$ mVNP (over the field $\mathbb{F}(\varepsilon)$) such that $\hat{f}_n(x) = f_n(x) + \varepsilon f_{n,1}(x) + \varepsilon f_{n,2}(x) + \ldots + \varepsilon^{t(n)} f_{n,t(n)}(x)$ for some $f_{n,1}, f_{n,2}, \ldots, f_{n,t(n)} \in \mathbb{R}[X]$. We know that $\hat{f}_n = \sum_{y_1, y_2 \ldots, y_{k'(n)} \in \{0,1\}} g_n(X, Y)$, where $k'(n)$ is polynomially bounded in n and $g_n(X, Y) \in$ mVP (over the field $\mathbb{F}(\varepsilon)$). Let $g_n(X, a_1, a_2, \ldots, a_{k'(n)})$ denote the evaluation of the polynomial $g_n(X, Y)$ at $y_1 = a_1$, $y_2 = a_2$, $\ldots$, $y_k = a_{k'(n)}$. Since the polynomial $\hat{f}_n$ is monotone and that it converges, each of the summand in $\hat{f}_n = \sum_{y_1, y_2 \ldots, y_{k'(n)} \in \{0,1\}} g_n(X, Y)$ must also necessarily converge. That is, for any boolean setting of variables $y_1 = a_2, y_2 = a_2, \ldots, y_{k'(n)} = a_{k'(n)}$, the polynomial $g_n(X, a_1, a_2, \ldots, a_{k'(n)})$ must converge. Therefore, $g_n(X, Y)$ must also converge. By Theorem 1, there exists a circuit C_{g_n} which computes $g_n(X, Y)$ such that there are no negative exponents of ε in any of its labels. Let $g_n(X, Y)_{\varepsilon=0}$ denote the polynomial obtained after substituting $\varepsilon = 0$ in $g_n(X, Y)$. We know that $f_n(x) = \sum_{y_1, y_2 \ldots, y_{k'(n)} \in \{0,1\}} g_n(X, Y)_{\varepsilon=0}$. By the definition of mVNP, the result follows. $\square$

We now state the final lemma of this section.

Lemma 6. *mVF $= \overline{\mathsf{mVF}}$*

We use the proof of mVBP $= \overline{\mathsf{mVBP}}$ from [4] to prove Lemma 6. We skip the details of this proof.

4 Complexity of $\mathrm{Det}_n^{\neq k}(X)$

It is not very hard to see that $\mathrm{Det}_n^{\neq k}(X)$ is VBP complete for $k = 1$. The proof idea is similar to the proof which shows that $\mathrm{Det}_n(X)$ is VBP complete. In the rest of this section, we prove that $\mathrm{Det}_n^{\neq k}(X)$ is VNP complete for all $k \geq 2$.

Proof Idea. Before going into the details of the complexity of $\mathrm{Det}_n^{\neq k}(X)$ ($k \geq 2$), we will look at the various ingredients required to show that $\mathrm{Det}_n^{\neq k}(X)$ ($k \geq 2$) is VNP-complete. In Sect. 4.1, we discuss the gadget construction $\mathcal{H}_k$ and its properties. In Sect. 4.2, we discuss the rosette construction and its properties. For any $f_n(X) \in$ VNP, we use the gadgets $\mathcal{H}_k$ and rosettes to construct the graph T_m ($m = \mathrm{poly}(n)$) such that $\mathrm{Det}_m^{\neq k}(X)$ defined over T_m computes $f_n(X)$.

4.1 Gadget Construction $\mathcal{H}_k$ and Its Properties

Gadget $\mathcal{H}_k$. For every $k > 1$, we construct a *partial iff gadget* $\mathcal{H}_k$.

- Let $V(\mathcal{H}_k) = \{a_i | 1 \leq i \leq 2k - 1\}$.
- Let $E(\mathcal{H}_k) = \{(a_t, a_{t+1}) | 1 \leq t \leq 2k - 2)\} \cup \{(a_1, a_{k+1}), (a_{k+1}, a_2), (a_{2k-1}, a_1)\}$
  $\cup \{(a_m, a_m) | 3 \leq m \leq 2k - 1\}$. We call the edges in $E(\mathcal{H}_k)$ as the gadget edges.
- The weights of (a_1, a_2) and the self-loop on a_{k+1} are -1. All the other edge weights are 1.

Consider a directed graph G with two distinct edges (u, v) and (u', v'). We say that the gadget $\mathcal{H}_k$ is placed between the edges (u, v) and (u', v'), if we delete both the edges (u, v) and (u', v') in G and we add the following directed edges, $\{(u, a_1), (a_1, v), (u', a_2), (a_2, v')\}$. We set $w(u, a_1) = w(u, v)$ and $w(u', a_2) = w(u', v')$. We set $w(a_1, v) = w(a_2, v') = 1$.

Before getting into the details of the properties of the gadget $\mathcal{H}_k$, we first define *perceived sign* and *perceived monomial* of a cycle cover.

Let G be a directed graph with edge labelling function $\Phi : E(G) \to X \cup \mathbb{F}$, where $X = \{x_i | 1 \leq i \leq n\}$. Let $\mathcal{C} = \{C_1, \ldots, C_k\}$ be a cycle cover of G. Let $\{e_1, \ldots, e_t\}$ be the edges in $\mathcal{C}$. The sign of the cycle cover, which we denote as $sign(\mathcal{C})$ is defined as $(-1)^k$. The monomial of $\mathcal{C}$, denoted as $mon(\mathcal{C})$, is defined as the product of the labels of edges in $\mathcal{C}$, i.e. $\prod_{i=1}^{t} \Phi(e_i)$.

Note that the coefficient of this monomial can be either positive or negative based on the number of negatively signed edge labels in $\mathcal{C}$. We denote this sign by $s(m_{\mathcal{C}})$. Based on this, we define *perceived sign* $\hat{s}(\mathcal{C})$ and *perceived monomial* $\hat{m}_{\mathcal{C}}$ as follows.

Let $\hat{s}(\mathcal{C})$ be $(-1)^k \cdot s(m_{\mathcal{C}})$. And let $\hat{m}_{\mathcal{C}} = m_{\mathcal{C}} \cdot s(m_{\mathcal{C}})$. Note that

$$s(\mathcal{C}) \cdot m_{\mathcal{C}} = \hat{s}(\mathcal{C}) \cdot \hat{m}_{\mathcal{C}}. \tag{4}$$

Properties of $\mathcal{H}_k$. We now state some important properties of $\mathcal{H}_k$. Consider a directed graph $G = (V, E)$ with two distinct edges (u, v) and (u', v'). Let G' be the graph obtained by placing the gadget $\mathcal{H}_k$ between the edges (u, v) and (u', v') in G. Let $\mathcal{C}^{\neq k}$ be a cycle cover (without any cycle of length k) with *perceived sign* $\hat{s}$ which uses either both or none of the edges (u, v) and (u', v') in G, then there exists another cycle cover $\mathcal{C}'^{\neq k}$ with *perceived sign* $\hat{s}'$ in graph G' such that the *perceived monomial* associated with both $\mathcal{C}^{\neq k}$ and $\mathcal{C}'^{\neq k}$ are same and $\hat{s} = \hat{s}'$. We also show that if there exists a cycle cover $\mathcal{C}^{\neq k}$ in G such that it uses exactly one of the edges from (u, v) and (u', v'), then such a cycle cover does not survive in G'.

1. If $\mathcal{C}^{\neq k}$ does not use any of the edges (u, v) and (u', v') then the $\mathcal{C}'^{\neq k}$ consists of all the cycles in $\mathcal{C}^{\neq k}$ and the gadget vertices are covered within themselves by a single cycle $(a_1, a_2, \ldots, a_{2k-1}, a_1)$. The total number of cycles in cycle cover $\mathcal{C}'^{\neq k}$ is one more than the number of cycles in cycle cover $\mathcal{C}^{\neq k}$, but since the weight of edge (a_1, a_2) is -1, the *perceived sign* of cycle cover $\mathcal{C}'^{\neq k}$ is same as the *perceived sign* of cycle cover $\mathcal{C}^{\neq k}$.

2. If $\mathcal{C}^{\neq k}$ uses both the edges (u, v) and (u', v') then $\mathcal{C}'^{\neq k}$ has all the cycles in $\mathcal{C}^{\neq k}$ except that the edges (u, v) and (u', v') are now replaced by directed path (u, a_1, v) and (u', a_2, v') respectively. The vertices a_3, a_4, $\ldots$, a_{2k-1} are covered by self-loops (of weight 1) in $\mathcal{C}'^{\neq k}$. Since the total number of cycles covered by self-loops in the gadget is always odd and the self-loop on the vertex a_{k+1} has weight -1, the *perceived sign* is preserved.

3. If $\mathcal{C}^{\neq k}$ uses the edge (u, v) but the not the edge (u', v') then there is only one way to cover the vertices $a_2, a_3, \ldots, a_{k+1}$ with a cycle $(a_2, a_3, \ldots, a_{k+1}, a_2)$ of length k, which is not a valid cycle cover.

4. If $\mathcal{C}^{\neq k}$ uses the edge (u', v') but not the edge (u, v) then there is only one way to cover the vertices $a_{k+1}, a_{k+2}, \ldots, a_{2k-1}$ with a cycle $(a_{k+1}, a_{k+2}, \ldots, a_{2k-1}, a_{k+1})$ of length k, which is not a valid cycle cover.

5. Note that after placing the gadget $\mathcal{H}_k$, there could be new cycle covers that arise in the graph G' which were not present in graph G. There are only two possible cases. Out of these two cases, in one of the cases, the contribution of all such cycle covers in the overall sum is 0 whereas in the other case, the contribution is not 0. We explain both the cases in detail below.

   (a) Let $\mathcal{C}^{\neq k}$ be a cycle cover consisting of a cycle, say $\mathcal{C}_1$ starting with vertex u followed by the edge (u, a_1) and then the edge path $\mathcal{P}_1 = (a_1, a_2)$ followed by an edge (a_2, v'). Since there are two paths from a_1 and a_2 (within the gadget vertices), there exists another cycle cover $\mathcal{C}'^{\neq k}$ consisting of a cycle $\mathcal{C}'_1$ starting with vertex u followed by the edge (u, a_1) and then the path $\mathcal{P}'_1 = (a_1 - a_{k+1} - a_2)$ followed by an edge (a_2, v') such that the perceived monomials of both $\mathcal{C}^{\neq k}$ and $\mathcal{C}'^{\neq k}$ are same. Moreover, since the number of cycles in cycle cover $\mathcal{C}'^{\neq k}$ is one less than the number of cycles in cycle cover $\mathcal{C}^{\neq k}$, their perceived signs are different. In other words, there exists a bijection Ψ between the set of cycle covers using $\mathcal{P}_1$ in one of its cycles and the set of cycle covers using $\mathcal{P}'_1$ in one of its cycles such that for any $\mathcal{C}$, $\Psi(\mathcal{C})$ and $\mathcal{C}$ have same perceived monomials but with opposite perceived signs. Therefore, the contribution of such cycle covers to the overall sum is 0.

   (b) Let $\mathcal{C}^{\neq k}$ be a cycle cover consisting of a cycle, say $\mathcal{C}_1$ with a path $\mathcal{P}_1$ in it starting with vertex u' followed by the edge (u', a_2) and then a path from vertex a_2 to vertex a_1 (within the gadget $\mathcal{H}_k$) followed by an edge (a_1, v). There are no cancellations possible in this case and therefore, such cycle covers survives.

Remark 3. Unlike the Valiant's iff gadget, in this gadget, we do not guarantee that the contribution of cycle covers (which may arise due to the placing of this gadget) is 0 (see Point 5(b) above). Therefore, we call $\mathcal{H}_k$ *partial iff gadget*. For the proof to work, we exploit the property of the graph on which these gadgets are placed such that the contribution of such new cycle covers is 0.

4.2 Rosette Construction $R(\ell, \mathcal{I})$

In this section, we describe the rosette construction $R(\ell, \mathcal{I})$ for every $\ell > k$ and $\mathcal{I} \subseteq [\ell]$. The construction of $R(\ell, \mathcal{I})$ is very similar to the construction of $R(\ell)$ as stated in [15], except for some modifications to incorporate the restriction about the length of the cycle in the cycle cover. Formally, we consider a directed cycle $\mathcal{C}$ of length ℓ with vertex set $|V(\mathcal{C})| = \{u_1, u_2, \ldots, u_\ell\}$ and $E(\mathcal{C}) = \{e_i = (u_i, u_{i+1}) | 1 \leq i \leq \ell - 1\} \cup \{e_\ell = (u_\ell, u_1)\}$. We call the edges in $|E(\mathcal{C})|$ as *connector edges* and the vertices in $|V(\mathcal{C})|$ as *connector vertices*. Consider a set $\mathcal{S}(\mathcal{I}) = \{e_i | i \in \mathcal{I}\}$. For every edge (u_i, u_j) in $\mathcal{S}(\mathcal{I})$, we add a new vertex $t_{i,j}$ and add edges $(u_i, t_{i,j})$ and $(t_{i,j}, u_j)$. We call the edges in set $\mathcal{S}(\mathcal{I})$ as *participating edges*. We add self-loops on all the vertices of our graph. We arbitrarily pick one of the connector vertices and set the weight of the self-loop on it to 1, whereas the weights of all the other self-loops are set to -1. The weights of all the edges in rosette $R(\ell, \mathcal{I})$ (except the self-loops) are set to 1. This completes the construction of $R(\ell, \mathcal{I})$. It is easy to observe that every $R(\ell, \mathcal{I})$ contains a unique longest cycle. We denote such a cycle by $\mathcal{Z}$. The rosette $R(\ell, \mathcal{I})$ satisfies the following four properties.

1. There is no cycle cover in $R(\ell, \mathcal{I})$ that contains a cycle of length k.
2. For any subset $\phi \neq X \subseteq \mathcal{S}(\mathcal{I})$, there exists exactly one cycle cover of $R(\ell, \mathcal{I})$ which, among the participating edges, contains exactly the edges in X. Such a cycle cover always contains a single cycle which is not a loop and all other remaining vertices are covered with self loops.
3. There are only two cycle covers of $R(\ell, \mathcal{I})$ which contain no participating edges. The first cycle cover consist of only self-loops on each of the vertices in $R(\ell, \mathcal{I})$. The other cycle cover consists of the unique longest cycle $\mathcal{Z}$.
4. There are no other cycle covers in $R(\ell, \mathcal{I})$.

Note that for any cycle cover C of rosette $R(\ell, \mathcal{I})$, the *perceived sign* is -1 and the *perceived monomial* is 1.

4.3 Construction of Graph T_m from $f_n(X) \in$ VNP

Let $f_n(X) \in$ VNP. From the definition of VNP we know that $f_n(X) = \sum_{y_1, y_2, \ldots, y_{p(n)} \in \{0,1\}} g'_n(X, Y)$, where $p : \mathbb{N} \longrightarrow \mathbb{N}$ is polynomially bounded function in n and $g'_n(X, Y)$ is in VP. Moreover, for any $f_n(X) \in$ VNP, $f_n(X) = \sum_{y_1, y_2, \ldots, y_{p(n)} \in \{0,1\}} g_n(X, Y)$, where $g_n(X, Y)$ is in VF [12].

1. Since, $g_n(X,Y) \in \mathsf{VF}$ and $\mathsf{VF} \subseteq \mathsf{VBP}$, there exists an algebraic branching program of size $s = \mathrm{poly}(n)$ to compute $g_n(X,Y)$, say $\mathcal{B}_n$. Let s_0 and t_0 be the source and sink of $\mathcal{B}_n$, respectively. Without loss of generality assume that the length of the longest path from s_0 to t_0 in $\mathcal{B}_n$ is at least k. We add a special vertex α and add directed edges from α to s_0 and from t to α. We set the weights of both the edges (α, s_0) and (t_0, α) to 1. We add self-loops on all the vertices in our constructed graph except α. We set the weight of all self-loops to be 1. We call this graph $\hat{\mathcal{B}}_n$.

2. We know that $Y = \{y_i | 1 \leq i \leq p(n)\}$. Let $occ(i)$ denote the total number of edges in $\hat{\mathcal{B}}_n$ which are labelled with variable y_i. Let $Y_1 = \{y_i | y_i \in Y, occ(i) > k\}$ and $Y_2 = \{y_i | y_i \in Y, occ(i) \leq k\}$. Cleary, $Y = Y_1 \cup Y_2$. For every $y_i \in Y_1$, we consider a rosette $R(occ(i), \mathcal{I} = [occ(i)])$. Similarly, for every $y_j \in Y_2$, we consider a rosette $R(occ(j) + k, \mathcal{I} \subset [occ(j)])$, where the set $|\mathcal{S}(\mathcal{I})| = occ(j)$. We call the graph constructed so far as the *partial graph* denoted by $\tilde{T}_m$.

3. For any $y_i \in Y_1$, let $c_{y_i,1}, c_{y_i,2}, \dots, c_{y_i,occ(i)}$ be the connector edges in $R(occ(i), \mathcal{I} = [occ(i)])$. Let $e_{y_i,1}, e_{y_i,2}, \dots, e_{y_i,occ(i)}$ be the distinct edges in $\hat{\mathcal{B}}_n$ which are labelled with variable y_i. We place $\mathcal{H}_k$ between the edges $c_{y_i,t}$ and $e_{y_i,t}$ for all $1 \leq t \leq occ(i)$.

4. For any $y_j \in Y_2$, let $c_{y_j,1}, c_{y_j,2}, \dots, c_{y_j,occ(j)}, \dots, c_{y_j,occ(j)+k}$ be the connector edges in $R(occ(j) + k, \mathcal{I} \subset [occ(j) + k])$. Without loss of generality, let us assume that $\mathcal{S}(\mathcal{I}) = \{c_{y_j,1}, c_{y_j,2}, \dots, c_{y_j,occ(j)}\}$. Let $e_{y_j,1}, e_{y_j,2}, \dots, e_{y_j,occ(j)}$ be the distinct edges in $\hat{\mathcal{B}}_n$ which are labelled with variable y_j. We place "partial iff gadget" $\mathcal{H}_k$ between the edges $c_{y_j,t}$ and $e_{y_j,t}$ for all $1 \leq t \leq occ(j)$.

This completes the construction of graph T_m. It is easy to note that $m = \mathrm{poly}(n)$.

4.4   Proof of **VNP** Hardness of $\mathrm{Det}_n^{\neq k}(X)$ for $k \geq 2$

Let $f_n(X) \in \mathsf{VNP}$. We have $f_n(X) = \sum_{y_1,y_2,\dots,y_{p(n)} \in \{0,1\}} g_n(X,Y)$, where $p(\cdot)$ is polynomially bounded function in n and $g_n(X,Y)$ is in VF. We know that $\mathcal{B}_n$ computes $g_n(X,Y)$. Let $\mathcal{P} = \{\mathcal{P}_i | 1 \leq i \leq \mu\}$ be the set of all $s_0 - t_0$ paths in $\mathcal{B}_n$. Let m_i be the monomial associated with $\mathcal{P}_i$ formed by multiplying all the edges of $\mathcal{B}_n$ participating in path $\mathcal{P}_i$. Let $m_j' := m_j(a_1, \dots, a_{p(n)})$ be an evaluation of m_j for a specific Boolean setting of Y variables.

Recall that T_m is the graph constructed from $f_n(X)$ as stated in Sect. 4.3. We will prove that for every non-zero m_j' there is a unique cycle cover $\mathcal{C}$ in T_m such that the product of the perceived sign of $\mathcal{C}$ and the perceived monomial of $\mathcal{C}$ is exactly equal to m_j'. Moreover, there are no other cycle covers in T_m. This will prove the VNP hardness. We skip the other details of the proof in the conference version.

Acknowledgement. We would like to thank Shourya Pandey, Radu Curticapean, and the anonymous reviewers who gave useful comments on the earlier draft of the paper.

References

1. Allender, E., Wang, F.: On the power of algebraic branching programs of width two. Comput. Complex. **25**(1), 217–253 (2016)
2. Bini, D.: Relations between exact and approximate bilinear algorithms. Applications. Calcolo **17**(1), 87–97 (1980)
3. Bini, D.A., Capovani, M., Romani, F., Lotti, G.: $O(n^{2.7799})$ complexity for $n \times n$ approximate matrix multiplication (1979)
4. Bläser, M., Ikenmeyer, C., Mahajan, M., Pandey, A., Saurabh, N.: Algebraic branching programs, border complexity, and tangent spaces. In: 35th Computational Complexity Conference, CCC 2020. LIPIcs, vol. 169, pp. 21:1–21:24 (2020)
5. Bringmann, K., Ikenmeyer, C., Zuiddam, J.: On algebraic branching programs of small width. J. ACM (JACM) **65**(5), 1–29 (2018)
6. Chaugule, P., Limaye, N., Varre, A.: Variants of homomorphism polynomials complete for algebraic complexity classes. ACM Trans. Comput. Theory **13**(4), 21:1–21:26 (2021)
7. Durand, A., Mahajan, M., Malod, G., de Rugy-Altherre, N., Saurabh, N.: Homomorphism polynomials complete for VP. Chicago J. Theor. Comput. Sci. **2016** (2016)
8. Grochow, J.A., Mulmuley, K.D., Qiao, Y.: Boundaries of VP and VNP. In: 43rd International Colloquium on Automata, Languages, and Programming, ICALP. LIPIcs, vol. 55, pp. 34:1–34:14 (2016)
9. Ikenmeyer, C., Sanyal, A.: A note on VNP-completeness and border complexity. Inf. Process. Lett. **176**, 106243 (2022)
10. Kumar, M.: On the power of border of depth-3 arithmetic circuits. ACM Trans. Comput. Theory (TOCT) **12**(1), 1–8 (2020)
11. Mahajan, M., Vinay, V.: Determinant: combinatorics, algorithms, and complexity. Chicago J. Theor. Comput. Sci. (1997)
12. Malod, G., Portier, N.: Characterizing Valiant's algebraic complexity classes. In: Královič, R., Urzyczyn, P. (eds.) MFCS 2006. LNCS, vol. 4162, pp. 704–716. Springer, Heidelberg (2006). https://doi.org/10.1007/11821069_61
13. Mulmuley, K.D., Sohoni, M.: Geometric complexity theory I: an approach to the P vs. NP and related problems. SIAM J. Comput. **31**(2), 496–526 (2001)
14. Raz, R.: Elusive functions and lower bounds for arithmetic circuits. In: Proceedings of the Fortieth Annual ACM Symposium on Theory of Computing, pp. 711–720. ACM (2008)
15. Saurabh, N.: Algebraic models of computation. M.S. thesis (2012)
16. Shpilka, A., Yehudayoff, A.: Arithmetic circuits: a survey of recent results and open questions. Found. Trends® Theor. Comput. Sci. **5**(3–4), 207–388 (2010)
17. Valiant, L.G.: Completeness classes in algebra. In: Proceedings of the Eleventh Annual ACM Symposium on Theory of Computing, STOC 1979, pp. 249–261 (1979)

MCSP is Hard for Read-Once
Nondeterministic Branching Programs

Ludmila Glinskih[1] and Artur Riazanov[2,3(✉)]

[1] Department of Computer Science, Boston University, Boston, USA
[2] EPFL, Lausanne, Switzerland
artur.riazanov@epfl.ch
[3] The Henry and Marylin Taub Faculty of Computer Science, Technion, Haifa, Israel

Abstract. We show that every read-once nondeterministic branching program computing the Minimum Circuit Size Problem on inputs of length N has size $\Omega(N^{\log \log(N)})$. This is the first superpolynomial lower bound on the size of **1-NBP** computing MCSP. This lower bound is tight for the version of MCSP restricted to a linear circuit size parameter.

To show this result we adapt a conditional lower bound of Ilango [10] on the deterministic Turing Machine time complexity of computing MCSP^*, the generalization of MCSP to partial functions. In contrast, our lower bound is unconditional and holds even for the total MCSP function.

En route, we get two results that may be of independent interest:
- The size of the minimal **1-NBP** computing MCSP equals, up to a constant factor, the size of the minimal **1-NBP** computing MCSP^*;
- The size of any **1-NBP** computing $(2n \times 2n)$-Bipartite Independent Set is $\Omega(n!)$.

1 Introduction

Branching programs[1] are a non-uniform model of computation that capture time-space trade-offs. Polynomial size deterministic and non-deterministic branching programs exactly recognize the non-uniform analogues of complexity classes **L** and **NL**, respectively [5]. Branching program size is polynomially related to Boolean formula size [17], and every branching program can be converted to a Boolean circuit of roughly the same size.

Restricted branching programs have been studied extensively. Read-once branching programs provide a way to analyze the power of linear-time non-uniform computations with logarithmic memory with an extra restriction that each variable is read at most once.

The Minimum Circuit Size Problem (MCSP) has been a central problem of study in computational complexity in recent years. This problem asks: given as input the truth-table of a Boolean function, and a size parameter s, determine whether the function can be computed by a Boolean circuit of size s. As the

---

[1] Sometimes are also called decision diagrams.

© Springer Nature Switzerland AG 2022
A. Castañeda and F. Rodríguez-Henríquez (Eds.): LATIN 2022, LNCS 13568, pp. 626–640, 2022.
https://doi.org/10.1007/978-3-031-20624-5_38

input length of MCSP is exponential in the input length of the function defined by the given truth table, it is clear that MCSP is contained in **NP**. Although it is not known whether MCSP is an **NP**-complete problem, it is widely believed to be hard to compute. Starting from the work of Kabanets and Cai [12] there has been mounting evidence that MCSP $\notin$ **P** [11,18], for example MCSP $\in$ **P** implies that widely believed cryptographic assumptions do not hold. On the other hand, there are complexity-theoretic barriers to proving **NP**-hardness of MCSP [15,16].

Recently [10] Ilango showed that, assuming the Exponential Time Hypothesis (ETH), the partial function variant of MCSP, denoted MCSP*, requires superpolynomial time to compute. Hence, ETH implies MCSP* $\notin$ **P**. MCSP is a special case of MCSP*, so the former is not harder than the latter. But an existence of a Turing reduction from MCSP* to MCSP is not known, even if we allow for a polynomial number of calls to an MCSP oracle.

Stronger lower bounds on MCSP are known for other models of computation. In recent years lower bounds that match the best lower bounds in corresponding models of computation were shown for DeMorgan [3], **CNF**, and **DNF** formulas [3], and for $\mathbf{AC}^0$ [3] and $\mathbf{AC}^0[p]$ [9] circuits computing MCSP. Polynomial lower bounds were also shown for deterministic and non-deterministic general branching programs [4]. Using local hitting set generators in [4] the authors proved exponential lower bounds on read-once **co**-non-deterministic branching programs computing MCSP. However, for non-deterministic read-once branching programs there are no known superpolynomial lower bounds against MCSP. In Sect. 1.2 we discuss why the technique from [4] is unlikely to work for non-deterministic read-once branching programs. Intuitively, as MCSP is naturally solved by nondeterministic guessing and checking, it is harder to find "weaknesses" of nondeterministic models of computations in order to obtain strong lower bounds. In this work we address this question and obtain a tight lower bound for read-once non-deterministic branching programs in a specific parameter regime of MCSP.

1.1 Results and Techniques

We study the complexity of computing MINIMUM CIRCUIT SIZE PROBLEM by read-once non-deterministic branching programs. Our main result is the following.

Theorem 1. *Every* 1-NBP *computing* MCSP *on inputs of length* N *requires size* $N^{\Omega(\log\log(N))}$. *Moreover, the size of the minimal* 1-NBP *computing* $\text{MCSP}_{s=n-1} : \{0,1\}^N \to \{0,1\}$ *(MCSP with the size parameter set to* $n-1$*) is* $N^{\Theta(\log\log(N))}$, *where* $N = 2^n$.

Our proof of the lower bound consists of three steps. First, we show a tight exponential lower bound on the size of read-once nondeterministic branching programs for a certain graph problem that is known to be ETH-hard. Second, we use a constructive reduction from the work of Ilango [10] that allows us to lift our lower bound for the graph problem to MCSP* with a linear size parameter. Finally, we show that the sizes of the minimal read-once nondeterministic branching programs computing MCSP and MCSP* are polynomially related.

The first step is the most technical one. To prove Theorem 3 we use bottleneck counting, adapting methods of classical lower bound for CLIQUE by Borodin, Razborov, and Smolensky [1] to show that $(2n \times 2n)$-BIPARTITE PERMUTATION INDEPENDENT SET $((2n \times 2n)$-BPIS) problem is not in 1-**NBP**$[2^{o(n \log n)}]$.

The second step is implemented in Theorem 4, where we show that Ilango's reduction can be implemented as a transformation of 1-**NBP**s.

Finally, in Theorem 6, we show that the size of the minimal 1-**NBP** computing MCSP is insensitive to whether we compute the total or partial MCSP function:

$$1\text{-}\mathbf{NBP}[\text{MCSP}] = \Theta(1\text{-}\mathbf{NBP}[\text{MCSP}^*]).$$

The proof of the upper bound is an application of the simple guess-and-check strategy that is natural for non-deterministic computations, and is provided for completeness in Lemma 4.

1.2   Related Work

Strong exponential lower bounds on MCSP are known for various model of computations such as $\mathbf{AC}^0$ circuits, and $\mathbf{AC}^0[p]$ circuits. Note, that $\mathbf{AC}^0$ circuits, and $\mathbf{AC}^0[p]$ circuits are incomparable with 1-**NBP**. Tseitin formulas on a d-regular expander graphs require 1-**NBP** of exponential size, though they can be computed by a **CNF** of a polynomial size [8]. Mod q function is hard for $\mathbf{AC}^0$, and $\mathbf{AC}^0[p]$ (for $p \neq q$) formulas, but has linear 1-**NBP** representation [20].

In contrast with other lower bounds against MCSP for weak computational models, which are based on the intuition that a random function is hard, our lower bound is based on explicit linear circuit lower bounds inherited from Ilango's technique [10]. Instead of relying on ETH, we prove an unconditional lower bound on $(2n \times 2n)$-BPIS for 1-**NBP**s.

The complexity of MCSP for branching programs was studied in [4], where the authors prove $N^{1.5-o(1)}$ lower bounds on MCSP and MKTP and an exponential lower bound on the size of 1-**coNBP**. The latter is the most relevant to this work. The proof of this lower bound goes roughly as follows. The authors construct a local hitting set (LHS) against 1-**NBP**s. The LHS is a function $H : \{0,1\}^s \rightarrow \{0,1\}^{2^n}$ such that $H(\{0,1\}^s) \subseteq \text{MCSP}_s$ (locality), and for every *small enough* 1-**NBP** D such that $D^{-1}(1) \geq \epsilon 2^{2^n}$ we have that there exists $y \in \{0,1\}^s$ such that $D(H(y)) = 1$. The existence of this function yields a lower bound for 1-**coNBP** computing MCSP.[2]

The part of this proof in [4] that seems to fail for an 1-**NBP** lower bound is a transition from Forbes-Kelley-based [7] LHS against Ordered Binary Decision

---

[2] See Lemma 14 in [4].

Diagrams[3] (OBDDs) to a generator against **1-NBP**s. The authors use a result of [1] which says that if f is computable by a small **1-NBP** then it can be represented as a disjunction of OBDD-computable functions, i.e. $f^{-1}(1)$ is a union of 1-preimages of OBDDs. Now, if we have a LHS that hits all OBDDs with relative 1-preimage size at least $1/2t$, where t is the number of OBDDs in the disjunction representation of f, then it must hit at least one of the OBDDs in the representation of f whenever the relative size of $f^{-1}(1)$ is at least $1/2$.

In order to get a lower bound on **1-NBP** computing MCSP in this setting one needs to construct a LHS against **1-coNBP**s. The main problem we face here is that a function computable by a **1-coNBP** is a *conjunction* of OBDD-computable functions. Thus hitting any of the OBDDs does not guarantee hitting some **1-coNBP** in the intersection.

1.3 Organization

The structure of the paper is as follows:

- In Sect. 2 we give definitions of the branching programs and various versions of MCSP we are working with, together with an overview of a reduction from Ilango's paper [10].
- In Sect. 3 we prove a lower bound on the size of **1-NBP** computing the ETH-hard problem $(2n \times 2n)$-BIPARTITE PERMUTATION INDEPENDENT SET.
- In Sect. 4 we show a reduction from the $(2n \times 2n)$-BIPARTITE PERMUTATION INDEPENDENT SET problem to MCSP*, together with a proof of equivalence between the sizes of **1-NBP** for MCSP* and MCSP.
- In Sect. 5 we provide a proof of an upper bound on the sizes of **1-NBP** computing MCSP and MCSP*, and prove the main theorem of this work, Theorem 1.
- In Sect. 6 we conclude with open problems related to our work.

2 Preliminaries

Throughout this work we denote by $[n]$ a set of n elements $\{1, \ldots, n\}$. Let a and b be partial assignments. If the supports of a and b do not intersect, denote by $a \cup b$ a partial assignment that coincides with a on the support of a and with b on the support of b. For a graph G we denote its set of vertices by $V(G)$ and its set of edges by $E(G)$. Throughout the work a *circuit* is a de Morgan circuit ($\wedge$, $\vee$ and $\neg$ gates) with gates of arity 2.

Definition 1. *A* Branching Program (BP) *is a form of representation of functions. A n-ary function[4] $f : D^n \to \{0,1\}$, where $|D| = d$, is represented by a directed acyclic graph with exactly one source and two sinks. Each non-sink*

---

[3] OBDD is a 1-BP in which variables in every path from the source to a sink appear in the same order.

[4] We will instantiate this definition for $D = \{0,1\}$ and $D = \{0,1,*\}$.

node is labeled with a variable; every internal node has exactly d outgoing edges: labeled with elements of D. One of the sinks is labeled with 1 and the other is labeled with 0. We say that a node queries x if it is labeled with a variable x.

The value of the function for given values of the variables is evaluated as follows: we start a path from the source such that for every node on this path we go along the edge that is labeled with the value of the corresponding variable. This path terminates in a sink. The label of this sink is the value of the function. We will refer to a path that terminates in a σ-labeled sink as σ-path.

Definition 2. *A* Nondeterministic Branching Program (NBP) *is a branching program that additionally has nondeterministic 'guessing' nodes that are unlabeled and have two outgoing unlabeled edges.*

The result of a function represented by a nondeterministic branching program for given values of variables equals 1, if there exists at least one path from the source to the sink labeled with 1 such that for every node labeled with a variable on its path we go along an edge that is labeled with the value of the corresponding variable.

A deterministic or nondeterministic branching program is a (syntactic) *read-once (1-BP or 1-NBP respectively)* if every path from the source to a sink contains at most a single occurrence of each variable.

Size of a deterministic or nondeterministic branching program is a number of deterministic (non-guessing) nodes in it. For a function f we define its *read-once nondeterministic branching program complexity*, denoted as **1-NBP**$[f]$, as a size of minimal non-deterministic read-once branching program that computes f. When we write **1-NBP**[MCSP] we denote a read-once nondeterministic branching program complexity of a characteristic function corresponding to the MCSP language.

Definition 3. *The* Minimum Circuit Size Problem (MCSP(f, s)) *gets as input the truth-table of a Boolean function* $f : \{0,1\}^n \to \{0,1\}$ *of length* $N = 2^n$ *and a size parameter s, and outputs 1 if there exists a Boolean circuit of size at most s that computes f. We use notation* MCSP$_{s'(n)}(f) = $ MCSP$(f, s'(n))$.

Definition 4. *The* Partial Minimum Circuit Size Problem (MCSP*(f^*, s)) *gets as input the partial truth-table of a function* $f^* : \{0,1\}^n \to \{0,1,*\}$ *of length* $N = 2^n$ *and a size parameter s, and outputs 1 if there exists a substitution of each * in the truth-table to a 0/1, transforming f^* to a Boolean function f, such that* MCSP$(f, s) = 1$. *We use notation* MCSP$^*_{s'(n)}(f) = $ MCSP$^*(f, s'(n))$. *Sometimes we abuse notation by using a partial assignment as an argument of* MCSP*.

2.1 Bipartite Permutation Independent Set Problem

In [10] Ilango showed that Partial MCSP is ETH-hard. The main idea of the proof is to show reduction from the $(2n \times 2n)$-Bipartite Permutation Independent Set problem that was previously shown to be ETH-hard in [13].

Definition 5. *In* $(2n \times 2n)$-BIPARTITE PERMUTATION INDEPENDENT SET PROBLEM (BPIS) *we are given an adjacency matrix of an undirected graph* G *over the vertex set* $[2n] \times [2n]$ *where every edge is between the sets of vertices* $J_1 = \{(i,j)|i,j \in [n]\}$ *and* $J_2 = \{(n+i, n+j)|i,j \in [n]\}$. *A graph* G *is a* YES-*instance iff it contains an independent set* $S \subseteq J_1 \cup J_2$ *of size* $2n$ *such that the coordinates of vertices in* S *define a permutation of* $[2n]$, *i.e.* $\forall i \in [2n]\, \exists j, k \in [2n] : v = (i,j), w = (k,i), v, w \in S.$

Note that a permutation of $[2n]$ *corresponding to a* YES-*instance of* $(2n \times 2n)$-BPIS *can be viewed as two permutations on disjoint n-element subsets, one permutation is defined by n vertices from* J_1 *(corresponds to a permutation of elements* $\{1, \ldots, n\}$*), and another by n vertices from* J_2 *(corresponds to a permutation of elements* $\{n+1, \ldots, 2n\}$*). We use each of these interpretations interchangeably.*

Theorem 2 ([13]). $(2n \times 2n)$-BPIS *cannot be solved in deterministic time* $2^{o(n \log n)}$ *unless ETH fails.*

Ilango in [10] shows a $2^{O(n)}$-time reduction from $(2n \times 2n)$-BPIS to MCSP*, hence, showing that MCSP* cannot be solved in deterministic time $2^{o(n \log n)}$ unless ETH fails.

3   1-NBP Lower Bound for Bipartite Permutation Independent Set Problem

Before proving the lower bound, we state one result that we use in our proof. We show that it is sufficient to prove a lower bound on the size of a **1-NBP** computing the $(2n \times 2n)$-BIPARTITE PERMUTATION CLIQUE PROBLEM $((2n \times 2n)$-BPC). This problem is very similar to $(2n \times 2n)$-BPIS, but it asks whether there are two sets of vertices, that both form a permutation on $[n]$, and also form a bipartite clique. Formally we define $(2n \times 2n)$-BPC as follows:

Definition 6. *In* $(2n \times 2n)$-BPC: $\{0,1\}^{([2n] \times [2n])^2} \rightarrow \{0,1\}$ *we are given an adjacency matrix of an undirected graph* G *over the vertex set* $[2n] \times [2n]$ *where every edge is between the sets of vertices* $J_1 = \{(i,j)|i,j \in [n]\}$ *and* $J_2 = \{(n+i, n+j)|i,j \in [n]\}$. *A graph* G *is a* YES-*instance iff it contains a subset* S *of* $2n$ *vertices from* $J_1 \cup J_2$ *that is a bipartite clique in the bipartite graph* $(J_1, J_2, E(G) \cap J_1 \times J_2)$, *and the coordinates of vertices in* S *define a permutation of* $[2n]$, *i.e. for each i there is exactly one vertex with the first coordinate equal to i, and exactly one vertex with the second coordinate equal to i.*

Throughout this section we consider branching programs solving $(2n \times 2n)$-BPIS and $(2n \times 2n)$-BPC. In both cases the input is an adjacency matrix of a graph. We slightly abuse the notation by saying that a node of a branching program queries an edge of a graph. Besides that, we use the following convention to distinguish graphs in $(2n \times 2n)$-BPC and underlying graphs of branching programs: the former has *vertices* and the latter has *nodes*.

Lemma 1. *If there exists a* **1-NBP** *of size* $f(n)$ *computing* $(2n \times 2n)$-BPIS: $\{0,1\}^{([2n] \times [2n])^2} \rightarrow \{0,1\}$, *then exists a* **1-NBP** *of size* $f(n)$ *computing* $(2n \times 2n)$-BPC: $\{0,1\}^{([2n] \times [2n])^2} \rightarrow \{0,1\}$.

Proof. Consider a **1-NBP** B of size s computing $(2n \times 2n)$-BPIS. For all edges in B we change labels from 0 to 1, and from 1 to 0 respectively, getting a **1-NBP** B'. Now, B' accepts only those graphs that are the complements of graphs with two permutations of a set $[n]$ that form a bipartite independent set. Therefore, each such graph has two permutations of $[n]$ that make a bipartite cliques on $2n$ vertices. Hence, we get that B' is a **1-NBP** of size s computing $(2n \times 2n)$-BPC.
□

Now we are ready to show a **1-NBP** lower bound for $(2n \times 2n)$-BPIS.

Theorem 3. *The size of every* 1-NBP *computing* $(2n \times 2n)$-BPIS *is* $2^{\Omega(n \log n)}$.

Proof. First, we show the lower bound on the size of a 1-NBP computing $(2n \times 2n)$-Bipartite Permutation Clique. Then, the same lower bound applies to the $(2n \times 2n)$-Bipartite Permutation Independent Set problem by Lemma 1.

Let B be the smallest **1-NBP** that decides $(2n \times 2n)$-Bipartite Permutation Clique. Consider the set G of all graphs on $2n \cdot 2n$ vertices that are exactly $(2n \times 2n)$-bipartite permutation cliques and have no other edges. On every such graph $x \in G$, the function $(2n \times 2n)$-BPC$(x) = 1$.

For each graph $g \in G$ there is at least one path, from the source to the 1-sink in B that is consistent with g. Pick any of such paths arbitrarily and denote it by π_g. We find the first node v on this path such that:

- v queries the value of the variable x_e;
- π_g assigns x_e to 1;
- v is the first node on the path π_g, such that after setting x_e to 1, for each of $2n$ vertices of the exactly bipartite permutation clique g, π_g assigns 1 to an edge incident to that vertex.

Such node should exist for every graph $g \in G$, as otherwise the 1-NBP would also accept inputs that do not contain a bipartite permutation clique. We call such node v a *red node* corresponding to g in B. We define a mapping $r : G \rightarrow N$, where N is the set of nodes of B and $r(g)$ is the red node corresponding to the graph g.

The plan of our proof is as follows.

1. Consider the set of graphs G. $|G| = n! \cdot n!$ (Lemma 2). For each graph $g \in G$ consider its red node $r(g)$.
2. Show that at most $2 \cdot n!$ graphs in G may share the same red node defined by mapping r (Lemma 3).
3. Since $|G| = n! \cdot n!$ and for each node in N its preimage under r has size at most $2 \cdot n!$, we get that $|N| \geq \frac{n!}{2}$.

Lemma 2. *Set* G *contains exactly* $n! \cdot n!$ *graphs.*

Proof. For each permutation on n vertices exists a unique set of n vertices in each part of a bipartite graph. As we need to choose n vertices in both parts, overall there are $n! \cdot n!$ choices of such vertices. Therefore, there are $n! \cdot n!$ graphs that contain the minimal amount of edges to form a permutation bipartite clique.

Lemma 3. *At most $2 \cdot n!$ graphs in G share the same red node.*

Proof. Consider a node r_0 querying x_e and let $e = uv$. Cover the set $r^{-1}(r_0)$ of all graphs in G that share the same red node r_0 with two (not necessary disjoint) subsets. The first is the set of graphs where u is the endpoint of e that has no incident edges that have been queried before. The second set is the set of cliques where v is the vertex with this property. We prove that each of these subsets has size at most $n!$.

Let g_1 and g_2 be two exact cliques from $r^{-1}(r_0)$ such that u is the vertex with no incident edges queried before e at r_0 (for v the proof is analogous). Let $p_i := \pi_{g_i}$ for $i \in \{1, 2\}$ (see Fig. 1). g_1 and g_2 are determined by the sets of nodes spanned by their edges. Let $U_1, U_2 \subseteq [n] \times [n]$ be the sets of nodes in the first part spanned by g_1 and g_2 respectively and $V_1, V_2 \subseteq ([2n] \setminus [n])^2$ be such sets in the second part. That is, $E(g_i) = U_i \times V_i$ for $i \in \{1, 2\}$. W.l.o.g $u \in U_i$.

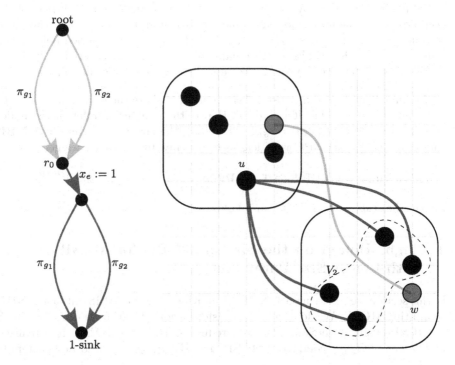

Fig. 1. The paths π_{g_1} and π_{g_2} **Fig. 2.** The graph h_1

Let us color the edges of g_1 and g_2 in two colors: the edges that have been queried along the beginning of the path p_i until r_0 are colored in yellow and the edges that have been queried after r_0 are colored in blue.

Consider the edge $e = uv$ in g_i for $i \in \{1, 2\}$. By the construction of r_0 it is blue and one of the vertices u has exactly n incident blue edges and no incident yellow edge. Then the blue neighbors of u in g_i are exactly the vertices V_i.

Suppose for the contradiction that $V_1 \neq V_2$.

Consider two graphs h_1 and h_2 where h_i consists of yellow edges of g_i and blue edges of g_{3-i}. By the structure of 1-**NBP** B, h_1 and h_2 both contain a permutation clique. Observe that $|E(h_1)| + |E(h_2)| = |E(g_1)| + |E(g_2)| = 2n^2$ thus we can assume w.l.o.g. that $|E(h_1)| \leq n^2$. Since h_1 contains a bipartite permutation clique $|E(h_1)| = n^2$, it has exactly n vertices of degree n in each part and all other vertices have degree zero. However, there are at least $n+1$ vertices of non-zero degree among $([2n] \setminus [n])^2$. Indeed, since $V_1 \neq V_2$ and $|V_1| = |V_2|$ there exists a vertex $w \in V_1 \setminus V_2$. w is not incident to e, so by the construction of r_0 it must have a yellow edge from g_1 incident to it, so its degree in h_1 is at least 1 (see Fig. 2). The vertices from V_2 are all connected to u in h_1 which sums up to $n + 1$. Therefore V_1 must be equal to V_2.

Therefore each graph from the set $r^{-1}(r_0)$ is uniquely determined by the endpoint of e that has no incident yellow edges and by the set of vertices in one of the parts. Since there are at most $n!$ permutation subsets in each part of the graph, we have the upper bound on the number of graphs in $r^{-1}(r_0)$ where u is the endpoint of e that has no incident edges that have been queried before. The same is true for the graphs where v happens to be the last-covered endpoint of e. As $r^{-1}(r_0)$ is covered by these two sets, we have our upper bound $2n!$.

Lemma 3 yields that at most $2n!$ elements of G share the same red node in B. By Lemma 2 we get that $|G| = n! \cdot n!$. Hence, the number of different red nodes in B is at least $|G|/(2 \cdot n!) = \frac{1}{2}n!$. Therefore, the size of the smallest 1-**NBP** computing $(2n \times 2n)$-BPC is at least $\frac{1}{2}n!$. Finally, by Lemma 1 we get

$$\text{1-}\mathbf{NBP}[(2n \times 2n)\text{-BPIS}] \geq \text{1-}\mathbf{NBP}[(2n \times 2n)\text{-BPC}] \geq n!/2 = 2^{\Omega(n \log n)}.$$

Hence, the statement of Theorem 3 holds. $\square$

4   Tight Bound on the Size of 1-NBP for MCSP with Linear Size Parameter

In this section we construct a 1-**NBP** computing $(2n \times 2n)$-BPIS out of a 1-**NBP** computing MCSP. We do it via two transformations. First, from $(2n \times 2n)$-BPIS to MCSP* using a version of Ilango's reduction [10] for 1-**NBP**s, it is realized in Theorem 4. The second from MCSP* to MCSP, it is realized in Theorem 6

Theorem 4. *If* MCSP*$: \{0, 1\}^{2^n} \times [2^n] \to \{0, 1\}$ *is in* 1-**NBP**$[f(n)]$, *then* $(2n \times 2n)$-BPIS *is in* 1-**NBP**$[f(n)]$.

The proof of this theorem amounts to checking that Ilango's reduction from [10] can be implemented as a **1-NBP** transformation.

Let us start with recalling how Ilango's reduction $\mathcal{R}: \{0,1\}^{\binom{2n \times 2n}{2}} \to \{0,1,*\}^{2^{6n}} \times [2^{6n}]$ works[5]. Let $G = ([2n] \times [2n], E)$ be an instance of $(2n \times 2n)$-BPIS. Then the reduction outputs the pair $(t, 6n - 1)$ where t is the truth table of a partial function $\gamma: \{0,1\}^{2n} \times \{0,1\}^{2n} \times \{0,1\}^{2n} \to \{0,1,*\}$ defined as

$$\gamma(x_1, \ldots, x_{2n}, y_1, \ldots, y_{2n}, z_1, \ldots, z_{2n}) =$$

$$
\begin{cases}
\bigvee_{i \in [2n]} (y_i \wedge z_i), & \text{if } x = 0^{2n} \\
\bigvee_{i \in [2n]} z_i, & \text{if } x = 1^{2n} \\
\bigvee_{i \in [2n]} (x_i \vee y_i), & \text{if } z = 1^{2n} \\
0, & \text{if } z = 0^{2n} \\
\bigvee_{i=1}^{n} x_i & \text{if } z = 1^n 0^n \text{ and } y = 0^{2n} \\
\bigvee_{i=n+1}^{2n} x_i & \text{if } z = 0^n 1^n \text{ and } y = 0^{2n} \\
1 & \text{if } \exists((j,k), (n+j', n+k')) \in E \text{ such that } (x,y,z) = (\overline{e_k e_{k'}}, 0^{2n}, e_j e_{j'}) \\
* & \text{otherwise}
\end{cases}
$$

where $x = x_1, \ldots, x_{2n}$, $y = y_1, \ldots, y_{2n}$, $z = z_1, \ldots, z_n$, $e_i \in \{0,1\}^n$ is the vector with 1 in the ith entry and zeroes in all the others, and $\overline{e_i}$ is such that $e_i + \overline{e_i} = 1^n$.

We make use of the following result by Ilango [10]:

Theorem 5 ([10]). *The reduction $\mathcal{R}$ as defined above is such that $G \in (2n \times 2n)-\text{BPIS}$ iff $\mathcal{R}(G) \in \text{MCSP}^*$.*

Proof of Theorem 4. Consider a **1-NBP** D solving MCSP* on $6n$-bit inputs. Let $f_{x,y,z}$ be a propositional variable encoding the function value in $(x,y,z) \in (\{0,1\}^{2n})^3$, i.e. a node of D querying the bit of the truth table corresponding to (x,y,z) is labeled with $f_{x,y,z}$. Let S be the set of $6n$ variables encoding the size threshold. Let $W := \{w_e\}_{e \in \binom{[2n] \times [2n]}{2}}$ be a family of propositional variables encoding an instance of $(2n \times 2n)$-BPIS (in the same sense: it is simply the names of the labels in the nodes of an **1-NBP**).

Let $F_0 = \{f_{x,y,z} \mid \exists j, k, j', k' \in [n]: (x,y,z) = (\overline{e_k e_{k'}}, 0^{2n}, e_j e_{j'})\}$ be a set of variables and $F_1 := S \cup (\{f_{x,y,z} \mid x, y, z \in \{0,1\}^{2n}\} \setminus F_0)$ be its complement. We argue that a **1-NBP** computing $(2n \times 2n)$-BPIS can be obtained from D by relabeling the nodes and edges, and removing some of the nodes. To construct a **1-NBP** computing the composition of MCSP* with $\mathcal{R}$ we need to simulate querying for a variable from $F_0 \cup F_1$ via queries to the variables from W.

1. Observe that $\mathcal{R}$ assigns values to the variables from F_1 independently of its argument, So there exists a partial assignment $\alpha: F_1 \to \{0,1,*\}$ that agrees with $\mathcal{R}$ on F_1.[6] Then for each node labeled with $x \in F_1$ we reroute all the edges incoming to this node to the endpoint of the edge labeled with $\alpha(x)$.

---

[5] We ignore all the edges except for ones between the sets $[n] \times [n]$ and $\{n+1, \ldots 2n\} \times \{n+1, \ldots, 2n\}$, the second argument is represented in binary form.

[6] Notice that here $*$ is a value of a variable and does *not* indicate that a variable is unassigned.

2. For each node u labeled with a variable $f_{(\overline{e_k e_{k'}}, 0^{2n}, e_j e_{j'})}$ from F_0 we apply a syntactic substitution

$$f_{(\overline{e_k e_{k'}}, 0^{2n}, e_j e_{j'})} := \begin{cases} 1 & \text{if } w_{(j,k),(n+j',n+k')} = 1 \\ * & \text{if } w_{(j,k),(n+j',n+k')} = 0 \end{cases}$$

in the following sense. First, replace the label of the node with $w_{(j,k),(n+j',n+k')}$, then remove the 0-labeled edge going out of u, and, finally, replace the labels on 1-labeled and $*$-labeled edges with 1 and 0 respectively.

The resulting **1-NBP** returns 1 iff $\mathcal{R}(w) \in \text{MCSP}^*$ which by Theorem 5 is equivalent to $(2n \times 2n)$-BPIS. Observe that the resulting **1-NBP** is read-once since the relabeling function $f_{(\overline{e_k e_{k'}}, 0^{2n}, e_j e_{j'})} \mapsto w_{(j,k),(n+j',n+k')}$ is injective. □

Now we show that MCSP and MCSP* are have equal **1-NBP** complexity. The same argument works for any non-deterministic model that can guess the extension of the function.

Theorem 6. *The size of the minimal **1-NBP** computing* MCSP *equals the size of the minimal **1-NBP** computing* MCSP*.

Proof. Consider an **1-NBP** D computing MCSP: $\{0,1\}^N \times \{0,1\}^n \to \{0,1\}$. Recall that the first argument encodes a truth table of a function with n-bit input and the second argument encodes the size threshold between 0 and $2^n - 1 = N - 1$. D returns 1 iff the given function has a circuit of size not exceeding the given size parameter.

In contrast, MCSP* : $\{0, 1, *\}^N \times \{0, 1\}^n \to \{0, 1\}$ equals 1 iff there exists a function f that extends the given partial function (that is, for all the $*$'s in the input the value of f in the corresponding point can be arbitrary) and has a circuit of size not exceeding the given size threshold. It is clear that

$$\text{MCSP}^*(p, s) = \bigvee_{\substack{f \in \{0,1\}^N \\ f \text{ extends } p}} \text{MCSP}(f, s).$$

We construct a **1-NBP** D' computing MCSP* in the following way: for each node u of D with successors v_0 and v_1, such that u queries a variable describing the function, add to the diagram a guessing node u'. Then add a new $*$-labeled edge from u to u', and then add two unlabeled edges from u' to v_0 and v_1. The size of the diagram after this operation does not change, as we do not count guessing nodes. We claim that D' constructed that way computes MCSP*.

Consider an arbitrary partial function $p : \{0,1\}^n \to \{0, 1, *\}$ and size threshold $s \in \{0, \ldots, N - 1\}$. Suppose there exists a 1-path ρ in D' that corresponds to an input (p, s). Let uu' be a $*$-labeled edge in ρ. By the construction of D' we have that u' is a guessing node and the successors of u' are exactly the 0-successor and the 1-successor of u. Then u' is followed in ρ by a node v_i which is i-successor of u for $i \in \{0, 1\}$. Let us then replace the edges uu' and $u'v_i$ in ρ

with the edge uv_i. We repeat this process until there are no *-labeled edges in ρ and denote the resulting path as ρ'. Observe that ρ' can be viewed as a 1-path in D as all nodes it contains are from D. Let (f, s) be the input to D that corresponds to ρ'. It is easy to see that by our construction of ρ', f extends p, so $\mathrm{MCSP}^*(p, s) \geq \mathrm{MCSP}(f, s) = D(f, s) = 1$. Therefore $D'(p, s) \leq \mathrm{MCSP}^*(p, s)$.

To prove the opposite inequality, suppose $\mathrm{MCSP}^*(p, s) = 1$ i.e. there exists f extending p such that $\mathrm{MCSP}(f, s) = 1$. Consider a 1-path ρ in D corresponding to f, let us view it as a path in D'. For each node u in ρ that queries a variable x such that $p(x) = *$ we replace the edge $uv_{f(x)}$ with edges uu' and $u'v_{f(x)}$, where v_0 and v_1 are the 0-successor and the 1-successor of u respectively, and u' is *-successor of u. It follows that after the applications of this operation the resulting 1-path ρ' corresponds to the input (p, s) which implies $\mathrm{MCSP}^*(p, s) \leq D'(p, s)$ as needed.   □

Corollary 1. *If MCSP on an n-bit function is in* 1-NBP$[f(n)]$, *then* $(2n \times 2n)$-BPIS *is in* 1-NBP$[O(f(n))]$.

5   Proof of Theorem 1

First, we show a simple upper bound on the size of **1-NBP** computing MCSP^*. The same upper bound then follows for MCSP, as it is a partial case of MCSP^*.

Lemma 4. $\mathrm{MCSP}_s^* \in$ **1-NBP** $\left[s^{O(s)} \cdot 2^n\right]$.

Proof. First observe that

$$\mathrm{MCSP}_s^*(p) = \bigvee_{C \text{ a circuit of size } \leq s} \underbrace{\bigwedge_{x \in \{0,1\}^n} [(p(x) \in \{0,1\}) \implies C(x) = p(x))]}_{=:F_C(p)}$$

$F_C(p)$ checks whether a circuit C computes an extension of a partial function p. There are $s^{\Theta(s)}$ circuits of size at most s: each gate is determined by the indices of int input ($O(\log s)$ bits) and its type (2 bits). Thus there are at most $s^{O(s)}$ many elements in the outer disjunction. Observe that $F_C \in$ **1-NBP**$[2^n]$. That is so since C is a constant, we query the values of the partial function one-by-one, and check that all non-star values coincide with the ones computed by C. For any functions f and g, $f \in$ **1-NBP**$[a] \wedge g \in$ **1-NBP**$[b] \implies f \vee g \in$ **1-NBP**$[a + b]$ (simply unite two diagrams, add a guessing node with edges to their roots and make it a new root). The last implication applied repeatedly to the outer disjunction in the representation of MCSP_s^* yields the statement of the lemma.   □

Now we have all ingredients to prove the Theorem 1:

Theorem 1. *Every* 1-NBP *computing* MCSP *on the inputs of length* N *requires size* $N^{\Omega(\log \log(N))}$. *Moreover, the size of the minimal* 1-NBP *computing* MCSP$(f, n - 1))$ *is* $N^{\Theta(\log \log(N))}$, *where* $f : \{0,1\}^n \to \{0,1\}, N = 2^n$.

Proof. By Theorem 3 we get that $\mathbf{1\text{-}NBP}[(2n \times 2n)\text{-BPIS}] = 2^{\Omega(n \log n)}$. By Theorem 4 we get that the same lower bound holds for MCSP^*, even if we fix size parameter to be linear in the length of the input. Hence we get $\mathbf{1\text{-}NBP}[\mathrm{MCSP}^*_{s(n)=n}] = 2^{\Omega(n \log n)}$. Finally, by Theorem 6 we get that $\mathbf{1\text{-}NBP}$-complexities of MCSP^* and MCSP are the same with respect to multiplicative constant factor. Therefore, $\mathbf{1\text{-}NBP}[\mathrm{MCSP}_{s(n)=n}] = 2^{\Omega(n \log n)}$.

By Lemma 4 we get that $\mathrm{MCSP}^*_{s(n)=n}$ can be decided by a $\mathbf{1\text{-}NBP}[n^{O(n)} \cdot 2^n]$ that, together with the lower bound for $N = 2^n$ gives us a tight bound on $\mathbf{1\text{-}NBP}$-complexity of $\mathrm{MCSP}^*_{s(n)=n}$ and $\mathrm{MCSP}_{s(n)=n}$:

$$\mathbf{1\text{-}NBP}[\mathrm{MCSP}^*_{s(n)=n}] = \mathbf{1\text{-}NBP}[\mathrm{MCSP}_{s(n)=n}] = 2^{\Theta(n \log n)} = N^{\Theta(\log \log N)}.$$

$\square$

6 Future Work

Our lower bound is tight for MCSP with a linear size parameter. One of the main open problems is to improve the lower bound for $\mathbf{1\text{-}NBP}$ from superpolynomial to exponential for MCSP with an exponentially large size-parameter. To obtain such lower bounds using similar methods, we would need a better reduction from an ETH-hard problem and an explicit construction of the truth-table of a function that has higher than linear circuit complexity. Here we reach an obstacle: the best circuit complexity lower bound for an explicit function is linear [6]. And, the recent results on hardness magnification in [2] witness that proving better than linear circuit lower bound may imply a major breakthrough in structural complexity. For example, a $n^{1+\varepsilon}$ circuit lower bound for a sparse language imply that $\mathbf{NP} \not\subset \mathbf{SIZE}[n^k]$ for all k. Therefore, to obtain a better lower bounds on read-once nondeterministic branching programs for MCSP, we would likely have to use much different techniques.

Stronger lower bounds for $\mathbf{1\text{-}NBP}$ are known for many explicit Boolean functions [1]. Though in this work we also show a strongly exponential lower bound on the size of $\mathbf{1\text{-}NBP}$ for explicit function $(2n \times 2n)$-BPC, the same approach doesn't work for MCSP directly. The crux of virtually every $\mathbf{1\text{-}NBP}$ lower bound is some form of analysis of a rectangle cover of the 1-preimage of the function (it is not explicit in Theorem 3, but manifests itself in the recombinations of two intersecting paths). Suppose the input variables of a function are partitioned into two sets X and Y. Then a rectangle is a Cartesian product of a set of assignments to X and a set of assignments to Y. To analyze the size of these rectangles one needs to reason about recombination of inputs: if (x_1, y_1) and (x_2, y_2) belong to a rectangle, then (x_1, y_2) and (x_2, y_1) must as well. In our proof of Theorem 3 we argue about the clique sizes of graphs recombined in this way, deriving that the size of rectangles are small enough. But given $\mathrm{MCSP}_s(x_1 \cup y_1) = \mathrm{MCSP}_s(x_2 \cup y_2) = 1$ what can we say about $\mathrm{MCSP}_s(x_1 \cup y_2)$ and $\mathrm{MCSP}_s(x_2 \cup y_1)$? Clearly, the minimal circuit size of such recombined truth tables can change drastically in either direction, which prohibits explicit combinatorial arguments like ours to work in this situation.

One possibility is to use a similar reduction as in [9], where the authors show hardness of MCSP against $\mathbf{AC}^0[p]$-circuits by constructing a reduction to MCSP from the coin problem.

Another possible direction is extending our lower bound for **1-NBP** to other computational models.

Observation 7. If for a computational model $\mathcal{C}$ the following holds:

1. $(2n \times 2n)$-BPIS is hard for $\mathcal{C}$;
2. Reduction $\mathcal{R}$ is efficiently computable in $\mathcal{C}$,

Then MCSP_n^* cannot be efficiently computed in $\mathcal{C}$. Moreover, if $\mathcal{C}$-complexity of MCSP_n^* and MCSP_n are polynomially related, which holds for reasonable nondeterministic models, then we get that MCSP_n cannot be efficiently computed in $\mathcal{C}$.

We also believe that a similar set of reductions can help us prove hardness of MINIMUM **1-NBP** SIZE PROBLEM, that is not known to be **NP**-hard yet. In this problem we are given a truthtable of a Boolean function f and a size parameter s, and we need to check, whether there is a **1-NBP** of size at most s deciding f. The **NP**-hardness of minimization is already known for **DNF**s [14] and $\mathbf{AC}^0$ [10]. There are weaker results for ordered binary decision diagrams (OBDDs) [19], which are restricted versions of 1-BPs. Can we extend this result to show hardness of minimization 1-BP or **1-NBP**?

Acknowledgements. We thank Mark Bun, Marco Carmosino, and the anonymous reviewers for their helpful comments and suggestions.

Ludmila Glinskih was supported by NSF grants CCF-1947889 and CCF-1909612. Artur Riazanov received funding from the European Union's Horizon 2020 research and innovation programme under grant agreement No. 802020-ERC-HARMONIC.

References

1. Borodin, A., Razborov, A.A., Smolensky, R.: On lower bounds for read-K-times branching programs. Comput. Complex. **3**, 1–18 (1993)
2. Chen, L., Jin, C., Ryan Williams, R.: Hardness magnification for all sparse NP languages. In: FOCS 2019, pp. 1240–1255. IEEE Computer Society (2019)
3. Cheraghchi, M., Hirahara, S., Myrisiotis, D., Yoshida, Y.: One-tape turing machine and branching program lower bounds for MCSP. In: Bläser, M., Monmege, B. (eds.) 38th International Symposium on Theoretical Aspects of Computer Science (STACS 2021), vol. 187, pp. 23:1–23:19. Leibniz International Proceedings in Informatics (LIPIcs). Schloss Dagstuhl - Leibniz-Zentrum für Informatik, Dagstuhl (2021). ISBN 978-3-95977-180-1. https://doi.org/10.4230/LIPIcs.ICALP.2019.39. https://drops.dagstuhl.de/opus/volltexte/2021/13668
4. Cheraghchi, M., Kabanets, V., Lu, Z., Myrisiotis, D.: Circuit lower bounds for MCSP from local pseudorandom generators. In: Baier, C., Chatzigiannakis, I., Flocchini, P., Leonardi, S. (eds.) 46th International Colloquium on Automata, Languages, and Programming, ICALP 2019, Patras, Greece, July 2019, vol. 132, pp. 9–12 (2019). LIPIcs, Schloss Dagstuhl - Leibniz-Zentrum für Informatik **39**(1–39), 14 (2019). https://doi.org/10.4230/LIPIcs.STACS.2021.23

5. Cobham, A.: The recognition problem for the set of perfect squares. In: 7th Annual Symposium on Switching and Automata Theory (SWAT 1966), pp. 78–87. https://doi.org/10.1109/SWAT.1966.30

6. Find, M.G., Golovnev, A., Hirsch, E.A., Kulikov, A.S.: A better-than-3n lower bound for the circuit complexity of an explicit function. In: FOCS 2016, pp. 89–98. IEEE Computer Society (2016)

7. Forbes, M.A., Kelley, Z.: Pseudorandom generators for readonce branching programs, in any order. In: 2018 IEEE 59th Annual Symposium on Foundations of Computer Science (FOCS), pp. 946–955. IEEE (2018)

8. Glinskih, L., Itsykson, D.: Satisfiable Tseitin formulas are hard for nondeterministic read-once branching programs. In: Larsen, K.G., Bodlaender, H.L., Raskin, J.-F. (eds.) 42nd International Symposium on Mathematical Foundations of Computer Science (MFCS 2017), vol. 83. Leibniz International Proceedings in Informatics (LIPIcs) (2017). Schloss Dagstuhl-Leibniz-Zentrum fuer Informatik, Dagstuhl **26**(1–26), 12 (2017). ISBN 978-3-95977-046-0. https://doi.org/10.4230/LIPIcs.MFCS.2017.26. https://drops.dagstuhl.de/opus/volltexte/2017/8076

9. Golovnev, A., Ilango, R., Impagliazzo, R., Kabanets, V., Kolokolova, A., Tal, A.: AC0[p] lower bounds against MCSP via the coin problem. In: ICALP, vol. 132, pp. 66:1–66:15. LIPIcs. Schloss Dagstuhl - Leibniz-Zentrum für Informatik (2019)

10. Ilango, R.: Constant depth formula and partial function versions of MCSP are hard. In: FOCS 2020, pp. 424–433. IEEE (2020)

11. Ilango, R., Ren, H., Santhanam, R.: Hardness on any samplable distribution suffices: new characterizations of one-way functions by meta-complexity. In: Electronic Colloquium on Computational Complexity, p. 82 (2021)

12. Kabanets, V., Cai, J.: Circuit minimization problem. In: STOC, pp. 73–79. ACM (2000)

13. Lokshtanov, D., Marx, D., Saurabh, S.: Slightly superexponential parameterized problems. SIAM J. Comput. **47**(3), 675–702 (2018)

14. Lukas, S., Czort, S.L.A.: The complexity of minimizing disjunctive normal form formulas (1999)

15. McKay, D.M., Murray, C.D., Ryan Williams, R.: Weak lower bounds on resource-bounded compression imply strong separations of complexity classes. In: Proceedings of the 51st Annual ACM SIGACT Symposium on Theory of Computing, STOC 2019, pp. 1215–1225. Association for Computing Machinery, Phoenix (2019). ISBN 9781450367059

16. Murray, C.D., Ryan Williams, R.: On the (non) NP-hardness of computing circuit complexity. Theory Comput. **13**(4), 1–22 (2017)

17. Pratt, V.R.: The effect of basis on size of boolean expressions. In: FOCS, pp. 119–121. IEEE Computer Society (1975)

18. Santhanam, R.: Pseudorandomness and the minimum circuit size problem. In: ITCS, vol. 151, pp. 68:1–68:26. LIPIcs. Schloss Dagstuhl - Leibniz-Zentrum für Informatik (2020)

19. Sieling, D.: The complexity of minimizing and learning OBDDs and FBDDs. Discret. Appl. Math. **122**(1), 263–282 (2002). ISSN 0166-218X. https://doi.org/10.1016/S0166-218X(01)00324-9. https://www.sciencedirect.com/science/article/pii/S0166218X01003249

20. Smolensky, R.: On representations by low-degree polynomials. In: Proceedings of 1993 IEEE 34th Annual Foundations of Computer Science, pp. 130–138 (1993). https://doi.org/10.1109/SFCS.1993.366874

Bounds on Oblivious Multiparty Quantum Communication Complexity

François Le Gall[ID] and Daiki Suruga[(✉)][ID]

Department of Mathematics, Nagoya University, Furocho, Chikusa-ku,
Nagoya 464-860, Japan
m19023e@math.nagoya-u.ac.jp

Abstract. The main conceptual contribution of this paper is investigating quantum multiparty communication complexity in the setting where communication is *oblivious*. This requirement, which to our knowledge is satisfied by all quantum multiparty protocols in the literature, means that the communication pattern, and in particular the amount of communication exchanged between each pair of players at each round is fixed *independently of the input* before the execution of the protocol. We show, for a wide class of functions, how to prove strong lower bounds on their oblivious quantum k-party communication complexity using lower bounds on their *two-party* communication complexity. We apply this technique to prove tight lower bounds for all symmetric functions with AND gadget, and in particular obtain an optimal $\Omega(k\sqrt{n})$ lower bound on the oblivious quantum k-party communication complexity of the n-bit Set-Disjointness function. We also show the tightness of these lower bounds by giving (nearly) matching upper bounds.

Keywords: Quantum complexity theory · Quantum communication complexity · Multiparty communication

1 Introduction

1.1 Background

Communication Complexity. Communication complexity, first introduced by Yao in a seminal paper [30] to investigate circuit complexity, has become a central concept in theoretical computer science with a wide range of applications (see [16,22] for examples). In its most basic version, called two-party (classical) communication complexity, two players, usually called Alice and Bob, exchange (classical) messages in order to compute a function of their inputs. More precisely, Alice and Bob are given inputs $x_1 \in \{0,1\}^n$ and $x_2 \in \{0,1\}^n$, respectively, and their goal is to compute a function $f : (x_1, x_2) \mapsto \{0,1\}$ by communicating with each other, with as little communication as possible.

There are two important ways of generalizing the classical two-party communication complexity: one is to consider classical *multiparty* communication complexity and the other one is to consider *quantum* two-party communication complexity. In (classical) multiparty communication complexity, there are k

© Springer Nature Switzerland AG 2022
A. Castañeda and F. Rodríguez-Henríquez (Eds.): LATIN 2022, LNCS 13568, pp. 641–657, 2022.
https://doi.org/10.1007/978-3-031-20624-5_39

players P_1, P_2, ..., P_k, each player P_i is given an input $x_i \in \{0,1\}^n$. The players seek to compute a given function $f : (x_1, \ldots, x_k) \mapsto \{0,1\}$ using as few (classical) communication as possible.[1] The other way of generalizing the classical two-party communication complexity is *quantum* two-party communication complexity, where Alice and Bob are allowed to use *quantum* communication, i.e., they can exchange messages consisting of quantum bits. Since its introduction by Yao [29], the notion of quantum two-party communication complexity has been the subject of intensive research in the past thirty years, which lead to several significant achievements, e.g., [4,5,10,11,28,29].

In this paper, we consider both generalizations simultaneously: we consider quantum multiparty communication complexity for $k > 2$ parties. This generalization has been the subject of several works [6,17,18,27] but, compared to the two-party case, is still poorly understood.

Set-Disjointness. One of the most studied functions in communication complexity is Set-Disjointness. For any $k \geq 2$ and any $n \geq 1$, the k-party n-bit Set-Disjointness function, written $\text{DISJ}_{n,k}$, has for input a k-tuple $(x_1, \ldots, x_k)$, where $x_i \in \{0,1\}^n$ for each $i \in \{1, \ldots, k\}$. The output is 1 if there exists an index $j \in \{1, \ldots, n\}$ such that $x_1[j] = x_2[j] = \cdots = x_k[j] = 1$, where $x_i[j]$ denotes the j-th bit of the string x_i, and 0 otherwise. The output can thus be written as

$$\text{DISJ}_{n,k}(x_1, \ldots, x_k) = \bigvee_{j=1}^{n} (x_1[j] \wedge \cdots \wedge x_k[j]).$$

Set-Disjointness plays a central role in communication complexity since a multitude of problems can be analyzed via a reduction from or to this function (see [9] for a good survey). In the two party classical setting, the communication complexity of Set-Disjointness is $\Theta(n)$: while the upper bound $O(n)$ is trivial, the proof of the lower bound $\Omega(n)$, which holds even in the randomized setting, is highly non-trivial [15,23]. The k-party Set-Disjointness function with $k > 2$ has received much attention as well, especially since it has deep applications to distributed computing [12]. Proving strong lower bounds on multiparty communication complexity, however, is significantly more challenging than in the two-party model. After much effort, a tight lower bound for k-party Set-Disjointness was nevertheless obtained in the classical setting: recent works [2,25] were able to show a lower bound $\Omega(kn)$ for $\text{DISJ}_{n,k}$, which is (trivially) tight.

In the quantum setting, Buhrman et al. [5] showed that the two-party quantum communication complexity of the Set-Disjointness function is $O(\sqrt{n} \log n)$, which gives a nearly quadratic improvement over the classical case. The logarithmic factor was then removed by Aaronson and Ambainis [1], who thus obtained an $O(\sqrt{n})$ upper bound. A matching lower bound $\Omega(\sqrt{n})$ was then proved by Razborov [24]. For k-party quantum communication complexity, an

---

[1] This way of distributing inputs is called the number-in-hand model. There exists another model, called the number-on-the-forehead model, which we do not consider in this paper.

$O(k\sqrt{n}\log n)$ upper bound is easy to obtain from the two-party upper bound from [5].[2] An important open problem, which is fundamental to understand the power of quantum distributed computing, is showing the tightness of this upper bound. In view of the difficulty in proving the $\Omega(kn)$ lower bound in the classical setting, proving a $\Omega(k\sqrt{n})$ lower bound in the quantum setting is expected to be challenging.

1.2   Our Contributions

Our Model. The main conceptual contribution of this paper is investigating quantum multiparty communication complexity in the setting where communication is *oblivious*. This requirement means that the communication pattern, and in particular the amount of communication exchanged between each pair of players at each round is fixed *independently of the input* before the execution of the protocol. (See Sect. 2.1 for the formal definition.) This requirement is widely used in classical networking systems (e.g., [13,19,21]) and classical distributed algorithms (e.g., [7]), and to our knowledge is satisfied by all known quantum communication protocols (for any problem) that have been designed so far. It has also been considered in the quantum setting by Jain et al. [14, Result 3], who gave an $\Omega(n/r^2)$ bound on the quantum communication complexity of r-round k-party oblivious protocols for a promise version of Set-Disjointness.

Our Results. The main result of this paper holds for a class of functions which has a property that we call *k-party-embedding*. We say that a k-player function f_k is a k-party-embedding function of a two-party function f_2 if the function f_2 can be "embedded" in f_k by embedding the inputs of f_2 in *any* position among the inputs of f_k. Many important functions such as any k-party symmetric function (including as important special cases the Set-Disjointness function $\text{DISJ}_{n,k}$ and the k-party Inner-Product function) or the k-party equality function have this property. For a formal definition of the embedding property, we refer to Definition 2 in Sect. 3. Our main result is as follows.

Theorem 1 (informal). *Let f_k be a k-party function that is a k-party-embedding function of a two-party function f_2. Then the oblivious k-party quantum communication complexity of f_k is at least k times the two-party quantum communication complexity of f_2.*

Theorem 1 enables us to prove strong lower bounds on oblivious quantum k-party communication complexity using the quantum two-party communication complexity.[3] This is useful since two-party quantum communication complexity

---

[2] We will show later (in Theorem 3 in Sect. 5) how to obtain an improved $O(k\sqrt{n})$ upper bound based on the protocol from [1].

[3] Note that in the two-party setting, the notions of oblivious communication complexity and non-oblivious communication complexity essentially coincide, since any non-oblivious communication protocol can be converted into an oblivious communication protocol by increasing the complexity by a factor at most two. To see this, without loss of generality assume that each player sends only one qubit at each round.

is a much more investigated topic than k-party quantum communication complexity, and many tight bounds are known in the two-party setting. For example, we show how to use Theorem 1 to analyze the oblivious quantum k-party communication complexity of $\mathrm{DISJ}_{n,k}$ and obtain a tight $\Omega(k\sqrt{n})$ bound:

Corollary 1. *In the oblivious communication model, the k-party quantum communication complexity of $\mathrm{DISJ}_{n,k}$ is $\Omega(k\sqrt{n})$.*

More generally, Theorem 1 enables us to derive tight bounds for the oblivious quantum k-party communication complexity of arbitrary symmetric functions. Since symmetric functions play an important role in communication complexity [8,20,24,26,31], our results might thus have broad applications. Additionally, we also give lower bounds for non-symmetric functions that have the k-party-embedding property, such as the equality function. Our results are summarized in Table 1.

To complement our lower bounds, we show tight (up to possible poly-log factors) upper bounds for these functions. The upper bounds are summarized in Table 1 as well. Note that if we apply our generic $O(k\log n \cdot G_n(f))$ bound in Table 1 to $\mathrm{DISJ}_{n,k}$, we only get the upper bound $O(k\log n \cdot \sqrt{n})$. We thus prove directly an optimal $O(k\sqrt{n})$ upper bound (Theorem 3) by showing how to adapt the optimal two-party protocol from [1] to the k-party setting.

Table 1. Our results for oblivious quantum k-party communication complexity, along with known bounds for the two-party setting. For a symmetric function f, the notation $G_n(f)$ refers to the quantity defined in Eq. (1).

Functions	2-party protocols		k-party oblivious protocols	
	Lower	Upper	Lower	Upper
Symmetric functions	$\Omega(G_n(f))$ in [24]	$O(\log n \cdot G_n(f))$ in [24]	$\Omega(k \cdot G_n(f))$ Proposition 3	$O(k\log n \cdot G_n(f))$ Theorem 4
Set-Disjointness	$\Omega(\sqrt{n})$ in [24]	$O(\sqrt{n})$ in [1]	$\Omega(k\sqrt{n})$ Corollary 1	$O(k\sqrt{n})$ Theorem 3
Set-Disjointness in M-round $(M \leq O(\sqrt{n}))$	$\tilde{\Omega}(n/M)$ in [3]	$O(n/M)$ (folklore)	$\tilde{\Omega}(k \cdot n/M)$ Proposition 5	$O(k \cdot n/M)$ Corollary 2
Equality function	$\Omega(1)$ (trivial)	$O(1)$ e.g., [16]	$\Omega(k)$ Proposition 4	$O(k)$ Proposition 6

2   Models of Quantum Communication

Notations: All logarithms are base 2 in this paper. We denote $[k] = \{1, \ldots, k\}$. For any set $\mathcal{X}$ and $k \geq 1$, $\mathcal{X}^k := \underbrace{\mathcal{X} \times \cdots \times \mathcal{X}}_{k}$.

Here we formally define the quantum multiparty communication model. As mentioned in Sect. 1.2, this communication model satisfies the oblivious routing condition (or simply the oblivious condition), meaning that the number of qubits used in communication at each round is predetermined (independent of inputs, private randomness, public randomness and outcome of quantum measurements). Since details of the model are important especially when proving lower bounds, we explain the model in detail below.

2.1   Quantum Multiparty Communication Model

In k-party quantum communication model, at each round, players are allowed to send quantum messages[4] to all of the players but the number of qubits used in communication is predetermined. This condition is called *oblivious*. Therefore for any k-player M-round protocol Π, we define the functions $C_{P_i \to P_j} : [M] \to \mathbb{N} \cup \{0\}$ $(i, j \in [k])$ which represent the number of qubits $C_{P_i \to P_j}(m)$ transmitted at m-th round from i-th player to j-th player.

Procedure: Before the execution of the protocol, all players $P_1, \ldots, P_k$ share an entangled state or public randomness if needed. Each player P_i is then given an input. At each round $m \le M$, every player P_i performs some operations (such as unitary operations, measurements, coin flipping) onto P_i's register and send $C_{P_i \to P_1}(m)$ qubits to the player P_1, $C_{P_i \to P_2}(m)$ qubits to the player P_2, $\cdots$, and $C_{P_i \to P_k}(m)$ qubits to the player P_k. All messages from all players are sent simultaneously. This continues until M-th round is finished. Finally, each player P_i output the answer based on the contents of P_i's register.

We define the communication cost of this protocol as

$$\mathrm{QCC}(\Pi) := \sum_{m \in [M]} \sum_{\substack{i, j \in [k] \\ i \ne j}} C_{P_i \to P_j}(m).$$

2.2   Coordinator Model

Let us also describe the definition of the following coordinator model so that discussions on the upper bounds in Sect. 5 become simpler.

In k-party coordinator model, there are k-players, each is given an input, and another player called a coordinator who is not given any input. Each player can communicate only with the coordinator. Similar to the ordinary communication model, the number of qubits used in communication is predetermined. Therefore for any k-player M-round protocol Π in coordinator model, we define the functions $C_{P_i \to \mathrm{Co}} : [M] \to \mathbb{N} \cup \{0\}$ and $C_{\mathrm{Co} \to P_i} : [M] \to \mathbb{N} \cup \{0\}$ for $i \le k$. The value $C_{P_i \to \mathrm{Co}}(m)$ (resp. $C_{\mathrm{Co} \to P_i}(m)$) represent the number of qubits transmitted at m-th round from i-th player to the coordinator (resp. the coordinator to i-th player).

---

[4] Trivially, players can send classical messages using quantum communication in this communication model.

Procedure: Before the execution of the protocol, all players $P_1, \ldots, P_k$ and the coordinator share an entangled state or public randomness if needed. Each player P_i is then given input. At each round $m \leq M$, each player P_i performs some operations onto P_i's register and send $C_{P_i \to \text{Co}}(m)$ qubits to the coordinator. After that, the coordinator, who received $C_{P_1 \to \text{Co}}(m) + \cdots + C_{P_k \to \text{Co}}(m)$ qubits, performs some operations (such as unitary operations, measurements, coin flipping) onto the coordinator's register and sends back $C_{\text{Co} \to P_i}(m)$ qubits to each player P_i. This continues until the M-th round is finished. Finally, each player P_i outputs the answer based on the contents of P_i's register.

We define the communication cost of this protocol as $\text{QCC}_{\text{Co}}(\Pi) := \sum_{m \in [M]} \sum_{i \in [k]} C_{P_i \to \text{Co}}(m) + C_{\text{Co} \to P_i}(m)$.

2.3   Protocol for Computing a Function

We define a protocol computing a function as follows.

Definition 1. *We say a protocol Π computes $f : \mathcal{X}_1 \times \cdots \times \mathcal{X}_k \to \mathcal{Y}$ with error $\varepsilon \in [0, 1/2)$ if*

$$\forall i \in [k], \ \forall x = (x_1, \ldots, x_k) \in \mathcal{X}_1 \times \cdots \times \mathcal{X}_k, \quad \Pr(\Pi_{\text{out}}^i(x) \neq f(x)) \leq \varepsilon$$

where $\Pi_{\text{out}}^i(x)$ denotes P_i's output of the protocol on input x.

We denote by $\mathcal{P}_k(f, \varepsilon)$ the set of k-party protocols computing a function f with error ε in the quantum multiparty communication model. The quantum communication complexity of function f with error ε in the model is defined as $\text{QCC}(f, \varepsilon) := \min_{\Pi \in \mathcal{P}_k(f, \varepsilon)} \text{QCC}(\Pi)$.

We also define the bounded round communication complexity of function f as $\text{QCC}^M(f, \varepsilon) := \min_{\Pi \in \mathcal{P}_k^M(f, \varepsilon)} \text{QCC}(\Pi)$ where we use the superscript M to denote the set of M-round protocols $\mathcal{P}_k^M(f, \varepsilon)$. Regarding the coordinator model, we define $\mathcal{P}_k(f, \varepsilon)_{\text{Co}}$, $\text{QCC}_{\text{Co}}(f, \varepsilon)$, $\mathcal{P}_k^M(f, \varepsilon)_{\text{Co}}$, and $\text{QCC}_{\text{Co}}^M(f, \varepsilon)$ in similar manners as above.

As is easily seen[5], $\text{QCC}^{2M}(f, \varepsilon) \leq \text{QCC}_{\text{Co}}^M(f, \varepsilon) \leq 2\,\text{QCC}^M(f, \varepsilon)$ holds. This means the two models asymptotically have the same power even in bounded round setting.

2.4   Symmetric Functions

A function $f : \{0, 1\}^n \times \{0, 1\}^n \to \{0, 1\}$ is symmetric[6] if there exists a function $D_f : [n] \cup \{0\} \to \{0, 1\}$ such that $f(x, y) = D_f(|x \cap y|)$, where $x \cap y$ is the intersection of the two sets $x, y \subseteq [n]$ corresponding to the strings x, y. This means

---

[5] To show $\text{QCC}^{2M}(f, \varepsilon) \leq \text{QCC}_{\text{Co}}^M(f, \varepsilon)$, assign P_1 the role of the coordinator. To show $\text{QCC}_{\text{Co}}^M(f, \varepsilon) \leq 2\,\text{QCC}^M(f, \varepsilon)$, consider the coordinator only passes messages without performing any operation.

[6] Although a function $f : \{0, 1\}^n \to \{0, 1\}$ is generally said to be symmetric when any permutation on the input does not change the value of f, in this paper we focus on functions of the form $f : \{0, 1\}^n \times \{0, 1\}^n \to \{0, 1\}$, and use the same definition for symmetric functions (predicates) as in [24].

that the function f depends only on the Hamming weight of (the intersection of) the inputs. For any symmetric function $f : \{0,1\}^n \times \{0,1\}^n \to \{0,1\}$, let us write

$$G_n(f) = \sqrt{n l_0(D_f)} + l_1(D_f), \tag{1}$$

where

$$l_0(D_f) = \max \{l \mid 1 \leq l \leq n/2 \text{ and } D_f(l) \neq D_f(l-1)\},$$
$$l_1(D_f) = \max \{n - l \mid n/2 \leq l < n \text{ and } D_f(l) \neq D_f(l+1)\}.$$

Razborov [24] showed the lower bound $\Omega(G_n(f))$ on the quantum two-party communication complexity of any symmetric function f, and also obtained a nearly matching upper bound $O(G_n(f) \log n)$. We also note that for any function D_f, this function is constant on the interval $[l_0(D_f), n-l_1(D_f)]$ by the definitions of $l_0(D_f)$ and $l_1(D_f)$. In Sect. 5.2, we use this fact to prove a nearly matching upper bound on the oblivious quantum multiparty communication model.

Analogously, a k-party function $f : \{0,1\}^{n \cdot k} \to \{0,1\}$ is symmetric when represented as $f(x_1, \ldots, x_k) = D_f(|x_1 \cap \cdots \cap x_k|)$ using some function $D_f : [n] \cup \{0\} \to \{0,1\}$. The k-party n-bit Set-Disjointness function $\mathrm{DISJ}_{n,k}$ defined in Sect. 1 is a symmetric function. The k-party n-bit (generalized) Inner-Product function $\mathrm{IP}_{n,k}$, defined for any $x_1, \ldots, x_k \in \{0,1\}^n$ as

$$\mathrm{IP}_{n,k}(x_1, \ldots, x_k) = (x_1[1] \wedge \cdots \wedge x_k[1]) \oplus \cdots \oplus (x_1[n] \wedge \cdots \wedge x_k[n])$$

is also symmetric.

On the other hand, the k-party n-bit equality function $\mathrm{Equality}_{n,k}$, defined for any $x_1, \ldots, x_k \in \{0,1\}^n$ as

$$\mathrm{Equality}_{n,k}(x_1, \ldots, x_k) = \begin{cases} 1 & \text{if } x_1 = x_2 = \cdots = x_k, \\ 0 & \text{otherwise}, \end{cases}$$

is not symmetric.

3 Lower Bounds

Here we show Proposition 1, which relates the oblivious communication complexity of a k-party function $f_k : \mathcal{X}^k \to \mathcal{Y}$ to the oblivious communication complexity of a two-party function $\tilde{f}_2 : \mathcal{X} \times \mathcal{X} \to \mathcal{Y}$ when f_k is a k-party-embedding function of $\tilde{f}_2$ in the following sense.

Definition 2. *A function* $f_k : \mathcal{X}^k \to \mathcal{Y}$ *is a* k-party-embedding function of $\tilde{f}_2 : \mathcal{X} \times \mathcal{X} \to \mathcal{Y}$ *if for any* $i \in [k]$, *there is a map* $x_{-i} : \mathcal{X} \to \mathcal{X}^{k-1}$ *such that*

$$\forall x_1, x_2 \in \mathcal{X} \quad \tilde{f}_2(x_1, x_2) = f_k([x_{-i}(x_2), i, x_1])$$

holds, where $[y, i, x] := (y_1 \ldots, y_{i-1}, x, y_i, \ldots, y_{k-1})$ *for* $y = (y_i)_{i \leq k-1} \in \mathcal{X}^{k-1}$ *and* $x \in \mathcal{X}$.

For example, $\mathrm{DISJ}_{n,k}$ ($k \geq 2$) is a k-party-embedding function of $\mathrm{DISJ}_{n,2}$ because we can take $\mathcal{X} = \{0,1\}^n$, $\mathcal{Y} = \{0,1\}$ and $x_{-i}(x) = (x, 1^n, \ldots, 1^n)$.

Using this definition, we show the following proposition.

Proposition 1. *Let $f_k : \mathcal{X}^k \to \mathcal{Y}$ be a function and suppose f_k is a k-party-embedding function of $\tilde{f}_2 : \mathcal{X} \times \mathcal{X} \to \mathcal{Y}$. For any protocol $\Pi_k \in \mathcal{P}_k(f_k, \varepsilon)$, there is a two-party protocol $\tilde{\Pi} \in \mathcal{P}_2(\tilde{f}_2, \varepsilon)$ such that $\mathrm{QCC}(\tilde{\Pi}) \leq \frac{2\mathrm{QCC}(\Pi_k)}{k}$ holds.*

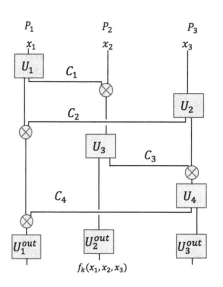

Fig. 1. Example of Π_k for f_k when $k = 3$. (Prior entanglement is omitted.) Assume $\mathrm{QCC}_1(\Pi_k) \leq \mathrm{QCC}(\Pi_k)/k$, i.e., $i_0 = 1$.

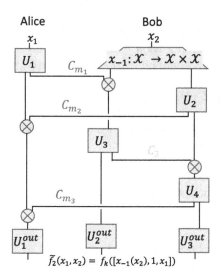

Fig. 2. Protocol $\tilde{\Pi}$ for $\tilde{f}_2$ created from Π_k when $i_0 = 1$. Here, the communication C_3 is internally computed by Bob and the entire communication cost is $\mathrm{QCC}(\tilde{\Pi}) = \mathrm{QCC}_1(\Pi_k)$.

Proof. Without loss of generality, we assume that at each round only one player sends a message in the protocol Π_k. Let $\mathrm{QCC}_i(\Pi_k)$ denote the communication cost of player i, which we define as the sum of the number of qubits exchanged, either sent or received, by player i. For example, in Fig. 1 showing an example[7] of the k-party protocol Π_k,

$$\mathrm{QCC}_1(\Pi_k) = C_1 + C_2 + C_4, \quad \mathrm{QCC}_2(\Pi_k) = C_1 + C_3, \quad \mathrm{QCC}_3(\Pi_k) = C_2 + C_3 + C_4.$$

where C_m denotes the number of qubits sent at the m-th round. This value satisfies the equation $2\mathrm{QCC}(\Pi_k) = \sum_{i \leq k} \mathrm{QCC}_i(\Pi_k)$ where the factor of two

---

[7] In Fig. 1, U_i and U_i^{out} denote classical or quantum operations and $\otimes$ denotes the operation of attaching registers. U_i^{out} usually includes measurement operations to output $f_k(x_1, x_2, x_3)$.

comes from the fact that for each communication C_m, there are two players, one sending C_m and one receiving C_m. This equation implies that there is $i_0 \in [k]$ such that $\mathrm{QCC}_{i_0}(\Pi_k) \leq 2\mathrm{QCC}(\Pi_k)/k$ (independent of the inputs). (This is where the oblivious condition is used. If the protocol is not oblivious, the coordinate i_0 usually varies depending on the player's inputs.)

For i_0, by the definition of the k-party-embedding property, there is a map $x_{-i_0} : \mathcal{X} \to \mathcal{X}^{k-1}$ such that $\tilde{f}_2(x_1, x_2) = f_k([x_{-i_0}(x_2), i_0, x_1])$ holds for any $x_1, x_2 \in \mathcal{X}$. Using the protocol Π_k, we then create a two-party protocol $\tilde{\Pi} \in \mathcal{P}_2(\tilde{f}_2, \varepsilon)$ with communication cost $\mathrm{QCC}_{i_0}(\Pi_k)$. We name the two players in the protocol $\tilde{\Pi}$ Alice and Bob. Each is given x_1, x_2 respectively. In the protocol $\tilde{\Pi}$, Alice plays the role of P_{i_0} and Bob plays the other $k - 1$ roles of $P_1, \ldots, P_{i_0-1}, P_{i_0+1}, \ldots, P_k$. Playing these roles, Alice and Bob simulate the original Π_k with the input $[x_{-i_0}(x_2), i_0, x_1]$. The communication cost of $\tilde{\Pi}$ is $\mathrm{QCC}_{i_0}(\Pi_k)$ because the communication between Alice and Bob is made only when the player P_{i_0} needs to communicate with others in the original protocol Π_k. The other communications are internally computed by Bob. (Figure 2 shows the two-party protocol $\tilde{\Pi}$ created from Π_k.) When the simulation is finished, Alice and Bob can output the answer with error $\leq \varepsilon$ because for the original protocol Π_k for any $i \in [k]$, $\Pr(\Pi_{\mathrm{out}}^i([x_{-i_0}(x_2), i_0, x_1]) \neq f_k([x_{-i_0}(x_2), i_0, x_1]) \leq \varepsilon$ holds. By the k-party-embedding property, we have $f_k([x_{-i_0}(x_2), i_0, x_1]) = \tilde{f}_2(x_1, x_2)$ which indicates $\tilde{\Pi} \in \mathcal{P}_2(\tilde{f}_2, \varepsilon)$ with communication cost $\mathrm{QCC}_{i_0}(\Pi_k) \leq \frac{2\mathrm{QCC}(\Pi_k)}{k}$. $\qquad\square$

We also show a proposition which considers the bounded round setting.

Proposition 2. *Let f_k and $\tilde{f}_2$ be the same as in Proposition 1. For any protocol $\Pi_k \in \mathcal{P}_k^M(f_k, \varepsilon)$, there is a protocol $\tilde{\Pi} \in \mathcal{P}_2^M(\tilde{f}_2, \varepsilon)$ such that $\mathrm{QCC}(\tilde{\Pi}) \leq \frac{\mathrm{QCC}(\Pi_k)}{k}$ holds.*

Proof. In a similar manner as in Proposition 1, we see that there is $i_0 \in [k]$ such that $\mathrm{QCC}_{i_0}(\Pi_k) \leq \frac{2}{k}\mathrm{QCC}(\Pi_k)$ holds. (Note that in this case, we do not restrict the number of players communicating at each round.) We create the desired two-party protocol $\tilde{\Pi}$ by Alice simulating P_{i_0} and Bob simulating all the other players, except for P_{i_0}. In the protocol $\tilde{\Pi}$, Alice and Bob need to communicate only if the player P_{i_0} need to communicate with other players in the original protocol Π_k. Therefore, the communication cost of the protocol satisfies

$$\mathrm{QCC}(\tilde{\Pi}) = \sum_{m \in [M]} \sum_{j \in [k] \setminus \{i_0\}} C_{P_{i_0} \to P_j}(m) + C_{P_j \to P_{i_0}}(m) \leq \frac{2\mathrm{QCC}(\Pi_k)}{k}.$$

We also observe the protocol $\tilde{\Pi}$ is M-round protocol, completing the proof.

Using Proposition 1, we next show the following theorem.

Theorem 1 (formal version). *Let $f_{n,k} : \{0,1\}^{n \cdot k} \to \{0,1\}$ be a k-party-embedding function of $\tilde{f}_n$. Then*

$$\forall n, k, \quad \mathrm{QCC}(f_{n,k}, \varepsilon) \geq \frac{k}{2} \cdot \mathrm{QCC}(\tilde{f}_n, \varepsilon).$$

Proof. Let $\Pi_{n,k}$ be an optimal protocol for $f_{n,k}$, i.e., $\mathrm{QCC}(\Pi_{n,k}) = \mathrm{QCC}(f_{n,k}, \varepsilon)$ $= \min_{\Pi \in \mathcal{P}_k(f_{n,k}, \varepsilon)} \mathrm{QCC}(\Pi)$. By Proposition 1, there is a two-party protocol $\tilde{\Pi} \in \mathcal{P}_2(\tilde{f}_n, \varepsilon)$ satisfying $\mathrm{QCC}(\tilde{\Pi}) \leq \frac{2\mathrm{QCC}(\Pi_{n,k})}{k}$. This yields

$$\mathrm{QCC}(\tilde{f}_n, \varepsilon) \leq \frac{2\mathrm{QCC}(\Pi_{n,k})}{k} = \frac{2\mathrm{QCC}(f_{n,k}, \varepsilon)}{k}$$

which means $\forall n, k, \quad \frac{k}{2}\mathrm{QCC}(\tilde{f}_n, \varepsilon) \leq \mathrm{QCC}(f_{n,k}, \varepsilon)$. □

We can also prove a similar proposition in the bounded round scenario using Proposition 2:

Theorem 2. *Let $f_{n,k}$ and $\tilde{f}_n$ be the same as Theorem 1. Then for any n, k, $\mathrm{QCC}^M(f_{n,k}, \varepsilon) \geq \frac{k}{2} \cdot \mathrm{QCC}^M(\tilde{f}_n, \varepsilon)$ holds.*

Proof. Note that in Proposition 2, the new protocol for $\tilde{f}_n$ preserves the round of the original protocol $\Pi_k \in \mathcal{P}_k^M(f_{n,k}, \varepsilon)$. Therefore in a similar manner as Theorem 1, we get

$$\forall n, k, \quad \frac{k}{2}\mathrm{QCC}^M(\tilde{f}_n, \varepsilon) \leq \mathrm{QCC}^M(f_{n,k}, \varepsilon).$$

□

4   Applications

Here we investigate the lower bounds of some important functions such as Symmetric functions, Set-disjointness and Equality.

We first apply Theorem 1 to symmetric functions. Recall that any k-party symmetric function f can be represented as $f(x_1, \ldots, x_k) = D_f(|x_1 \cap \cdots \cap x_k|)$ (each player is given $x_i(1 \leq i \leq k)$ as input) using some function $D_f : [n] \cup \{0\} \to \{0,1\}$.

Proposition 3. $\mathrm{QCC}(f_{n,k}, 1/3) \in \Omega\left(k\{\sqrt{nl_0(D_{f_{n,k}})} + l_1(D_{f_{n,k}})\}\right)$ *holds for any k-party n-bit symmetric function $f_{n,k}$.*

Proof. For $i \in [k]$, define $x_{-i}(x) := (x, 1^n, \ldots, 1^n) \in \{0,1\}^{n \cdot (k-1)}$. Then we have that for any $i \in [k]$ and any $x_1, x_2 \in \{0,1\}^n$, $f_{n,2}(x_1, x_2) = f_{n,k}([x_{-i}(x_2), i, x_1])$. This implies $f_{n,k}$ is a k-party-embedding function of $f_{n,2}$. Therefore, Theorem 1 yields $\mathrm{QCC}(f_{n,k}, 1/3) \in \Omega(k \cdot \mathrm{QCC}(f_{n,2}, 1/3))$. Applying the well known lower bound $\Omega(\sqrt{nl_0(D_{f_{n,2}})} + l_1(D_{f_{n,2}}))$ of the two-party function $f_{n,2}$ [24], we obtain

$$\mathrm{QCC}(f_{n,k}, 1/3) \in \Omega\left(k\{\sqrt{nl_0(D_{f_{n,k}})} + l_1(D_{f_{n,k}})\}\right).$$

This lower bound is so strong that we get the optimal $\Omega(n \cdot k)$ bound for Inner-Product function (as $l_0(D_{f_{n,k}}) = l_1(D_{f_{n,k}}) = \Theta(n)$ holds) and $\Omega(k\sqrt{n})$ lower bound for Set-disjointness function (as $l_0(D_{f_{n,k}}) = 1$ and $l_1(D_{f_{n,k}}) = 0$ holds), which turns out to be optimal in our setting as described in Sect. 5.

Next, we examine the lower bound of Equality function.

Proposition 4. $QCC(Equality_{n,k}, 1/3) \in \Omega(k)$.

Proof. For $i \in [k]$, define $x_{-i} : \{0,1\}^n \to \{0,1\}^{n \cdot (k-1)}$ as $x_{-i}(x) = (x, x, \ldots, x)$ (i.e., making $k - 1$ copies of x). Then we have that for any $i \in [k]$, any $x_1, x_2 \in \{0,1\}^n$, $Equality_{n,2}(x_1, x_2) = Equality_{n,k}([x_{-i}(x_2), i, x_1])$. Therefore by Theorem 1, the trivial lower bound $\Omega(1)$ of two-party n-bit Equality function yields $QCC(Equality_{n,k}, 1/3) \in \Omega(k)$. $\qquad\qquad\qquad\qquad\qquad\qquad\square$

We also prove a lower bound in bounded round scenario using Theorem 2.

Proposition 5. $QCC^M(DISJ_{n,k}, 1/3) \in \Omega\left(n \cdot k / (M \log^8 M)\right)$.

Proof. Since the two-party M-round Set-disjointness requires $\Omega\left(n/(M \log^8 M)\right)$ communication [3], we obtain $QCC^M(DISJ_{n,k}, 1/3) \in \Omega\left(n \cdot k / (M \log^8 M)\right)$. This is nearly tight as shown in Sect. 5. $\qquad\qquad\qquad\qquad\qquad\qquad\square$

5   Matching Upper Bounds

In this section, we show the upper bound $O(k\sqrt{n})$ for $DISJ_{n,k}$, the upper bound $O(k \log n(\sqrt{nl_0(D_f)} + l_1(D_f)))$ for symmetric functions and the upper bound $O(k)$ for $Equality_{n,k}$ by creating efficient protocols for each function. Without being noted explicitly, all of our protocols satisfy the oblivious routing condition. These are (sometimes nearly) matching upper bounds since we have the same lower bounds in Sect. 4.

5.1   Optimal Protocol for $DISJ_{n,k}$

Here, we adopt the arguments from [1, Section 7], which gives a two-party protocol for DISJ with $O(\sqrt{n})$-communication cost, and present the protocol with $O(k \cdot \sqrt{n})$ cost in coordinator model.

Let us first briefly describe the two-party protocol given in [1]. In the two-party protocol, inputs are represented as $(x_{ijk})_{(i,j,k)\in[n^{1/3}]^3} \in \{0,1\}^n$ to Alice and $(y_{ijk})_{(i,j,k)\in[n^{1/3}]^3} \in \{0,1\}^n$ to Bob.[8] They cooperate and communicate with each other to perform the following five operations (and their inverse operations) onto their registers:

Denoting the register for Alice (for Bob) as $|\psi\rangle_A$ ($|\psi\rangle_B$) and Alice holding an additional one qubit register $|z\rangle$,

- $O : |(i,j,k), z\rangle_A |(i,j,k)\rangle_B \mapsto |(i,j,k), z \oplus (x_{ijk} \wedge y_{ijk})\rangle_A |(i,j,k)\rangle_B$
- $W : |(i,j,k), z\rangle_A |(i,j,k)\rangle_B \mapsto (-1)^z |(i,j,k), z\rangle_A |(i,j,k)\rangle_B$
- S_V: For a subset $V \subset [n^{1/3}]^3$,

$$S_V : |(i,j,k), z\rangle_A |(i,j,k)\rangle_B \mapsto \begin{cases} (-1)^{\delta_{0z}} |(i,j,k), z\rangle_A |(i,j,k)\rangle_B & \text{if } (i,j,k) \in V . \\ |(i,j,k), z\rangle_A |(i,j,k)\rangle_B & \text{otherwise} \end{cases}$$

---

[8] If $n^{1/3}$ is not an integer, inputs are embedded to a larger cube of size $\lceil n^{1/3} \rceil^3$. In this case, for any coordinate $i \in \lceil n^{1/3} \rceil^3 \setminus [n]$, the i-th inputs x_i and y_i are set to 0.

- For $d = 1, 2, 3$,

$$Z_{\text{plus}}^d : |(i, j, k), z\rangle_A |(i, j, k)\rangle_B \mapsto \begin{cases} |(i+1, j, k), z\rangle_A |(i+1, j, k)\rangle_B & \text{if } d = 1, \\ |(i, j+1, k), z\rangle_A |(i, j+1, k)\rangle_B & \text{if } d = 2, \\ |(i, j, k+1), z\rangle_A |(i, j, k+1)\rangle_B & \text{if } d = 3. \end{cases}$$

- $Z_{\alpha,\beta}^d$ ($d = 1, 2, 3$; $\alpha, \beta \in \mathbb{C}$ s.t. $|\alpha|^2 + |\beta|^2 = 1$)
  For specific subsets V_1, V_2 and V_3 (defined in the original paper [1]),

$$Z_{\alpha,\beta}^1 : |(i, j, k), z\rangle_A |(i, j, k)\rangle_B \mapsto \begin{cases} (\alpha|i\rangle_{AB}^{\otimes 2} + \beta|i+1\rangle_{AB}^{\otimes 2})|z\rangle_A |j, k\rangle_{AB}^{\otimes 2} \\ \qquad\qquad\qquad\qquad \text{if } (i, j, k) \in V_1, \\ \\ |(i, j, k), z\rangle_A |(i, j, k)\rangle_B \quad \text{otherwise.} \end{cases}$$

$$Z_{\alpha,\beta}^2 : |(i, j, k), z\rangle_A |(i, j, k)\rangle_B \mapsto \begin{cases} (\alpha|j\rangle_{AB}^{\otimes 2} + \beta|j+1\rangle_{AB}^{\otimes 2})|z\rangle_A |i, k\rangle_{AB}^{\otimes 2} \\ \qquad\qquad\qquad\qquad \text{if } (i, j, k) \in V_2, \\ \\ |(i, j, k), z\rangle_A |(i, j, k)\rangle_B \quad \text{otherwise.} \end{cases}$$

$$Z_{\alpha,\beta}^3 : |(i, j, k), z\rangle_A |(i, j, k)\rangle_B \mapsto \begin{cases} (\alpha|k\rangle_{AB}^{\otimes 2} + \beta|k+1\rangle_{AB}^{\otimes 2})|z\rangle_A |i, j\rangle_{AB}^{\otimes 2} \\ \qquad\qquad\qquad\qquad \text{if } (i, j, k) \in V_3, \\ \\ |(i, j, k), z\rangle_A |(i, j, k)\rangle_B \quad \text{otherwise.} \end{cases}$$

As shown in [1], each operation is achieved by at most two qubits of communication: O and $Z_{\alpha,\beta}^d$ requires two qubits of communication and other operations W, S_V and Z_{plus}^d are achieved without any communication. In the two-party protocol, Alice and Bob use these operations $O(\sqrt{n})$ times to compute Set-Disjointness. Therefore in total $2O(\sqrt{n}) = O(\sqrt{n})$ communication is sufficient in two-party case.

In the following theorem, we explain how to extend these operations appropriately for the quantum multiparty communication model.

Theorem 3. $\text{QCC}(\text{DISJ}_{n,k}, 1/3) \in O(k\sqrt{n})$.

Proof. Without loss of generality, we assume that the communication model is the coordinator model. In our extension, the coordinator plays the role of Alice and k-players play the role of Bob. For example, the query operation O is extended to

$$O_k : |(i, j, l), z\rangle_{\text{Co}} |(i, j, l)\rangle_{P_1 \cdots P_k}^{\otimes k} \mapsto |(i, j, l), z \oplus (x_{ijl}^1 \wedge \cdots \wedge x_{ijl}^k)\rangle_{\text{Co}} |(i, j, l)\rangle_{P_1 \cdots P_k}^{\otimes k}.$$

Note that in this case each player $P_{i'}$ who is given an input $(x_{ijk}^{i'})$ holds the register $|i, j, l\rangle$. We now explain how to extend each operation to that of coordinator model and how many qubits are needed to perform these operations.

– $O_k : |(i,j,l), z\rangle_{\text{Co}}|(i,j,l)\rangle_{P_1\cdots P_k}^{\otimes k} \mapsto |(i,j,l), z \oplus (\wedge_{i' \leq k} x_{ijl}^{i'})\rangle_{\text{Co}}|(i,j,l)\rangle_{P_1\cdots P_k}^{\otimes k}$
First, each player $P_{i'}$ performs $|(i,j,l)\rangle|0\rangle \mapsto |(i,j,l)\rangle|x_{ijl}\rangle$ using an auxiliary qubit $|0\rangle$. Then they send the encoded qubits $|x_{ijl}^1\rangle \cdots |x_{ijl}^k\rangle$ to the coordinator who next performs

$$|(i,j,l), z\rangle|x_{ijl}^1, \ldots, x_{ijl}^k\rangle \mapsto |(i,j,l), z \oplus (\wedge_{i' \leq k} x_{ijl}^{i'})\rangle|x_{ijl}^1, \ldots, x_{ijl}^k\rangle$$

and return $|x_{ijl}^{i'}\rangle$ to each player $P_{i'}$. Finally, each player clears the register: $|x_{ijl}^{i'}\rangle \mapsto |0\rangle$. The total communication cost for this operation is $2k$ qubits.

– $Z_{\alpha,\beta}^1 : |(i,j,l), z\rangle_{\text{Co}}|(i,j,l)\rangle_{P_1\cdots P_k}^{\otimes k} \mapsto \alpha|(i,j,l), z\rangle_{\text{Co}}|(i,j,l)\rangle_{P_1\cdots P_k}^{\otimes k} + \beta|(i+1,j,l), z\rangle_{\text{Co}}|(i+1,j,l)\rangle_{P_1\cdots P_k}^{\otimes k}$ iff $(i,j,k) \in V_1$.
First, the coordinator creates $|0\rangle_C^{\otimes k} \mapsto \alpha|0\rangle_C^{\otimes k} + \beta|1\rangle_C^{\otimes k}$ from auxiliary qubits $|0\rangle_C^{\otimes k}$ and performs $|(i,j,l)\rangle_{\text{Co}}(\alpha|0\rangle_C^{\otimes k} + \beta|1\rangle_C^{\otimes k}) \mapsto \alpha|(i,j,l)\rangle_{\text{Co}}|0\rangle_C^{\otimes k} + \beta|(i+1,j,l)\rangle_{\text{Co}}|1\rangle_C^{\otimes k}$. Next, the coordinator sends the auxiliary qubits to players, each player is given the single qubit. On the received qubit $C_{i'}$ and the register $P_{i'}$, each player performs, for $a \in \{0,1\}$, $|a\rangle_{C_{i'}}|(i,j,l)\rangle_{P_{i'}} \mapsto |a\rangle_{C_{i'}}|(i+a,j,l)\rangle_{P_{i'}}$. They then return the auxiliary qubits to the coordinator who finally performs $|(i+1,j,l)\rangle_{\text{Co}}|1\rangle_C^{\otimes k} \mapsto |(i+1,j,l)\rangle_{\text{Co}}|0\rangle_C^{\otimes k}$ iff $(i,j,k) \in V_1$. The total communication for this operation is $2k$ qubits. Other operations $Z_{\alpha,\beta}^2, Z_{\alpha,\beta}^3$ are achieved similarly.

– The operations W, S, Z_{plus}^d are done without any communication.

Suppose in the two-party protocol, Alice and Bob finally create the state $\sum_{(i,j,l)} \alpha_{ijl}|(i,j,l), z_{ijl}\rangle_A|(i,j,l)\rangle_B$ applying the above operations $O(\sqrt{n})$ times. Then, with the same amount of steps, the coordinator and players can create the state $\sum_{(i,j,l)} \alpha_{ijl}|(i,j,l), z_{ijl}\rangle_{\text{Co}}|(i,j,l)\rangle_{P_1\cdots P_k}^{\otimes k}$ whose amplitude $\{\alpha_{ijl}\}$ is the same as of the state in two party protocol. Therefore, the coordinator can output the same answer as in the two-party protocol which implies that the success probability in the coordinator protocol is the same as in the two-party protocol. After the coordinator obtain the answer, he/she finally send it to all players.

Let us consider the communication cost needed to achieve this protocol. In the coordinator model, there are $O(\sqrt{n})$ steps and each step needs at most $2k$ communication. This shows $O(k\sqrt{n})$ upper bound of $\text{DISJ}_{n,k}$ in the coordinator model.   □

Using the protocol described in Theorem 3, we can create $O(M)$-round protocol for $\text{DISJ}_{n,k}$ with $O(n \cdot k/M)$ communication cost when $M \leq O(\sqrt{n})$. The important fact here is that in the protocol with $O(k\sqrt{n})$ cost, the coordinator and players interact only for $O(\sqrt{n})$ rounds. To create the desired protocol, let us now divide the input $x \in \{0,1\}^n$ into n/M^2 sub-inputs, each contains M^2 elements. We next apply the above protocol *in parallel* with the n/M^2 sub-inputs where each of sub-inputs uses $O(M)$ rounds and $O(kM)$ communication. The new protocol still uses $O(M)$ rounds although the communication cost grows up to $\frac{n}{M^2}O(kM) = O(n \cdot k/M)$. The success probability is still the same since the original protocol is a one-sided error protocol.

Therefore, this protocol has $O(M)$ rounds and the communication cost $O(n \cdot k/M)$ which nearly matches the lower bound $\Omega\left(n \cdot k/(M \log^8 M)\right)$ described in Sect. 4. By converting this M-round coordinator protocol to the ordinary protocol, we obtain the following corollary:

Corollary 2. $\mathrm{QCC}^M(\mathrm{DISJ}_{n,k}, 1/3) \in O(n \cdot k/M)$ *when* $M \leq O(\sqrt{n})$.

5.2 Symmetric Functions

Theorem 4. *For any k-party n-bit symmetric function $f_{n,k}$,*

$$\mathrm{QCC}(f_{n,k}, 1/3) \in O\left(k \log n \{ \sqrt{nl_0(D_{f_{n,k}})} + l_1(D_{f_{n,k}}) \} \right).$$

Proof. This proof is a generalization of [24, Section 4] which investigates only the two-player setting. Without loss of generality, we assume our model of communication to be the coordinator model.

Let us first describe some important facts based on the arguments in [24,26]. For any symmetric function $f_{n,k}$, the corresponding function $D_{f_{n,k}}$ is constant on the interval $[l_0(D_{f_{n,k}}), n - l_1(D_{f_{n,k}})]$. Without loss of generality, assume $D_{f_{n,k}}$ takes 0 on the interval. (If $D_{f_{n,k}}$ takes 1 on the interval, we take the negation of $D_{f_{n,k}}$.) Defining D_0 and $D_1 : [n] \cup \{0\} \to \{0,1\}$ as

$$D_0(m) = \begin{cases} D_{f_{n,k}}(m) & \text{if } m \leq l_0 \\ 0 & \text{else} \end{cases}, \quad D_1(m) = \begin{cases} D_{f_{n,k}}(m) & \text{if } m > n - l_1 \\ 0 & \text{else} \end{cases}$$

(abbreviating $l_0 := l_0(D_{f_{n,k}})$ and $l_1 := l_1(D_{f_{n,k}})$), $D_{f_{n,k}} = D_0 \vee D_1$ holds. Therefore, by defining $f_{n,k}^0(x_1, \ldots, x_k) := D_0(|x_1 \cap \cdots \cap x_k|)$ and $f_{n,k}^1(x_1, \ldots, x_k) := D_1(|x_1 \cap \cdots \cap x_k|)$, we get $f_{n,k} = f_{n,k}^0 \vee f_{n,k}^1$. This means, computing $f_{n,k}^0$ and $f_{n,k}^1$ separately is sufficient to compute the entire function $f_{n,k}$. As another important fact needed for our explanation, we note that the query complexity of $f_{n,k}^0$ equals to $O(\sqrt{nl_0(D_{f_{n,k}})})$ which is proven in [20].

Let us now explain a nearly optimal protocol for symmetric functions. In this protocol, a coordinator computes $f_{n,k}$ by computing $f_{n,k}^0$ and $f_{n,k}^1$ separately. By the query complexity $O(\sqrt{nl_0(D_{f_{n,k}})})$ of the function $f_{n,k}^0$, the coordinator can compute $f_{n,k}^0$ by performing the query $|i\rangle|y\rangle \mapsto |i\rangle|(x_1^i \wedge \cdots \wedge x_k^i) \oplus y\rangle$ $(1 \leq i \leq n)$ for $O(\sqrt{nl_0(D_{f_{n,k}})})$ times. We describe then how this query is implemented with $O(k \log n)$ communication. For an $|i\rangle|y\rangle$, the procedure goes as follows.

(Step 1) Coordinator creates k copies of $|i\rangle$: $|i\rangle|y\rangle \mapsto |i\rangle^{\otimes k+1}|y\rangle$ (using additional ancillary qubits to create $|i\rangle^{\otimes k}$) and sends each of them to k players.

(Step 2) Each player j $(1 \leq j \leq k)$ of the k players performs $|i\rangle|0\rangle \mapsto |i\rangle|x_j^i\rangle$ and sends the coordinator these qubits. Now the coordinator obtains $|i\rangle^{\otimes k+1}|y\rangle|(x_1^i, \ldots, x_k^i)\rangle$

(Step 3) Coordinator performs $|y\rangle|(x_1^i, \ldots, x_k^i)\rangle \mapsto |(\wedge_{j \le k} x_j^i) \oplus y\rangle|(x_1^i, \ldots, x_k^i)\rangle$
and return each $|i\rangle|x_j^i\rangle$ to player j.

(Step 4) Each player j clears the register $|i\rangle|x_j^i\rangle \mapsto |i\rangle|0\rangle$ and returns $|i\rangle$. Now
the coordinator's register is $|i\rangle^{\otimes k+1}|(x_1^i \wedge \cdots \wedge x_k^i) \oplus y\rangle$.

This is how the query is implemented.

Let us analyze how many qubits of communication is needed for this query.
Step 1 requires $k \cdot \log n$ communication since $i \in [n]$ is represented by $\log n$
qubits. Step 2 requires $k(\log n + 1)$ qubits by $\log n$ qubits for i and one qubit for
$x_j^i \in \{0, 1\}$. Step 3 requires the same $k(\log n+1)$ qubits and Step 4 requires $k \log n$
qubits. Therefore, in total, this query is implemented by $O(k \log n)$ qubits of
communication and this protocol requires $O(k \log n \sqrt{n l_0(D_{f_{n,k}})})$ communication
to compute $f_{n,k}^0$.

We next explain a protocol to compute $f_{n,k}^1$ which is simpler comparing to
the protocol for $f_{n,k}^0$. For the coordinator to compute $f_{n,k}^1$, each player j tells
the coordinator (1) if there are more than $(n - l_1(D_{f_{n,k}}))$ zeros and (2) where
are zeros in the input $(x_j^i)_{i \le n}$ when the first answer is YES (if the answer is
NO, the player send an arbitrary bit string). This takes one qubit for the first
question and $\log \left(\Sigma_{m=n-l_0(D_{f_{n,k}})+1}^n \binom{n}{m} \right) = O(l_1(D_{f_{n,k}}) \log n)$ qubits[9] for the
second question. This needs $O(k l_1(D_{f_{n,k}}) \log n)$ communication in total. With
the information from players, the coordinator compute $f_{n,k}^1$ as follows. First, if
there is NO answered in the first question, the coordinator determines $f_{n,k}^1 = 0$. If
every answer of the first question from players is YES, the coordinator calculates
how many zeros are in $x_1 \cap \cdots \cap x_k \in \{0, 1\}^n$ which in turn gives the value of
$|x_1 \cap \cdots \cap x_k|$. Therefore, the coordinator can compute the value of the function
$f_{n,k}^1(x_1, \ldots, x_k) = D_1(|x_1 \cap \cdots \cap x_k|)$ even when there is no NO answer from
players.

Combining these two protocols (one is for $f_{n,k}^0$ and the other is for $f_{n,k}^1$), the
coordinator computes $f_{n,k}$ with $O(k \log n \sqrt{n l_0(D_{f_{n,k}})}) + O(k l_1(D_{f_{n,k}}) \log n) = O(k \log n \{\sqrt{n l_0(D_{f_{n,k}})} + l_1(D_{f_{n,k}})\})$ communication. Finally, the coordinator
sends the output to all players with the negligible k bits of communication.   □

5.3   Optimal Protocol for Equality$_{n,k}$

Applying a public coin protocol with $O(1)$ communication cost for Equality$_{n,2}$
(see, e.g., [16]) to the k-party case, we obtain the following proposition.

Proposition 6. QCC(Equality$_{n,k}$, 1/3) $\in O(k)$.

Acknowledgement. FLG was partially supported by JSPS KAKENHI grants
Nos. JP16H01705, JP19H04066, JP20H00579, JP20H04139 and by MEXT Q-LEAP
grants Nos. JPMXS0118067394 and JPMXS0120319794. DS would like to take this
opportunity to thank the "Nagoya University Interdisciplinary Frontier Fellowship"
supported by JST and Nagoya University.

---

[9] Here we use the fact that for any $n_0 \le \frac{n}{2}$, $\log \left(\Sigma_{m=n-n_0+1}^n \binom{n}{m} \right) = O(n_0 \log n)$.

References

1. Aaronson, S., Ambainis, A.: Quantum search of spatial regions. In: 44th Annual IEEE Symposium on Foundations of Computer Science. pp. 200–209 (2003). https://doi.org/10.1109/SFCS.2003.1238194

2. Braverman, M., Ellen, F., Oshman, R., Pitassi, T., Vaikuntanathan, V.: A tight bound for set disjointness in the message-passing model. In: 54th Annual Symposium on Foundations of Computer Science, pp. 668–677 (2013). https://doi.org/10.1109/FOCS.2013.77

3. Braverman, M., Garg, A., Ko, Y.K., Mao, J., Touchette, D.: Near-optimal bounds on the bounded-round quantum communication complexity of disjointness. SIAM J. Comput. **47**(6), 2277–2314 (2018). https://doi.org/10.1137/16M1061400

4. Buhrman, H., Cleve, R., Watrous, J., de Wolf, R.: Quantum fingerprinting. Phys. Rev. Lett. **87**, 167902 (2001). https://doi.org/10.1103/PhysRevLett.87.167902

5. Buhrman, H., Cleve, R., Wigderson, A.: Quantum vs. classical communication and computation. In: 30th Annual ACM Symposium on Theory of Computing, pp. 63–68 (1998). https://doi.org/10.1145/276698.276713

6. Buhrman, H., van Dam, W., Høyer, P., Tapp, A.: Multiparty quantum communication complexity. Phys. Rev. A **60**, 2737–2741 (1999). https://doi.org/10.1103/PhysRevA.60.2737

7. Censor-Hillel, K., Kaski, P., Korhonen, J.H., Lenzen, C., Paz, A., Suomela, J.: Algebraic methods in the congested clique. Distrib. Comput. **32**(6), 461–478 (2016). https://doi.org/10.1007/s00446-016-0270-2

8. Chakraborty, S., Chattopadhyay, A., Høyer, P., Mande, N.S., Paraashar, M., de Wolf, R.: Symmetry and quantum query-to-communication simulation. In: 39th International Symposium on Theoretical Aspects of Computer Science, vol. 219, pp. 20:1–20:23 (2022). https://doi.org/10.4230/LIPIcs.STACS.2022.20

9. Chattopadhyay, A., Pitassi, T.: SIGACT news complexity theory column 67. SIGACT News **41**(3), 58 (2010). https://doi.org/10.1145/1855118.1886592

10. Cleve, R., Buhrman, H.: Substituting quantum entanglement for communication. Phys. Rev. A **56**(2), 1201 (1997). https://doi.org/10.1103/PhysRevA.56.1201

11. Cleve, R., van Dam, W., Nielsen, M., Tapp, A.: Quantum entanglement and the communication complexity of the inner product function. In: Williams, C.P. (ed.) QCQC 1998. LNCS, vol. 1509, pp. 61–74. Springer, Heidelberg (1999). https://doi.org/10.1007/3-540-49208-9_4

12. Drucker, A., Kuhn, F., Oshman, R.: The communication complexity of distributed task allocation. In: ACM Symposium on Principles of Distributed Computing, PODC 2012, pp. 67–76 (2012). https://doi.org/10.1145/2332432.2332443

13. Fiat, A., Woeginger, G.J. (eds.): Online Algorithms: The State of the Art. Lecture Notes in Computer Science, vol. 1442. Springer, Heidelberg (1998). https://doi.org/10.1007/BFb0029561

14. Jain, R., Radhakrishnan, J., Sen, P.: A lower bound for the bounded round quantum communication complexity of set disjointness. In: 44th Annual IEEE Symposium on Foundations of Computer Science, pp. 220–229 (2003). https://doi.org/10.1109/sfcs.2003.1238196

15. Kalyanasundaram, B., Schintger, G.: The probabilistic communication complexity of set intersection. SIAM J. Discret. Math. **5**(4), 545–557 (1992). https://doi.org/10.1137/0405044

16. Kushilevitz, E., Nisan, N.: Communication Complexity. Cambridge University Press, Cambridge (1996). https://doi.org/10.1017/CBO9780511574948

17. Le Gall, F., Nakajima, S.: Multiparty quantum communication complexity of triangle finding. In: 12th Conference on the Theory of Quantum Computation, Communication and Cryptography (2018). https://doi.org/10.4230/LIPIcs.TQC.2017.6

18. Lee, T., Schechtman, G., Shraibman, A.: Lower bounds on quantum multiparty communication complexity. In: 24th Annual IEEE Conference on Computational Complexity, pp. 254–262 (2009). https://doi.org/10.1109/ccc.2009.24

19. Ni, L.M., McKinley, P.K.: A survey of wormhole routing techniques in direct networks. Computer **26**(2), 62–76 (1993). https://doi.org/10.1109/2.191995

20. Paturi, R.: On the degree of polynomials that approximate symmetric Boolean functions (preliminary version). In: 24th Annual ACM Symposium on Theory of Computing, pp. 468–474 (1992). https://doi.org/10.1145/129712.129758

21. Räcke, H., Schmid, S.: Compact oblivious routing. In: 27th Annual European Symposium on Algorithms, vol. 144, pp. 75:1–75:14 (2019). https://doi.org/10.4230/LIPIcs.ESA.2019.75

22. Rao, A., Yehudayoff, A.: Communication Complexity: and Applications. Cambridge University Press, Cambridge (2020). https://doi.org/10.1017/9781108671644

23. Razborov, A.A.: On the distributional complexity of disjointness. Theoret. Comput. Sci. **106**(2), 385–390 (1992). https://doi.org/10.1016/0304-3975(92)90260-m

24. Razborov, A.A.: Quantum communication complexity of symmetric predicates. Izvestiya Math. **67**(1), 145 (2003). https://doi.org/10.1070/im2003v067n01abeh000422

25. Rosén, A., Urrutia, F.: A new approach to multi-party peer-to-peer communication complexity. In: 10th Innovations in Theoretical Computer Science, vol. 124, pp. 64:1–64:19 (2018). https://doi.org/10.4230/LIPIcs.ITCS.2019.64

26. Sherstov, A.A.: The pattern matrix method. SIAM J. Comput. **40**(6), 1969–2000 (2011). https://doi.org/10.1137/080733644

27. Tani, S., Nakanishi, M., Yamashita, S.: Multi-party quantum communication complexity with routed messages. IEICE Trans. Inf. Syst. **92**(2), 191–199 (2009). https://doi.org/10.1587/transinf.e92.d.191

28. Touchette, D.: Quantum information complexity. In: 47th Annual ACM Symposium on Theory of Computing, STOC 2015, pp. 317–326. Association for Computing Machinery, New York (2015). https://doi.org/10.1145/2746539.2746613

29. Yao, A.C.C.: Quantum circuit complexity. In: 34th Annual Foundations of Computer Science, pp. 352–361 (1993). https://doi.org/10.1109/sfcs.1993.366852

30. Yao, A.C.C.: Some complexity questions related to distributive computing (preliminary report). In: 11th Annual ACM Symposium on Theory of Computing, pp. 209–213 (1979). https://doi.org/10.1145/800135.804414

31. Zhang, Z., Shi, Y.: Communication complexities of symmetric xor functions. Quantum Inf. Comput. **9**(3), 255–263 (2009). https://doi.org/10.26421/qic9.3-4-5

Improved Parallel Algorithms
for Generalized Baumslag Groups

Caroline Mattes[(✉)] and Armin Weiß[iD]

Institut für Formale Methoden der Informatik, Universität Stuttgart,
Stuttgart, Germany
{caroline.mattes,armin.weiss}@fmi.uni-stuttgart.de

Abstract. The Baumslag group had been a candidate for a group with
an extremely difficult word problem until Myasnikov, Ushakov, and Won
succeeded to show that its word problem can be solved in polynomial
time. Their result used the newly developed data structure of power
circuits allowing for a non-elementary compression of integers. Later this
was extended in two directions: Laun showed that the same applies to
generalized Baumslag groups $\mathbf{G}_{1,q}$ for $q \geq 2$ and we established that the
word problem of the Baumslag group $\mathbf{G}_{1,2}$ can be solved in TC^2.

In this work we further improve upon both previous results by showing
that the word problems of all the generalized Baumslag groups $\mathbf{G}_{1,q}$ can
be solved in TC^1 – even for negative q. Our result is based on using
refined operations on reduced power circuits.

Moreover, we prove that the conjugacy problem in $\mathbf{G}_{1,q}$ is strongly
generically in TC^1 (meaning that for "most" inputs it is in TC^1). Finally,
for every fixed $g \in \mathbf{G}_{1,q}$ conjugacy to g can be decided in TC^1 for *all*
inputs.

Keywords: Algorithmic group theory · Power circuit · TC^1 · Word
problem · Conjugacy problem · Baumslag group · Parallel complexity

1 Introduction

In the early 20th century, Dehn [6] introduced the *word problem* as one of the
basic algorithmic problems in group theory: given a word over the generators
of a group G, the questions is whether this word represents the identity of G.
Already in the 1950s, Novikov and Boone constructed finitely presented groups
with an undecidable word problem [4,25]. Still, many natural classes of groups
have an (efficiently) decidable word problem – most prominently, the class of
linear groups (groups embeddable into a matrix group over some field): their
word problem is in $\mathsf{LOGSPACE}$ [15,27] – in particular, in NC, i.e., decidable
by Boolean circuits of polynomial size and polylogarithmic depth. There are
several other results on word problems of groups in small complexity classes
defined by circuits, for example for solvable linear groups in TC^0 (constant depth

This work has been partially supported by DFG Grant WE 6835/1-2.

© Springer Nature Switzerland AG 2022
A. Castañeda and F. Rodríguez-Henríquez (Eds.): LATIN 2022, LNCS 13568, pp. 658–675, 2022.
https://doi.org/10.1007/978-3-031-20624-5_40

with threshold gates) [13], for Baumslag-Solitar groups in LOGSPACE [29], and for hyperbolic groups in $SAC^1 \subseteq NC^1$ [16]. Nevertheless, there are also finitely presented groups with decidable, yet arbitrarily hard, word problems [26].

A *one-relator group* is a group that can be written as a free group modulo a normal subgroup generated by a single element (*relator*). A famous algorithm called the *Magnus breakdown procedure* [18] shows that one-relator groups have decidable word problems (see also [17,19]). Its complexity remains an open problem: while it is not even clear whether the word problems of one-relator groups are solvable in elementary time, [2] asks for polynomial-time algorithms.

In 1969 Gilbert Baumslag defined the group $\mathbf{G}_{1,2} = \langle a, b \mid bab^{-1}a = a^2bab^{-1} \rangle$ as an example of a one-relator group enjoying certain remarkable properties. It is infinite and non-abelian, but all its finite quotients are cyclic [3]. Moreover, Gersten showed that the Dehn function of $\mathbf{G}_{1,2}$ is non-elementary [9] making $\mathbf{G}_{1,2}$ a candidate for a group with a very difficult word problem. Indeed, when applying the Magnus breakdown procedure to an input word of length n, one obtains as intermediate results words of the form $v_1^{x_1} \cdots v_m^{x_m}$ where $v_i \in \{a, b, bab^{-1}\}$, $x_i \in \mathbb{Z}$, and $m \leq n$. The issue is that the x_i might grow up to $\tau_2(\log n)$ (with $\tau_2(0) = 1$ and $\tau_2(i+1) = 2^{\tau_2(i)}$ for $i \geq 0$ – the tower function). However, Myasnikov, Ushakov and Won succeeded to show that the word problem of $\mathbf{G}_{1,2}$ is, indeed, decidable in polynomial time [23]. Their crucial contribution were so-called *power circuits* in [24] for compressing the x_i in the above description.

Roughly speaking, a (base-2) *power circuit* is a directed acyclic graph with edges labelled by $\{-1, 0, 1\}$. One defines an evaluation of a vertex P as two raised to the power of the (weighted) sum of the successors of P. Hence, the value $\tau_2(n)$ can be represented by an $n + 1$-vertex power circuit – thus, power circuits allow for a non-elementary compression. The crucial feature for the application to the Baumslag group is that they not only efficiently support the operations $+$, $-$, and $(x, y) \mapsto x \cdot 2^y$, but also the test whether $x = y$ or $x < y$ for two integers represented by power circuits can be done in polynomial time. The main technical part of the comparison algorithm is to compute a so-called *reduced* power circuit.

Based on these striking results, Diekert, Laun and Ushakov [7] improved the running time for the word problem of the Baumslag group from $\mathcal{O}(n^7)$ down to $\mathcal{O}(n^3)$ and described a polynomial-time algorithm for the word problem of the Higman group H_4 [10]. Subsequently, more applications of power circuits to similar groups emerged: In [14] Laun gave a polynomial-time solution for the word problem of generalized Baumslag groups $\mathbf{G}_{1,q} = \langle a, b \mid bab^{-1}a = a^qbab^{-1} \rangle$ for $q \geq 1$ and also for generalized Higman groups. In order to do so, he generalized power circuits to arbitrary bases $q \geq 2$ and adapted the corresponding algorithms from [7,24]. Of particular interest here is the computation of so-called *compact markings*, which allow for a unique representation of integers; for arbitrary bases it is considerably more involved than for base two.

In [8] the conjugacy problem of the Baumslag group is shown to be strongly generically in P and in [1] the same is done for the conjugacy problem of the Higman group. Here "generically" roughly means that the algorithm works for most inputs – for a precise definition, see Sect. 1.1 below. The idea is that often the "generic-case behavior" of an algorithm is more relevant than its average-case or worst-case behavior. We refer to [11,12] where the foundations of this theory were developed and to [22] for applications in cryptography.

Finally, in [20], we studied the word problem of the Baumslag group $\mathbf{G}_{1,2}$ from the point of view of parallel complexity. We showed that it can be solved in the circuit class TC^2. The proof consists of two main steps: first, to show that for a power circuit of logarithmic depth a corresponding reduced power circuit can be computed in TC^1 (in contrast to the general case where computing reduced power circuits is P-complete [20, Theorem C]) and, second, to show that the Magnus breakdown procedure can be performed in a tree-shape manner leading to a logarithmic number of rounds with each individual round doable in TC^1.

Contribution. In this work we combine the results of [14] and [20] by considering the parallel complexity of the word problem of generalized Baumslag groups $\mathbf{G}_{1,q} = \langle a, b \mid bab^{-1}a = a^q bab^{-1} \rangle$. As a first step, we show how to compute compact base-q signed-digit representations in AC^0 (see Theorem 7). Moreover, we not only unify [14] and [20] but also prove improved complexity bounds:

Theorem A. *For every $q \in \mathbb{Z}$ with $|q| \geq 2$ the word problem of the generalized Baumslag group $\mathbf{G}_{1,q}$ is in TC^1.*

Note that for the first time we allow q to be negative. We do not consider the case $q = \pm 1$ since then the word problem can be solved even in TC^0 using a different approach; we refer to future work. The main ingredient to the improvement of the complexity from TC^2 (in [20]) to TC^1 (here) is that we succeed to perform all operations directly on reduced base-$|q|$ power circuits. For this we allow operations in a SIMD (single instruction multiple data) fashion: many operations of the same type are performed on the same power circuit in parallel in TC^0. Furthermore, we improve the algorithm to get power circuits of quasi-linear size – thus, close to the optimal size as in the sequential algorithms [7,14].

In the last part of our paper, we consider the conjugacy problem for $\mathbf{G}_{1,q}$. We use our results for the word problem to improve the complexity of the strongly generic algorithm from [8] by showing:

Theorem B. *For every $q \in \mathbb{Z}$ with $|q| \geq 2$ the conjugacy problem of the generalized Baumslag group $\mathbf{G}_{1,q}$ is strongly generically in TC^1.*

Moreover, for every fixed $g \in \mathbf{G}_{1,q}$, the problem to decide whether some input word w is conjugate to g is in TC^1.

Note that for the second part of Theorem B not even a polynomial-time algorithm has been described before. It seems to stand in contrast to the conjecture that the conjugacy problem of $\mathbf{G}_{1,2}$ cannot be solved in elementary time [8, Corollary 2]. The crucial point is that here $g \in \mathbf{G}_{1,q}$ is fixed. Due to space constraints most proofs are omitted; we refer to the full version on arXiv [21].

1.1   Notation and Preliminaries

The logarithm log is with respect to base two, while $\log_q$ denotes the base-q logarithm. Let $q \in \mathbb{N}$. Then the base-q *tower function* $\tau_q \colon \mathbb{N} \to \mathbb{N}$ is defined by $\tau_q(0) = 1$ and $\tau_q(i + 1) = q^{\tau_q(i)}$ for $i \geq 0$. It is primitive recursive, but already $\tau_2(6)$ written in binary cannot be stored in the memory of any conceivable real-world computer. We denote the support of a function $f \colon X \to \mathbb{R}$ by $\sigma(f) = \{x \in X \mid f(x) \neq 0\}$. Furthermore, the interval of integers $\{i, \ldots, j\} \subseteq \mathbb{Z}$ is denoted by $[i \mathbin{..} j]$. For $q, x \in \mathbb{Z}$, we write $q \nmid x$ if q does not divide x. Moreover, $\mathrm{sgn}(x)$ denotes the sign of $x \in \mathbb{Z}$. We write $\mathbb{Z}[1/q] = \{m/q^k \in \mathbb{Q} \mid m, k \in \mathbb{Z}\}$ for the set of fractions with powers of q as denominators.

Let Σ be a set. The set of words over Σ is denoted by $\Sigma^* = \bigcup_{n \in \mathbb{N}} \Sigma^n$. The length of $w \in \Sigma^*$ is denoted by $|w|$. A dag is a directed acyclic graph. For a dag Γ we write $\mathrm{depth}(\Gamma)$ for the length (number of edges) of a longest path in Γ.

Complexity. Throughout, we assume that languages L (resp. inputs to functions f) are encoded over the binary alphabet $\{0, 1\}$. A Boolean circuit is a dag where the vertices are either input gates $x_1, \ldots, x_n$, or NOT, AND, or OR gates. There are one or more designated output gates and there is an order given on the output gates. All gates may have unbounded fan-in (i.e., there is no bound on the number of incoming wires). Let $k \in \mathbb{N}$. A language $L \subseteq \{0, 1\}^*$ belongs to AC^k if there exists a family $(C_n)_{n \in \mathbb{N}}$ of Boolean circuits such that $x \in L \cap \{0, 1\}^n$ if and only if the (unique) output gate of C_n evaluates to 1 when assigning $x = x_1 \cdots x_n$ to the input gates. Moreover, C_n may contain at most $n^{\mathcal{O}(1)}$ gates and have depth $\mathcal{O}(\log^k n)$. Likewise AC^k-computable functions are defined.

The class TC^k is defined analogously also allowing MAJORITY gates (which output 1 if the input contains more 1s than 0s). For more details on circuits we refer to [28]. The classes TC^k are contained in P if we consider uniform circuits. Roughly speaking, a circuit family is *uniform* if the n-input circuit can be computed efficiently from the string 1^n. To not overload the presentation, we state our results only in the non-uniform case. We use two basic building blocks, iterated addition and sorting, which can be done in TC^0.

Example 1. Base-q iterated addition is as follows: on input of n base-q numbers $A_1, \ldots, A_n$ each having n digits, compute $\sum_{i=1}^n A_i$ (in base-q representation). For binary numbers this is well-known to be in TC^0 (see e.g. [28, Theorem 1.37]). The standard proof can be translated for other bases.

Generic Case Complexity. A set $I \subseteq \Sigma^*$ is called *strongly generic* if the probability to find a random string outside I converges exponentially fast to zero – more precisely, if $|\Sigma^n \setminus I|/|\Sigma^n| \in 2^{-\mathcal{O}(n)}$. Let $\mathcal{C}$ be some complexity class. A problem $L \subseteq \Sigma^*$ is called *strongly generically in* $\mathcal{C}$ if there is a strongly generic set $I \subseteq \Sigma^*$ and a (partial) algorithm (or circuit family) $\mathcal{A}$ running within the bounds of $\mathcal{C}$ such that $\mathcal{A}$ computes the correct answer for every $w \in I$; outside of I it provides either the correct answer or none (or outputs "unknown").

2   Compact Representations

Based on the concept of compact sums and power circuits with base 2 (as introduced in [24]), Laun [14] described so-called power sums: A power sum to base
$q \geq 2$ is a sum $\sum_{i \geq 0} a_i q^i$ with $a_i \in [-q+1 .. q-1]$ and only finitely many a_i
are non-zero. We are interested in *compact representations* of such power sums.
In [14, Proposition 2.18] it is shown that each power sum has a unique compact representation, which is obtained using a confluent rewriting system. Using
Boolean formulas for this construction we show that it is in AC^0. This will be
an important ingredient for our power circuit operations to be in TC^0. Observe
that in [20, Theorem 11] we gave a proof for base $q = 2$. Here, we fix $q \geq 2$.

Definition 2. *Let* $A = (a_0, \dots, a_{m-1})$ *be a sequence with* $a_i \in [-q+1 .. q-1]$.

- *We define* $\mathrm{val}_q(A) = \sum_{i=0}^{m-1} a_i \cdot q^i$.
- *We call* A *a (base-q) signed-digit representation (short sdr) of* $\mathrm{val}_q(A)$.
- *We call* A *compact if the following conditions hold for all* $i \in [0 .. m-2]$:
  (1) if $|a_i| = q-1$, *then* $|a_{i+1}| < q-1$,
  (2) if $|a_i| \neq 0$, *then* $a_{i+1} = 0$ *or* $\mathrm{sgn}(a_i) = \mathrm{sgn}(a_{i+1})$.

We set $a_i = 0$ for $i \geq m$. Note that, if $A = (a_0, \dots, a_{m-1})$ is an sdr with $a_i \in$
$[0 .. q-1]$, we have a usual base-q representation of an integer. Allowing negative
digits gives more flexibility when working with power circuits – however, with
the price that representations are no longer unique. This uniqueness property
can be regained by requiring the sdr to be compact. Moreover, compact base-q
signed-digit representations (for short base-q csdr) can be compared easily.

Lemma 3 ([14, Proposition 2.18]). *For every* $x \in \mathbb{Z}$ *there is a unique compact base-q signed-digit representation* $A = (a_0, \dots, a_{m-1})$ *with* $\mathrm{val}_q(A) = x$.
   Moreover, two base-q csdrs $A = (a_0, \dots, a_{m-1})$ *and* $B = (b_0, \dots, b_{m-1})$ *can
be compared using the lexicographical order – more precisely,* $\mathrm{val}_q(A) < \mathrm{val}_q(B)$
if and only if $a_{i_0} < b_{i_0}$ *for* $i_0 = \max\{i \in [0 .. m-1] \mid a_i \neq b_i\}$.

Lemma 4. *If* $A = (a_0, \dots, a_{m-1})$ *is a base-q csdr, then* $\mathrm{val}_q(A) \leq \left\lfloor \frac{q^{m+1}}{q^2-1} \right\rfloor$.

This lemma follows by an easy calculation using that in a base-q csdr of maximal
value always digits $q-1$ and $q-2$ alternate. Next, we construct the base-q csdr
of a given sdr $A = (a_0, \dots, a_{m-1})$. We start by restricting A to only non-negative
digits. In a first step we need the following formula for $i \geq 0$:

$$e_i = \bigvee_{j \in [1..i]} \left(a_j = q-1 \wedge a_{j-1} = q-1 \wedge \bigwedge_{k \in [j+1..i-1]} ((a_k = q-1 \vee a_{k+1} = q-1) \wedge a_k \geq q-2) \right)$$

Lemma 5. *For every sdr* $A = (a_0, \dots, a_{m-1})$ *with* $a_i \in [0 .. q-1]$ *the sequence*
$B = (b_0, \dots, b_m)$ *defined by* $b_i = a_i - q \cdot e_{i+1} + e_i$ *satisfies:* $\mathrm{val}_q(A) = \mathrm{val}_q(B)$,
$b_i \in [-1 .. q-1]$, $b_i = -1$ *implies* $b_{i+1} = 0$, *and* $b_i = q-1$ *implies* $b_{i+1} < q-1$.

For the second step – to make a signed-digit representation $B = (b_0, \ldots, b_m)$ as in Lemma 5 compact – we need the following formula:

$$f_i = \bigvee_{j \in [i..m]} \left(b_j = -1 \wedge \bigwedge_{\ell \in [i-1..j-1]} b_\ell > 0 \right)$$

Lemma 6. *If $B = (b_0, \ldots, b_m)$ is a sdr satisfying the conditions of the output of Lemma 5, then $C = (c_0, \ldots, c_m)$ defined by $c_i = b_i - q \cdot f_{i+1} + f_i$ satisfies $\mathrm{val}_q(B) = \mathrm{val}_q(C)$, $c_i \in [-q+1 .. q-1]$, and C is compact.*

Theorem 7. *The following is in AC^0:*

Input: *A base-q signed-digit representation $A = (a_0, \ldots, a_{m-1})$.*
Output: *A base-q csdr $B = (b_0, \ldots, b_m)$ such that $\mathrm{val}_q(A) = \mathrm{val}_q(B)$.*

Proof. Observe that there exist signed-digit representations $C = (c_0, \ldots, c_{m-1})$ and $D = (d_0, \ldots, d_{m-1})$ such that $c_i, d_i \in [0 .. q-1]$ and such that $\mathrm{val}_q(A) = \mathrm{val}_q(C) - \mathrm{val}_q(D)$ (we just collect the negative digits of A into D and the positive ones into C). Now, we compute $|\mathrm{val}_q(C) - \mathrm{val}_q(D)|$ in the usual base q representation (which is in AC^0 – see e.g. [28, Theorem 1.15] for base 2; the general case follows the same way) and make it compact by first applying Lemma 5 and then Lemma 6. If $\mathrm{val}_q(C) - \mathrm{val}_q(D) < 0$, we invert this number digit by digit. $\qquad\square$

3   Power Circuits

The original definition [24] is for power circuits with base 2. Here, following [14], we define power circuits with respect to an arbitrary base q – hence, from now on we fix $q \geq 2$.

Consider a pair (Γ, δ) where Γ is a set of n vertices and δ is a mapping $\delta \colon \Gamma \times \Gamma \to [-q+1 .. q-1]$. Note that $(\Gamma, \sigma(\delta))$ is a directed graph. Throughout we require that $(\Gamma, \sigma(\delta))$ is acyclic – i.e., it is a dag. In particular, $\delta(P, P) = 0$ for all vertices P. A *marking* is a mapping $M \colon \Gamma \to [-q+1 .. q-1]$. Each node $P \in \Gamma$ is associated in a natural way with a marking $\Lambda_P \colon \Gamma \to [-q+1 .. q-1]$, $Q \mapsto \delta(P, Q)$ called its *successor marking*. We define the *evaluation* $\varepsilon(P) \in \mathbb{R}_{>0}$ of a node ($\varepsilon(M) \in \mathbb{R}$ of a marking resp.) bottom-up in the dag by induction: nodes of out-degree zero evaluate to 1 and, in general,

$$\varepsilon(P) = q^{\varepsilon(\Lambda_P)} \quad \text{for a node } P, \qquad \varepsilon(M) = \sum_P M(P)\varepsilon(P) \quad \text{for a marking } M.$$

Definition 8. *A (base-q) power circuit is a pair (Γ, δ) with $\delta \colon \Gamma \times \Gamma \to [-q+1 .. q-1]$ such that $(\Gamma, \sigma(\delta))$ is a dag and all nodes evaluate to an integer in $q^{\mathbb{N}}$.*

The size of a power circuit is the number of nodes $|\Gamma|$. If M is a marking on Γ and $S \subseteq \Gamma$, we write $M|_S$ for the restriction of M to S. Let (Γ', δ') be a power circuit, $\Gamma \subseteq \Gamma'$, $\delta = \delta'|_{\Gamma \times \Gamma}$, and $\delta'|_{\Gamma \times (\Gamma' \setminus \Gamma)} = 0$. Then (Γ, δ) itself is a power circuit. We call it a *sub-power circuit* and denote this by $(\Gamma, \delta) \leq (\Gamma', \delta')$. If M is a marking on $S \subseteq \Gamma$, we extend M to Γ by setting $M(P) = 0$ for $P \in \Gamma \setminus S$. With this convention, every marking on Γ also can be seen as a marking on Γ' if $(\Gamma, \delta) \leq (\Gamma', \delta')$. For a list of markings $\vec{M} = (M_1, \ldots, M_n)$ we define $S(\vec{M}) = \sum_{i=1}^n |\sigma(M_i)|$ (and $S(M) = |\sigma(M)|$ for a single marking).

Example 9. We can represent every integer in the range $[-q^n + 1, q^n - 1]$ by some marking on a base q power circuit with nodes $\{P_0, \ldots, P_{n-1}\}$ with $\varepsilon(P_i) = q^i$ for $i \in [0 .. n - 1]$. Thus, we can convert the q-ary notation of an n-digit integer into a power circuit with n vertices, $\mathcal{O}(n \log_q n)$ edges (each successor marking requires at most $\lfloor \log_q n \rfloor + 1$ edges) and depth at most $\log_q^* n$ – see e.g. Fig. 1.

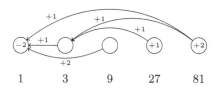

$$
\begin{array}{ccccc}
1 & 3 & 9 & 27 & 81
\end{array}
$$

Fig. 1. Each integer $z \in [-242 .. 242]$ can be represented by a marking on the following power circuit. The marking given in blue is representing the number 187. (Color figure online)

Definition 10. *We call a marking M compact if for all $P, Q \in \sigma(M)$ with $P \neq Q$ we have $\varepsilon(P) \neq \varepsilon(Q)$ and, if $|M(P)| = |M(Q)| = q - 1$ or $\operatorname{sgn}(M(P)) \neq \operatorname{sgn}(M(Q))$, then $|\varepsilon(\Lambda_P) - \varepsilon(\Lambda_Q)| \geq 2$. A reduced power circuit of size n is a power circuit (Γ, δ) with Γ given as a sorted list $\Gamma = (P_0, \ldots, P_{n-1})$ such that all successor markings are compact and $\varepsilon(P_i) < \varepsilon(P_j)$ whenever $i < j$. In particular, all nodes have pairwise distinct evaluations.*

Note that by [20, Theorem 37] it is crucial that the nodes in Γ are sorted by their values. Still, sometimes it is convenient to treat Γ as a set – we write $P \in \Gamma$ or $S \subseteq \Gamma$ with the obvious meaning.

Also note some slight differences compared to other literature: In [7,14], the definition of a reduced power circuit also contains a bit-vector indicating which nodes have successor markings differing by one. Moreover, in [14] the (successor) markings of a reduced power circuit do not have to be compact.

Definition 11. *Let (Γ, δ) be a reduced power circuit with $\Gamma = (P_0, \ldots, P_{n-1})$.*

(i) *A chain C of length $\ell = |C|$ in Γ is a sequence $(P_i, \ldots, P_{i+\ell-1})$ such that $\varepsilon(P_{i+j+1}) = q \cdot \varepsilon(P_{i+j})$ for all $j \in [0 .. \ell - 2]$.*

(ii) *We call a chain C maximal if it cannot be extended in either direction. We denote the set of all maximal chains by $\mathcal{C}_\Gamma$.*

(iii) *There is a unique maximal chain C_0 containing the node P_0 of value 1. We call C_0 the initial maximal chain of Γ and denote it by $C_0 = C_0(\Gamma)$.*

3.1   Operations on Reduced Power Circuits

We continue with fixed $q \geq 2$ and assume that all power circuits are with respect to base q. Following [20, Proposition 14], we can also compare compact markings on reduced base-q power circuits in AC^0. The proof is a straightforward application of Lemma 3.

Lemma 12. *The following problem is in* AC^0:

 Input: *A reduced power circuit* (Γ, δ) *with compact markings* L, M.
 Question: *Is* $\varepsilon(L) \leq \varepsilon(M)$?

The next lemma turns out to be quite versatile and of interest on its own. In particular, it allows to compare a marking on a reduced power circuit in TC^0 with some integer given in binary. Moreover, we use it to get rid of the technical condition $\mu \leq \lfloor 2^{|C_0(\Gamma)|+1}/3 \rfloor$ of [20, Lemma 20] leading to Lemma 15 below.

Lemma 13. *The following problem is in* TC^0:

 Input: *A reduced power circuit* (Γ, δ) *and* $\mu \in \mathbb{N}$ *given in unary.*
 Output: *A reduced power circuit* (Γ', δ') *such that* $|C_0(\Gamma')| \geq \mu$ *and*
           $(\Gamma, \delta) \leq (\Gamma', \delta')$ *(and* $|\Gamma'| \leq |\Gamma| + \mu$).

Proof (Sketch). For a node P with $\varepsilon(\Lambda_P) \leq \mu$ the marking Λ_P uses only the first $\nu + 1$ nodes for $\nu = \lceil \log_q \mu \rceil$. There are at most $2^{\nu+1}$ possibilities which of the first $\nu + 1$ nodes are missing in Γ. For each of these possibilities we can check in parallel whether it is the possibility which actually applies. Now, for each missing value we define a new node P using the compact signed digit-representation of $\varepsilon(\Lambda_P)$.   □

In [20] we proved that we can reduce a power circuit (Π, δ_Π) using TC circuits of depth $\mathcal{O}(\text{depth}(\Pi))$ ($\mathsf{LinDepParaTC}^0$ parametrized by $\text{depth}(\Pi)$). In this paper, the non-reduced power circuits are "almost reduced": $\Pi = (\Gamma \cup \Xi, \delta)$ with $\Gamma \cap \Xi = \emptyset$ and (Γ, δ_Γ) is a reduced power circuit. If $P \in \Xi$, then Λ_P is a compact marking on Γ. Reducing such a power circuit is possible in TC^0. We borrow the following two lemmas from [20]. While [20] treats only power circuits with base 2, here we allow an arbitrary base $q \geq 2$. The proof of the first lemma is actually verbatim the same for $q \geq 2$, for the second one we indicate the small differences.

Lemma 14 (UPDATENODES, [20, Lemma 19]). *The following is in* TC^0:

 Input: *A power circuit* $(\Gamma \cup \Xi, \delta)$ *as above.*
 Output: *A reduced power circuit* (Γ', δ') *such that for each* $Q \in \Xi$ *there is a*
           node $P \in \Gamma'$ *with* $\varepsilon(P) = \varepsilon(Q)$. *In addition,* $(\Gamma, \delta|_{\Gamma \times \Gamma}) \leq (\Gamma', \delta')$
           and $|\Gamma'| \leq |\Gamma| + |\Xi|$ *and* $|C_{\Gamma'}| \leq |C_\Gamma| + |\Xi|$.

Lemma 15 (EXTENDCHAINS, [20, **Lemma 20**]). *The following is in* TC^0:

Input: *A reduced power circuit* (Γ, δ) *and* $\mu \in \mathbb{N}$ *given in unary.*

Output: *A reduced power circuit* (Γ', δ') *such that for each* $P \in \Gamma$ *and each* $i \in [0 .. \mu]$ *there is a node* $Q \in \Gamma'$ *with* $\varepsilon(\Lambda_Q) = \varepsilon(\Lambda_P) + i$. *In addition,* $(\Gamma, \delta) \leq (\Gamma', \delta')$ *and* $|\Gamma'| \leq |\Gamma| + |\mathcal{C}_\Gamma| \cdot \mu$ *and* $|\mathcal{C}_{\Gamma'}| \leq |\mathcal{C}_\Gamma|$.

To prove Lemma 15, we use Lemma 13 to prolongate C_0 such that the last μ nodes are not already present in Γ. This replaces Step 1 in the proof of [20, Lemma 20]. Now we can continue with step 2 of [20, Lemma 20].

Addition is one of the basic operations on power circuits introduced in [24]. Here is our parallel version for addition of arbitrarily many markings on a reduced power circuit:

Lemma 16 (ADDITION). The following is possible in TC^0:

Input: A reduced power circuit (Γ, δ) with compact markings $L_j^{(i)}$ on Γ for $i \in [1 .. \ell]$, $j \in [1 .. k]$.

Output: A reduced power circuit (Γ', δ') such that $(\Gamma, \delta) \leq (\Gamma', \delta')$ and compact markings $M^{(i)}$ on Γ' with $\varepsilon(M^{(i)}) = \varepsilon(L_1^{(i)}) + \cdots + \varepsilon(L_k^{(i)})$ for $i \in [1 .. \ell]$ such that $|\mathcal{C}_{\Gamma'}| \leq |\mathcal{C}_\Gamma|$ and $|\Gamma'| \leq |\Gamma| + \lceil \log_q(k) \rceil \cdot |\mathcal{C}_\Gamma|$, and $\left| \sigma(M^{(i)}) \right| \leq \sum_{j=1}^{k} \left| \sigma(L_j^{(i)}) \right|$ for each $i \in [1 .. \ell]$.

To obtain the marking $M^{(i)}$ in Lemma 16, we add the markings on the maximal chains separately and use Theorem 7 to get a compact signed-digit representation. After applying EXTENDCHAINS($\lceil \log_q(k) \rceil$) this is well-defined.

To represent also numbers $r \in \mathbb{Z}[1/q]$ by markings in power circuits, we use a floating point representation. Observe that for each such $r \in \mathbb{Z}[1/q] \setminus \{0\}$ there exist unique $u, e \in \mathbb{Z}$ with $q \nmid u$ such that $r = u \cdot q^e$. The floating point representation of an integer valued marking can be obtained as follows:

Lemma 17. *The following is possible in* TC^0:

1. MULTBYPOWER:

   Input: *A reduced power circuit* (Γ, δ) *with compact markings* $K^{(i)}, L^{(i)}$ *on* Γ *for* $i \in [1 .. \ell]$ *such that* $\varepsilon(K^{(i)}) \cdot q^{\varepsilon(L^{(i)})} \in \mathbb{Z}$.

   Output: *A reduced power circuit* (Γ', δ') *with compact markings* $M^{(i)}$ *such that* $\varepsilon(M^{(i)}) = \varepsilon(K^{(i)}) \cdot q^{\varepsilon(L^{(i)})}$ *for* $i \in [1 .. \ell]$.

2. MAKEFLOATINGPOINT:

   Input: *A reduced power circuit* (Γ, δ) *with compact markings* $K^{(i)}$ *for* $i \in [1 .. \ell]$.

   Output: *A reduced power circuit* (Γ', δ') *with compact markings* $U^{(i)}, E^{(i)}$ *such that* $\varepsilon(K^{(i)}) = \varepsilon(U^{(i)}) \cdot q^{\varepsilon(E^{(i)})}$ *with* $\varepsilon(U^{(i)}) = 0$ *or* $q \nmid \varepsilon(U^{(i)})$ *for* $i \in [1 .. \ell]$.

In both cases we have $(\Gamma, \delta) \leq (\Gamma', \delta')$ *and*

- $|\Gamma'| \leq |\Gamma| + |\mathcal{C}_\Gamma| + \sum_{i=1}^{\ell} \left|\sigma(K^{(i)})\right|$
- $|\mathcal{C}_{\Gamma'}| \leq |\mathcal{C}_\Gamma| + \sum_{i=1}^{\ell} \left|\sigma(K^{(i)})\right|$
- $\left|\sigma(M^{(i)})\right| = \left|\sigma(K^{(i)})\right|$ (resp. $\left|\sigma(U^{(i)})\right| = \left|\sigma(K^{(i)})\right|$) for all $i \in [1..\ell]$.

Notice that the size of Γ' and $\left|\sigma(M^{(i)})\right|$ does not depend on $L^{(i)}$.

Proof (Sketch). MULTBYPOWER: First, for every i and every $P \in \sigma(K^{(i)})$ we define a new node $R_P^{(i)}$ with $\varepsilon(\Lambda_{R_P^{(i)}}) = \varepsilon(\Lambda_P) + \varepsilon(L^{(i)})$ using ADDITION(2). So $\Lambda_{R_P^{(i)}}$ is a compact marking on a reduced power circuit. Thus we can apply UPDATENODES to reduce the power circuit containing these newly defined nodes and define the marking $M^{(i)}$ on the resulting power circuit as follows: $M^{(i)}(P) = K^{(i)}(P)$ if there is a node $R_P^{(i)}$ with $\varepsilon(R_P^{(i)}) = P$ and $M^{(i)}(P) = 0$ otherwise.

MAKEFLOATINGPOINT: Let $\sigma(K^{(i)}) = \{Q_1, \ldots, Q_k\}$. If $k = 0$, then $\varepsilon(K^{(i)}) = 0$, so we set $\varepsilon(U^{(i)}) = \varepsilon(E^{(i)}) = 0$. Now let $k \geq 1$. Because (Γ, δ) is reduced, we know that $\varepsilon(Q_1) < \varepsilon(Q_j)$ for all $j \in [2..k]$. Therefore, $u = \varepsilon(K^{(i)}) \cdot q^{-\varepsilon(\Lambda_{Q_1})}$ is integral but not divisible by q. We set $E^{(i)} = \Lambda_{Q_1}$ and use MULTBYPOWER with input $K^{(i)}$ and $-E^{(i)}$ to compute a marking $U^{(i)}$ with $\varepsilon(U^{(i)}) = u$. By the first part of the lemma, we can do this for all $i \in [1..\ell]$ in parallel.   □

Definition 18. *Let (Γ, δ) be a reduced power circuit and $r \in \mathbb{Z}[1/q]$. We call $R = (U, E)$ a reduced power circuit representation (red-PC rep.) for r over (Γ, δ) if U and E are compact markings on Γ with $r = \varepsilon(U) \cdot q^{\varepsilon(E)}$ and $\varepsilon(U)$ is either zero or $q \nmid \varepsilon(U)$. We write $\varepsilon(R) = \varepsilon(U) \cdot q^{\varepsilon(E)}$ and define $S(R) = S(U)$ (recall that $S(U) = |\sigma(U)|$, i.e., we only count nodes in the support of the mantissa).*

Likewise, for $m \in \mathbb{Z}$ we call a compact marking M on Γ with $\varepsilon(M) = m$ a reduced power circuit representation (red-PC rep.) of m over (Γ, δ). Moreover, for $\vec{R} = ((U_1, E_1), \ldots, (U_\ell, E_\ell))$ we write $S(\vec{R}) = S((U_1, \ldots, U_\ell)) = \sum_{i=1}^{\ell} S((U_i, E_i))$.

The operations ADDITION, MULTBYPOWER and MAKEFLOATINGPOINT are our main ingredients for the Britton-reduction algorithm in $\mathbf{G}_{1,q}$. The next result combines these operations to work with floating point numbers (more precisely, their red-PC rep.s). The proof consists of several applications of Lemma 16 and Lemma 17.

For an analogous statement for floating point operations on non-reduced power circuits, see [20, Lemma 28]. Notice that, while [20, Lemma 28] deals only with a single operation, here we consider an unbounded number of operations of the same type on the same reduced power circuit. Moreover, the constructions are all in TC^0, while in [20] the depth of the TC circuit depends on the depth of the input power circuit.

Corollary 19. *The following constructions are in TC^0:*

a) Input: *A red-PC rep. $\vec{R} = (R^{(i)})_{i \in [1..\ell]}$ over (Γ, δ) for $r^{(i)} \in \mathbb{Z}[1/q]$ and compact markings $M^{(i)}$ on Γ for $i \in [1..\ell]$.*

   Output: *A red-PC rep. $\vec{S} = (S^{(i)})_{i \in [1..\ell]}$ for $r^{(i)} \cdot q^{\varepsilon(M^{(i)})}$ over a power circuit (Γ', δ') such that $S(S^{(i)}) = S(R^{(i)})$.*

b) Input: *Red-PC rep.s* $\vec{R} = \left(R^{(i)}\right)_{i \in [1..\ell]}$ *over* (Γ, δ) *for* $r^{(i)} \in \mathbb{Z}[1/q]$.

  Output: *A reduced power circuit* (Γ', δ') *and for each* $i \in [1..\ell]$ *the answer whether* $r^{(i)} \in \mathbb{Z}$ *and, if yes, a compact marking* $M^{(i)}$ *such that* $\varepsilon(M^{(i)}) = r^{(i)}$ *and* $\mathrm{S}(M^{(i)}) = \mathrm{S}(R^{(i)})$.

c) Input: *Red-PC rep.s* $\vec{R} = \left(R_j^{(i)}\right)_{i \in [1..\ell], j \in [1..k]}$ *over* (Γ, δ) *for* $r_j^{(i)} \in \mathbb{Z}[1/q]$.

  Output: *Red-PC rep.s* $\vec{S} = \left(S^{(i)}\right)_{i \in [1..\ell]}$ *over a power circuit* (Γ', δ') *for* $\sum_{j=1}^k r_j^{(i)}$ *such that* $\mathrm{S}(S^{(i)}) \leq \sum_{j=1}^k \mathrm{S}(R_j^{(i)})$.

In all cases we have $(\Gamma, \delta) \leq (\Gamma', \delta')$. *In addition, there is some constant* c *such that* $|\mathcal{C}_{\Gamma'}| \leq |\mathcal{C}_\Gamma| + c \cdot \mathrm{S}(\vec{R})$ *and* $|\Gamma'| \leq |\Gamma| + c \cdot (|\mathcal{C}_\Gamma| + \mathrm{S}(\vec{R}))$ *(in (in cases a) and b)), and* $|\Gamma'| \leq |\Gamma| + (c + \log_q k) \cdot (|\mathcal{C}_\Gamma| + \mathrm{S}(\vec{R}))$ *(in case c)).*

Modulo in Power Circuits. As usual for $a, b \in \mathbb{Z}$, we write $a \bmod b$ for the unique $x \in \mathbb{Z}$ with $x \equiv a \bmod b$ and $x \in [0..b-1]$. In [8, Theorem 1] it is shown that to compute a marking for $\varepsilon(M) \bmod \varepsilon(L)$ on input of markings M and L on a power circuit can lead to a non-elementary blow-up. Thus, in general, calculating modulo is certainly not in TC^0. Nevertheless, this changes for a small modulus.

Proposition 20. *For every fixed* $k \geq 2$ *the following problem is in* TC^0:

  Input: *A reduced power circuit* (Γ, δ) *and a compact marking* M *on* Γ.
  Output: $\varepsilon(M) \bmod k$.

For the proof of Proposition 20 we decompose $k = \ell \cdot r$ with $\gcd(\ell, q) = \gcd(\ell, r) = 1$. We calculate $\varepsilon(P) \bmod r$ (using comparison in power circuits) and $\varepsilon(P) \bmod \ell$ (using induction and $q^{\varphi(\ell)} \equiv 1 \bmod \ell$, where φ is Euler's totient function). The proof of the next lemma uses Proposition 20 and applies a similar decomposition.

Lemma 21. *Let* $r \in \mathbb{Z}$ *be a constant. The following problem is in* TC^0:

  Input: *A reduced power circuit* (Γ, δ) *with compact markings* K, L.
  Output: *A reduced power circuit* (Γ', δ') *with a compact marking* M *such that* $\varepsilon(M) = \varepsilon(L) \bmod q^{\varepsilon(K)} \cdot r$.

4 The Word Problem of $G_{1,q}$

First let us fix our notation from group theory. Let G be a group and $\eta \colon \Sigma^* \to G$ a surjective monoid homomorphism. We treat words over Σ both as words and as their images under η. We write $v =_G w$ with the meaning that $\eta(v) = \eta(w)$. The word problem of G is as follows: given a word $w \in \Sigma^*$, is $w =_G 1$? For further background on group theory, we refer to [17].

The Baumslag-Solitar Group and the Baumslag Group. Let $q \in \mathbb{Z}$ with $|q| \geq 2$. The Baumslag-Solitar group is defined by $\mathbf{BS}_{1,q} = \langle a, t \mid tat^{-1} = a^q \rangle$. We have $\mathbf{BS}_{1,q} \cong \mathbb{Z}[1/q] \rtimes \mathbb{Z}$ via the isomorphism $a \mapsto (1,0)$ and $t \mapsto (0,1)$. The multiplication in $\mathbb{Z}[1/q] \rtimes \mathbb{Z}$ is defined by $(r, m) \cdot (s, n) = (r + q^m s, m + n)$. In the following we use $\mathbf{BS}_{1,q}$ and $\mathbb{Z}[1/q] \rtimes \mathbb{Z}$ as synonyms. The Baumslag group $\mathbf{G}_{1,q}$ can be understood as an HNN extension (for a definition see [17] – this is how the Magnus breakdown procedure works) of the Baumslag-Solitar group:

$$\mathbf{G}_{1,q} = \langle \mathbf{BS}_{1,q}, b \mid bab^{-1} = t \rangle = \langle a, t, b \mid tat^{-1} = a^q, bab^{-1} = t \rangle.$$

Note that the letter t can be seen as an abbreviation for bab^{-1}; by removing it, we obtain exactly the presentation $\langle a, b \mid bab^{-1}a = a^q bab^{-1} \rangle$. Moreover, $\mathbf{BS}_{1,q}$ is a subgroup of $\mathbf{G}_{1,q}$ via the canonical embedding. We have $b(k, 0)b^{-1} = (0, k)$, so a conjugation by b "flips" the two components of the semi-direct product if possible (i.e. if $k \in \mathbb{Z}$). Henceforth, we will use the alphabet $\Sigma = \{1, a, a^{-1}, t, t^{-1}, b, b^{-1}\}$ to represent elements of $\mathbf{G}_{1,q}$ (the letter 1 represents the group identity).

Britton Reductions. Britton reductions are a standard way to solve the word problem in HNN extensions. Let $\Delta_q = \mathbf{BS}_{1,q} \cup \{b, b^{-1}\}$ be an infinite alphabet (note that $\Sigma \subseteq \Delta_q$). A word $w \in \Delta_q^*$ is called *Britton-reduced* if it is of the form $w = (s_0, n_0)\beta_1(s_1, n_1) \cdots \beta_\ell(s_\ell, n_\ell)$ with $\beta_i \in \{b, b^{-1}\}$ and $(s_i, n_i) \in \mathbf{BS}_{1,q}$ for all i (i.e., w does not have two successive letters from $\mathbf{BS}_{1,q}$) and there is no factor of the form $b(k, 0)b^{-1}$ or $b^{-1}(0, k)b$ with $k \in \mathbb{Z}$. If w is not Britton-reduced, one can apply one of the rules $(r, m)(s, n) \to (r + q^m s, m + n)$, $b(k, 0)b^{-1} \to (0, k)$, or $b^{-1}(0, k)b \to (k, 0)$ in order to obtain a shorter word representing the same group element. The following lemma is well-known (see also [17, Section IV.2]).

Lemma 22 (Britton's Lemma for $\mathbf{G}_{1,q}$ [5]). *Let $w \in \Delta_q^*$ be Britton-reduced. Then $w \in \mathbf{BS}_{1,q}$ as a group element if and only if w does not contain any letter b or b^{-1}. In particular, $w =_{\mathbf{G}_{1,q}} 1$ if and only if $w = (0,0)$ or $w = 1$ as a word.*

Example 23. Let $q \geq 2$ and define words $w_0 = t$ and $w_{n+1} = b w_n a w_n^{-1} b^{-1}$ for $n \geq 0$ with $w_n \in \Delta_q^*$ for all $n \geq 0$. Then we have $|w_n| = 2^{n+2} - 3$ but $w_n =_{\mathbf{G}_{1,q}} t^{\tau_q(n)}$. While the length of the word w_n is only exponential in n, the length of its Britton-reduced form is $\tau_q(n)$.

4.1 Conditions for Britton-Reductions in $\mathbf{G}_{1,q}$

The following lemma was already used in [20] to find a maximal suffix of u which cancels with a prefix of v on input of two Britton-reduced words u and v. The proof is exactly the same for arbitrary q with $|q| \geq 2$, just replace 2 by q.

Lemma 24 ([20, Lemma 31]). *Let $w = \beta_1(r, m)\beta_2 \, x \, \beta_2^{-1}(s, n)\beta_1^{-1} \in \Delta_q^*$ with $\beta_1, \beta_2 \in \{b, b^{-1}\}$ such that $\beta_1(r, m)\beta_2$ and $\beta_2^{-1}(s, n)\beta_1^{-1}$ both are Britton-reduced and $\beta_2 x \beta_2^{-1} =_{\mathbf{G}_{1,q}} (g, k) \in \mathbf{BS}_{1,q}$ (in particular, $k = 0$ and $g \in \mathbb{Z}$, or $g = 0$).*
  Then $w \in \mathbf{BS}_{1,q}$ if and only if the respective condition in the table below is satisfied. Moreover, if $w \in \mathbf{BS}_{1,q}$, then $w =_{\mathbf{G}_{1,q}} \hat{w}$ according to the last column.

β_1	β_2	Condition		$\hat{w}$
b	b	$r + q^{m+k}s$	$\in \mathbb{Z},\ m+n+k=0$	$(0,\, r+q^{-n}s)$
b	b^{-1}	$r + q^m(g+s)$	$\in \mathbb{Z},\ m+n=0$	$(0,\, r+q^m(g+s))$
b^{-1}	b	$r + q^{m+k}s$	$= 0$	$(m+n+k,\, 0)$
b^{-1}	b^{-1}	$r + q^m(g+s)$	$= 0$	$(m+n,\, 0)$

Note that in the second and third case, the outcome $\hat{w}$ depends on (g,k). We consider these cases in Lemma 25 below. Let us fix the following notation for elements $u, v \in \mathbf{G}_{1,q}$ written as words over Δ_q:

$$u = (r_h, m_h)\beta_h \cdots (r_1, m_1)\beta_1(r_0, m_0), \quad v = (s_0, n_0)\tilde{\beta}_1(s_1, n_1) \cdots \tilde{\beta}_\ell(s_\ell, n_\ell) \quad (1)$$

with $(r_j, m_j), (s_j, n_j) \in \mathbb{Z}[1/q] \rtimes \mathbb{Z}$ and $\beta_j, \tilde{\beta}_j \in \{b, b^{-1}\}$. We write $|u|_\beta = |u|_b + |u|_{b^{-1}} = h$. Moreover, we define

$$uv[i,j] = \beta_{i+1}(r_i, m_i) \cdots \beta_1(r_0, m_0)\,(s_0, n_0)\tilde{\beta}_1 \cdots (s_j, n_j)\tilde{\beta}_{j+1}. \quad (2)$$

Note that, to simplify some notation, here we start with β_{i+1} while in [20] we had a similar notation starting with β_i. Further notice that as an immediate consequence of Britton's Lemma we obtain that, if u and v as in (1) are Britton-reduced and $uv[i,i] \in \mathbf{BS}_{1,q}$ for some i, then also $uv[j,j] \in \mathbf{BS}_{1,q}$ for all $j \leq i$.

On input of red-PC rep.s of Britton-reduced words u and v we want to construct a red-PC rep. for a Britton-reduced word $w =_{\mathbf{G}_{1,q}} uv$ using Corollary 19. Note that in [20], in the third case of Lemma 24, we used the log-operation to compute the outcome. As this approach does not allow for good bounds on the size of the power circuits, we use a different approach based on iterated addition as shown in the next lemma which is proved by a straightforward induction:

Lemma 25. *Let $u, v \in \mathbf{G}_{1,q}$ be Britton-reduced and denoted as in Eq. (1) and $j \leq \lfloor \frac{i}{2} \rfloor$ with $i + 1 \leq \min\{|u|_\beta, |v|_\beta\}$. Further assume that*

$$uv[i,i] = b(r_i, m_i)b^{-1} \cdots b(r_{i-2j}, m_{i-2j})b^{-1}yb(s_{i-2j}, n_{i-2j}) \cdots b(s_i, n_i)b^{-1} \in \mathbf{BS}_{1,q}$$

with $b^{-1}yb = (g, 0) \in \mathbf{BS}_{1,q}$. Then, for $\kappa_\theta = \sum_{\varsigma=0}^{\theta} m_{i-2\varsigma}$ and $h_\theta = i - (2\theta + 1)$ we have

$$uv[i,i] = \left(0, r_i + q^{\kappa_j} \cdot (g + s_{h_j+1}) + \sum_{\theta=0}^{j-1} q^{\kappa_\theta} \cdot (m_{h_\theta} + n_{h_\theta} + r_{h_\theta-1} + s_{h_\theta+1})\right).$$

4.2 The Algorithm for $\mathbf{G}_{1,q}$

Definition 26. *Let (Γ, δ) be a reduced base-$|q|$ power circuit and $w = w_1 \cdots w_n \in \Delta_q^*$. A red-PC rep. of w over (Γ, δ) is a list $\mathcal{W} = ((B_i, U_i, E_i, M_i))_{i \in [1..n]}$ with $B_i \in \{b, b^{-1}, \$\}$ and U_i, E_i, M_i compact markings on Γ such that for $i \in [1..n]$*

- if $w_i \in \{b, b^{-1}\}$, then $B_i = w_i$ and U_i, E_i, M_i are the zero marking,
- if $w_i = (r_i, m_i) \in \mathbf{BS}_{1,q}$, then $B_i = \$$ and (U_i, E_i) is a red-PC rep. for r_i and M_i a red-PC rep. for m_i (as in Definition 18).

We write $|\mathcal{W}|_\beta = |w|_\beta$ and $\mathrm{S}(\mathcal{W}) = \mathrm{S}((U_1, \ldots, U_n)) + \mathrm{S}((M_1, \ldots, M_n)) = \sum_{i=1}^{n}(|\sigma(U_i)| + |\sigma(M_i)|)$ and call n the length of $\mathcal{W}$.

Note that we always use power circuits with a positive base – even when q is negative! In the following, we do not specify the base of the power circuit – it is always $|q|$. Recall that we assume $|q| \geq 2$.

Be aware that for $\mathrm{S}(\mathcal{W})$ we do not count the markings in $\vec{E}$. In ADDITION or MULTBYPOWER, the number of new nodes we insert in the worst case only depends on the number of maximal chains and the markings M_i and U_i, but not on the markings E_i. Moreover, $\mathrm{S}(\mathcal{W})$ does not increase by any of our operations.

Lemma 27. *There is a constant c such that the following problem is in* TC^0:

Input: *Red-PC rep.s* $\mathcal{U}^{(i)}, \mathcal{V}^{(i)}$ *for Britton-reduced words* $u^{(i)}, v^{(i)} \in \Delta_q^*$ *over* (Γ, δ) *for* $i \in [1 .. \nu]$.

Output: *Red-PC rep.s* $\mathcal{W}^{(i)}$ *over* (Γ', δ') *for Britton-reduced words* $w^{(i)} \in \Delta_q^*$ *with* $w^{(i)} =_{\mathbf{G}_{1,q}} u^{(i)} v^{(i)}$ *for* $i \in [1 .. \nu]$ *and*

- $\sum_{i=1}^{\nu} \mathrm{S}(\mathcal{W}^{(i)}) \leq \mathcal{S}$,
- $|C_{\Gamma'}| \leq |C_\Gamma| + c \cdot \mathcal{S}$,
- $|\Gamma'| \leq |\Gamma| + c \cdot \log(n) \cdot (|C_\Gamma| + \mathcal{S})$,

where $n = \max\limits_{i \in [1 .. \nu]} |\mathcal{U}^{(i)}|_\beta + |\mathcal{V}^{(i)}|_\beta$ *and* $\mathcal{S} = \sum_{i=1}^{\nu} \mathrm{S}(\mathcal{U}^{(i)}) + \mathrm{S}(\mathcal{V}^{(i)})$.

Proof (Sketch). Let us first describe the idea for $\nu = 1$ writing $u = u^{(1)}$, $v = v^{(1)}$. The outline is similar to [20], but the way we compute the red-PC rep.s differs. First for each $i < \min\{|u|_\beta, |v|_\beta\}$ we compute a red-PC rep. for $g_i, k_i \in \mathbb{Z}$ such that $uv[i-1, i-1] =_{\mathbf{G}_{1,q}} (g_i, k_i)$ if $uv[i-1, i-1] \in \mathbf{BS}_{1,q}$. This can be done using a constant number of applications of Lemma 24 (depending on the β_j) and at most one time Lemma 25 (with the maximal j possible). We use these g_i, k_i to compute for each i whether the implication $uv[i-1, i-1] \in \mathbf{BS}_{1,q} \implies uv[i, i] \in \mathbf{BS}_{1,q}$ hold. We only use operations in Corollary 19, Lemma 16 and 17. If $q < 0$, we have to take special care: to obtain $q^\kappa r$ we compute $z = \kappa \bmod 2$ using Proposition 20 and then either $-|q|^\kappa r$ or $|q|^\kappa r$ (depending on z) using Corollary 19.

Next, let i_0 be maximal such that the implication $uv[i-1, i-1] \in \mathbf{BS}_{1,q} \implies uv[i, i] \in \mathbf{BS}_{1,q}$ holds for all $i \in [-1 .. i_0]$. Since $uv[-1, -1] = 1 \in \mathbf{BS}_{1,q}$, it follows that $uv[i, i] \in \mathbf{BS}_{1,q}$ for all $i \leq i_0$. Moreover, $uv[j, j] \notin \mathbf{BS}_{1,q}$ for $j \geq i_0 + 1$. The red-PC rep. for $uv[i_0, i_0]$ is computed as above and allows us to output the Britton reduction for uv.

By Corollary 19, Lemma 16 and 17 we can apply the above process to the words $u^{(i)} v^{(i)}$ for $i \in [1 .. \nu]$ independently in parallel. Thus, all the red-PC rep.s $\mathcal{W}^{(i)}$ of the Britton-reduced words $w^{(i)} =_{\mathbf{G}_{1,q}} u^{(i)} v^{(i)}$ are computed using constantly many TC^0 steps. Moreover, $\mathrm{S}(\mathcal{W}^{(i)}) \leq \mathrm{S}(\mathcal{U}^{(i)}) + \mathrm{S}(\mathcal{V}^{(i)})$ since each application of Lemma 24 or Lemma 25 uses different $r_\theta, m_\theta, s_\theta, n_\theta$ and in the sum in Lemma 25 none of them appears twice – if we ignore the exponents κ_θ (we do not consider them for $S(\mathcal{W}^{(i)})$). Finally, in each step we add at most $(c + \log_{|q|} n) \cdot (|\mathcal{C}_\Gamma| + \mathcal{S})$ new nodes and $c \cdot \mathcal{S}$ new chains. This leads to the size conditions in Lemma 27. □

Theorem 28. *There exists a constant c such that the following is in TC^1:*

> Input: *A red-PC rep. $\mathcal{W}$ for a word $w \in \Delta_q^*$ over (Γ, δ) with $n = |w|$.*
> Output: *A red-PC rep. $\mathcal{W}'$ for a Britton-reduced word $w_{red} \in \Delta_q^*$ over (Γ', δ') such that $w_{red} =_{\mathbf{G}_{1,q}} w$ and $|\Gamma'| \leq |\Gamma| + c \cdot n \cdot \log(n)^3 \cdot |\Gamma|$.*

Note that the size of the power circuits in Theorem 28 is close to the optimum $\mathcal{O}(n)$ (for $|\Gamma| = 1$) for the sequential algorithm in [7] and much better than the rough polynomial bound in [20]. To prove Theorem 28, we apply Lemma 27 $\lceil \log(n) \rceil$ many times: We first Britton-reduce all factors of length two, then all factors of length four and so on. The size conditions in Lemma 27 imply that for each intermediate result $(\Gamma^{(k)}, \delta^{(k)})$, we have that $|\Gamma^{(k)}| \leq |\Gamma| + c' \cdot n \cdot \log(n)^3 \cdot |\Gamma|$ – also proving that the inputs of all subsequent stages are of polynomial size and, thus, the composition is in TC^1. By Britton's Lemma, we obtain the following consequences, which also comprise Theorem A from the introduction.

Corollary 29.(a) *The word problem in $\mathbf{G}_{1,q}$ and the subgroup membership problem for $\mathbf{BS}_{1,q}$ in $\mathbf{G}_{1,q}$ (given $w \in \Sigma^*$, decide whether w presents some element in $\mathbf{BS}_{1,q}$) are in TC^1.*

(b) *The word problem in $\mathbf{G}_{1,q}$ and the subgroup membership problem for $\mathbf{BS}_{1,q}$ in $\mathbf{G}_{1,q}$ with the input word given as a red-PC rep. are in TC^1.*

4.3   Conjugacy

Let $u \in \Delta_q^*$. A word $v \in \Delta_q^*$ is called a *cyclic permutation* of u if we can write $u = xy$ and $v = yx$ for some $x, y \in \Delta_q^*$. A word $u \in \Delta_q^*$ is called *cyclically Britton-reduced* if all its cyclic permutations are Britton reduced. For $g, h \in \mathbf{G}_{1,q}$ and $A \subseteq \mathbf{G}_{1,q}$ we write $g \sim_A h$ if there exists some $z \in A$ with $g =_{\mathbf{G}_{1,q}} z^{-1}hz$.

Corollary 30. *The following is in TC^1:*

> Input: *Words $u, v \in \Sigma^*$.*
> Output: *Is u conjugate to some element in $\mathbf{BS}_{1,q}$? If no, is $u \sim_{\mathbf{G}_{1,q}} v$?*

In particular, the conjugacy problem for $\mathbf{G}_{1,q}$ is strongly generically in TC^1.

Proof (Sketch). By Theorem 28, we can Britton-reduce u and v in TC^1. By [29, Lemma 25] it suffices to apply one cyclic permutation to each u and v and Britton-reduce them again using Lemma 27, to arrive at cyclically Britton-reduced words, which we also call u and v. Now, by Collins' Lemma (see [17, Theorem IV.2.5]), can be u conjugated into $\mathbf{BS}_{1,q}$ if and only if u is a single letter from $\mathbf{BS}_{1,q}$. Thus, from now on, we assume that u cannot be conjugated into $\mathbf{BS}_{1,q}$. Now, we follow [8, Theorem 3]. Its proof shows how to apply Collins' Lemma and distinguishes three cases (note that it is only for $\mathbf{G}_{1,2}$; however, the proof for $\mathbf{G}_{1,q}$ is a verbatim repetition with 2 replaced by q). In each of the three cases it highlights a unique $\alpha \in \mathbf{G}_{1,q}$ such that $u \sim_{\langle t \rangle \cup \langle a \rangle} v$ if and only if $u =_{\mathbf{G}_{1,q}} \alpha^{-1} v' \alpha$. Moreover, this α can be easily computed from the red-PC rep.s for u and v. Then it remains to solve the word problem (Corollary 29). Finally, by [8, Theorem 4] the set of words $u \in \Delta_q^*$ representing elements of $\mathbf{G}_{1,q} \setminus \mathbf{BS}_{1,q}$ is strongly generic. Hence, the second part of the corollary and the first part of Theorem B follows. $\qquad\square$

Proposition 31. *For every fixed $g \in \mathbf{G}_{1,q}$ the problem, given a word $w \in \Sigma^*$, decide whether $g \sim_{\mathbf{G}_{1,q}} w$ is in TC^1.*

Proof (Sketch). By Corollary 30, we only need to consider the case that $g = (r, m) \in \mathbf{BS}_{1,q}$. As before, we can assume that $g = (s, n)$ is already cyclically Britton-reduced. Now, if $m \neq 0$ and $(r, m) \not\sim_{\mathbf{BS}_{1,q}} (0, m)$, by [8, Proposition 5], it suffices to test for conjugacy in $\mathbf{BS}_{1,q}$. W.l.o.g. $m > 0$. By [8, Equation (5)], we have $(r, m) \sim_{\mathbf{BS}_{1,q}} (s, n)$ if and only if $m = n$ and there is some $k \in [0 \mathinner{..} m - 1]$ with $r \cdot q^k \equiv s \bmod (q^m - 1)$. Trying all constantly many possibilities for k, by Proposition 20, this can be checked in TC^0. The other case is straightforward using [8, Proposition 6] – however, the test whether $(s, n) \sim_{\mathbf{G}_{1,q}} (0, n)$ is quite involved: It follows from [8, Proposition 6] that $(s, n) \sim_{\mathbf{G}_{1,q}} (0, n)$ and $(s, n) \sim_{\mathbf{G}_{1,q}} g$ can only be the case if $n = q^k r$. Again we may assume $n > 0$. By [8, Equation (5)] we have $(s, q^k r) \sim_{\mathbf{G}_{1,q}} (0, q^k r)$ if and only if there is some $x \in [0 \mathinner{..} q^k r - 1]$ with $s \cdot q^x \equiv 0 \bmod q^{q^k r} - 1$. Next we show that $x = 0$. After that the last condition can be evaluated using some calculation based on Lemma 21. This also shows the second part of Theorem B. $\qquad\square$

Conclusion. We have shown the word problem of generalized Baumslag groups $\mathbf{G}_{1,q}$ to be in TC^1. The same complexity applies to the conjugacy problem for elements outside $\mathbf{BS}_{1,q}$ or if one element is fixed. TC^1 seems to be the best possible using the current approach of tree-like Britton reductions. We conclude with some open questions: What is the "actual" complexity of the word problem of $\mathbf{G}_{1,q}$? Are there any better lower bounds other than that it contains a non-abelian free subgroup? Can our methods be generalized to the Higman group H_4? This is closely related to the growth of its Dehn function, which, to the best of our knowledge, is not known to be in $\tau(\mathcal{O}(\log n))$ like for the Baumslag group.

References

1. Baker, O.: The conjugacy problem for Higman's group. Int. J. Algebra Comput. **30**(6), 1211–1235 (2020). https://doi.org/10.1142/S0218196720500393
2. Baumslag, G., Myasnikov, A.G., Shpilrain, V.: Open problems in combinatorial group theory. In: Combinatorial and Geometric Group Theory. Contemporary Mathematics, 2nd edn, vol. 296, pp. 1–38. American Mathematical Society (2002)
3. Baumslag, G.: A non-cyclic one-relator group all of whose finite quotients are cyclic. J. Aust. Math. Soc. **10**(3–4), 497–498 (1969)
4. Boone, W.W.: The word problem. Ann. Math. **70**(2), 207–265 (1959)

5. Britton, J.L.: The word problem. Ann. Math. **77**, 16–32 (1963)
6. Dehn, M.: Ueber unendliche diskontinuierliche Gruppen. Math. Ann. **71**, 116–144 (1911). https://doi.org/10.1007/BF01456932
7. Diekert, V., Laun, J., Ushakov, A.: Efficient algorithms for highly compressed data: the word problem in Higman's group is in P. Int. J. Algebra Comput. **22**(8), 1240008 (2013). https://doi.org/10.1142/S0218196712400085
8. Diekert, V., Myasnikov, A.G., Weiß, A.: Conjugacy in Baumslag's group, generic case complexity, and division in power circuits. Algorithmica **76**(4), 961–988 (2016). https://doi.org/10.1007/s00453-016-0117-z
9. Gersten, S.M.: Isodiametric and isoperimetric inequalities in group extensions (1991, preprint)
10. Higman, G.: A finitely generated infinite simple group. J. London Math. Soc. **26**, 61–64 (1951)
11. Kapovich, I., Miasnikov, A.G., Schupp, P., Shpilrain, V.: Generic-case complexity, decision problems in group theory and random walks. J. Algebra **264**, 665–694 (2003)
12. Kapovich, I., Myasnikov, A., Schupp, P., Shpilrain, V.: Average-case complexity and decision problems in group theory. Adv. Math. **190**, 343–359 (2005)
13. König, D., Lohrey, M.: Evaluation of circuits over nilpotent and polycyclic groups. Algorithmica **80**(5), 1459–1492 (2017). https://doi.org/10.1007/s00453-017-0343-z
14. Laun, J.: Efficient algorithms for highly compressed data: the word problem in generalized Higman groups is in P. Theory Comput. Syst. **55**(4), 742–770 (2013). https://doi.org/10.1007/s00224-013-9509-5
15. Lipton, R.J., Zalcstein, Y.: Word problems solvable in logspace. J. ACM **24**, 522–526 (1977)
16. Lohrey, M.: Decidability and complexity in automatic monoids. Int. J. Found. Comput. Sci. **16**(4), 707–722 (2005)
17. Lyndon, R., Schupp, P.: Combinatorial Group Theory. Classics in Mathematics, Springer, Heidelberg (2001). https://doi.org/10.1007/978-3-642-61896-3First edition (1977)
18. Magnus, W.: Das Identitätsproblem für Gruppen mit einer definierenden Relation. Math. Ann. **106**, 295–307 (1932)
19. Magnus, W., Karrass, A., Solitar, D.: Combinatorial Group Theory. Dover, New York (2004)
20. Mattes, C., Weiß, A.: Parallel algorithms for power circuits and the word problem of the Baumslag group. In: Proceedings of MFCS 2021. LIPIcs, vol. 202, pp. 74:1–74:24. Schloss Dagstuhl - Leibniz-Zentrum für Informatik (2021). https://doi.org/10.4230/LIPIcs.MFCS.2021.74
21. Mattes, C., Weiß, A.: Improved parallel algorithms for generalized Baumslag groups. arXiv eprints abs/2206.06181 (2022). https://arxiv.org/abs/2206.06181
22. Myasnikov, A., Shpilrain, V., Ushakov, A.: Group-based cryptography. Advanced Courses in Mathematics. CRM Barcelona, Birkhäuser Basel (2008). http://books.google.de/books?id=AbPZJf1wcKcC
23. Myasnikov, A.G., Ushakov, A., Dong-Wook, W.: The word problem in the Baumslag group with a non-elementary Dehn function is polynomial time decidable. J. Algebra **345**, 324–342 (2011). http://www.sciencedirect.com/science/article/pii/S0021869311004492
24. Myasnikov, A.G., Ushakov, A., Dong-Wook, W.: Power circuits, exponential algebra, and time complexity. Int. J. Algebra Comput. **22**(6), 3–53 (2012)

25. Novikov, P.S.: On the algorithmic unsolvability of the word problem in group theory. Trudy Mat. Inst. Steklov 1–143 (1955). (in Russian)
26. Sapir, M.V., Birget, J.C., Rips, E.: Isoperimetric and isodiametric functions of groups. Ann. Math. **156**(2), 345–466 (2002)
27. Simon, H.U.: Word problems for groups and contextfree recognition. In: Proceedings of Fundamentals of Computation Theory (FCT 1979), Berlin/Wendisch-Rietz (GDR), pp. 417–422. Akademie-Verlag (1979)
28. Vollmer, H.: Introduction to Circuit Complexity. Springer, Heidelberg (1999). https://doi.org/10.1007/978-3-662-03927-4
29. Weiß, A.: A logspace solution to the word and conjugacy problem of generalized Baumslag-Solitar groups. In: Algebra and Computer Science, Contemporary Mathematics, vol. 677, pp. 185–212. American Mathematical Society (2016)

Computational Geometry

Piercing Pairwise Intersecting Convex Shapes in the Plane

Saman Bazargani[1(✉)], Ahmad Biniaz[2], and Prosenjit Bose[3]

[1] Department of Computer Science and Electrical Engineering, University of Ottawa, Ottawa, Canada
sbaza033@uottawa.ca
[2] School of Computer Science, University of Windsor, Windsor, Canada
abiniaz@uwindsor.ca
[3] School of Computer Science, Carleton University, Ottawa, Canada
jit@scs.carleton.ca
https://cglab.ca/saman/, https://cglab.ca/biniaz/,
https://cglab.ca/jit/

Abstract. Let C be any family of pairwise intersecting convex shapes in a two dimensional Euclidean space. Let $\tau(C)$ denote the piercing number of C, that is, the minimum number of points required such that every shape in C contains at least one of these points. Define a shape to be α-fat when the ratio of the radius of the smallest disk that encloses the shape over the radius of the largest disk that is enclosed in the shape is at most α. Define $\alpha(C)$ to be the minimum value where each shape in C is $\alpha(C)$-fat. We prove that $\tau(C) \leq 43.789\alpha(C) + 4 = O(\alpha(C))$ for any set C consisting of pairwise intersecting convex α-fat shapes. This improves the previous best known upper-bound of $O(\alpha(C)^2)$. This result has a number of implications on other results concerning fat shapes, such as designing data structures with less complexity for 3-D vertical ray shooting and computing depth orders. Additionally, our results reduce the time complexity of the query time of these data structures. We also get better bounds for some restricted families of shapes. We show that $(5\sqrt{2} + 2)\alpha(C) + 25 + 5\sqrt{2} \leq 9.072\alpha(C) + 32.072 = O(\alpha(C))$ piercing points are sufficient to pierce a set of arbitrarily oriented α-fat rectangles. We also prove that $\tau(C) = 2$ when C is a set of pairwise intersecting homothets of regular hexagons. We show that the piercing number of a set of pairwise intersecting homothets of an arbitrary convex shape is at most 15. This improves the previous best upper-bound of 16. We also give an algorithm to calculate the exact location of the piercing points.

1 Introduction

Let H be a set of convex shapes in d−dimensions such that every subset of $d+1$ shapes in H has a non-empty intersection. In 1923, Helly [11] proved that the intersection of all shapes in H is non-empty. This result is known as the *Helly's theorem*. For example, if H is a set of convex shapes in $\mathbb{R}^2$ such that every three of them have a common intersection, then by Helly's theorem all shapes in H

Research supported in part by NSERC.

A. Castañeda and F. Rodríguez-Henríquez (Eds.): LATIN 2022, LNCS 13568, pp. 679–695, 2022.
https://doi.org/10.1007/978-3-031-20624-5_41

have a common intersection. In the other words, all shapes in H can be pierced with one point.

Consider the following fundamental geometric problem: What is the minimum number of points that is sufficient to pierce a given set of pairwise intersecting shapes in the plane? In the case of homothetic triangles, three points are sufficient, as was shown by Chakerian et al. (1967) [5]. In the case of disks, four points are sufficient. The proof of the existence of four piercing points was independently shown by Danzer (1956, 1986) [6] and Stacho (1981) [23,24]. To pierce a set of n pairwise intersecting line segments, $\Omega(n)$ points are sometimes required. This huge gap between the number of points required, from a constant to linear, to pierce different sets of pairwise intersecting shapes gives rise to many interesting problems. Notice that the linear lower-bound to pierce a set of pairwise intersecting line segments comes from the fact that line segments are essentially "thin". How round or fat an object is plays a vital role in the number of points needed to pierce the set. The main problem that we study in this paper is the following: How many points are sufficient to pierce a set of pairwise intersecting shapes in terms of their fatness parameter?

In the literature, the main approach used by researchers to pierce a set C of pairwise intersecting α-fat shapes is by constructing a grid whose resolution is quadratic in the fatness parameter [2,14,18,19]. In this article, we are able to reduce the number of points to linear with respect to the fatness parameter by placing points near the perimeter of a shape that has a non-empty intersection with every other shape in the set. In essence, we show that it is possible to pierce the set by focusing on the perimeter of an object as opposed to filling an area with points. The details of our approach are given in Sect. 2.

1.1 Preliminaries

Informally the fatness of a shape is a parameter that tries to capture how close a shape is to a disk. There are many different definitions and variations of the fatness of a shape [1,7,15,17,19,20,25]. Most of them share some similarities. In this paper we use the following measure of fatness. The fatness of a shape c is the ratio of the radius of the smallest disk that encloses c over the radius of the largest disk that is enclosed in c. This measure of fatness will be denoted by α. We say that a shape c is α-fat if its fatness is at most α. A set C of shapes is referred to as $\alpha(C)$-fat if $\alpha(C)$ is the smallest value such that $\forall c_i \in C$, the fatness of c_i is lesser than or equal to $\alpha(C)$. We note that a set of disks is 1-fat, since a disk is perfectly fat according to our fatness definition.

The *piercing number* of a family of sets $\mathcal{F}$ is the smallest integer k for which it is possible to partition $\mathcal{F}$ into subfamilies $\mathcal{F}_1, \ldots, \mathcal{F}_k$ such that the sets in each $\mathcal{F}_i$ have a non-empty intersection for every i such that $1 \leq i \leq k$ [8]. We say that a set of points P pierces a set of shapes C if every shape in C contains at least one point of P.

A shape B is a *homothet* of a shape A if B can be obtained by scaling and translating the shape A. Two geometric figures are *homothetic* if one is a homothet of the other. If every pair of shapes in a set C is homothetic we call the set C *homothets*. In this paper we only consider positive homotheties. A set of shapes C is *unit* if all the shapes in C have the same area.

1.2   Our Contributions

In this paper we prove the following results in 2-dimensions:

- Any set C of pairwise intersecting arbitrary convex shapes with fatness $\alpha(C)$ can be pierced with less than or equal to $43.789\alpha(C) + 4 \in O(\alpha(C))$ points.
- Any set C of pairwise intersecting rectangles of arbitrary orientation with fatness $\alpha(C)$, can be pierced by $(5\sqrt{2}+2)\alpha(C) + 25 + 5\sqrt{2} \in O(\alpha(C))$ points.
- Any set of pairwise intersecting convex homothets can be pierced by 15 points.
- A set of pairwise intersecting homothets of regular hexagons can be pierced by 2 points.

Known results for piercing sets of pairwise intersecting convex sets.

Family of convex shapes	Known results	Our results
Homothetic Triangles	3 Points [5]	–
Homothetic Rectangles	1 Point [folklore]	–
Homothetic regular Hexagons	Not known	2 Points, Theorem 4
Disks	4 Points [6,21,23,24]	–
Centrally symmetric	7 Points [9]	–
Unit Shapes	3 Points [12]	–
Convex Homothets	16 [16]	15 Points, Theorem 3
α-fat Rectangles	$O(\alpha^2)$ [2,14,18,19]	$\leq 9.072\alpha + 32.072$, Theorem 2
α-fat Convex shapes	$O(\alpha^2)$ [2,14,18,19]	$\leq 43.789\alpha + 4$, Theorem 1

1.3   Previous Results

Overmars et al. (1994) [19] proved that for a set of disjoint convex α-fat objects and a restricted range query (with diameter $h \times p$ where h is a constant and p is the radius of the minimal enclosing hyper-sphere among the objects in the set) in d-dimensions, $O((\alpha d^d h)^d)$ points are enough to pierce all the shapes. They use a grid of points inside and around the range query to pierce such a set. Agarwal et al. (1995) [2], Katz (1996) [14] and Nielsen (2000) [18] among other results proved that $O(\alpha^2)$ points can pierce a set of pairwise intersecting α-fat shapes in 2-dimension. The definitions of fatness that they use are similar. The unifying theme among these proofs is to cover the area around and inside the smallest shape with a grid of $\Theta(\alpha^2)$ piercing points. To find the piercing points Nielsen (2000) [18] uses Fredman's sampling technique [14]. Agarwal et al. (1995) [2], Katz (1996) [14] use a similar gridding technique.

Let F be a family of pairwise intersecting and centrally symmetric convex homothets. Grünbaum (1959) [9] showed that $\tau(F)^1 \leq 7$. He transforms all the shapes from Euclidean space into Minkowski space. The reason behind this transformation is that any centrally symmetric shape in Euclidean space can be treated as a disk in Minkowski space. This transformation maintains the pairwise intersecting property of the set. The fact that 4 points pierces a set of pairwise intersecting disks applies [6,21,23,24]. Grünbaum (1959) [9] also showed that $\tau(F) = 3$ when F is a family of pairwise intersecting and centrally symmetric convex unit-shapes. He conjectured that $\tau(F) = 3$ for any family of pairwise intersecting convex unit-shapes. This conjecture was proved by Karasev (2000) [12]. Karasev (2001) [13] subsequently showed an upper-bound of $d + 1$ on the number of points sufficient to pierce a family of d-wise intersecting homothets of a simplex in $\mathbb{R}^d$. He also gave an upper-bound for a family F of d-wise intersecting spheres which is the following: $\tau(F) \leq 3(d+1)$ when $d \geq 5$ and $\tau(F) \leq 4(d+1)$ when $d \leq 4$.

In case of pairwise intersecting disks, Danzer (1956, 1986) [6] and Stacho (1981) [23,24] were the first to give a proof of the existence of 4 piercing points. However, both of their proofs are essentially non-constructive. Har-Peled et al. [10] were the first to present a deterministic and constructive algorithm. They find 5 piercing points, in $O(n)$ expected time, that pierces a set of n pairwise intersecting disks. Biniaz, Bose and Wang [3] gave a linear algorithm that finds 5 piercing points given a set of pairwise intersecting disks that does not use an LP-*type* framework unlike Har-Peled's algorithm. Carmi, Katz and Morin [21] gave a linear time algorithm to compute 4 piercing points which also uses LP-*type* machinery.

2   General Convex Shapes

2.1   Piercing a Set of Fat Shapes

In this section we prove the following theorem which is our main result.

Theorem 1. *Any set C of pairwise intersecting arbitrary convex shapes on a plane with fatness $\alpha(C)$ can be pierced with $(12 + 6\sqrt{2} + 2^{\frac{15}{4}}\sqrt{3})\alpha(C) + 4 \leq 43.789\alpha(C) \in O(\alpha(C))$ points.*

Proof (Proof of Theorem 1).
    Let $S = \{S_0, S_1, \ldots, S_{n-1}\}$ be a set of pairwise intersecting convex shapes with fatness at most α. For all i, let α_i be the fatness of S_i. Let o be the smallest disk that has a non empty intersection with every shape in the set S. Let δ be the radius of o. Define sq_1 to be an axis-parallel square that is concentric with o. Let the side length of sq_1 be $2c\delta$ for a constant c. And let sq_2 be an axis-parallel

---

[1] Let $\tau(C)$ denote the *piercing number* of C, that is, the minimum number of points required such that every shape in C contains at least one of these points.

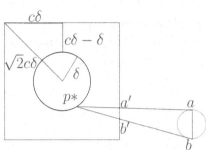

Fig. 1. outer case

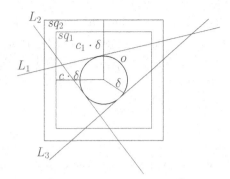

Fig. 2. Initial setup and information required for the proof

square concentric with o with side length $2c_1\delta$ for a constant c_1 $(c_1 > c)$. We specify the exact values of c and c_1 at the end of the proof (Fig. 2).

If all the shapes in S have a common intersection we can pierce the whole set with one point and as a result o will have radius zero. Otherwise, o is tangent to at least three shapes, say S_1, S_2, S_3. Let L_1, L_2, L_3 be the three tangent lines to o where S_1, S_2, S_3 intersect o. Notice that no two tangent line can be parallel, otherwise, either the intersection of every shape in S is non-empty or it contradicts the fact that two corresponding shapes intersect. Moreover, L_1, L_2, L_3 form a triangle, otherwise it contradicts with the minimality of o (See Fig. 2).

We partition the set S into two groups, S^{gp1} and S^{gp2}. A shape $S_i \in S$ will be in S^{gp1} if the center of the largest enclosed disk in S_i or at least one of the largest enclosed disks in S_i (in case S_i has multiple largest enclosed disks) is located completely outside of sq_1. Otherwise, S_i will be in S^{gp2}.

Piercing S^{gp1}: By the definition of o, every shape in S^{gp1} intersects o. Every shape S_i in S^{gp1} is convex, intersects o and has at least one of the largest disk(s) enclosed in S_i centered outside of sq_1. These three facts plus the fact that sq_1 encloses o implies that S_i intersects a continuous portion of the boundary of sq_1. We now show how to place a set of points on the boundary of sq_1 to pierce all the shapes in S^{gp1}. Let S_i be an arbitrary shape in S^{gp1}. Let p^* be an arbitrary point in the intersection of S_i and the boundary of o. Let o' be the largest disk enclosed in S_i and centered outside of sq_1 (in the case of multiple disks satisfying these conditions, pick an arbitrary one). Without loss of generality, assume that S_i intersects the right vertical side of sq_1. Let ab be the diameter of o' parallel to the y-axis. Since S_i is convex, there exists a triangle p^*ab that is contained in S_i. Let the boundary of the triangle p^*ab cross the boundary of sq_1 at points a' and b'. Now the minimum possible length of the segment $a'b'$ gives us the required resolution of points to put on the boundary of sq_1 to pierce S^{gp1}. Recall that α_i is the fatness of S_i. The smallest disk that encloses S_i has a radius greater than or equal to $\frac{|p^*b|}{2}$ since the segment p^*b is in S_i and any disk with diameter less

than $|p^*b|$ cannot have a segment of length $|p^*b|$ in it. Thus, $\alpha_i \geq \frac{\frac{|p^*b|}{2}}{\frac{|ab|}{2}} = \frac{|p^*b|}{|ab|}$.
Moreover, since the two triangles p^*ab and $p^*a'b'$ are similar we get following
equation: $\frac{|a'b'|}{|p^*b'|} = \frac{|ab|}{|p^*b|} \implies |a'b'| = \frac{|ab|\cdot|p^*b'|}{|p^*b|} \implies |a'b'| \geq \frac{|p^*b'|}{\alpha_i}$. Furthermore,
note that $(c-1)\delta \leq |p^*b'| \leq 2\sqrt{2}c\delta$ and $\alpha(S) \geq \alpha_i$. ("Therefore, $|a'b'|$ is at least
$\frac{(c-1)\delta}{\alpha(S)}$. See Fig. 1").

Exceptional Case: The only exceptional case in this scenario is when a' is not
located on the same side of sq_1 as b'. Considering the fact that the disk o' is
centered outside of sq_1, the convexity of S_i implies that S_i contains a corner of
sq_1. To pierce such shapes we put points on the 4 corners of sq_1.

The perimeter of sq_1 is $8c\delta$, therefore the number of points placed on the
perimeter of sq_1 to pierce all the shapes in S^{gp1} is $4 + \frac{8c\delta}{\frac{(c-1)\delta}{\alpha(S)}} = 4 + \frac{8c}{c-1}\alpha(S)$

Piercing S^{gp2}: Let S_i be an arbitrary shape in S^{gp2}. Let L'_i be a line through
the center of circle o and parallel to L_i for $i \in \{1,2,3\}$. We call a point p *proper*
with respect to $L_i, i \in [1,3]$ if it is located inside sq_2, and p is located on the
same side of L_i and L'_i but p is closer to L'_i.

Lemma 1. *Any point inside sq_2 is proper with respect to some $L_i, i \in [1,3]$.*

Proof. Let $H_i, i \in [1,3]$ be the halfspace that is tangent to L'_i and does not
contain L_i. Since H_1, H_2, H_3 intersect at a point and the union of the angle
that they cover is 2π (otherwise it contradicts with the fact L_1, L_2, L_3 form a
triangle). Using the result of Bose et al. [4] we have that the $\cup H_i$, for $i \in [1,3]$
covers the entire plane. Thus, they cover any point in sq_1 as well.   □

The number of points sufficient to pierce the set
Without loss of generality, assume that the center of at least one of the largest
disks enclosed in S_i is a proper point with respect to L_1. Such a disk exists, since,
at least one of the largest disks enclosed in S_i is centered in sq_1.

("We analyze two cases, namely when $S_i \cap L_1 \cap sq_2 \neq \emptyset$ and when $S_i \cap L_1 \cap sq_2 = \emptyset$. See Fig. 3 and 4")

Case 1. S_i has an intersection with L_1 inside sq_2.

Let p^* be an arbitrary point in the intersection of L_1 and S_i interior to sq_2. Let
o' be a largest disk enclosed in S_i centered inside sq_1. Let ab be the diameter of
o' parallel to L_1. Since the center of o' is proper point with respect to L_1, the
triangle p^*ab intersects L'_1 at two points a' and b'. Recall that α_i is the fatness
of S_i, the smallest disk that encloses S_i has a radius greater than or equal
to $\frac{|p^*b|}{2}$, since the segment p^*b is in S_i and any disk with diameter less than
$|p^*b|$ cannot have a chord of length $|p^*b|$. Thus, $\alpha_i \geq \frac{\frac{|p^*b|}{2}}{\frac{|ab|}{2}} = \frac{|p^*b|}{|ab|}$. Moreover,
since the triangles p^*ab and $p^*a'b'$ are similar we get the following equation:
$\frac{|a'b'|}{|p^*b'|} = \frac{|ab|}{|p^*b|} \implies |a'b'| = \frac{|ab|\cdot|p^*b'|}{|p^*b|} \implies |a'b'| \geq \frac{|p^*b'|}{\alpha_i}$. By definition and the
relation between L_1 and L'_1, $\delta \leq |p^*b'| \leq 2\sqrt{2}c\delta$ and $\alpha(S) \geq \alpha_i$. Therefore,
$|a'b'| \geq \frac{\delta}{\alpha(S)}$.

The length of $L_1' \cap sq_2$ is at most $2\sqrt{2}c_1\delta$, which is the diameter of sq_2. Hence, we place $\frac{2\sqrt{2}c_1\delta}{\frac{\delta}{\alpha(S)}} = 2\sqrt{2}c_1\alpha(S)$ points on L_1'. So, in total $6\sqrt{2}c_1\alpha(S)$ points on L_1', L_2', L_3' are sufficient to pierce shapes in this case.

Case 2. S_i only intersects L_1 outside of sq_2.

As a result S_i intersects with the boundary of sq_2. Let p^* be a point in the intersection of S_i and the boundary of sq_2. Without loss of generality, assume that p^* is located on the right side of sq_2. Let o' be a largest disk enclosed in S_i centered inside sq_1. Let ab be the diameter of o' parallel to the y-axis. Recall that α_i is the fatness of S_i. The smallest disk that encloses S_i has a radius greater than or equal to $\frac{|p^*b|}{2}$. By an identical argument as in Case 1, we have $|a'b'| = \frac{|ab| \cdot |p^*b'|}{|p^*b|} \implies |a'b'| \geq \frac{|p^*b'|}{\alpha_i}$. By definition of sq_1, sq_2 and $P^*a'b'$; $(c_1 - c)\delta \leq |p^*b'| \leq 2\sqrt{2}c_1\delta$ and $\alpha(S) \geq \alpha_i$. Therefore, the minimum length of $|a'b'|$ is $\frac{(c_1-c)\delta}{\alpha(S)}$. The perimeter of sq_1 is $8c\delta$, therefore the number of points required to put on the perimeter of sq_1 is $\frac{8c\delta}{\frac{(c_1-c)\delta}{\alpha(S)}} = \frac{8c}{c_1-c}\alpha(S)$.

Let $m = max(\frac{c}{c_1-c}, \frac{c}{c-1})$. The number of points sufficient to pierce the set S using the placements described above is $4 + (8m + 6\sqrt{2}c_1)\alpha(S)$. The minimum value is roughly 43.789 when we set $c = 1.6866$ and $c_1 = 2.3732$. $\qquad \square$

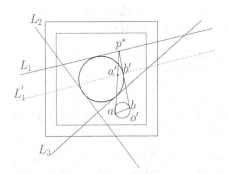

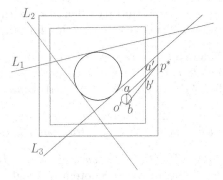

Fig. 3. A convex shape with the largest enclosed disk centered in sq_1 and intersecting L_1 inside sq_2

Fig. 4. A convex shape with the largest enclosed disk centered in sq_1 and intersecting L_1 only outside of sq_2

2.2 Implications

Our result has a number of implications on other research problems concerning sets of fat objects, such as computing depth orders, 3-D vertical ray shooting, 2-D point enclosure, range searching, and arc shooting for convex fat objects. The following are some Corollaries where the asymptotic complexity is improved from $O(\alpha^2)$ to $O(\alpha)$ using our results:

1. **Piercing a set of pairwise intersecting c-oriented convex polygons [18]**

Corollary 1. *The piercing number $\tau(\beta)$ when β is a set of pairwise intersecting c-oriented α-fat polygons is $O(\alpha)$.*

2. **Computing depth order for fat objects [2]**

Corollary 2. *The time complexity of 2-dimensional linear-extension problem is of $O(\alpha n \lambda_s^{1/2}(n) \log^4 n)$.*

3. **3-D vertical ray shooting and 2-D point enclosure, range searching, and arc shooting amidst convex fat objects [14]**

Corollary 3. *For a given query point p, the object of C lying immediately below p (if such an object exists) can be found in $O(\alpha \log^4 n)$ time.*

Corollary 4. *For a given query point p, the k objects of C containing p can be reported in $O(\alpha \log^3 n + k \log^2 n)$ time.*

2.3   Piercing Fat Rectangles

In this section we demonstrate how to pierce a set C of pairwise intersecting rectangles of arbitrary orientation with fatness $\alpha(C)$. This theorem is a generalization of pairwise intersecting line segments in terms of fatness.

Theorem 2. *Any set C of pairwise intersecting rectangles of arbitrary orientation of fatness $\alpha(C)$ can be pierced with $(5\sqrt{2} + 2)\alpha(C) + 25 + 5\sqrt{2} \leq 9.072\alpha(C) + 32.072 = O(\alpha(C))$ points.*

Lemma 2. *The maximum area of a square that does not have any lattice point in it is less than 2. A lattice point is a point with integer coordinates.*

Proof (Proof of Theorem 2). Let $R = \{r_0, r_1, ..., r_{n-1}\}$ be a set of pairwise intersecting rectangles. Denote a rectangle r of width w and height h as (w, h). We assume that the longer side of an arbitrary oriented rectangle is the height of the rectangle ($h \geq w$). Without loss of generality, Let $r_1 = (w_1, h_1)$ be the rectangle with the minimum width among all rectangles in the set R. Without loss of generality, assume that r_1 is axis parallel.

Structure of the grid points: Let $r_i = (w_i, h_i)$ be an arbitrary rectangle in R, let p^* be one of the intersection points of the boundary of r_1 with r_i. By definition of r_1 we have the following two inequalities: $w_i \geq w_1$ and $h_i \geq w_1$. For every $r_i \in R$ there exists a square s_i located inside r_i of side length w_1 such that, s_i contains the point p^*. Suppose that such a square does not exist. It implies that any square of side length w_1 that contains p^* intersects with the boundary of r_i. Thus, either $w_i < w_1$ or $h_i < w_1$, both of which contradicts with the minimality of w_1.

Let G be a grid of points whose resolution is 1. Let G' be a grid of points whose resolution is $\frac{w_1}{\sqrt{2}}$. By Lemma 2 we see that any square of side length at least w_1 must contain at least one point of G' (To see the argument simply scale down the squares and G' by factor of $\frac{w_1}{\sqrt{2}}$). By definition, every s_i intersects r_1, and, the distance from any point in any $s_i, i \in [0, n-1]$ to the boundary of r_1 is at most $\sqrt{2}w_1$ (in the worst case the point can be on the opposite side of the diameter of a square). Therefore, we cover an axis-parallel rectangle centered with r_1, whose distance to r_1 is at most $\sqrt{2}w_1$, with a grid of points. See Fig. 5 for illustration. That rectangle has width $w_1 + 2\sqrt{2}w_1$ and height $h_1 + 2\sqrt{2}w_1$. Therefore, The number of the grid points $(\frac{w_1 + 2\sqrt{2}w_1}{\frac{w_1}{\sqrt{2}}} + 1) \times (\frac{h_1 + 2\sqrt{2}w_1}{\frac{w_1}{\sqrt{2}}} + 1) = (\sqrt{2} + 5) \times (\sqrt{2}\frac{h_1}{w_1} + 5) \le (5\sqrt{2} + 2)\alpha(C) + 25 + 5\sqrt{2}$. Therefore, the sufficient number of points on the grid to pierce the set C is $(5\sqrt{2} + 2)\alpha(C) + 25 + 5\sqrt{2}$ that is less than or equal to $9.072\alpha(C) + 32.072$. □

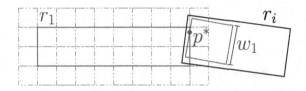

Fig. 5. How an arbitrary rectangle in R gets pierced by a point on the grid.

3 Refined Results for Specific Shapes

In this section we study the number of points sufficient to pierce more specific sets of shapes. First we study sets of pairwise intersecting homothets and design an algorithm that computes the exact location of the points that pierce the set. Next, we show that 2 points are sometimes necessary and always sufficient to pierce a set of pairwise intersecting homothets of a regular hexagon.

3.1 Homothets of a Convex Shape

In this subsection, we show how one can pierce any set of pairwise intersecting homothetic shapes with a constant number of points. More precisely, we give an upper-bound of 15 piercing points. Kim et al. [16] proved that 16 points are sufficient to pierce any set of pairwise intersecting homothetic convex shapes. Kim's proof [16][Lemmas 4,13] requires the existence of two homothetic parallelotopes p_A and P_A such that $p_A \subseteq A \subseteq P_A$ where A is a convex shape. In this paper, our parallelotopes of choice are the pair of rectangles provided by Schwarzkopf et al.'s [22] Algorithm. This pair of rectangles satisfies the required conditions for Kim's [16] proof. Let S be a set of pairwise intersecting homothetic shapes. We prove that 15 points are enough by eliminating one of the 16 points. Finally, given a set S of n k-gons, we give an algorithm of complexity $O(n + \log^2 k)$ to find the exact location of 16 piercing points and $O(n \log k + \log^2 k)$ to find 15 piercing points.

Let $S = \{S_0, S_1, \ldots, S_{n-1}\}$ be a set of pairwise intersecting homothetic convex shapes in the plane. We transform every shape $S_i \in S$ into a pair of homothetic orthogonal rectangles (r_i, R_i), with each pair satisfying the following three conditions: First, r_i is enclosed in S_i, and, R_i encloses S_i. Second, the side length of r_i is at least half of the side length of R_i. Third, the vertices of r_i are located on the boundary of S_i.

For a shape S_i, define C_i to be a *cross-shaped* polygon with edges parallel to the edges of r_i, with $r_i \subseteq S_i \subseteq C_i \subseteq R_i$. Let V_i be the vertices of C_i which include the vertices of r_i (Fig. 6).

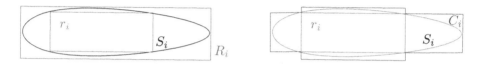

Fig. 6. r_i and C_i are enclosed in S_i and R_i

The existence of such a pair of enclosed and enclosing rectangles for any convex shape was shown by Schwarzkopf et al. [22]. They designed an algorithm to compute such a pair for a convex polygon in time $O(\log^2 k)$ when the k vertices of the polygon are given in an array and sorted in a lexicographic order.

Let S^* be the smallest shape homothetic to the shapes in S that intersects every shape in S. Assume that $\cap_{i=0}^{n-1} S_i = \emptyset$. Minimality of S^* implies that there exist at least 3 shapes in S, say S_1, S_2, S_3, that are tangent to S^* at points x_1, x_2, x_3. Let L_1, L_2, L_3 be the tangent lines to S^* at x_1, x_2, x_3. These three lines form a triangle. For simplicity we assume that S^* is an element of S.

Theorem 3. *Any set of pairwise intersecting convex homothets can be pierced by 15 points.*

Before proving this theorem, we prove a few helper lemmas. According to Kim et al. [16] the 16 piercing points form a grid (see Fig. 7). We label the points from 1 to 16 starting at the top left point. We show that a corner point can be removed from this set of piercing points. Let (r^*, R^*) be the corresponding rectangles to S^*. According to Kim et al. [16] points $\{6, 7, 10, 11\}$ are vertices of r^*.

Every shape in S contains at least one of these 16 piercing points. If there is no shape in the set S that contains only one corner piercing point (points $1, 4, 13, 16$) we can simply remove one of the corner points and reduce the piercing number to 15. Otherwise, without loss of generality, let S_4(resp. S_5, S_6, S_7) $\in S$ be a shape that only contains the point 13 (resp. $1, 4, 16$).

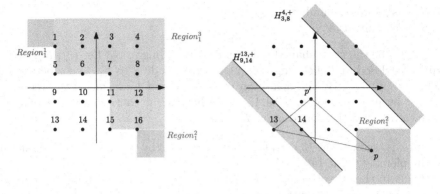

Fig. 7. Dividing the space into different regions and the area that the intersection of S_4 and S_6 cannot be.

Let $H_{i,j}^{k,+}$ (resp. $H_{i,j}^{k,-}$) be a halfspace defined by the line that goes through the piercing points i, j and, contains the piercing point k (resp. does not contain the point k).

Let $Region_1$ (resp. $Region_2$, $Region_3$, $Region_4$) be the area defined as $H_{5,7}^{4,+} \cup H_{7,15}^{4,+}$ (resp. $H_{7,15}^{4,+} \cup H_{10,12}^{13,+}$, $H_{2,10}^{13,+} \cup H_{10,12}^{13,+}$, $H_{5,7}^{4,+} \cup H_{2,10}^{13,+}$). Let $Region_1^1$ be $H_{1,5}^{4,+} \cap H_{1,2}^{13,-}$. Let $Region_1^2$ be $H_{15,16}^{4,-} \cap H_{12,16}^{13,-}$. Let $Region_1^3$ be $Region_1 \cap (H_{1,5}^{4,+} \cap H_{15,16}^{4,+})$ and $Region_2^3$ be $Region_2 \cap (H_{1,2}^{13,+} \cap H_{12,16}^{13,+})$.

Lemma 3. $S_4 \cap S_6 \nsubseteq Region_1^1 \cup Region_1^2$.

Proof. Let $p \in S_4 \cap S_6$. Suppose, for the sake of a contradiction, $p \in Region_1^2$. Let $p' \in S_4 \cap S^*$ and let $p'' \in S_6 \cap S^*$.

- Suppose that p is not located in $Region_1^2 \cap H_{3,8}^{4,+}$.
  In this case the triangle formed by 4, p'' and p will contain point 8, which is a contradiction to the definition of S_6.
- Suppose that p is not located in $Region_1^2 \cap H_{9,14}^{13,+}$.
  Similarly, in this case the triangle formed by 13, p' and p will contain point 14, which is a contradiction to the definition of S_4.

Since $H_{3,8}^{4,+}$ and $H_{9,14}^{13,+}$ have an empty intersection, it implies that the point p cannot be in $Region_1^2$. The same argument holds for $Region_1^1$. Thus, S_4 and S_6 cannot intersect in $Region_1^1$ either.

$\square$

Lemma 4. $S_4 \cap Region_1^3 = \emptyset$.

Proof. Let p be a point on the boundary of $Region_1^3 \cap S_4$. Let S_4' be a shape homothetic to S_4 with the following conditions:

1. S_4' has the same size as S^*.

2. S_4' has p on its boundary.
3. S_4' is contained in S_4.

Let C_4' be the cross shaped polygon corresponding to S_4'. Let (r_4', R_4') be the pair of enclosed and enclosing rectangle corresponding to S_4'. Let v be the top right vertex of r_4'. By the definition of S_4', v cannot be inside of the rectangle defined by piercing points $9, 10, 13, 14$. Otherwise, it contradicts C_4' having an intersection with the boundary of $Region_1^3$.

- ("If v is below point 13 then the triangle defined by point $13, p$ and v contains the piercing point 14. See Fig. 8").
- If v is to the left of point 13 then the triangle defined by point $13, p$ and v contains the piercing point 9.
- If v is to the right and above the piercing point 13, then since the resolution of the piercing points and resolution of the vertices that define r_4' (size of r_4') are equal it implies that r_4' as well as S_4 contains another piercing point beside point 13. See Fig. 8.

□

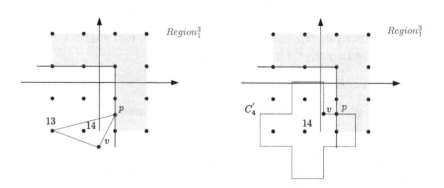

Fig. 8. S_4 does not have an intersection with $Region_1^3$

A similar argument holds for $S_6 \cap Region_2^3$. Lemmas 3 and 4 imply that $S_4 \cap S_6$ is located in the rectangle formed by piercing points $6, 7, 10, 11$.

Lemma 5. $S_4 \cap Region_1$ *has an empty intersection.*

Proof. Let p be a point from the intersection of the boundary of $Region_1$ and S_4. Notice that $Region_1 = Region_1^3 \cup (H_{11,15}^{4,+} \cap H_{15,16}^{4,-}) \cup (H_{1,5}^{4,-} \cap H_{5,6}^{4,+})$. Suppose that S_4 intersects $Region_1$. We analyze the following three cases:

1. $p \in Region_1^3$: According to Lemma 3, p cannot be in $Region_1^3$

2. $p \in H_{11,15}^{4,+} \cap H_{15,16}^{4,-}$: Let p' be a point from the intersection of S_4 and S_6. $S_4 \cap S_6$ is located in the rectangle formed by vertices $6, 7, 10, 11$. This means that p' is to the right of point 14. And it implies that the triangle formed by points 13, p and p' contains point 14 which is a contradiction to the definition of S_4.

3. $p \in H_{1,5}^{4,-} \cap H_{5,6}^{4,+}$: Let p' be a point from the intersection of S_4 and S_6. $S_4 \cap S_6$ is located in the rectangle formed by vertices $6, 7, 10, 11$. This means that p' is above point 9. And it implies that the triangle formed by points 13, p and p' contains point 9 which is a contradiction to the definition of S_4.

Thus, S_4 has an empty intersection with $Region_1$.   □

For a region Reg, let $\overline{Reg}$ be the complement of the region Reg. More precisely, $\overline{Reg} = \{x | x \notin Reg, \forall x \in \mathbb{R}^2\}$.

Notice that S_1, S_2, S_3 are tangent to S^*. Also, S_4 intersects with S_1, S_2, S_3 and S^*. These two facts imply that S_4 intersects with L_1, L_2 and L_3. Moreover, Lemma 5 implies that the intersection of S_4 with each L_1, L_2, L_3 should be located in $\overline{Region_1}$. By symmetry S_5 (resp. S_6, S_7) intersects with L_1, L_2, L_3. This intersection is located in $\overline{Region_2}$ (resp. $\overline{Region_3}, \overline{Region_4}$).

Lemma 6. *Each $\overline{Region_i}, i \in [1,4]$ has a non-empty intersection with at least one of L_1, L_2, L_3.*

Proof. Notice that the bottom-left vertex of r^* is in $\overline{Region_1}$ and the triangle defined by the intersection of L_1, L_2, L_3 encloses r^*. Suppose that $\overline{Region_1}$ has empty intersection with all L_1, L_2, L_3. This implies that the triangle defined by the intersection of L_1, L_2, L_3 does not enclose r^*, which is a contradiction. Similarly this argument can be applied for $\overline{Region_2}, \overline{Region_3}$ and $\overline{Region_4}$.   □

Lemma 7. *At least two of the regions in $\{\overline{Region_i}, i \in [1,4]\}$ do not have an intersection with all three of L_1, L_2, L_3.*

Proof. Observe that each $L_i, i \in [1,3]$ can intersect with at most three regions of $\{\overline{Region_i} | i \in [1,4]\}$. Thus, we have at most 9 pairs of $(L_i, Region_j)$ when L_i intersects with $Region_j$ for $i \in [1,3], j \in [1,4]$. According to Lemma 6 and the Pigeonhole theorem at least two of the regions in $\{\overline{Region_i} | i \in [1,4]\}$ do not intersect with all three of L_1, L_2, L_3.   □

Proof (Proof of Theorem 3). Lemma 7 implies that the regions corresponding to at least two of the shapes S_4, S_5, S_6, S_7 do not intersect with all three of L_1, L_2, L_3. This contradicts the fact that the shapes in the set S are pairwise intersecting. Thus, at least one piercing point can be removed from our piercing point set.   □

Algorithm to find the exact location of the piercing points: First, we find the smallest shape, S_1, of the set in $O(n)$. Next we apply the Schwarzkopf et al.'s [22] algorithm on S_1 to compute the vertices of $r_1 = (w_1, h_1)$ in $O(\log^2 k)$ time. This allows us to compute the 16 points outlined in Kim's [16] proof in a constant time. Thus the time complexity of finding 16 piercing points is $O(n + \log^2 k)$. Next, we can determine in $O(n \log k)$ time which of the 4 corner points can be removed. Thus, we can find 15 piercing points in $O(n \log k + \log^2 k)$ time.

3.2   Hexagons

In this subsection we determine the piercing number of a set of pairwise inter-secting homothets of a regular hexagon. We show that two points are always sufficient and sometimes necessary to pierce such a set. For a hexagon s with an edge parallel to the x-axis, we refer to its edges by *Bottom, BottomRight, TopRight, Top, TopLeft, BottomLeft* edges. We denote the *Bottom* edge of s by s^B. Respectively we refer to *BottomRight, TopRight, Top, TopLeft, BottomLeft* edges of s by $s^{BR}, s^{TR}, s^T, s^{TL}, s^{BL}$.

Theorem 4. *Any set of pairwise intersecting homothets of a regular hexagon can be pierced by two points.*

Proof (Proof of Theorem 4). Let $C = \{C_0, C_1, \ldots C_{n-1}\}$ be a set of pairwise intersecting homothets of a regular hexagon. Without loss of generality, assume that the bottom edge of every hexagon in C is parallel with the x-axis. Let $TL = \{C_i^{TL} | \forall C_i \in C\}$, and $TR = \{C_i^{TR} | \forall C_i \in C\}$, and $BE = \{C_i^B | \forall C_i \in C\}$. Each element of these sets is a line segment. Segments of each set are associated with the same side of hexagons in C. $\forall C_i^{TL} \in TL$ let TL_i^+ be the halfspace defined by C_i^{TL} that does not contain the corresponding hexagon to C_i^{TL}. Let $TL^+ = \{TL_0^+, TL_1^+ \ldots, TL_{n-1}^+\}$. $\forall C_i^{TR} \in TR$ let TR_i^+ be the halfspace defined by C_i^{TR} that does not contain the corresponding hexagon to C_i^{TR}. Let $TR^+ = \{TR_0^+, TR_1^+ \ldots, TR_{n-1}^+\}$. Let tl^* be the halfspace in TL^+ that contains all other halfspaces in TL^+. Such a halfspace exists since all of the halfspaces in TL^+ are parallel. Let tr^* be the halfspace in TR^+ that contains all other halfspaces in TR^+. Such a halfspace exists since all of the halfspaces in TR^+ are parallel.

Let tl be the corresponding segment in TL to tl^*. Let tr be the corresponding segment in TR to tr^*. Let be be the top most segment in BE.

Define L_1 the line that is parallel to and goes through tl. Similarly, define R_1 to be the line that is parallel to and goes through tr, and B_1 to be the line that is parallel to and goes through be.

Assume that L_1, R_1, B_1 do not intersect at a point p. Otherwise, the two piercing points will be on top of each other at p, thus, p pierces the whole set. Observe that, the intersection points of L_1, R_1, B_1 form a equilateral triangle $T_h = ABD$. Let point B be the intersection point of lines L_1 and R_1. Let point D be the intersection point of lines L_1 and B_1 and let point A be the intersection point of lines B_1 and R_1. This triangle can have one of the two following possible shapes.

Case 1. In the first case, the point B is located above the segment AD. There-fore, the left top side of any hexagon in C is to the left of L_1. Similarly, the right top side of any hexagon in C is to the right of R_1 and any bottom side of any hexagon in the set is below B_1.

Case 2. In the second case, the point B is located below the segment AD. Therefore, the bottom right side of any hexagon in C is to the right of L_1 since, each pair of hexagons have to intersect. The top side of any hexagon in C is above B_1 and, the bottom left side of any hexagon is to the left of

R_1, otherwise it contradicts the fact that each pair of hexagons intersects. Observe that this case is symmetric to the first case. Therefore, giving the proof for the first case is sufficient.

Proof for the case 1: Let M_L, M_R and M_B be the mid points corresponding to sides AB, BD and AD of the triangle $T_h = ABD$. We prove that every hexagon in C contains either M_B or B.

Take an arbitrary hexagon s from C. If s contains M_B then we are done. Suppose that s does not contain the point M_B. The point M_B can be either to the right of s^{BR} or to the left of s^{BL} and both cases are symmetric. Without loss of generality, assume that the point M_B is to right of the s^{BR}. By the definition of R_1 the top right edge of any hexagon in C is to the right of R_1. Similarly, the bottom edge of any hexagon in C is to the bottom of B_1. This implies that the right bottom side of s should cross the lines B_1 and R_1. Let i_1 be the intersection point of s^{BR} and B_1, and i_2 be the intersection point of s^{BR} and R_1.

The point i_1 is to the left of M_B and i_2 is to the left of M_R. Observe that the triangle $i_1 i_2 D$ is similar to $M_B M_R D$ and $|Di_2| > |DM_B|$. Therefore, the segment $i_1 i_2$ is greater than $M_R M_B$. Moreover, the side length of s is greater than or equal to the length of the segment $i_1 i_2$, and it is greater than $M_B M_R$ ($|s^B| = |s^{BR}| \geq |i_1 i_2| > |M_R M_B| = |M_R B| = |M_B A|$). Considering the facts that the side length of s is greater than $|M_R M_B|$, and s^{TR} is parallel to R_1 and to the right of R_1. Thus, by convexity of s, s^{TR} crosses L_1, and similarly s^B crosses L_1. Thus s contains the segment AB and in particular the point B.

□

4   Conclusion

In this paper we showed that pairwise intersecting convex shapes of fatness α with arbitrary orientation can be pierced by a linear number of points with respect to the fatness parameter of the shapes in the set. The main idea to achieve our results is to avoid covering an area with a grid of high resolution but rather focusing on the perimeter of a specific shape. By using this idea we reduce the number of points sufficient to pierce any pairwise intersecting convex α-fat shapes from $O(\alpha^2)$ to $O(\alpha)$. Our theorem is an improvement over Fredman's sampling algorithm to find piercing points.

Moreover, for a set of pairwise intersecting homothets we showed that the piercing number is at most 15 points. The piercing number of a set of pairwise intersecting set of homothets of regular hexagons is 2 which is tight. We leave as an open problem to improve our upper bounds.

References

1. Agarwal, Pankaj K.., Katz, Matthew J.., Sharir, Micha: Computing depth orders and related problems. In: Schmidt, Erik M.., Skyum, Sven (eds.) SWAT 1994. LNCS, vol. 824, pp. 1–12. Springer, Heidelberg (1994). https://doi.org/10.1007/3-540-58218-5_1

2. Agarwal, P.K., Katz, M.J., Sharir, M.: Computing depth orders for fat objects and related problems. Comput. Geom. Theory App. **5**(4), 187–206 (1995)
3. Biniaz, A., Bose, P., Wang, Y.: Simple linear time algorithms for piercing pairwise intersecting disks. In: He, M., D., Sheehy, M., (eds.) Proceedings of the 33rd Canadian Conference on Computational Geometry, CCCG 2021, 10–12 August 2021, Dalhousie University, Halifax, Nova Scotia, Canada, pp. 228–236 (2021)
4. Bose, P., et al.: The floodlight problem. Int. J. Comput. Geom. Appl. **7**(1/2), 153–163 (1997)
5. Chakerian, G.D., Stein, S.K.: Some intersection properties of convex bodies. Proc. Am. Math. Soc. **18**(1), 109–112 (1967)
6. Danzer, L.: Zur Lösung des Gallaischen Problems über Kreisscheiben in der Euklidischen Ebene. Stud. Sci. Math. Hung. **21**(1–2), 111–134 (1986)
7. Efrat, A., Rote, G., Sharir, M.: On the union of fat wedges and separating a collection of segments by a line. In: Lubiw, A., Urrutia, J., (eds.), Proceedings of the Fifth Canadian Conference on Computational Geometry, Waterloo, pp. 115–120 (1993)
8. Goodman, J.E., O'Rourke, J. (eds.): Handbook of Discrete and Computational Geometry. CRC Press Inc., USA (1997)
9. Grünbaum, B.: On intersections of similar sets. Portugal. Math. **18**, 155–164 (1959)
10. Har-Peled, S., et al.: Stabbing pairwise intersecting disks by five points. In: 29th International Symposium on Algorithms and Computation (ISAAC 2018). Schloss Dagstuhl-Leibniz-Zentrum fuer Informatik (2018)
11. Ed. Helly. Über mengen konvexer körper mit gemeinschaftlichen punkte. Jahresbericht der Deutschen Mathematiker-Vereinigung. **32**, 175–176 (1923)
12. Karasev, R.N.: Transversals for families of translates of a two-dimensional convex compact set. Disc. Comput. Geom. **24**(2), 345–354 (2000)
13. Karasev, R.N.: Piercing families of convex sets with the d-intersection property in $\mathbb{R}^d$. Disc. Comput. Geom. **39**(4), 766–777 (2008)
14. Katz, M.J.: 3-d vertical ray shooting and 2-d point enclosure, range searching, and arc shooting amidst convex fat objects. Comput. Geom. **8**(6), 299–316 (1997)
15. Katz, M.J., Overmars, M.H., Sharir, M.: Efficient hidden surface removal for objects with small union size. Comput. Geom. **2**(4), 223–234 (1992)
16. Kim, S.-J., Nakprasit, K., Pelsmajer, M.J., Skokan, J.: Transversal numbers of translates of a convex body. Disc. Math. **306**(18), 2166–2173 (2006)
17. Matousek, J., Pach, J., Sharir, M., Sifrony, S., Welzl, E.: Fat triangles determine linearly many holes. SIAM J. Comput. **23**, 154–169 (1994)
18. Nielsen, F.: On point covers of c-oriented polygons. Theo. Comp. Sci. **265**(1–2), 17–29 (2001)
19. Overmars, M.H., van der Stappen, F.A.: Range searching and point location among fat objects. J. Algorithms **21**(3), 629–656 (1996)
20. Pach, J., Safruti, I., Sharir, M.: The union of congruent cubes in three dimensions. In Proceedings of the Seventeenth Annual Symposium on Computational Geometry, SCG 2001, pp. 19–28. Association for Computing Machinery, New York, NY, USA (2001)
21. Carmi, P., Morin, P., Katz, M.J.: Stabbing pairwise intersecting disks by four points (2020)
22. Schwarzkopf, O., Fuchs, U., Rote, G., Welzl, E.: Approximation of convex figures by pairs of rectangles. Comput. Geom. **10**(2), 77–87 (1998)
23. Stachó, L.: Über ein Problem für Kreisscheibenfamilien. Acta Scientiarum Mathematicarum (Szeged) **26**, 273–282 (1965)

24. Stachó, L.: A solution of Gallai's problem on pinning down circles. Mat. Lapok. **32**(1–3), 19–47 (1981/1984)
25. van Kreveld, M.J.: On fat partitioning, fat covering and the union size of polygons. Comput. Geom. **9**(4), 197–210 (1998)

Local Routing Algorithms on Euclidean Spanners with Small Diameter

Nicolas Bonichon[1(✉)], Prosenjit Bose[2], and Yan Garito[3]

[1] LaBRI, U-Bordeaux, Bordeaux, France
`bonichon@labri.fr`
[2] Carleton University, Ottawa, Canada
`jit@scs.carleton.ca`
[3] ENS Rennes, Rennes, France
`yan.garito@ens-rennes.fr`

Abstract. Given a set of n points in the plane, we present two constructions of geometric r-spanners with $r \geq 1$ based on a hierarchical decomposition. These graphs have $\mathrm{O}(n)$ edges and diameter $\mathrm{O}(\log n)$. We then design online routing algorithms on these graphs.

The first construction is based on Θ_k-graphs (with $k > 6$ and $k \equiv 2 \bmod 4$). The routing algorithm is memoryless and local (i.e. it uses information about the closed neighborhood of the current vertex and the destination). It has routing ratio $1/(1 - 2\sin(\pi/k))$ and finds a path with $\mathrm{O}(\log^2 n)$ edges.

The second construction uses a TD-Delaunay triangulation, which is a Delaunay triangulation where the empty regions are homothets of an equilateral triangle. The associated routing algorithm is local and memoryless, has a routing ratio of $5/\sqrt{3}$, finds a path consisting of $\mathrm{O}(\log^2 n)$ edges and requires the pre-computation of vertex labels of $\mathrm{O}(\log^2 n)$ bits (assuming the nodes are placed on a grid of polynomial size).

We have examples that show when using either of our routing algorithms, in the worst case, the paths returned by the algorithm can consist of $\Omega(\log^2 n)$ edges.

1 Introduction

We focus on two fundamental problems in networking: network design and online routing in these networks. These two problems go hand in hand; we build our network and then design efficient online algorithms to route in these networks. We focus on *geometric* networks, which are graphs whose vertices are points in the plane and whose edges are line segments connecting these points. The edges are weighted by the Euclidean distance between their endpoints. Geometric spanners have been studied extensively [2, 19, 26] since their introduction by Chew [13]. In this paper, we are particularly interested in geometric spanners with low hop-diameter. The literature is vast and many constructions of such

Research of P. Bose and Y. Garito supported in part by NSERC.

A. Castañeda and F. Rodríguez-Henríquez (Eds.): LATIN 2022, LNCS 13568, pp. 696–712, 2022.
https://doi.org/10.1007/978-3-031-20624-5_42

spanners exist [1–4,12,17,24,25]. However, most are probabilistic constructions with no online routing algorithm (such as in [3,4,17]). In some cases, the proof of the spanning ratio is constructive but does not lend itself to a local routing algorithm since knowledge of the whole graph is required or paths are computed by working inward from both the source and destination vertices. For the constructions that do have online routing algorithms (such as in [1,12]), the networks store routing tables at the vertices and the routing algorithms require $\Theta(\log n)$ bits of overhead memory, where n is the size of the network. In contrast, both our constructions are deterministic and our main contribution is the design of online routing algorithms that only need to know the destination and do not use any additional memory overhead (i.e. they are memoryless).

We present two constructions of geometric r-spanners with $r \geq 1$ that have low diameter on which we design simple efficient local routing algorithms that return r-spanning paths consisting of $O(\log^2 n)$ edges. The key obstacle in designing local routing algorithms is that the routing algorithm does not have the whole graph at its disposal. When the whole graph is available, then standard shortest path algorithms, such as Dijkstra's algorithm [16], can be used to efficiently find short paths. In the online setting, the routing algorithm must simultaneously explore the graph while trying to find a short path to the destination, which is the challenge. Our routing algorithms are deterministic, *memoryless* and *local*. A routing algorithm is considered memoryless and local if the only information available to the algorithm, prior to deciding which edge to follow out of its current vertex, consists only of the information stored at the current vertex (which is typically the closed neighborhood of the vertex) and the label of the destination vertex. So, for example, even simple graph exploration algorithms such as Depth-First Search cannot be executed in a memoryless manner. In fact, a deterministic, memoryless, local routing algorithm will not know if it has visited a vertex multiple times which means that these algorithms are prone to cycling. As such, additional properties of the graph must be used to route successfully.

In order to achieve a low diameter, our spanners are built in a hierarchical manner where the hierarchy has logarithmic height. It is the hierarchy that provides a low diameter to these graphs. However, the difficulty for memoryless routing in such spanners is that locally it is unclear when one should go up or down the hierarchy to find a short path with few edges. We summarize our results below.

We first present a hierarchical construction using Θ_k-graphs at each level (with $k > 6$ and $k \equiv 2 \mod 4$) [14,21,22]. The routing algorithm is memoryless, and local (i.e. the local information consists of its current position, the positions of its neighbors and the position of the destination). It has routing ratio[1] $1/(1 - 2\sin(\pi/k))$ and finds a path with $O(\log^2 n)$ edges. We also show that their exist configurations such that the path obtained with our routing algorithm has $\Omega(\log^2 n)$ edges.

---

[1] The routing ratio of a routing algorithm is the spanning ratio of the path returned by the algorithm.

The second hierarchical construction uses a TD-Delaunay triangulation at each level, which is a Delaunay triangulation where the empty regions are homothets of an equilateral triangle. The advantage here is that each level consists of a planar graph, which is the main graph class used for guaranteed delivery of packets in wireless networks [11,20]. The associated routing algorithm is memoryless, local, has a routing ratio of $5/\sqrt{3}$, and finds a path consisting of $O(\log^2 n)$ edges. The local information in addition to storing the information about the closed neighborhood of the current vertex also requires the pre-computation of vertex labels of $O(\log^2 n)$ bits (assuming the nodes are placed on a grid of polynomial size). We show when using either of our routing algorithms, in the worst case, the paths returned by the algorithm can consist of $\Omega(\log^2 n)$ edges.

2   Preliminaries

The graph theoretic terminology we use is standard [15]. The terminology of spanners we use is from Narasimhan and Smid [26]. We focus on *geometric* graphs which are weighted graphs whose vertices are points in the plane and whose edges are segments weighted by the Euclidean distance between their endpoints. We use $V(G)$ and $E(G)$ to denote the vertex set and edge set of a graph G. Given an edge $uv \in E(G)$, we use $|uv|$ to denote its length. A geometric graph G is an r-spanner if for all $u, v \in V(G)$, there exists a path Π from u to v such that the sum of the weights of the edges of Π is at most $r|uv|$. The path Π is called an r-*spanning path* (or simply spanning path when r is understood from the context) from u to v. The smallest value of r for which G is an r-spanner is its *spanning ratio*. A D-*independent set* in a graph G is an independent set of vertices whose maximum degree is D.

The class of Θ_k-graphs play an important role in both our constructions [14, 22]. A Θ_k-graph is constructed in the following way: the plane around each vertex u is partitioned into k cones with apex u and cone angle $\theta = 2\pi/k$. The k cones around u are labeled in clockwise order as $C_0^u, \ldots, C_{k-1}^u$ with all index manipulation modulo k (see Fig. 1a). Each cone C_i^u is defined by two lines L_i^u, L_{i+1}^u through u. In each cone, u is joined to the point, v, whose projection on the bisector of the cone is closest to u. The *canonical triangle* Δ_{uv} between u and v is the triangle defined by the two rays of u's cone containing v and the line through v perpendicular to the cone bisector. Without loss of generality, we assume that no point lies on a cone boundary. Note that Δ_{uv} is empty if and only if there is an edge from u to v. Although a Θ_k-graph is naturally directed, we focus on the underlying undirected graph. We assume that our points are vertices of a polynomial size grid with $O(\log n)$ bit coordinates.

When $k \equiv 2 \bmod 4$, we define the *half-Θ_k-graph*. The construction is virtually identical to the Θ_k-graph except that for each point u, edges are only computed in the even cones, which we call *positive* cones. The cones with odd index are called *negative* cones. Bonichon et al. [5] showed that the half-Θ_6 graph is identical to the TD-Delaunay graph. For the half-Θ_6-graph, we relabel the cones as in Fig. 1b. Note that if v is in a negative cone of u, then u is in a positive cone of v.

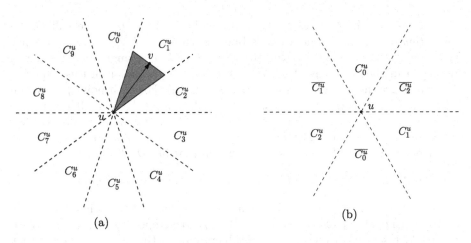

Fig. 1. (a): The cones around u for Θ_{10}. u has an edge to v as the triangle Δ_{uv} is empty (b): The relabeled cones in the case of the half-Θ_6 graph: each cone C_i has an "opposite" cone $\overline{C_i}$

3   The Hierarchical Construction

In this section, we present our hierarchical construction. Our hierarchy is similar to the hierarchy defined by Arya et al. [4]. Their hierarchy is also based on Θ_k-graphs but they use a randomized Skip-List-like strategy to build the levels of their hierarchy [29]. To make our hierarchy deterministic, we use an idea similar to Kirkpatrick's point location structure [23] by repeatedly removing large independent sets of bounded degree, which has been used in the context of geometric spanners before by Hoedemakers [18].

Before describing our hierarchy, we prove a simple property about sparse graphs. A graph G is κ-sparse if $|E(G)| \leq \kappa n$ where $n = |V(G)|$. We assume throughout that all our graphs are connected.

Lemma 1. *There are at least* $n(d + 1 - 2\kappa)/d$ *vertices of degree at most d for $d \geq 2\kappa$ in an n-vertex κ-sparse graph G.*

Proof. Let V_d be the set of vertices of degree at most d in G. We note that $\sum_{v \in V(G)} \deg(v) = 2|E(G)| \leq 2\kappa n$. Since $\sum_{v \in V(G)} \deg(v) \geq |V_d| + (n - |V_d|)(d + 1)$, the result follows. $\qquad\square$

Lemma 2. *There is an independent set of size at least* $n(d+1-2\kappa)/(d(d+1))$ *of vertices of degree at most d for $d \geq 2\kappa$ in an n-vertex κ-sparse graph G.*

Proof. Follows immediately from Lemma 1. $\qquad\square$

Hierarchy Construction: Let P be a set of n points in the plane. Let $\mathcal{A}$ be a function that computes a κ-sparse geometric graph given P as input. A *hierarchy of graphs* built from the graph $G = \mathcal{A}(P)$ is a finite sequence of graphs

$G_0 = G, G_1 = (V_1 \subset V(G), E_1), ..., G_h = (V_h \subset V_{h-1}, E_h)$. We call G the *foundation* on which the hierarchy is built. Each $G_i = \mathcal{A}(V_i)$ is called a *layer*, with i being the *level* of G_i within the hierarchy. G_h is a single vertex and h is the height of the hierarchy. A *Hierarchical graph* is the union of this sequence of κ-sparse graphs into one graph $H = (V_0, \cup_{i=0}^h E_i)$. For each point $p \in V_0$, we define $\ell(p)$, the *level* of p, to be the largest i such that $p \in V_i$. We call a hierarchy *compact*, if $|V_i \setminus V_{i+1}| \geq c|V_i|$ for a fixed constant c independent of n. This implies that compact hierarchies have height $O(\log n)$. We now show how to construct a compact hierarchy H from a κ-sparse graph that has $|E(H)| = O(n)$.

Lemma 3. *A κ-sparse graph G admits a compact hierarchical graph H.*

Proof. Let $c = 1/(2\kappa + 1)^2$. By Lemma 2, we have that each G_i contains a 2κ-independent set of size at least $c|V(G_i)|$. By repeatedly removing these independent sets from one level of the hierarchy to the next, we ensure that $|V_{i+1}| \leq (1-c)|V_i|$. Since G_h has size 1, the height h is $\log_{1/(1-c)} n = O(\log n)$. Moreover, $\sum_{i=0}^{h} |E_i| \leq \sum_{i=0}^{h} \kappa|V(G_i)| \leq \sum_{i=0}^{h} \kappa(1-c)^i n = O(n)$.   □

Henceforth, in this article, all hierarchies are constructed to be compact.

Lemma 4. *Let H be a Hierarchical graph built from G with height h. If every level of H is connected, then H has diameter $O(h)$.*

Proof. Let p be the vertex of H with level h. A vertex u at level $i < h$ always has a neighbor whose level is greater than i. Therefore, every vertex in H has a path to p of length $O(h)$ by simply going up the hierarchy.   □

Observation 1. *Let H be a Hierarchical graph based on Θ_k-graphs or half-Θ_6-graphs. Since at each level we remove an independent set, $\ell(u) \neq \ell(v)$ for every edge uv of H.*

4   The Θ_k-Hierarchical Graph

We begin by defining a class of graphs which we call the *augmented Θ_k-graph*. Throughout this section we assume that $k \equiv 2 \mod 4$, $k > 6$ and all index manipulation is done mod k. Note that we are using the directed version of Θ_k graphs, although we can forget direction if we want to.

The crucial property of Θ_k-graphs when $k \equiv 2 \mod 4$ is that the edges of Δ_{uv} are parallel to L_i, L_{i+1} and $L_{i+\lceil k/4 \rceil}$. In other words, each edge of the canonical triangle is parallel to a cone boundary, which is not the case when $k \not\equiv 2 \mod 4$. When computing an edge uv in C_i^u, one can view this process as sweeping a line parallel to $L_{i+\lceil k/4 \rceil}$ starting at u and moving away from u along the bisector of C_i until reaching the first point, v. We now define a Θ_k^j-graph, which has the same vertex set as the Θ_k-graph. However, to compute an edge uv in C_i^u, we sweep a line parallel to L_{i+j} starting at u and moving away from u along the bisector of C_i, where $j \in \{1, \ldots, (k-4)/2\}$ ((see Fig. 2a). The augmented Θ_k-graph is defined as the union of Θ_k^j-graphs for all values of j. We will refer to the canonical triangle defined by u and v in Θ_k^j as Δ_{uv}^j. We drop the superscript j when referring to the canonical triangle in the Θ_k-graph.

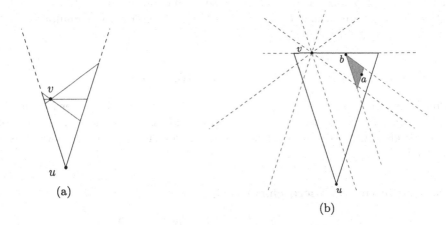

(a)

(b)

Fig. 2. (a): The different canonical triangles of Θ_k^j-graphs when $k = 10$. The blue lines represent top edges of the $(k-4)/2 = 3$ different canonical triangles. (b): The points a and b inside Δ_{uv} must have a canonical triangle that is completely inside Δ_{uv}. In this case, it is the red triangle (Color figure online)

Let $G_0, G_1, \ldots, G_h$ be a compact hierarchy of graphs built on a foundation that is an augmented Θ_k-graph on a set P of n points. Since an augmented Θ_k-graph has $O(k^2 n)$ edges, by Lemma 3, we know that such a hierarchy exists. The only reason that we considered augmented Θ_k-graphs is to extract the sequence of points $P = P_0 = V(G_0) \supset P_1 = V(G_1) \supset \ldots \supset P_h = V(G_h)$. These points have a special geometric property that we exploit in our routing algorithm. We construct a compact hierarchy by building a Θ_k-graph on this sequence of points. The Θ_k-*Hierarchical* graph H is the union of $\Theta_k(P_i)$ for $i \in \{0, \ldots, h\}$. It is important to note that the hierarchy is built with Θ_k-graphs in each layer and **not** augmented Θ_k-graphs. The following lemma is the key geometric property of the Θ_k-Hierarchical graph that aids in the routing.

Lemma 5. *Let* $\ell \in \{0, 1, \ldots, h\}$. *Let* u, v *be two vertices of* H *such that* Δ_{uv} *is empty at level* ℓ. *Then* Δ_{uv} *contains at most one point at level* $\ell - 1$.

Proof. Assume there are two points a and b inside Δ_{uv} at level $\ell - 1$. Without loss of generality, we can assume that $a \in \Delta_{ub}$. Let $v \in C_j^u$ and let L_d, L_{d+1} be the two lines defining this cone. Let C be the cone of b that contains a.

If $C \cap \Delta_{uv}$ is a triangle then this triangle must be bounded by the cone lines defining C and one of the lines defining C_j^u. Therefore, one of Δ_{ba}^d or Δ_{ba}^{d+1} must be contained in $C \cap \Delta_{uv}$ (see Fig. 2b). Otherwise, $C \cap \Delta_{uv}$ is not a triangle. This can only happen if C is defined by both L_d and L_{d+1} as $k \equiv 2 \mod 4$. In other words, $b \in C_j^a$. Thus, $\Delta_{ab} \subseteq \Delta_{uv}$.

In both cases, we have some canonical triangle Δ defined by a and b that is contained in Δ_{uv}. Since $\ell(a) = \ell(b)$, by construction of the Θ_k-Hierarchical graph, ab cannot be an edge of the augmented Θ_k-graph at level $\ell - 1$. There must be a point c in Δ at level $\ell - 1$ such that one of ac or bc is an edge. However, this implies that $\ell(c) > \ell - 1$ which contradicts the assumption that Δ_{uv} is empty at level ℓ.

$\square$

Lemma 6. *If $ab \in E(H)$ and $u \in \Delta_{ab}$, then $\ell(u) < \ell(a)$.*

Proof. Assume otherwise: $\ell(u) \geq \ell(a)$. Then $u \in \Delta_{ab}$ is present in levels $0, 1, ..., \ell(a)$. By construction of the Θ_k-graph, ab cannot be an edge in any of those levels. Therefore, by construction of H, ab cannot be an edge, which is a contradiction. $\square$

Routing in the Θ_k-Hierarchical Graph

In a Θ_k-graph, for $k > 6$, the *greedy* routing algorithm always reaches the destination and the routing ratio is $1/(1 - 2\sin(\pi/k))$ [8,30]. The greedy routing algorithm is simple: let u be the current vertex and t be the destination. Move to the vertex v adjacent to u in Δ_{ut} such that Δ_{uv} is empty. We slightly modify this algorithm when routing on a Θ_k-Hierarchical graph and refer to this algorithm as the *hierarchical greedy (routing)* algorithm. The only difference from the standard greedy algorithm is that we follow the edge $uv \in \Delta_{ut}$ such that v is the vertex of highest level **inside** Δ_{ut} among all vertices w adjacent to u.

We first show that this simple algorithm always finds a path to t and the path has spanning ratio at most $1/(1 - 2\sin(\pi/k))$. We then show that this path consists of $O(\log^2 n)$ edges.

Lemma 7. *This algorithm always reaches t and has spanning ratio at most $1/(1 - 2\sin(\pi/k))$.*

Proof. Consider the Θ_k-graph constructed on the points of the path generated by the hierarchical greedy routing algorithm. By construction, the greedy routing algorithm on this graph generates the same path. The result follows from the result in [30]. $\square$

When the hierarchical greedy algorithm follows an edge uv with $\ell(u) < \ell(v)$, we say that it is *going up* the hierarchy, otherwise ($\ell(u) > \ell(v)$ by Remark 1) we say it is *going down*. We fix a source s and a destination t. Let $s = u_0, u_1, ..., u_m = t$ be the points visited while routing from s to t using this algorithm. What remains to be shown is that $m = O(\log^2 n)$. Notice that the hierarchical greedy algorithm may go up and down the hierarchy multiple times. We need to bound the number of times this can happen. We first examine what happens when we are going down the hierarchy. We assume that $p \in \{0, 1, ..., m - 1\}$.

Lemma 8. *If $\ell(u_{p+1}) < \ell(u_p)$ then $\Delta_{u_p t}$ is empty at level $\ell(u_{p+1}) + 1$.*

Proof. Assume there is some v in the interior of $\Delta_{u_p t}$ at level $\ell(u_{p+1}) + 1$. Since $\Delta_{u_p t}$ is not empty at this level, and $\ell(u_{p+1}) < \ell(u_p)$, u_p must have an edge to a point inside $\Delta_{u_p t}$ by construction of the Θ_k-graph. Without loss of generality, we can assume v is this point. This means $u_p v$ is an edge in the Θ_k-Hierarchical graph with $\ell(u_{p+1}) < \ell(v)$ and v is inside $\Delta_{u_p t}$. The hierarchical greedy algorithm would then have followed the edge to v rather than to u_{p+1}, which is a contradiction. $\qquad\square$

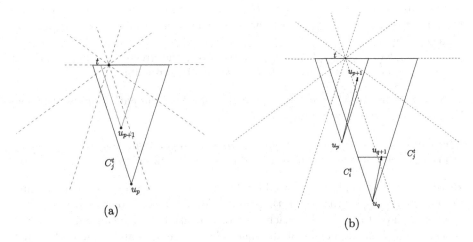

Fig. 3. (a): If $u_p \in C_j^t$ and $u_{p+1} \in C_j^t$, then there is an inclusion between their canonical triangles (b): Illustration of the situation of Lemma 10 with two transitional edges from C_i^t to C_j^t. We are at u_p and are going to u_{p+1}. Previously, we were at u_q and went down to u_{q+1}.

Lemma 9. *Let $u_p \in C_j^t$. If $\ell(u_{p+1}) < \ell(u_p)$ and $u_{p+1} \in C_j^t$, then $\Delta_{u_{p+1} t}$ is empty at level $\ell(u_{p+1})$.*

Proof. Let v be a point in $\Delta_{u_{p+1} t}$. Lemma 8 implies that $\Delta_{u_p t}$ is empty at the level $\ell(u_{p+1}) + 1$. Since $u_{p+1} \in C_j^t$, we know that $\Delta_{u_{p+1} t} \subseteq \Delta_{u_p t}$ (see Fig. 3a). By Lemma 5 we have that u_{p+1} is the only point inside $\Delta_{u_p t}$ at its level. In particular, $\Delta_{u_{p+1} t}$ must be empty. $\qquad\square$

When routing, the path can "zigzag" between different cones of t as in Fig. 3b; that is, the path can go from a cone i of t to a cone j, then back to cone i, then to cone j, etc. We call an edge uv a *transitional* edge, if u and v lie in distinct cones of t. Otherwise, when u and v lie in the same cone, we call it a *stable* edge. If the routing algorithm only follows stable edges, then the path can only consist of at most $2h$ edges where h is the height of the hierarchy. In this case, the path first goes up the hierarchy, then goes down the hierarchy to reach t. Lemmas 8 and 9 guarantee that we cannot go up and down the hierarchy more than once. Thus, if the path is long, then it must consist of many transitional edges. The next lemma is the key to bounding the number of transitional edges.

Lemma 10. *Let $u_p u_{p+1}$ be a transitional edge from C_i^t to C_j^t. Suppose there exists a $q \in \{0, 1, ..., p-2\}$ such that $u_q u_{q+1}$ is also a transitional edge from C_i^t to C_j^t. If $\ell(u_{q+1}) < \ell(u_q)$, then $\ell(u_{p+1}) < \ell(u_{q+1})$*

Proof. There are two cases to consider: $\Delta_{u_p t} \cap C_j^t \subseteq \Delta_{u_q t}$ or $\Delta_{u_p t} \cap C_j^t \nsubseteq \Delta_{u_q t}$.

We consider the former case first. By Lemmas 5 and 8, we know that u_{q+1} is the only point in $\Delta_{u_q t}$ at level $\ell(u_{q+1})$. Therefore, since $u_{p+1} \in \Delta_{u_q t}$, we must have $\ell(u_{p+1}) < \ell(u_{q+1})$, otherwise we contradict that fact that $u_q u_{q+1}$ is a greedy edge.

In the latter case, $|u_p t| > |u_{q+1} t|$. However, the greedy routing algorithm always moves closer to t with each step, so this is impossible. □

We now have all the tools to bound the length of the path generated by the hierarchical greedy routing algorithm.

Theorem 1. *The hierarchical greedy routing algorithm finds a path using $O(\log^2 n)$ edges and has routing ratio at most $1/(1 - 2\sin(\pi/k))$.*

Proof. Let Π be the path generated by the hierarchical greedy routing algorithm. Since there are k cones around t, there are $\binom{k}{2}$ different types of transitional edges in Π. Lemma 10 implies that each type of transitional edge can only appear h times, where h is the height of the hierarchy. Lemmas 8 and 9 guarantee that there are at most $2h$ consecutive stable edges in Π. Thus, the length of Π is at most $2\binom{k}{2}h^2$ which is $O(\log^2 n)$.

The routing ratio of the algorithm follows from Lemma 7.

□

Theorem 2. *The hierarchical greedy routing algorithm may take $\Omega(\log^2 n)$ steps on a Θ_k-hierarchical graph.*

Proof (sketch). The key idea is to use the pattern highlighted by Lemma 10. We repeat the following pattern $\Omega(\log n)$ times. $u \in C_i^t$ goes down to $v \in C_{i-1}^t$. Then v goes up to $w \in C_i^t$. We have no information about $\Delta_{wt} \setminus \Delta_{ut}$ so we may go down $\Omega(\log n)$ times inside it before reaching $u' \in C_i^t$.

In total, we would take $\Omega(\log^2 n)$ steps by following this path. The exact details are ommited due to space constraints.

5   The TD-Delaunay-Hierarchical Graph

In this section, we present the construction of the TD-Delaunay-Hierarchical graph (TDH for short) on a set of n points in the plane. Then we show how to compute labels of size $O(\log^2 n)$ bits for each vertex to aid in routing. Finally, given these labels, we provide a local memoryless routing algorithm that finds a path between two vertices s, t in a TDH that has spanning ratio $5/\sqrt{3}$ and $O(\log^2 n)$ edges. The main draw to the TD-Delaunay-Hierarchical graph over the Θ_k-Hierarchical graph is that each level of the hierarchy is a connected planar graph.

Let P be a set of n points in the plane. A *Delaunay triangulation* T on point set P is a triangulation such that every triangle Δ of T satisfies the *empty circle property*, that is that the circumcircle of Δ contains no point of P in its interior. Delaunay triangulations have been studied extensively (see [27] for a survey), and variations have been introduced by changing the empty circle property for different shapes [7,13,28]. The *TD-Delaunay* graph [28] replaces the empty circle property with an empty *triangle* property. We fix Δ_0 an equilateral triangle. A triangle Δ is said to satisfy the empty triangle property for point set P if there is a homothet of Δ_0 circumscribing Δ, and this homothet contains no points of P in its interior. Note that three points in the plane may not have a circumscribing Δ_0, so this graph is not necessarily a triangulation. To remedy that, we assume hereafter that any point set we consider has three points v_0^*, v_1^* and v_2^* so that $P \backslash \{v_0^*, v_1^*, v_2^*\} \subset \cup_{i=0}^2 \overline{C_i^{v_i}}$, in essence, a large triangle that contains our point set. This ensures that the TD-Delaunay graph is a triangulation.

Since the TD-Delaunay graph is sparse, using Lemma 3, we build a compact Hierarchical graph H with height $h = O(\log n)$ whose foundation is a TD-Delaunay graph on a set P of n points in the plane. For ease of analysis, we set the level of v_0^*, v_1^* and v_2^* to be above the level of all vertices of P. We call this graph a *TD-Delaunay Hierarchical graph*, which we abridge to TDH graph.

In [10], a routing algorithm on the TD-Delaunay graph is presented. We describe it formally in Algorithm 1. We use this algorithm as a blackbox by providing the neighborhood of u as input. This routing algorithm is known to have routing ratio $5/\sqrt{3}$ [10, Corollary 4.1].

Routing in the TD-Delaunay-Hierarchical Graph

In order to route in a memoryless and local fashion in a TDH, we need to encode some information into the vertex labels of the TDH. For each node u and each cone number i, we build a path P_i^u inside each of its positive cones i (see Algorithm 2). The path P_i^u finds the node v adjacent to u in cone C_i such that v is the vertex of highest level adjacent to u in C_i^u. Then the path continues from v until it reaches v_i^*. The label $L(u)$ for each vertex u contains its level and the coordinates of the vertices of the three paths P_i^u. Since each path has size $O(\log n)$, and the vertices are placed on a grid of polynomial size, the label size of each node is $O(\log^2 n)$ bits.

We define some notation used to define our routing algorithm. To simplify notation, when there is no ambiguity, we write P_j instead of P_j^t. Let $P_*^t = P_0^t \cup P_1^t \cup P_2^t$. Let $\Delta_{P_j^t} := \cup_{ab \text{ an edge of } P_j^t} \Delta_{ab} \cup P_j^t$ and $\Delta_{P_*} := \Delta_{P_1} \cup \Delta_{P_2} \cup \Delta_{P_3}$. If we remove Δ_{P_*} from the triangle v_0^*, v_1^*, v_2^*, we get three connected regions,

---

Data: u the current point, t the destination, N the neighborhood of u

if $t \in C_i^u$ *for some* i **then**

   | Follow the edge in C_i^u;

end

if $t \in \overline{C_i^u}$ *for some* i **then**

    $X_0^u \leftarrow \Delta_{tu} \cap \overline{C_i^u}$; $X_1^u \leftarrow \Delta_{tu} \cap C_{i-1}^u$; $X_2^u \leftarrow \Delta_{tu} \cap C_{i+1}^u$;

    if X_1^u *is empty* **then**

      | Follow the clockwise first edge which leads inside Δ_{tu};

    else if X_2^u *is empty* **then**

      | Follow the clockwise last edge which leads inside Δ_{tu};

    else if *There is an edge uv leading inside* X_0^u **then**

      | Follow uv;

    else

      | Follow the edge leading into the smaller of X_1^u or X_2^u;

    end

end

Algorithm 1: TD Routing [10]

Data: u the starting point, i a cone number, H the TDH graph

$v \leftarrow u$;

$P \leftarrow \{u\}$;

while $v \neq v_i^*$ **do**

    $v \leftarrow$ the point a inside C_i^v of highest level such that va is an edge in the graph;

    $P \leftarrow P \cup \{v\}$;

end

return P;

Algorithm 2: Computation of the path P_i^u

R_0, R_1 and R_2 that respectively contain the interior of the cones $\overline{C_0^t}, \overline{C_1^t}$ and $\overline{C_2^t}$. See Fig. 4a.

We now show how to route from a vertex s to a vertex t in a given TDH. The routing algorithm is formally described by Algorithm 3. We consider the neighborhood of the current point to be an implicit argument since this information is stored at the current point. Essentially, the routing algorithm operates in three phases:

1. In the first phase, we are at $u \in R_i$. We follow an edge in C_i^u and ensure that this edge either remains in R_i or moves into Δ_{P_*}. We reach the second phase once we leave R_i.

2. In the second phase, we have that $u \in \Delta_{P_*}$. Note that either $u \in P_j$, in which case we move on to phase three, or there is an $ab \in P_j$ such that $u \in \Delta_{ab}$. We then follow the TD-routing algorithm defined in [10] to route to a.

3. In the third phase, we are at $u \in P_j$ for some j. We then follow P_j backwards to t.

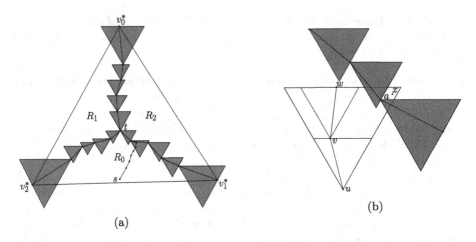

(a)   (b)

Fig. 4. (a): The three regions R_0, R_1 and R_2. The blue shaded triangles represent Δ_{P_*}. (b): Illustration of Lemma 11. If we go down from u to v and up from v to w, it must be that there is $a \in P_* \cap \Delta_{uw}$. The point x is the neighbor of u at its level. Note that $x \in \Delta_{uw}$ and thus prevents the edge uw from existing. (Color figure online)

We now prove the correctness and the efficiency of Algorithm 3. We start with the following lemma that will help bound the number of steps during Phase 1.

Lemma 11. *Let uv and vw be two consecutive steps taken during phase 1. If $\ell(v) < \ell(u)$ and $\ell(v) < \ell(w)$, then there exists $a \in P_*$ with $a \in \Delta_{uw}$.*

Proof. Assume $\ell(v) < \ell(u)$, $\ell(v) < \ell(w)$ and for all $a \in P_*$, $a \notin \Delta_{uw}$. Since the algorithm went to v instead of w despite $\ell(w) > \ell(v)$, there must be some point $x \in \Delta_{uw}$ with $\ell(x) > \ell(w)$ such that ux is an edge in the graph that crosses a triangle Δ_{ab} of Δ_{P_*} as x cannot be inside the region $R_i \cup \Delta_{P_*} \cup P_*$. Since $a, b \notin \Delta_{uw} \supseteq \Delta_{ux}$, x must be inside Δ_{ab}, and that is a contradiction as $x \notin \Delta_{P_*}$. $\square$

Data: u the current point, $L(t)$ the label of the destination
if $u \in R_i$ *(Phase 1)* **then**
    $v \leftarrow$ neighbor of u in $C_i^u \cap (R_i \cup \Delta_{P_*} \cup P_*)$ with maximum level;
    Follow the edge uv;
end
if $u \in \Delta_{ab}$, ab *is an edge of* P_j^t *for some* j *(Phase 2)* **then**
    $N_u \leftarrow$ neighborhood of u at its level;
    TDRouting(u, a, N_u);
end
if $u \in P_j^t$ *for some* j *(Phase 3)* **then**
    Go to the predecessor of u in P_j^t;
end

Algorithm 3: Routing in the TD-Delaunay-Hierarchical graph

Lemma 12. *There exists an edge $ab \in P_*$ such that all edges of the second phase of Algorithm 3 are included in $a \cup \Delta_{ab}$ and are going up. Moreover the last step of phase 2 ends at point a.*

Proof. Let Π be a path computed by Algorithm 3. First observe that if Π doesn't contain any edge of phase 2, then the lemma is true.

Let $u = u_0, u_1, u_2, \ldots$ be the (possibly infinite) sequence of points such that $u_i u_{i+1}$ is an edge of Π of Phase 2. By the definition of Phase 2, u_0 must be in a triangle of Δ_{P_*}. Let ab be an edge of P_* such that $u_0 \in \Delta_{ab}$.

By construction of Algorithm 1, for all i, $u_{i+1} \in \Delta_{au_i} \cup a$. By Lemma 6, this implies that for all i, $\ell(u_i) < \ell(a)$. Furthermore, by construction of Algorithm 3, for all i, $u_{i+1} \in N_{u_i}$ where N_{u_i} is the neighborhood of u_i at its level. By Observation 1, we then know that for all i, $\ell(u_i) < \ell(u_{i+1})$. The sequence $u_0, u_1, u_2, \ldots$ must then be finite so let v be the last point of this sequence. Assume $v \neq a$. We know $\ell(v) < \ell(a)$. Furthermore, since the algorithm stopped at v, it must be because we do not have an edge to a or an edge leading inside Δ_{av} at the level of v. This implies that Δ_{av} is empty at the level of v. Therefore, by construction of the half-Θ_6 graph, av should be an edge at this level, which is a contradiction. Therefore, $v = a$. $\qquad\square$

Lemma 13. *Let Π be a path computed by Algorithm 3. This path is finite and goes from s to t.*

Proof. During phase 1, by construction, the path Π cannot go from a region R_i to a region R_j with $i \neq j$. Hence during this phase, we only take steps in a cone C_i, thus never backtracking. Hence this phase is finite and we must reach a point of $\Delta_{P_*} \cup P_*$ and move on to phase 2 or phase 3 at some point. By Lemma 12, phase 2 reaches a point $a \in P_*$. Phase 3 reaches t by construction of the paths P_0, P_1, P_2. Hence Π is finite and goes from s to t. $\qquad\square$

We introduce some notation for the next lemmas. Let uv be an edge in a half-Θ_6 graph such that $v \in C_i^u$. We call i the *color* of the edge uv. Note that since $u \in \overline{C_i^v}$, the color of an edge is never ambiguous. Furthermore, we denote the length of the side of Δ_{uv} as $|\Delta_{uv}|$. We also denote the maximum (resp. minimum) distance between v and a corner $w \neq u$ of Δ_{uv} as $|\Delta_{uv}|_+$ (resp. $|\Delta_{uv}|_-$). Note that $|\Delta_{uv}| = |\Delta_{uv}|_+ + |\Delta_{uv}|_-$. If $v \in C_i^u$ for some i, for convenience, we define $\Delta(u,v)$ as Δ_{uv} and otherwise $\Delta(u,v)$ is defined as Δ_{vu}.

The next lemmas will be used to show that Algorithm 3 has routing ratio $5/\sqrt{3}$. We first give some intuition as to why Algorithm 3 has constant routing ratio. Let s, t be two points in the TDH graph. We execute Algorithm 3 to route from s to t. Let u be the point reached at the end of phase 1 and a be the point reached at the end of phase 2. Note that the TD-Delaunay graph of $\{s, u, a, t\}$ is exactly the path s, u, a, t. Since the TD-Delaunay graph is a 2-spanner [28], we know $|su| + |ua| + |at| \leq 2|st|$. Since the paths during phases 1 and 3 consist of only one color, by using projections, we can easily show that these paths have spanning ratio $2/\sqrt{3}$. Since phase 2 routes to a using Algorithm 1, we know [9] that the path built during phase 2 has spanning ratio $5/\sqrt{3}$. Let Π be the full path travelled by Algorithm 3. We then have $|\Pi| \leq (2|su| + 5|ua| + 2|at|)/\sqrt{3} \leq 5(|su| + |ua| + |at|)/\sqrt{3} \leq 10/\sqrt{3}|st|$.

With a more careful analysis, we prove that $|\Pi| \leq 5|st|/\sqrt{3}$.

Lemma 14. *Let Π be a path built by Algorithm 3 from s to t. Let a the first point of Π in P_*. If there exists i such that $s \in R_i$ then $|\Pi| \leq |\Delta(s,t)| + |\Delta(s,t)|_+$ otherwise $|\Pi| \leq 2|\Delta_{as}| + |\Delta_{as}|_- + |\Delta_{ta}|$.*

Proof. Let Π_1 (resp. Π_2) be the set of points of Π visited during phase 1 (resp. phase 2). Note that $|\Pi_1 \cap \Pi_2| = 1$. Let $N_{\Pi_2} = \cup_{u \in \Pi_2}(N_u \cap \Delta_{au})$ where N_u is the neighborhood of u at its level. We consider the TD-Delaunay graph T on $\Pi_1 \cup \Pi_2 \cup N_{\Pi_2}$.

Let us first assume $s \in R_i$. This implies $a \in C_i^s$. If Algorithm 1 routes in T from s to a, the path it follows will have length bounded by $|\Delta_{sa}| + |\Delta_{sa}|_+$ by Theorem 4.1 of [9]. We claim the path it follows this way is exactly the path followed during phase 1 and phase 2 by the algorithm. At first, it must follow the edges followed during phase 1 as a lies in a positive cone of the current node. We then reach u_0 the first point seen during phase 2 after having followed all of the edges of phase 1.

Let $u \in \Pi_2$. Let $v \in (N_u \cap \Delta_{au})$. Assume uv is not an edge in T. There must be $w \notin N_u$ such that $w \in \Delta_{uv}$ or $w \in \Delta_{vu}$. However, by Lemma 12, it must be that $\ell(w) > \ell(u)$. This implies that w is present at the level of u; thus, $w \in N_u$ and $v \notin N_u$, which is a contradiction. Now, let $v \in \Delta_{au} \backslash (N_u \cap \Delta_{au})$ be a vertex of T. By the same reasoning, uv cannot be an edge in T. Therefore, the neighborhood of u in T is exactly N_u.

By construction of Algorithm 3, we use Algorithm 1 to route to a during phase 2. In T, when Algorithm 1 reaches $u \in C_j^a$, it makes its decision based only on its neighborhood inside Δ_{au}. This neighborhood is exactly $N_u \cap \Delta_{au}$. Since it reaches u_0, it must follow every edge followed during phase 2.

Therefore, $|\Pi|$ is bounded by $|\Delta_{sa}| + |\Delta_{sa}|_+ + |\Pi_3|$ where Π_3 is the part of Π corresponding to phase 3. Since this part is monocolored, it is bounded by $|\Delta_{ta}|$ as shown in [10]. Putting it all together, we have $|\Pi| \leq |\Delta_{sa}| + |\Delta_{sa}|_+ + |\Delta_{ta}| \leq |\Delta(s,t)| + |\Delta(s,t)|_+$

Assume now that $s \in \Delta_{P_j}$. During phase 2, we use Algorithm 1 to route. When we are at u, we consider the subset N_u of its neighbors which corresponds to the neighbors it has at its level. By construction, if $v \in N_u$, then uv is a valid TD-Delaunay edge. Therefore, at each u reached during phase 2, Algorithm 1 makes decisions as if in a TD-Delaunay graph. By Theorem 4.1 of [10], we have that $\delta_{sa} \leq 2|\Delta_{as}| + |\Delta_{as}|_-$ where δ_{sa} is the distance travelled during phase 2. During phase 3, the path is mono-colored so we know $\delta_{at} \leq |\Delta_{ta}|$ where δ_{at} is the distance travelled during phase 3. Thus, in this case we have, $|\Pi| \leq \delta_{sa} + \delta_{at} \leq 2|\Delta_{as}| + |\Delta_{as}|_- + |\Delta_{ta}|$.

Theorem 3. *The TDH routing algorithm always reaches its destination. Its routing ratio is $5/\sqrt{3}$. Its hop length is $O(\log^2 n)$. It uses $O(\log^2 n)$ bit labels for the vertices. This overhead is fixed at the start and no additional memory is used throughout the routing.*

Proof. Let Π be the path followed by Algorithm 3 from s to t. We first show that the routing ratio of Algorithm 3 is $5/\sqrt{3}$. Lemma 14 gives us bounds on the length of the path in all cases.

In the case, $s \in R_i$, $|P| \leq |\Delta_{st}| + |\Delta_{st}|_+ \leq 2|st|$ by Corollary 4.1 of [10].

Otherwise, when $s \notin R_i$, we have $|P| \leq 2|\Delta_{as}| + |\Delta_{as}|_- + |\Delta_{ta}|$. Since $|\Delta_{as}| + |\Delta_{ta}| = |\Delta_{ts}|$. This further implies $|\Delta_{as}| + |\Delta_{ta}|_+ \leq |\Delta_{ts}|$. Since $\Delta_{as} \subset \Delta_{ts}$ and both are equilateral, $|\Delta_{as}|_- \leq |\Delta_{ts}|_-$. Finally, $|P| \leq 2|\Delta_{ts}| + |\Delta_{ts}|_-$. Therefore, by Corollary 4.1 of [10], $|P| \leq 5|st|/\sqrt{3}$.

The routing ratio of Algorithm 3 is at most $5/\sqrt{3}$.

Finally, we show that the number of edges in the path is $O(\log^2 n)$. We consider the three phases of the algorithm: During phase 1, we may go up or down TDH. By Lemma 11, we can only go down and then up by moving *past* a point a on some P_*. Let $u, w \in R_i$ be two points reached during phase 1 such that there is $a \in P_* \cap \Delta_{uw}$. Then $a \notin C_i^w$. This implies that any pair of points u', v' reached during phase 1 inside C_i^w is such that $a \notin \Delta_{u'v'}$. Therefore, each time we go past a point $a \in P_*$, we cannot move past it again during phase 1. This implies we can only alternate going down and up at most $O(\log n)$ times (once for each point of each P_j). Each sequence of points going up or down cannot be longer than $O(\log n)$ steps, so phase 1 takes at most $O(\log^2 n)$ steps in total.

During phase 2, by Lemma 12 we only go up, so we take at most $O(\log n)$ steps. During phase 3, we take at most $O(\log n)$ steps by construction of P_j. In total, we take at most $O(\log^2 n)$ steps from s to t.   □

Theorem 4. *Algorithm 3 may take $\Omega(\log^2 n)$ steps.*

Proof (sketch). The main idea is to use the pattern highlighted by Lemma 11 to build a path of size $\Omega(\log^2 n)$ steps. We describe a pattern to be repeated $\Omega(\log n)$ times. We start at $u \in C_0^t$. We add $a \in C_1^u$ and $v \in C_1^u \cap \overline{C_2^a}$. We put the next point on the path to follow $w \in \Delta_{uv} \cap C_2^a$ at a lower level. We then add $u' \in C_1^w \backslash \Delta_{uv}$ the start of the next pattern. We then add $\Omega(\log n)$ points to be followed in $\Delta_{wu'} \backslash \Delta_{uv}$. These steps will all be going up. This gives us $\Omega(\log^2 n)$ steps. The exact details are ommited due to space constraints.

6   Conclusion

We presented two graphs, the Θ_k-Hierarchical graphs and the TD-Delaunay-Hierachical graphs. These two graphs have a bounded spanning ratio and a diameter $O(\log n)$. We also give local routing algorithms for these graphs that find paths of bounded stretch and consisting of $O(\log^2 n)$ edges. Moreover, we have examples where $\Omega(\log^2 n)$ can be attained. We leave open the problem of finding a sparse graph together with a memoryless local routing algorithm that finds paths of at most $O(\log n)$ edges and bounded spanning ratio.

TD-Delaunay triangulations are also the starting point of 6-spanners with constant maximum degree [6]. These graphs can naturally be used to build a Hierachical graph of small diameter. Is it possible to route efficiently in these hierarchies also?

References

1. Abraham, I., Malkhi, D.: Compact routing on euclidean metrics. In: Proceedings of the Annual ACM Symposium on Principles of Distributed Computing, vol. 23, Jul 2004
2. Ahmed, A.R., et al.: Graph spanners: A tutorial review. Comput. Sci. Rev. **37**, 100253 (2020)
3. Arya, S., Das, G., Mount, D., Salowe, J., Smid, M.: Euclidean spanners: Short, thin, and lanky. In: Proceedings of the 27th ACM STOC, Mar1996
4. Arya, S., Mount, D.M., Smid, M.H.M.: Dynamic algorithms for geometric spanners of small diameter: Randomized solutions. Comput. Geom. **13**(2), 91–107 (1999)
5. Bonichon, N., Gavoille, C., Hanusse, N., Ilcinkas, D.: Connections between theta-graphs, delaunay triangulations, and orthogonal surfaces. In: Thilikos, D.M. (ed.) WG 2010. LNCS, vol. 6410, pp. 266–278. Springer, Heidelberg (2010). https://doi.org/10.1007/978-3-642-16926-7_25
6. Bonichon, N., Gavoille, C., Hanusse, N., Perković, L.: Plane spanners of maximum degree six. In: Abramsky, S., Gavoille, C., Kirchner, C., Meyer auf der Heide, F., Spirakis, P.G. (eds.) ICALP 2010. LNCS, vol. 6198, pp. 19–30. Springer, Heidelberg (2010). https://doi.org/10.1007/978-3-642-14165-2_3
7. Bose, P., Carmi, P., Collette, S., Smid, M.: On the stretch factor of convex delaunay graphs. J. Comput. Geo. **1**(1), 41–56 (2010)
8. Bose, P., Carufel, J.D., Morin, P., van Renssen, A., Verdonschot, S.: Towards tight bounds on theta-graphs: More is not always better. Theor. Comput. Sci. **616**, 70–93 (2016)
9. Bose, P., Fagerberg, R., van Renssen, A., Verdonschot, S.: Competitive routing in the half-Θ_6-graph. In: Rabani, Y. (ed.) Proceedings of the Twenty-Third Annual ACM-SIAM Symposium on Discrete Algorithms, SODA 2012, Kyoto, Japan, 17–19 Jan 2012, pp. 1319–1328. SIAM (2012)
10. Bose, P., Fagerberg, R., van Renssen, A., Verdonschot, S.: Optimal local routing on delaunay triangulations defined by empty equilateral triangles. SIAM J. Comput. **44**(6), 1626–1649 (2015)
11. Bose, P., Morin, P., Stojmenovic, I., Urrutia, J.: Routing with guaranteed delivery in ad hoc wireless networks. Wirel. Networks **7**(6), 609–616 (2001)
12. Chan, T.H., Gupta, A., Maggs, B.M., Zhou, S.: On hierarchical routing in doubling metrics. ACM Trans. Algorithms **12**(4), 55:1–55:22 (2016)
13. Chew, P.: There is a planar graph almost as good as the complete graph. In: Proceedings of the Second Annual Symposium on Computational Geometry, SCG 1986, New York, USA, pp. 169–177. Association for Computing Machinery (1986)
14. Clarkson, K.: Approximation algorithms for shortest path motion planning. In: Proceedings of the Nineteenth Annual ACM Symposium on Theory of Computing, pp. 56–65 (1987)
15. Diestel, R.: Graph Theory, 4th edn., vol. 173, Graduate texts in mathematics. Springer (2012). https://doi.org/10.1007/978-1-4612-9967-7
16. Dijkstra, E.W.: A note on two problems in connexion with graphs. Numer. Math. **1**, 269–271 (1959)
17. Elkin, M., Solomon, S.: Optimal euclidean spanners: Really short, thin, and lanky. Proceedings of the Annual ACM Symposium on Theory of Computing, vol. 62, Jul 2012
18. Hoedemakers, C.: Geometric spanner networks master thesis (2015)

19. Kao, M., (ed.) Encyclopedia of Algorithms - 2016 edn. Springer (2016). https://doi.org/10.1007/978-0-387-30162-4

20. Karp, B., Kung, H.T.: GPSR: greedy perimeter stateless routing for wireless networks. In: MobiCom, pp. 243–254. ACM (2000)

21. Keil, J.M.: Approximating the complete euclidean graph. In: Karlsson, R., Lingas, A. (eds.) SWAT 1988. LNCS, vol. 318, pp. 208–213. Springer, Heidelberg (1988). https://doi.org/10.1007/3-540-19487-8_23

22. Keil, J.M., Gutwin, C.A.: Classes of graphs which approximate the complete euclidean graph. Dis. Comput. Geometry **7**(1), 13–28 (1992). https://doi.org/10.1007/BF02187821

23. Kirkpatrick, D.: Optimal search in planar subdivisions. SIAM J. Comput. **12**, 28–35 (1983)

24. Le, H., Solomon, S.: Truly optimal euclidean spanners. In: 2019 IEEE 60th Annual Symposium on Foundations of Computer Science (FOCS), Los Alamitos, CA, USA, pp. 1078–1100. IEEE Computer Society, Nov 2019

25. Le, H., Solomon, S.: Truly optimal euclidean spanners. SIAM J. Comput. FOCS 19-135 (2022)

26. Narasimhan, G., Smid, M.: Geometric Spanner Networks. Cambridge University Press, New York, NY, USA (2007)

27. Okabe, A., Boots, B., Sugihara, K., Chiu, S.N., Kendall, D.G.: Spatial Tessellations: Concepts and Applications of Voronoi Diagrams, Second edn.. Wiley Series in Probability and Mathematical Statistics. Wiley (2000)

28. Paul Chew, L.: There are planar graphs almost as good as the complete graph. J. Comput. Syst. Sci. **39**(2), 205–219 (1989)

29. Pugh, W.W.: Skip lists: A probabilistic alternative to balanced trees. Commun. ACM **33**(6), 668–676 (1990)

30. Ruppert, J., Seidel, R.: Approximation algorithms for shortest path motion planning. In: Proceedings of the 3rd Canadian Conference on Computational Geometry, pp. 207–210 (1991)

On r-Guarding SCOTs – A New Family of Orthogonal Polygons

Vasco Cruz[(✉)] and Ana Paula Tomás

CMUP, Departamento de Ciência de Computadores, Faculdade de Ciências,
Universidade do Porto, Rua do Campo Alegre s/n, 4169-007 Porto, Portugal
vasco.j.r.cruz@gmail.com, aptomas@fc.up.pt

Abstract. We define a new family of orthogonal polygons, the SCOTs, which are made up of rectangular rooms linked by rectangular corridors, mimicking properties of real-world buildings. We prove that, if a SCOT P is simple or r-*independent*, meaning that the extensions of all pairs of adjacent corridors with the same direction are either coincident or disjoint, a minimum-cardinality guard set for guarding P under r-visibility can be computed in polynomial time. To this end, we propose three methods: a linear-time algorithm for simple SCOTs, for both vertex- or point-guards; an $\mathcal{O}(c\sqrt{c})$-time algorithm for an r-independent SCOT with c corridors for vertex-guards; and another one running in time $\mathcal{O}\left(c^3 \log c\right)$ for point-guards in an r-independent SCOT. For SCOTs with holes and not r-independent, we show that the problem becomes NP-hard – proving that the decision problem is NP-complete for the case of point-guards.

Keywords: Art gallery problem · Orthogonal polygons · Computational complexity

1 Introduction

The Art Gallery Problem is not only appealing, but also one of the most well-known and thoroughly studied problems in Computational Geometry. The original formulation, posed in 1973 by Klee [16], asks about the minimum number of points, called guards, that are required to watch over the interior of any simple polygon P (that is, one without holes or self-intersections), where two points $p, q \in P$ *see* each other if the closed line segment pq lies completely within P. Chvátal [6] proved, in 1975, that $\lfloor n/3 \rfloor$ point-guards are always sufficient, and sometimes necessary, to guard any simple polygon with n vertices. For orthogonal polygons – those whose internal angles all measure 90° or 270° –, a tighter bound of $\lfloor n/4 \rfloor$ guards was shown [12,19]. Finding a minimum-cardinality guard

This work was partially supported by CMUP, member of LASI, which is financed by national funds through FCT – Fundação para a Ciência e a Tecnologia, I.P., under the project with reference UIDB/00144/2020.

A. Castañeda and F. Rodríguez-Henríquez (Eds.): LATIN 2022, LNCS 13568, pp. 713–729, 2022.
https://doi.org/10.1007/978-3-031-20624-5_43

set for a given polygon has been proved to be NP-hard, even for simple polygons [15] and orthogonal polygons [17,18]. Variations of the problem have been studied through the years, concerning different visibility models, polygon families, subsets of the polygon that we wish to guard and valid positions for the guards. If guards may be placed anywhere in P, they are called *point-guards*; if we force them to lie at vertices or anywhere on the boundary, they are called *vertex-guards* and *boundary-guards* [13]. In this paper, we focus on the r-visibility (rectangular visibility) model for orthogonal polygons, according to which two points $p, q \in P$ *see* each other (or *r-see* each other, or *are visible* from each other) if the minimal axis-aligned rectangle that contains p and q does not intersect the exterior of P. If the polygon is simple, an optimal r-guard set can be found in time $\widetilde{\mathcal{O}}\left(n^{17}\right)$, where $\widetilde{\mathcal{O}}(\cdot)$ hides a polylogarithmic factor [20], but for polygons with holes the problem becomes NP-hard [3,11]. A linear-time algorithm has also been proposed for orthogonal polygons with bounded treewidth [3].

Although the Art Gallery Problem is hard for generic polygon instances, buildings and art galleries that we find in real life are not *random*: they show a very specific structure and a spatial organization that is characteristic of the human way of architecting. In this paper, we define a new family of orthogonal polygons, the SCOTs, which prove to be useful models for mimicking properties that are close to those one can find in real-world rectangular galleries. We study the algorithmic complexity of guarding SCOTs under the r-visibility model – which we denominate MINIMUM SCOT r-GUARD –, describe subfamilies that can be solved in polynomial time by providing algorithms for them, in Sects. 2 and 3, provide hardness results for the general case, in Sect. 4, and leave some open problems in the conclusions.

1.1 Preliminaries

A SCOT[1] is a connected orthogonal polygon that is made up of rectangular rooms linked by rectangular corridors. A corridor C links two rooms R_1 and R_2 in a SCOT if a side of C is strictly contained in an edge of R_1 and the opposite side is strictly contained in an edge of R_2. By "strictly contained" we mean that a corridor C is always narrower than the rooms R_1 and R_2 that are incident to it and none of the four vertices of C coincide with either of the four corners of R_1 or R_2. A SCOT has no "dangling" corridors, so every corridor is incident to exactly two rooms. Each corridor has an implicit direction, *horizontal* or *vertical*, depending on whether it connects two rooms to its right/left or above/below it, respectively. There may exist cycles that involve more than two rooms, in which case the SCOT is called *cyclic* (Figs. 1b and 1c). The same pair of rooms may be linked by multiple, non-intersecting corridors (Fig. 1c). If there exist *room cycles* or *multiple corridors* connecting the same pair of rooms, the SCOT has holes; if it does not have holes, it is said to be *simple* (Fig. 1a).

---

[1] In Portuguese, *"salas e corredores ortogonais"*, which means *"orthogonal rooms and corridors"*.

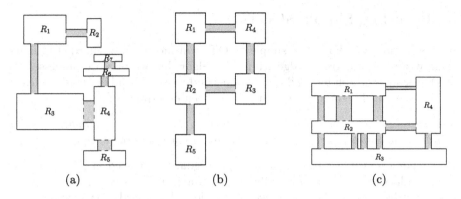

Fig. 1. (a) Simple $(\mathcal{R},\mathcal{C})$-SCOT, with $|\mathcal{R}| = 7$ and $|\mathcal{C}| = 6$. Rooms are represented as white rectangles and corridors as shaded rectangles. The corridor connecting rooms R_1 and R_2 is *horizontal* and the corridor connecting rooms R_1 and R_3 is *vertical*. **(b)** SCOT with a cycle involving rooms R_1, R_2, R_3 and R_4. **(c)** SCOT with multiple corridors between the same pairs of rooms.

For convenience, we use $(\mathcal{R},\mathcal{C})$-SCOT to refer to a SCOT with set of rooms $\mathcal{R}$ and set of corridors $\mathcal{C}$. The number of vertices is given by $n = 4|\mathcal{R}| + 4|\mathcal{C}|$, since each room and each corridor contribute with four new vertices to the total count. An obvious way to guard a SCOT is to place an individual guard in each room and in each corridor. So, we have a trivial upper bound of $|\mathcal{R}| + |\mathcal{C}| = n/4$ for the number of guards needed under r-visibility, both for vertex-guards and for point-guards. This solution, however, may in general not be optimal. The algorithm of [20] with time complexity $\widetilde{\mathcal{O}}\left(n^{17}\right)$ establishes that MINIMUM SCOT r-GUARD can be solved in polynomial time for *simple* SCOTs with point-guards. As this is not a very gracious running time, we are interested in exploiting the structure of SCOTs for developing more efficient algorithms for them, and understanding whether the problem remains polynomial for SCOTs with holes. A first observation towards a better algorithm is given by Lemma 1, which states that reflex vertices are always better choices to place guards at, meaning that, for finding a minimum-cardinality vertex-guard set for a SCOT, one could opt to ignore its convex vertices altogether without risking losing optimality.

Lemma 1. *Let P be an $(\mathcal{R},\mathcal{C})$-SCOT with $|\mathcal{R}| \geq 2$. For any convex vertex u of P, there is a reflex vertex v of P that r-sees more than u, that is, v r-sees a superset of what u r-sees.*

Proof. Since P has at least two rooms and is connected, every room has at least two reflex vertices. The only convex vertices in P are the four corners of each room and these only r-see the room they belong to, given that corridors are narrower than incident rooms. A reflex vertex is shared between a room and a corridor and therefore r-sees that room and (at least) that corridor.   □

2   Guarding Simple SCOTs

In this section, we show that simple SCOTs can indeed be r-guarded in linear time by means of a greedy algorithm. The algorithm we propose processes the SCOT as a tree graph induced by its rooms and corridors and detaches its leaves one by one, until a complete guard set has been determined.

Theorem 1. *Let P be a simple $(\mathcal{R}, \mathcal{C})$-SCOT. The minimum number of point-guards (and, indeed, vertex-guards) for* r*-guarding P is exactly $|\mathcal{R}|$.*

Proof. For necessity, it is mandatory that a guard be placed in the interior or the boundary of each room for (fully) r-guarding it. Otherwise, the room corners would not be covered. Therefore, we need at least $|\mathcal{R}|$ guards. We now provide a greedy algorithm as a constructive proof that $|\mathcal{R}|$ guards are sufficient. Figure 2 illustrates its idea. Build an undirected graph $T = (V, E)$, with $|V| = |\mathcal{R}|$ and $|E| = |\mathcal{C}|$. Each node in V represents a room of P and there is an edge in E linking two nodes if the corresponding rooms in P are connected by a corridor. As P has no holes, T is a tree and has at least one leaf (a node with degree 0 or 1). While T is not empty, select an arbitrary leaf u corresponding to a room R. If u has degree 0, place a guard anywhere in R (namely, at one of the corners), remove u from T and terminate. Otherwise (i.e., if u has degree 1), place the guard at one of the reflex vertices that are shared between R and its incident corridor C and remove both u and the edge corresponding to C. The tree T will become empty (i.e., P will become r-guarded) after exactly $|\mathcal{R}|$ steps and in each step we have placed a vertex-guard, so $|\mathcal{R}|$ guards are enough. $\square$

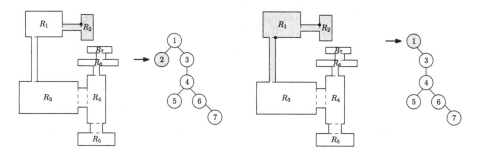

Fig. 2. The first two steps of the algorithm for finding a minimum vertex-guard set for a simple SCOT by pruning leaves in increasing order of the node identifier.

3   Guarding r-independent SCOTs

In this section, we define a subfamily of SCOTs, the r-independent SCOTs, and prove that they can also be guarded in polynomial time, although by means of conceptually different algorithms regarding vertex-guards or point-guards.

We define the *stretch* of a corridor C in a SCOT as the infinite extension of C along its direction (horizontal or vertical), that is, as the unbounded corridor defined by the two supporting lines of C (Fig. 3). For *adjacent corridors* C_1 and C_2, i.e., corridors incident to the same room R, when C_1 and C_2 are both horizontal or both vertical, we say that C_1 and C_2 are *disjoint* if their stretches do not intersect and that C_1 and C_2 are *aligned* if their stretches coincide (Fig. 3b). An *r-independent SCOT* is a SCOT in which every pair of adjacent corridors with the same direction is either disjoint or aligned.

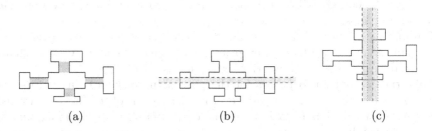

Fig. 3. (a) A simple SCOT. (b) Two aligned horizontal corridors. Their stretches match perfectly. (c) Two vertical corridors that are not aligned nor disjoint, although their stretches intersect.

3.1 Super-Corridors

Let P be an r-independent $(\mathcal{R}, \mathcal{C})$-SCOT. For every pair (C_1, C_2) of aligned adjacent corridors, C_1 is r-seen if and only if C_2 is r-seen. This induces an equivalence relation between corridors (r-equivalence), according to which two adjacent corridors are equivalent if and only if they are aligned. By transitivity, two nonadjacent corridors that belong to the same succession of aligned adjacent corridors are also r-equivalent.

We now describe a general strategy that we will employ several times for dealing with r-independent SCOTs. The idea is to replace each maximal succession of aligned adjacent corridors $C^{(1)}, C^{(2)}, \ldots$ by a *super-corridor* C' that may cross multiple rooms. So, C' stands for a single long corridor that represents that equivalence class. By this transformation, the problem is reduced to the case where all pairs of adjacent (super-)corridors are disjoint. We denote by $\mathcal{C}'$ the set of all super-corridors in P, so that $|\mathcal{C}'|$ is the number of equivalence classes of corridors. Because the decomposition into super-corridors is so important for developing algorithms specific to r-independent SCOTs, we will often denote by $(\mathcal{R}, \mathcal{C}, \mathcal{C}')$-SCOT an r-independent SCOT whose set of super-corridors is $\mathcal{C}'$.

Lemma 2. *Let P be an r-independent $(\mathcal{R}, \mathcal{C}, \mathcal{C}')$-SCOT. There are at most $2|\mathcal{C}|$ pairs (R, C'), with $R \in \mathcal{R}$ and $C' \in \mathcal{C}'$, such that R and C' are incident.*

Proof. Let Q be the set of all incident pairs of rooms and corridors in P. Each corridor is incident to exactly two rooms and a room R is incident to a super-corridor C' if and only if it is incident to some corridor C that belongs to C'. Therefore, $|\{(R, C') : R \in \mathcal{R}, C' \in \mathcal{C}', R \text{ is incident to } C'\}| \leq |Q| = 2|\mathcal{C}|$. □

3.2 Algorithm for Vertex-Guards

We now explain how to find an optimal vertex-guard set for an r-independent $(\mathcal{R}, \mathcal{C}, \mathcal{C}')$-SCOT in polynomial time. Once more, we need one guard per room, otherwise the corners of some room will not be covered. Since a vertex-guard can r-see at most one room and one super-corridor of P, it is intuitive that, when placing a guard g in a room R, we would better try to also cover one of the incident super-corridors C', by placing g at one of the reflex vertices shared by R and C', so as to avoid wasting extra guards for C' later.

We thus consider two phases for the positioning of guards in P. First, we place one guard in every room, also covering exactly one of the (not yet covered) incident super-corridors in each; then, and only after that, we spend a guard for separately watching over each of the remaining uncovered rooms and/or super-corridors. In the first phase, we find a maximum matching between rooms and incident super-corridors. For that, we build a bipartite graph $H = (V, E)$ with $V = \mathcal{R} \cup \mathcal{C}'$ and $E = \{(R, C') : R \in \mathcal{R} \text{ is incident to } C' \in \mathcal{C}'\}$. Note that a separate edge should be added for each room crossed by the super-corridor.

A concrete selection of pairs of rooms and super-corridors that may be chosen to be r-seen collectively, each with a single guard, can be modeled as a matching in graph H; in particular, an optimal selection of guards includes a maximum cardinality matching (Fig. 4). The proof of Lemma 3 explains how we can extract a vertex-guard set for P from a given maximum matching $M^\star$ in H.

Lemma 3. *The graph H has a matching M of size $k \in \mathbb{Z}_0^+$ if and only if the SCOT P has a vertex-guard set $\mathcal{G}$ of size $|\mathcal{R}| + |\mathcal{C}'| - k$.*

Proof. ($\Rightarrow$) Let M be a matching in H of cardinality k and $\mathcal{G}$ be an empty vertex-guard set. For each edge (R, C') in M, add to $\mathcal{G}$ a guard at a reflex vertex that is shared between the room R and a corridor in the super-corridor C'; there are either two or four possibilities for choosing that vertex (depending on whether C' contains one or two corridors incident to R, respectively), but it can in fact be picked arbitrarily, because all these shared vertices r-see the exact same region in P. Finally, place a guard in each of the rooms and/or super-corridors that may eventually still be left unguarded. $\mathcal{G}$ is feasible and has size $|M| + (|\mathcal{R}| - |M|) + (|\mathcal{C}'| - |M|) = |\mathcal{R}| + |\mathcal{C}'| - k$.

($\Leftarrow$) Let $\mathcal{G}$ be a vertex-guard set for P, with $|\mathcal{G}| = |\mathcal{R}| + |\mathcal{C}'| - k$ and let $\mathcal{G}'$ be a subset of $\mathcal{G}$ with $|\mathcal{G}'| = |\mathcal{C}'|$ that covers all the super-corridors. $\mathcal{G}'$ exists because each super-corridor requires a different vertex-guard. By hypothesis, $\mathcal{G}'$ leaves at most $|\mathcal{R}| - k$ rooms uncovered and, thus, covers at least k different rooms. As we have placed each guard in a different super-corridor, there is a matching of size k between room and super-corridor nodes in H. □

$M^\star$ may, for instance, be found using Hopcroft-Karp's algorithm [10], which runs in time $\mathcal{O}(\sqrt{|V|}|E|)$ for any bipartite graph $H = (A \cup B, E)$ – indeed in $\mathcal{O}(\sqrt{\min\{|A|, |B|\}}|E|)$. Since, by Lemma 2, $|E| \leq 2|\mathcal{C}|$, Theorem 2 and the bound $\mathcal{O}(|\mathcal{C}|\sqrt{|\mathcal{C}|})$ follow.

Theorem 2. *A minimum-cardinality vertex-guard set for an* $(\mathcal{R}, \mathcal{C}, \mathcal{C}')$-*SCOT P that is r-independent can be determined in time* $\mathcal{O}(|\mathcal{C}|\sqrt{\min\{|\mathcal{R}|, |\mathcal{C}'|\}})$.

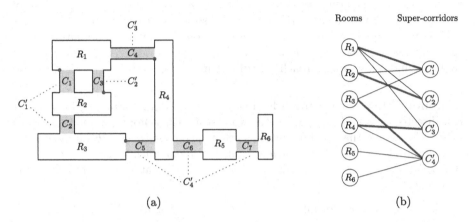

(a)   (b)

Fig. 4. (a) An r-independent SCOT P, with four super-corridors: $C_1' = \{C_1, C_2\}$, $C_2' = \{C_3\}$, $C_3' = \{C_4\}$ and $C_4' = \{C_5, C_6, C_7\}$. The four marked vertices form a partial guard set that covers P except for rooms R_5 and R_6. Each guard covers exactly one room and one super-corridor. (b) The associated bipartite graph H. The four highlighted edges define the $\mathcal{C}'$-perfect matching $M^\star$ that yields that partial guard set, which can be extended with two guards, one in R_5 and another in R_6, to optimally cover the entirety of P.

3.3   Algorithm for Point-Guards

A point-guard placed in a room R which is incident to a horizontal and a vertical super-corridor, respectively C_h' and C_v', can actually r-see both super-corridors if it is placed in the intersection of the room and the stretches of C_h' and C_v'. Hence, for point-guards, Theorem 2 cannot be applied. Nevertheless, we will show that, for r-independent SCOTs with point-guards, the problem is still polynomial-time solvable. We start by presenting some useful notions and lemmas.

Lemma 4. *For any r-independent SCOT P, there is a minimum-cardinality point-guard set for P in which every guard lies inside a room.*

Proof. Suppose P admits a minimum point-guard set $\mathcal{G}$ with at least one guard g being placed strictly inside a corridor C and only r-seeing the super-corridor corresponding to C. Shift g along the direction of C until it becomes strictly contained in a room, call g' its new position and let $\mathcal{G}' = (\mathcal{G} \setminus \{g\}) \cup \{g'\}$ be the new guard set, with $|\mathcal{G}'| = |\mathcal{G}|$. Now g' still sees the super-corridor corresponding to C, but it does also see a room. Therefore, the new position for the guard g' has not made $\mathcal{G}'$ worse than $\mathcal{G}$ in terms of covered area. Repeat the same argument while there are guards strictly outside rooms in $\mathcal{G}'$.   □

The algorithm we propose starts by transforming the problem into one of network flows, called MINIMUM FLOW WITH DEMANDS. This problem, sometimes referred to as *circulation problem with lower bounds* [14], is a generalization of MAXIMUM FLOW in a network and can be solved in polynomial time. Before delving into the reduction, we take a short detour so that we can focus on what the problem of MINIMUM FLOW WITH DEMANDS is about, recall some relevant concepts and detail an approach for solving it.

Minimum Flow with Demands. Like in the standard setting, a *flow network* is a directed graph $G = (V, E)$ with two distinguished nodes, a *source* s and a *sink* t. We extend the usual definition so that each edge $(u, v) \in E$ has a *capacity* $c(u, v)$ and also a lower bound $d(u, v)$ called *demand*, with $0 \leq d(u, v) \leq c(u, v)$. A *feasible flow* in G is a function $f : V \times V \to \mathbb{R}$ that satisfies the following properties:

- Flow conservation: for all $u \in V \setminus \{s, t\}$, $\sum_{v \in V} f(u, v) = \sum_{v \in V} f(v, u)$.
- Edge constraints: for all $u, v \in V$, $d(u, v) \leq f(u, v) \leq c(u, v)$.

MINIMUM FLOW WITH DEMANDS asks for a feasible flow $f^\star$ of minimum value for the given network G – or to report that no feasible flow exists. For solving MINIMUM FLOW WITH DEMANDS, we reduce it to MAXIMUM FLOW, closely following the method described in [9]. For that, we construct a new network $G' = (V', E')$, without demands on edges, satisfying $|V'| = |V| + 2$ and $|E'| \leq 2|V| + |E| - 1$, that, in addition to the same nodes as G, has a new source s' and target t' and satisfies the following capacity function $c' : E' \to \mathbb{R}$:

- $c'(s', v) = \sum_{u \in V} d(u, v)$ and $c'(v, t') = \sum_{w \in V} d(v, w)$ for every vertex $v \in V$.
- $c'(u, v) = c(u, v) - d(u, v)$ for every edge $(u, v) \in E$.
- $c'(t, s) = \infty$.

From [9], a function $f : E \to \mathbb{R}$ is a feasible flow from s to t in G if and only if the function $f' : E' \to \mathbb{R}$ satisfying $f'(u, v) = f(u, v) - d(u, v)$ (when restricted to the original set of edges, E) is a saturating flow from s' to t' in G', meaning that its value equals the sum of the capacities of the edges leaving s'. Also, the entire flow of G flows along the edge (t, s) in G', that is, $f'(t, s) = |f|$. Hence, we may determine a minimum feasible flow $f^\star$ in G by binary searching the smallest value one can assign to the capacity $c'(t, s)$ (instead of ∞) such that the corresponding flow f' in the new network G' is still saturating – which in turn we can decide by running any maximum flow algorithm on G'. This method runs in time $\mathcal{O}\left(T(G') \log |f^\star|\right)$, where $T(G')$ is the time complexity of the algorithm used for computing a maximum flow in G'.

Network Construction. Having defined the problem MINIMUM FLOW WITH DEMANDS, we are ready to describe the reduction from MINIMUM SCOT r-GUARD for r-independent SCOTs with point-guards, that is, how a network G can be built from a given r-independent $(\mathcal{R}, \mathcal{C}, \mathcal{C}')$-SCOT P. The idea is that, in our construction, every guard will be placed in a room (which is plausible by Lemma 4) and will r-see *exactly* one room, one horizontal super-corridor and one vertical super-corridor – perhaps these super-corridors being mock (artificial), as we explain below. First, we define two types of gadgets. A *real-gadget* is a set of two nodes u and v that are connected by a directed edge (u, v) with demand 1 and capacity ∞. We call node u the *in-node* of the gadget and the other one, v, its *out-node*. A *pseudo-gadget* is defined identically to a real-gadget, but the edge linking the in-node to the out-node has demand 0. A real-gadget and a pseudo-gadget are illustrated on Fig. 5 (above and below the network, respectively). The edges in real-gadgets are the only ones in G that have positive demand. Every edge in G has capacity ∞.

Each room, horizontal super-corridor and vertical super-corridor in P is represented by a corresponding real-gadget in G. Let us denote the gadgets corresponding to a given room R, to a horizontal super-corridor C'_h and to a vertical super-corridor C'_v as $\Gamma(R)$, $\Gamma(C'_h)$ and $\Gamma(C'_v)$, respectively. Also, add to G two pseudo-gadgets, $\Gamma(C'_{hf})$ and $\Gamma(C'_{vf})$, representing two mock super-corridors, one horizontal and one vertical, respectively. Connect the out-node of $\Gamma(C'_{hf})$ to the in-node of the real-gadget $\Gamma(R)$ of every room R with demand 0 and connect the out-node of the real-gadget $\Gamma(R)$ of every room to the in-node of $\Gamma(C'_{vf})$. Connect s to $\Gamma(C'_{hf})$ and $\Gamma(C'_{vf})$ to t with demands 0. For each pair (C'_h, R) such that C'_h and R are incident in P, connect the out-node of $\Gamma(C'_h)$ to the in-node of $\Gamma(R)$ with an edge of demand 0. For each pair (R, C'_v) such that R and C'_v are incident in P, connect the out-node of $\Gamma(R)$ to the in-node of $\Gamma(C'_v)$, also with demand 0. Connect the source node s of G to the in-node of the real-gadget $\Gamma(C'_h)$ of each horizontal super-corridor with a demand of 0. Finally, connect the out-node of every real-gadget $\Gamma(C'_v)$ corresponding to a vertical super-corridor to the sink t, again with demand 0. See Fig. 5 for an illustration of the whole reduction from a SCOT to a network of flows with demands.

Proposition 1. *The size of the network G and the time it takes to construct it are linear in the SCOT size, $|\mathcal{R}| + |\mathcal{C}|$. Specifically, $|V| = 2(|\mathcal{R}| + |\mathcal{C}'|) + 2 \times 2 + 2 \leq 2(|\mathcal{R}| + |\mathcal{C}|) + 6$ and $|E| \leq 2(|\mathcal{C}'_h| + 1) + 2(|\mathcal{C}'_v| + 1) + 3|\mathcal{R}| + 2|\mathcal{C}| \leq 3|\mathcal{R}| + 4|\mathcal{C}| + 4$, where $\mathcal{C}'_h$ and $\mathcal{C}'_v$ are the sets of horizontal and vertical super-corridors in P.*

The meaning that we assign to one unit of flow on G is a point-guard in P; for each unit that flows from s to t, one guard is placed at some room in P. The network G always admits a feasible flow; in particular, one with $|f^\star| = \infty$. In practice, we can replace ∞ by an upper bound for the number of guards needed for covering P entirely, such as $|\mathcal{R}| + |\mathcal{C}|$. Intuitively, the edges in real-gadgets, which were assigned a positive demand, require any feasible flow on G to pass through them and, thus, force every room and (real) super-corridor to have at least one guard covering it. Edges in pseudo-gadgets have demand 0

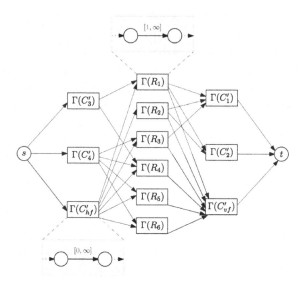

Fig. 5. The network G corresponding to the r-independent SCOT P of Fig. 4a. It has a total of 26 nodes and 41 edges (including those on the gadgets).

because we do not explicitly require any guard to see them; their only importance is to ensure a correspondence between the network construction and the polygon. Given that only edges in the real-gadgets of G contribute with positive capacities to G', we have $|E'| = |E| + 2(|\mathcal{R}| + |\mathcal{C}'|) + 1 \leq 5|\mathcal{R}| + 6|\mathcal{C}| + 5$. This reduction to MINIMUM FLOW WITH DEMANDS then implies a method for determining a minimum point-guard set for r-independent SCOTs. Its overall time complexity depends on the exact MAXIMUM FLOW algorithm that is used as subroutine. For instance, if Dinic's algorithm [8] is used, the running time that one obtains is $\mathcal{O}(|V'|^2|E'|\log|f^\star|) = \mathcal{O}(|\mathcal{C}|^3 \log |\mathcal{C}|)$. We remark that [5] solves MAXIMUM FLOW in a network $H = (V_H, E_H)$ in almost-linear time $|E_H|^{1+o(1)}$. By Proposition 1, this result speeds up our algorithm to $\mathcal{O}\left((5|\mathcal{R}| + 6|\mathcal{C}| + 5)^{1+o(1)} \log |\mathcal{C}|\right) = \mathcal{O}\left(|\mathcal{C}|^{1+o(1)} \log |\mathcal{C}|\right)$. Fig. 6 exemplifies the parallelism between $f^\star$ in G and the optimal point-guard set in P. Lemma 5 establishes the connection between MINIMUM SCOT r-GUARD and MINIMUM FLOW WITH DEMANDS, which allows us to conclude Theorem 3.

Lemma 5. *The $(\mathcal{R}, \mathcal{C}, \mathcal{C}')$-SCOT P has a point-guard set $\mathcal{G}$ of size $k \leq |\mathcal{R}| + |\mathcal{C}|$ if and only if the network with demands G has a feasible flow f of value $|f| = k$.*

Proof. ($\Rightarrow$) Let $\mathcal{G}$ be a point-guard set for P with $|\mathcal{G}| = k$. By Lemma 4, assume every guard in $\mathcal{G}$ is placed inside a room. For each guard $g \in \mathcal{G}$, let $R \in \mathcal{R}$, $C'_h \in \mathcal{C}'$ and $C'_v \in \mathcal{C}'$ be the room, horizontal super-corridor and vertical super-corridor that g r-sees; if it does not see any horizontal and/or vertical super-corridor, take C'_h and/or C'_v as mock super-corridors. Send 1 unit of flow from s to t through the path $s \to \Gamma(C'_h) \to \Gamma(R) \to \Gamma(C'_v) \to t$. This can always be done because all the edge capacities are ∞. This yields a flow of value k. Since the guard set $\mathcal{G}$ covers every room, every horizontal super-corridor and every vertical super-corridor in P, the demand of every edge in G will be satisfied and therefore the flow is feasible.

($\Leftarrow$) Let $\mathcal{G} = \emptyset$. Do what follows while there is some path from s to t in G that passes only through edges (u, v) with $f(u, v) > 0$. Let $\gamma = s \to \Gamma(C_h') \to \Gamma(R) \to \Gamma(C_v') \to t$ be such a path, which can be found by a breadth-first search from s. Place a guard anywhere in the intersection of R, the stretch of C_h' and the stretch of C_v'; if C_h' and/or C_v' are mock super-corridors, place the guard in an arbitrary y-coordinate and/or an arbitrary x-coordinate, respectively, within room R. Insert this guard into $\mathcal{G}$. Decrement in 1 the value of $f(u, v)$ for every edge (u, v) along γ. Since the flow f satisfied every edge demand in G, for any room and corridor in P there will be at least one guard in $\mathcal{G}$ watching over it and $|\mathcal{G}| = k$. The number of processed paths γ is $|f|$, being $|f| \leq |\mathcal{R}| + |\mathcal{C}|$, and, since every considered path γ has 4 edges, the cost per iteration is $\mathcal{O}(1)$, leading to an overall linear-time reduction. □

Theorem 3. *A minimum-cardinality point-guard set for an $(\mathcal{R}, \mathcal{C}, \mathcal{C}')$-SCOT P that is r-independent can be determined in time $\mathcal{O}(|\mathcal{C}|^{1+o(1)} \log |\mathcal{C}|)$.*

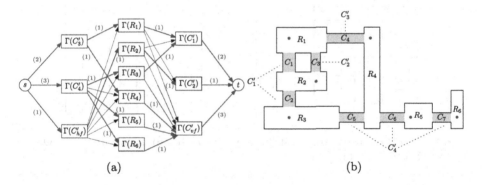

Fig. 6. (a) Minimum feasible flow f^* in the network G, with $|f^*| = 6$. The flow has not been annotated on edges (u, v) with $f^*(u, v) = 0$ to avoid clutter. (b) SCOT P, with an optimal set of 6 point-guards determined by flow f^*.

4   NP-Completeness with Point-Guards

In Sects. 2 and 3, we have presented efficient, polynomial-time algorithms for solving the cases where the SCOT instance has no holes or is r-independent. These were majorly based upon the specific structure of the polygon, which allowed for determining guard placements that we can combinatorially prove optimal. We now prove the hardness of the general version of the problem – the one where the SCOT simultaneously has holes and is not r-independent.

Theorem 4. MINIMUM SCOT r-GUARD *with point-guards is* NP-*hard*.

For the proof, we present a polynomial reduction from MINIMUM POLYOMINO r-GUARD, which is the problem of finding a minimum-cardinality point-guard set for a polyomino under r-visibility. An *m-polyomino* P_m is a polyform that results from the finite union of m unit squares, called *cells*, edge to edge [1,4]. Deciding whether a polyomino with m cells can be r-guarded with up to k point-guards, for a positive integer k, is NP-hard [11]. At a high level, the SCOT B we construct for a given m-polyomino P is composed of a sequence of identical gadgets replicated side by side, horizontally. Each two consecutive gadgets in the sequence are connected by one or more horizontal corridors as we will describe in due course. For an overview of the reduction, see Fig. 7. We define as $\Delta(P)$ the minimal unit grid that contains the polyomino P. Suppose its dimensions are N rows $\times$ M columns and assume that the top-left corner of $\Delta(P)$ has coordinates $(0,0)$.

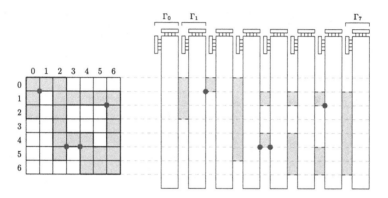

Fig. 7. Reduction from MINIMUM POLYOMINO r-GUARD to MINIMUM SCOT r-GUARD, not presented to scale.

Anchor Gadget. We begin by defining the *anchor gadget* Γ that will be instantiated several times in the construction of our SCOT (Fig. 8). It consists of an $(N+7) \times 5$ room (R_c) whose top and bottom walls are contained, respectively, in the lines $y = 0$ and $y = N + 7$ (for convenience, we assume that the y-axis grows downwards). Two tiny rooms are appended onto the walls of R_c by means of corridors. On the left there is a 5×1 room (R_l), whose top edge also satisfies $y = 0$. R_l is connected to the big room R_c by means of two 1×1 corridors: one of the corridors (C_γ) has its top edge contained in $y = 1$ and the other corridor (C_δ) has its own contained in $y = 3$. Similarly, above R_c there is a 1×5 room (R_t) whose top wall is contained in $y = -2$. R_t is also connected to R_c through two 1×1 corridors whose left walls obey $x = 1$ (C_α) and $x = 3$ (C_β), respectively.

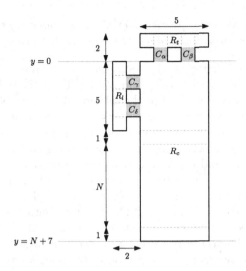

Fig. 8. Anchor gadget Γ.

Instance Transformation. SCOT B is formed by $M + 1$ identical anchor gadgets $\Gamma_0, \Gamma_1, \ldots, \Gamma_M$ arranged horizontally side by side, numbered 0 through M from left to right. The big rooms of each two consecutive anchor gadgets Γ_i and Γ_{i+1} $(0 \leq i < M)$ are separated by a horizontal space of 3 units and the line $y = 0$ passes through the top edges of all the big rooms. All the corridors connecting two consecutive anchor gadgets in B have fixed length 3 but variable width. For placing corridors in B, sweep through the grid $\Delta(P)$, column by column, from left to right. When processing the i-th column $(0 \leq i < M)$, consider all the vertical maximal connected components of cells that belong to that column in P. For each connected component spanning rows $a, a+1, \ldots, b$ of the i-th column, add a corridor with length 3 and width $b - a + 1$ connecting big rooms i and $i+1$ of B. That corridor has to be placed so that its top and bottom edges are contained, respectively, in the lines $y = a+6$ and $y = (b+1)+6 = b+7$.

Lemma 6. *The reduction is polynomial on m, the number of cells in P.*

Proof. $N, M \leq m$ is a suitable upper bound for the dimensions $N \times M$ of the minimal unit grid $\Delta(P)$ containing P, so building $M+1$ anchor gadgets is done in $\mathcal{O}(m)$ time. Sweeping through the columns of the grid $\Delta(P)$, determining maximal vertical connected components and adding the corresponding corridors to the SCOT B can be done in time linear on the grid size, $\mathcal{O}(m^2)$. Every anchor gadget takes 7 horizontal and $N+9$ vertical units of space, each two consecutive anchor gadgets are connected by horizontal corridors of length 3 and all the coordinates of vertices in B are defined by integers bounded by $\mathcal{O}(m)$. □

Lemma 7. *Let Γ be an anchor gadget, possibly with extra horizontal corridors attached to its left and right walls that result from the presented construction.*

Exactly 3 *point-guards are required for guarding* Γ, *ignoring the extra corridors. Moreover, we can construct such a guard set in which every guard satisfies* $y \leq 5$.

Proof. We first show that exactly 3 point-guards are required for guarding Γ. For the necessity, the tiny rooms R_t and R_l of Γ lie in the half-plane defined by $y \leq 5$ and we must place a guard in the interior of the big room and of the two tiny rooms. We now prove sufficiency. First, we choose one corridor among C_α and C_β and also choose one corridor among C_γ and C_δ. Suppose, without loss of generality, that we picked C_α and C_γ. Place a guard on one of the reflex vertices shared by R_t and C_α and another guard on one of the reflex vertices shared by R_l and C_γ. So far we have covered R_t, R_l, C_α and C_γ. The remaining parts of Γ can be covered with a third guard: simply place it somewhere in the intersection of the stretches of C_β and C_δ. The horizontal corridors that remain can then be guarded independently from the gadget with extra guards.

We now show that there is always an optimal solution for guarding Γ for which every guard satisfies $y \leq 5$. Suppose, by contradiction, that there is a better strategy, that is, one could instead choose to place the third guard in the intersection of the big room R_c and the stretch of another horizontal corridor $C^\star$ (other than C_γ and C_δ) incident in the big room to try to r-see $C^\star$. We would then still need a fourth guard somewhere in the stretch of C_δ to cover up C_δ: since all the remaining horizontal corridors incident in Γ are, by construction, positioned in the half-plane $y \geq 6$, they are r-independent from C_δ and therefore could not possibly be exploited to guard C_δ using less guards. Given that C_α has already been covered, we have nothing to lose by also placing the fourth guard in the stretch of C_β; so the fourth guard could be placed in the intersection of the stretch of C_β and the stretch of C_δ. But then one could swap the third and fourth guards to obtain precisely the solution described before.   □

Lemma 8. *[4] Under the* r-*visibility model, for any polyomino P there exists an optimal solution for guarding P whose guards are placed only at cell corners.*

Lemma 9. *The polyomino P can be* r-*guarded by* $k \in \mathbb{Z}^+$ *point-guards if and only if the SCOT B can be* r-*guarded by* $3(M+1) + k$ *point-guards.*

Proof. We show that we can map any solution for P to an analogous one for B which has 3 extra guards per anchor gadget (and vice-versa).

($\Rightarrow$) Let $\mathcal{G}_B = \emptyset$ and let $\mathcal{G}_P$ be a guard set for P with $|\mathcal{G}_P| = k$ in which every guard is placed at a cell corner by Lemma 8. For each guard $g \in \mathcal{G}_P$ that is placed at the intersection of the line $y = q$, for some $0 \leq q \leq N$, and the left (resp. right) cell corner of the i-th column of $\Delta(P)$ ($0 \leq i < M$), place a guard in $\mathcal{G}_B$ at the intersection of $y = q + 6$ and the left (resp. right) side of a corridor connecting gadgets i and $i + 1$ in B. Note that, if $i < M - 1$, the guard g also belongs to the $(i+1)$-th column of $\Delta(P)$, but we do not place an extra, redundant guard in $\mathcal{G}_B$. Next, insert $3(M+1)$ guards satisfying $y \leq 5$ into $\mathcal{G}_B$ to cover every gadget in P, according to Lemma 7. We have that $|\mathcal{G}_B| = 3(M+1) + k$. By the assumption that $\mathcal{G}_P$ covers up the entirety of P, and since rooms in B are wider than the incident corridors and distances do not affect r-visibility, $\mathcal{G}_B$ covers up all the corridors connecting adjacent anchor gadgets in B and, thus, B is covered entirely as well.

($\Leftarrow$) Let $\mathcal{G}_P = \emptyset$ and let $\mathcal{G}_B$ be a guard set for SCOT B with $|\mathcal{G}_B| = 3(M+1)+k$ in which every guard covering an anchor gadget satisfies $y \leq 5$ by Lemma 7. Assume that every guard in $\mathcal{G}_B$ other than those $3(M+1)$ that see anchor gadgets are placed in corridors connecting consecutive anchor gadgets Γ_i and Γ_{i+1} ($0 \leq i < M$). We can assume that because, by moving a guard g from a big room R_c of an anchor gadget to an incident corridor, the only region g stops seeing completely is R_c, which is still covered by one of the $3(M+1)$ guards we have discriminated for watching over anchor gadgets. For each guard $g \in \mathcal{G}_B$ that is placed at the intersection of the line $y = q + 6$, for some $0 \leq q \leq N$, and a corridor connecting gadgets Γ_i and Γ_{i+1} in B ($0 \leq i < M$), let d be the horizontal distance between g and Γ_i and place a guard in $\mathcal{G}_P$ at the point $(i + d, q)$. Big rooms R_c of anchor gadgets in B do not block r-visibility from incident corridors. Therefore, guards in $\mathcal{G}_B$ covering up all the corridors imply that the corresponding guards in $\mathcal{G}_P$ will also cover up all the cells of P and, thus, $\mathcal{G}_P$ is a valid guard set for P, with $|\mathcal{G}_P| = k$. $\square$

From [2] it follows that we can reduce candidate guard positions in a SCOT to a finite set with polynomial size by partitioning it into rectangular pieces. Being any such solution verifiable in time polynomial in the instance size, we are finally set to state our main result.

Theorem 5. *The decision version of* MINIMUM SCOT r-GUARD *with point-guards, i.e., deciding whether $k \in \mathbb{Z}^+$ point-guards are sufficient, is* NP-*complete.*

5   Conclusion

In this paper, we define a new family of orthogonal polygons, the SCOTs. Their structure encompasses many properties which enable us to develop efficient algorithms for r-guarding them. We prove that, if the SCOT is simple or r-independent, the MINIMUM SCOT r-GUARD problem is in P. A linear-time algorithm is given for simple SCOTs, for both vertex- and point-guards, based on a tree decomposition of the polygon. For r-independent SCOTs with vertex-guards, we develop an algorithm that runs in time $\mathcal{O}(|\mathcal{C}|\sqrt{\min\{|\mathcal{R}|, |\mathcal{C}'|\}})$, based on bipartite matchings. Finally, a third one is given for r-independent SCOTs with point-guards, based on a reduction to MINIMUM FLOW WITH DEMANDS, which runs in time $\mathcal{O}\left(T(G') \log |f^\star|\right)$, where $T(G')$ is the time it takes to compute a maximum flow in a network G' that we define and $|f^\star|$ is the optimal number of guards. On the contrary, should the SCOT have holes and not be r-independent, we show that the problem becomes NP-hard – indeed NP-complete for the case of point-guards – by reducing from MINIMUM POLYOMINO r-GUARD. We observe that MINIMUM SCOT r-GUARD with vertex-guards is also in NP, because any solution is a subset of the vertices of the SCOT, and it can be approximated with factor $\mathcal{O}(\log |\mathcal{C}|)$ in polynomial time [7]. However, whether it remains NP-hard or even APX-hard is left as an open question. As a final note, we remark that, for standard visibility, along the lines of [18], we can prove that even finding a minimum-cardinality guard set for the vertices of a simple SCOT is APX-hard.

Acknowledgments. The authors would like to thank anonymous reviewers for constructive comments.

References

1. Biedl, T., Irfan, M.T., Iwerks, J., Kim, J., Mitchell, J.S.B.: The art gallery theorem for polyominoes. Discrete Comput. Geom. **48**(3), 711–720 (2012). https://doi.org/10.1007/s00454-012-9429-1

2. Biedl, T., Mehrabi, S.: On r-Guarding thin orthogonal polygons. arXiv preprint arXiv:1604.07100 (2016), (extended version of ISAAC'2016)

3. Biedl, T., Mehrabi, S.: On orthogonally guarding orthogonal polygons with bounded treewidth. Algorithmica **83**(2), 641–666 (2020). https://doi.org/10.1007/s00453-020-00769-5

4. Biedl, T.C., Irfan, M.T., Iwerks, J., Kim, J., Mitchell, J.S.B.: Guarding polyominoes. In: Proceedings of the 27th ACM Symposium on Computational Geometry, SoCG 2011. pp. 387–396. ACM (2011). https://doi.org/10.1145/1998196.1998261

5. Chen, L., Kyng, R., Liu, Y.P., Peng, R., Gutenberg, M.P., Sachdeva, S.: Maximum flow and minimum-cost flow in almost-linear time. arXiv preprint arXiv:2203.00671 (2022). https://doi.org/10.48550/arXiv.2203.00671

6. Chvátal, V.: A combinatorial theorem in plane geometry. J. Comb. Theory Series B **18**(1), 39–41 (1975)

7. Cruz, V.: Algorithms for art gallery problems. Master's thesis, Universidade do Porto (2022), https://hdl.handle.net/10216/142907

8. Dinitz, Y.: Dinitz' algorithm: the original version and Even's version. In: Goldreich, O., Rosenberg, A.L., Selman, A.L. (eds.) Theoretical Computer Science. LNCS, vol. 3895, pp. 218–240. Springer, Heidelberg (2006). https://doi.org/10.1007/11685654_10

9. Erickson, J.: Algorithms - Lecture 25: Extensions of maximum flow, lecture notes (2015). https://courses.engr.illinois.edu/cs498dl1/sp2015/notes/25-maxflowext.pdf. Accessed Aug 2022

10. Hopcroft, J.E., Karp, R.M.: An $n^{5/2}$ algorithm for maximum matchings in bipartite graphs. SIAM J. Comput. **2**(4), 225–231 (1973). https://doi.org/10.1137/0202019

11. Iwamoto, C., Kume, T.: Computational complexity of the r-visibility guard set problem for polyominoes. In: Akiyama, J., Ito, H., Sakai, T. (eds.) JCDCGG 2013. LNCS, vol. 8845, pp. 87–95. Springer, Cham (2014). https://doi.org/10.1007/978-3-319-13287-7_8

12. Kahn, J., Klawe, M., Kleitman, D.: Traditional galleries require fewer watchmen. SIAM J. Algebraic Discrete Methods **4**(2), 194–206 (1983)

13. Katz, M.J., Roisman, G.S.: On guarding the vertices of rectilinear domains. Comput. Geom. **39**(3), 219–228 (2008)

14. Kleinberg, J., Tardos, E.: Algorithm Design. Addison-Wesley, Pearson Education (2006)

15. Lee, D., Lin, A.: Computational complexity of art gallery problems. IEEE Trans. Inf. Theory **32**(2), 276–282 (1986). https://doi.org/10.1109/TIT.1986.1057165

16. O'Rourke, J.: Art Gallery Theorems and Algorithms, vol. 57. Oxford, New York (1987)
17. Schuchardt, D., Hecker, H.D.: Two NP-hard art-gallery problems for ortho-polygons. Math. Log. Q. **41**(2), 261–267 (1995). https://doi.org/10.1002/malq. 19950410212
18. Tomás, A.P.: Guarding thin orthogonal polygons is hard. In: Gasieniec, L., Wolter, F. (eds.) FCT 2013. LNCS, vol. 8070, pp. 305–316. Springer, Heidelberg (2013). https://doi.org/10.1007/978-3-642-40164-0_29
19. Urrutia, J.: Sixth Proof of the Orthogonal Art Gallery Theorem. Technical report TR-97-03. Department of Computer Science, University of Ottawa, February 1997. http://www.math.unam.mx/~urrutia/online_papers/IllumOrt.pdf. Accessed Oct 2022
20. Worman, C., Keil, J.M.: Polygon decomposition and the orthogonal art gallery problem. Int. J. Comput. Geom. Appl. **17**(02), 105–138 (2007). https://doi.org/ 10.1142/S0218195907002264

Complexity Results on Untangling Red-Blue Matchings

Arun Kumar Das[1], Sandip Das[1], Guilherme D. da Fonseca[2],
Yan Gerard[3], and Bastien Rivier[3]([✉])

[1] Indian Statistical Institute, Kolkata, India
arund426@gmail.com, sandipdas@isical.ac.in
[2] Aix-Marseille Université and LIS, Marseille, France
guilherme.fonseca@lis-lab.fr
[3] Université Clermont Auvergne and LIMOS, Clermont-Ferrand, France
{yan.gerard,bastien.rivier}@uca.fr

Abstract. Given a matching between n red points and n blue points by line segments in the plane, we consider the problem of obtaining a crossing-free matching through flip operations that replace two crossing segments by two non-crossing ones. We first show that (i) it is NP-hard to α-approximate the shortest flip sequence, for any constant α. Second, we show that when the red points are colinear, (ii) given a matching, a flip sequence of length at most $\binom{n}{2}$ always exists, and (iii) the number of flips in any sequence never exceeds $\binom{n}{2}\frac{n+4}{6}$. Finally, we present (iv) a lower bounding flip sequence with roughly $1.5\binom{n}{2}$ flips, which shows that the $\binom{n}{2}$ flips attained in the convex case are not the maximum, and (v) a convex matching from which any flip sequence has roughly $1.5\,n$ flips. The last four results, based on novel analyses, improve the constants of state-of-the-art bounds.

Keywords: Reconfiguration · Matching · Planar geometry · NP-hard

1 Introduction

We consider the problem of untangling a perfect red-blue matching drawn in the plane with straight line segments. We are given a set of $2n$ points in the plane, partitioned into a set R of n red points, and a set B of n blue points, in general position (no three colinear points, unless they have the same color).

In combinatorial reconfiguration, a flip is an operation changing a configuration into another [9,20]. In our case, a *configuration* is a set of n line segments where each point of R is matched to exactly one point of B, i.e. a perfect straight-line red-blue matching (a *matching* for short), and a *flip* replaces two crossing segments by two non-crossing ones (Fig. 1).

This work is partially supported by the IFCAM project Applications of Graph Homomorphisms (MA/IFCAM/18/39), and by the French ANR PRC grant ADDS (ANR-19-CE48-0005). The full version is available at arxiv.org/abs/2202.11857.

A. Castañeda and F. Rodríguez-Henríquez (Eds.): LATIN 2022, LNCS 13568, pp. 730–745, 2022.
https://doi.org/10.1007/978-3-031-20624-5_44

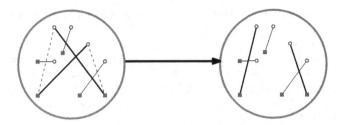

Fig. 1. Matchings before and after a flip. Solid squares are red points and hollow circles are blue points. (Color figure online)

The *reconfiguration graph* is the directed simple graph whose vertices $\mathcal{V}$ are the configurations, and such that there is a directed edge from a configuration M_1 to another one M_2 whenever a flip transforms M_1 into M_2. Note that, in our case, the reconfiguration graph is acyclic [8]. Let $\mathcal{S} \subseteq \mathcal{V}$ be the set of sinks, which corresponds to the crossing-free matchings. Given two configurations $u, v \in \mathcal{V}$, let $\mathcal{P}(u, v)$ be the set of directed paths from u to v. Given a path P, let the *length* of P, denoted $|P|$, be the number of edges in P. We are interested in two parameters of this reconfiguration graph:

$$\mathbf{d}(R, B) = \max_{u \in \mathcal{V}} \min_{v \in \mathcal{S}} \min_{P \in \mathcal{P}(u,v)} |P| \quad \text{and} \quad \mathbf{D}(R, B) = \max_{u \in \mathcal{V}} \max_{v \in \mathcal{S}} \max_{P \in \mathcal{P}(u,v)} |P|.$$

This leads to the definitions of $\mathbf{d}(n)$ and $\mathbf{D}(n)$ respectively as the maximum of $\mathbf{d}(R, B)$ and $\mathbf{D}(R, B)$ over all sets R, B with $|R| = |B| = n$. An *untangle sequence* is a path in the reconfiguration graph ending in $\mathcal{S}$. Intuitively, $\mathbf{d}$ corresponds to the minimum length of an untangle sequence in the worst case, while $\mathbf{D}$ corresponds to the longest untangle sequence.

We also consider a more specific version of the problem where the red points are colinear [6], say, on the x-axis. As the flips on each half-plane defined by the x-axis are independent, we additionally suppose all blue points to lie on the upper half-plane without loss of generality. The matchings in this case are called *red-on-a-line* matchings.

Related Work. The parameters $\mathbf{d}$ and $\mathbf{D}$ have been studied in several different contexts with similar definitions of a flip, but considering other configurations.

In 1981, an $O(n^3)$ upper bound on $\mathbf{D}(n)$ was stated in the context of optimizing a TSP tour [22] (the configurations are polygons). This upper bound should be compared to the exponential lower bound on $\mathbf{D}(n)$ when the flips are not restricted to crossing segments, as long as they decrease the Euclidean length of the tour [12]. The convex case (i.e. the case where the points are in convex position) has been studied in [25,28].

In the non-bipartite version of the straight-line perfect matching problem, there are two possible pairs of segments to replace a crossing pair. This additional choice yields an $n^2/2$ upper bound on $\mathbf{d}(n)$ [8].

Table 1. Lower and upper bounds on $\mathbf{d}(n)$ and $\mathbf{D}(n)$ for red-blue matchings.

	$\mathbf{d}(n)$ bounds		$\mathbf{D}(n)$ bounds	
	Lower	Upper	Lower	Upper
General	$\frac{3}{2}n-2$, Theorem 5*	$\binom{n}{2}(n-1)$, [8,22]	$\frac{3}{2}\binom{n}{2}-\frac{n}{4}$, Theorem 4*	$\binom{n}{2}(n-1)$, [8,22]
Convex	$\frac{3}{2}n-2$, Theorem 5*	$2n-2$, [6]	$\binom{n}{2}$, [8]	$\binom{n}{2}$, [6]
Red-on-a-line	$n-1$, [8]	$\binom{n}{2}$, Theorem 2	$\frac{3}{2}\binom{n}{2}-\frac{n}{4}$, Theorem 4*	$\binom{n}{2}\frac{n+4}{6}$, Theorem 3

* For even n.

It is also possible to relax the flip definition to all operations that replace two segments by two others with the same four endpoints, whether they cross or not, and generalize the configurations to multigraphs with the same degree sequence [14,15,20]. In this context, finding the shortest path from a given configuration to another in the reconfiguration graph is NP-hard, yet 1.5-approximable [4,5,13,27]. If we additionally require the configurations to be connected graphs, the same problem is NP-hard and 2.5-approximable [10].

Reconfiguration problems in the context of triangulations are widely studied [24]. A flip consists of removing one edge and adding another one while preserving a triangulation. It is know that $\Theta(n^2)$ flips are sufficient and sometimes necessary to obtain a Delaunay triangulation [18,21]. Determining the flip distance between two triangulations of a point set [23,26] and between two triangulations of a simple polygon [1] are both NP-hard.

Considering perfect matchings of an arbitrary graph (instead of the complete bipartite graph on R, B), a flip amounts to exchanging the edges in an alternating cycle of length four. It is then PSPACE-complete to decide whether there exists a path from a configuration to another [7]. There is, actually, a wide variety of reconfiguration contexts derived from NP-complete problems where this same accessibility problem is PSPACE-complete [19]. Many other reconfiguration problems are presented in [17].

Getting back to our context of straight-line red-blue matchings, the values of $\mathbf{d}$ and $\mathbf{D}$ have been determined almost exactly in the convex case (see Table 1). Notice that the $n-1$ lower bound on $\mathbf{d}(n)$ carries to both the general and red-on-a-line cases [8]. It is notable that the upper bound on $\mathbf{D}(n)$ is also the best known bound on $\mathbf{d}(n)$ and has not been improved since 1981 [22].

As a final side note, given a red-blue point set, a crossing-free red-blue matching can be computed in $O(n \log n)$ time [16] (improving over an $O(n \log^2 n)$ algorithm [3]). The algorithm is based on semi-dynamic convex hull data structures and does not use flips. The problem has also been considered in higher dimensions [2].

Contributions. We show in Sect. 2 that it is NP-hard to α-approximate the shortest untangle sequence starting at a given matching, for any fixed $\alpha \geq 1$.

The following results are summarized in Table 1. An improved lower bound on $d(n)$ in the convex case is presented in Sect. 5.2. The remainder of the paper considers the red-on-a-line case. In Sect. 3, we slightly improve the former $2\binom{n}{2}$

upper bound on $\mathbf{d}(n)$ [6], using a simpler algorithm and a novel analysis. In Sect. 4, we asymptotically divide by 6 the historical $\binom{n}{2}(n-1)$ upper bound on $\mathbf{D}(n)$ [8,22], using a different potential argument.

In Sect. 5.1, we present a counter-example to the intuition that the longest untangle sequence is attained in the convex case (where the number of crossings is maximal). We take advantage of points that are not in convex position to increase the lower bound by a factor of $\frac{3}{2}$. This red-on-a-line lower bound on $\mathbf{d}(n)$ carries over to the general case (and even to non-bipartite perfect matchings). The conjecture that $\mathbf{D}(n)$ is quadratic [8] remains open, though.

2   NP-Hardness

In this section, we consider the proof of the NP-hardness of the following problem. Let $d(M)$ denote the minimum path length from a matching M to $\mathcal{S}$, the set of crossing-free matchings, in the reconfiguration graph. The proof is presented in the full version.

Problem 1. Let $\alpha \geq 1$ be a constant.
Input: M, a red-blue matching with rational coordinates.
Output: An untangle sequence starting at M of length at most α times $d(M)$.

We have the following theorem.

Theorem 1. *Problem 1 is NP-hard for all $\alpha \geq 1$.*

De Berg and Khosravi [11] showed that the rectilinear planar monotone 3-SAT problem (*RPM 3-SAT*) is NP-hard. The RPM 3-SAT problem is a special case of the classic 3-SAT problem in which the clauses consist only of either all positive or all negative literals and the layout is planar (Fig. 2). We reduce RPM 3-SAT to Problem 1. The key elements of the reduction are described next.

Given a planar embedding of an RPM 3-CNF formula Φ (Fig. 2), we construct a matching M_Φ of polynomial size. The property of this matching M_Φ is that its shortest untangle sequence has a length below a certain constant if Φ is satisfiable and above α times this constant otherwise.

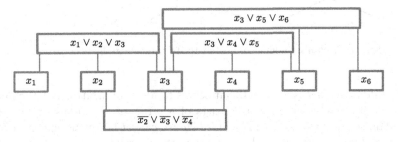

Fig. 2. A planar embedding of an RPM 3-CNF formula.

The aforesaid matching M_Φ is built using two types of gadgets. The variable rectangles are replaced by variable gadgets (Fig. 3). The clause rectangles together with the corresponding edges are replaced with padded clause gadgets (Fig. 6).

A *variable gadget* is a three-segment matching with two crossings. It allows for two possible flips, either of which produces a crossing-free matching, as shown in Fig. 3. The flip generating the topmost segment stands for *false* ($x = 0$ in Fig. 3), while the flip generating the bottom segment stands for *true* ($x = 1$).

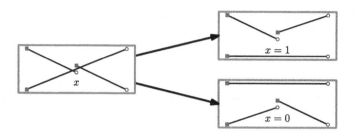

Fig. 3. A variable gadget and its two untangle sequences.

A *clause gadget* is an OR gate with three inputs (Fig. 4). The RPM 3-CNF clauses are either positive or negative. We describe the gadget for a positive clause, but the gadget for a negative clause can be defined analogously (by a vertical reflection). Three variable gadgets are the inputs of a clause gadget. In the crossing-free matching obtained for the clause gadget, the presence of the topmost segment (dashed in Fig. 4, 5, and 6) stands for a false output.

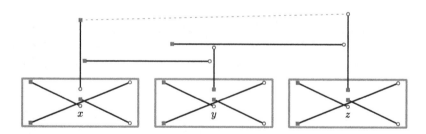

Fig. 4. A clause gadget connected to its variable gadgets x, y, and z.

A *padding gadget* is a gadget that serves to force an arbitrarily large number k of flips if a clause is false. It consists of a series of k non-crossing segments (the plain segments in Fig. 5). A *padded clause gadget* is a clause gadget coupled with a padding gadget in such a way that the presence of the output segment triggers k extra flips (Fig. 6).

Fig. 5. A padding gadget.

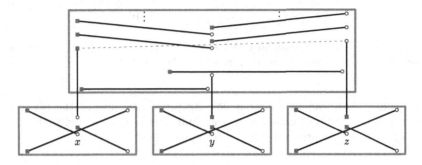

Fig. 6. A padded clause gadget connected to its variable gadgets x, y, and z.

Let c be the number of clauses and v be the number of variables of the formula Φ. If Φ is *satisfiable*, then the shortest untangle sequence of M_Φ has at most 5 flips per clause plus 1 flip per variable. In this case, we have $d(M_\Phi) \leq 5c + v$. We choose the size of the padding gadget so that a non-satisfied clause triggers $k = \alpha(5c + v) + 1$ flips. If the formula Φ is *not satisfiable*, then at least one of the padding gadgets is triggered and $d(M_\Phi) > \alpha(5c + v)$.

3   Upper Bound on d(n)

In this section, we prove the following upper bound.

Theorem 2. *In the red-on-a-line case,* $\mathbf{d}(n) \leq \binom{n}{2}$.

The proof consists of the analysis of the number of flips performed by the recursive algorithm described next. Throughout, we assume general position (no two blue points with the same y-coordinate). Let the *top* segment of a red-on-a-line matching be the segment with the topmost blue endpoint (Fig. 7(a)).

While the top segment s_1 of the matching crosses another segment s_2, we flip s_1 and s_2. If multiple segments cross s_1, then we choose s_2 as the top segment among the segments crossing s_1.

The previous loop stops when the top segment s_1 has no crossings. At this point, we have that s_1 splits the matching into at most two non-empty submatchings, one to each side of s_1. We recursively call the algorithm on these submatchings (Fig. 7(b)).

The correctness of the algorithm follows from the next lemma.

Lemma 1 ([8]). *If a matching admits a partition of submatchings whose convex hulls are all disjoint, then, any sequence of flips in one of the submatchings never affects the other submatchings (Fig. 8).*

Fig. 7. (a) A red-on-a-line matching with s_1 as the top segment. (b) The matching just before the first recursive calls of the algorithm, where s_1 is free. (Color figure online)

Proof. This result can be found in [8]. Its proof amounts to the observation that the flip operation leaves the convex hull unchanged (in Fig. 8, the dashed segments are the results of possible flip sequences).   □

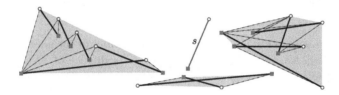

Fig. 8. A partition of 4 submatchings whose convex hulls are all disjoint. The segment s is the only free segment.

We say that a segment s is *free* if the matching admits a partition of submatchings whose convex hulls are all disjoint, and one of the submatchings consists of the segment s alone. In Fig. 8, the segment s is the only free segment. It is easy to see that the algorithm always makes the top segment free before recursive calls. The correctness of the algorithm then follows from Lemma 1.

Flip Complexity. The analysis of the number of flips performed by the algorithm stems from the following observations. We define three possible *states* for a pair of segments (Fig. 9).

- State **X**: the segments are crossing.
- State **H**: the segments are not crossing and their endpoints are in convex position.
- State **T**: the endpoints are not in convex position.

In the convex case, there are no **T**-states and a flip increases the number of **H**-pairs by at least 1 unit, and decreases the number of **X**-pairs as well. Hence, counting either **X** or **H**-pairs yields the $\binom{n}{2}$ upper bound on $\mathbf{D}(n)$. However, when the points are not in convex position, counting **H** and **X**-pairs is fundamentally different. We will see that counting **H**-pairs is more useful to prove the desired bounds.

Fig. 9. The three different states of pairs of segments.

When the points are not in convex position, a flip may decrease the number of **H**-pairs. Fig. 10 shows two such situations where flipping s_1, s_2 does not increase the number of **H**-pairs. There is one **H**-pair involving segment s before the flip, and none after the flip. Notice that, if we added multiple segments close to s, the number of **H**-pairs would actually decrease. However, the algorithm avoids these situations by choosing to flip top segments. The full proof involves state tracking, a novel approach to analyse flip sequences, which is described next.

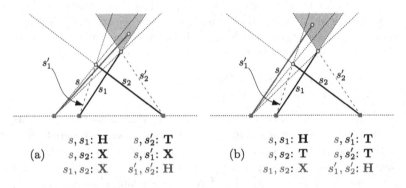

$$
\begin{array}{cc}
s, s_1:\textbf{H} & s, s_2':\textbf{T} \\
(a) \quad s, s_2:\textbf{X} & s, s_1':\textbf{X} \\
s_1, s_2:\textbf{X} & s_1', s_2':\textbf{H}
\end{array}
\qquad
\begin{array}{cc}
s, s_1:\textbf{H} & s, s_1':\textbf{T} \\
(b) \quad s, s_2:\textbf{T} & s, s_2':\textbf{T} \\
s_1, s_2:\textbf{X} & s_1', s_2':\textbf{H}
\end{array}
$$

Fig. 10. Two cases where flipping s_1, s_2 does not increase the number of **H**-pairs. The upper cone of s_1, s_2' is shaded.

State Tracking. We have $\binom{n}{2}$ pairs of segments before and after a flip. Each pair has an associated state. However, since two segments change in the matchings, there is no clear correspondence between the state of each pair before and after the flip. State tracking establishes this correspondence by making choices of which pair of segments in the initial matching corresponds to which pair of segments in the resulting matching. These choices are performed deliberately to obtain certain state transitions instead of others and prove the desired bounds.

The following notations will be used throughout the rest of this section and are summarized in Fig. 11. Let r_1, r_2 be two red points and b_1, b_2 be two blue points. Let s_1, s_2, s_1', s_2' be the following four segments respectively: $r_1 b_1$, $r_2 b_2$, $r_1 b_2$, $r_2 b_1$. We consider a flip that replaces the pair of segments s_1, s_2 by s_1', s_2'. Let M denote the matching before the flip and M' denote the resulting matching after the flip.

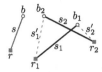

Fig. 11. Notations for a flip and for a variable segment s.

We order the $\binom{n}{2}$ pairs of segments of M in a column vector. There are three types of pairs of segments in M with respect to the flip: the *unaffected* pairs (involving neither s_1 nor s_2), the *flipping* pair s_1, s_2, and the *affected* pairs (involving exactly one of s_1 or s_2). We choose the new order of the $\binom{n}{2}$ pairs of segments of M' in a way that satisfies the following properties with respect to the previous vector. The unaffected pairs keeps the same indices. The pair s_1', s_2' gets the index of s_1, s_2. Next, we describe the remaining indices.

Let s be a segment of M distinct from s_1 and s_2. Let r and b be the red and blue endpoints of s. Let i_1 and i_2 be the indices of s, s_1 and s, s_2, and let $\mathbf{S_1}$ and $\mathbf{S_2}$ be their respective states. Let $\mathbf{S_1'}$ and $\mathbf{S_2'}$ be the respective states of s, s_1' and s, s_2'. We restrict our choice to the following two options:

– index s, s_1' with i_1, and s, s_2' with i_2, or
– index s, s_1' with i_2, and s, s_2' with i_1.

We call such a choice a *tracking choice*. We say that a pair of segments in M *turns into* a pair in M' when they have the same index. We denote $\mathbf{S} \to \mathbf{S'}$ to specify that the pairs of segments with a given index go from the state $\mathbf{S}$ to the state $\mathbf{S'}$. In the following figures, we use $\mathbf{S_1}, \mathbf{S_2} \to \mathbf{S_1'}, \mathbf{S_2'}$ as a shorthand notation to say that we have the two following tracking choices: either $\mathbf{S_1} \to \mathbf{S_1'}$ and $\mathbf{S_2} \to \mathbf{S_2'}$ or $\mathbf{S_1} \to \mathbf{S_2'}$ and $\mathbf{S_2} \to \mathbf{S_1'}$.

There are 3^2 possible such *transitions* $\mathbf{S} \to \mathbf{S'}$. Yet, the next two lemmas ensure that some transitions can be ruled out by tracking choices. Lemma 2 actually holds for any (possibly non-bipartite) matching, while Lemma 3 is specific to the red-on-a-line case. Both lemmas are proved analyzing the tracking choices of each possible position of a segment s relatively to the flipping pair.

Lemma 2. *There always exists a tracking choice avoiding the $\mathbf{H} \to \mathbf{X}$ transition.*

Proof. There clearly exists a tracking choice avoiding the $\mathbf{H} \to \mathbf{X}$ transition unless we have either a transition (i) $\mathbf{H}, \mathbf{H} \to \mathbf{X}, \mathbf{S}$ or (ii) $\mathbf{H}, \mathbf{S} \to \mathbf{X}, \mathbf{X}$, where $\mathbf{S} \in \{\mathbf{X}, \mathbf{H}, \mathbf{T}\}$. We show that these two cases are not possible.

(i) $\mathbf{H}, \mathbf{H} \to \mathbf{X}, \mathbf{S}$: If both the pairs s, s_1 and s, s_2 are $\mathbf{H}$ while at least one of the two pairs s, s_1' and s, s_2' is $\mathbf{X}$, then the final $\mathbf{X}$ state implies that s crosses s_1 or s_2, which contradicts the two initial $\mathbf{H}$ states.

(ii) $\mathbf{H}, \mathbf{S} \to \mathbf{X}, \mathbf{X}$: If one of the two pairs s, s_1 and s, s_2 is $\mathbf{H}$ while both pairs s, s_1' and s, s_2' are $\mathbf{X}$, then the two final $\mathbf{X}$ states imply that s crosses s_1' and s_2'. It follows that s also crosses s_1 and s_2, which is again a contradiction. □

If there are neither $\mathbf{H} \to \mathbf{X}$ nor $\mathbf{H} \to \mathbf{T}$ transitions (as in the convex case), then an upper bound of $\binom{n}{2}$ on the number of flips follows immediately. Unfortunately, it is not always possible to avoid the $\mathbf{H} \to \mathbf{T}$ transition. However, the following lemma shows that, in the red-on-a-line case we can avoid $\mathbf{H} \to \mathbf{T}$ if we carefully choose which flip to perform. The proof is presented in the full version. We define the *upper cone* of two segments $r_3 b_3, r_4 b_3$ as the locus of the points that are separated from the horizontal line $r_3 r_4$ by the two lines $r_3 b_3$ and $r_4 b_3$. In Fig. 10, the upper cone of s_1, s_2' is shaded.

Lemma 3. *In the red-on-a-line case, if the blue point b of s is not in any of the two upper cones of s_1, s_2' and s_2, s_1', then there always exists a tracking choice that avoids $\mathbf{H} \to \mathbf{T}$ for the pairs s, s_1 and s, s_2 while still avoiding $\mathbf{H} \to \mathbf{X}$.*

We are now ready to prove Theorem 2.

Proof. Let $f(M)$ be the total number of flips performed by the algorithm on an n-segment input matching M and let $g(M)$ be the number of flips performed by the algorithm before the recursive calls. Let M_r denote the matching before the recursive calls. The recursive calls take two submatchings of M_r that we call M_1 and M_2, yielding the following recurrence relation.

$$f(M) = f(M_1) + f(M_2) + g(M)$$

Let $\bar{h}(M)$ be the number of $\mathbf{X}$-pairs plus the number of $\mathbf{T}$-pairs in a matching M, that is, the number of pairs that are not $\mathbf{H}$-pairs. Lemma 3 ensures that

$$g(M) \le \bar{h}(M) - \bar{h}(M_r) \le \bar{h}(M) - \bar{h}(M_1) - \bar{h}(M_2).$$

Clearly, $f(\emptyset) = 0$. We suppose that, for all M' with less than n segments, we have $f(M') \le \bar{h}(M')$. Then by induction we get

$$f(M) \le \bar{h}(M_1) + \bar{h}(M_2) + \bar{h}(M) - \bar{h}(M_1) - \bar{h}(M_2) = \bar{h}(M).$$

Theorem 2 follows since $\bar{h}(M) \le \binom{n}{2}$.   □

4   Upper Bound on $\mathbf{D}(n)$

In this section we prove the following theorem.

Theorem 3. *In the red-on-a-line case, $\mathbf{D}(n) \le \binom{n}{2} \frac{n+4}{6}$.*

To prove Theorem 3, we define a potential function Φ that maps a red-on-a-line matching to an integer from 0 to $\binom{n}{2} \frac{n+4}{3}$. Since Φ decreases by at least 2 units at each flip, the theorem follows. We first give the definitions needed to present Φ. Then, we prove four lemmas yielding Theorem 3.

Let M be a red-on-a-line matching. Let $r_1, \ldots, r_n$ be the red points, from left to right. Let $<$ be a line, parallel to the line of the red points and above them. For each k in $\{1, \ldots, n\}$, we project the blue points onto $<$, using r_k as a focal

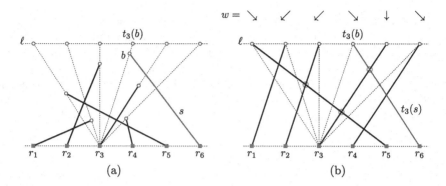

Fig. 12. (a) The projection t_k for $k = 3$. (b) The segments $t_3(\cdot)$. The three 3-observed crossing 3-pairs are circled.

point. More precisely, each blue point b maps to a point $t_k(b)$, the intersection between the ray $r_k b$ and the line $<$ (Fig. 12(a)). We also define the function t_k of a red-blue segment rb as the segment $t_k(rb) = rt_k(b)$ (Fig. 12(b)).

We may abbreviate a pair of segments $r_i b, r_j b'$ as $\langle i, j \rangle$ when the points b and b' can be deduced from the underlying matching. Let k be an integer in $\{1, \ldots, n\}$. We say that a pair of segments $\langle i, j \rangle$ is a *k-pair* if $i \leq k \leq j$. A *k-flip* is then a flip of a k-pair. We say that two segments are *k-observed crossing* if the extended projection $t_k(\cdot)$ maps them to crossing segments (Fig. 12(b)). We have the following lemma.

Lemma 4. *A crossing k-pair is necessarily k-observed crossing.*

Proof. Let $r_i b, r_j b'$ be a crossing k-pair. We suppose, without loss of generality, that $i < j$ (e.g. $i = 2$, $k = 3$, and $j = 4$ in Fig. 12).

The fact that the k-pair $r_i b, r_j b'$ is crossing means that the four points are in convex position, and that they appear as r_i, r_j, b, b' on their convex hull in counter-clock-wise order. Since $i \leq k \leq j$, the point r_k is also on the boundary of the convex hull of the four points. Therefore, the projection $t_k(\cdot)$ will not change the convex-hull order and the segments $r_i t_k(b)$ and $r_j t_k(b')$ will cross. □

We define $\Phi_k(M)$, the *k-th potential* of M, as the number of k-observed crossing k-pairs (Fig. 12(b)). Lemma 5 shows that the k-th potential Φ_k is at most $(k-1)(n-k) + n - 1$. Lemma 6 shows that Φ_k never increases, and decreases by at least 1 unit at each k-flip.

Lemma 5. *The k-th potential Φ_k takes integer values from 0 to $k(n+1) - k^2 - 1$.*

Proof. The k-th potential $\Phi_k(M)$ is at most the number of k-pairs in M, crossing or not. There are exactly $(k-1)(n-k)$ k-pairs of the form $\langle i, j \rangle$ with $i < k < j$. There are exactly $k-1$ k-pairs of the form $\langle i, k \rangle$ with $i < k$. There are exactly $n-k$ k-pairs of the form $\langle k, j \rangle$ with $k < j$. In total, there are $(k-1)(n-k) + k - 1 + n - k$ k-pairs in M. □

Lemma 6. *The k-th potential Φ_k never increases, and decreases by at least 1 unit at each k-flip.*

Proof. We order the projected blue points on ℓ from left to right. We then map each projected blue point $t_k(b)$ to an element in $\{\nearrow, \downarrow, \searrow\}$:

- $t_k(b)$ is mapped to $\nearrow$ if b is matched to a red point on the left of r_k,
- $t_k(b)$ is mapped to $\downarrow$ if b is matched to r_k,
- $t_k(b)$ is mapped to $\searrow$ if b is matched to a red point on the right of r_k.

Let $w = w_1 \ldots w_n$ be the word on the alphabet $\{\nearrow, \downarrow, \searrow\}$ induced by the order of the projected blue points and the map. For instance, in Fig. 12 with $k = 3$, $w = \searrow\nearrow\nearrow\searrow\downarrow\searrow$.

Let the total order of the symbols be $\nearrow \prec \downarrow \prec \searrow$. An *inversion* in w is a pair w_i, w_j with $i < j$ and $w_j \prec w_i$. The inversions in w are in bijection with the k-observed crossing k-pairs in M. Thus, by definition, $\Phi_k(M)$ is the number of inversions in w. Lemma 6 follows from the following two observations.

(i) Any flip which is not a k-flip swaps two $\nearrow$ or two $\searrow$ in w, resulting in word w' identical to w.

(ii) Lemma 4 ensures that a crossing k-pair corresponds to an inversion in w. Thus, a k-flip exchanges the two symbols of an inversion in w, resulting in word w' with at least one inversion less than in w. $\qquad\square$

We now define $\Phi(M)$, the *potential* of M, as the sum of $\Phi_k(M)$, for k in $\{1, \ldots, n\}$. The following lemma presents the key properties of Φ.

Lemma 7. *The potential Φ takes integer values from 0 to $\binom{n}{2}\frac{n+4}{3}$, and decreases by at least 2 units at each flip.*

Proof. We know that Φ takes non-negative integer values by definition and, by Lemma 5, an upper bound on Φ is $\sum_{k=1}^{n}\left(k(n+1) - k^2 - 1\right) = \binom{n}{2}\frac{n+4}{3}$.

Finally, Lemma 6 ensures that Φ decreases by at least 2 units at each flip. Indeed, a flip of a pair $\langle i, j\rangle$ is counted at least twice: once in Φ_i as an i-flip, and once in Φ_j as a j-flip. $\qquad\square$

Theorem 3 follows from Lemma 7.

5   Lower Bounds

In this section, we consider the following two lower bounds. The proofs are presented in the full version.

Theorem 4. *In the red-on-a-line case, for even n, $\mathbf{D}(n) \geq \frac{3}{2}\binom{n}{2} - \frac{n}{4}$.*

Theorem 5. *In the convex case, for even n, $\mathbf{d}(n) \geq \frac{3n}{2} - 2$.*

To prove Theorem 4, it suffices to present a long untangle sequence. The initial matching of the sequence is represented in Fig. 13(a). To prove Theorem 5, we need to show that every untangle sequence starting at a given configuration (represented in Fig. 13(b)) is long enough. We do so, by showing that every flip reduces the number of crossings by exactly one unit.

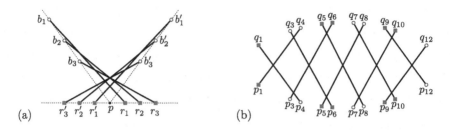

Fig. 13. (a) A 3-butterfly to lower bound $\mathbf{D}(6)$. (b) A 5-fence to lower bound $\mathbf{d}(10)$.

5.1 Lower Bound on $D(n)$

We provide a $2m$-segment red-on-a-line matching which we call an m-*butterfly*. There exists an untangle sequence starting at an m-butterfly of length $\frac{3}{2}\binom{2m}{2} - \frac{m}{2}$. Next, we give the precise definition of an m-butterfly and some intuition.

For an integer m, we define an m-*butterfly* as the following matching with $n = 2m$ segments. For i from 1 to m we have red points $r_i = (i/(m+1), 0)$ and $r'_i = (-i/(m+1), 0)$ as well as blue points $b_i = (i - (m+1), (m+1) - i)$ and $b'_i = ((m+1) - i, (m+1) - i)$. We match r_i to b_i and r'_i to b'_i. Next, we discuss important properties of an m-butterfly.

We call a red-on-a-line *convex* matching an n-*star* if all the $\binom{n}{2}$ pairs of segments cross. We say that an n-star *looks* at a point p if the blue points are all on a common line, and if p is the intersection of this line with the line of the red points. We also say that two red-blue point sets R, B and R', B' are *fully crossing* if all the pairs of segments of the form $\{rb, r'b'\}$ cross, where $(r, b, r', b') \in R \times B \times R' \times B'$. Two matchings are fully crossing if their underlying red-blue point sets are fully crossing. An m-butterfly is a red-on-a-line matching consisting of two fully crossing m-stars both looking at the same point $p = (0, 0)$ (Fig. 13(a) represents these properties but it is not drawn to scale).

In the following, we use the state tracking framework from Sect. 3 to describe how to come up with an untangle sequence starting at an m-butterfly with more than $\binom{2m}{2}$ flips. We consider a sequence of tracking choices with no $\mathbf{H} \rightarrow \mathbf{X}$ transition (Lemma 2) for the long untangle sequence we build. We take advantage of the non-convex position of the blue points to create flip situations such as in Fig. 10(a), where an $\mathbf{H}$-pair is turned into a $\mathbf{T}$-pair.

For instance, let us consider an $\mathbf{X}$-pair of one of the m-stars composing the m-butterfly. At some point of the untangle sequence, we flip this $\mathbf{X}$-pair, turning it into an $\mathbf{H}$-pair. Latter on, we turn this $\mathbf{H}$-pair into a $\mathbf{T}$-pair, as in Fig. 10(a). Still latter on, we turn this $\mathbf{T}$-pair into an $\mathbf{X}$-pair again, similarly to the pairs involving the horizontal segment in Fig. 1. This $\mathbf{X}$-pair will be flipped again.

We manage to carry out this whole process to flip twice all the $2\binom{m}{2}$ pairs of the two m-stars composing the m-butterfly while still having one flip for every other pair. In total, we reach $\frac{3}{2}\binom{2m}{2} - \frac{m}{2}$ flips.

5.2   Lower Bound on d(n)

We provide a convex red-blue matching which we call an *m-fence*, with $2m$ segments and $3m - 2$ crossings (Fig. 13(b)). All untangle sequences starting at an m-fence have length $3m - 2$, that is, each flip reduces the number of crossings by exactly one unit. Next, we give the precise definition of an m-fence and the idea of the proof.

Let $q_{2m+2}, q_{2m}, q_{2m-1}, \ldots, q_4, q_3, q_1, p_1, p_3, p_4, \ldots, p_{2m-1}, p_{2m}, p_{2m+2}$ be $4m$ points in convex position, ordered counter-clockwise, and with colors alternating every two points (Fig. 13(b)). More precisely, points p_i, q_i are red if $i \equiv 1, 2$ mod 4 and blue otherwise. We deliberately avoid using the indices 2 and $2m + 1$ to simplify the description. The segments of an m-fence are the $p_i q_{i+3}$ and the $q_i p_{i+3}$ where i is odd and varies between 1 and $2m - 1$.

For $1 \leq k \leq m + 1$, the *k-th column* consists of the at most 4 points with indices $2k - 1$ and $2k$. We say that a convex red-blue matching with the same point set as an m-fence is a *derived m-fence* if, for all $k \in \{2, \ldots, m\}$, for all $w \in \{p, q\}$, one of the following statements holds:

1. w_{2k-1} is matched to a point of the $(k - 1)$-th column, and w_{2k} is matched to a point of the $(k + 1)$-th column, or
2. w_{2k-1} is matched to a point of the $(k + 1)$-th column, and w_{2k} is matched to a point of the $(k - 1)$-th column.

Two examples of derived m-fences are presented in Fig. 14. Note that an m-fence is in particular a derived m-fence. To prove Theorem 5, we first show that a flip changes a derived m-fence into another derived m-fence. Finally, we show that a flip of a derived m-fence reduces its number of crossings by exactly one unit.

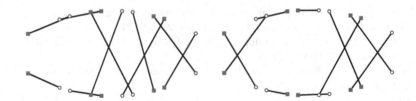

Fig. 14. Two examples of derived 5-fences.

6   Concluding Remarks

Untangle sequences of TSP tours have been investigated since the 80s, when a cubic upper bound on $D(n)$ has been discovered [22]. This bound also holds for matchings (even non-bipartite ones) and has not been improved ever since. Except for the convex case, there are big gaps between the lower and upper bounds, as can be seen in Table 1. Experiments on tours and matchings have shown that, in all cases tested, the cubic upper bound is not tight and the lower bounds seem to be asymptotically tight.

Untangle sequences have many unexpected properties which make the problem harder than it seems at first sight. The following questions remain open.

1. If we add a new segment to a crossing-free matching, what is the maximum length of an untangle sequence? Notice that an $o(n^2)$ bound would lead to an $o(n^3)$ bound for $\mathbf{d}(n)$.
2. Is it always possible to find an untangle sequence that does not flip the same pair of segments twice? Using a balancing argument, we can show that the number of *distinct* flips in any untangle sequence is $O(n^{8/3})$.
3. What is the maximum number of flips involving a given point? The cubic potential provides a quadratic bound which leads again to an $O(n^3)$ bound for $\mathbf{D}(n)$.

We proved the NP-hardness of computing the shortest untangle sequence for a red-blue matching. What is the complexity of computing the shortest untangle sequence for a TSP tour, for a red-on-a-line matching, or even for a convex instance? What about the longest untangle sequence?

References

1. Aichholzer, O., Mulzer, W., Pilz, A.: Flip distance between triangulations of a simple polygon is NP-complete. Discrete Comput. Geom. **54**(2), 368–389 (2015)
2. Akiyama, J., Alon, N.: Disjoint simplices and geometric hypergraphs. In: Third International Conference on Combinatorial Mathematics, pp. 1–3 (1989)
3. Atallah, M.J.: A matching problem in the plane. J. Comput. Syst. Sci. **31**(1), 63–70 (1985)
4. Bereg, S., Ito, H.: Transforming graphs with the same degree sequence. In: Computational Geometry and Graph Theory, pp. 25–32 (2008)
5. Bereg, S., Ito, H.: Transforming graphs with the same graphic sequence. J. Inf. Process. **25**, 627–633 (2017)
6. Biniaz, A., Maheshwari, A., Smid, M.: Flip distance to some plane configurations. Comput. Geom. **81**, 12–21 (2019)
7. Bonamy, M., et al.: The perfect matching reconfiguration problem. In: 44th International Symposium on Mathematical Foundations of Computer Science. LIPIcs, vol. 138, pp. 80:1–80:14 (2019)
8. Bonnet, É., Miltzow, T.: Flip distance to a Non-crossing perfect matching. Computing Research Repository abs/1601.05989 (2016)
9. Bose, P., Hurtado, F.: Flips in planar graphs. Comput. Geom. **42**(1), 60–80 (2009)
10. Bousquet, N., Joffard, A.: Approximating shortest connected graph transformation for trees. In: Chatzigeorgiou, A., Dondi, R., Herodotou, H., Kapoutsis, C., Manolopoulos, Y., Papadopoulos, G.A., Sikora, F. (eds.) SOFSEM 2020: Theory and Practice of Computer Science: 46th International Conference on Current Trends in Theory and Practice of Informatics, SOFSEM 2020, Limassol, Cyprus, January 20–24, 2020, Proceedings, pp. 76–87. Springer International Publishing, Cham (2020). https://doi.org/10.1007/978-3-030-38919-2_7

11. De Berg, M., Khosravi, A.: Optimal binary space partitions for segments in the plane. Int. J. Comput. Geom. Appl. **22**(03), 187–205 (2012)
12. Englert, M., Röglin, H., Vöcking, B.: Worst case and probabilistic analysis of the 2-Opt algorithm for the TSP. Algorithmica **68**(1), 190–264 (2014)
13. Erdős, P.L., Király, Z., Miklós, I.: On the swap-distances of different realizations of a graphical degree sequence. Comb. Prob. and Comput. **22**(3), 366–383 (2013)
14. Hakimi, S.L.: On realizability of a set of integers as degrees of the vertices of a linear graph i. J. Soc. Ind. Appl. Math. **10**(3), 496–506 (1962)
15. Hakimi, S.L.: On realizability of a set of integers as degrees of the vertices of a linear graph ii. uniqueness. J. Soc. Ind. Appl. Math. **11**(1), 135–147 (1963)
16. Hershberger, J., Suri, S.: Applications of a semi-dynamic convex hull algorithm. BIT Numer. Math. **32**(2), 249–267 (1992)
17. van den Heuvel, J.: The complexity of change. Surv. Comb. **409**, 127–160 (2013)
18. Hurtado, F., Noy, M., Urrutia, J.: Flipping edges in triangulations. Discrete Comput. Geom. **22**(3), 333–346 (1999)
19. Ito, T., et al.: On the complexity of reconfiguration problems. Theor. Comput. Sci. **412**(12), 1054–1065 (2011)
20. Joffard, A.: Graph domination and reconfiguration problems. Ph.D. thesis, Université Claude Bernard Lyon 1 (2020)
21. Lawson, C.L.: Transforming triangulations. Discrete Math. **3**(4), 365–372 (1972)
22. van Leeuwen, J.: Untangling a traveling salesman tour in the plane. In: 7th Workshop on Graph-Theoretic Concepts in Computer Science (1981)
23. Lubiw, A., Pathak, V.: Flip distance between two triangulations of a point set is NP-complete. Comput. Geom. **49**, 17–23 (2015)
24. Nishimura, N.: Introduction to reconfiguration. Algorithms **11**(4) (2018)
25. Oda, Y., Watanabe, M.: The number of flips required to obtain non-crossing convex cycles. In: Kyoto International Conference on Computational Geometry and Graph Theory, pp. 155–165 (2007)
26. Pilz, A.: Flip distance between triangulations of a planar point set is apx-hard. Comput. Geom. **47**(5), 589–604 (2014)
27. Will, T.G.: Switching distance between graphs with the same degrees. SIAM J. Discrete Math. **12**(3), 298–306 (1999)
28. Wu, R., Chang, J., Lin, J.: On the maximum switching number to obtain non-crossing convex cycles. In: 26th Workshop on Combinatorial Mathematics and Computation Theory, pp. 266–273 (2009)

On Vertex Guarding Staircase Polygons

Matt Gibson-Lopez[1], Erik Krohn[2], Bengt J. Nilsson[3], Matthew Rayford[2],
Sean Soderman[1(✉)], and Paweł Żyliński[4]

[1] The University of Texas at San Antonio, San Antonio, TX, USA
{matthew.gibson,sean.soderman}@utsa.edu
[2] University of Wisconsin - Oshkosh, Oshkosh, WI, USA
krohne@uwosh.edu
[3] Malmö University, Malmö, Sweden
bengt.nilsson.TS@mau.se
[4] University of Gdańsk, Gdańsk, Poland
zylinski@inf.ug.edu.pl

Abstract. In this paper, we consider the variant of the art gallery problem where the input polygon is a staircase polygon. Previous works on this problem gave a 2-approximation for point guarding a staircase polygon (where guards can be placed anywhere in the interior of the polygon and we wish to guard the entire polygon). It is still unknown whether this point guarding variant is NP-hard. In this paper we consider the vertex guarding problem, where guards are only allowed to be placed at the vertices of the polygon, and we wish to guard only the vertices of the polygon. We show that this problem is NP-hard, and we give a polynomial-time 2-approximation algorithm.

1 Introduction

A polygon that does not have any holes is said to be a *simple polygon*. The boundary of a simple polygon is a Jordan curve that separates the plane into two regions: *inside* the polygon and *outside* the polygon. For any two points u, v inside a simple polygon, we say u sees v if and only if the line segment connecting u and v does not go outside the polygon.

In the famous *art gallery problem*, we are given a simple polygon P, and we wish to compute a minimum number of *guards* that collectively see P. There are several variants of the art gallery problem that have been considered. In the *point guarding* variant, any point inside P is eligible to be chosen as a guard. In the *vertex guarding* variant, guards are only allowed to be the vertices of the polygon. Some variants require that the guards see the entire polygon, and other variants require that the guards see some subset of the polygon (e.g., just the vertices of the polygon).

The art gallery problem for simple polygons has proven to be difficult to obtain tight bounds on polynomial-time approximation algorithms. There have been several hardness results [1,9], and the problem is known to be APX-hard [3]. There is an $O(\log OPT)$ approximation algorithm for point guarding [2].

Due to the difficulty with the simple polygon setting, more restricted versions of polygons have been considered. A simple polygon such that a vertical line

© Springer Nature Switzerland AG 2022
A. Castañeda and F. Rodríguez-Henríquez (Eds.): LATIN 2022, LNCS 13568, pp. 746–760, 2022.
https://doi.org/10.1007/978-3-031-20624-5_45

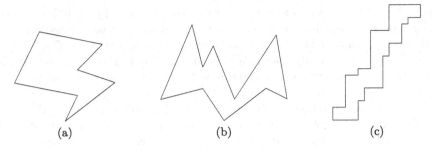

Fig. 1. (a) A simple polygon that is not x-monotone. (b): An x-monotone polygon. (c) A staircase polygon.

intersects the boundary of the polygon in at most two points is said to be an *x-monotone polygon*, or simply, a monotone polygon. See Fig. 1 (b). We call the leftmost vertex of a monotone polygon $\mathcal{L}$ and the rightmost point of a monotone polygon $\mathcal{R}$. Then the boundary of the monotone polygon can be partitioned into two x-monotone polygonal chains: the *ceiling* and the *floor*. The ceiling (resp. floor) is defined as the boundary of P when walking clockwise (resp. counterclockwise) from $\mathcal{L}$ to $\mathcal{R}$. Note that a vertical line can intersect the ceiling at most once and the floor at most once. For monotone polygons, both the vertex guarding and point guarding variants are known to be NP-hard [7,8], and there is an $O(1)$-approximation algorithm for point guarding a monotone polygon [8]. For vertex guarding a monotone polygon, there is no $O(1)$-approximation algorithm currently known.

The art gallery problem has also been considered in the special case where we wish to guard an x-monotone polygonal chain by itself, commonly referred to as *terrains*. In this context, we say two points u and v on the terrain see each other if the line segment connecting them does not go under the terrain. In this setting, the problem is known to be NP-hard [6] and there is a PTAS [4], so the complexity of guarding terrains is settled.

In this paper, we consider the art gallery problem in the context where the polygon we wish to guard is a *staircase polygon*. Similarly to monotone polygons, we can partition the boundary of a staircase polygon into a ceiling and a floor, but instead of the ceiling and floor being x-monotone polygonal chains, they are "staircases". See Fig. 1 (c). Abusing notation, we let $\mathcal{L}$ denote the lower left vertex and let $\mathcal{R}$ denote the upper right vertex. These vertices are on both the ceiling and the floor, and every other vertex will either be only on the floor or on the ceiling. Let $f_1, f_2, \ldots, f_k$ denote the vertices of the floor (not including $\mathcal{L}$ and $\mathcal{R}$). For any point p, we let $p.x$ denote the x-coordinate of p and let $p.y$ denote the y-coordinate of p. Then $f_1, \ldots f_k$ are such that $f_1.y = \mathcal{L}.y$ and $f_1.x > \mathcal{L}.x$, f_i for even i satisfies $f_i.x = f_{i-1}.x$ and $f_i.y > f_{i-1}.y$, and f_i for odd $i > 1$ satisfies $f_i.y = f_{i-1}.y$ and $f_i.x > f_{i-1}.x$. Additionally f_k satisfies $f_k.x = \mathcal{R}.x$ and $f_k.y < \mathcal{R}.y$ (note k must be odd). Intuitively, if we walk along the floor from $\mathcal{L}$ to $\mathcal{R}$, we alternate between walking horizontally to the right and then straight up; each "turn" is a 90 degree angle. We call f_i such that i is odd *convex* vertices and we call f_i such that i is even *reflex* vertices. The ceiling vertices (other than $\mathcal{L}$ and $\mathcal{R}$) can be denoted $c_1 \ldots, c_{k'}$ and satisfy symmetric properties to the floor except the walk from $\mathcal{L}$ to $\mathcal{R}$ along the ceiling starts vertically upward and finishes horizontally to the right. We also have that c_i for odd i are convex vertices and c_i for even i are reflex vertices.

There is one previous paper [5] on guarding staircase polygons which presented a 2-approximation for point guarding a staircase polygon. The point guarding variant is not known to be NP-hard, and the $O(1)$-approximation for point guarding monotone polygons can also be applied to the staircase setting, so the main contribution here was to reduce the approximation ratio for the staircase setting.

1.1   Our Results

In this paper, we consider the vertex guarding variant of the art gallery problem when the polygon to be guarded is a staircase polygon. Unlike for the point guarding variant, there is no previously-known $O(1)$-approximation algorithm for monotone polygons that can be applied here, so previously the best known solution for this is $O(\log OPT)$ from simple polygons. In this paper, we present a polynomial-time 2-approximation for this problem as well as a prove the problem is NP-hard. Our algorithm applies to when we want to guard the entire polygon as well as when we want to guard only the vertices of the polygon.

Theorem 1. *There is a polynomial-time 2-approximation algorithm for vertex guarding any staircase polygon.*

Theorem 2. *Given a staircase polygon P and an integer k, it is NP-hard to determine if there is subset of vertices of P of size at most k that sees all of the vertices of P.*

2   Preliminaries

In this section we state some observations about staircase polygons that will be used in the paper. The first is about the visibility of a convex vertex. First note that we do not consider $\mathcal{L}$ and $\mathcal{R}$ as convex vertices in this analysis. Consider any convex vertex f on the floor of a staircase polygon P. It is adjacent to a vertical edge on the boundary of P that extends up from it. This edge will prevent f from seeing any point $p \in P$ such that $p.x > f.x$. Similarly, f is adjacent to a horizontal edge that extends to its left. This edge will prevent f from seeing any $p \in P$ such that $p.y < f.y$. And finally, it is easy to see that for any $p \in P$ that satisfies $p.x \leq f.x$ and $p.y \geq f.y$, then f *must* see p. A symmetric argument holds for ceiling convex vertices. So we have the following observations.

Observation 3. *A convex vertex f on the floor of a staircase polygon P sees a point $p \in P$ if and only if $p.x \leq f.x$ and $p.y \geq f.y$.*

Observation 4. *A convex vertex c on the ceiling of a staircase polygon P sees a point $p \in P$ if and only if $p.x \geq c.x$ and $p.y \leq c.y$.*

Another observation that we will use involves points that are blocked from the ceiling by a floor point or vice versa. Let p be any point in a staircase polygon P, and let c be any point on the ceiling of P that p does not see. Without loss of generality, assume that $p.x < c.x$. If there is a floor vertex f that is over the line segment $\overline{p,c}$, then p will not be able to see any point q such that $q.x \geq c.x$ or $q.y \geq c.y$. Indeed, if the line segment $\overline{p,q}$ passes under f, then f blocks p from q. If $\overline{p,q}$ passes over q, then c will be under $\overline{p,q}$. The symmetric argument for when a ceiling vertex blocks a point from some floor point holds as well.

3   Approximation Algorithm

In this section, we give a polynomial-time 2-approximation algorithm for vertex guarding the vertices of any staircase polygon P. In our algorithm, all of the guards that we pick will be convex vertices of P. We will not consider the lower left vertex $\mathcal{L}$ or the upper right vertex $\mathcal{R}$ to be convex vertices here. Let $F = \{f_1, f_3, f_5, \ldots, f_{k-2}, f_k\}$ denote the convex vertices on the floor from left-to-right, and let $C = \{c_1, c_3, \ldots, c_{k'-2}, c_{k'}\}$ denote the convex vertices on the ceiling from left-to-right. Note that there are no reflex vertices in either F or C. Also note that since all of the indices of vertices in F and C are odd, the "next" vertex after f_i in F (resp. after c_i in C) is f_{i+2} (resp. c_{i+2}). To aid in the analysis, our algorithm computes two disjoint sets of guards: red guards R and blue guards B. Our final guard set is then $R \cup B$. The algorithm is formally stated in Algorithm 1.

3.1   Algorithm Correctness

Let us index the guards in R as $\{r_1, r_2, r_3, \ldots\}$ where the index corresponds to the order we added the guard to R. Note that the algorithm alternates adding vertices in F and C, so each r_j such that j is odd is from F, and each r_j such that j is even is from C. We will argue that whenever we add a guard r to R, that $R \cup B$ sees every $p \in P$ such that $p.x \leq r.x$ or $p.y \leq r.y$. We will argue inductively. In the base case when we have $r_1 = f_1$, we get that this holds true by Observation 3 as there are no $p \in P$ such that $p.y < f_1.y$. So now suppose that the claim holds true for the last red guard r_j we added to R, and we will show that it will hold true after we add the next red guard r_{j+1} to R. There are two symmetric cases to consider depending on if r_j is on the floor or is on the ceiling.

Floor to Ceiling. Suppose r_j is on the floor and the next red guard, r_{j+1} is on the ceiling. We know that all points p in the region with $p.x \leq r_j.x$ and $p.y \geq r_j.y$ are covered by r_j. Likewise, we know that all points q such that $q.x \geq r_{j+1}.x$ and $q.y \leq r_{j+1}.y$ are covered. Now consider the points unseen by the red guards, u, with $u.x > r_j.x$ and $u.x < r_{j+1}.x$. These points have a maximum y value of $v.y$, where v is the ceiling reflex vertex incident to r_{j+1}

```
R = ∅;
B = ∅;
r ← f₁ ;
R ← R ∪ {r};
while R ∪ B doesn't guard all of P do
    if r ∈ F then
        cᵢ ← rightmost vertex c ∈ C s.t. c.x ≤ r.x;
        B ← B ∪ {cᵢ};
        if i < k' then
            r ← cᵢ₊₂;
            R ← R ∪ {r};
        end
    else
        if r ∈ C then
            fᵢ ← highest vertex f ∈ F s.t. f.y ≤ r.y;
            B ← B ∪ {fᵢ};
            if i < k then
                r ← fᵢ₊₂;
                R ← R ∪ {r};
            end
        end
    end
end
return R ∪ B
```

Algorithm 1: Guarding algorithm

with $v.x = r_{j+1}.x$. This is because the points to the right of r_{j+1} are covered by r_{j+1}. Now consider the blue guard b that was placed along with r_{j+1}. The b guard is on the rightmost convex ceiling vertex c_i such that $c_i.x \leq r_j.x$. As c_{i+2} is where we placed r_{j+1}, c_i is incident to v, so $b.y = v.y$. Since b is on a convex ceiling vertex and $u.y \leq v.y = b.y$, the u points are covered by b (Fig. 2).

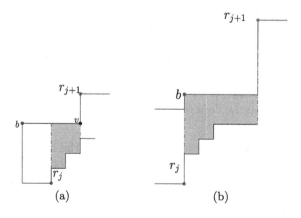

Fig. 2. (a): Illustration of b covering the blue region left unseen by the red guards. (b): The same as (a), except when $r_j.x = b.x$. (Color figure online)

Ceiling to Floor. Suppose r_j is on the ceiling and the next red guard, r_{j+1} is on the floor. There is a region of unseen points between these guards, where any such point u satisfies $u.y > r_j.y$ and $u.y < r_{j+1}.y$. For such u, $u.x \leq v.x$, where v is the floor reflex vertex incident to r_{j+1} with $v.y = r_{j+1}.y$. Since our blue guard, b, is placed on the highest floor convex vertex f_i such that $b.y \leq r_j.y$, and f_{i+2} is where r_{j+1} sits, b is incident to v. Thus, $u.x \leq v.x = b.x$ and any such point u is seen by b.

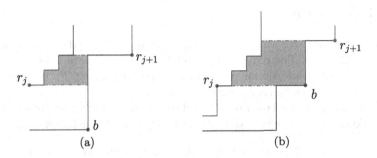

Fig. 3. (a): Illustration of b covering the blue region left unseen by the red guards, (b): The same as (a) except when $r_j.y = b.y$. (Color figure online)

Termination. Consider the last red guard r we place. By the inductive argument, we know we see every point $p \in P$ such that $p.x \leq r.x$ or $p.y \leq r.y$. But there may be $p \in P$ such that $p.x > r.x$ and $p.y > r.y$. We argue that the last blue guard we add will see all such points.

We can see this by noting the the last blue guard b must be the last blue convex vertex on its side of P. For example, if b is on the ceiling, then there cannot be another convex vertex on the ceiling or it would have been another red guard. This means that b must be adjacent to $\mathcal{R}$ whether it is on the ceiling or the floor. If b is on the ceiling then there are no $p \in P$ with $p.y > b.y$. If b is on the floor, then there are no $p \in P$ with $p.x > b.x$. It follows that b sees the rest of the polygon (Fig. 3).

3.2 Approximation Ratio

Let OPT be an optimal solution for guarding P. We will show that our solution $R \cup B$ contains at most twice the number of guards of OPT.

The key observation is that every point in P sees at most one red guard from our solution. Indeed, it follows from the algorithm that if r_i and r_{i+1} are two consecutive red guards from our algorithm, then it must be that $r_i.x < r_{i+1}.x$ and $r_i.y < r_{i+1}.y$. Then by our observations on the visibilities of convex vertices, there is no point that can see r_i and r_{i+1} at the same time. It then follows that no point can see any pair of red guards at the same time, and therefore it must be that $|OPT| \geq |R|$.

We then complete the argument by showing that $|B|$ is either $|R|$ or $|R| - 1$. When we place a blue guard, we can charge this guard to the red guard we just placed before it. Every red guard will be charged exactly once, perhaps with the exception of the very last red guard placed. If the last red guard resulted in the entire P to be guarded, then we do not add a blue guard later. Either way, our final solution either has size $2|R|$ or $2|R| - 1$. Therefore our algorithm is a 2-approximation. This completes the proof of Theorem 1.

4   NP-hardness

In this section we will show that vertex guarding the vertices of a staircase polygon is NP-hard. The reduction is from vertex guarding a monotone polygon that satisfies some additional properties. In particular, let P_1 be a monotone polygon with vertices $v_1, \ldots v_n$ indexed such that $v_1.x < v_2.x < \cdots < v_n.x$. Note that the indices are based only on x-coordinate and not how they are encountered when walking around the boundary. Suppose P_1 satisfies the following properties:

1. v_i such that i is odd are on the ceiling and v_i such that i is even are on the floor.
2. For every $i \in \{2, \ldots, n\}$, we have $v_i.x - v_{i-1}.x = 1$.

It is already known that vertex guarding a monotone polygon (where the entire polygon must be seen) is NP-hard [8], and we show in the appendix that we can extend this to vertex guarding the vertices of a monotone polygon that satisfies the additional properties by adding additional vertices to the boundary of the construction used in [7] so that the polygon satisfies properties 1 and 2 above.

The first step of our reduction is to apply a linear transformation to the x-monotone polygon P_1 to obtain a polygon P_2 that is *both* x-monotone *and* y-monotone in a manner such that the visibilities are preserved. That is, two vertices u and v of P_2 see each other if and only if their corresponding vertices of P_1 see each other.

4.1   Transforming Any X-Monotone Polygon into One that Is X-Monotone and Y-Monotone

Let P denote any monotone polygon and let G be its *visibility graph*. That is, the vertices of G correspond to the vertices of P, and an edge connects two vertices of G if and only if their corresponding vertices of P see each other. We will show how to convert P into a polygon $T(P)$ that is both x-monotone and y-monotone, and such that G also is the visibility graph of $T(P)$.

The transformation is as follows. Let M be any scalar such that $M > \frac{v_{i-1}.y - v_i.y}{v_i.x - v_{i-1}.x}$ for every $i \in \{2, \ldots, n\}$. We construct $T(P)$ by taking each vertex v_i of P and creating a vertex t_i such that $t_i.x := v_i.x$ and $t_i.y := M \cdot v_i.x + v_i.y$. The sequence of vertices on the boundary of $T(P)$ is the same as in P. See Fig. 4 for an illustration.

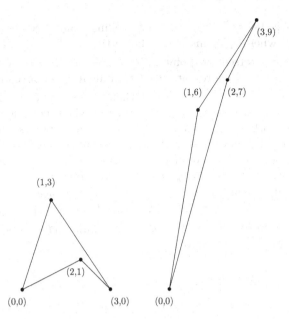

Fig. 4. Left: an x-monotone polygon P. Right: a transformed polygon $T(P)$ that is x-monotone and y-monotone. We used $M = 3$ for $T(P)$.

Lemma 1. *For any monotone polygon P with visibility graph G, $T(P)$ is x-monotone and y-monotone and G is also the visibility graph of $T(P)$.*

Proof. Let $t_1, \ldots, t_n$ denote the vertices of $T(P)$ such that $t_{i-1}.x < t_i.x$ for each $i \in \{2, \ldots, n\}$. Clearly $T(P)$ is x-monotone as the x-coordinates of the vertices of $T(P)$ are exactly the same as those in P. To see that $T(P)$ is y-monotone, we will show that $t_i.y - t_{i-1}.y > 0$.

$$
\begin{aligned}
t_i.y - t_{i-1}.y &= (M \cdot v_i.x + v_i.y) - (M \cdot v_{i-1}.x + v_{i-1}.y) \\
&= M \cdot (v_i.x - v_{i-1}.x) + v_i.y - v_{i-1}.y \\
&> \frac{v_{i-1}.y - v_i.y}{v_i.x - v_{i-1}.x} \cdot (v_i.x - v_{i-1}.x) + v_i.y - v_{i-1}.y \\
&= v_{i-1}.y - v_i.y + v_i.y - v_{i-1}.y \\
&= 0
\end{aligned}
$$

Therefore $T(P)$ is both x-monotone and y-monotone.

We now show that this mapping preserves the visibilities. Consider two vertices v_i and v_k of P such that $i < k$. In a monotone polygon, whether v_i and v_k see each other or not is only depends upon the vertices v_j such that $i < j < k$. In particular, v_i and v_k will see each other if and only if every floor point between them is below $\overline{v_i, v_k}$ and every ceiling point between them is above $\overline{v_i, v_k}$. Consider where $\overline{v_i, v_k}$ intersects the vertical line $x = v_j.x$ for a j satisfying $i < j < k$. This intersection point is $(v_j.x, \frac{(v_k.x - v_j.x) \cdot v_i.y + (v_j.x - v_i.x) \cdot v_k.y}{v_k.x - v_i.x})$. We can then subtract the y-coordinates to determine if v_j is above or below $\overline{v_i, v_k}$. If $\frac{(v_k.x - v_j.x) \cdot v_i.y + (v_j.x - v_i.x) \cdot v_k.y}{v_k.x - v_i.x} - v_j.y > 0$, then v_j is below $\overline{v_i, v_k}$. If the difference is negative then v_j is above $\overline{v_i, v_k}$. If the difference is 0 then the three points are colinear. Some work assumes that with three colinear points, the middle point blocks the outer two and some work assumes that it does not block them; our transformation and hardness reduction works for either definition. So this vertical distance between v_j and $\overline{v_i, v_k}$ is critical to determining if v_i sees v_k. Let us call this vertical distance $d_1(i, j, k)$. Similarly, the vertical distance between t_j and $\overline{t_i, t_k}$ in $T(p)$ is $\frac{(t_k.x - t_j.x) \cdot t_i.y + (t_j.x - t_i.x) \cdot t_k.y}{t_k.x - t_i.x} - t_j.y$. Let us call this distance $d_2(i, j, k)$. We claim that $d_1(i, j, k) = d_2(i, j, k)$, implying v_j is below $\overline{v_i, v_k}$ if and only if t_j is below $\overline{t_i, t_k}$. We have

$$
\begin{aligned}
d_2(i, j, k) &= \frac{(t_k.x - t_j.x) \cdot t_i.y + (t_j.x - t_i.x) \cdot t_k.y}{t_k.x - t_i.x} - t_j.y \\
&= \frac{(v_k.x - v_j.x)(M v_i.x + v_i.y) + (v_j.x - v_i.x)(M v_k.x + v_k.y)}{v_k.x - v_i.x} - (M v_j.x + v_j.y) \\
&= \frac{(v_k.x - v_j.x) v_i.y + (v_j.x - v_i.x) v_k.y}{v_k.x - v_i.x} - v_j.y \\
&= d_1(i, j, k)
\end{aligned}
$$

Then this implies that v_i sees v_k if and only if t_i sees t_k. Indeed, if v_i sees v_k then every floor point between them is below $\overline{v_i, v_k}$ (and the corresponding floor point in $T(P)$ will be below $\overline{t_i, t_k}$) and every ceiling point between them is above $\overline{v_i, v_k}$ (and the corresponding ceiling point in $T(P)$ will be above $\overline{t_i, t_k}$). Similarly, if v_i does not see v_k, then there is some floor point that is above $\overline{v_i, v_k}$ or a ceiling point that is below $\overline{v_i, v_k}$ which will block v_i from v_k. The same blocking point will block t_i and t_k in $T(P)$.

We remark that this transformation applies to terrains as well. Both the NP-hardness proofs for monotone polygons [7,8] and terrains [6] featured a steep canyon where gadgets passed information back and forth across the canyon. The canyon seemed to be a key feature of the reduction, but as this transformation shows, the hardness constructions could have been drawn in a y-monotone manner as well.

4.2   Our Reduction

We are given a monotone polygon P_1 that satisfies the two properties listed at the beginning of this section, and we are given an integer k. We want to know if there is a subset of vertices O of P_1 of size at most k such that every vertex of P_1 sees at least one vertex of O. This problem is NP-hard (proof omitted due to lack of space). We will reduce this problem to the staircase polygon vertex guarding problem. By translating the polygon, we can assume without loss of generality that the largest y-coordinate of P_1 is α, the smallest y-coordinate of P_1 is $-\alpha$, and $v_i.x = i$ for each vertex. We apply the transformation to P_1 to obtain a polygon P_2 that is both x-monotone and y-monotone. Note that since the difference in x-coordinates of v_i and v_{i-1} is 1, we then have that any $M > 2\alpha$ will suffice to make P_2 be y-monotone. We pick M to be 180α. Since the visibilities are preserved, we have that P_2 can be guarded with at most k guards if and only if P_1 can be guarded with at most k guards. Moreover, P_2 satisfies the same two properties that we assumed P_1 satisfies.

The next step is to create a staircase polygon S that uses vertices at the exact same locations as the vertices of P_2. We call the vertices of S that correspond to vertices of P_2 the *gold* vertices. All of the gold vertices will be reflex vertices of S. We then add additional vertices to the gold vertices so that the boundary satisfies the conditions of a staircase polygon and the added vertices do not alter the visibilities of any two gold vertices. In particular, if two vertices in P_2 see each other, then their corresponding gold vertices will not be blocked by any of the added vertices and therefore will still see each other. To achieve this, let's make an observation. The entire polygon P_1 is contained between the horizontal lines $y = \alpha$ and $y = -\alpha$. Then P_2 will be contained between the lines $y = (180\alpha) \cdot x + \alpha$ and $y = (180\alpha) \cdot x - \alpha$ (seen by applying the transformation to the points on the lines $y = \alpha$ and $y = -\alpha$ respectively). We call the region between these lines the *strip*. Note that the line segment connecting two gold vertices that see each other stays inside the strip. We ensure that all of the extra vertices we add to S are outside the strip, and therefore they cannot block two gold vertices who see each other in P_2. Note that it suffices to describe just the reflex vertices of S, as if we know two consecutive floor reflex vertices of S, then we know exactly where the convex vertex between them must be.

Consider some v_i floor gold vertex. Due to the properties P_2 satisfies, it follows that v_{i-1} exists and is a ceiling gold vertex. We will now describe a gadget that we will place on the ceiling between v_{i-1} and v_i that is above the strip, as well as a symmetric gadget that we will place on the floor between v_{i-1} and v_i that is below the strip. The intuition of the gadgets is that it allows us to turn this portion of the boundary of S into a staircase, and we will show that we can "force" a guard placement of S to have to include a specific vertex on the floor gadget that will see all of the vertices in the ceiling gadget without seeing any of the gold vertices, and similarly we will force a guard placement of S to include a specific vertex on the ceiling gadget that will see all of the vertices we add to the floor gadget without seeing any of the gold vertices. The intuition then is that after we place guards at all of the forced locations, the only unseen vertices will be the gold vertices which will essentially require us to find a set of gold vertices of size k to guard just the gold vertices (the technical details are a bit more complicated than this as there will have to be some non-gold vertices that can see some gold vertices, but this is the high level idea).

High Level Overview of the Gadget. There is one gadget that we use to convert P_2 into a staircase polygon S. Between any two gold vertices v_{i-1} and v_i, we will have one gadget on the floor, and a symmetric gadget on the ceiling. Each copy of the gadget contains six key points that are carefully placed to ensure that they see exactly what we want them to see. The six points can be broken up into three pairs of points that we color code for ease of description. We have two green points on the "outside" of the gadget, then nested between the green points are two purple points, then nested between the purple points are two blue points. See Fig. 5 for an illustration. All 6 of these points will be reflex vertices of S. Between the purple and blue vertices are a set of $\Theta(n)$ staircase steps that we call C_i^1 and C_i^2, where n is the number of vertices of P_1. Finally there is a convex vertex f_i^* between the two blue vertices that plays a critical role in the reduction.

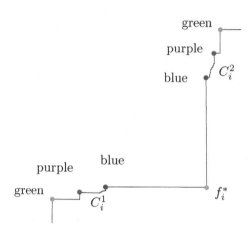

Fig. 5. The floor gadget. The coordinates shown here are not exact. (Color figure online)

Now we describe the interaction between the floor gadget and the ceiling gadget that are between the same two gold vertices. See Fig. 6 (a). The ceiling gadget is the same as the floor gadget but "mirrored". That is, the x-coordinate of the rightmost green vertex of the floor gadget (relative to v_i) is the same as the leftmost green vertex of the ceiling gadget (relative to v_{i-1}). In Fig. 6 (a), the thick diagonal lines are the bounding lines of the strip. The vertical dotted lines show the x-coordinates of v_{i-1} and v_i, and the horizontal dotted lines show the y-coordinates of v_{i-1} and v_i if they were in the exact middle of the strip (they could be up to α above or below those lines). The dashed lines are the "midpoint" lines of the dotted lines, breaking the plane up into four quadrants. Let C_i^3 and C_i^4 denote the staircase steps on the ceiling gadget. C_i^1 and C_i^3 are in the bottom left quadrant, and C_i^2 and C_i^4 are in the upper right quadrant. The convex vertices between the blue vertices are the only vertices in the top left and bottom right quadrants. They are positioned left-to-right so that C_i^3 is entirely to the left of C_i^1, which is entirely to the left of C_i^4, which is entirely to the left of C_i^2. Then from bottom-to-top they are positioned so that C_i^1 is entirely below C_i^3, which is entirely below C_i^2, which is entirely below C_i^4. What this placement does is ensure that a guard placed at any vertex of C_i^1 will see the convex vertices of C_i^3, but it will not see *any* convex vertices of C_i^4 (same vice versa). Similarly, a guard placed at any vertex of C_i^2 will see all of the convex vertices of C_i^4, but it will not see *any* convex vertices of C_i^3 (same vice versa). However, the convex vertex f_i^* will see the *entire* ceiling gadget. The intuition is that this "forces" a solution to place a guard at f_i^* to guard the ceiling gadget, and symmetrically we must place a guard at the corresponding ceiling gadget convex vertex to guard the entire floor gadget. f_i^* will then see every vertex on the ceiling from the convex vertex above v_{i-1} through the convex vertex above the right green vertex of the ceiling gadget.

It becomes important to keep track of which gold vertices can be seen from gadget vertices. We show that the only gadget vertices that can see gold vertices are the green vertices as well as the convex vertices "outside" the green vertices (no vertices between the green vertices can see any gold vertices). However we show that the gold vertices that can be seen from such a vertex is a subset of what a nearby gold vertex sees, so it ends up not being a problem (it is here where we use the fact that P_2 satisfies properties 1 and 2).

Now we overview how we connect all the gadgets together with the gold vertices to obtain our staircase polygon S. Consider a floor gadget between two gold vertices v_{i-1} and v_i. One of these gold vertices is a floor vertex (they could be $\mathcal{L}$ and/or $\mathcal{R}$ which can be viewed as both being ceiling and floor vertices here). Without loss of generality, assume v_i is a floor vertex. Then the rightmost green vertex of the gadget connects to v_i with a single convex vertex in between, and the leftmost green vertex of the gadget connects to the rightmost green vertex of another floor gadget between v_{i-2} and v_{i-1} (again with a convex vertex between them). The floor gadgets between v_{i-2} and v_{i-1} are mirrored for symmetry. The ceiling gadgets are connected similarly. See Fig. 6 (b). The black vertices in the figure represent the f_i^* vertices where we are "forced" to place guards. These black vertices collectively will see every vertex of S except for the gold vertices, which essentially forces us to have to find a solution that will guard P_2.

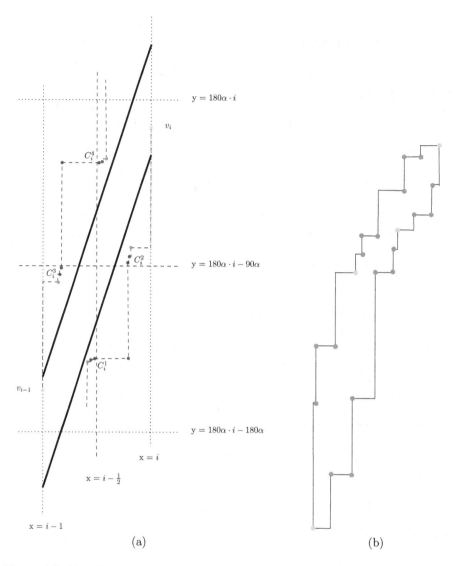

Fig. 6. (a) The relative positioning of the floor and ceiling gadgets. Note that the coordinates are not exact and the visibilities of the gold vertices and the gadget vertices are not correct here. In the actual construction, given the width of the strip, the picture would be much steeper. (b) Connecting gadgets to each other and gold vertices. For simplicity, only green points are shown in the gadget. Coordinates are not exact here. The actual construction is much steeper. (Color figure online)

Green Vertex Placement. The gold vertex v_{i-1} is such that $v_{i-1}.x = i - 1$, and $v_{i-1}.y \in [180\alpha \cdot i - 181\alpha, 180\alpha \cdot i - 179\alpha]$. The gold vertex v_i is such that $v_i.x = i$, and $v_i.y \in [180\alpha \cdot i - \alpha, 180\alpha \cdot i + \alpha]$. We first will describe the floor gadget (the ceiling gadget is symmetric). The first reflex vertex on the floor to the right of v_{i-1} is the point $g_i^1 = (i - \frac{11}{20}, 180\alpha \cdot i - 104\alpha)$. The last reflex vertex of the floor gadget to the left of v_i is the point $g_i^2 = (i - \frac{2}{5}, 180\alpha \cdot i - 77\alpha)$. We call g_i^1 and g_i^2 the *green* vertices of the floor gadget. These two green vertices will be the only reflex vertices on the floor gadget that can see a gold vertex, but as we show, the gold points they can see is restricted to just a few "nearby" gold vertices. The proof is omitted due to lack of space.

Lemma 2. *The green vertices g_i^1 and g_i^2 cannot see any gold vertex v_j such that $j \leq i - 3$ or $j \geq i + 3$.*

A corollary of the lemma is that the green vertices can see at most 5 gold vertices: $v_{i-2}, v_{i-1}, v_i, v_{i+1}$, and v_{i+2}. Note that v_i must also see all 5 of these gold vertices. v_i must see v_{i-2} and v_{i+2} because they are adjacent vertices on the floor of P_2. v_i must see v_{i-1} because there are no vertices between v_i and v_{i-1} to block them (v_{i+1} is argued similarly). So while the green points can see some gold vertices, they will see a subset of the gold vertices seen by v_i; this fact will play an important role in the correctness of the reduction.

Purple Vertex Placement. Next we place the *purple* vertices p_i^1 and p_i^2, which respectively will be the first reflex vertex on the floor to the right of g_i^1 and the first reflex vertex on the floor to the left of g_i^2. The coordinates of $p_i^1 = (i - \frac{8}{15}, 180\alpha \cdot i - 103.5\alpha)$, and the coordinates of p_i^2 are $(i - \frac{201}{500}, 180\alpha \cdot i - 80\alpha)$. We argue that these purple vertices do not see any gold vertices. The proof is omitted due to lack of space.

Lemma 3. *The purple vertices p_i^1 and p_i^2 do not see any gold vertices. More specifically, g_i^1 blocks p_i^1 and p_i^2 from all gold vertices v_j such that $j \leq i - 1$, and g_i^2 blocks p_i^1 and p_i^2 from all gold vertices v_j such that $j \geq i$.*

Blue Vertex Placement. We next place two blue vertices b_i^1 and b_i^2. The coordinates of b_i^1 are $(i - \frac{31}{60}, 180\alpha \cdot i - 103.25\alpha)$, and the coordinates of b_i^2 are $(i - \frac{202}{500}, 180\alpha \cdot i - 82\alpha)$. We claim that the blue points do not see any gold points. The line through p_i^1 and b_i^1 is "almost horizontal" which makes it easy to see that b_i^1 will not see any gold vertices v_j for $j \leq i - 1$, and to see that it will not see any gold vertices v_j such that $j \geq i$, we can use the fact that we already know g_i^2 blocks p_i^1 from all such points. Indeed, the slope of the line through p_i^1 and b_i^1 is less than the slope of the line through p_i^1 and g_i^2. This implies b_i^1 is under this line segment and therefore g_i^2 will also block b_i^1 from all gold vertices to the right of b_i^1. The argument for b_i^2 is symmetric.

The blue vertices both connect to the critical convex vertex f_i^* that will then have the coordinates $(i - \frac{202}{500}, 180\alpha \cdot i - 103.25\alpha)$.

The Ceiling Gadget. We now give the coordinates for the gadget on the ceiling between v_{i-1} and v_i. The gadget is symmetric to the floor gadget. We give the coordinates for two green points g_i^3 and g_i^4 where g_i^3 will be to the right of g_i^4.

The coordinates of g_i^3 are symmetric to g_i^2 on the floor, and the coordinates of g_i^4 is symmetric to g_i^1. The purple and blue points p_i^3, p_i^4, b_i^3 and b_i^4 are defined similarly. So $g_i^3 = (i - \frac{3}{5}, 180\alpha \cdot i - 103\alpha)$, $g_i^4 = (i - \frac{9}{20}, 180\alpha \cdot i - 76\alpha)$, $p_i^3 = (i - \frac{299}{500}, 180\alpha \cdot i - 100\alpha)$, $p_i^4 = (i - \frac{7}{15}, 180\alpha \cdot i - 76.5\alpha)$, $b_i^3 = (i - \frac{298}{500}, 180\alpha \cdot i - 98\alpha)$, and $b_i^4 = (i - \frac{29}{60}, 180\alpha \cdot i - 76.75\alpha)$. With respect to seeing gold vertices, the arguments for the floor also apply to the ceiling gadget, and with the coordinates precisely defined, one can verify that the ceiling and floor gadgets are arranged in the layout as described in the high level overview with regards to the left-to-right and bottom-to-top ordering of the vertices.

4.3   Correctness

We are now ready to prove that our reduction is correct. We set $k' := 2(n-1)+k$ where n is the number of vertices of P_1, and we argue that there is a cover for P_1 of size k if and only if there is a cover of size k' of S. The proof is omitted due to lack of space. This completes the proof of Theorem 2.

Lemma 4. *There is a subset of vertices of size at most k that sees every vertex of P_1 if and only if there is a subset of vertices of size at most k' of S that sees every vertex of S.*

References

1. Aggarwal, A.: The art gallery theorem: its variations, applications and algorithmic aspects. Ph.D. thesis, The Johns Hopkins University (1984)
2. Bonnet, É., Miltzow, T.: An approximation algorithm for the art gallery problem. In: 33rd International Symposium on Computational Geometry, SoCG 2017, Brisbane, Australia, 4–7 July 2017, pp. 20:1–20:15 (2017). https://doi.org/10.4230/LIPIcs. SoCG.2017.20
3. Eidenbenz, S.: Inapproximability results for guarding polygons without holes. In: ISAAC, pp. 427–436 (1998)
4. Gibson, M., Kanade, G., Krohn, E., Varadarajan, K.R.: Guarding terrains via local search. J. Comput. Geom. 5(1), 168–178 (2014). https://doi.org/10.20382/ jocg.v5i1a9
5. Gibson, M., Krohn, E., Nilsson, B.J., Rayford, M., Zylinski, P.: A note on guarding staircase polygons. In: Friggstad, Z., Carufel, J.D. (eds.) Proceedings of the 31st Canadian Conference on Computational Geometry, CCCG 2019, 8–10 August 2019, pp. 105–109. University of Alberta, Edmonton, Alberta, Canada (2019)
6. King, J., Krohn, E.: Terrain guarding is NP-hard. SIAM J. Comput. 40(5), 1316–1339 (2011)
7. Krohn, E., Nilsson, B.J.: The complexity of guarding monotone polygons. In: Proceedings of 24th Canadian Conference on Computational Geometry (CCCG), pp. 167–172 (2012)
8. Krohn, E., Nilsson, B.J.: Approximate guarding of monotone and rectilinear polygons. Algorithmica 66(3), 564–594 (2013). https://doi.org/10.1007/s00453-012-9653-3
9. Lee, D.T., Lin, A.K.: Computational complexity of art gallery problems. IEEE Trans. Inf. Theory 32(2), 276–282 (1986)

On the Complexity of Half-Guarding Monotone Polygons

Hannah Miller Hillberg, Erik Krohn$^{(\boxtimes)}$ ⓘ, and Alex Pahlow

University of Wisconsin-Oshkosh, Oshkosh, WI 54901, USA
{hillbergh,krohne}@uwosh.edu, apahlow22@alumni.uwosh.edu

Abstract. We consider a variant of the art gallery problem where all guards are limited to seeing to the right inside a monotone polygon. We call such guards: half-guards. We provide a polynomial-time approximation for point guarding the entire monotone polygon. We improve the best known approximation of 40 to 8. We also provide an NP-hardness reduction for point guarding a monotone polygon with half-guards.

1 Introduction

An instance of the original *art gallery problem* takes as input a simple polygon P. A polygon P is defined by a set of points $V = \{v_1, v_2, \ldots, v_n\}$. There are edges connecting (v_i, v_{i+1}) where $i = 1, 2, \ldots, n - 1$. There is also an edge connecting (v_1, v_n). If these edges do not intersect other than at adjacent points in V (or at v_1 and v_n), then P is called a simple polygon. The edges of a simple polygon give us two regions: inside the polygon and outside the polygon. For any two points $p, q \in P$, we say that p sees q if the line segment $\overline{pq}$ does not go outside of P. The art gallery problem seeks to find a guarding set of points $G \subseteq P$ such that every point $p \in P$ is seen by a point in G. In the point guarding problem, guards can be placed anywhere inside of P. In the vertex guarding problem, guards are only allowed to be placed at points in V. The optimization problem is defined as finding the smallest such G.

1.1 Previous Work

There are many results about guarding art galleries. Several results related to hardness and approximations can be found in [3,9,10,13,20,21]. Whether a polynomial time constant factor approximation algorithm can be obtained for vertex guarding a simple polygon is a longstanding and well-known open problem, although a claim for one was made in [5].

Additional Polygon Structure. Due to the inherent difficulty in fully understanding the art gallery problem for simple polygons, much work has been done guarding polygons with additional structure, see [4,6,15,19]. In this paper we consider monotone polygons.

α**-Floodlights.** Motivated by the fact that many cameras and other sensors often cannot sense in 360°, previous works have considered the problem when

© Springer Nature Switzerland AG 2022
A. Castañeda and F. Rodríguez-Henríquez (Eds.): LATIN 2022, LNCS 13568, pp. 761–777, 2022.
https://doi.org/10.1007/978-3-031-20624-5_46

guards have a fixed sensing angle α for some $0 < \alpha \leq 360$. This problem is often referred to as the α-*floodlight problem* and was first studied in [2]. Some of the work on this problem has involved proving necessary and sufficient bounds on the number of α-floodlights required to guard (or illuminate) an n vertex simple polygon P, where floodlights are anchored at vertices in P and no vertex is assigned more than one floodlight, see for example [22, 23]. Orthogonal polygons can always be guarded with 90°-floodlights using any of the illumination rules defined in [12], Lemma 1 there changing NE for NW, or SW. It follows that there is a rotation of an orthogonal polygon that can be illuminated with $\lfloor \frac{3n-4}{8} \rfloor$ orthogonal floodlights, and the bound is tight. Computing a minimum cardinality set of α-floodlights to illuminate a simple polygon P is APX-hard for both point guarding and vertex guarding [1]. For more problems that have been studied for floodlights see [8, 24]. More specifically, 180°-floodlights, or *half-guards*, see only in one direction. Half-guarding may have the ability to help with full-guarding. A *full-guard* can see 360°. In [11, 17], the authors use half-guarding to show a 4-approximation for terrain guarding using full-guards. A constant factor approximation for half-guarding a monotone polygon was shown in [15] and NP-hardness for vertex half-guarding a monotone polygon was shown in [14].

1.2   Definitions

A simple polygon P is *x-monotone* (or simply *monotone*) if any vertical line intersects the boundary of P in at most two points. In this paper, we define *half-guards* as guards that can see only in one direction. If we assume half-guards can only see right, then we redefine *sees* as: a point p sees a point q if the line segment $\overline{pq}$ does not go outside of P and $p.x \leq q.x$, where $p.x$ denotes the x-coordinate of a point p. In a monotone polygon P, let l and r denote the leftmost and rightmost point of P respectively. Consider the "top half" of the boundary of P by walking along the boundary clockwise from l to r. We call this the *ceiling* of P. We obtain the *floor* of P by walking counterclockwise along the boundary from l to r. A vertical line that goes through a point p is denoted l_p. Given two points p, q in P such that $p.x < q.x$, we use (p, q) to denote the points s such that $p.x < s.x < q.x$, (resp. $[p, q]$ for $p.x \leq s.x \leq q.x$). If referring to only boundary points in the range, we use the subscript c (resp. f) to denote the ceiling (resp. floor), for example, $(p, q]_c$ represents ceiling points c where $p.x < c.x \leq q.x$.

1.3   Our Contribution

Krohn and Nilsson [19] give a constant factor approximation for point guarding a monotone polygon using full-guards. There are monotone polygons P that can be completely guarded with one full-guard that require $\Omega(n)$ half-guards considered in this paper [15]. Due to the restricted nature of half-guards, new observations are needed to obtain the approximation given in this paper. A 40-approximation for this problem was presented in [15]. The algorithm in [15] places guards in 5 steps: guard the ceiling vertices, then the floor vertices, then the entire ceiling boundary, then the entire floor boundary, and finally any missing portions of the interior. We propose a new algorithm that requires only 3 steps: guarding the entire ceiling, then the entire floor, and lastly the interior. By providing improved analysis of the new algorithm, we obtain an 8-approximation.

In addition, we show that point guarding a monotone polygon with half-guards is NP-hard. An NP-hardness proof for vertex guarding a monotone polygon with half-guards was presented in [14]. However, if a guard was moved off a vertex, it would see too much of the polygon and the reduction would fail. Thus, new insights were needed for point guarding.

The remainder of the paper is organized as follows. Section 2 gives an algorithm for point guarding a monotone polygon using half-guards. Section 3 provides an NP-hardness proof for point guarding a monotone polygon using half-guards. Finally, Sect. 4 gives a conclusion and possible future work.

2   8-Approximation for Point Guarding a Monotone Polygon with Half-Guards

We start with an algorithm for point guarding the boundary of a monotone polygon P with half-guards. We first give a 2-approximation algorithm for guarding the entire ceiling. A symmetric algorithm works for guarding the entire floor giving us a 4-approximation for guarding the entire boundary of the polygon. Finally, even though the entire boundary is seen, portions of the interior may be unseen. We show that by doubling the number of guards, we can guarantee the entire polygon is seen giving an 8-approximation.

We first provide a high level overview of the algorithm for guarding the entire ceiling boundary. Any feasible solution must place a guard at the leftmost vertex l where the ceiling and floor come together (or this vertex would not be seen). The algorithm begins by placing a guard here. The algorithm iteratively places guards from left to right, letting S denote the guards the algorithm has already placed. When placing the next guard, we let p denote the rightmost point on the ceiling such that for every ceiling point $x \in [l, p]_c$, there exists a guard $g' \in S$ that sees x. In other words, the point on the ceiling directly to the right of p is not seen by any guard in S. Note that p may be a ceiling vertex or any point on the ceiling. The next guard g that is placed will lie somewhere on the vertical line l_p. The algorithm initially places g at the intersection of l_p and the floor, and it slides g upwards vertically along l_p. The algorithm locks in a final position for the guard g by sliding it upwards along l_p until moving it any higher would cause g to no longer see some unseen point on the ceiling (i.e. a point not seen by some $g' \in S$); let r_g be the leftmost such point, see, for example, Fig. 1. In this figure, when g is initially placed on the floor, it does not see r_g, but as the algorithm slides g up the l_p line, r_g becomes a new point that g can see. If g is slid any higher up than as depicted in the figure, then g would no longer see r_g, and therefore g is locked in that position. The algorithm then adds g to S, and repeats this procedure until the entire ceiling is guarded. The ceiling guarding algorithm is formally described in Algorithm 1.

Algorithm 1. Ceiling Guard

1: **procedure** CEILING GUARD(monotone polygon P)
2:     $S \leftarrow \{g\}$ such that g is placed at the leftmost point l.
3:     **while** there is a point on the ceiling that is not seen by a guard in S **do**
4:         Let p be the rightmost ceiling point such that for each ceiling point $x \in [l,p]_c$, there exists a guard $g' \in S$ where g' sees x. Place a guard g where l_p intersects the floor and slide g up. Let r_g be the first ceiling point not seen by any guard in S that g would stop seeing if g moved any further up. Place g at the highest location on l_p such that g sees r_g.
5:         $S \leftarrow S \cup \{g\}$.
6:     **end while**
7:     **return** S
8: **end procedure**

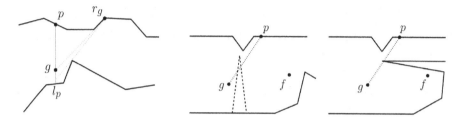

Fig. 1. A guard g slides up l_p and sees a point r_g. If g goes any higher, it will stop seeing r_g.

Fig. 2. Assume $g.x < p.x < f.x$ and that p is on the ceiling. (Left) If the floor blocks g from a ceiling point p and the polygon is monotone, then g will not see any point f. (Right) If g sees the point f and the floor blocks g from p, then the polygon is not monotone.

2.1   Sliding Analysis

All steps, except the sliding step, can be trivially done in polynomial time. The analysis of [15] uses a sliding step but only considers guarding the vertices. When considering an infinite number of points on the ceiling, it is not immediately clear that the sliding can be done in polynomial time since each time a guard moves an ϵ amount upwards, it will see a different part of the boundary. We prove that there are at most $O(n^3)$ potential locations on l_p that must be considered. We use the following lemma to prove the bound on the number of locations a guard g must consider on l_p.

Lemma 1. *Consider points $g, f \in P$ such that $g.x < f.x$. If g sees f, then the floor (resp. ceiling) cannot block g from seeing any ceiling (resp. floor) point in $(g, f)_c$ (resp. $(g, f)_f$).*

Proof. Consider a point on the ceiling p such that $g.x < p.x < f.x$. In order for the floor to block g from p, the floor must pierce the $\overline{gp}$ line segment. If g is being blocked from seeing p because of a floor vertex, since the polygon is monotone, then the floor must pierce the $\overline{gf}$ line segment, a contradiction that g sees f, see Fig. 2 (left). If the floor pierces the $\overline{gp}$ line segment and g sees f, then the polygon is not monotone, see Fig. 2 (right).

Corollary 1. *If a point g is blocked by the floor (resp. ceiling) from seeing a point p on the ceiling (resp. floor), then g does not see any points in $(p, r]$.*

Each guard g placed by the algorithm starts on the floor and is slid upwards. The approximation analysis in Sect. 2.2 relies on the fact that g has the largest y-coordinate on l_p as possible such that if g is moved any further upwards, then there is a point on the ceiling (r_g) that would not be seen. This ceiling point (r_g) could be (A) a vertex, blocked by another vertex, (B) a point on an edge that marks the endpoint of sight from a previously-placed guard, or (C) a point seen specifically from where l_p intersects the floor or the ceiling. These possible slide-stopping ceiling points correspond to points on l_p that are the potential guard locations, denoted PGL, that the algorithm must consider.

PGL(A): Vertex-Vertex. To represent vertices that cause other vertices to no longer be seen as a guard slides upwards, PGL(A) are given by rays shot from a vertex through another vertex until they hit l_p. This includes the rays extending each edge of the polygon. For example, in Fig. 3, the guard g_{j+1} is placed where it would stop seeing $r_{g_{j+1}}$. With n vertices, there are $O(n^2)$ PGL(A) on l_p.

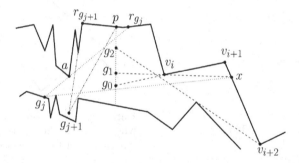

Fig. 3. When placing guard g_{j+2} on l_p, g_0 and g_2 are given by PGL(A), and g_1 is given by PGL(B). The algorithm places the guard at g_1 to ensure the leftmost previously unseen portion of edge $[v_{i+1}, v_{i+2}]$ from x and to the right is seen. (When previously placing guards g_j and g_{j+1}, g_j location was given by PGL(C), and g_{j+1} by PGL(A).)

PGL(B): Vertex-Unseen Edge Points. It is possible that only part of a ceiling edge is seen from previously placed guards. For example, in Fig. 3, g_j sees $[v_{i+1}, x]$, but $(x, v_{i+2}]$ is unseen. PGL(B) represent lines of sight to points marking the start of unseen portions of the ceiling. Let X_g be the set of points on the ceiling that are hit by rays shot from some guard $g \in S$ through all vertices. For example, in Fig. 3: $x, r_{g_j} \in X_{g_j}$. For all $g \in S$, $X_S = \bigcup X_g$. For any edge that is only partially covered, these points, along with the vertices of the polygon, represent all of the potential endpoints of these partially seen edges.

For all points $x \in X_S$, shoot a ray from x through all vertices until they hit l_p. For example in Fig. 3, a ray from x through vertex v_i intersects l_p at g_1. These locations on l_p are denoted as G_{X_S} and are the PGL(B). In Fig. 3, if location $g_1 \in G_{X_S}$ was not considered, it would miss the leftmost unseen points $(x, x+\epsilon]_c$ on $[v_{i+1}, v_{i+2}]$.

In order to bound $|G_{X_S}|$, consider that when a guard is being placed, at most n guards have been placed already. To illustrate, if a guard g is placed in the polygon in the range of $[v_i, v_{i+1})$, where v_i and v_{i+1} are ceiling vertices, then the final location of g must see vertex v_{i+1}. If g does not see v_{i+1} when g started on the floor, then a floor vertex must have blocked g from seeing v_{i+1}. However, visibility of g to the ceiling is not lost if g is pushed up such that it sees over the floor blocker and sees v_{i+1}. When g is placed, the subsequent guard will be placed at or beyond $l_{v_{i+1}}$. Therefore, when a guard is being placed, at most n guards have been placed already.

With n vertices and n previously placed guards $g \in S$, a ray shot from every previously placed g through every vertex gives an upper bound of: $|X_S| = O(n^2)$. Finally, a ray shot from each $x \in X_S$ through every vertex yields a total of $|G_{X_S}| = O(n^3)$ PGL(B) on l_p.

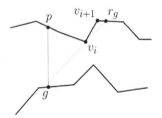

Fig. 4. Ceiling visibility is maximal when g is on the floor. If g is moved up, then r_g would not be seen.

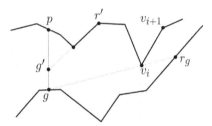

Fig. 5. Ceiling visibility is maximal when g is on the floor. The guard g moves up until it stops seeing r'.

PGL(C): l_p Floor and Ceiling Intersection Points. Assume g is on the floor and there exists some vertex v_i such that g sees vertex v_i, but v_i is blocking g from seeing v_{i+1}, see Fig. 4. If there are multiple v_i candidates, choose the leftmost v_i candidate. Shoot a ray from g through v_i and let r_g be the point on the boundary that is hit. If r_g is on the ceiling, then no guard to the left of g is able to see r_g. If g is slid up, then g would no longer see r_g. Thus, g stays where l_p hits the floor.

If r_g is on the floor, then moving g upwards will not cause g to see any more ceiling points to the right of r_g because some v_i is blocking g from seeing them (Corollary 1), see Fig. 5. By Lemma 1, the floor cannot be blocking g from seeing any ceiling points to the left of r_g. Therefore, moving g upwards will not result in seeing any more ceiling points. However, the approximation analysis relies on g being as high as possible on l_p such that g sees r_g and r_g is a ceiling point. Consider guards S placed by the algorithm up to this point. Now consider ceiling points R that are not seen by any guard $g' \in S$. We slide g upwards until it would have stopped seeing some point $r' \in R$ using PGL(A) and PGL(B). We set r_g to be r'. In the case where there are no PGL(A) or PGL(B) on l_p, then any point on l_p sees all of the unseen ceiling points of $[p, v_i]$, we place g on the ceiling at point p and we assign r_g to be any point in $(p, v_i]$ that is also in R.

A trivial analysis of the number of locations on l_p that PGL(A), PGL(B) and PGL(C) could generate gives a polynomial upper bound on the number of locations the algorithm has to consider at $O(n^2) + O(n^3) + 2 = O(n^3)$.

Necessity of PGL(A), PGL(B) and PGL(C): If PGL(A) are not included, then the g_1 guard location from Fig. 6 (left) would not be considered and g_2 would not be able to be placed to see the remainder of the ceiling. If PGL(B) are not included, then the g_2 guard location in Fig. 6 (middle) would not be considered and g_3 would not be able to be placed to see the remainder of the ceiling. If the ceiling point of PGL(C) is not included, then the g_2 guard location in Fig. 6 (right) would not be considered and g_3 would not be able to be placed to see the remainder of the ceiling. A similar figure can be drawn for the necessity of the floor point of PGL(C).

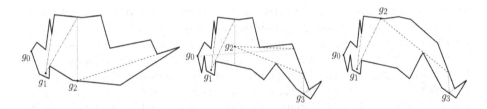

Fig. 6. Necessity of PGL(A), PGL(B) and PGL(C).

The following lemma is used to help show sufficiency. The proof is omitted due to lack of space.

Lemma 2. *Consider points x, y on a vertical line such that x is strictly below y and there is a point q that is strictly to the right of the vertical line. If the floor (resp. ceiling) blocks y (resp. x) from seeing q, then x (resp. y) does not see q.*

Sufficiency of PGL(A), PGL(B) and PGL(C): Assume some set of guards S has been placed. Consider 2 PGL x and y from PGL(A), PGL(B) and PGL(C) on l_p such that there are no PGL between x and y, and x is below y. Assume that the algorithm placed a guard at location y. Now consider a ceiling point q not seen by any guard $g \in S$. Neither x nor y sees q but there is some location z on l_p that sees q where z is above x and below y, see Fig. 7. In other words, z is a location that the algorithm does not consider as a PGL. By Lemma 2, the floor must block x from seeing q and the ceiling must block y from seeing q. Let v_i be the rightmost ceiling vertex that blocks y from seeing q. If a ray shot from v_{i+1} through v_i hits the floor to the right of l_p, then the guard would have been placed in the lowest guard location on l_p to see v_i (i.e. from PGL(A)). Such a guard location is below the y location, see Fig. 7 (left), a contradiction that the algorithm placed a guard at y.

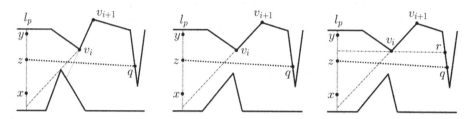

Fig. 7. A guard is placed at location y by the algorithm. A guard location z that lies between x and y that sees an unseen ceiling point $q \in R$ does not exist.

If a ray shot from v_{i+1} through v_i goes through l_p, then one needs to consider multiple cases. Before considering the cases, note that this ray must intersect l_p below y. If the ray were to have intersected above y, then either y would have seen v_{i+1}, contradicting the fact that v_i is the rightmost vertex that blocks y from q. Or, if y did not see v_{i+1}, then the floor would have blocked y from v_{i+1}. By Corollary 1 and Lemma 2, z would not have seen q, a contradiction.

If v_{i+1} was not seen by a previously placed guard, then the algorithm would have placed a guard at this intersection point because of the PGL(A), see Fig. 7 (middle). This contradicts the algorithm choosing guard location y. If v_{i+1} was seen by a previously placed guard, then consider the rightmost ceiling point r between $[v_{i+1}, q]_c$ that was seen by a guard in S. Such a location must exist since some guard $g' \in S$ sees v_{i+1} and no guard in S sees q. PGL(B) would have shot a ray from r through v_i. If this ray hits l_p below y, then the algorithm would have chosen that point as our guard location, contradicting that the algorithm chose y. If this ray hits l_p above y, then v_i could not have blocked y from seeing q. Therefore, it is not possible for guard location z to exist.

2.2 Approximation Analysis

We will now prove why Algorithm 1 will place no more than 2 times the number of guards in the optimal solution. The argument is similar to the argument presented in [15]. An optimal solution $\mathcal{O}$ is a minimum cardinality guard set such that for any point p on the ceiling of P, there exists some $g \in \mathcal{O}$ that sees p. The argument will be a charging argument; every guard placed will be charged to a guard in $\mathcal{O}$ in a manner such that each guard in $\mathcal{O}$ will be charged at most twice. First, charge the leftmost guard placed to the optimal guard at the same location, call this guard g_1. All optimal guardsets include a guard at the leftmost vertex in the polygon else the leftmost vertex is unseen. Now consider two consecutive guards $g_i, g_{i+1} \in S$ returned by the algorithm. Assuming the g_i

guard has already been charged to some guard in the optimal solution, we must find an optimal guard to charge g_{i+1} to.

Case 1: If at least one optimal guard is in $(g_i, g_{i+1}]$, then we charge g_{i+1} to any optimal guard, chosen arbitrarily, in that range.

Case 2: If there is no optimal guard in $(g_i, g_{i+1}]$, then consider the point on the ceiling directly above g_{i+1}, call this point p. The g_{i+1} guard was placed on l_p because no previously placed guard saw the ceiling point to the right of p and the entire ceiling from $[l, p]_c$ is seen by guards in S. Let p_ϵ be a point on the ceiling that is an ϵ amount to the right of p. By assumption, g_i does not see p_ϵ.

If the Floor Blocks g_i From Seeing p_ϵ: By Corollary 1, g_i does not see any ceiling point to the right of p_ϵ. The reason that g_i stopped moving upwards is because it saw some point r_{g_i} on the ceiling that no previously placed guard saw. Since g_i cannot see to the right of p_ϵ, $r_{g_i}.x \leq p.x$. By assumption, the optimal guard o' that sees r_{g_i} must be to the left of, or on, l_{g_i}. If g_i were to have moved any higher up, it would have missed r_{g_i}. Any guard that sees r_{g_i} must be "below" the $\overline{g_i r_{g_i}}$ line, see the gray shaded region of Fig. 8. Any point on the ceiling that o' sees to the right of r_{g_i}, g_i will also see (Lemma 1, [15]). We charge g_{i+1} to o' and o' cannot be charged again. The o' guard will never be charged in Case 1 again since any subsequent g_k and g_{k+1} guards are strictly to the right of o'. The o' guard will not be charged in Case 2 again since any point on the ceiling to the right of g_i that o' sees, g_i will also see. In other words, no future g_k guard, where $k > i + 1$, will be charged to o' in Case 2. The g_k guard will stop moving up because of some previously unseen point r_{g_k}. If o' saw r_{g_k}, then g_i would have also seen r_{g_k} and thus r_{g_k} is not an unseen point.

If the Ceiling Blocks g_i From Seeing p_ϵ: Consider the optimal guard o' that sees p_ϵ. If $o'.x \leq g_i.x$, then the ray $\overrightarrow{p_\epsilon o'}$ must cross the l_{g_i} vertical line. If the ray crosses below g_i's final location, then at some point, g_i saw p_ϵ since a ray shot from p through some v_i, where $g_i.x < v_i.x < p.x$, would have crossed l_{g_i} below g_i. The g_i guard would not have continued above this PGL(B) as it saw p_ϵ. The ray $\overrightarrow{p_\epsilon o'}$ cannot cross above g_i on l_{g_i} since the ceiling blocks g_i from p_ϵ and will thus block o' from p_ϵ. Therefore, an optimal guard that sees p_ϵ must lie in $(g_i, g_{i+1}]$ and we revert back to Case 1.

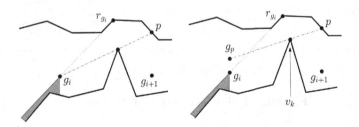

Fig. 8. If no optimal guard exists in $(g_i, g_{i+1}]$, then o' must exist to see r_{g_i} such that $o'.x \leq g_i.x$. Let g_p be the lowest point on l_{g_i} that sees p. The g_i guard must be below g_p otherwise g_i would have seen p and also the ceiling point to the right of p.

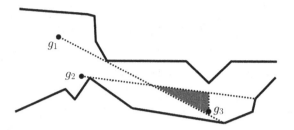

Fig. 9. The g_1, g_2 and g_3 guards see all of the boundary to the right of g_1. However, the shaded region is unguarded.

The entire ceiling can be guarded with at most $2 \cdot |\mathcal{O}|$ guards. A similar algorithm is applied to the floor to give at most $4 \cdot |\mathcal{O}|$ guards to guard the entire boundary. Finally, even though the entire boundary is guarded, it is possible that a portion of the interior is unseen, see Fig. 9. Let us assume that a guardset $G = \{g_1, g_2, \ldots g_k\}$ guards the entire boundary of a monotone polygon such that for all i, $g_i < g_{i+1}$. In [18], they prove that between any consecutive guards of G, a (potentially disconnected) region can exist that is unseen by any of the guards in G. To see this possible disconnect, imagine a g_0 guard in Fig. 9 that has a narrow beam of visibility that splits the shaded region in two. However, they prove that this (potentially disconnected) region is convex and can be guarded with 1 additional guard. In our setting, a half-guard placed at the leftmost point of the unseen convex region covers the entire unseen portion. If $|G| = k$, then there are at most $k - 1$ guards that need to be added to guard these unseen interior regions. This doubles the approximation to give us the following theorem:

Theorem 1. *There is a polynomial-time 8-approximation algorithm for point guarding a monotone polygon with half-guards.*

3 NP-Hardness for Point Guarding a Monotone Polygon with Half-Guards

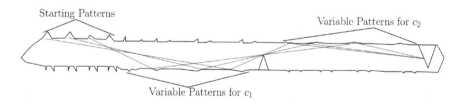

Fig. 10. A high level overview of the reduction.

In this section, we show that point guarding a monotone polygon with half-guards is NP-hard. NP-hardness for vertex guarding a monotone polygon with half-guards was shown in [14]. However, moving the guards a small amount off of the vertex causes the entire reduction to fail. Additional insight and additional gadgets/patterns were needed to show hardness for point guarding a monotone polygon with half-guards. We show the following theorem:

Theorem 2. *Finding the smallest point guard cover for a monotone polygon using half-guards is NP-hard.*

The reduction is from *SAT*. A SAT instance (X, C) contains a set of n Boolean variables, $X = \{x_1, x_2, \ldots, x_n\}$ and a set of m clauses, $C = \{c_1, c_2, \ldots, c_m\}$. A SAT instance is satisfiable if a satisfying truth assignment for X exists such that all clauses c_i are true. We show that any SAT instance is polynomially transformable to an instance of point guarding a monotone polygon using half-guards. We construct a monotone polygon P from the SAT instance such that P is guardable by $K = n \cdot (1 + 2m) + 1$ or fewer guards if and only if the SAT instance is satisfiable.

The high level overview of the reduction is that specific potential guard locations represent the truth values of the variables in the SAT instance. Starting patterns are placed on the ceiling on the left side of the polygon, see Fig. 10. In these starting patterns, one must choose one of two guardset locations in order to guard distinguished edges for that particular pattern. A *distinguished edge* is an edge that is only seen by a small number of specific guard locations. Then, to the right of the starting patterns, variable patterns are placed on the floor, then the ceiling, then the floor, and so on for as many clauses as are in the SAT instance. In each variable pattern, similar to a starting pattern, certain guard locations will represent a truth assignment of true/false for a variable (x_i). This Boolean information is then "mirrored rightward" such that there is a consistent choice of all true x_i locations or all false $\overline{x_i}$ locations for each variable. Most previous results had Boolean information mirrored from the "left side" of the polygon/terrain to the "right side" of the polygon/terrain and then back to the left side [7, 16, 18], with [14] being an exception. A *distinguished clause vertex* is placed to the right of each sequence of variable patterns such that only the guard locations representing the literals in the specific clause can see the distinguished clause vertex. A high level example of the entire reduction is shown in Fig. 10.

3.1 Hardness Details

Starting Pattern: The starting pattern, shown in Fig. 11, appears along the ceiling of the left side of the monotone polygon a total of n times, each corresponding to a variable from the SAT instance, in order from left to right $(x_1, x_2, \ldots, x_n)$. In each pattern, there are 2 distinguished edges with labels: $\{A = \overline{v_2 v_3}, B = \overline{v_5 v_6}\}$. These edges are seen by a specific, small range of points in each starting pattern. It is important to note that no other point in any other starting pattern sees these distinguished edges.

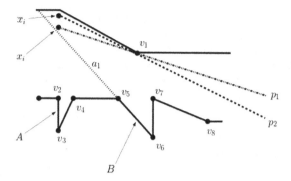

Fig. 11. A starting pattern for variable x_i, with distinguished edges $\{A, B\}$.

In order to see A, a guard must be placed on the vertical line that goes through v_2. We can make the A edge be almost vertical such that any guard placed to the left of the vertical line through A will miss v_3. In order to guard the entire pattern with 1 guard, that guard must see B as well. We are left with a small range of potential guard locations as seen in Fig. 11. Any guard placed above a_1 on the l_{v_2} line will see these distinguished edges. Foreshadowing the variable mirroring, a guard placed at x_i will represent a true value for x_i. Any guard placed at $\overline{x_i}$ will represent a false value for x_i.

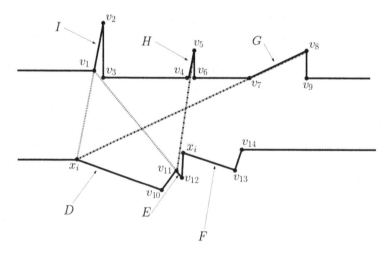

Fig. 12. A variable pattern with distinguished edges $\{D, E, F, G, H, I\}$. Critical lines of sight are shown. Guards must be placed at either $\{x_i, v_{11}\}$ (if F is previously seen) or $\{\overline{x_i}, v_1\}$ (if D is previously seen).

Variable Pattern: On the floor of the polygon to the right of all the n starting patterns are the first n variable patterns, one for each variable, that verify and propagate the assigned truth value of each variable. The variables are in reverse order from the initial starting pattern $(x_n, x_{n-1}, \ldots, x_1)$. When the variables are "mirrored" rightward again to the ceiling, the ordering will again reverse.

A single variable pattern is shown in Fig. 12. There are 6 distinguished edges located at $\{D = \overline{x_i v_{10}}, E = \overline{v_{11} v_{12}}, F = \overline{x_i v_{13}}, G = \overline{v_7 v_8}, H = \overline{v_4 v_5}, I = \overline{v_1 v_2}\}$. Of those, E, G, H and I are only visible within the variable pattern. Further, no guard that sees I can also see H. Therefore, at least 2 guards are required to see the vertices in a variable pattern. However, with D or F seen by a previously placed guard, then 2 guards are sufficient for guarding the variable pattern.

If neither D nor F is seen by a previously placed guard, then 3 guards are required to see all of $\{D, E, F, G, H, I\}$. Assume that no guard in a previous pattern saw either D or F in the variable pattern. If this is the case, then a guard must be placed on (or to the left of) the l_{x_i} line in order to see D. No guard that sees D also sees H. If a guard that sees D does not see I, then 3 guards are required since no guard sees both H and I. We assume that the guard that sees D also sees I. The only guard that sees both D and I is at x_i. If the guard is moved left or right, it will miss part of D. If it is moved up, it will miss I. Therefore, a guard is placed at x_i and we now also see G. This leaves E, F and H to be guarded. A guard that sees E and H must lie on v_{11}. If the guard is pushed higher, it will not see H. If it is pushed left or right, it will not see E. The vertex $\overline{x_i}$ blocks v_{11} from seeing F. Therefore, the F edge goes unseen and a third guard is required. In summary, if a previously placed guard does not see D nor F, 3 guards are required to guard the variable pattern.

Let us assume that F is seen by a previously placed guard. If this happens, then $\{D, E, G, H, I\}$ must still be guarded. To see all of D, a guard must be placed on the vertical line through (or to the left of) x_i, see Fig. 12. If the guard is placed above x_i, then I would remain unguarded. If this happens, then 2 guards are still required to see I and H. Therefore, a guard that sees D must also see I, and the only such location is x_i. Placing a guard at x_i sees edges D, G and I, leaving E and H to be seen. The only guard that sees all of E and H is at vertex v_{11}. Therefore, two guards suffice and they are placed at x_i and v_{11}.

Now let us assume that D is seen by a previously placed guard. If this happens, then $\{E, F, G, H, I\}$ must still be guarded. In order to see I, a guard must be on (or to the left of) l_{v_1}. Edge E is angled in such a way that it does not see any of the vertical line below v_1.

If a guard is placed at v_1, then edges E and I are guarded and we need to place a second guard that sees F, G and H. It should be noted that a guard placed at v_1 does not see to the right of $\overline{x_i}$ since v_3 is blocking v_1 from seeing too far to the right. A ray shot from v_5 through v_4 goes just above the $\overline{x_i}$ vertex, so placing a guard too far above $\overline{x_i}$ will not see edge H. Thus, a guard must be placed arbitrarily close to $\overline{x_i}$ to see F, G and H. Therefore, two guards suffice in this instance and guards are placed arbitrarily close to both $\overline{x_i}$ and v_1.

It is important to note that a ray shot from v_{13} through $\overline{x_i}$ is above a ray shot from v_8 through v_7 when they cross l_{v_1}. In other words, no guard that sees I can see both F and G. The following 3 cases assume that the guard that sees I does not see E and it will lead to a contradiction in each case. This will show that when D is seen by a previously placed guard, one must put guards arbitrarily close to $\overline{x_i}$ and at v_1.

If a guard is placed on l_{v_1} that sees neither E, F nor G. In this case, edges E, F, G and H are unseen. A guard that sees E and H must lie on v_{11}. If the guard is pushed higher, it will not see H. If it is pushed left or right, it will not see E. The vertex $\overline{x_i}$ blocks v_{11} from seeing F and G. Therefore, the F and G edges go unseen and a third guard is required. If a guard placed on l_{v_1} (resp. l_{v_1}) to see I and also sees F (resp. G), then edges E, G (resp. F) and H are unseen. Similar to the previous case, a guard that sees E and H must lie on v_{11}. The vertex $\overline{x_i}$ blocks v_{11} from seeing G (resp. F). Therefore, the G (resp. F) edge goes unseen and a third guard is required. To summarize, if D or F is seen, then in order to guard the entire variable pattern with two guards, there are exactly two sets of potential guard locations: $\{x_i, v_{11}\}$, and $\{v_1, \overline{x_i}\}$.

Connecting Starting/Variable Patterns to subsequent Variable Patterns: To ensure truth values are mirrored correctly, one must ensure that choosing the x_i location in a starting/variable pattern forces the choice of x_i in the subsequent variable pattern. This is done by ensuring that x_i (resp. $\overline{x_i}$) in the starting/variable pattern sees distinguished edge F (resp. D) in the subsequent variable pattern. By doing this, the remaining edges: $\{E, F, G, H, I\}$ or $\{D, E, G, H, I\}$ are seen by locations described in the previous section. Full details of the mirrorings between starting/variable patterns and between variable/variable patterns have been omitted due to lack of space.

Clauses: For each clause c in the SAT instance, there is a sequence of variable patterns $x_1, \ldots, x_n$ along either the floor or ceiling of the polygon. Immediately to the right of one such sequence of variable patterns exists a clause pattern. A clause pattern consists of one vertex such that the vertex is only seen by the variable patterns corresponding to the literals in the clause; see Fig. 13. The distinguished vertex of the clause pattern is the c_3 vertex. This vertex is seen only by specific vertices in its respective sequence of variable patterns.

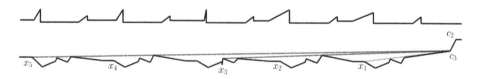

Fig. 13. High level overview of a clause point c_3 being placed for clause $c_i = x_1 \lor \overline{x_3} \lor x_5$.

To see how a clause distinguished point is placed in the polygon, consider Fig. 13 that represents the clause $x_1 \lor \overline{x_3} \lor x_5$. Let's assume the clause point is on the floor. The ceiling has a symmetric argument. The potential guard locations in the variable patterns of v_1 and v_{11} do not see c_3. Referring to Fig. 12, the v_1

vertex is blocked from seeing v_{13} using the ceiling vertex of v_3. By Corollary 1, v_1 cannot see anything to the right of v_{13}, including c_3. In a similar manner, v_{11} is blocked from seeing v_8 using $\overline{x_i}$ and therefore, v_{11} cannot see c_3. Initially, all variable patterns on the floor have their x_i and $\overline{x_i}$ vertices see c_3. The lines of sight go just over the v_{14} vertex. If neither x_i nor $\overline{x_i}$ is supposed to see c_3, then v_{14} is raised a small amount to block them from seeing c_3. The v_{14} vertex is only blocking these vertices from the floor and does not affect the mirroring of x_i to the subsequent variable pattern on the ceiling. Since the $\overline{x_i c_3}$ line must be above the $\overline{x_i c_3}$ line, if x_i is supposed to see c_3, then v_{14} is raised just enough to block $\overline{x_i}$ from seeing c_3. Lastly, if only $\overline{x_i}$ is supposed to see c_3, then x_i must be pushed a small amount down-and-to-the-right on the D edge. In this instance, $\overline{x_i}$ blocks x_i from seeing c_3. The remaining vertices of the variable pattern, namely v_1, v_3 and v_7 are adjusted locally to account for this new x_i location. Since x_i was pushed down-and-to-the-right on the D edge, no tweaks in any previous patterns need to be made. The polygon will be drawn from left-to-right and therefore, no future patterns will need to be tweaked either since they have yet to be drawn. The mirroring of the x_i variable is not affected.

Putting it all Together: We choose our truth values for each variable in the starting variable patterns. The truth values are then mirrored in turn between variable patterns on the floor and the ceiling. Consider the example of Fig. 13 the SAT clause corresponds to $c_i = (x_1 \vee \overline{x_3} \vee x_5)$. Hence, a guard placement that corresponds to a truth assignment that makes c_i true, will have at least one guard on or near the $x_1, \overline{x_3}$ or x_5 vertex and can therefore see vertex c_3 without additional guards. Neither x_2 nor x_4 sees the vertex c_3. They are simply there to transfer their truth values in case these variables are needed in later clauses.

The monotone polygon we construct consists of $11n + (16n + 3)m + 2$ vertices where n is the number of variables and m is the number of clauses. Each starting variable pattern has 11 vertices, each variable pattern 16 vertices, the clause pattern has 3 vertices, plus 2 vertices for the leftmost and rightmost points of the polygon. Exactly $K = n \cdot (1 + 2m) + 1$ guards are required to guard the polygon. A guard is required to see the distinguished edges of the starting patterns and 2 guards are required at every variable pattern, of which there are (mn) of them. Lastly, since a starting pattern cannot begin at the leftmost point, a guard is required at the leftmost vertex of the polygon. If the SAT instance is satisfiable, then $K = n \cdot (1 + 2m) + 1$ guards are placed at locations in accordance with whether the variable is true or false in each of the patterns. Each clause vertex is seen since one of the literals in the associated clause is true and the corresponding c_3 clause vertex is seen by some guard.

4   Conclusion and Future Work

In this paper, we present an 8-approximation for point guarding a monotone polygon with half-guards. We also show that point guarding a monotone polygon with half-guards is NP-hard. Future work might include finding a better approximation for both the point guarding and vertex guarding versions of this problem. Insights provided in this paper may help with guarding polygons where the guard can choose to see either left or right, or in other natural directions. One may also be able to use these ideas when allowing guards to see 180° but guards can choose their own direction (180°-floodlights).

References

1. Abdelkader, A., Saeed, A., Harras, K.A., Mohamed, A.: The inapproximability of illuminating polygons by α-floodlights. In: CCCG, pp. 287–295 (2015)
2. Abello, J., Estivill-Castro, V., Shermer, T.C., Urrutia, J.: Illumination of orthogonal polygons with orthogonal floodlights. Int. J. Comput. Geom. Appl. **8**(1), 25–38 (1998). https://doi.org/10.1142/S0218195998000035
3. Aggarwal, A.: The art gallery theorem: its variations, applications and algorithmic aspects. Ph.D. thesis, The Johns Hopkins University (1984)
4. Bhattacharya, P., Ghosh, S., Roy, B.: Approximability of guarding weak visibility polygons. Discrete Appl. Math. **228**, 109–129 (2017)
5. Bhattacharya, P., Ghosh, S.K., Pal, S.P.: Constant approximation algorithms for guarding simple polygons using vertex guards. CoRR (2017)
6. Bhattacharya, P., Ghosh, S.K., Roy, B.: Vertex guarding in weak visibility polygons. In: Ganguly, S., Krishnamurti, R. (eds.) Algorithms and Discrete Applied Mathematics, pp. 45–57. Springer International Publishing, Cham (2015). https://doi.org/10.1007/978-3-319-14974-5
7. Bonnet, É., Giannopoulos, P.: Orthogonal terrain guarding is np-complete. J. Comput. Geom. **10**(2), 21–44 (2019). https://doi.org/10.20382/jocg.v10i2a3
8. Bose, P., Guibas, L.J., Lubiw, A., Overmars, M.H., Souvaine, D.L., Urrutia, J.: The floodlight problem. Int. J. Comput. Geom. Appl. **7**(1/2), 153–163 (1997). https://doi.org/10.1142/S0218195997000090
9. Efrat, A., Har-Peled, S.: Guarding galleries and terrains. Inf. Process. Lett. **100**(6), 238–245 (2006)
10. Eidenbenz, S.: Inapproximability results for guarding polygons without holes. In: Chwa, K.-Y., Ibarra, O.H. (eds.) ISAAC 1998. LNCS, vol. 1533, pp. 427–437. Springer, Heidelberg (1998). https://doi.org/10.1007/3-540-49381-6_45
11. Elbassioni, K.M., Krohn, E., Matijevic, D., Mestre, J., Severdija, D.: Improved approximations for guarding 1.5-dimensional terrains. Algorithmica **60**(2), 451–463 (2011). https://doi.org/10.1007/s00453-009-9358-4
12. Estivill-Castro, V., O'Rourke, J., Urrutia, J., Xu, D.: Illumination of polygons with vertex lights. Inf. Process. Lett. **56**(1), 9–13 (1995). https://doi.org/10.1016/0020-0190(95)00129-Z
13. Ghosh, S.K.: On recognizing and characterizing visibility graphs of simple polygons. Discrete Comput. Geom. **17**(2), 143–162 (1997). https://doi.org/10.1007/BF02770871
14. Gibson, M., Krohn, E., Rayford, M.: Guarding monotone polygons with vertex half-guards is np-hard. In: European Workshop on Computational Geometry (2018). https://conference.imp.fu-berlin.de/eurocg18/program
15. Gibson, M., Krohn, E., Rayford, M.: Guarding monotone polygons with half-guards. In: Gudmundsson, J., Smid, M.H.M. (eds.) Proceedings of the 29th Canadian Conference on Computational Geometry, CCCG 2017, 26–28 July 2017, Carleton University, Ottawa, Ontario, Canada, pp. 168–173 (2017)

16. King, J., Krohn, E.: Terrain guarding is np-hard. SIAM J. Comput. **40**(5), 1316–1339 (2011)
17. Krohn, E.: Survey of terrain guarding and art gallery problems. In: Comprehensive Exam, University of Iowa (2007). http://faculty.cs.uwosh.edu/faculty/krohn/papers/comprehensive.pdf
18. Krohn, E., Nilsson, B.J.: The complexity of guarding monotone polygons. In: CCCG, pp. 167–172 (2012)
19. Krohn, E., Nilsson, B.J.: Approximate guarding of monotone and rectilinear polygons. Algorithmica **66**(3), 564–594 (2013). https://doi.org/10.1007/s00453-012-9653-3
20. Lee, D.T., Lin, A.K.: Computational complexity of art gallery problems. IEEE Trans. Inform. Theory **32**(2), 276–282 (1986)
21. O'Rourke, J.: Art Gallery Theorems and Algorithms, vol. 57. Oxford New York, NY, USA (1987)
22. Speckmann, B., Tóth, C.D.: Allocating vertex π-guards in simple polygons via pseudo-triangulations. Discrete Comput. Geom. **33**(2), 345–364 (2005). https://doi.org/10.1007/s00454-004-1091-9
23. Tóth, C.D.: Art galleries with guards of uniform range of vision. Comput. Geom. **21**(3), 185–192 (2002)
24. Urrutia, J.: Art gallery and illumination problems. In: Sack, J., Urrutia, J. (eds.) Handbook of Computational Geometry, pp. 973–1027. North Holland / Elsevier (2000). https://doi.org/10.1016/b978-044482537-7/50023-1

Author Index

Printed in the United States
by Baker & Taylor Publisher Services